W9-ADE-098

Source Books in the History of the Sciences

Edward H. Madden, *General Editor*

From Frege to Gödel: A Source Book in Mathematical Logic, 1879–1931
Jean van Heijenoort

A Source Book in Animal Biology
Thomas S. Hall

A Source Book in Astronomy and Astrophysics, 1900–1975
Kenneth R. Lang and Owen Gingerich

A Source Book in Chemistry, 1900–1950
Henry M. Leicester

A Source Book in Classical Analysis
Garrett Birkhoff

A Source Book in Geography
George Kish

A Source Book in Geology, 1400–1900
Kirtley F. Mather and Shirley L. Mason

A Source Book in Greek Science
Morris R. Cohen and I. E. Drabkin

A Source Book in Mathematics, 1200–1800
Dirk J. Struik

A Source Book in Medieval Science
Edward Grant

A Source Book in the History of Psychology
Richard J. Herrnstein and Edwin G. Boring

A Source Book in Astronomy and Astrophysics, 1900–1975

Edited by Kenneth R. Lang and Owen Gingerich

Harvard University Press
Cambridge, Massachusetts,
and London, England
1979

Copyright © 1979 by the President and Fellows of Harvard College
All rights reserved
Printed in the United States of America

Library of Congress Cataloging in Publication Data

Main entry under title:

A Source book in astronomy and astrophysics, 1900–1975.

(Source books in the history of the sciences)
Includes bibliographical references and index.
1. Astronomy—Addresses, essays, lectures.
2. Astrophysics—Addresses, essays, lectures.
I. Lang, Kenneth R. II. Gingerich, Owen. III. Series.
QB51.S67 520'.8 78-9463
ISBN 0-674-82200-5

0288661

General Editor's Preface

The *Source Books* in this series are collections of classical papers that have shaped the structure of the various sciences. Some of these classics are not readily available and many of them have never been translated into English, thus being lost to the general reader and in many cases to the scientist himself. The point of this series is to make these texts easily accessible and to provide good translations of the ones that have not been translated at all, or only poorly.

The series was planned to include volumes in all the major sciences from the Renaissance through the nineteenth century. It has been extended to include ancient and medieval science and the development of the sciences in the twentieth century. Many of these books have been published already and others are in various stages of preparation.

The Carnegie Corporation originally financed the series by a grant to the American Philosophical Association. The History of Science Society and the American Association for the Advancement of Science have approved the project and are represented on the Editorial Advisory Board. This Board at present consists of the following members:

Marshall Clagett, History of Science, Institute for Advanced Study, Princeton
I. Bernard Cohen, History of Science, Harvard University
Thomas A. Goudge, Philosophy, University of Toronto
Gerald Holton, Physics, Harvard University
Ernst Mayr, Zoology, Harvard University
Ernest Nagel, Philosophy, Columbia University
Dorothy Needham, Chemistry, Cambridge University
Harry Woolf, History of Science, Institute for Advanced Study, Princeton

I am indebted to the members of the Advisory Board for their indispensable aid in guiding the course of the *Source Books*.

Edward H. Madden
Department of Philosophy
State University of New York at Buffalo

Preface

The 132 selections that make up this *Source Book* represent the seminal contributions to twentieth-century astronomy and astrophysics through the year 1975. One of the first things that will strike an observant browser is the wide variation of level, ranging from Hale's popular article on a proposed giant telescope in *Harper's Magazine* to the tensor calculus of Schwarzschild and Einstein. Our book is not necessarily made for general reading, although we hope that readers will find numerous inviting and informative articles. Rather, we have tried to illuminate the vigorous development of our celestial science by giving something of its entire fabric—from Forest Moulton's account of the collision hypothesis in his elementary textbook and Edwin Hubble's identification of the Crab nebula with the supernova of 1054 in a *Leaflet of the Astronomical Society of the Pacific*, to Ralph Fowler's application of degenerate gas statistics to white dwarfs in the *Monthly Notices of the Royal Astronomical Society*, and Jan Oort's demonstration of galactic rotation in the *Bulletin of the Astronomical Institutes of the Netherlands*.

It is more disappointing than astonishing that astronomers so quickly forget the exciting moments of yesterday's science. Today it seems obvious that the universe is predominantly hydrogen; yet this was a landmark discovery made less than fifty years ago, when astronomers still believed that the universe, like the earth, was predominantly composed of the heavier metallic elements. For better or for worse, it is all too easy to accept well-formulated theories without remembering the conflicting, embryonic ideas from which they arose. We hope that our selections will stimulate a renewed interest in the sources and history of twentieth-century astronomy.

In this volume readers can rediscover A. S. Eddington's elegant proof of the virial theorem for star clusters, Walter Baade and Fritz Zwicky's 1932 proposal for the existence of neutron stars, and Thomas Gold's forecast of collapsed stars as radio sources. They can reread, in the words of the original observers, the discoveries of the cosmic rays, the Van Allen belts, Martian volcanoes and canyons, pulsars, interstellar hydrogen, cosmic magnetic fields, quasars, and the remnant background radiation of the primeval big bang.

Here will also be found a dozen new translations of articles originally in German, French, or Dutch. For example, we include the first English translations of Heinrich Vogt's article on stellar interiors; C. F. von Weizsäcker's discovery of the CNO cycle; Karl Schwarzschild's original derivation of the space-time metric of a static, spherical mass; Albert Einstein's explanation of the motion of Mercury; and Aleksandr Friedmann's description of the curvature of space.

We have grouped together sequences of key articles to tell the story of the development of basic ideas. Readers can, for example, find the changing answers to such fundamental questions as: How did the solar system originate? What makes the stars shine? What lies in the vacuous space between the stars? Are the spiral nebulae distant "island universes"? Do the intense radio sources lie outside our galaxy? Will the universe continue to expand forever?

Within each chapter we have adopted a generally chronological arrangement, although occasionally certain thematic subgroups have their own characteristic ordering.

We fully realize that the selections presented here cannot take the place of the hundreds of important articles on which modern astronomy and astrophysics are based. Instead, we have tried to choose articles that present an important idea or a significant new observation. A concept or measurement need not be right to be influential, and our selections document a few erroneous observations and theories. Nevertheless, being right generally helps make an article memorable. Being first also increased the likelihood of an article's inclusion, but we have not felt bound to include the first appearance of each idea. Often the second or third presentation is so much more lucid, complete, or recognizably significant that the clear choice for an original "source" rests with the later publication. We have resisted the temptation to document the progress of twentieth-century astronomy through review articles, excellent though they may be. In this respect our selections differ from those of Harlow Shapley's earlier *Source Book of Astronomy, 1900–1950*; our articles present the fundamental ideas in their original statements, rather than in more polished or popularized forms.

We have, with each selection, tried to state why it is important. When related material has appeared earlier, we have attempted to cite the relevant references. For example, Karl Jansky's pioneering work on extraterrestrial radio sources appeared in three articles between 1930 and 1935, but the astronomical connection is only just hinted at in the first two, and therefore we have chosen part of the third. Again, Carl Wirtz anticipated Edwin Hubble in finding the redshift-magnitude relation for galaxies, but Hubble's results were so much more thorough that they must be considered the primary source. In each case, however, our introductory note mentions the earlier research. Similarly, we have also attempted to reference closely related subsequent work.

Our original goal was to present these papers in uncut versions, but as we sifted through the 600 articles specifically considered for inclusion, we quickly realized that such a policy would force us to exclude many important contributions. Even with substantial abridgment of the longer selections, our original selection of 180 articles or groups of articles was simply too expensive to print. In order to keep this volume within the price range of the individual consumer, we were forced to eliminate 48 more selections. During this final cutting we tended to eliminate mathematical works available in other sources, together with those articles whose recent publication makes their long-range importance difficult to assess. Thus, for example, Karl Schwarzschild's article on the equilibrium of the stellar atmosphere and Albert Einstein's article on the cosmological implications of his relativity theory were excluded because they appear, respectively, in *Selected Papers on the Transfer of Radiation* and *The Principle of Relativity* (both Dover paperbacks).

At least one extremely important type of astronomical publication has, by its nature, been excluded from this compendium. These are the catalogues—of stars and galaxies and radio sources, of positions and magnitudes and spectral types—basic for the progress of astronomy yet impossible to present here. HD, FK3, GC, 3C—these are but four of the household codes in the astronomical vocabulary, indicating the Henry Draper Catalogue of stellar spectral types, the Dritter Fundamental Katalog of stellar positions, the Boss General Catalogue of positions and proper motions of some 33,342 stars, and the third Cambridge survey of radio sources. These laborious compilations are the truly fundamental sources for our contemporary science. We saw no way to include them, yet we realize that with their omission a significant slice of twentieth-century astronomy passes unheralded in our selections.

In collecting important contributions from a variety of sources and putting them within the covers of a single book, the editors and Press found that a number of compromises had to be made in matters of editorial consistency and style. In some cases these decisions can perhaps be criticized, but in the interest of economy and expeditiousness they have been accepted even when a better solution later presented itself. We have retained the original numbering of equations, but both figures and tables have been renumbered, and we have added identifying captions or titles when these were absent in the original sources. Figures have frequently been redrawn for clearer reproduction, and the Press has edited the tables to attain greater aesthetic unity. We have renumbered the references and added citations to articles originally published as "in press," but we have made no attempt to force the references into a uniform style. The systems of units and their abbreviations reproduce those of the original articles, including the early labeling of radial velocities in "km" rather than "km/sec." We hope that none of the editorial changes made by ourselves or by the Press have affected the information contained in the selections. Incidentally, although we have both carefully proofread the entire volume, we know from the comparative statistics that errors undoubtedly remain; we console ourselves with the conviction that we have proofread more accurately than many of the original authors!

We have profited from the advice of many persons in assembling this collection. Among those who contributed useful suggestions are Ludwig Biermann, Alastair Cameron, David Dewhirst, William Fowler, Jesse Greenstein, Icko Iben, William McCrea, Donald Menzel, Jan Oort, Carl Sagan, and Sydney van den Bergh. As our editing reached its final stages, we asked a number of our colleagues to check one or more of the introductions. Foremost among these friendly critics have been Bart J. Bok and Cecilia Payne-Gaposchkin, and the list includes Hannes Alfvén, Ralph Alpher, Viktor Ambartsumian, Robert d'E. Atkinson, John Bahcall, Alan Barrett, Eric Becklin, Hans Bethe, Ludwig Biermann, Adriaan Blaauw, John Bolton, Hermann Bondi, Geoffrey Burbidge, Margaret Burbidge, Bernard Burke, Alastair Cameron, George Clark, Theodore Dunham, Bengt Edlén, Farouk El-Baz, Harold Ewen, George Field, William Fowler, Kenneth Franklin, Riccardo Giacconi, Vitallii Ginzburg, Thomas Gold, Jesse Greenstein, Herbert Gursky, Robert Hanbury Brown, Chushiro Hayashi, George Herbig, Robert Herman, Antony Hewish, James Hey, Gerald Holton, Michael Hoskin, James Kemp, Frank Kerr, Robert Kraft, C. C. Lin, Per Olof Lindblad, Ursula Marvin, Thomas Matthews, Cornell Mayer, Stanley Miller, William Morgan, Philip Morrison, Gerry Neugebauer, Jan H. Oort, Ernst Öpik, Franco Pacini, Arno Penzias, Edward Purcell, Grote Reber, John Rogerson, Martin Ryle, Carl Sagan, Edwin Salpeter, Maarten Schmidt, Martin Schwarzschild, Irwin Shapiro, Iosif Shklovskii, Lyman Spitzer, Albrecht Unsöld, Hendrik van de Hulst, Carl F. von Weizsäcker, Gart Westerhout, John Wheeler, Fred Whipple, Charles Whitney, Gerald Whitrow, and Robert Wilson.

We gratefully acknowledge the generous permission to reprint selections from the following sources: *Arkiv für matematik, astronomi och fysik*; *Astronomical Journal* (American Astronomical Society); *Astronomicheskii zhurnal* (Soviet Astronomy—American Institute of Physics); *Astronomische Nachrichten*; *Bulletin of the Astronomical Institutes of the Netherlands*; *Doklady akademii nauk USSR* (Soviet Physics Doklady—American Institute of Physics); Dover Publications; *Harper's Magazine*; *Icarus* (Academic Press); International Astronomical Union; *Journal of Atmospheric Sciences* (American Meteorological Society); *Journal of Geophysical Research* (American

Geophysical Union); *Lick Observatory Bulletin*; *Monthly Notices of the Royal Astronomical Society* (Blackwell Scientific Publications); *Nature* (Macmillan Journals); *Naturwissenschaften*; *Nuovo cimento* (Società Italiana di fisica); *Observatory*; *Philosophical Transactions of the Royal Society*; *Physical Review* and *Physical Review Letters* (American Institute of Physics); *Physikalische Zeitschrift*; *Physikalishe Zeitschrift der Sowjetunion*; *Proceedings of the Institute of Radio Engineers*; *Proceedings of the National Academy of Sciences*; *Proceedings of the Royal Society*; *Publications de l'Observatoire de Tartu*; *Publications of the Astronomical Society of Japan*; *Publications of the Astronomical Society of the Pacific* (California Academy of Sciences); *Reviews of Modern Physics* (American Physical Society); *Science* (American Association for the Advancement of Science); *Space Research* (North Holland Publishing Company); *Sky and Telescope*; *Vistas in Astronomy* (Pergamon Press); *Zeitschrift für Astrophysik* and *Zeitschrift für Physik* (Springer-Verlag).

We also wish to acknowledge permission to reprint a number of articles from the *Astrophysical Journal* (American Astronomical Society and the University of Chicago Press).

We are especially indebted to the staff of the scientific periodicals library of the Cambridge Philosophical Society and the library of the Harvard-Smithsonian Center for Astrophysics for their aid in locating references. We would like to thank the American Philosophical Society and the Tufts University Faculty Research Committee for financial aid.

Kenneth R. Lang
Tufts University

Owen Gingerich
Harvard-Smithsonian
Center for Astrophysics

Contents

CHAPTER I

New Windows on the Universe

1. On the Application of Interference Methods to Astronomical Measurements

Albert Abraham Michelson

(*Philosophical Magazine 30*, 1–21 [1890])

In this paper Albert Abraham Michelson describes the use of interference methods in measuring the angular size and the one-dimensional brightness distribution of sources that are too small to be resolved by a single telescope. This fundamental technique has found widespread application both in optical and in radio astronomy. At optical wavelengths, radiation from a source is received by two mirrors and combined on the focal plane of a telescope. If the separation of the mirrors, D, is not too great, the source is unresolved and the coherent light will produce interference fringes of alternating light and dark bands. As the two mirrors are gradually separated, the fringes will disappear when the source is resolved. In this case, the angular diameter of a source is 1.22 λ/D, where λ is the wavelength of the observation. The first successful measurement of the angular diameter of a star was made by Michelson and Francis G. Pease on December 13, 1920, by using two mirrors separated by 20 ft and placed at the end of the open tube of the 100-in Hooker telescope.[1] They found that the angular diameter of the supergiant star α Orionis (Betelgeuse) was 47×10^{-3} sec of arc. Although the angular diameters of six giant stars were measured with this instrument,[2] subsequent efforts to extend the measurements to the smaller main-sequence stars failed. This was due partly to the effects of atmospheric scattering and partly to the practical difficulty of constructing large mirror separations. These problems were finally overcome with the development of the intensity interferometer described in our next selection.

When extraterrestrial radiation was found at radio wavelengths, Michelson interferometers were constructed to measure the angular sizes and the brightness distributions of the sources. In this case, radio frequency signals received at two telescopes were transmitted to some central location for correlation, and coherence was maintained by the transmission of a common local oscillator signal to the two telescopes. In this way, the angular sizes of several bright radio sources were found to be a few minutes of arc; later detailed maps showed that many of the radio objects consisted of two components.[3] The Michelson type of interferometer has been since used to give angular resolutions up to 0.1 sec of arc at a radio wavelength of 6 cm;[4] and the very long baseline (V.L.B.I.) adaptation using signals recorded simultaneously at two independent radio telescopes has given angular resolutions up to 0.001 sec of arc.

R. Hanbury Brown and his colleagues suggested that mutually coherent local oscillators were not necessary at the two telescopes, and showed that post-detection correlation of signals recorded at two different telescopes could be used to measure the angular sizes of radio sources (see selection 2). This method has vastly extended the interferometer baselines at both optical and radio wavelengths.

1. *Astrophysical Journal 53*, 249 (1921).

2. *Ergebnisse der exakten Naturwissenschaften 10*, 84 (1931).

3. M. Ryle, *Proceedings of the Royal Society* (London) *A211*, 351 (1952); B. Y. Mills, *Australian Journal of Physics 6*, 452 (1953).

4. A historical account of the use of interferometry in radio astronomy is given by J. S. Hey in *The Evolution of Radio Astronomy* (New York: Neale Watson Academic Publications, 1973).

In a recent paper on "Measurement by Light-Waves"[1] it was shown that the limitation of the effective portions of an objective to the extreme ends of a diameter converted the instrument into a refractometer; and although definition and resolution are thereby sacrificed, the accuracy may be increased ten to fifty fold.

The simplest way of effecting this in the case of a telescope is to provide the cap of the objective with two slits adjustable in width and distance apart. If such a combination be focused on a star, then, instead of an image of the star, there will be a series of coloured interference-bands with white centre, the bands being arranged at equal distances apart and parallel to the two slits. The position of the central white fringe can be marked from ten to fifty times as accurately as can the centre of the telescopic image of the star.

One of the most promising applications of the method is the measurement of the angular magnitudes of small sources of light.

This may be accomplished by taking advantage of the well-known principle that in order to obtain clear interference-bands from two pencils diverging from the same source (width a) at an angle β, it is necessary that either β or a be very small.

Thus let us take

$a = ee_1$ = width of the source.
d = distance of source from the objective.
$b = SS_1$ = distance between the slits.

Also put $S_1P = \Delta$, $a/d = \alpha$, $b/d = \beta$.

The distance $b = SS_1$ is the distance between the two slits that are placed in a plane perpendicular to the telescope axis. The distance $\Delta = S_1P$ is the extra distance between S_1 and the wave-front that lies along the line extending from point S to point P.

Then the usual statement is that the interference-fringes vanish when $Se_1 - Se = \beta a/2 = b\alpha/2 = \lambda/2$, or when

$$\alpha = \frac{\lambda}{b}.$$

But λ/b is the "limit of resolution" of the telescope of aperture b, and if this be denoted by α_0 we have

$$\alpha = \alpha_0.$$

Or, in words, the fringes disappear when the source subtends an angle which can just be resolved by the telescope.

The experiment was first tried with an objective of 45 millim. effective diameter (distance between the slits) at a distance of ten metres from an adjustable slit which served as the source.

It was found that the first indication of indistinctness occurred when a was 0.08 millim. wide, and at 0.14 millim. the fringes almost vanished.

But on continuing to widen the slit they again became clearly visible, to disappear and reappear at regular intervals.

Now, though it might with truth be urged that the observation of the indefinite vanishing of interference-fringes depends so much on the attendant circumstances, and especially on the condition of the observer, that it can scarcely be called a precise measurement, yet the statement applies no longer when the disappearance depends on the existence of well-marked *minima of distinctness*; and, as will appear below, it is possible to measure, with accuracy, by the observation of these minima the width of a source of light, which in a telescope can with difficulty be ascertained to have an appreciable size.

The theory of these successive appearances and disappearances is as follows:

Let x be the distance of any element of the source from the axis of the telescope, dx the width of the element, and $y = \phi(x)$ the length.

Then the difference in the two paths xS and xS_1P terminating at the wave-front P, which makes an angle γ with the plane perpendicular to the axis of the telescope, will be $\beta x - \gamma b$, and the resulting intensity in the direction γ for the whole source will therefore be

$$I = \int \phi(x)\left[1 + \cos\frac{2\pi}{\lambda}(\beta x - \gamma b)\right]dx. \tag{1}$$

Case I. Uniformly Illuminated Slit

If the source be a slit whose centre is in the axis, and whose length is parallel to the slits SS_1, and whose width is a, then

$$I = a + \frac{\lambda}{\pi\beta}\sin\frac{\pi\beta}{\lambda}a\cos\frac{2\pi\gamma}{\lambda}b. \tag{2}$$

If I_1 be the intensity at the centre of a bright fringe, and I_2 that at the centre of a dark fringe, then the visibility of the fringes may be expressed by

$$V = \frac{I_1 - I_2}{(I_1 + I_2)}. \tag{3}$$

But

$$I_1 = a + \frac{\lambda}{\pi\beta}\sin\frac{\pi\beta}{\lambda}a,$$

$$I_2 = a - \frac{\lambda}{\pi\beta}\sin\frac{\pi\beta}{\lambda}a;$$

$$\therefore\quad V = \frac{\sin\frac{\pi\beta}{\lambda}a}{\frac{\pi\beta}{\lambda}a}$$

or finally,

$$V = \frac{\sin \pi \frac{\alpha}{\alpha_0}}{\pi \frac{\alpha}{\alpha_0}}. \tag{4}$$

Hence the fringes will disappear whenever α is a multiple of α_0. They will be clearest when

$$\alpha = \frac{\alpha_0}{\pi} \tan \pi \frac{\alpha}{\alpha_0}. \tag{5}$$

Case II. Uniformly Illuminated Disk,
Case III. Illumination Not Uniform

Here Michelson derives the formulae for the visibility function for a uniformly illuminated disk and the nonuniform disk of the sun. Michelson then presents a plot of equation (4), together with plots of the visibility function for a uniformly illuminated disk and the sun's disk. The three plots differ only slightly from each other and are well represented by the plot of equation (4), which is given in figure 1.1.

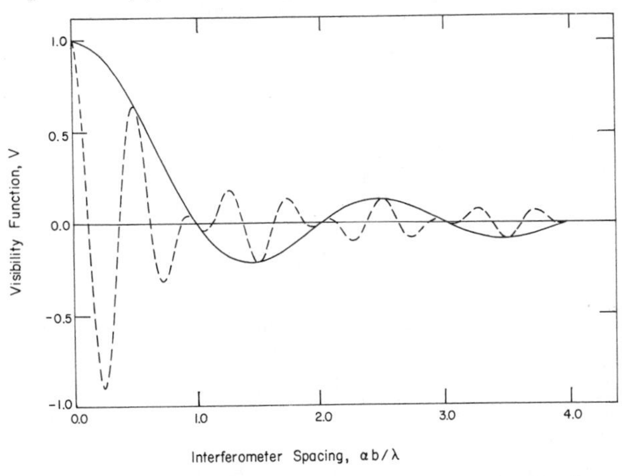

Fig. 1.1 A plot of the interference visibility, V, as a function of α/α_0 for a uniformly illuminated source of angular extent α (solid curve and equation [4]), and for two equal, uniformly illuminated sources of angular separation 4α and individual angular extents $\alpha_1 = 1.00\ \alpha$ (dashed curve and equation [16]). For a two-element interferometer with a linear spacing between elements of b and a wavelength λ, the parameter $\alpha_0 = \lambda/b$.

Case IV. Double Source

Michelson considers two equal symmetrical sources of light centered at C and C_1 with a linear separation $CC_1 = 2s$. The radii r of the two sources are given by $r = OC = CB = DC_1 = C_1E$, where the linear extents of the two sources are OB and DE, respectively, for the sources centered at C and C_1. A point on any source is given the rectangular coordinates x and y, with the origin of the coordinate system centered at C.

Then the intensity of the interference-fringes will be

$$I = \int y\left(1 + \cos\frac{2\pi\Delta}{\lambda}\right)dx \ldots \text{as in (1).}$$

Integrating from O to B and from D to E, we have

$$I = \int_0^{2r} y_1\left(1 + \cos\frac{2\pi\Delta}{\lambda}\right)dx + \int_{2s}^{2s+2r} y_2\left(1 + \cos\frac{2\pi\Delta}{\lambda}\right)dx; \tag{11}$$

where

$$y_1 = f(x - r) \quad \text{and} \quad y_2 = f[x - (2s + r)]$$

and

$$\Delta = \beta x - \gamma b.$$

In the first integral put $w_1 = (x - r)$.

In the second integral put $w_2 = x - (2s + r)$.

Then we obtain

$$I = \int_{-r}^{+r} f(w_1)\left[1 + \cos\frac{2\pi}{\lambda}(\beta w_1 + \beta r - \gamma b)\right]dw_1 + \int_{-r}^{+r} f(w_2)\left[1 + \cos\frac{2\pi}{\lambda}(\beta w_2 + \beta r - \gamma b + 2\beta s)\right]dw_2. \tag{12}$$

Expanding the first of these integrals we obtain:

$$\int_{-r}^{+r} f(w_1)\,dw_1 + \cos\frac{2\pi}{\lambda}(\beta r - \gamma b)\int_{-r}^{+r} f(w_1)\cos\frac{2\pi}{\lambda}\beta w_1\,dw_1 - \sin\frac{2\pi}{\lambda}(\beta r - \gamma b)\int_{-r}^{+r} f(w_1)\sin\frac{2\pi}{\lambda}(\beta w_1)\,dw_1;$$

in which the first term is half the area of the aperture, and the last term (since $f(w_1)$ is a symmetrical function) is 0. The same is also true of the expansion of the second integral.

Michelson then carries out the appropriate integrations to obtain

$$I = Q + \tfrac{1}{2}QA \cos\frac{2\pi}{\lambda}(\beta r - \gamma b) + \tfrac{1}{2}QA \cos\frac{2\pi}{\lambda}(\beta r - \gamma b + 2\beta s)$$

or

$$I = Q + QA\left(\cos\frac{2\pi}{\lambda}[\beta s + \beta r - \gamma b]\cos\frac{2\pi}{\lambda}\beta s\right); \quad (13)$$

whence for the visibility

$$V = \frac{I_1 - I_2}{I_1 + I_2} = A\cos\frac{2\pi}{\lambda}\beta s = A\cos\pi\frac{\alpha}{\alpha_0}, \quad (14)$$

where

$$\int_{-r}^{+r} f(w)\,dw = \tfrac{1}{2}Q,$$

$$\int_{-r}^{+r} f(w)\cos\frac{2\pi}{\lambda}\beta w\,dw = \tfrac{1}{2}QA$$

and $f(w)$ is the intensity distribution across the source of radiation.

When the sources are two equal uniformly illuminated slits of height $2h$,

$$f(w) = h, \quad \text{and} \quad Q = 4rh.$$

Hence

$$A = \frac{1}{r}\int_0^r \cos\frac{2\pi\beta}{\lambda}w\,dw = \frac{\sin\frac{2\pi}{\lambda}\beta r}{\frac{2\pi}{\lambda}\beta r}. \quad (15)$$

Putting $(2\pi/\lambda)\beta r = \pi(\alpha_1/\alpha_0)$, and substituting for A in equation (14),

$$V = \frac{\sin\pi\frac{\alpha_1}{\alpha_0}}{\pi\frac{\alpha_1}{\alpha_0}}\cos\pi\frac{\alpha}{\alpha_0}. \quad (16)$$

For the case of two equal uniformly illuminated circular apertures of radius r,

$$f(w) = \sqrt{r^2 - w^2}, \quad \text{and} \quad Q = \pi r^2.$$

Hence

$$A = \frac{4}{\pi r^2}\int_0^r \sqrt{r^2 - w^2}\cos\frac{2\pi}{\lambda}\beta w\,dw.$$

Putting $w/r = z$, this reduces to the form already given for Airy's integral, and the expression for the visibility of the fringes is

$$V = A_1\cos\pi\frac{\alpha}{\alpha_0}. \quad (17)$$

When the distance between the sources is more than five or six times the width, the periodicity of the second term alone is of importance, the term A_1 representing the amplitude of its variation.

The fringes vanish whenever $\alpha = (2n - 1)(\alpha_0)/2$.

If, on the other hand, the two sources coincide the expression reduces to that previously found for a single circular source.

Michelson then discusses various methods of measuring stellar interference fringes and proposes an instrument consisting of two mirrors a and b, which are separated but which both reflect onto an auxilliary mirror m.

The instrument would be used as follows: The mirrors a and b being moved as close together as possible, and the auxiliary mirror m being in place, the mirror a is adjusted by the screws SS till the two images of a source of light L as viewed in the telescope T appear to coincide. Next the mirror m is adjusted in azimuth by the screw K till the paths of the two pencils are equal. (As before mentioned, this angular motion does not affect the mutual inclination of the pencils, and therefore the breadth of the fringes is unaltered.) The telescope T is then adjusted till the two images of its illuminated cross-hairs coincide with the cross-hairs themselves, and is then clamped.

The mirror m is now detached and the instrument is ready for use.

Suppose the object to be measured is a minor planet or satellite. The whole instrument, which would have to be placed on an equatorial mounting, is pointed so that the image of the body is exactly on the cross-hairs. The interference-fringes will at once appear if the adjustment has been properly made.

Next, by means of a right-and-left-hand screw, the mirrors a and b are separated *until the fringes disappear*. If this disappearance is due to an accident, it will immediately become evident by observing any star in the neighbourhood. If the examination of the star shows the fringes while they are absent in the case of the planet, it may be considered certain that the cause of the disappearance in the latter case is its appreciable disk.

The angular diameter of the latter can be found (on the supposition of uniform illumination) by the formula

$$\alpha = 1.22 \frac{\lambda}{b},$$

where λ is the wave-length of the light, and b the distance between the centres of the mirrors a and b.[2]

From what has gone before, it will be inferred that the chief object of the method proposed is the measurement of the apparent size of minute telescopic objects, such as planetoids, satellites, and possibly star disks, and also double stars too close to be resolved in the most powerful telescope. But it is clear that interference methods may also be employed for the measurement of star-places.

Thus, in observations for right ascension, the slits would be placed parallel to the meridian, and the instant of passage of the central white fringe across the spider-lines noted; and in observations for declination, the slits would be horizontal, the cross-hair being brought to coincide with the centre of the white fringe.

The increase in accuracy to be expected from this method would, however, be limited by the imperfections of our present means of measuring time and angles; still, it would appear that by its use a one-inch glass may be made to do the work now required of a ten-inch.

CONCLUSION

(1) Interference phenomena produced under appropriate conditions from light emanating from a source of finite magnitude become indistinct as the size increases, finally vanishing when the angle subtended by the source is equal to the smallest angle which an equivalent telescope can resolve, multiplied by a constant factor depending on the shape and distribution of light in the source and on the order of the disappearance.

(2) The vanishing of the fringes can ordinarily be determined with such accuracy that single readings give results from fifty to one hundred times as accurate as can be obtained with a telescope of equal aperture.

(3) The principal applications of the methods herein described are the measurement of the apparent magnitudes of very small or very distant sources of light such as planetoids and satellites (though larger bodies are not excluded), and of the angular distances between very close double stars.

(4) On account of the narrowness of the interference-fringes when a very minute body is under examination, the method of obtaining these fringes (by a pair of adjustable slits in front of the objective of a telescope) is open to objection, from which the refractometer method is entirely free. Further, this last modification makes it possible to extend the effective aperture of the equivalent telescope without limit. Thus, while it would be manifestly impracticable to construct objectives much larger than those at present in use, there is nothing to prevent increasing the distance between the two mirrors of the refractometer to even ten times this size. If among the nearer fixed stars there is any as large as our sun, it would subtend an angle of about one hundredth of a second of arc; and the corresponding distance required to observe this small angle is ten metres, a distance which, while utterly out of question as regards the diameter of a telescope-objective, is still perfectly feasible with a refractometer. There is, however, no inherent improbability of stars presenting a much larger angle than this; and the possibility of gaining some positive knowledge of the real size of these distant luminaries would more than repay the time, care, and patience which it would be necessary to bestow on such a work.

1. *American Journal of Science 39* (Feb. 1890).
2. Better, the distance between the centres of the apertures in front of these mirrors.

2. A Test of a New Type of Stellar Interferometer on Sirius

Robert Hanbury Brown and Richard Q. Twiss

(*Nature 178*, 1046–48 [1956])

Early in this century Albert Michelson and Francis Pease,[1] using an optical interferometer mounted at the front of the Mount Wilson 100-in reflector, measured the angular diameter of half a dozen giant and supergiant stars. Further attempts to measure directly ordinary main-sequence stars failed because of the size and clumsiness of the equipment, although many stellar diameters were deduced from the study of eclipsing binaries. In 1952 Robert Hanbury Brown and his colleagues first demonstrated that mutually coherent oscillators are not necessary for the separated antennae of a radio-frequency interferometer.[2] Out of this demonstration grew the idea that if the intensity fluctuations of starlight could be recorded independently at two optical telescopes, they might be correlated later. In this way Hanbury Brown and Richard Twiss overcame the practical difficulties of obtaining optical interferometers wide enough to measure the angular diameters of ordinary stars and of making reliable measurements of angular diameter through the turbulent atmosphere of the earth.

The mathematical theory describing the intensity interferometer, as applied to radio waves, was first developed by Hanbury Brown and Twiss in 1954, and by 1957 they had applied the theory to light waves.[3] In the paper reproduced here, the first tests of the optical instrument are described.[4] The intensity interferometer eventually became the first instrument to measure the diameter of a main-sequence star and a hot star (stellar angular diameters measured with the Narrabri intensity interferometer are given here as an appendix).

Radio wavelength interferometers that use independent local oscillators, but which differ from the intensity interferometer, have now been used with baselines nearly equal to the earth's diameter to give radio frequency angular resolutions as small as one thousandth of a second of arc. In this way individual components of the quasi-stellar objects have been resolved, and angular resolutions comparable to that of the optical intensity interferometer have been achieved at radio wavelengths.

1. *Astrophysical Journal 53*, 249 (1921); *Ergebnisse der exakten Naturwissenschaften 10*, 84 (1931).
2. *Nature 170*, 1061 (1952).
3. *Philosophical Magazine 45*, 663 (1954); *Proceedings of the Royal Society* (London) *A242*, 300 (1957), *A243*, 291 (1957), *A248*, 199 (1958).
4. Hanbury Brown has described the first tests of both the radio and the optical instruments in *The Intensity Interferometer* (New York: Halsted Press, 1974).

We have recently described[1] a laboratory experiment which established that the time of arrival of photons in coherent beams of light is correlated, and we pointed out that this phenomenon might be utilized in an interferometer to measure the apparent angular diameter of bright visual stars.

The astronomical value of such an instrument, which might be called an 'intensity' interferometer, lies in its great potential resolving power, the maximum usable base-line being governed by the limitations of electronic rather than of optical technique. In particular, it should be possible to use it with base-lines of hundreds, if not thousands of feet, which are needed to resolve even the nearest of the *W*-, *O*- and *B*-type stars. It is for these stars that the measurements would be of particular interest since the theoretical estimates of their diameters are the most uncertain.

The first test of the new technique was made on Sirius (α Canis Majoris A), since this was the only star bright enough to give a workable signal-to-noise ratio with our preliminary equipment.

The basic equipment of the interferometer consisted of two mirrors M_1, M_2, which focused light on to the cathodes of the photomultipliers P_1, P_2 and which were guided manually on to the star by means of an optical sight mounted on a remote-control column. The intensity fluctuations in the anode currents of the photomultipliers were amplified over the band 5–45 Mc./s., which excluded the scintillation frequencies, and a suitable delay was inserted into one or other of the amplifiers to compensate for the difference in the time of arrival of the light from the star at the two mirrors. The outputs from these amplifiers were multiplied together in a linear mixer and, after further amplification in a system where special precautions were taken to eliminate the effects of drift; the average value of the product was recorded on the revolution counter of an integrating motor. The readings of this counter gave a direct measure of the correlation between the intensity fluctuations in the light received at the two mirrors; however, the magnitude of the readings depended upon the gain of the equipment, and for this reason the r.m.s. value of the fluctuations at the input to the correlation motor was also recorded by a second motor. Since the readings of both revolution counters depend in the same manner upon the gain, it was possible to eliminate the effects of changes in amplification by expressing all results as the ratio of the integrated correlation to the r.m.s. fluctuations, or uncertainty in the final value. The same procedure was also followed in the laboratory experiment described in a previous communication.[1]

There is no necessity in an 'intensity' interferometer to form a good optical image of the star. It is essential only that the mirrors should focus the light from the star on to a small area, so that the photocathodes may be stopped down by diaphragms to the point where the background light from the night sky is relatively insignificant. In the present case, the two mirrors were the reflectors of two standard searchlights, 156 cm. in diameter and 65 cm. in focal length, which focused the light into an area 8 mm. in diameter. However, for observations of Sirius, the circular diaphragms limiting the cathode areas of the photomultipliers (*R.C.A.* type 6342) were made as large as possible, namely, 2.5 cm. in diameter, thereby reducing the precision with which the mirrors had to be guided.

The first series of observations was made with the shortest possible base-line. The searchlights were placed north and south, 6.1 metres apart, and observations were made while Sirius was within 2 hr. of transit. Since the experiments were all carried out at Jodrell Bank, lat. 53° 14′ N., the elevation of the star varied between $15\frac{1}{2}°$ and 20°, and the average length of the base-line projected normal to the star was 2.5 metres; at this short distance Sirius should not be appreciably resolved.

Throughout the observations the average d.c. current in each photomultiplier was recorded every 5 min., together with the readings of the revolution counters on both the integrating motors. The small contributions to the photomultiplier currents due to the night-sky background were measured at the beginning and end of each run. The gains of the photomultipliers were also measured and were found to remain practically constant over periods of several hours.

In order to ensure that any correlation observed was not due to internal drifts in the equipment, or to coupling between the photomultipliers or amplifier systems, dummy runs of several hours duration were made before and after every observation; for these runs the photomultiplier in each mirror was illuminated by a small lamp mounted inside a detachable cap over the photocathode. In no case was any significant correlation observed.

In this initial stage of the experiment, observations were attempted on every night in the first and last quarters of the Moon in the months of November and December 1955; the period around the full moon was avoided because the background light was then too high. During these months a total observation time of 5 hr. 45 min. was obtained, an approximately equal period being lost due to failure of the searchlight control equipment. The experimental value for the integrated correlation $C(d)$ at the end of the observations was recorded. The value of $C(d)$ is the ratio of the change in the reading of the counter on the correlation motor to the associated r.m.s. uncertainty in this reading.

In the second stage of the experiment the spacing between the mirrors was increased and observations were carried out with east–west base-lines of 5.6, 7.3 and 9.2 metres. These measurements were made on all possible nights during the period January–March 1956, and a total observing time of $12\frac{1}{4}$ hr. was obtained.

As a final check that there was no significant contribution to the observed correlation from any other source of light in the sky, such as the Čerenkov component from cosmic rays,[2] a series of observations was made with the mirrors close together and exposed to the night sky alone. No significant correlation was observed over a period of several hours.

The results have been used to derive an experimental value for the apparent angular diameter of Sirius. The four measured values of $C(d)$ were compared with theoretical values for uniformly illuminated disks of different angular sizes, and the best fit to the observations was found by minimizing the sum of the squares of the residuals weighted by the observational error at each point. In making this comparison, both the angular diameter of the disk and the value of $C(o)$, the correlation at zero base-line, were assumed to be unknown, and account was taken of the different light flux and observing time for each point. Thus the final experimental value for the diameter depends only on the relative values of $C(d)$ at the different base-lines, and rests on the assumption that these relative values are independent of systematic errors in the equipment or in the method of computing $C(d)$ for the models. The best fit to the observations was given by a disk of angular diameter 0.0068″ with a probable error of ±0.0005″.

The angular diameter of Sirius, which is a star of spectral type *A*1 and photovisual magnitude −1.43, has never been measured directly; but if we assume that the star radiates like a uniform disk and that the effective black body temperature[3,4] and bolometric correction are 10,300° K. and −0.60, respectively, it can be shown that the apparent angular diameter is 0.0063″, a result not likely to be in error by more than 10 per cent. (In this calculation the effective temperature, bolometric magnitude and apparent angular diameter of the Sun were taken as 5,785° K., −26.95, and 1,919″, respectively.) Thus it follows that the experimental value for the angular diameter given above does not differ significantly from the value predicted from astrophysical theory.

A detailed comparison of the absolute values of the observed correlation with those expected theoretically has also been made, and the results are given in figure 2.1. In making this comparison, it is convenient to define a normalized correlation coefficient $\Gamma^2(d)$, which is independent of observing time, light flux and the characteristics of the equipment, where $\Gamma^2(d) = C(d)/C(o)$ and $C(d)$ is the correlation with a base-line of length d, and $C(o)$ is the correlation which would be observed with zero base-line under the same conditions of light flux and observing time. The theoretical values of $\Gamma^2(d)$ for a uniformly illuminated disk of diameter 0.0063″ are shown as the dashed line in figure 2.1.

For monochromatic radiation it is simple to evaluate $\Gamma^2(d)$, since it can be shown[5] that it is proportional to the square of the Fourier transform of the intensity distribution across the equivalent strip source; however, in the present case, where the light band-width is large, the values of $\Gamma^2(d)$ were calculated by numerical integration.

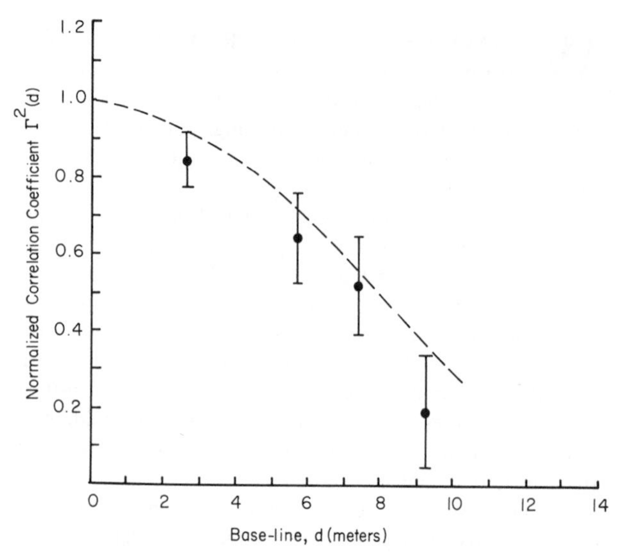

Fig. 2.1 Comparison between the values of the normalized correlation coefficient $\Gamma^2(d)$ observed from Sirius and the theoretical values for a star of angular diameter 0.0063 sec of arc. The errors shown are the probable errors of the observations.

Here Hanbury Brown and Twiss discuss the determination of $C(o)$ as specified in their paper[1] and show that scintillation effects on measurements of $C(d)$ are negligible.

Thus, despite their tentative nature, the results of this preliminary test show definitely that a practical stellar interferometer could be designed on the principles described above. Admittedly such an instrument would require the use of large mirrors. Judging from the results of this test experiment, where the peak quantum efficiency of the phototubes was about 16 per cent and the overall bandwidth of the amplifiers was about 38 Mc./s., one would need mirrors at least 3 metres in diameter to measure a star, near the zenith, with an apparent photographic magnitude +1.5. Mirrors of at least 6 metres in diameter would be required to measure stars of mag. +3, and an increase in size would also be needed for stars at low elevation because of atmospheric absorption. However, the optical properties of such mirrors need be no better than those of searchlight reflectors, and their diameters could be decreased if the overall band-width of the photomultipliers and the electronic apparatus could be increased, or if photocathodes with higher quantum efficiencies become available. It must also be noted that the technique of using two mirrors, as described here, would probably be restricted to stars of spectral type earlier than G, since cooler stars of adequate apparent magnitude would be partially resolved by the individual mirrors.

The results of the present experiment also confirm the theoretical prediction[6] that an 'intensity' interferometer should be substantially unaffected by atmospheric scintillation. This expectation is also supported by experience with a radio 'intensity' interferometer[5,7,8] which proved to be virtually independent of ionospheric scintillation. It is also to be expected that the technique should be capable of giving an extremely high resolving power. Without further experience it is impossible to estimate the maximum practical length of the base-line; however, it is to be expected that the resolving power could be at least one hundred times greater than the highest value so far employed in astronomy, and that almost any star of sufficient apparent magnitude could be resolved.

1. R. Hanbury Brown and R. Q. Twiss, *Nature 177*, 27 (1956).

2. W. Galbraith and J. V. Jelley, *J. Atmos. Terr. Phys. 6*, 250 (1955).

3. G. P. Kuiper, *Astrophys. J.*, *88*, 429 (1938).

4. P. C. Keenan and W. W. Morgan, *Astrophysics*, ed. J. A. Hynek (New York: McGraw-Hill, 1951).

5. R. Hanbury Brown and R. Q. Twiss, *Phil. Mag. 45*, 663 (1954).

6. R. Hanbury Brown and R. Q. Twiss, *Proceedings of the Royal Society* (London) *A242*, 300 (1957), *A243*, 291 (1957).

7. R. Hanbury Brown, R. C. Jennison, and M. K. Das Gupta, *Nature 170*, 1061 (1952).

8. R. C. Jennison and M. K. Das Gupta, *Phil. Mag. 48*, 55 (1956).

Appendix. Stars whose angular diameters have been measured by the intensity interferometer. (The radii are in units of the sun's radius, which has the value of 6.959 × 10^{10} cm.)

Star number[1]	Star name	Type	Angular diameter $\times 10^{-3}$ sec of arc	Temperature (K)	Radius ($R_\odot^{-1}$)
472	α Eri	B3 (Vp)	1.85 ± 0.07	13,700 ± 600	
1713	β Ori	B8 (Ia)	2.43 ± 0.05	11,500 ± 700	
1790	γ Ori	B2 (III)	0.70 ± 0.04	20,800 ± 1,300	
1903	ε Ori	B0 (Ia)	0.67 ± 0.04	24,500 ± 2,000	
1948	ζ Ori	O9.5 (Ib)	0.47 ± 0.04	26,100 ± 2,200	
2004	κ Ori	B0.5 (Ia)	0.44 ± 0.03	30,400 ± 2,000	
2294	β CMa	B1 (II–III)	0.50 ± 0.03	25,300 ± 1,500	
2326	α Car	F0 (Ib–II)	6.1 ± 0.7	7,500 ± 250	42 ± 22
2421	γ Gem	A0 (IV)	1.32 ± 0.09	9,600 ± 500	4.2 ± 0.7
2491	α CMa	A1 (V)	5.60 ± 0.15	10,250 ± 150	1.69 ± 0.05
2618	ε CMa	B2 (II)	0.77 ± 0.05	20,800 ± 1,300	
2693	δ CMa	F8 (Ia)	3.29 ± 0.46	— —	
2827	η CMa	B5 (Ia)	0.72 ± 0.06	14,200 ± 1,300	
2943	α CMi	F5 (IV–V)	5.10 ± 0.16	6,500 ± 200	2.1 ± 0.1
3165	ζ Pup	O5 (f)	0.41 ± 0.03	30,700 ± 2,500	15.6 ± 2.2
3207	γ^2 Vel	WC8 + 0.9 (I)	0.43 ± 0.05	29,000 ± 3,000	16.3 ± 2.9
3685	β Car	A1 (IV)	1.51 ± 0.07	9,500 ± 350	4.5 ± 1.8
3982	α Leo	B7 (V)	1.32 ± 0.06	12,700 ± 800	3.8 ± 1.0
4534	β Leo	A3 (V)	1.25 ± 0.09	9,050 ± 450	1.9 ± 0.2
4662	γ Crv	B8 (III)	0.72 ± 0.06	13,100 ± 1,200	
4853	β Cru	B0.5 (III)	0.702 ± 0.022	27,900 ± 1,200	
5056	α Vir	B1 (IV)	0.85 ± 0.04	22,400 ± 1,000	7.9 ± 0.7
5132	ε Cen	B1 (III)	0.47 ± 0.03	26,000 ± 1,800	
5953	δ Sco	B0.5 (IV)	0.45 ± 0.04	— —	
6175	ζ Oph	O9.5 (V)	0.50 ± 0.05	— —	
6556	α Oph	A5 (III)	1.53 ± 0.12	8,150 ± 400	3.1 ± 0.5
6879	ε Sgr	A0 (V)	1.37 ± 0.06	9,650 ± 400	
7001	α Lyr	A0 (V)	3.08 ± 0.07	9,250 ± 350	2.8 ± 0.2
7557	α Aql	A7 (IV, V)	2.78 ± 0.13	8,250 ± 250	1.65 ± 0.09
7790	α Pav	B2.5 (V)	0.77 ± 0.05	17,100 ± 1,400	
8425	α Gru	B7 (IV)	0.98 ± 0.07	14,800 ± 1,200	2.2 ± 0.6
8728	α PsA	A3 (V)	1.98 ± 0.13	9,200 ± 500	1.6 ± 0.2

Source: Adapted from R. Hanbury Brown, *The Intensity Interferometer* (New York: Halsted Press, 1974).
[1] *Yale University Observatory Bright Star Catalogue* (New Haven, 1964).

3. Concerning Observations of Penetrating Radiation on Seven Free Balloon Flights

Victor Franz Hess

TRANSLATED BY BRIAN DOYLE

(*Physikalishe Zeitschrift 13*, 1084–91 [1912])

When Victor Hess, an ardent amateur balloonist, began the observations summarized in this paper, it was already known that the ionizing radiation believed to originate from radioactive materials at the earth's surface did not decrease with increasing altitude as fast as had been expected. Hess designed airtight ionization chambers that would withstand the pressure differentials in the flight of an open balloon gondola and that would be sensitive only to very penetrating γ rays. By using these detectors together with an intense radium source, Hess showed that γ rays must be almost completely absorbed by the atmosphere at a height of 500 m. Nevertheless, at altitudes of about 5,000 m the ionization rate increased to levels as high as 6 times that observed on the ground. By observing at night and even during a solar eclipse, Hess was able to show that the observed radiations did not come from the sun. The observed results could only be explained by assuming that radiation of great penetrating power impinges upon our atmosphere from above.

Further balloon observations by Werner Kolhörster showed that the ionization continued to increase with height to values as large as fifty times that at sea level, but the variations in ionization at different latitudes seemed to discredit Hess's bold hypothesis. Thirteen years later, however, Robert Millikan convinced everyone that the radiation did come from beyond the earth's atmosphere, and incidentally gave the radiation its present name of cosmic rays.[1]

Because cosmic ray particles are the only form of matter known to reach the earth from outside the solar system, they also have important astrophysical consequences. Both the cosmic ray electrons and the more abundant cosmic ray protons may be directly accelerated in such cosmic ray sources as supernovae explosions, and the observed abundances of the heavier cosmic rays have consequences for theories of stellar nucleosynthesis. As the cosmic ray electrons travel through interstellar space they will be deflected by the interstellar magnetic field and may give rise to the radio frequency radiation of our galaxy. Furthermore, because the energy density of cosmic rays is comparable to the energy density of starlight, the turbulent kinetic energy of the interstellar gas, and the energy density of the interstellar magnetic field, the cosmic rays are one of the prominant factors in determining the physical state of the interstellar medium.

1. R. A. Millikan and G. H. Cameron, *Physical Review 28*, 851 (1926).

BY LAST YEAR I had already had the opportunity to undertake two balloon flights to investigate the penetrating radiation; I have reported on the first flight at the scientific congress in Karlsruhe.[1] In both flights no significant change in radiation from ground level to 1,100-m altitude was found. In two balloon flights Gockel[2] also had not been able to find the expected decrease in radiation with altitude. The inference was drawn that, in addition to the γ radiation of radioactive substances in the earth's crust, there must exist still another source of penetrating radiation.

A subvention from the Kaiserliche Akademie der Wissenschaften in Vienna made it possible this year for me to carry out a sequence of seven further balloon flights, whereby more extensive and, in several respects more extended observational material was obtained.

To observe the penetrating radiation two Wulf radiation apparatuses of 3-mm wall thickness served in the first line. These were closed in a completely airtight way and were also made to withstand all the pressure changes occurring on balloon flights.

Apparatus 1 had an ionization chamber of 2,039 cm^3; the capacitance was 1.597 cm. Apparatus 2 had a volume of 2,970 cm^3; the capacitance was 1.097 cm.

An ionization strength of $q = 1.56$ ions cm^{-3} sec^{-1} thus corresponded to a charge leakage of 1 v hr^{-1} in apparatus 1. In apparatus 2 such a charge leakage corresponded to $q = 0.7355$ ions cm^{-3} sec^{-1}.

On the suggestion of Dr. Bergwitz, both apparatuses were galvanized electrochemically on the inside to achieve the greatest possible reduction of the proper radiation of the receptacle walls. After this treatment apparatus 1 showed a normal ionization of about 16 ions cm^{-3} sec^{-1}; apparatus 2, about 11 ions cm^{-3} sec^{-1}. (The firm of Günther and Tegetmeyer in Braunschweig has brought about a further significant improvement: formerly the focus on the filaments resulted from a displacement of the ocular alone, which was connected with a not inconsiderable change in magnification. This caused reading differences up to 0.5 upon repeated adjustment. Günther and Tegetmeyer has now constructed, in the ocular tube, a sliding negative lens which achieves the focus on the different separations of the filaments without a noticeable change in magnification. By this means the precision of adjustment is considerably increased.)

Because the wall thickness was 3 mm in both apparatuses 1 and 2, only γ radiation could affect the inside.

For the purpose of studying the behavior of β radiation at the same time, I utilized yet a third apparatus. This, not built airtight, was a common Wulf two-filament electrometer; inverted on it was a cylindrical ionization receptacle of 16.7-l volume, made out of the thinnest sheet zinc commercially available (wall thickness 0.188 mm). Thus, soft radiation, such as β radiation, could still be to some extent effective. A zinc peg set on the filament bearer of the electrometer at a height of 20 cm served as a dissipation body [Zerstreuungskörper]. The capacitance was 6.57 cm.

For the thick-walled Wulf radiation apparatuses 1 and 2, the determination of the insulation leakage was made in the usual way, by means of a lowered shield pipe. The hourly charge leakage amounted to 0.2 v for apparatus 1 and 0.7 v for apparatus 2. An insulation discharge has never occurred, even in the most humid weather.

For the thin-walled apparatus, the ionization chamber must be taken off to determine the insulation leakage. The voltage leakage of the electrometer alone is corrected for the capacitance of the completely assembled apparatus and is then subtracted from the observed total dissipation.

According to all observers of the penetrating radiation on towers, a steady decrease in radiation had been confirmed, while Gockel and I had not been able to find, with certainty, such a decrease in free balloons. Thus, there was a need for longer flights at low altitudes to carry out measurements and thereby obtain reliable mean values. Parallel observations with the thin-walled apparatus 3 should show whether the softer radiation behaves in the same way as the γ rays.

Further attention was to be given to the simultaneous variations in the radiation. Pacini[3] has established certain variations in the discharge rate by means of parallel observations on two Wulf radiation apparatuses at hourly read-off intervals. These were seen on land as well as above the sea. The cause of the variations, therefore, lies manifestly outside the apparatus, in the radiation itself. It was now very important to establish whether such simultaneous variations in the radiation are also observable in a balloon in several apparatuses. Because these measurements are most irreproachably accomplished by means of lengthy flights at the same altitude, I have carried out the predominant part of the observations on night flights.

The last and most important point of my investigation was the measurement of the radiation at the greatest possible altitudes. On the six flights undertaken with Vienna as starting point, the small carrying capacity of the gas used, as well as meteorological chance, did not permit measurement at very high altitudes; but I did succeed in taking measurements up to 5,350-m altitude on an ascent begun at Aussig on the Elbe.

Before each flight, control observations were made for several hours with all three apparatuses. Here the apparatuses were fastened, exactly as on a flight itself, by means of brackets to the balloon basket. The observations before the ascents were carried out at the clubhouse of the Austrian Aeroclub, on a flat lawn in the Prater in Vienna. L. V. King[4] has expressed the conjecture that the balloon observations could be disturbed by the proximity of the possibly radioactive ballast sand, but I have never found a heightening of the radiation in the immediate vicinity of greater supplies of ballast sand.

Inside the ionization chamber in apparatuses 1 and 2 the same air density as at the point of departure is steadily maintained (an average 750 mm). In the thin-walled apparatus 3 the same air pressure is maintained as in the surroundings. Thus, especially for observations at higher altitudes, a correction to the directly observed values is necessary. Under the assumption that the ionization produced by the penetrating radiation is proportional to the pressure, the actually observed radiation values are multiplied by the ratio of the normal pressure 750 mm to the mean pressure prevailing during the observation interval. It must be noted that this correction conceals a definite uncertainty within itself, because it is thereby tacitly assumed that the residual radiation of the chamber walls changes proportionately with air pressure in the chamber, which need not be the case if this residual radiation has only as weak a penetrating power as that of α rays. Therefore, especially for measurements at greater altitudes, the uncorrected values obtained from apparatus 3 will also be discussed.

In the following tables q_1, q_2, q_3 denote the penetrating radiation observed in apparatus 1, 2, 3 in ions cm^{-3} sec^{-1}. The elementary quantity of charge is hereby accepted to be $e = 4.65 \times 10^{-10}$ ESU.

The mean altitude of the balloon during the observation intervals in question (as a rule about 1 hr) was inferred by a graphical procedure from the barograph curve. From the height above sea level of the actual place flown above, a mean value was then calculated for the relative height. The hours of the day stated in the tables are given in the 24-hr classification.

The detailed report giving all the observations made in the balloon has been presented to the Kaiserliche Akademie der Wissenschaften in Vienna and appears in their Sitzungsberichte. Reported here in detail are only the two most important flights; for the other flights quotations of the mean values are given.

Flight 1

Flight 1 took place during the very strong partial solar eclipse in lower Austria on April 17, 1912. Observations were made from 11 to 1 o'clock around midday at 1,900–2,750-m absolute altitude over an almost completely solid cumulus bank. There was no diminution in the penetrating radiation with increased darkening. For example, apparatus 2 showed before the ascent an ionization of 10.7 ions, at 1,700-m relative altitude 11.1 ions, from 1,700 to 2,100 m during the first phase of the eclipse 14.4 ions, later at about 50% totality 15.1 ions. The balloon was compelled to descend because of the cooling of the gases, and further measurements were not possible.

A magnification of the radiation was found, then, at about 2,000 m. Because no influence of the eclipse on the radiation was observed, I conclude that, if a part of the radiation should be of cosmic origin, it can hardly come from the sun, at least if one pictures a γ ray moving in a straight line directly to the eye. This interpretation is further strengthened by the fact that, on the later balloon flights, I never found a pronounced difference in radiation between day and night.

Flight 2

Table 3.1 gives an overview of all the observations of the second flight. Ascent began at 11:00 P.M. on April 26, 1912. Through careful guidance we were able to hold the balloon at about the same height for almost 6 hr, which was important for establishing the variations in the radiation. The flight headed south over the Prater, then was becalmed, and finally turned back north and passed over Florisdorf, Stockerau, and Guntersdorf toward Mähren. We landed about 10:30 A.M., April 27, after reaching a maximum altitude of 2,100 m at Pausram, south of Brünn. The sky was cloudless during the entire $11\frac{1}{2}$-hr flight.

First of all, it follows from table 3.1 that at low altitudes above the ground the radiation is, in fact, lower than on the ground itself. Let us construct the mean values (q values in ions cm^{-3} sec^{-1}):

Apparatus 1	Apparatus 2	Apparatus 3
	Before the ascent	
$q_1 = 17.5$	$q_2 = 11.55$	$q_3 = q_{3\,corr} = 20.2$
	At 140–190 m over the ground	
$q_1 = 14.9$	$q_2 = 9.8$	$q_3 = 18.2\, q_{3\,corr} = 18.7$

The ion differences are correspondingly 2.6, 1.8, 2.0, and 1.5. On the average the difference is about 2 ions. This decrease in the radiation by about 2 ions manifestly originates in the absorption of the γ rays of radioactive substances of the earth's crust by the air. According to King[5] the γ radiation is already weakened to 24% at 160 m. Therefore, the difference of 2 ions corresponds to about three-quarters of the total ionizing strength produced by the radioactive γ radiation of the earth's crust. *The total γ radiation of the earth's crust may thus be set at about 3 ions cm^{-3} sec^{-1} in a zinc container.* (At the low height of 160 m a component of the radiation coming from above can be seen from the decrease in the induction capacity and the eventually possible increase.)

Also, the variations in the radiation become very strikingly evident in table 3.1. First, we already see from the observations on the ground that the measured values of several apparatuses placed next to one another are not exactly parallel. This variation is caused by read-off errors in the first approximation. The hourly decrease of the filaments amounted, on the average, to 6 scale divisions on apparatus 1, 9 scale divisions on apparatus 2, and 15 scale divisions on apparatus 3. If we accept as the possible error for each single reading 0.1 scale divisions, in the extreme case 0.2 scale divisions, then an error of 0.4, at worst 0.8, divisions can arise from the initial and final readings of both filaments. This gives as possible errors, for the read-off in hourly intervals: apparatus 1—7%;

Table 3.1 Observed radiation, flight 2 (q values in ions cm^{-3} sec^{-1})
Balloon: "Excelsior" (1600 cbm coal gas). Pilot: Captain W. Hoffory. Observer: V. F. Hess.

		Mean altitude		Observed radiation			
		Absolute	Relative	Apparatus 1	Apparatus 2	Apparatus 3	
No.	Time	m	m	q_1	q_2	q_3	q_3 (corrected)
1[a]	16h40–17h40	156	0	15.6	11.5	—	—
2[a]	17h40–18h40	156	0	18.7	11.8	21.0	21.0
3[a]	18h40–21h00	156	0	17.8	11.6	19.5	19.5
4[a]	21h30–22h30	156	0	17.8	11.3	20.0	20.0
5	23h26–0h26	300	140	14.4	9.6	19.4	19.8
6	0h26–1h26	350	190	16.2	9.9	17.4	17.9
7	1h26–2h26	300	140	14.4	10.1	17.7	18.1
8	2h26–3h32	330	160	15.0	9.6	18.2	18.7
9	3h32–4h32	320	150	14.4	9.8	18.5	19.0
10	4h32–5h35	300	70	17.2	13.2	20.6	21.0
11	5h35–6h35	540	240	17.8	11.8	19.6	20.8
12	6h35–7h35	1,050	800	17.6	10.0	18.1	20.3
13	7h35–8h35	1,400	1,200	12.2	8.8	17.3	20.3
14	8h35–9h35	1,800	1,600	17.5	10.9	17.3	21.3

a. Before the ascent at the clubhouse, Vienna.

apparatus 2—4.4%; and apparatus 3—2.7% (at worst 14%, 9%, 5%, respectively).

We will be able to claim with certainty as real variations in the radiation only those changes which are indicated by all three apparatuses simultaneously and which exceed in magnitude the possible read-off errors.

After studying the table we notice variations of such a kind, for example, at observation 13 the value $q_1 = 12.2$ in apparatus 1 remains about 40% behind the mean. At the same time apparatuses 2 and 3 also register a decrease of 2 ions. Therefore, a variation in the radiation may be taken to lie here, although the decrease of about 5 ions indicated by apparatus 1 may be partly incorrect because of read-off errors. That apparatus 3 shows no greater decrease may be explained by the perhaps somewhat irregular behavior of the additional radiation in the heights, which will be discussed later.

Without doubt, a real oscillation of the radiation is also to be noted at observation 10 between 4:30 and 5:30 A.M. Simultaneously, in all three apparatuses, an increase in the radiation is found of 2.8, 3.4, and 2.0 ions, respectively. In no way is this change connected with the closeness of the earth's surface at 70 m, because such a connection would oppose all the experience I have acquired on the later flights.

Because the variations were accompanied by no kind of meteorological change, one can hardly attribute them to a change in the distribution of radioactive substances in the atmosphere.

At higher balloon altitudes (observations 11–14), a weak increase in the radiation was generally observed. With the exclusion of observation 13, influenced by the variation in the radiation, $q_1 = 17.6$, $q_2 = 10.5$, and $q_3 = 20.8$ were the mean values produced at altitudes 800–1,600 m. The values are about as large as those which had been found before the ascent.

Flight 3

The flight 3 ascent began on May 20, 1912, at 10:10 P.M. The balloon flew rapidly over Korneuburg, Neuhäusel on the Thaya, toward the NNW. Toward morning it became partly cloudy. At 4:00 A.M. we reached Kuttenberg in Bohemia. Because of a ballast shortage we landed at 5:30 A.M. after attaining a maximum altitude of 1,020 m at Sadowa-Dohalice, north of Königgrätz.

The mean values of the radiation, with q values in ions cm^{-3} sec^{-1}, were

Apparatus 1	Apparatus 2	Apparatus 3
Before the ascent in Vienna		
$q_1 = 16.9$	$q_2 = 11.4$	$q_3 = 19.7$
During the night, 10:30–2:30, at 150–340-m altitude		
$q_1 = 16.9$	$q_2 = 11.1$	$q_{3\,corr} = 19.2$
2:30–4:30 at about 500-m relative altitude		
$q_1 = 14.7$	$q_2 = 9.6$	$q_{3\,corr} = 17.6$

The observations at 150–340 m deviated very little from those on the ground. Probably the normal decrease in γ radiation from the earth's surface was masked by a chance enhancement of the residual radiation. In contrast, the decrease in the measurements at 500 m stands out clearly: the ion difference compared to the values before the ascent amounted to 2.2, 1.8, and 2.1 ions at the three apparatuses, respectively. Between 11:30 P.M. and 12:30 A.M. a radiation variation, indeed an enhancement, was found in the three apparatuses, of 2.8, 2.0, and 1.0 ions respectively.

Flight 4

The fourth ascent took place on June 3 at 9:45 P.M. in a 2,200-cbm balloon, with which great altitudes could be obtained. Unfortunately, a closely approaching thunderstorm compelled us to land at 1:30 A.M. The maximum altitude was 1,900 m (absolute). I made observations only between 10:30 and 12:30 P.M. at altitudes of 800 to 1,100 m above the ground. The mean values attained at this altitude were $q_1 = 15.5$, $q_2 = 11.2$, and $q_{3\,\mathrm{corr}} = 21.8$; while before the ascent the values were $q_1 = 15.5$, $q_2 = 11.7$, and $q_{3\,\mathrm{corr}} = 21.3$. The values at 800–1,100 m depart very little from the first values.

Flight 5

Flight 5 took place on June 19, beginning at 5:00 P.M. Because I ascended alone and thus had to attend to the guidance of the balloon, I took along just one radiation apparatus. At 850–950-m relative altitude I found $q_2 = 9.8$ up to $q_2 = 10.7$ ions, while I had observed $q_2 = 12.3$ to 14.5 ions in the 2 hr before ascent. The mean value at 900-m altitude was about 3 ions smaller than on the ground.

From the results of flights 4 and 5 we will conclude that *the verifiable decrease in radiation up to 1,000 m above the ground may be such that it can be hidden under circumstances like those of flight 4 through a chance increase in the residual radiation.*

Flight 6

On flight 6 the radiation in the immediate vicinity of the earth was to be investigated again. I ascended on June 28 at 11:30 P.M. and was able to hold the balloon at altitudes from 280 to 360 m for 5 hr. I had taken along only the two thick-walled apparatuses, which gave completely parallel readings all night long. Between 1:30 A.M. and 2:30 A.M. an enhancement in the radiation by 1.9 and, respectively, 1.6 ions was noted without any altitude change in the balloon or any kind of meteorological change to note. At 360 m above the ground the radiation decrease relative to the ground values was 2.1 ions in apparatus 1 and 2.4 ions in apparatus 2. These results confirm the conclusion drawn earlier.

Flight 7

We ascended at 6:12 A.M. on August 7, 1912, from Aussig on the Elbe. We flew over the Saxony border at Peterswalde,

Table 3.2 Observed radiation, flight 7 (q values in ions cm^{-3} sec^{-1}).
Balloon: "Böhmen" (1680 cbm hydrogen). Pilot: Captain W. Hoffory Meteorological observer: E. Wolf.
Electrical observer: V. F. Hess.

No.	Time	Mean altitude: Absolute m	Mean altitude: Relative m	Observed radiation: Apparatus 1 q_1	Apparatus 2 q_2	Apparatus 3 q_3	Apparatus 3 $q_{3\,\mathrm{corr}}$	Temp.	Rel. humidity %
1[a]	15h15–16h15	156	0	17.3	12.9	—	—	—	—
2[a]	16h15–17h15	156	0	15.9	11.0	18.4	18.4	—	—
3[a]	17h15–18h15	156	0	15.8	11.2	17.5	17.5	—	—
4	6h45–7h45	1,700	1,400	15.8	14.4	21.1	25.3	+6.4°	60
5	7h45–8h45	2,750	2,500	17.3	12.3	22.5	31.2	+1.4°	41
6	8h45–9h45	3,850	3,600	19.8	16.5	21.8	35.2	−6.8°	64
7	9h45–10h45	4,800	4,700	40.7	31.8	—	—	−9.8°	40
—	—	(4,400–5,350)		—	—	—	—	—	—
8	10h45–11h15	4,400	4,200	28.1	22.7	—	—	—	—
9	11h15–11h45	1,300	1,200	(9.7)	11.5	—	—	—	—
10	11h45–12h10	250	150	11.9	10.7	—	—	+16°	68
11[b]	12h25–13h12	140	0	15.0	11.6	—	—	—	—

a. In Vienna, $1\frac{1}{2}$ days before the ascent.
b. After landing in Pieskow, Brandenburg.

Struppen bei Pirna, Bischofswerda, and Kottbus. In the vicinity of the Schwielochsee we reached 5,350-m altitude. At 12:15 P.M. we landed at Pieskow, 50 km east of Berlin. Unfortunately, no observations at the take-off site could be made before the flight. However, after the landing, under the yet-to-be-dismantled balloon, measurements were taken in order to see whether the balloon, which had just descended from 5,000 m, had become covered with radioactive inductions and would thus be emitting radiations itself. However, as is apparent from table 3.2 (observation 11), no trace of an enhancement of radiation under the landed balloon was observed.

On this flight the weather was not completely clear. A barometric depression approaching from the west made itself noticeable through the onset of cloudiness. Yet let it be expressly noted that we never found ourselves in a cloud, indeed not once in the vicinity of a cloud; because, at the time when the cumulus clouds appeared scattered in isolated balls over the whole horizon, we were already at altitudes above 4,000 m. When we traveled at the maximum altitude, there was a thin cloud layer, still much higher, above us. Its underside must have been at least 6,000-m altitude. The sun shimmered through only very weakly.

At 1,400–2,500-m mean altitude the radiation was approximately as strong as it is usually found to be on the ground. Then, however, *a clearly noticeable rise in radiation began in both thick-walled apparatuses, 1 and 2, with increasing altitude*; at 3,600 m above the ground the values already were 4–5 ions higher than on the ground.

In the thin-walled apparatus, the rise in radiation is apparent at even lower altitudes. Because of the uncertainty previously discussed, arising from the correction of the values of this apparatus to normal pressure, one may view this conclusion as not completely certain.[6] Moreover, the qualitative rise in the uncorrected values of q_3 should also be recognized. The readings for apparatus 3 found an unintended end at 10:45 A.M.; an unexpected shock just before the reading at maximum height caused the ionization cylinder to loosen itself, and when it touched the center post the apparatus discharged itself.

For the two γ-ray apparatuses the values at the maximum altitude are from 20 to 24 ions higher than on the ground. On the descent the values $q_1 = 28.1$ and $q_2 = 22.7$, which are also very high, were still found at 4,400 m average absolute height. These extend beyond the normal values by about 12, respectively 11, ions. On the ensuing very rapid descent (2 m $\sec^{-1}$) the very much lower value 9.7 ions was measured in apparatus 1, while in apparatus 2 the normal value 11.5 ions was registered. I believe it possible that in apparatus 1, which had very thick filaments, a definite stiffness of the filaments made itself disturbingly noticeable. As already noted, for both apparatuses the results obtained after landing, under the still-filled balloon, were entirely normal.

In order to obtain an overview of the change in the penetrating radiation with height, as represented in the mean values, I have assembled in table 3.3 all eighty-eight of the radiation values I observed on the balloon, arranged according to the corresponding altitude range. Because, for each altitude range, mean values are built up from several single values, obtained under different conditions and perhaps influenced by the already noted variations in time, one may not expect to obtain an entirely exact picture of the change in radiation with increasing altitude.

From table 3.3 we notice that directly above the earth the total radiation decreases a little. In mean values these decreases amount to 0.8 to 1.4 ions. Because, however, for the single flights, a decrease up to 3 ions has several times been found, for many measurements over 2 ions, we will claim something like 3 ions as the maximum value of the decrease. This decrease reaches to approximately 1,000 m above the ground. As mentioned previously, it manifestly originates in the absorption of γ rays which emanate from the earth's surface. I conclude that *the γ rays from the earth's surface and the uppermost lower layers of the earth excite in zinc containers an ionization of about 3 ions cm^{-3} sec^{-1}.*

Already at altitudes of 2,000 m, a marked increase in radiation appears. In both thick-walled apparatuses, at 3,000–4,000 m the increase reaches the amount of 4 ions, at 4,000–5,200 m the amount of 16 to 18 ions. For the thin-walled apparatus 3 the increase appears much earlier and more strongly, if one reduces the values to normal atmospheric pressure.

What is the cause of this increase in penetrating radiation with altitude, which has been observed several times and simultaneously in all three apparatuses? If one restricts oneself to the point of view that only the well-known radioactive substances in the earth's crust and in the atmosphere emit a radiation with the character of γ rays and produce an ionization in a closed container, then great difficulties face any explanation.

According to the direct determination of the absorption coefficient of γ rays in air by me[7] and by Chadwick,[8] the absorption of the radiation leaving the earth's surface is quite strong, so that at 500 m above the ground hardly 10% of the radiation can still be detected. As mentioned, I have also succeeded in proving this decrease experimentally. It follows, of course, that the radioactive substances in the earth's surface play a much less predominant role in the total radiation than many authors have believed: the actual portion was determined to be 3 ions cm^{-3} sec^{-1}.

At high altitudes the destruction products of the emanations still remain as ionizers with γ rays. Because of their short lifetimes, the thorium and actinium emanations, as well as their decay products, will not explain the observed ionization. Only the radium emanation with a half-life of almost 4 days will be able to be borne aloft by air currents to great heights.

Table 3.3 Mean values (q values in ions cm^{-3} sec^{-1}).

Mean altitude over the earth (m)	Observed radiation			
	Apparatus 1	Apparatus 2	Apparatus 3	
	q_1	q_2	$q_{3\,corr}$	q_3
0	16.3(18[a])	11.8(20)	19.6(9)	19.7(9)
to 200	15.4(13)	11.1(12)	19.1(8)	18.5(8)
200–500	15.5(6)	10.4(6)	18.8(5)	17.7(5)
500–1,000	15.6(3)	10.3(4)	20.8(2)	18.5(2)
1,000–2,000	15.9(7)	12.1(8)	22.2(4)	18.7(4)
2,000–3,000	17.3(1)	13.3(1)	31.2(1)	22.5(1)
3,000–4,000	19.8(1)	16.5(1)	35.2(1)	21.8(1)
4,000–5,200	34.4(2)	27.2(1)	—	—

a. The parenthetical numbers denote the number of observations from which the corresponding mean values were constructed.

Nevertheless, in general, the concentration of the emanations and, with that, the proportion of RaC in the air will soon decrease with height. An increase in the radiation with altitude could only arise from a chance accumulation of RaC of a purely local character: it would be conceivable, for example, that in stable layers with temperature inversion, or in cumulus clouds, or in fogs, such accumulations occur, because it is known that the RaC atoms frequently function as condensation nuclei; but a uniform increase with altitude in the penetrating radiation, as seen in my observations, can probably not be explained in this way. Also, I have observed no enhancement in radiation on flights 2 and 6, in which the balloon traveled near the earth for hours on a stable layer with temperature inversion, although in the vicinity of the earth the proportion of RaC in the air must be greater. At a height of 5,000 m the RaC proportion generally will not suffice to effect as great an enhancement of the radiation as I found.

Also, the variations in the radiation found by Pacini[9] and Gockel[10] on the land and sea and by me in a balloon above the ground cause great difficulties for an explanation of the penetrating radiation based exclusively on radioactive theory. I have observed such variations repeatedly in the middle of the night in a quiescent atmosphere. Because of the lack of any meteorological alteration, there is no reason for attributing the variations to changes in the distribution of radioactive substances in the atmosphere.

The results of the preceding observations may most easily be explained on the assumption that a radiation of very great penetrating power impinges on our atmosphere from above, and still evokes in the lowest layers a part of the ionization observed in closed vessels. The intensity of this radiation seems to be subject to oscillations in time which are still recognizable in hour-long read-off intervals. Because I did not find any decrease in radiation either at night or during a solar eclipse, one can hardly view the sun as the cause of this hypothetical radiation, at least as long as one thinks only of a direct γ ray with straight line propagation.

That the increase in the radiation first becomes noticeably strong at 3,000 m is not so very surprising: in the first 1,000 m the decrease in the γ radiation from the earth's surface is predominant, and thereafter the decrease in induction capacity, which is first clearly noticeable above 3,000 m, appears. In that case the absorption of the radiation coming from above develops according to an exponential curve: the increase in the radiation becomes steeper and steeper at the higher altitudes.

Finally, consider the experiment that Wright, Simpson, McLennon, and Wulf, among others, have carried out to demonstrate that the penetrating radiation on the earth's surface originates almost exclusively from radioactive substances in the earth, and not from the atmosphere. Universally, Wright et al. found that over water or ice, even at a short distance from land, the radiation is 4–6 ions smaller. They found, therefore, a greater difference than I have established between the values on the earth's surface and at a few hundred meters altitude (2–3 ions). In best agreement with my results is an observation by Professor Wulf, who has found on the peak of the Eiffel Tower in Paris a radiation about 2–3 ions smaller than on the ground. It is well-known that γ rays excite secondary β and γ rays on impact with matter. The secondary radiation excited in hard bodies is, however, much stronger than the secondary radiation produced in a sheet of water.[11] The radiation coming from above will surely be able to excite secondary radiation on the earth's surface. Over land, however, the secondary radiation produced will be greater than that over water. By this means the difference in the evaluation of the earth's radiation from balloon experiments over land and from experiments over water can be explained.

The hitherto existing investigations have shown that the penetrating radiation observed in closed vessels has a very complex origin. A part of the radiation originates in the earth's surface and in the uppermost layers of the earth and is altered relatively little. A second portion, influenced by meteorological factors, originates in radioactive substances in the air, most significantly RaC. My balloon observations seem to prove that there exists still a third component in the total radiation. This component increases with altitude and also produces noteworthy intensity variations on the ground. The greatest attention will have to be paid to this component in any further researches.

1. *Physikalishe Zeitschrift 12*, 998–1001 (1911); *Wien. Sitz-Ber. 120*, 1575–85 (1911).
2. *Physikalishe Zeitschrift 12*, 595–597 (1911).
3. *Le radium 8*, 307–312 (1911).
4. *Philosophical Magazine* (6) *23*, 242 (1912).
5. Ibid. (6) *23*, 247 (1912).
6. I have disregarded applying a correction for the variation in absolute temperature; on flights 1–6 this change never reached a significant amount. Also, on the just-discussed measurements this correction was not important: the temperature in the measuring chamber of the apparatus, because of the solar radiation, could not be nearly as low as the exterior temperature measured with an aspiration thermometer.
7. *Wien. Sitz.-Ber. 120*, 1205–1212 (1911).
8. *Le radium 9*, 200–202 (1912).
9. Ibid. *8*, 307–312 (1911).
10. *Jahrbuch der Radium u. Elektron. 9*, 1–15 (1912).
11. A. Brommer, *Wien. Sitz.-Ber. 121*, 583 (1912).

4. The Possibilities of Large Telescopes

George Ellery Hale

(*Harper's Magazine 156*, 639–646 [1928])

Copyright © 1928 by Harper's Magazine

Probably the greatest scientific entrepreneur of the twentieth century was the solar astronomer George Ellery Hale. He was one of the first of a long series of astronomers who have become publicists for a popular form of science that appeals to private industrialists, governments, and foundations. In the 1890s Hale persuaded a wealthy Chicagoan, C. T. Yerkes, to provide over $349,000 to build the world's largest refractor—a 40-in—for the University of Chicago. Early in the 1900s Hale migrated to southern California, and by 1904 he had founded a solar observatory on Mount Wilson with funds from the Carnegie Institution of Washington, a private organization founded by Andrew Carnegie and devoted to the advancement of scientific research. By 1908 Hale had installed a 60-in reflector superseding the 40-in refractor as the largest telescope. After persuading John D. Hooker, a Los Angeles businessman, to provide $45,000 for a 100-in reflector, Hale obtained $600,000 from the Carnegie Institution to install the instrument on Mount Wilson.

From 1917 to 1948 the 100-in reflector was the world's largest telescope, but in 1928 Hale wrote this stirring call for a 200-in telescope. Early in 1928 Hale asked the editor of *Harper's* to send a copy of his article to the Rockefeller Foundation, where it had a catalytic effect in bringing forth six million dollars for building a 200-in telescope. In this case, the grant was given to the California Institute of Technology, which was to work closely with the Carnegie Institution and the existing Mount Wilson Observatory. Under active construction at the time of Hale's death in 1938, the Palomar 200-in telescope was finally finished and dedicated in his honor ten years later. For more than three decades this instrument has reigned as the unrivaled leader in optical astronomy.

Like buried treasures, the outposts of the universe have beckoned to the adventurous from immemorial times. Princes and potentates, political or industrial, equally with men of science, have felt the lure of the uncharted seas of space, and through their provision of instrumental means the sphere of exploration has rapidly widened. If the cost of gathering celestial treasure exceeds that of searching for the buried chests of a Morgan or a Flint, the expectation of rich return is surely greater and the route not less attractive. Long before the advent of the telescope, pharaohs and sultans, princes and caliphs built larger and larger observatories, one of them said to be comparable in height with the vaults of Santa Sophia. In later times kings of Spain and of France, of Denmark and of England took their turn, and more recently the initiative seems to have passed chiefly to American leaders of industry. Each expedition into remoter space has made new discoveries and brought back permanent additions to our knowledge of the heavens. The latest explorers have worked beyond the boundaries of the Milky Way in the realm of spiral "island universes," the first of which lies a million light-years from the earth while the farthest is immeasurably remote. As yet we can barely discern a few of the countless suns in the nearest of these spiral systems and begin to trace their resemblance with the stars in the coils of the Milky Way. While much progress has been made, the greatest possibilities still lie in the future.

I have had more than one chance to appreciate the enthusiasm of the layman for celestial exploration. Learning in August, 1892, that two discs of optical glass, large enough for a forty-inch telescope, were obtainable through Alvan Clark, I informed President Harper of the University of Chicago, and we jointly presented the opportunity to Mr. Charles T. Yerkes. He said he had dreamed since boyhood of the possibility of surpassing all existing telescopes, and at once authorized us to telegraph Clark to come and sign a contract for the lens. Later he provided for the telescope mounting and ultimately for the building of the Yerkes Observatory at Lake Geneva, Wisconsin.

In 1906 Mr. John D. Hooker of Los Angeles, a business man interested in astronomy, agreed to meet the cost of making the optical parts for an 84-inch reflecting telescope in the shops of the Mount Wilson Observatory in Pasadena, where a 60-inch mirror had recently been figured by Ritchey. Before the glass could be ordered he increased his gift to provide for a still larger mirror. Half a million dollars was still needed for the mounting and observatory building, and Mr. Carnegie, who was greatly taken with the project during his visit to the Observatory in 1910, wanted the Carnegie Institution of Washington to supply it. The entire income of the Institution was required, however, to provide for the annual expenses of its ten departments of research, of which the Observatory is one. Nearly a year later I was on my way to Egypt. At Ventimiglia, on the Italian frontier, I bought a local newspaper, in which an American cable had caught my eye. Mr. Andrew Carnegie, by a gift of ten million dollars, had doubled the endowment of the Carnegie Institution of Washington. A paragraph in his letter to the Trustees especially appealed to me: "I hope the work at Mount Wilson will be vigorously pushed, because I am so anxious to hear the expected results from it. I should like to be satisfied before I depart, that we are going to repay to the old land some part of the debt we owe them by revealing more clearly than ever to them the new heavens."

I hope that the 100-inch Hooker telescope, thus named at Mr. Carnegie's special request, has justified his expectations. Its results, described in part in *The New Heavens*, *The Depths of the Universe*, and *Beyond the Milky Way* have certainly surpassed our own forecasts. They have given us new means of determining stellar distances, a greatly clarified conception of the structure and scale of the Galaxy, the first measures of the diameter of stars, new light on the constitution of matter, new support for the Einstein theory, and scores of other advances. They have also made possible new researches beyond the boundaries of the Milky Way in the region of the spiral nebulae. Moreover, they have convinced us that a much larger telescope could be built and effectively used to extend the range of exploration farther into space. Lick, Yerkes, Hooker, and Carnegie have passed on, but the opportunity remains for some other donor to advance knowledge and to satisfy his own curiosity regarding the nature of the universe and the problems of its unexplored depths.

El Karakat, an Arabian astronomer who built a great observatory at Cairo in the twelfth century, once exclaimed to the Sultan, "How minute are our instruments in comparison with the celestial universe!" In his day the amount of light received from a star was merely that which entered the pupil of the eye, and large instruments were constructed, not with any idea of discovering new celestial objects, but in the hope of increasing the precision of measuring the positions of those already known. Galileo's telescope, which suddenly expanded the known stellar universe at the beginning of the seventeenth century, had a lens about $2\frac{1}{4}$ inches in diameter, with an area eighty times that of the pupil of the eye. This increase in light-collecting power was sufficient to reveal nearly half a million stars (over the entire heavens), as compared with the few thousands previously within range. The 100-inch mirror of the Hooker telescope, which collects about 160,000 times as much light as the eye, is capable of recording photographically more than a thousand million stars.

While the gain since Galileo's time seems enormous, the possibilities go far beyond. Starlight is falling on every square mile of the earth's surface, and the best we can do at present is to gather up and concentrate the rays that strike an area 100 inches in diameter. From an engineering standpoint our telescopes are small affairs in comparison with modern battleships and bridges. There has been no such increase in size since Lord Rosse's six-foot reflector, completed in 1845, as engineering advances would permit, though advantage has been taken of the possible gain in precision of workmanship. The time thus seems to be ripe for an examination of present opportunities, which must be considered in the light of recent experience.

II

I have never liked to predict the specific possibilities of large telescopes, but the present circumstances are so different from those of the past that less caution seems necessary. The astronomer's greatest obstacle is the turbulence of the earth's atmosphere, which envelops us like an immense ocean, agitated to its very depths. The crystal-clear nights of frosty winter, when celestial objects seem so bright, are usually the very worst for observation. Watch the excessive twinkling of the stars, and you will appreciate why this is true. In a perfectly quiet and homogeneous atmosphere there would be no twinkling, and star images would remain sharp and distinct even when greatly magnified. Mixed air of varying density means irregular refraction, which causes twinkling to the eye and boiling images, blurred and confused, in the telescope. Under such conditions a great telescope may be useless.

This is why Newton wrote in his *Opticks*:

> If the Theory of making Telescopes could at length be fully brought into practice, yet there would be certain Bounds beyond which Telescopes could not perform. For the Air through which we look upon the Stars, is in a perpetual Tremor; as may be seen by the tremulous Motion of Shadows cast from high Towers, and by the twinkling of the fix'd stars. The only remedy is a most serene and quiet Air, such as may perhaps be found on the tops of the highest Mountains above the grosser Clouds.

Even at the best of sites, in a climate marked by long periods of great tranquility, unbroken by storms, the atmosphere remains the chief obstacle. For this reason we could not be sure how well the 60-inch and 100-inch reflecting telescopes would work on Mount Wilson until we had rigorously tested them. Large lenses or mirrors, uniting in a single image rays which have traveled through widely separated paths, are more sensitive than small ones to atmospheric tremor. So it has always been a lottery, as we frankly told the donors of the instruments, whether the next increase in size might not fail to bring the advantages we sought.

Fortunately we have found, after several years of constant use, that on all good nights the gain of the 100-inch Hooker telescope over the 60-inch is fully in proportion to its greater aperture. The large mirror receives and concentrates in a sharply defined image nearly three times as much light as the smaller one, with consequent immense advantages. But the question remains whether we could now safely advance to an aperture of 200 inches, or, better still, to 25 feet.

Our affirmative opinion is based not merely upon the performance of the Hooker telescope, but also upon tests of the atmosphere made with apertures up to 20 feet. The Michelson stellar interferometer, with which Pease has succeeded in measuring the diameters of several stars, is attached to the upper end of the tube of the Hooker telescope. When its two outer mirrors are separated as far as possible, they unite in a single image beams of starlight entering in paths 20 feet apart. By comparing these images with those observed when the mirrors are 100 inches or less apart, Pease concludes that an increase of aperture to 20 feet or more would be perfectly safe. For the first time, therefore, we can make such an increase without the uncertainties that have been unavoidable in the past.

Other reasons that combine to assure the success of a larger telescope are the remarkable opportunities for new discoveries revealed by recent astronomical progress and the equally remarkable means of interpreting them afforded by recent advances in physics.

III

These new possibilities are so numerous that I must confine myself to three general examples, bearing upon the structure of the universe, the evolution of stars, and the constitution of matter. A 200-inch telescope would give us four times as much light as we now receive with the 100-inch, while a 300-inch telescope would give nine times as much. How would this help in dealing with these questions?

The first advantage that strikes one is the immense gain in penetrating power and the means thus afforded of exploring remote space. The spiral structure of nebulae beyond the Milky Way was unknown until Lord Rosse discovered it with his six-foot reflector in 1845. The Hooker telescope, greatly aided by optical and mechanical refinements and by the power of photography, can now record many thousands of these remarkable objects. Moreover, in the hands of Hubble it has shown that they are in fact "island universes," perhaps similar in structure to the Galaxy, of which our solar system is an infinitesimal part.

Our present instruments are thus powerful enough to give us this imposing picture of a universe dotted with isolated systems, some of them probably containing millions of stars brighter than our sun. It is also possible to measure the distance of the Great Nebula in Andromeda and one or two other spirals that lie about a million light-years from the earth. Much larger telescopes are needed, however, to continue the analysis of these nearest spirals, now only just begun, and to extend it to some of those at greater distances. Needless to say, the greater power of larger telescopes would also give us a far better understanding than we now possess of the structure and nature of the Galaxy, of which we still have much to learn. For example, we cannot yet say whether it shares the characteristic form of the spiral nebulae, nor do we even know with certainty whether it rotates about its center at the enormous velocity that seems equally characteristic of the "island universes." In fact, our own stellar system offers countless opportunities for productive research, as the important advances in our knowledge of the Galaxy recently made by Seares with the 60-inch Mount Wilson reflector so clearly indicate.

If our ideas of the structure of the universe are thus in a very early stage, the same may be said of our knowledge of the evolution of the stars. Recent discoveries in physics have greatly modified our conception of stellar evolution, affording a rational explanation of scores of questions formerly unanswered, but raising many new and fascinating problems. Giant stars with diameters several hundreds of times that of the sun, expanded by internal pressure to gossamer tenuity, lie near one end of our present stellar vista, with dwarfs of a density more than fifty thousand times that of water near the other. The sun, a condensing dwarf, 1.4 times as dense as water, stands on the downward slope of stellar life. The continual radiation that marks the transition from giant to dwarf is now attributed to the transformation of stellar mass into radiant energy, thus harmonizing with Einstein's views and accounting for the decrease in mass observed with advancing age. Surface temperatures ranging from about 3,000°C. in the earlier stage of stellar life to about 100,000° at its climax, and internal temperatures perhaps reaching hundreds of millions of degrees are among the incidents of stellar existence. But here again, while theory and observation have recently joined in painting a new and surprising picture of celestial progress, important differences of opinion still exist and many of these await a more powerful telescope to discriminate between them. For while theories based on modern physics have been our chief guide in recent years, the final test is that of observation, and often our present instruments are insufficient to meet the demand.

So much in brief for the questions of celestial structure and evolution, though I have had to pass over the greatest of these problems: that of determining with certainty the successive stages in the development of the spiral nebulae, a phase of evolution vastly transcending that involved in the birth, life, and decline of a particular star. I have space to add only a word regarding the role of great telescopes in the study of the constitution of matter.

The range of mass, temperature, and density in the stars and nebulae is of course incomparably greater than the physicist can match in the laboratory. It is, therefore, not surprising that some of the most fundamental problems of modern physics have been answered by an appeal to experiments performed for us in these cosmic laboratories. For example, one of the most illuminating tests of Bohr's theory of the atom has just been made at the Norman Bridge Laboratory by Bowen in a study of the characteristic spectrum of the nebulae, where the extreme tenuity of the gas permits hydrogen and nitrogen to exist in a state harmonizing with the theory but unapproachable in any vacuum tube. Similarly, Adams' observations of the companion of Sirius with the Hooker telescope confirmed Eddington's prediction that matter can exist thousands of times denser than any terrestrial substance. In fact, things have reached such a point that a far-sighted industrial leader, whose success may depend in the long run on a complete knowledge of the nature of matter and its transformations, would hardly be willing to be limited by the feeble range of terrestrial furnaces. I can easily conceive of such a man adding a great telescope to the equipment of a laboratory for industrial research if the information he needed could not be obtained from existing observatories.

IV

The development of new methods and instruments of research is one of the most effective means of advancing science. In hundreds of cases the utilization of some obvious principle, long known but completely neglected, has suddenly multiplied the possibilities of the investigator by opening new highways into previously inaccessible territory. The telescope, the microscope, and the spectroscope are perhaps the most striking illustrations of this fact, but new devices are constantly being found, and the result has been a complete transformation of the astronomical observatory.

From our present point of view the chief question is the bearing of these developments on the design of telescopes. To Galileo a telescope was a slender tube, three or four feet in length, with a convex lens at one end for an object glass, and a concave lens at the other for an eyepiece. With this "optic glass" the surprising discoveries described in the *Sidereus Nuncius* were made, which shifted the sun from its traditional position as a satellite of the earth to the center of the solar system, and greatly enlarged the scale of the universe. After his time the telescope grew longer and longer, finally reaching the ungainly form of a lens supported on a pole as much as two or three hundred feet from the eyepiece. The invention of the achromatic lens brought the telescope back to manageable dimensions and permitted the use of an equatorial mounting, equipped with driving-clock to keep the celestial object at rest in the field of view. With the improvement of optical glass the aperture steadily increased, finally reaching 36 inches in the Lick and 40 inches in the Yerkes telescope.

Meanwhile it had become clear that the reflecting telescope, designed by Newton to avoid the defects of single lenses, possessed many advantages over the refractor. Chief among these are its power of concentrating light of all colors at the same focus and the fact that the light does not pass through the mirror, but is reflected from its concave front surface. Speculum metal, a highly polished alloy of tin and copper, was used for the early reflectors, reaching a maximum size in Lord Rosse's six-foot telescope. Mirrors of glass, silvered on the front surface, were then introduced, and proved greatly superior in lightness and reflecting power. Moreover, optical glass perfect enough for lenses cannot be obtained in very large sizes, and even if it could, the loss of light by absorption in transmission through the glass would prevent its use for objectives materially exceeding that of the Yerkes telescope.

Therefore, our hopes for the future must lie in some form of reflector.

It is evident that a lens, through which the starlight passes to the eye, must be mounted in a very different way from a concave mirror, which receives the light on its surface and reflects it back to the focus. The large concave mirror lies at the bottom of the telescope tube, which is usually of light skeleton construction, open at the top. The surface of the mirror is figured to a paraboloidal form, which differs somewhat from a sphere in curvature, and has the power of concentrating the parallel rays from a star in a point at the focus. This focus is near the top of the tube, opposite the center of the mirror.

For some classes of work it is desirable to place the photographic plate, small spectroscope, or other accessory instrument at this principal focus, centrally within the tube. Some starlight is thus cut off from the large mirror, but the loss is small and is less than with other arrangements. Newton interposed a plane mirror, fixed at an angle of 45°, which reflected the light to the side of the tube, where he placed the eyepiece. Cassegrain substituted a convex mirror for Newton's plane. Supported centrally at right angles to the beam, it changes the convergence of the rays and brings them to a focus near the large mirror. An inclined plane mirror may be used to intercept them, thus bringing the secondary focus at the side of the tube, or the large mirror may be pierced with a hole, allowing the rays to come to a focus close behind it.

In a third arrangement, the rays may be sent through the hollow polar axis of the telescope to a secondary focus at a fixed point in a constant temperature laboratory. This arrangement, first suggested by Ranyard, was embodied with both the Newtonian and Cassegrain methods in the mountings of the 60-inch and 100-inch telescopes of the Mount Wilson Observatory. By these means we may obtain any desired equivalent focal length (which varies with the curvature and position of the small convex mirrors) and thus photograph celestial objects on a large or small scale, as required by the problem in hand. Furthermore, we can use to the best advantage all types of spectroscope, photometer, interferometer, thermocouple, radiometer, photo-electric cell, and the many other accessories developed in recent years.

These accessory instruments and devices have made possible most of the discoveries of modern astrophysics. The stellar spectroscope, originally merely a small laboratory instrument attached to a telescope, has grown to the dimensions of the powerful fixed spectrograph of 6 inches aperture and 15 feet in length, recently used with splendid success by Adams in photographing the spectra of some of the brightest stars. The development of this method of high dispersion stellar spectroscopy, initiated in the early days of the Yerkes Observatory, was one of my chief incentives in endeavoring to obtain large reflecting telescopes for the Mount Wilson Observatory. The recent advances in our knowledge of the atom and the consequent complete transformation of spectroscopy from an empirical to a rational basis greatly increase the possibilities of analyzing starlight. In most of the small-scale spectra photographed with ordinary stellar spectrographs the lines are so closely crowded together that they cannot be separately measured. With a larger telescope we could push the dispersion to the point attained by Rowland in his classic studies of the solar spectrum, and thus take full advantage of the great possibilities of discovery offered us by recent advances in physics.

V

These details are important because they point directly to the type of telescope required. It is true that in some cases lenses may be used instead of convex mirrors for enlarging the image; but in our judgment the design should permit observations to be made in the principal focus of the large mirror, at a secondary focus just below the (pierced) mirror, and at another secondary focus in a fixed laboratory. It should also be possible to attach to the tube a large Michelson stellar interferometer, arranged for rotation in position angle and thus suitable for the measurement of very close double stars.

A mounting designed by Pease of the Mount Wilson Observatory meets these requirements and is worthy of careful consideration. It is large enough to carry a mirror 25 feet in diameter, collecting nine times as much light as the 100-inch Hooker telescope. It would thus enlarge our sphere of observation to three times its present diameter and increase the total number of galactic stars to three or four times that now within range.

This, of course, is a tentative design, subject to modification in the light of an exhaustive study. Of all the optical and mechanical problems involved only one presents real difficulties, but there is no reason to think that these cannot be readily surmounted. This is the manufacture of the glass for the large mirror.

Our chief difficulty in the case of the Hooker telescope was to obtain a suitable glass disc. The largest previously cast was that for the 60-inch mirror of our first large reflector. This is 8 inches thick and weighs a ton. The 100-inch disc, 13 inches thick, weighs nearly five tons. To make it three pots of glass were poured in quick succession into the mold. After a long annealing process, to prevent the internal strains that result from rapid cooling, the glass was delivered to us. Unlike the discs previously sent by the French makers, it contained sheets of bubbles, doubtless due in part to the use of the three pots of glass, while but one had sufficed before. Any considerable lack of homogeneity would result in unequal expansion or contraction under temperature changes, and experiments were, therefore, continued at the glass factory in the Forest of St. Gobain in the hope of producing a flawless disc. As they did not succeed, the disc containing the bubbles was given a spherical figure and tested optically under a wide range of temperature. Its performance convinced us that the disc could safely be given a paraboloidal figure for use in the telescope, where it has served ever since for a great variety of visual and photographic observations.

Recently, important advances have been made in the art of glass manufacture, and mirror discs much larger and better than the 100-inch can now undoubtedly be cast. Pyrex glass, so useful in the kitchen and the chemical laboratory because it is not easily cracked by heat, is also very advantageous for telescope mirrors. Observations must always be made through the widely opened shutter of the dome, at temperatures as nearly as possible the same as that of the outer air. As the temperature rises or falls the mirror must respond. The small expansion or contraction of Pyrex glass means that mirrors made of it undergo less change of figure and, therefore, give more sharply defined star images—a vitally important matter in all classes of work, especially in the study of the extremely faint stars in the spiral nebulae, for which Pease's design is especially adapted.

Dr. Arthur L. Day of the Carnegie Institution of Washington, working in association with the Corning Glass Company, has succeeded in producing glass with a higher silica content than Pyrex and, therefore, with a lower coefficient of expansion. Moreover, Dr. Elihu Thomson and Mr. Edward R. Berry of the General Electric Company have recently made discs up to 12 inches in diameter of transparent fused quartz (pure silica), which is superior to all other substances for telescope mirrors. The chief difficulty in the manufacture of fused quartz has been the elimination of bubbles. These would do no harm whatever within a large telescope mirror, provided its upper surface were freed from them by a method proposed by Dr. Thomson. In fact, the presence of a great number of bubbles would be a distinct advantage in reducing the weight of the disc. As there is every reason to believe that a suitable Pyrex or quartz disc could be successfully cast and annealed, and as the optical and engineering problems of figuring, mounting, and housing it present no serious difficulties, I believe that a 200-inch or even a 300-inch telescope could now be built and used to the great advantage of astronomy.

Limitations of space have prevented mention of many interesting matters of detail. It goes without saying that all questions relating to the optical as well as the engineering design should be thoroughly investigated by a group of competent authorities, who should also include those best qualified to deal with related problems involving the design of spectroscopes and the many other accessory instruments required. As for photographic plates, it is well known that the power of photographic telescopes could be materially increased by improving their quality, so that no effort in this direction should be spared.

Perhaps a word as to procedure may be added. The first step should be to determine by experiment how large a mirror disc, preferably of fused quartz, can be successfully cast and annealed. Meanwhile all questions as to the final design of the mounting and accessories could be settled. With the completion of the mirror disc the only uncertainty would vanish and the optical and mechanical work could begin.

5. A Rapid Coma-Free Mirror System

Bernhard V. Schmidt

TRANSLATED BY NICHOLAS U. MAYALL

(*Zentralzeitung für Optik und Mechanik 52*, 25–26[1931]
Trans. in *Publications of the Astronomical Society of the Pacific 58*, 285–290[1946])

In this article Bernhard Schmidt describes a new type of optical system characterized by a wide angular field of view and a high speed. These advantages have made it possible to survey large areas of the sky in a reasonable time, something impossible with the narrow fields of the conventional parabolic reflectors. The basic principle of the Schmidt camera is that a single concave spherical mirror has no unique axis and therefore yields equally good images at all points of its field. However, the images formed by a spherical mirror are badly distorted by spherical aberration. To overcome this distortion, Schmidt introduced a thin, nonspherical glass corrector plate at the center of curvature of the mirror. Schmidt built such a system only after he conceived an "elegant" scheme for generating the complex optical figure on the plate; he bent the corrector plate over a partial vacuum and then polished a flat surface on the stressed glass. He also first discussed the detailed theory of figuring the fourth-degree curves of the corrector plate.[1]

The advantages of this telescope for survey work were quickly appreciated at Palomar Observatory, where a 48-in Schmidt camera was erected. In 1949 the National Geographic Society and the Mount Wilson and Palomar Observatories began a program to map the entire northern sky with this instrument. The field of view of this camera is so large that a single plate covers 44 square degrees of sky area. The ambitious sky survey program is now complete, and the entire sky north of declination -33° is available on 935 pairs of plates.[2] Each pair of plates covers the same field of view, but the O plate is sensitive to the blue and the E plate to the red light. Photographic contact prints of the original plates are available, with an angular scale of 1 mm = 67.19 sec of arc, a resolution limit of 2 sec of arc, and a magnitude limit of 21.1 for the blue and 20.0 for the red. This sky atlas, made possible by Schmidt's invention, is now routinely used by most major observatories for purposes ranging from the discovery of unusual interacting galaxies to the identification of radio sources with galaxies or quasars.

1. See Ira S. Bowen, "Schmidt Cameras," in *Stars and Stellar Systems I: Telescopes:* (Chicago: University of Chicago Press, 1960), pp. 43–61.

2. Albert G. Wilson, "The National Geographic Society—Palomar Observatory Sky Survey," in *Transactions of the International Astronomical Union 8*, 335–336 (1952).

If losses of light of a mirror and of a lens system are compared with each other, then for the same aperture ratio the mirror shows a smaller loss of light than the lens system. A freshly silvered mirror reflects at least 90 per cent of the incident light, while a two-lens system transmits at most 80 per cent, and a three-lens system at most 70 per cent of the incident light. In the case of large lenses, the situation is still more unfavorable because of the stronger absorption of short wave lengths by the glass.

In large telescopes, the parabolic mirror thus would be more advantageous, in general, than a lens system, but unfortunately with large aperture ratios the usable field of view is very limited by coma. For an aperture ratio of 1 to 3, the spreading due to coma amounts to 37 seconds of arc for a field diameter of only 1 degree; moreover, the spreading due to astigmatism becomes 5 seconds of arc. Coma increases in direct proportion to the field diameter, astigmatism quadratically. As a result, astigmatism in the vicinity of the axis is negligibly small and almost pure coma is present, while at greater distances from the axis it is modified by astigmatism.

Nevertheless, a parabolic mirror of aperture ratio 1 to 8 or 1 to 10 surpasses the ordinary two-lens objective as regards image sharpness, which is due to the fact that chromatic aberration is entirely absent in the mirror. But it is a disadvantage that the light-distribution in the aberration disk of the mirror image is one-sided, for this condition can produce systematic radial displacements in measurements of such images.

But it will be shown below that even a purely spherical mirror of aperture ratio 1 to 8 or 1 to 10 is still quite usable. If the aperture stop were brought directly in front of the mirror, then there would be no advantage over the parabolic mirror, since the spherical mirror has exactly the same aberrations; besides, spherical aberration would be present. which increases the existing aberrations over the whole field of view. However, if the aperture stop is brought to the center of curvature, the spherical mirror no longer has any but longitudinal aberrations, for coma and astigmatism are zero. The image surface lies on a sphere whose radius is the focal length and which is concentric with the mirror surface, so that the image surface is turned with the convex side to the mirror.

The aberration of a spherical mirror of 1 to 8 or 1 to 10 ratio amounts to 12.5 or 6.4 seconds of arc at the paraxial image point, and the smallest possible aberrations are only one-fourth of that, 3.1 or 1.6 seconds of arc.[1] In practice, even sharper pictures can be obtained if the focus is set between these two positions. Under normal conditions these aberrations are smaller than the spreading inherent in the photographic layer. Therefore, even with the use of flat plates, the image quality at the edge of the field is better than with a parabolic mirror of corresponding aperture ratio; the star images are round everywhere, with a symmetrical light-distribution.

Moreover, if a round, flat film is curved by pressing it with a ring over a spherical surface corresponding to the image surface, which is easily possible without wrinkling, then the confusion disks are of the same size over the entire field. The same thing can be accomplished with a sharp-edged plano-convex condenser lens in front of a flat photographic plate (plane side of the lens toward the plate).

If the aperture ratio is greatly increased, however, then the spherical aberration becomes very large, since it increases with the third power of the aperture ratio. For 1 to 3, or 1 to 2, the aberrations at the paraxial image point are 240 or 800 seconds of arc. The smallest possible confusion disk has a diameter of 60 or 200 seconds of arc. With a focal length of 1 meter [39.4 in.], the paraxial disks then would be 1.2 or 4 mm [0.047 or 0.157 in.], or the smallest possible ones 0.3 or 1 mm [0.012 or 0.039 in.]. In this case, therefore, the spherical mirror no longer would be useful.

I shall now show how completely sharp images can be obtained with a spherical mirror of large aperture ratio.

In order to produce a parabolic mirror from a spherical mirror, the latter's edge must be flattened, that is, be given a greater radius of curvature. However, a concentric curved glass plate (of the same thickness everywhere) can be placed on the spherical mirror, and one of its surfaces deformed. But now the curvatures must be reversed, and its edge must be more strongly curved than its center. Also, the amount of the deformation must be about twice as great, because now the deviation results from refraction. In general, in order to obtain the same deviation by refraction as by reflection, about four times as great inclinations have to be given, but in this case, since the rays go through the glass surfaces twice, only twice as large deformations are necessary.

This plate can also be optically "sagged" to such an extent that one side becomes plane again, while the other then has a pure deformation curve. That is to say, a plane-parallel plate, instead of a zero-power meniscus, can be deformed just as well from the beginning. Almost the same effect is thus obtained optically with this correction plate as with a parabolic mirror.

A suitably shaped cover plate of this kind for a spherical mirror[2] also has the practical advantage that the silver coat of the mirror is well protected. It is a disadvantage in that, owing to the passage of light twice through the glass plate, the loss of light reaches about 20 per cent.

The correcting plate can also be placed in another position in the optical path. If it is located beyond the focal surface, then the light goes through the plate only once. The plate then obviously must have twice as much deformation as in the first case. The loss of light is then only 10 per cent.

If the correcting plate is now brought to the center of curvature of the mirror, then there results the same relationships as before in the case of the spherical mirror with aperture stop in the center of curvature, but with the difference that now the spherical aberration is abolished, even over the whole field. Thus it is possible to use aperture ratios of 1 to 3 or 1 to 2, and to obtain freedom from coma, astigmatism, and spherical aberration.

If the inclination of the incident rays is very large, then the correcting plate is projected as an ellipse, and the deformation is not projected on the correct places of the mirror, so that the correction is variable and even introduces an overcorrection in the radial direction.

However, large inclinations do not need to be considered at all, since the photographic plate soon would become greater than the clear aperture. In practice, photographic plates greater than one-fourth to one-third the aperture can hardly be used, the inclination aberrations then being negligibly small.

The case is somewhat different for the chromatic aberrations of the correcting plate. In order to keep these as small as possible, the correcting plate is so shaped that the central part acts like a weak condensing lens, and the outer parts have a divergent effect. If the neutral zone is placed at 0.866 of the diameter, then the chromatic aberration is a minimum. If the point of inflection of the curve is at 0.707, then the thickness of the edge is equal to the central thickness. The remaining difference in thickness between the thickest and thinnest parts of the plate is very small, only several hundredths of a millimeter, so that a disturbing color effect does not occur; in any case, the effect usually is much smaller than the secondary spectrum of a corresponding objective.

This chromatism is identical with the so-called "chromatic difference of the spherical aberration."

If the mirror has the same diameter as the correcting plate, then the incident cylinder of rays for outer images falls eccentrically on the mirror, and there is a part of it left out, so that the outer portions of the plate obtain somewhat less light. If this is to be avoided, the mirror must have a greater diameter than the clear aperture, and of course it must be greater by about twice the plate diameter. In a mirror of 50 cm [19.7 in.] clear aperture (diameter of the correcting plate) and of 1 m [39.4 in.] focal length, the photographic plate for a field of 6 degrees has a diameter of 10.5 cm [4.1 in.], and accordingly the mirror must have a diameter of 71 cm [28 in.].

The rapid coma-free mirror system described here offers, according to the preceding considerations, great advantages in regard to light-gathering power and aberration-free imagery. There is assumed, however, a technically complete understanding of the correcting plate.

1. Translator's note: The pair of larger figures refers to the size of the aberration disk at the focus for paraxial rays, the pair of smaller ones to the size of disk at the focus for rays from the outermost zone.

2. Translator's note: One form of this optical system is known as the Mangin mirror; it is described in Czapski-Eppenstein, *Grundzüge der Theorie der Optischen Instrumente*, 3d ed. (1924), pp. 110–11.

6. On the Discovery of Extraterrestrial Radio Waves

A Note on the Source of Interstellar Interference

Karl G. Jansky

(*Proceedings of the Institute of Radio Engineers 23*, 1158–63 [1935])

Copyright © 1935 by the Institute of Radio Engineers
(now the Institute of Electrical and Electronics Engineers)

Cosmic Static

Grote Reber

(*Astrophysical Journal 100*, 279–287 [1944])

These two papers report the first exploration of the heavens at wavelengths far beyond those normally visible to the eye. Before the discovery of cosmic radio sources, astronomical investigations had depended solely on observations with telescopes operating at optical wavelengths of about 10^{-5} cm.

In the early 1930s radio waves were, however, being used extensively for communications across distances as large as the Atlantic Ocean; and at that time the Bell Telephone Laboratories assigned to Karl Jansky the task of investigating the atmospheric static that interferes with radio communications. For this purpose he built a rotating "merry-go-round" array, which detected signals coming at the long radio wavelengths of 14.6 m. In his first report on these studies[1] Jansky documented a steady hiss of static of unknown origin that seemed comparable in intensity to the radio noise produced by lightning discharges in distant thunderstorms. By observing the variation of the intensity of this "unknown" radio signal as a function of direction and time of arrival, Jansky established in the following year that the radio source must lie outside the solar system.[2] The most intense radio emission was found to be coming from the direction of the center of our galaxy, and further analysis of the data showed that weaker radio waves were coming from all directions in the Milky Way. In this paper he concludes that the source is in the stars themselves or in the interstellar material in the Milky Way.

Jansky's radio techniques were so much outside the traditional methods of astronomy that no observatory started an investigation of his discovery. The next steps were taken by an amateur, Grote Reber, who built a parabolic antenna in his back yard to pursue these findings. By 1940 Reber had confirmed Jansky's discovery of cosmic static from the Milky Way,[3] and in this paper he gives the first contour maps of the radio emission at a wavelength of 1.87 m. He also detected discrete sources of radio emission that lie in the direction of the galactic center, Cygnus, and Cassiopeia.

Jansky's experiment, which led to the discovery of an entirely new radio universe, was the first of several unexpected discoveries that have characterized radio astronomy. Thirty years later Arno Penzias and Robert W. Wilson, also working at the Bell Telephone Laboratories, accidentally discovered the microwave remnant of the primeval big-bang explosion of our universe. Similarly, the pulsars were accidentally discovered by a group of Cambridge radio astronomers while they were searching for scintillations of extragalactic radio sources in the interplanetary medium.

1. *Proceedings of the Institute of Radio Engineers 20*, 1920–32 (1932).
2. Ibid. *21*, 1387–98 (1933).
3. Ibid. *28*, 68–70 (1940).

Jansky, The Source of Interstellar Interference

Summary—Further consideration of the data obtained during observations on interstellar interference has shown that these radiations are received any time the antenna system is directed towards some part of the Milky Way system, the greatest response being obtained when the antenna points towards the center of the system. This fact leads to the conclusion that the source of these radiations is located in the stars themselves or in the interstellar matter distributed throughout the Milky Way.

Because of the similarity in the sound produced in the receiver headset, it is suggested that these radiations might be due to the thermal agitation of charged particles.

In former papers, [1,2] it was explained how interstellar interference was first observed with an automatic field strength recording system making use of a highly directional rotating antenna. It was pointed out that the directions of arrival were fixed in space and that there seemed to be only a single direction of arrival having a right ascension of eighteen hours and a declination of −20 degrees.

Some of the data did not check very accurately the theory of a single direction of arrival and it was suggested that a possible explanation of the discrepancies might be found in refraction of the waves during their passage through the ionized layers of the atmosphere.

Since the publication of the above papers further consideration of the data has led to some very interesting conclusions and speculations. The data obtained from the system are in the form of a continuous record of the output for all hours of the day and, since the antenna rotates continuously, for all directions as well. If we examine a typical day's record of the disturbances, the following facts are evident. Besides varying gradually in height throughout the day the peaks obtained for each revolution of the antenna also change decidedly in shape. In figure 6.1 from 12:00 M to 3:00 P.M. the peaks are very broad, in fact one peak covers the time taken up by one complete revolution of the antenna. From 3:00 P.M. on, the peaks gradually get narrower and narrower until at 10:00 P.M. they are only one quarter the previous width. During the same time a much smaller and weaker peak begins to appear on the record. At 9:20 P.M. it is very clear.

Upon determining the direction towards which the antenna system points in space at these times, it is discovered that when the peaks are broad, the antenna is so located in space that it sweeps along the Milky Way and the maximum response is obtained when it points in the direction of the center of the Milky Way. When the large sharp peaks and the alternate small peaks are obtained the antenna is so located in space that it sweeps across the Milky Way, the large peak being obtained when that section of the Milky Way nearest the center is crossed and the small peak when that section farthest from the center is crossed.

If we consider the belief now held by astronomers that the Milky Way is a large galaxy of stars having the same general shape as a huge discus or grindstone with the solar system, and therefore the earth, located at some distance from the center and almost in the galactic plane, then the phenomena described above would seem to indicate that the disturbances recorded are due to radiations emanating from the stars themselves. The various heights and widths of the peaks obtained on the record would then be explained in the following manner.

If the axis of rotation of the antenna were perpendicular to the plane of the Milky Way the antenna would rotate so that it always pointed at some part of the Milky Way and therefore would always receive some energy. This energy should reach a maximum value when the antenna points in the direction of the center of the Milky Way System, for the greatest number of stars would then be included within the angle of reception of the antenna. As the antenna rotates the number of stars included within this angle would very gradually decrease until the antenna points in just the opposite direction when the number of stars within the angle would be a minimum. As the antenna rotates further the number of stars within the angle would again increase until the maximum was again reached, etc. Thus the energy received at such a time would show a gradual decrease and increase with one maximum and one minimum for a single rotation of the antenna.

Actually the axis of rotation of the antenna is never exactly perpendicular to the plane of the Milky Way, but approaches that condition closely when the meridian of the receiving location has a right ascension of twelve hours and forty minutes. Translated in terms of the time on the records this occurs about five hours and twenty minutes before the azimuth of the direction of arrival of the disturbances is south. For the record shown in the figures this time would be about 1:50 P.M. at which time the angular distance between the axis of rotation and the perpendicular to the plane of the Milky Way is twelve degrees and twenty minutes. However, due to the facts that the Milky Way has a very appreciable width, and the vertical directional characteristic of the antenna is very broad, the discussion given above is still applicable and explains the type of record that should be and is obtained at this time. Furthermore, at this time, the center of the Milky Way System as seen from the receiving location has an azimuth very slightly south of east which checks exactly the direction of the maximum disturbance as obtained from the curves.

Since the direction of the axis of rotation of the antenna changes as the earth rotates, the above condition exists for only a short time. Thus, after seven hours and forty-seven and one-third minutes (when the right ascension of the meridian of the receiving location is twenty-two hours and twenty-seven and one-third minutes) the axis of rotation will lie in the plane of the Milky Way instead of being perpendicular to it. In this position the antenna will sweep across the Milky Way twice for every revolution and we would expect two peaks on the record where previously we had only one, a large peak when the antenna sweeps across that part of the Milky Way nearest the center and a smaller one when it crosses that part farthest from the center. These peaks should be relatively sharp because the number of stars within the angle of reception of the antenna changes rapidly. The first case occurs when the antenna points in a southwesterly direction and the second when it points north-northeast.

Turning now to figure 6.1, we see that at $9{:}37\frac{1}{3}$ P.M. (seven hours and forty-seven and one-third minutes after 1:50 P.M.) we have two definite peaks on the record for every revolution of the antenna, the larger of which is obtained when the antenna is pointing southwest and the smaller when it is pointing north-northeast, checking the predictions exactly.

After another eight hours and twenty-five and one-third minutes, or at $6{:}02\frac{1}{3}$ A.M. on the record given, the axis of rotation of the antenna again lies in the Milky Way, but this time the two points where the antenna sweeps across the Milky Way are both some distance from the center so that neither peak should be very large. At this time the two points have directions of northwest by north and southeast by south, the former being the nearest to the center of the Milky Way. Turning again to the figures we find that at $6{:}02\frac{1}{3}$ A.M. the peaks are much weaker than at $9{:}37\frac{1}{3}$ P.M. and that they have the directions predicted.

A more detailed analysis of the data has shown that every time the antenna points towards some part of the Milky Way the record shows an increase in the energy received, and also every time the record shows an increase of energy received the antenna is found to be pointing towards some part of the Milky Way.

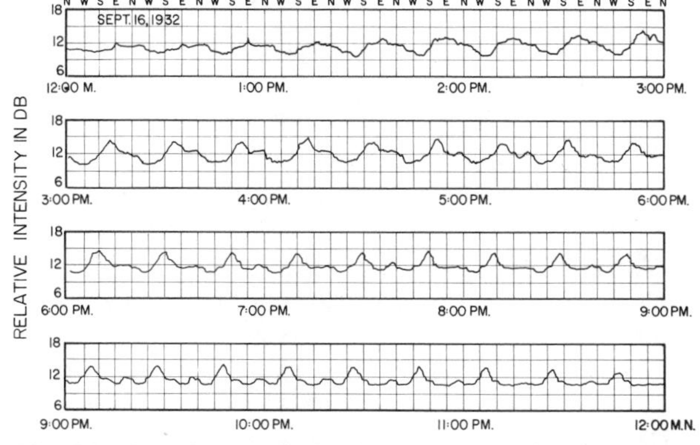

Fig. 6.1 Sample record of radio waves of interstellar origin taken with a rotating antenna and showing a peak in received intensity every time the antenna beam sweeps across the Milky Way. The wavelength of the radio emission is about 15 m; the day of observation was September 16, 1932.

As said before, the most obvious explanation of these phenomena is one that assumes that the stars themselves are sending out these radiations and that the direction of arrival at the receiving location, instead of being confined to a single direction as was formerly intimated, includes all directions, a greater indication being obtained for those directions confined to the Milky Way because of the greater star density there.

Another plausible explanation is one based on an hypothesis previously suggested,[2] that the waves which reach the antenna are secondary radiations caused by some form of bombardment of the atmosphere by high speed particles which are shot off by the stars.

Upon examining the characteristics of these radiations for further clues as to their source, one is immediately struck by the similarity between the sounds they produce in the receiver headset and that produced by the thermal agitation of electric charge.[3] In fact the similarity is so exact that it leads one to speculate as to whether or not the radiations might be caused by some sort of thermal agitation of charged particles. Such particles are found not only in the stars, but also in the very considerable amount of interstellar matter that is distributed throughout the Milky Way, which matter, according to Eddington[4] has an effective temperature of 15,000 degrees centigrade. If the radiations come from such particles one would expect the response obtained to depend upon the directional characteristic and gain of the antenna and the way it is pointed relative to the Milky Way, an expectation which agrees with the observed facts.

Attempting to explain the radiations in question on the basis of any of the above hypotheses immediately raises a serious question as to the effect of the sun. Since the sun is a star, although in comparison with some, a rather insignificant star, and since it is much closer to the earth than the stars of the Milky Way, so close in fact that the energy received from it in the form of light and heat is many times the combined energy received from all the stars and all the interstellar matter of the Milky Way, it would naturally be expected that radiations similar to those in question would be found coming from the sun with an intensity much greater than the intensity of those coming from the other sections of the Milky Way. So far no such solar radiations have been detected, and, although a possible explanation for their absence might be based on a supposition that the temperature of the sun is such that the ratio of the energy radiated by it on the wavelengths studied to that radiated in the form of light and heat is much less than for some other classes of heavenly bodies found in the Milky Way, the question will have to remain unanswered until more data have been taken.

1. Karl G. Jansky, "Directional Studies of Atmospherics at High Frequencies," *Proc. I.R.E. 20*, 1920 (Dec. 1932).

2. Karl G. Jansky, "Electrical disturbances apparently of Extraterrestrial Origin," *Proc. I.R.E. 21*, 1387 (Oct. 1933).

3. F. B. Llewellyn, "A Study of Noise in Vacuum Tubes and Attached Circuits," *Proc. I.R.E. 18*, 243 (Feb. 1930).
4. Arthur Stanley Eddington, *Stars and Atoms* (New Haven: Yale University Press, 1927), pp. 66–69.

Reber, Cosmic Static

Abstract—Cosmic static is a disturbance in nature which manifests itself as electromagnetic energy in the radio spectrum arriving from the sky. The results of a survey at a frequency of 160 megacycles per second show the center of this disturbance to be in the constellation of Sagittarius. Minor maxima appear in Cygnus, Cassiopeiae, Canis Major, and Puppis. The lowest minimum is in Perseus. Radiation of measurable intensity is found coming from the sun.

EXPERIMENTS ON THE MEASUREMENT of electromagnetic energy at radio wave lengths arriving from the sky have been conducted at Wheaton, Illinois, for a number of years. Preliminary results have already been published.[1] During the year 1943 new and improved apparatus was put into operation and considerably better data were obtained.

The electromagnetic energy is captured by a parabolic mirror and is directed to the mouth of the drum at the focal point of the mirror. Within the drum are a pair of cone antennae. These convert the electromagnetic energy into alternating current. This current is fed from the tips of the cones up a parallel-wire transmission line to the receiver-mounting on the end of the drum. On this mounting is attached a five-stage amplifier of about 90-decibels gain over the frequency range 156–164 megacycles per second.[2] It uses type 954 acorn tubes and a circuit arrangement involving transmission-line elements. The output of the amplifier is rectified by a type 9006 diode, and the resulting direct-current voltage is fed down a concentric cable into the house, where the associated power supply and recorder equipment are located.

It is well known that considerable amounts of random-charge voltage are generated in the grid and plate circuits of the first tube of a multistage amplifier. This background random-charge voltage will be rectified by the diode and appears as a constant direct-current voltage, V_{on}, at the end of the concentric cable. Its magnitude has been the point of much investigation and serves as a reference level for the calibration of the intensity of cosmic static phenomena. The value of V_{on} is noted on a meter before and after a set of data is taken. During the actual taking of data, V_{on} is canceled out by a small battery so that the cosmic static may be more easily perceived.

The mirror is mounted on an east-west axis so that it may be pointed to any angle of declination between the limits of $-32\overset{\circ}{.}5$ and $+90°$ along the north-south meridian. One of the circular tracks is calibrated in degrees. The mirror is set to point at the desired declination; and then, as the earth rotates, the mirror sweeps out a band in the sky along this particular declination. Whenever the mirror passes over a cosmic static disturbance, energy is collected, and the voltage at the end of the concentric cable increases by an amount ΔV_{on}. The recorder is set to a chart speed of 6 inches per hour. If no cosmic static is intercepted, the recorder will draw a straight line on the chart. If cosmic static is encountered, the pen will move up by an amount ΔV_{on}.

Here Reber reproduces a chart recording in which a pen has drawn a slow rise and fall while the Milky Way has passed through the antenna beam. Sharp spikes caused by the opening and closing of switches are also recorded, together with small rises caused by the ignition noise of passing automobiles.

Referring to a previous paper,[3] the intensity of cosmic static is

$$I = \frac{2U}{E_d E_m A \phi_e \phi_m}, \qquad (1)$$

where

$U = 15.1 \times 10^{-14}\,\Delta$ watts/M.C. band,
$E_d = 1.00$ = Drum efficiency,
$E_m = 0.85$ = Mirror efficiency,
$A = 7 \times 10^5$ sq. cm. = Mirror area,
$\phi_e = 8°$ = Resolving-power in plane of electric vector,
$\phi_m = 6°$ = Resolving-power in plane of magnetic vector,

and

$$\Delta = \frac{\Delta V_{on}}{V_{on}}.$$

Assembling these values gives

$$I = 10.6 \times 10^{-21}\Delta \text{ watts/sq. cm., cir. deg., M.C. band.} \qquad (2)$$

Several charts are taken at each declination under consideration. All charts of like declination are then worked up, the right ascension for a given intensity is determined, and the results are averaged to give mean right ascension for this intensity. About two hundred charts were obtained in 1943. The final results plotted on a flattened globe are shown in figure 6.2, *a* and *b*, for the two hemispheres of the sky. The constant-intensity lines are in terms of 10^{-22} watts/sq. cm., cir. deg., M.C. band. The small numbers at the centers of major maxima are the top values in this direction. The

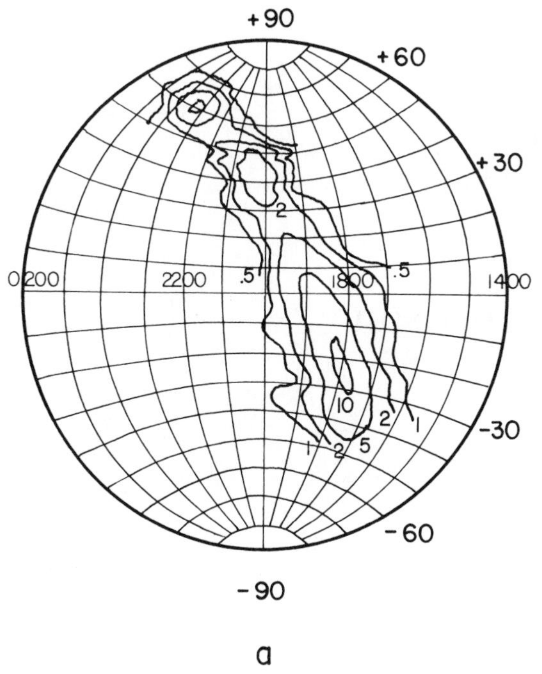

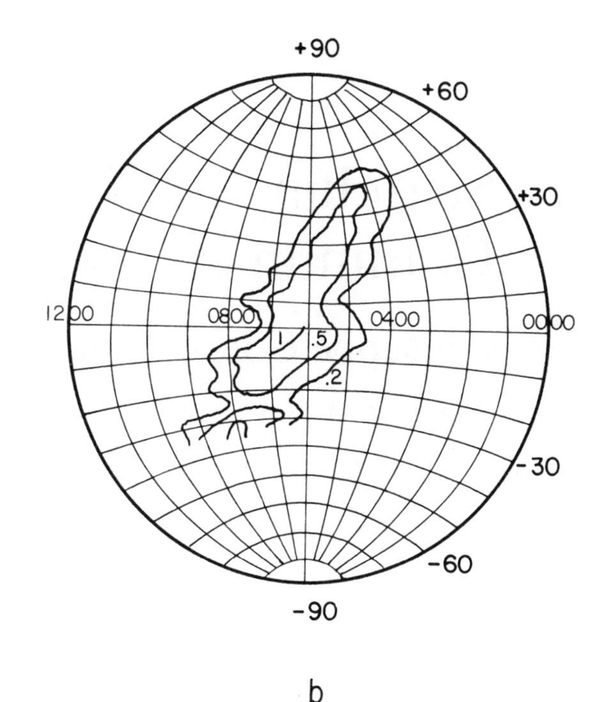

Fig. 6.2 Constant intensity lines of radio-frequency radiation at a wavelength near 1.87 m. The units are in terms of 10^{-22} w per cm^2 per circular degree per MHz bandwidth. The numbers on the central line of each circle denote right ascension; the numbers along the circle perimeters denote declination.

points at declination $-48°$ in figure 6.2, *a* are beyond the normal range of the collector machine. They were obtained as the result of an accident when the machine was run off the end of the track in a heavy snowstorm. It lodged with the mirror nearly vertical and the drum somewhat out of focus and resting in the service tower. These points show the general trend of the phenomena but are not of a high order of accuracy.

Too little is known about the cause of this phenomenon to read a great deal from figure 6.2. However, it is suggested that this disturbance is in some way connected with the amount of material in space. Since the wave length is long (1.87 meters), the absorption caused by dust is small. Therefore, the intensity is roughly indicative of the amount of material between us and the edge of the Milky Way. On this basis the various maxima point to the directions of projections from the Milky Way. These projections may be similar to the arms often photographed in other spiral nebulae. In the case of the Milky Way this general picture would call for the center toward Sagittarius, and arms in the directions of Cygnus, Cassiopeiae, and Canis Major. A minimum occurs in Perseus, indicating that we are nearest the edge of the galaxy in that direction. The maximum in lower Puppis is possibly a general rise toward the center. This region from Puppis to Scorpius is out of reach at the latitude of Wheaton. However, it is deserving of study, as it may contain other minor maxima.

It has been suggested that this long-wave radiation could be set up in the corona of the sun. Until recently no positive evidence was available. However, the new apparatus was sufficiently stable to allow some data to be taken during the day in spite of severe automobile noise. A series of charts was taken over several months as the sun passed through the center of the Milky Way. V_{on} is different on the various charts; hence they are not strictly comparable. In any case the sun had the rather surprising center intensity of 10×10^{-22} watts/sq. cm., cir. deg., M.C. band. In spite of the apparent great strength from the sun, this source must be greatly discounted when explaining the origin of cosmic static. If it were the source and the Milky Way were made of average stars like the sun, a very large area in Sagittarius would have a visible intensity equal to that of the sun. Since this is not the case, some other cause must be found to make up the difference of 20 or 30 mag.

A few other objects, such as the moon, the Pleiades, the Orion Nebula, and the spiral in Andromeda have been tested without conclusive results. The last object seems to be at the threshold of sensitivity of the present apparatus, namely, 0.2×10^{-22} watt/sq. cm., cir. deg., M.C. band. Further improved sensitivity will probably pick up a variety of other objects, and an increase in resolving-power will probably show more detail in the galactic structure.

1. G. Reber, *Proc. I.R.E. 28*, 68 (1940); *Ap. J. 91*, 621 (1940); *Proc. I.R.E. 30*, 367 (1942).
2. G. Reber, *Electronic Industries 3*, 89–92 (1944).
3. *Proc. I.R.E. 30*, 376, eq. (20) (1942).

7. Searching for Interstellar Communications

Giuseppe Cocconi and Philip Morrison

(*Nature 184*, 844–846[1959])

Current interpretations of the origin of our solar system picture the formation of planetary systems as a typical event in the process of star formation. In fact, planetary companions may have been detected for several nearby dwarf stars. If these planets have an earth-like composition and reasonable surface temperatures, then proteins and nucleic acids may form in the primitive planetary atmospheres. Living organisms might well evolve, and it is imaginable that some of the planets contain civilizations superior to our own in technological development.

Around 1959 it became clear that communication over galactic distances is possible using radio waves and that existing radio wavelength transmitters could be used to send messages to the nearby stars. In the paper provided here, Giuseppe Cocconi and Philip Morrison show that interstellar communication is best carried out at radio wavelengths. They also provide an initial stimulus for considering a search for extraterrestrial civilizations by radio waves. At about the same time, Frank Drake initiated Project Ozma, in which he unsuccessfully searched for 21-cm signals from two solar-type stars, τ Ceti and ε Eridani.[1]

Since the presentation of this pioneering paper, several books have been written on the general subject of the existence of, and possible communication with, intelligent extraterrestrial life. During the 1970s several prominent radio astronomers have participated in routine searches for signals from the nearby stars, and a concerted effort has been made to design an optimum system for radio contact with extraterrestrial systems.[2] The start of a systematic search for signals from stars is proposed in 1978 by centers of the National Aeronautics and Space Administration.[3]

1. *Physics Today 14*, 40 (1961).
2. B. M. Oliver, ed., *Project Cyclops: A Design Study of a System for Detecting Extraterrestrial Life*, NASA report no. CR-114445 (Washington, 1973).
3. P. Morrison, J. Billingham, and J. Wolfe, eds., *The Search for Extraterrestrial Intelligence SETI*, NASA SP-419 (Washington, 1977).

NO THEORIES yet exist which enable a reliable estimate of the probabilities of (1) planet formation; (2) origin of life; (3) evolution of societies possessing advanced scientific capabilities. In the absence of such theories, our environment suggests that stars of the main sequence with a lifetime of many billions of years can possess planets, that of a small set of such planets two (Earth and very probably Mars) support life, that life on one such planet includes a society recently capable of considerable scientific investigation. The lifetime of such societies is not known; but it seems unwarranted to deny that among such societies some might maintain themselves for times very long compared to the time of human history, perhaps for times comparable with geological time. It follows, then, that near some star rather like the Sun there are civilizations with scientific interests and with technical possibilities much greater than those now available to us.

To the beings of such a society, our Sun must appear as a likely site for the evolution of a new society. It is highly probable that for a long time they will have been expecting the development of science near the Sun. We shall assume that long ago they established a channel of communication that would one day become known to us, and that they look forward patiently to the answering signals from the Sun which would make known to them that a new society has entered the community of intelligence. What sort of a channel would it be?

THE OPTIMUM CHANNEL

Interstellar communication across the galactic plasma without dispersion in direction and flight-time is practical, so far as we know, only with electromagnetic waves.

Since the object of those who operate the source is to find a newly evolved society, we may presume that the channel used will be one that places a minimum burden of frequency and angular discrimination on the detector. Moreover, the channel must not be highly attenuated in spaces or in the Earth's atmosphere. Radio frequencies below ~ 1 Mc./s., and all frequencies higher than molecular absorption lines near 30,000 Mc./s., up to cosmic-ray gamma energies, are suspect of absorption in planetary atmospheres. The bandwidths which seem physically possible in the near-visible or gamma-ray domains demand either very great power at the source or very complicated techniques. The wide radio-band from, say, 1 Mc./s. to 10^4 Mc./s., remains as the rational choice.

In the radio region, the source must compete with two backgrounds: (1) the emission of its own local star (we assume that the detector's angular resolution is unable to separate source from star since the source is likely to lie within a second of arc of its nearby star); (2) the galactic emission along the line of sight.

Let us examine the frequency dependence of these backgrounds. A star similar to the quiet Sun would emit a power which produces at a distance R (in metres) a flux of:

$$10^{-15} f^2/R^2 \quad \text{W.m.}^{-2}(\text{c./s.})^{-1}$$

If this flux is detected by a mirror of diameter l_d, the received power is the above flux multiplied by l_d^2.

The more or less isotropic part of the galactic background yields a received power equal to:

$$\left(\frac{10^{-12.5}}{f}\right)\left(\frac{\lambda}{l_d}\right)^2 (l_d)^2 \quad \text{W.(c./s.)}^{-1}$$

where the first factor arises from the spectrum of the galactic continuum, the second from the angular resolution, and the third from the area of the detectors. Thus a minimum in spurious background is defined by equating these two terms. The minimum lies at:

$$f_{\text{min.}} \approx 10^4 \left(\frac{R}{l_d}\right)^{0.4} \quad \text{c./s.}$$

With $R = 10$ light years $= 10^{17}$ m. and $l_d = 10^2$ m., $f_{\text{min.}} \approx 10^{10}$ c./s. The source is likely to emit in the region of this broad minimum.

At what frequency shall we look? A long spectrum search for a weak signal of unknown frequency is difficult. But, just in the most favored radio region there lies a unique, objective standard of frequency, which must be known to every observer in the universe: the outstanding radio emission line at 1,420 Mc./s. ($\lambda = 21$ cm.) of neutral hydrogen. It is reasonable to expect that sensitive receivers for this frequency will be made at an early stage of the development of radio-astronomy. That would be the expectation of the operators of the assumed source, and the present state of terrestrial instruments indeed justifies the expectation. Therefore we think it most promising to search in the neighbourhood of 1,420 Mc./s.

POWER DEMANDS OF THE SOURCE

The galactic background around the 21-cm. line amounts to:

$$\frac{dW_b}{dS\, d\Omega\, df} \approx 10^{-21.5} \quad \text{W.m.}^{-2}\ \text{ster.}^{-1}\ (\text{c./s.})^{-1}$$

for about two-thirds of the directions in the sky. In the directions near the plane of the galaxy there is a background up to forty times higher. It is thus economical to examine first those nearby stars which are in directions far from the galactic plane.

If at the source a mirror is used l_s metres in diameter, then the power required for it to generate in our detector a signal as large as the galactic background is:

$$\frac{dW_s}{df} = \frac{dW_b}{dS\,d\Omega\,df}\left(\frac{\lambda}{l_s}\right)^2\left(\frac{\lambda}{l_d}\right)R^2$$

$$= 10^{-24.2}\,R^2/l_s^2 l_d^2 \quad \text{W.(c./s.)}^{-1}$$

For source and receiver with mirrors like those at Jodrell Bank ($l = 80$ m.), and for a distance $R \simeq 10$ light years, the power at the source required is $10^{2.2}$ W.(c./s.)$^{-1}$, which would tax our present technical possibilities. However, if the size of the two mirrors is that of the telescope already planned by the U.S. Naval Research Laboratory ($l = 200$ m.), the power needed is a factor of 40 lower, which would fall within even our limited capabilities.

We have assumed that the source is beaming towards all the sun-like stars in its galactic neighbourhood. The support of, say, 100 different beams of the kind we have described does not seem an impossible burden on a society more advanced than our own. (Upon detecting one signal, even we would quickly establish many search beams.) We can then hope to see a beam toward us from any suitable star within some tens of light years.

Signal Location and Band-Width

In all directions outside the plane of the galaxy the 21-cm. emission line does not emerge from the general background. For stars in directions far from the galactic plane search should then be made around that wave-length. However, the unknown Doppler shifts which arise from the motion of unseen planets suggest that the observed emission might be shifted up or down from the natural co-moving atomic frequency by $\pm \sim 300$ kc./s. (± 100 km. s.$^{-1}$). Closer to the galactic plane, where the 21-cm. line is strong, the source frequency would presumably move off to the wing of the natural line background as observed from the direction of the Sun.

So far as the duration of the scanning is concerned, the receiver band-width appears to be unimportant. The usual radiometer relation for fluctuations in the background applies here, that is:

$$\frac{\Delta B}{B} \propto \sqrt{\frac{1}{\Delta f_d \cdot \tau}}$$

where Δf_d is the band-width of the detector and τ the time constant of the post-detection recording equipment. On the other hand, the background accepted by the receiver is:

$$B = \frac{dW_b}{df}\Delta f_d \quad \text{and} \quad \tau \propto \frac{\Delta f_d}{(\Delta B)^2}$$

If we set ΔB equal to some fixed value, then the search time T required to examine the band F within which we postulated the signal to lie is given by:

$$T = \frac{F\tau}{\Delta f_d} \propto \frac{F}{(\Delta B)^2}$$

independent of receiver band-width Δf_d.

Of course, the smaller the band-width chosen, the weaker the signal which can be detected, provided $\Delta f_d \geq \Delta f_s$. It looks reasonable for a first effort to choose a band-width Δf_d normal in 21 cm. practice, but an integration time τ longer than usual. A few settings should cover the frequency range F using an integration time of minutes or hours.

Nature of the Signal and Possible Sources

No guesswork here is as good as finding the signal. We expect that the signal will be pulse-modulated with a speed not very fast or very slow compared to a second, on grounds of band-width and of rotations. A message is likely to continue for a time measured in years, since no answer can return in any event for some ten years. It will then repeat, from the beginning. Possibly it will contain different types of signals alternating throughout the years. For indisputable identification as an artificial signal, one signal might contain, for example, a sequence of small prime numbers of pulses, or simple arithmetical sums.

The first effort should be devoted to examining the closest likely stars. Among the stars within 15 light years, seven have luminosity and lifetime similar to those of our Sun. Four of these lie in the directions of low background. They are τ Ceti, o_2 Eridani, ε Eridani, and ε Indi. All these happen to have southern declinations. Three others, α Centauri, 70 Ophiuci and 61 Cygni, lie near the galactic plane and therefore stand against higher backgrounds. There are about a hundred stars of known spectral type within some fifty light years. All main-sequence dwarfs between perhaps $G0$ and $K2$ with visual magnitudes less than about $+6$ are candidates.

The reader may seek to consign these speculations wholly to the domain of science-fiction. We submit, rather, that the foregoing line of argument demonstrates that the presence of interstellar signals is entirely consistent with all we now know, and that if signals are present the means of detecting them is now at hand. Few will deny the profound importance, practical and philosophical, which the detection of interstellar communications would have. We therefore feel that a discriminating search for signals deserves a considerable effort. The probability of success is difficult to estimate; but if we never search, the chance of success is zero.

8. The Photoelectric Photometry of the Stars

The Measurement of the Light of Stars with a Selenium Photometer, with an Application to the Variations of Algol

Joel Stebbins

(*Astrophysical Journal 32*, 185–214[1910])

Fundamental Stellar Photometry for Standards of Spectral Type in the Revised System of the Yerkes Spectral Atlas

Harold L. Johnson and William W. Morgan

(*Astrophysical Journal 117*, 313–352[1953])

The luminosities and colors of stars are the basic data upon which physical interpretations are based, and these are the data used to infer stellar temperatures, radii, and evolutionary sequences. By the beginning of the twentieth century photographic techniques were well established as the standard procedure for measuring both magnitudes and colors. Early in this century two other energy-measuring devices were introduced: the thermocouple and the photoresistive cell. In the hands of F. Coblentz and especially of Edison Pettit and Seth B. Nicholson,[1] the thermocouple was employed to measure a wide spectral range of heat energy arriving from the stars at the earth's surface.

It was the photoelectric cell, however, that eventually furnished the sensitivity and accuracy to become a major tool for the measurements of stellar magnitudes and colors. Joel Stebbins pioneered these techniques in 1907, when he first demonstrated that a photoresistive selenium cell responded to moonlight.[2] By 1910 Stebbins had discovered that the sensitivity of the selenium cell was improved when the cell was kept at low temperatures in an ice pack and when the exposures to light were shorter than the periods required for cell recovery. In the paper given here, Stebbins reports the use of these techniques to measure, for the first time, the secondary minimum of the variable star Algol, and hence to deduce the brightness of the invisible companion. This was possible because Stebbins's selenium cell measurements recorded magnitudes quickly, with a probable error of only 0.02 mag—at that time a technique considerably more accurate than the use of photographic plates.

In addition to the problem of refrigeration, the chief disadvantage of the photoresistive selenium cell was its low sensitivity, which meant that only very bright objects could be measured. For this reason selenium cell photometry was quickly abandoned when more sensitive photoelectric cells were developed and applied to astronomy. Attempts to construct practical photoelectric cells were begun in Germany in 1893 by J. Elster and H. Geitel, and by 1912 P. Guthnick had installed one of Elster and Geitel's successful cells on the 30-cm refractor of the Berlin-Babelsberg Observatory. The following year Guthnick was able to report on photoelectric observations of the variable star β Cephei. In the meantime, news of the photoelectric cell reached Stebbins when Jakob Kunz visited the University of Illinois in 1911. During the following year Stebbins visited Guthnick in Berlin, while Kunz and W. F. Shultz carried on at Illinois in his absence. In fact, an account of photoelectric astronomy at Illinois was published in the *Astrophysical Journal* a few months before Guthnick's pioneering paper.[3]

During the next three decades astronomers concentrated on improving the sensitivity of photoelectric cells. Among the important developments during this period were the use of fused quartz for the cell tubes in 1916, the application of the Lindemann electrometer in 1927, and the perfection of the thermionic amplifier with the FP-54 tube by Albert Whitford in 1932. Each new increase in sensitivity closely paralleled commercial developments, which received a major impetus with the advent of the "talkies"—sound motion picture projectors, which used the photoelectric cells for decoding the film soundtrack. Shortly after World War II, the Radio Corporation of America developed the 1P-21 photomultiplier tube, whose reliability and greatly increased sensitivity made feasible the measurement of the brightnesses and colors of thousands of fainter stars.

Because color measurements depend upon the spectral sensitivity of the photomultiplier and upon the transmission curves of the filters used, a standard set of observational wavelengths and filter bandwidths had to be developed if different observers were to compare their results. A foremost photometrist, H. L. Johnson, teamed up with a spectroscopist, W. W. Morgan, to produce the widely used *UBV* system of photometric standards; their work is reported in the second paper given here. Relatively broad spectral regions were included in order to measure the weaker sources, and the center wavelength of each region was chosen partially for historical reasons. The inclusion of *U* (ultraviolet) wavelengths made it possible to discriminate between the intrinsic colors and the space-reddening of stars.

To facilitate the calibration of observations, Johnson and Morgan used the 1P-21 tube together with their standard three-color filters to provide a catalogue containing the visual magnitudes, V, and the color indices $B - V$ and $U - B$ of 290 stars that include a broad range of spectral classes and absolute magnitudes. In this way, they were able to show that supergiants, main-sequence stars, and white dwarfs describe different paths on the color-color plot of $B - V$ and $U - B$. Appended to their paper is the impressive color-color diagram of over 20,000 stars compiled at the

Geneva Observatory, clearly showing the diagonal scatter caused by interstellar reddening, a feature quantitatively analyzed by Johnson and Morgan.

Johnson and Morgan not only used the *UBV* colors to sort out stars of different luminosities; they also correlated this research with previous results on the two-dimensional MKK spectral classification (which included both spectral type and luminosity class). Their paper defined the new MK classification, a modification of the earlier Morgan-Keenan-Kellman scheme.[4] A sample page from the earlier MKK atlas is also appended.

1. E. Pettit and S. B. Nicholson, *Astrophysical Journal 68*, 279 (1928).
2. J. Stebbins and F. C. Brown, *Astrophysical Journal 26*, 326 (1907).
3. P. Guthnick, *Astronomische Nachrichten 196*, 356 (1913); W. F. Schulz, *Astrophysical Journal 38*, 187 (1913).
4. W. W. Morgan, P. C. Keenan, and E. Kellman, *Atlas of Stellar Spectra* (Chicago: University of Chicago Press, 1943).

Stebbins, Measurement with a Selenium Photometer

SOME THREE YEARS AGO I became interested in the possible application of selenium to astronomical photometry, and with Dr. F. C. Brown, I began to experiment on selenium cells, with the hope of using them for accurate measures of the light of stars. The idea is an attractive one: the proposed method being to expose a selenium surface to the light of a star, focused by a large lens, and to note the change of resistance by means of a galvanometer: the brighter the star, the larger the galvanometer-deflection. In theory this is very simple, but at the outset we met some of the difficulties which confront everyone who tries to work with selenium. Other agencies than light affect the resistance, and apparently no experimenter has solved, to his own satisfaction, the mysteries of this peculiar element.

After some futile attempts to obtain results from starlight at the focus of a 12-inch refractor, we found that a selenium cell gave large effects when exposed directly to the moon. This led to a study of the variation of the moon's light throughout a lunation,[1] and it was also found that various selenium cells have widely different curves of color-sensibility.[2] One reason that physicists and electricians have not used these cells in photometry is that selenium is relatively most sensitive to red light, but in certain kinds of astronomical work, such as the study of eclipsing variable stars, it is quite immaterial what part of the spectrum is used.

During several years of experimenting I have tried the selenium cells of different makers, whose processes are secret, also a number of cells made by Dr. Brown and myself. The best cells known to me are those by Giltay of Delft, Holland. In the form which he uses, two wires are wound in a double spiral about a flat insulator, and the spaces on one face are filled with selenium which is properly treated to make it sensitive. In response to my inquiry, Mr. Giltay has said that his method of sensitizing is indeed a secret, but he adds that it is often a secret even to himself, for after thirty years of experience in making cells he frequently has surprises, and usually of the disagreeable kind.

In working with faint lights, we found that a small variation in the temperature of a cell gives a greater change of resistance than is produced by the light of the brightest stars. I made numerous attempts to maintain a constant temperature with thermostats, but although some improvement was noted, the irregularities remained almost hopelessly large. One of the fundamental obstacles is the heating effect of the current which is necessarily passed through the selenium to measure the resistance. It is unnecessary to enumerate the various experiments which were made, but after a long time I learned that there are three simple precautions necessary to insure success, and they are given in what seems to be the relative order of importance:

1. The selenium should be kept at a uniform, low temperature, 0° C., or lower.
2. The current should pass continually through the selenium.
3. Exposures to light should be short, say 10 seconds, with longer intervals for recovery.

Here Stebbins notes that the resistance of a selenium cell increases at lower temperatures and that if the cell is surrounded with an ice pack the effects of variations in room temperature on the detected signal are minimized. He also argues that short exposure times with long intervals between exposures are needed because of the long recovery time of selenium from light action. Starlight is measured by connecting the selenium cell to one arm of a Wheatstone bridge, and the strength of the current passing through the cell is measured by a galvanometer of the d'Arsonval type. Meter readings were recorded by one observer while another observer controlled exposure times by manually opening and closing a shutter in time to the beats of a chronometer.

APPLICATION TO THE LIGHT-VARIATIONS OF ALGOL

As soon as I began experimenting with selenium, I had it in mind as one of the problems to take up, when it became possible to measure starlight with accuracy, to investigate the light-curve of Algol, especially for a secondary minimum, which, if present, would give us valuable information concerning the companion. The history of the investigations of Algol is so well known, even to the general reader, that it is superfluous to mention the names of the astronomers who have contributed to our knowledge of this star. In fact, we may say that if any star in the whole sky has been subjected to an exhaustive study, Algol is that one. For this very reason, however, it seemed advisable to try the selenium photometer, for this star would furnish an excellent test of the improvement of our work over visual methods.

It was found on trial that with good conditions a 10-second exposure of our best selenium cell to Algol at maximum light gave a scale deflection of 8.0 mm. The probable error of a single deflection was of the order of 0.16 mm, or about 2 per cent, and as this was a better accordance than can be secured in visual work, I felt ready to adopt a program for observing this star during several months. In illustration of the accuracy which was insisted upon, it may be said that when the cell happened to become so irregular that the probable error of one deflection became higher than 5 per cent, observations were not undertaken.

On account of the atmospheric absorption, there is of course no method of measuring the light of a star, which does not include observations of one or more comparison stars. On consideration, I decided to use α Persei and δ Persei. After making this decision, I found that these same stars had been chosen by Müller,[3] whose determination of the light-curve is perhaps the best that has ever been made visually. The normal plan with selenium was to make 4 exposures on α Persei, then 8 on Algol, and finally 4 on α Persei. Such a series of 8 readings on each star will be called a set. When Algol was near minimum, both α and δ were used, and a set of readings usually consisted of 4 on α, 4 on δ, 8 on Algol, 4 on δ, and 4 on α. When only two stars were observed, the 16 deflections of a set were secured in about 20 minutes, including the time for moving the dome and telescope. Ordinarily it was considered that 4 to 6 sets of measures were sufficient for one night, but when the phase of Algol was near principal minimum, or the supposed secondary minimum, the observations were continued all night, or as long as conditions would permit.

Stebbins presents a sample of the recorded meter readings and other observational parameters for one night of observation. He then describes the corrections for the progressive drift of the galvanometer during a given night and for the effects of differential atmospheric absorption. The difference in magnitude between α and δ Persei was found to be 1.10 mag, and this value was used to compute the drift. For most cases the correction was less than 0.01 mag. Stebbins presents a table of the difference in magnitude between Algol and α Persei at different times and a table giving the residuals between the computed light curve for Algol and the observed magnitude differences. From these residuals he finds that the probable error in the magnitude is ± 0.023 mag and ± 0.006 mag, respectively, near principal minimum and secondary minimum.

One peculiarity of the selenium photometer is that the errors, expressed in magnitudes, increase for faint stars, but in light-units the accordance is practically the same for any intensity. In visual observations, the residuals in magnitudes are about the same over a wide range of brightness. Where a variable star has a range of 1 magnitude and eye observations are expressed in light-units, the probable error near minimum is only $\frac{2}{5}$ that at maximum, and according to the usual method the weights should be assigned as 6 to 1. So far as I know, no computer has taken this fact into account, although there have been many elaborate calculations of the elements of different variable stars. In this connection, I may say that I am quite unable to see the justification for publishing the elements of a system like that of Algol to 4 or even 5 significant figures, when the original data are often not exact to 2 places, and when even the first figure of some of the results may be in error.

The normal magnitudes are shown graphically in figure 8.1. The first peculiarity of the light-curve which will attract those familiar with Algol is the existence of a secondary minimum. This has been sought for in vain by visual observers, though in my opinion if Algol were not so bright, the variation of 0.06 magnitude might have been detected. Of perhaps equal interest is the continuous variation in the light between minima, showing this star to be a distant relative of β Lyrae, though the maximum brilliancy does not seem to occur midway between minima, the curve being highest just before and after secondary minimum. The fact that a smooth, symmetrical light-curve fits in with the observations is perhaps sufficient reason for believing this continuous variation to be real.

Theory of the System of Algol

On the basis of the observations which have been tabulated, let us now consider what conditions in the system of Algol will account for the variations in light. Since Vogel's classic determination of the radial velocities before and after light-minimum, the eclipse theory has been regarded as established. The presence of the secondary minimum now proves that the

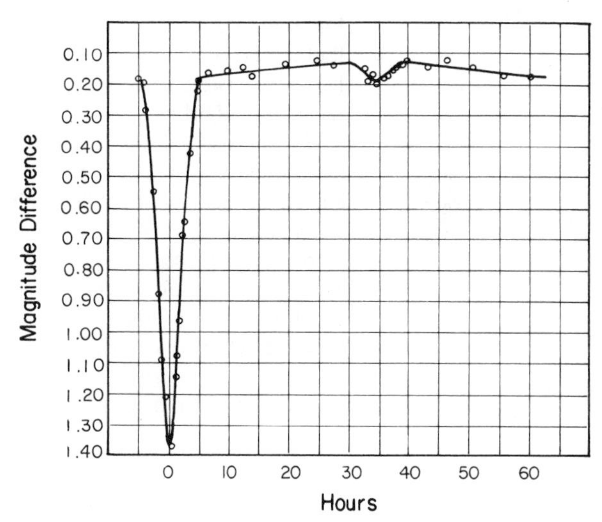

Fig. 8.1 The light curve for the variable star Algol as measured with a selenium cell.

companion is not wholly dark, and it is evident that we have to deal with a bright and a relatively faint body, which in what follows I shall designate as Algol and the companion.

As a first approximation let us assume that the orbit is circular, that both bodies are spheres, and further that they are devoid of extensive atmospheres which would cause a decrease of intensity from center to limb, such as we know does exist in the case of the sun. The most obvious explanation of the continuous variation of light between minima is that the companion keeps one face toward Algol, and is brighter on that side due to radiation received from the primary. I therefore assume that the companion rotates uniformly once in the period of revolution, and that it is divided into two hemispheres each uniformly intense. While this last assumption is probably far from the truth, it is sufficient for the accuracy of the observations.

Here Stebbins gives equations that relate the observed light curve to the orbital parameters and the radii of Algol and its companion (following Pickering[4] and Harting[5]). He then derives the elements of Algol by a method of successive approximations, starting with preliminary values given by Schlesinger in the *Publications of the Allegheny Observatory 1*, 123 (1909). The calculated elements are given in the summary, together with the theoretical and observed light curves.

By using a parallax of 0.07 sec of arc Stebbins shows that the total light from the bright and weak companions of Algol are, respectively, 3.0 and 1.7 times that of the sun. In a note added to the proof he uses a more recent parallax value of 0.01 sec of arc to show that the total light from Algol is 240 times that of the sun. By assuming that the densities and the masses of the two members of Algol are related by $d = d_s$ or $m = 2m_s$, where d denotes density and m denotes mass, he derives the elements given in the summary.

SUMMARY

1. It has been demonstrated that for bright stars the selenium photometer, attached to a 12-inch telescope, yields results which are considerably more accurate than have ever been obtained by visual or photographic methods.
2. An application of this new device to observations of Algol has led to the discovery that the companion, far from being dark, gives off more light than our sun, and in addition is much brighter on the side which is turned toward the primary.
3. A discussion of the photometric and other results for Algol gives the principal facts concerning this system.

From the Light-curve:

—	Radius of Algol	1
κ	Radius of companion	1.14 ± 0.05
r	Distance between centers	4.77 ± 0.05
i	Inclination of orbit	$82^{\circ}.3 \pm 0^{\circ}.3$
—	Surface-intensity of Algol	1
λ	Surface-intensity of faint hemisphere of companion	0.050 ± 0.010
$\lambda + \lambda_1$	Surface-intensity of bright hemisphere of companion	0.088 ± 0.012
—	Total period	$68^{h}.816$
—	Duration of eclipse	$9^{h}.80$
—	Mean density of the system	0.07 ☉
—	Limiting density of Algol	0.18 ☉
—	Limiting density of companion	0.12 ☉

From $\pi = 0''.07$ (sun $= -26.6$ mag., Algol $= 2.2$ mag.):

		☉ = 1	Stellar Magnitude
l	Total light of Algol	26	2.2
$l\kappa^2\lambda$	Total light of faint hemisphere of companion	1.7	5.2
$l\kappa^2(\lambda + \lambda_1)$	Total light of bright hemisphere of companion	3	4.6

Two assumptions from $a \sin i = 1{,}600{,}000$ km:

		$d = d_s$	$m = 2m_s$
R	Radius of Algol	0.81 ☉	1.45 ☉
$R\kappa$	Radius of companion	0.92	1.66
m	Mass of Algol	0.04	0.37
m_s	Mass of companion	0.06	0.18
d	Density of Algol	0.07	0.12
d_s	Density of companion	0.07	0.04
σ	Surface-intensity of Algol	40	12
$\sigma\lambda$	Surface-intensity of faint hemisphere of companion	2	0.6
$\sigma(\lambda + \lambda_1)$	Surface-intensity of bright hemisphere of companion	3.5	1.1

Stebbins then thanks his assistants and the Rumford Committee of the American Academy of Arts and Sciences for successive grants of $200 and $350, which enabled him to carry out his work.

1. *Astrophysical Journal 26*, 326 (1907).
2. Ibid. *27*, 183 (1908).
3. *Astronomische Nachrichten 156*, 178 (1901).
4. *Proceedings of the American Academy of Arts and Sciences 16*, 1 (1880).
5. *Untersuchungen über den Lichtwechsel des Sternes β Persei* (Munich, 1889).

Johnson and Morgan, Fundamental Stellar Photometry

Abstract—A system of photoelectric photometry is outlined which utilizes the revised zero point of the visual magnitude scale of the North Polar Sequence and which returns to the original definition for the zero point of color indices in terms of main-sequence stars of class A0; the interval A0-gK0 is 1 mag. The revised Yerkes *Atlas* system (MK) of spectral classification is taken as standard. The latter is described briefly, and a list of standard stars is included.

Magnitudes and color indices from measures in three wavelength bands are given for stars selected by spectral type and luminosity class to be representative of the principal regions of the H-R diagram. A few white dwarfs are also included.

A standard main sequence is defined for the new color-absolute magnitude diagram by the use of stars of large parallax, together with the galactic clusters NGC 2362, the Pleiades, the Ursa Major nucleus, and Praesepe. A standard main sequence is also defined for the relationship between the two systems of color index.

A purely photometric method for determining spectral types and space reddening for B stars in galactic clusters is described.

TERMINOLOGY

y. Deflection through yellow filter, corrected for sky.

b. Deflection through blue filter, corrected for sky.

u. Deflection through ultraviolet filter, corrected for sky.

C_y. Observed blue-yellow color index, reduced to outside the earth's atmosphere.

C_u. Observed ultraviolet-blue color index, reduced to outside the earth's atmosphere.

V. Observed magnitude through yellow filter, reduced to outside the earth's atmosphere. This is approximately equivalent to the photovisual magnitude on the International System.

B. Observed magnitude through blue filter, reduced to outside the earth's atmosphere and including a zero-point correction to satisfy the condition

$$B - V = 0$$

for main-sequence stars of class A0 on the MK system.

U. Observed magnitude through ultraviolet filter, reduced to outside the earth's atmosphere and including a zero-point correction to satisfy the condition

$$U - B = 0$$

for main-sequence stars of class A0 on the MK system.

E_y. Color excess on $B - V$ system.

E_u. Color excess on $U - B$ system.

MKK. The system of spectral classification outlined in: W. W. Morgan, P. C. Keenan, and Edith Kellman, *An Atlas of Stellar Spectra.*

MK. The revised MKK system. The MK system is outlined in the present paper.

Introduction

Because of nonlinear (and sometimes multi-valued) relationships between various systems of color indices,[1] the definition of a fundamental system of magnitudes and colors becomes difficult; in particular, for regions like that of the North Polar Sequence, where no early-type stars are available, a system for stars bluer than class A0 (and for reddened B stars in general) can hardly be considered to exist. The principal difficulty here is that it is not justifiable to extrapolate a color equation determined from A–K stars to those of class B; in addition, the spectrum of a reddened B star does not have the same energy distribution as does that of a later-type star;[2] therefore, a different color equation may be necessary for reddened and unreddened stars.

A fundamental photometric system for stars should therefore include:

1. Magnitudes and color indices for unreddened stars from all parts of the H-R diagram; these should include white dwarfs and subdwarfs, as well as supergiants, giants, and main-sequence stars.
2. The same photometric data for stars having interstellar reddening and of known spectral type and luminosity class.
3. A series of color indices extending from the ultraviolet to the infrared, so that reductions to the standard color system can be made by a process of interpolation rather than extrapolation; the "six-color photometry" of Stebbins and Whitford satisfies this condition extremely well.
4. A determination of the zero point of the color indices in terms of a certain *kind* of star which can be accurately defined spectroscopically; that is, in terms of a kind of star whose spectral energy distribution can be predicted accurately from its spectral type and luminosity class; in addition, the stars used for the zero point should be plentiful.

The above requirements cannot be satisfied by using small selected regions of the sky; the standard stars are, of necessity, scattered over the sky, and many of them must be very bright if condition 1 is to be satisfied.

It is, therefore, of importance to supplement the scattered standards with regional secondary standards which fulfil some of the conditions and which are located in a small region. The most useful of the regional standards would be accurately observed main-sequence stars in open clusters, together with some yellow giants.

An observing program was arranged to include a number of bright and near-by stars to satisfy most of conditions 1, 2, and 4, and also several open clusters for defining a standard main sequence. The observations of the bright and near-by stars are included herewith. Three open clusters (Praesepe, the Pleiades, and IC 4665) were selected as regional secondary standards; the measures for Praesepe have been published (*Ap. J.*, *116*, 640, 1952), the Pleiades are given in the present paper, and IC 4665 has yet to be completed. All the photometric observations were made by Johnson at the McDonald Observatory in the winter of 1950–1951 and the summer of 1951.

Certain details of the photoelectric photometry are given in the next section, including estimates of the probable errors. The following section tabulates the minor changes that had been incorporated into the MKK spectral classification system in the decade since the publication of the Yerkes *Atlas of Stellar Spectra*, thereby defining a revised scheme, the so-called MK system of spectral classification. Table 2 (omitted here) gives 339 standard stars arranged according to luminosity class.

The Zero Point for the Color Indices

The original zero point of the International System was set by the condition that the color index be zero for A0 stars near the sixth magnitude.[3] Various circumstances resulted in the observed mean values for main-sequence A0 stars ranging from around -0.14 to -0.04.[4]

Recently, the magnitudes and colors of nine stars of the North Polar Sequence as determined by Stebbins, Whitford, and Johnson[5] have been adopted as standards for photoelectric work. Since we here adopt a different zero point for the color systems, some justification would appear to be called for.

For this, it is necessary to anticipate some of the results of the present paper and a following one. The fundamental conclusion arrived at is that, to specify a photometric system with the highest accuracy, it is necessary to know the spectral types and luminosity classes of the stars concerned, as well as the photometric data. We cannot reduce one system of colors to another by a single relation, unless we are satisfied to make a sacrifice in accuracy. For two color systems, where the observations are affected by hydrogen absorption, we will usually have one relation for main-sequence stars, another for the yellow giants, and still another for reddened O and B stars. Examples of these relations will be found in a later paper.

This complexity in ordinary blue-yellow color systems makes imperative the definition of a zero point for colors in a more specific way than is afforded by the nine standard Polar Sequence stars. Two of the highly accurate color systems have adopted different zero points. The "six-color photometry" of Stebbins and Whitford has a zero point set by the mean of ten main-sequence stars of types G and K, and averaging G6. The Greenwich gradients depend for zero point on nine specified stars of HD type A0. We have followed a procedure similar to the latter; the zero point of the present color systems has been set by the mean values for six stars of class A0 V on the MK system; the stars are: α Lyr, γ UMa, 109 Vir, α CrB, γ Oph, and HR 3314. For the mean of these stars

$$U - B = B - V = 0.$$

The color indices were derived from the observed colors by the following formulae:

$$B - V = C_y + 1.040, \qquad U - B = C_u - 1.120.$$

After the zero point for $B - V$ has been set according to the above definition, it is found that, for the K0 III stars listed in Table 3, $B - V = +1.01 \pm 0.007$ (p.e.). Therefore, both the zero point and the scale for $B - V$ satisfy the original definition of the International System.

Omitted here are table 3, giving the photometric observations for 290 stars, table 4, giving an approximate reduction of Kuiper's spectral types to the MK system, and tables 5–7, containing the magnitudes and color indices for the Pleiades, M36, and NGC 2362.

COLOR-MAGNITUDE DIAGRAMS

A number of the stars listed in table 3 have accurately known parallaxes, through which the absolute magnitudes may be determined. Table 8 contains all the stars in table 3 having trigonometric parallaxes of 0″.100 or greater; three yellow gaints from the Hyades cluster and four A and F stars from the Ursa Major cluster for which accurate cluster parallaxes are known; α Boo and α Tau, the two yellow giants later than K0 having the largest parallaxes; and all the degenerate stars (white dwarfs) contained in table 3 for which parallaxes have been measured. The first two columns of table 8 contain the same serial numbers and names of the stars as table 3; the third, the parallax; and the fourth the absolute visual magnitude. The parallaxes are, in general, from the Yale 1935 catalogue.

Here we omit table 8, which gives the parameters of stars with accurately known parallaxes, and table 9, which compares the absolute magnitudes of B stars with those obtained by Keenan and Morgan.

We can construct a color-luminosity diagram from these stars, but the number of bluish stars having accurately known absolute magnitudes is very small, and no stars earlier than A0 can be included. We cannot, therefore, form any idea about the main sequence earlier than A0, and much of the detail for later-type stars is lost, especially in the A to G stars. We can, however, fill in many of the gaps by using the color-magnitude diagrams from galactic clusters to interpolate between and extrapolate beyond the stars with known absolute magnitudes. In this way we can show the slope and some of the details of the main sequence. We do, of course, make the assumption that the main sequences of the clusters may legitimately be fitted to near-by stars.

For these processes of interpolation and extrapolation we shall use the following galactic clusters: The Pleiades (we omit the brighter stars, which have luminosity classes III and IV, and a strongly reddened A star); Praesepe,[6] and NGC 2362. The spectral types of the stars in these clusters range from B1 to M0, and adequate fits to the near-by stars are possible. Two of the clusters (the Pleiades and NGC 2362) are reddened, while the third is not. We shall correct the Pleiades for +0.04 mag. reddening and NGC 2362 for +0.11 mag. (on the $B - V$ scale); the determination of these values is described later.

We first fit the Pleiades to the near-by stars (correcting for +0.04 mag. reddening in the process) over the range of spectral type A0 to G0, corresponding to $0.00 < B - V < +0.60$. The resultant apparent distance modulus is $5^m.9 \pm 0^m.1$ (p.e.), and there is no significant systematic difference in shape between the main sequence of the Pleiades and that of the near-by stars. When we correct for the ratio of total to selective absorption, we find that the true distance modulus of the Pleiades is $5^m.8 \pm 0^m.1$ (p.e.), corresponding to a distance of 144 ± 7 (p.e.) parsecs.

We next fit the Praesepe main sequence to the Pleiades main sequence for $+0.30 < B - V < +0.60$. Since there is no systematic difference between the Pleiades main sequence and that of the near-by stars, this process is equivalent to fitting Praesepe to the near-by stars over the same range of color. The resultant distance modulus is $6^m.2 \pm 0^m.1$ (p.e.), corresponding, since Praesepe is not reddened, to a distance of 174 ± 8 (p.e.) parsecs. As a check, let us fit the Praesepe main sequence to the near-by stars for $B - V > +0.60$, the range of color that was not used in the above fit. We find, after rounding off to the nearest tenth, the same distance modulus as before—6.2 mag. In this computation Gmb 1830 has been omitted because it is known to lie well below the main sequence. As in the case of the Pleiades, there is no significant systematic difference in shape between the main sequence of Praesepe and that of the near-by stars.

Finally, we fit NGC 2362 to the upper portion of the Pleiades main sequence, taking into account the reddening of +0.11 mag. The fit is not so certain as the other two, because of the appreciable width of the Pleiades main sequence in this region and the lack of any method, independent of the photometry, for choosing cluster members for NGC 2362. We find an apparent distance modulus of $11^{\mathrm{m}}.9 \pm 0^{\mathrm{m}}.2$ (p.e.). When we correct for the ratio of total to selective absorption, we find a true distance modulus of $11^{\mathrm{m}}.6 \pm 0^{\mathrm{m}}.2$ (p.e.), corresponding to a distance of 2,090 ± 190 (p.e.) parsecs.

This last distance, 2,090 parsecs, seems rather large for the amount of reddening, +0.11 mag., of the cluster. A satisfactory check on the reliability of the distance modulus for NGC 2362 can be had by comparing the absolute magnitudes from table 7 with those given by Keenan and Morgan[7] for the same spectral types. This comparison is shown in table 9, where it is evident that the agreement of the cluster magnitudes with the independent ones of Keenan and Morgan is quite good. The discordance at B9 V can be removed if NGC 2362, No. 16, which lies far above the main sequence of the cluster, is omitted.

We are now prepared to plot the composite color-magnitude diagram given in figure 8.2., which shows the main sequence from B1 V to M6 V, the yellow giant branch from K0 III to K5 III, and a few white dwarfs. Several stars that fall somewhat below the main sequence (as defined by the Praesepe stars) will be noted around $B - V = +0.7$. In order to avoid confusing the diagram with too many points, the Praesepe stars bluer than $B - V = +0.60$ have been omitted. The Pleiades stars adequately represent this portion of the diagram.

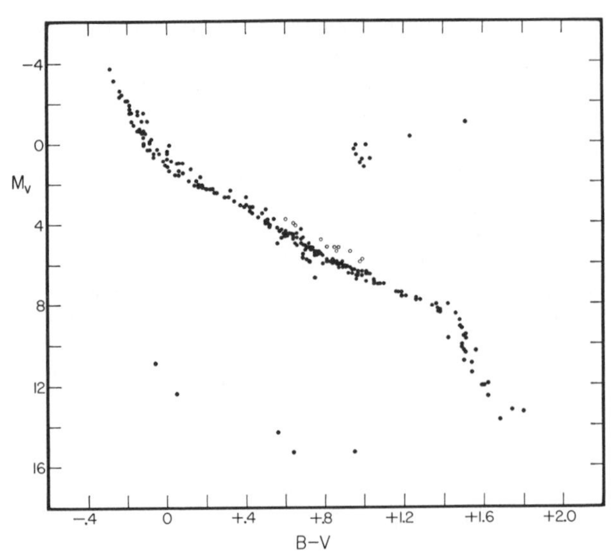

Fig. 8.2 A standard main sequence for the color system $B - V$ and the absolute magnitude system M_V. The stars plotted include main-sequence objects that have trigonometric parallaxes >0.100 sec of arc; the Pleiades, corrected for a mean interstellar reddening (one highly reddened A star omitted); Praesepe; and NGC 2362, corrected for a mean interstellar reddening. In addition, five white dwarfs, three yellow giants from the Hyades, and several other yellow giants of large parallax are included. The open circles refer to a few stars lying above the main sequence in Praesepe that may be binaries. (Courtesy W. W. Morgan.)

Here Johnson and Morgan present a plot of M_V versus $U - B$ for the stars used in figure 8.2. This plot, not reproduced here, has more scatter than that of figure 8.2. This scatter is attributed to differential reddening in the Pleiades star cluster, which does not affect figure 8.2.

The Relations Between $U - B$ and $B - V$

For the discussion of the relations between $U - B$ and $B - V$, all the main-sequence stars in table 3 that can reasonably be expected to be unreddened, the Praesepe stars, and the stars in NGC 2362 have been used. The Pleiades have been omitted because of the differential reddening within the cluster. Also we have assumed that reddening in $U - B$ is 0.72 of that in $B - V$, as will be shown below. Corrections for the reddening of NGC 2362 can therefore be made in both colors.

The mean relation for the main sequence as defined by the stars mentioned above is shown in figure 8.3. Again the dip caused by the hydrogen absorption in the A stars is evident. The few stars falling above the main sequence (as defined by the Praesepe stars) at $B - V \sim +0.7$ are the same stars—μ Cas, Gmb 1830, τ Cet, and χ Dra—that fell below the main sequence in figure 8.2. It may be, therefore, that there is a luminosity effect here, in the sense that stars below the main sequence in luminosity have stronger ultraviolet than do main-sequence stars. If this effect is universal, it would seem to remove from cluster membership the five stars below the main sequence in Praesepe,[8] since all five stars fall along with the main-sequence stars in figure 8.3.

Large scatter among the M dwarfs ($B - V > +1.40$) is evident. This scatter is much too large to be explained by observational errors. There appears to be some correlation between $U - B$ and the strength of hydrogen emission in these stars: Wolf 47[9] has extremely strong hydrogen emission lines on a spectrogram obtained at the McDonald Observatory by W. P. Bidelman, and it also has the strongest ultraviolet of all the M dwarfs observed. Barnard's star and Gmb 34B have no appreciable emission and have the weakest ultraviolet of the M dwarfs observed.

Johnson and Morgan compare figure 8.3 with the relation between $U - B$ and $B - V$ for weakly

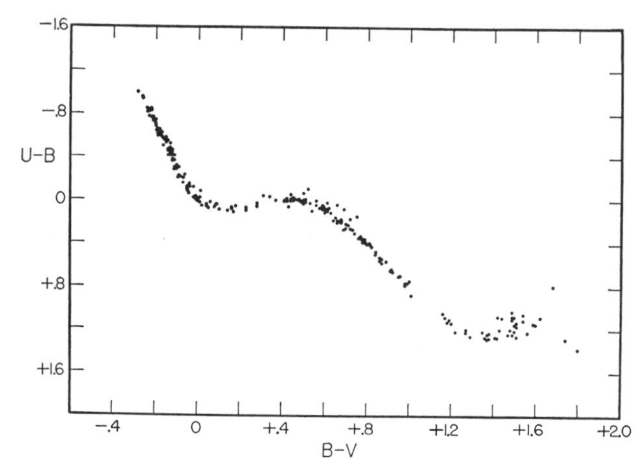

Fig. 8.3 The relation between the color systems $U - B$ and $B - V$ for unreddened main-sequence stars. (Courtesy W. W. Morgan.)

reddened supergiants, yellow giants, and white dwarfs. White dwarfs with $B - V$ lying between -0.2 and $+0.2$ have a $U - B$ lying between 0.5 and 1.0. These white dwarfs and the early-type supergiants are systematically stronger in the ultraviolet than the main-sequence stars of the same $B - V$. The part of the supergiant diagram for stars with $B - V$ lying between -0.4 and $+0.4$ is shown in figure 8.4.

One notices immediately that the early-type supergiants have systematically stronger ultraviolet than do the main-sequence stars with the same values of $B - V$; in addition, the hydrogen absorption in the early A supergiants is not so pronounced as in the main-sequence stars. The K and M giants and supergiants ($B - V > +0.80$) appear to run pretty well together, both deviating systematically in $U - B$ from the dwarfs among the late K and M stars ($B - V > +1.40$). It is evident that the white dwarfs for which $B - V \sim 0$ have much stronger ultraviolet than do the main-sequence stars of the same color. This fact makes the identification of white dwarfs easy, since the ultraviolet deflection is so much larger than for a normal star that, if the possibility of the star's being a supergiant can be eliminated, one knows immediately whether or not a white dwarf is being observed. Among the later-type white dwarfs the same systematic excess in the ultraviolet compared with the normal stars appears, although here it is not so pronounced. In the case of the white dwarfs, there appears to be little or no continuous hydrogen absorption.

These figures demonstrate very effectively the nonlinear and multivalued relationships between two color systems, one of which contains significantly more ultraviolet short of 3800 A than does the other. Not only is the relationship between the two colors for the main-sequence stars nonlinear, but it obviously does not apply to the supergiants, white dwarfs, and yellow giants.

It will be noticed that in figure 8.4 the lines connecting reddened and unreddened stars of the same spectral type have very nearly the same slope. The color excesses for $U - B$ and $B - V$ are, respectively, E_u and E_y; if we compute a quantity $Q = (U - B) - (E_u/E_y)(B - V)$, this quantity will be independent of interstellar reddening. We can therefore separate certain intrinsic properties of the stars, among which are, as we shall show, the spectral type and intrinsic (unreddened) colors. This approach is very nearly the same as W. Becker's[10] "color-difference" method, except that Becker has chosen his effective wave lengths so that the ratio of the color excess is unity.

The remainder of the paper demonstrates explicitly how $B - V$ and Q used in conjunction with a spectral type can be used to determine the intrinsic colors of B-type stars.

1. *Ap. J. 116*, 272 (1952).
2. See Frances Sherman and W. W. Morgan, *Ap. J. 89*, 515 (1939); Joel Stebbins and A. E. Whitford, *Ap. J. 98*, 20 (1943).
3. *Trans. I.A.U. 1*, 79 (1922).
4. W. W. Morgan and W. P. Bidelman, *Ap. J. 104*, 245 (1946).
5. *Ap. J. 112*, 469 (1950).
6. Only main-sequence and yellow giant stars are used.
7. *Astrophysics*, ed. J. A. Hynek (New York: McGraw-Hill Book Co., 1951).
8. *Ap. J. 116*, 640 (1952).
9. Wolf 47 "flared" during one of the observations.
10. *Ap. J. 107*, 278 (1948).

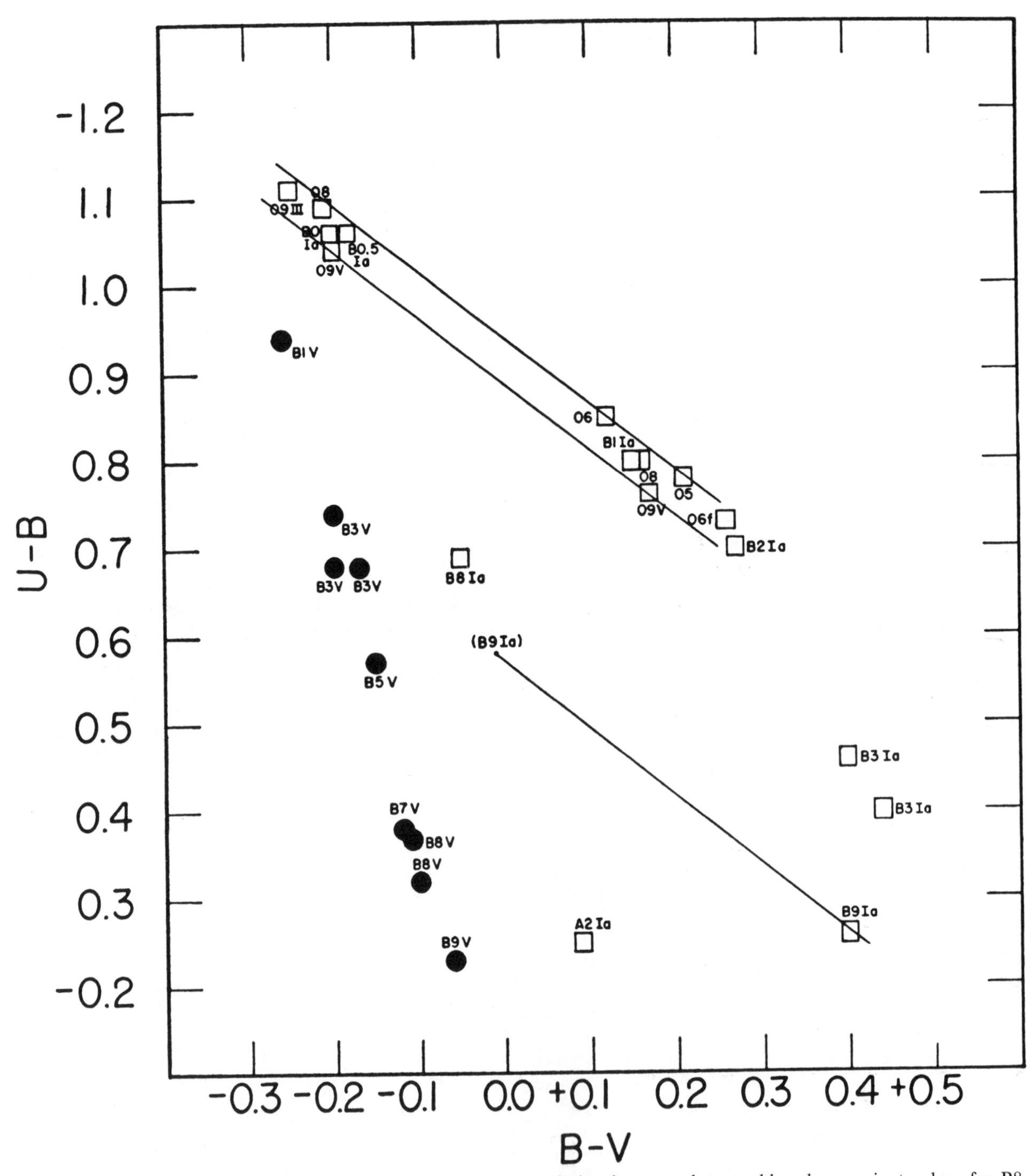

Fig. 8.4 Two-color ($B-V$, $U-B$) diagram for main-sequence B1–B9 stars (*filled circles*) and O stars and B–A2 supergiants (*squares*). The reddening path for the O stars is marked. The point (B9 Ia) was determined by linear interpolation between the unreddened supergiant values for B8 Ia and A2 Ia. The approximate reddening line for B9 Ia is illustrated. (Courtesy W. W. Morgan.)

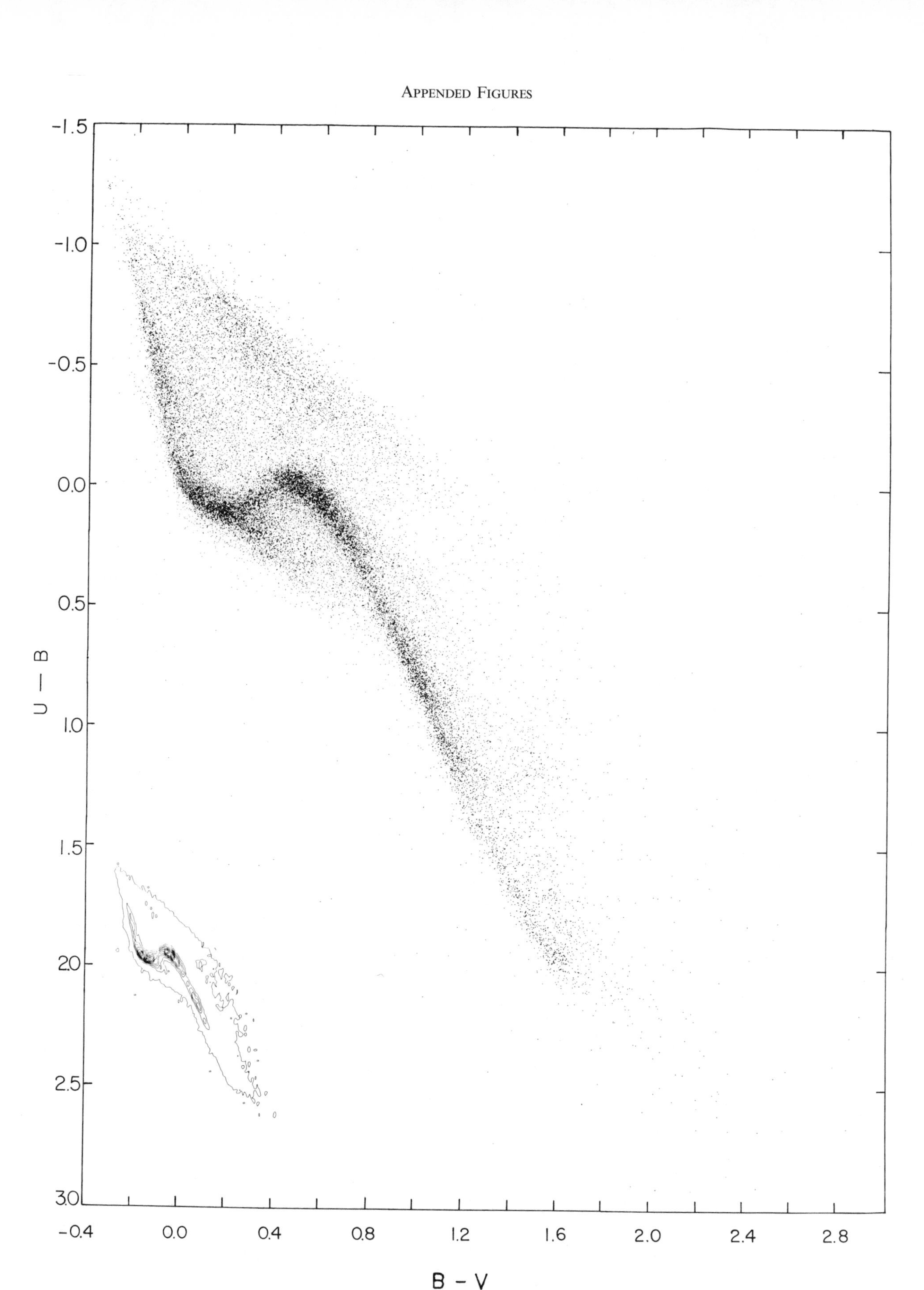

Fig. 8.5 The two-color ($B - V$, $U - B$) diagram for over 29,000 stars. This diagram exhibits a large amount of diagonal scatter not present in figure 8.3, which is caused by interstellar reddening of the starlight. (Courtesy of the Geneva Observatory.)

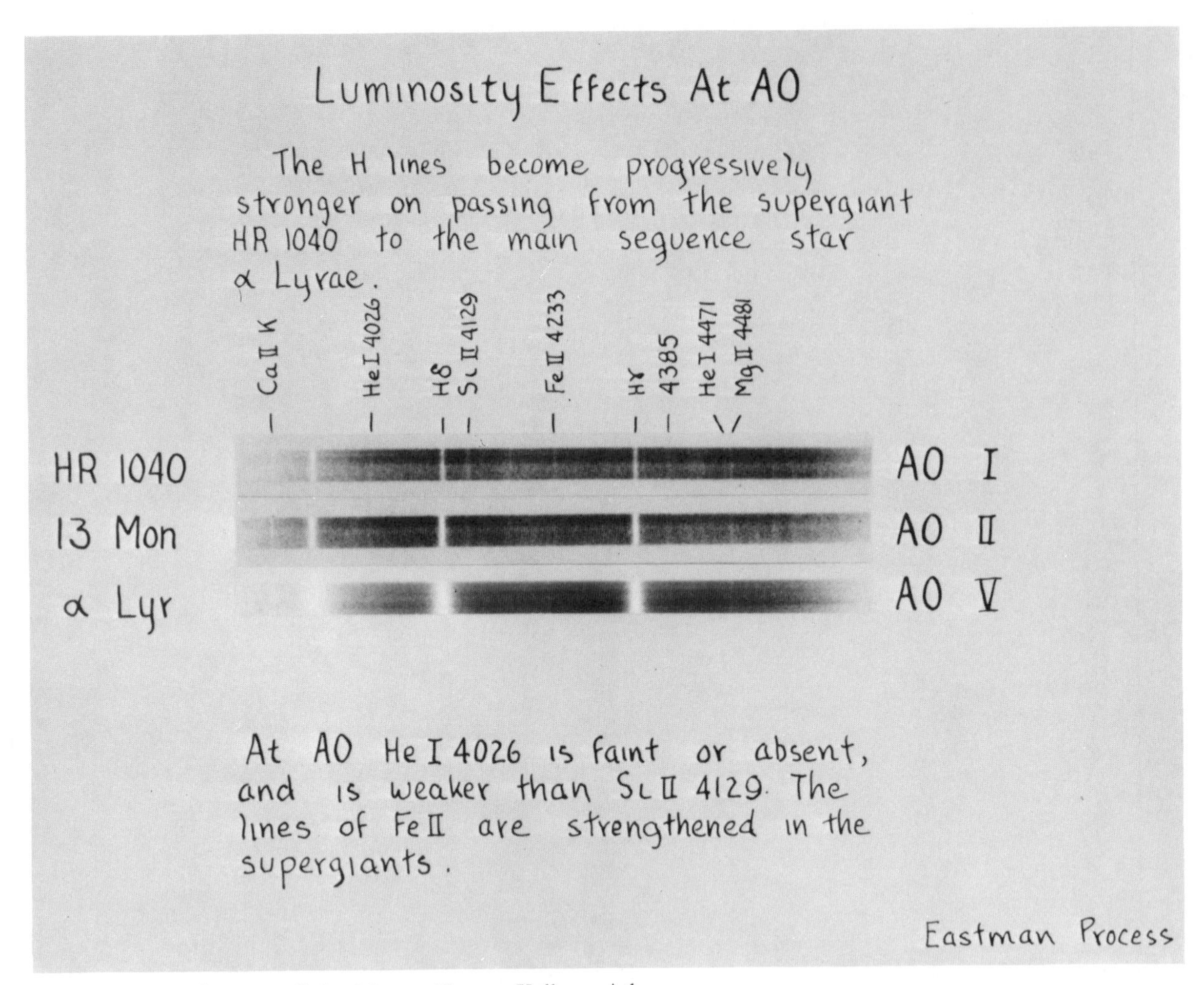

Fig. 8.6 A sample page of the Morgan-Keenan-Kellman Atlas.

9. The First Results Obtained from Photographs of the Invisible Side of the Moon

N. P. Barabashov and Yu. N. Lipskii

(*Doklady academii nauk SSSR 129*, 288–289 [1959]; trans. in *Soviet Physics* [*Doklady*] *4*, 1165–69 [1960])

When the Russians launched the first artificial earth satellite, *Sputnik I*, on October 4, 1957, a new era of direct experimental study of the terrestrial environment was introduced. Within three months of the Russian launch, the Americans had their first artificial satellite, *1958α*, in orbit. In just two years the American satellite program had mapped the geomagnetic field, discovered the Van Allen radiation belts that girdle the earth, and directly determined the harmonic coefficients of the earth's gravitational field.

On October 4, 1959, exactly two years after the launch of the world's first artificial satellite, the Russians launched a space rocket, *Luna III*, that inaugurated another new era for astronomy—that of remote sensing and direct exploration of the moons, the planets, and the interplanetary space. In the paper given here, two Russian academicians describe the first photographs of the hitherto invisible side of the moon. As well as noting a lower incidence of maria on the moon's far side and giving names to the most prominent far-side lunar features, the authors of this paper also made the important suggestion that it is theoretically feasible to deliver instruments close enough to the planets to photograph them. From this epoch-making beginning, extraordinary technological developments have allowed men to explore the moon, to sample the interplanetary spaces, and to guide spacecraft so accurately that even the moons and volcanoes of Mars have been photographed (see selections 29 and 30). Furthermore, in the sixteen years since 1959, more than fifty spacecraft have flown near or landed on Earth's moon, twelve astronauts have walked on and sampled the lunar surface, and twenty-eight interplanetary spacecraft have been launched in order to probe the atmospheres and surface features of Mercury, Mars, Venus, Jupiter, and Saturn.

As has already been reported, the third cosmic rocket was successfully launched on October 4, 1959, and placed an automatic interplanetary station (AIS) into a preassigned orbit. The launching of the AIS was undertaken to investigate a number of problems connected with space research, and to obtain photographs of the reverse side of the moon and its boundary regions. The moon was photographed starting from a distance of 65,200 km (start of photographing) to 68,400 km (end of photographing), when the AIS was approximately on a straight line joining the sun and the moon. A camera with two objectives was used for taking photographs: the first had a relative aperture of 1:5.6 and $F_1 =$ 200 mm; the other had a relative aperture of 1:9.5 and $F_2 =$ 500 mm. Various exposure times were provided for by the program of observation. During the 40 minutes of operation of the camera, the reverse side of the moon was photographed many times. The time chosen for taking photographs made it possible for the AIS to obtain photographs of a large portion of the lunar surface invisible from the earth, and a small region with features already known. The presence of the latter region in the photographs made it appreciably easier to relate the new features on the reverse side of the moon to those already known. The methods used for processing the film, the television equipment, the techniques for telemetering the images of the moon to earth, and other data, have been published elsewhere.[1]

A large number of frames obtained by means of the objective with a focal length of 200 mm, as well as with the objective of 500 mm focal length, were available to us for investigation. First of all, each original negative was examined by us on a special projector under conditions of optimal magnification. In this way, we selected for a preliminary investigation those negatives which contained the smallest number of flaws in the form of lines of the television raster crossing the lunar image.

For the preservation of the unique negatives, further work was carried out, partly with enlarged positive prints, as well as positive prints made without a change of scale. In later work, the original negatives were only used for purposes of checking.

A study of the photographs obtained in various frames led to the immediate identification of Mare Crisium, and then, also, of Mare Foecunditatis, Mare Humboldtianum, Mare Marginis, Mare Undarum, Mare Spumans, Mare Smithii, and Mare Australe. These features are clearly marked in most of the photographs studied.

As was to be expected, the familiar outlines of Mare Foecunditatis and Mare Crisium were altered, because of the unusual angle of the photographs. The border regions are particularly affected by this effect. The well-investigated Mare Australe, in which, to the south of crater Hanno there are crater maria, was found to lie, for the greater part, on the reverse side of the moon. On this side, the boundary of Mare Australe has an irregular and sinuous shape. Mare Smithii, in contrast to Mare Australe, is more regular in outline and, from the south, it is penetrated by a mountainous region. Most of Mare Smithii is also situated on the invisible side of the moon. Mare Marginis, in which the region visible from the earth is characterized by a very low albedo, has an elongated shape in the photographs studied, and has a depression on the side away from Mare Crisium. Most of Mare Marginis situated on the reverse side of the moon apparently also has a very low albedo. It should be noted that Mare Humboldtianum, which from the earth appears to lie in a depression in the border region, has an unusual pear-shaped outline with a sinuous edge.

On the basis of a preliminary investigation of the available photographs, it is possible to conclude that the invisible part of the lunar surface is rich in mountainous regions. On the reverse side, there are very few maria similar to those situated on the side of the moon visible from the earth. Crater maria, lying in the southern and equatorial regions, are prominent features. It should be pointed out that all the region on the reverse side of the moon adjoining the western border region has an albedo which is intermediate in magnitude between that of maria and mountainous regions. In reflecting power, this region is similar to the region of the moon situated between the craters Tycho, Petavius and Mare Nectaris.

To the south-southeast of Mare Humboldtianum, near the above-mentioned region, there is a mountain range with a total length of over 2,000 km which runs, with breaks in some places, across the equator and extends into the southern hemisphere. Behind the mountain range, there is an extensive continent with a higher reflecting power. The name "Soviet" has been given to this mountain range.

In the region with the coordinates latitude $+20^\circ$ to $+30^\circ$ and longitude $+140^\circ$ to 160°, there lies a crater mare with a diameter of about 300 km, which has been called Sea of Moscow. The southern part of this sea contains a sinus, which has been named Bay of Astronauts. In the southern hemisphere, in the region with the coordinates latitude -20 to -30° and longitude $+130^\circ$, there is a large crater with a diameter of over 100 km, having a dark inner surface and a bright central peak surrounded by a wide and lightly colored bank; this crater has been named Tsiolkovskii.

A group of four craters is situated at latitude $+30^\circ$ between the Soviet mountain range and Mare Marginis. The largest of this group of craters, with a diameter of about 70 km and a noticeable central peak, has been named crater Lomonosov, while the crater furthest removed from the mountain range has been named crater Joliot-Curie. The shapes of the remaining two craters of this group require clarification. Near the equator, at a longitude of about $+110^\circ$ there is a separate crater with a circular outline. In the southern hemisphere near the edge of the lunar disc, and approximately at the same longitude as Sea of Moscow, there is a large sea, Sea of Dreams, which apparently extends to the other side of the disc. Above Sea of Dreams in the equatorial region there is a

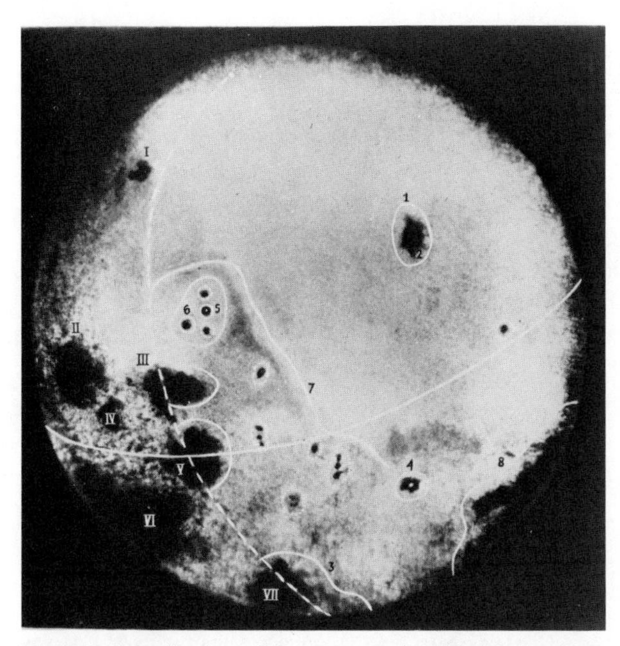

Fig. 9.1 Distribution of features on the side of the moon invisible from the earth, derived from a preliminary analysis of photographs obtained from the Automatic Interplanetary Station: (*1*) a large crater sea with a diameter of about 300 km—Sea of Moscow; (*2*) Bay of Astronauts in the Sea of Moscow; (*3*) continuation of Mare Australe on the reverse side of the moon; (*4*) a crater with a central peak—Tsiolkovskii; (*5*) a crater with a central peak—Lomonosov; (*6*) crater Joliot-Curie; (*7*) Soviet mountains; (*8*) Sea of Dreams. Roman numerals denote features on the visible side of the moon: (*I*) Mare Humboldtianum; (*II*) Mare Crisium; (*III*) Mare Marginis, which has a continuation on the invisible side; (*IV*) Mare Undazum and Mare Spumans; (*V*) Mare Smithii, which has a continuation on the invisible side; (*VI*) Mare Foecunditatis; (*VII*) Mare Australe, which has a continuation on the invisible side. (Courtesy *Sky and Telescope.*)

large region near the edge of the disc with a markedly lower reflecting power.

In addition to those listed above, the photographs contain regions with a slightly increased or decreased reflecting power and numerous small features. The nature of these features, their shapes and sizes, can be found from a detailed study of the available photographs. The distribution of features established during the preliminary analysis of photographs obtained on board the AIS is presented in figure 9.1. The full line running across the picture is the lunar equator, the broken line is the boundary between the lunar regions visible and invisible from the earth. Full lines are also used to indicate the features on the reverse side which have been established reliably during the preliminary analysis of the photographs. The features which require a further analysis have been encircled by a broken line. Regions whose classification is being established are surrounded by a series of dots. The material obtained is being analyzed, and the remaining regions are being investigated.

In conclusion, we express our unbounded admiration of the group of Soviet scientists, engineers, and workers who have realized a great scientific achievement—the launching into a preassigned orbit of the first automatic interplanetary station, which has created a revolution in the methods for the investigation of celestial bodies.

We express our deep gratitude for the honor given us—the chance to analyze the first photographs of the moon's side invisible from the earth—and for the help which has been given us by the specialists of the various sections.

1. The first photographs of the reverse side of the moon [in Russian] (*Izd. AN SSSR*, 1959).

10. X-Ray and Extreme Ultraviolet Observations of the Sun

Herbert Friedman

(*Space Research II*, ed. H. C. van de Hulst, C. de Jager, and A. F. Moore [Amsterdam: North-Holland Pub. Co., 1961], pp. 1021–35)

Because a hot gas radiates its maximum energy at a wavelength that varies inversely with its temperature, the hottest astronomical sources will shine most powerfully at short wavelengths. However, the earth's atmosphere blocks out wavelengths shorter than 2,900 Å; so direct observations of celestial ultraviolet and X-ray radiation had to await the advent of the space age. The first such spectrograms were obtained by the United States Naval Research Laboratory (NRL), using captured German V-2 rockets. In this paper, Herbert Friedman summarizes the chief results of these rocket observations of the sun. The main motivation for the very first rocket observations was a desire to understand the solar control of the earth's ionosphere. An early V-2 rocket flight showed that the brightest line of the solar ultraviolet spectrum, the Lyman-α radiation at 1,216 Å, produces the ionospheric D region, by ionization of the trace constituent nitric oxide, and later flights disclosed that the E region can be attributed to soft X rays, which ionize all of the atmospheric constituents.

These preliminary results also led to the first descriptive accounts of the solar surface at short wavelengths. The first X-ray image of the sun is shown here, and this photograph illustrates the association of soft X-ray emission with condensations in the plage regions of the sun. This primitive picture set the stage for the detailed soft X-ray pictures of the sun recently taken with the X-ray telescope aboard the *Skylab* satellite.[1] The loops, arches, and other features of these photographs provide striking illustrations of the interaction between the solar magnetic field and the ionized gas in the solar corona.

In the early 1950s, when rocket astronomy was in its infancy, it was already well known that the sun had an extensive million-degree corona (see selection 21), and the rocket spectra revealed numerous emission lines from highly ionized gases. Because different temperature layers give rise to emission from different ions, the various atomic lines serve as effective probes of different physical levels in the solar atmosphere. As a result, the far ultraviolet spectrum maps out the physical conditions of the solar atmosphere from the photosphere (effective temperature of 6,000 K), through the temperature minimum (about 4,000 K), through the chromosphere and the corona (upwards of 2,000,000 K). From the elementary beginnings reported in this paper, the stage was set for the extreme ultraviolet spectra taken during the NASA Orbiting Solar Observatory program. One result of this program has been a detailed map of electron density and electron temperature as a function of height above the solar corona.[2]

1. G. S. Vaiana, J. M. Davis, R. Giacconi, A. S. Krieger, J. K. Silk, A. F. Timothy, and M. Zombeck, *Astrophysical Journal 185*, L47 (1973).
2. R. W. Noyes, *Annual Reviews of Astronomy and Astrophysics 9*, 209–236 (1971).

Solar Spectroscopy

The first ultraviolet spectrogram of the sun showing emission lines other than Lyman-α was obtained by Johnson, Malitson, Purcell and Tousey of NRL in 1955.[1] In 1959, using the same type of spectrograph but with improved optical coatings to reduce stray light, approximately 100 emission lines (including 8 members of the Lyman series) and the Lyman continuum were recorded. A description and analysis of this 1959 spectrogram was presented at the first COSPAR Symposium in Nice (1960) by Purcell, Packer and Tousey.[2]

Here Friedman describes the new ultraviolet spectrograph, which is essentially a combination of the normal incidence spectrograph and a predisperser grating external to the slit in place of a collecting mirror. The predisperser grating was used to reduce the stray light background and was mechanically deformed to neutralize the astigmatism of the second grating. An example of the spectra obtained is shown in figure 10.1.

Examination of these new spectrograms shows that the solar continuum can be measured down to about 1,000 Å, even though the spectrum to the short wavelength side of 1,550 Å is predominately composed of emission lines. Furthermore, the continuum below 1,550 Å is free of Fraunhofer lines. Between 1,000 Å and 912 Å the continuum can be seen very faintly in the brightest plage. The Lyman continuum stands out clearly between 912 and 800 Å. Unfortunately, the spectrum below 800 Å was marred by a stray light leak in the instrument, but the strongest emission lines are still discernible. With decreasing wavelength the spectrum shows progressively the variation in emission with increasing altitude in the solar atmosphere. The Fraunhofer lines disappear when the emission level is near the top of the photosphere. Below 1,550 Å, the continuum must arise in the lower chromosphere

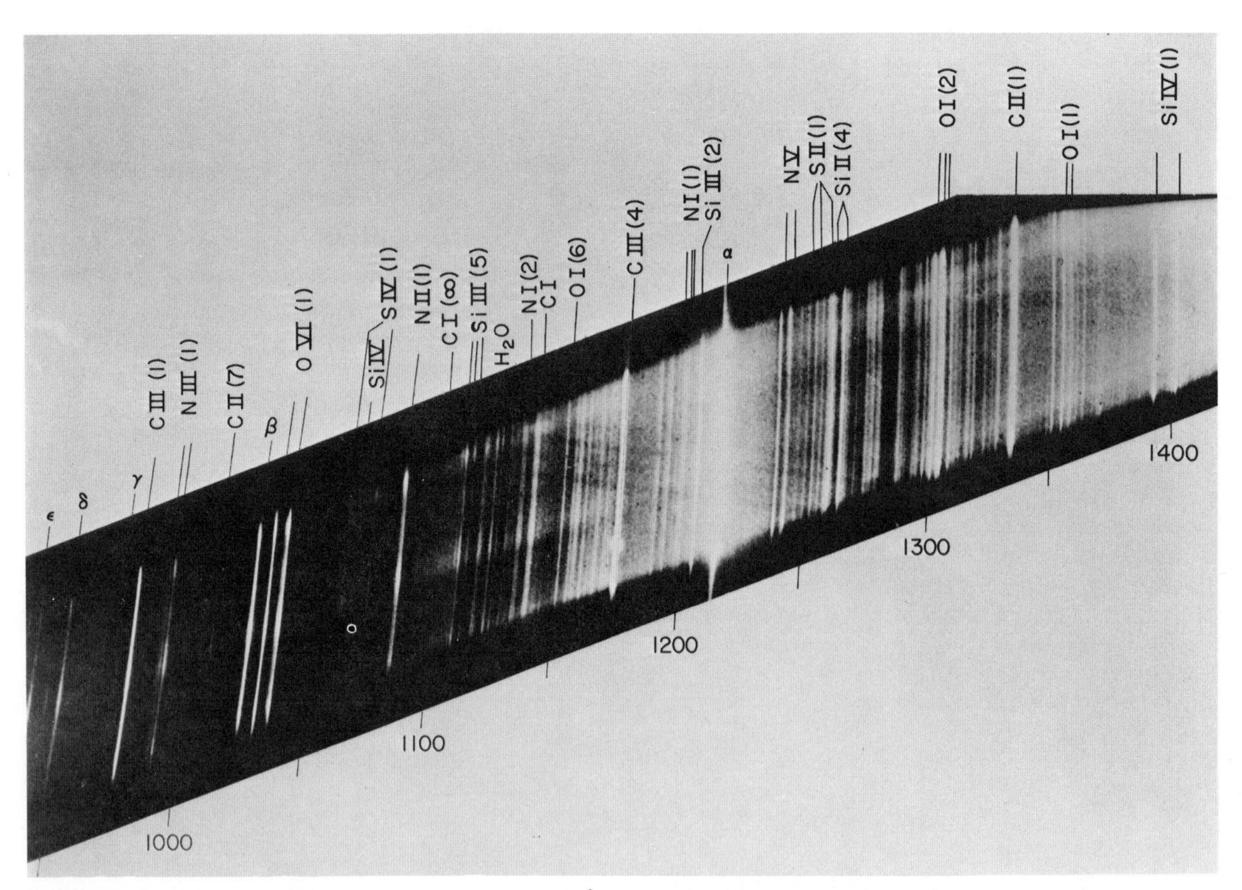

Fig. 10.1 The solar spectrum between 950 and 1,400 Å, photographed from an altitude of 148–200 km on April 19, 1960, with DC-3 film and 60-sec exposure.

and may be attributable to the superposed recombination continua of silicon, iron and hydrogen. Almost all of the nearly 200 emission lines have been identified and a detailed analysis is contained in a paper by Detwiler, Purcell and Tousey, to be published in the Proceedings of the Tenth International Astrophysical Symposium, Liège, July 1960.

The absorption bands appearing between 1,130 Å and 1,040 Å are due to water vapor carried with the rocket as a contaminant. The absorption details within the Lyman continuum, however, are attributable to N_2. Molecular nitrogen also accounts for the weakness of Lyman-γ, even at a height of 220 km because of a close coincidence with the N_2 bandhead at 972.1 Å.

The intensity of the Lyman continuum from 912 Å to 820 Å is about 0.24 erg cm^{-2} s^{-1}, which corresponds to a black body temperature of 6,600 deg K. For comparison with the photoelectric spectrometer measurements made by Hinteregger[3] of GRD, the NRL data have been averaged over 10A wide intervals. More recent data obtained by Hinteregger bring his results into fairly close agreement with the NRL data.

X-ray Solar Disk Photography

The same rocket that carried the above described spectrograph experiment also carried a pinhole camera to photograph the sun in its X-ray emission. The experiment was prepared by R. L. Blake and A. Unzicker. The pinhole was 0.013 cm in diameter and the length of the camera was 16 cm, giving a resolution of about one-tenth of the solar diameter. Covering the pinhole was a film of Parlodion, a form of cellulose nitrate, which transmitted fairly efficiently up to about 60 Å. This served as the substrate for an evaporated aluminum coating of sufficient thickness (2,500 Å) to make the system opaque to visible as well as ultraviolet light. The camera was mounted on a biaxial pointing control, built by the University of Colorado, which kept it directed toward the sun to within less than 1 minute of arc. However, the pointing control did not compensate for motion about the principal axis of the pinhole camera resulting from the rocket's precession. The images of discrete X-ray sources were therefore smeared through about 170 degrees of arc.

Comparison of the X-ray photograph (figure 10.2) with CaK line photographs shows a remarkable correlation when allowance is made for the rotation of the camera. Two bright plage areas on the west limb, one which is clearly evident a day later on the east limb and a bright plage in the northeast quadrant, account for virtually all of the X-ray emission. Densitometry of the X-ray image indicates that the intensity corresponding to the brightest plage region is about 12 times that from the weakest region of the background. When the effect of the rotation of the camera is taken into account, it appears that the X-ray emission from the brightest plage area may have been at least 70 times as intense as from the quiet background.

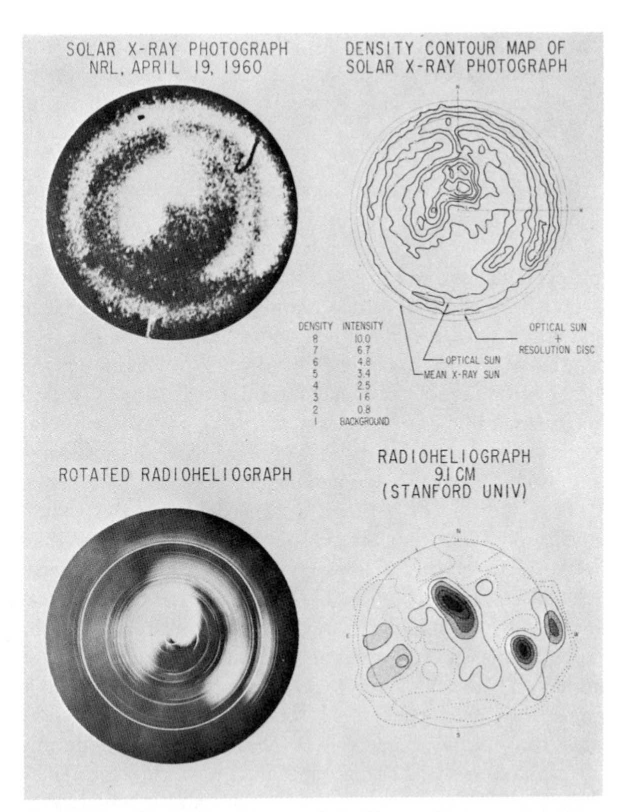

Fig. 10.2 *Clockwise from top left*: X-ray photograph of the sun obtained on April 19, 1960, with a pinhole camera flown in an Aerobee rocket; density contour map of an NRL X-ray photograph prepared from measurements of Orin Mohler, University of Michigan; radioheliogram of the sun at 9.1 cm by R. N. Bracewell and G. Swarup, Stanford University; photograph of a radioheliogram with rotation introduced during exposure to match rocket X-ray camera rotation.

Strong similarities are also evident when the X-ray image is compared with a radioheliogram made on the same date. Figure 10.2 includes a radioheliogram at 9.1 cm obtained by Bracewell and Swarup at Stanford University. The contour intervals were shaded by R. L. Blake to match the density range of the X-ray photograph. If the radioheliogram is photographed while being rotated in the same fashion as the original X-ray exposure, it yields a blurred image closely resembling the X-ray photograph.

The slowly varying component of radio decimeter wave emission appears to originate in coronal condensations at temperatures ranging up to a cutoff of about 1.6×10^6 deg K with a most typical value being about 0.6×10^6 deg K. Waldmeier[4] estimated densities 20 times normal from the observed brightness of the condensations in white light. The radio emitting regions cover areas approximately equal to the calcium plages and lie above them at heights from several tens of thousands to 100,000 km. These dense regions appear to extend in a columnar form above the photosphere and may be constrained in the shape of streamers by the actions of local magnetic fields.

Since both the X-ray emission and the decimeter waves are generated by thermal processes at temperatures of the order of 10^6 deg K and since the intensities of both radiations are proportional to the square of the electron density, it is not surprising that the X-ray disk photograph resembles the plage pattern as closely as does the radio source distribution. The sources of the slowly varying component have lifetimes of the order of three months. It has been noted that the E-region has a similar 27-day periodicity persisting over several rotations corresponding probably to the life time of X-ray emission from a particular coronal condensation.

The X-ray image of figure 10.2 was produced by the portion of the solar spectrum comprising wavelengths shorter than 60 Å. Eighty percent of this radiation was contained in local sources comparable in size to the plage areas. If the sources of solar X-ray emission in the corona are distributed in the same manner as the sources of radio microwave emission, the radiation below 60 Å must represent only the highly variable portion of the total coronal X-ray emission. It is possible that wavelengths from 60 Å to 200 Å are more uniformly distributed, as is the background microwave emission. These longer wavelengths may contribute a flux at least equal to the variable shorter than 60 Å radiation. Since the measured solar flux below 60 Å approached 1 erg/cm^2/sec at solar maximum, the total flux below 200 Å may have been of the order of 2 erg/cm^2/sec.

Here Friedman reports that an unsuccessful attempt to measure helium resonance radiation in the night sky indicates that the night helium glow must be less than 10% of the intensity of the Lyman-α glow. The low intensity lends additional support to the view that the night Lyman-α glow arises almost entirely from a telluric hydrogen geocorona. Measurements of the extreme ultraviolet flux from individual stars show that they appear as point sources and are not surrounded by nebulous emission.

1. F. S. Johnson, H. H. Malitson, J. D. Purcell, and R. Tousey, *Astrophys. J. 127*, 80 (1959).

2. J. D. Purcell, D. M. Packer and R. Tousey, *Space Research, Proc. of the First International Space Science Symposium, Nice, 1960*, ed. H. Kallmann-Bijl (Amsterdam: North-Holland Publishing Co., 1960), p. 581.

3. H. E. Hinteregger, *Astrophys. J. 132*, 801 (1960).

4. M. Waldmeier, *J. Terr. Mag. Atmos. Elec. 52*, 333 (1947).

Appended Figure

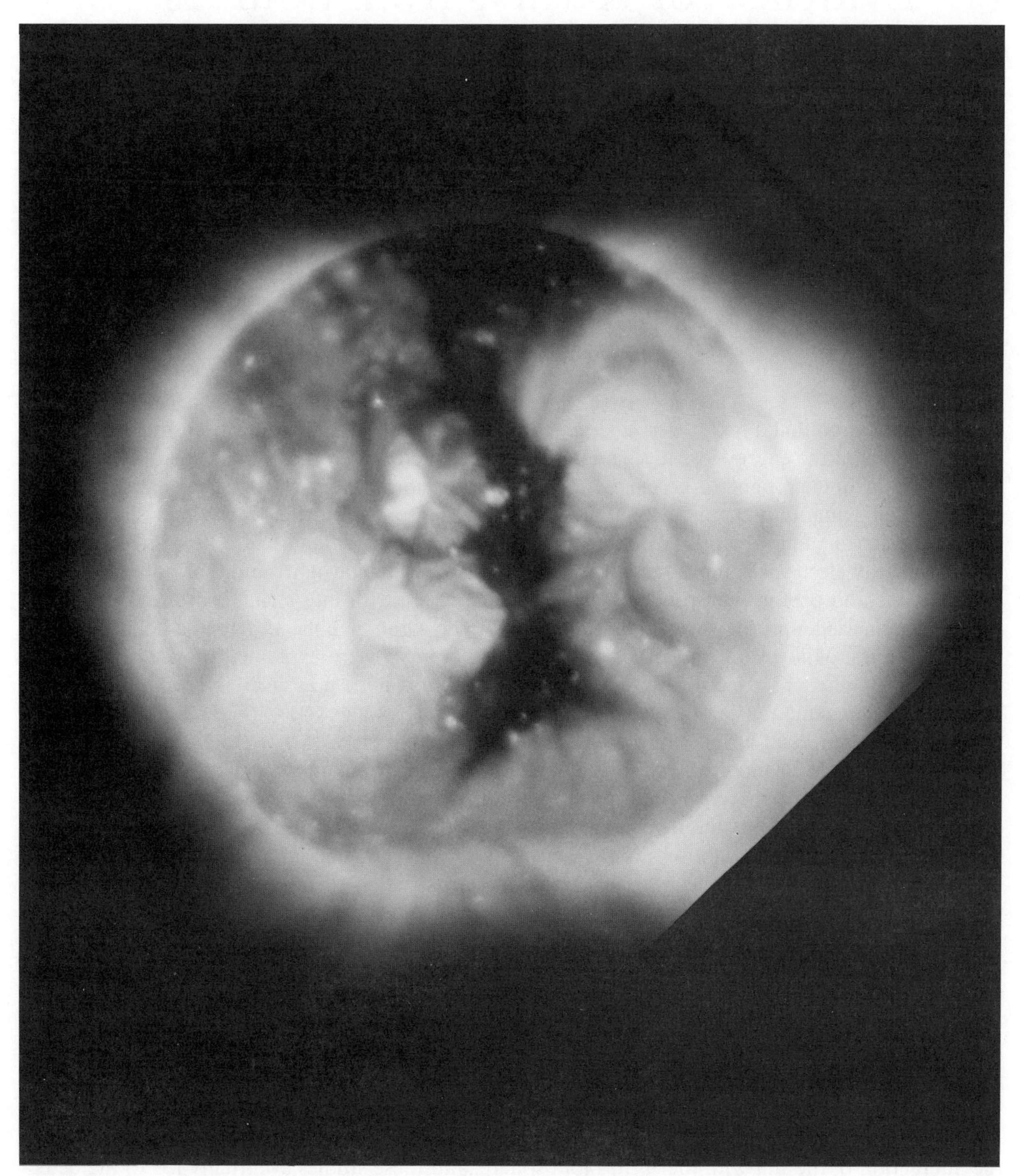

Fig. 10.3 An X-ray photograph (wavelengths 1–60 Å) of the solar corona, from a satellite of the American Science and Engineering Co. The map shows active regions (large bright features distributed along the solar equator), small bright points nearly uniformly distributed over the solar surface, and a coronal hole seen as a large dark area extending from the pole down through the middle of the disk. These coronal holes are regions where plasma might easily escape, giving rise to high-velocity streams in the solar wind. (After G. Vaiana et al., Meeting of the American Geophysical Union, April 8, 1974).

11. Evidence for X-Rays from Sources outside the Solar System

Riccardo Giacconi, Herbert Gursky, Frank R. Paolini, and Bruno B. Rossi

(*Physical Review Letters* 9, 439–443 [1962])

The first indication that a flux of very energetic radiation might be present in the cosmos came from considerations of the energy loss mechanisms of the cosmic rays. Ironically, the cosmic rays themselves were at first thought to be energetic γ rays (radiation at very short wavelengths of about 10^{-10} cm with quantum energies of about 1 MeV or 10^{-6} ergs), but they were later shown to be energetic electrons and protons (see selection 3). In the late 1940s and early 1950s it was shown that cosmic ray electrons can produce γ rays by interacting with starlight[1] and that both the cosmic ray electrons and protons can produce γ rays by interacting with interstellar matter.[2] During this time the polarized optical wavelength radiation of the Crab nebula and the radio wavelength radiation of the Milky Way and other discrete radio sources were explained by the synchrotron radiation of high-energy electrons interacting with weak magnetic fields (see selections 72, 99, and 115), and this explanation led to speculations that a variety of energetic processes might give rise to observable sources of γ radiation.[3]

In the meantime Bruno Rossi, who had previously played an important role in the study of cosmic rays, adopted a more cautious approach and reasoned that important astronomical consequences would result from the detection of X rays, whose wavelengths of about 10^{-7} cm are a thousand times longer than those of γ rays but still a thousand times shorter than those detected by conventional optical wavelength telescopes. In retrospect, it seems that astronomy at X-ray wavelengths has been extraordinarily successful, because detectable X-ray radiation is easily generated by a very hot gas and by the interaction of energetic electrons with either magnetic fields or radio wavelength radiation; in contrast, only a few celestial objects seem to generate enough γ rays to be detected.

Nevertheless, this situation was not well understood in the early days of X-ray astronomy. As a result of discussions with Rossi, Riccardo Giacconi at American Science and Engineering Company (AS & E) investigated possible celestial sources of X-ray radiation and concluded that conventional stellar objects other than the sun would be too faint to be detected with existing instruments.[4] During this period scientists at the Naval Research Laboratory (NRL) were using rockets to study the interaction of the sun with the earth's ionosphere, and these studies eventually led to the detection of X rays from the million-degree solar corona (see selection 10). Rossi and Giacconi reasoned that solar radiation might produce fluorescent X rays on the moon's surface; and the AS & E group designed the sensitive equipment needed to detect the moon's X rays, as well as searching for other sources of X-ray radiation.

As shown in the paper given here, both discrete X-ray sources and a diffuse X-ray background were discovered. Although the discrete sources had to lie outside the solar system, the observed X-ray radiation was nearly a hundred million times more intense than that expected from the ordinary thermal radiation from a nearby star. Perhaps because the X-ray sources were in the general direction of well-known radio sources in the galactic center, Cygnus, and Cassiopeia, Giacconi et al. made the argument given here that the X-ray emission might be caused by a synchrotron radiation process similar to that which accounts for the nonthermal radiation observed at radio wavelengths. In the following year, it became clear that the X-ray source Sco X-1 was not coincident with the galactic center, and this explanation for the X-ray emission became untenable. At the same time X-ray emission was detected from the direction of the Crab nebula.[5] The NRL group later used a stabilized rocket to observe this source during its occultation by the moon and concluded that the X-ray emission is extended (about 1 min of arc) and centered on the optically visible nebula.[6] Ironically, the X-ray emission from the Crab nebula is caused by synchrotron radiation, and its existence could have been inferred a decade earlier from Iosif Shklovskii's explanation of the radio and visible light emission from the Crab (see selection 72). The X-ray source Sco X-1 was, however, later found to be a rapidly varying, starlike object, and the subsequent discovery of nonperiodic, rapid X-ray pulsations in the Cygnus source played an important role in the discovery of a candidate black hole (see selection 67).

The initial results presented in this paper set the stage for a host of rocket and satellite observations of the X-ray background and of discrete X-ray sources. Over 160 discrete X-ray sources have now been detected, including starlike variable X-ray sources, X-ray supernovae, X-ray clusters of galaxies, X-ray radio galaxies, and X-ray quasars; these X-ray sources are all listed in the UHURU catalogue.[7] Perhaps the two most significant results of X-ray astronomy have been the detection of pulsating X-ray stars with optically visible companions and the discovery of X-ray emission from clusters of galaxies. The orbital parameters of the binary star systems have been used to infer the masses and moments of inertia of the neutron stars that are thought to account for the pulsating X-ray emission, whereas the X-ray emission from clusters of galaxies is thought by many to come from a hot (100 million degrees) intergalactic gas representing a mass equal to that contained in the optically visible galaxies in the clusters.[8]

1. E. Feenberg and H. Primakoff, *Physical Review 73*, 449 (1948).

2. S. Hayakawa, *Progress in Theoretical Physics* (Kyoto) *8*, 517 (1952); G. R. Hutchinson, *Philosophical Magazine 43*, 847 (1952).

3. P. Morrison, *Nuovo cimento 1*, 858 (1958).

4. An interesting historical account of developments during this period is given by R. Giacconi in *X-ray Astronomy*, ed. R. Giacconi and H. Gursky (Boston: D. Reidel Publishing Co., 1974). Speculations about the existence of celestial X-ray sources appeared in a variety of places, ranging from technical reports of AS & E to Herbert Friedman's report of possible extraterrestrial X-rays at the International Astronomical Union in Moscow in 1958. (See also H. Friedman's account of rocket astronomy in *Annals of the New York Academy of Science 198*, 267 [1972]).

5. H. Gursky, R. Giacconi, F. R. Paolini, and B. B. Rossi, *Physical Review Letters 11*, 530 (1963). See also S. Bowyer, E. T. Byram, T. A. Chubb, and H. Friedman, *Nature 201*, 1307 (1964).

6. S. Bowyer, E. T. Byram, T. A. Chubb, and H. Friedman, *Science 146*, 912 (1964).

7. R. Giacconi, S. Murray, H. Gursky, E. Kellogg, E. Schreier, and H. Tananbaum, *Astrophysical Journal 178*, 281 (1972).

8. H. Tananbaum, H. Gursky, E. M. Kellogg, R. Levinson, E. Schreier, and R. Giacconi, *Astrophysical Journal 174*, L143 (1972); E. M. Kellogg, *Astrophysical Journal 197*, 689 (1975).

DATA FROM an Aerobee rocket carrying a payload consisting of three large area Geiger counters have revealed a considerable flux of radiation in the night sky that has been identified as consisting of soft x rays.

The entrance aperture of each Geiger counter consisted of seven individual mica windows comprising 20 cm^2 of area placed into one face of the counter. Two of the counters had windows of about 0.2-mil mica, and one counter had windows of 1.0-mil mica. The sensitivity of these detectors for x rays was between 2 and 8 Å, falling sharply at the extremes due to the transmission of the filling gas and the opacity of the windows, respectively. The mica was coated with lamp-black to prevent ultraviolet light transmission. The three detectors were disposed symmetrically around the longitudinal axis of the rocket, the normal to each detector making an angle of 55° to that axis. Thus, during flight, the normal to the detectors swept through the sky, at a rate determined by the rotation of the rocket, forming a cone of 55° with respect to the longitudinal axis. No mechanical collimation was used to limit the field of view of the detectors. Also included in the payload was an optical aspect system similar to one developed by Kupperian and Kreplin.[1] The axes of the optical sensors were normal to the longitudinal axis of the rocket. Each Geiger counter was placed in a well formed by an anticoincidence scintillation counter designed to reduce the cosmic-ray background. The experiment was intended to study fluorescence x rays produced on the lunar surface by x rays from the sun and to explore the night sky for other possible sources. On the basis of the known flux of solar x rays, we had estimated a flux from the moon of about 0.1 to 1 photon cm^{-2} sec^{-1} in the region of sensitivity of the counter.

The rocket launching took place at the White Sands Missile Range, New Mexico, at 2359 MST on June 18, 1962. The moon was one day past full and was in the sky about 20° east of south and 35° above the horizon. The rocket reached a maximum altitude of 225 km and was above 80 km for a total of 350 seconds. The vehicle traveled almost due north for a distance of 120 km. Two of the Geiger counters functioned properly during the flight; the third counter apparently arced sporadically and was disregarded in the analysis. The optical aspect system functioned correctly. The rocket was spinning at 2.0 rps around the longitudinal axis. From the optical sensor data it is known that the spin axis of the rocket did not deviate from the vertical by more than 3°; for purposes of analysis, the spin axis is taken as pointing to zenith. The angle of rotation of the rocket corresponds with the azimuth Φ and is measured from north as zero and increasing to the east. The data were reduced by using the optical aspect information to determine the azimuth as a function of time. Each complete rotation of the rocket was divided into sixty equal intervals, and the number of counts in each of these intervals was recorded separately.

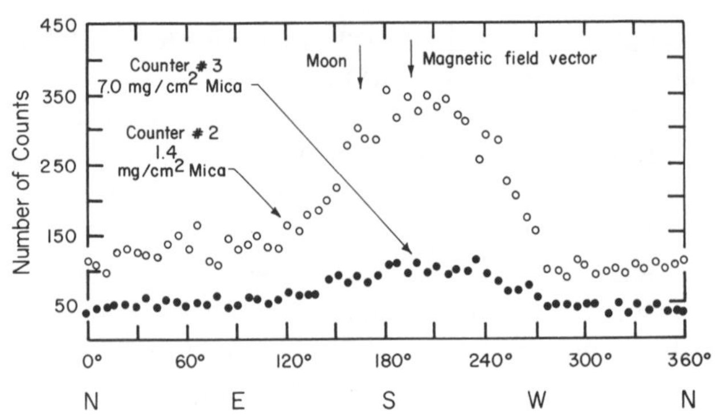

Fig. 11.1 Number of counts versus azimuth angle. The numbers represent counts accumulated in 350 sec in each 6° angular interval.

The total data accumulated in this manner during the entire flight are shown in figure 11.1 for the operating Geiger counters. The observed region of the sky is shown in figure 11.2.

Here Giacconi et al. discuss the apparent symmetry in the maximum peak of the observed radiation and note the large difference in the counting rates for the two windows. They argue

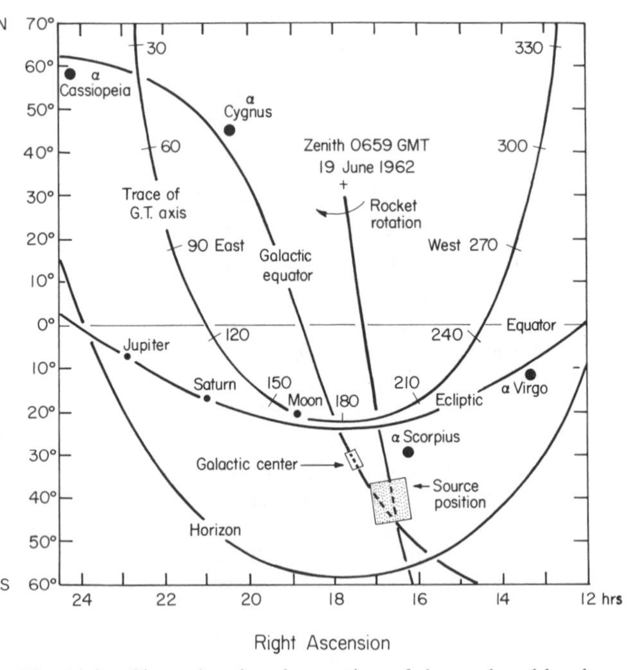

Fig. 11.2 Chart showing the portion of sky explored by the counters.

that high-energy-charged particles would give rise to a double-peaked distribution (because of the rocket roll and the axial symmetry of a charged particle spiraling about a magnetic field). They conclude that the bulk of the observed radiation is not corpuscular but electromagnetic in nature. The electromagnetic radiation is then shown to consist of soft X rays whose wavelength is about 3 Å. The intense source of emission lying near the galactic center (see figure 11.2 for source position) has a measured flux of 5.0 photons cm^{-2} sec^{-1}.

The diffuse character of the observed background radiation does not permit a positive determination of its nature and origin. However, the apparent absorption coefficient in mica and the altitude dependence is consistent with radiation of about the same wavelength as that responsible for the peak. Assuming the source lies close to the axis of the detectors, one obtains the intensity of the x-ray background as 1.7 photons cm^{-2} sec^{-1} sr^{-1} and of the secondary maximum (between 102° and 18°) as 0.6 photon cm^{-2} sec^{-1}. In addition, there seems to be a hard component to the background of about 0.5 cm^{-2} sec^{-1} sr^{-1} which does not show an altitude dependence and which is not eliminated by the anticoincidence.

The question arises whether the source of the observed x radiation could be associated with the earth's atmosphere and ascribed to some form of auroral activity. The rarity of occurrence of auroras of the magnitude required to account for the observed intensities at the latitude of the measurement makes this possibility very unlikely.

In addition, the following comments are apropos. The bulk of the measurements were obtained between altitudes of 100 to 225 km and over a range of distances from the firing point of the rocket of 15 to 100 km. The variation of the measured intensity within these limits is consistent with a source at infinity. A lower limit on the position of the source places it at a distance greater than 1,000 km. This number, combined with the measured elevation angle of 10°, places a lower limit on the altitude of the source which is about 400 km above the earth's surface. Auroral electrons of energy sufficient to produce the observed radiation would, on the other hand, penetrate down to an altitude of about 100 km and would expend the bulk of their energy at altitudes lower than 200 km. We conclude that the hypothesis of an auroral source for the observed radiation is not consistent with the data.

From figure 11.2, showing the locations of the source as well as of the moon and planets, it is clear that the observed source does not coincide with any obvious scattering body belonging to our solar system. Further, the intensity of solar x radiation at the observed wavelength is much too low in this period of the solar cycle to account for the observed intensities of the peak or of the background on the basis of back-scattered solar radiation. It would thus appear that the radiation does not originate in our solar system.

From figure 11.2 we see that the main apparent source is in the vicinity of the galactic center at a G.T. azimuthal angle of about 195°. We also see that the trace of the G.T. axis lies close to the galactic equator for a value of the azimuthal angle near 40°, which is the region where the background radiation is recorded with greater intensity. This apparent maximum of the background radiation is the general region of the sky where two peculiar objects—Cassiopeia *A* and Cygnus *A*—are located. It is perhaps significant that both the center of the galaxy where the main apparent source of x rays lies, and the region of Cassiopeia *A* and Cygnus *A* where there appears to be a secondary x-ray source, are also regions of strong radio emission. Clark[2] has pointed out that the probable mechanism for the production of the nonthermal component of the radio noise, namely, synchrotron radiation from cosmic electrons in the galactic magnetic fields, can also give rise to the x rays we observe.

In the cosmic-ray air shower experiment presently being carried out in Bolivia,[3] tentative evidence has been obtained for the existence of cosmic γ rays in the energy region of 10^{14} eV at a rate of $10^{-3} - 10^{-4}$ of the charged cosmic-ray flux at the same energy with an indication of enhanced emission in the galactic plane. Clark has shown that cosmic electrons must be produced along with γ rays by the decay of mesons that arise in the interactions of cosmic rays with interstellar matter. Since electrons at these energies lose their energy predominantly via synchrotron radiation in the galactic magnetic field, one should observe roughly the same total energy in synchrotron radiation at the earth as in γ-ray energy. For electrons of 2×10^{14} eV in a field of 3×10^{-6} gauss, the peak of the synchrotron emission is at 3 Å; in a stronger field this will happen at lower electron energies. It has been shown[4] that x rays in this wavelength region are not appreciably absorbed over interstellar distances.

With this one experiment it is impossible to completely define the nature and origin of the radiation we have observed. Even though the statistical precision of the measurement is high, the numerical values for the derived quantities and angles are subject to large variation depending on the choice of assumptions. However, we believe that the data can best be explained by identifying the bulk of the radiation as soft x rays from sources outside the solar system. Synchrotron radiation by cosmic electrons is a possible mechanism for the production of these x rays. Ordinary stellar sources could also contribute a considerable fraction of the observed radiation.

1. J. E. Kupperian and R. W. Kreplin, *Rev. Sci. Instr. 28*, 19 (1957).

2. G. W. Clark, *Nuovo cimento 30*, 727 (1963).

3. Joint Program, MIT, University of Tokyo, University of Michigan, and University at La Paz. The MIT Cosmic Ray Group has graciously made available to us the preliminary results of the air shower experiment. We have also profited considerably from discussions with Professor M. Oda of the University of Tokyo and with Professor G. W. Clark of MIT.

4. S. E. Strom and K. M. Strom, *Publs. Astron. Soc. Pacific 73*, 43 (1961).

12. Infrared Observations of the Galactic Center

Eric E. Becklin and Gerry Neugebauer

(*Astrophysical Journal 151*, 145–161 [1968])

Infrared radiation was originally discovered by an astronomer, William Herschel, who noted the rise in temperature in a thermometer placed beyond the visible red light cast by a prism. As early as 1909 V. M. Slipher had shown that photographic emulsions could record the infrared spectrum, and he found that the major planets exhibited infrared absorption lines not present in sunlight. Subsequently, these bands were identified with ammonia and methane. It was not until the 1960s, however, that the increase in the sensitivity of detectors made "infrared astronomy" a major enterprise.

Among the advantages offered by the longer infrared wavelengths is their greater penetrating power in regions of space filled with interstellar dust. In 1918 Harlow Shapley (see selection 79) had proposed that the center of our Milky Way system lay at a considerable distance from the sun in the direction of Sagittarius, but by the time this structure was commonly accepted, astronomers also realized that the galactic nucleus was completely hidden from direct view by layers of gas and dust. Infrared radiation offered a chance to detect the nucleus, and in 1947 Joel Stebbins and A. E. Whitford[1] reported on their search at 10,000 Å. They found a previously invisible nucleus concentrated at the galactic center, extending about 8° in galactic longitude and about 5° in latitude. These observations were soon confirmed, and by the early 1950s it was generally accepted that the central region of our galaxy contains an extended infrared nucleus.

Meanwhile, radiation at radio wavelengths was discovered along the Milky Way, with the most intense source coming from the direction of Sagittarius. By 1951 J. H. Piddington and H. C. Minnett[2] had shown that a sharp maximum in the radio noise appears near the galactic center. If we assume that the center of our galaxy is similar to that of M 31, it should have a nucleus about 20 pc in diameter imbedded in a larger nuclear bulge about 1,000 pc across. Because the radio evidence strikingly agrees with this model, the International Astronomical Union adopted the sharp intense radio source Sagittarius A as the position of the center of our galaxy.

Nevertheless, this narrow discrete radio source had not been observed at either optical or infrared wavelengths until Eric Becklin and Gerry Neugebauer showed in the paper given here that the galactic nucleus is observable in the infrared at 22,000 Å. They found that the near infrared flux was over 1,000 times that expected from extrapolations of the radio flux. Furthermore, it was soon realized that the galactic nucleus is even brighter at longer infrared wavelengths.[3] The nuclei of some normal galaxies, including our own, and most Seyfert galaxies, radiate infrared emission that is orders of magnitude larger than their optical or radio emission, and all these objects also exhibit emission line spectra which suggest very large internal motions.[4] These discoveries suggest that the high infrared luminosity of our galaxy might reflect violent nuclear activity.

In an extensive review published in 1977 Jan Oort summarized various observations which suggest the expulsion of matter from the galactic center, and in addition he discussed the gravitational field and the mass distribution of the center.[5] An infrared core containing a remarkable concentration of discrete infrared sources is found within one pc of the center, whereas the actual center of the galaxy is presumably an ultracompact radio source smaller than ten astronomical units (or 0.00005 pc) with a mass that may be as large as five million solar masses. Within a few thousand pc of the galactic center, regions of neutral hydrogen are moving away from the center at speeds of around a hundred kilometers per second, and within a few hundred pc of the center massive molecular clouds move away from the center with comparable speeds. The total mass in the molecular complexes is on the order of one hundred million times that of the sun. Because of the ramifications to a wide variety of astrophysical problems, the study of the galactic nucleus remains a very active branch of research.

1. J. Stebbins and A. E. Whitford, *Astrophysical Journal 106*, 235 (1947).

2. J. H. Piddington and H. C. Minnett, *Australian Journal of Scientific Research 4A*, 459 (1951). See also selection 6.

3. W. F. Hoffmann and C. L. Frederick, *Astrophysical Journal 155*, L9 (1969). E. E. Becklin and G. Neugebauer, *Astrophysical Journal 157*, L31 (1969).

4. F. J. Low and D. E. Kleinmann, *Astronomical Journal 73*, 868 (1968). F. J. Low and H. H. Aumann, *Astrophysical Journal 162*, L79 (1970).

5. J. H. Oort, *Annual Reviews of Astronomy and Astrophysics 15*, 295 (1977). Also see *I.A.U. Symposium No. 84: Large-Scale Characteristics of the Galaxy*, ed. W. B. Burton (D. Reidel, 1978).

Abstract—Infrared radiation from the nucleus of the Galaxy has been detected at effective wavelengths of 1.65, 2.2, and 3.4 μ with angular resolutions from 0.'08 to 1.'8. The structure consists of: (1) a dominant source 5′ in diameter; (2) a pointlike source centered on the dominant source; (3) an extended background; and (4) additional discrete extended sources. Contour maps of the 2.2-μ brightness distribution of the galactic center region are given for resolutions of 1.'8, 0.'8, and 0.'25.

A comparison of the infrared and radio observations shows that the dominant infrared source and the radio source Sagittarius A have the same coordinates and similar sizes.

An analysis of the observed infrared radiation predicts about 25 mag of visual absorption between the Sun and the galactic center if the source of infrared radiation is stellar. A comparison is also made between the infrared radiation from the galactic center and that from the nucleus of M31 which shows agreement in both the apparent structure and infrared luminosity of the two nuclei.

INTRODUCTION

THE DYNAMICAL CENTER of the Galaxy lies 10 kpc from the Sun in the direction of the constellation Sagittarius (Oort and Rougoor 1960; *I.A.U. Bull.*, No. 11, p. 11, 1963). Some details about the galactic center have been obtained through observations at radio wavelengths; for example, data in the decimeter and centimeter range, recently summarized by Downes and Maxwell (1966), show a discrete 10-pc diameter non-thermal source, Sagittarius A, which is believed to coincide with a small galactic nucleus.

Indications as to the nature of the galactic center have also been obtained from visual and near-infrared observations in selected areas where obscuration by interstellar dust is low (Stebbins and Whitford 1947; Baade 1951; Dufay 1960; Arp 1965). Further information has been derived from optical observations of galaxies thought to be similar to the Milky Way; specifically, the spiral galaxy M31 has a central starlike region with a diameter less than 10 pc (Hubble 1929, Baade 1955).

Observations of the galactic nucleus are not possible at visible wavelengths because of strong obscuration by intervening dust; it is well known, however, that the amount of obscuration decreases at longer wavelengths. Stebbins and Whitford (1947), using a photocell, scanned across the galactic equator at an effective wavelength of 1.03 μ but detected no small discrete source. Moroz (1961) made scans at an effective wavelength of 1.7 μ in the vicinity of Sagittarius A, but no radiation was detected. In August, 1966, one of us (E. B.) scanned the region of Sagittarius A using the wavelength band from 2.0 to 2.4 μ. On these scans infrared radiation was discovered which agrees both in position and extent with the radio source Sagittarius A.

INSTRUMENTATION

The data were obtained using an infrared photometer designed to measure extended objects. The photometer has a mirrored 5 Hz chopper which admits radiation alternately from the field to be measured and from an out-of-focus reference field 5 cm away. A reference field in the sky is normally used to cancel fluctuations in the background emission from the telescope and atmosphere, but under special conditions a constant-temperature black-body reference can be substituted for the sky. The field of view of the detector is determined by a circular aperture placed in the focal plane of the telescope; interference filters are used to determine the wavelength response.

The infrared detector consists of a liquid nitrogen-cooled $\frac{1}{2} \times \frac{1}{2}$ mm PbS cell placed behind an 8-mm f/1 field lens. The noise equivalent power of the total system mounted on a telescope is about 10^{-13} watt in the wavelength range 2.0–2.5 μ. A cooled photomultiplier makes simultaneous measurements of the signal field in the visual or near infrared.

OBSERVATIONS

Observations of the galactic center region were made in the wavelength bands 0.8–1.1 μ ($\lambda_{\text{effective}} = 0.9\ \mu$), 1.5–1.8 μ ($\lambda_{\text{effective}} = 1.65\ \mu$), 2.0–2.4 μ ($\lambda_{\text{effective}} = 2.2\ \mu$), and 3.1–3.8 μ ($\lambda_{\text{effective}} = 3.4\ \mu$); the most extensive and highest-quality observations are those made in the 2.0–2.4-μ band. A sufficient number of right-ascension and declination scans were made at 2.2 μ to permit construction of brightness distribution contour maps with resolutions of 1.'8, 0.'8, and 0.'25. Scans were also made at 1.65 μ, but for these the signal-to-noise ratios were not sufficiently large to permit construction of contour maps; however, the signal was high enough to measure the average surface brightness over many regions. At 3.4 μ, sky fluctuations were so large that scans of the general region were not possible although the flux density at 3.4 μ has been measured over small areas. No radiation was detected around 0.9 μ, and only an upper limit to the surface brightness has been obtained.

All of the data were obtained with the photometer mounted at the f/16 Cassegrain foci of the 24-, 60-, and 200-inch reflecting telescopes of the Mount Wilson and Palomar Observatories. This range of telescope apertures, when used with either a 6-, 4-, or 2-mm focal-plane diaphragm, resulted in resolutions ranging from 1.'8 to 0.'08 diameter. The data were reduced to either a flux density or surface brightness by using stars measured by Johnson (1964) as standards. The energy calibration, which agrees with that of Johnson (1965*b*), is probably accurate to better than 20 per cent. Because the emission from the telescope and sky varies with the position of the telescope, the true zero of energy can be determined only if it is possible to observe a region in the sky which is close to

Fig. 12.1 A 2.2-μ map of the galactic center in photographic form (top), together with a red photograph (bottom) of the same region. (Courtesy Eric Becklin.)

the source and which is known to have no emission. The extended nature of the center of the Galaxy precluded such a measurement; therefore, for each map a local zero of energy was chosen near the edge of that map.

Omitted here are figures that describe the infrared photometer, strip chart recordings, and contour maps of the 2.2-μ brightness distribution at 0.8 and 0.2 min of arc resolution. Included, however, is figure 12.1, which compares a 2.2-μ map of the galactic center in picture form with a red photograph of the same region. We also omit tables 1, 2, and 3, which respectively record the observation parameters, the observed flux densities, and the observed color ratios. The 2.2-μ contour map of the galactic center is compared with a 1.9-cm map in figure 12.2. Becklin and Neugebauer's comparison of the radio frequency and infrared flux densities of Sagittarius A is not reproduced. At about 14 μ the flux density of this source is comparable to the radio frequency flux (about 5×10^{-24} W m^{-2} Hz^{-1} for a 3.5-min of arc source), whereas the radio frequency data show a flux density that falls off with increasing frequency. Extrapolations of the radio-frequency data suggest a flux density of about 5×10^{-27} W m^{-2} Hz^{-1} at 14 μ.

Summary of Data

DESCRIPTION Radiation from the galactic center region at 2.2 μ can conveniently be discussed in four parts: (1) a dominant source which agrees in position and extent with the radio source Sagittarius A; (2) a pointlike source of radiation located within the dominant source (1) and near the position of its maximum brightness; (3) a general background radiation distributed predominantly along the galactic plane; (4) several smaller extended sources.

1. The dominant source has a full width at half-maximum of 3′–5′ when observed with 1.′8 resolution, and has a total extent of 5′–10′ diameter with a definite elongation along the galactic plane.

The flux densities at 2.2 μ observed with different-sized circular apertures centered on the point of maximum brightness were also computed. For diameters greater than 1.′8 it was necessary to integrate the data numerically; in all cases the energy contained in the pointlike source (2) was subtracted. It is seen that the spatial distribution is non-Gaussian but approximately follows a power law such that the flux density within a circular area of diameter D is proportional to $D^{1.2 \pm 0.1}$ corresponding to a mean surface brightness which is proportional to $D^{-0.8 \pm 0.1}$.

The power law described above holds over the entire range of the dominant source. At the finest resolutions used there is evidence that the distribution of flux density departs from the power-law dependence and flattens into a core of radiation with a full width at half-maximum of approximately 0.′3. Although the surface brightness of the core is larger than that of the surroundings, the total flux density in the core at 2.2 μ is about 4 per cent of the total flux density in the dominant source (1). The determination of the largest diameter for which the power law holds is complicated by secondary sources and the low surface brightness of the dominant source away from its center.

The color ratio $B_\nu(1.65)/B_\nu(2.2)$ is almost constant over the dominant source with a mean value of 0.20 ± 0.03; within the central 0.′25 the color ratio $B_\nu(3.4)/B_\nu(2.2)$ equals 3.2 ± 0.8 while $B_\nu(0.9)/B_\nu(2.2)$ is less than 0.001. The observed energy distribution is similar to that of a 900° K black body.

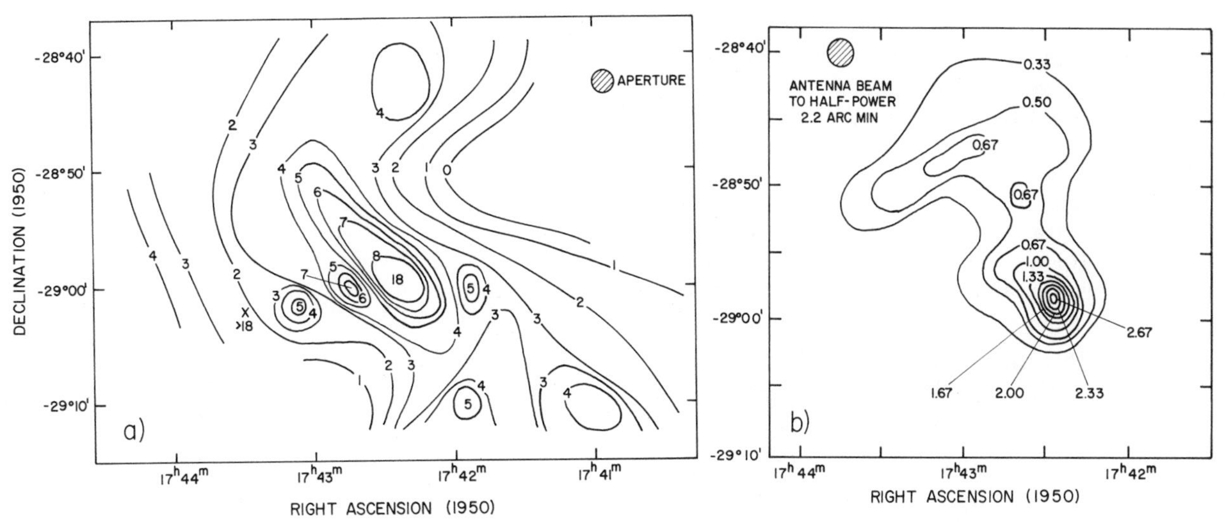

Fig. 12.2 Contour maps of the galactic center region taken at (*left*) 2.2 μ and (*right*) 1.9 cm. The 2.2-μ scans were taken with an aperture of 1.8 min of arc in diameter, and the contour lines are separated by 8.5×10^{-20} W m^{-2} Hz^{-1} sterad^{-1}. An X marks a pointlike source. The 1.9-cm map is taken from Downes, Maxwell, and Meeks (1965), and the contours represent antenna temperature in degrees Kelvin.

2. The data also show a pointlike source displaced 10″ from the centroid of the bright core of radiation. At 2.2 μ the source has a diameter less than 5″, a K-magnitude of 6.7, and a flux density of 1.4×10^{-26} W m^{-2} Hz^{-1}. The color ratio $B_\nu(1.65)/B_\nu(2.2)$ of this source is 0.15 ± 0.03, slightly smaller than the average color ratio of the dominant source (1).

3. The extent and brightness of the background is not well determined, although it appears that it may be an extension of the dominant source. The total flux density in the background is approximately 10^{-23} W m^{-2} Hz^{-1} at 2.2 μ over the area $\frac{1}{3}$ square degree surrounding the source, but is subject to large uncertainties since the level of zero flux is unknown.

4. Figure 12.2 shows seven additional localized sources of radiation. The 2.2-μ flux density in each secondary source is about one fifth that measured in the dominant source. Sources B, C, and D all have color ratio $B_\nu(1.65)/B_\nu(2.2)$ approximately three times larger than the dominant source. Only sources C and D were scanned with a resolution finer than 1.′8. With 0.′8 resolution source C remained extended, while source D was resolved into two point-like sources plus an extended source.

COMPARISON WITH DYNAMICAL CENTER The coordinates of the dynamical center of the Galaxy have been measured by Oort and Rougoor (1960) to be:

Right ascension (1950)	$17^h42^m.7 \pm 0^m.4$;
Declination (1950)	$-28°56' \pm 5'$.

As seen from figure 12.2, these coordinates lie within 4′ of the position of maximum brightness of the dominant source.

COMPARISON OF INFRARED AND RADIO OBSERVATIONS Radio observations of the center of the Galaxy (Downes and Maxwell 1966) show (*a*) a bright discrete source 3′–4′ in diameter–Sagittarius A; (*b*) several weaker secondary sources a few arc min in diameter, and (*c*) an extended background about 1° in diameter.

The dominant infrared source (1) agrees in position with Sagittarius A; the 1950 coordinates of the position of maximum infrared brightness with 0.′25 resolution and the mean position of Sagittarius A (Downes and Maxwell 1966) are:

	Right Ascension	Declination
Radio	$17^h42^m28^s \pm 2^s$	$-28°58'.5 \pm 0'.5$
Infrared	$17^h42^m30^s \pm 1^s$	$-28°59'.4 \pm 0'.1$

When observed with 1.′8 resolution, the dominant source has a full width at half-maximum of 3′–5′, which agrees with the width measured by Downes, Maxwell, and Meeks (1965) at 1.9 cm using 2.′2 resolution. Radio observations with pencil-beam resolutions finer than 2′ have not been published, so a comparison of the fine structure within the source cannot be made. The agreement in the positions and sizes of the radio source Sagittarius A and the dominant infrared source strongly suggests that the two sources are spatially coincident.

An extrapolation of the radio flux density into the infrared predicts a 3.4-μ flux density a factor of 10^3 less than that observed; therefore the mechanisms for generating the infrared and radio energy require further discussion.

Except for the dominant infrared source (1) there seems to be little correlation between the localized radio and infrared sources. However, infrared scans have been made over only a few of the radio sources in the galactic center region.

The data on the infrared background radiation are not sufficient to make a definitive comparison with the radio background. However, both are concentrated along the galactic plane, and extend about $\frac{1}{2}°$ from the dominant source.

COMPARISON WITH M31 Here Becklin and Neugebauer show that at 14 pc resolution the 2.2-μ profiles of the galactic center and of the nuclear region of M31 (Neugebauer and Becklin 1967) are comparable. At 3.4 μ, where interstellar absorption is low, the observed mean surface brightness of M31 (Johnson 1966*a*) agrees within a factor of 2 with the extrapolated mean surface brightness of a similar area of the galactic center.

Discussion

We now consider the nature of the source of the infrared radiation from the galactic center. For the most part we shall assume that the infrared radiation is of stellar origin and that the observed 900° K black-body spectrum arises from the effects of interstellar absorption upon the spectra of ordinary stars. The "point" source will be considered after the extended sources are discussed.

INTERSTELLAR ABSORPTION The energy spectrum observed in the dominant infrared source is a function of both the emitted energy spectrum and the total amount and spectral characteristics of the interstellar absorption. In the following discussion, the intrinsic color of the galactic center and the interstellar reddening curve are assumed known; the total amount of obscuration and the infrared energy emitted by the source are then calculated.

The intrinsic energy spectrum of the galactic center region is taken to be that measured in the nucleus of M31 by Johnson (1966*a*) and by Neugebauer and Becklin (1967); it approximates a 4,000° K black body. The infrared interstellar absorption law is averaged from the results of Johnson and Borgman (1963) and Johnson (1965*a*) for the solar neighborhood in the direction of the galactic center; this average (Table 12.1) agrees with the observations of Whitford (1948) and calculations by van de Hulst (1949) for dielectric grains (Curve No. 15).

The total absorption is found from the equation

$$(B_{\nu 1}/B_{\nu 2})_{\text{obs}} = (B_{\nu 1}/B_{\nu 2})_{\text{emit}} \exp[-\tau_0(A_1 - A_2)],$$

Table 12.1
Extinction coefficient

$\lambda(\mu)$	A	$\lambda(\mu)$	A
0.55	1.00	1.65	0.17
0.90	0.46	2.2	0.10
1.25	0.30	3.4	0.045

Table 12.2 Calculated absorption to galactic center

$\lambda(\mu)$	ΔM	$(B_\nu)_{\text{obs}}/(B_\nu)_{\text{emitted}}$
0.55	27.0	10^{-10}
0.90	13.0	10^{-5}
1.65	4.6	0.02
2.2	2.7	0.1
3.4	1.2	0.3

where $B_{\nu 1}$ is the surface brightness (W m^{-2} Hz^{-1} sterad^{-1}) at ν_1, A_1 is the extinction coefficient at ν_1 normalized to 1.00 at 0.55 μ (table 12.1), and τ_0 is the total optical depth at 0.55 μ. The data at the wavelength pairs 2.2 and 1.65 μ, and 2.2 and 3.4 μ, give $\tau_0 = 25 \pm 4$, and 30 ± 10, respectively; a lower limit $\tau_0 = 20$ is established from the data at 2.2 and 0.9 μ. The agreement of all three values gives confidence that the initial assumptions about the intrinsic energy spectrum and the absorption law are correct. For convenience, the value $\tau_0 = 25$ will be adopted for further calculations.

Further confidence in the assumptions is provided by the calculation that the mean intrinsic surface brightness of the galactic center at 2.2 μ out to a diameter of 14 pc is 10^{-17} W m^{-2} Hz^{-1} sterad^{-1}, while the comparable quantity for M31 is 0.4×10^{-17} W m^{-2} Hz^{-1} sterad^{-1} (Neugebauer and Becklin 1967). In comparison, the mass of the galactic nucleus as determined by Rougoor and Oort (1960) and the mass of the nucleus of M31 as determined by Kinman (1965) from the data of Lallemand, Duchesne, and Walker (1960) agree within a factor of 3.

The total calculated absorption in magnitudes and the ratio of measured to emitted brightness are given as a function of wavelength in table 12.2. The total absorption of 27 mag at 0.55 μ is in reasonable agreement with the estimate of Münch (1952) of 2 mag of absorption per kpc for the region within 1 kpc of the Sun.

The absorption curve used above included no wavelength-independent absorption, although Moroz (1961) explained his failure to detect radiation as most probably caused by the existence of 1 or 2 mag of neutral absorption. If this amount of neutral absorption were present, then at 2.2 μ the galactic center would be intrinsically about 10 times brighter than a similar region of M31.

For selected areas of the sky, Johnson (1965*a*, 1966*b*) has derived a reddening curve which, for wavelengths less than 3 μ, has a much larger extinction coefficient than shown in table 12.1. If the reddening of the galactic center radiation were given by the average of the extinction curves recently summarized by Johnson (1966*b*, Figs. 24–37), then the galactic nucleus would be intrinsically 75 times brighter than the nucleus of M31.

In summary, if the luminosities of the nucleus of M31 and of the galactic nucleus are equal within a factor of 3 and both nuclei have similar colors, then the interstellar reddening curve extrapolates to $1/\lambda \rightarrow 0$ in a manner shown in table 12.1 and the extinction coefficient at 2.2 μ cannot be greater than 15 per cent of the visual extinction coefficient.

SECONDARY EXTENDED SOURCES If the infrared radiation of the dominant source is of stellar origin, it is important to ask whether the secondary sources are extensions of the central source viewed through relatively little obscuration or whether they are separate clusters of stars.

Evidence that the secondary sources could be extensions of the central source comes from their observed relatively blue colors and their brightnesses. When the brightnesses of the secondary sources are corrected using an optical depth derived purely from the observed colors, the secondary sources no longer stand conspicuously above the background. Unfortunately, the uncertainties in the background levels are too large to make the conclusion quantitative.

The sources could also be discrete star clouds, in which case the observed blue colors imply they are probably between the Sun and the galactic center. If the close angular separations between the center of the Galaxy and the secondary sources indicate these sources are in fact separated by about 100 pc, there must be 10 to 15 mag of absorption within the region of the galactic center; extragalactic systems do not exhibit such a dense concentration of absorption toward their nuclear regions (Baade 1951).

PHYSICAL PROPERTIES OF THE GALACTIC CENTER Some inferences about the physical nature of the galactic center can be made using the results of the previous sections. The 2.2-μ intrinsic brightnesses of the dominant source averaged over circular areas centered on the position of maximum brightness have been calculated from the derived optical depths. In addition, the total luminosities have been estimated assuming an energy distribution equal to that measured by Johnson (1966*a*) for the nuclear region of M31. The results of these calculations are presented in table 12.3 as a function of diameter; the uncertainty in the interstellar reddening curve introduces a factor of 2 uncertainty in these results.

The total mass can be found from the calculated luminosity if a mass-to-luminosity ratio for the stars in the galactic center is estimated. For M31, the following values of M/L_{pg} have been determined: 23 (Schmidt 1957), 16 (Spinrad 1962), and 3.6 (Lallemand, Duchesne, and Walker 1960). Schmidt's determination is for the whole of M31, while the others are for the nuclear region only. The total mass producing the observed radiation within a circular area is given in table 12.3 for $M/L_{pg} = 10$; since the surface brightness increases rapidly toward the center, the observed mass within a given projected diameter is effectively equal to the mass within a sphere of the same diameter. The mean density in the 1 pc diameter central core is approximately 10^7 times the density in the neighborhood of the Sun. The power-law dependence of the flux density implies that the mass density of material falls from the center approximately as $R^{-1.8}$.

Rougoor and Oort (1960) have determined the mass within 70 pc of the galactic center and have estimated, by extrapolation, that the mass within a radius of 20 pc is $0.7 \times 10^8 \, M_\odot$. According to table 12.3, the mass inside a radius of 20 pc is $2.3 \times 10^8 \, M_\odot$; this agreement gives added confidence in the assumptions.

Table 12.3 Calculated intrinsic physical properties

Diameter (pc)	Mean intrinsic 2.2-μ surface brightness (10^{-18} W m^{-2} Hz^{-1} sterad^{-1})	Intrinsic luminosity ($10^6 \, L_\odot$)[a]	Mass ($10^6 \, M_\odot$)[b]
1	78	1	3
2	45	2	6
5	21	7	20
10	12	15	45
20	7	35	100
40	4	80	230
60	3	130	370

[a] $L_\odot = 4 \times 10^{26}$ w. [b] $M_\odot = 2 \times 10^{30}$ kg.

The high stellar density obtained for the bright 1 pc diameter core suggests that stellar collisions should be considered. Using the results tabulated by Spitzer and Saslaw (1966), the average time before a star of 1 solar mass is involved in a physical collision is about 10^{10} years in the 1 pc core and the rate of collisions is about 10^{-4} yr^{-1}.

OTHER MODELS FOR THE GALACTIC CENTER Although the assumption that the observed infrared radiation arises from a stellar population similar to that present in the nuclear region of M31 is consistent with the data, other models for the source of radiation which cannot be ruled out at present will be considered.

Infrared radiation could originate from a non-thermal source at the center of the Galaxy. In fact, if 15 mag of visual absorption are assumed between the Sun and the galactic center, the corrected spectral energy distribution of the galactic center has features strikingly like those of the energy distribution measured from NGC 1068 (Howard and Maran 1965; Epstein 1967; Pacholczyk and Wisniewski 1967). The energy distribution of 3C 273 also has qualitatively similar features (Low and Johnson 1965), but more measurements of the galactic center and of non-thermal sources are needed at frequencies between 10^{10} and 10^{14} Hz before meaningful comparisons can be made.

If the infrared radiation originates in an optically thin H II region, the correction for interstellar absorption would be 20 mag at 0.55 μ, and the average emission measure for a region 10 pc in diameter would be 2×10^7 pc cm^{-6}. Although this emission measure is consistent with the measured radio flux density, the spectral index of -0.7 indicates that emission from an H II region cannot account for all of the observed radio radiation. Downes and Maxwell (1966) estimate an emission measure of about 10^6 pc cm^{-6} which would account for only 5 per cent of the observed brightness at 2.2 μ.

It is, of course, possible that the observed radiation arises from a stellar population dominated by infrared stars. If the observed energy distribution is corrected for a minimum of 10 mag of absorption at 0.55 μ the source of radiation must have a color temperature of 1,200° K. No other stellar grouping of this nature has been observed.

It should be emphasized that any of the above models contradicts the assumption that the galactic center is similar to the nucleus of M31.

Becklin and Neugebauer discuss the pointlike infrared source near the galactic center. If the source is a single luminous star obscured by 27 mag of visual absorption, its 2.2-μ absolute mag of -11.0 would be similar to that of α Orionis, a Population I star. Spitzer and Saslaw (1966) have suggested that these young stars may form at the dynamical center of the Galaxy from matter dispersed during collisions.

SUMMARY

We have observed an extended source of infrared radiation which we believe to be at the nucleus of the Galaxy. The infrared source is similar in shape and luminosity to the nucleus of M31, and its coordinates agree with those of the dynamical center of the Galaxy.

We consider the further agreement between the position and extent of the infrared source and the radio source Sagittarius A as conclusive evidence that Sagittarius A also lies at the dynamical center of the Galaxy.

We believe that the infrared radiation most likely originates from stars with a mean infrared black-body temperature greater than 4,000° K located at the center of the Galaxy. The visual extinction is on the order of 25 mag; the physical properties deduced are similar to those observed in the nucleus of M31. However, a non-thermal origin for all or part of the observed radiation cannot be excluded.

The nature of the pointlike source observed near the center of the dominant source is unclear. Whether it is stellar or non-thermal in nature must await further measurements, especially at longer wavelengths.

Arp, H. 1965, *Ap. J. 141*, 43.
Baade, W. 1951, *Pub. Obs. Univ. Mich. 10*, 7.
———. 1955, *Mitt. Astr. Ges.*, p. 51.
Downes, D., and Maxwell, A. 1966, *Ap. J. 146*, 653.
Downes, D., Maxwell, A., and Meeks, M. L. 1965, *Nature 208*, 1189.
Dufay, J. 1960, *Ann. d'ap. 23*, 451.
Epstein, E. 1967 (private communication) [See Schorn, R. A., Epstein, E. E., Oliver, J. P., Soter, S. L., and Wilson, W. J. 1968, *Ap. J. 151*, L27 and Low, F. J., Kleinmann, D. E., Forbes, F. F. and Aumann, H. H. 1969, *Ap. J. 157*, 197.]
Howard, W. E., and Maran, S. P. 1965, *Ap. J. Suppl. 10*, 1.
Hubble, E. 1929, *Ap. J. 69*, 103.
Hulst, H. C. van de. 1949, *Rech. Astr. Obs. Utrecht 11*, Pt. 2.
Johnson, H. L. 1964, *Bol. Obs. Tonantzintla y Tacubaya 3*, 305.
———. 1965*a*, *Ap. J. 141*, 923.
———. 1965*b*, *Com. Lunar and Planet. Lab. 3*, No. 53.
———. 1966*a*, *Ap. J. 143*, 187.
———. 1966*b*, *Stars and Stellar Systems*: vol. 7, *Nebulae and Interstellar Matter*, ed. L. H. Aller and B. M. Middlehurst (Chicago: University of Chicago Press).
Johnson, H. L., and Borgman, J. 1963, *B.A.N. 17*, 115.
Kinman, T. D. 1965, *Ap. J. 142*, 1376.
Lallemand, A., Duchesne, M., and Walker, M. F. 1960, *Pub. A.S.P. 72*, 76.
Low, F. J., and Johnson, H. L. 1965, *Ap. J. 141*, 336.
Moroz, V. I. 1961, *Astr. Zh. 38*, 487.
Münch, G. 1952, *Ap. J. 116*, 575.
Neugebauer, G., and Becklin, E. 1967, Pub. by Sandage, A. R., Becklin, E., and Neugebauer, G. 1969, *Ap. J. 157*,
Oort, J. H., and Rougoor, G. W. 1960, *M.N.R.A.S. 121*, 171.
Pacholczyk, A. G., and Wisniewski, W. Z. 1967, *Ap. J. 147*, 394.
Rougoor, G. W., and Oort, J. H. 1960, *Proc. Nat. Acad. Sci. 46*, 1.
Schmidt, M. 1957, *B.A.N. 14*, 17.
Spinrad, H. 1962, *Ap. J. 135*, 715.
Spitzer, L., and Saslaw, W. C. 1966, *Ap. J. 143*, 400.
Stebbins, J., and Whitford, A. E. 1947, *Ap. J. 106*, 235.
Whitford, A. E. 1948, *Ap. J. 107*, 102.

13. Interstellar Deuterium Abundance in the Direction of Beta Centauri

John B. Rogerson and Donald G. York

(*Astrophysical Journal* (*Letters*) *186*, L95–L98[1973])

The ultraviolet portion of the spectrum is of particular interest to astronomers both because the energy of very hot objects radiates primarily in this region and because the resonance lines of many neutral and ionized atoms emit these wavelengths. In particular, the Lyman series of neutral hydrogen, from 1,215 Å to 912 Å, and the vibrational transitions of molecular hydrogen around 1,100 Å fall in the ultraviolet. Until the advent of rocket, balloon, and satellite astronomy about two decades ago, the spectral region below 3,000 Å was inaccessible for astronomical investigations.

The most sophisticated of the new space hardware devices for ultraviolet studies are the Orbiting Solar Observatories[1] and the Orbiting Astronomical Observatories.[2] The third OAO, named the *Copernicus* satellite, which was successfully launched in the fall of 1972, carried a high-resolution (0.05 Å) spectrometer specially designed by a team of Princeton University astronomers for observations of the narrow interstellar absorption lines at ultraviolet wavelengths. Within a year the Princeton group had published five papers on far ultraviolet absorption lines.

Perhaps the most significant result of the *Copernicus* mission was the detection of interstellar deuterium, the report of which is selected here as representative of space research in the ultraviolet. The deuterium was presumably created in the big-bang explosion that produced the expansion of the universe. As John B. Rogerson and Donald G. York point out, the abundance of deuterium has considerable significance for cosmology; their observations provide one of the most important pieces of evidence for an open, hyperbolic universe that will forever continue to expand.

1. L. Goldberg, R. W. Noyes, W. H. Parkinson, E. M. Reeves, and G. L. Withbroe, *Science 162*, 95 (1968).
2. J. B. Rogerson, L. Spitzer, J. F. Drake, K. Dressler, E. B. Jenkins, D. C. Morton, and D. G. York, *Astrophysical Journal* (*Letters*) *181*, L97 (1973).

Abstract—Interstellar absorption lines due to the Lyman series transitions in hydrogen and deuterium have been observed in the spectrum of β Cen. From these, a ratio of deuterium to hydrogen, by number, of 1.4 ± 0.2 (m.e.) $\times 10^{-5}$ has been obtained. If one assumes that the present deuterium abundance is a relic of the big-bang element synthesis, a value of 1.5×10^{-31} g cm^{-3} for the present density of the Universe is derived.

Introduction

A NUMBER of investigations have been undertaken to determine the interstellar deuterium abundance, but completely satisfying results have not yet been obtained. The deuterium analog of the 21-cm hydrogen hyperfine transition at 91.6 cm has been carefully observed by Weinreb (1962), who obtained an upper limit on N(D I)/N(H I) of 7.7×10^{-5}, and more recently by Cesarsky, Moffet, and Pasachoff (1973) who report a probable detection of this line yielding values between 3.3×10^{-5} and 5.0×10^{-4}. Radio emission lines of DCN, measured in the Orion Nebula by Jefferts, Penzias, and Wilson (1973), imply a value between 1.2×10^{-7} and 1.0×10^{-5} (Solomon and Woolf 1973). Black and Dalgarno (1973), using measurements of N(HD)/N(H_2) by Spitzer *et al.* (1973), find that N(D)/N(H) lies between 5×10^{-6} and 2×10^{-4} in the ζ Oph cloud.

In this *Letter* we report on a direct measurement of the interstellar deuterium abundance ratio using the *Copernicus* spectrometer. The Lyman series lines of hydrogen and deuterium from Lβ to Lζ have been scanned at 0.05 Å resolution in the spectrum of the B1 III star β Cen.

Data Analysis

The Lβ through Lε lines of deuterium are all visible, but the weak Lε line was excluded from the analysis due to a poor signal-to-noise ratio. The wavelengths and measured equivalent widths are given in table 13.1.

Theoretical curves of growth were computed for each line and for b-values of 4, 6, and 8 km s^{-1} ($b = 2^{1/2}$ times the atomic velocity dispersion in the line of sight to the star). The measured equivalent widths have been analyzed with these curves of growth and yield a mean column density of 4.8 ± 0.3 (m.e.) $\times 10^{14}$ cm^{-2} and a mean b-value of 6.6 ± 0.3 (m.e.) km s^{-1}.

Table 13.1 Data on interstellar hydrogen and deuterium lines

Source	λ_{lab}	$\lambda_{obs}{}^{a} - \lambda_{lab}$	W_λ (Å)
H	1,025.722	−0.010	0.739 (+.031, −.055)[b]
D	1,025.442	−0.005	0.075
H	972.537	+0.005	0.323 (+0.008, −0.011)[c]
D	972.272	−0.002	0.051
H	949.743	−0.003	0.235 ± 0.020
D	949.485	-0.010 ± 0.015	0.034 ± 0.001
H	937.804	$+0.001 \pm 0.015$	0.191 ± 0.012
H	930.748	-0.004 ± 0.008	0.177 ± 0.037

[a] Based on one determination except for 949.485 (2 measures), 937.805 (4 measures), and 930.748 (2 measures).

[b] Obtained by fitting a damping profile.

[c] Obtained by fitting a Voigt profile (see text). Unless otherwise noted, equivalent widths were measured directly from the plotted data.

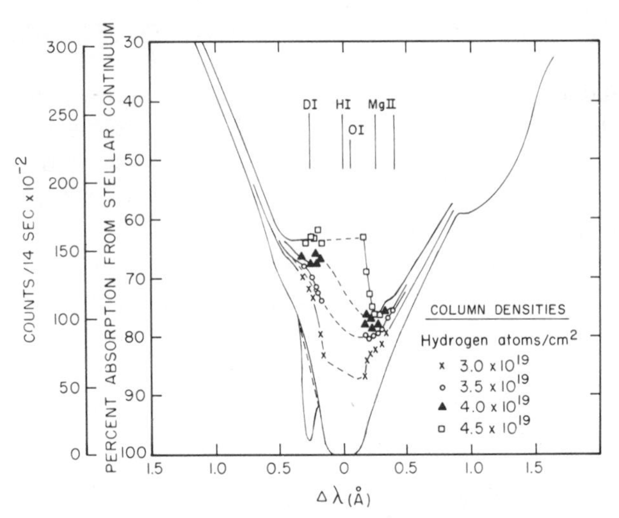

Fig. 13.1 Examples of continua derived from profile fitting (see text) for different values of N(H I) for the observed Lyman-β line. Wavelengths of interstellar lines that probably affect the long-wavelength wing are noted. Two continua are shown at the deuterium line. The solid line is the line used in the fitting; the dashed line is the lowest continuum that can reasonably be drawn for the line. The dashed line between the derived continua for hydrogen (blue wing) for N(H I) equal to 3.0 and 3.5×10^{19} shows the effect of using a continuum intermediate between the two continua shown at the deuterium Lyman-β line.

The hydrogen column density can be derived directly from the Lβ line which is on the damping part of the curve of growth. Figure 13.1 shows the continuum derived by multiplying the observed counts by exp($+\tau_\lambda$), where τ_λ is the optical depth of a purely damped Lβ line at a given τ_λ, for a variety of column densities. The red wing of Lβ is deeper than the blue wing, due to blends of interstellar lines of O I and Mg II (2 lines) (and possibly to very weak, hydrogen components) and probably to other stellar features. The true continuum for the interstellar line consists of the core of the broad stellar Lβ line plus whatever other lines are present; and without detailed model fits, the true continuum cannot

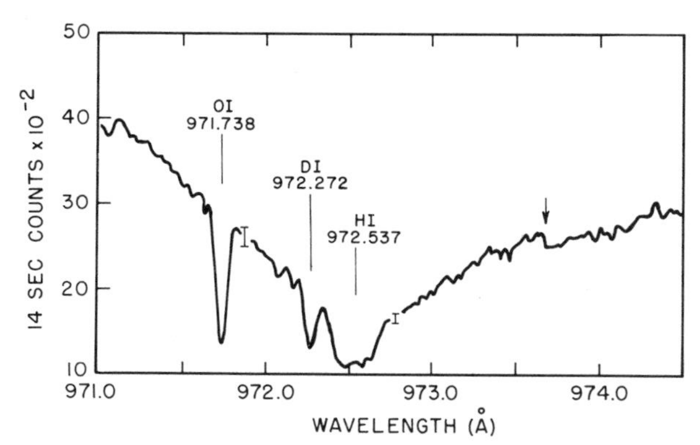

Fig. 13.2 The wavelength region near Lyman-γ in the far-ultraviolet spectrum of β Centauri. The stellar Lyman-γ line dominates the figure. Wavelengths of interstellar lines are marked. Interpolation between points in the raw data yielded a point for every 0.012 Å, and two scans have been averaged in some places. The arrow marks a discontinuity in baseline for data taken in separate orbits.

be determined. However, as shown in figure 13.1, the derived continua for $N(\text{H I}) > 4.5 \times 10^{19}$ and $N(\text{H I}) < 3 \times 10^{19}$ imply structure that would in fact be smeared out by the 160 km s^{-1} rotational velocity of β Cen (Uesugi and Fukuda 1970). The adopted value is $N(\text{H I}) = 3.5 \pm 0.5 \times 10^{19}$. The distance to β Cen, based on the spectroscopic parallax, is 81 pc, yielding a mean volume density of 0.14 cm^{-3}.

Table 13.1 shows the wavelengths, equivalent widths, and method of determination of the equivalent widths. Fitting the quoted equivalent width of Lε to the relevant curve of growth yields a b-value of 8.8 ± 1.0 km s^{-1}, for $N = 3.5 \times 10^{19}$. Using a Voigt profile, the best fit to the Lγ line, assuming $N(\text{H I}) = 3.5 \times 10^{19}$ gives $b = 10.2\,(+0.5, -1.0)$ km s^{-1}. The average of these two b-values, $b = 9.5\,(+1.1, -1.4)$ km s^{-1}, is in good agreement with the deuterium b-value quoted above, if the Doppler velocity is assumed to be entirely thermal, since then $b(\text{deuterium}) = b(\text{hydrogen})/\sqrt{2}$. The deuterium value then scales to a hydrogen b-value of 9.4 ± 0.4. The inferred temperature of 5,400° (+1,300°, −1,500°) K may be regarded as an upper limit for this line of sight.

DISCUSSION

The results of the previous two sections yield a value for $N(\text{D I})/N(\text{H I})$ of 1.4 ± 0.2 (m.e.) $\times 10^{-5}$. Since lines of H_2 are not observed in this spectrum, and since the degree of ionization is likely to be the same for the two isotopes, the measured ratio is essentially identical to the deuterium abundance, $N(\text{D})/N(\text{H})$. Assuming this value to be typical of the interstellar gas, we compare it with solar-system measurements.

Deuterium has not been detected in the solar photosphere or in the solar wind, but under the assumption that most if not all the ^{3}He measured in the solar wind results from the nearly complete burning of deuterium into ^{3}He in the Sun, Geiss and Reeves (1972) estimate a deuterium abundance for the protosolar gas of 2.5×10^{-5}. They further point out that chemical isotope fractionation during formation of at least the inner solar system can increase this value to the observed ocean-water deuterium abundance of 15.7×10^{-5} (Craig 1961). Trauger *et al.* (1973) report a deuterium abundance of $2.1 \pm 0.4 \times 10^{-5}$ based on measurements of HD in the spectrum of Jupiter. These results should not be compared directly with the value reported here since the interstellar deuterium has presumably decreased between the time the Sun formed and the present due to the dilution of the interstellar gas by deuterium-free matter released by evolved stars. According to the galactic evolutionary model of Truran and Cameron (1971) this decrease amounts to a factor of about 1.8, which allows us to predict a deuterium abundance for the protosolar gas of 2.5×10^{-5} in satisfactory agreement with the above results.

The deuterium abundance is of special interest since it is not produced in stellar nuclear processes nor (in significant quantities) by cosmic-ray spallation processes (see Reeves *et al.* 1973). Colgate (1973) has suggested that deuterium can be formed in abundance during supernova explosions, but this point of view has been strongly criticized by Reeves (1973). We infer that whatever deuterium we now see is a relic of a pregalactic phase of deuterium creation. If this phase of deuterium creation is identified with the big bang, then an estimate can be made of the present density of the Universe. Wagoner (1973) has calculated the results of element synthesis during the high-temperature phase of the big bang and (assuming zero leptonic number) gives the present density (when the temperature has fallen to 2.7° K) as a function of the mass fraction created of each element species. From the model of Truran and Cameron (1971) we find that the present mass fraction of deuterium is a factor 6.4 lower than it was at the end of the big bang. Thus the mass fraction of big-bang deuterium is 1.8×10^{-4}; and with Wagoner's results, we find a value of 1.5×10^{-31} g cm^{-3} for the present density of the Universe. Under the extreme assumption of no evolutionary deuterium depletion, the density would be 4.7×10^{-31} g cm^{-3}.

This value of the matter density does not conflict with the lower limit found by Shapiro (1971) of 5×10^{-32} g cm^{-3} (assuming a Hubble constant of 50 km s^{-1} Mpc^{-1}), but it falls a factor 27 short of the critical density for closing the Universe (4×10^{-30} g cm^{-3}).

Black, J. H., and Dalgarno, A. 1973, *Ap. J.* (*Letters*) *184*, L101.

Cesarsky, D. A., Moffet, A. T., and Pasachoff, J. M. 1973, *Ap. J.* (*Letters*) *180*, L1.

Colgate, S. A. 1973, *Ap. J.* (*Letters*) *181*, L53.

Craig, H. 1961, *Science 133*, 1833.

Geiss, J., and Reeves, H. 1972, *Astr. and Ap. 18*, 126.
Jefferts, K. R., Penzias, A. A., and Wilson, R. W. 1973, *Ap. J.* (*Letters*) *179*, L57.
Reeves, H. 1973, in *Supernovae and Supernova Remnants*, ed. C. B. Cosmovici (Boston: Reidel), p. 381.
Reeves, H., Audouze, J., Fowler, W. A., and Schramm, D. N. 1973 *Ap. J. 179*, 909.
Shapiro, S. L. 1971, *A.J. 76*, 291.
Solomon, P. M., and Woolf, N. J. 1973, *Ap. J.* (*Letters*) *180*, L89.
Spitzer, L., Drake, J. F., Jenkins, E. B., Morton, D. C., Rogerson, J. B., and York, D. G. 1973, *Ap. J.* (*Letters*) *181*, L116.
Trauger, J. T., Roesler, F. L., Carleton, N. P., and Traub, W. A. 1973, *Ap. J.* (*Letters*) *184*, L137.
Truran, J. W., and Cameron, A. G. W. 1971, *Ap. and Space Sci. 14*, 179.
Uesugi, A., and Fukuda, I. 1970, *Cont. Inst. Ap. and Kwasan Obs., Univ. Kyoto*, no. 189.
Wagoner, R. V. 1973, *Ap. J. 179*, 343.
Weinreb, S. 1962, *Nature 195*, 367.

CHAPTER II

The Solar System

14. The Moon's Face: A Study of the Origin of Its Features

Grove Karl Gilbert

(*Bulletin of the Philosophical Society of Washington 12*, 240–292 [1892])

In this paper Grove Karl Gilbert discusses the origin of the main surface features of the moon, most of which were first observed by Galileo in 1610. Lunar craters consist of a circular plain with a surrounding raised wreath of material; the central plain is a few thousand feet lower than the general surface of the moon. As Gilbert points out, most volcanic craters are shallow depressions perched on top of conical mountains, and this difference in form suggests that the lunar craters are not volcanic in origin. The circular rims and depressed central plains can be explained, however, as the natural result of meteoritic impact.

Gilbert develops his impact theory into the moonlet theory for the origin of the moon itself. According to this hypothesis, meteoritic material gathered together into aggregates called moonlets, concentrated in the plane of the earth's equator. The moon was slowly built up as the moonlets collided, and the surface features we now see are the remnants of the last moonlets to hit. Because the moonlets must have also collided with the earth, Gilbert concludes that the observed lunar features are the result of impacts that occurred before the process of erosion began on the earth.

The modern space age has confirmed most of Gilbert's views. The crust of the moon is now believed to be a well-preserved museum of impact scars more than 3.9 billion years old. Shock effects on the rocks attest to the violent impact history, and examination of basin and crater rocks shows that they are not of volcanic origin. A large part of the lunar surface is covered by ejecta from the Imbrium basin, and the white rays are the ejecta from craters. We even find in the Apollo summary a mention that "as the solid anorthositic crust was formed it kept a record of moonlets that struck it leaving numerous scars." (see selection 30).

THE FACE which the moon turns ever toward us is a territory as large as North America, and, on the whole, it is perhaps better mapped. As its surveyor, even if armed with the most powerful of telescopes, is still practically several hundred miles away, his map does not represent the smallest features; but as all parts are equally accessible and as he has labored industriously these many years, there is no remaining space on which to write the legend "unexplored." Upon his map are a score of great plains with dark floors, which he calls *maria*; there are a score of mountain chains; there are a few trough-like valleys remarkable for their straightness; there are many thousand circular valleys with raised rims, which it is convenient this evening to call craters[1] although for the purposes of detailed description he has found it convenient to give them many distinctive names; there are thousands of bright streaks, which are neither ridges nor hollows, but mere bands of color; there are many hundred narrow linear depressions, which he calls rills.

Despite the persistent enthusiasm, the patience, and the industry with which he has studied his field, it must nevertheless be admitted that he has rarely satisfied himself and never satisfied his fellow-workers with the explanations he has suggested as to the origin of the features his map delineates. But selenographers are not the only students of the moon's face. There are also selenologists, who use the telescope comparatively little but cogitate much, and who have evolved theories of great ingenuity and variety. Far be it from me to say aught to their disparagement, for this evening I join myself to their ranks; but, again, it must be confessed that the selenographers do not look upon the teachings of the selenologists with favor. So, despite all that has been done, the field of theory is still open and this is my excuse for putting forth ideas founded neither on protracted observation nor on protracted study—this and the further plea that the problem is largely a problem of the interpretation of form, and is therefore not inappropriate to one who has given much thought to the origin of the forms of terrestrial topography.

Crater Characters

In the study of lunar physiography—or physiognomy, if you prefer—interest naturally centers in the craters, for these are the dominant features. All theories begin with them; and, before examining the theories, it will be well to place clearly in view the characteristics of the lunar craters. The range in size is great, extending from a maximum of about 800 miles diameter to a minimum of less than one mile. The size of the smallest ones is not known, as they are beyond the present power of the telescope. Within this range are several varieties, more or less correlated with size, but their intergradation is so perfect that they are all regarded as phases of a single type.[2] Those of medium size will be first described.

Picture to yourself a circular plain ten, twenty, fifty, or one hundred miles in diameter, surrounded by an acclivity which everywhere rises steeply but irregularly to a rude terrace, above which is a circular cliff likewise facing inward toward the plain. This cliff is the inner face of a rugged, compound, annular ridge, composed of shorter ridges which overlap one another, but all trend concentrically. Seen from above, this ridge calls to mind a wreath, and it has been so named. From the outer edge of the wreath a gentle slope descends in all directions to the general surface of the moon, which it is convenient to call here the outer plain. The outer slope of the crater may be identical in surface character with the outer plain, or it may be radially and somewhat delicately ridged, as though by streams of lava. The inner slope, from the base of the cliff to the margin of the inner plain, is broken by uneven and discontinuous terraces, which have the peculiar habit of land-slip terraces as one sees them about the flanks of a plateau capped by a heavy sheet of basalt. From the center of the inner plain rises a hill or mountain, sometimes symmetric but usually irregular and crowned by several peaks. From the outer plain to the base of the wreath the ascent is 1,000 or 2,000 feet, and the ascent thence to the top of the wreath may be as much more. The descent from the wreath to the inner plain is ordinarily from 5,000 to 10,000 feet, and the height of the central hill is 1,000 to 5,000 feet. With rare exceptions, the inner plain is several thousand feet lower than the outer plain.

Here Gilbert notes that the central hill and the wreath of craters are diminished or missing for the larger craters.

The craters are more abundant in some regions than in others, and there are comparatively few upon the maria. Usually the small craters are far more numerous than those of medium size, but in certain districts well covered by craters those of small size are less abundant. The craters overlap one another with every conceivable relation, except that the overlapping is never reciprocal. It is in every case possible to distinguish the newer from the older, the older being partially effaced by the newer. Small craters occur on all parts of larger ones, not excepting the wreath and the steep inner slope.

Volcanic Theory

By the majority of writers the craters are assumed to be volcanic, and as they differ in size, abundance, and form from terrestrial volcanoes, it is thought that they represent some special type of volcanism determined by physical conditions peculiar to the moon. Let us compare the lunar and terrestrial craters and see how far their differences can be explained as dependent on differences of physical condition.

Gilbert notes that lunar craters are 10 times more abundant than volcanic craters on the earth, for a similar surface area, but that erosion

Fig. 14.1 Lunar crater Clavius showing grouping of craters. The overall diameter is 230 km, and the depth is 3 km.

processes could have demolished many terrestrial volcanoes. He then shows that the largest lunar craters have linear extents over 25 times those of the largest terrestrial volcano craters. The lower surface gravity of the moon, however, could allow a lunar crater to be 6 times as broad or deep as a crater on the earth if both were formed in the same way.

In vertical dimensions there is no important discrepancy. Lunar craters of the first rank range from 8,000 to 15,000 feet in depth; terrestrial, probably from 2,000 to 4,000. Dividing the lunar measures, as before, by six, we obtain 2:3 as the ratio of lunar depth to terrestrial; but as few terrestrial craters have been measured, this result cannot claim high precision.

The contrasts as to form are of greater importance. To set them forth fully it is necessary to give separate consideration to several types of terrestrial craters. These may be called the ordinary or Vesuvian, the Hawaiian, and the maar types. Craters of the Vesuvian type—and these include nineteen-twentieths of all terrestrial volcanoes—are formed of lavas containing a considerable amount of water, and usually result from extravasation and explosion in alternation. As the lava rises in its conduit the contained water is converted into steam, by which the lava is torn to fragments and thrown into the air. That which falls back into the vent is again thrown upward, and that which falls outside the vent builds the crater rim. From time to time drier lava wells up and overflows the rim, or else forces a way to the surface at some lower level. In this manner there is accumulated a conical mountain with a funnel-shaped cavity at the top. Eruption is not continuous, but is interrupted by periods of quiescence, and sometimes, after a long interval of quiet, operations are again initiated by a great explosion of steam, the upper portion of the cone being blown out and an immense cavity left in its place. Eventually the reissue of lava builds a new cone inside the great crater, and this cone, which always carries a crater at top, may grow so as to bury completely the wreck of the great explosion.

With the forms resulting from this process, or alternation of processes, the lunar craters have little in common. Ninety-nine times in one hundred the bottom of the lunar crater lies lower than the outer plain; ninety-nine times in a hundred the bottom of the Vesuvian crater lies higher than the outer plain. Ordinarily the inner height of the lunar crater rim is more than double its outer height; ordinarily the outer height of the Vesuvian crater rim is more than double its inner height. The lunar crater is sunk in the lunar plain; the Vesuvian is perched on a mountain top. The rim of the Vesuvian crater is not developed, like the lunar, into a complex wreath, but slopes outward and inward from a simple crest-line. If the Vesuvian crater has a central hill, that hill bears a crater at summit and is a miniature reproduction of the outer cone; the central hill of the lunar crater is entire, and is distinct in topographic character from the circling rim. The inner cone of a Vesuvian volcano may rise far higher than the outer; the central hill of the lunar crater never rises to the height of the rim and rarely to the level of the outer plain. The smooth inner plain characteristic of so many lunar craters is either rare or unknown in craters of Vesuvian type. Thus, through the expression of every feature the lunar crater emphatically denies kinship with the ordinary volcanoes of the earth. If it was once nourished by a vital fluid, that fluid was not the steam-gorged lava of Vesuvius and Etna.

Craters of the Hawaiian type are produced by lavas containing so little moisture that its conversion into steam does not cause violent explosions. Craters of this sort are somewhat rare, but their rarity does not affect their value as interpreters of extra-telluric phenomena. As long ago pointed out by Dana, they resemble the moon's craters much more closely than do those of ordinary volcanoes. They agree with lunar craters in the possession of inner plains, and to a certain extent in the terracing of their inner walls. They differ in the fact that they occupy the tops of mountains; in the absence of the wreath; in the absence of the central hill, and usually in the presence of level terraces due to the formation of successive crusts. In my judgment the differences far outweigh the resemblances, and I have not succeeded in imagining such peculiarities of local condition as might account for the divergence in form.

Before passing to the examination of other theories, it is well to bring together the results of our inquiry into the adequacy of the volcanic. The comparative abundance of lunar craters is readily accounted for without prejudice to the theory. Their greater maximum width, though partly referable to a gravitational factor, constitutes a real difficulty, especially as volcanoes appear to have a definite size limit, while lunar craters do not. Form differences effectually bar from consideration all volcanic action involving the extensive eruption of lavas, whether dry or saturated with water. They also exclude the maar process (single explosion) as an explanation of medium and large craters, but not as an explanation of small craters. The volcanic theory, as a whole, is therefore rejected, but a limited use may be found for the maar phase of volcanic action in case no other theory proves broad enough for all the phenomena.

Gilbert discusses the tidal theory of crater origin, according to which a thin lunar crust atop a rapidly rotating, liquid lunar core was pulled apart by the earth's tidal force, causing the interior liquid to pour out. He then argues that the stresses of the observed crater rims would rend a crust of granite 100 mi thick and that a thin crust therefore cannot support the observed craters. The theory assuming that craters are pools of water in a snow- or ice-covered moon cannot account for the observed crater morphology.

Meteoric Theories

All other theories which I have been able to discover appeal in one way or another to the collision of other bodies with the moon's surface, and for want of a better term I shall call them meteoric. If a pebble be dropped into a pool of pasty mud, if a raindrop fall upon the slimy surface of a sea marsh when the tide is low, or if any projectile be made to strike any plastic body with suitable velocity, the scar produced by the impact has the form of a crater. This crater has a raised rim, suggestive of the wreath of the lunar craters. With proper adjustment of material, size of projectile, and velocity of impact, such a crater scar may be made to have a central hill. Thus scars of impact may simulate in many ways the scars of the moon's face, and a number of theories have accordingly been broached which agree in regarding the craters as due to the bombardment of the moon by projectiles coming from without. As the present study is primarily physiographic, these similitudes of form have been considered with great care, and it is my belief that all features of the typical lunar crater and of its varieties may be explained as the result of impact. The special considerations presently to be adduced are along this line.

Long ago it was suggested that the projectiles might have been fired from terrestrial volcanoes, but the speed actually acquired by the ejecta of volcanoes falls so far short of that necessary to carry them beyond the sphere of the earth's attraction that this view is no longer entertained. All other suggestions have regarded the material as cosmic. Every shooting star records by its brief coruscation the collision with our atmosphere of a particle of star dust; and though the number of these which can be seen by a single observer in one night is not great, it has been computed that no less than 400 millions are captured by the earth in the course of twenty-four hours. So minute are they in general that their ashes do not contribute to the earth's surface an appreciable layer of dust; but a few have such size that they are not completely consumed in traversing the atmosphere and fall to the earth as aërolites weighing grains, ounces, pounds, or even tons. For the most part they strike the atmosphere with a velocity far higher than could be induced by the earth's attraction, and we must believe that they are speeding through space in all directions in numbers that defy the imagination. They must collide with all planetary bodies in numbers depending chiefly on the area of surface exposed, and the moon, of course, receives its share.[3]

As the moon either is without atmosphere or has one of extreme tenuity, the mechanical effect of this bombardment may be important, for the average velocity of the meteors is from fifty to one hundred times as great as that with which the swiftest ball leaves the cannon, and the energy of a projectile is measured by the square of its velocity. Nevertheless it is incredible that even the largest meteors of which we have direct knowledge should produce scars comparable in magnitude with even the smallest of the visible lunar craters. Recognizing this difficulty, advocates of meteoric theories have assumed that at some earlier period the meteors encountered by our solar system were of greater size than now, and as no evidence has been found that the earth was subjected to a similar attack, there is assigned to the lunar bombardment an epoch more remote than all the periods of geologic history, any similar scars produced on the earth having been obliterated by the processes which continually reconstruct and remodel its surface.

Another difficulty has been found in imagining a condition of lunar surface which should admit at the same time of plastic molding and of the preservation of the resulting forms.

Gilbert then notes that the high slopes of lunar craters require strong material for their stability but that the energy of impacting meteorites will suffice to heat volcanic rock to 3,500° F.

The average velocity of shooting stars is estimated at 45 miles per second, or thirty times that of a body falling freely to the moon, and it is easy to understand that the heat developed by the sudden arrest of a fragment of rock traveling with such speed might serve not only to melt the fragment itself, but also to liquefy a considerable tract of the rock mass by which its motion was arrested.

The third difficulty is found in the relation of the volume of the rim to the capacity of the hole.

Gilbert summarizes observational details relating to the volumes of crater rims and hollows: the rim volume is sometimes larger and sometimes smaller than the hollow volume.

Though the imperfection of the data gives a large probable error to the determinations, there can be no question of the general fact that in many instances the rims of large craters are quite inadequate to fill the cavities they surround. This is an important fact, but it is not necessarily inimical to the impact theory. In the course of a series of laboratory experiments, in which craters were produced by throwing projectiles of various plastic materials against targets of similar materials, it was occasionally found that the rim when pared away would not fill the hollow, and the cause of this result was discovered. When target and projectile were of uniform consistency throughout, there was no defect of rim; but when the general mass of the target was softer than the portion at the surface, the uplift consequent on the production of the hollow was only partly localized about its periphery, the remaining part being widely distributed through flow of the softer material below. It is possible, therefore, to interpret the quantitative relations discovered on the moon in terms of local physical condition without rejecting the impact theory.

A fourth difficulty is connected with the circular contours of the craters. If a ball of mud be allowed to fall vertically upon a horizontal surface of the same material, the resulting crater is circular; but if instead it be thrown obliquely, the resulting crater has an oval contour. Except for irregularities which may be counted as details of form, some of the lunar craters are as nearly circular as can be determined by measurement; others are slightly elliptic; a few only are notably elongate. It is inferred that the predominant direction of the incident bodies supposed to have formed them was vertical to the lunar surface or nearly so; but it can be shown from simple geometric considerations that the predominant angle of incidence of swift-moving meteoric bodies approaching from all directions would be 45 degrees, and the scars produced by such collisions would be predominantly oval instead of predominantly circular.

Moonlet Theory

Besides the nomadic and apparently individual meteors of space, there are certain groups symmetrically arranged and moving in a systematic and orderly way. One of these groups is arranged in the form of a ring and encircles the planet Saturn. This ring is broad and thin, and all parts of it lie nearly in one plane. The meteors which constitute it are so numerous that portions of the ring appear continuous and solid. They are too small to be individually perceived, but there can be little question that they all travel about the planet in a system of parallel orbits and with correspondingly adjusted velocities. It is my hypothesis that before our moon came into existence the earth was surrounded by a ring similar to the Saturnian ring; that the small bodies constituting this ring afterward gradually coalesced, gathering first around a large number of nuclei, and finally all uniting in a single sphere, the moon. Under this hypothesis the lunar craters are the scars produced by the collision of those minor aggregations, or moonlets, which last surrendered their individuality.

This change of conception yields a material difference in the law of the directions in which minor bodies approach the moon, the difference depending on the fact that all the minor bodies colliding with the greater body have initial orbits lying approximately in the same plane. To render this clear it is necessary to amplify the statement already made with reference to the predominant angle at which cosmic meteors encounter the surface of the moon. Their velocities are so high, as compared with the acceleration due to lunar attraction, that their courses in the vicinity of the moon may, without sensible error, be regarded as straight. The angle at which each one strikes the moon's surface depends upon the nearest distance of its produced orbit from the moon's center, and is entirely independent of the direction from which it approaches. We may therefore simplify the discussion of incidence angles by assuming that the meteors all come from the same direction and move along parallel lines. The number of meteorites being indefinitely large and their distribution entirely independent of the moon, we may for this purpose conceive them as an evenly distributed rain, of which the moon receives a certain portion.

Here Gilbert develops his moonlet theory and computes the general expression for the angle of incidence for moonlets colliding with the moon in the absence of the earth's gravitational attraction. He finds that the number of colliding moonlets whose angle of incidence is less than i varies with the square root of the sin of i. This means that 70% deviate less than 30° from the vertical to the lunar surface.

In fine, the hypothesis of the Saturnian ring, by restricting the colliding bodies to a single plane, by substituting a low initial velocity and thus rendering the moon's attraction the dominant influence, and by introducing a system of directions controlling, and therefore adjusted to, the moon's rotation, relieves the meteoric theory of its most formidable difficulty. It also explains in a simple way the abundance of colliding bodies of a different order of magnitude from ordinary meteorites and aërolites.

Since the area of the moon's surface directly struck by the moonlet is a function of the square of the diameter of the moonlet, while the energy applied to that area, being measured by the mass of the moonlet, is a function of the cube of its diameter, more energy would be applied to a unit of space in the case of large moonlets than in the case of small, and the temperatures caused by large moonlets would therefore be greater. To this relation I ascribe the restriction of inner plains, indicative of fusion, to the larger craters, and the same explanation applies less directly to the limited distribution of central hills.

In the production of small craters by small moonlets I conceive that the bodies in collision either were crushed or were subjected to plastic flow, and in either case were molded into cups in a manner readily illustrated by laboratory experiments with plastic materials. The material displaced in the formation of the cup was built into a rim, partly by overflow at the edges of the cup, but chiefly by outward mass movement in all directions, resulting in the uplifting of the surrounding plain into a gentle conical slope. This outward and upward movement was accentuated, possibly through the agency of heat, about the immediate edge of the cup, occasioning the special elevation called the wreath. The cups thus formed, having dimensions commensurate with the strength of the lunar material, were stable and permanent. The impact of a larger moonlet produced a larger cup, and at the same time fused a portion of the material and softened other portions. The walls of this cup were so lofty that they could not sustain their own weight, and they were further weakened by the effects of heating; consequently they settled downward and their lower portions flowed inward toward the center of the cup.

Gilbert then develops his moonlet theory in greater detail; portions of that development follow here.

SCULPTURE

The rims of certain craters are traversed by grooves or furrows, which arrest attention as exceptions to the general configuration. In the same neighborhood such furrows exhibit

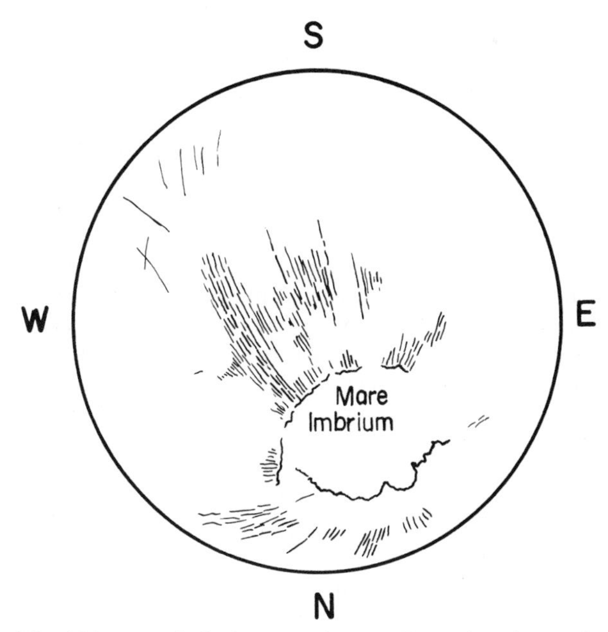

Fig. 14.2 Trends in lunar sculpture. General sculpture is represented by shading and great furrows by heavy lines. Irregular lines show crests of uplands surrounding Mare Imbrium.

parallelism of direction. Similar furrows appear on tracts between craters, and are there associated with ridges of the same trend, some of which seem to have been added to the surface. Elsewhere groups of hills have oval forms with smooth contours and parallel axes, closely resembling the glacial deposits known as drumlins, but on a much larger scale. Tracing out these sculptured areas and plotting the trend lines on a chart of the moon, I was soon able to recognize a system in their arrangement, and this led to the detection of fainter evidences of sculpture in yet other tracts. The trend lines converge toward a point near the middle of the plain called Mare Imbrium, although none of them enter that plain. Associated with the sculpture lines is a peculiar softening of the minute surface configuration, as though a layer of semiliquid matter had been overspread, and such I believe to be the fact; the deposit has obliterated the smaller craters and partially filled some of the larger. These and allied facts, taken together, indicate that a collision of exceptional importance occurred in the Mare Imbrium, and that one of its results was the violent dispersion in all directions of a deluge of material—solid, pasty, and liquid. Toward the southwest the deluge reached nearly to the crater Theophilus, a distance of 900 or 1,000 miles, and southward it extended nearly to the latitude of Thebit. Northward and northeastward it probably extended to the limb.

Furrows

In strong contrast with all other features of the moon's surface are a series of gigantic furrows. In general direction they are remarkably straight, but their sides and bottoms, with a single exception, are jagged, abounding in acute salients and reëntrants. If one thinks only of their apparent size instead of their real magnitude as he examines them through the telescope, he is reminded of the rude grooves sometimes seen on glaciated surfaces where the corner of a hard boulder, dragged forward by the ice, has plowed its way through a brittle rock. Despite the enormous disparity in size—a disparity no less than that of a mountain to a molehill—I believe that this resemblance is more than accidental, and that the lunar furrows were really formed by the forceful movement of a hard body; but the graving tool in this case, instead of being slowly pushed forward by a matrix of ice, moved with high velocity and was controlled only by its own inertia. It was my first idea that the furrows are the tracks left by solid moonlets whose orbits at the instant of collision were nearly tangent to the surface of the moon, and for some of them I have still no better explanation to suggest; but when they came to be plotted on a chart of the moon's face it was found that more than half of them accord in direction with the trend lines of the outrush, a relation which can be seen in figure 14.2, where they are represented by heavy lines. It thus appears possible, if not probable, that they were produced simultaneously with the Imbrian deluge, and the implication of power is thereby rendered even more impressive. What must have been the violence of a collision whose scattered fragments, after a trajectory of more than a thousand miles, scored valleys comparable in magnitude with the Grand Canyon of the Colorado!

White Streaks

The only remaining great group of features are the white streaks. These are bands of color, sometimes faint, sometimes brilliant, but always indefinitely outlined, like the tail of a comet, and some of them stretch for long distances across the moon's surface. Their courses are independent of the configuration. They pass up and down the slopes of craters without either modifying their forms or being interrupted by them. The more prominent of them, and probably all, occur in systems, and those of each system radiate from some crater. This crater is itself lined with white and is usually more resplendent than the radiating streaks.

Gilbert quotes from a letter from one of his students, Mr. Würdemann:

> "The most remarkable appearance on the moon, for which nothing on earth furnishes an example, is presented by those immense radiations from a few of the larger craters—perfectly straight lines, as though marked with chalk along a ruler—starting from the center of the crater and extending to great distances over every obstruction. My explanation is that a meteorite, striking the moon with great force, spattered some whitish matter in various directions. Since gravitation is much feebler on the moon than with us and atmospheric obstruction of consequence does not exist, the great distance to which the matter flew is easily accounted for."

Retrospect

The analytic examination of volcanic processes left the possibility that the small craters of the moon are maars, the results of explosion without eruption of lava; the tidal process might perhaps make large craters, but could not make small ones. These are the only suggested reactions originating in the moon itself which appear competent to produce the crater forms actually observed. Taken together, they cover all the craters, but they cannot be applied as a joint theory without arbitrarily dividing a series the gradation of which is complete as to both size and form. The impact theory applies a single process to the entire series (excepting only the rill pits), correlating size variation with form variation in a rational way. Specialized by the assumption of an antecedent ring of moonlets, it accounts also for the great size of many craters. It brings to light the history of a great cataclysm, whose results include the remodeling of vast areas, the flooding of crater cups, the formation of irregular maria, and the conversion of mere cracks to rills with flat bottoms. It explains the straight valleys and the white streaks. In fine, it unites and organizes as a rational and coherent whole the varied strange appearances whose assemblage on our neighbor's face cannot have been fortuitous.

Gilbert concludes by arguing that during its whole period of growth the moon was cold and that the absence of impact craters on the earth suggests the formation of the moon before erosion processes began on the earth.

1. The word *crater*, derived from the Greek name of a kind of bowl, is used chiefly to designate the bowl-shaped cavities of volcanoes. In this paper, as in most selenographic writings, it designates a topographic form without implication as to the origin of the form.

2. The only exceptions to the type that I have noted are associated with certain of the rills. They are so small that I could not determine their characters with certainty, but they seemed to lack rims and to be hopper-shaped.

3. I have discovered no published statement of meteoric theories more than twenty years old, but the idea is older and various obscure allusions indicate that it was earlier in print. Proctor makes a meteoric suggestion in 1873 (*The Moon*, p. 346), and advocates it in 1878 (*Belgravia*, vol. 36, p. 153). A meteoric theory is said to be contained in *Die Physiognomie des Mondes*, by "Asterios," (Nordlingen, 1879). A. Meydenbauer advances another in *Sirius*, for February 1882, and he includes bodies other than cosmic.

15. Of Atmospheres upon Planets and Satellites

George Johnstone Stoney

(*Astrophysical Journal* 7, 25–55 [1898])

In this paper George Johnstone Stoney lays the physical foundation for all subsequent considerations of the evolution of planetary atmospheres. He shows that the stability of an atmosphere of a given composition depends upon the thermal motions of the individual molecules as well as upon the gravitational pull of the planet. If a molecule is to escape the gravitational pull of a planet, its velocity must exceed the planet's escape velocity. Because the molecule's thermal velocity increases with its temperature but decreases with its mass, the lighter molecules are more likely to escape. Stoney notes that hydrogen and helium are not observed in the earth's atmosphere, whereas water vapor and other heavier molecules are. This means that molecules with molecular weights somewhere between 4 and 18 are just sufficiently massive to be retained in the earth's atmosphere.

Twenty years later James Jeans put Stoney's arguments in a more rigorous mathematical framework by calculating the rate of escape of an atmosphere under the assumption of a Maxwellian velocity distribution of particles in an isothermal atmosphere.[1] He found that molecules will escape in less than a billion years if the planet's escape velocity is less than 5 times the thermal velocity of the individual molecules. Lyman Spitzer subsequently extended this result for the special case when the temperature of the upper atmosphere differs appreciably from that of the lower atmosphere.[2]

Of special interest in this paper is the discussion of the probable constituents of the other planets' atmospheres. Stoney argues that no water vapor can be retained in the Martian atmosphere and suggests that the Martian polar caps and snows must be produced by carbon dioxide. He also argues that because of Jupiter's greater gravity and distance from the sun, its atmosphere must contain every element known to man. In particular, the lightest element, hydrogen, could and should be present in the Jovian atmosphere. Thirty-two years after Stoney's paper, Donald Menzel used the same dynamic arguments to infer the presence of hydrogen in the atmospheres of the giant planets.[3] By that time, however, it was known that hydrogen is many times more abundant than all the other elements combined, at least in the outer layers of the sun and stars. It was natural for Menzel to suppose that if the sun and planets were all formed from the same primeval gas, the giant planets would have hydrogen atmospheres. When Menzel wrote his article it was widely believed that the low densities of the giant planets were caused by extensive atmospheres of hot gas, but Menzel argued that an abundance of hydrogen produced the low densities. Radiometric surface measurements soon showed that the giant planets were quite cold, and fifty years after Stoney's original prediction, hydrogen was found in the Jovian atmosphere.

The only faulty aspects of Stoney's logic were his assumptions that the atmospheres of sufficiently massive planets will resemble the composition of the earth's atmosphere and that all the planets now retain their primeval atmospheres. We now suspect that the atmospheres of Venus, Earth, and Mars are the products of the secondary outgassing from volcanoes, and the atmospheres of Venus and Mars are known to be primarily composed of carbon dioxide. However, the more massive major planets do retain the lighter hydrogen compounds—methane, ammonia, and molecular hydrogen—that must have made up the primeval terrestrial atmosphere.

1. J. H. Jeans, *The Dynamical Theory of Gases* (Cambridge: Cambridge University Press, 1916).

2. L. Spitzer, "The Terrestrial Atmosphere above 300 Kilometers," in *The Atmospheres of the Earth and Planets*, ed. G. P. Kuiper (Chicago: University of Chicago Press, 1949).

3. D. H. Menzel, *Publications of the Astronomical Society of the Pacific 42*, 228 (1930).

Introduction

The present writer began early in the sixties to investigate the phenomena of atmospheres by the kinetic theory of gas, and in 1867 communicated to the Royal Society a memoir,[1] which pointed out the conditions which limit the height to which an atmosphere will extend, and in which it was inferred that the gases of which an atmosphere consists attain elevations depending on the masses of their molecules, the lighter constituents overlapping the others. This was disputed at the time, on account of its supposed conflict with Dalton's law of the equal diffusion of gases;[2] but physical astronomers now recognize its truth.

On December 19, 1870, the author delivered a discourse before the Royal Dublin Society, which was the first of the series of communications, of which an account is given in the following pages. One of the topics of that discourse was the absence of atmosphere from the Moon. This was accounted for by the kinetic theory of gas; inasmuch as the potential of gravitation on the Moon is such that a free molecule moving in any outward direction with a velocity of 2.38 kilometers per second[3] would escape, and, accordingly, the Moon is unable to retain any gas, the molecules of which can occasionally reach this speed at the highest temperature that prevails on the surface of the Moon.

Shortly after, a second communication was made to the Royal Dublin Society, at one of its evening scientific meetings, based on the supposition that the Moon would have had an atmosphere consisting of the same gases as those of the Earth's atmosphere, were it not for the drifting away of the molecules. It was shown that if the molecules of these gases can escape from the Moon, it necessarily follows that the Earth is incompetent to imprison free hydrogen; and this was offered as explaining the fact that, though hydrogen is being supplied in small quantities to the Earth's atmosphere by submarine volcanoes and in other ways, it has not, even after the lapse of geological ages, accumulated in the atmosphere to any sensible extent. This communication was followed at intervals by others, in which the investigation was extended to other bodies in the solar system, in which an endeavor was made to trace what becomes of the molecules that filter away from these several bodies, and in which it was suggested that the gap in the series of terrestrial elements between hydrogen and lithium may be accounted for by the intermediate elements (except helium) having escaped from the Earth at a remote time, when the Earth was hot.

In one of the earlier of these communications, it was pointed out that it is probable that no water can remain on Mars—a probability which is now raised to a certainty by the recent discovery, that helium (with a molecular mass twice that of hydrogen) is being constantly supplied in small quantities to the Earth's atmosphere by hot springs, and probably in other ways, and that nevertheless there is no sensible accumulation of it in the Earth's atmosphere after the infiltration has been going on for cosmical ages of time. In the absence of water, carbon dioxide was suggested as, with some probability, the substance that produces the polar snows upon Mars. Moreover, on the Earth, snow, rain, and cloud are produced by the lightest constituent of our atmosphere; but if the atmosphere of Mars consists of nitrogen and carbon dioxide, snow, frost, and fog on that planet are being produced by the heaviest constituent. An attempt was made to follow out the consequences of this state of things, and to refer to it those recurring appearances upon Mars which, though very imperfectly seen owing to the great distance from which we observe them, have been (perhaps too definitely) mapped and described under the name of canals.

Of this series of communications, though known to many, only imperfect printed accounts have appeared; and it is the object of the present communication to present the subject in a more complete form. The opportunity will be taken of substituting better numerical results for those originally given, by basing them on the fact which has recently come to our knowledge, that not only hydrogen, but helium also, with a density twice that of hydrogen, can escape from the Earth. The most notable change that this makes is, that what was before probable is now certain—that water cannot in any of its forms, be present upon Mars.

Chapter 1. Of the Fundamental Facts

In order to see why neither hydrogen nor helium remains in the Earth's atmosphere, and why there is neither air nor water on the Moon, it is necessary to understand the conditions which determine the limit of an atmosphere. These were investigated under the kinetic theory of gas by the present writer in a memoir communicated to the Royal Society in May, 1867: see his paper "On the Physical Constitution of the Sun and Stars," in the *Proceedings of the Royal Society*, No. 105, 1868, from page 13 of which it will be convenient to make the following extract:[4]

> Let us consider what it is that puts a limit to the atmosphere. Let us first suppose that it consists of but one gas, and let us conceive a layer of this gas between two horizontal surfaces of indefinite extent, so close that the interval between them is small compared with the mean distance to which molecules dart between their collisions, but yet thick enough to have, at any moment, several molecules within it. Molecules are constantly flying in all directions across this thin stratum. Some of them come within the sphere of one another's influence while within the layer, and therefore pass out of it with altered direction and speed. Let us call them the molecules emitted by the layer. If the same density and pressure prevail above and below the layer, the molecules which strike down into it will, on account of gravity, arrive with somewhat more speed on the average than those which rise into it. Hence those

molecules which suffer collision within the stratum will not scatter equally in all directions, but will have a preponderating downward motion, so that of the molecules emitted by the stratum more will pass downwards than upwards. This state of things is unstable, and will not arrive at an equilibrium until either the density or the temperature is greater on the underside of the layer. If the density be greater, more molecules will fly into the stratum from beneath than from above; and if the temperature be greater the molecules will strike up into it, both more frequently and with greater speed. In the Earth's atmosphere it is by a combination of both these that the equilibrium is maintained; both the temperature and the density decrease from the surface of the Earth upwards.

We have hitherto taken into account only those molecules which, after a collision, have arrived at the stratum from the side on which the collision took place. But besides these there will be a certain number of molecules which, having passed through the stratum from beneath, fall back into it without having met with other molecules, either by reason of the nearly horizontal direction of their motion, or because of their low speed. The number of molecules that will thus fall back into the stratum will be a very inconsiderable proportion of the whole number passing through the stratum, so long as the temperature and density are at all like what they are at the surface of the Earth. In the lower strata of the atmosphere, therefore, the law by which the temperature and density decrease will not be appreciably affected by molecules thus falling back. But in those regions where the atmosphere is both cold and very attenuated, where accordingly the distance between the molecules is great and the speed with which they move feeble, the number of cases in which ascending molecules become descending without having encountered others will begin to be sensible. From this point upwards the density of the atmosphere will decrease by a much more rapid law, which will, within a short space, bring the atmosphere to an end.

It appears, then, that the atmosphere round any planet or satellite will, *cæteris paribus*, range to a greater height the less gravity upon that body is; and that if the potential of gravitation be sufficiently low, and the speed with which the molecules dart about sufficiently great, individual molecules will stream away from that body, and become independent wanderers throughout space.

Thus, we shall presently see that, in the case of the Earth, a velocity of about eleven kilometers per second (nearly seven miles per second) would be enough to carry a molecule at the boundary of our atmosphere off into space, if the Earth were alone and at rest; and a somewhat less velocity of projection (about 10.5 kilometers per second) is sufficient, on account of the rotation of the Earth, and because westerly winds sometimes blow in the upper regions of the atmosphere. The modification introduced by these subsidiary causes will be examined in Chapter IV, and the amount of their effect will be determined. The behavior of molecules is also slightly affected by the Moon, which is near enough sensibly to alter the orbits of molecules if shot up in some directions.

Here Stoney argues that although free hydrogen can interact with other atmospheric constituents, there would be some present in the earth's atmosphere if it were not able to escape because of its thermal motion. The evidence for the escape of gas from the earth's atmosphere is even more conspicuous in the case of helium, which does not interact with other atmospheric constituents. Stoney then notes that heavier elements, such as water vapor, nitrogen, oxygen, and carbon dioxide, are retained in the earth's atmosphere.[5]

We may infer from this that *the boundary between those gases that can effectually escape from the Earth and those which cannot, lies somewhere between gas consisting of molecules with twice the mass of molecules of hydrogen and gas with molecules whose mass is nine times*[6] *the mass of molecules of hydrogen.*

This we may take to be one fact which we can ascertain by observing what occurs upon the Earth, and the telescope has been able to reveal to us another fact of a like kind, viz., that there is either no atmosphere upon the Moon, or excessively little—a fact which has been made certain by the application of very delicate tests.

Chapter II. Interpretation by the Kinetic Theory

In order to make these facts the starting-point for fresh advances, we must study their precise physical meaning when interpreted by the kinetic theory of gas.

Here Stoney gives the [root] mean square velocity of the individual molecules of a gas, V_T, first given by Rudolf Clausius (*Philosophical Magazine 14*, 124 [1857]) and by James Clerk Maxwell (*Philosophical Magazine 19*, 19 [1860]).

$$V_T = \left(\frac{3RT}{\mu}\right)^{1/2} = \left(\frac{3kT}{m}\right)^{1/2} = 0.157(T/\mu)^{1/2}\ \text{km sec}^{-1},$$

where T is the temperature in degrees Kelvin; the molecular weight, μ, is the mass of the individual molecule in units of the mass of the hydrogen atom; R is the universal gas constant; k is Boltzmann's constant; and m is the mass of the individual gas molecule. Stoney then shows that the mean square velocity of hydrogen at $-66°$ C is 1.6 km sec^{-1}.

The actual velocities of the molecules are, of course, some of them considerably more and others considerably less than this mean, even if the hydrogen or helium be unmixed with other gases; and the divergences of some of the individual velocities from the mean will become exaggerated when the encounters to which the molecules of these lighter gases are subjected are sometimes with molecules many times more massive, and which may, when the encounter takes place, be moving with more than their average speed, as must often happen in our atmosphere. Under these circumstances we should be prepared to find that a velocity several times the foregoing mean is not unfrequently reached; and the evidence (see Chapter IV) goes to show that *a velocity which is between nine and ten times the velocity of mean square*, a velocity which is able to carry molecules of either hydrogen or helium away from the Earth, *is sufficiently often attained to make the escape of gas effectual.*

We are now in a position to aim at making our results so definite that they may be extended to other bodies in the solar system.

Chapter III. Dynamical Equations

In making our calculations with reference to the planets and satellites of the solar system, it will simplify the work, and be sufficient for our purpose, to treat them as spherical bodies, consisting of layers each of which is a spherical shell of uniform density. In that case, let B be one of these bodies.

Stoney derives the velocity of escape V_E, given by the condition that the kinetic energy per unit mass equals the potential energy per unit mass on the surface of the planet.

$$V_E = \left(\frac{2GM}{R}\right)^{1/2} = 11.3\left(\frac{M}{M_e}\frac{R_e}{R}\right)^{1/2} \text{ km sec}^{-1},$$

where M and R denote, respectively, the planet mass and radius and M_e and R_e denote, respectively, the mass and radius of the earth.

If a missile were projected from a planet B with this speed, it would just be able to reach infinity, *i.e.*, this speed is the least which would enable a molecule to get completely away from B. We may, therefore, call it *the minimum speed of escape* from B when B is at rest. If B rotates, a less velocity relatively to the surface of B will suffice, provided that the missile is shot off in the direction towards which the station from which it starts was being carried by the rotation at the instant of projection.

Chapter IV. Of the Earth

Under the assumption that the outer atmosphere of the earth is at a temperature of $-66°$ C, Stoney derives a thermal velocity of $V_T = 1.6/\sqrt{\mu}$ km sec^{-1} for a molecule whose molecular weight is μ. He then computes the velocity of escape at the top of the earth's atmosphere as 11.0 km sec^{-1}. When the rotation velocity of the earth is taken into account, the escape velocity can be as low as 10.54 km sec^{-1}.

Now, we found above that, in order that any gas may cease to be imprisoned by the Earth, its molecules must now and then be able to attain at least a speed of 10.5 kilometers per second. Whenever this happens to a molecule favorably circumstanced it escapes. Hence, since hydrogen succeeds in leaking away from the Earth, its molecules must in sufficient number attain this speed, which is 6.55 times the velocity of mean square in that gas at a temperature 66° below zero; and since helium can escape, its molecules must sufficiently often reach a speed equal to or exceeding 9.27 times what we have found to be the velocity of mean square in helium at a temperature of $-66°$ C.

On the other hand, in order that a molecule of water may escape from the Earth, it has to get up a speed of 19.66, nearly twenty times the velocity of mean square in that vapor at the above temperature: and the fact that water does not drain away from the Earth in sensible quantities shows that this seldom happens.

We are now in a position to make a very important deduction in molecular physics from these facts, which is that *in a gas a molecular speed of 9.27 times the velocity of mean square is reached sufficiently often to have a marked effect upon the progress of events in nature*; while, on the other hand, a molecular speed of twenty times the velocity of mean square is an event which occurs so seldom that it exercises no appreciable influence over the cosmical phenomena which we have been considering. We must remember, however, that there are other events in nature—in chemistry, and especially in biology—which may be, and probably are, determined by conditions that occur far more rarely.

The separation of the swiftest moving molecules from the boundary of our atmosphere is, of necessity, accompanied by a lowering *pro tanto* of the temperature of the atmosphere left behind. It is one of the many operations carried on by nature to which the second law of thermodynamics does not apply. We must remember that this law is only a law of molecular averages, and therefore is not a law of nature where, as in this case, nature separates one class of molecules (those moving fastest) from the rest.

Chapter V. Extension of the Inquiry to Other Bodies

Stoney then gives the criteria that $9.3V_T \geq V_E$ for the escape of a molecule and $19.7V_T \leq V_E$ for its retention.

Chapter VI. Of the Moon

When we turn to the Moon, we find the conditions to be such that it can rid itself of an atmosphere with much ease.

Stoney finds the escape velocity of the moon to be 2.38 km sec^{-1}, its presently accepted value. He then argues that the moon will retain an atmosphere at $-66°$ C only if the molecules have a molecular weight greater than 40. Because the lunar surface is expected to be hotter than $-66°$ C, it is not expected to retain any atmosphere.

Chapter VII. Of Mercury

Stoney obtains an escape velocity of 4.64 km sec^{-1} for Mercury, whereas the presently accepted value is 4.2 km sec^{-1}. He then argues that gas molecules with molecular weights smaller than 11 will escape a Mercurian atmosphere whose temperature is $-66°$ C.

The general conclusion then is:

1. That water with a density of 9 certainly cannot exist upon Mercury. Its molecules would very promptly fly away.
2. That it is in some degree probable that both nitrogen and oxygen, with densities of 14 and 16, would more gradually escape.

It is, therefore, not likely that there are, in whatever atmosphere Mercury may be able to retain, any of the constituents of the Earth's atmosphere except perhaps argon and carbon dioxide.

Chapter VIII. Of Venus

The state of Venus' atmosphere need not detain us long. The potential of gravitation is so nearly the same on this planet as on the Earth that its atmosphere almost certainly retains and dismisses the same gases as does the atmosphere of the Earth. The only element of uncertainty arises from its period of rotation being imperfectly known, but the nearly globular form of the planet assures us that its rotation cannot be swift enough seriously to affect the problem.

The similarity of the two atmospheres is confirmed by the appearance of the planet. Venus is presumably a much younger planet than the Earth, and its temperature is consequently what the Earth's was many ages ago, when through excessive evaporation water was the largest constituent of our atmosphere, and when clouds were present everywhere and without intermission.

The conditions upon Venus are so nearly akin to those on the Earth that we cannot be mistaken in regarding the vapor which forms the abundant cloud we see on that planet as none other than the vapor of water. If we may assume this, we can advance a step farther than the statements made in Chapter IV.

We omit Stoney's argument that the presence of water on Venus means that molecules do not attain a velocity 18 times the mean square often enough to enable appreciable quantities to escape. He concludes that:

1. A velocity of 9.27 times that of mean square is attained by the molecules of a gas sufficiently often to enable helium to escape from the Earth.
2. A velocity 18 times that of mean square is so seldom attained that Venus has been able to retain its stock of water.
3. Since Venus can prevent the escape of water, the Earth, with its larger potential, is competent to retain its hold upon a gas of somewhat less density, viz., one whose density is $\rho = 7.43$.

Accordingly, as regards the Earth, we may come to the following conclusions: (1) Gases with a density of 2 or less than 2 can certainly escape from the Earth; (2) a gas with a density of 7.43, and all denser gases,[1] are effectually imprisoned by the Earth; (3) the information supplied by Venus, supplemented by our present chemical knowledge, does not determine what would be the fate of a gas, if there be such, whose density lies between 2 and 7.43.

Chapter IX. Of Mars

Stoney gives the escape velocity of Mars as 5.0 km sec^{-1}, its presently accepted value, and argues that water would quit Mars at $-78°$ C as easily as helium escapes from the earth at $-66°$ C.

It appears here to be worth reviewing the state of things that must prevail if the atmosphere of Mars consists mainly of nitrogen and carbon dioxide. Without water, there can be no vegetation upon Mars, at least not such vegetation as we know; and, in the absence of vegetation, it is not likely that there is much free oxygen. Under these circumstances, the analogy of the Earth suggests that the atmosphere of Mars consists mainly of nitrogen, argon, and carbon dioxide.

Carbon dioxide, the most condensible gas of such an atmosphere, would behave very differently from the way in which water behaves on the Earth. Water in the state of vapor is so much lighter than the other constituents of our atmosphere that it hastens upward through the atmosphere; and, accordingly, its condensation into clouds, whether of droplets of water or spicules of ice, takes place usually at very sensible elevations. There would be no such hurry to rise on the part of carbon dioxide, it would, on the contrary, show great sluggishness in diffusing upward through an atmosphere of nitrogen. When brought to the ground in the form of snow or frost (for there would probably be no rain), and when subsequently evaporated, the carbon dioxide gas would crawl along the surface, descending into valleys, occupying plains and pushing its way under the nitrogen, mixing only slowly with the nitrogen; and, as a result, only a very small proportion of the whole stock would be at any one time found elsewhere in the atmosphere than near the ground. It is suggested that the fogs, the snows, the frosts, and the evaporation of such a constituent of the atmosphere may account for the peculiar and varying appearances upon Mars, which, though recorded in our maps as if they were definite, are in reality very imperfectly seen from our distant Earth. In fact, Mars, when nearest the Earth, which unfortunately seldom happens, is still 140 times farther off than the Moon. Fogs over the low-lying plains which on Mars correspond to the bed of our ocean, with mountain chains projecting through the fog, and a border of frost along either flank of these ranges, would perhaps account for some of the appearances which have been glimpsed; and extensive displacements of the vapor, consequent upon its distillation towards the two poles alternately, would perhaps account for the rest.

CHAPTER X. OF JUPITER

In the case of Jupiter, Stoney obtains an escape velocity of 61 km sec^{-1}, the presently accepted value, and argues that a gas one-tenth the density of hydrogen would be retained by Jupiter if its temperature were $-66°$ C.

Jupiter is accordingly able to imprison all gases known to chemists. His atmosphere may therefore, so far as can be determined by the present inquiry, have in it all the constituents of the Earth's atmosphere, with the addition of helium and hydrogen, and any elements between hydrogen and lithium which the Earth may have lost; except that, if the hydrogen is sufficiently abundant, there can be no free oxygen. Owing to the chemical reaction that would then take place, the oxygen will have been used up in adding to the stock of water.

CHAPTER XI. OF SATURN, URANUS, AND NEPTUNE

Stoney obtains escape velocities of 34.9, 21.6 and 22.6 km sec^{-1}, respectively, for Saturn, Uranus, and Neptune; the presently accepted values are 37, 22, and 25 km sec^{-1}.

Thus the information we gain with reference to these three planets amounts to this—that we have no definite information as regards hydrogen; that Saturn is able to detain helium, but that we do not know whether the other two planets can or cannot; that all other gases known to chemists would be more firmly imprisoned by any one of these planets than they are by the Earth; and that, if there be gases lighter than hydrogen, it is certain that Saturn cannot detain any of which the density falls as low as one-third of that of hydrogen, Neptune cannot hold any as light as two-thirds, nor Uranus any lighter than three-quarters of the density of hydrogen. On the whole, the probability seems to be that the atmosphere of Saturn is the same as that of Jupiter; while the atmospheres of Uranus and Neptune more nearly approximate to that of the Earth, with perhaps the addition of any gases with densities less than 7.43 that may possibly have left the Earth when the Earth was hotter, and whose withdrawal from the Earth is perhaps what has left the gaps in the series of terrestrial elements which appear to exist between hydrogen and helium, and between helium and lithium.

CHAPTER XII. OF THE SATELLITES AND MINOR PLANETS

We have no sufficient information as to the densities of any of these bodies. But the asteroids, or minor planets, which lie between the orbits of Mars and Jupiter, are all of them bodies so small that, even if they were as dense as osmium, iridium, or platinum, they could not retain their hold upon an atmosphere. The same may be said of the two satellites of Mars, of the two satellites of Jupiter, of most of the satellites of Saturn, and of the small bodies that make up the rings of Saturn. None of these can condense any atmosphere upon them. If there are molecules of gases traveling in their neighborhood, they also are, each of them, an independent satellite.

One satellite of Saturn and three of Jupiter are larger than our Moon; and one other of Saturn and one of Jupiter, though smaller than the Moon, are not much smaller. We should need to know the densities of these bodies before we could speak with confidence about them. The presumption, however, is that, as their primaries are very much less dense than the Earth, so these satellites are probably less dense than the

Moon. If so, they also, as well as the smaller satellites, must be devoid of atmosphere.

We know too little about the satellites of Uranus and Neptune to venture upon any conclusion about them. The satellite of Neptune appears to be a body of considerable size, and, with some probability, it may have an atmosphere.

Chapter XIII. What Becomes of the Molecules That Escape

The speed of the Earth in its orbit is about 30 km./sec. Now it follows, from the dynamics of potential, that the potential of the Sun at the distance of the Earth is represented by the square of this number if the Sun's mass be measured in gravitational units.

Stoney then argues that the combined escape velocity of the sun and the earth is 43.83 km sec^{-1} at the earth's orbit. Because the earth's escape velocity is only 11.2 km sec^{-1}, he argues that nearly all the molecules that have left the earth have remained in the solar system and are in fact now traveling as independent planets around the sun.

We have taken the special case of a molecule leaving the Earth's atmosphere. A similar treatment applies to molecules leaving the atmospheres of other planets and satellites. In every case the velocity required to enable a molecule to quit the solar system is markedly in excess of that which enables it to escape from its own atmosphere. Accordingly, almost all such wandering molecules are still denizens of the solar system.

Chapter XIV. Former Size of the Sun

The Sun is contracting, and therefore in past time was larger than it now is. The question then arises, how much larger may it have been while it was still globular? We can place a limit on its possible size *if we assume that it was then, as now, able to prevent the escape of free hydrogen*, and if we assign a temperature below which its outer boundary did not fall.

In order to arrive at definite results, let us suppose this temperature to be 0° C. Here we might take into consideration the probability that, at a sufficiently remote period, the planets formed part of the Sun. But it is needless to do this, as the addition to be then made to its present mass would be only about $\frac{1}{750}$ part, which is too slight an increase sensibly to affect our present computation.

The surface of the Sun would need to have been about $6\frac{1}{4}$ times farther from the Sun than the Earth now is, in order that hydrogen at 0° C. should escape from it as freely as helium does from the Earth at −66° C. And it would need to have been 1.64 times farther than the Earth to imprison the hydrogen as firmly as water is held by Venus.

Hence, the *greatest* size which the Sun can have had since it became a sphere, consistently with its not allowing hydrogen at 0° C. to escape, is an immense globe extending to some situation intermediate between the orbits of Mars and Jupiter. From some such vast size it may have been ever since slowly contracting.

1. See an extract from this Memoir below. The *Astrophysical Journal* selection is reprinted from an advance copy of a paper in the *Transactions of the Royal Dublin Society*, Vol. VI, Part 13. Communicated by the author.

2. According to the Kinetic theory, Dalton's law will be true of mixtures of gases if the free paths of the molecules between their encounters are straight. This is the case, to an excessively close approximation, in all laboratory experiments; but the law ceases to hold at elevations in the atmosphere where the longer and more slowly pursued free paths are sensibly bent by gravity.

3. It is very desirable that the names of metric measures should be made English words, and pronounced as such. Thus kilometer, hektometer, and dekameter should be pronounced with the accent on the second syllable, as in thermometer, barometer, etc. This would have the further useful effect of better distinguishing these names from decimeter, centimeter, and millimeter, which have accents on the first and third syllables.

4. Further information on this subject will be found in sections 22, 24, 25, 26, and in the footnote to section 93, of the paper here quoted.

5. We need not suppose that there is absolutely no escape of the molecules of the denser gases, but only that the event is an excessively rare one. Thus, if the molecules of a gas escape so very seldom that only a million succeed in leaving the entire atmosphere of the Earth in each second, then a simple computation will show that it would take rather more than 30 millions of years for a uno-twenty-one (the number represented by 1 with 21 ciphers after it) of these molecules to have escaped. Now a uno-twenty-one is about the number of molecules which are present within every cubic centimeter of the gas at such temperatures and pressures as prevail at the bottom of our atmosphere. An escape of molecules of the denser constituents of the atmosphere on this excessively small scale, or even on a scale considerably larger, may be and probably is going on. See a paper on the "Internal Motions of Gases" in the *Philosophical Magazine* for August 1868, where the number of molecules in a gas is estimated. Readers of that paper are requested to correct a mistake at the end of the third paragraph, where 16^2 was by an oversight inserted instead of $\sqrt{16}$.

6. We shall find in the chapter on Venus that the presence of water on that planet enables us to somewhat lower the upper of these two limits.

16. On the Probable Existence of a Magnetic Field in Sun-Spots

George Ellery Hale

(*Astrophysical Journal 28*, 315–343 [1908])

Twentieth-century solar astronomy, inaugurated by George Ellery Hale, differs from that of the previous century because of the invention of specialized observing instruments, the development of a theoretical understanding of the mechanisms involved in the production of atomic spectra, and the realization that laboratory experiments can be used to test interpretations of solar spectra. Hale developed the first modern solar observatory at Mount Wilson in the early 1900s. Using high-dispersion spectroscopy, made possible by the development of the spectroheliograph, he confirmed Norman Lockyer's observations that in sunspots many spectral lines are strengthened, whereas others are weakened or absent. Hale and his colleagues then used laboratory experiments to demonstrate that the great majority of the strengthened lines in the sunspots are also enhanced in the laboratory when the temperature of the gas is decreased.[1]

As described in this paper, photographs of the sun's surface using the hydrogen Hα line suggest vortical currents that surround sunspots, and the direction of the whirling currents seems to change with the part of the solar surface observed. Hale hypothesized that the ionized particles of the solar atmosphere are whirled at high velocities into the sunspots and that these moving charges then create a magnetic field, much as a rotating, electrically charged disk produces a magnetic field. In order to test this idea, Hale looked for the Zeeman effect, in which a magnetic field along the line of sight splits a line into two circularly polarized lines. Not only was he successful in detecting this effect, but he was also able to use laboratory measurements to show that the magnetic fields in sunspots are as strong as 2,900 gauss.

By 1913 Hale presented a general review of the solar magnetic field, in which solar storms were illustrated by photographs of solar prominences, and a general dipole magnetic field for the sun was inferred from the polar streamers seen in the solar corona during solar eclipses.[2] The illustrations appended to the paper given here show that solar prominences reflect the curved lines of force of the magnetic fields which join sunspot pairs and that eclipse photographs do suggest a general solar field extending out into interplanetary space. Although Hale argued in his 1913 paper that the general magnetic field of the sun has a strength of 50 gauss at the solar poles, subsequent measurements of the sun's polar field suggest that it varies and that the strength is a few gauss.

Surprisingly, when J. Evershed attempted to use the Doppler effect to detect the vortical motion that by Hale's assumption gave rise to the magnetic fields, he found no evidence for vortical motion about an axis perpendicular to the solar surface. Evershed's observations were consistent instead with a radial outflow of gases along the axis of the spot vortex.[3] Theoretical models suggest, furthermore, that the sunspot magnetic fields are actually produced deep within the solar interior by the rotational and convective motions of its ionized gases. Nevertheless, Hale had correctly established the existence of strong magnetic fields in sunspots; and his work on sunspot structure, completed after the paper given here, was so thorough that no significant new information has been obtained since. In this later work Hale and his colleagues found that the majority of sunspots occur in pairs with opposite magnetic polarity. The preceding and following spots of binary groups are of opposite polarity, and the corresponding spots of such groups in the northern and southern hemispheres are also opposite in sign. Furthermore, all of the spots reverse their polarity every 22 years, at twice the 11-year period of the variation in the number of sunspots.[4]

1. G. E. Hale, W. S. Adams, and H. G. Gale, *Astrophysical Journal 24*, 185 (1906).

2. G. E. Hale, *Smithsonian Report for 1913*, pp. 145–158; reproduced in *Early Solar Physics*, ed. A. J. Meadows (London: Pergamon Press, 1970).

3. J. Evershed, *Monthly Notices of the Royal Astronomical Society 69*, 454 (1909).

4. G. E. Hale, F. Ellerman, S. B. Nicholson, and A. H. Joy, *Astrophysical Journal 49*, 153 (1919); G. E. Hale, *Annual Report to the Director of the Mount Wilson Observatories* (1924).

THE DISCOVERY of vortices surrounding sun-spots, which resulted from the use of the hydrogen line Hα, for solar photography with spectroheliograph,[1] disclosed possibilities of research not previously foreseen. Photographs taken daily on Mount Wilson with this line suggest that all sun-spots are vortices, and provide material for a discussion of spot theories which will soon be undertaken. Revealing, as they do, the existence of definite currents and whirls in the solar atmosphere, they afford the requisite means of testing the operation in the sun of certain physical laws previously applied only to terrestrial phenomena. The present paper describes an attempt to enter one of the new fields of research opened by this recent work with the spectroheliograph.

Electric Convection

In 1876 Rowland discovered that an electrically charged ebonite disk, when set into rapid rotation, produced a magnetic field, capable of deflecting a magnetic needle suspended just above the disk.[2] It thus appeared, in accordance with Maxwell's anticipation, that a rapidly moving charged body gives rise to just such effects as are caused by an electric current flowing through a wire. Rowland's whirling disk therefore corresponds to a wire helix, within which a magnetic field is produced when a current is passed through it.

Recent studies of the discharge of electricity in gases prove that gases and vapors, when ionized by one of several means, contain electrically charged particles. Moreover, at high temperatures carbon and many other elements which occur in the sun emit negatively charged corpuscles in great numbers; the complementary positively charged particles must also be present, more or less completely separated from the negative corpuscles.[3] Thus electromagnetic disturbances on a vast scale may result from the rapid motions of charged particles produced by eruptions or other solar disturbances.

Soon after the discovery of the vortices associated with sun-spots, it occurred to me that if a preponderance of positive or negative ions or corpuscles could be supposed to exist in the rapidly revolving gases, a magnetic field, analogous to that observed by Rowland in the laboratory, should be the result. An equal number of positive and negative ions, when whirled in a vortex, would produce no resultant field,[4] since the effect of the positive charges would exactly offset that of the negative charges. But Thomson's statement regarding the possible copious emission of corpuscles by the photosphere, and the tendency of negative ions to separate themselves, by their greater velocity, from positive ions, led to the belief that the conditions necessary for the production of a magnetic field might be realized in the solar vortices.

Thanks to Zeeman's discovery of the effect of magnetism on radiation it appeared that the detection of such a magnetic field should offer no great difficulty, provided it were sufficiently intense. When a luminous vapor is placed between the poles of a powerful magnet the lines of its spectrum, if observed along the lines of force, appear in most cases as doublets, having components circularly polarized in opposite directions. The distance between the components of a given doublet is directly proportional to the strength of the field. As different lines in the spectrum of the same element are affected in different degree, it follows that in a field of moderate strength many of the lines may be simply widened, while others, which are exceptionally sensitive, may be separated into doublets.

The Sun-Spot Spectrum

It has long been known that the spectrum of a sun-spot differs from the ordinary solar spectrum in several particulars. If, for example, we examine the iron lines in a spot, we find some of them are more intense than in the solar spectrum, while others are weaker. Again, we perceive that many of the spot lines are widened, and that the degree of widening varies for different lines. Finally, if the observations are made with an instrument of high dispersion, it will be seen that some of the iron lines, which are single in the solar spectrum, are double in the spot spectrum. Such double lines were first seen by Young in 1892 with a large spectroscope attached to the 23-inch Princeton refractor. Walter M. Mitchell, who subsequently observed them with the same instrument, described the doublets as "reversals," which they closely resemble. Mitchell's papers contain a valuable series of observations of these "reversals" and other sunspot phenomena.[5]

Our previous investigations in this field on Mount Wilson may be summarized as follows:

1. The application of photography to the study of sun-spot spectra. A Littrow or auto-collimating spectrograph of 18 feet (5.5 m) focal length, used with the Snow telescope, gave good results, and permitted a great number of spot lines and bands, not previously known, to be recorded.[6] On the completion of the tower telescope last autumn, these observations were continued with a vertical spectrograph of 30 feet (9.1 m) focal length.[7] Although the only grating available for work in the higher orders is a 4-inch (10 cm) Rowland, having 14,438 lines to the inch (567 to the mm), employed in my experiments in photographing sun-spot spectra at the Kenwood and Yerkes Observatories,[8] the results secured with this instrument are very satisfactory, greatly surpassing those obtained with the 18-foot spectrograph. They give the first photographic records of the "reversals" or doublets seen visually by Young and Mitchell, and reveal thousands of faint lines beyond the reach of visual observation.

2. The preparation of a photographic map of the sun-spot spectrum and a catalogue of all the lines. A preliminary map, consisting of 26 sections of 100 Ångströms each, covering the region λ 4,600–7,200, was prepared last

year by Mr. Ellerman from negatives made with the 18-foot spectrograph, and supplied to visual observers taking part in the sun-spot work of the International Solar Union. A much better map, to be made from negatives obtained with the tower telescope and 30-foot spectrograph, will be ready, it is hoped, within a year. The catalogue of lines, which involves a great amount of measurement for the determination of wave-lengths, is well advanced, and one section has been published by Mr. Adams.[9]

3. The identification of the numerous lines which constitute the flutings in the spot spectrum. Photographs of the spectra of titanium oxide, magnesium hydride, and calcium hydride,[10] made in our laboratory by Dr. Olmsted, have furnished the material for this purpose. The measurement of the lines in these flutings is well advanced.

4. The interpretation of the change of the relative intensity of lines observed in passing from the solar spectrum to the spot spectrum. Investigations on the spectra of iron, manganese, chromium, titanium, vanadium, and other metals conspicuous in spots, made with the arc, spark, and flame, indicated that this change is due to a reduction of the temperature of the spot vapors.[11] Subsequent work with a new electric furnace by Dr. King,[12] the details of which have not yet been published, seems to leave little doubt that this explanation is correct. It is supported by the presence in the spot of compounds which appear to be dissociated at the higher temperature outside the spot, and by the resemblance of spot spectra to the spectra of red stars.[13]

While our investigations have thus furnished a plausible explanation of some of the characteristic phenomena of sun-spot spectra, the widening of lines and the presence of doublets are among the remaining peculiarities that demanded consideration. As we have seen, however, these very peculiarities are precisely what would be expected if a magnetic field were present. Prompted by the theoretical considerations outlined above, and encouraged by their apparent agreement with the facts of observation, I decided to test the components of the spot doublets for evidences of circular polarization and to seek for other indications of the Zeeman effect.

Method of Observation

In Hale's tower telescope, an image of the sun about 17 cm in diameter was formed on the slit of a vertical spectrograph after passing through a Fresnel rhomb and a Nicol prism. The rhomb transforms circularly polarized radiation into plane polarized radiation, and rotation of the Nicol prism permits the detection of this plane polarized radiation. Partially covering the slit made it possible to photograph only the sunspot spectrum, and comparison spectra were taken on other regions of the sun.

Circular Polarization along the Lines of Force

A magnetic field oriented along the line of sight to a sunspot causes the light of the two components of a doublet to be circularly polarized in opposite directions; this effect was detected when the relative intensities of the components of the spot doublets showed a reversal during a rotation of the Nicol prism through 90°. Omitted here is a photograph showing this effect for the vanadium doublet at λ5,940.87.

Reversed Polarities of Right and Left-Handed Vortices

A second test, which also bears upon the hypothesis that the field is produced by the revolution of electrically charged particles in the spot-vortex, may now be described. If a Nicol is set so as to cut off the violet component of a doublet observed along the lines of force of a magnetic field, reversal of the current will cause the red component to disappear and the violet component to become visible. Reversal of the direction of the current in a magnet corresponds to reversal of the direction of revolution in a solar vortex. If it could be shown, by an independent method, that in two sun-spot vortices the charged particles are revolving in opposite directions, the red components of the doublets should appear in the spectrum of one spot, and the violet components in that of the other, the position of the rhomb and Nicol remaining unchanged.

Fortunately the spectroheliograph plates indicate the direction of revolution in the solar vortices. The vortices are constantly changing in appearance, and the stream lines are not always clearly defined.

Here Hale presents the Hα photograph given in figure 16.1, which shows right- and left-handed vortices for two sunspots located, respectively, in the northern and southern hemispheres of the sun.[14] We omit here portions of the spectra of these spots showing only the red component of

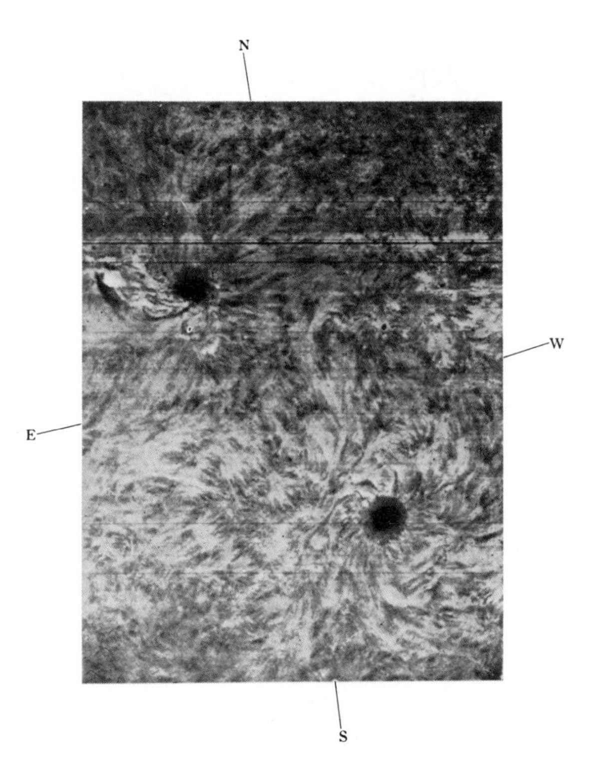

Fig. 16.1 Sunspots and hydrogen flocculi showing right- and left-handed vortices. Picture taken at the Hα wavelength at 6^h20^m A.M. on September 9, 1908. (From *Astrophysical Journal 28*, 323 [1908].)

the doublet in the southern spot and only the violet component of the doublet in the northern spot. When the Nicol was rotated by 90° the colors were reversed.

This result has been confirmed by other photographs, which indicate that the direction of the displacement always depends upon the direction of the revolution in the vortex. If this relation is found by future observations to hold generally, we may conclude that the field is always produced by the revolution of particles carrying charges of like sign.

PLANE POLARIZATION ACROSS THE LINES OF FORCE

So far we have confined our attention to polarization phenomena observed along the lines of force. But it is well known that the doublets are, in general, transformed into triplets, when observed in a magnetic field at right angles to the lines of force. The components of the triplets are plane polarized, the central line in a plane at right angles to the plane of polarization of the side components. It should be possible to detect similar phenomena in spot spectra, if they are produced in a magnetic field.

Mitchell has recorded cases when spot doublets were seen as triplets,[15] and Hale reports polarization measurements of triplets in spots seen near the limb of the sun. As expected theoretically, with one orientation of the Nicol, only the central component of the triplet is visible, whereas a 90° rotation of the Nicol causes the central component to disappear and the outer two lines of the triplet to appear.

LABORATORY TESTS

If the widened lines and doublets in spot spectra are produced by a magnetic field, an equal degree of widening and an equal separation of the components of doublets should be found in the laboratory when the same lines are observed in a field of equal strength.

The iron and chromium doublets at $\lambda \approx 6{,}300$ and $\lambda \approx 5{,}780$ Å, respectively, were sufficiently strong to be detected in the laboratory spectra of a spark in the presence of a magnetic field 15,000 gauss in strength. The laboratory measurements show that the doubling of these lines in sunspots is caused by a magnetic field of about 2,900 gauss. Some titanium lines support this conclusion, but others give ambiguous results. Hale concludes that the titanium lines produced by absorption at higher levels in the atmosphere reflect a weaker magnetic field than those produced at lower levels. That the D lines of sodium are only slightly displaced by Nicol rotation suggests that the field strength is reduced to a fraction of its maximum value at elevations of about 5,000 mi in the chromosphere.

If a magnetic field is the principal cause of the widening of lines in spots, their widths should be roughly proportional to the separation of the components of the corresponding doublets observed in a field of equal strength. Bearing in mind the differences in the character of the lines, and the probable effect of variations in the mean level of absorption, we can hardly expect a very close agreement. But some evidences of relationship should appear, if a magnetic field is present.

Hale compares the widths of various sunspot lines with the separations of the doublets observed in the laboratory in the presence of a magnetic field of 15,000 gauss. There is a progressive decrease toward the violet for the mean width of spot lines and for the separation of the corresponding doublets in the spark. The iron doublets follow very closely the law $\Delta\lambda/\lambda^2 =$ constant.

SIGN OF THE CHARGE THAT PRODUCES THE FIELD IN SUN-SPOTS

If the evidence presented in this paper renders probable the existence of a magnetic field in sun-spots, it is of interest to inquire concerning the sign of the charge which, according to our hypothesis, produces the field. In Lorentz's theory of the Zeeman effect in its simplest form, the motion of a single electron in a molecule of a luminous source is discussed.[16] This electron is supposed to be capable of displacement in all directions from its position of equilibrium, toward which it is drawn by an elastic force, which is proportional to the displacement but independent of its direction. Let e be the charge of the particle, m its mass, fr the elastic force caused by a displacement r, f being a positive constant. The frequency of the vibrations, whether they be linear, elliptical, or circular, will be

$$n_0 = \sqrt{\frac{f}{m}}.$$

We may now suppose the light-source to be placed in a homogeneous magnetic field of intensity H. A particle carrying a charge e, and moving with velocity v, will be subjected to a force perpendicular to the field and to the direction of motion of the particle, the magnitude of which may be represented by $ev\text{H} \sin (v,\text{H})$. It is evident that the electron may have three different motions, each with its own frequency. Linear vibrations parallel to the lines of force, having the frequency n_0, will not be affected by the magnetic field. Circular vibrations in a plane perpendicular to the lines of force will be affected differently, depending upon whether they are right-handed or left-handed. If r is the radius of a circular orbit and n the frequency, the velocity of the electron will be $v = nr$ and the centripetal force will have the value mn^2r. We may now consider the effect on the motion of the electron of the elastic force fr and of an electromagnetic force

$$ev\text{H} = enr\text{H}.$$

For a positive charge the latter force is directed toward the center if the motion is clockwise, as seen by an observer toward whom the lines of force are directed. We then have

$$mn^2r = fr + enr\text{H}.$$

This frequency n differs very slightly from the frequency n_0; thus the last term of the equation must be much smaller than the term fr, so that we may write

$$n = n_0 + \frac{e\text{H}}{2m}. \qquad (1)$$

This expression gives the frequency of the right-handed (clockwise) vibrations. For the left-handed vibrations we have

$$n = n_0 - \frac{e\text{H}}{2m}. \qquad (2)$$

As seen along the lines of force a single line in the spectrum is thus transformed into a doublet, the components of which are circularly polarized. An observer toward whom the lines of force are directed will find that the light of the component of greater wave-length, whose frequency has been decreased by the field, is circularly polarized in the right-handed or clockwise direction. Hence (2) is greater than (1), and it follows that the charge e of the electron which produces the spectral lines must be negative.

In the case of the solar vortices we have to consider two sets of charged particles, which may be entirely distinct from one another: (1) those whose vibrations give rise to the lines in the spectra of spots, and (2) those that carry the charge which, by the hypothesis, produces the magnetic field. The Zeeman effect supplies the means of determining the direction of the lines of force of the sun-spot fields, and photographs of the vortices, made with the spectroheliograph, indicate the direction of their rotation. Thus we are in a position to determine the sign of the charge carried by the particles which produce the fields. As pointed out independently by König and Cornu, the violet component of a magnetic doublet observed along the lines of force is formed by circular vibrations, having the direction of the current flowing through the coils of the magnet.[17] From observations of circularly polarized light, made in our Mount Wilson laboratory by Dr. St. John and confirmed by myself, it appears that when the Nicol prism of the tower spectrograph stands at 60° E. it transmits the violet component of a doublet produced in a magnetic field directed toward the observer. From Biot and Savart's law the direction of the current causing such a field is counterclockwise, as seen by the observer. In the same position the Nicol also transmits the violet component of a doublet produced in a sun-spot surrounded by a vortex rotating clockwise. As a negative charge rotating clockwise produces a field of the same polarity as an electric current flowing counterclockwise, we may conclude that the magnetic field in spots is caused by the motion of negative ions or electrons.

PROBABLE SOURCE OF THE NEGATIVE CORPUSCLES

We may now consider the probable source of a sufficient number of negative corpuscles to produce a field of about 2,900 gausses in sun-spots.

In his *Conduction of Electricity through Gases*, p. 164, J. J. Thomson writes as follows:

> We thus are led to the conclusion that from an incandescent metal or glowing piece of carbon "corpuscles" are projected, and though we have as yet no exact measurements for carbon, the rate of emission must, by comparison with the known much smaller rate for platinum, amount in the case of a carbon filament at its highest point of incandescence to a current equal to several amperes per square centimeter of surface. This fact may have an important application to some cosmical phenomena, since, according to the generally received opinion, the photosphere of the sun contains large quantities of glowing carbon; this carbon will emit corpuscles unless the sun by the loss of its corpuscles at an earlier stage has acquired such a large charge of positive electricity that the attraction of this is sufficient to prevent the negatively electrified particles from getting right away from the sun; yet even in this case, if the temperature were from any cause to rise above its average value, corpuscles would stream away from the sun into the surrounding space.

On another page (168) Thomson also remarks: "The emission of the negative corpuscles from heated substances is not, I think, confined to the solid state, but is a property of the atom in whatever state of physical aggregation it may occur, including the gaseous." After illustrating this in the case of sodium vapor, Thomson adds (p. 168):

> The emission of the negatively electrified corpuscles from sodium atoms is conspicuous as it occurs at an exceptionally low temperature; that this emission occurs in other cases although at very much higher temperatures is, I think, shown by the conductivity of very hot gases (or at any rate by that part of it which is not due to ionization occurring at the surface of glowing metals), and especially by the very high velocity possessed by the negative ions in the case of these gases; the emission of negatively electrified corpuscles from atoms at a very high temperature is thus a property of a very large number of elements, possibly of all.

Thus the chromosphere, as well as the photosphere, may be regarded as copious sources of negatively electrified corpuscles. The part played by these corpuscles in the sun-spots cannot be advantageously discussed until the nature of the vortices is better understood.[18] At present it is enough to recognize that the supply of negative electricity appears amply sufficient to account for the magnetic fields.

Let n be the number of corpuscles per unit cross-section passing a given point in unit time and e the charge on each corpuscle. Then we have, for the current carried by the corpuscles, $c = ne$. H. A. Wilson found that in a vacuum tube, at pressures up to 8.5 mm, the current at the cathode was $0.4\,p$ milliamperes per sq. cm, where p is the pressure in millimeters.[19] If $p = 8.5$, we have $c = 3.4 \times 10^{-3}$ amperes. Assume the velocity of the corpuscles in this case to be of the order of 10^4 km per sec. In a solar vortex (if the charged particles are carried with it) the velocity may be taken as of the order of 100 km per sec.[20] Then if the number of corpuscles per sq. cm were the same in the two cases, the current in the sun would be of the order of 3.4×10^{-5} amperes per sq. cm at the same pressure.

We may now assume the corpuscles to be moving at a velocity of 100 km per second in an annulus 25,000 km wide, 1,000 km deep, and 100,000 km in diameter surrounding a sun-spot. Taking the current strength to be as above, 3.4×10^{-5} amperes per sq. cm, the intensity of the resulting magnetic field comes out 1,000 gausses.

Such a calculation is of little value, except for the purpose of indicating that a magnetic field of the observed order of magnitude might conceivably be produced on the sun.[21]

External Field of Sun-Spots

We have already seen that the strength of the field in spots apparently changes very rapidly along a solar radius, and is small at the upper level of the chromosphere.

If subsequent work proves this to be the case, it will appear very improbable (as indicated by theory) that terrestrial magnetic storms are caused by the direct effect of the magnetic fields in sun-spots. Their origin may be sought with more hope of success in the eruptions shown on spectroheliograph plates in the regions surrounding spots.

Conclusion

Although the combined evidence presented in this paper seems to indicate the probable existence of a magnetic field in sun-spots, the weak points of the argument should be clearly recognized. Among these are the following:

1. The failure of our photographs to show the central line of spot triplets before the spots are very close to the limb.
2. The presence in the spot spectrum of at least one triplet, which appears as a doublet when observed along the lines of force in the laboratory.
3. The absence of evidence to support the hypothesis that the imperfect agreement between spot and laboratory results is due to differences in the mean level of absorption.
4. The apparent constancy of the field strength, as indicated by the nearly uniform width of the doublets in different spots.
5. The difficulty of explaining, on the basis of our present fragmentary knowledge of solar vortices, the observed strength of field in the umbra and penumbra, and especially its variation with level.

As the resolving power of the 30-foot spectrograph is sufficient to resolve completely only the wider spot doublets, the central line could not be separately distinguished in other cases, even if it were present. Hitherto it has been possible to photograph the spectra of only the largest spots, because the images of other spots, as given by the tower telescope, are too small. The need of a telescope giving a much larger image of the sun, and a spectrograph of greater resolving power and focal length, which has been felt in previous work, is strongly emphasized by this investigation. Such apparatus would also permit the spectrum of the chromosphere, and many other solar phenomena, to be studied to great advantage.

As regards the nature of the vortices, the principal question is whether the gyratory motion primarily concerned in the production of the magnetic field is outside the boundaries of the spot or within the umbra. In the former case we must face various difficulties, such as the apparent constancy of the field in different spots, and the fact that its intensity rapidly decreases upward. If a spot vortex may be considered analogous to an anti-cyclone, and the assumption be made that the gyratory motion of the low-level vapors produces the field, these difficulties may be lessened. The view that the field is produced by the gyratory motion of vapors within the umbra raises other difficulties which may also be serious. Fortunately there is reason to hope that observations now in progress may throw light on several of these questions.

Hale states in an addendum that laboratory photographs show certain iron and titanium lines appearing as doublets when the magnetic field is either parallel or perpendicular to the line of sight. These and other measurements provide reasons for the failure of triplets to appear when spots are as much as 60° from the sun center, and they leave no doubt in Hale's mind that the doublets and triplets observed in the sunspot spectra are actually caused by a magnetic field.

1. Hale, "Solar Vortices," *Contributions from the Mount Wilson Solar Observatory*, no. 26; *Astrophysical Journal 28*, 100 (1908).

2. Rowland, "On the Magnetic Effect of Electric Convection," *American Journal of Science* (3) *15*, 30 (1878).

3. J. J. Thomson, *Conduction of Electricity through Gases*, p. 165.

4. Unless separated by centrifugal force, as suggested by Professor Nichols.

5. Walter M. Mitchell, "Reversals in the Spectra of Sun-Spots," *Astrophysical Journal 19*, 357 (1904); "Researches in the Sun-Spot Spectrum, Region F to *a*," ibid *22*, 4 (1905); "Results of Solar Observations at Princeton, 1905–1906," ibid *24*, 78 (1906).

6. Hale and Adams, "Photographic Observations of the Spectra of Sun-Spots," *Contributions from the Mount Wilson Solar Observatory*, no. 5; *Astrophysical Journal 23*, 11 (1906).

7. Hale, "The Tower Telescope of the Mount Wilson Solar Observatory," *Contributions from the Mount Wilson Solar Observatory*, no. 23; *Astrophysical Journal 27*, 204 (1908).

8. Hale, "Solar Research at the Yerkes Observatory," *Astrophysical Journal 16*, 211 (1902).

9. Adams, "Preliminary Catalogue of Lines Affected in Sun-Spots, Region λ 4000 to λ 4500," *Contributions from the Mount Wilson Solar Observatory*, no. 22; *Astrophysical Journal 27*, 45 (1908).

10. Olmsted, "Sun-Spot Bands Which Appear in the Spectrum of a Calcium Arc Burning in the Presence of Hydrogen," *Contributions from the Mount Wilson Solar Observatory*, no. 21; *Astrophysical Journal 27*, 66 (1908).

11. Hale, Adams, and Gale, "Preliminary Paper on the Cause of the Characteristic Phenomena of Sun-Spot Spectra," *Contributions from the Mount Wilson Solar Observatory*, no. 11; *Astrophysical Journal 24*, 185 (1906); Hale and Adams, "Second Paper on the Cause of the Characteristic Phenomena of Sun-Spot Spectra," *Contributions from the Mount Wilson Solar Observatory*, no. 15; *Astrophysical Journal 25*, 75 (1907).

12. King, "An Electric Furnace for Spectroscopic Investigations, with Results for the Spectra of Titanium and Vanadium," *Contributions from the Mount Wilson Solar Observatory*, no. 28; *Astrophysical Journal 28*, 300 (1908).

13. Hale and Adams, "Sun-Spot Lines in the Spectra of Red Stars," *Contributions from the Mount Wilson Solar Observatory*, no. 8; *Astrophysical Journal 23*, 400 (1906); Adams, "Sun-Spot Lines in the Spectrum of *Arcturus*," *Contributions from the Mount Wilson Solar Observatory*, no. 12; *Astrophysical Journal 24*, 69 (1906).

14. Right- and left-handed vortices have also been found in the same hemisphere.

15. *Astrophysical Journal 24*, 80 (1906).

16. The following outline of the theory is taken from Lorentz's "Theorie des phénomènes magnéto-optiques récemment découverts," *Rapports, Congrés international de physique 3*, 1 (1900).

17. See Cotton, *Le phénomène de Zeeman*, chap. vii; König, *Wied. Ann. 62*, 240 (1897).

18. For this reason a discussion of the very interesting suggestion of Professor E. F. Nichols, that the positively and negatively charged particles are separated by centrifugal action in the spot vortex, is reserved for a subsequent paper.

19. *Philosophical Magazine* (6) *4*, 613 (1902).

20. *Solar Vortices*, p. 13.

21. See a similar calculation by Zeeman in *Nature* for August 20, 1908.

Appended Figures

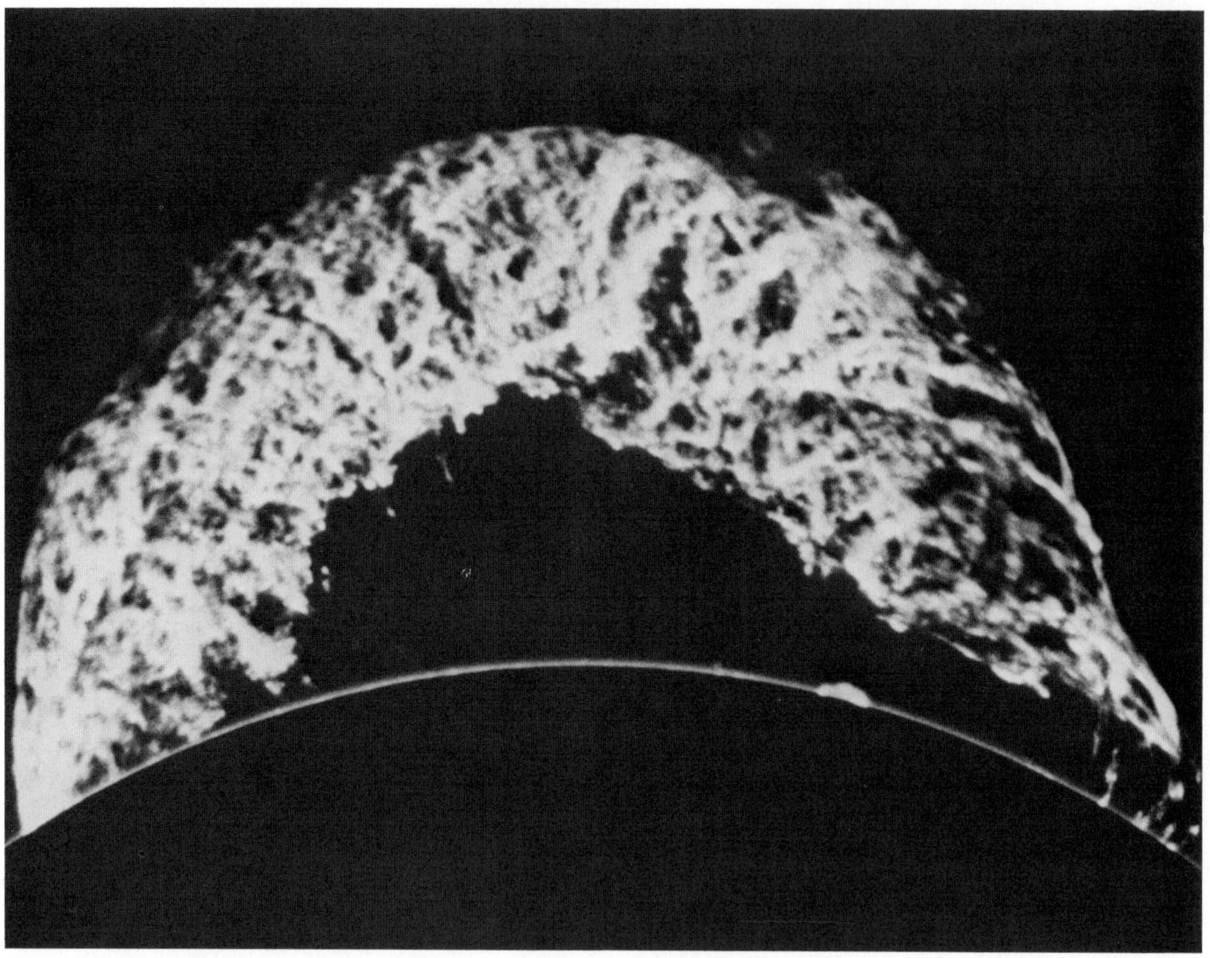

Fig. 16.2 An arch prominence photographed on June 4, 1946, with the coronagraph at Climax, Colorado. The height of this prominence is over 320,000 km. (Courtesy *Sky and Telescope.*)

Fig. 16.3 The solar corona, photographed at the solar eclipse of June 30, 1973, in Loyengalani, Kenya, with a filter that compensates for the sharp decrease in coronal brightness with distance from the sun. (Courtesy High Altitude Observatory.)

17. How Could a Rotating Body such as the Sun Become a Magnet?

Joseph Larmor

(*Report for the British Association for the Advancement of Science*, 159–160[1919])

In this short report, Joseph Larmor addresses the problem of producing the large-scale magnetic fields of the earth and the sun. Because of the high internal temperatures of both earth and sun, Larmor found the hypothesis of a permanent magnet unacceptable. His solution was a self-exciting dynamo with magnetic fields generated by internal motions of conducting material. Later T. G. Cowling[1] showed that rotational motion alone, in the axially symmetric case, was insufficient, but in 1939 Walter Elsasser[2] demonstrated that the addition of convective motions could maintain the magnetic field. Besides reviving the dynamo theory for the origin of magnetic fields, Elsasser's paper marked the beginning of a wide range of applications of magnetohydrodynamics to cosmic phenomena.

1. T. G. Cowling, *Monthly Notices of the Royal Astronomical Society 94*, 39 (1933).
2. W. M. Elsasser, *Nature 143*, 374 (1939), and *Physical Review 55*, 489 (1939).

THE OBVIOUS SOLUTION by convection of an electric charge, or of electric polarisation is excluded; because electric fields in and near the body would be involved, which would be too enormous. Direct magnetisation is also ruled out by the high temperature, notwithstanding the high density. But several feasible possibilities seem to be open.

1

In the case of the sun, surface phenomena point to the existence of a residual internal circulation mainly in meridian planes. Such internal motion induces an electric field acting on the moving matter: and if any conducting path around the solar axis happens to be open, an electric current will flow round it, which may in turn increase the inducing magnetic field. In this way it is possible for the internal cyclic motion to act after the manner of the cycle of a self-exciting dynamo, and maintain a permanent magnetic field from insignificant beginnings, at the expense of some of the energy of the internal circulation. Again, if a sunspot is regarded as a superficial source or sink of radial flow of strongly ionised material, with the familiar vortical features, its strong magnetic field would, on these lines, be a natural accompaniment: and if it were an inflow at one level compensated by outflow at another level, the flatness and vertical restriction of its magnetic field would be intelligible.

2

Theories have been advanced which depend on a hypothesis that the force of gravitation or centrifugal force can excite electric polarisation, which, by its rotation, produces a magnetic field. But, in order to obtain a sensible magnetic effect, there would be a very intense internal electric field such as no kind of matter could sustain. That, however, is actually got rid of by a masking distribution of electric charge, which would accumulate on the surface, and in part in the interior where the polarisation is not uniform. The circumstance that the two compensating fields are each enormous is not an objection; for it is recognised, and is illustrated by radioactive phenomena, that molecular electric fields are, in fact, enormous. But though the electric masking would be complete, the two distributions would not compensate each other as regards the magnetic effects of rotational convection: and there would be an outstanding magnetic field comparable with that of either distribution taken separately. Only rotation would count in this way; as the effect of the actual translation, along with the solar system, is masked by relativity.

3

A crystal possesses permanent intrinsic electric polarisation, because its polar molecules are orientated: and if this natural orientation is pronounced, the polarisation must be nearly complete, so that if the crystal were of the size of the earth it would produce an enormous electric field. But, great or small, this field will become annulled by masking electric charge as above. The explanation of pyro-electric phenomena by Lord Kelvin was that change of temperature alters the polarisation, while the masking charge has not had opportunity to adapt itself: and piezo-electric phenomena might have been anticipated on the same lines. Thus, as there is not complete compensation magnetically, an electrically neutralised crystalline body moving with high speed of rotation through the æther would be expected to produce a magnetic field: and a planet whose materials have crystallised out in some rough relation to the direction of gravity, or of its rotation, would possess a magnetic field. But relativity forbids that a crystalline body translated without rotation at astronomical speeds should exhibit any magnetic field relative to the moving system.

The very extraordinary feature of the earth's magnetic field is its great and rapid changes, comparable with its whole amount. Yet the almost absolute fixity of length of the astronomical day shows extreme stability of the earth as regards its material structure. This consideration would seem to exclude entirely theories of terrestrial magnetism of the type of (2) and (3). But the type (1), which appears to be reasonable for the case of the sun, would account for magnetic change, sudden or gradual, on the earth merely by change of internal conducting channels: though, on the other hand, it would require fluidity and residual circulation in deepseated regions. In any case, in a celestial body residual circulation would be extremely permanent, as the large size would make effects of ordinary viscosity nearly negligible.

18. Polarization of the Moon and of the Planets Mars and Mercury

Bernard Lyot

TRANSLATED BY OWEN GINGERICH

(*Comptes rendus de l'Académie des Sciences* (Paris) *178*, 1796–98 [1924])

The light rays of the sun that illuminate the planets may be reflected, refracted, diffracted, scattered, or absorbed by the planetary atmosphere or surface. Hence, observations of the intensity and polarization of the planet's light can provide considerable physical information about the planet. The photographic magnitudes of the planets can be used to infer their albedo, the ratio of the total light reflected from the planet to the total incident on it. As early as 1916 Henry Norris Russell had determined the photographic albedo of most of the planets; he found that the values for the moon, Mars, and Mercury were about 0.06 and that those for Earth, Venus, Jupiter, Saturn, Uranus, and Neptune were about 0.60. Because observations of basalt rocks and terrestrial clouds gave respective albedos of 0.06 and 0.6, the planetary results seemed to confirm "the familiar view that the reflecting power of Venus and the outer planets is comparable to that which might be expected from cloud surfaces, while that of Mercury, Mars, and the moon, which have little or no atmospheres, is similar to that of ordinary rock."[1] By observing the polarization of planetary light as a function of the angle between the directions of observation and of illumination, Bernard Lyot was eventually able to show that the polarization curves of Venus and Jupiter differ completely from those of solid materials, whereas those for the moon, Mercury, and Mars are very similar to those of solid materials.[2] The earlier Lyot paper given here, one written while he was still a graduate student, shows the close similarity of the moon and Mercury. Because the planet Mars also has a polarization curve similar to that of the moon, Lyot concluded that the moon, Mercury, and Mars have surfaces with analogous physical composition.

After studying many substances in the laboratory, Lyot concluded that the moon, Mercury, and Mars are covered almost entirely with a powdery material resembling volcanic ashes found on the earth, although this covering could be thin.[2] As early as December 1924 Lyot had suggested the presence of dust storms on Mars, after he observed that a considerable reduction in polarization occurred when a large portion of the planet's surface was covered with a dense yellow veil. The presence of a powdery lunar soil has been established by the Apollo missions, and the Mariner missions to Mars have confirmed the existence of windblown dust storms on its surface.

The similarity of Mercury's crater-pocked surface to that of the moon was dramatically confirmed by the *Mariner X* photographs taken in 1974; appended here is the photomosaic from that space mission.

1. H. N. Russell, *Astrophysical Journal 43*, 173 (1916).
2. B. Lyot, *Annales de l'Observatoire de Paris* (Meudon) *8*, 37–161 (1929).

I have described in an earlier note[1] how the polarization of Venus varies when the phase angle between the directions of illumination and of observation successively takes on all the values for which the equipment can measure this weak polarization. The polarizations of the moon, Mercury, and Mars, studied with the same equipment, vary in an entirely different manner.

The measurements of the moon, made with a refractor of 0.175 m aperture, were carried out from the start principally on twenty-three regions distributed over the entire surface and ranging from very dark gray, such as Sinus Aestuum, to the most brilliant white of the white spot situated to the southeast of Furnerius. The dark regions are represented by the curve *D*, the light regions by the curve *E* (figure 18.1).

All these regions yielded similar curves, exhibiting for a phase angle of about 100° a maximum degree of polarization lying between 0.045 and 0.180, being higher where the color of the surface is darker.

The positions had no influence on the plane of polarization, which was always either parallel or normal to the plane of illumination and observation; and the position seemed to affect the proportion of the light polarized only in the vicinity of the terminator, where the polarization was a little stronger.

Beginning with the first quarter, the polarization of all these points, parallel to the plane of illumination and observation, diminished according to curves that blended into curve G.

About 15 hr before the full moon, for precisely the phase angle of 24°, the polarization vanished, first on the eastern side and then on the western, reappearing less than 2 hr later in a perpendicular plane.

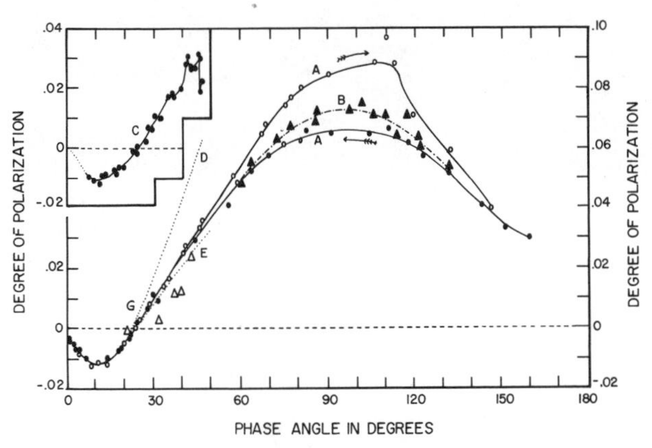

Fig. 18.1 The degree of polarization as a function of phase angle for the waning (*open circles*) and waxing (*filled circles*) moon (*curve A*); for Mercury (*curve B*) during the day (*open triangles*) and night (*filled triangles*); and for Mars (*curve C*). All the curves exhibit the same negative polarization in the region *G*, and the dark and light regions of the moon are represented, respectively, by curves *D* and *E*.

After this inversion, the degree of polarization, which we now for convenience call negative, attains for an angle of 10–11° a maximum of −0.011 and −0.012 for the continental areas and of −0.013 to −0.014 for the maria; it then finally vanishes at the exact opposition.

Another series of measurements, carried out for the light of the entire moon, has been made with the same polarimeter by forming a small lunar image on the pupil of the observer; the values thus obtained, represented by the points on curve *A*, belong to many lunations, and consequently the libration has caused a scatter for some points near the quadrature.

The differences between the eastern and western halves are clearly shown by the much stronger polarization during the waning phase than during the waxing phase. The abrupt drop in polarization observed after the last quarter is undoubtedly caused by the disappearance from the terminator of the darkest regions of the Sea of Storms.

For Mars the observations, numbering twenty-eight, began with the opposition of 10 June 1922. The polarization, null at that moment, varied as shown in curve *C*, which is almost identical to that for the moon.

The polarizations of the continental areas, and thereafter those of the maria, which were a little larger in absolute value, changed signs successively during the nights of July 9 and 10 for a phase angle of 24°, a value identical to that found for the moon.

The limbs of the planet, the polar caps, and the terminator were a little more polarized than the central region. The oscillations seen on the curve can be attributed to the rotation of the planet; they are in fact especially noticeable in the straight part, where the phase angle changes very slowly.

The observations of Mercury were made primarily at dusk near the quadratures of May 1922 and April 1924 with the 0.83 m equatorial of the Meudon Observatory. They furnish a series of points (curve *B*) that fall between the curves of the waxing and the waning moons. In order to carry the phase angles to small angles, a daytime series of observations was necessary. The arrangement previously used for Venus was used again to compensate for the atmospheric polarization, but the smallness of the disk and especially the faintness of the light of its surface, compared to the atmosphere in the vicinity of the sun, have greatly diminished the precision of the measurements. The points obtained show, however, that the polarization varies in essentially the same way as those of the moon and of Mars and must vanish at nearly the same phase angle.

The curves described testify to the close relationship among the three planets: their surfaces probably have analogous physical compositions. Furthermore, if Mercury has an atmosphere, it cannot be thick enough to hide the surface from us.

1. *Comptes rendus de l'Académie des sciences* (Paris) *177*, 1015 (1923).

Appended Figure

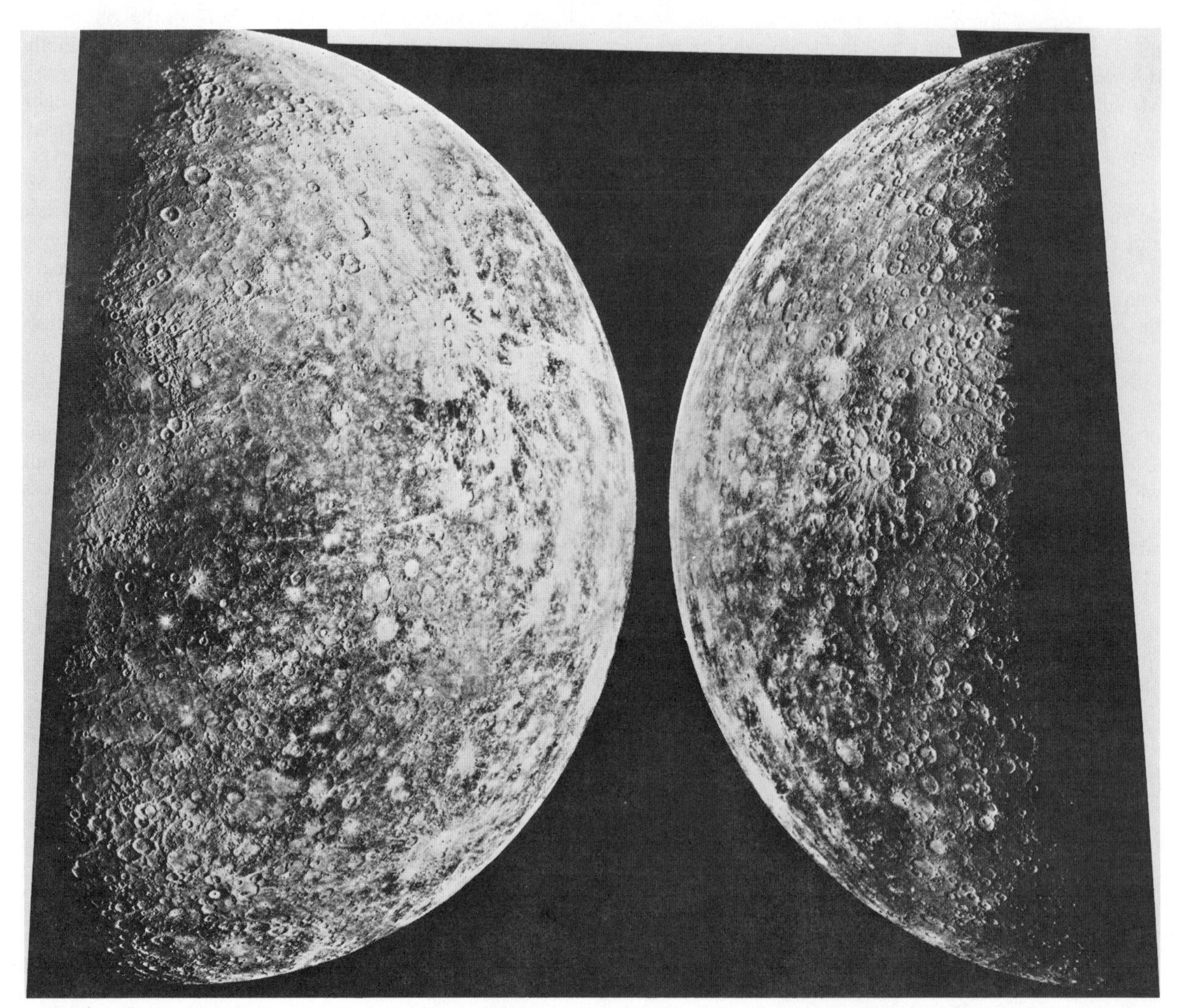

Fig. 18.2 *Right*: photomosaic of Mercury made from nine computer-enhanced pictures taken at 234,000 km, 6 hr before closest approach to the planet. North is at the top, and the sun's illumination is from the left. The evening terminator is at about 20° W longitude, and the bright limb is at 110° W longitude. This part of Mercury is heavily cratered, the craters ranging in size up to almost 200 km. Apparently the freshest craters are those with bright rays, for example, the bright crater (diameter about 40 km) near the center of the picture. This picture and that on the left show the whole of the illuminated portion of the planet; adjoining areas lie along the bright limb of both pictures. *Left*: photomosaic made from twelve computer-enhanced pictures taken at 210,000 km, $5\frac{1}{2}$ hr after closest approach. Half of a 1,300-km multiringed basin appears near the center of the disk on the terminator. The inner portion and parts of the surrounding area are relatively poorly cratered plains materials, similar in aspect to the lunar mare units. Several craters with bright rays are evident; a bright ray pattern, which appears to connect several young craters, forms a smooth hyperbolic curve. The North Pole is at the top, and the equator extends from left to right about two-thirds of the way down the photograph. The bright limb is at 110° W and the terminator at 200° W. (Courtesy National Space Science Data Center.)

19. The Theory of Continental Drift

Alfred Wegener
TRANSLATED BY JOHN BIRAM

(*Die Enstehung der Kontinente und Ozeane* [Germany: Friedrich Vieweg & Sons, 1929], trans. in *The Origin of Continents and Oceans* [New York: Dover Publications, 1966], pp. 5–21)

In this selection Alfred Wegener presents evidence to support his view that the continents wander across the surface of the earth. He suggests that at one time all the continents were joined together into one huge landmass, Pangaea, which occupied half the earth's surface while the other half was covered by the Pacific Ocean. At a later time Pangaea began to rift into fragments, causing the continents to move apart like icebergs in water. To support this view, Wegener calls attention to the correspondence between the Atlantic coasts of Africa and South America, to the matching distribution of mountain ranges truncated by coastlines, and to the presence of similar fossil assemblages and geological features on different continents.

Unfortunately, Wegener offers no satisfactory mechanism either to move the continents apart or to uplift the mountains, and his suggestion that the earth's rotation is responsible was quickly dismissed as inadequate. In particular, Harold Jeffreys showed that the Eötvös force that tends to displace a floating body toward regions of lower gravity is insufficient to move continents.[1] It is only in the past few years that new discoveries in geophysics and oceanography have compelled the acceptance of Wegener's ideas and have created the new study of "plate tectonics." Energy from the decay of radioactive elements in the earth's mantle provides heat to generate a slow but large-scale convective circulation of the rocks beneath the oceans and continental crust. Molten lava pours out from a system of mid-oceanic ridges at the astonishing rate of about 50 billion tons per year.[2] The ocean floor spreads away from the ridge-crests at rates of 1–5 cm per year. To compensate for the spreading the rigid lithosphere plunges deep into the earth's mantle along the oceanic trenches. Both the continents and ocean floors ride on a mosaic of large crustal plates that move as rigid units and suffer deformation only at their margins, from contact with neighbouring plates. As a result, the plate boundaries are marked by the major mountain ranges, the areas of volcanic activity, and the earth's seismically active zones.

1. H. Jeffreys, *The Earth: Its Origin, History, and Physical Constitution* (Cambridge: Cambridge University Press, 1924; 5th ed., 1970).

2. E. C. Bullard, *Nature 170*, 200 (1952); R. Revelle and A. E. Maxwell, *Nature 170*, 199 (1952); G. J. F. MacDonald, *Science 143*, 921 (1964).

It is a strange fact, characteristic of the incomplete state of our present knowledge, that totally opposing conclusions are drawn about prehistoric conditions on our planet, depending on whether the problem is approached from the biological or the geophysical viewpoint.

Palæontologists as well as zoo- and phytogeographers have come again and again to the conclusion that the majority of those continents which are now separated by broad stretches of ocean must have had land bridges in prehistoric times and that across these bridges undisturbed interchange of terrestrial fauna and flora took place. The palæontologist deduces this from the occurrence of numerous identical species that are known to have lived in many different places, while it appears inconceivable that they should have originated simultaneously but independently in these areas. Furthermore, in cases where only a limited percentage of identities is found in contemporary fossil fauna or flora, this is readily explained, of course, by the fact that only a fraction of the organisms living at that period is preserved in fossil form and has been discovered so far. For even if the whole groups of organisms on two such continents had once been absolutely identical, the incomplete state of our knowledge would necessarily mean that only part of the finds in both areas would be identical and the other, generally larger, part would seem to display differences. In addition, it is obviously the case that even where the possibility of interchange was unrestricted, the organisms would not have been quite identical in both continents; even today Europe and Asia, for example, do not have identical flora and fauna by any means.

Comparative study of *present-day* animal and plant kingdoms leads to the same result. The species found today on two such continents are indeed different, but the genera and families are still the same; and what is today a genus or family was once a species in prehistoric times. In this way the relationships between present-day terrestrial faunas and floras lead to the conclusion that they were once identical and that therefore there must have been exchanges, which could only have taken place over a wide land bridge. Only after the land bridge had been broken were the floras and faunas subdivided into today's various species. It is probably not an exaggeration to say that if we do not accept the idea of such former land connections, the whole evolution of life on earth and the affinities of present-day organisms occurring even on widely separated continents must remain an insoluble riddle.

Here Wegener quotes various scientists who support the view that ancient land bridges must be inferred from the present geographical distribution of mammals, plants, and other organisms. These hypothetical bridges are postulated for regions now under water.

Since it was previously taken for granted that the continental blocks—whether above sea level or inundated—have retained their mutual positions unchanged throughout the history of the planet, one could only have assumed that the postulated land bridges existed in the form of intermediate continents, that they sank below sea level at the time when interchange of terrestrial flora and fauna ceased and that they form the present-day ocean floors between the continents.

This assumption of sunken intermediate continents was in fact the most obvious so long as one based one's stand on the theory of the contraction or shrinkage of the earth, a viewpoint we shall have to examine more closely in what follows. The theory first appeared in Europe. It was initiated and developed by Dana, Albert Heim and Eduard Suess in particular, and even today dominates the fundamental ideas presented in most European textbooks of geology. The essence of the theory was expressed most succinctly by Suess: "The collapse of the world is what we are witnessing." Just as a drying apple acquires surface wrinkles by loss of internal water, the earth is supposed to form mountains by surface folding as it cools and therefore shrinks internally. Because of this crustal contraction, an overall "arching pressure" is presumed to act over the crust so that individual portions remain uplifted as horsts. These horsts are, so to speak, supported by the arching pressure. In the further course of time, these portions that have remained behind may sink faster than the others and what was dry land can become sea floor and vice-versa, the cycle being repeated as often as required. This idea, put forth by Lyell, is based on the fact that one finds deposits from former seas almost everywhere on the continents. There is no denying that this theory provided historic service in furnishing an adequate synthesis of our geological knowledge over a long period of time. Furthermore, because the period was so long, contraction theory was applied to a large number of individual research results with such consistency that even today it possesses a degree of attractiveness, with its bold simplicity of concept and wide diversity of application.

Ever since our geological knowledge was made the subject of that impressive synthesis, the four volumes by Eduard Suess entitled *Das Antlitz der Erde*, written from the standpoint of contraction theory, there has been increasing doubt as to the correctness of the basic idea. The conception that all uplifts are only apparent and consist merely of remnants left from the general tendency of the crust to move towards the centre of the earth, was refuted by the detection of absolute uplifts. The concept of a continuous and ubiquitous arching pressure, already disputed on theoretical grounds for the uppermost crust has proved to be untenable because the structure of eastern Asia and the eastern African rift valleys have, on the contrary, enabled one to deduce the existence of tensile forces over large portions of the earth's crust. The concept of mountain folding as crustal wrinkling due to internal shrinkage of the earth led to the unacceptable result that pressure would have to be transmitted inside the earth's crust over a span of 180 great-circle degrees. Many authors have opposed this quite rightly, claiming that the surface of the earth would have to undergo regular overall wrinkling, just as the drying apple does. However, it was particularly the discovery of the scale-like "sheet-fault structure" or overthrusts in the Alps which made the shrinkage theory of mountain formation, which presented enough difficulties in any case, seem more and more inadequate. Since the present-day width of the chain is about 150 km, a stretch of crust from 600 to 1,200 km wide (5–10 degrees of latitude) must have been compressed in this case. Yet in the most recent large-scale synthesis on Alpine sheet-faults, R. Staub agrees with Argand that the compression must have been even greater. He concludes:

"The Alpine orogenesis is the result of the northward drift of the African land mass. If we smooth out only the Alpine folds and sheets over the transverse section between the Black Forest and Africa, then in relation to the present-day distances of about 1,800 km, the original distance separating the two must have been about 3,000 to 3,500 km, which means an alpine compression (in the wider sense of the word Alpine) of around 1,500 km. Africa must have been displaced relative to Europe by this amount. What is involved here is a true continental drift of the African land mass and an extensive one at that."

Moreover, even the apparently obvious basic assumption of contraction theory, namely that the earth is continuously cooling, is in full retreat before the discovery of radium. This element, whose decay produces heat continuously, is contained in measurable amounts everywhere in the earth's rock crust accessible to us. Many measurements lead to the conclusion that even if the inner portion had the same radium content, the production of heat would have to be incomparably greater than its conduction outwards from the centre, which we can measure by means of the rise of temperature with depth in mines, taking into account the thermal conductivity of rock. This would mean, however, that the temperature of the earth must rise continuously. Of course, the very low radioactivity of iron meteorites suggests that the iron core of the earth presumably contains much less radium than the crust, so that this paradoxical conclusion can perhaps be avoided. In any case, it is no longer possible, as it once was, to consider the thermal state of the earth as a temporary phase in the cooling process of a ball that was formerly at a higher temperature. It should be regarded as a state of equilibrium between radioactive heat production in the core and thermal loss into space. In fact, the most recent investigations into this question, which will be discussed in more detail later on, imply that actually, at least under the continental blocks, more heat is generated than is conducted away, so that here the temperature must be rising, though in the ocean basins conduction exceeds production. These two processes lead to equilibrium between production and loss rate, taking the earth as a whole. In any case, one can see that through these new views the foundation of the contraction theory has been completely removed.

There are still many other difficulties which tell against the contraction theory and its mode of thinking. The concept of an unlimited periodic interchange between continent and sea floor, which was suggested by marine sediments on present-day continents, had to be strictly curtailed. This is because more precise investigation of these sediments showed with increasing clarity that what was involved was coastal-water sediments, almost without exception. Since coastal shallows must be counted, geophysically, as part of the continental blocks, the nature of these marine fossils implies that these blocks have been "permanent" throughout the history of the earth and have never formed ocean floors. Are we then still to assume that today's sea floors were ever continents? The justification for this conclusion is obviously removed by establishing that the marine sediments found on continents were formed in shallows. But more than this, the conclusion now leads to an open contradiction. If we reconstruct intercontinental bridges filling up a large part of today's ocean basins without having the possibility of compensating for this by submergence of present-day continental regions to the sea-floor level, there would be no room for the volume of the world's oceans in the now much reduced deep-sea basins. The water displacement of the intercontinental bridges would be so enormous that the level of the world's oceans would rise above that of the whole continental area of the earth and all would be flooded, today's continents and the bridges alike. The reconstruction would not therefore achieve the desired end, i.e., dry land bridges between continents.

Of the many objections to the contraction theory, one more only will be emphasised; it has very special importance. Geophysicists have decided, mainly on the basis of gravity determinations, that the earth's crust floats in hydrostatic equilibrium on a rather denser, viscous substrate. This state is known as *isostasy*, which is nothing more than hydrostatic equilibrium according to Archimedes' principle, whereby the weight of the immersed body is equal to that of the fluid displaced. The introduction of a special word for this state of the earth's crust has some point because the liquid in which the crust is immersed apparently has a very high viscosity, one which is hard to imagine, so that oscillations in the state of equilibrium are excluded and the tendency to restore equilibrium after a perturbation is one which can only proceed with extreme slowness, requiring many millennia to reach completion. Under laboratory conditions, this "liquid" would perhaps scarcely be distinguishable from a "solid." However, it should be remembered here that even with steel, which we certainly consider a solid, typical flow phenomena occur, just before rupture, for example.

Wegener illustrates the equilibrium process of isostasy through recorded postglacial elevations of Finland, Sweden, Norway, and North America.

One can see immediately that this result runs quite counter to the ideas of contraction theory and that it is very hard to combine one with the other. In particular, it seems impossible, in view of the isostatic principle, that a continental block the size of a land bridge of required size could sink to the ocean bottom without a load or that the reverse should happen. Isostasy is therefore in contradiction not only to contraction theory, but in particular also to the theory of sunken land bridges as derived from the distribution of organisms.

Here Wegener mentions the theory of permanence, in which the ocean basins are assumed to be permanent, unchanging features of the earth's surface, and the isostasy theory, which proves it impossible to regard the present-day ocean floors as sunken continents.

So we have the strange spectacle of two quite contradictory theories of the prehistoric configuration of the earth being held simultaneously—in Europe an almost universal adherence to the idea of former land bridges, in America to the theory of the permanence of ocean basins and continental blocks.

Wegener then argues that a long tradition of European geology has led to the present-day acceptance of the contraction theory by European scientists and that the absence of any such tradition in American geology has provided American scientists with no temptation to adopt this theory. He concludes that it is impossible to overlook the existence of land bridges and also impossible to overlook the grounds on which the exponents of the permanence theory deny the existence of sunken continents.

This is the starting point of displacement or drift theory. The basic "obvious" supposition common to both land-bridge and permanence theory—that the relative position of the continents, disregarding their variable shallow-water cover, has never altered—must be wrong. The continents must have shifted. South America must have lain alongside Africa and formed a unified block which was split in two in the Cretaceous; the two parts must then have become increasingly separated over a period of millions of years like pieces of a cracked ice floe in water. The edges of these two blocks are even today strikingly congruent. Not only does the large rectangular bend formed by the Brazilian coast at Cape Sao Roque mate exactly with the bend in the African coast at the Cameroons, but also south of these two corresponding points every projection on the Brazilian side matches a congruent bay on the African, and conversely. A pair of compasses and a globe will show that the sizes are precisely commensurate.

In the same way, North America at one time lay alongside Europe and formed a coherent block with it and Greenland, at least from Newfoundland and Ireland northwards. This block was first broken up in the later Tertiary, and in the north as late as the Quaternary, by a forked rift at Greenland, the sub-blocks then drifting away from each other. Antarctica, Australia and India up to the beginning of the Jurassic lay alongside southern Africa and formed together with it and South America a single large continent, partly covered by shallow water. This block split off into separate blocks in the course of the Jurassic, Cretaceous and Tertiary, and the sub-blocks drifted away in all directions. Our three world maps (figure 19.1) for the Upper Carboniferous, Eocene, and Lower Quaternary show this evolutionary process. In the case of India the process was somewhat different: originally it was joined to Asia by a long stretch of land, mostly under shallow water. After the separation of India from Australia on the one hand (in the early Jurassic) and from Madagascar on the other (at the transition from Tertiary to Cretaceous), this long junction zone became increasingly folded by the continuing approach of present-day India to Asia; it is now the largest folded range on earth, i.e., the Himalaya and the many other folded chains of upland Asia.

There are also other areas where the continental drift is linked causally with orogenesis. In the westward drift of both Americas, their leading edges were compressed and folded by the frontal resistance of the ancient Pacific floor, which was deeply chilled and hence a source of viscous drag. The result was the vast Andean range which extends from Alaska to Antarctica. Consider also the case of the Australian block, including New Guinea, which is separated only by a shelf sea: on the leading side, relative to the direction of displacement, one finds the high-altitude New Guinea range, a recent formation. Before this block split away from Antarctica, its direction was a different one, as our maps show. The present-day east coastline was then the leading side. At that time New Zealand, which was directly in front of this coast, had its mountains formed by folding. Later as a result of the change

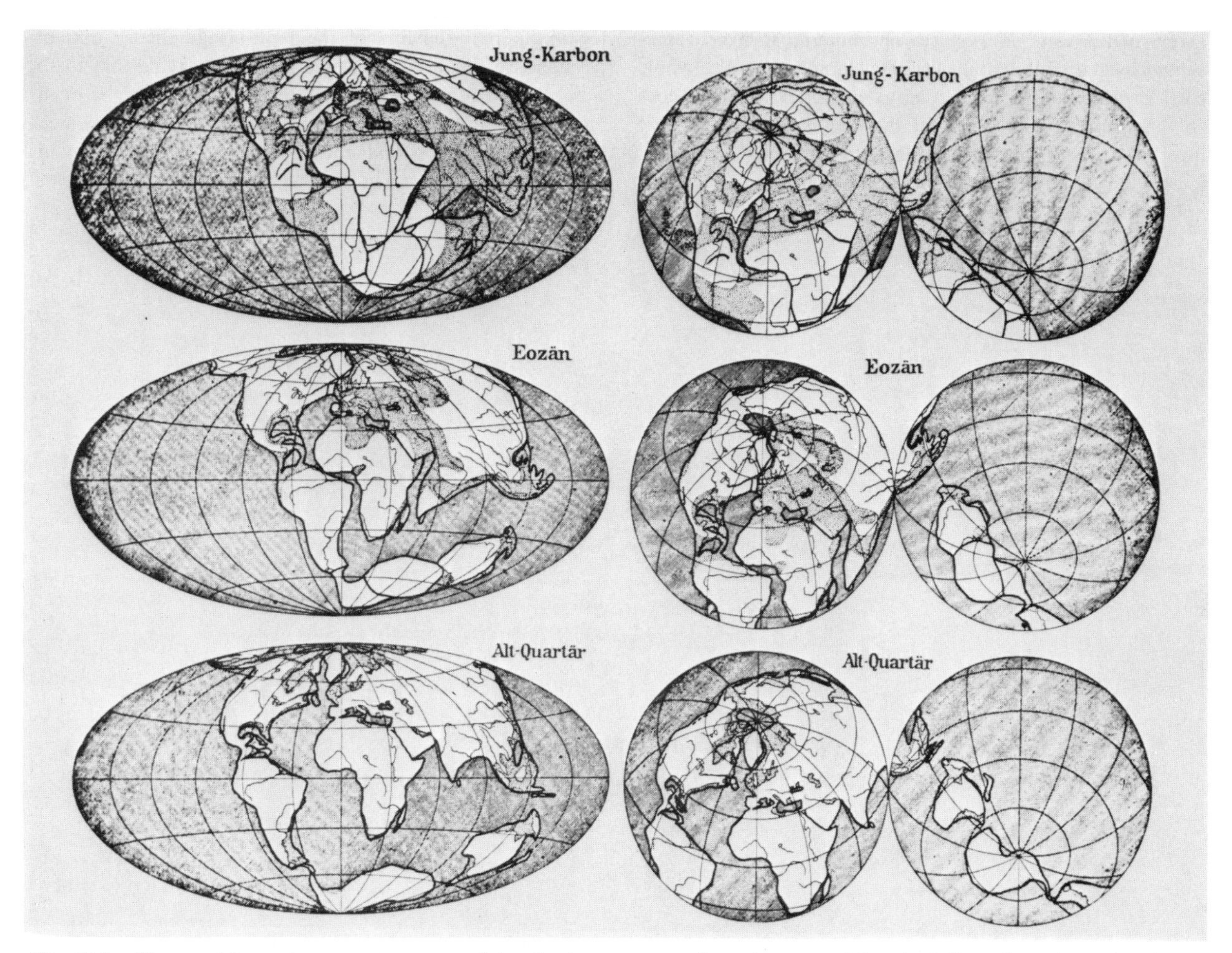

Fig. 19.1 The world continent as it appeared in the late Carboniferous, Eocene, and Early Quaternary periods, according to Wegener, who first used these maps in the third German edition (1922) of his book. The solid lines denote the boundaries of the continents as we know them now, the stippling marks shallow seas, the white areas denote uncovered continents, and the wiggly lines show the major rivers. According to Wegener, all the continents were clustered together into a single supercontinent named Pangaea a few hundred million years ago, during the late Carboniferous period.

in direction of displacement, the mountains were cut off and left behind as island chains. The present-day cordilleran system of eastern Australia was formed in still earlier times; it arose at the same time as the earlier folds in South and North America, which formed the basis of the Andes (precordilleras), at the leading edge of the continental blocks, then drifting as a whole before dividing.

We have just mentioned the separation of the former marginal chain, later the island chain of New Zealand, from the Australian block. This leads us to another point: smaller portions of blocks are left behind during continental drift, particularly when it is in a westerly direction. For instance, the marginal chains of East Asia split off as island arcs, the Lesser and Greater Antilles were left behind by the drift of the Central American block, and so was the so-called Southern Antilles arc (South Shetlands) between Tierra del Fuego and western Antarctica. In fact, all blocks which taper off towards the south exhibit a bend in the taper in an easterly direction because the tip has trailed behind: examples are the southern tip of Greenland, the Florida shelf, Tierra del Fuego, the Graham Coast and the continental fragment Ceylon.

It is easy to see that the whole idea of drift theory starts out from the supposition that deep-sea floors and continents consist of different materials and are, as it were, different layers of the earth's structure. The outermost layer, represented by the continental blocks, does not cover the whole earth's surface, or it may be truer to say that it no longer does so. The ocean floors represent the free surface of the next layer inwards, which is also assumed to run under the blocks. This is the geophysical aspect of drift theory.

If drift theory is taken as the basis, we can satisfy all the legitimate requirements of the land-bridge theory and of permanence theory. This now amounts to saying that there were land connections, but formed by contact between blocks now separated, not by intermediate continents which later sank; there is permanence, but of the area of ocean and area of continent as a whole, but not of individual oceans or continents.

20. The Spectra of Venus, Mars, Jupiter, and Saturn under High Dispersion

Theodore Dunham

(*Publications of the Astronomical Society of the Pacific 45*, 202–204 [1933])

When Joseph Fraunhofer first discovered absorption lines in the spectrum of the sun, he also noticed that the spectra of Venus and Mars contain the same fixed lines. Although this is just what would be expected if the planets shine by reflected sunlight, there should be additional absorption lines present if the sunlight passes through a planetary atmosphere. The first spectroscopic indication that the major planets might contain atmospheres came in the 1860s, when an absorption line not present in the solar spectrum was observed in the spectra of Jupiter and Saturn.

A new era of planetary spectroscopy began when V. M. Slipher photographed the spectra of the major planets. Early in the twentieth century Slipher extended the spectra to the red and infrared wavelengths and showed that the major planets exhibit strong absorption lines at a variety of wavelengths.[1] As shown in the illustrations appended to the paper given here, the observed bands grow in number, breadth, and intensity as one goes from Jupiter through Saturn and Uranus and on to Neptune. Slipher's work was discussed by Rupert Wildt, who interpreted some of the bands of Jupiter as absorptions by ammonia and methane.[2] It remained for Theodore Dunham and Walter Adams to take the high dispersion spectra that confirmed Wildt's speculation. The identifications of ammonia and methane were verified when Dunham filled a long pipe with each gas and measured absorption lines at the same wavelengths as those found in the spectra of the planets.

The results of the observations given in this paper may be summarized by stating that large planets have atmospheres containing hydrogen compounds, middle sized planets have atmospheres containing oxygen compounds, and small objects, such as the earth's moon, have no atmosphere at all. The simple explanation for these differences had already been set forth by George Johnstone Stoney in 1898 (see selection 15). The smaller bodies do not have sufficient gravitational pull to retain any atmosphere; medium-sized bodies can only retain the heavier, slower-moving molecules; and only the larger, major planets have sufficient gravitational pull to retain the lightest molecule, hydrogen.

1. V. M. Slipher, *Lowell Observatory Bulletin no. 42* (1909).
2. R. Wildt, *Naturwissenschaften 19*, 109 (1931). *Veröffentlichungen der Universitäts-Sternwarte zu Göttingen 2*, no. 22, 171 (1932).

INFRA-RED SPECTRA of Venus taken by Dr. Adams and the present author show no evidence of planetary bands due to oxygen or water vapor. Three new bands well resolved into individual lines have been detected. The spacing of lines in these bands leads to a moment of inertia in close agreement with that determined for carbon dioxide by Houston from the Raman spectrum. The identification has been verified by photographing the head of the strongest band in the absorption spectrum of 40 meters of carbon dioxide at a pressure of ten atmospheres.

With new plates prepared by Dr. Mees, the B-band of oxygen in the spectrum of Mars has been photographed in collaboration with Dr. Adams, both before and after the recent opposition. Measurements on ten plates fail to show any displacement of the telluric lines which can be attributed to planetary oxygen. The shapes of the lines when the planet was approaching the Earth have been compared with the shapes of the same lines when the planet was receding. No evidence for an asymmetry due to Martian lines in the wings of the telluric lines has been detected. Both methods indicate that if Martian oxygen exists its amount must be less than one per cent of that present in the Earth's atmosphere.

Grating spectra of Jupiter and Saturn have been obtained with various spectrographs ranging from the 9-foot Littrow arrangement for Jupiter in the visible red to a 9-foot collimator combined with a 9-inch camera for Saturn beyond λ9,000. Wildt has shown that the mean wave-lengths of some unresolved bands in the spectra of the major planets agree with those of bands due to ammonia and methane and that several partially resolved lines in Slipher's photographs of the λ6,450 band in the spectrum of Jupiter agree in position with lines measured by Badger in a weak ammonia band. On the present photographs a number of bands between λ6,450 and λ9,100 are clearly resolved, so that it has been possible to compare the positions of individual lines with those of band lines in the absorption spectra of ammonia and methane obtained at Mount Wilson with a 40-meter path. More than sixty lines due to ammonia have been identified in the bands near λ6,450 and λ7,920 in the spectrum of Jupiter. The identification of methane has been established with certainty in the spectra of both Jupiter and Saturn by the agreement in position of eighteen lines in the absorption spectrum of methane with lines in the nearly unblended planetary band near λ8,640. There is a narrow band of intense absorption near λ9,750 in the spectrum of Jupiter, and four other intense but wider bands between that wave-length and λ10,000.

A rough comparison between the intensities of ammonia band lines in the spectrum of Jupiter and in the laboratory absorption spectrum of that gas with known pressure and length of path indicates that above the reflecting level in the atmosphere of Jupiter there is the equivalent of a path about five to ten meters thick of ammonia at atmospheric pressure. A minimum temperature of about 170° K for the reflecting level may be inferred, since at lower temperatures the vapor pressure of ammonia would be insufficient to support the observed amount of gas on a planet whose surface gravity is that of Jupiter.

Appended Figures

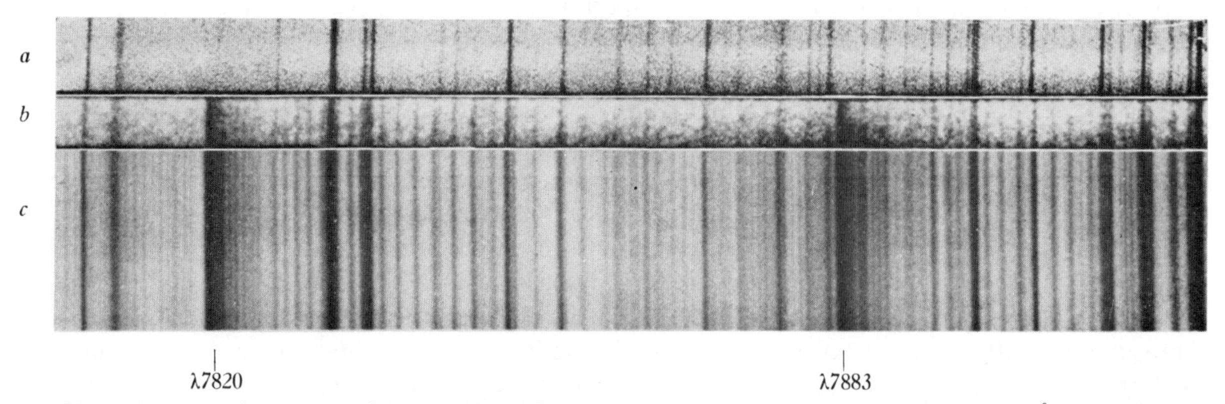

Fig. 20.1 The absorption spectra of the sun (*a*) and Venus (*b* and *c*), showing the well-known Fraunhofer lines and the lines of carbon dioxide at 7,820 and 7,883 Å. (From *Publications of the Astronomical Society of the Pacific 44*, 243 [1932].)

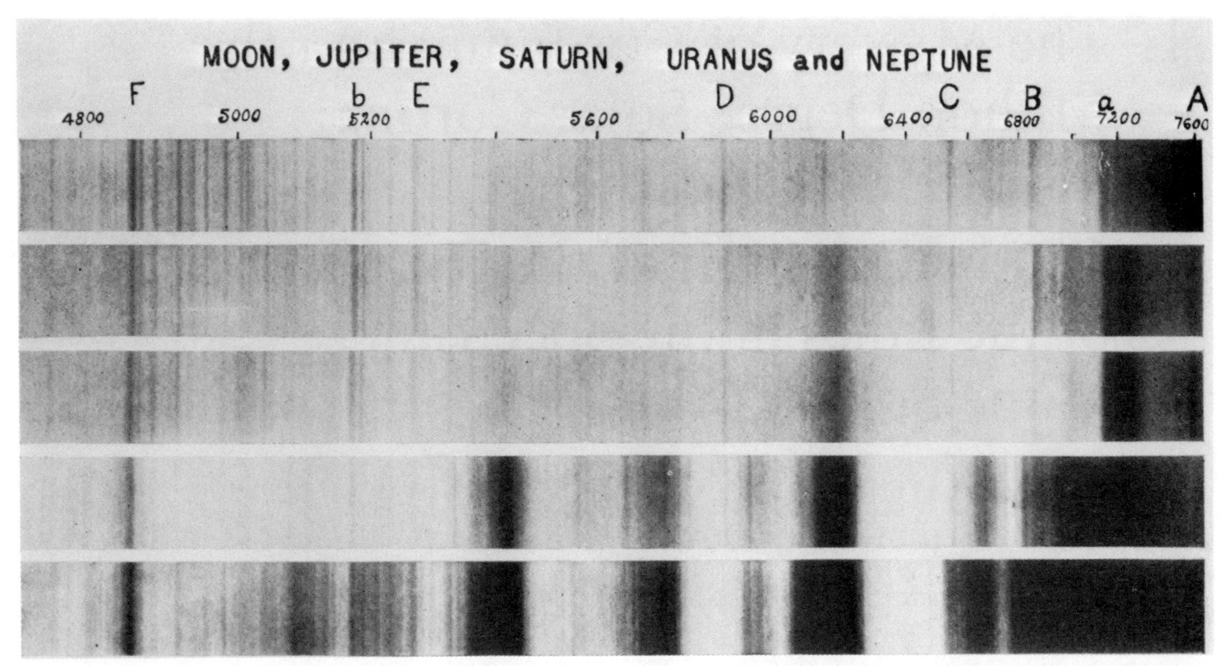

Fig. 20.2 The absorption spectra of the moon, Jupiter, Saturn, Uranus, and Neptune (*top to bottom*), showing the well-known Fraunhofer lines and the fifth, sixth, seventh, and eighth harmonics of the fundamental methane vibration frequency. The Fraunhofer lines are denoted by the letters at the top of the photograph, along which the wavelengths in Ångstroms are also numbered; the methane vibration lines are the dark features becoming more intense and wider as one progresses to the more distant planets. (From *Monthly Notices of the Royal Astronomical Society*, *93*, 661 [1933].)

21. The Mystery of Coronium and the Million-Degree Solar Corona

On the Question of the Significance of the Lines in the Spectrum of the Solar Corona

Walter Grotrian
TRANSLATED BY BRIAN DOYLE

(*Naturwissenschaften 27*, 214 [1939])

An Attempt to Identify the Emission Lines in the Spectrum of the Solar Corona

Bengt Edlén

(*Arkiv för matematik, astronomi och fysik 28B*, 1–4 [1941])

During the solar eclipse of August 7, 1869, both Charles Young and William Harkness first found that the spectrum of the solar corona was characterized by a conspicuous green emission line. By the beginning of the 20th century, eclipse observations revealed at least ten coronal emission lines, none of which had been observed to come from terrestrial substances. For the want of a better explanation, astronomers concluded that the solar corona consisted of some mysterious substance, which they named "coronium"; but it was strongly suspected that the large number of emission lines did not all come from the same substance.

Meanwhile, observations of gaseous nebulae also disclosed bright, unidentified emission lines, which were attributed to the otherwise unknown element "nebulium." The element helium had been discovered and named from its appearance in the solar chromospheric spectrum over twenty years before it was detected terrestrially, and astronomers hoped eventually to identify coronium and nebulium with new elements. By 1930, however, the atomic periodic table had been fairly well established, and there remained little possibility that these solar and nebular lines could be identified with new elements. Hence, theoreticians began to search for new spectral guises of familiar atoms in unusual environments. In 1928 Ira Bowen (see selection 85) demonstrated that the nebulium lines arose from ionized oxygen and nitrogen in a gas of such low density that collisions were extremely rare, making them lines from so-called forbidden transitions.

The solution of the coronium puzzle proved more difficult because of the unexpectedly high temperature of the solar corona, which created an astonishingly high degree of ionization. The coronium lines identified by Walter Grotrian and Bengt Edlén in these two selections turned out not only to be forbidden lines but to come from atoms deprived of 10–15 electrons. This meant that the solar corona must have an excitation temperature of a few million degrees, and conclusive evidence for a hot solar corona was provided a few years later by solar observations at radio wavelengths.[1] The existence of the forbidden lines proves that the coronal gas is exceedingly tenuous, however, which means that comparatively little energy from the sun is required to maintain the hot corona.

1. D. F. Martyn and J. L. Pawsey, *Nature 158*, 632 (1946). Also see V. L. Ginzburg, *Comptes rendus (Doklady) de l'Académie des sciences de l'URSS 52*, 487 (1946) and selection 10.

Grotrian, The Significance of the Lines in the Spectrum of the Solar Corona

RECENTLY I. S. Bowen and B. Edlén have shown that forbidden lines of Fe VII have appeared in the spectrum of Nova RR Pictoris,[1] and the appearance of five corona lines in the spectrum of RS Ophiuchi, during a phase of the nova-like outbreak in the year 1933, has been demonstrated beyond a doubt by W. S. Adams and A. H. Joy.[2] Also, the indications have become more and more conclusive that, in the outer zones of the solar atmosphere, conditions are present for the excitation of spectral lines that far exceed those to be expected for thermal equilibrium. Therefore, it no longer seems misguided to discuss the question whether the coronal lines are to be explained as forbidden lines of highly ionized atoms.

In this connection it seems legitimate to point to the following numerical agreements. B. Edlén has identified some lines and terms of the [Fe X] spectrum.[3] For the difference in ground terms of the [Fe X] spectrum $\Delta\nu = 3p^2P_2 - 3p^2P_1$, the values $\Delta\nu = 15.66 \times 10^3$ cm^{-1} and 15.71×10^3 cm^{-1} are produced by two independent determinations. The mean is therefore $\Delta\nu = 15.68 \times 10^3$ cm^{-1}. The frequency corresponding to the well-known red corona line at $\lambda = 6{,}374.51$ A is $\nu = 15.683 \times 10^3$ cm^{-1}. This agreement is as good as one can expect from the precision with which the difference of terms is determined; nonetheless, the agreement can of course be fortuitous, and a conclusive decision can only be brought about through a more precise determination of the term difference in question.

For correctness of the identification, the following can be cited. Because the single forbidden line of the [Fe X] spectrum corresponds to the transition $3p^2P_1 - 3p^2P_2$, we expect this line to stand with no other coronal lines in a close connection. In fact, the observations show that the line $\lambda 6{,}374$ deviates considerably in intensity from the other lines, particularly, for example, from the known green line $\lambda 5{,}302.8$,[5] but an agreement in intensity is found with the line $\lambda 7{,}891.94$, recently measured by B. Lyot.[6] After the term analysis by B. Edlén,[7] we could presume that this line is to be explained as the forbidden transition between the ground terms of the [Fe XI] spectrum $3p^3P_1 - 3p^3P_2$. The term difference is $\Delta\nu = 12.675 \times 10^3$ cm^{-1}, and the frequency of the line is $\nu = 12.668 \times 10^3$ cm^{-1}. If this identification is correct, the agreement in intensity of the lines 6,374 and 7,892 would be, after all, understandable.

1. *Nature 143*, 374 (1939).
2. *Publications of the Astronomical Society of the Pacific 45*, 301 (1933).
3. *Zeitschrift für Physik 104*, 407 (1937).
4. *Comptes rendus de l'Académie des sciences* (Paris) *203*, 1327 (1937).
5. *Zeitschrift für Astrophysik 7*, 26 (1933).
6. *L'Astronomie* 201 (1938).
7. *Zeitschrift für Physik 104*, 188 (1937).

Edlén, An Attempt to Identify the Emission Lines in the Spectrum of the Solar Corona

EVER SINCE the first emission lines in the solar corona were discovered some 70 years ago, many suggestions of the most different kind have been put forward to explain their origin. So far, however, none of these attempts has led to an acceptable result. Even a critical discussion of the matter by Swings finished with the conclusion that the origin of the coronal lines remains as mysterious as ever.

In spite of the well-known difficulties of coronal observations, there are at present somewhat more than 20 relatively well-measured lines, which can be considered as certainly belonging to the specific coronal spectrum (table 21.1). None of these lines has ever been observed either in a laboratory light-source or in any other astronomical object, except for a short appearance at a nova outburst of RS Ophiuchi[1]. In the present article an identification is proposed which accounts for the main part of the coronal lines and gives the explanation for the unique character of the coronal spectrum. At the same time the reason for the failure of all previous attempts to connect it with any known atomic or molecular spectrum becomes evident.

Grotrian's recent observation[2] that the wave-numbers of the coronal lines 6,374 and 7,892 coincide with the separations $^2P_{3/2} - {}^2P_{1/2}$ and $^3P_2 - {}^3P_1$ in the ground terms of Fe X and Fe XI, gave the impulse to a systematic search for analogous term separations. Hereby two other coronal lines were found to coincide with the separations $^2P_{3/2} - {}^2P_{1/2}$ and $^3P_2 - {}^3P_1$ of Ca XII and Ca XIII as determined by unpublished measurements of the author. This fact increased the probability that the coincidences did not happen by chance and that the whole coronal spectrum might consist of "forbidden lines" analogous to those of the gaseous nebulae but from atoms much more highly ionised than had previously been considered at all.

As the absence of any more coincidences might well be attributed to the impossibility of analyzing the corresponding extreme ultraviolet spectra, an attempt has been made to predict by extrapolation such term separations as could be expected in the corona according to those already found. We have then to consider the configurations s^2p, s^2p^2, s^2p^4, and s^2p^5 of cosmically abundant atoms in appropriate ionisation stages. There were good reasons to expect intense lines especially of Fe XIII and Fe XIV because of (1) the rich abundance of iron as judged from the intensity of Fe X and Fe XI, (2) the very high degree of ionisation indicated by the nonappearance of the well-known forbidden lines of Fe VII, and the presence of Ca XII and Ca XIII with much higher ionisation potentials than Fe XIII and XIV. It is thus of great significance indeed that the extrapolated separation in the

Table 21.1 The emission lines in the spectrum of the solar corona

λ^a A	cm^{-1}	Intensity[b]			Identification	I. P.
3,328.1	30,039	8	1.0	—	Ca XII $^2P_{3/2}$—$^2P_{1/2}$	589
3,388.10	29,506.6	20	16	—	Fe XIII $^3P_2$—$^1D_2$	325
3,454.13	28,942.6	8	2.3	—	—	—
3,600.97	27,762.4	12	2.1	—	Ni XVI $^2P_{1/2}$—$^2P_{3/2}$	455
3,642.87	27,443.1	—	—	—	Ni XIII $^3P_1$—$^1D_2$	350
3,800.77	26,303.0	1	—	—	—	—
3,986.88	25,075.2	5	0.7	—	Fe XI $^3P_1$—$^1D_2$	261
4,086.29	24,465.2	6	1.0	—	Ca XIII $^3P_2$—$^3P_1$	655
4,231.4	23,626.2	8	2.6	—	Ni XII $^2P_{3/2}$—$^2P_{1/2}$	318
4,311.5	23,187.3	1	—	—	—	—
4,359	22,935	1	—	—	—	—
4,567	21,890	3	1.1	—	—	—
5,116.03	19,541.0	5	4.3	2.6	Ni XIII $^3P_2$—$^3P_1$	350
5,302.86	18,852.5	20	100	120	Fe XIV $^2P_{1/2}$—$^2P_{3/2}$	355
5,694.42	17,556.2	—	—	1.5	—	—
6,374.51	15,683.2	6	8.1	28	Fe X $^2P_{3/2}$—$^2P_{1/2}$	233
6,701.83	14,917.2	4	5.4	3.3	Ni XV $^3P_0$—$^3P_1$	422
7,059.62	14,161.2	—	—	4	—	—
7,891.94	12,667.7	—	—	29	Fe XI $^3P_2$—$^3P_1$	261
8,024.21	12,458.9	—	—	1.3	Ni XV $^3P_1$—$^3P_2$	422
10,746.80	9,302.5	—	—	240	Fe XIII $^3P_0$—$^3P_1$	325
10,797.95	9,258.5	—	—	150	Fe XIII $^3P_1$—$^3P_2$	325

a. Wavelengths from S. A. Mitchell, *Hb. der Astroph.*, vol. 7 (1935), p. 400, and B. Lyot, *M. N. 99*, 580 (1939).

b. First column estimated by Grotrian, *Zs. f. Astroph. 2*, 106 (1931); second column measured by Grotrian, *Zs. f. Astroph. 7*, 26 (1933); third column measured by Lyot, *M. N. 99*, 580 (1939).

ground term of Fe XIV definitely points to the strongest line of the coronal spectrum 5,303, and that the three next strongest coronal lines 10,747, 10,798 and 3,388 permit a spontaneous identification with the transitions in Fe XIII as given in table 21.1. Besides, the fainter 3,987 can be attributed to a transition in Fe XI. Together with 6,374 and 7,892 these identifications of Fe X, XI, XIII, and XIV comprise more than $\frac{9}{10}$ of the total intensity of the coronal line emission. The fact that no other ionisation stages of iron appear is quite naturally explained, as the extrapolated Fe XII-transitions are unobservable and the neighbouring Fe VIII, IX, XV, XVI and XVII have no deep metastable levels. The absence of $^3P_2$—$^3P_0$ of Fe XI and Fe XIII as well as the comparatively very low intensity of Fe XI $^3P_1$—$^1D_2$ is in agreement with the theoretical transition probabilities recently calculated by Pasternack.[3]

If now the iron identifications mentioned above are accepted, one can predict with considerable accuracy the position of the corresponding transitions in the elements nearby. It then appears that within the observable range the complete set of transitions extrapolated to Ni is found among the remaining coronal lines with roughly the same relative intensities as for iron. In this way six more coronal lines have been identified. The average intensity ratio of Fe to Ni lines is approximately 10:1, in accordance with the usual cosmical abundance of these elements. The proposed identifications for both Fe and Ni seem furthermore to compare well with all reliable physical classifications of the coronal lines as shown, for instance, by Lyot's "groups" in table 21.1.

Comparing the intensities of lines from different ions, one finds that for Fe the two higher stages of ionisation are the more frequent, while for Ni the opposite is true. This indicates a maximum abundance for ions with ionisation potentials of about 400 volts. In this connection we may remind that Lyot,[4] assuming the profiles observed for some coronal lines as due to a thermal Doppler effect in oxygen atoms, found an equivalent temperature of 660,000° which becomes 2,300,000° when recalculated to iron atoms.

With the present identifications we have reached a fairly complete explanation of the corona spectrum, leaving unidentified only some very faint lines forming less than 3% of the total line emission. So far, only the three elements: calcium, iron, and nickel, have been documented. It is quite possible that still others will be found through the identification of the remaining faint lines. It seems fairly certain, however, that the elements K, Cr, Mn, and Co, for which possible transitions can be accurately predicted, do not account for any of the coronal lines as yet observed. This gives an important information about the composition of the coronal matter, for if present in the same ratios as in the solar atmosphere, these elements should give quite observable lines. On the other hand the absence of K, Cr, Mn, and Co and the proportion of Fe to Ni agrees well with the average composition of meteorites, which suggests a rather plausible origin for the coronal matter. It may be remarked that all the other elements which are abundant in meteorites: O, Mg, Al, Si, and S, have at the expected ionisation stages all their strong lines outside the observable range.

The more detailed argumentation underlying the explanation of the coronal spectrum which has been proposed in this preliminary note will be given in a later report.[5]

1. W. S. Adams and A. H. Joy, *Publications of the Astronomical Society of the Pacific 45*, 301 (1933).
2. W. Grotrian, *Naturwissenschaften 22*, 214 (1939).
3. S. Pasternak, *Astrophysical Journal 92*, 129 (1940).
4. B. Lyot, *L'Astronomie 51*, 203 (1937).
5. B. Edlén, *Monthly Notices of the Royal Astronomical Society 105*, 323 (1945).

22. Corpuscular Influences upon the Upper Atmosphere

Sydney Chapman

(*Journal of Geophysical Research 55*, 361–372 [1950])

(Copyright © 1950 American Geophysical Union)

Carl Störmer and Kristian Birkeland first considered the impact of charged solar particles on the earth's dipole magnetic field and concluded that their trajectories could account for the horseshoe-shaped configurations of some aurorae and also for the appearance of the aurorae only at the poles of the earth.[1] As Sydney Chapman explains in this paper, a solar emission mechanism also provides a persuasive explanation for the 27-day recurrence period that had been observed for the geomagnetic disturbances. We can imagine that the magnetic storms are produced by corpuscular streams that travel radially outward from the center of the sunspots. By virtue of the sun's rotation, however, the geometrical form of these streams will be an Archimedian spiral. A particular stream will sweep past the earth at intervals of about 27 days, corresponding to the apparent rotational period of the sun as viewed from the earth. Furthermore, because magnetic storms follow the emission of solar flares by an interval of about 1 day, the corpuscles must travel to the earth at a relatively slow velocity, on the order of 1,000 km sec^{-1}.

How do the moving charged particles interact with the magnetic field of the earth? Chapman and his colleague V. C. A. Ferraro had been able to show that the pressure of the electron-ion gas on the geomagnetic field would result in the sudden commencement and initial phase of a geomagnetic storm.[2] A cavity is formed in the gas stream by the deflection of the positively and negatively charged particles, and the earth's magnetic field is confined in this cavity, a region now known as the magnetosphere. Charged solar particles are deflected from the magnetosphere except possibly at the polar areas, and in this way nearly all the earth's surface is protected from the possibly lethal solar corpuscles.

1. C. Störmer, *Archives des sciences physiques et naturelles* (Geneva) *24*, 5, 113, 221, 317 (1907); K. Birkeland, *Comptes rendus de l'Académie des sciences* (Paris) *152*, 345, *153*, 938 (1911).
2. S. Chapman and V. C. A. Ferraro, *Monthly Notices of the Royal Astronomical Society 89*, 470–479 (1929); *Terrestrial Magnetism and Atmospheric Electricity 36*, 77, 171 (1931), *37*, 147, 421 (1932), *38*, 79 (1933), *45*, 245 (1940).

Abstract—The evidence for the corpuscular theory of magnetic storms and aurorae is briefly reviewed. No certain success has attended the efforts made to detect the solar corpuscles during their emission from the sun or their passage to the earth; recent observations of atomic hydrogen lines in the auroral spectrum, which are broadened and displaced by Doppler effect, give the most direct support yet available to the corpuscular theory.

Introduction

It is generally considered that the earth, and especially its upper atmosphere, is affected by particles coming from the sun, the most visible effect being the aurora polaris, with which are associated magnetic and ionospheric disturbances.

This view has been (and from time to time still is) contested. It is urged that the sun's (wave-) radiation can account for these phenomena.[1]

The most serious, interesting, and sustained challenge of this kind was that made by E. O. Hulburt (later in conjunction with H. B. Maris) in 1928-29. Its enunciation, development, and discussion enriched the literature of geophysics, and I think the challenge was completely met.

Other writers have more recently proposed (ultraviolet) radiation hypotheses in a vaguer way.[2,3]

It seems worth while to re-examine and restate the case for the solar corpuscular theory, to indicate its strength and its deficiencies, and to try to develop it in somewhat more detail than has yet been done.

It continues to seem to me that the onus rests with those who would urge the abandonment of the corpuscular theory for some alternative hypothesis, to show clearly (if it be possible) that the evidence here adduced can be reinterpreted, and the arguments outweighed by other arguments, so as to shift the balance of judgment in favor of their view.

The Solar Associations of Magnetic Storms and Aurorae

In many ways magnetic storms and aurorae show strong statistical connections with the changing phenomena on the sun, of which spots, prominences, and faculae are the most evident. The annual mean frequency curves for the solar and terrestrial phenomena, over a period of many sunspot cycles, show much general parallelism, as well as many systematic or detailed differences. The correlation coefficients between the annual means of the sunspot numbers and of appropriate indices of geomagnetic activity exceed 0.8. The association between great magnetic storms and the presence of large sunspot groups on the sun is far closer than can be accounted for by chance.

This indicates that the sun is the cause of the magnetic storms and aurorae, but further consideration is needed before concluding by what means it acts.

Ultraviolet radiation certainly plays a great part in magnetic and ionospheric phenomena. As is shown by the solar eclipse changes in the ionosphere, which I shall discuss later, it is the main cause of the ionized layers in which flow the currents that produce the daily magnetic variations. Radio measurements and these daily magnetic variations show that the ultraviolet radiation waxes and wanes with the sunspot cycle, and therefore varies in general parallelism with magnetic storms and aurorae.

Moreover, bursts of ultraviolet light of exceptional intensity are occasionally observed on the sun, many of them from small regions in active sunspot areas. These bursts, called solar flares, whose duration is commonly about an hour or less, produce terrestrial effects simultaneous with our observation of the flares, so that the cause must be wave-radiation traveling with the speed of light. These effects consist of a downward extension of the lowermost ionospheric layer *D*, causing abrupt radio fade-outs on certain wavelengths, together with an amplification of the electric currents in the ionosphere, augmenting the daily magnetic variations during the brief period of the flare and fade-out. These terrestrial effects, like the regular daily process of ionization of the upper atmosphere, are confined to the sunlit hemisphere. The magnetic changes are quite different from those called magnetic storms.

When the flare is very intense, and occurs within about 45° from the centre of the sun's disk, it is often succeeded, after an interval of about a day, by a great magnetic storm, which commences suddenly, and simultaneously over the whole earth to less than a minute.

The question arises, is this also an effect of the sun's radiation that reached the earth during the period of visibility of the flare—an effect taking about a day to develop and manifest itself?

The only mechanism of this kind yet proposed in any detail proved subject to a fatal objection. A more general objection is the unlikelihood that an effect produced by radiation falling only on one hemisphere, and requiring about a day to develop, should finally mature so suddenly and with such world-wide simultaneity.

The corpuscular hypothesis offers an alternative interpretation, as follows: It proposes that the visible flare of light is accompanied by the ejection of solar gas, and that much, probably nearly all, of the delay between the flare and the storm represents the time taken by the gas to travel from the sun to the distance of the earth. This implies a speed of the order of 1,000 km or miles a second.

Let us now consider the evidence for the corpuscular hypothesis of aurorae and magnetic storms, both for the great storms and for those that are moderate and weak.

The Directive Property and Lateral Limitation

As has been stated, a magnetic storm follows a day or so after a flare only if the flare occurs within about 45° from the centre of the sun's disk. The radio fade-out and augmentation of the daily magnetic variation, which are clearly due to solar wave-radiation, may occur when the flare is anywhere on the disk. This indicates that the storm-producing influence of the flare is limited within a core of rather wide angular diameter, about 90°, and that it does not extend, like the wave-radiation, in almost all directions outward from the flare. This directive property and lateral limitation of the flare-influence suggest that it has a material, corpuscular character.

In the case of the weaker magnetic storms, this directive property and lateral limitation reveal themselves in another way, by the 27-day recurrence tendency. This is the tendency shown by geomagnetic disturbance and auroral activity to recur after about 27 days, or an integral multiple thereof; at these intervals, after a day of marked character (whether stronger or weaker than normal), the activity tends to depart from the average in the same sense. It is definitely not a true periodicity.

This tendency is revealed most clearly, and in a way that involves no pre-conception as to the length of the recurrence interval, by Chree's pulse diagram, obtained by the method of superposed epochs.

A century ago there was some controversy as to whether the recurrence interval, of about 27 days, indicated a solar or a lunar influence. It is surprising to see this question recently raised anew. Nothing in lunar tidal theory seems to warrant the revival, in view of the strict periodicity in lunar effects, and the absence of any true periodicity in the recurrence tendency; and in view of the enormous difference between the capacities of the sun and the moon to influence the earth's physical state; and in view also of the many independent indications of the solar causation of magnetic storms.

The natural interpretation, due to Maunder, identifies the approximately 27-day recurrence interval with the average rotation period of the sun relative to the moving earth; the sun's rotational speed varies with latitude, and the actual recurrence interval roughly fits the period for the sunspot zones.

Having established the recurrence tendency and determined the recurrence interval by the pulse diagram, recourse can further be had to the 27-day time pattern, as developed by Maunder, Chree and Stagg, and Bartels; this indicates the recurrence and other characteristics of the phenomena with much greater richness and detail. Particular interest attaches to the comparison of the 27-day time patterns of the daily magnetic-activity character-figures, and of the daily sunspot numbers. The differences between the two patterns are at least as remarkable as the resemblances.

The feature I wish now particularly to emphasize is the succession of black or white patches (representing magnetic disturbance or quiet) in the same columns of successive rows of the magnetic time pattern. These indicate the recurrence tendency, which naturally is manifested also on the sunspot pattern, where, however, the recurrences are shown by much broader bands; a recurrent spot as it crosses the sun's disk influences the sunspot number and pattern continuously for up to 13 or 14 days. On the magnetic pattern, the width of the blacker parts of the recurrence columns is more like two or three days.

The simplest and most unforced explanation of the recurrence tendency—which is not manifested by the great magnetic storms—is that the solar cause of the moderate and weak storms and aurorae is propagated along a narrow stream, which continues from the same limited region of the sun for two or more solar rotations, and produces a magnetic storm and aurora when it impinges on the earth. Breaks in the recurrence, or its termination, can be explained by intermittence or cessation of the stream, or changes in its direction.

The conical angle of the stream may be inferred to bear the same proportion to 360° as does the breadth of the recurrence columns in the time pattern to 27 days; taking the breadth to be from two to three days, the conical angular diameter will be from about 30° to 40°, considerably less than for the wide-angle burst from solar flares. There seems to be no possibility of a solar beam of *light* being so limited, so that the beam must be corpuscular, formed of solar gas.

The Time Lag

Next consider the *speed* of the supposed corpuscular emissions that produce magnetic storms and aurorae. In the case of the flare-produced storms, the flare and the storm commencements are both well-marked events, and the time lag is therefore definite. If the storm is supposed to develop rapidly when the stream reaches the earth (as the suddenness and simultaneity of the commencement would suggest), the time of travel is definite to within the duration of the flare, that is, half an hour or so; hence the speed is well determined, being of the order of 1,000 km per second or 1,000 miles per second.

During weaker storms there is no detectable flare in the sunspots, and it is therefore difficult to measure a time lag. Nevertheless, if the weaker storms are correlated with the simple presence of sunspots, velocities on the order of several hundreds of kilometers per second are inferred.

The Search for Direct Evidence of Corpuscular Emissions from the Sun

Solar gas is continually observed in violent vertical or oblique motion in the sun's atmosphere, both at the limb (or edge of the disk), and in absorption on the disk, with Doppler displacements indicating motion along the line of sight.

These motions are clearly not free motions, subject only to gravity after the original impulsion. They show rapid accelerations, by changes of speed and direction.

The speeds range from tens to hundreds of km/sec. They rarely attain the minimum required, with upward vertical projection, to escape from the sun's gravitational pull; in the absence of all other forces, this speed is 615 km/sec. Only a few cases have been observed of gas with upward speeds greater than this, and therefore destined to escape from the sun. Up to 1937 the highest recorded speed was 728 km/sec; in this case, the gas was rising with increasing, not diminishing, speed.

In 1926 Milne indicated a mechanism by which atoms could be continuously accelerated away from the sun, approaching limiting speeds of the order of 1,000 miles a second.

In the few cases where solar gas has been seen to be traveling upward with speed sufficient to escape from the sun, it has been in the form of short whisps, very different from either the continuous narrow-angle beams inferred for weak magnetic storms, or the wide-angle short bursts inferred for solar flares.

The observation of bursts of radio noise connected with strong solar flares has been found to suggest in some cases that the source of the noise is rising through the sun's atmosphere with a speed of several hundred km/sec;[4] there is also some indication[5] that the radio noise may suffer absorption by solar corpuscular streams.

There is, however, a regrettable scantiness of positive facts of direct solar observation and theory supporting the case for these supposed escaping masses of gas from the sun that affect the earth. The escape of the gas would seem to be in general invisible.

If so, it is similar to the upward motion of the major part of the moving gas commonly observed near the sun's edge, for this gas is mostly falling. It must previously have ascended, but invisibly.

Mention may also be made, at this point, of the occurrence of a very few great magnetic storms, and of several minor storms, at times when for some days previously there had been no sunspots or visible sign of solar disturbance. This is not evidence that tells in favour of either radiative or corpuscular theories of storms or aurorae. It merely indicates that earth-influencing solar emissions can occur without the usual association with visible spots. It may be recalled that in like manner local magnetic fields on the sun are occasionally found in spotless regions, though ordinarily such local fields are associated with spots.

In 1929 I suggested that during a magnetic storm the solar gas approaching to impinge on the earth might possibly be detectable by its absorption of some of the sunlight passing through it. If the solar gas is a normal sample of the sun's atmosphere, the absorption produced should resemble that in the Fraunhofer spectrum, in which two of the strongest lines are the H and K lines of singly ionized calcium. The absorption lines due to the gas streaming towards the earth, however, should show a Doppler displacement to the violet, amounting for example to 14 Å if the speed of approach is 1,000 km per sec, or 22 Å if it is 1,000 miles per sec.

In 1944 R. S. Richardson of Mount Wilson,[6] and in 1946 H. A. Brück and F. Ruttlant at Cambridge, England,[7] tentatively reported success in this observation, during different great magnetic storms.

It is very desirable to continue such attempts to detect the corpuscular bursts and streams, to confirm or (at least negatively) to discount these observations, which are certainly difficult. F. D. Kahn and I have since discussed the probability that such detection will be feasible, on the basis of the scanty available knowledge or expectation as to what is likely to happen.[8] We infer that the chance of feasibility is greatest in the case of flare bursts, within the first hour or so after the onset of the flare; but that even then the calcium atoms may be too predominantly doubly ionized to give detectable H and K lines, which are produced by singly ionized calcium.

Magnetic and ionospheric storms have a wide range of intensity, however, and positive evidence of this kind, even in a very few cases, where the corpuscular streams were particularly dense, would be of great value.

Attempts are being made also by radio methods to detect the streams when outside the atmosphere but near the earth. Success has not yet been achieved, and it is uncertain whether or not it will be possible.

It is also of interest to point out that electron densities of the order of 100 per cc, in the stream or burst when near the earth, are not likely to reveal themselves by their scattering of sunlight. F. L. Whipple and J. L. Gossner[9] have lately inferred that the observed 15 per cent polarization of the zodiacal light sets an upper limit of 1,000 per cc to the density of electrons near the earth—though they consider that the probable *value* of the electron density may be much less than this, the polarization being due to reflexion by dust particles rather than to electron scattering. Hence, even if the solar streams much increase the electron density temporarily, near the earth, their density and lateral limitation would keep their contribution to the zodiacal polarization below observable limits, except possibly in rare cases of storms of exceptional intensity.

Corpuscular Ionospheric Solar Eclipses

In 1932 I showed that if the ionosphere was partly ionized by the entry of *neutral* particles into the atmosphere, traveling along straight lines like rays of light, the eclipse of the beam of such particles, by the passage of the moon between the sun and the earth, would differ greatly, both in time and position, from the "optical" eclipse of the sun's (wave-) radiation. Radio observations soon showed that a major ionospheric effect coincides with the optical eclipse, and that therefore ultraviolet radiation is the main ionizing agent in the formation of the ionosphere.

Here Chapman suggests that several of his assumptions in predicting the detection of a neutral corpuscular eclipse effect might be in error, including those asserting that the flow of neutral corpuscles is continual, that the speed of the neutral corpuscles is constant, and that the flux of the neutral corpuscles is unaffected by the presence of the charged particles intermingling with them.

The Course of the Charged Corpuscles near the Earth

One outstanding feature of magnetic storms is that they do not show the marked difference of intensity as between the sunlit and night hemispheres, that characterizes the ionosphere and the daily magnetic variations. The magnetic and auroral activity over the dark hemisphere, and the strong localization of aurorae and intense magnetic activity along and within the auroral zones, are generally regarded as among the clearest pointers to the corpuscular causation of these phenomena. Experiments such as those by Birkeland and Brüche, and the theoretical discussion by Störmer and (in connection with cosmic rays) by Lemaître and Vallarta, indicate how great can be the geomagnetic influence on the motion of rapidly moving charged particles approaching the earth from a great distance.

However, these experiments and theories are not applicable to a cloud of neutral but ionized gas (unless its density is far less than one particle per cc). The motion of an ionized neutral gas near the earth has been considered by Ferraro and myself, with only limited success. We show that the earth's magnetic field will carve a hollow in the solar gas, but we have *not* been able to infer in any detail how some of the gas will stream off from the inside of the hollow, to reach the earth on both the light and dark hemispheres, and especially in high magnetic latitudes. It seems to me true to say that as yet we have no theory of aurorae, but nevertheless the polar and nocturnal (as well as daytime) incidence of magnetic storms and aurorae is in my view a powerful argument for their extra-terrestrial origin.

Other cogent auroral evidence is the determination by L. Harang[10] of the luminosity distribution along auroral rays; this fits well with the conception of rapidly moving particles entering the atmosphere from outside, along the lines of magnetic force, and being slowed down as they lose their energy by exciting and ionizing the atmosphere along their path. Harang's work is also of great interest in indicating the upward increase of atmospheric scale height.

The complexity of auroral forms and motions shows that no theory of the aurora can be simple, but the extreme rapidity with which auroral curtains sometimes sweep across the sky suggests a cause such as the changing direction of impact of streams of particles coming from outside the atmosphere, whose paths are influenced by weak variable electromagnetic fields over a considerable distance around the earth.

The alternative view, on the ultraviolet radiation hypothesis, is that aurorae are electric discharges occurring and produced entirely within the atmosphere, by hydrodynamic and electrodynamic causes energized by solar wave radiation. Many observers have reported the visual impression of electrical discharges occurring during aurorae, but this may be an illusion, and meanwhile the case for an electric discharged interpretation of aurorae has scarcely begun to be developed seriously.

It would be improper, however, to omit mention of the difficulty long felt in the corpuscular theory, namely, that the speeds required by atomic particles to penetrate the atmosphere to the levels of the lowest aurorae materially exceed the estimated speed of travel of the streams and bursts from the sun. No detailed explanation of this discrepancy has yet been found, but as has appeared through conversations between Dr. D. F. Martyn and myself, the probable solution of the difficulty is contained in Part I of the Chapman-Ferraro series of papers on magnetic storms. There we showed, by consideration of particular simple cases, that at the surface of a neutral ionized stream of gas, moving with speed v transverse to a magnetic field H, there is a charged layer from which the charges are repelled with a force of order vH/c esu (c = speed of light). In the hollow round the earth, this force will be of the order vH_0/cZ^3, where H_0 denotes the earth's field (0.3 gauss) at the surface at the equator, and Za (a = earth's radius) the distance from the earth's centre to the point considered on the surface of the hollow. If the outer particles of the surface layer move under a force of this order of magnitude, over a distance of order Za, the gain of energy will be of order evH_0a/cZ^2, or somewhat less than this, but probably not by a factor more than 10. Taking v as 10^8 cm/sec, Z as 5, the gain of energy per particle is 1.2×10^{-5} erg, or, reduced by the suggested factor of 10, 1.2×10^{-6}. The kinetic energy of a hydrogen atom with the speed v is 8.3×10^{-9} erg. The electrostatic increase of energy by the amount 1.2×10^{-6} erg would increase the speed of the hydrogen atom to about 10^9 cm/sec, thus giving a reasonable explanation of the degree of penetration usually observed in aurorae.

Geomagnetic Effects

The aurora is visible only at night, though twilight observations show that it occurs also in the sunlit atmosphere outside the earth's shadow. Geomagnetic studies indicate the growth, soon after the onset of a magnetic storm, of polar systems of electric currents around both magnetic axis-poles; the currents flow in high concentration (both eastwards and westwards, in different parts) along the auroral zones, and mainly complete their flow across the polar caps. The type of current system is remarkably similar, except for differences of orientation, from one storm to another.

It is natural to infer from the visible phenomenon of the aurora, and from the associated radio observation of exceptional ionization at and below the level of the *E* layer, at times of magnetic storms and intense aurorae, that at such times the auroral zone is a channel of abnormally high electrical conductivity; as the channel includes the sunlit as well as the dark part of the auroral zones, the corpuscular penetration may be inferred to take place all round the zone. It may be hoped that in the future the occurrence of aurorae in full daylight will be proved by direct observation, either from rockets or from the ground, by some refinement of photographic methods comparable with (though different from) that by which Lyot has enabled the solar corona to be observed without an eclipse.

The electromotive forces that impel the polar electric currents may well be continually present, their effects being determined by the degree of corpuscular precipitation and ionization along and within the auroral zones.

These polar currents seem to take some time to develop, after the sudden commencement of a magnetic storm. The first phase of the storm, with its suddenness and high degree of simultaneity over the earth, receives a natural explanation in terms of the corpuscular hypothesis. As shown by Ferraro and myself, a neutral ionized stream, of density of the order of 100 protons and electrons per cc, would be very rapidly retarded on entry into the earth's magnetic field, and electric currents would be induced over the surface of the hollow carved in the stream by the earth's field. These currents would shield the interior of the stream from the field, and would increase the earth's horizontal field, as is observed during the first phase of the storm.

At the same time, currents will be induced in the ionosphere, tending to shield the space below from the magnetic changes outside. In this way, the corpuscular stream may affect the ionosphere powerfully though indirectly.

In the succeeding main phase of the storm, these induction effects on the ionosphere will continue with changing character. A. A. Ashour and A. T. Price[11] and M. Sugiura[12] have made a beginning in the theory of these effects, which can be studied even though the cause of the main phase itself is still only partly understood.

The Evidence of the Auroral Spectrum

The strongest present direct evidence for the corpuscular theory now seems to me to be provided by the hydrogen lines in the auroral spectrum; this evidence came to my notice only after the preceding part of this paper was written. C. W. Gartlein[13] has obtained some remarkably excellent auroral spectra (which I have lately had the privilege of seeing) in 1946 and at other times; on these spectra the atomic hydrogen lines $H\alpha$, $H\beta$, and $H\gamma$ clearly appear, but are broader than would be expected for single atomic lines. This broadening he attributes to Doppler effect; the spread of the $H\beta$ and $H\gamma$ lines corresponds to a *range* of speed of 800 km/sec. He does not give the actual displacements from the true positions of the lines, the dispersion presumably being considered inadequate; though attributing the broadening to motion in the line of sight, he therefore did not indicate whether the motion is towards or away from the observer, or upward or downward.

The idea that the hydrogen already existing in the upper atmosphere (in extremely low concentration) should acquire great upward speeds, not shared even in moderate degree by the oxygen and nitrogen that contribute the bulk of the auroral spectrum, is sufficiently improbable to permit the conclusion that the hydrogen is coming downwards from outside, thus constituting a most important confirmation of the corpuscular theory.

Vegard,[14] with his new spectrograph of high light-power and large dispersion, also finds $H\beta$ on a recent spectrogram of date February 23, 1950, and states that it "is broad and diffuse and has its maximum somewhat displaced by Doppler effect towards shorter waves, indicating rapid downward motions of the hydrogen atoms." Thus he establishes that the motion is downward.

Neither author states the inclination of the line of sight to the direction of the auroral rays (or the direction of the geomagnetic vector at the location of the aurora). It is important that in future such communications this be mentioned so that the speed of motion along the auroral rays, the true corpuscular speed, may be inferred. Gartlein's spectrum was probably taken at an angle very oblique to the auroral rays, so that the true atomic hydrogen speed may have been 1,000 km/sec or more. As Vegard's spectrum shows molecular oxygen bands, it probably refers to an aurora below about 150 km. At this level, the incoming particles will already have lost some of their initial speed (probably approaching 10,000 km/sec, in view of the observed penetration), though the most rapid loss is suffered near the end of their path.

Vegard's lack of mention of diffuse $H\alpha$ and $H\gamma$ lines on his spectrum requires explanation, which may be the presence of other lines or bands near their wavelengths.

In high geomagnetic latitudes, where the aurorae are viewed at only a small inclination to their ray direction, the spread of the decreasing speeds along the rays, and the resultant diffuse Doppler broadening, may render the hydrogen lines so diffuse that they may escape observation except by very refined means.

It seems clear that these auroral observations open a new and fascinating chapter of auroral research.

Later note: During the intense auroral storm of 18–20 August 1950, A.B. Meinel[15] observed the spectral region of the $H\alpha$ line with a high resolution spectrograph, both with the spectrograph pointed towards the magnetic zenith, and also towards the magnetic horizon. In the latter case, the $H\alpha$ line was strong and diffuse but not displaced. The zenith observations, however, gave a very unsymmetrical $H\alpha$ profile; the maximum was displaced 10 Å, and the violet wing was shifted 71 Å. This shows that protons were entering the atmosphere with a speed of at least 3,200 km/sec. These observations give direct support to the corpuscular theory of aurorae and magnetic storms.

1. See *Geomagnetism*, by S. Chapman and J. Bartels (Oxford: Clarendon Press, 1940); or *Terrestrial Magnetism and Electricity*, ed. J. A. Fleming (New York: McGraw-Hill Book Co., 1939); for references prior to 1939.

2. O. R. Wulf, *Terr. Mag. 50*, 185–197, 259–278 (1945); with S. B. Nicholson, *Pub. Astr. Soc. Pacific 60*, 37–53, 259–262 (1948); *Phys. Rev. 73*, 1204–05 (1948); with M. W. Hodge, *J. Geophys. Res. 55*, 1–20 (1950).

3. R. H. Woodward, *Terr. Mag. 53*, 1–25 (1948).

4. R. Payne-Scott, et al., *Nature 160*, 256 (1947).

5. J. S. Hey, S. J. Parsons, and J. Phillips, *Mon. Not. R. Astr. Soc. 108*, 354–371 (1948); cf. also A. Unsöld and S. Chapman, *Observatory 69*, 219–221 (1949).

6. R. S. Richardson, *Trans. Amer. Geophys. Union*, 25th annual meeting, pt. 4, 558–560 (1944).

7. H. A. Brück and F. Ruttlant, *Mon. Not. R. Astr. Soc. 106*, 130 (1946).

8. F. D. Kahn, *Mon. Not. R. Astr. Soc. 109*, 324–336 (1949).

9. F. L. Whipple and J. L. Gossner, *Astroph. J. 109*, 380–390 (1949).

10. L. Harang, *Terr. Mag. 51*, 381–400 (1946).

11. A. A. Ashour and A. T. Price, *Proc. R. Soc., A 195*, 198–224 (1948).

12. M. Sugiura, *Geophys. Notes No. 19* (Tokyo Univ.), *2*, no. 2 (1949).

13. C. W. Gartlein, *Trans. Amer. Geophys. Union 31*, no. 1, 18–20 (1950).

14. L. Vegard, *Nature 165*, 1012 (1950).

15. A. B. Meinel, *Phys. Rev. 80*, 1096 (1950); *Astroph. J. 113*, 50, 583 (1951).

23. The Origin and Nature of Comets

The Structure of the Cloud of Comets Surrounding the Solar System and a Hypothesis Concerning Its Origin

Jan H. Oort

(*Bulletin of the Astronomical Institutes of the Netherlands 11*, 91–110 [1950])

A Comet Model I: The Acceleration of Comet Encke

Fred L. Whipple

(*Astrophysical Journal 111*, 375–394 [1950])

Compared to the planets, with their stately motion in nearly circular orbits, the comets are the jetsam of the solar system. Their masses are small and their orbits almost always elongated; so they are subject to major gravitational perturbations from the planets. In his *Méchanique céleste*, Pierre Simon Laplace argued that comets come from interstellar space and that close encounters with Jupiter occasionally divert them into smaller orbits, thus accounting for the short-period comets. The capture theory of comets was extensively studied by H. A. Newton,[1] who showed that out of a billion comets entering a sphere described about the sun with a radius equal to Jupiter's orbit, less than 1,000 will have their orbits changed into ellipses with an orbital period less than Jupiter's.

In 1948 A. J. J. van Woerkom[2] demonstrated that if the comets originally belonged to the planetary system they would long ago have been diffused out of it by the action of Jupiter. He also concluded that small orbital perturbations by Jupiter could account for the observed distribution of the axes of the long-period comets. Van Woerkom argued that Laplace's interstellar origin theory was highly improbable, because of the lack of hyperbolic orbits among the observed comets.

Building on van Woerkom's result, Jan H. Oort postulated the existence of a vast cometary cloud belonging to the outer parts of the solar system. The extract of Oort's paper given here argues that the comets form a cloud extending to distances as large as 150,000 A.U. from the sun. Here he supports Ernst Öpik's earlier evidence that such a cloud is statistically stable against stellar perturbation.[3] In addition, Oort shows that passing stars throw enough comets into the central parts of the solar system to replenish those lost by disintegration and by outward diffusion through Jovian perturbations.

Oort's unique contribution was his provision of a detailed account of the various perturbations, which he compared to the observational data. These considerations made it possible, from the observed frequency of perihelion passages of new comets within 1.5 A.U. of the sun, to estimate the total number of comets contained in the reservoir between 30,000 and 100,000 A.U. This number is on the order of 200,000 million comets. He also showed that a cloud of exactly the observed dimensions could automatically have formed in the early history of the solar system from the smaller bodies left over from the formation of the planets.

Today the comets are generally viewed as frozen vestiges of the same nebula that produced the sun and its planetary system. In fact, it is likely that the comets were the building blocks for the great outer planets Uranus and Neptune. However, in 1950, when Fred L. Whipple described the icy conglomerate model in the second paper given here, comets were commonly regarded as loose aggregations of meteoritic material. By this time Whipple's photographic studies of meteors had begun to reveal the great difference between the fragile meteoric material associated with comets and the firm, even steely, meteorites that could survive the atmospheric flight to the earth's surface. He also knew, from the studies of Karl Wurm and Pol Swings, that the radicals observed in cometary spectra could arise from such parent molecules as water and carbon dioxide, which could have remained frozen in a cometary nucleus if comets are stored in the distant, cold comet cloud envisaged by Oort.

Whipple recognized that if the frozen comet was rotating as it approached perihelion, the ejection of gases resulting from the sublimation of the ices would create nongravitational forces on the comet. He used these effects on Comet Encke as a test of his theory.

In 1964 Ludwig Biermann and Eleanor Trefftz[4] considered the resonance fluorescence radiation of comets in the ultraviolet and predicted that one consequence of Whipple's icy conglomerate model would be the existence of a hydrogen halo that could be detected by its Lyman-α radiation at ultraviolet wavelengths. When the first ultraviolet observation of a comet was obtained in 1970, using the Orbiting Astronomical Observatory (OAO-2), it revealed an extensive Lyman-α halo.[5] Furthermore, recent Orbiting Geophysical Observatory space observations of Comet Bennett (1970 II) and Aerobee rocket observations of Comet Kohoutek (1973 XII) (see the figure we have appended to this selection) have shown that the hydrogen envelope is caused primarily by photodissociation of water. For every atom of hydrogen generated there is approximately another radical of OH, and together they account for the loss of about a million tons of water per day when the comet is near the sun. The observations of the hydrogen halo have reaffirmed Whipple's view that a comet is fundamentally a dirty snowball.

1. H. A. Newton, *Memoirs of the National Academy of Sciences* (Washington, D.C.) *6*, 7 (1893).
2. A. J. J. van Woerkom, *Bulletin of the Astronomical Institutes of the Netherlands 10*, 445 (1948).
3. E. J. Opik, *Proceedings of the American Academy of Arts and Sciences 67*, 169 (1932).
4. L. Biermann and E. Trefftz, *Zeitschrift für Astrophysik 59*, 1 (1964).
5. A. D. Code, T. E. Houck, and C. E. Lillie, "Ultraviolet Observations of Comets," in *The Scientific Results from the Orbiting Astronomical Observatory* (*OAO-2*), ed. A. D. Code, NASA report no. SP-310 (Washington, 1972).

Oort, The Cloud of Comets Surrounding the Solar System

Abstract—The combined effects of the stars and of Jupiter appear to determine the main statistical features of the orbits of comets.

From a score of well-observed original orbits it is shown that the "new" long-period comets generally come from regions between about 50,000 and 150,000 A.U. distance. The sun must be surrounded by a general cloud of comets with a radius of this order, containing about 10^{11} comets of observable size; the total mass of the cloud is estimated to be of the order of 1/10 to 1/100 of that of the earth. Through the action of the stars fresh comets are continually being carried from this cloud into the vicinity of the sun.

The article indicates how three facts concerning the long-period comets, which hitherto were not well understood, namely the random distribution of orbital planes and of perihelia, and the preponderance of nearly-parabolic orbits, may be considered as necessary consequences of the perturbations acting on the comets.

The theoretical distribution curve of $1/a$ following from the conception of the large cloud of comets is shown to agree with the observed distribution, except for an excess of observed "new" comets. The latter is taken to indicate that comets coming for the first time near the sun develop more extensive luminous envelopes than older comets. The average probability of disintegration during a perihelion passage must be about 0.014. The preponderance of direct over retrograde orbits in the range from a 25 to 250 A.U. can be well accounted for.

The existence of the huge cloud of comets finds a natural explanation if comets (and meteorites) are considered as minor planets escaped, at an early stage of the planetary system, from the ring of asteroids, and brought into large, stable orbits through the perturbing actions of Jupiter and the stars.

The investigation was instigated by a recent study of van Woerkom on the statistical effect of Jupiter's perturbations on comet orbits. Action of stars on a cloud of meteors has been considered by Öpik in 1932.

Sketch of the Problem

Among the so-called long-period comets there are 22 for which, largely by the work of Elis Strömgren, accurate calculations have been made of the orbits followed when they were still far outside the orbits of the major planets.[1] Approximate calculations of the original orbits by Fayet[2] are available for 8 other comets with well-determined osculating orbits. For the present limiting ourselves to the comets for which the perturbations were rigorously determined, and excluding 3 for which the mean error of the reciprocal major axis, $1/a$, is larger than 0.000100, the values of $1/a$ for the remaining 19 comets are distributed as shown in table 23.1.

Table 23.1 Distribution of original semi-major axes (a in A.U.)

$1/a$	n
<0.00005	10
0.00005— 0.00010	4
0.00010— 0.00015	1
0.00015— 0.00020	1
0.00020— 0.00025	1
0.00025— 0.00050	1
0.00050— 0.00075	1
>0.00075	0

The mean errors of $1/a$ are all smaller than 0.000061; their average is $\pm$.000027. The steepness of the maximum for small values of $1/a$ indicates that the real mean errors of the original $1/a$ cannot greatly exceed these published mean errors. The 22 comets do not form a representative sample of the long-period comets; there has been a selection for small values of $1/a$, so that the real proportion of comets with $1/a >.00050$ is much larger than indicated in the table. It can be shown, however, that the selection has not appreciably influenced the relative numbers in the rest of the table. Among the comets in the first division there are two with negative values of $1/a$, viz. $-.000007$ and $-.000016$, probably due to observational errors.

It is evident from table 23.1 that the frequency curve of $1/a$ shows a steep maximum for very small values. The average for the 10 orbits in the first interval is .000018, thus corresponding to a major axis of 110,000 A.U. We may conclude that a sensible fraction of the long-period comets must have come from a region of space extending from a distance $2a =$ 20,000 to distances of at least 150,000 A.U. from the sun; that is, almost to the nearest star. This does not mean that they are interstellar. They belong very definitely to the solar system, because they share accurately the sun's motion. Yet, the prevalence of these very large major axes has led several astronomers to investigate the question whether the comets could not be of interstellar origin. It is evident that they cannot *directly* come from interstellar space, for in that case there would have to be many more outspoken hyperbolic orbits than nearly parabolic ones. So far, no comet has been found for which the eccentricity exceeds 1 by an amount large enough to be considered as real. It is conceivable, however, that comets would be caught from an interstellar field by the action of the major planets, and would then move for a long time in orbits of large dimensions, so that the number of comets caught would gradually become far larger than the number of hyperbolic comets passing through the solar system. This suggestion has recently been studied by Dr van Woerkom.[3] He concludes that this possibility must be ruled out, because the action of the major planets which causes the comets to be captured would at the same time result in a distribution of the values $1/a$ which is constant over a considerable range of negative as well as positive values. There would again be a large preponderance of hyperbolic comets, which is contradicted by observations. For a more exhaustive discussion of this problem I may refer to section 4 of van Woerkom's article.

There is no reasonable escape, I believe, from the conclusion that the comets have always belonged to the solar system. They must then form a huge cloud, extending, according to the numbers cited above, to distances of at least 150,000 A.U., and possibly still further. It is not necessary at this point to enter upon the question how this cloud has originated. It might conceivably be considered as part of the remnants of a disrupted planet (see section 6). An alternative hypothesis, repeatedly put forward, according to which comets would be formed by eruptions from Jupiter and the other planets, does not appear to be likely (cf. van Woerkom, *l.c.* p. 464 a.f.).

Accepting this existence of a huge cloud of comets we are still faced with a difficulty that has been put into full light by van Woerkom's study. Jupiter, and to a lesser extent the other planets, exert a diffusing action on the long-period comets. According to van Woerkom's calculations the small perturbations by Jupiter suffered by an observable comet during its passage through the "inner" part of the planetary system will on the average change the reciprocal major axis by about 0.0005; positive and negative changes are equally probable. By these perturbations the long-period comets will gradually disappear, partly into interstellar space, partly into the families of short-period comets. In addition, the comets may gradually diminish in brightness through the sun's action, or be dissolved. It is evident from van Woerkom's study that within one or two million years after their first perihelion passage practically all long-period comets will have disappeared. As it is highly improbable that the comets we observe have only originated within the last two million years we are led to conclude that comets already existing outside the region where they are subject to the perturbing action of sun and planets are continually being brought into this region.

A direct indication of the probable escape of a considerable fraction of the comets of very long period has been given by Fayet.[4] Among 36 comets for which he has made approximate calculations of the orbits which they must have described after they passed out of the action of Jupiter he found 7 for which this orbit was hyperbolic. A more complete calculation for a similar case (comet 1898 VII) where the final orbit is definitely hyperbolic, has recently been made by Sinding.[5]

If we assume that at the start the velocity distribution of the comets in the huge cloud surrounding the planetary system was a random distribution, there must have been comets, even in the outer parts of the cloud, whose velocities were so nearly directed towards the sun that these comets would eventually pass through the "observable region" (i.e. the region within about 2 A.U. from the sun). Even if the radius of the cloud was 150,000 A.U. all the comets which could come into the vicinity of the earth would have done so within roughly 20 million years. All these comets will diffuse into space or be disintegrated. No new comets would come in after this period unless they were made to do so by some perturbation. Van Woerkom's discussions make it clear that perturbations by *planets* cannot be effective in bringing comets into the observable region: their influence on the major axis and the period is always much more important than on the perihelion distance. Their perturbations will diffuse the comets out of the long-period range long before they have caused a change of any importance in the perihelion distance.

Two alternative types of perturbations offer themselves, namely resistance by an interplanetary medium, and influence of passing stars. It seems extremely unlikely that the former mechanism could have an observable influence on the perihelia of comets. For a general influence of this kind to be effective a density of interplanetary gas would be required that is quite inadmissible on dynamical grounds. Moreover, a resisting medium would in the first place tend to decrease the major axes, while for the nearly parabolic comets the perihelion distances would appear to be practically unaffected, so that it could never solve our problem.

The purpose of the present paper is to investigate the second possibility, the action of passing stars.

Oort first examines the general influence of stars on the comet cloud and shows that their perturbations limit the cloud to about 200,000 A.U.

It is important to see what effects the passing stars will have had on the *shape* of the cloud, and on the distribution of orbital eccentricities. The fact that neither the orbital planes nor the aphelia of the long-period comets show a distinct preference for the ecliptic, has often been taken as an indication that they are of interstellar origin.[6] However, if we take account of the influence of the other stars on the cloud of comets, we see that, even if this cloud had originally been strongly concentrated towards the plane of the ecliptic, not much trace of this could have remained.

As regards the eccentricities, although *observations* cannot tell us whether the distant regions, into which we have seen that many of the elongated orbits extend, contain also comets with less eccentric orbits, we may infer from consideration of the action of stellar perturbations that this is probable. We may even conclude that, unless comets are only a recent phenomenon, their velocity distribution in these regions must be nearly isotropic.

In the next section I have attempted to give a working model for a cloud extending to 200,000 A.U. In this model the average square velocity at $r = 50{,}000$ is $1.60 \cdot 10^8$ or in one co-ordinate $0.53 \cdot 10^8$. Now $\frac{1}{3}\overline{\Delta V^2} = 0.35 \cdot 10^8$. So, even if originally the velocities had all been directed along the radius vector to the sun, or had all been parallel to the ecliptic, the comets would by now have acquired velocities of practically the same amount in the other co-ordinates. The argument applies still more strongly to the comets at larger distances from the sun. It does *not* apply to smaller distances. If originally there had been a flattening of the part of the cloud within, say, 40,000 A.U., the flattening should still be visible. But it is probable that most comets come originally from distances larger than 40,000 A.U.; for these we cannot expect to find a sensible deviation from random distribution.

After investigating the velocity distribution within the comet cloud, Oort turns next to the effect of the stellar perturbations on the velocity distribution, first for comets at 50,000 A.U. from the sun and then for comets with aphelion q less than that distance.

We see from table 23.2 that practically all "new" comets must have had orbits with semi-major axes larger than 25,000 A.U. This agrees very well with observation (see table 23.1 and the next section) and gives us a clear insight into the reason why all new comets appear to come exclusively from such very large distances.

The numbers in table 23.2 enable us also to compute the density $\nu(50{,}000)$. As we shall see below, we may estimate that on the average 97 observable new comets pass per century through perihelion with $q < 1.5$. According to table 23.2 this

Table 23.2 Number of comets, n, passing per century through a perihelion within 1.5 A.U. from the sun (ν = no. of comets per cubic A.U., and u denotes the radial velocity in cm sec^{-1} at 50,000 A.U.).

q	a	$u(50{,}000)$	n
200,000	100,000	$1.637 \cdot 10^4$	
			$2.31 \cdot 10^5\ \nu(50{,}000)$
141,000	70,700	$1.518 \cdot 10^4$	
			$3.10 \cdot 10^5\ \nu(50{,}000)$
100,000	50,000	$1.336 \cdot 10^4$	
			$4.56 \cdot 10^5\ \nu(50{,}000)$
70,700	35,400	$1.022 \cdot 10^4$	
			$6.32 \cdot 10^5\ \nu(50{,}000)$
50,000	25,000	0	
			$0.5 \cdot 10^5\ \nu(50{,}000)$
35,400	17,700	—	
			$0.14 \cdot 10^5\ \nu(50{,}000)$
25,000	12,500	—	
			$0.07 \cdot 10^5\ \nu(50{,}000)$
17,700	8,840	—	
			$0.03 \cdot 10^5\ \nu(50{,}000)$
12,500	6,250	—	

number should be equal to $1.7 \cdot 10^6\ \nu(50{,}000)$. We find therefore that

$$\nu(50{,}000) = 5.7 \cdot 10^{-5} \text{ per (A.U.)}^3.$$

The number of comets between $r = 25{,}000$ and $r = 200{,}000$ is then equal to

$$4\pi \cdot 50{,}000^2 \cdot 5.7 \cdot 10^{-5} \int_{25{,}000}^{200{,}000} \frac{N(r)}{N(50{,}000)}\, dr.$$

The values of $N(r)/N(50{,}000)$ can be inferred from my working model. The total number of comets in the cloud is thus found to be $1.9 \cdot 10^{11}$.

There are no good estimates of the average mass of a comet, except that it must probably be larger than about 10^{14}, and smaller than 10^{20} grams. A plausible estimate is perhaps about 10^{16} g (cf. also van Woerkom, *l.c.* p. 462, footnote). With such an average mass the total mass of the cloud of comets would be 10^{27}, or about 1/10th of the earth's mass. This estimate is uncertain by one or two factors of 10.

Oort then compares the observed distribution of major axes with the stationary distribution that would be produced by the action of Jupiter.

The present theory seems capable of accounting satisfactorily for all statistical data concerning the long-period comets down to periods of about a century, viz. the remarkable form of the distribution curve of $1/a$, the random distribution of inclinations and of the directions of perihelion, and the decreasing inclinations for orbits with a between 25 and 250 A.U. The relation of the short-period comets, in particular those of the Jupiter "family", to the long-period ones has been extensively investigated in the past, especially by H. A. Newton,[7] H. N. Russell,[8] and most recently by van Woerkom.[9] The work on this intricate problem is still far from complete, but the evidence available appears to be in at least approximate agreement with what one would expect if the population of the family is kept up by the captures by Jupiter from the field of long-period comets.[10] The Jupiter family would thus be in equilibrium with the long-period comets.

It may be noted that, except for the incompleteness in the discussion of the Jupiter and Saturn families, we have now a fairly comprehensive theoretical picture of the distribution of cometary orbits. The picture is not confined to the comets that come within the observable region, but may be extended to any perihelion distance, because for the long-period comets, as H. A. Newton has already remarked, the number passing through a perihelion between $q - \frac{1}{2}$ and $q + \frac{1}{2}$ may be expected to be independent of q. For the short-period comets the conditions are vastly more complicated, but nevertheless it would seem possible to work out the theory of these orbits in a statistical way.

Oort next assumes a common origin for meteorites and comets and, observing that this speculative conclusion in no way affects the considerations set forth in the rest of the article, argues that the comets have diffused out to the comet cloud from an origin in the asteroid belt.

1. A list of these is given by Sinding, *Danske Vidensk. Selsk., Mat.-Fys. Medd. 24*, no. 16 (1948), or *Publ. o. Mindre Medd. Köbenhauns Obs.* no. 146. Van Biesbroeck's orbit for comet 1908 III has been added to this list.

2. *Thèse* (Paris, 1906); also in *Ann. Paris, Mém. 26A* (1910).

3. *B.A.N.* no. 399 (1948).

4. *Ann. Bur. Longitudes 10B; Comptes Rendus 189*, 1122 (1929).

5. This was communicated to me before publication by the kindness of Mr Sinding and Prof. Strömgren.

6. Statistics of the distribution of perihelia have been given, among others, by Oppenheim, *Festschrift für H. von Seeliger* (1924), p. 131, and by Bourgeois and Cox, *B.A. 8*, 271, and *9*, 349 (1934).

7. "On the Capture of Comets by Planets, Especially their Capture by Jupiter," *Mem. Nat. Ac. Washington 6*, 7 (1893).

8. "On the Origin of Periodic Comets," *A.J. 33*, 49 (1920).

9. *L.c.* section 5.

10. The essential difference between these captures and the small perturbations considered in the present paper is that the captures are due to one, or a very few, quite large perturbations. The comets in the Jupiter group may therefore well be "younger" on the average than the comets which have come down to orbits with major axes of the order of 100 A.U. by successive, small perturbations. The family is partly made up of comets whose original orbits happened to come exceptionally close to the orbit of Jupiter.

Whipple, A Comet Model

Abstract—A new comet model is presented that resolves the chief problems of abnormal cometary motions and accounts for a number of other cometary phenomena. The nucleus is visualized as a conglomerate of ices, such as H_2O, NH_3, CH_4, CO_2 or CO, (C_2N_2?), and other possible materials volatile at room temperature, combined in a conglomerate with meteoric materials, all initially at extremely low temperatures ($<50°$ K). Vaporization of the ices by externally applied solar radiation leaves an outer matrix of nonvolatile insulating meteoric material. Quantitative and qualitative study shows that heat transfer through thin meteoric layers in a vacuum is chiefly by radiation, that the heat transfer is inversely proportional to the effective number of layers, and that an appreciable time lag in heat transfer can occur for a rotating cometary nucleus. Because of the time lag, such a cometary nucleus rotating in the "forward" sense will emit its vaporized ices with a component toward the antapex of motion. The momentum transfer from the kinetic velocity of the emitted gas will propel the nucleus in the forward sense, reduce the mean motion, and increase the eccentricity of the orbit. Such orbital effects occur for Comet D'Arrest; the mean daily motion of Comet Wolf I also appears to be decreasing.

Retrograde rotation can produce an acceleration in mean motion and a decrease in eccentricity, as observed for Comet Encke. If the decelerating force component is taken as one-quarter its maximum theoretical value, the present observed acceleration in the mean motion of Comet Encke can be produced by a loss of 0.002 of its mass per revolution. The corresponding mass loss for Comet D'Arrest is 0.005. For both comets the observed changes in eccentricity are obtained if the force acts proportionately to the solar energy flux but is cut off at a solar distance of about 2 A.U.

A second paper (Part II) soon forthcoming will be concerned with the physical problems of comet structure, loss of meteoric and gaseous material, and correlations with observed meteoric phenomena.

INTRODUCTION

THE RECENT and valuable contribution by A. J. J. van Woerkom[1] on the origin of comets has strengthened the growing confidence in the concept that comets are ancient members of the solar system. He has shown that the process of replenishment of the periodic comets from the extremely long-period comets via capture by Jupiter appears to be the most plausible, if not the only plausible, process for maintaining the supply of periodic comets. The remaining question as to the statistical stability of a solar family of comets with semimajor axes up to about 10,000 A.U. has been studied by E. Öpik.[2] The losses to such a system by the gravitational action of passing stars is not serious over a period of even 3×10^9 years. Only a close approach of a passing star to the sun would remove the sun from control of a large fraction of the comets. The loss by stellar attraction for individual comets with periods up to a million years is statistically unimportant over such a long interval of time; but comets with longer periods will suffer statistical increases in perihelion distance that will tend appreciably to place them beyond Jupiter's reach.

There still remains, however, a discrepancy of approximately 20 between van Woerkom's calculated number of

Jupiter "captures" of long-period comets into short-period orbits (one per 650 years) and the estimated loss of three periodic comets per century. This discrepancy might be removed by assuming a greater number of comets with perihelion in the neighborhood of Jupiter, by establishing more certainly the rate of disintegration of comets, or by accepting G. Fayet's[3] suggestion that, statistically, several short-period comets arise from each parent-comet.

In any case, the lifetime of a short-period comet must lie generally in the range of from 3,000 (one hundred comets being lost at a rate of three per century) to possibly 60,000 years. Probably more important is the number of small perihelion passages that can be weathered by a comet—of the order of several hundreds, at least, for perihelion distances as small as 0.5–1.0 A.U. For considerably greater perihelion distances the number probably increases to several thousand, thus permitting comets with periods up to 10^6 years to persist throughout all or most of the past history of the solid earth.

Even though parts of the preceding discussion are somewhat conjectural, we must certainly accept the conclusion that individual short-period comets cannot exist indefinitely in their present orbits and also that they must previously have existed at great distances from the sun, where their temperature throughout remained at extremely low values. K. Wurm[4] has discussed certain effects of such low temperatures in cometary phenomena. N. T. Bobrovnikoff[5] and P. Swings[6] have pointed out that certain possible parent-compounds, such as CO_2, H_2O, and NH_3, may be responsible for the observed radicals, such as CO, OH, and NH, in cometary spectra; Swings has suggested that these materials would exist in the solid state within the nuclei of comets.

In the present discussion I propose to investigate the possibility that the molecules responsible for most of the light of comets near perihelion arise primarily from gases long frozen in the nuclei of comets. Furthermore, I propose that these primitive gases constitute an important, if not a predominant, fraction of the mass of a "new" or undisintegrated comet.

On the basis of these assumptions, a model comet nucleus then consists of a matrix of meteoric material with little structural strength, mixed together with the frozen gases—a true conglomerate. Since no meteorites are known certainly to arise from cometary debris, we know very little about the physical structure of the meteoric material except that the pieces seem generally to be small. Hence we assume that the larger pieces are perhaps a few centimeters in radius and the smallest are perhaps molecular. As a convenience in terminology, the term "ices" will be used in referring to substances with melting points below about 300° C and "meteoric material" to substances with higher melting points.

Our only chemical knowledge of the meteoric material comes from the spectra of meteors,[7] which tell us that Fe, Ca, Mn, Mg, Cr, Si, Ni, Al, and Na, at least, are present. Physically the meteoric material is strong enough to withstand some shock in the atmosphere, but more than 3 per cent of the Harvard photographic meteors are observed to break into two or more pieces. A much larger percentage show flares in brightness, an indirect evidence of breaking. The high altitude of the disappearance of the photographic Giacobinid meteors of October 9, 1946, as observed by P. M. Millman and analyzed by L. Jacchia and Z. Kopal,[8] suggests that those meteoric bodies may have been unusually fragile or porous. It is difficult to defend the hypothesis that, as a whole, the bodies producing photographic meteors possess great physical rigidity or strength.

A careful determination of the relative abundances of the primitive ices in the nucleus of a comet and their physical properties will require an exhaustive study of the theory of cometary spectra and related phenomena, including evolutionary hypotheses. Only a few comments will be made here. The observed gases CH, CH^+, CH_2, CO, NH, NH_2, OH, and OH^+ can be accounted for by four possible parent-molecules of great stability, viz., CH_4, CO_2, NH_3, and H_2O. Photodissociation appears capable of producing the various radicals from these parent-molecules, although Wurm prefers CO instead of CO_2. Only the very important observed C_2, N_2^+, and CN molecules are unaccounted for above. The choice of C_2N_2 as a parent-molecule does not seem desirable because the dissociation of C_2N_2 is exothermic; nevertheless, I shall include it in the present discussion for lack of a better substitute (C_2H_2, N_2, HCN?). Quite possibly some of the radicals can exist permanently at very low temperatures.

The metals Na, Fe, Ni, and Cr—all observed in meteor spectra—have been observed in comet spectra at small solar distances. They require the presence of metals or, more generally, meteoric materials in molecular or atomic forms within the comet nucleus.

Some physical data for five of the possible parent-gases are given in table 23.3.

As our model comet nucleus approaches perihelion, the solar radiation will vaporize the ices near the surface. Meteoric material below some limiting size will blow away (Part II, forthcoming) because of the low gravitational attraction of the nucleus and will begin the formation of a meteor stream. Some of the larger or denser particles may be removed by shocks (see below), but the largest particles or matrix will remain on the surface, to produce an insulating layer. After a short time (probably in the geologic past for all known comets) the loss of gas will be reduced materially by the insulation so provided.

A comet such as Encke's, if made of a nonvolatile solid, would come eventually to a temperature of approximately 140° K at aphelion.[9] Thus CH_4 would melt and vaporize quickly, while the other ices of table 23.3 would vaporize more slowly. At perihelion temperatures, all the substances in table 23.3 would be gaseous, even under high pressures.

Table 23.3 Properties of certain molecules.

Properties	Molecule				
	CH_4	CO_2	NH_3	C_2N_2	H_2O
Melting point (°K)	90	217	198	239	273
Heat of fusion (cal/gm)	50	45	108	—	80
Boiling point at 1 atm. (°K)	111	195	240	252	373
Heat of vaporization from solid (cal/gm)	188+	138+	435+	103++	670+
Vapor pressure at 191° K (atm.)	45.8	0.74	0.038	8.0×10^{-3}	3.7×10^{-7}

The quasi-equilibrium state arising from a slow external heating of the extremely cold ices can be visualized qualitatively as follows: at the base of the meteoric layer, only the ice with the lowest vapor pressure will still remain; hence this layer will consist only of "rotten" ice (H_2O) and meteoric material; the next layer will contain, in addition, C_2N_2 (if present), etc.

It is important to note that practically all heat reaching the ice (H_2O) at temperatures above about 180° K will be used in vaporization. The H_2O vapor that does not escape the comet will condense on the cooler layers beneath, producing evaporation of these materials.

The outer icy layers will arrive at a quasi-equilibrium state in the order of the vapor pressures of the ices at low temperatures. The thicknesses of the various layers will depend upon the temperature gradient, which, in turn, will depend upon the effective heat conductivities of the layers and the temperature of the outer layer. It will be shown below that the effective conductivities of such a conglomerate must be very low. The deep interior of the cometary nucleus will remain extremely cold, not only because of the low heat conductivity but also because the available heat will be used in vaporization, an extremely effective cooling mechanism in a vacuum. Hence the comparative rates of escape of the various primitive gases from the nucleus will depend primarily not upon their physical-chemical properties but upon their abundances. Some second-order effects may occur because of the variation of temperature gradient in the upper layers with the external heating; but such effects should not be manifest, for example, in correlations of cometary emission spectra with age except in the most extreme stages of disintegration.

The weakening of the upper layers of the icy core by selective vaporization of the ices may be expected to produce cometary activity of considerable intensity, especially near the sun. The surface gravities of cometary nuclei are certainly extremely low; hence surprisingly weak structures can persist over rather large areas of the nucleus. At irregular intervals collapses must occur. The heated meteoric material will then fall into the ices and produce rapid vaporization. The dust and smaller particles held in the upper layers will be shaken out and blown away, so that insulation produced by this material will be much reduced. Solar heat, consequently, will be much more effective in vaporizing the ices in the pit until equilibrium is again established. Such "cave-ins" might spread over appreciable areas. Other effects might occur if "pockets" of an ice with low melting points exist within an ice of higher melting point. Phenomena of mildly explosive, jet, or cracking types may occur, forcing out pieces of material much larger than those carried normally by the outgoing gas. Hence the type of nuclear activity that is observed for large comets with small perihelion distances would be expected from this type of comet model.

If the primitive ices constitute a large percentage of the total mass, the comet truly disintegrates with time. Its actual substance vaporizes; the surface gravity decreases; and, finally, all activity ceases as the last of its ice reservoir is exhausted. The observed sequence of phenomena in dying comets is entirely consistent with this picture. In the later stages, only a very small nucleus of the largest meteoric fragments remains (note the asteroid Hidalgo as a possible example).

The period of rotation of a comet with a single spheroidal nucleus would generally remain constant with age, so that the comet might dissipate slowly and uniformly. If, however, the nucleus were multiple or irregular in shape, the vaporization of ices could materially affect the rotation. Suppose, for example, a part of the surface were nearly in a plane passing through the center of gravity of the nucleus, while the remaining surfaces were generally smooth and approximately oval in shape. Meteoric material would fall from the vertical surface, exposing it to the full action of sunlight. Hence the excess of gas evolved from this surface would exert a force moment on the nucleus as a whole.

The effect of the resulting rotation, depending upon the initial circumstances, might easily produce rotational instability, permitting the sun's tidal action to complete the splitting of the nucleus. If the larger parts of the separated nucleus were unstable, the comet might disappear quickly. On the other hand, the pieces might be large enough to persist for a long period of time as individual comets. In fact, the phenomenon of splitting has occurred for several comets and has been followed by disappearance in some cases, but not in others. Either possibility may be expected on the basis of the present comet model, depending upon the mass, shape, and rotation of the nucleus.

It is clear that the answers to certain of the problems concerning the proposed comet model can best be determined in the laboratory rather than by theory. A pertinent experiment would involve the making of conglomerates of the various ices and meteoric materials, submitting them to small pressures and then observing the vaporization when the conglomerates were exposed to radiation from one surface, the other surfaces remaining insulated or refrigerated. An exhaust pump could simulate the conditions of low gas pressure in space, while even the action on meteoric material might be studied by utilizing fine powders to compensate the model for the large terrestrial gravity.

Even without such experiments, however, the suggested model of a comet nucleus is subject to a number of tests, both by theory and by observation. Certain phases of the problem will be discussed in the following sections, and the conclusions will be applied to demonstrate possible mechanisms for the observed acceleration of the mean motion of Comet Encke (and similar phenomena for other comets) and to explain in more detail the ejection of meteoric material from comet nuclei.

The Problem of Heat Transfer

It is obvious that the total solar radiation falling upon our model comet nucleus would be sufficient, in a relatively short time, to melt and vaporize quite sizable masses. In 1 year at 1 A.U. from the sun, a layer of ice some 4 meters thick would be vaporized from the surface of a spherical body, if all the solar radiation were absorbed. A much thicker layer of the other ices in table 23.3 would be lost. Hence a cometary nucleus, 1 km in diameter and made of such ices, would scarcely persist for a hundred perihelion passages within 1 A.U. Also Minnaert[10] has shown that the temperature rise within the solid nucleus of a periodic comet is relatively rapid if the material conducts heat like ordinary stone. The surface layers of meteoric material, however, will greatly reduce the rate of heat transfer as compared to that of a solid body, and vaporization will maintain a low internal temperature. Let us consider, therefore, the likely processes of heat transfer.

If the meteoric layer is a coarse aggregate, very poorly cemented, direct conduction of heat by solids will be very slow because of the small areas of contact between discrete particles on the surface of the nucleus. The coefficient of heat conduction for the compact solid may be reduced by the order of ten thousand times, making this form of heat transfer negligible except within the particles themselves.

Here Whipple computes the effective conductivity of a layer of particles of a given thickness at a given temperature. He finds that 1% of the incident solar radiation is transmitted to the bottom of a layer between 30 and 400 times the thickness of the individual particles.

Time Lags in Heat Transfer

The question of time lags in heat transfer through the meteoric layers of a rotating cometary nucleus is of great importance in the discussion of the acceleration of Comet Encke. Since the classical theories of periodic heat flow are not applicable to the general case of a rotating cometary nucleus if heat transfer is primarily by radiation, this problem must be investigated by methods of numerical integration. Such an investigation is under way and will be presented in a later paper. Nevertheless, some fragmentary information can be obtained from classical heat theory if constant values of the conductivity and diffusivity are assumed.

Whipple then derives a relation between the lag in maximum heat transfer at a given depth and the rotational period of the comet. In general, the lag increases with increasing depth, with increasing distance from the sun, and with increasing rotational period. For a comet at a distance of 1 a.u. from the sun, the lag becomes zero at a thickness of 10 particle diameters if the rotational period is 6 days.

The Luminosity Law

Here Whipple discusses the dependence of the total luminosity of a comet on its distance, *r*, from the sun. Observational results for 45 comets give a luminosity varying as $r^{3.32 \pm 0.16}$, but the observed fluctuations in the exponent are enormous from comet to comet. The observed exponent fits various models in which the light from a comet is assumed to arise from the emission of gases escaping from the head of the comet.

The Accelerations of Comet Encke and Other Comets

The systematic increase in mean motion and the decrease in the eccentricity of Encke's Comet, conspicuous during the first half of the nineteenth century and smaller but definite since then, has long been a subject of speculation. Theories of a resisting medium stumble upon the lack of other evidence for the medium, upon the variation in the rate of acceleration with time, and upon the lack of uniformity or nonoccurrence of such effects for other comets. We will first discuss some of the observations, both for Comet Encke and for other comets.

Whipple next discusses observations of Encke's Comet and comes to the conclusion that the acceleration is variable and does not ever become negative. The Periodic Comet D'Arrest does, however, exhibit deceleration. In general, the angular change in the mean daily motion per revolution is approximately $5-10 \times 10^{-5}$ times the value of the mean daily motion.

The proposed comet model provides a possible mechanism for accelerating or decelerating the motion of a rotating comet. The gas escaping from the nucleus will leave with a velocity corresponding roughly to the mean speed of the gas molecules at the temperature of the surface layer of the meteoric blanket. The momentum of the escaping gas will exert a force on the nucleus. If there is an appreciable lag between the time of gas escape and the meridian passage of the sun with respect to the rotating nucleus, this force will possess a component perpendicular to the radius vector of the comet's orbit. The force may act in any direction, depending upon the direction of the axis and the sense of rotation of the nucleus.

Here Whipple assumes a spherical, rotating cometary nucleus and computes the mass loss per revolution that will explain the accelerations under the assumption of a lag in heat transfer. For Comet Encke he finds that the observed acceleration can be accounted for by a mass loss not exceeding one-fifth of 1% per revolution.

Effective Solar Attraction for Comets

If comets are losing material in the manner proposed in the preceding discussion, there is little question that the component of ejection will be greater statistically along the radius vector toward the sun than normal to it. Certainly, this will be true whenever the time lag in vaporization corresponds to less than an eighth period of cometary rotation, when the rotation is extremely slow or when irregular ejection occurs because of "cave-ins." In the case of extremely great time lags the H_2O ice may tend to freeze on the night hemisphere of the nucleus, while all gases will be adsorbed to a greater or lesser extent on the night hemisphere. Hence, statistically, the cometary nuclei will tend to emit material toward the sun. The phenomenon of the sunward ejection of material from cometary nuclei has long been recognized for the bright comets. Early drawings show the effect strikingly.

The sunward component of the ejection momentum will reduce the solar attraction for a cometary nucleus and affect its orbital motion. We may adopt the rough approximation that the quantity of gas ejected toward the sun by vaporization is proportional to the solar radiation flux and, therefore, inversely proportional to the solar distance. If we neglect the small variation with solar distance of the average speed of the ejected gas particles, the resultant force on the cometary nucleus varies according to the inverse-square law of solar distance.

Whipple derives an expression relating the expected effective reduction in the gravitational constant for comets to the rate of mass loss. He ends his article by noting that F. W. Bessel long ago showed that material streams out of Halley's Comet, that this mass loss must affect the orbital motion of the comet, and that the acceleration of Comet Encke might thereby be explained (*Astronomische Nachrichten 13*, nos. 3, 185, 345 [1836]). Whipple also mentions the work of A. D. Dubiago (*Astronomical Journal Soviet Union 25*, 361 [1948]), who attempted to explain the secular acceleration of five comets by the force of expulsion of solid particles.

1. *B.A.N. 10*, 445 (1948); see also H. N. Russell, *A.J. 33*, 49–61 (1921); and H. A. Newton, *Mem. Nat. Acad. Sci. Washington 6*, 1 (1893).
2. *Proc. Amer. Acad. Arts and Sci. 67*, 169 (1932). In a recent and extremely important paper, J. H. Oort (*B.A.N. 11*, 91 [1950]) has independently expanded the work of Öpik to demonstrate that an extended cloud of comets about the sun would be sufficiently stable, yet disturbed enough by passing stars to provide comets for Jupiter's capture after 3×10^9 years. The postulated cloud extends about 1 parsec about the sun.
3. *Bull. astr. 28*, 168 (1911).
4. *Mitt. Hamburger Sternw., Bergedorff*, vol. 8, no. 51 (1943).
5. *Rev. Mod. Phys. 14*, 164–178 (1942).
6. *Ann. d'ap. 11*, 124 (1948).
7. See, e.g., P. M. Millman, *Harvard Ann. 82*, 113 (1932), and *82*, 149 (1935); F. G. Watson, *Between the Planets* (Philadelphia: Blakiston Co., 1941), p. 108.
8. Private communication.
9. M. G. J. Minnaert, *Kon. Ned. Akad. Wet. Amsterdam 50*, 826 (1947).
10. Ibid.

Appended Figure

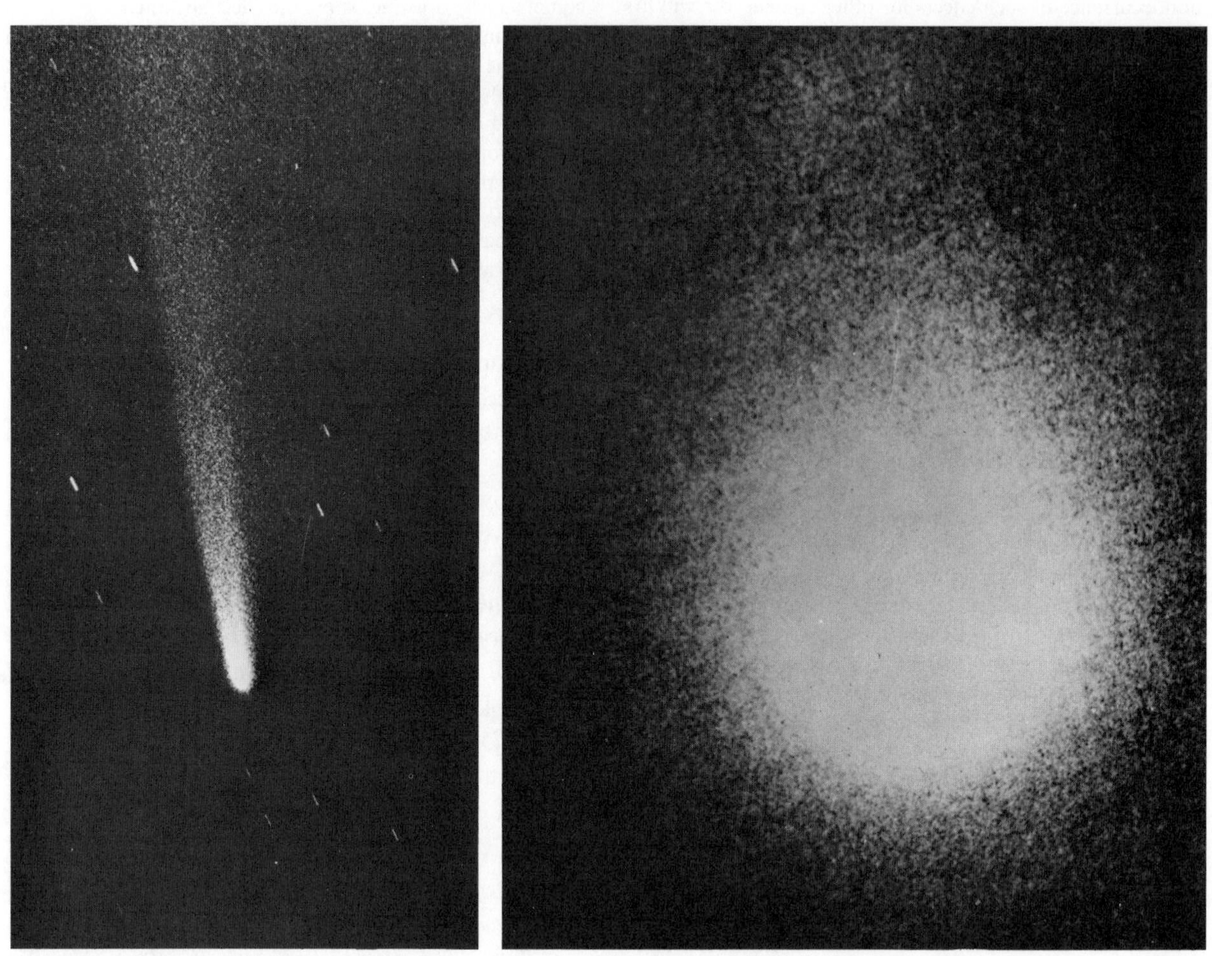

Fig. 23.1 A comparison of the visible light image (*left*) of Comet Kohoutek with a Lyman-α image (*right*) on the same scale. Both pictures were taken from an Aerobee rocket on January 5, 1974. The Lyman-α image shows the presence of a gigantic halo of hydrogen, nearly 10 million km in size, which is being fed by the cometary nucleus at the rate of 5×10^{29} atoms of hydrogen every second. Because the halo contains comparable abundances of hydrogen (H) and hydroxyl (OH), it is thought that the halo is the result of the photodissociation of water (H_2O) in the comet's nucleus. (Courtesy F. Whipple.)

24. Observations of a Variable Radio Source Associated with the Planet Jupiter

Bernard F. Burke and Kenneth L. Franklin

(*Journal of Geophysical Research 60*, 213–217[1955])

The discovery of intense radio emission from Jupiter is one of the many examples of the accidental discovery of an unexpected phenomenon while looking for something else. For example, thirteen years after the discovery of Jupiter's radio emission, Jocelyn Bell, Antony Hewish, Jack Pilkington, Paul Scott, and Robin Collins accidentally found the pulsars (see selection 74); in both instances the radio astronomers were observing radio wavelength radiation from other radio sources, and in both cases the periodic radio bursts were first attributed to interference.

What radiation mechanism accounts for the unexpected radio bursts from Jupiter? Hannes Alfvén and Nicolai Herlofson had argued that discrete radio sources shine by the synchrotron radiation of high-speed electrons spiraling about a magnetic field, and Iosif Shklovsky had argued that the Crab nebula would exhibit polarized radiation if it emitted synchrotron radiation (see selections 115 and 72). Polarized radio emission was subsequently detected from both the Crab nebula and Jupiter, suggesting the presence of magnetic fields in both objects. Consequently, George Field, Frank Drake, and S. Hvatum suggested that Jupiter's radio emission is caused by the synchrotron radiation of magnetically trapped electrons surrounding Jupiter.[1] Because of its fast rotation period and large size, Jupiter should have larger radiation belts than the terrestrial Van Allen belts, and early measurements of Jupiter's radio radiation suggested a magnetosphere which extends to fifty Jupiter radii.[2] Three findings fully vindicated the trapped electron hypothesis: interferometric studies showing that Jupiter's equatorial radio diameter is about 3 times that of the planet; the correlation of the occurrence of the radio bursts with the orbital position of Jupiter's satellite Io; and the demonstration by the *Pioneer X* and *XI* spacecraft that Jupiter has a strong dipole magnetic field with a surface field strength of about 14 gauss. Electrons with energies greater than 3 MeV were found to be trapped within Jupiter's magnetosphere at less than 20 Jupiter radii, lending additional credence to the view that the radio emission originates from the synchrotron radiation of high-energy electrons trapped within Jupiter's magnetic field.[3]

1. G. B. Field, *Journal of Geophysical Research 64*, 1169 (1959); F. D. Drake and S. Hvatum, *Astronomical Journal 64*, 329 (1959).
2. J. Warwick, *Astrophysical Journal 137*, 41 (1963).
3. V. Radhakrishnan and J. A. Roberts, *Physical Review Letters 4*, 493 (1960); E. K. Bigg, *Nature 203*, 1008 (1964); A. G. Opp, *Science 183*, 302 (1974).

Abstract—A source of variable 22.2-Mc/sec radiation has been detected with the large "Mills Cross" antenna of the Carnegie Institution of Washington. The source is present on nine records out of a possible 31 obtained during the first quarter of 1955. The appearance of the records of this source resembles that of terrestrial interference, but it lasts no longer than the time necessary for a celestial object to pass through the antenna pattern. The derived position in the sky corresponds to the position of Jupiter and exhibits the geocentric motion of Jupiter. There is no evident correlation between the times of appearance of this phenomenon and the rotational period of the planet Jupiter, or with the occurrence of solar activity. There is evidence that most of the radio energy is concentrated at frequencies lower than 38 Mc/sec.

DURING THE MONTHS of January, February, and March 1955, the large 22-Mc/sec "Mills Cross" of the Carnegie Institution of Washington was used for detailed observations in declination strips containing the Crab Nebula (M1). This antenna system, inspired by a design of Mills and Little[1] and utilizing the phase-switching principle first described by Ryle[2] will be discussed more fully in a subsequent paper.

The antenna may be briefly described as a pair of 64-dipole linear arrays, each 2,047 feet in length, arranged to form a slightly flattened X. The intersection of the fan-beams of the two arrays defines the effective beam position. By phasing the dipoles of the arrays, the direction of the fan-beams and hence the direction of the effective pencil-beam can be changed. Normally, the two arrays are phased identically, in which case the antenna is used as a meridian transit instrument. By tapering the dipole feeds in each array, all side-lobe responses have been reduced to roughly one one-hundreth that of the main beam. A slightly elliptical beam is produced, measuring $1^\circ\!.6 \times 2^\circ\!.4$ at half-power points, with the larger dimension in declination.

Here Burke and Franklin present a figure that illustrates the received signal when the radio sources IC 443 and M1 (the Crab nebula) pass through the antenna beam. This figure is not reproduced here because the traces on the right-hand side of figure 24.1 illustrate similar data for the Crab nebula.

Inspection of a number of records obtained at or close to the declination of the Crab Nebula revealed that on a number of occasions interference appeared for a brief time, the duration of which was about the same as the length of time required for a point-source at this declination to go through the antenna pattern. Six such occurrences are given in figure 24.1. In all cases, the bursts occurred during the night hours, at times when terrestrial interference was rare. In fact, of the 31 records obtained in this region of the sky, nine show the bursts at approximately the same sidereal time, while only one record showed interference of this intensity at any other time during the night. In view of such absence of interference at other times, it appears unlikely that terrestrial interference, approximately at the same sidereal time, should persist for nearly three months.

The noise bursts exhibited one puzzling feature: The dependence on sidereal time was not exact, but showed a slow drift in right ascension. The short vertical lines indicate our estimate of the position where the burst amplitude reaches approximately twice the noise level. In one case, March 10, which appears in figure 24.1 it was difficult to assign a position for the beginning. Not all side-lobes of the antenna have been suppressed, and it may be that the initial negative bursts represent unusually intense events occurring just before the source enters the main beam. We have chosen the time at which all bursts are clearly positive as the significant position, but the March 10 example is not as clearly defined as the other instances.

If a variable point-source passing through the beam is responsible for the observed signals, its true position should lie about half-way between beginning and ending of the disturbance, although the uncertainty would be greater than for a steady source of noise such as M1.

Here Burke and Franklin present a plot comparing the right ascension and duration of the bursts with the positions of Jupiter, Uranus, NGC 2420, and NGC 2392. This plot, not reproduced here, shows that the mean position of the disturbance coincides with Jupiter and that it cannot be caused by Uranus or any celestial object with a fixed right ascension.

The character of the bursts is interesting, for large fluctuations in intensity occur during the brief period while the source is in the antenna beam. The larger, more persistent peaks may result from the superposition of many shorter bursts, but this has not been clearly demonstrated as yet. The integrating time-constant of the receiver is 15 seconds, and, since several bursts sometimes occur within a minute, it appears that the duration of each smaller burst may be shorter than 15 seconds. Measurements with receivers having shorter time-constants should establish the single burst duration.

Although the peak intensity of the burst cannot be determined until we are certain that the receiver time-constant is considerably shorter than the burst duration, an order of

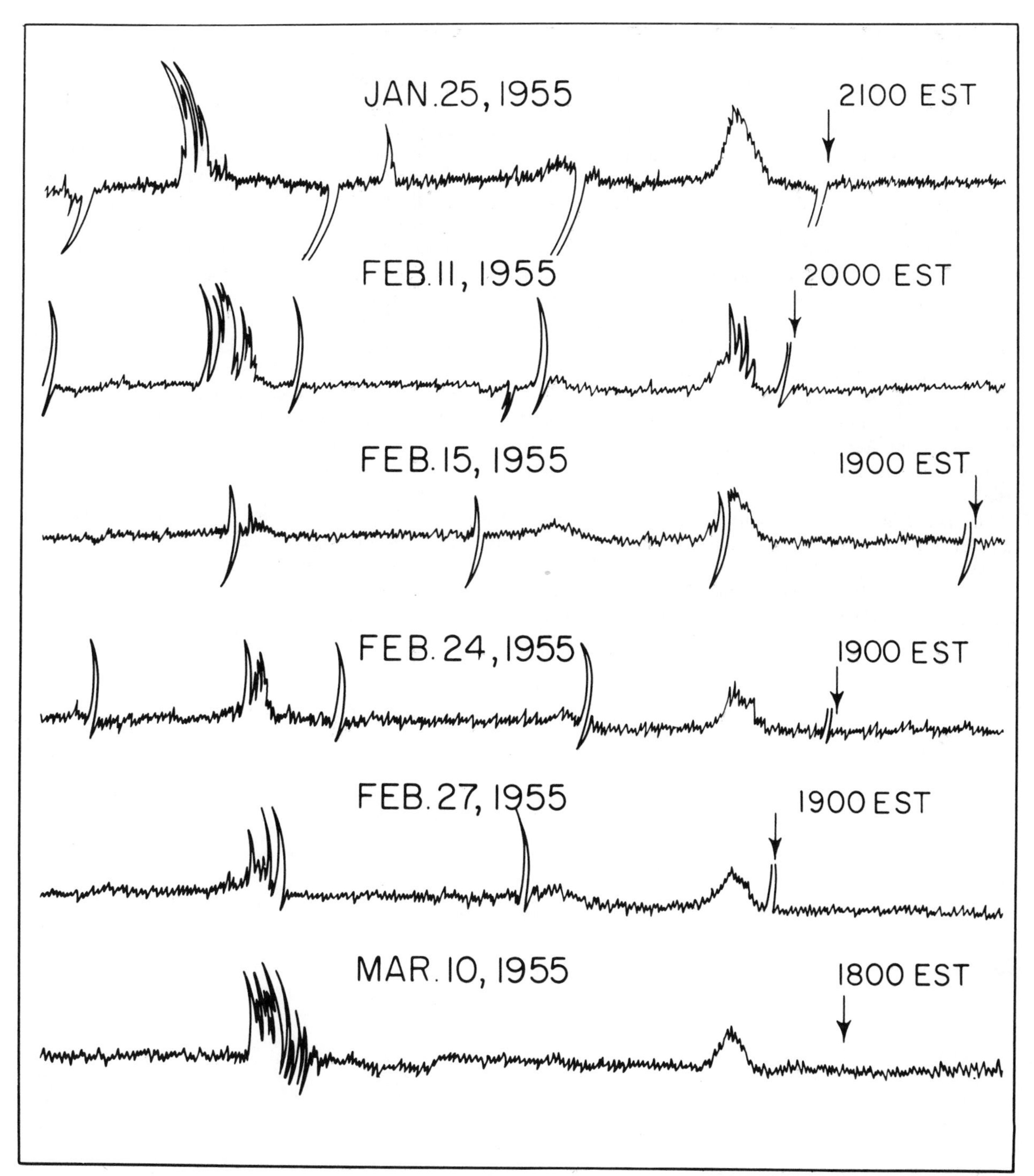

Fig. 24.1 Phase-switching records of the radio intensity at 22 MHz as a function of time for different days in 1955. The sporadic fluctuations at the left of each trace are attributed to Jupiter; the symmetric deflection at the right of each trace is caused by the Crab nebula. That the deflections at the left have the same total duration as those of the Crab nebula suggests that the source giving rise to the deflections is extraterrestrial; the changing time interval between the left- and right-hand deflections means that the unknown source cannot be a celestial object with a fixed right ascension. (Courtesy Bernard Burke.)

magnitude estimate can be made, comparing the apparent burst height to the trace of the Crab Nebula. Of the nine events seen so far, seven reach apparent peak intensities that exceed M1, and on two occasions, February 11 and February 15, the peak intensities remain less than M1 for the entire period of observation. It appears that rather violent outbursts are favored, reaching a peak intensity at least as great as that of the Crab Nebula. Our determination of the flux per unit bandwidth for M1 at this frequency (22.2 Mc/sec) is 5.2×10^{-23} w.m.$^{-2}$ sec, including both polarization components. If we assume that the bursts we associate with Jupiter are also randomly polarized, the peak intensities are at least as great as this, and occasionally are several times larger. If the bursts originate from Jupiter, which was between 4.25 and 4.67 astronomical units (AU) from the earth during the period of observation, and if we assume that the source is an isotropic radiator, we can estimate the peak power per unit bandwidth generated at this frequency. Taking a distance of 4.5 AU, for example, the peak power per burst must be at least as great as 300 watts per cps of bandwidth.

The possibility that the occurrence of the bursts was correlated with a region on the surface of Jupiter was investigated. No obvious periodicity related to the rotational period of the planet has been found. Furthermore, there does not appear to be any clear correlation with either radio or optical solar activity.

Information about the spectrum of the bursts is limited to a single observation obtained with interferometers having much lower antenna gain. During the month of June 1954, simultaneous interferometer records were obtained at frequencies of 22.2 and 38.7 Mc/sec. On one occasion, June 6, 1954, an intense radio source of fluctuating intensity was observed with the 22.2-Mc/sec interferometer, which consisted of two 4-dipole arrays, spaced 67.4 wavelengths along an east-west line. For nearly two hours, the source was prominent, moving through the interferometer lobes at a sidereal rate, with an intensity that was occasionally more than ten times greater than that of the Crab Nebula. From the rate of passage of the source through the interferometer lobes, and from the change of this rate with time, an approximate position was deduced which agreed with the position of Jupiter at that time, within the experimental uncertainty of 5°. At the same time, the 38.7-Mc/sec interferometer, which consisted of a pair of antennas having twice the gain of the lower-frequency arrays, failed to show any detectable trace of such a source. Since the Crab Nebula gave a trace having an amplitude 15 times the noise level, it would appear, from this single observation, that most of the radio noise is concentrated in the low-frequency range, below 38 Mc/sec.

1. B. Y. Mills and A. G. Little, *Aust. J. Phys. 6*, 272 (1953).
2. M. Ryle, *Proc. R. Soc.*, *A 211*, 351 (1952).

25. Solar Corpuscular Radiation and the Interplanetary Gas

Ludwig F. Biermann

(*Observatory 77*, 109–110[1957])

By the late 1940s it had been generally accepted that a hot, dense, million-degree solar corona pervades at least the innermost parts of interplanetary space and that corpuscular radiation from the sun causes the geomagnetic storms and the polar aurorae. Noting the general character of the observed motions in the ion tails of comets, which always stream away from the sun with velocities many times larger than could be caused by solar radiation pressure, Ludwig Biermann suggested that the ion tails are accelerated by a moving plasma of solar origin and proposed that the sun emits a continuous flow of solar corpuscles of the same type as those causing the geomagnetic storms, but weaker by some orders of magnitude and in all directions. He showed that the deviation of the ion tails' direction from the antisolar direction, until then unexplained, could be understood by this theory.[1]

In the paper given here Biermann argues that similar interactions must take place continuously between the solar corpuscular radiation and any component of the interplanetary plasma, to the extent that any observed interplanetary ionized gas outside comets must be identical with the solar corpuscular radiation. A year later Eugene Parker, considering the dynamic consequences of Biermann's suggestion that ionized gas is streaming outward in all directions from the sun at hundreds of kilometers per second, discovered that such a solar wind follows from the hydrodynamic equations of a million-degree solar corona.[2]

Owing to an erroneous but rather generally accepted identification of the particles causing the zodiacal light's polarization with electrons, the true density of the interplanetary plasma became known only several years later. Observations with interplanetary space probes have, however, now confirmed Biermann's basic idea. A continuous solar wind flows radially outward from the sun with velocities between 350 and 800 km sec^{-1}. The solar wind also contains a magnetic field whose average strength is about 0.00006 gauss; the flux amounts to some 10^8 ions cm^{-2} sec^{-1}, which is adequate to accelerate the ion tails.

1. L. Biermann, *Zeitschrift für Astrophysik 29*, 274 (1951), and *Mémoires de la société royale des sciences de Liège, 4éme sér. 13*, 291 (1953) (in English).
2. E. N. Parker, *Astrophysical Journal 128*, 664 (1958).

THE OBJECT of this letter is to draw attention to the fact that there are strong reasons against the assumption, made in some recent papers, that a stationary interplanetary gas exists which contributes, for example, to the polarized component of the zodiacal light.

1

The acceleration of the ion tails of comets (type I of Bredichin) has been recognized as being due to the interaction between the corpuscular radiation of the Sun and the tail plasma.[1,2] The observations of comets indicate that there is practically always a sufficient intensity of solar corpuscular radiation to produce an acceleration of the tail ions of at least about twenty times solar gravity (the mean value[3] being ~ 100, the maximum value several thousand times solar gravity). The mass density of the tail plasma should not be different by more than a few powers of ten from that of the interplanetary gas ($\sim 10^{-21}$ g/cm^3, at 1 A.U. near the plane of the ecliptic, from observations[4] of the polarization of the zodiacal light); near the head of the comet the mass density of the cometary plasma should be larger than 10^{-21} g/cm^3 in typical comets.[5] There is no reason why similar interactions should not be expected between the solar corpuscular radiation and a stationary interplanetary plasma; it follows that an assumed interplanetary cloud (which near 1 A.U. could only be ionized) would not remain stationary, since an additional acceleration of only a fraction of solar gravity would remove it before long.

2

The particle density of the solar corpuscular radiation is not very well known. Unsöld and Chapman[6] estimated 10^5 cm^{-3} at 1 A.U. in a very strong magnetic storm; values of the order of some 10^2 cm^{-3} for normal geomagnetically quiet conditions (when activity is observed only at polar stations) would be consistent with their estimate, and such values have been used in the work on comets quoted above.[1] They are also in agreement with the evidence from the polarization of the zodiacal light.[4] An independent check is provided by the circumstance that the ionization of the CO and N_2 molecules in the heads of comets is almost certainly produced by the solar corpuscular radiation,[2] since the alternative formerly discussed,[3] namely ionization by ultraviolet light, can now be strictly ruled out on the basis of the recent observations of solar ultraviolet radiation with rockets. The rate of development of ionization has been observed for some comets, and these observations are compatible with known collision cross sections and the density of solar corpuscular radiation quoted above.

3

Owing to the existence of large scale magnetic fields on the Sun—especially the polar-type field of about 2 Gauss at the poles observed by H. W. and H. D. Babcock[7]—the corona, including its invisible low-density extensions, must rotate together with the Sun, as was first pointed out by Alfvén[8] in 1942. Lüst and Schlüter's calculations of force-free magnetic fields[9] indicate, in a general way, the geometry of the magnetic fields around the Sun and the conditions to be fulfilled near the outer boundary of the co-rotating atmosphere (possibly near but inside Mercury's orbit). It follows from this, that in fact the co-rotating atmosphere in a certain sense belongs to the Sun, and may have a temperature of the order of that of the corona (10^6 degrees), whereas no considerable flow of heat by conduction should be expected across the boundary of the co-rotating atmosphere, owing to the influence of the magnetic field. The conditions of this atmosphere are locally, of course, continually disturbed by the corpuscular radiation, but their large scale features should restore themselves automatically.

4

If the identity of the solar corpuscular radiation and the interplanetary gas is accepted, one should expect fluctuations in the polarized component of the zodiacal light. Up to now the observations do not seem to allow any conclusion as to whether this expectation is confirmed or not. Suitable new observations would therefore seem to be highly desirable.

1. L. Biermann, *Zs. f. Ap. 29*, 274 (1951); and *Zs. f. Naturf. 7a*, 127 (1952).
2. L. Biermann, *Mémoires Soc. R. Sc. Liège*, 4th ser. *13*, 291 (1953).
3. K. Wurm, *Mitt. der Sternw. Hamburg-Bergedorf 8*, no. 51 (1943).
4. A. Behr and H. Siedentopf, *Zs. f. Ap. 32*, 19 (1953); and H. C. van de Hulst, *Mémoires Soc. R. Sc. Liège*, 4th ser. *15*, 89 (1954).
5. *C.f.* reference 3, above. Since the *f*-values were overestimated at that time by between one and two powers of ten, Wurm's values of the density have to be increased in that proportion.
6. A. Unsöld and S. Chapman, *The Observatory 69*, 219 (1949).
7. H. W. Babcock and H. D. Babcock, *Ap. J. 121*, 349 (1955).
8. H. Alfvén, *Ark. Mat. Astron. Fysik*, ser. A *28*, no. 6 (1942).
9. R. Lüst and A. Schlüter, *Zs. f. Ap. 34*, 263 (1954); *38*, 190 (1955).

26. Radiation Observations with Satellite 1958ε

James A. Van Allen, Carl E. McIlwain, and George H. Ludwig

(*Journal of Geophysical Research 64*, 271–286 [1959])

(Copyright © 1959 American Geophysical Union)

The existence of charged particles trapped in orbits in the earth's magnetic field was anticipated by Kristian Birkeland and Carl Störmer in their explanations for the aurorae and the geomagnetic disturbances.[1] In the Birkeland-Störmer theory, charged particles flowing from the sun are captured by the geomagnetic field and forced into magnetic reservoirs, where they remain trapped. These trapped particles then cause the visible aurorae. This basic hypothesis provided the framework for similar theories developed by S. Chapman and V. C. A. Ferraro, H. Alfvén, S. B. Treiman, and S. Singer.

The paper given here provided the first conclusive evidence for geomagnetically trapped, high-energy electrons and protons, in what is now called the inner Van Allen belt. These satellite observations showed an unexpectedly high flux of high-energy charged particles at altitudes above 2,000 km from the surface of the earth. Extrapolations of the observed data suggested a doughnut-shaped region of trapped particles, located in the earth's magnetic equator at a height of 1 terrestrial radius above the surface of the earth.

In order to determine the structure and extent of the radiation belt, Van Allen arranged for a rocket flight in the following year. The particle detectors aboard this rocket confirmed the existence of this radiation belt and also found another radiation belt about 2.5 terrestrial radii above the surface of the earth.[2] Both belts girdle the earth around the same geomagnetic equator, and they have roughly the same flux of high-energy particles. These belts are in turn enclosed within the magnetosphere—the cavity that the earth's magnetic field hollows out in the solar wind. On the side facing the sun, the magnetosphere is separated from the solar wind by a bow-shaped shock wave; a long magnetic tail flows away from the sun in the opposite direction.

1. K. Birkeland, *Comptes rendus de l'Académie des sciences* (Paris) *157*, 275 (1913); C. Störmer, *Archives des sciences physiques et naturelles* (Geneva) *24*, 317 (1907).

2. J. A. Van Allen and L. A. Frank, *Nature 183*, 430 (1959). For a detailed discussion see Van Allen's invited discourse in the *Transactions of the International Astronomical Union 11B*, 99–136 (1962).

Introduction

THE EXISTENCE of a high intensity of corpuscular radiation in the vicinity of the earth was discovered by apparatus carried by Satellite 1958α, launched at 0348 UT on February 1, 1958. The discovery was confirmed and knowledge of the distribution of radiation was greatly extended with more elaborate equipment in Satellite 1958γ, launched at 1738 UT on March 26, 1958. A preliminary account of this work has been given.[1]

The data from 1958α and 1958γ showed that:

(a) The intensity of radiation up to some 700 km altitude was in good accord with that to be expected for cosmic rays only, when proper account was taken of the increasing opening angles of geomagnetically allowed cones with increasing altitude and of the concurrent shrinking of the solid angle subtended at the observing point by the solid earth.

(b) Above some 1,000 km (this transition altitude being longitude and latitude dependent) the intensity of radiation increased very rapidly with increasing altitude, in a way totally inconsistent with cosmic ray expectations.

(c) At the higher altitudes ($\sim$ 2,000 km) the true counting rate of a Geiger tube with a geometric factor[2] $G_0 = 17.4\ \text{cm}^2$ and with total shielding of about 1.5 g/cm^2 of stainless steel (extrapolated range for electrons of energy 3 Mev or range for protons of 30 Mev) exceeded 25,000 counts per second. Hence the omnidirectional intensity exceeded 1,700 cm^{-2} sec^{-1} if the radiation consisted wholly of penetrating charged particles ($\varepsilon = 0.85$); or it exceeded some 10^8 cm^{-2} sec^{-1} if the radiation consisted wholly of electrons whose range was less than 1.5 g/cm^2 ($E < 3$ Mev) but whose bremsstrahlung was sufficiently energetic to penetrate the absorber with little attenuation ($E \gtrsim 50$ kev).

(d) Since the atmospheric path length between altitudes of, for example, 1,000 km and 700 km was negligible compared with the effective wall thickness of the counter, it was evident that the primary radiation was restrained from reaching lower altitudes by the earth's magnetic field and must therefore consist of *charged* particles.

It was proposed in our May 1, 1958, report[1] that the radiation was corpuscular in nature, was presumably trapped in Stoermer-Treiman[3] lunes about the earth, and was likely intimately related to that responsible for aurorae. On the basis of these tentative beliefs it was thought likely that the observed trapped radiation had originally come from the sun in the form of ionized gas which may or may not have been subjected to acceleration in the outer reaches of the earth's magnetic field, in some such manner as discussed by Chapman and Ferraro.

The existence of such radiation had been presaged by our earlier rocket observations in the northern and southern auroral zones.[4,1] Indeed, further specific experiments on the arriving auroral radiations has been planned by us[1] and by Bennett[5] for satellites in high inclination orbits; but it had not been anticipated that the "auroral soft radiation," as we originally called it, would be encountered at such low altitudes and low latitudes as was found with 1958α and 1958γ to be the case.

Here Van Allen, McIlwain, and Ludwig describe the satellite apparatus, which included plastic and crystal scintillators and Geiger tubes. The satellite was tracked by radio interferometric methods and by optical methods, providing data that enabled the position of the satellite to be specified to within 10 km. Sample sets of data are given, together with contours of constant counting rates as functions of altitude and latitude for the various detectors. Contours for the crystal scintillation detector are presented; figure 26.1 presents similar data from a later flight in the same year. Two detectors looking in dia-

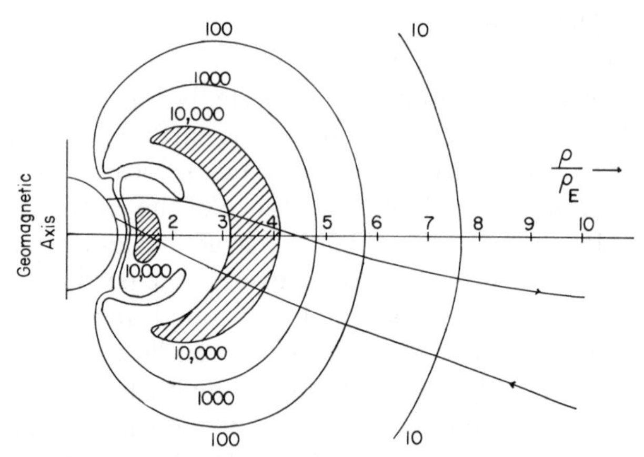

Fig. 26.1 Original diagram of the intensity structure of the trapped radiation around the earth. The diagram is a section in a geomagnetic meridian plane of a three-dimensional figure of revolution around the geomagnetic axis. Contours of intensity are labeled with the numbers 10, 100, 1,000, 10,000, which are the true counting rates of an Anton 302 Geiger tube carried by *Explorer IV* and *Pioneer III*. The linear scale of the diagram is relative to the radius of the earth—6,371 km. The outbound and inbound legs of the trajectory of *Pioneer III* are shown by the slanting, undulating lines. (From J. A. Van Allen and L. A. Frank, *Nature 183*, 430 [1959].)

metrically opposite directions had counting rates whose maxima and minima were in phase. This situation suggested that the angular distribution of the radiation is disklike in nature.

REMARKS ON INTERPRETATION

On the basis of the evidence presented above, we regard it as established that the great radiation belt around the earth consists of charged particles, temporarily trapped in the earth's magnetic field. The rapid diminution of intensity at the lower altitudes is almost certainly due to atmospheric scattering and absorption. The dominant loss mechanism for electrons is scattering. Thus, if an electron is trapped along a specified magnetic line-of-force, the 'guiding center' of its corkscrew paths spends most of its time near the points at which the velocity of the particle is nearly orthogonal to the magnetic line (the 'mirror' points). The cumulative effect of multiple coulomb scattering is to drive the mirror point to lower altitudes where the lifetime is rapidly reduced. If one adopts as a crude criterion for loss from the trapping region an rms scattering of one radian, then the 'lifetime' in the reservoir can be estimated as a function of mirror altitude.

Van Allen, McIlwain, and Ludwig estimate that the mean lifetime of a trapped 1-MeV electron is between 10^4 and 10^6 sec for mirror altitudes between 400 and 1,000 km and that the lifetime is proportional to the square of the electron's energy. If the volume of the trapping region is equal to that of a sphere of radius of 4 earth radii and if the average number density of energetic particles is 1 cm^{-3}, then about 10^{29} particles are present in the reservoir. Van Allen, McIlwain, and Ludwig then propose that the radiation belt is the reservoir whose leakage of particles is the direct cause of the visible aurorae, going on to argue that the radiation belt may well be the seat of a "ring" current encircling the earth and that perturbations of the belt because of the arrival of solar plasma may be directly responsible for magnetic storms.

1. J. A. Van Allen, "Study of the Arrival of Auroral Radiations," chap. 21, pp. 188–193, of *Scientific Uses of Earth Satellites* (Ann Arbor: University of Michigan Press, 1956). J. A. Van Allen, "Direct Detection of Auroral Radiation with Rocket Equipment," *Proc. Natl. Acad. Sci. U.S. 43*, 57–62 (1957). J. A. Van Allen, G. H. Ludwig, E. C. Ray, and C. E. McIlwain, "Observation of High Intensity Radiation by Satellites 1958 Alpha and Gamma," *IGY Satellite Rep. Ser no. 3: Some Preliminary Reports of Experiments in Satellites 1958 Alpha and Gamma*, pp. 73–92 (Washington, D.C.: Natl. Acad. Sci., 1958). Also *IGY Bull. Trans. Am. Geophys. Union 39*, 767–769 (1958a). J. A. Van Allen, G. H. Ludwig, E. C. Ray, and C. E. McIlwain, "Observation of High Intensity Radiation by Satellites 1958 Alpha and Gamma," *Jet Propulsion 28*, 588–592 (1958b).

2. V. Vouk, "Projected Area of Convex Bodies," *Nature 162*, 330–331 (1948).

3. S. B. Treiman, "The Cosmic-Ray Albedo," *Phys. Rev. 91*, 957–959 (1953).

4. L. H. Meredith, M. B. Gottlieb, and J. A. Van Allen, "Direct Detection of Soft Radiation above 50 Kilometers in the Auroral Zone," *Phys. Rev. 97*, 201–205 (1955).

5. Willard H. Bennett, "Proposed Measurement of Solar Stream Protons," chap. 22, pp. 194–197, of *Scientific Uses of Earth Satellites* (Ann Arbor: University of Michigan Press, 1956).

27. The Hot Surface Temperature of Venus

Observations of Venus at 3.15 cm Wave Length

Cornell H. Mayer, Timothy P. McCullough, and Russell M. Sloanaker

(*Astrophysical Journal 127*, 1–10 [1958])

Soft Landing of Venera 7 on the Venus Surface

V. S. Avduevsky, M. Ya. Marov, M. K. Rozhdestvensky, N. F. Borodin, and V. V. Kerzhanovich

(*Journal of Atmospheric Science 28*, 263–269 [1971])

(Copyright © 1971 American Meteorological Society)

Early evidence of an atmosphere on Venus included its high albedo, the absence of apparent surface features, the extension of the horns of the crescent into a complete ring when seen near inferior conjunction, and the polarization curve of its reflected sunlight. Eventually spectroscopic measurements established carbon dioxide as the principal constituent of the Venus atmosphere, and by 1924 the temperature of its outer layers had been measured at about the freezing point of water (273 K).[1] With remarkable foresight, Rupert Wildt noticed that carbon dioxide is so highly opaque to the surface radiation that it would produce a considerable greenhouse effect on Venus; the actual surface temperature could therefore be considerably higher. Here Wildt was building upon Frank Very's earlier suggestion[2] that the atmospheres of the major planets would allow optically visible sunlight to pass through to the ground, which would heat up and reradiate at infrared wavelengths. Because the atmospheres are opaque to the infrared spectrum, this radiation would be trapped beneath the atmosphere, where it could further heat the planetary surface.

As shown in the first paper given here, Wildt's assertion that the Venus surface might be hotter than its outer atmosphere was indeed correct. In 1958 Cornell Mayer and his colleagues reported that Venus has a microwave brightness at a wavelength of 3 cm corresponding to a temperature of around 600 K. Subsequent measurements of the microwave spectrum and other properties of Venus showed that this radiation could only be caused by emission from the hot surface of the planet.[3] Such a high surface temperature required a more efficient greenhouse than Wildt's pure carbon dioxide atmosphere provided, and Carl Sagan first suggested that small amounts of water vapor might accelerate the greenhouse effect.[4]

In late 1967 the Russians successfully launched a series of space probes that were to provide *in situ* measurements of the Venus atmosphere by entry probes. The first three successful Russian probes, *Veneras IV*, *V*, and *VI*, obtained direct measurements of the principal constituents of the atmosphere of Venus and found that it consists of $97 \pm 4\%$ carbon dioxide with 1.1% water vapor at a pressure level of 0.6 atm.[5] Both the *Venera IV* and the United States *Mariner* flyby mission gave additional evidence that the surface temperature of Venus was well above 500 K. This suggestion, which was consistent with the thermal interpretations of the microwave emissions, was dramatically confirmed by the *Venera VII* spacecraft. As explained in the second paper given here, an entry probe was parachuted into the hot, dense Venus atmosphere to measure the temperature and pressure profiles down to the Venus surface. Finding the temperature and pressure at the surface equal to 747 ± 20 K and 90 ± 15 kg cm^{-2}, the probe demonstrated that the carbon dioxide–water vapor greenhouse effect is a correct explanation.

1. E. Pettit and S. B. Nicholson, *Publications of the Astronomical Society of the Pacific 36*, 227 (1924).
2. F. W. Very, *Philosophical Magazine 16*, 462 (1908).
3. B. G. Clark and A. D. Kuz'min, *Astrophysical Journal 142*, 23 (1965). See also C. Sagan, *Nature 216*, 1198 (1967).
4. C. Sagan, *Astronomical Journal 65*, 352 (1960).
5. A. P. Vinogradov, Yu. A. Surkov, and B. M. Andreichikov, *Soviet Physics* (*Doklady*) *15*, 4 (1970).

Mayer, McCullough, and Sloanaker, Observations of Venus

Abstract—The observations of radiation from Venus at 3.15-cm wave length on 34 days in May-June, 1956, are described. The apparent black-body temperature for Venus derived from the measurements changed from about 620° ± 110° K (m.e.) in early May to about 560° ± 73° K (m.e.) near inferior conjunction. Two single observations at 9.4-cm wave length are described, which suggest that the radiation follows a thermal spectrum.

Introduction

THERMAL RADIATION from planets has not previously been investigated at radio wave lengths because of the small flux density of radiation at the earth. The strong bursts of long-wave-length radio noise which were first identified with Jupiter by Burke and Franklin (1955, 1956) and subsequently by others (Kraus 1956*a*; Shain 1956) and also later identified with Venus by Kraus (1956*b*, 1957) are presumably associated with some electrical phenomena in the atmospheres of the planets. The flux density of this impulsive radiation from Jupiter apparently falls off rapidly with decreasing wave length and has not yet been measured at wave lengths shorter than about 11 meters (Smith 1955). If the impulsive radiation from Venus has similar spectral characteristics, it should not be an important factor compared to thermal radiation at centimeter wave lengths.

Preliminary estimates indicated that thermal radiation from Venus at inferior conjunction should be easily detectable with the Naval Research Laboratory 50-foot reflector and a sensitive radiometer at a wave length of 3.15 cm. As a result, a series of observations was made during May and June, 1956, a period just prior to and including inferior conjunction, with the object of measuring the flux density of the radiation and the corresponding apparent black-body radiation temperature of Venus at this wave length. The observations were made on 34 days spread over the period May 2 to June 23, so that a variation of the radiating properties with time could also be tested.

Whether the radiation measured at 3.15 cm is all of thermal origin and whether the radiation originates in the atmosphere of Venus or at the surface cannot be deduced from the present measurements without additional data. Two single subsidiary observations of the radiation from Venus at 9.4-cm wave length are described in the appendix. These measurements suggest that the bulk of the received radiation has a spectrum similar to that of thermal radiation, but the precision of the spectral determination is low because of the extremely weak flux densities and the small number of observations at the longer wave length.

Apparatus

Here Mayer, McCullough, and Sloanaker explain that their receiving system is a superhetrodyne system that is switched between Venus and a comparison part of the sky at the rate of 30 times a second. (cf. Dicke 1946 and Mayer 1954, 1956). The temperature of the received signal was calibrated by the use of thermal noise sources. A 50-ft diameter antenna was used to give a beamwidth of about 0.14° at 3.15-cm wavelength. The aperture efficiency was measured to have the value of 0.56.

Observations

The observations of Venus were made by pointing the antenna in a fixed direction and allowing the rotation of the earth to scan the antenna beam through the position of Venus. As the distance between Venus and the earth decreased, the measured flux density increased. The observations are summarized in figure 27.1, where both the flux density, S, and the effective temperature, T, are given.

The flux density of unpolarized radiation, S, from a point source is related to the corresponding change in antenna temperature, Δt_a, during a drift-curve by the following formula, where k is Boltzmann's constant and A is the effective area of the antenna for reception:

$$S = \frac{2k\Delta t_a}{A}. \qquad (1)$$

The flux density in two polarizations at a wave length, λ, from a uniformly bright black-body radiator at a temperature, T, and subtending a solid angle, Ω, is given with sufficient accuracy for this experiment by the Rayleigh-Jeans approximation:

$$S = \frac{2kT\Omega}{\lambda^2}. \qquad (2)$$

The results of the radio observations of Venus and the results of other observations may be compared with reservation. Recently reported results of infrared radiometric observations by Pettit and Nicholson (1955) and by Sinton (cited from Menzel and Whipple 1955) indicate an apparent black-body temperature of about 235° or 240° K. These measurements probably refer to the top of the cloud layer on Venus. The rotational temperature derived from the CO_2 bands of Venus has recently been reported by Chamberlain and Kuiper (1956) as 285° K. The maximum temperature at the surface of Venus has been estimated by Wildt (1940) from solar heating considerations as 408° K.

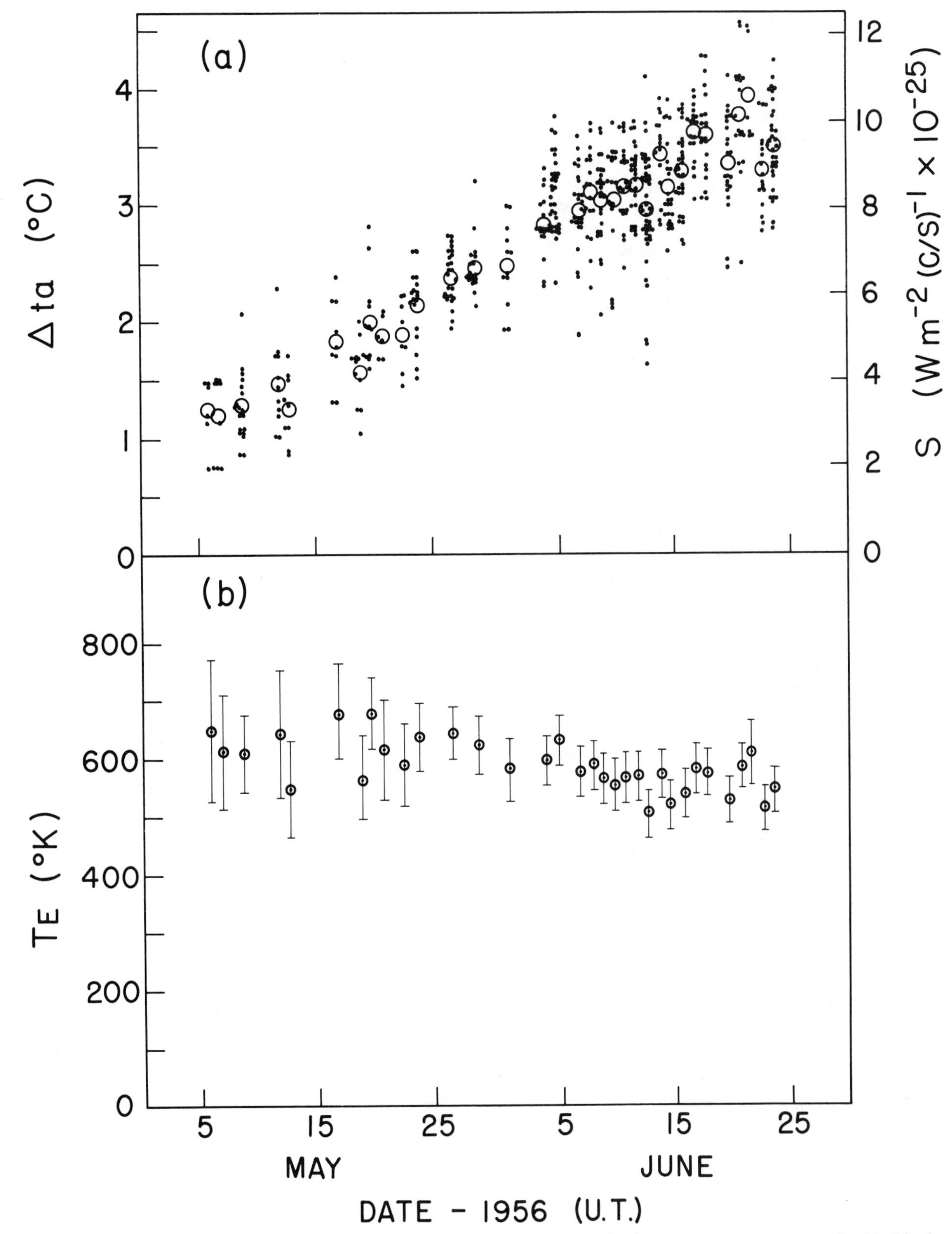

Fig. 27.1 Summary of the observations of Venus at a wavelength of 3.15 cm. In (*a*) individual measurements and daily averages of antenna temperature, $\Delta t\hat{a}$, and flux density, S, are given. In (*b*) the apparent temperature of a blackbody subtending the same solid angle as Venus is derived from the measured daily averages. (Courtesy Cornell Mayer.)

The radio observations indicate black-body temperatures considerably higher than these. Because of the uncertain nature of the atmosphere of Venus, no attempt will be made in the present discussion to assign the region of emission of the radio radiation, but it is undoubtedly different from that of the infrared. At least part of the 3.15-cm radiation may be emitted at the surface of the planet, and it is perhaps possible that part may arise in a heavily ionized atmosphere layer, although this seems unlikely, as an electron-density orders of magnitude higher than those found in the earth's atmosphere would probably be required. Also, the radio measurements are not yet complete enough to rule out the possibility that part of the observed radiation might be of nonthermal origin. Two single measurements at 9.4-cm wave length were an attempt to put rough limits on the radio spectrum, and the results tend to confirm thermal radiation.

Conclusions

The observed radio radiation from Venus at 3.15-cm wave length over a period covering nearly 2 months showed the characteristics of steady radiation with no apparent linearly polarized component to within the limits of accuracy. The measured flux density of the radiation approximated the inverse-square-law variation as the distance between Venus and the earth decreased but suggests a slight decrease in the radiation level during the period. The apparent black-body temperature of Venus deduced from the measurements was about 620° ± 110° K (m.e.) near the beginning of the period and about 560° ± 73° K (m.e.) near the end of the period. Two single observations at 9.4-cm wave length suggest that the bulk of the radiation follows a thermal spectrum, but, the accuracy of these measurements is low. A more complete series of measurements of the radiation from Venus at different wave lengths will be needed to define the spectrum.

Appendix

Measurements at 9.4-cm wavelength are described. Apparent blackbody temperatures of 430 K and 740 K were obtained at two different times.

Burke, B. F., and Franklin, K. L. 1955, *J. Geophys. Res. 60*, 213.

———. 1956, *A. J. 61*, 177.

Chamberlain, J. W., and Kuiper, G. P. 1956, *Ap. J. 124*, 399.

Dicke, R. H. 1946, *Rev. Sci. Instr. 17*, 268.

Kraus, J. D. 1956*a*, *A. J. 61*, 182.

———. 1956*b*, *Nature 178*, 33, 103, 159.

———. 1957. *A. J. 62*, 21.

Mayer, C. H. 1954. *J. Geophys. Res. 59*, 188.

———. 1956. *I. R. E. Trans. PGMTT 4*, 24.

Menzel, D. H., and Whipple, F. L. 1955, *Pub. A. S. P. 67*, 161.

Pettit, E., and Nicholson, S. B. 1955, *Pub. A. S. P. 67*, 293.

Shain, C. A. 1956, *Australian J. Phys. 9*, 61.

Smith, F. G. 1955. *Observatory 75*, 252.

Wildt, R. 1940, *Ap. J. 91*, 266.

Avduevsky et al., Soft Landing of Venera 7

Abstract—A soft landing on the planet Venus was successfully accomplished by the automatic interplanetary station Venera 7. The temperature of the Venus atmosphere was measured during the descent and at the surface after landing. The variation of temperature and pressure with altitude on Venus was determined down to the surface by combining the temperature measurements with descent velocity derived from the Doppler shift data during the descent, and by considering the data collected previously during the flights of Veneras 4, 5 and 6.

On 15 DECEMBER 1970 the automatic interplanetary station Venera 7 reached the planet Venus. At 08:37:32 Moscow time after a 35-min descent, the apparatus landed on the planet's surface, on the night side, ~2,000 km from the sunrise terminator.

The first radio signals transmitted by the descending apparatus were received at 08:02:50. Telemetry information was acquired throughout the entire period of descent through the atmosphere, from 08:02:50 to 08:37:32 and after landing on the surface. The apparatus continued to operate at the surface for ~23 min until 09:00:30.

The design and general structure of Venera 7 are essentially the same as the earlier stations, Veneras 4, 5 and 6, with, however, some important modifications (*Pravda*, 1971). Since the principal aim of Venera 7 was to land and operate on the Venus surface, the descending apparatus was designed as a landing vehicle. While the total weight of Venera 7 was unchanged (1,180 kg), the descending apparatus was about 100 kg heavier than those of Veneras 5 and 6. Taking into account the results obtained in the previous Venera flights and the subsequently developed model of the Venus atmosphere (Avduevsky *et al.*, 1970; Marov, 1971), the Venera 7 descent apparatus was designed to withstand a maximum external temperature of 800 K and a maximum pressure of 180 atm.

In addition, since ordinary thermal insulating materials lose their protective properties at such high pressure, new thermal insulation was developed which possessed the necessary qualities. This insulation also served as a damping device to decrease the shock and facilitate landing on the surface.

The descent apparatus of Venera 7 is oblong in shape. Inside of the outer aerodynamic body is a hermetically sealed, spherical container holding the radio-telemetry and measuring devices, the automatic control and thermal control systems, and the power supplies. The capsule is protected by both thermal insulation and by heat protection systems. The heat protection acts as a shield, protecting the apparatus from the high temperatures produced by the aerodynamic drag of the atmosphere. The thermal insulation maintains a favorable heat balance inside the apparatus during its operation in the hot atmosphere and on the planetary surface.

In the upper portion of the descent apparatus a parachute container is housed. For the Venera 7 station a single-cascade parachute system with an extraction parachute was used. The parachute was made of a thermostable material able to withstand temperatures of up to 800 K. To better withstand the heat inflow of the ambient atmosphere near the surface, the capsule was cooled near the end of the interplanetary phase of the mission so that at the moment of departing from the orbital module, the temperature inside the descent apparatus was between −6 and 8 C.

The descent apparatus was equipped with instruments to measure the temperature and pressure of the Venus atmosphere. Resistance thermometers and aneroid manometers, with a range of 25 to 540 C and 0.5 to 150 atm, respectively, were installed to measure these parameters.

The carrier frequency of the radio transmitters were stabilized by a thermostatted crystal oscillator. This assured the accuracy of the Doppler frequency shift determination and allowed the measurement of components of the velocity of the apparatus in the Venus-Earth direction. This direction corresponded to an angle equal to 10°20′ with the local vertical on Venus in the region of descent. Apparatus velocity measurements as a function of time provided the information necessary to determine height increments during the entire descent phase. This procedure is more accurate than the procedure used earlier to estimate wind velocities in the Venus atmosphere (Kerzhanovich *et al.*, 1970).

Measurements

Analysis of the telemetry data has shown that the on-board telemetry commutator, which was to scan the outputs of the measurement devices during the descent and after landing, remained in a fixed position. For this reason only the data on ambient temperature were transmitted. The rms error of temperature measurements, allowing for discrete telemetry readings, was ±20 K for the whole range.

The results of the temperature-time measurements with an indication of the error limits are shown in figure 27.2 along with the averaged vertical velocity. The latter was calculated

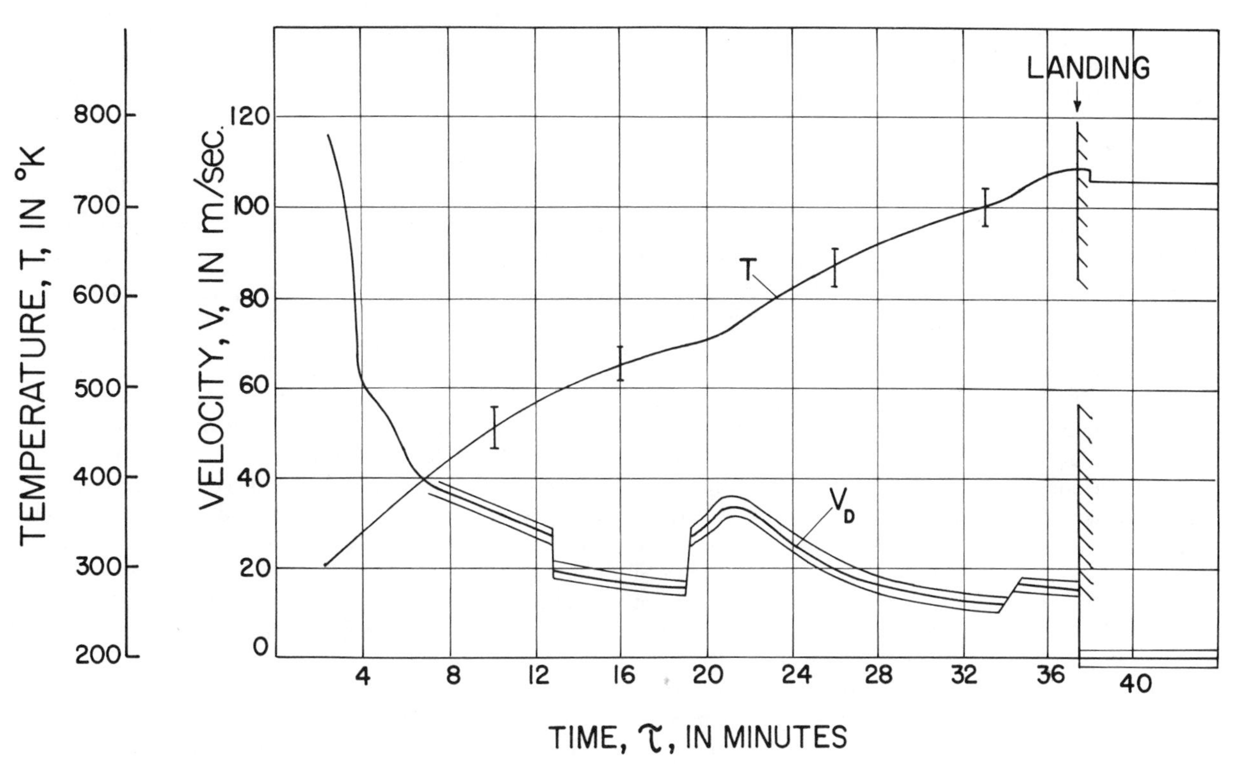

Fig. 27.2 The atmospheric temperature of Venus and the probe descent velocity (Doppler frequency shift) as a function of time during probe descent. Tolerance ±1.5 m sec^{-1}.

on the assumption that the horizontal components of the apparatus descent velocity were negligible during the descent. The thin lines bracketing the velocity curve correspond to the possible errors in the descent velocity (V_d) determination. These do not exceed ± 1.5 m sec^{-1}.

The open-loop method was used to measure the descent velocity of the apparatus in the Venus atmosphere. The calculated velocity resulted from the deviation of the frequency of the signal received on Earth from the basic frequency f_0 of the on-board crystal oscillator. The frequency f_0 was calculated by extrapolating frequency calibration data obtained during the interplanetary phase of the mission. A comparison of velocity measurements made by both open- and closed-loop methods was used for the calibration. Corrections for the change in the frequency of the crystal oscillator produced by the temperature increase in the apparatus during descent and the frequency change produced by the refractivity of the Venus atmosphere were taken into account in deriving the values of Doppler frequency f_D. The refractivity effect on the velocity values was estimated at less than 0.3 m sec^{-1} to the end of descent.

The velocity of the apparatus with respect to the surface was obtained by subtracting all the components associated with the relative motion of Venus and the receiving station on earth from the measured values of total velocity.

Here a figure, not reproduced, gives a record of the frequency of the received signal.

The parachute deployment at 08:02:50, when Venera 7 began its descent through the atmosphere, is indicated in figure 27.2 by the onset of a steady increase in temperature accompanied by a decrease in velocity. According to the program, at a temperature of ~ 470 K (at 08:13:03) the parachute opened, causing a reduction of the descent velocity from 27 to 19 m sec^{-1}. A constant parachute descent associated with increasing aerodynamic drag persisted for 6 min.

At 08:19:08, a sudden increase of velocity from 15 to 26 m sec^{-1} occurred, followed by a decrease, with the mean velocity exceeding that anticipated. This increase in velocity gave rise to a more rapid increase in the measured temperature with time than had been anticipated. Because of the velocity increase, the total descent time was 35 min, rather than the predicted time of 60 min. The descending apparatus touched the Venus surface at 08:37:32, travelling at a velocity of ~ 16.5 m sec^{-1}. On landing, the frequency shift of the received signal due to the apparatus descent velocity dropped to zero. The fact of the landing was also established by the equality of the apparatus velocity with the velocity obtained independently by means of the frequency jump at the moment of touching the Venus surface.

After the surface had been reached, the power of the radio signal decreased by about a factor of 100, probably due to the inclination of the apparatus and the resulting orientation of the antenna axis with respect to the direction of earth. The reduced-power signal received from the surface was subject to a harmonic analysis to determine its instantaneous spectrum, and then decoded according to a special program. The pertinent description of this procedure is given elsewhere.

Signals were received until 09:00:30, with telemetry data on atmospheric temperature from the Venus surface being transmitted for ~ 20 min, until 8:57:00. The temperature measurement upon landing was 747 K.

A constant temperature of 747 K was transmitted during the first 50 sec after landing, and a temperature of 730 K during the subsequent time. The difference between these two values is within the limits of telemetry sampling errors. The value recorded at the moment of touchdown should be taken as the most reliable figure. From this information, the temperature of the surface of Venus in the region of the Venera 7 landing is 747 K $\pm$ 20 K.

Here Avduevsky et al. use the descent velocity as a function of time to obtain the spacecraft altitude as a function of time. The temperature distribution as a function of altitude is then determined from the data shown in figure 27.2. The resulting profile is characteristic of an atmosphere that obeys the adiabatic law and is represented by a straight line whose slope is 8.6 K km^{-1}. Given the value of pressure at one point, chosen to be 10.1 kg cm^{-2} at the 500 K level, the variation of pressure with height is determined from the height distribution of temperature. An atmospheric abundance of 97% carbon dioxide and 3% nitrogen is assumed.

CONCLUSION

As a result of the Venera 7 flight, the first landing of a spacecraft on the planet Venus was achieved, and the temperature of the atmosphere was measured from a height of 55 km to the planetary surface. Temperature measurements continued to be taken for 20 min after the landing. The temperature-height profile $T(h)$ in the Venus atmosphere was obtained by combining temperature measurements with apparatus descent velocity measurements, the latter defined by means of Doppler shift frequency. The height-pressure profile $p(h)$ was then obtained by using the distribution of $T(h)$ indicated by Venera 7 and the results of measurements taken by Veneras 4, 5 and 6.

The distribution of temperature with height above the surface is close to adiabatic, apparently strengthening the hypothesis of strong convective mixing in the Venus atmosphere, and reducing the probability of the greenhouse mechanism for trapping solar radiation near the surface. Temperature on the surface of Venus in the region of the Venera 7 landing (at a distance of 2,000 km from the sunrise terminator) is 747 K $\pm$ 20 K. The height distribution of pressure was calculated by two methods, verifying that atmospheric pressure at the surface is between 86 and 97 kg cm^{-2}. Taking into account the errors of the T and V_D measurements and the $p(h)$ calculations, the most probable value of pressure at the surface is 90 $\pm$ 15 kg cm^{-2}.

The data obtained in the present experiment in the overlapping region agrees with the results of previous measurements made by Veneras 4, 5 and 6. The model of the Venus atmosphere that was calculated on the basis of the former measurements, assuming an adiabatic lapse rate in the regions of extrapolation to the planetary surface, is also in agreement with the data obtained. The surface level in the region of the Venera 7 landing coincides within 1–2 km of the mean surface level in the region of the Venera 5 and 6 probes. The results of Veneras 4–6 agree within 2–3 km with Mariner 5 atmospheric profiles (Kliore *et al.*, 1969) deduced from refractivity measurements above the level of critical refraction in the Venus atmosphere ($p \approx 4.5$ kg cm^{-2}). This comparison leads to a mean value of the Venus radius that is in reasonable agreement with the radar data [$R_♀ = 6050 \pm 5$ km (Ash *et al.*, 1968)].

From the analysis of the signal transmitted by Venera 7 at the moment of impact, it seems fairly certain that the surface of Venus is hard enough to bring a spacecraft to an abrupt stop.

Ash, M. E., Ingalls, R., Pettengill, G., Shapiro, I., Smith, W., Slade, M., Campbell, D., Dyce, R., Jurgens, R., and Thompson, T. 1968, "The Case for the Radar Radius of Venus," *J. Atmos. Sci. 25*, 560–563.

Avduevsky, V. S., Marov, M. Ya., and Rozhdestvensky, M. K. 1970, "A Tentative Model of the Atmosphere of the Planet Venus Based on the Results of Measurements of Space Probes Venera 5 and Venera 6," *J. Atmos. Sci. 27*, 561–568.

Kerzhanovich, V. V., Andreev, B. N., and Gotlib, B. M. 1970, "Investigation of Venus Atmosphere Dynamics by the Automatic Interplanetary Stations Venera 5 and Venera 6," *Dokl. Akad. Nauk SSSR 194*, No. 2.

Kliore, A., Cain, D., Fjeldbo, G., and Rasool, S. I. 1969, "Structure of the Atmosphere of Venus Derived from Mariner V S-band Measurements," *Space Research IX* (Amsterdam: North-Holland Publ. Co.).

Marov, M. Ya. 1971, "A Model of the Venus Atmosphere," *Dokl. Akad. Nauk SSSR 196*, no. 1.

Pravda. 1971, "Soviet Interplanetary Station on the Surface of Venus," no. 27, 27 January.

28. Radar Determinations of the Rotations of Venus and Mercury

Rolf B. Dyce, Gordon H. Pettengill, and Irwin I. Shapiro

(*Astronomical Journal 72*, 351–359 [1967])

World War II led to the accelerated development of radar (an acronym for radio detection and ranging). At the end of the war several radar groups, finding themselves with excess equipment and time, undertook studies of natural radar echoes. Proof that short scatter echoes came from meteor trails, for example, was provided by J. S. Hey and G. S. Stewart in 1946. Other war surplus radar equipment was used to bounce radio waves off the moon. In 1946 the United States Army Signal Corps Laboratory first recorded lunar echo signals, approximately 2.5 sec after transmission.[1] These crude investigations set the stage for a wide variety of radar measurements of the moon, culminating in a detailed map of the lunar surface. As first outlined by Paul Green and Gordon Pettengill[2] (and illustrated here in figure 28.2), the lunar surface can be divided into a circular annulus defined by a short echo pulse of a given time delay or range and into strips of different frequencies corresponding to different Doppler shifts. By analyzing the reflected signal for range and Doppler shift, the echoes from different parts of the surface can be mapped in detail.

The most spectacular results of the delay-Doppler mapping technique has been its extension to the nearby planets Mercury and Venus. These results, summarized in the paper given here, include the measurements of axial rotation rates, which show that Venus is rotating with a long retrograde period of about 245 days and that Mercury has a direct rotation in about 59 days. Because optical observations could not penetrate the featureless cloud cover of Venus, the rotation rate of Venus could not be measured before the advent of radar techniques. Most surprising was the discovery that Venus, unlike most of the other planets, does not rotate in the direction in which it orbits the sun. Optical observations of Mercury had purported to show that it always faced the sun throughout its orbit, apparently locked into synchronous rotation, much as the moon is forced by tidal torque always to face the earth. Once the radar results had shown that Mercury spins on its axis with a period two-thirds its orbital period, Giuseppe Colombo suggested that tidal torques might produce the commensurate rotation rate that would cause opposite sides of Mercury to face the sun on alternate perihelion passages.[3] (The optical observations used for recording surface markings of Mercury had, in fact, been made at alternate passages, and observations at other times had been systematically discarded!)

1. J. Mofenson, *Electronics Magazine 19*, 92 (1946). See also Z. Bay, *Hungarica Acta Physica 1*, 1 (1946).

2. P. E. Green, *Journal of Geophysical Research*, *65*, 1108 (1960). G. H. Pettengill, *Proceedings of the Institute of Radio Engineers 48*, 933 (1960).

3. G. Colombo, *Nature 208*, 575 (1965). G. Colombo and I. I. Shapiro, *Astrophysical Journal 145*, 296 (1966). Also see P. Goldreich and S. J. Peale, *Nature 209*, 1078 (1966).

Abstract—By measuring the frequency dispersion of time-gated radar echoes from Venus and Mercury at the Arecibo Ionospheric Observatory, the apparent rotation rates of these planets have been unambiguously determined. The combination of many such measurements made over a period of time has permitted a separation of the intrinsic rotation from the contribution of the relative orbital motion of the earth and the target planet. From observations over a three-month period surrounding the inferior conjunction of 1964, we find that Venus has a solar rotation period (i.e., with respect to the planet-sun line) of 117 ± 1 days and a sidereal rotation period of 245.1 ± 2 days (retrograde); its rotation axis has a declination of −66.7 ± 1 deg and a right ascension of 90.3 ± 1 deg (1950.0). The inclination of the axis is −89.8 ± 1 deg with respect to the ecliptic and −86.7 ± 1 deg with respect to the orbital plane of Venus. These results are consistent with Venus making four complete axial rotations as seen by an earth-based observer between inferior conjunctions. For such a retrograde rotation a sidereal period of 243.16 days is required. Data obtained near the inferior conjunctions of Mercury in April and August 1965 indicate that the axial rotation of Mercury is direct and has an average solar period of 176 ± 9 days and a sidereal period of 59 ± 3 days. The direction of the rotation axis, although not well determined by these data, seems to be inclined to the normal to Mercury's orbital plane by less than 28°. These results are consistent with Mercury's axial rotation being locked to its orbital motion such that the sidereal period of the former is exactly two-thirds of that of the latter.

Enhanced scattering from distinct parts of the Venus surface have also been detected and a corresponding contour map of the surface radar reflectivity is presented.

INTRODUCTION

OPTICAL OBSERVATIONS have not provided useful information on the rotation of Venus because the surface is always hidden behind a largely opaque and featureless cloud cover. The only reliable quantitative determinations of the direction and rate of rotation of Venus have resulted from radar measurements. The first successful Venus radar experiments, conducted in 1961 by the Jet Propulsion Laboratory (Victor and Stevens 1961) and by MIT Lincoln Laboratory (Staff, Millstone Hill Radar 1961), indicated that the rotation period was very long, but were not accurate enough to determine the sense of rotation. Smith (1963) did, however, note the possibility of retrograde rotation.

More sensitive radar observations of Venus performed at JPL in 1962 resulted in two separate determinations of the rotation rate. One (Carpenter 1964) was based on following from day to day the spectral position of a "feature" on the surface of Venus; the other (Goldstein 1964; Carpenter 1964) made use of frequency dispersion techniques. The results were consistent and showed the rotation to be retrograde with a period of about 250 ± 40 days and an axis nearly normal to the ecliptic. Although omitting details, Kotelnikov *et al.* (1963) also reported a retrograde rotation with a period between 200 and 300 days, based on their 1962 measurements.

The 1964 measurements at JPL were consistent with the rotation rate results obtained there in 1962 but were considerably more accurate; Goldstein (1965) reported a period of 250 ± 9 days with the axis direction having a declination $\delta = -62 \pm 4$ deg and a right ascension $\alpha = 81 \pm 6$ deg. Carpenter (1966) concluded that the period lay between 244 and 254 days with the direction of the axis having coordinates: $\delta = -68 \pm 4$ deg and $\alpha = 75(+10, -4)$ deg. Measurements made at Jodrell Bank in 1964 also confirmed the retrograde aspect of the rotation and yielded a rotation period between 100 and 300 days (Ponsonby *et al.* 1964). Similarly, the 1964 Lincoln Laboratory spectral measurements are consistent with a retrograde rotation of Venus (Evans *et al.* 1965). The 1964 radar measurements of Venus made at the Arecibo Ionospheric Observatory of Cornell University, which are reported here in detail, were presented in preliminary form by one of us (Shapiro 1964) to the International Astronomical Union. The more refined results indicate a rotation period of 245.1 ± 2 days with the axis having a declination of −66.7 ± 1 deg and a right ascension of 90.3 ± 1 deg (mean equinox and equator of 1950.0), in reasonably good agreement with the JPL results.

The planet Mercury is not encumbered with a veil of clouds, as is Venus, and optical observations have been carried out from the early 19th century. The first sustained observations by Schröter were analyzed by Bessel (1813), who claimed that the rotational period was in the vicinity of 24 h. Later, however, Schiaparelli (1889) concluded from his extensive observations that Mercury was rotating slowly with a period most likely synchronous with its orbital period of approximately 88 days. Antoniadi (1934), Dollfus (1953) and others also found the period of rotation to be 88 days from analyzing their visual and photographic observations.

Early radar measurements by Carpenter and Goldstein (1963) seemed to confirm the 88-day rotational period. These measurements, however, were based on spectra integrated over all ranges, and, in view of the limited signal-to-noise ratio were possibly subject to considerable systematic error in interpretation. The 1965 radar measurements of Mercury, discussed below in detail, indicate that its rotation is direct with a period of 59 ± 3 days. (A preliminary report, based on some of these data, was published previously by Pettengill and Dyce 1965.)

After the disclosure that Mercury's axial period was substantially shorter than its orbital period, several groups undertook a re-examination of the optical measurements (McGovern, Gross, and Rasool 1965; Colombo and Shapiro 1965). Except for some of Schiaparelli's data, all groups of optical observations published had been made at intervals separated by very nearly even multiples of Mercury's orbital period and hence were unable to distinguish between a rotation value of 88 days and the new radar-determined value of 59 days (i.e., two-thirds of the orbital period), as pointed out by Colombo and Shapiro (1965). More recently Dollfus (1966) has revealed that some previously unpublished optical observations are definitely inconsistent with an 88-day rotation period but are consistent with 59 days, apparently confirming the radar value.

Theory of Rotation Determination

Before discussing the data reduction in detail we review briefly the fundamental principles involved in the radar determination of planetary rotations. The mathematical theory is given by Shapiro (1967).

Most of the echo energy returned by a planet to the observing radar results from quasi-specular reflection from areas on the planet that are perpendicular to the incident radar waves. For a smooth sphere these areas are located near the intersection with the planetary surface of the line joining the radar site and the center of the planet. (This intersection is commonly called the subradar point on the target.) The rough elements on the surface, which are comparable in size to the illuminating wavelength, give rise to weaker but detectable echoes at greater delays because of their contribution to the energy that is scattered back in the direction of the radar even at positions where the mean surface of the planet is not normal to the direction of incidence. Each element of area on the surface will, therefore, return an echo characterized by a particular value of time delay and also by a particular value of Doppler frequency shift, the latter depending on the component of velocity of the element along the direction of propagation of the incident wave. These facts underlie the fundamental principle of delay-Doppler (often called "range-Doppler") mapping (Green 1960).

This principle may, perhaps, be pictured more clearly by considering a very short radar pulse reflected from a distant, rotating planet. At any particular instant slightly later than its time of arrival at the subradar point on the planet, the pulse illuminates a particular annular region centered on the subradar point. Reflections from each such annulus, or ring, are, therefore, associated with a particular value of time delay. To determine the association between Doppler shifts and surface points, we note that for a rigid target all surface points that lie in a plane parallel to both the target's apparent axis of rotation (see below) and the radar line of sight have the same component of velocity in the direction of the radar. Each strip where such a plane intersects the target's surface is, therefore, associated with a unique Doppler shift. Furthermore, the difference between the Doppler shift associated with the subradar point and that which characterizes each strip is proportional to the plane's displacement from the target's center. By combining these two ways of associating points on the surface of the target with energy received at a particular time delay and with a particular Doppler shift, we can produce a delay-Doppler map of the surface. This technique was employed in the work described in the following sections.

A twofold ambiguity in mapping from delay-Doppler coordinates on to the actual surface exists because pairs of points symmetrically placed with respect to the target's apparent equator have identical delay and Doppler coordinates (see figure 28.2). For this reason, along the equator the Jacobian of the transformation from usual surface coordinates to delay-Doppler coordinates is singular. This peculiarity proves to be of enormous benefit for the application to the rotation rate determinations. The power associated with the instantaneous backscattered reflections from a given annulus will be greatest (as will the signal-to-noise ratio) at the maximum and minimum frequencies represented since these correspond to the two regions (on the apparent equator of the target) where the Doppler strip and the delay annulus are tangent to each other, and, therefore, enclose the greatest surface area within the bounds of the given delay-Doppler resolution. The bandwidth (i.e., the difference between the values of the maximum and minimum frequencies represented) is directly proportional to the rate of rotation of the planet as seen from the radar. This apparent angular velocity of rotation is composed of two parts: (1) the contribution of the intrinsic, or sidereal, rotation of the planet, and (2) the contribution attributable to the relative motion of the radar site and the center of the planet. The bandwidth is thus proportional to the projection on a plane perpendicular to the radar line-of-sight of the vector sum of the intrinsic angular velocity of rotation of the planet and the angular velocity associated with the relative orbital motion. As the orbital positions change, not only does the relative orbital angular velocity vector change, but the vector sum of it and the intrinsic angular velocity is projected on a plane of differing orientation. By making a series of bandwidth measurements from different relative orbital positions, all ambiguities are resolvable (in principle, at least) and all three scalar parameters associated with the axial rotation of the planet can be estimated by using the method of weighted-least-squares.

Radar Characteristics and Method of Data Reduction

Here Dyce, Pettengill, and Shapiro describe the characteristics of the 1,000-ft-diameter Arecibo antenna, which at the signal frequency of 430 MHz transmitted a peak power of 2 million watts into a beam whose half-width was 9 min of arc. The power was transmitted in a train of pulses, each of between 0.1 and 1.0 millisec duration, for a time period slightly longer than the round-trip travel time to the planet, and the receiving system was then turned on for a 30-sec time period. More complete descriptions of the data-taking procedures are given by

Pettengill, Dyce, and Campbell (1967), Pettengill (1960), and Pettengill and Henry (1962). The spectral dispersions, or bandwidths, of the received pulses were then used to specify the rotation period of the planet, using the technique discussed by Shapiro (1967).

Venus Observations

Omitted here is a figure showing that the residual difference between the observed bandwidth and the "best-fit" rotation period for Venus amounts to at most ± 0.4 c/sec.

The rotation period for this solution is 245.1 ± 0.7 days, while the direction of the rotation axis has $\delta = -66.7 \pm 0.4$ deg and $\alpha = 90.3 \pm 0.6$ deg (referred to the mean equinox and equator of 1950.0). The axis is almost normal to the ecliptic having an inclination to this plane of -89.8 ± 0.4 deg. The corresponding inclination to the orbital plane of Venus is -86.7 ± 0.4 deg. The errors quoted here are the standard deviations obtained formally from the covariance matrix that follows from the measurement error standard deviations. Since a known, although small, source of systematic error [i.e., the convolutional effects of the finite pulse length and the receiver filter characteristics (Shapiro 1967)] was not included, we consider it prudent to increase the error associated with the period to 2 days and those associated with the angle determinations to about 1 deg.

Our result for the rotation period is close to the value of 243.16 days for which Venus would make on average four complete revolutions between inferior conjunctions as seen by an earth-based observer. We therefore computed the residuals based on this value and found them to be clearly larger than those shown, although the differences, as intimated above, may not be significant.

The presentation does not exhibit the time variation of the bandwidths caused by the relative orbital motion of the radar site and Venus. Therefore we show in figure 28.1, as a function of date, the extrapolations of the measured bandwidths to ones that would have been measured had each of the annuli been at the limb of Venus. The corresponding best-fit curve is also included. The dip near inferior conjunction implies that the contribution to the apparent angular velocity arising from the relative orbital motions of earth and Venus must be opposing that of the sidereal spin angular velocity of Venus and indicates clearly that Venus is rotating in a retrograde direction.

The theoretical model used in the above analyses assumed the radius ρ of Venus to be 6,089 km (de Vaucouleurs and Menzel 1960), and all motions were considered, including the movement of the radar site with respect to the center of the earth. Any error in ρ does not affect the result greatly since the result depends essentially only on the square root of ρ. Of course, ρ can also be estimated from the bandwidth data; a preliminary attempt indicated that the error would exceed 200 km, and hence the estimate would not be useful.

Mercury Observations

During August, both 0.5 and 0.1 msec pulses were used; with the latter it was possible to observe at delays up to 0.4 msec. Although the frequency spreads in this case are smaller, they are nevertheless well resolved and their greater number offered an excellent opportunity to verify internal

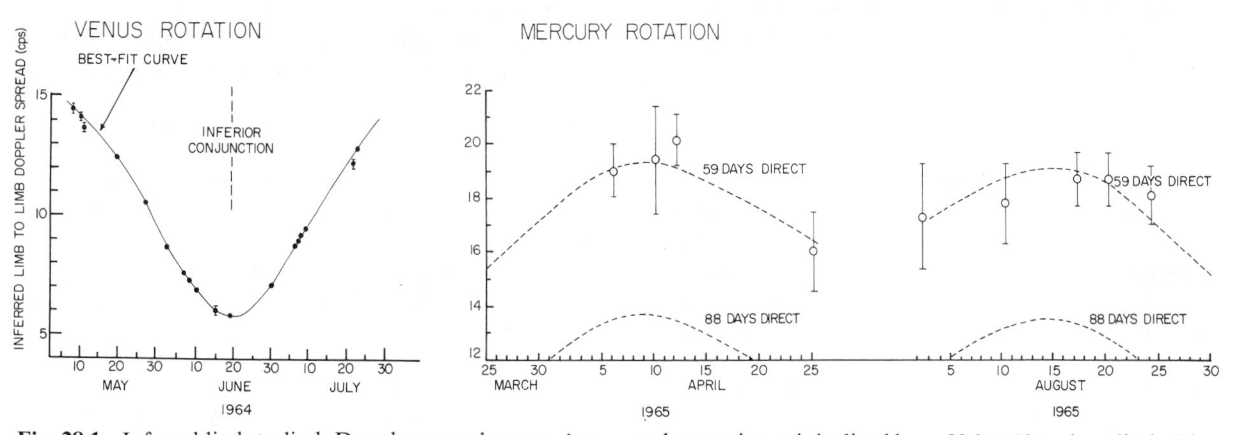

Fig. 28.1 Inferred limb-to-limb Doppler spread versus date for observations of Venus and Mercury taken at the Arecibo Ionospheric Observatory during 1964 and 1965. The solid curve represents the least-mean-square fit to the data for Venus and corresponds to a rotation period of 245.1 ± 2 days and a rotation axis inclined by $-89.8 \pm 1°$ to the ecliptic. The dashed curves show the theoretical variation for Mercury with direct sidereal periods of rotation of 59 and 88 days, under the assumption that the axis of rotation is normal to the orbital plane of Mercury.

consistency. Data obtained using both pulse widths agreed well, as shown in figure 28.1, where the extrapolated values for limb-to-limb Doppler bandwidth are plotted vs date of measurement for all the 1965 observations at AIO. In the calculations the radius of Mercury was assumed to be 2,420 km (Allen 1964). The measurement errors indicated were obtained from a consideration of the internal consistency of the results for a given day and the estimated accuracy in determining the positions of the wings of the spectra.

Mercury's rotation vector could be estimated from the inferred limb-to-limb Doppler bandwidths as mentioned before. Because of the smaller arc of the orbit over which observations were made, however, the determination of the axis direction would not be nearly so accurate for Mercury as for Venus. Therefore, at first the axis was constrained to lie perpendicular to the plane of Mercury's orbit. The sum of the squares of the differences between the measured and calculated values was then determined for a number of possible values for the rotation period. This sum was found to be a minimum for an assumed period of 59 days and to double for periods of 56 and 62 days. With the constraint on the axis direction removed, a period of 56 days and an axis direction 28° from the normal to the orbital plane were obtained; the sum of the squares of the residuals was nearly the same as for the constrained 59-day case. Thus, we have concluded from these data that the axial rotation period of Mercury is 59 ± 3 days and that its axis is pointed within 28° of the normal to its orbital plane.

Venus Surface Markings

Dyce, Pettengill, and Shapiro then mention that nonuniformities in echo strength suggest Venusian features extending to 3,800 km. We have appended a recent radar map of the surface of Venus as figure 28.3.

Discussion and Conclusions

The rotation period found for Venus suggests that its axial motion may be controlled by the earth since retrograde rotation with a period of 243.16 days implies that Venus presents the same "face" to the earth at every inferior conjunction. It is hard to understand both how such a rotational state was reached and how it could be dynamically stable. Several groups (e.g., Goldreich and Peale 1966; Shapiro 1966; and Bellomo *et al.* 1966) have studied the dynamics of the rotation but none has found a satisfactory solution.

The value of 59 ± 3 days for the axial rotation period of Mercury, although not nearly so well determined as that of Venus, appears to be quite consistent with almost all of the optical data (McGovern, Gross, and Rasool 1965; Colombo and Shapiro 1965; Dollfus 1966) and with theoretical analyses (e.g., Peale and Gold 1965; Colombo and Shapiro 1965). In fact, Colombo (1965) pointed out the possibility that Mercury's rotation period might be 58.65 days, i.e., exactly two-thirds of its orbital period. This value of the rotation period would apparently make Mercury unique in the solar system in having its axial motion locked to its orbital motion in such a manner. The dynamic stability and the possible evolution of Mercury's axial rotation are actively being investigated by many groups. The key element in the preferred status of the two-to-three relation is the large eccentricity of Mercury's orbit. This eccentricity allows the solar tidal torque to change sign during an orbital revolution (Peale and Gold 1965) and leads to an asymptotically stable spin state (Bellomo *et al.* 1966).

Note added in proof: Using a radius of 6,055 km for Venus, as found by Ash *et al.* (1967), we obtain a sidereal rotation period of 244.3 ± 2 days with an axis direction specified by $\delta = -66.4 \pm 1$ deg and $\alpha = 90.9 \pm 1$ deg (1950.0).

Allen, C. W. 1964. *Astrophysical Quantities* (London: Athlone Press), p. 142.

Antoniadi, E. M. 1934, *La planète Mercure et la rotation des satellites* (Paris: Gauthier-Villars).

Ash, M. E., Shapiro, I. I., and Smith, W. B. 1967, *Astron. J. 72*, 338.

Bellomo, E., Colombo, G., and Shapiro, I. 1966, "Theory of the Axial Rotations of Mercury and Venus," paper presented to Conference on the Mantles of the Earth and Terrestrial Planets, Newcastle, England (March).

Bessel, F. W. 1813, *Berl. Astron. Jahrb.*, p. 253

Carpenter, R. L. 1964, *Astron. J. 69*, 2.

———. 1966, ibid. *71*, 142.

Carpenter, R. L., and Goldstein, R. M. 1963, *Science 142*, 381.

Colombo, G. 1965, *Nature 208*, 575.

Colombo, G., and Shapiro, I. I. 1965, *Smithsonian Astrophys. Obs. Spec. Rept.* no. 188; see also 1966, *Astrophys. J. 145*, 296.

de Vaucouleurs, G., and Menzel, D. H. 1960, *Nature 188*, 28.

Dollfus, A. 1953, *Bull. Soc. Astron. France 67*, 61.

———. 1966, unscheduled paper presented to COSPAR, Vienna (May).

Evans, J. V., Brockelman, R. A., Henry, J. C., Hyde, G. M., Kraft, L. G., Reid, W. A., and Smith, W. W. 1965, *Astron. J. 70*, 486.

Goldreich, P., and Peale, S. J. 1966, *Nature 209*, 1117.

Goldstein, R. M. 1964, *Astron. J. 69*, 12.

———. 1965, *J. Res. Natl. Bur. Stds. Radio Sci. 69D*, 1623.

Green, P. E., Jr. 1960, *J. Geophys. Res. 65*, 1103.

Kotelnikov et al. 1963, *Dokl. Akad. Nauk 151*, 532.

McGovern, W. E., Gross, S. H., and Rasool, S. I. 1965, *Nature 208*, 375.

Peale, S. J., and Gold, T. 1965, ibid. *206*, 1240.

Pettengill, G. H. 1960, *Proc. IRE 48*, 933.

Pettengill, G. H., and Dyce, R. B. 1965, *Nature 206*, 1240.

Pettengill, G. H., Dyce, R. B. and Campbell, D. 1967, *Astron. J. 72*, 330.

Pettengill, G. H., and Henry, J. C. 1962, *J. Geophys. Res. 67*, 4881.

Ponsonby, J. E. B., Thomson, J. H., and Imrie, K. S. 1964, *Nature 204*, 63.

Schiaparelli, G. V., 1889, *Astron. Nachr. 123*, 241.

Shapiro, I. I. 1964, "Radar Determinations of Planetary Motions," paper presented to XII General Assembly IAU, Hamburg (Sept.).

———. 1966, "Planetary Radar Astronomy," Proc. VIIth International Space Science Symposium, Vienna (May).

———. 1967, *Astron. J. 72*, 1309.

Smith, W. B. 1963, ibid. *68*, 15.

Staff, Millstone Radar. 1961, *Nature 190*, 592.

Victor, W. K., and Stevens, R. 1961, *Science 134*, 46.

Appended Figures

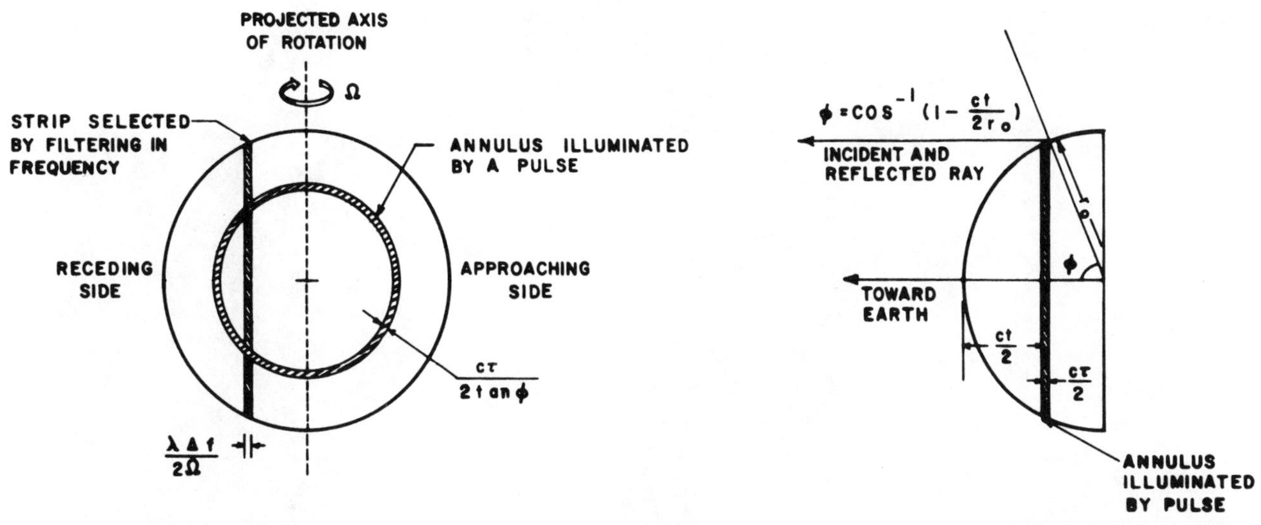

Fig. 28.2 The disk and near hemisphere of a planet, illustrating the annulus of constant range delay t for a pulse of duration τ; the strip selected by filtering in frequency with a resolution Δf at a frequency determined by the wavelength, λ; and the instantaneous apparent angular velocity, Ω, of the planet. The planet radius is r_0, and the angle of incidence and reflection is $\phi = \cos^{-1}[1 - ct/(2r_0)]$. The range delay, t, is the time after the pulse first strikes the surface. (From K. R. Lang, *Astrophysical Formulae* [New York: Springer Verlag, 1974].)

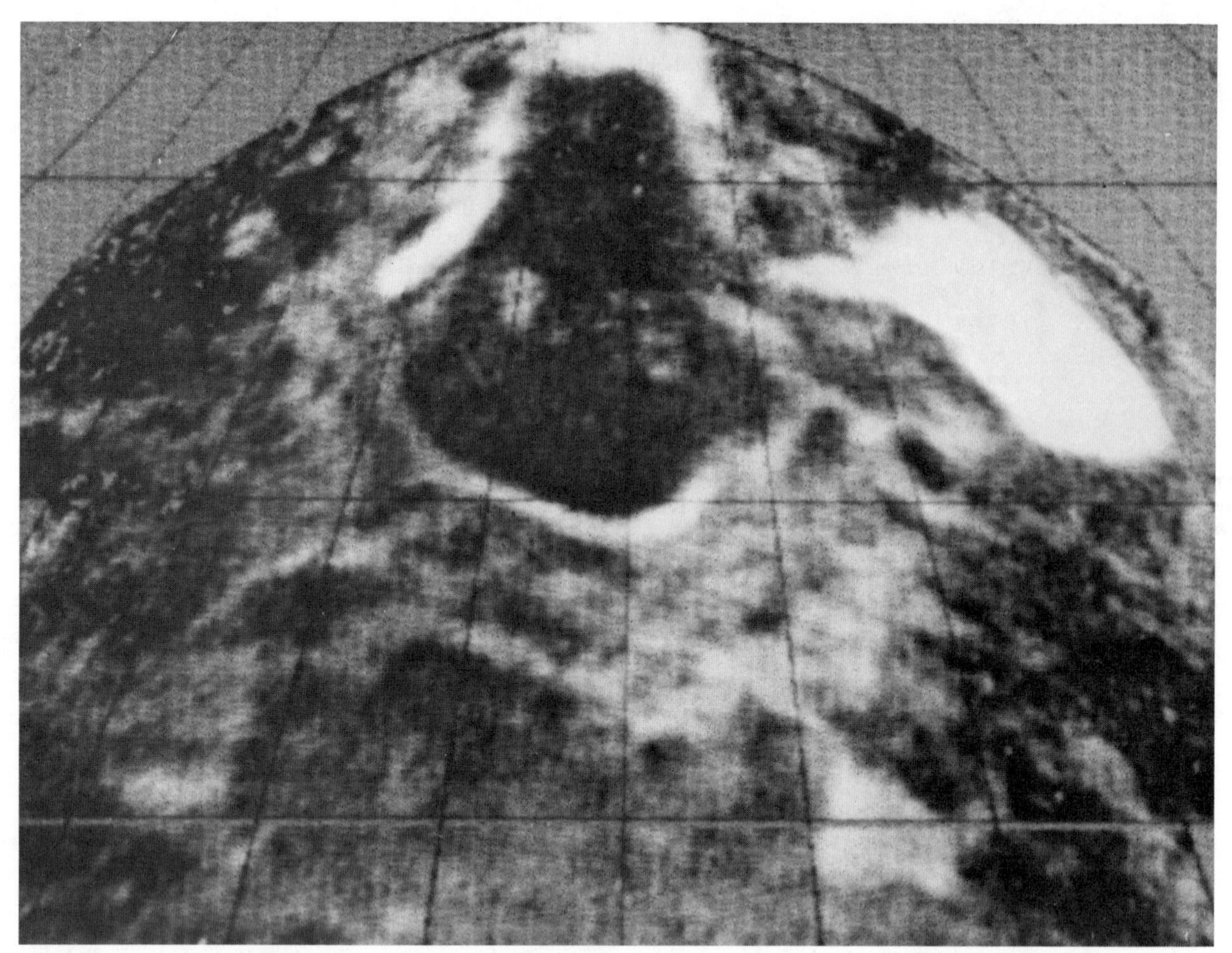

Fig. 28.3 Radar photograph of the surface of Venus between latitudes 36° and 75°. The meridians of longitude are spaced at 10°, with the 0° meridian passing through the center of the large bright feature on the upper right of the photograph. This photograph, which has a resolution of 20 km, utilized the new radar system at the Arecibo Observatory of the U.S. National Astronomy and Ionosphere Center, operating at a wavelength of 12.6 cm. Two features seem to dominate this region. One is the low-contrast area with a well-defined rim on its southern side and with two high-contrast features forming its northern and northwest edges. This region, called the northern basin, extends about 1,500 km in the north-south direction and has an average width of about 1,000 km. The other dominant region consists of the same two small high-contrast features on the northern edge of the basin and the large, very high-contrast feature, tentatively named Maxwell, that straddles 0° meridian and extends from 60° to 70° latitude. (Courtesy Donald Campbell.)

29. Mars as Viewed from Mariner 9

Harold Masursky, Bradford A. Smith, Carl Sagan, Conway B. Leovy, Bruce C. Murray, John F. McCauley, and James B. Pollack

(*Icarus 17*, 289–407 [1972])

Copyright © Academic Press, N.Y. (1972)

Perhaps the most exciting and longest-lived astronomical question concerns the possibility of life on Mars. Soon after Giovanni Schiaparelli reported the observation of narrow, dark, straight lines crisscrossing the Martian surface, Percival Lowell argued that these dark markings were actually irrigation canals, which marked "the lines of communication for water purposes between the polar caps on the one hand, and the centers of population on the other."[1] Although this view appealed to the public imagination, Alfred Russel Wallace soon argued that such an enormous network of straight canals could not exist under Martian conditions. Furthermore, the surface gravity of Mars was shown to be insufficient to retain water vapor, which led George Johnstone Stoney to argue that the Martian polar caps are actually composed of frozen carbon dioxide.[2] In Lowell's time it was well known that the dark areas of the Martian surface wax and wane with the Martian seasons, and these dark areas were imagined to be covered with lichenlike vegetation that survives the cold Martian temperature and flourishes in the increased humidity of the local Martian spring.

Beginning in November 1971 the *Mariner IX* spacecraft began to map Mars's surface from a Martian orbit with resolutions between 100 and 1,000 m. These photographs reveal the nature of the Martian features in rich detail but fail to find the canals, which in retrospect are seen to have been optical illusions or figments of Schiaparelli's and Lowell's imaginations. As shown in this selection, some of the small dark spots and larger dark basins on the Martian surface are actually the scars of ancient meteoritic impacts, and the seasonal variations of the Martian features are caused by the alternate deposition and displacement of windblown dust. This last result was actually anticipated by Dean B. McLaughlin in 1954, when he argued that the morphology of the Martian features recalls the terrestrial trade winds and that the time-variable features reflect deposits of windblown dust rather than plant life.[3] The cameras of *Mariner IX* revealed a profusion of unexpected and exotic detail, including towering volcanic mountains, huge canyons, and enigmatic, sinuous valleys that may have been caused by running water at previous epochs. Furthermore, the south polar cap appears to be composed of a thin layer of carbon dioxide frost covering a more substantial mass of water ice. The presence of water ice in the polar region, together with features that look like dry river beds, suggests the presence of liquid water and possibly the existence of organic substances during the past history of Mars.

The kaleidoscopic results of *Mariner IX* have been subsequently confirmed and extended by the Viking missions to Mars.[4] When the Viking spacecraft flew over the northern polar cap of Mars, as much as 99 μ of water vapor were detected, and the average surface temperature of the cap was found to be about 100° above the freezing point of carbon dioxide. These results mean that the Martian poles are composed of substantial amounts of water ice, seasonally covered with a carbon dioxide frost. Detailed Viking photographs provide additional evidence for water flow in the past. The descent of the Viking lander to the surface of Mars also permitted measurement of the principal constituents of the Martian atmosphere. The first analysis indicates that the atmosphere is composed of about 95% carbon dioxide, 2–3% nitrogen, 1–2% argon, 0.3–0.4% oxygen, and traces of krypton and xenon. Perhaps the most exciting aspect of the Viking landing was the performing of experiments to test for life in the Martian soil. Although no organic molecules were found, it was unexpectedly discovered that the Martian soil is highly oxidizing.

1. G. V. Schiaparelli, *Osservazioni astronomiche e fisiche sull'asse di rotazione e sulla topografia del pianeta Marte* (Rome. 1881, 1896); P. Lowell, *Mars and Its Canals* (London: Macmillan and Co., 1906).

2. A. R. Wallace, *Is Mars Habitable?* (London: Macmillan and Co., 1907). See also selection 15.

3. D. B. McLaughlin, *Publications of the Astronomical Society of the Pacific 66*, 161, 221 (1954).

4. *Science 194*, 57, 819, 924 (1976), *Icarus 32*, 131–251 (1977).

THE MAIN TEXT of this selection has been omitted, and only the introductions, discussions, summaries, and conclusions of each section of this 118 page article have been included.

THE GEOLOGY OF MARS

Mariner 9 pictures indicate that the surface of Mars has been shaped by impact, volcanic, tectonic, erosional and depositional activity. The moonlike cratered terrain, identified as the dominant surface unit from the Mariner 6 and 7 flyby data, has proven to be less typical of Mars than previously believed, although extensive in the mid- and high-latitude regions of the southern hemisphere. Martian craters are highly modified but their size-frequency distribution and morphology suggest that most were formed by impact. Circular basins encompassed by rugged terrain and filled with smooth plains material are recognized. These structures, like the craters, are more modified than corresponding features on the Moon and they exercise a less dominant influence on the regional geology. Smooth plains with few visible craters fill the large basins and the floors of larger craters; they also occupy large parts of the northern hemisphere where the plains lap against higher landforms. The middle northern latitudes of Mars from 90 to 150° longitude contain at least four large shield volcanoes each of which is about twice as massive as the largest on Earth. Steep-sided domes with summit craters and large, fresh-appearing volcanic craters with smooth rims are also present in this region. Multiple flow structures, ridges with lobate flanks, chain craters, and sinuous rilles occur in all regions, suggesting widespread volcanism. Evidence for tectonic activity post-dating formation of the cratered terrain and some of the plains units is abundant in the equatorial area from 0 to 120° longitude. Some regions exhibit a complex semiradial array of graben that suggest doming and stretching of the surface. Others contain intensely faulted terrain with broader, deeper graben separated by a complex mosaic of flat-topped blocks. An east-west-trending canyon system about 100–200 km wide and about 2,500 km long extends through the Coprates–Eos region. The canyons have gullied walls indicative of extensive headward erosion since their initial formation. Regionally depressed areas called chaotic terrain consist of

Fig. 29.1 Typical cratered terrain on the Martian surface, showing both old, smooth-rimmed craters and younger sharp-rimmed ones. The large crater at the top is 165 km across, with a conspicuous flat floor and slumped walls. Note the small doublet craters with central peaks (*lower left*).

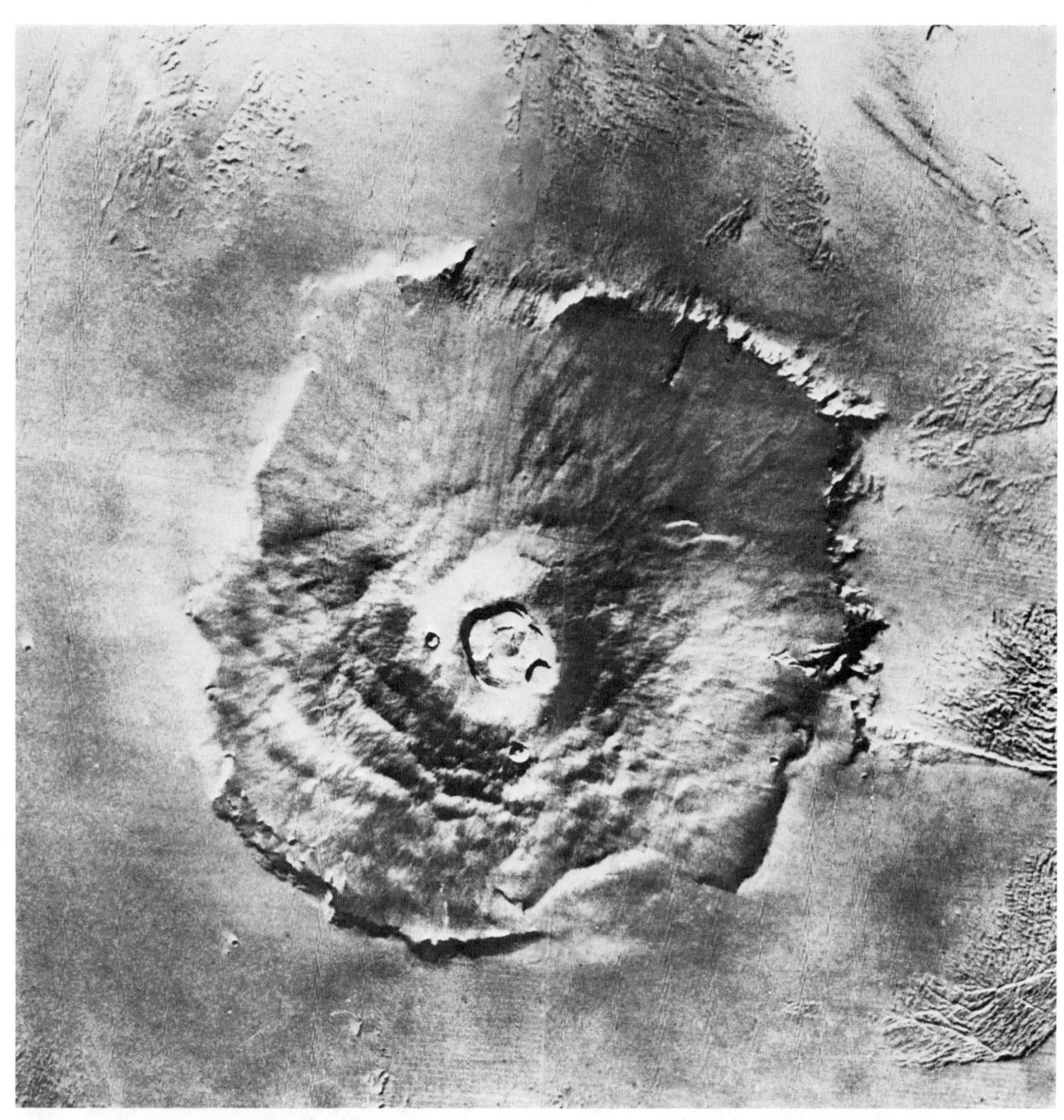

Fig. 29.2 Photomosaic of Olympus Mons, the largest of the Mars volcanic mountains. The volcanic structure is 500 km across and about 29 km high, with a complex summit caldera about 70 km across. These dimensions make it the largest volcanic structure known, much larger than the island of Hawaii, which at 200 km across (on the ocean floor) and 9 km high is the largest volcanic pile on the earth. The scarp around the base of Olympus Mons stands 1–4 km high and may have been produced by wind erosion. Originally the volcanic pile probably graded smoothly into the surrounding plain.

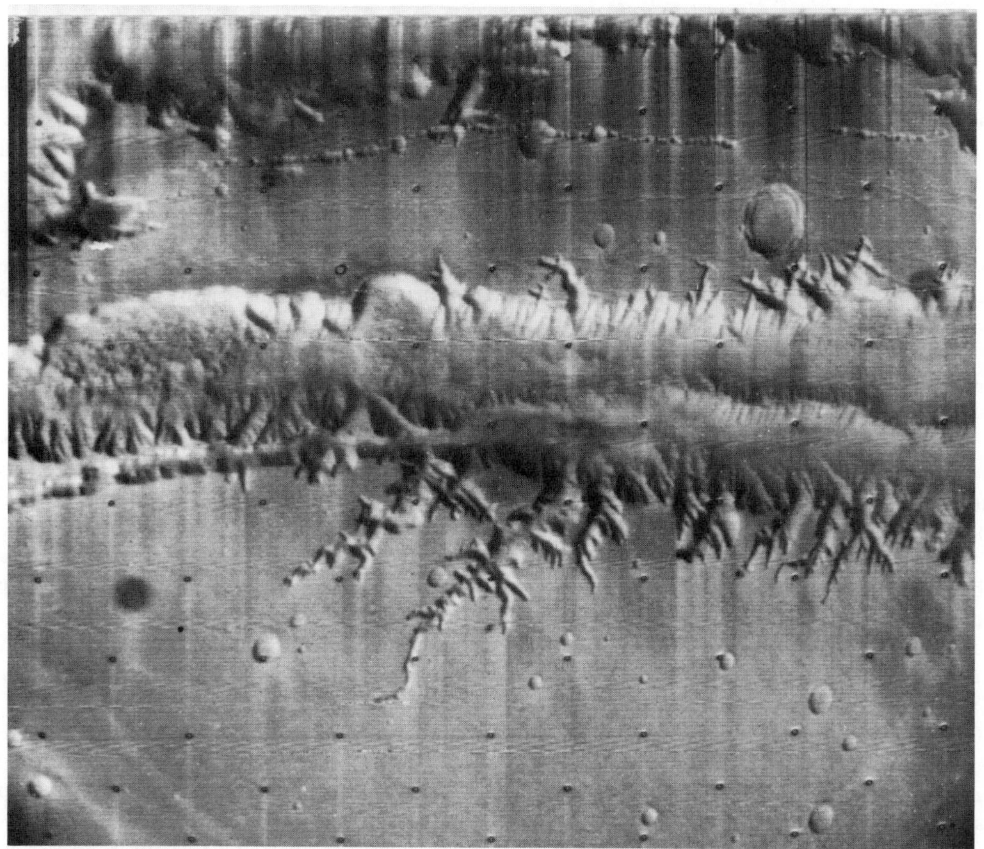

Fig. 29.3 The great Martian canyon system, Valles Marineris, more than 5,000 km long and at least 6 km deep, dwarfs any stream valleys on the continents of the earth. A minor side canyon is similar in length and depth to the Grand Canyon of Arizona. Features on the earth most closely comparable in size to Valles Marineris are the Rift Valley system of Africa and related rift valleys on ocean floors. Like the earth's rift valleys, Valles Marineris may have been formed where the crust of the planet has pulled apart. Pieces of the crust forming the floor of the canyon have probably subsided along faults. Subsequently, the rim of the valley has been sculptured by mysterious processes of erosion. The intricate system of canyons extending back from the rim may have been developed during melting and evaporation of subsurface ice.

intricately broken and jumbled blocks and appear to result from breaking up and slumping of older geologic units. Compressional features have not been identified in any of the pictures analyzed to date. Plumose light and dark surface markings can be explained by eolian transport. Mariner 9 has thus revealed that Mars is a complex planet with its own distinctive geologic history and that it is less primitive than the Moon.

MAJOR CONCLUSIONS Mariner 9 has shown that Mars is geologically far more heterogeneous than previously suspected from the earlier flyby missions. The analyses to date indicate convincingly that it has a geological style of its own different from that of either the Earth or Moon, and suggest that it represents a body intermediate in its evolutionary sequence somewhere between the Earth and Moon. It is now tempting to consider Mars as a planet that has partly made the transition from a relatively primitive impact-dominated (but not primordial) body like the Moon to an orogenically mobile, volcanically active, water-dominated planet like the Earth.

Like the Moon, Mars shows extensively cratered regions as well as numerous large circular basins. The basins, some of which are larger than any lunar basin, seem to exercise less control on the regional topography and distribution of the volcanic units than they do on the Moon.

The crater and basin terrains and the plains material that fills depressions in it are reminiscent of the Moon. But over much of the planet, the later parts of the Martian geological record are punctuated by huge and spectacular tectonic and volcanic features, not moonlike and only partly earthlike, that have destroyed or covered its earlier crater and basin aspect.

Extensive tectonic activity has occurred in huge regions of Mars. Much of this can be ascribed to circumferential tension in the upper parts of the lithosphere and to local doming. No shear or compressional features have been identified to date.

Volcanism has also played an important role in shaping the surface of Mars and probably has contributed in great part to its tenuous atmosphere. Martian volcanism is dramatically more varied and may span a larger part of the planet's history than lunar volcanic activity. Preliminary crater frequency studies point to the possibility that the major shield volcanos, calderas, plains, and other volcanic features could be relatively young.

Erosion and sedimentation, neither of them related to impact cratering, has occurred on a planetwide scale and surface modification processes are more wide-spread than previously envisioned. Erosion channels and depositional features abound, commonly related geographically and probably genetically to terrain that has collapsed chaotically. Extensive transport of materials has occurred in these channels, and moving fluids probably in episodic surges seem to be the only possible mechanism by which this can be explained, that is consistent with the present observational data.

Mars has clearly undergone a different proportionate mix of major surface-shaping processes than the Earth; the interplay between impact, volcanism, tectonism, and various erosion and sedimentation processes is clearly distinctive.

THE SOUTH POLAR REGION OF MARS

The first 4 months of Mariner 9 photography of the south polar region are discussed. Three major geological units have been recognized, separated by erosional unconformities. From oldest to youngest they are: cratered terrain, pitted plains, and laminated terrain. The latter unit is unique in occurrence to the polar region, volatiles are probably involved in its origin, and may still be present within the laminated terrain as layered ice.

The residual south polar cap has been observed to survive the disappearance of the thin annual CO_2 frost deposit and to last virtually unchanged in outline through the southern summer. That exposed deposit is inferred to be composed of water-ice. The residual cap appears to lie at the apex of an unusual quasicircular structure composed of laminated terrain; a similar structure also appears to exist near the north pole.

DISCUSSION Neither the pictures nor other measurements made by Mariner 9 unambiguously indicate whether the residual cap is solid CO_2 or H_2O. However, the Leighton/Murray model indicates the evaporation rate of a CO_2 deposit during the Martian southern summer would be so high as to make a permanently exposed CO_2 deposit there unlikely. Water-ice, on the other hand, once collected into extensive surface deposits at either pole, should be characterized by a sufficiently low evaporation rate to permit indefinite survival, even if subjected to direct summer sunshine. If water has been abundantly supplied in the vapor form at some time in the history of the planet, much of it may have migrated to the polar regions, since these areas likely constitute "cold traps." It is significant, in this regard, that Mariner 9 has detected the presence of large constructional volcanic features which may be relatively young in terms of Mars history. Indeed, possible constructional volcanic features have been detected in the south polar region itself (lat. 65° S, long. 10° W). The formation of such volcanic features may have provided an unusually large flux of water vapor to the atmosphere at various times, either directly through magmatic exhalation or through sublimation of permafrost and ground ice encountered in the subsurface. Accordingly, we prefer the concept of a residual water-ice core, exposed each Martian year following the disappearance of the thin annual CO_2 cap.

At the present time, the pitted plains in general cannot convincingly be attributed in origin to uniquely polar processes. Future work, however, may lead to subdivisions within this loosely defined unit whose origin and evolution are unique to the geological history of the Martian south polar region.

The origin of laminated terrains very probably involved solid carbon dioxide and/or water-ice, because their occurrence is restricted to the polar regions. It seems possible that fine dust or volcanic ash, circulated by frequent storms over the planet, is trapped by CO_2 or H_2O seasonal ices and accumulated into stratified deposits. The ice subsequently may be entirely volatilized leaving stratified deposits consisting solely of wind-transported detritus. Alternatively, water-ice (or possibly CO_2 ice) may still be present, and the laminations may reflect varying ratios of ice and dust. This second possibility is consistent with the peculiar convex and smoothed appearance of the upper edges of the laminated deposits. Such rounded edges might well have originated by evaporation and wind action. Also, the near-circular symmetry of the erosional edges of the "plates" about the rotational pole is suggestive of ablation of volatiles. If this is the case, it would mean a very much larger amount of condensates is present on the surface of Mars than otherwise indicated.

The regularity of the laminae is suggestive of periodic variations in conditions of stratification. Annual accumulations of dust or volatiles would be several orders of magnitude smaller than the observed laminae. However, one strong periodic effect in the atmospheric conditions has been hypothesized, which could result in laminae of the order of meters or tens of meters in thickness. Because the orbit of Mars is eccentric, the planet is usually closer to the Sun during winter in one hemisphere than it is during winter in the other hemisphere. Precession of both the perihelion and the axis of rotation causes the pole with the colder average climate to alternate with a period of about 50,000 years. The laminae may reflect such climatic alternations. The larger scale structure of the plates themselves (as indicated by the dark bands in the Mariner 9 photographs) must correspond to some other process of a longer period.

Currently, the northern climate is the more severe with cooler summers. It is conceivable that another lamination now may be forming there, while in the southern polar region similar laminated deposits (or even large-scale plates) are undergoing ablation and wind erosion.

We cannot offer at this time an explanation for the offset of the residual ice deposit from the geometric south pole. It may well be that further topographic information will indicate that the stacked plates indeed are sufficiently concave (bowl shaped) to provide the topographic control adequate to offset the point of minimal insolation from the geometric south pole. In this regard, the final disappearance of a large outlier of frost north of the south residual cap is of interest. Assuming this to have been an annual CO_2 deposit, a gentle poleward slope would best account for its survival so late into the southern summer. Such a regional slope is consistent with the concave plate concept. In addition, it may be that there are other factors involved in determining the initial locus of formation of the residual cap. For example, wind circulation patterns and surface roughness may be important as well as simple absorption and reradiation of solar energy.

VARIABLE FEATURES ON MARS

Systematic Mariner 9 photography of a range of Martian surface features, observed with all three photometric angles approximately invariant, reveals three general categories of albedo variations: (1) an essentially uniform contrast enhancement due to the dissipation of the dust storm; (2) the appearance of splotches, irregular dark markings at least partially related to topography; and (3) the development of both bright and dark linear streaks, generally emanating from craters.

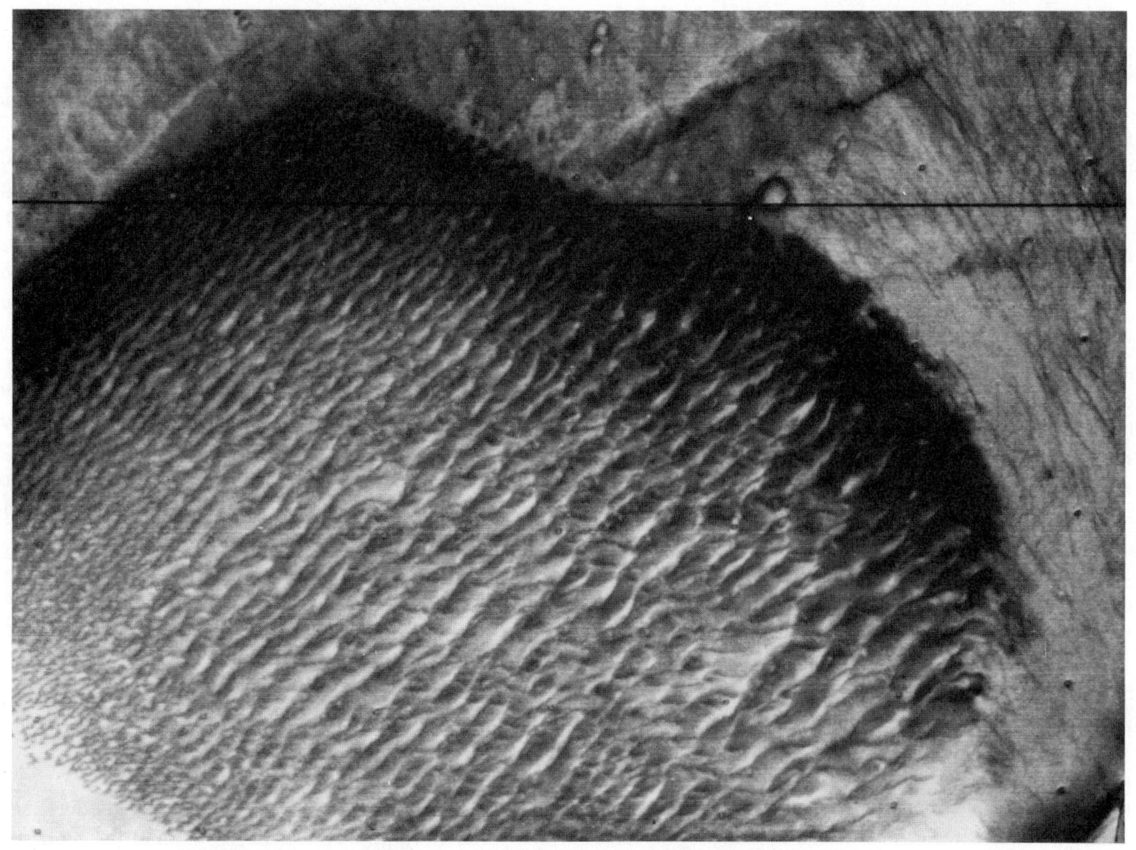

Fig. 29.4 An elliptical, Martian dune field about 130 by 65 km. The dune field consists of a series of subparallel ridges, 1–2 km apart, closely resembling terrestrial transverse dunes. Many of the ridges appear to have rounded crests with similar slopes on either side, which suggests that, although the wind here generally blows at right angles to the transverse ridges, it may intermittently reverse its direction and thereby even out the slopes on the windward and lee sides of the dunes. The unusually dark appearance on what appears to be the more windward side of the dunes may be concentrations of such dark heavy minerals as ilmenite and small dark lithic fragments. On Mars concentrations of such heavy minerals may have become preferentially trapped in crater floors through wind action.

Some splotches and streaks vary on characteristic time-scales ~ 2 weeks; they have characteristic dimensions of kilometers to tens of kilometers. The loci of these features appear in some cases to correspond well to the ground-based albedo markings, and the integrated time variation of splotches and streaks is suggested to produce the classical "seasonal" and secular albedo changes on Mars. The morphology and variability of streaks and splotches, and the resolution of at least one splotch into an extensive dune system, implicates windblown dust as the principal agent of Martian albedo differences and variability.

DISCUSSION A variety of classical time-variable bright and dark areas on Mars are found to be constituted of, and have their variations controlled by, the distribution and appearance of two categories of features: splotches and streaks. There are apparently only dark splotches, but both bright and dark streaks. Both are connected with topography, particularly craters, ridges, scarps, and faults. Some cases, possibly intermediate between streaks and splotches, also exist, and it may be that the essential distinction between the two types of features has to do with wind velocity. The parallel arrangement and streamlined shape of the streaks points very strongly to an aeolian origin. A given variation in a streak or splotch requires the deposition or removal of only a thin layer of material, easily sequestered in craters or elsewhere.

The bright streaks are most readily understood if we assume that, in the waning stages of the dust storm, fine particles, settling through the atmosphere, are preferentially trapped in topographic features such as craters by lateral transport. Deflation of craters by subsequent high-velocity winds produces the parallel array of streaks. The appearance of bright craters towards the end of the dust storm is consistent with this idea, although some contribution to higher albedo follows from the longer slant path through a dusty atmosphere above craters than above neighboring terrain.

Wind velocities ~ 100 m/sec are deduced from lee wave clouds and from the IRIS atmospheric temperature structure via the thermal wind equation, as well as from a variety of theoretically expected wind regimes. Such high-velocity sandblasting in the thin Martian atmosphere should result in aeolian erosional processes much more efficient than those common on the Earth. Because of the lower atmospheric temperature and the larger mean molecular weight of the Martian atmosphere the velocity of sound on Mars is less than on Earth. Velocities required to just initiate particle motion, correspond to velocities of ~ 70 m/sec above the boundary layer, or Mach 0.4. Considerably higher velocities may be expected, at least at some places and times. Mars therefore presents an unusual possibility: that wind velocities in the transonic range (between Mach 0.9 and 1.0) occasionally exist. The erosional, depositional, and deflationary effects of transonic meteorology, if it exists, remain entirely unexplored.

CONCLUSIONS We have found splotches and streaks as the albedo fine structure of Martian bright and dark areas. The superposition of splotches and streaks seems to determine the configuration of ground-based albedo markings. Both splotches and streaks vary on characteristic time scales of the order of 2 weeks. The integration of such variations is suggested to be the cause of the classical seasonal and secular variations, although no clear dependence of changes on latitude or season is found in these preliminary results. Aeolian transport of fine particles appears clearly to be implicated in the morphology and variation of splotches and streaks. We suggest that both bright and dark material is laterally transported on Mars.

The Martian Atmosphere

Atmospheric phenomena appearing in the Mariner 9 television pictures are discussed in detail. The surface of the planet was heavily obscured by a global dust storm during the first month in orbit. Brightness data during this period can be fitted by a semi-infinite scattering and absorbing atmosphere model with a single-scattering albedo in the range 0.70–0.85. This low value suggests that the mean radius of the particles responsible for the obscuration was at least 10 μm. By the end of the second month, this dust storm had largely dissipated, leaving a residual optical depth ~ 0.1. Much of the region north of 45° N was covered by variable clouds comprising the north polar hood. The cloud structures revealed extensive systems of lee waves generated by west-to-east flow over irregular terrain. Extensive cloud systems in this region resembled baroclinic wave cyclones. Clouds were also observed over several of the large calderas; these clouds are believed to contain water ice. Several localized dust storms were seen after the global dust storm cleared. These dust clouds appeared to be intensely convective. The convective nature of these storms and the stirring of large dust particles to great heights can be explained by vertical velocities generated by the absorption of solar radiation by the dusty atmosphere.

SUMMARY We summarize here the significant observations described above.

(1) During the global storm, dust was stirred to heights in excess of 30 km.

(2) This dust storm could be characterized by a single-scattering albedo, between 0.7 and 0.85. These low values suggest that the mean particle radius was comparable to that of bright areas, i.e., a radius in excess of 10 μm.

(3) A persistent layer of condensates was observed near the 0.02 mb pressure level (55 km) indicating a tem-

perature minimum there. At least over the South Polar region, water ice is the probable constituent of this layer.

(4) The polar hood consisted of clouds of CO_2 ice, and perhaps also water ice, which show large day-to-day variations in the 45–65° N latitude zone. Some of the variations were characteristic of active baroclinic waves.

(5) Gravity waves generated by flow over irregular topography were much more prominent than they are in the earth's atmosphere, and they indicated persistent westerly flow north of 45° N during this northern late winter season. Wavelengths of about 30 km were predominant, and persisted far downstream from the generating obstacle. Recurrent brightening observed from the earth in the Arcadia and Tempe regions at this Mars season is due to prominent lee wave systems.

(6) Contrary to our expectations, there was little evidence in the pictures of CO_2 ice or other condensate on the surface up to latitudes as high as 75° N. Only isolated frost deposits around the rims of craters were apparent between 45 and 75° N.

(7) There is a remarkable correlation between very high volcanic mountains and recurrent afternoon clouds. Clouds observed over the South Spot caldera probably consisted of water ice.

(8) Localized dust storms occurred rather frequently. On one occasion the mechanism appears to have been a strong southward movement of cold air following a cold front. The dust storms were highly convective, and dust was carried rapidly to heights of 15–20 km. Extensive growth of a dark area following a local dust storm can be attributed to the removal of fine dust by wind.

(9) The great height to which dust from the major storm was raised and the highly convective appearance of the local dust storms can be attributed to strong vertical velocities generated by absorption of solar radiation in the dusty air.

THE MARTIAN SATELLITES, PHOBOS AND DEIMOS

Mariner 9 photographs of Phobos and Deimos have yielded new information about the orbits, rotation periods, sizes, shapes, and surface characteristics of the satellites. Both satellites appear to be in synchronous rotation. They are irregular, heavily cratered bodies whose shapes appear to have been determined largely by impact fragmentation and spalling. The surfaces of both satellites have crater densities close to saturation and nearly identical, very low albedos. Lower limits on the tensile and yield strengths are estimated, and it is concluded that both satellites may consist of well-consolidated, though possibly highly fractured material.

DISCUSSION Some preliminary corrections to the ephemerides of the satellites were previously presented. In this section we summarize some of the other important results of this paper and discuss their implications. Both Phobos and Deimos are very irregular, heavily cratered bodies whose shapes seem to have been determined by impact fragmentation and spalling. Both satellites are most likely in synchronous rotation, which implies that tidal locking has occurred and that 10^6 years have elapsed since the last major impact on Phobos. The corresponding figure for Deimos is 10^8 years.

Both satellites have similar albedos. This suggests a similar surface composition and perhaps origin. The unusually low value of the albedo is somewhat diagnostic of composition, since basalts and carbonaceous chondrites are the only two sufficiently dark materials which are likely to be abundant in the solar system. The evolutionary histories for the satellites implied by these compositions are very different. We hope eventually to be able to distinguish between these two possibilities.

The surfaces of both satellites have similar crater densities which are close to saturation. From this we infer that these surfaces are probably as old as the solar system, although there are several uncertainties in such an estimate.

Nevertheless, it is certain that the surfaces of Phobos and Deimos are at least 2×10^6 years old, and, if the best estimate is used, the age is found to be greater than that of the solar system assuming a constant cratering rate. This suggests that the surfaces of the satellites date from the earliest aeon of solar system history, when the cratering rate was probably higher.

We cannot, at the present time, distinguish between a capture origin for the satellites and an origin as debris left over from the time of formation of Mars.

Lower limits of 6×10^6 dyn/cm^2 and 2×10^6 dyn/cm^2 are estimated for the tensile and yield strengths of Phobos and Deimos, respectively, based on Baldwin's energy scaling. These values imply a well-consolidated though possibly fractured and jointed material. Therefore, the satellites may have been originally parts of larger objects in which gravitational forces were sufficient to form well-consolidated material. We have also speculated that the energy of the impact which produced the largest crater on Phobos may have been within an order of magnitude of the satellite's binding energy. If so, parts of the satellite may have broken off, resulting in its present angular appearance.

Both Phobos and Deimos appear to be the remnants of a larger body (or bodies) evolved through a complex collisional natural selection. Their appearance may be typical of many asteroids and other small solar system objects.

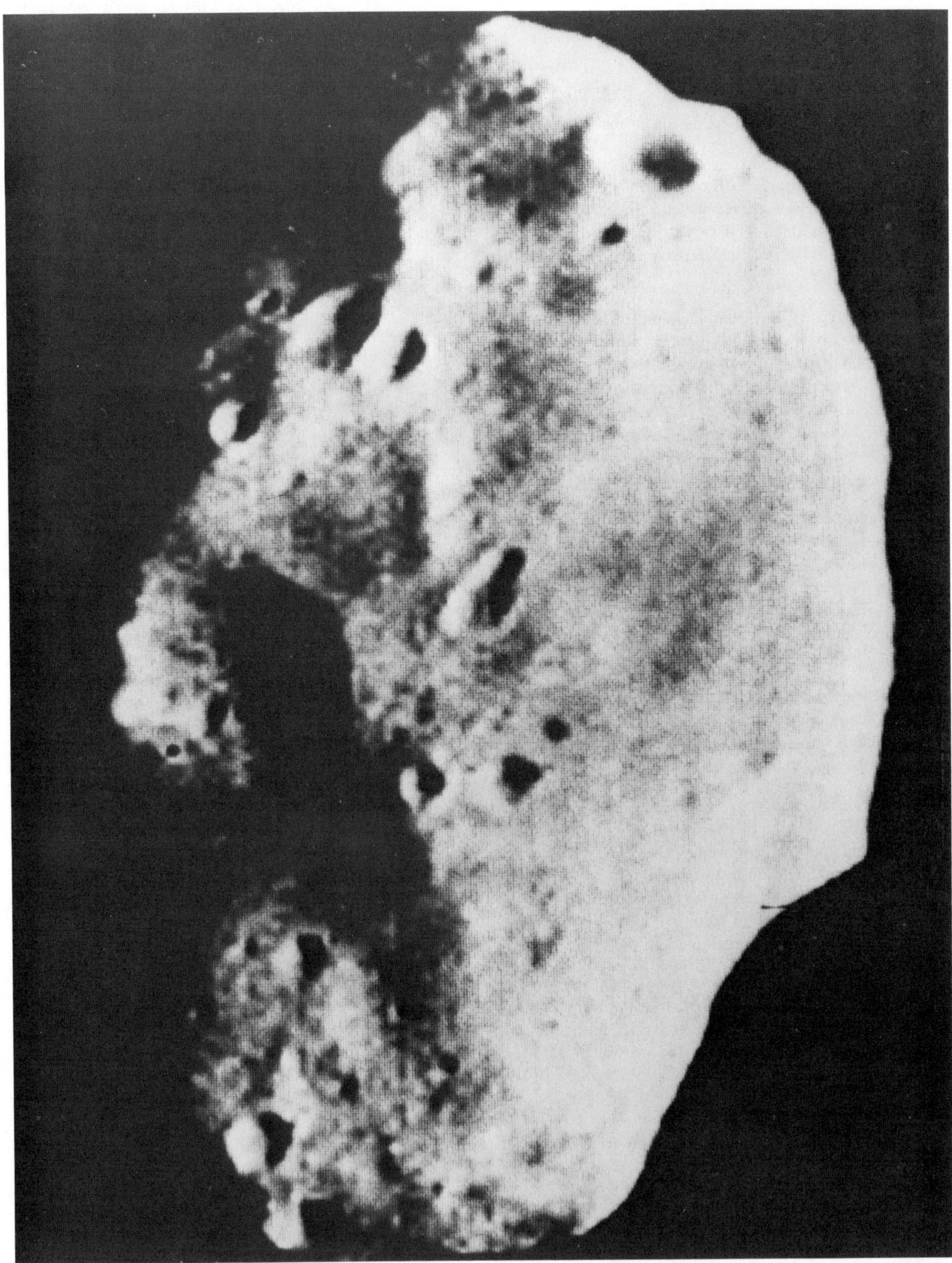

Fig. 29.5 Phobos, the inner and larger of the Martian moonlets, which orbits at an average distance of 6,100 km above the surface of Mars. It proves to be an oblong mass about 20 by 25 km in its major dimensions. Deimos orbits roughly 20,000 km above Mars and is 10 by 16 km in size. Because the Martian moons are so small, their gravity fields are too weak to force them into spherical shape. As with our moon, each keeps the same side turned toward the planet. In this photograph of Phobos, the larger crater at middle right appears to have at least one small crater on its rim, and more than a dozen other small craters are visible. The irregular edges of Phobos strongly suggest fragmentation.

30. The Moon after Apollo

Farouk El-Baz

(*Icarus 25*, 495–537 [1975])

Copyright © Academic Press, N.Y. (1975)

One of the most widely debated astronomical questions has been the nature of the lunar surface features. For decades a fierce controversy raged over whether most craters were volcanic or meteoritic in origin. As early as 1892, Grove Karl Gilbert had advocated the now accepted view that meteoritic impact caused the craters (see selection 14). Around 1950 Ralph Baldwin, Gerard Kuiper, and Harold Urey espoused this idea and extended it to interpret the lunar maria as lava flows resulting from the impact of planetesimals.

Nevertheless, the great complexity of the lunar history remained unknown as long as astronomers relied solely on earth-based observations. The great proliferation of data from space, beginning with the *Luna III* back views of the moon in 1959 (see selection 9) has not only multiplied our knowledge of the moon manyfold but has found every theory of the lunar origin inadequate in at least one significant detail.

The initial space exploration of the moon culminated on June 20, 1969, when an Apollo vehicle first landed men on the moon. During the subsequent three and a half years five Apollo vehicles landed men on the moon and two unmanned Soviet Lunas returned lunar material to the earth. The scientific results from such an extensive cooperative enterprise are scattered in scores of jointly-authored publications, so we give here Farouk El-Baz's summary of the more than fifty scientific spacecraft observations of the moon. His account does more than explain the origin of the lunar features. The available data are sufficient to map out a detailed evolutionary history of the moon.

Apollo Missions

HERE WE GIVE in succession the summaries that follow much longer discussions of each mission.

MARE TRANQUILLITATIS In summary, the Apollo 11 rock samples provided significant new information from which the following are emphasized.

1. The majority of Mare Tranquillitatis samples are volcanic rocks of basaltic composition. This is a confirmation of pre-Apollo photogeologic interpretations, as well as interpretations of data transmitted by the Surveyor missions.
2. The rocks are remarkably similar in composition except for the concentrations of a few minor and trace elements. By terrestrial standards, they are rich in iron and titanium, and poor in sodium, carbon and water. They also do not contain any evidence that life once existed on the Moon.
3. The volcanic rocks are derivatives of differentiation in a melt at depth (rather than being impact-produced, near-surface melts). There is also ample chemical evidence that the original magmas had low oxidation levels.
4. Samples of fine dust and soil are in most part derived from the basaltic rocks. The soil is chemically similar to the rock samples except for minor components which are probably derived from the nearby (anorthositic) highlands.
5. Rocks, rock fragments, and fine dust are most likely the product of repeated impacts on solid bed rock surfaces. There is ample evidence of modifications of the original rock structures by shock metamorphism, including the formation of glass and breccias.
6. The lunar volcanic rocks at the Apollo 11 site are extremely old; they crystallized from melts 3.7 billion years ago. (This particular fact contradicts some earlier suggestions that the Moon has been volcanically active in the recent past.) The model age of the soil, 4.6 billion years, probably represents the age of the Moon.

OCEANUS PROCELLARUM Among the most important conclusions that could be drawn from the findings of Apollo 12 are the following.

1. The chemistry of the two mare landing sites, Apollo 11 and 12, resembles the findings of Surveyor V and VI. This chemistry distinguishes lunar basalts from basaltic meteorites and terrestrial basalts.
2. At least two, probably three or more, mare units were sampled on Apollo 12 based on petrologic and petrographic evidence. This confirms photogeologic interpretations that suggest the presence of more than one lava flow in the landing site area.
3. The age of the Apollo 12 basalts, which is approximately 3.2 billion years, attests to the fact that the lunar maria are geologically very old. The wide difference between the ages of Mare Tranquillitatis and Oceanus Procellarum basalts (600 million years) indicates a prolonged episode of mare filling on the Moon.
4. The high albedo material collected in the fine samples as well as rock 13 appear to be foreign to the Procellarum basalts. These light-colored materials seem to be part of the Copernicus ray. KREEP components in the Apollo 12 site may, therefore, be part of buried Fra Mauro Formation materials. If so, KREEP basalts should be an important component in the Apollo 14 rocks, which they are.

FRA MAURO FORMATION In summary, important results of the study of the Apollo 14 samples are:

1. The complex history of shock metamorphism of all the rocks, for which there are three probable causes: (a) excavation by the Imbrium impact of preexisting brecciated materials; (b) formation of breccia within breccia during the Imbrium event, and the base-surge that resulted from it; and (c) local cratering events superposed on the Imbrium event.
2. The composition of the Fra Mauro Formation rocks is different from the mare rocks of Apollo 11 and 12. The highland or KREEP basalts of Apollo 14 are characterized by a lower iron oxide content and higher aluminum oxide and trace element concentrations than those of mare basalts.
3. Since the recrystallization age of the Apollo 14 rocks is about 3.95 billion years, and if the Fra Mauro Formation is indeed ejecta of the Imbrium basin, then we have established the absolute age of the Imbrium event which shaped much of the near side of the Moon.

APENNINE-HADLEY Among the most significant results of the Apollo 15 mission are the following.

1. The basalts of the Apollo 15 site are similar chemically and texturally to the Apollo 12 basalts. Both mare basalts are also about 3.2 billion years old. They constitute the youngest group of sampled basalts. The Luna 16 and Apollo basalts are 3.5 and 3.7 billion years old, respectively. The oldest basalt is one of the Apollo 14 rocks, number 53, which is 3.95 billion years old.

2. Rocks and soil samples of Hadley δ are in general terms anorthositic gabbros and gabbroic anorthosites. Some of these are crystalline rocks, which indicates that brecciation was not as extensive at the rim of the Imbrium scar as it was farther out on the ejecta blanket (Apollo 14 breccias).
3. The lunar highlands are indeed older than the maria, which is in agreement with photogeologic interpretations. The fact that "Genesis Rock" is 4.1 billion years old is very significant, in that the rock is older than the Imbrium basin, and was not metamorphosed by that event.
4. Although samples and close up photographs were taken of Rima Hadley, no new information was gathered on the origin of that sinuous rille. All existing theories of sinuous rille formation are still competing, and there may be many modes of origin for these rilles.

DESCARTES HIGHLANDS From studies of the Apollo 16 samples, the following results are most significant.

1. Photogeologic interpretations by many investigators, including myself, of landforms in the Descartes region were proved incorrect. The hilly and grooved topography in that area is probably unrelated to terra volcanism and most likely to the accumulation of basin ejecta.
2. Although the Imbrium basin appears to be the most likely source of the Descartes area deposits, differences between the Apollo 16 and 14 rocks are still not fully understood.
3. The ages of Apollo 16 rocks (about 3.9 billion years) are consistent with an Imbrium origin, although some samples dated by argon 40-argon 39 give crystallization ages of approximately 4.1 to 4.2 billion years.
4. The high aluminum and calcium rocks of the Apollo 16 site are crushed anorthosite breccias, probably derived from cataclysmic metamorphism of an anorthositic complex containing 70 to 90 percent plagioclase.
5. Although the lunar environment is completely void of water and is characterized by exceedingly low oxygen pressure, goethite was formed in the Apollo 16 rocks as a mineral reaction rim. The water vapor may have been part of the impacting (cometary) projectile that formed North Ray crater.

TAURUS–LITTROW In summary, among the most important findings of the Apollo 17 surface exploration are the following.

1. The basalts contained in the Taurus–Littrow Valley are grossly similar to those from Mare Tranquillitatis in their high content of titanium, as well as in their age. The age of these basalts, about 3.7 billion years, contradicts earlier expectations that the Apollo 17 basalts would be young.
2. Orange, brown and black spheres of glass that are probably the product of volcanism are also 3.7 billion years old. They may represent a more complex history of volcanism in that region. High concentrations of volcanic glass spheres may be responsible for the apparent smoothness of the dark mare unit.
3. The highland rocks from the massifs show more petrologic variety than any other group collected on previous sites. These rocks might represent both materials uplifted at the rim of the Serenitatis basin as well as ejecta from both Serenitatis and Imbrium.
4. The age of the highland rocks of Apollo 17 (about four billion years) lends support to the idea of a cataclysmic event or events which affected much of the near side of the Moon around 3.9 to 4 billion years ago. However, discovery of older dunite fragments (about 4.48 billion years old) also indicates that earlier differentiates may still be preserved in the lunar crust.

GLOBAL CHEMISTRY

Mare materials form only about one-fifth of the lunar surface. The distribution of maria is an obvious indication of the asymmetry of the Moon, more mare material on the near side than on the far side.

Large exposures of mare materials can be divided into two types of occurrences: (1) the probably thick (tens of kilometers) inner fill of circular basins; and (2) the relatively thin (a few kilometers) cover of basin troughs and other low-lands. As first discovered by Muller and Sjogren, the first occurrences display large positive gravity anomalies or mass concentrations (mascons). The gravimetric data, which are available for the nearside and limb regions, show mascons underlying the maria of Smythii, Crisium, Nectaris, Serenitatis, Imbrium, Humorum, and Orientale. However, large mare expanses of irregular shape, such as Oceanus Procellarum, do not show a positive gravity anomaly; both topography and gravimetrics indicate that the mare material in this region formed only a shallow veneer.

The basaltic composition of mare materials, although suggested by telescopic observations and photogeologic interpretations, was first indicated by the α-scattering elemental analyses made by Surveyors V and VI in Mare Tranquillitatis and Sinus Medii, respectively. The widely spaced locations of the Surveyor V and VI sampling sites, as well as Apollo missions 11, 12, 15, 17, and Luna 16 sites, leave no doubt that basalts are the major component of all the lunar maria. The basic difference between lunar and terrestrial basalts is that the former are depleted in sodium, along with all other volatile elements.

Highland materials, which make up about four-fifths of the lunar surface are chemically and petrologically more complex than the mare materials. In general, the sampled highlands are feldspathic breccias. The common brecciation, recrystallization, and apparent partial melting is consistent with patterns of overlapping ejecta from numerous impacts.

All the lunar rock and soil returned by Apollo and Luna missions originated from differentiates; they have the specialized chemical compositions that are characteristically produced by magmatic differentiation. This indicates that the part of the Moon that was sampled had been molten at one time or another. The basic pattern of primary differentiation may have been relatively simple, but the picture has been greatly complicated by the tendency of large cratering events and lava eruptions to transport and mix chemical entities during and after the original differentiation.

EVOLUTION OF THE SURFACE

Before the return of lunar samples, geologists were limited to establishment of a relative time scale to describe events and major provinces on the Moon. This relative age time scale can effectively be applied to the entire Moon, most of which will not be sampled. The absolute time scale, which is based on age dating measurements of the lunar rocks, can now be used to calibrate the relative time scale, as will be discussed below.

RELATIVE AGES Global morphology of the Moon is controlled by the large features of impact origin known as multiring circular basins. From the time of its formation, the crust of the Moon has probably been repeatedly shaped by basin and crater formation. Materials of those basins and smaller craters compose most of the lunar highlands. As discussed above, mare materials, of later volcanic origin, are concentrated in the basins and their peripheral troughs, and the nearby lowlands.

Thus, the main surface sculpting mechanisms on the Moon appear to be meteorite bombardment and the downslope movement of fine-grained debris from higher to lower levels, aided by tectonic and/or seismic shaking.

ABSOLUTE AGES In a nutshell, the Moon is assumed to have formed 4.6 billion years ago; global planetary differentiation processes took place between 4.6 and 4.3 billion years ago. As the solid anorthositic crust was formed it kept a record of moonlets that struck it leaving numerous scars (pre-Nectarian time). The last of the impact basins were formed on the near side about 4.0 to 3.9 billion years ago (Imbrian System, starting about 3.95). Volcanic eruptions filled the basins and low lands with basalts from 3.7 to 3.2 billion years ago (Imbrian and Eratosthenian time). Post-basalt history included transport and mixing of the lunar materials, mostly by impact craters, that continues to the present day (Copernican System).

LAYERS OF THE MOON

1. Crust: a layer enriched in aluminum and calcium (anorthositic gabbros), approximately 60 km thick in the region of the Apollo seismic network, as compared to the 5 to 35 km thick crust on Earth. In this layer there are local mass concentrations of iron-rich rocks, and a rapid increase of velocity down from the surface due to pressure effects on dry rocks. There is also a minor discontinuity at a depth of about 25 km which may be related to intensive fracturing of the upper crust and/or to petrological differences. The seismic velocity is nearly constant (6.8 km sec^{-1}) between 25 and 65 km.

2. Mantle: at the major discontinuity at about 65 km depth, there is a large increase in the velocity to about 9.0 km sec^{-1}. This interface is comparable to the rise of velocity at the Mohorovičić discontinuity on Earth. This layer appears to continue to the depth of 1,100 km. Compositions of the surface rocks, the measured seismic velocities, and constraints imposed by the Moon's mean density and moments of inertia support a pyroxene-rich composition for the lunar mantle.

3. Core: the moment of inertia and the overall density of the Moon place an upper limit of about 500 km on the radius of the lunar core, if a core exists. A molten or partially molten silicate zone or core below about 1,100 km from the surface may be responsible for deep-seated moonquakes, which are focused between 800 and 1,100 km deep. The data however, do not rule out the possibility of a small, 400 km radius, iron-rich (Fe–FeS?) core.

QUESTIONS OF ORIGIN

There are three basic theories of origin of the Moon. She is either Earth's wife, captured from some other orbit; daughter, fissioned directly from "proto" Earth; or sister, accreted from the same binary system.

Before Apollo, it was widely believed that the lunar exploration missions would provide the necessary information for a final answer to the question of origin. This did not happen. As discussed above, the primitive lunar crustal materials have been subjected to much metamorphism, due to large impacts.

Thus, the three basic theories of lunar origin are still competing, especially since all three have been modified to accommodate the new findings. Each theory has its share of positive points as well as shortcomings, and these vary depending on whose paper one reads. However, one of the most important constraints on the theories of origin is the chemical makeup of the Moon. Any theory will have to account for the lower metal and volatile content, and the higher refractory element content in the lunar rocks, as compared to the compositions of terrestrial rocks and meteorites.

SUMMARY

To me, the most important findings from Apollo lunar exploration are the following.

1. The lunar samples indicate that the Moon does not now, nor did it ever, host any water or life forms of any kind.
2. The Moon is made up of the same chemical elements as is the Earth, but in varying proportions. The lunar rocks are enriched in refractory elements and depleted in volatile elements.
3. The Moon is a differentiated body: extensive melting may have occurred in the upper layers of the Moon, to about 100 km depth, at the time of accretion. As the hot magma cooled, light plagioclase crystals floated to form the low-density crust leaving denser materials below.
4. The low-density highland rocks are exposed everywhere on the lunar crust except where covered by denser basalts. These are the products of internally generated volcanic melts which spread on the surface during a period of 600 million years (between 3.7 and 3.2 billion years ago).
5. There are many indications of the shaping of the Moon by large impacts, basin and crater formation being the major sculpting mechanism on the Moon. Shock effects on the rocks attest to the violent impact history. Although the bombardment is now continuing, the size and frequency of impacting projectiles were much larger in the early history of the Moon.
6. As confirmed by the orbiting geochemical experiments, samples collected at the landing sites most probably represent the whole Moon.
7. The figure of the Moon shows pronounced asymmetry; the far side is higher, and the near side is lower, relative to a mean lunar radius of 1,738 km. Also, the center of mass is shifted about 2 km towards the Earth relative to the center of the figure.
8. There is lateral asymmetry in natural γ-ray radiation on the Moon. Pockets of radioactive materials are concentrated in the western near side; only a small anomaly in the vicinity of the crater Van de Graaff exists on the lunar far side.
9. The moon is probably layered into a crust, about 65 km on the near side and probably thicker on the far side; a solid mantle, about 1,100 km thick; and a silicate core, about 500 km thick, that is in part molten or contains a partially molten zone.
10. The data so far collected do not provide enough proof for final conclusions regarding the origin of the Moon.

31. The Encounter Theories of the Origin of the Solar System

The Spiral Nebula Hypothesis

Forest Ray Moulton

(*An Introduction to Astronomy* [New York: Macmillan Co., 1906, pp. 463–472])

The Dissipation of Planetary Filaments

Lyman Spitzer

(*Astrophysical Journal 90*, 675–688 [1939])

At the close of the eighteenth century, Pierre Simon Laplace developed the so-called nebular theory for the origin of the solar system, a scheme suggested earlier by Immanuel Kant. According to this theory, the planets formed as minor condensations in a cloud of gas and dust in interstellar space, whose gravitational contraction also led to the formation of the primitive sun. As a rotating protosun contracts, it must spin faster in order to conserve angular momentum and must shed mass in an equatorial plane; doing so, our protosun may have given rise to the rotating nebular disk from which the planets subsequently formed. As apparently first pointed out by Jacques Babinet in 1861 and by Maurice Fouché in 1884,[1] according to the nebular theory the sun should contain most of the angular momentum of the solar system, but in fact the sun possesses only 2% of this momentum, the planets possessing the remaining 98%.

Recognizing the serious angular momentum difficulty in the solar system, Thomas C. Chamberlin and Forest Ray Moulton proposed an alternative encounter theory, according to which a star, passing close to the sun, raised a solar tide, so that material was lost from the sun. If the sun and other stars were to impart sufficient angular momentum to the ejected material, the angular momentum distribution of the solar system might be explained. Here Chamberlin and Moulton were building upon the previous proposal of Georges-Louis Buffon, who in 1745 suggested that a comet hitting the sun might have torn off sufficient material to produce planets. The first selection given here comes from Moulton's elementary astronomy textbook, which closely parallels his

Astrophysical Journal article on the same topic.[2] He here argues that if a nearby star were exerting tidal forces, material normally ejected from the sun (which would ordinarily fall back in) would be drawn out into elongated filaments projecting outward from the near and far sides of the sun. The material in these filaments would quickly cool and solidify, creating small chunks of matter called planetesimals, which would subsequently coalesce to form the planets.

Harold Jeffreys and James Jeans advanced another variety of the encounter theory in the period between 1916 and 1919.[3] They argued that tidal action itself would suffice to explain the origin of the solar system, without calling upon intermittent eruptions from the sun, and they postulated that the material drawn out of the sun was in the form of a long cigar-shaped filament, which later condensed at various points along its length to give the planets. The cigar shape would account for the more massive planets in the center of the system.

One of the earliest criticisms of the encounter theories was based on the extreme rarity of stellar collisions. The sun, for example, would take 10^5 years to travel to its nearest neighbor. The probability of its striking this particular star is, however, only one in 10^{16}. This case can be extrapolated to the entire collection of 10^{11} stars in our galaxy: the probability of any given star colliding with any other star during the entire lifetime of the galaxy is only 10^{-11}. Accordingly, only one or two star collisions can ever have occurred in our galaxy, but various observational evidence suggests that planetary systems are considerably more common than a few collisions could explain.

In the early 1930s, Friedrich Nölke and Henry Norris Russell advanced other serious objections to the encounter theory for the origin of the solar system.[4] Nölke showed that if a collision did result in the extraction of matter from the sun, the shearing force of the solar gravitational field would be too strong to allow the matter to condense by its own gravitational contraction; Russell showed that if the encounter were responsible for the present distribution of angular momentum, the planets closest to the sun would have been the largest. In the second of these two papers, Lyman Spitzer delivers the death blow to the encounter theories, noting that material ejected from the sun will dissipate rather than condense into planets. This is essentially because the ejected gas will have a kinetic energy of thermal motion exceeding its gravitational potential energy. As we now realize, a hot origin for planetary material conflicts with the observed abundance of deuterium, lithium, beryllium, and boron on the earth. In the sun these elements are consumed by nuclear reactions at temperatures on the order of 10^6 K, the approximate temperature of the solar material. If the earth were made of material torn from the hot sun, the observed abundances of these elements could not be explained.

1. J. Babinet, *Comptes rendus de l'Académie des sciences (Paris) 52*, 481 (1861), M. Fouché, *Comptes rendus de l'Académie des sciences (Paris) 99*, 903 (1884).
2. T. C. Chamberlin, *Astrophysical Journal 14*, 17 (1901); F. R. Moulton, *Astrophysical Journal 22*, 165 (1905).
3. J. H. Jeans, *Monthly Notices of the Royal Astronomical Society 77*, 186 (1916), *Memoirs of the Royal Astronomical Society 62*, 1 (1917), *Problems of Cosmogony and Stellar Dynamics* (Cambridge: Cambridge University Press, 1919); H. Jeffreys, *Monthly Notices of the Royal Astronomical Society 77*, 84 (1916), *78*, 424 (1918).
4. F. Nölke, *Monthly Notices of the Royal Astronomical Society 93*, 159 (1932); H. N. Russell, *The Solar System and Its Origin* (New York: Macmillan Co., 1935).

Moulton, The Spiral Nebula Hypothesis

Hypothesis Respecting the Antecedents of our Present System

THE SOLAR SYSTEM exists and is in the midst of an evolution; the problem is to trace out this evolution. The historical theories have been seen to be untenable, and the question arises whether at the present time an hypothesis can be formulated whose implications are in agreement with observed phenomena. An attempt is now being made to work out a theory along somewhat new lines, and a sketch of its main features will be given.

Instead of supposing that the solar system started from a vast gaseous mass in equilibrium under the law of gravitation and the laws of gaseous expansion, the spiral hypothesis postulates that the matter of which the sun and planets are composed was, at a previous stage of its evolution, in the form of a great spiral swarm of discrete particles whose positions and motions were dependent upon their mutual gravitation and their velocities. Gaseous expansion preserved the dimensions of the Laplacian nebula, while in this the orbital motions were the dominant factor. Because of the fact that every particle is supposed to have moved nearly independently like a planet, Chamberlin calls the theory the *Planetesimal Hypothesis.*

Before discussing the possible origin of a spiral swarm of particles, and the details and merits of the planetesimal theory, attention should be called to the fact that there is not an example of a Laplacian ring nebula among the thousands of nebulas which are known. On the other hand, spirals are very numerous, particularly among the smaller and fainter nebulas. The photographs which Keeler made at the Lick Observatory shortly before his death led him to the conclusion that the spiral is the normal type. He said:

> 1. Many thousands of unrecorded nebulas exist in the sky. A conservative estimate places the number within the reach of the Crossley reflector at about 120,000. The number of nebulas in our catalogues is but a small fraction of this.
>
> 2. The nebulas exhibit all gradations of apparent size from the great nebula in Andromeda down to an object which is hardly distinguishable from a faint star disk.
>
> 3. Most of these nebulas have a spiral structure.... While I must leave to others an estimate of the importance of these conclusions, it seems to me that they have a very direct bearing on many, if not all, questions concerning the cosmogony. If, for example, the spiral is the form normally assumed by a contracting nebulous mass, the idea at once suggests itself that the solar system has been evolved from a spiral nebula, while the photographs show that the spiral is not, as a rule, characterized by the simplicity attributed to the contracting mass in the nebular (Laplacian) hypothesis. This is a question which has already been taken up by Chamberlin and Moulton of the University of Chicago.

The spiral nebulas have dark lanes down between their arms, and it is evident that, if the distribution of matter in them is even approximately as it appears to be, their forms are preserved almost entirely by the motions of their separate parts instead of by gaseous pressure.

A Possible Origin of Spiral Nebulas

The theory of the evolution of the system from a spiral nebula is largely independent of any hypothesis about the origin of the spiral. However, a possible, and even probable, mode of generation of these remarkable forms has been suggested by Chamberlin; and for the sake of having a definite theory to work on, it will be assumed, at least provisionally, that the solar spiral nebula was developed in this way.

The stars are moving with respect to one another, often with very great velocities, and apparently in every direction. It follows that in the course of a time which may be extremely long indeed they will pass very near other stars, or possibly collide with them. If a collision occurs, the chances are very great that it will be oblique rather than central, and a spiral may be formed; but the chances of simply a near approach are enormously greater, and only this case will be treated here.

When two large bodies come near each other they are subject to great tidal strains, which, according to the researches of Roche, entirely break them up if their distance apart is less than 2.44 . . . times their radii. Suppose, however, that this limit is not reached. Then they do not disintegrate under tidal strains alone, but when these forces are added to the eruptive tendencies of highly heated gaseous bodies, it is almost certain that masses of matter burst out and recede to great distances. It is like the eruptive prominences on the sun, only on a vastly greater scale. These eruptions occur in the directions of the greatest disturbing forces. It follows from the character of the tide-raising forces that they are straight toward and from the tide-raising body. If this matter were undisturbed it would fall straight back on the body from which it burst forth, but the tide-raising body changes its orbit into an ellipse, as will be shown.

Here Moulton considers the gravitational attractions between two stars, S and S', moving relative to one another, and finds that material ejected from one star will always be drawn toward the line joining the two stars. If matter is continuously ejected as one star moves past another, spiral arms will be drawn out from the other star.

When we see a spiral nebula we do not see the paths which the separate masses have described, but the positions which they occupy at the time. In the present case if a smooth curve is drawn through the regions where the matter is densest, it will form a sort of double spiral as represented by the full lines. There will be nuclei here and there along the arms of the spiral where large masses have been ejected, and the whole space will be more or less filled with finely divided and nebulous material. It must be remembered that *the matter does not move along the arms of the spiral*, but in orbits which cross them at large angles. The particles in the smaller orbits will move faster than the outer ones, and the spiral will become more and more coiled with age until its spiral character can no longer be discerned. In the photographs of spiral nebulas the *two arms* can nearly always be easily made out, and it is significant that no other number is certainly found. But it is almost certain that the spiral nebulas which have been photographed are much greater than the one from which our system may have developed.

The Development of the Solar System from a Spiral Nebula

The consequences of the hypothesis that the solar system has developed from a spiral nebula of the type just considered will now be worked out, and the results will be compared with the known condition of the system.

For simplicity in the exposition, the statements will be made positively as though there were no further question about the evolution of the system, and as if this were simply an account of its development. All those phenomena which were enumerated as supporting or opposing the Laplacian hypothesis will be reviewed in connection with the spiral theory.

The Origin of Planets

The various planets have grown out of the original nuclei by the gradual accretion of the smaller particles whose orbits crossed, or passed near, their orbits. The larger nuclei, especially those that passed through regions rich in the finer material, gave rise to the larger planets. Marked irregularities in the masses are to be expected rather than the opposite. The planetoids have formed from a mass of material with no large dominating nucleus.

The Origin of Satellites

When the planetary nuclei left the sun they were accompanied by smaller secondary nuclei. Those secondary nuclei which had large velocities with respect to their primary nucleus escaped from its gravitative control and became independent bodies. When their relative velocities were very small, they fell upon the primary nucleus. When their velocities were moderate, they became satellites.

The Planes of the Planetary Orbits

The planes of the orbits of the planets are very nearly the same as the plane of the orbit of S' when it passed by S. It is clear that small differences in the planes are to be expected because of the various conditions which control the direction of projection, and that the greatest difference may well be in the nearest planet whose material was for a shorter time under the influence of S'. Moreover, the process of sweeping up the scattered material, which in general would be distributed symmetrically with respect to the plane of the orbit of S', would tend to reduce the planes of the orbits of the various bodies to coincidence. In general, the more a nucleus grew by sweeping up the scattered material, the more nearly the plane of its orbit coincided with the general plane of the system.

Turning to the observational data we actually find the orbit of Mercury more highly inclined to the average plane of the system than that of any other planet. The orbits of all the great planets are nearly in the same plane. On the other hand, the planes of the orbits of the planetoids are often highly inclined. For example, the orbit of Eros is inclined about 10° to the orbits of the earth and Mars, an unexplainable condition under the Laplacian theory.

Rotation and Equatorial Acceleration of the Sun

The present rotation of the sun is the resultant of its rotation before the appearance of S', and of the disturbances produced by this body. Its original axis of rotation is unknown, but it is very improbable that it was exactly perpendicular to the plane of motion of S'. S' disturbed the rotation of S in two ways. First it raised an enormous tide in S which it pulled around S in the direction of its revolution. In this way it contributed a large moment of momentum around an axis perpendicular to the plane of its orbit, and this moment of momentum has been constant ever since. In the second place, a considerable quantity of the material which was ejected and had its straight line orbit changed to an elliptical orbit would still have its perihelion distance less than the radius of S. Consequently, it would be precipitated again on the sun, and inevitably in such a way as to increase the moment of momentum of the sun around the same axis. This cause may have been more efficient in determining the character of the sun's rotation than the other. Hence, since the sun's present rotation is the resultant of its original rotation and these disturbing factors, we should expect to find its equator near, but not exactly in, the average plane of the planetary orbits.

Both of the factors which have just been noted were more important in the equatorial zone than in any other. Consequently there was an original equatorial acceleration which has not yet been worn out by friction on the lower parts. The spots are in those zones where the relative motions of the different layers are the greatest, and they are probably due in some way to these relative motions.

Moulton next describes how nuclear concentrations acquire nearly circular orbits and direct rotations as they gradually capture small solid planetesimals and grow into planets. When any two bodies with the same orbital axes and eccentricities are subject only to their mutual perturbations and unite by collision, the orbit of the combined mass is less eccentric than the original orbits were. For this reason, the originally eccentric orbits of the nuclear condensations are gradually rounded out by the accretion of planetesimals. Interactions with the resisting medium of small solid planetesimals is also shown to account for the direct rotations of the planets. Although satellites may revolve around their planet in any direction, the chance of surviving as independent bodies during interaction with the planetesimals is greatest for satellites which have direct orbits. The orbits of these satellites are also rendered less eccentric by their impact with planetesimal material.

The Moment of Momentum of the System

The moment of momentum of the system is nearly all possessed by the planets, Jupiter having over 95 per cent of that which is within Saturn's orbit. This condition of affairs is precisely that which the postulated origin of the spiral nebula would lead to, and it is entirely in agreement with the theory which has been outlined. The total moment of momentum of the system is the measure of the disturbances produced by the sun S'.

The Zodiacal Light

The zodiacal light is probably the light reflected from the vast number of particles moving in orbits so nearly parallel to that of the earth that they have been swept up very slowly. Many of them are the remains of the original material scattered by S', though some of them are the remains of disintegrated comets. The large meteorites which fall upon the earth, often containing large quantities of occluded gases, are probably part of the material ejected from the sun, though some of them give very strong evidence of having been once parts of a large solid body. They may have come from planets which existed before the advent of S', and which were broken up and destroyed by this body.

The Evolution of the Planets

The evolutions of the small and large planetary nuclei have been quite different. They were all very hot at the time of their ejection. The small nuclei did not have sufficient gravitative control to retain their lighter gases. In a comparatively short time they had no appreciable atmospheres, and they speedily cooled until they became solid. The meteoric matter which fell in upon them was also in a solid state. The relative velocity was in general so small that no great amount of heat was generated by the impact, and what was produced speedily radiated away. After the masses began to assume earth-like dimensions the interior pressure became very great and they diminished in volume. This shrinking produced *interior* heat just as it does in the case of the sun. Computation shows that if the earth shrank so that its density increased from the average density of known meteorites to its present density, enough heat would be generated to increase its temperature (assuming a specific heat of 0.2, which is about that of rock) more than 10,000° Fahrenheit. This heat would be conducted to the surface and lost very slowly, and it is much more than sufficient to account for all the known igneous action in the case of the earth. But the earth as a whole has been solid throughout its history.

The earth acquired its atmosphere chiefly after it became about as large as Mercury. The atmospheric gases came from the interior, squeezed, as it were, out of the heated and compressed material. Bodies much smaller than Mercury have never retained any real atmospheres. This applies to most of the satellites and to all of the planetoids.

On the other hand, the large planetary nuclei were so massive that they never lost their light gaseous envelopes. Because of this their original heat was largely retained, and they have not yet contracted to any great extent. They are less dense than the smaller planets both because they retained nearly all of the original light elements, and also because the conditions have been unfavorable to their cooling and contracting.

The Age of the Solar System

No certain answer can be given to the question of the length of time that has been required for this evolution of our system to take place. It is certainly very great. The greatest difficulty is in accounting for the apparently undiminished vigor of the sun. If it was in its maturity at the time of the visitation of S', apparently it ought to be far in its decline now. The only explanation available, and the one which had to be evoked also in the Laplacian theory, is that probably the contraction theory of the sun's heat accounts for only a small part of the energy which becomes available in this form.

Spitzer, The Dissipation of Planetary Filaments

Abstract—An investigation is made of the forces acting on a cylindrical filament of stellar matter produced by an encounter between two stars. It is shown that under rather general assumptions such a filament will expand under its own internal pressure much more rapidly than it will lose energy either by radiation or by turbulent convection. A similar conclusion may be deduced for a thin ribbon-like filament produced by a grazing collision. It seems probable that a stellar encounter will simply produce an extended gaseous nebula around one or more of the stars involved.

SINCE LYTTLETON'S[1] important paper there has been a renewed interest in encounter theories of the origin of the solar system. The papers by Luyten[2] and by others which have followed have dealt primarily with the stellar dynamics of the proposed encounter and have been largely concerned with difficulties of energy and momentum.

It is also relevant to investigate the conditions under which a filament of stellar matter, presumably torn out of both stars by the encounter, can condense into planets. Early work[3] on this phase of the problem involved the assumption that the temperatures of the filaments were of the same order as those of the solar photosphere. As Russell[4] has stressed, however, this assumption is certainly incorrect, since a filament with as much mass as Jupiter must come from layers in which the stellar temperature is at least 10^6 degrees. This fact had also been discussed by Jeffreys,[5] who showed that the planets would expand to radii comparable with the distances of their satellites. Jeffreys discussed only equilibrium configurations, however, and did not investigate either the rate of expansion or the rate of loss of internal energy by radiation.

As will be shown hereinafter, the gases within a filament at a temperature of hundreds of thousands of degrees will accelerate outward and within a few hours will reach the velocity of escape. The filament will then presumably dissipate. This fate might perhaps be avoided by the radiation of energy at a sufficiently rapid rate. The optical depth of any filament with appreciable mass, however, will be at least several millions, and the time required to radiate half the internal energy of the filament would be at least several months if there were no expansion. Since no convection currents with velocities less than the escape velocity can dispose of so much energy in a few hours, the filament must therefore explode.[6]

In section 1 of this paper it is shown that a filament produced by an encounter has, in general, a positive energy, i.e., more than enough to dissipate the entire filament to infinity. Section 2 presents the formulae for the rate of expansion of a cylindrical filament of circular cross-section on the assumption that no radiative loss of energy occurs. In section 3 the possibility of convection or turbulence is considered, and a solution of the equation of radiative transfer is given for the case in which the intensity of radiation is a function of the time; it is shown that the rate of energy loss is less than a hundredth of that required to avert the dissipation of the filament. Section 4 extends the analysis to a ribbon-like filament. Section 5 discusses the relevance of these results to the encounter theory of the origin of the solar system.

1

It is evident that, if a gaseous filament has a positive energy, the filament will expand to infinity (provided γ exceeds 4/3) unless it loses energy at a sufficiently rapid rate. If a filament is to be unstable, then, it is a necessary condition that the ratio of the internal positive energy to the absolute value of the negative gravitational energy exceed unity. A lower limit for this quantity is easily calculated.

One may assume that the matter in the filament, assumed to be of total mass M, was originally in one of the two stars involved in the supposed encounter, and that this matter had originally a maximum density ρ_{st} and a maximum temperature T_{st}. Since, in general, the temperature within a star decreases outward much less rapidly than the density, the mean stellar temperature of the matter in the filament may be taken as not less than $(1/2)T_{st}$. (In the standard model, for instance, the mean temperature for the entire star is 0.585 times the central temperature.)

The total internal energy W of the filament will then be not less than $3kM[(1/2)T_{st}]/2\mu m_0$ ergs, where μ is the mean molecular weight, m_0 is the mass of unit atomic weight, and k is the usual Boltzmann constant. If γ, the ratio of the specific heats, is less than 5/3, W is, of course, considerably greater than the foregoing lower limit.

The self-gravitational energy $-\Phi$ of the matter in the filament will have an absolute value not greater than that of a sphere of uniform density ρ_{st} with a radius S equal to $(3M/4\pi\rho_{st})^{1/3}$, for which the potential energy is $-3GM^2/5S$, where G is the gravitational constant. If we use the well-known formula for T^3/ρ in a gaseous star, assume that radiation pressure is negligible, and use the Eddington mass-luminosity relationship, a minimum value of W/Φ is given by

$$W/\Phi > 0.283\left(\frac{M_0}{M}\right)^{2/3}, \quad (1)$$

where M_0 is the mass of the parent-star. This is greater than unity when M is less than $0.15M_0$.

We see that W/Φ is greater than unity for small masses. If M is comparable with M_0, a separate investigation is required. Let us consider the most favorable case, in which a star is split by an encounter into two spheres of equal mass. From the virial theorem it follows that, if radiation pressure is negligible, W/Φ for the parent-star is 0.50—almost twice as large as the lower limit found above when M is set equal to M_0. After the encounter the original kinetic energy will be divided equally between the two spheres. The potential energy of each, however, will be $(1/2)^{5/3}$ times the former value of $-\Phi$, and hence W/Φ for each sphere will be 0.79. Unless the stellar matter consists entirely of hydrogen and helium, the additional energy of ionization and excitation would suffice to raise W/Φ well above unity.

In addition, the encounter itself may be expected to increase W/Φ even further. If the passing star is sufficiently massive and energetic to split the original star into two equal masses, it will presumably give gravitational energy to each component, decreasing Φ. Physically this means that, if a star is torn in two by tidal action, one may expect each component to be enormously elongated and hence to have much less gravitational energy. The internal energy W could not be decreased by the encounter, since superelastic collisions seem impossible.

In other words, a gaseous star A in equilibrium has a reserve of internal energy equal in magnitude to at least one-half its gravitational energy. A catastrophic encounter with star B will divert energy to A from the mutual motion of A and B. This will result in an enormous decrease of the gravitational energy of A, reflected partly in the disruption of A and partly in the elongation of each component of A. If, moreover, the encounter is a grazing one, the thermal energy of A will be increased. As a result, the internal energy will exceed the value of the negative gravitational energy, and either or both components will explode.

One may conclude, therefore, that any mass of stellar matter torn loose from one star by the passage of another will have more than enough energy to dissipate itself to infinity. Radiation of energy, however, will decrease the value of W; whether or not a filament formed by an encounter would actually disrupt completely therefore depends on the rates both of expansion and of radiation.

2

In order to calculate the rate of expansion, certain assumptions of uniformity are apparently necessary; these simplifications should not alter the order of magnitude of the results. One may consider an idealized cylindrical filament of circular cross-section stretched between two stars, one of which is receding with a velocity V relative to the other. Let M, R, and z denote the mass, radius, and total length of the filament; M', R', and z' will be used to denote the same quantities in units of the solar mass and radius. The masses and radii of the two stars are left unspecified, as they play no primary part in the expansion of the filament.

The material in the filament will be acted upon by the pressure of the gases in the filament and by the gravitational attraction both of the filament itself and of the two stars. Only those forces will be considered which act in the radial direction in the filament, perpendicularly to the axis of the filament. The equation of motion in this simplified case is

$$a = -\frac{1}{\rho}\frac{\partial p}{\partial r} - \frac{\partial \varphi}{\partial r}, \tag{2}$$

where r is the perpendicular distance from the axis of the filament, a is the radial acceleration, p the pressure, and φ the gravitational potential. If we assume that all quantities are functions of r and t alone, the mass average of the acceleration, which will be denoted by $\bar{a}$, becomes

$$\bar{a} = -\frac{2\pi z}{M}\int_0^R \left(\frac{\partial p}{\partial r} + \rho\frac{\partial \varphi}{\partial r}\right) r\,dr. \tag{3}$$

The integral of the first term on the right-hand side of (3) will be denoted by $\bar{a}_1$, that of the second, by $\bar{a}_2$. If now we integrate by parts for $\bar{a}_1$ and substitute $\pi R^2 z\rho_m$ for M, where ρ_m is the mean density, we have

$$\bar{a}_1 = \frac{2}{R^2\rho_m}\int_0^R P\,dr = \frac{2}{R\rho_m}\bar{P}. \tag{4}$$

P_m, the average of P over the volume, will be less than $\bar{P}$ as defined in (4), since the large values of P for small r are weighted much more heavily in $\bar{P}$ than in P_m. We may therefore find a lower limit for $\bar{a}_1$ by setting $\bar{P}$ equal to P_m. The value of T does not vary appreciably throughout most of the filament, and we may legitimately assume that T is constant. With these two assumptions, we have

$$\bar{a}_1 = \frac{2kT}{\mu m_0 R}, \tag{5}$$

where the symbols have the same meanings as before. If the filament has been assumed to expand adiabatically, we know that

$$T = T_{st}\left(\frac{\rho}{\rho_{st}}\right)^{\gamma-1}, \tag{6}$$

where γ is the ratio of the specific heats. Formula (6) is rigorous only for the portion of the filament which was initially at the maximum temperature T_{st} and had the initial maximum density ρ_{st}. To an adequate approximation, however, this formula may be applied to the mean temperature and density of the gases in the filament.

To evaluate $\bar{a}_2$ we appeal to the virial theorem. This theorem states that a gaseous system expands when its thermal kinetic energy W is greater than $\frac{1}{2}\Phi$, where $-\Phi$ is again the gravitational energy. Since in the previous section it was shown that W is actually greater than Φ, one may deduce that $\bar{a}_1$ is at least twice as great as that part of $\bar{a}_2$ which arises from the gravitational attraction of the filament on itself. It is therefore possible to neglect this gravitational attraction in the computation of the rate of expansion; the rate so calculated will be too great by a factor of less than 2.

The gravitational attraction which the two stars exert on the filament is difficult to discuss in detail but is not likely, in any case, to prevent the expansion. Only the radial component of this attraction is effective in this connection, and since the force perpendicular to the line connecting the two stars varies as the inverse cube of z, it is evident that as soon as z' is much greater than unity such a force will be negligible.

We may let t_0 be the time at which this stellar gravitational field becomes negligible and the filament begins to expand. Similarly, ρ_0 and z_0 will be used to denote values of ρ_m and z at the time t_0. If t_1 denotes the value of t for which the radial velocity outward, v, equals the velocity of escape from the filament, v_∞, then we have

$$v_\infty = \int_{t_0}^{t_1} a(t)\,dt. \tag{7}$$

If we neglect $\bar{a}_2$, assume that a_1 is constant through time, and replace v_∞ by its value for a spherical mass M of radius R, we find from (5), (6), and (7) that

$$t_1^* - t_0 = \left(\frac{\rho_{st}}{\rho_0}\right)^{\gamma-1} \frac{\mu m_0}{kT_{st}} \left(\frac{MGR}{2}\right)^{1/2}. \tag{8a}$$

Substituting $M/\pi R^2 z_0$ for ρ_0, we have

$$t_1^* - t_0 = 1.70 \cdot 10^{10} (\rho_{st} z_0')^{2/3} R'^{11/6} M'^{-1/6} T_{st}^{-1} \text{ sec}, \tag{8b}$$

where R', z', and M' again denote $R/R_\odot$, $z/R_\odot$, and $M/M_\odot$, respectively; t_1^* denotes the value of t_1 derived on the assumption that ρ is constant with time; and γ in (8*b*) has been set equal to 5/3, its value for a monatomic gas. We assume throughout that μ is unity. If T_{st} is 10^6 degrees, M' is 1/500, ρ_{st} is 1/10, and other quantities are of order unity, $t_1^* - t_0$ will equal approximately three hours. Since v_∞ is of the order of 30 km/sec, the change in R during this time will be $\frac{1}{4}R_\odot$; the change in z during the same period will, of course, be considerably larger.

More realistically, we may assume that ρ is proportional to $1/z$ and that z increases uniformly at a rate V, the velocity of recession of the one star relative to the other when t is equal to t_0. The time origin is chosen so that z equals Vt. In this case t_0 must be at least as great as the value of t for which the filament ceases to acquire additional mass, the filament expanding along its axis as the stars recede. With this assumption we obtain from (7)

$$t_1 = t_0 \left\{1 + (2-\gamma)\frac{t_1^* - t_0}{t_0}\right\}^{1/(2-\gamma)}, \tag{9}$$

where $t_1^* - t_0$ is given in (8) above. The quantity t_0 is equal to z_0/V; if V is 10^8 cm/sec and z_0 is $2R_\odot$, t_0 equals $1.4 \cdot 10^3$ sec, or about half an hour.

3

One may consider next the loss of energy from the filament. It will be shown that neither convection currents nor radiative transfer can be responsible for an appreciable loss of energy from the filament in the short space of a few hours.

Convection currents are easily disposed of. No current of matter can reach the surface of the filament from the far interior before the time t_1^* unless the velocity of the current is greater than the velocity of escape from the filament. More quantitatively, if the condition that $(t_1^* - t_0)v_\infty$ be less than R is combined with (7) and with the formula for v_∞ in the spherical case, this gives roughly the condition that $\bar{a}_1$ be greater than $2\bar{a}_2$, which, as we have already seen, is fulfilled in all relevant cases. Hence, if such currents are responsible for any important transfer of heat, they must leave the filament entirely and will promote the dissipation of the filament rather than hinder it.

A number of approximations reduce the problem of radiative transfer to a tractable form in the present case. As a first approximation we assume that the filament is static.

Spitzer proceeds to solve the equations of radiative transfer and radiative equilibrium in the cylindrical case. He defines ξ as the ratio of the energy density of radiation to the total energy density within the filament, and t_2 as the time at which the energy density equals one-half the value it had at t_0 when the filament started to expand. With the opacity k at the value for electron scattering and M' at least 10^{-3}, $t_2^* - t_0$ is scarcely less than a year.

If the gases in the filament are to condense into planets, they must radiate more than half their internal energy before the time t_1, and the ratio of $t_2 - t_0$ to $t_1 - t_0$ must be considerably less than unity. When ρ and k are assumed constant through time, this ratio is given by (8*b*) and (16*b*), which yield

$$\frac{t_2^* - t_0}{t_1^* - t_0} = \frac{6.42 M'^{7/6} k T_{st}}{z_0'^{5/3} R'^{11/6} \rho_{st}^{2/3} \xi}. \tag{17}$$

Since k should be at least unity, k/ξ may be given a minimum value of 10. We may equate M to one-half the solar mass exterior to the radius at which the density and temperature before the encounter were ρ_{st} and T_{st}, and determine these quantities from the usual standard model. If z' and R' are assumed to be less than 10 and unity, respectively, we find that $(t_2^* - t_0)/(t_1^* - t_0)$ is at least 80,000.

Spitzer next considers a uniform longitudinal expansion of the filament, and he derives a modified formula to find an extreme minimum value of the foregoing ratio. If t_1^* and t_2^* are both considerably greater than t_0, this lower limit becomes

$$\frac{t_2 - t_0}{t_1 - t_0} = 38.2 t_0^{5/2} \frac{(t_2^* - t_0)^{1/2}}{(t_1^* - t_0)^3}, \qquad (18a)$$

$$= \frac{3.27 \cdot 10^3 M' k^{1/2} T_{st}^3}{\xi^{1/2} \rho_{st}^2 R'^{11/2} V^{5/2}}. \qquad (18b)$$

Even if V is as great as 10^8 cm/sec, formula (18*b*) still gives a value greater than a hundred, provided values of M, ρ_{st}, and T_{st} are again determined from the standard polytropic model for the sun, while R' and k/ξ are taken to be not greater than unity and not less than 10, respectively. Thus, we see that in general the idealized filament discussed here is unable to radiate an appreciable fraction of its energy in a sufficiently short time to avert complete dissipation.

4

The preceding arguments have been developed for the case of filaments with circular cross-section. It has been suggested also that a very close encounter or grazing collision might produce a thin ribbon or sheet of stellar matter connecting the two stars. Essentially the same analysis may be applied in this case as well.

There is no need, however, to derive another set of formulae applicable to this case. It is evident from (18*b*) that for a filament of given mass M and given length z, the least value of $(t_2 - t_0)/(t_1 - t_0)$ is attained for maximum radius R. It also follows from general principles that the value of this ratio will increase as the initial cross-section of the filament is deformed from a circle of radius R to an ellipse with semi-major axis R and minor axis h, keeping the mass of the filament constant.

Spitzer next shows that the ratio $(t_2 - t_1)/(t_1 - t_0)$ will increase slightly as h is decreased. The escape velocity is relatively unchanged during the deformation, and if convection currents convey enough matter to the surface to produce an appreciable effect their velocity must be greater than the escape velocity of the filament. From these considerations and from the conclusions reached in section 3, Spitzer infers that a cylindrical filament will be unable to avert expansion and dissipation, regardless of what its cross section may be.

It is of interest to note the magnitude of these effects. For a filament of length $R_\odot$, of width $2R_\odot$, of thickness $R_\odot/50$, of mass $M_\odot/500$, and for which $\rho_{st}^{2/3}/T_{st}$ equals 10^{-6}, the acceleration is so great that less than two minutes are sufficient for the gases to acquire the escape velocity of roughly 35 km/sec; during this time the thickness of the filament increases, on the average, not more than 15 per cent.

5

The foregoing analyses are based on several approximations, but there is a considerable margin of safety in the conclusions. Even under extreme conditions there is a difference of two orders of magnitude between the short time required to produce macroscopic velocities sufficient to disrupt the filament and the long interval necessary to radiate half the internal energy. It is not likely that any or all of the various approximations made could introduce so large a factor.

It is difficult, therefore, to see how the filament as a whole can avoid dissipation. Although only the average acceleration has been discussed, it is unlikely that any part of the filament can remain after most of it has been forced off into space. The radial acceleration near the center of the filament is small, to be sure, but the velocity of escape decreases linearly with r for a uniform filament. Furthermore, the central gases will remain at the highest temperature for the longest time and will thus have more chance for expansion.

Nor is it probable that the outer portions of the filament might condense into planets as the filament expanded. Any sphere of gas in the expanding filament will itself be expanding in a direction perpendicular to the radius vector of the filament with a velocity directly proportional to the radius of the sphere. But the velocity of escape from any sphere will also vary directly as the radius. Thus, all spheres within the filament have the same chance of condensing, independently of their radii. The filament as a whole, however, and hence a sphere with a radius equal to that of the filament, cannot condense after the time t_1 has been reached. It follows that, as soon as the gases of the filament have reached the velocity of escape from the filament, no large portion of the filament can condense, independently of the temperature to which it may fall.

If, then, such a filament would not be likely to condense into planets, what would be the fate of the matter so ejected? Such of it as did not fall back into the stars or dissipate into free space would form an enormously extended atmosphere of some sort around one of the two stars directly involved in the encounter or around their possible companions. With an assumed mass of $2 \cdot 10^{30}$ grams and a radius of 30 astronomical units, such an atmosphere would have a mean density of $5.4 \cdot 10^{-15}$ gm/cm^3 and would be comparable with the solar chromosphere. Such an atmosphere is reminiscent of the Laplace nebular hypothesis, except that in this case there need be no lack of angular momentum. The validity of the encounter theory as an explanation of the origin of the solar system rests apparently on whether or not a nonuniformly rotating atmosphere could condense into solid bodies.

1. *M.N. 96*, 559 (1936).

2. W. J. Luyten and E. L. Hill, *Ap. J. 86*, 470 (1937); *M. N. 99*, 692 (1939); R. A. Lyttleton, *M. N. 98*, 536 (1938).

3. H. Jeffreys, *The Earth* (Cambridge: Cambridge University Press, 1929), p. 27.

4. *The Solar System and Its Origin* (New York: Macmillan, 1935), p. 112.

5. *M. N. 89*, 731 (1929).

6. This result is reminiscent of A. W. Bickerton's theory of partial impact (*Trans. N. Zealand Inst. 11*, 125 [1878]), in which the appearance of a nova was attributed to the explosion of such a filament. The origin of the planets, however, was also attributed to this process. Bickerton deserves mention for what is probably the first detailed statement of the encounter theory of the formation of the solar system.

The expansion of such a filament has also been discussed by F. Nölke in his thorough investigation of encounter theories (*Der Entwicklungsgang unseres Planetensystems* [1930], pp. 188–191). His analysis is primarily intended, however, to show that the filament will leave the gravitational field of the sun, and he considers the motion of a point mass in the gravitational field of the two stars, neglecting both the mass and the internal pressure of the filament. Nölke's investigation is thus concerned with the same dynamical difficulties that Lyttleton and Luyten have more recently discussed, and has no direct connection with the present analysis.

32. The Nebular Theory of the Origin of the Solar System

The Stability of a Spherical Nebula

James H. Jeans

(*Philosophical Transactions of the Royal Society of London 199*, 1–53 [1902])

The Origin of the Solar System

Carl Friedrich von Weizsäcker

TRANSLATED BY BRIAN DOYLE

(*Naturwissenschaften 33*, 8–14 [1946])

The contemporary view that the solar system was formed from a cloud of interstellar gas goes back to Immanuel Kant in 1755 and Pierre Simon Laplace in 1796. Yet neither Kant nor Laplace made a detailed mathematical analysis of the problem, and in our first selection here James Jeans provides the first quantitative framework for the nebular theory. Jeans considers the conditions under which a spherical nebula will become unstable, for it is only under the onset of some instability that a primeval nebular gas will contract to form a star. If a single gaseous mass has no tendency to break up into condensations, we would have no reason to expect that stars would form in galaxies or that galaxies themselves would exist. As Jeans shows, only objects that exceed a certain critical size, called the Jeans wavelength, and a certain critical mass, called the Jeans mass, will become gravitationally unstable and begin to contract.

Although the Jeans criteria for gravitational instability provided a mathematically rigorous background for the nebular theory of the origin of stars, the nebular theory could not account for the observed distribution of angular momentum of the solar system. The central, compact sun should contain most of the angular momentum of the solar system, but the planets, with only 0.14% of the mass of the solar system, actually possess 98% of the angular momentum. This dilemma led to the abandonment of the nebular hypothesis in the early part of the twentieth century.

It was the German physicist C. F. von Weizsäcker who developed one of the first plausible modifications of the nebular theory, and as the second of these selections, we provide a translation of Weizsäcker's short, nontechnical account of his theory.[1] Weizsäcker applied the physical theory of turbulence to the primeval nebula and argued that the planets contain most of the angular momentum of the solar system because turbulent motions remove angular momentum from the sun. By introducing turbulent motions, vortex cells, and whirlpool eddies into a rotating gaseous envelope, Weizsäcker provided physical interpretations for several hitherto unexplained features of the solar system, including the spacing of the planets in an orderly sequence (the Titius-Bode law).

By presenting Weizsäcker's paper we do not mean to suggest that his theory provides the most likely explanation for the origin of the solar system; rather, we select it as a representative pioneering work. Hannes Alfvén and Otto Schmidt have, for example, both proposed interesting theories in which the planets were formed when a well-developed sun passed through an interstellar gas cloud. Schmidt argued that the inner sun would naturally collect most of the material with less angular momentum, and Alfvén argued that the sun's magnetic forces can help explain the distribution of mass and angular momentum in the solar system. Similarly, Fred Hoyle and others have invoked a solar magnetic field for the transfer of angular momentum from the sun to the protoplanets. Raymond Lyttleton has developed the theory of accretion of matter in a Schmidt-like gas cloud, whereas William McCrea has put forward a modified nebular theory in which minor condensations are subsequently captured by the protosun.[2]

1. For a technical account see C. F. von Weizsäcker, *Zeitschrift für Astrophysik 22*, 319 (1943).

2. For a recent review of these theories see M. M. Woolfson, *Reports of Progress in Physics 32*, 135 (1969). Also see O. Y. Schmidt, *A Theory of Earth's Origin* (Moscow: Foreign Languages Publishing House, 1958).

Jeans, The Stability of a Spherical Nebula

INTRODUCTION

THE OBJECT of the present paper can be best explained by referring to a sentence which occurs in a paper by Professor G. H. Darwin.[1] This is as follows:

"The principal question involved in the nebular hypothesis seems to be the stability of a rotating mass of gas; but, unfortunately, this has remained up to now an untouched field of mathematical research. We can only judge of probable results from the investigations which have been made concerning the stability of a rotating mass of liquid."

In so far as the two cases are parallel, the argument by analogy will, of course, be valid enough, but the compressibility of a gas makes possible in the gaseous nebula a whole series of vibrations which have no counterpart in a liquid, and no inference as to the stability of these motions can be drawn from an examination of the behaviour of a liquid. Thus, although there will be unstable vibrations in a rotating mass of gas similar to those which are known to exist in a rotating liquid, it does not at all follow that a rotating gas will become unstable, in the first place, through vibrations which have a counterpart in a rotating liquid: it is at any rate conceivable that the vibrations through which the gas first becomes unstable are vibrations in which the compressibility of the gas plays so prominent a part, that no vibration of the kind can occur in a liquid. If this is so, the conditions of the formation of planetary systems will be widely different in the two cases.

With a view to answering the questions suggested by this argument, the present paper attempts to examine in a direct manner the stability of a mass of gravitating gas, and it will be found that, on the whole, the results are not such as could have been predicted by analogy from the results in the case of a gravitating liquid. The main point of difference between the two cases can be seen, almost without mathematical analysis, as follows:

Speaking somewhat loosely, the stability or instability may be measured by the resultant of several factors. In the case of an incompressible liquid we may say that gravitation tends to stability, and rotation to instability; the liquid becomes unstable as soon as the second factor preponderates over the first. The gravitational tendency to stability arises in this case from the surface inequalities caused by the displacement: matter is moved from a place of higher potential to a place of lower potential, and in this way the gravitational potential energy is increased. As soon as we pass to the consideration of a compressible gas the case is entirely different.

Suppose, to take the simplest case, that we are dealing with a single shell of gravitating gas, bounded by spheres of radii r and $r + dr$, and initially in equilibrium under its own gravitation, at a uniform density ρ_0.

Suppose, now, that this gas is caused to undergo a tangential compression or dilatation, such that the density is changed from

$$\rho_0 \text{ to } \rho_0 + \Sigma \rho_n S_n,$$

where ρ_n is a small quantity, and S_n is a spherical surface harmonic of order n.

It will readily be verified that there is a decrease in the gravitational energy of amount

$$4\pi r^3 (dr)^2 \Sigma \frac{\rho_n^{\,2}}{(2n+1)} \iint S_n^{\,2} \sin\theta \, d\theta \, d\phi.$$

As this is essentially a positive quantity, we see that any tangential displacement of a single shell will decrease the gravitational energy.

This example is sufficient to show that when the gas is compressible, the tendency of gravitation may be towards instability. The gravitation of the surface inequalities will as before tend towards stability, but when we are dealing with a gaseous nebula, it is impossible to suppose that a discontinuity of density can occur such as would be necessary if this tendency were to come into operation. Rotation as before will tend to instability, and the factor which makes for stability will be the elasticity of the gas.

We can now see that there is nothing inherently impossible, or even improbable, in the supposition that for a gaseous nebula the symmetrical configuration may become unstable even in the absence of rotation. The question which we shall primarily attempt to answer is, whether or not this is, in point of fact, a possible occurrence, and if so, under what circumstances it will take place. To investigate this problem, it will be sufficient to consider the vibrations of a non-rotating nebula about a configuration of spherical symmetry.

Unfortunately, the stability of a gaseous nebula of finite size is not a subject which lends itself well to mathematical treatment. The principal difficulty lies in finding a system which shall satisfy the ordinarily assumed gas equations, and shall at the same time give an adequate representation of the primitive nebula of astronomy.

If we begin by supposing a nebula to consist of a gas which satisfies at every point the ordinarily assumed gas equations, and to be free from the influence of all external forces, then the only configuration of equilibrium is one which extends to an infinite distance, and is such that the nebula contains an infinite mass of gas. The only alternative is to suppose the gas to be totally devoid of thermal conductivity, and in this case there is an equilibrium configuration which is of finite size and involves only a finite mass of gas. But the assumption that a gas may be treated as nonconducting finds no justification in nature. When we are dealing, as in the present case, with changes extending through the course of thousands of years, we cannot suppose the gas to be such a bad conductor of heat, that any configuration, other than one of thermal equilibrium, may be regarded as permanent.

Professor Darwin has pointed out that a nebula which consists of a swarm of meteorites may, under certain limitations, be treated as a gas of which the meteorites are the "molecules."[2] In this quasi-gas the mean time of describing a free path must be measured in days, rather than (as in the case of an actual gas) in units of 10^{-9} second. The process of equalisation of temperature will therefore be much slower than in the case of an actual gas, and it is possible that the conduction of heat may be so slow that it would be legitimate to regard adiabatic equilibrium as permanent.[3]

Except for this the mathematical conditions are identical, whether we assume the gaseous or meteoritic hypothesis. The present paper deals primarily with a nebula in which the equilibrium is conductive, but it will be found possible from the results arrived at, to obtain some insight into the behaviour of a nebula in which the equilibrium is partially or wholly convective.

Whether we suppose the thermal equilibrium of the gas to be conductive or adiabatic, we are still met by the difficulty that the gas equations break down over the outermost part of the nebula, through the density not being sufficiently great to warrant the statistical methods of the kinetic theory. This difficulty could be avoided by supposing that the nebula is of finite size, and that equilibrium is maintained by a constant pressure applied to the outer surface of the nebula. If this pressure is so great that the density of gas at the outer surface of the nebula is sufficiently large to justify us in supposing that the gas equations are satisfied everywhere inside this surface, then the difficulty in question will have been removed. On the other hand, this pressure can only be produced in nature by the impact of matter, this matter consisting either of molecules or meteorites, so that we are now called upon to take account of the gravitational forces exerted upon the nebula by this matter. This whole question is, however, deferred until a later stage; for the present we turn to the purely mathematical problem of finding the vibrations of a mass of gas which is in equilibrium in a spherical configuration. We shall consider two distinct cases. In the first, equilibrium is maintained by a constant pressure applied to the outer surface of the nebula, this surface being of radius R_1. In the second, the nebula extends to infinity, and it is assumed that the ordinary gas equations are satisfied without limitation. We suppose for the present that the gas is in thermal equilibrium throughout. It is not, however, supposed that the gas is all at the same temperature; to make the question more general, and to give a closer resemblance to the state of things which may be supposed to exist in nature, it will be supposed that the gas is collected round a solid spherical core of radius R_0, and the temperature will be supposed to fall off as we recede from this core to the surface, the equation of conduction of heat being satisfied at every point. We shall also suppose that the gas is acted upon by an external system of forces, this system being, like the nebula, spherically symmetrical. The reason for these generalisations will be seen later; it will at any time be possible to pass to less general cases.

The Criterion of Stability

Here Jeans sets up the equations describing a gaseous nebula that is symmetrical about the origin of the coordinate system. In general, the total potential is denoted by V, the gravitational constant by γ, the mass density by ρ, the pressure by ϖ, the temperature by T, and the gas constant by λ. The last of these is given by the ideal gas law

$$\varpi = \lambda \mathrm{T} \rho,$$

or in more familiar notation

$$\mathrm{P} = \varpi = \frac{k}{\mu m_H} \mathrm{T} \rho,$$

where $\lambda = R/\mu = k/(\mu m_H)$, the universal gas constant $R = 8.314 \times 10^7$ erg mole^{-1} K^{-1}, the mean molecular weight is μ, the mass of the hydrogen atom is $m_H = 1.67 \times 10^{-24}$ grams, and Boltzmann's constant $k = 1.3806 \times 10^{-16}$ erg K^{-1}.

Jeans initially assumes an equilibrium configuration in which the density, ρ_0, is a constant at the coordinate x. This configuration is next assumed to be perturbed so that the density $\rho = \rho_0 + \rho_1$ at the displaced coordinate $x + \xi$, and $\rho = \rho_0 + \rho'$ at the coordinate x. The subscripts 0 and 1 and the superscript $'$ are assumed to have the same general meaning for other perturbed quantities. As well as the ideal gas law, the continuity equation and Poisson's equation are assumed to hold. A consequence of the continuity equation is

$$\rho_1 = -\rho_0 \Delta,$$

or in modern terminology

$$\frac{\partial \rho_1}{\partial t} = -\rho_0 \nabla v_1,$$

where t denotes the time variable and v_1 is the velocity of the perturbation.

PROBLEMS OF COSMIC EVOLUTION

INFINITE SPACE FILLED WITH MATTER A limiting solution of the equations of equilibrium gives a nebula in which the density is constant everywhere. This solution may be supposed to represent infinite space filled with matter distributed at random. If space has no boundary there is presumably no need to satisfy a boundary-equation at infinity, so that ρ may have any value; if, however, this equation must be satisfied the only solution is $\rho = 0$.

Let us consider the former case. Space is filled with a medium of mean density ρ and of mean temperature T. Since the space under consideration is infinite, we may measure linear distances on any scale we please, and, by taking this scale sufficiently great, we can cause all irregularities in density and temperature to disappear. We may, therefore, suppose at once that the density and temperature have the constant values ρ and T.

The equations of motion for small displacements referred to rectangular axes are, in the old notation, since V_0 and ϖ_0 are constants,

$$\frac{d^2\xi}{dt^2} = \frac{dV'}{dx} - \frac{1}{\rho_0}\frac{d\varpi'}{dx}, \text{\&c.,} \qquad (137)$$

or, operating with d/dx, d/dy, d/dz, and adding

$$\frac{d^2\Delta}{dt^2} = \nabla^2 V' - \frac{1}{\rho_0}\nabla^2\varpi'. \qquad (138)$$

Since V′ is the gravitational potential of a distribution of density $-\Delta\rho$, we have

$$\nabla^2 V' = 4\pi\Delta\rho, \qquad (139)$$

while if we suppose, for the sake of simplicity, that the motion is adiabatic, so that the ratio of pressure to density changes at a constant rate κ, we have

$$\nabla^2\varpi' = \kappa\nabla^2\rho' = -\kappa\rho_0\nabla^2\Delta.$$

Hence equation (138) becomes

$$\frac{d^2\Delta}{dt^2} - 4\pi\rho\Delta - \kappa\nabla^2\Delta = 0. \qquad (140)$$

The simplest solution of this is of the form

$$\Delta = \frac{1}{r}e^{i(pt \pm qr)}, \qquad (141)$$

where

$$q^2 = \frac{p^2 + 4\pi\rho}{\kappa} \qquad (142)$$

and the general solution can be built up by superposition of such solutions.

[In modern nomenclature, equation (141) is a plane wave solution for the perturbed density ρ_1, in which the frequency $\omega = p$ is related to the wavelength $\lambda = 2\pi/q$ by the relation

$$\omega^2 = \left(\frac{2\pi}{\lambda}\right)^2 s^2 - 4\pi G\rho_0,$$

where $s = [kT/(\mu m_H)]^{1/2}$ is the velocity of sound, and the gravitation constant is $G = 6.673 \times 10^{-8}$ dyn cm^{-2} gm^{-2}. Any fluctuation with a wavelength greater than the critical value $\lambda_J = s[\pi/(G\rho_0)]^{1/2} = 6 \times 10^7[T/(\mu\rho_0)]^{1/2}$ cm will grow exponentially with increasing time, and the waves are unstable. Masses larger than the critical mass $M_J = \pi\rho_0\lambda_J{}^3/6 = 10^{23}(T/\mu)^{3/2}/\rho_0{}^{1/2}$ gm will be gravitationally unstable and will contract continuously.]

Now solution (141) gives $\Delta = 0$ at infinity, provided q is real, and therefore provided $p^2 + 4\pi\rho$ is positive, a condition which admits of p being imaginary. There is therefore a possible motion, which consists of a concentration of matter about some point, the amount of this concentration vanishing at infinity, and the amount at any point increasing, in the initial stages, exponentially with the time.

We conclude, therefore, that a uniform distribution in space will be unstable, independently of the mean temperature or density of this distribution.[4]

THE EVOLUTION OF NEBULÆ We can also see that a distribution of matter which is symmetrical about a single point will be equally unstable. For, if this distribution of matter were perfectly homogeneous, the whole mass of matter would form a spherical nebula of literally infinite extent, and would therefore be in neutral equilibrium. The introduction of even the smallest irregularities into this structure is equivalent to the application of an external field of force. This, as has already been seen, will destroy the spherical symmetry, and it can easily be seen that the motion from spherical symmetry is such as to lead to a concentration of matter about points of maximum density.

It appears, therefore, that the configuration which will naturally be assumed by an infinite mass of matter in the gaseous or meteoritic state consists of a number of nebulæ (*i.e.*, clusters round points of maximum density). We may either suppose the outer regions of these nebulæ to overlap, each nebula satisfying the gas-equations by being of infinite extent, or we may suppose the nebulæ to be distinct and of finite size, the interstices being filled by meteorites or other matter, which by continual bombardment upon the surfaces of the nebulæ supply the pressure which is required at these surfaces by the equations of equilibrium.

What, we may inquire, will determine the linear scale upon which these nebulæ are formed? Three quantities only can be concerned: γ the gravitational constant, ρ the mean density, and λT the mean elasticity. Now these quantities can combine in only one way so as to form a length, namely, through the expression

$$\sqrt{\frac{\lambda T}{\gamma \rho}},$$

of which the dimensions will be readily verified to be unity in length, and zero in mass and time. We conclude, then, that the distance between adjacent nebulæ will be comparable with the above expression.

Now the value of γ is 65×10^{-9}, and if we assume the primitive temperature to be comparable with 1,000° (absolute) we may take $\lambda T = 10^9$ (corresponding accurately to an absolute temperature of 350° for air, 2,800° for hydrogen). If we take the sun's diameter as a temporary unit of length, the earth's orbit is (roughly) of diameter 200. If we suppose the fixed stars to be at an average parallactic distance of 0.5″ apart, measured with respect to the earth's orbit, we find for their mean distance apart, about 4×10^7 sun's radii. The density of the sun being, in C.G.S. units, roughly equal to unity, we may, to the best of our knowledge, suppose the mean density of the primitive distribution of matter to be about $(4 \times 10^7)^{-3}$, or say 10^{-23}. Substituting these values for γ, λT and ρ, we find as the scale of length a quantity of the order of $10^{19.5}$ centims. The distance which corresponds to a parallax of 0.5″ would be about $10^{18.6}$ centims. It will therefore be seen that we are dealing with distances which are of the astronomical order of magnitude.

THE EVOLUTION OF PLANETARY SYSTEMS Let us now regard a single centre, together with the matter collected round it, as the spherical nebula which is the subject of discussion. On account of the way in which it has been formed, this nebula will, in general, be endowed with a certain amount of angular momentum. We have seen that a primitive nebula of this kind may be supposed, under certain conditions, to become unstable. We have also seen that the motion, when the nebula becomes unstable, is such as to strongly suggest the ejection of a satellite.

As a nebula cools the rotation increases, owing to the contraction of the nebula, and the surface velocity $\Omega = \omega r$ also increases. Thus the quantity $\Omega^2/3\lambda^2 T^2$, which measures the rotational tendency to instability, has a double cause of increase; firstly owing to the increase in Ω, and secondly owing to the decrease in T. We can accordingly imagine the primitive nebula becoming unstable time after time, throwing off a satellite each time.

In the usually accepted form of the nebular hypothesis, the rotation is supposed to be the sole cause of instability, so that the system resulting from a single nebula ought theoretically to be entirely symmetrical about an axis. On the view of the present paper, there is no reason for expecting this symmetry. For large rotations of the primitive nebula, the configuration of the resultant planetary system will approximate to perfect symmetry, but for small rotations, a slight irregularity occurring at the critical moment, at a point out of the equatorial plane, may produce a satellite of which the orbit is far removed from the equatorial plane.

In conclusion, two particular cases of "irregularities" may be referred to. If the nebula is penetrated by a wandering meteorite, at a moment at which it is close to a state of instability, the presence of the meteorite will constitute an irregularity, and may easily result in the formation of a satellite. And if a quasi-tide is raised in the nebula by the presence of a distant mass, the same result may be produced. In the former case, the plane of the satellite would, if the rotation is sufficiently small, be largely determined by the path of the meteorite; in the second case, by the position (or path) of the attracting mass. It would not, in either case, depend much upon the axis of rotation of the nebula.

1. "On the Mechanical Conditions of a Swarm of Meteorites, and on Theories of Cosmogony," *Phil. Trans, A 180*, 1 (1888).
2. G. H. Darwin, ibid.
3. Ibid., p. 64.
4. An interesting field of speculation is opened by regarding the stars themselves as molecules of a quasi-gas. If space were Euclidean and unbounded, there would be no objection to this procedure, and we should be led to the conclusion that the matter of the universe must become more and more concentrated in the course of time. If space is non-Euclidean, this concentration might reach a limit as soon as the coarse-grainedness of the structure attained a value so great that the distance between individual units became comparable with the radii of curvature of space. In any case, it may reach a limit as soon as an appreciable fraction of the space in question becomes occupied by matter.

Weizsäcker, The Origin of the Solar System

The Structure of the Solar System

In introducing his article[1] Weizsäcker provides a historical summary of explanations for the wandering paths of the planets across the sky. For Copernicus the question of the planetary system was a problem of structure and not of mechanical causality. For Kepler the sun was the seat of power for the motion of the planets, but his basic guiding principle was one of perfect form. Newton provided a mechanical explanation for the present behavior of the planetary system but inferred the intervention of God in explaining its origin. Weizsäcker argues that a mechanical explanation for the origin of the solar system must be sought and that such an explanation is a reflection or first indication of God. Questions other than this causal one remain, including: what is matter? why are there laws of nature? and is there a first cause? Weizsäcker then states that he does not want to say more about these questions, for they will be implicitly present in the thoughts of those who are interested in them.

Kant's Theory

As an introduction to the problem of the origin of the planets, I must briefly describe for you the theory of Kant. Kant begins with an observation: besides the zodiac we see a second distinguished great circle in the sky, the Milky Way. Let us conceive of it as a plane in which we ourselves are located. Its white lustre is the collective light of countless stars, which are concentrated in this plane. Kant asked himself how this immense star system, in which our sun itself plays as undistinguished a role as an asteroid plays in the solar system, would appear seen from outside. He believed that certain elliptically shaped nebulae in the sky were distinct Milky Ways. Today we know that he was correct in this conjecture. One of these nebulae is the well-known nearby spiral galaxy in Andromeda.

Why do the great assemblages of matter in the universe have this form? Kant explains their structure by means of their rotation: they rotate around an axis and flatten themselves thereby, like every rotating body. At the same time, they are acted upon by gravity. All their parts tend toward the common center. Now, it follows from the law for the conservation of angular momentum that for a rotating object, the nearer its parts get to the axis of rotation, the faster they must orbit. By this means the flattening also increases. The final state will be reached when each part of the nebula describes a circular orbit about the center with an exact balance between gravity and inertia, just like a planet that orbits the sun; in this state the whole nebula must become flat like a plate. Moreover, we should notice that the nebula cannot rotate like a rigid plate, because then its outermost parts would have to complete their orbits about the center in the same time as the innermost, whereas the balance between gravity and inertia demands, according to Kepler's third law, that the inner parts orbit quickly and the outer parts slowly.

With such a nebula, but a millionfold smaller, Kant now compares the earliest state of our solar system. Just as the Milky Way, which was probably originally a diffuse gas mass, has decomposed into single stars while retaining its form, so also should the solar system have been built out of its gaseous state by degrees into discrete bodies: a great central body—the sun—and smaller, peripheral bodies—the planets, comets, and meteors.

This notion explains in one stroke countless things. Let us recall what the Kepler laws encompass and what they do not!

It is a part of the second Kepler law and an elementary consequence of mechanics that a planet's orbit must lie in a plane. However, it does not follow from mechanics that all planetary orbits must lie in one and the same plane. Mechanically, it would be equally possible for the path of the planet Jupiter to lie in a plane perpendicular to that of the planet Saturn, and so on. Then there would be no zodiac. However, if we conceive the planets as the remnants of a mass that originally had a uniform rotation, then we immediately understand that their orbital planes are identical and that they all revolve in the same direction about the sun. Also, the rotation of the planets about their own axes takes place in the same direction (if with a certain variation in angle), and similarly, the moons orbit their planets in the plane of the zodiac. Only the outermost planets. Uranus and Neptune, which perhaps have suffered a subsequent disturbance, are exceptions to these rules. On the whole we may say that in the actual planetary system a close connection exists among the orbits of different planets, although mechanics treats each planet chiefly as if it moved independently of all the other planets. This conception now is, according to Kant, not a mechanical but a historical fact: it is the visible remnant of the original physical unity of the whole system.

Also, the first Kepler law still leaves too much freedom. It only establishes that the orbital curve is an ellipse, but, in fact, for all the major planets it is nearly a circle. This, too, is explained by the nebular hypothesis: the parts of the nebula had to orbit in circles because of the uniform rotation, and so also must the planets that arose from them.

Finally, Kepler's third law states how quickly a planet orbits the sun, if its distance from the sun is given, but this law says nothing about what distances actually occur. You will recall that the distances of the planets from the sun form a rather regular sequence. This regularity is expressed mathematically in a rule established by Titius and Bode in the eighteenth century. In a simplified version that is valid only for the planets from Mars out, the rule holds that each planet is about twice as distant from the sun as the preceding one. Indeed, this rule is not very precise. It has become famous, however, because through it a new planet was predicted. Between Mars and Jupiter a gap exists, according to the rule. Jupiter is not twice but four times as far as Mars from the sun and, therefore, between the two there should be another planet. In the year 1800 a planet (later defined as a multitude of planetoids) was in fact discovered there.

However, the Kantian theory now finds an obstacle in the Titius-Bode rule. Indeed, according to the Kantian theory, it is not implausible for the distances of the planets from the sun to arrange themselves according to some law or other. But the exact form of the law could not be explained, either then or by the later theories. It seems especially strange that a law holds for the distances of the planets but not for their sizes. Directly next to Jupiter, the largest of all the planets, which contains more material than all the others together, stand the asteroids, which collectively have a smaller mass than Mercury, the smallest of the major planets. If one considers the planets to have arisen out of a fairly uniformly divided gas mass, then a connection should exist between the relative distances of the planets and their sizes. A very large planet should be far removed from its neighbors; a very extensive one should be followed in a small distance by a very small one. Nevertheless, although the outer planets, whose relative distances are large, also are on the whole more massive than the inner ones, individually the rule is not valid.

This is not the only difficulty of the Kantian theory. The orbital direction of the moons offers a problem. In fact, the moons orbit their planets in the direction in which the planets orbit the sun. According to Kantian theory, the direction should be reversed for the moons. The moons are built from material that the planets drew in from the surrounding gas mass. This material, according to Kepler's third law, does not now generally have the same orbital velocity about the sun as the planet. The material that was originally closer than the planet to the sun must move more quickly than this planet; the material that was originally further from the sun must move more slowly. If the moons circling the planets are made from such material, then they obviously must continually overtake the planets on the inside the lag behind them on the outer side (that is, when they are further than the planet from the sun). In fact, however, the moons reverse this theoretical process.

A much more fundamental difficulty is the Kantian theory's inability to show that a diffuse mass circling the sun consolidates into discrete, large planetary bodies. These considerations are quite complicated, and indeed the different versions that Kant, Laplace, and their successors suggested have led not to solutions but always to new problems.

Finally, it is not an inner contradiction but an unsolved problem that the theory does not tell us from where the initial rotation of the system came.

I cannot here discuss the numerous newer hypotheses proposed to replace Kant's. I now want only to portray how, in my opinion, one can remove the difficulties in the Kantian theory, if one applies the present astrophysical knowledge consistently.

New Version of the Theory[2]

Kant did not know how the sun and the planets were chemically composed. Today we can say something like this: all bodies in the universe of which we know consist of the same chemical elements; only the ratios of the components show certain variations. These variations occur mainly in the proportions of the lightest elements, above all hydrogen and helium, to the remaining heavier elements, whereas these heavier elements show, among themselves, about the same distribution throughout the universe. The planets consist almost entirely of these heavier elements. The sun, in contrast, is about one-half hydrogen by weight, with the other half consisting almost entirely of the light gases, up to oxygen. The elements heavier than oxygen account for about one-half the earth's weight but less than 1% of the weight of the sun.

Unfortunately, I can only indicate to you how this knowledge has been obtained. We can directly investigate the earth's surface. We can draw inferences from geology about the earth's interior. The appearance of the planetary surfaces likewise allows inferences to be made. Furthermore, we know the specific weights of the planets, because we can determine their spatial volumes directly and their total mass from their gravitational interactions with other bodies. The chemical composition of the solar surface follows from the spectrum of the light the sun emits. We cannot see freely into the interior of the sun, but a sequence of strong theoretical conclusions, chiefly based on atomic physics, indicates that the great predominance of hydrogen shown in the spectrum of the surface is also found in the interior. For example, we know today that the enormous amount of heat continually radiated by the sun must be produced by the conversion (in the deepest part of the solar interior) of nuclei of hydrogen into those of helium. The manner in which this energy is transported outward also depends strongly on the sun's hydrogen content. Finally, the most telling argument may be the consideration that in the interior of the sun, which is completely agitated because of the high temperatures at which gaseous matter exists and because of the rotation of the sun, a permanent, thorough mixing of the matter must take place by means of convection, so that a difference of composition between the interior and the surface could not be maintained.

What now follows from this purely mechanical distinction between the sun and the planets? It would be strange if the undivided gas mass out of which all the bodies in the system were built should have produced such differences in composition. It is very probable that the difference arose first in the formation of the planets, and this conclusion follows from a closer consideration of the formation process of the planets.

All we know about the predominance of hydrogen in the universe indicates that the solar composition is normal and the planets are the exception. We want therefore to accept also that the gas mass out of which the planets arose consisted mainly of light gases. This mass has to have been located in the vicinity of the sun, at a temperature that prevails on the planets, about like that on earth. In what physical and chemical state would it be at these temperatures? All the elements that are gaseous on the earth, that is, hydrogen, nitrogen, oxygen, and all the inert gases, must have been gaseous then. On the other hand, the elements we know as solid bodies or as liquids must have condensed into little fragments and drops. These fragments and drops were swirled around, collided with one another, and united themselves into always bigger bodies. One can estimate that during this process, which may have lasted between 10 and 100 million years, bodies of planet size must have formed. Because we can determine from radioactive minerals that the earth is at least 2 billion years old, a time of 100 million years can indeed be a suitable formation time.

We now understand why the planets contain only the heavier elements: only these condense at the prevailing temperature. By means of chemical bonding to the condensing matter, a little hydrogen, nitrogen, and oxygen may have been blended with the planets. On the other hand, the inert gases, which enter into no chemical unions, had to remain entirely outside the planetary bodies. Practical experience confirms this conclusion. The inert gases are found in the earth only in traces, and yet at least two of them, helium and neon, are among the most predominant elements in the universe.

If, however, the entire remainder of the gas shell did not condense to planets, why is it not on hand today? One can show that it necessarily had to destroy itself with time, because its state of motion was not stable. This instability depends on the viscosity of the gas. According to Kepler's third law, the outer parts of the gas move more slowly than the inner ones. Now, the gas clouds that move with different velocities are not separated from one another but touch and rub against one another. The viscosity seems to equalize their velocity difference. By that means the inner parts of the gas slow down and the outer parts speed up. The inner parts now move too slowly. Inertia no longer has the power to exert the necessary resistance against gravity, and gradually they sink in toward the sun. Conversely, the outer parts now move too quickly. The force of gravity no longer dominates their motion, and they escape further and further into space. I believe that each rotating gas mass must gradually decompose in this manner. The shape of the spiral nebula also discloses a decomposition into a concentrated nucleus and a very flattened shell dispersing to the outside. An evaluation of the time necessary to destroy the gas shell of the sun completely gives again around 100 million years.

According to our hypothesis, it is clear that planets should have formed. Can we, however, also remove the other difficulties of the Kantian theory?

The two questions about the Titius-Bode law, for the distances of the planets and the direction of rotation of the moons, prove to be connected. Their explanation is, of course, somewhat complicated. This is the part of my report that I myself consider still the most hypothetical. We have to investigate the inner motions of the gas shell somewhat more precisely. I said previously that the gas shell rotates so that each of its parts moves on a circular orbit in accordance with Kepler's third law. Viscosity causes one part of the mass to move inward and the other part to stream outward. Viscosity affects all parts at the same time, however; therefore, motions (both inward and outward) must arise everywhere simultaneously, and manifestly, a turbulent motion arises. I believe, however, that this motion proceeds in a not entirely unordered way and that a regular system of turbulent eddies develops. A schematic diagram of these hypothetical vortices is shown in figure 32.1.

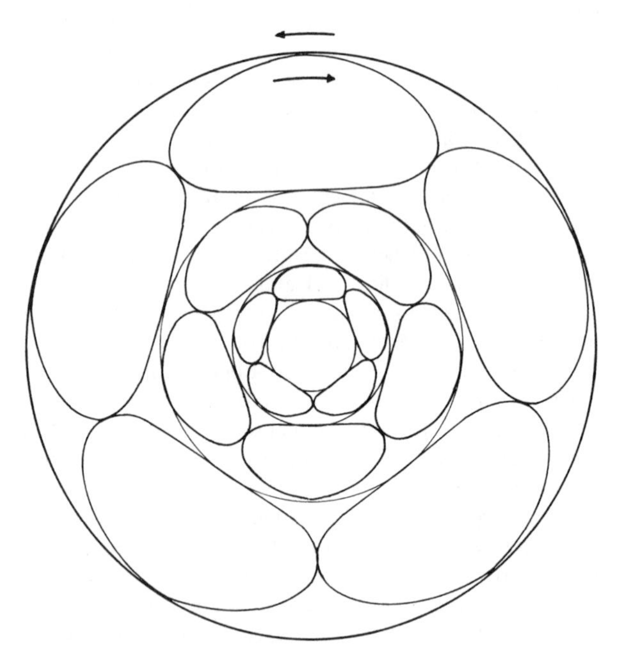

Fig. 32.1 Diagram illustrating vortex cells formed by turbulent motions in the primeval solar nebula. Planets form in whirlpool eddies between the vortices that appear within the circular rings. In this diagram the direction of rotation of the nebula and the eddies is counterclockwise, and that of each of the vortices is clockwise.

The outer arrow indicates the rotation direction of the gas mass on the whole. The vortex systems drawn in all move, therefore, in this direction around the sun, which is to be considered as a small point at the common center of the circle. Each bean-shaped figure now represents a boundary of a gas mass region; these regions jointly circle the sun and do not decompose as long as they are not disturbed from outside. Thereby, viewed from their own centers, which also orbit the sun, they have an inner rotation represented by the inner arrow. Of course, they rotate around themselves exactly once during one revolution about the sun. All vortices within a circular zone move with equal speed around the sun and thus do not alter their relative positions. In contrast, the vortices of the inner rings move more quickly than those of the outer rings, according to Kepler's third law. Only three rings of vortices are shown, but the figure can be continued, in an analogous manner, much further inward and also much further outward.

The distinguishing thing about this eddy motion is that it can be realized under the influence of gravity and inertia alone. The separate gas clouds inside the vortex describe, of course, precise Keplerian ellipses as seen from the sun, and the vortex persists permanently. Inside the vortex the motion is thus perfectly steady and there is practically no viscosity; all the friction is found at the outer boundary. I would like to conjecture that this is the final state of the gas shell, which has the least viscosity, and that therefore it is stable. If now, at the inner and outer borders of the gas shell, matter permanently streams across, on the inside to the sun and on the outside toward space, then this matter can be replenished from the neighboring vortices; each vortex acts like a ferris wheel scooping up matter.

For the construction of the planetary bodies the boundary regions between the vortices now are important. Because the motion is so uniform inside the vortex, the fragments of condensed matter are thrown together to a very small extent and thus have very little opportunity to collide and thereby to grow. In the boundary zones, however, in which the vortices touch while in motion in opposed directions, a strongly turbulent motion must prevail, and here the fragments are often hurled together and grow quickly. So countless planet-like bodies are built at each contact circle between two vortex rings. The asteroids represent one end point of this state. The union of these countless bodies into a single large planet probably first takes place when one of them has grown so large that it can pull the others into it by means of its gravity. With the asteroids, none grew large enough: their large number is probably a consequence of their small total mass.

The Titius-Bode law can now nearly be inferred from figure 32.1. The number of vortices on a ring has to be almost always the same, no matter how far the ring is from the sun. Therefore, the diameter of the vortex and the distance of the rings from one another are both proportional to the distance from the sun, just as observation shows. I have not succeeded in deducing theoretically how many vortices must lie on a ring. If we take the empirical distances as a basis, it must be for the region between Mars and Uranus exactly 5, and it is so drawn in the figure. Inside the orbit of Mars the number of vortices must be somewhat larger. We can now understand why a law exists for the distances of the planets but not for their sizes. The location of the formation of the planets is established by the vortex system, and how much condensed matter comes together there can depend on chance. In the neighborhood of a very large planet like Jupiter, a special material-poor zone like that of the asteroids may lie: Jupiter, in forming, has presumably "sucked dry" the neighborhood of the asteroids.

Now, how can we explain the common directions of motion of the planets and their moons? The internal rotation of a vortex is opposite in direction to the overall rotation of the system; this follows necessarily from the condition that each part of the vortex orbits around the sun on a Keplerian ellipse. On the boundary circles between vortices the matter is now turned like a ball bearing between two oppositely turning wheels; therefore, it is turned in the direction of the overall rotation. Here, however, planets and moons are built up. They necessarily have the direction of rotation of the overall system, just as in fact is observed. For Uranus, whose moon moves in a plane perpendicular to the zodiac, the schematization of the motion as a flat arrangement is probably no longer sufficient.

In conclusion I turn to the often-asked question about the origin of the rotation of the system. I cannot, of course, answer this question completely, but I can push the uncertainty one step further back. The whole solar system is thought to have arisen out of the originally diffuse matter of the great Milky Way, just like the other fixed stars. The Milky Way system rotates as a whole, but with the inner parts moving more quickly than the outer ones; a strong turbulent inner motion had to form in it in consequence of these differences in motion. Because of this motion, the matter did not subside evenly from all sides into the forming sun; instead, vigorous side motions arose at the same time. These motions may have been disturbed according to the law of chance. If so, on the average they will not have equalized completely, and there must have remained some resulting unpredictable rotational motion.

The question is often asked whether there are other planetary systems in the universe besides our own. We cannot see the planets of other stars, because of their great distances, and we are therefore directed to a theoretical conclusion. Now, in the considerations I have presented to you, there is nothing that could not also be applied to other stars. I would thus like to believe that there are many other planetary systems in the universe. In many cases, perhaps, a planet grew so large that it became a second sun, and instead of a system of planets, a double star arose; we do, indeed, see many double stars in the sky. Perhaps in other cases the matter sufficed only to form meteors and comets. There are probably still a great number of cases in which planet systems similar to our own were formed.

If we wanted to investigate the question about the origin of the rotation further, we would now have to turn to the development of the Milky Way system and to the genetic connection among the spiral nebulae. I do not want to raise this question today.

1. This article was a lecture first given in Munich in winter 1943–1944.

2. The new theory is given in *Zeitschrift für Astrophysik 22*, 319 (1944).

33. A Production of Amino Acids under Possible Primitive Earth Conditions

Stanley L. Miller

(*Science 117*, 528–529 [1953])

The theory of the evolution of species implies that we can extrapolate backward in time to the point when the first living organism was formed. In the 1930s Aleksandr Oparin opened the argument that the first organism could have been generated spontaneously if sufficient quantities of organic compounds were present in the shallow seas of the primitive earth.[1] Under this assumption, a key problem in explaining the origin of life becomes the one of determining how the organic constituents of living matter arose in the first place. Many astronomers believe that the primitive atmosphere of the earth was rich in hydrogen, quite unlike the present oxygen-rich atmosphere. Harold Urey has persuasively argued that the cold, original atmosphere of the earth must have been composed of the stable molecules of methane, ammonia, water, and hydrogen.[2] This view is partly a result of the old observation that Jupiter, Saturn, and Titan contain ammonia and methane and partly a result of the fact that the most abundant element of the primitive solar nebula must have been hydrogen.

In the paper given here, Stanley Miller describes how an electrical discharge in a simulation of the primitive earth atmosphere can give rise to one of the basic building blocks of living organisms, the amino acids. Since that time, a number of experiments have been performed in which the simple molecules of the earth's primitive atmosphere are converted into more complex molecules by the action of electric discharge, ultraviolet light, and ionizing radiation.[3] As a result of these experiments, we believe that the carbon and hydrogen in methane, the hydrogen and oxygen in water, and the hydrogen and nitrogen in ammonia, can be liberated by various radiation processes and can combine to produce all the complex molecules that subsequently form the living species.

Recently, an alternative possibility has been introduced, according to which organic compounds are formed in the solar nebula by reactions involving no external energy sources. Straight-chained hydrocarbons (thought by many to be indigenous components of meteorites) can be produced in the terrestrial laboratory by the Fischer-Tropsch reaction, in which carbon monoxide mixes with molecular hydrogen in the presence of an iron catalyst to form both water and hydrocarbons. In addition, when ammonia is added to the gas mixture, intermediates are generated that lead to amino acids and other organic compounds that have been found in meteorites. These developments point to the possibility that some of the basic components of biological life could have been formed through the interaction of abundant organic molecules in the primitive solar nebula or on the primitive earth. Nevertheless, special physical conditions are required for the Fischer-Tropsch reaction, and it now appears that a variety of reactions, including those involving electrical discharges and an aqueous environment, are needed to account for all the organic compounds synthesized on the primitive earth.[4]

1. A. I. Oparin, *The Origin of Life* (New York: Macmillan Co., 1938; 3rd ed., New York: Academic Press, 1957).
2. H. C. Urey, *The Planets: Their Origin and Development* (New Haven: Yale University Press, 1952).
3. S. L. Miller and H. C. Urey, *Science 130*, 245 (1959); S. L. Miller and L. E. Orgel, *The Origins of Life on Earth* (Englewood Cliffs, N. J.: Prentice Hall, 1974).
4. S. L. Miller, H. C. Urey, and J. Oro, *Journal of Molecular Evolution 9*, 59 (1976).

THE IDEA that the organic compounds that serve as the basis of life were formed when the earth had an atmosphere of methane, ammonia, water, and hydrogen instead of carbon dioxide, nitrogen, oxygen, and water was suggested by Oparin[1] and has been given emphasis recently by Urey[2] and Bernal.[3]

In order to test this hypothesis, an apparatus was built to circulate CH_4, NH_3, H_2O, and H_2 past an electric discharge. The resulting mixture has been tested for amino acids by paper chromatography. Electrical discharge was used to form free radicals instead of ultraviolet light, because quartz absorbs wavelengths short enough to cause photo-dissociation of the gases. Electrical discharge may have played a significant role in the formation of compounds in the primitive atmosphere.

Water is boiled in one flask, mixes with the gases in another flask, circulates past the electrodes, condenses and empties back into the boiling flask. The U-tube prevents circulation in the opposite direction. The acids and amino acids formed in the discharge, not being volatile, accumulate in the water phase. The circulation of the gases is quite slow, but this seems to be an asset, because production was less in a different apparatus with an aspirator arrangement to promote circulation. The discharge, a small corona, was provided by an induction coil designed for detection of leaks in vacuum apparatus.

The experimental procedure was to seal off the opening in the boiling flask after adding 200 ml of water, evacuate the air, add 10 cm pressure of H_2, 20 cm of CH_4, and 20 cm of NH_3. The water in the flask was boiled, and the discharge was run continuously for a week.

During the run the water in the flask became noticeably pink after the first day, and by the end of the week the solution was deep red and turbid. Most of the turbidity was due to colloidal silica from the glass. The red color is due to organic compounds adsorbed on the silica. Also present are yellow organic compounds, of which only a small fraction can be extracted with ether, and which form a continuous streak tapering off at the bottom on a one-dimensional chromatogram run in butanol-acetic acid. These substances are being investigated further.

At the end of the run the solution in the boiling flask was removed and 1 ml of saturated $HgCl_2$ was added to prevent the growth of living organisms. The ampholytes were separated from the rest of the constituents by adding $Ba(OH)_2$ and evaporating *in vacuo* to remove amines, adding H_2SO_4 and evaporating to remove the acids, neutralizing with $Ba(OH)_2$, filtering and concentrating *in vacuo*.

The amino acids are not due to living organisms because their growth would be prevented by the boiling water during the run, and by the $HgCl_2$, $Ba(OH)_2$, H_2SO_4 during the analysis.

In figure 33.1 is shown a paper chromatogram run in *n*-butanol-acetic acid-water mixture followed by water-saturated phenol, and spraying with ninhydrin. Identification of an amino acid was made when the R_f value (the ratio of the distance traveled by the amino acid to the distance traveled by the solvent front), the shape, and the color of the spot were

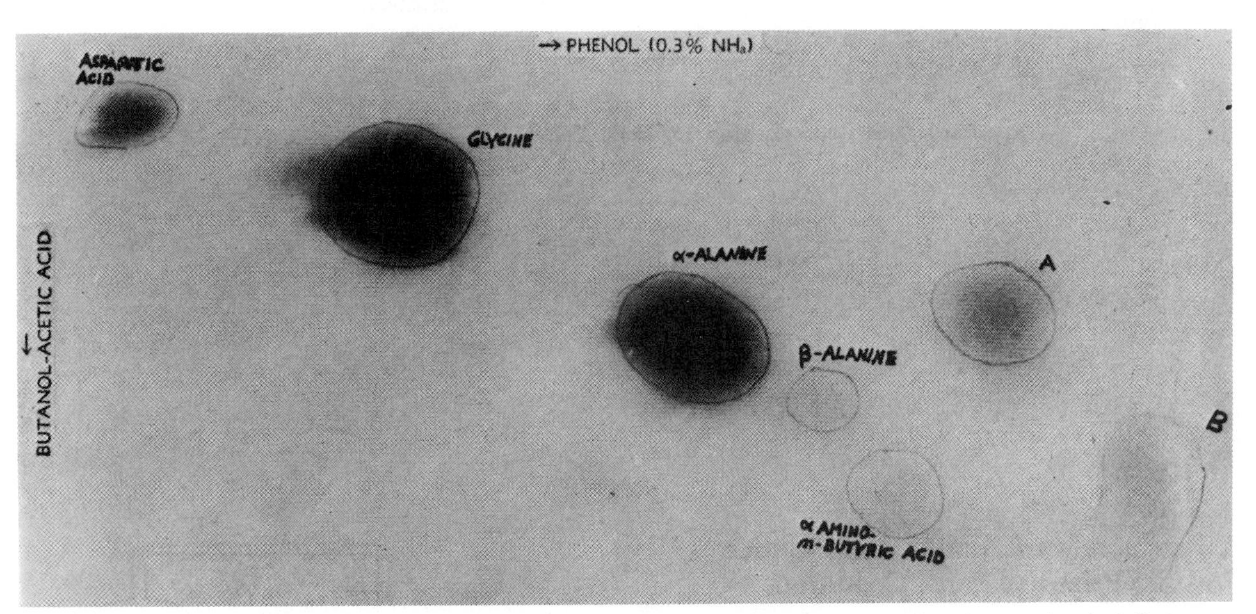

Fig. 33.1 A paper chromatogram showing the presence of amino acids in a mixture of methane, ammonia, water, and hydrogen that has been subjected to an electrical discharge. (Courtesy Stanley Miller.)

the same on a known, unknown, and mixture of the known and unknown; and when consistent results were obtained with chromatograms using phenol and 77% ethanol.

On this basis glycine, α-alanine and β-alanine are identified. The identification of the aspartic acid and α-amino-*n*-butyric acid is less certain because the spots are quite weak. The spots marked A and B are unidentified as yet, but may be beta and gamma amino acids. These are the main amino acids present, and others are undoubtedly present but in smaller amounts. It is estimated that the total yield of amino acids was in the milligram range.

In this apparatus an attempt was made to duplicate a primitive atmosphere of the earth, and not to obtain the optimum conditions for the formation of amino acids. Although in this case the total yield was small for the energy expended, it is possible that, with more efficient apparatus (such as mixing of the free radicals in a flow system, use of higher hydrocarbons from natural gas or petroleum, carbon dioxide, etc., and optimum ratios of gases), this type of process would be a way of commercially producing amino acids.

1. A. I. Oparin, *The Origin of Life* (New York: Macmillan Co., 1938).

2. H. C. Urey, *Proc. Natl. Acad. Sci. U.S. 38*, 351 (1952); *The Planets* (New Haven: Yale University Press, 1952), chap. 4.

3. J. D. Bernal, *Proc. Phys. Soc.* (London) *62A*, 537 (1949), *62B*, 597 (1949); *Physical Basis of Life* (London: Routledge and Kegan Paul, 1951).

CHAPTER III

Stellar Atmospheres and Stellar Spectra

34. On the Radiation of Stars

Ejnar Hertzsprung

Translated by Harlow Shapley and Vincent Icke

(*Zeitschrift für wissenschaftliche Photographie 3*, 429–442[1905])

In the 1860s Lewis Rutherford and Angelo Secchi first noted that stars of different colors exhibit different spectral lines, and Secchi visually classified about 4,000 stars into four main groups. Later in the century Antonia C. Maury and Annie Jump Cannon, working on photographic plates under the direction of Edward C. Pickering, extended these early investigations.[1] In her lifetime Cannon eventually classified over 500,000 stars for the *Henry Draper Catalogue* and its extension. She distinguished the stars on the basis of the absorption lines in their spectra and subsequently arranged most of them in a smooth and continuous spectral sequence: O, B, A, F, G, K, M. Eventually, Maury examined about as many stars as Cannon, and in the process discovered that representation of the different spectra required a two-dimensional scheme. During the same period, the parallaxes of a large number of nearby stars and the proper motions of distant stars were measured, and these results allowed determinations of the distances and absolute luminosities of individual stars and groups of stars. In the paper given here, Ejnar Hertzsprung shows that the absolute luminosities of stars of Maury's subclasses *a* and *b* exhibit a systematic decrease with the progression of spectral class from O to M. Hertzsprung also notes that stars with unusually narrow and sharp spectral lines, Maury's subclass *c*, have much higher luminosities than those of subclasses *a* and *b*, but that there is comparatively little variation in mass between the stars of all subclasses. In this way Hertzsprung first recognizes what have come to be called dwarf and giant stars.

Hertzsprung suggests here that the class *c* stars have lower densities than those of classes *a* and *b*. Two years later he concluded that the bright red stars of class *c* are "rare per unit volume of universal space, and that those which belong to the normal solar sequence form the vast majority. The bright red stage is therefore rapidly traversed, or the relevant stars belong to a parallel sequence. Whether there is a gap between this possible parallel sequence and the solar sequence is a matter yet to be decided." In that same paper, Hertzsprung argued that the high luminosity of the red class *c* stars is caused by their low densities and not by larger mass.[2]

1. A. C. Maury and E. C. Pickering, *Annals of the Harvard College Observatory 28* (1897), A. J. Cannon and E. C. Pickering, *Annals of the Harvard College Observatory 91–99* (1918–24).
2. E. Hertzsprung, *Zeitschrift für wissenschaftliche Photographie 5*, 89 (1907).

IN VOLUME 28 of the "Annals of the Astronomical Observatory of Harvard College" a detailed survey of the spectra is given for northern and southern bright stars by Antonia C. Maury and Annie J. Cannon, respectively.

The first two columns of table 34.1 give a short summary of the spectral class designation used by the two authors. In the last two columns are listed characteristic stars along with their spectral types. For a more detailed description of the characteristics used we must refer to the original papers cited above. Here we can find room for only a few words concerning the three sub-classifications *b*, *a*, and *c*. The *b* stars have broader lines than those of "division" *a*. The relative intensities of the lines seem, however, to be equal for *a*- and *b*-stars "so that there appears to be no decided difference in the constitution of the stars belonging, respectively, to these two divisions." As the most important characteristics of subclass *c* we can mention, first, that the lines are unusually narrow and sharp; second, that among the "metallic" lines others occur which are not identifiable with any solar lines, and the relative intensities of the remainder do not correspond with the intensities observed in the solar spectrum. "In general, division *c* is distinguished by the strongly defined character of its lines, and it seems that stars of this division must differ more decidedly in constitution from those of division *a* than is the case with those of division *b*." Antonia C. Maury suspects that the *a*- and *b*-stars on the one hand and the *c*-stars on the other, belong to collateral series of development. That is to say not all stars have the same spectral development. What determines such a differentiation (e.g., differences in mass and constitution) is a question that remains unanswered.

The question arises how great the systematic differences of the brightness, reduced to a common distance, of stars of the different groups will be. For this purpose I have used the proper motions of the stars in the following simple manner.

For each group a value was determined above and below which lies, respectively, one-half of the proper motions expressed in arc of a great circle, and reduced to magnitude 0. These values are listed in column V of table 34.1. In column VI are found the corresponding magnitudes reduced to a proper motion of 1″ in a hundred years. (Reduced to 1″ annual proper motion the stars would be 10 magnitudes brighter.) In column VIII are the mean reduced stellar magnitudes for somewhat large groups, and in the following two columns the values above and below which 15% of the total lies. These values will be, therefore, the mean deviation from the median. Finally there are listed in column XI the mean errors of the medians.

Table 34.1 contains only stars of subclasses *a* and *b* for which I have found proper motions based on the latest determinations of the Fundamental stars (Newcomb precession constants). Also in addition to the *c*-stars, all stars are omitted which are recognized as variable or the spectra of which were described as "peculiar." The total number of the *a* and *b* stars found in Antonia C. Maury's catalogue are given in column III, and in column IV the number of stars remaining after these omissions. I have also attempted to bring together all stars brighter than the 5th magnitude for which spectral class (according to the above-named authors, or to the Draper Catalogue) as well as proper motion could be found, and I come to the same result as that which appears in table 34.1. In spite of the small number (308) of stars taken into consideration in table 34.1, I consider the picture they give us as more reliable than would be that from a larger number of much more uncertainly classified spectra used in connection with a too great value for the small proper motions (Orion stars).

The radial velocities found for about 60 stars have an approximately typical distribution with a mean deviation from zero of some ± 20 km/sec. It is therefore probable that the projection of the absolute proper motions on a randomly chosen direction would also have a typical distribution. We have, however, also considered the projection of the apparent proper motions on a plane at right angles to the line of sight; and we ask which mean deviation in the stellar magnitudes, reduced to equal apparent proper motions, would uniquely result from this procedure only (corresponding to the assumption that all stars have the same absolute magnitude). The values are about $+1.12$ and -1.57 magnitudes. Comparing these with those in columns IX and X in table 34.1, we find that especially the stars which were put together in the A-class cannot differ very much among themselves in absolute magnitude. According to this result, combined with the fact that membership in spectral A-class is easily recognized, I have assembled for 100 A-stars of magnitude 4.62–5.00 the proper motions in declination only. If one arranges these according to magnitude, the value $-.''008$ lies in the middle, and respectively 15% of the total is over $+.''0325$ and under $-.''0575$. From this can be calculated the mean deviation $\pm.''0448$ annually, which would correspond to a speed of ± 20 km/sec, or 4 astronomical units per year. According to this, we find for the 100 A-stars of mean magnitude 4.84 the mean parallax of $.''0112$. In table 34.1 the magnitudes are reduced to a mean annual proper motion of $.''01$ in arc of a great circle, corresponding to a parallax of some $.''002$. For the 100 A-stars we compute with the parallax the mean stellar magnitude of 8.6, in fair agreement with the value 8.05 from table 34.1. The proper motion in right ascension is treated in the same way, after multiplication by the cosine of the declination.

Further I have in column XIII, table 34.1, inserted values which can be taken as a sort of color-equivalent and which were derived in the following way from the visual magnitudes taken from the revised Harvard Photometry (H.P.) and the photographic magnitudes (corresponding to G-line light of wave length $.432\mu$) taken from the Draper Catalogue (D.C.). Within each group, for the number of stars in column XII, both magnitudes m_H and m_D were brought together, and, on

Table 34.1 Proper motions and magnitudes of stars of different spectral class

Spectral class according to AJC I	Spectral class according to ACM II	Number in ACM III	Number Adjusted IV	Mean annual proper motion for $m_H = 0$ V	Mean magnitude reduced to the annual proper motion of ″.01 VI	Number VII	VIII	Mean deviation from the mean IX	X	Mean error of the mean XI	Number XII	Star color m_D for $m_H = 4.5$ XIII	Star color Δm_H : Δm_D XIV	Characteristic star Name XV	Characteristic star Spectrum XVI
Oe5B	I	7	3	0″.067	(4.13)						4	(4.66)	—	*S* Monocerotis	I*b*
B to B3A	II	14	11	0.059	3.84						9	4.73	1.36	ε Orionis	II*a*
	III	17	11	0.069	4.20	38	4.37	−3.02	+2.08	±0.41	13	4.65	1.35	α Virginis	III*b*
	IV	30	9	0.069	4.19						26	4.58	1.35	γ Orionis	IV*a*
	IV'	20	7	0.127	5.51						18	4.59	1.17	π_4 Orionis	IV'*a*
B5A to B9A	V	22	9	0.199	6.49						19	4.55	1.32	τ Orionis	V*b*
	VI	22	10	0.411	8.07	21	7.25	−1.40	+1.22	±0.29	19	4.58	1.30	α Leonis	VI*b*
	VI'	3	2	0.520	(8.58)						3	(4.49)	—	η Aquarii	VI'*b*
A	VII	43	21	0.386	7.93	47	8.05	−1.38	+1.48	±0.21	38	4.56	1.27	α Canis maj.	VII*a*
	VIII	49	26	0.439	8.21						43	4.65	1.34	α Geminorum	VIII*a*
AF and F	IX	29	16	0.547	8.69						26	4.73	1.22	δ Ursae maj.	IX*b*
	X	19	9	0.646	9.05	34	9.06	−2.03	+1.77	±0.33	15	4.82	1.36	α Aquilae	X*b*
	XI	13	7	0.721	9.29						13	4.86	1.21	δ Aquilae	XI*a,b*
	XI'	3	2	0.568	(8.77)						3	(4.80)	—	ζ Leonis	XI'*ab*
FG	XII	30	18	1.197	10.39	30	11.23	−2.25	+2.06	±0.40	24	5.03	1.36	α Canis min.	XII*a*
	XIII	21	12	3.251	12.56						21	5.02	1.41	χ_1 Orionis	XIII*a*
G and GK	XIII'	1	1	2.466	(11.96)						1	(5.16)	—	θ Persei	XIII'*a*
	XIV	19	11	0.481	8.41	24	7.93	−2.46	+4.01	±0.31	16	5.23	1.34	α Aurigae, Sun	XIV*a*
	XIV'	21	12	0.321	7.53						20	5.29	1.53	κ Geminorum	XIV'*a*
K	XV_1	25	21	1.081	10.17						24	5.39	1.46	α Bootis	XV_1*a*
	XV	49	21	0.745	9.36	74	9.38	−2.06	+1.68	±0.22	42	5.44	1.49		
	XV_2	38	32	0.658	9.09						36	5.46	1.58	α Cassiopeiae	XV_2*a*
KM	XV'	6	5	0.908	4.79	21	7.77	−3.07	+2.01	±0.56	6	(5.75)	—	β Cancri	XV'*a*
	XVI	23	16	0.449	8.26						22	5.68	1.75	α Tauri	XVI*a*
Ma	XVII	18	9	0.378	7.89	15	8.28	−1.12	+1.15	±0.30	15	5.81	1.41	β Andromedae	XVII*a*
	XVIII	20	6	0.479	8.40						16	5.84	1.45	α Orionis	XVIII*a*
	XIX	6	1	0.891	(9.75)						5	5.81	—	ρ Persei	XIX*a*

the approximately correct assumption that a linear relation exists between them, that value of m_D was calculated, using least squares, which corresponds to $m_H = 4.5$. Further we have in column XIV for each group the computed ratios $\Delta m_H : \Delta m_D$. Actually they should be constant with the value 1. That they increase from white through yellow to red may be due to the Purkinje phenomenon. That they all lie appreciably above 1 can be due to the circumstance that the normal intensity scale, which was used for the determination of the D.C. magnitudes through comparison of the spectral darkening in the neighborhood of the G-line ($\lambda = .432\mu$), was established not in pure G-light but by means of the Carcel-lampe, where a possible different sensitivity of the utilized emulsion in the different colors—the Purkinje effect of the plate—must appear.

The minimum shown in column XIII in the neighborhood of the A-group appears to be real. Accordingly the Orion stars would be somewhat yellower than the A-stars.

Here Hertzsprung gives a table (not reproduced by us) of the spectral class, magnitudes, and proper motions (reduced to zero magnitude) for the fundamental *c* and *ac* stars of Antonia C. Maury's catalogue.

In any case we may say that the annual proper motion of an average *c*-star, reduced to magnitude 0, amounts to only a few hundredths of a second. With the relatively large errors of these small values, a dependence on spectral class cannot be recognized. In other words, the *c*-stars are at least as bright as the Orion stars. In both of the spectroscopic binaries *o* Andromedae and β Lyrae the brightness of the *c*-star and of the companion star of the Orion type appear to be of the same order of brightness. The proper motions (not here given) are all small, according to the Auwers-Bradley Catalogue. (The star e Navis has an annual proper motion of 0″.5 when reduced to zero magnitude). For the stars in Annie J. Cannon's listing that have narrow sharp lines, I can also find only small proper motions. This result confirms the assumption of Antonia C. Maury that the *c*-stars are something unique.

When the *c*- and *ac*-stars are looked at in summary fashion one sees that with increasing Class number [advancing toward redder spectra] the *c*-characteristic diminishes, and that these stars stop exactly where the bright K-stars begin.

Hertzsprung presents tables of the measured parallaxes of the brighter stars, showing that stars in the G spectral class tend to have the larger parallaxes. He then gives a formula for calculating a star's magnitude reduced to unit mass and parallax and presents a table of these magnitudes for 53 double stars. We do not reproduce this table here, but we do reproduce Hertzsprung's conclusions.

There should be an explanation for the fact that the brightness of two stars of not very different mass (γ Leonis at 6.5 solar masses and 70 Ophiuchi at 2.5 solar masses), after reduction to unit mass and parallax—stars that even belong to the same subdivision of a spectral class (XV_1a)—differ by less than 5.75 mag. With the same surface brightness, γ Leonis should have a mass density less than one three-thousandth that of 70 Ophiuchi.

For our sun, $m_r = -0.33$; in comparison, the reddish state of the star 70 Ophiuchi, with $m_r = +0.87$, seems a naturally later phase. It is not likely, however, that the sun would be similar, upon further cooling, to γ Leonis. Therefore, γ Leonis must either represent an earlier phase or belong to a collateral series. [Hertzsprung is here using absolute magnitudes on the old system in which the standard distance is only 1 parsec.]

Hertzsprung explains the division of stars by brightness and spectral class by the hypothesis of two collateral evolutionary series, in which the light strength decreases with the increasing redness of the stars in each series. Representative stars for the two series are then listed:

Star series I	Star series II
S Monocerotis	χ_2 Orionis
ε Orionis	β Orionis
γ Orionis	α Cygni
α Leonis	δ Canis maj.
α Canis maj.	α Bootis
α Aquilae	α Orionis
α Canis min.	Vogel's Type IV
Sun	
70 Ophiuchi	
61 Cygni	
o_2 Eridani B. C.	
"black" stars	

The division of stars by brightness and spectral class is further explained if stars spend more time in the redder parts of the series.

35. Relations between the Spectra and Other Characteristics of Stars

Henry Norris Russell

(*Popular Astronomy 22*, 275–294 [1914])

Henry Norris Russell first described what has come to be called the "Hertzsprung-Russell diagram" at a meeting of the Royal Astronomical Society in June 1913, and in December of the same year, at the American Astronomical Society meeting in Atlanta, he presented the amplified version given here. Russell's collection of the increasingly available data on stellar parallax enabled him to calculate the luminosities of stars, which he plotted against spectral type to show two well-defined classes of stars. These he called dwarf and giant stars, designations he erroneously attributed to Ejnar Hertzsprung. For many years this figure was called the Russell diagram, but eventually astronomers realized that Hertzsprung had already plotted such diagrams for the Pleiades and Hyades in 1911.[1]

Always a much more speculative thinker than Hertzsprung, Russell looked on the diagram primarily as a device for understanding stellar evolution. Earlier, Augustus Ritter[2] had argued that there could be three classes of stars, a cool category with rising temperatures, an intermediate division of hot stars near the maximum temperature, and another cool category with falling temperatures. Norman Lockyer had taken up this scheme[3] and had tried to identify certain spectral types with the red stars on the rising temperature branch and others with the descending branch. Russell was greatly influenced by Lockyer's attempt, and his new distinction of giant and dwarf stars seemed to fit into such an evolutionary scheme far more naturally than Lockyer's classes.

In sorting out the various physical parameters of stars, Russell noticed that there is relatively little variety in mass, and he failed to find any correlation of mass with luminosity. Since mass seems not to be involved, the difference between dwarf and giant stars appears to be entirely one of size and density. Russell concluded that the giant stars are the youngest and are in the process of gravitational collapse, which was their source of energy and which caused them to grow smaller and hotter. Accordingly, the A-type stars are of intermediate age, and the sun is even more advanced in the evolutionary sequence. This idea contrasted with the then commonly accepted belief that the A-type stars were the youngest and hence were "early-type" stars; this terminology still persists, although its evolutionary premise has been forgotten. Today the Hertzsprung-Russell diagram remains a primary tool for tracing the paths of stellar evolution, but the routes are more complex than Russell imagined. The giant stars in fact represent later rather than earlier stages in a star's life cycle.

1. E. Hertzsprung, *Publikationen des astrophysikalischen Observatoriums zu Potsdam 22*, no. 63 (1911).
2. A. Ritter, *Wiedemann Annalen 20*, 137, 897 (1883), *Astrophysical Journal 8*, 293 (1898).
3. J. N. Lockyer, *Proceedings of the Royal Society* (London) *43*, 117 (1887), *44*, 1 (1888).

INVESTIGATIONS INTO THE NATURE of the stars must necessarily be very largely based upon the average characteristics of groups of stars selected in various ways—as by brightness, proper motion, and the like. The publication within the last few years of a great wealth of accumulated observational material makes the compilation of such data an easy process; but some methods of grouping appear to bring out much more definite and interesting relations than others, and, of all the principles of division, that which separates the stars according to their spectral types has revealed the most remarkable differences, and those which most stimulate attempts at a theoretical explanation.

In the present discussion, I shall attempt to review very rapidly the principal results reached by other investigators, and shall then ask your indulgence for an account of certain researches in which I have been engaged during the past few years.

Thanks to the possibility of obtaining with the objective prism photographs of the spectra of hundreds of stars on a single plate, the number of stars whose spectra have been observed and classified now exceeds one hundred thousand, and probably as many more are within the reach of existing instruments. The vast majority of these spectra show only dark lines, indicating that absorption in the outer and least dense layers of the stellar atmospheres is the main cause of their production. Even if we could not identify a single line as arising from some known constituent of these atmospheres, we could nevertheless draw from a study of the spectra, considered merely as line-patterns, a conclusion of fundamental importance.

The spectra of the stars show remarkably few radical differences in type. More than ninety-nine per cent of them fall into one or other of the six great groups which, during the classic work of the Harvard College Observatory, were recognized as of fundamental importance, and received as designations, by the process of "survival of the fittest," the rather arbitrary series of letters B, A, F, G, K, and M. That there should be so few types is noteworthy; but much more remarkable is the fact that they form a continuous series. Every degree of gradation, for example, between the typical spectra denoted by B and A may be found in different stars, and the same is true to the end of the series, a fact recognized in the familiar decimal classification, in which B5, for example, denotes a spectrum half-way between the typical examples of B and A. This series is not merely continuous; it is *linear*. There exist indeed slight differences between the spectra of different stars of the same spectral class, such as A0; but these relate to minor details, which usually require a trained eye for their detection, while the difference between successive classes, such as A and F, are conspicuous to the novice. Almost all the stars of the small outstanding minority fall into three other classes, denoted by the letters O, N, and R. Of these O undoubtedly precedes B at the head of the series, while R and N, which grade into one another, come probably at its other end, though in this case the transition stages, if they exist, are not yet clearly worked out.

From these facts it may be concluded that the principal differences in stellar spectra, however they may originate, arise in the main from variations in a single physical condition in the stellar atmospheres. This follows at once from the linearity of the series. If the spectra depended, to a comparable degree, on two independently variable conditions, we should expect that we would be obliged to represent their relations, not by points on a line, but by points scattered over an area. The minor differences which are usually described as "peculiarities" may well represent the effects of other physical conditions than the controlling one.

The first great problem of stellar spectroscopy is the identification of this predominant cause of the spectral differences. The hypothesis which suggested itself immediately upon the first studies of stellar spectra was that the differences arose from variations in the chemical composition of the stars. Our knowledge of this composition is now very extensive. Almost every line in the spectra of all the principal classes can be produced in the laboratory, and the evidence so secured regarding the uniformity of nature is probably the most impressive in existence. The lines of certain elements are indeed characteristic of particular spectral classes; those of helium, for instance, appear only in Class B, and form its most distinctive characteristic. But negative conclusions are proverbially unsafe. The integrated spectrum of the Sun shows no evidence whatever of helium, but in that of the chromosphere it is exceedingly conspicuous. Were it not for the fact that we are near this one star of Class G, and can study it in detail, we might have erroneously concluded that helium was confined to the "helium stars". There are other cogent arguments against this hypothesis. For example, the members of a star-cluster, which are all moving together, and presumably have a common origin, and even the physically connected components of many double stars, may have spectra of very different types, and it is very hard to see how, in such a case, all the helium and most of the hydrogen could have collected in one star, and practically all the metals in the other. A further argument—and to the speaker a very convincing one—is that it is almost unbelievable that differences of chemical composition should reduce to a function of a single variable, and give rise to the observed linear series of spectral types.

I need not detain you with the recital of the steps by which astrophysicists have become generally convinced that the main cause of the differences of the spectral classes is difference of temperature of the stellar atmospheres. There is time only to review some of the most important evidence which, converging from several quarters, affords apparently a secure basis for this belief.

Those metallic lines that are strongest in class M and weakest in class K stars are those that are present in the spectra seen in the cooler parts of laboratory flames; this observation suggests that M stars are cooler than K stars (see Fowler, *Proceedings of the Royal Society* (London) *72*, 219 [1904]). A comparison of stellar brightness measured visually and photographically shows that the color index (the relative photographic magnitude of stars of equal visual brightness) increases with increasing spectral type (See table 35.1 and Parkhurst, *Astrophysical Journal 36*, 218 [1912]).

Table 35.1 Spectra, color indices and surface temperatures of stars

	Color-index			
Spectrum	King	Parkhurst	Schwarzschild	Temperature[a]
B0	−0.32	—	—	20,000:
B5	−0.17	−0.21	−0.20	14,000
A0	0.00	0.00	0.00	11,000
A5	0.19	0.23	0.20	9,000
F0	0.30	0.43	0.40	7,500
F5	0.52	0.65	0.60	6,000
G0	0.71	0.86	0.84	5,000
G5	0.90	1.07	1.10	4,500
K0	1.16	1.30	1.35	4,200
K5	1.62	1.51	1.80	3,200
M	1.62	1.68	—	3,100
N	—	2.5	—	2,300:

a. This column gives the effective temperature at which a black-body would emit light of the same color as that observed.

It should be expressly stated that the "temperatures" here spoken of are the effective "black body" temperatures corresponding to the spectral distribution of the radiation. Unless the surfaces of the stars possess decided selective emissivity for certain wave-lengths, these effective temperatures should also indicate with tolerable accuracy the energy density of the flux of radiation which escapes from them. This tells us little about the temperature of the deeper regions; but it must be the main, if not the only, factor in determining the temperature of those outer and nearly transparent layers of the atmospheres in which the characteristic line absorption takes place. If we further assume, in accordance with Abbot's studies of the solar atmosphere, that the absorption is nearly complete in so small a thickness of the atmosphere that wide variations in its depth and density would modify its total absorption but little, it becomes easy to see how the influence of its temperature (which presumably determines the relative strengths of absorption in different lines) may predominate so greatly over that of all other factors in determining the spectral type.

More than half the stars brighter than 6.25 mag (those visible to the naked eye) are classes A and K, whereas the remaining stars are divided fairly evenly among the other four principal classes. Stars of classes O and B are strongly concentrated in the galactic plane; those of the other classes are not. Russell summarizes data on the proper motions, parallaxes, and velocities of stars of different spectral types and then notes that short-period binary stars are confined to classes B, A, and F.

Having thus made a rapid survey of the general field, I will now ask your attention in greater detail to certain relations which have been the more special objects of my study.

Let us begin with the relations between the spectra and the real brightness of the stars. These have been discussed by many investigators—notably by Kapteyn and Hertzsprung—and many of the facts which will be brought before you are not new; but the observational material here presented is, I believe, much more extensive than has hitherto been assembled. We can only determine the real brightness of a star when we know its distance; but the recent accumulation of direct measures of parallax, and the discovery of several moving clusters of stars whose distances can be determined, put at our disposal far more extensive data than were available a few years ago.

Figure 35.1 shows graphically the results derived from all the direct measures of parallax available in the spring of 1913 (when the diagram was constructed). The spectral class appears as the horizontal coordinate, while the vertical one is the absolute magnitude, according to Kapteyn's definition—that is, the visual magnitude which each star would appear to have if it should be brought up to a standard distance, corresponding to a parallax of $0''.1$ (no account being taken of any possible absorption of light in space.) The absolute magnitude −5, at the top of the diagram, corresponds to a luminosity 7,500 times that of the Sun, whose absolute magnitude is 4.7. The absolute magnitude 14, at the bottom, corresponds to 1/5,000 of the Sun's luminosity. The larger dots denote the stars for which the computed probable error of the parallax is less than 42 per cent of the parallax itself, so that the probable error of the resulting absolute magnitude is less than $\pm 1^m.0$. This is a fairly tolerant criterion for a "good parallax," and the small dots, representing the results derived from the poor parallaxes, should hardly be used as a basis for

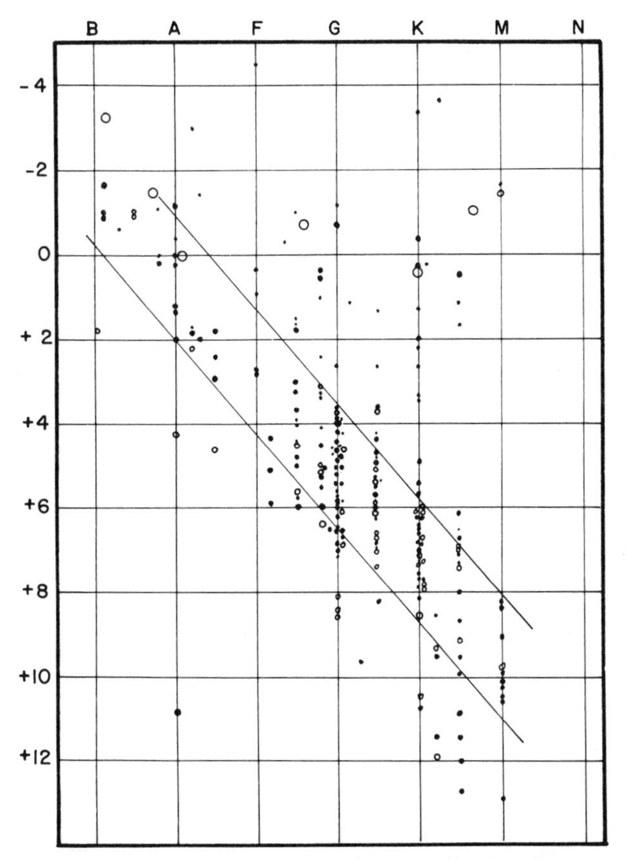

Fig. 35.1 The spectral class-absolute luminosity diagram for individual bright stars whose distances have been measured. The spectral class appears as the horizontal coordinate, the absolute magnitude appears as the vertical coordinate, the larger dots denote stars whose absolute magnitudes have a probable error of 1 mag, and the small dots represent results derived from poor parallaxes. The large open circles denote mean results for about 120 bright stars of small proper motion; the smaller open and filled circles denote, respectively, stars whose parallaxes have been measured once or more than once. The two diagonal lines mark the boundaries of Hertzsprung's main sequence (called the dwarf series by Russell), and the exceptional point in the lower left-hand corner is the faint companion of the double star system σ^2 Eridani. [This faint companion is now known to be a white dwarf star.]

any argument. The solid black dots represent stars whose parallaxes depend on the mean of two or more determinations; the open circles, those observed but once. In the latter case, only the results of those observers whose work appears to be nearly free from systematic error have been included, and in all cases the observed parallaxes have been corrected for the probable mean parallax of the comparison stars to which they were referred. The large open circles in the upper part of the diagram represent mean results for numerous bright stars of small proper motion (about 120 altogether) whose observed parallaxes hardly exceed their probable errors. In this case the best thing to do is to take means of the observed parallaxes and magnitudes for suitable groups of stars, and then calculate the absolute magnitudes of the typical stars thus defined. These will not exactly correspond to the mean of the individual absolute magnitudes, which we could obtain if we knew all the parallaxes exactly, but they are pretty certainly good enough for our purpose.

Upon studying figure 35.1, several things can be observed.

1. All the white stars, of Classes B and A, are bright, far exceeding the Sun; and all the very faint stars—for example, those less than 1/50 as bright as the Sun—are red, and of Classes K and M. We may make this statement more specific by saying, as Hertzsprung does, that there is a certain limit of brightness for each spectral class, below which stars of this class are very rare, if they occur at all. Our diagram shows that this limit varies by rather more than two magnitudes from class to class. The single apparent exception is the faint double companion to o^2 Eridani, concerning whose parallax and brightness there can be no doubt, but whose spectrum, though apparently of Class A, is rendered very difficult of observation by the proximity of its far brighter primary.

2. On the other hand, there are many red stars of great brightness, such as Arcturus, Aldebaran and Antares, and these are as bright, on the average, as the stars of Class A, though probably fainter than those of Class B. Direct measures of parallax are unsuited to furnish even an estimate of the upper limit of brightness to which these stars attain, but it is clear that some stars of all the principal classes must be very bright. The range of actual brightness among the stars of each spectral class therefore increases steadily with increasing redness.

3. But it is further noteworthy that all the stars of Classes K5 and M which appear on our diagram are either very bright or very faint. There are none comparable with the Sun in brightness. We must be very careful here not to be misled by the results of the methods of selection employed by observers of stellar parallax. They have for the most part observed either the stars which appear brightest to the naked eye or stars of large proper motion. In the first case, the method of selection gives an enormous preference to stars of great luminosity, and, in the second, to the nearest and most rapidly moving stars, without much regard

to their actual brightness. It is not surprising, therefore, that the stars picked out in the first way (and represented by the large circles in figure 35.1) should be much brighter than those picked out by the second method (and represented by the smaller dots). But if we consider the lower half of the diagram alone, in which all the stars have been picked out for proper motion, we find that there are no very faint stars of Class G, and no relatively bright ones of Class M. As these stars were selected for observation entirely without consideration of their spectra (most of which were then unknown) it seems clear that this difference, at least, is real, and that there is a real lack of red stars comparable in brightness to the Sun, relatively to the number of those 100 times fainter.

The appearance of figure 35.1 therefore suggests the hypothesis that, if we could put on it some thousands of stars, instead of the 300 now available, and plot their absolute magnitudes without uncertainty arising from observational error, we would find the points representing them clustered principally close to two lines, one descending sharply along the diagonal, from B to M, the other starting also at B, but running almost horizontally. The individual points, though thickest near the diagonal line, would scatter above and below it to a vertical distance corresponding to at least two magnitudes, and similarly would be thickest near the horizontal line, but scatter above and below it to a distance which cannot so far be definitely specified, so that there would be two fairly broad bands in which most of the points lay. For Classes A and F, these two zones would overlap, while their outliers would still intermingle in Class G, and probably even in Class K. There would however be left a triangular space between the two zones, at the right-hand edge of the diagram, where very few, if any, points appeared; and the lower left-hand corner would be still more nearly vacant.

We may express this hypothesis in another form by saying that there are two great classes of stars—the one of great brightness (averaging perhaps a hundred times as bright as the Sun) and varying very little in brightness from one class of spectrum to another; the other of smaller brightness, which falls off very rapidly with increasing redness. These two classes of stars were first noticed by Hertzsprung, who has applied to them the excellent names of *giant* and *dwarf* stars. The two groups, on account of the considerable internal differences in each, are only distinctly separated among the stars of Class K or redder. In Class F they are partially, and in Class A thoroughly intermingled, while the stars of Class B may be regarded equally well as belonging to either series.

In addition to the stars of directly measured parallax, represented in figure 35.1, we know with high accuracy the distances and real brightness of about 150 stars which are members of the four moving clusters whose convergent points

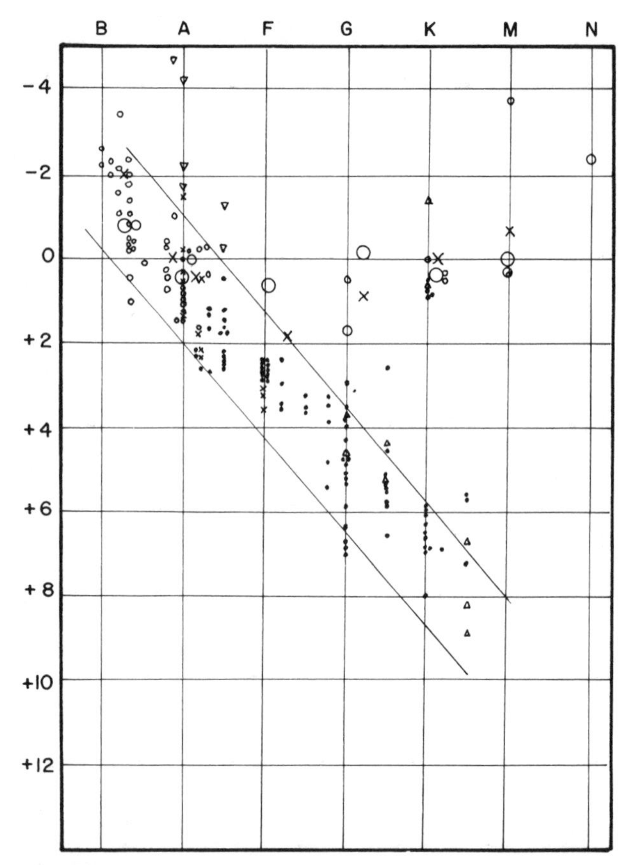

Fig. 35.2 The spectral class-absolute luminosity diagram for four moving clusters: *black dots*, the Hyades; *small crosses*, the Ursa Major group; *small open circles*, the large group in Scorpius; and *triangles*, the 61 Cygni group. The large circles and crosses represent points calculated from the mean parallaxes and magnitudes of other groups of stars.

are known—namely, the Hyades, the Ursa Major group, the 61 Cygni group, and the large group in Scorpius, discovered independently by Kapteyn, Eddington, and Benjamin Boss, whose motion appears to be almost entirely parallactic. The data for the stars of these four groups are plotted in figure 35.2, on the same system as in figure 35.1. The solid black dots denote the members of the Hyades; the open circles, those of the group in Scorpius; the crosses the Ursa Major group, and the triangles the 61 Cygni group. Our lists of the members of each group are probably very nearly complete down to a certain limiting (visual) magnitude, but fail at this point, owing to lack of knowledge regarding the proper motions of the fainter stars. The apparently abrupt termination of the Hyades near the absolute magnitude 7.0, and of the Scorpius group at 1.5 arises from this observational limitation.

The large circles and crosses in the upper part of figure 35.2, represent the absolute magnitudes calculated from the mean parallaxes and magnitudes of the groups of stars investigated by Kayteyn, Campbell, and Boss. The larger circles represent Boss's results, the smaller circles Kapteyn's, and the large crosses Campbell's.

It is evident that the conclusions previously drawn from figure 35.1 are completely corroborated by these new and independent data. Most of the members of these clusters are dwarf stars, and it deserves particular notice that the stars of different clusters, which are presumably of different origin, are similar in absolute magnitude. But there are also a few giant stars, especially of Class K (among which are the well known bright stars of this type in the Hyades); and most remarkable of all is Antares, which, though of Class M, shares the proper motion and radial velocity of the adjacent stars of Class B, and is the brightest star in the group, giving out about 2,000 times the light of the Sun. It is also clear that the naked eye stars, studied by Boss, Campbell and Kapteyn, are for the most part giants. With this in mind, we are now in a position to explain more fully the differences between the results of these investigators.

Here Russell argues that Kapteyn and Boss obtained different mean proper motions and parallaxes because Boss considered only dwarf stars and Kapteyn considered both giants and dwarfs. Empirical formulae are then established for determining the absolute magnitudes of dwarf stars of different spectral classes.

The questions now arise:

What differences in their nature or constitution give rise to the differences in brightness between the giant and dwarf stars, and why should these differences show such a systematic increase with increasing redness, or "advancing" spectral type?

We must evidently attack the first of these questions before the second. The absolute magnitude (or the actual luminosity) of a star may be expressed as a function of three physically independent quantities—its mass, its density, and its surface brightness. Great mass, small density, and high surface brightness make for high luminosity, and the giant stars must possess at least one of these characteristics in a marked degree, while the dwarf stars must show one or more of the opposite attributes.

A good deal of information is available concerning all these characteristics of the stars. The masses of a considerable number of visual and spectroscopic binaries are known with tolerable accuracy, the densities of a larger number of eclipsing variable stars have recently been worked out; and the recent investigations on stellar temperatures lead directly to estimates of the relative surface brightness of the different spectral classes (subject, of course, to the uncertainty whether the stars really radiate like black bodies, as they are assumed to do). We will take these matters up in order.

There appears to be some correlation between mass and luminosity, but stellar masses are remarkably similar, in comparison to the million-fold range in stellar luminosities. By assuming that the total mass of the two components of visual binary stars is equal to twice that of the sun, Russell obtains values for their dynamical parallaxes. The distances of these stars are then obtained from the parallaxes and from measurements of the relative motion of the pair of stars. The spectrum-luminosity diagram is then given for the binary stars and is shown to be very similar to the previous diagrams.

But this new evidence does much more than to confirm that which we have previously considered; it proves that the distinction between the giant and dwarf stars, and the relations between their brightness and spectral types, do not arise (primarily at least), from differences in mass. Even when reduced to equal masses, the giant stars of Class K are about 100 times as bright as the dwarf stars of similar spectrum, and for Class M the corresponding ratio is fully 1,000. Stars belonging to the two series must therefore differ greatly either in surface brightness or in density, if not in both.

There is good physical reason for believing that stars of similar spectrum and color-index are at least approximately similar in surface brightness, and that the surface brightness falls off rapidly with increasing redness. Indeed, if the stars radiate like black bodies, the relative surface brightness of any two stars should be obtainable by multiplying their relative color-index by a constant (which is the ratio of the mean effective photographic wave-length to the difference of the mean effective visual and photographic wave-lengths, and lies usually between 3 and 4, its exact value depending upon the systems of visual and photographic magnitude adopted as standards). Such a variation of surface brightness with redness will evidently explain at least the greater part of the change in absolute magnitude among the dwarf stars (as Hertzsprung and others have pointed out); but it makes the problem of the giant stars seem at first sight all the more puzzling.

The solution is however very simple. If a giant star of Class K, for example, is 100 times as bright as a dwarf star of the same mass and spectrum, and is equal to it in surface brightness, it must be of ten times the diameter, and 1/1,000 of the density of the dwarf star. If, as in Class M, the giant star is 1,000 times as bright as the dwarf, it must be less than 1/30,000 as dense as the latter. Among the giant stars in general, the diminishing surface brightness of the redder stars must be compensated for by increasing diameter, and therefore by rapidly decreasing density (since all the stars considered have been reduced to equal mass).

But all this rests on an assumption which, though physically very probable, cannot yet be said to be proved, and its consequences play havoc with certain generally accepted ideas. We will surely be asked—Is the assumption of the existence of stars of such low density a reasonable or probable one? Is there any other evidence that the density of a star of Class G or K may be much less than that of the stars of Classes B and A? Can any other evidence than that derived from the laws of radiation be produced in favor of the rapid decrease of surface brightness with increasing redness?

The densities of stars of different spectral classes are computed from eclipsing binary data. The great majority of stars of spectral classes A and B have very similar densities, lying between 1/3 and 1/45 that of the sun. The stars of classes G and K exhibit a wide range of densities, between twice and one-millionth that of the sun. The giant stars are shown to have the lowest densities.

We may now answer decisively, and in the affirmative, the first two questions which were put a few moments ago. Some stars actually have densities quite as low as any that might be required to explain the great brightness of the reddest giant stars; and these stars of low density show a very marked preference for the "later" spectral classes, while practically all the stars of "earlier" type are far denser.

We can answer the third question as well, and in a quantitative fashion, if we are willing to assume that the eclipsing binaries, and also the telescopic double stars, of the various spectral classes are typical of the stars of these classes as a whole.

Giant and dwarf stars of widely different spectral classes are shown to have very similar masses.

We may now summarize the facts which have been brought to light as follows:

1. The differences in brightness between the stars of different spectral classes, and between the giant and dwarf stars of the same class, do not arise (directly at least), from differences in mass. Indeed, the mean masses of the various groups of stars are extraordinarily similar.
2. The surface brightness of the stars diminishes rapidly with increasing redness, changing by about three times the difference in color-index, or rather more than one magnitude, from each class to the next.
3. The mean density of the stars of Classes B and A is a little more than one-tenth that of the Sun. The densities of the dwarf stars increase with increasing redness from this value through that of the Sun to a limit which cannot at present be exactly defined. This increase in density, together with the diminution in surface brightness, accounts for the rapid fall in luminosity with increasing redness among these stars.
4. The mean densities of the giant stars diminish rapidly with increasing redness, from one-tenth that of the Sun for Class A to less than one twenty-thousandth that of the Sun for Class M. This counteracts the change in surface brightness, and explains the approximate equality in luminosity of all these stars.
5. The actual existence of stars of spectra G and K, whose densities are of the order here derived, is proved by several examples among the eclipsing variables—all of which are far less dense than any one of the more numerous eclipsing stars of "early" spectral type, with the sole exception of Beta Lyrae.

These facts have evidently a decided bearing on the problem of stellar evolution, and I will ask your indulgence during the few minutes which remain for an outline of the theory of development to which it appears to me that they must inevitably lead.

Of all the propositions, more or less debatable, which may be made regarding stellar evolution, there is probably none that would command more general acceptance than this—that as a star grows older it contracts. Indeed, since contraction converts potential energy of gravitation into heat, which is transferred by radiation to cooler bodies, it appears from thermodynamic principles that the general trend of change must in the long run be in this direction. It is conceivable that at some particular epoch in a star's history there might be so rapid an evolution of energy, for example—of a radioactive nature—that it temporarily surpassed the loss by radiation, and led to an expansion against gravity; but this would be at most a passing stage in its career, and it would still be true in the long run that the order of increasing density is the order of advancing evolution.

If now we arrange the stars which we have been studying in such an order, we must begin with the giant stars of Class M, follow the series of giant stars, in the reverse order from that in which the spectra are usually placed, up to A and B, and then, with density still increasing, though at a slower rate, proceed down the series of dwarf stars, in the usual order of the spectral classes, past the Sun, to those red stars (again of

Class M), which are the faintest at present known. There can be no doubt at all that this is the order of increasing density; if it is also the order of advancing age, we are led at once back to Lockyer's hypothesis that a star is hottest near the middle of its history, and that the redder stars fall into two groups, one of rising and the other of falling temperature. The giant stars then represent successive stages in the heating up of a body, and must be more primitive the redder they are; the dwarf stars represent successive stages in its later cooling, and the reddest of these are the farthest advanced. We have no longer two separate series to deal with, but, a single one, beginning and ending with Class M, and with Class B in the middle—all the intervening classes being represented, in inverse order, in each half of the sequence.

The great majority of the stars visible to the naked eye, except perhaps in Class F, are giants; hence for most of these stars the order of evolution is the reverse of that now generally assumed, and the terms "early" and "late" applied to the corresponding spectral types are actually misleading.

This is a revolutionary conclusion; but, so far as I can see, we are simply shut up to it with no reasonable escape. If stars of the type of Capella, Gamma Andromedae, and Antares represent later stages of development of bodies such as Delta Orionis, Alpha Virginis, and Algol, we must admit that, as they grew older and lost energy, they have expanded, in the teeth of gravitation, to many times their original diameters, and have diminished many hundred—or even thousand—fold in density. For the same reason, we cannot regard the giant stars of Class K as later stages of those of Class G, or those of Class M as later stages of either of the others, unless we are ready to admit that they have expanded against gravity in a similar fashion. We may of course take refuge in the belief that the giant stars of the various spectral classes have no genetic relations with one another—that no one class among them represents any stage in the evolution of stars like any of the others—but this is to deny the possibility of forming any general scheme of evolution at all.

Here Russell argues that as a star loses energy it will contract and its temperature will rise, following Lane's law, which states that the temperature varies inversely with the radius.

The resemblance between the characteristics that might thus be theoretically anticipated in a mass of gas of stellar dimensions, during the course of its contraction, and the actual characteristics of the series of giant and dwarf stars of the various spectral classes is so close that it might fairly be described as identity. The compensating influences of variations in density and surface brightness, which keep all the giant stars nearly equal in luminosity, the rapid fall of brightness among the dwarf stars, and the ever increasing difference between the two classes, with increasing redness, are all just what might be expected. More striking still is the entire agreement between the actual densities of the stars of the various sorts and those estimated for bodies in the different stages of development, on the basis of the general properties of gaseous matter. The densities found observationally for the giant stars of Classes G to M are such that Lane's Law must apply to them and they must grow hotter if they contract; that of the Sun (a typical dwarf star), is so high that the reverse must almost certainly be true; and the mean density of the stars of Classes B and A (about one-ninth that of the Sun, or one-sixth that of water) is just of the order of magnitude at which a contracting mass of gas might be expected to reach its highest surface temperature.

Russell then argues that less massive stars will be colder and that this is the reason why no luminous stars of mass less than one-tenth that of the sun are known. It is also the reason why Jupiter and Saturn are dark even though they have densities comparable to dwarf stars.

One further application of the theory may be very briefly mentioned. If we have a large number of contracting masses of gas, endowed with various moments of momentum, more and more of them will split up into pairs as they grow denser, and the pairs latest formed will have the shortest periods. A large percentage of spectroscopic binaries, especially of short period, is therefore direct evidence of a fairly advanced state of evolution, and the occurrence of this condition among the stars of Classes B and A supports—indeed, almost by itself compels—the view that they are far removed from a primitive condition. Most of the stars which have been investigated for radial velocity are giants, and the absence of spectroscopic binaries of short period among the redder ones is in agreement with the view that they are in earlier stages of evolution.

Russell argues that the number of stars per unit volume decreases rapidly with increasing luminosity because bright stars radiate their energy more rapidly than less luminous stars.

I have endeavored in the past hour to set before you the present state of knowledge concerning the real brightness, masses, densities, temperatures and surface brightness of the stars, and to sketch the theory of stellar evolution to which the study of these things has led me. This theory is inconsistent with the generally accepted view. Its fundamental principle is identical with that of Lockyer's classification, but it differs radically from the latter in the principles according to which it assigns individual stars, and even whole classes of stars, to the series of ascending or descending temperature.

(For example, Lockyer puts such conspicuously giant stars as Canopus, Capella, Arcturus, and Beta Cygni, and all the stars of Class N, into the descending series, and places β Hydri and δ Pavonis (which are clearly dwarf stars) in the ascending series.)

Two things have gone farthest to convince me that it may be a good approximation to the truth; the way in which it explains and coördinates characteristics of the different spectral types which previously appeared to be without connection or reason, and the way in which a number of apparent exceptions to its indications have disappeared, one by one, as more accurate information concerning spectra, orbits of double stars, and the like, became available, till only one doubtful case remains.

36. Some Spectral Criteria for the Determination of Absolute Stellar Magnitudes

Walter S. Adams and Arnold Kohlschütter

(*Astrophysical Journal 40*, 385–398 [1914])

"If we could only determine the absolute luminosity of a star from its spectrum, we would be in a position to compute its parallax from the apparent magnitude of the star together with its spectrum ... The finding of such spectral equivalents of luminosity will therefore be a particularly rewarding task."[1]

So wrote Ejnar Hertzsprung in 1911. The task was accomplished by the American Walter Adams and the German Arnold Kohlschütter working with the spectrograph on the 60-in telescope at Mount Wilson Observatory. They showed that the relative intensities of certain neighboring spectral lines can be used to determine the absolute luminosities of both dwarf and giant stars, a conclusion that was later confirmed and extended to a study of over 4,000 stars by Adams and his colleagues.[2] By showing that stars of the same temperature but different luminosities exhibit spectral differences, Adams and Kohlschütter set the stage for a two-dimensional stellar classification scheme that includes both spectral type and luminosity class. Such a classification scheme has been incorporated in the *Atlas of Stellar Spectra*, illustrated in selection 8.

1. E. Hertzsprung, *Publikationen des astrophysikalischen Observatoriums zu Potsdam 22*, no. 63 (1911).
2. W. S. Adams, A. H. Joy, M. L. Humason, and A. M. Brayton, *Astrophysical Journal 81*, 188 (1935).

In the course of a study of the spectral classification of stars whose spectra have been photographed for radial velocity determinations some interesting peculiarities have been observed. The stars investigated are of two kinds: first, those of large proper motion with measured parallaxes; second, those of very small proper motion, and hence, in general, of great distance. The apparent magnitudes of the large proper motion, or nearer stars, are somewhat less on the average than those of the small proper motion stars, so that the difference in absolute magnitude must be very great between the two groups. The spectral types range from A to M.

The principal differences in the spectra of the two groups of stars are:

1. The continuous spectrum of the small proper motion stars is relatively fainter in the violet as compared with the red than is the spectrum of the large proper motion stars. The magnitude of this effect appears to depend on the spectral type, and increases with advancing type between F0 and K0.
2. The hydrogen lines are abnormally strong in a considerable number of the small proper motion stars. Thus six stars which show the well developed titanium oxide bands characteristic of type M have hydrogen lines which would place them in types G4 to G6, and many others which show the bands strongly would be classified under type K from their hydrogen lines. That the spectra of these stars are not composite is shown by their radial velocities. The hydrogen lines in the spectra of the large proper motion stars which show the titanium oxide bands are without exception very weak.
3. Certain other spectrum lines are weak in the large proper motion stars, and strong in the small proper motion stars, and conversely. It is with the possibility of applying this fact to the determination of absolute magnitudes that the results given in this communication mainly have to deal.

Intensity of the Continuous Spectrum

Here Adams and Kohlschütter describe their method of obtaining the continuum intensity of stars at different wavelengths by using spectral photographs compared with standard plates of α Tauri. Points at violet and red wavelengths were selected in regions as free from lines as possible.

The values for the groups of stars are given in table 36.1. The densities for the three violet wave-lengths have been combined to form a mean at λ 4,220, and similarly for the four wave-lengths near λ 4,955.

Table 36.1 A comparison of stellar intensities

	Number of stars	Average μ	Average type	Density at λ 4,220	Density at λ 4,955
A0–A9	15	0″.020	A2	0.30	0.32
	16	.13	A3	.29	.32
F0–F9	10	.012	F4	.25	.37
	23	.66	F6	.32	.37
G0–G4	8	.009	G3	.22	.41
	30	.64	G2	.33	.41
G5–G9	14	.011	G7	.25	.48
	22	.64	G7	.40	.48
K0–K4	24	.011	K2	.16	.44
	22	.70	K1	.31	.44

The features of note in these results are:

a) The small proper motion stars of types F to K are decidedly weaker in the violet part of the spectrum than the large proper motion stars.

b) The difference is inappreciable for two groups of A-type stars for which the ratio of proper motions is 1:6.5.

c) The difference increases with advancing type from F to K, being twice as great for the latter. The ratio of proper motions for the groups of small and of large proper motion stars is nearly the same for the stars between F and K. Hence if interpreted in terms of distance the ratio of distances should be nearly the same, and it would appear that at least a part of the absorption in the violet part of the spectrum of the distant stars must be ascribed, not to scattering of light in space, but to conditions in the stellar atmospheres. In the case of the A-type stars the results are inconclusive, since the ratio of the proper motions shows that the negative result found may be due to the fact that the difference of distance between the two groups of stars is insufficient to produce a measurable amount of scattering.

The Hydrogen Lines

The abnormal strength of the hydrogen lines in the spectra of certain of the small proper motion stars is of peculiar interest because of the possibility of selective absorption by hydrogen gas in interstellar space. The radial velocity affords a means of determining the origin of the additional absorption since it is highly improbable that the hydrogen in space would have the motion of the stars observed. Accordingly we have given especial attention to the determination of the radial velocities of these stars from the hydrogen lines as compared with other selected lines in the spectrum. The results obtained indicate that within the limits of error of measurement the hydrogen lines give essentially the same values as the other lines, and no differences have been found of an order to correspond to the abnormal intensity of the lines.

Adams and Kohlschütter give a table of the differences in velocities based on the metallic and hydrogen lines of 15 stars. Within the accuracy of the measurements, the velocities are equal, and therefore all the hydrogen absorption must occur in the stellar atmospheres.

The Relation of Line Intensity to Absolute Magnitude

Systematic differences of intensity for certain lines between stars of large and stars of small proper motion soon became evident in the course of the study of the spectral classification of these stars. In order to secure an accurate system of classification as well as to investigate these differences the following method was adopted. Pairs of lines were selected not far from one another in the spectrum and of as nearly as possible the same intensity and character, and estimations were made of their relative intensities. For classification purposes a line decreasing in intensity with advancing type, such as a hydrogen line, was combined with a line increasing in intensity with advancing type, such as an ordinary metallic line. In addition to these pairs used for classification purposes several pairs were selected which included all lines suspected of systematic deviations in certain stars.

The estimations were made on an arbitrary scale extending from 1 to about 12, 1 being the smallest difference in intensity which could be detected. The method, therefore, is analogous to the *Stufenmethode* of Argelander used in estimations of variable stars; hence, for physiological reasons, our scale will be approximately proportional to the logarithm of the intensity differences of the two lines. In general three plates were used for each star, and the photographs of the large and the small proper motion stars were intermingled in order that systematic effects on the estimation scale might be avoided.

Here Adams and Kohlschütter list the pairs of lines showing the most significant change with spectral type. When spectral type was assigned by using these pairs, the remaining pairs of lines were used to investigate changes with absolute magnitude. Absolute magnitudes were obtained from apparent magnitudes and the measured parallaxes. The ratios of line intensities as a function of two mean absolute magnitudes are given in table III, not reproduced here.

The most prominent cases of lines where systematic differences are seen to exist between the stars of high and of low luminosity are, for stars of high luminosity:

Strong	Weak
	4,325 Sc
4,216 Sr	4,435 Ca
4,395 Ti, V, Zr	4,456 Ca
4,408 V, Fe	4,535 Ti

The Sr line at λ 4,216 is an extremely prominent chromospheric line, and the same is true in less degree of the enhanced Ti line at λ 4,395. The line at λ 4,408 is a blend, and as given by Rowland consists of V and Fe. Some other element may perhaps contribute to the stellar line. All four of the lines which are relatively weak in the high luminosity stars are well known sun-spot lines, being greatly strengthened in the umbrae of spots.

The following five pairs of lines were selected from table III as the basis for an investigation of the individual stars:

$$\frac{4{,}216}{4{,}250} \quad \frac{4{,}395}{4{,}415} \quad \frac{4{,}408}{4{,}415} \quad \frac{4{,}456}{4{,}462} \quad \frac{4{,}456}{4{,}495}$$

The results given in table III, estimated value – normal value, for these five pairs of lines were combined into means. By assuming a linear relationship between these mean values D, and the absolute magnitude M, we then derived the formulae:

$$\text{F8–G6 stars: } M = +5.6 - 1.6\,D$$
$$\text{G6–K9 stars: } M = +6.8 - 1.8\,D.$$

Adams and Kohlschütter use the measured intensities of these lines to specify absolute magnitudes and compare these with the values obtained from the apparent magnitudes and measured parallaxes. The average difference between the two sets of data was 1.5 mag.

Summary

Including the results described here, we have found as a product of our investigations of the spectra of large and of small proper motion stars three phenomena which appear to have a distinct bearing upon the problem of the determination of the absolute magnitudes of stars.

1. The continuous spectrum of the small proper motion stars is decidedly less intense in the violet region relative to the red than the spectrum of the nearer and smaller stars. This effect appears to be a function of the spectral type, and so must be ascribed in part, at least, to conditions in the stellar atmospheres.
2. A considerable number of the small proper motion stars show hydrogen lines of abnormally great intensity. Measures of the radial velocity show the source of the additional absorption to be mainly, if not wholly, in the stars themselves.
3. Certain lines are strong in the spectra of the small proper motion stars, and others in the spectra of the large proper motion stars. The use of the relative intensities of these lines gives results for absolute magnitudes in satisfactory agreement with those derived from parallaxes and proper motions.

It seems very probable from physical considerations that the spectra of stars of quite different mass and size would differ considerably in certain respects even when the main spectral characteristics were the same. If the depth of the atmosphere for stars of similar spectral type is at all in proportion to the linear dimensions of the stars, we should expect the deeper reversing layers of the larger stars to produce certain modifications of the spectrum lines. Owing to the small scale of the stellar spectrum photographs, only the most marked changes could be distinguished, and among these the effect of the deep atmosphere upon the violet end of the spectrum should be especially prominent.

A case of somewhat similar nature is that found in observations of the center and the limb of the sun. The length of path through the solar atmosphere is much greater at the limb, and greater relatively for the lower and lower strata. On large-scale solar photographs the differences between the center and the limb spectra are very marked, but on very small-scale photographs, no doubt, only the most prominent differences could be observed. The difference, however, in the relative intensity of the violet portion of the continuous spectrum at center and limb as compared with the red portion, which is so marked a feature of the observations, would appear equally well on photographs taken with high and low dispersion.

37. On the Radiative Equilibrium of Stars

Arthur Stanley Eddington

(*Monthly Notices of the Royal Astronomical Society* 77, 16–35, 596–612[1916–1917])

In 1916 A. S. Eddington announced, in the first part of this selection, that radiation pressure must stand with gravitation and gas pressure as the third major factor in maintaining the equilibrium of a star. All previous investigators had assumed that convection currents transferred the heat from the center of a star to its surface, but all had encountered the same difficulty: if the inner temperature remained high enough to produce a gas pressure sufficient to counterbalance gravitation, then the outflow of energy would greatly exceed that actually observed. Ten years earlier Karl Schwarzschild had shown that a radiative (rather than convective) transport of energy could prevail in the atmosphere of a star, and Eddington here extended the theory to the interior of a star. He showed that for giant stars the radiant energy will almost balance the entire downward pull of gravity.

In the 1916 paper Eddington adopted 54 as the average weight of the gas particles, indicating an iron star. He was quickly convinced by James Jeans that "a rather extreme state of distintegration is possible" under the high temperature of the stellar interior, and in a second paper, written a few months later as a continuation of the first, he accepted a mean molecular weight of 2, corresponding to the complete ionization of atoms (and to the assumption that hydrogen plays a negligible role in the composition of stars). The likelihood of complete ionization meant that the equations could be applied also to dwarf stars; this paper therefore provided the first indication that main-sequence stars as well as giants could be treated as gaseous spheres. Eddington had found in his first paper that the luminosity of giant stars depended only on their mass, and a few years later he generalized this mass-luminosity relation for other stars as well (see selection 46).

Concerning the 1917 paper, Eddington's biographer has written, "This second paper represents a truly remarkable performance. Not only did he recapitulate in revised, compact form the mathematical arguments establishing his formulae for radiative equilibrium but he developed them farther, computed difficult numerical evaluations, and drew seven important conclusions."[1] Deciding that contraction does not provide a reasonably large supply of stellar energy, Eddington wrote "If contraction is the only source of energy, the giant stage of a star's existence can scarcely exceed 100,000 years." Hence, he turned to the annihilation of matter as an inexhaustible store of energy, and he also pointed out that the current theories of stellar evolution were without foundation, since contraction played no major role in the development of stars.

1. A. Vibert Douglas, *Arthur Stanley Eddington* (London: Thomas Nelson and Sons, 1956), p. 59.

1. Outline of the Investigation

THE THEORY of radiative equilibrium of a star's atmosphere was given by K. Schwarzschild in 1906.[1] He did not apply the theory to the interior of a star; but the necessary extension of the formulæ (taking account of the curvature of the layers of equal temperature) is not difficult. It is found that the resulting distribution of temperature and density in the interior follows a rather simple law.

Taking a star—a "giant" star of low density, so that the laws of a perfect gas are strictly applicable—and calculating from its mass and mean density the numerical values of the temperature, we find that the temperature gradient is so great that there ought to be an outward flow of heat many million times greater than observation indicates. This contradiction is not peculiar to the radiative hypothesis; a high temperature in the interior is necessary in order that the density may have a low mean value notwithstanding the enormous pressure due to the weight of the column of material above.

There is a way out of the difficulty, however, if we are ready to admit that the radiation-pressure due to the outward flow of heat may under calculable conditions of temperature, density, and absorption nearly neutralise the weight of the column, and so reduce the pressure which would otherwise exist in the interior. For the giant stars it is necessary that only a small fraction of the weight should remain uncompensated. (For the dwarf stars, on the other hand, radiation-pressure is practically negligible.)

We thus arrive at the theory that a rarefied gaseous star adjusts itself into a state of equilibrium such that the radiation-pressure very approximately balances gravity at interior points. This condition leads to a relation between mass and density on the one side and effective temperature on the other side, which seems to correspond roughly with observation. The laws arrived at differ considerably from those of Lane and Ritter.[2]

The principal results are given in §§7–10. The theory enables us to estimate the average densities of the giant stars of different spectral types; it shows that the average luminosity will be roughly the same for the different types, and determines this luminosity as compared with the Sun; it determines the maximum effective temperature which a star can attain; and it indicates the extent to which the masses of individual stars are likely to deviate from the mean mass. It is scarcely necessary to say that the conclusions here given are tentative, being based on analysis which is only concerned with obtaining a probable approximation; but there seems to be a satisfactory accordance with observation, so far as is known. The present results also remove an objection which might be urged against the theories of Lane and Ritter, viz. that they require the heat-energy retained in the star to be much greater than that generated by contraction.

The outermost layers of the star are outside the scope of this investigation, and the formulæ here given do not apply to them. In speaking of conditions at the boundary of the star, I refer to a depth negligible compared with the radius, but deep from the point of view of the spectroscopist. Hence the theory has no bearing on the interpretation of spectroscopic results. Frequent reference is made to the effective temperature, because it affords a measure of the total outflow of radiation; we can use this measure without discussing the conditions of the layer which actually possesses the effective temperature.

It is clear that we cannot arrive at much certainty with regard to the conditions in a star's interior, except in so far as the treatment can be based on the most general laws of nature. There are some physical laws so fundamental that we need not hesitate to apply them even to the most extreme conditions; for instance, the density of radiation varies as the fourth power of the temperature, the emissive and absorbing powers of a substance are equal, the pressure of a gas of given density varies as its temperature, the radiation-pressure is determined by the conservation of momentum—these provide a solid foundation for discussion. The weak link in the present investigation is that I have assumed without much justification that a certain product $k\varepsilon$ is constant throughout a star. I have given some evidence that if it is variable the general character of the results would not be greatly altered; and, as a step towards the elucidation of the problem of stellar temperatures, I plead to be allowed provisionally one rather artificial assumption.

2. Radiative Equilibrium in the Case of Spherical Symmetry

Schwarzschild's treatment deals with the case when the surfaces of equal temperature are parallel planes. We have to consider the case when they are concentric spheres. The fundamental principle is that the transfer of energy takes place by radiation; convection and conduction are considered negligible.

Let ξ be the distance of a point P from the centre O of a star. At P, let the intensity of radiation travelling outwards in a direction making an angle θ with OP be expressed as a series of zonal harmonics,

$$(\mathrm{A} + \mathrm{B}\cos\theta + \mathrm{C}P_2(\cos\theta) + \mathrm{D}P_3(\cos\theta) + \cdots)\, d\mathrm{S}\, d\omega,$$

where $d\mathrm{S}$ is the cross-section of the stream of radiation, and $d\omega$ is the solid angle containing directions near to θ.

Consider a small cylinder of matter of cross-section $d\mathrm{S}$, and with axis ds in the direction θ. The fraction of the radiation absorbed in this cylinder will be proportional to the density ρ, and may be set equal to

$$k\rho\, ds,$$

where k is a coefficient of absorption and represents the absorption by a cylinder of unit *mass* and unit cross-section.

Assuming that each molecule absorbs independently of the others, it will make no difference whether the cylinder is long and of low density, or short and of high density.

Hence the loss of the beam of radiation in traversing the cylinder is

$$k\rho \, ds(A + B\cos\theta + CP_2(\cos\theta) + \cdots)\, dS\, d\omega. \quad (1)$$

The matter in the cylinder emits energy equally in all directions at a rate proportional to (1) the mass contained, (2) the fourth power of the temperature T, and (3) the specific emissive power of the substance. The last quantity may be set equal to k, since the absorbing and emissive powers of a substance are necessarily equal. Hence the energy emitted from the cylinder in directions included in $d\omega$ is

$$\mu k T^4 \rho \, ds\, dS\, d\omega, \quad (2)$$

where μ is an absolute constant of nature, connecting the units of energy and temperature.

The loss (1) and gain (2) are equated in order to specify the change in intensity of the radiation traversing a length ds. By using geometry to convert ds into the coordinate θ and the radius ξ, and by noticing that A, B, and C are functions of ξ only, Eddington equates the coefficients in the combined loss-gain equation to obtain the results given below.

The constant terms give

$$\frac{1}{3}\frac{dB}{d\xi} + \frac{2}{3}\frac{B}{\xi} = k\rho(-A + \mu T^4). \quad (5)$$

The coefficients of $\cos\theta$ give

$$\frac{dA}{d\xi} = -k\rho B. \quad (6)$$

The coefficients of $P_2(\cos\theta)$ give

$$\frac{2}{3}\frac{dB}{d\xi} - \frac{2}{3}\frac{B}{\xi} = -k\rho C,$$

which verifies the statement made above that C is small compared with B. We shall therefore neglect C henceforth.

Equation (5) can be written

$$\frac{1}{\xi^2}\frac{d}{d\xi}(B\xi^2) = 3k\rho(-A + \mu T^4). \quad (7)$$

To interpret the quantity B, we remark that the total radiation flowing outwards across unit surface transverse to the radius is obtained by integrating $(A + B\cos\theta)\cos\theta\, d\omega$ over the corresponding hemisphere. The factor $\cos\theta$ is required, since the cross-section of a beam of radiation flowing obliquely through a surface dS is equal to $dS\cos\theta$. The result is

$$\pi A + \tfrac{2}{3}\pi B. \quad (8)$$

Similarly the amount flowing inwards is

$$\pi A - \tfrac{2}{3}\pi B. \quad (8a)$$

Thus the net flow outwards is $(4/3)\pi B$ per unit area, or $(16/3)\pi^2 B\xi^2$ across the sphere of radius ξ.

In a strictly steady state the total energy between two boundaries must be constant, and therefore the net outward flow across all boundaries must be the same. Thus $B\xi^2$ is constant, and by (7)

$$A = \mu T^4.$$

In an actual star the stream of energy flowing outwards is supplied by slow changes occurring within the star. The simplest theory results if we suppose that the energy is produced by radioactive processes. Let the amount thus liberated per unit mass be $4\pi\varepsilon$. Then in the steady state the outward flow of heat across the outer boundary of a spherical shell will exceed that across the inner boundary by the amount of heat generated in the shell. That is,

$$d(\tfrac{16}{3}\pi^2 B\xi^2) = 4\pi\varepsilon \cdot 4\pi\rho\xi^2\, d\xi.$$

Hence

$$\frac{1}{\xi^2}\frac{d}{d\xi}(B\xi^2) = 3\rho\varepsilon. \quad (9)$$

Imagine the successive spherical layers to be expanded or contracted until the whole star has a uniform density τ. Let x be the radius in the uniform star, which corresponds to ξ in the original star. Then

$$\rho\xi^2\, d\xi = \tau x^2\, dx \quad (10)$$

Hence from (6) and (9)

$$\frac{1}{\tau}\frac{dA}{dx} = -kB\frac{x^2}{\xi^2} \quad (11)$$

$$\frac{1}{\tau}\frac{d}{dx}(B\xi^2) = 3\varepsilon x^2 \quad (12)$$

Integrating (12),

$$B = \tau\varepsilon \frac{x^3}{\xi^2} \tag{13}$$

and

$$\frac{1}{\tau}\frac{dA}{dx} = -k\tau\varepsilon\frac{x^5}{\xi^4}. \tag{14}$$

If the outflowing energy is produced by contraction instead of by radioactivity, it is not so easy to give a precise statement, because, strictly speaking, the conditions are changing with the time. We may write down equation (13) as a definition of ε; and it will be seen that $4\pi\varepsilon$ is then the net flow of radiation across the spherical surface divided by the mass within the surface, *i.e.* the average energy per unit mass generated by contraction. I shall generally take ε to be constant, representing roughly a state of affairs such that the energy of stellar radiation comes from processes going on in all parts of the star, and not from a singularity at the centre. It must be noticed that, except near the surface, ε is extremely small compared with μT^4, so that in any case $A = \mu T^4$ approximately.

Let g be the value of gravity at ξ, and G the constant of gravitation, then

$$g = \tfrac{4}{3}\pi G\tau x^3/\xi^2,$$

since the numerator gives the total mass within a sphere of radius ξ.

The pressure equation is

$$\frac{1}{\rho}\frac{dp}{d\xi} = -g,$$

whence by (10)

$$\frac{1}{\tau}\frac{dp}{dx} = -g\frac{x^2}{\xi^2}$$

$$= -\tfrac{4}{3}\pi G\tau\frac{x^5}{\xi^4}$$

$$= \frac{4}{3}\frac{\pi G}{k\varepsilon}\frac{1}{\tau}\frac{dA}{dx} \quad \text{by (14).} \tag{15}$$

If k and ε are constants, integrating and setting $A = \mu T^4$,

$$p = \frac{4\pi\mu G}{3k\varepsilon}T^4. \tag{16}$$

Strictly speaking, there is a small constant of integration, but it can be neglected in comparison with the large values of p in the interior of the star.

Up to this point we have not used the gas-equation. For a *gaseous* star obeying Boyle's Law, $p = RT\rho$. Hence from (16), $\rho \propto T^3$ and $p \propto \rho^{4/3}$.

The distribution of density, temperature, and pressure for radiative equilibrium is identical with that occurring in the adiabatic equilibrium of a mass of gas for which the ratio of the specific heats is $\tfrac{4}{3}$.[3]

A gas for which $\gamma = \tfrac{4}{3}$ can be at the same time in radiative and adiabatic equilibrium; if γ alters, adiabatic equilibrium alters but radiative equilibrium does not.

3. Intensity of the Outward Radiation

The net outward radiation has been found to be $(16/3)\pi^2 B\xi^2$

$$= \tfrac{16}{3}\pi^2\tau\varepsilon x^3 \quad \text{by (13).}$$

If the effective temperature of the star is T_1, the outflow is equivalent to that due to isotropic radiation of intensity $A_1 = \mu T_1{}^4$. By (8) this outflow is πA_1 per unit area. Hence

$$\pi\mu T_1{}^4 \cdot 4\pi\xi^2 = \tfrac{16}{3}\pi^2\tau\varepsilon x^3 \tag{17}$$

where x and ξ have their surface values.

Eliminate ε between (16) and (17), we obtain

$$kT_1{}^4 = \tfrac{4}{3}\pi G\tau\frac{x^3}{\xi^2}\frac{T^4}{p}$$

$$= \tfrac{4}{3}g_1\frac{T^4}{p} \tag{18}$$

where g_1 is the value of gravity at the surface.

This equation gives the relation between the constant of absorption and the effective temperature. T and p are any corresponding values of the temperature and pressure. It is usually most convenient to use the central values, T_0 and p_0, so that

$$k = \frac{4}{3}\frac{g_1}{p_0}\left(\frac{T_0}{T_1}\right)^4. \tag{19}$$

4. Numerical Example

Eddington assumes that a star has a radius of $r = 7 \times 10^{11}$ cm, a mass density of $\rho = 0.002$ gm cm^{-3} (comparable to that of giant stars), and a molecular weight of 54 (monatomic iron vapor) to obtain a central temperature of 1.52×10^8 °C, a central pressure of $p_0 = 2.5 \times 10^{13}$ dyn cm^{-2}, and a surface gravity of $g_1 = 390$ cm sec^{-2}. The necessary tables and formulae come from Emden's *Gaskugeln*. Using these constants together with an effective temperature of $T_1 = 6{,}500$ K in equation (19), a mass absorption coefficient of $k = 6.2 \times 10^6$ is obtained. Such a high opacity is inconsistent with the known properties of matter.

A further indication that something is wrong is given by a calculation of the total amount of radiant energy contained in the star. The radiant energy per unit volume is aT^4 ergs, where $a = 7.1 \times 10^{-15}$ and T is reckoned in degrees. On integrating throughout the volume, we find the total imprisoned radiant energy is

$$H = \frac{1}{2}\frac{aT_0{}^4GM^2}{\rho_0 r} = 5.85 \times 10^{52} \text{ ergs},$$

whereas the whole energy so far generated by contraction is

$$\Omega = \frac{3}{2}\frac{GM^2}{r} = 1.18 \times 10^{48} \text{ ergs}.$$

One naturally asks. Where has all the radiant energy come from?

The radiant energy probably possesses electromagnetic mass; but this does not give rise to any difficulty. The mass is found by dividing H by the square of the velocity of light

$$H/c^2 = 6.5 \times 10^{31} \text{ grams},$$

which is less than (1/40)th of the whole mass of the star.

5. Discussion of the Assumptions

Eddington examines the assumptions of the theory that has led to the improbable results and finds them sound.

6. Radiation-Pressure

It appears then that no modification of the assumptions mentioned could affect the main result, viz. that the interior temperature is so high as to cause much too great an outflow of radiation, unless we imagine the material to be almost perfectly opaque. The high temperature in the interior is inevitable if the gas—necessarily of moderate density in a giant star—is to support the weight of the enormous column of material above.

This points to a way out of the discrepancy—the weight of the column may be partly supported by radiation-pressure. In introducing radiation-pressure at this stage we do so not as a *hypothesis* to explain the discrepancy, but because in the conditions we have found radiation-pressure would be extremely powerful. Whether radiation-pressure is important or not depends on the value of k. If we had arrived at a low value of k, our neglect of radiation-pressure would have been justified. With the high value found radiation-pressure would far out-balance gravity, and our neglect of it is clearly illegitimate.

As there seems to be a rather widespread impression that gases are not subject to radiation-pressure, it may be advisable to state the theory briefly. The pressure is simply a consequence of absorption or scattering. A beam of radiation carries a certain forward-momentum proportional to its intensity; after passing through a sheet of absorbing medium a weaker beam emerges carrying proportionately less momentum; the difference of incident and emergent momentum is retained by the medium and constitutes the pressure. The medium, in fact, absorbs the momentum of the beam in the same proportion as it absorbs the energy. The calculations of radiation-pressure on small solid particles are simply calculations of absorption and scattering by these particles; it is not possible to apply such methods to atoms and molecules, which absorb by some internal mechanism. But the relation between absorption and pressure is a perfectly general one, depending only on the conservation of momentum.

Consider a small disc of thickness $d\xi$, and let radiation carrying momentum h fall on it, travelling at an angle θ to the normal. The length of path is $d\xi \sec\theta$. Hence the momentum absorbed is $k\rho h\, d\xi \sec\theta$. Resolving along the normal, the normal outward momentum absorbed is $k\rho h\, d\xi$, which is independent of θ.

In the present problem we have energy $\pi A + (2/3)\pi B$ flowing outwards and $\pi A - (2/3)\pi B$ inwards across unit surface; hence the net outward momentum absorbed is $c \cdot (4/3)\pi B k\rho\, d\xi$, where c is the factor relating the momentum and energy of a beam. Now the pressure on a black body is numerically equal to the density of the energy (assumed isotropic). The density of the energy at temperature T is aT^4, where a is the universal constant 7.06×10^{-15}. The outward flow of isotropic energy across unit surface is $\pi A = \pi\mu T^4$, and the outward flow of momentum is $c \cdot \pi\mu T^4$. Hence

$$aT^4 = c\pi\mu T^4,$$

so that

$$c = a/\pi\mu.$$

It follows that the force of radiation-pressure on an element $d\xi$ is

$$\frac{4}{3}\frac{a}{\mu} Bk\rho\, d\xi \tag{23}$$

$$= -\frac{4}{3}\frac{a}{\mu} dA \quad \text{by (6)}$$

$$= -\tfrac{4}{3}a\,d(T^4). \tag{24}$$

Equation (24) is quite general. Assuming now the constancy of $k\varepsilon$ so that $T^4 \propto p$, the radiation-force is

$$-\tfrac{4}{3}a \frac{T_0^{\,4}}{p_0}\, dp.$$

For the star discussed the values of T_0 and p_0 in § 4 give

$$\tfrac{4}{3}a \frac{T_0^{\,4}}{p_0} = 2.0 \times 10^5,$$

so that radiation-pressure would be 200,000 times stronger than gravitation. Of course, the result does not mean that the existing radiation-pressure is so strong as this; it shows that our conclusions contradict our premises, viz. that radiation-pressure was negligible.

We must now form the modified equations in which radiation-pressure is included. The weight of an element $g\rho\, d\xi$ will be partly counterbalanced by the radiation-force (23).

Eddington then replaces this force in the pressure equation and lets $1 - \beta$ represent the ratio of radiation pressure to gravitation. The effect of radiation pressure is equivalent to an alteration of the constant of gravitation to βG. Substituting this alteration into equation (16), Eddington obtains

$$1 - \beta = \tfrac{4}{3}a\beta^4 \left(\frac{T_0^{\,4}}{p_0}\right)_c, \tag{27}$$

where $(T_0^{\,4}/p_0)_c$ refers to the values calculated in §4 without taking account of radiation-pressure.

Equation (29) is given more succinctly in section 14 as equation (42).

7. Results for a Giant Star

Substituting in (27) the values calculated in §4, we have

$$1 - \beta = [5.2980]\beta^4.$$

I have given the logarithm of the coefficient. Solving this equation

$$\beta = 0.0468,$$

so that less than one-twentieth of the weight is left uncompensated. It will be seen that $(T_0^{\,4}/p_0)_c$ does not depend on the density; thus β is the same for all giant stars of the same mass.

Using this value of β in (42) we obtain

$$k = 29.5.$$

This value should presumably be the same for all stars. It indicates that the radiation would be reduced in the ratio $1/e$ after passing through a column containing 1/30 gm. per sq. cm. section (about 25 cm. of air). This seems possible, since it must be remembered that at the high temperatures concerned the bulk of the radiation is of very short wave-length.

The new values of the central temperature and pressure are

$$T_0 = 7.12 \times 10^6 \text{ degrees}$$

$$p_0 = 1.19 \times 10^{12} \text{ dynes cm.}^{-2}$$

Surface temperature gradient = 0°. 30 per kilometre.

The total radiant energy imprisoned is

$$H = 2.81 \times 10^{47} \text{ ergs}$$

$$= 0.238\ \Omega,$$

so that it is no longer an excessive amount.

8. The Relation between Density and Effective Temperature

Under the assumption that radiation pressure balances the inward force of gravity, Eddington shows that the effective temperature, T_1, is proportional to the sixth root of the mass density. Although the effective temperature depends only on the twelfth root of the stellar mass under these assumptions, the total radiation of a star is independent of its mass density and depends only on its mass. In this way, it is thought possible to relate the small observed range of stellar masses to the fact that the absolute magnitudes of the giant stars are nearly the same. Eddington later states that his conclusions apply for gaseous stars like the giant stars and not for dwarf stars like the sun. The effective temperature ceases to increase with density for the dwarf stars because of a failure of the gas law at high densities.

9. 10. 11. 12. Energy of a Star, Dwarf Stars, Molecular Weight, and the Constant of Absorption

Eddington then reiterates that his theory brings the differences between radiated energy and the energy supplied by gravitational contraction into closer accord. He modifies the gas law to take into account the high-density dwarf stars and concludes that for the sun the radiation pressure is practically negligible in comparison to gravitation. The difference in luminosity of the dwarf types is thought to be caused mainly by differences in effective temperature. Eddington argues that the importance of radiation pressure in determining stellar equilibrium is reduced if the molecular weight is smaller but that, even if a star is composed of helium, its radiation pressure can still be intense enough to counterbalance half its gravitation. Finally, the exact expected value for the mass absorption coefficient is shown to be very uncertain.

13. Ionisation in the Stars

In the first paper calculations were made on the assumption of an average molecular weight 54 for the material of a star. In the present paper the average molecular weight will be taken to be 2.

The first value fairly represents the hypothesis that the ultimate particles, which move independently, are atoms. It will be shown that the value 2 represents the hypothesis that the atoms are highly ionised, so that most or all of the electrons outside the nucleus have been broken off and move as independent particles. The suggestion that at these high temperatures we are concerned with particles smaller than the atom was made to me independently by Newall, Jeans, and Lindemann. By an argument which now appears insufficient, I had supposed that atomic disintegration, though undoubtedly occurring, could not have proceeded very far; but Jeans has convinced me that a rather extreme state of disintegration is possible, and indeed seems more plausible. In any case, the prevalence of this view of the interior state of a star makes it desirable to present the theory of stellar temperatures and densities on this alternative basis.

The principal result of the first paper—that the effective temperature of a giant star varies as the sixth root of the density—is not changed by taking account of ionisation. The main difference in the results is that the influence of the star's *mass* on the temperature is considerably increased by ionisation. The change seems on the whole to improve the agreement with observation in minor details; but since the theory is necessarily only a rough approximation, it is not desirable to lay too much stress on this removal of small discrepancies. A more important confirmation is that the theory, when thus modified by introducing ionisation, provides an explanation of why stellar masses are of the order of magnitude actually found.

It will be remembered that the temperatures within the stars are chiefly from 10^6 to 10^7 degrees. The radiation at these temperatures consists mainly of waves a little longer than X-rays, having strong ionising power. But since we do not know how fast recombination of the ions takes place, it is difficult to predict what proportion of the atoms are ionised at any moment. For many elements the first electron is removed by an ionisation potential of about 4 volts; and, in accordance with the theory of the photoelectric effect, this separation can be effected by radiation of wave-length 3,200 Å or less (since a quantum of this frequency carries sufficient energy). At 9,000° the maximum radiation is of the required wave-length. Although the theory gives no certain guidance, it seems reasonable to suppose that when the material is in equilibrium with radiation at temperature 9,000° a fair proportion of the atoms must be in the ionised state, and that the proportion will increase rapidly as the temperature rises. We should expect then the loss of one electron to take place at temperatures of the order 10,000°; succeeding electrons will be lost at higher temperatures which can scarcely be predicted; but we can estimate on the same basis the temperature at which the last electron is removed, since the nucleus attended by a single electron forms the simple system considered by Bohr. The necessary frequency is given by the formula $2\pi^2 N^2 me^4/h^3$, where N is the atomic number, m and e the mass and charge of a negative electron, and h Planck's constant. The wave-length is, in fact, $\frac{3}{4}$ that of one of the characteristic X-rays of the element. For aluminium it is 5.4 Å; for copper 1.1 Å. The maximum radiation at 10^7 degrees is of wave-length 2.9 Å. It seems probable therefore that, though disintegration is nearing completion, it will scarcely be complete even in the hottest part of the stars, and that all except the lightest elements will retain a small remnant of their electrons.

If N electrons break away from each atom, they will act like independent particles in causing pressure, so that for the same temperature and density the pressure is increased $(N + 1)$fold. Using the term molecules to denote the ultimate particles, whether atoms or electrons, the average molecular weight will be $A/(N + 1)$, where A is the atomic weight; and it is this average which occurs as the molecular weight in the pressure-temperature-density equation of a gas. (I assume that the electric charges of the particles do not affect the calculation of the pressure.) Now the atomic number, or number of electrons outside the nucleus, is very near to half the atomic weight for all elements except hydrogen. Hence, if disintegration is complete, $A/(N + 1)$ is approximately equal to 2, whatever may be the composition of the star. Although 2 is thus the extreme lower limit for the molecular weight, the retention of the last few electrons would make comparatively little difference; thus iron (atomic number 26) might retain 8 electrons and gold (79) might retain 14 without the molecular weight increasing above 3.

It may be suggested that the nucleus could also be broken up, leading to a further diminution of the average molecular weight. But it seems probable that this would require much greater energy; and it could scarcely be accomplished by radiation of less frequency than the characteristic X-rays.

14. Summary of the Theoretical Formulæ

The theory given in the first paper will here be collected into a form more convenient for reference and development. I wish especially to show which formulae are general, and which depend on particular assumptions.

The notation may be summarised as follows:

ξ is the distance from the centre of the star.

p, ρ, T, the pressure, density, and temperature of the material.

k is the mass-coefficient of absorption.

$\frac{4}{3}\pi B$ is the net outward flow of radiation per unit area.

a is the constant 7.06×10^{-15} such that aT^4 is the density of the radiant energy.

μ is a constant equal to $ac/4\pi$ where c is the velocity of light.

The two fundamental equations are

$$d(p + \tfrac{4}{3}aT^4) = -g\rho\, d\xi. \tag{34}$$

$$d(T^4) = -\frac{B}{\mu} \cdot k\rho\, d\xi. \tag{35}$$

Equation (34) is the usual hydrostatic equation with the correction for radiation-pressure, $d[(4/3)aT^4]$, as found in (24). Equation (35) is the equation of radiative equilibrium, expressing that the flow of energy, B, is proportional to the gradient of the energy-density, $dT^4/d\xi$, and inversely proportional to the opacity of the medium, $k\rho$; it is equivalent to (6). Both equations are perfectly general, and hold for any kind of material; they could only fail if some additional forces or effects were present, *e.g.* electric forces or convection currents.

Eliminating $d\xi$, we have,

$$d(p + \tfrac{4}{3}aT^4) = \frac{g\mu}{B} \cdot \frac{1}{k}\, d(T^4). \tag{36}$$

The surface-value of B is given by the condition that the net outward flow of radiation near the boundary is $\pi\mu T_1{}^4$ per unit area, T_1 being the effective temperature.[4] It follows that the surface-value of $(g\mu)/B$ is $(4/3)(g_1/T_1{}^4)$, where g_1 is the value of g at the surface. Let the ratio B/g at any interior point be η times the surface-value, then

$$d(p + \tfrac{4}{3}aT^4) = \frac{4}{3}\frac{g_1}{T_1{}^4} \cdot \frac{1}{\eta k}\, d(T^4). \tag{37}$$

Now g is proportional to the mass within a sphere of radius ξ, divided by ξ^2; B is proportional to the energy liberated per unit time in the same sphere, divided by ξ^2. Hence η varies as the energy liberated per unit mass, averaged through the part of the star interior to the point considered. Owing to this averaging, η varies very little even when the energy is liberated at very different rates in different parts of the star. For example, if the radiated energy comes entirely from contraction, and the star passes through a series of homologous states of gradually increasing density, the energy liberated in any volume per unit time will be proportional to the energy present in that volume. In that case, with the law of temperature considered in the first paper ($p \propto T^4$), I find that η rises slowly from 1 at the surface to 1.7 at the centre.

Eddington then assumes that $\eta = 1$ and introduces the average absorption coefficient k_0, which differs little from k but represents a temperature-weighted function of k.

Integrating (37), we have

$$p = \frac{4}{3}\left(\frac{g_1}{k_0 T_1{}^4} - a\right)T^4. \tag{39}$$

One useful result follows immediately from this formula. Since p cannot be negative,

$$k_0 < \frac{g_1}{aT_1{}^4}. \tag{40}$$

If ξ_1 is the radius of the star, $g_1\xi_1{}^2$ gives the star's mass, and $aT_1{}^4\xi_1{}^2$ gives the total radiation emitted, *i.e.* the bolometric magnitude. Hence the right-hand side of (40) can be estimated from observation. With the observational data previously used,

$$k_0 < 31.$$

This has to hold for all temperatures up to the highest interior temperature in the star.

Introduce a quantity β defined by

$$\frac{p}{\frac{4}{3}aT^4} = \frac{\beta}{1-\beta} \tag{41}$$

where β is in general variable.

We then obtain from (39)

$$T_1{}^4 = \frac{1}{k_0 a} g_1(1-\beta). \tag{42}$$

Equations (41) and (42) are formally equivalent to (26*a*) and (29).

If k_0 *is constant throughout the star,* β is constant by (42), and hence $p \propto T^4$. Thus (for a perfect gas) the temperature and density distribution is of a well-known type studied by Emden. We find also from (41) that $1 - \beta$ is the ratio of the radiation-pressure to gravitation, since dp represents the excess of gravitation over radiation-pressure on an element, and $d[(4/3)aT^4]$ represents the radiation-pressure. Radiation-pressure can thus be allowed for by replacing the constant of gravitation G by βG.

For a gaseous star in which $p \propto T^4$, the central temperature and pressure are related to the mass M, mean density ρ_m, molecular weight m, and effective constant of gravitation βG, in the ratios[5]

$$\left.\begin{aligned} T_0 &\propto \rho_m{}^{1/3}M^{2/3}\beta m \\ p_0 &\propto \rho_m{}^{4/3}M^{2/3}\beta \end{aligned}\right\}. \tag{43}$$

Hence (41) leads to an equation for β of the form

$$\frac{\beta}{1-\beta} = \frac{1}{m^4M^2\beta^3} \times \text{constant}$$

or

$$1 - \beta = y\beta^4 \tag{44}$$

where y varies as M^2m^4.

This equation enables us to find β for any giant star, provided k is constant.

In default of information as to how k varies with the temperature, I make this assumption of the constancy of k in what follows. It has been seen that it leads to the law that the bolometric magnitude of a giant star is independent of the density, which seems to be verified by observation; and it is difficult to see how this could happen unless the assumption were approximately true.

In section 15 Eddington evaluates the radiation pressure for low molecular weight. The ratio of radiation pressure to gas pressure, $1 - \beta$, is a much more sensitive function of a change in stellar mass at low molecular weight. At low molecular weight, radiation pressure no longer counterbalances the greater part of gravitation, its place in sustaining stellar equilibrium being taken by the pressure of free electrons. In section 16 Eddington calls attention to the striking coincidence between the actual masses of stars of low molecular weight and the critical range of values of β for which radiation pressure increases from negligible to predominant influence. This coincidence points to the reason why stars are so nearly of one standard size: stars of high radiation pressure probably become unstable. Even at low molecular weight, the temperatures of giant stars still vary as the sixth root of the density, and the bolometric magnitude is independent of the stage of evolution as long as the star behaves as a perfect gas (section 17).

18. The Temperatures of Dwarf Stars

With a high molecular weight it was found that the whole dwarf series must be comprised within densities 0.7 to 1.5; but with molecular weight 2 the range is extended in both directions. Maximum temperature occurs at densities 0.2 to 0.4 according to mass, and it is no longer necessary or possible to assume that the sun is compressed to the maximum possible density. Thus we can have dwarf stars with densities up to 3 or 4. The temperature-density curve becomes more open, and it is possible to follow the relations some way down the dwarf series.

I have thought it worth while to work out in detail the temperatures near the maximum, and as far as practicable down the dwarf series. It is true that in this (as in some other places) I am pressing the theory further than it could be expected to go. But it may be interesting and suggestive to follow out fully the behavior of a theoretical model; and since a comparison with observation is often possible, we may seek to find out whether the model is defective.

I have represented the deviations from a perfect gas by van der Waal's equation

$$p = R\rho T(1 - \rho/\rho_0)^{-1},$$

the term $a\rho^2$ usually added to p being negligible at the high pressures here considered. The equation is, of course, not accurately satisfied by any gas, but it seems likely to represent sufficiently well for our purpose the kind of deviation from Boyle's law which most concerns us. In order to obtain the correct temperature for the sun, it is found necessary to take ρ_0 to be 3.9. The constants have thus been adjusted to agree with observation at two points, viz. to give a temperature of about 6,000° for the sun, and 6,500° for a star of mass 1.5 and density 0.002.

The method of calculation was explained in §10, but I have now found it much simpler to proceed from the centre of the star outwards; this avoids a series of adjustments by trial and error. Starting with an arbitrary central density and an arbitrary value of β, we can always build up a possible star; its mass and mean density are found at the end of the calculation. The density is unalterable; but we can afterwards change the mass to whatever value we please, changing β so as to keep $(1 - \beta)/\beta^4M^2$ constant. Thus β can be found for stars of that particular density and of any assigned mass. In this way I have obtained table 37.1.

Table 37.1 Effective temperatures of stars

Mass[a] / Density[a]	0·2	0·5	1·0	1·5	3·0	4·5
2·11	—	2,590°	3,880°	4,920°	7,350°	9,300°
1·53	—	3,580	5,360	6,790	10,080	12,640
0·97	2,720	4,650	6,960	8,770	12,810	15,690
0·65	3,140	5,350	7,950	9,960	14,210	17,000
0·33	3,520	5,980	8,820	10,900	14,930	17,320
0·13	3,710	6,250	9,060	10,980	14,370	16,230
0.035	3,540	5,910	8,350	9,900	12,460	13,830
0·002	2,410	4,000	5,550	6,500	8,020	8,820

[a] The masses are measured in terms of the sun as unit; the densities in terms of water.

Omitted here is a plot of the results given in table 37.1.

Table 37.1 illustrates H. N. Russell's contention that the small stars cannot rise to the high temperatures characteristic of the so-called earlier types. The maximum temperature attained varies nearly as the square-root of the mass. The following tabulation (derived by interpolation) shows the mass necessary to attain a specified temperature:

Effective temperature	Minimum mass
14,000°	2·5
12,000	1·8
10,000	1·2
8,000	0·8
6,000	0·46
4,500	0·29
3,000	0·14

I had not anticipated so rapid an increase of maximum temperature with mass; but I believe that the observational evidence on the whole favours it.

It is easy to calculate from the curves the absolute luminosity of a star of mass M and effective temperature T. Take the density ρ from the curve; then the volume of the star is to the volume of the sun as $M/\rho : 1/1.38$; their surfaces are in the ratio $(1.38M/\rho)^{2/3}$ and their total radiations in the ratio $(1.38M/\rho)^{2/3}(T/6{,}000)^4$. Converting this into magnitudes and applying the correction from table III [not reproduced], we obtain the difference between the absolute visual magnitude of the star and that of the sun ($+5^m.1$).

It may be noted that whereas the luminosities of the giants of given type vary approximately as M^2, the luminosities of the dwarfs cannot vary more widely than $M^{2/3}$, and even this variation is partly compensated by the corresponding change of density. In the same spectral class the range of luminosity of the dwarfs ought therefore to be much less than that of the giants; but unless we use very fine divisions of spectral classification this uniformity of the dwarfs is likely to be masked by the great change in luminosity from type to type.

Here we omit section 19, which compares the temperature-density data of the previous section with observational data. There is no evidence for any visible star with a mass as small as one-tenth the sun's, because a star with a mass less than this limit cannot attain an effective temperature of 3,000°.

20. Energy of a Star

It is well known that the hypothesis that the sun's energy is derived from contraction leads to a value of the sun's age which is much too small to be reconciled with estimates of the earth's age based on studies of radioactive minerals and other geological evidence. Perhaps it has not been realised that the contraction hypothesis involves a rapidity of evolution of the giant stars which is from some points of view even more startling. I give a numerical instance worked out on the present theory, but on any other theory the order of magnitude of the results could scarcely be altered.

Consider the time taken for a giant star of mass 1.5 to rise from effective temperature 3,000° to 6,000°, *i.e.* from type M to G. At 6,500° the energy of contraction Ω was found to be 1.18×10^{48} ergs, and since $\Omega \propto T^2$ we find $\Omega_{6,000} - \Omega_{3,000} = 7.54 \times 10^{47}$ ergs. Of this, the fraction $(1+\beta)/4$ is required for ætherial and translational molecular energy in the star, and therefore, taking the new value of β (0·826), we have at most 4.1×10^{47} ergs available for radiation. The actual amount is less than this, since some energy must be used in increasing the ionisation and internal atomic energy.

Taking the solar constant to be 1.93 gram-calories per sq. cm. per minute, we find that the sun radiates $1.20 \cdot 10^{41}$ ergs per year. The giant star in question is 5.4 magnitudes brighter (bolometrically), and therefore radiates 145 times as fast. Hence the time taken to rise from 3,000° to 6,000° is not greater than $4.1 \cdot 10^{47} \div (145 \times 1.20 \cdot 10^{41})$, or 24,000 years!

We can only avoid a short time-scale by supposing that the star has some unknown supply of energy. Probably the simplest hypothesis of this kind is that there may be a slow process of annihilation of matter (through positive and negative electrons occasionally annulling one another). This would set free an almost inexhaustible store of energy. But the defect of all such processes is that they supply energy at some fixed rate—unlike the energy of contraction, which is drawn on as required and automatically replaces the amount radiated. A star would soon expand or contract until the rate of radiation balanced the fixed rate of supply. But in that case it is difficult to understand why the rate should be so very different in different stars of similar mass.

The usual view that evolution takes place in the direction of continually increasing density has practically no foundation, unless we believe that the energy comes mainly from contraction.

On the contraction hypothesis the question whether in the giant stage evolution proceeds more rapidly in heavy stars or light stars depends (according to the present theory) on the molecular weight. If r is the radius of the star—

$$\text{Energy of contraction} \propto \frac{M^2}{r}.$$

$$\text{Rate of radiation} \propto M(1-\beta).$$

$$\text{Effective temperature} \propto M^{1/4} r^{-1/2}.$$

Hence, assuming that a constant fraction of the energy of contraction is available for radiation,[6]

$$\text{Time for a given rise of effective temperature} \propto M^{1/2}/(1-\beta).$$

This increases with M when the molecular weight is high, but decreases with increasing M (up to a certain limit) for molecular weight 2.

There is thus a possibility that during the giant stage heavy stars may change type more rapidly than light ones; but, nevertheless, the light stars have, as it were, not so far to go, and may still reach the dwarf stage first.

Summary of Conclusions

1. If the mass-coefficient of absorption is the same for all stars, the effective temperature varies as the sixth-root of the density, and the total radiation is independent of the stage of evolution, so long as the star is in a perfect gaseous condition. Conversely, the approximate confirmation of these laws by observation seems to show that the absorption is nearly the same for all stars.

2. Owing to the compensatory effect of radiation-pressure the distribution of temperature in the interior depends only to a very small degree on the molecular weight of the material or the amount of ionisation. For a giant star of type G the central temperature is of the order 5,000,000°.

3. Assuming the highest possible degree of ionisation in the interior, the mass-coefficient of absorption is 5.4 C.G.S. units. If the coefficient depends on the temperature, this value refers to approximately the highest temperature in the interior. The absorption cannot anywhere exceed 30 C.G.S. units, except possibly for a very limited range of temperature.

4. The actual masses of stars correspond to the range in which radiation-pressure increases from trifling to predominant influence. Assuming that a gas-sphere in which gravitation is largely counter-balanced by radiation-pressure would be in a state approaching instability, and would readily break up under small rotational velocity or other causes, it is evident that the matter of the universe must aggregate into bodies of mass agreeing quantitatively with the observed masses of stars.

5. With low ionisation the total radiation of a giant star varies approximately as the mass, with high ionisation as the square of the mass. In either case the observed range of luminosity of the giants indicates a very close uniformity of mass of the majority of the stars.

6. Determinations of the temperatures of stars too dense to behave as a perfect gas involve more doubtful assumptions. Assuming a high degree of ionisation, a theoretical model has been found which appears to represent the chief results of observation satisfactorily. According to this, a body must have at least one-seventh the sun's mass in order to attain an effective temperature of 3,000°; stars with mass less than 2.2 times the sun's mass will not reach the B stage. The maximum temperature is attained when the mean density is between 0.2 and 0.5. The distribution of the luminosities of the B stars may be expected to be strongly asymmetrical, the maximum frequency corresponding to minimum luminosity.

7. If contraction is the only source of energy, the giant stage of a star's existence can scarcely exceed 100,000 years.

1. K. Schwarzschild, *Nachrichten von der königlichen Gesellschaft der Wissenschaften zu Göttingen 195*, 41 (1906); [trans. in D. H. Menzel, ed., *Selected Papers on the Transfer of Radiation* (New York: Dover Publications, 1966)].

2. *Astrophysical Journal 8*, 307 (1898).

3. It may be well to restate the restriction, viz., that $k\varepsilon$ must be constant.

4. I take as definition of effective temperature the temperature of a blackbody, which gives the same total radiation per unit area.

5. R. Emden, *Gaskugeln*, (Leipzig and Berlin: B. G. Teubner, 1907), pp. 69, 97.

6. The fraction available for radiation is $(3-\beta)/4$, *less* the unknown internal atomic energy. Our ignorance of the latter renders the conclusion somewhat uncertain.

38. Ionization in the Solar Chromosphere

Meghnad Saha

(*Philosophical Magazine 40*, 479–488 [1920])

Stellar spectroscopy originated in 1817 when Joseph von Fraunhofer discovered that the continuum spectrum of the sun is crossed by an enormous number of dark lines. These observations received a physical explanation in 1859, when Gustav Kirchhoff used laboratory measurements to show that dark line spectra are caused by the absorption of a cool gas overlying a hot gas.[1] By the late 1860s Kirchhoff, his associate Robert Bunsen, and the Swedish spectroscopist Anders Ångström had shown that the solar spectrum contains lines of hydrogen, sodium, calcium, barium, strontium, magnesium, copper, iron, chromium, nickel, cobalt, zinc, and gold. During the subsequent decade Norman Lockyer showed experimentally that elements exhibit different spectra under varying conditions of temperature and pressure and that, in particular, the arc and higher temperature spark spectra of the same element differed.[2]

In order to achieve a quantitative analysis of stellar spectra, it was necessary to calculate the distribution of atoms in their various states, but before this could be done, a better understanding of the nature of the atom itself was required. Perhaps most fundamental were Joseph Thomson's discovery of the electron in 1897 and Ernest Rutherford's 1911 discovery that most of the atom's mass and all its positive charge are concentrated in a nucleus near its center. The first successful attempt at explaining atomic spectra came with Niels Bohr, who pictured the atom in terms of specific electron orbits surrounding the massive atomic nucleus.

Shortly after Bohr had proposed his atomic model, a young physicist in Calcutta, Meghnad Saha, began a systematic study of the previous twenty-five years of the *Monthly Notices of the Royal Astronomical Society*. Thus, when he read J. Eggert's 1919 paper on the dissociation of gases in *Physikalische Zeitschrift*, he was well prepared to begin the work on thermal ionization that can be considered the starting point of modern quantitative astrophysics. In 1919 Saha visited Alfred Fowler's laboratory at Imperial College, London, and later Hermann Nernst's laboratory in Berlin, so there is some question where he wrote the fundamental paper presented here. In any event, it was Edward Milne and Ralph Fowler who subsequently developed important extensions to the work and provided a quantitative theory of the spectral sequence.[3]

In his analysis Saha, seeing the analogy between the dissociation of molecules and the ionization of atoms, merely replaced the mass of the atom with the mass of the electron in the well known expression for the degree of dissociation of a molecule to obtain his famous ionization equation.

Saha's equation relates the degree of ionization of an atom to temperature and pressure, and therefore indicates that the relative intensities of different spectral lines are caused, in part, by differences in the pressure and temperatures of stellar atmospheres. For example, the application of Saha's equation explained for the first time the apparent absence of rubidium and caesium in the sun: these elements, with low ionization potentials, are almost entirely ionized at the temperature of the solar photosphere.

Within a year after this paper, Saha had used his ionization equation to specify the temperatures of stars of different spectral types.[4] In addition to confirming that the spectral sequence of stars is a temperature sequence, Saha's result indicated that differences in stellar spectra are caused by differences in excitation rather than differences in chemical composition. It thus paved the way for the work of Cecilia Payne, who showed that stars with different spectra have essentially the same composition.

1. J. Fraunhofer, *Denkschriften der königlichen Akademie der Wissenschaften zu München 5*, 193 (1817); trans. by J. S. Ames in *Prismatic and Diffraction Spectra* (New York: Harper, 1898) and reproduced by H. Shapley and H. E. Howarth in *A Source Book in Astronomy* (New York: McGraw-Hill, 1929); G. R. Kirchhoff, *Monatsberichte der königlichen Preussischen Akademie der Wissenschaften zu Berlin*, 662 (1859). Also see *Philosophical Magazine 19*, 193 (1860), *21*, 185, 241 (1861), *22*, 329, 498 (1861).
2. J. N. Lockyer, *The Chemistry of the Sun* (London: Macmillan and Co., 1887).
3. R. H. Fowler and E. A. Milne, *Monthly Notices of the Royal Astronomical Society 83*, 403 (1923), *84*, 499 (1924).
4. M. Saha, *Proceedings of the Royal Society* (London) *99A*, 135 (1921). Also see E. A. Milne, *Observatory 44*, 261 (1921).

It has been known for a long time that the high-level chromosphere is generally distinguished by those lines which are relatively more strengthened in the spark than in the arc, and which Lockyer originally styled as enhanced lines.

Here Saha presents a table showing that the lines of calcium, strontium, barium, scandium, and titanium present in the chromosphere are more intense in the laboratory spark than in the laboratory flame.[1]

It appears that no satisfactory explanation of this fact, as well as of the extraordinary height reached by these lines, has yet been offered. It is intimately connected with the physical mechanism of the arc and the spark. In this connexion, it is well to recall Lockyer's original hypothesis, which, however, does not seem to have been, at any time, much in favour with the physicists. According to Lockyer, the passage from the arc to the spark means a great, though localised, increase of temperature, to which mainly the enhancement of the lines was to be ascribed. But, apart from its physical incompleteness, Lockyer's theory launches us amidst great difficulties as far as the interpretation of solar phenomena is concerned. It would lead us to the hypothesis that the outer chromosphere is at a substantially higher temperature than the photosphere, and the lower chromosphere; and that the temperature of the sun increases as we pass radially outwards. This hypothesis is, however, quite untenable and is in flagrant contradiction to all accepted theories of physics.

A much more plausible explanation is that the lines in question are not due to radiations from the normal atom of the element, but from "an ionized atom, *i.e.*, one which has lost an electron." The high-level chromosphere is, according to this view, the seat of very intense ionization. Let us see briefly how this hypothesis has grown up.

Modern theories of atomic structure and radiation leave little doubt that the "enhanced lines" are due to the ionized atom of the element. As a concrete example, let us take the case of the calcium H, K, and g lines. The "H, K" lines are of the enhanced type, while "g" is of the normal type. The "H, K" are the leading members of the principal pair-series of the system of double lines of Calcium, while the "g-" line is the first member of the system of single lines of Calcium. Lorenser and Fowler[2] have shown that the series formula of the double lines is of the type

$$\nu = 4\mathrm{N}\left[\frac{1}{\{f(m)\}^2} - \frac{1}{\{\phi(n)\}^2}\right],$$

while the series formula of the single lines is of the type

$$\nu = \mathrm{N}\left[\frac{1}{\{f'(m)\}^2} - \frac{1}{\{\phi'(n)\}^2}\right],$$

where $f(m)$, $\phi(n)$ are functions of the form $m + \alpha$, according to Rydberg, and $m + \alpha + \beta[t(m)]$, according to Ritz, $t(m)$ being a function of m which vanishes with increasing values of m.

In other words, in the series formula of the enhanced lines, the spectroscopic constant is 4N instead of the usual Rydberg number N. In the light of Bohr's theory, this is to be understood in the sense that, during the emission of the enhanced lines, the nucleus, and the system of electrons (excluding the vibrating one) taken together behave approximately as a double charge, so that the spectroscopic constant, $= (2\pi^2e^2E^2m)/h^3$, becomes 4N, as $\mathrm{E} = 2e$. This means that if the nuclear charge is n, the total number of electrons is $(n - 1)$, and the system has been produced by the removal of one electron from the normal atom.

What has been said of the Calcium lines H and K is also true of the Strontium pair 4,216 and 4,078, and the Barium

Table 38.1 Levels at which different chromospheric lines originate

Element	Lines due to the ionized atom[a]	Chromospheric level	Lines due to the normal atom[a]	Chromospheric level
Ca	(H) 3,968 (K) 3,933	14,000	(g) 4,227	5,000
Sr	4,216 4,078	6,000	4,607	350
Ba	4,934 4,554	750 1,200	5,536	400

[a] The lines chosen are the fundamental lines or the first lines of the principal series, having the symbolic formula $\nu = (1.\mathrm{S}) - (2.\mathrm{P})$.

pair 4,934 and 4,554, *i.e.*, they are due to the ionized atom of these elements. The principal lines of the system of single lines of these elements also occur in the flash spectrum, but table 38.1 shows that they reach a much lower level.

No satisfactory series formula are known for the other high-level chromospheric elements, viz., Titanium, Scandium, Iron, and other elements. But the recent remarkable work of Kossel and Sommerfeld[3] makes it quite clear that the spark-lines of these elements are due to the ionized atom. The spark-lines of alkalies have not been much investigated and lie in the ultraviolet beyond 3,000, so that, even if they are present in the high-level chromosphere, we shall have no means of detecting them.[4]

As regards Hydrogen, ionized Hydrogen would mean simply the hydrogen core, and this probably by itself would be incapable of emitting any radiation. But as H_α and H_β lines occur high in the chromosphere, we have to admit that hydrogen probably is not much ionized in the chromosphere.

The case of helium is very interesting. It is well known that the Fraunhofer spectrum does not contain any helium lines, which are obtained only in the flash spectrum. But these lines are all due to normal helium, and the highest level reached by the second line of the so-called principal series is some 8,500 kms., while the better-known D_3 reaches a level of 7,500 kms. The lines due to ionized helium are represented by the general series formula

$$\nu = 4N\left[\frac{1}{m^2} - \frac{1}{n^2}\right],$$

and the best known of them, in the visible range, are the Rydberg line 4,686 and the Pickering system

$$\nu = N\left[\frac{1}{2^2} - \frac{1}{(m+\frac{1}{2})^2}\right]$$

once ascribed to "cosmic hydrogen." Mitchell[5] states that 4,686 occurs in the flash spectrum, and reaches a level of 2,000 kms. If the identification be all right, helium would present a seemingly anomalous case, for, whereas other elements are ionized in the upper strata, it is ionized in the lower strata of the chromosphere.

The above sketch embodies, in short, the problems before us. The alkaline earths and the heavier elements are ionized throughout the whole of the solar atmosphere, but the ionization is complete in the chromosphere, which seems to contain no normal atom at all. But hydrogen and helium are probably unionized throughout the whole chromosphere, and in the case of helium we have probably some slight ionization in the lower parts—a rather anomalous case.

The explanation of these problems, and some other associated problems of solar physics, will be attempted in this paper. The method is based upon a recent work of Eggert[6]—"On the State of Dissociation in the Inside of fixed Stars." In this problem, Eggert has shown that by applying Nernst's formula of "Reaction-isobar,"

$$K = \frac{p_M{}^{\nu_m} p_N{}^{\nu_n} \ldots}{p_A{}^{\nu_A} p_B{}^{\nu_B} \ldots},$$

to the problems of gaseous equilibrium in the inside of stars, it is possible to substantiate many of the assumptions made by Eddington[7] in his beautiful theory of the constitution of stars. These assumptions are that in the inside of stars the temperature is of the range of 10^5 to 10^6 degrees and the pressure is about 10^7 Atm., and the atoms are so highly ionized that the mean atomic weight is not much greater than 2. This method is directly applicable to the study of the problems sketched above. The equation of the Reaction-isobar is

$$\log K = \log \frac{p_M{}^{\nu_m} p_N{}^{\nu_n} \ldots}{p_A{}^{\nu_A} p_B{}^{\nu_B} \ldots} = -\frac{U}{4.571\,T} + \frac{\sum \nu C_p}{R} + \sum \nu C, \qquad (1)$$

where K = the Reaction-isobar,
U = heat of dissociation,
C_p = specific heat at constant pressure,
C = Nernst's Chemical constant,

and the summation is extended over all the reacting substances. The present case is treated as a sort of chemical reaction, in which we have to substitute ionization for chemical decomposition. The next section shows how U is to be calculated. The equation will be resumed in §3.

2

We may regard the ionization of a calcium atom as taking place according to the following scheme, familiar in physical chemistry,

$$Ca \rightleftarrows Ca_+ + e - U. \qquad (2)$$

Where Ca is the normal atom of calcium (in the state of vapour), Ca_+ is an atom which has lost one electron, U is the quantity of energy liberated in the process. The quantity considered is 1 gm. atom.

The value of U in the case of alkaline earths, and many other elements, can easily be calculated from the value of the ionization potential of elements as determined by Franck and Hertz, MacLennan[8] and others. Let V = ionization potential. Then, to detach one electron from the atomic system, we must add to each atom an amount of energy equivalent to that acquired by an electron falling through a potential difference V, where V (in volts) is given by the quantum relation,

$$\frac{eV}{300} = h\nu_0, \qquad (3)$$

ν_0 being the convergence frequency of the principal series, *i.e.*, (1, *s*) in Paschen's notation.[9] If this quantity be multiplied by the Avogadro number N, and expressed in calories, we obtain U.

Thus if V = 1 volt, we have

$$\mathrm{U} = \frac{e\mathrm{V}\cdot\mathrm{N}}{\mathrm{J},300} = \frac{9645\cdot 10^8}{4.19 \times 10^7} = 2.302\cdot 10^4 \text{ calories.}$$

Table 38.2 contains for future use the values of the ionization potentials as far as known, and the calculated value of U. Here I wish to remark that an element may have more than one ionization potential, depending upon the successive transfer of the outer electrons one by one to infinity, or the simultaneous existence of two more constitutions of the normal atom (*e.g.* helium and parhelium). The ionization potential given in the table corresponds to the case when only one electron is transferred to infinity leaving an excess of unit positive charge in the atom. We have made it clear in the introduction that the high-level alkaline earth-lines are due to the atoms with one plus charge in excess.

Table 38.2 Values of the ionization potentials and of U (U = heat of dissociation)

Element	Ionization potential	U in calories
Mg	7.65	1.761×10^5
Ca	6.12	1.409×10^5
Sr	5.7	1.313×10^5
Ba	5.12	1.178×10^5
Ra	?	?
Na	5.112	1.177×10^5
K	4.318	0.994×10^5
Rb	4.155	0.957×10^5
Cs	3.873	0.892×10^5
Zn	9.4	2.164×10^5
Cd	9	2.072×10^5
Hg	10.45	2.406×10^5

The cases of hydrogen and helium will be taken up later on.

3. Equation of the Reaction-Isobar for Ionization

As mentioned in the introduction, the equation of gaseous equilibrium proceeds according to the equation,

$$\log \mathrm{K} = -\frac{\mathrm{U}}{4.571\ \mathrm{T}} + \frac{\sum \nu \mathrm{C}_p}{\mathrm{R}} \log \mathrm{T} + \sum \nu \mathrm{C}, \qquad (1)$$

where the reaction proceeds according to the scheme,

$$\nu_a\mathrm{A} + \nu_b\mathrm{B} + \cdots = \nu_m\mathrm{M} + \nu_n\mathrm{N} + \cdots$$

and K is the "Reaction-isobar,"

$$\frac{p_{\mathrm{M}}{}^{\nu_m} p_{\mathrm{N}}{}^{\nu_n} \ldots}{p_{\mathrm{A}}{}^{\nu_a} p_{\mathrm{B}}{}^{\nu_b} \ldots}$$

$p_{\mathrm{M}}{}^{\nu_m}$, $p_{\mathrm{N}}{}^{\nu_m} \ldots$ being the partial pressures of the reacting substances—M, N, etc.

In the present cases, viz., for a reaction of the type,

$$\mathrm{Ca} \rightleftarrows \mathrm{Ca}_+ + e - \mathrm{U}, \qquad (2)$$

we have

$$\sum \nu \mathrm{C}_p = (\mathrm{C}_p)_{\mathrm{Ca}+} + (\mathrm{C}_p)_e - (\mathrm{C}_p)_{\mathrm{Ca}}.$$

We can take

$$(\mathrm{C}_p)_{\mathrm{Ca}} = (\mathrm{C}_p)_{\mathrm{Ca}+},$$

and $(\mathrm{C}_p)_e = (5/2)\mathrm{R}$, the electron being supposed to behave like a monatomic gas.

Eggert calculates the chemical constant from the Sackur-Tetrode-Stern relation,

$$\mathrm{C} = \log \frac{(2\pi \mathrm{M})^{3/2} k^{5/2}}{h^3 \mathrm{N}^{3/2}} = -1.6 + \tfrac{3}{2} \log \mathrm{M}, \qquad (4)$$

where M = molecular weight, the pressure being expressed in atmospheres.

Now C has the same value for Ca and Ca_+. For the electron $\mathrm{M} = 5.5 \times 10^{-5}$, and $\mathrm{C} = -6.5$.

We have thus

$$\sum \nu \mathrm{C} = -6.5. \qquad (5)$$

To calculate the "Reaction-isobar" K, let us assume that P is the total pressure, and a fraction x of the Ca-atoms is ionized.

Then we have

$$\log \mathrm{K} = \log \frac{x^2}{1 - x^2}\mathrm{P} = -\frac{\mathrm{U}}{4.571\ \mathrm{T}} + 2.5 \log \mathrm{T} - 6.5. \qquad (1')$$

This is the equation of the "reaction-isobar" which is throughout employed for calculating the "electron-affinity" of the ionized atom.

Ionization of Calcium, Barium, and Strontium

With the aid of formula (1), the degree of ionization for any element, under any temperature and pressure, can be calculated when the ionization potential is known. As a concrete example, we may begin with Calcium, Strontium, and Barium.

Saha notes that pressure has a great influence on the degree of ionization and presents tables showing the percentages of ionization of calcium, strontium, and barium under varying conditions of pressure and temperature. As an example, an extract from the table for calcium is given below. Saha then assumes that the temperature of the photosphere and the reversing layer are, respectively, 7,500 K and 6,000 K. The partial pressure in the reversing layer is supposed to vary from 10 atm in the reversing layer to 10^{-12} atm in its outermost layers.

Table 38.3 Percentage ionization of calcium

Pressure (atm.) Temp.	10	1	10^{-1}	10^{-2}	10^{-3}	10^{-4}
4,000°	0	0	0	3	9	26
5,000	0	2	6	20	55	90
6,000	2	8	26	64	93	99
7,000	7	23	68	91	99	100
8,000	16	46	84	98.5	100	100
10,000	46	85	98.5	100	100	100
12,000	76	96.5	100	100	100	100

An examination of tables IV, V, VI, [not reproduced] shows that, under the above-mentioned assumptions, about 34 per cent. of the Ca-atoms are ionized on the photosphere. When the pressure falls to 10^{-4} atmosphere, almost all the atoms get ionized, so that up to this point in the solar atmosphere, we shall get combined emission of the H, K, and the *g*-line, but above this point, we shall have only the H, K lines. This is in very good agreement with observed facts.

In the case of strontium and barium, owing to their comparatively low ionization potential, ionization at 6,000° is practically complete at 10^{-3} atmosphere, and the heights shown by the lines of the unionized atoms of these elements are still lower.

The results of the flash-spectrum observations are thus seen to be very satisfactorily accounted for on the basis of our theory.

Laboratory experiments also, as far as they go, are in qualitative agreement with our theory. It is well known that in the flame, the [lines] due to the ionized atom either do not occur at all, or even if they do occur they are extremely faint compared with the lines of the unionized atom. As the temperature is increased, the "enhanced lines" begin to strengthen, until at the temperature of the arc they are comparable in intensity to the lines of the normal atom.

Here Saha presents the intensity of enhanced and ordinary lines in furnace spectra as a function of temperature[10] and compares these lines with the photosphere lines, with the chromosphere lines, and with his theory.

The tables show that an increase of temperature causes an increase of ionization and the proportion of emission centres of the enhanced lines. The increasing intensities of the double lines are mainly to be ascribed to this fact. These become comparable in intensity to the principal lines of the normal atom only when the degree of ionization is rather large. Comparing the relative intensities of the corresponding lines of the calcium and barium group, we find that for the same temperature the enhanced lines of barium are relatively stronger than the calcium lines; and this, according to our theory, is due to the comparatively lower ionization potential of barium.

4. HYDROGEN IN THE SUN

Saha notes that the hydrogen in the sun has not been detected in ionized or molecular form. The dissociation of the hydrogen molecule into atoms is shown to be complete in the conditions prevailing in the sun. Consideration of the ionization of hydrogen shows that, at a temperature of 6,000 K, hydrogen can be completely ionized at the pressure of 10^{-11} atm. Thus, only at the highest point of the chromosphere, where the partial pressure falls to 10^{-11} atm, can the ionization be complete and the vanishing of the atomic hydrogen lines be expected. Helium is then shown to be unionized under conditions in the solar atmosphere, because of its high ionization potential.

SUMMARY

1. In the present paper it has been shown from a discussion of the high-level chromospheric spectrum that this region is chiefly composed of ionized atoms of Calcium, Barium, Strontium, Scandium, Titanium, and Iron. In the lower layers both ionized and neutral atoms occur.

2. An attempt has been made to account for these facts from the standpoint of Nernst's theorem of the "Reaction-isobar," by assuming that the ionization is a sort of reversible chemical process taking place according to the equation $Ca \rightleftarrows Ca_{+} + e - U$. The energy of

ionization U can be calculated from the ionization-potential of elements as determined by Franck and Hertz, and MacLennan. For determining Nernst's chemical constant and the specific heat, the electron has been assumed to be a monatomic gas having the atomic weight of 1/1836.

3. The equation shows the great influence of pressure on the relative degree of ionization attained. The almost complete ionization of Ca, Sr, and Ba atoms in the high-level chromosphere is due to the low pressure in these regions. The calculated values are in very good accord with observational data and the laboratory experiments of King.

4. Hydrogen has been shown to be completely dissociated into atoms at all points in the solar atmosphere.

5. It has also been shown that the greater the ionization potential of an element, the more difficult ionization will be for that element under a given thermal stimulus. Calculations have been made in the case of hydrogen (V = 13.6 volts) and helium (V = 20.5 volts), which show that these elements cannot get ionized anywhere in the Sun to an appreciable extent. Helium can have appreciable ionization only in stars having the highest temperature (> 16,000° K.), which only are therefore capable of showing the Rydberg line 4,686 and the Pickering lines $\nu = N[(1/2^2) - 1/(m + 1/2)^2]$.

1. Mitchell, *Astrophysical Journal 38*, 424 (1913).
2. Fowler, *Phil. Trans. 214*, 225 (1914).
3. Kossel and Sommerfeld, *Ber, d. d. Phys. Gesellschaft*, Jahrgang *21*, 240.
4. Ibid., p. 250.
5. Mitchell, pp. 490–491.
6. Eggert, *Phys. Zeitschrift 20*, 570 (1919).
7. Eddington, *M.N.R.A.S. 77*, 16, 596.
8. McLennan, *Proceedings of the Physical Society of London, 31*, 1, 30 (1918).
9. Paschen uses the symbol (1.5, *s*), but following Sommerfeld (p. 243), I have taken off .5 and used (1, *s*).
10. King, *Astrophysical Journal 48*, 13 (1918).

39. The Abundances of the Chemical Elements in Stellar Atmospheres

The Relative Abundances of the Elements

Cecilia H. Payne

(*Stellar Atmospheres*, Harvard Observatory Monograph no. 1 [Cambridge, Mass.: Harvard University Press, 1925], chap. 13)

On the Composition of the Sun's Atmosphere

Henry Norris Russell

(*Astrophysical Journal 70*, 11–82 [1929])

Spectroscopic analyses in the nineteenth and early twentieth centuries led to the conclusion that the earth and sun contain the same chemical elements with the same relative abundances. Stars of other spectral types showed conspicuous lines of other elements, however, suggesting that different stars have different chemical compositions. When the theory of ionization in stellar atmospheres was developed, it became clear that the presence or absence of specific spectral lines did not necessarily indicate the chemical composition of a star's atmosphere. By the early 1920s Meghnad Saha had realized that the spectral lines of different elements are excited under different conditions of temperature and pressure, and Ralph H. Fowler and Edward Milne had shown that the number of atoms or ions responsible for the production of a spectral line can be estimated from the line intensity, once the temperature and pressure of the stellar atmosphere are known (see selection 38). In 1925 Cecilia Payne extended Fowler and Milne's work and demonstrated that the abundances of the elements were similar in virtually all bright stars. Her work, presented as a doctoral dissertation at Harvard College Observatory, was called by Otto Struve "the most brilliant PhD thesis ever written in astronomy." We give here a short section of her conclusions.

Fowler and Milne had calculated the fractional concentration of the ions or neutral atoms of an element at different temperatures and the same pressure. Payne assumed that the number of effective atoms required to make a spectral line barely visible in a stellar spectrum is the same for all lines of all elements and that the reciprocals of these fractional concentrations could be used to give the relative abundances of the elements. The results showed that the relative abundances of the elements in the sun and stars are similar to those in the earth's crust.

The single outstanding discrepancy between the compositions of the sun and of the earth was the apparent predominance of hydrogen in the solar atmosphere. Because hydrogen is terrestrially rare, Payne assumed that the intensity of the hydrogen lines in the sun is caused by some abnormal behavior of hydrogen rather than its high solar abundance. Within three years, however, Albrecht Unsöld had used a completely independent method to show that hydrogen is in fact the most abundant element in the sun.[1] He used the observed contours of resonance lines to determine the number of atoms involved in the production of the line. Using this method, Unsöld established that the observed lines of sodium, aluminum, calcium, strontium, and barium require about 10^6 atoms cm^{-2} for their production, whereas the hydrogen lines require a million times more atoms for their production. Unsöld's conclusion that hydrogen was a million times more abundant than any other element in the sun was soon substantiated by William H. McCrea, who used measurements of the absolute and relative intensities of the hydrogen lines appearing in the flash spectrum to show that the number density of hydrogen atoms at the base of the chromosphere is 10^{12} atoms cm^{-3}.[2]

In the same year that McCrea presented his paper on the hydrogen chromosphere, Henry Norris Russell published the analysis of the Fraunhofer absorption lines that is given as the second of these two papers. Russell calibrated his revised intensity scale of Henry Rowland's atlas of solar absorption lines with Unsöld's results to show that it takes 6×10^{12} atoms cm^{-2} to produce a line of Rowland intensity zero. He thus confirmed Unsöld's and McCrea's conclusion that hydrogen is the predominant element in the solar atmosphere. In the course of his investigation, Russell visited Lick Observatory, where Donald Menzel showed him his work on the flash spectrum of the sun. This work, not published until 1931,[3] apparently convinced Russell that hydrogen had to be very abundant in the sun. Russell's work, in addition to confirming this important fact, also provided values for the relative abundances of hydrogen, carbon, nitrogen, oxygen, silicon, and iron that are within a factor of 2 of modern determinations.

Once it was established that hydrogen was the most abundant element in the solar atmosphere, Bengt Strömgren[4] extended A. S. Eddington's work on the interiors of stars by calculating their hydrogen content. Under the assumption that stellar material is thoroughly mixed and therefore chemically homogeneous, Strömgren confirmed Eddington's conjecture that the observed luminosities could be explained if stellar interiors had a hydrogen content of about one third by weight. Hence, it was Strömgren's calculations that established the high hydrogen abundance for entire stars and not just their atmospheres.

1. A. Unsöld, *Zeitschrift für Physik 46*, 765 (1928).
2. W. H. McCrea, *Monthly Notices of the Royal Astronomical Society 89*, 483 (1929).
3. D. H. Menzel, *Publications of the Lick Observatory 17*, 1 (1931). After calibration of the Rowland scale in 1928, Russell still regarded the apparent large abundance of hydrogen as abnormal and possibly related to radiation pressure on the light element and to departures from thermodynamic equilibrium in stellar atmospheres (W. S. Adams and H. N. Russell, *Astrophysical Journal 68*, 9[1928]).
4. B. Strömgren, *Zeitschrift für Astrophysik 4*, 118 (1932), *7*, 222 (1933).

Payne, The Relative Abundance of the Elements

THE RELATIVE FREQUENCY of atomic species has for some time been of recognized significance. Numerous deductions have been based upon the observed terrestrial distribution of the elements; for example, attention has been drawn to the preponderance of the lighter elements (comprising those of atomic number less than thirty), to the "law of even numbers," which states that elements of even atomic number are far more frequent than elements of odd atomic number, and to the high frequency of atoms with an atomic weight that is a multiple of four.

The existence of these general relations for the atoms that occur in the crust of the earth is in itself a fact of the highest interest, but the considerations contained in the present chapter indicate that such relations also hold for the atoms that constitute the stellar atmospheres and therefore have an even deeper significance than was at first supposed. Data on the subject of the relative frequency of the different species of atoms contain a possible key to the problem of the evolution and stability of the elements. Though the time does not as yet seem ripe for an interpretation of the facts, the collection of data on a comprehensive scale will prepare the way for theory, and will help to place it, when it comes, on a sound observational basis.

The intensity of the absorption lines associated with an element immediately suggests itself as a possible source of information on relative abundance. But the same species of atom gives rise simultaneously to lines of different intensities belonging to the same series, and also to different series, which change in intensity relative to one another according to the temperature of the star. The intensity of the absorption line is, of course, a very complex function of the temperature, the pressure, and the atomic constants.

The observed intensity can therefore be used *directly* for only a crude estimate of abundance. Roughly speaking, the lines of the lighter elements predominate in the spectra of stellar atmospheres, and probably the corresponding atoms constitute the greater part of the atmosphere of the star, as they do of the earth's crust. Beyond a general inference such as this, few direct conclusions can be drawn from line-intensities. Russell[1] made the solar spectrum the basis of a discussion in which he pointed out the apparent similarity in composition between the crust of the earth, the atmosphere of the star, and the meteorites of the stony variety. The method used by him should be expected, in the light of subsequent work, to yield only qualitative results, since it took no account of the relative probabilities of the atomic states corresponding to different lines in the spectrum.

Uniformity of Composition of the Stellar Atmosphere

The possibility of arranging the majority of stellar spectra in homogeneous classes that constitute a continuous series, is an indication that the composition of the stars is remarkably uniform—at least in regard to the portion that can be examined spectroscopically. The fact that so many stars have *identical* spectra is in itself a fact suggesting uniformity of composition; and the success of the theory of thermal ionization in predicting the spectral changes that occur from class to class is a further indication in the same direction.

If departures from uniform distribution did occur from one class to another, they might conceivably be masked by the thermal changes of intensity. But it is exceedingly improbable that a lack of uniformity in distribution would *in every case* be thus concealed. It is also unlikely, though possible, that a departure from uniformity would affect equally and solely the stars of one spectral class. Any such departure, if found, would indicate that the presence of abnormal quantities of certain elements was an effect of temperature. This explanation appears, however, to be neither justified nor necessary; there is no reason to assume a sensible departure from uniform composition for members of the normal stellar sequence.

Marginal Appearance of Spectrum Lines

Fowler and Milne[2] pointed out that the "marginal appearance," when the line is at the limit of visibility, is a function of the abundance of the corresponding atom. For this reason their own theory, which dealt not with the marginal appearance but with the maximum of an absorption line, was capable of a more satisfactory observational test than Saha's. It is possible, as shown below, to extend the Fowler-Milne considerations and to use the observed marginal appearances as a measure of relative abundance.

The conditions for marginal appearance must first be formulated. When a strong absorption line is at maximum, the light received from its center comes from the deepest layer that is possible for the corresponding frequency. The actual depth depends upon the number of absorbing atoms per unit volume, and upon the atomic absorption coefficient for the frequency in question. The suggestions indicate that different lines, at their maxima, arise from different "effective levels," the more abundant atoms appearing, other things being equal, at higher levels.

As an absorption line is traced through the classes adjacent to the one at which it attains maximum, it begins to diminish in intensity, owing to the decrease in the number of suitable atoms. If the line is very intense, the first effect of the fall in the number of suitable atoms is a reduction in the width and wings. As the number of suitable atoms per unit volume decreases further, a greater and greater thickness of atmosphere is required to produce the same amount of absorption, and accordingly the line originates deeper and deeper in the atmosphere of the star. As the "effective level" falls, the temperature of the layer that gives rise to the line increases, owing to the temperature gradient in the stellar reversing layer. The observed fall in the intensity of the line is caused both by the reduction in the number of suitable atoms, and by the decreased contrast between the line and the background. The former cause predominates for strong (saturated) lines, and the latter for weak (unsaturated) lines.

As the atoms suitable to the absorption of the line considered decrease in number, the effective level from which the line takes its origin falls, and ultimately coincides with the photosphere (the level at which the *general* absorption becomes great enough to mask the *selective* absorption due to individual atoms). The line then disappears owing to lack of contrast. Immediately before the line merges into the photosphere (the approximate point estimated as "marginal appearance"), *all* the suitable atoms above the photosphere are clearly contributing to the absorption; in other words the *line* is unsaturated. The position in the spectral sequence of the marginal appearance of a line must then depend directly upon the *number of suitable atoms above the photosphere*; considerations of effective level are eliminated.

The conditions at maximum and marginal appearance of a line in the spectral sequence are to some extent reproduced for an individual absorption line at the center of the line and at the edge of its wing. A hydrogen line displays wings that may extend to thirty Ångström units on either side of the center. The energy contributing to the wings is evidently light coming from hydrogen atoms with a frequency that deviates somewhat from the normal. Atoms with small deviations are more numerous than atoms with large deviations, and therefore the light received from them originates in a higher effective level. The line center corresponds to the highest level of all. At points far out upon the wings, lower and lower levels are represented, until, where the line merges into

Table 39.1 Observed marginal appearances of all available lines

Atomic number	Atom	Series	Line	Classes			Atomic number	Atom	Series	Line	Classes		
1	H	1S – 2P	4,340	—	A3	—	20	Ca+	$1^2S - 1^2P$	3,933	—	—	B0
2	He	$1^2P - 3^2D$	4,471	B9	B3	O	22	Ti	1F – F	3,999	[a]	[a]	A2
2	He	1S – 2P	5,015	B9	B3	O	22	Ti	1F – G	4,862	[a]	[a]	A2
2	He	1P – 4D	4,388	B9	B3	O	22	Ti	—	4,867	[a]	[a]	A2
2	He+	4F – 9G	4,542	O	O	—	22	Ti	—	4,856	[a]	[a]	A2
3	Li	$1^2S - 1^2P$	6,707	[a]	[a]	—	22	Ti	$1^5F - {}^5F$	4,536	—	—	A5
6	C+	$2^2D - 3^2F$	4,267	B9	B3	O	22	Ti	—	4,535	—	—	A5
11	Na	$1^2S - 1^2P$	5,889	[a]	[a]	A0[b]	23	V	$1^6G - {}^6G$	4,333	[a]	[a]	F0
11	Na	—	5,896	[a]	[a]	A0[b]	23	V	—	4,330	[a]	[a]	F0
12	Mg	$1^3P - 1^3D$	5,184	—	?	A0[b]	24	Cr	$1^7S - 1^7P$	4,290	[a]	[a]	A2
12	Mg	—	5,173	—	?	A0[b]	24	Cr	—	4,275	[a]	[a]	A2
12	Mg	—	5,167	—	?	A0[b]	24	Cr	—	4,254	[a]	[a]	A2
12	Mg	$1^3P - 2^3D$	3,838	—	?	A0	24	Cr	$1^5S - 1^5P$	4,497	—	M1	A7
12	Mg	—	3,832	—	?	A0	25	Mn	$1^6S - 1^6P$	4,034	[a]	[a]	A2
12	Mg	—	3,829	—	?	A0	25	Mn	—	4,033	[a]	[a]	A2
12	Mg+	$2^2D - 3^2F$	4,481	—	A3	B0	25	Mn	—	4,030	[a]	[a]	A2
13	Al	$1^2P - 1^2S$	3,962	[a]	[a]	A0	25	Mn	$1^6D - {}^6D$	4,084	—	K2	A3
13	Al	—	3,944	[a]	[a]	A0	25	Mn	—	4,041	—	K2	A3
14	Si	—	3,905	—	G0	A2	26	Fe	$1^3F - {}^3G$	4,325	—	K2	A2
14	Si+	—	4,128	F0	A0	O	26	Fe	$1^3F - {}^3F$	4,072	—	K0	A0
14	Si+	—	4,131	F0	A0	O	30	Zn	$1^3P - 1^3S$	4,811	G5	G0	A7[c]
19	K	$1^2S - 1^2P$	4,044	[a]	[a]	F8	30	Zn	—	4,722	G5	G0	A7[c]
19	K	—	4,047	[a]	[a]	F8	38	Sr	1S – 1P	4,607	[a]	[a]	F0
20	Ca	1S – 1P	4,227	[a]	[a]	B9	38	Sr+	$1^2S - 1^2D$	4,078	—	K2	A0
20	Ca	$1^3P - 2^3D$	4,455	—	K2	F0	54	Ba+	$1^2S - 1^2P$	4,555	—	?	A2

a. The ultimate lines of the neutral atom, which are strongest at low temperature and have no maximum.
b. Estimates from dyed plates made with slightly smaller dispersion.
c. Estimates by Menzel.

the continuous background, the level from which it originates coincides with the photosphere, and the "marginal appearance" of the line (if it may so be called) is reached. Accurate photometry of the centers and wings of strong absorption lines would seem to have an important bearing on the structure of the stellar atmosphere, as it would provide an immediate measure of the factor that produces the deviations from normal frequency. The success of parallel work in the laboratory[3] indicates that intensity distribution should be amenable to observation and to theory.

Observed Marginal Appearances

The spectral class at which a line is first or last seen is obviously, to some extent, a function of the spectroscopic dispersion used, for, with extremely small dispersion, many of the fainter lines fail to appear at all. A line will also probably appear somewhat later, and disappear somewhat earlier, with small than with large dispersion. It is therefore a matter of some difficulty to obtain measures of marginal appearance that shall be absolute, but the present discussion neither assumes nor requires them. The method used is designed for the estimation of *relative* abundances, and all that is required of the data is that they shall be mutually consistent.

In order to attain the maximum degree of consistency, the estimates used in this chapter were derived chiefly from two series of plates. All the plates used were made with the same dispersion (two 15° objective prisms) and were of comparable density, and of good definition. The data furnished by the writer's own measures were supplemented by some estimates derived by Menzel[4] from a similar series of plates, of the same dispersion and comparable quality. The estimate of the marginal appearance of potassium was very kindly suggested by Russell from solar observations.

The observed marginal appearances of all the lines that are available are summarized in table 39.1. Successive columns contain the atomic number and atom, the series relations, the wave-length of the line used, and the Draper classes at which the line is observed, respectively, to appear, to reach maximum, and to disappear.

Method of Estimating Relative Abundances

If the physical conception of marginal appearance above outlined is correct, the *number of atoms* of a given kind above the photosphere will practically determine the class at which the corresponding line is last seen.[5] Now at marginal appearance the number of suitable atoms is only a small fraction of the total amount of the corresponding element that is present in the reversing layer, and this fraction is precisely the "fractional concentration" evaluated by Fowler and Milne. If then it be assumed that the number of atoms required for marginal appearance is the same for all elements, the reciprocals of the computed fractional concentrations at marginal appearance should give directly the relative abundances of the atoms.

In applying the theory of line intensities to obtain the relative abundances of atoms, Payne assumes that the stellar atmospheres are of uniform composition, that at marginal appearance all lines are unsaturated, and that the same number of atoms is represented at the marginal appearance of a line, whatever the element. The last assumption implies the equality of the atomic absorption coefficients under the conditions involved, and this assumption seems to be applicable to stellar atmospheres.

As stated above, the relative abundances of the atoms are given directly by the reciprocals of the respective fractional concentrations at marginal appearance. The values of the relative abundance thus deduced are contained in table 39.2.

Table 39.2 Relative abundances, a_r, of the atoms

Atomic number	Atom	Log a_r
1	H	11
2	He	8.3
2	He+	12
3	Li	0.0
6	C+	4.5
11	Na	5.2
12	Mg	5.6
12	Mg+	5.5
13	Al	5.0
14	Si	4.8
14	Si+	4.9
14	Si+++	6.0
19	K	3.5
20	Ca	4.8
20	Ca+	5.0
22	Ti	4.1
23	V	3.0
24	Cr	3.9
25	Mn	4.6
26	Fe	4.8
30	Zn	4.2
38	Sr	1.8
38	Sr+	1.5
54	Ba+	1.1

Comparison of Stellar Atmosphere and Earth's Crust

The preponderance of the lighter elements in stellar atmospheres is a striking aspect of the results, and recalls the similar feature that is conspicuous in analyses of the crust of the earth.[6] A distinct parallelism in the relative frequencies of the atoms of the more abundant elements in both sources has already been suggested by Russell,[7] and discussed by H. H. Plaskett.[8]

Here Payne presents data that confirm and amplify the similarity of the relative abundances of the atoms in the earth's crust, the meteorites, and the stellar atmospheres. This data is presented in a table not reproduced here.

The most obvious conclusion that can be drawn from table 39.2 is that all the commoner elements found terrestrially, which could also, for spectroscopic reasons, be looked for in the stellar atmosphere, are actually observed in the stars. The twenty-four elements that are commonest in the crust of the earth, in order of atomic abundance, are oxygen, silicon, hydrogen, aluminum, sodium, calcium, iron, magnesium, potassium, titanium, carbon, chlorine, phosphorus, sulphur, nitrogen, manganese, fluorine, chromium, vanadium, lithium, barium, zirconium, nickel, and strontium.

The most abundant elements found in stellar atmospheres, also in order of abundance, are silicon, sodium, magnesium, aluminum, carbon, calcium, iron, zinc, titanium, manganese, chromium, potassium, vanadium, strontium, barium, (hydrogen, and helium). All the atoms for which quantitative estimates have been made are included in this list. Although hydrogen and helium are manifestly very abundant in stellar atmospheres, the actual values derived from the estimates of marginal appearance are regarded as spurious.

The absence from the stellar list of eight terrestrially abundant elements can be fully accounted for. The substances in question are oxygen, chlorine, phosphorus, sulphur, nitrogen, fluorine, zirconium, and nickel, and none of these elements gives lines of known series relations in the region ordinarily photographed.

The outstanding discrepancies between the astrophysical and terrestrial abundances are displayed for hydrogen and helium. The enormous abundance derived for these elements in the stellar atmosphere is almost certainly not real. Probably the result may be considered, for hydrogen, as another aspect of its abnormal behavior, already alluded to; and helium, which has some features of astrophysical behavior in common with hydrogen, possibly deviates for similar reasons. The lines of both atoms appear to be far more persistent, at high and at low temperatures, than those of any other element.

The uniformity of composition of stellar atmospheres appears to be an established fact. The quantitative composition of the atmosphere of a star is derived from estimates of the "marginal appearance" of certain spectral lines, and the inferred composition displays a striking parallel with the composition of the earth.

The observations on abundance refer merely to the stellar atmosphere, and it is not possible to arrive in this way at conclusions as to internal composition. But marked differences of internal composition from star to star might be expected to affect the atmospheres to a noticeable extent, and it is therefore somewhat unlikely that such differences do occur.

1. Russell, *Science 39*, 791 (1914).
2. R. H. Fowler and Milne, *M.N.R.A.S. 83*, 403 (1923).
3. Harrison, unpub.
4. *H. C. 258* (1924).
5. Payne, *Proc. N. Ac. Sci. 11*, 192 (1925).
6. Clarke and Washington, *Proc. N. Ac. Aci. 8*, 108 (1922).
7. Russell, *Science 39*, 791 (1914).
8. *Pub. Dom. Ap. Obs. 1*, 325 (1922).

Russell, The Composition of the Sun's Atmosphere

Abstract—The *energy of binding* of an electron in different quantum states by *neutral* and *singly ionized* atoms is discussed with the aid of tables of the data at present available. The *structure of the spectra* is next considered, and *tables* of the *ionization potentials* and *the most persistent lines* are given. The *presence* and *absence* of the lines of different elements in the solar spectrum are then simply explained. The *excitation potential*, *E*, for the strongest lines in the observable part of the spectrum is the main factor. Almost all the elements for which this is small show in the sun. There are *very few solar lines* for which *E exceeds* 5 *volts*; the only strong ones are those of hydrogen.

The *abundance* of the various elements in the solar atmosphere is calculated with the aid of the calibration of Rowland's scale developed last year and of Unsöld's studies of certain important lines. The *numbers of atoms* in the more important energy states for each element are thus determined and found to decrease with increasing excitation, but a little more slowly than demanded by thermodynamic considerations.

The *level of ionization* in the solar atmosphere is such that atoms of *ionization potential 8.3 volts are 50 per cent ionized.*

Tables are given of the *relative abundance* of *fifty-six elements* and *six compounds*. These show that six of *the metallic elements*, Na, Mg, Si, K, Ca, and Fe, contribute 95 per cent of the whole mass. The whole number of metallic atoms above a square centimeter of the surface is 8×10^{20}. Eighty per cent of these are ionized. Their mean atomic weight is 32 and their total mass 42 mg/cm^2. The well-known difference between elements of *even* and *odd* atomic number is conspicuous—the former averaging *ten times as abundant* as the latter. The *heavy metals*, from *Ba* onward, are but little less abundant than those which follow Sr, and the hypothesis that the heaviest atoms sink below the photosphere is not confirmed. *The metals from Na to Zn, inclusive, are far more common* than the rest. The *compounds* are present in but small amounts, cyanogen being rarer than scandium. Most of those elements which *do not appear* in the solar spectrum should not show observable lines unless their abundance is much greater than is at all probable. There is a chance of finding faint lines of some additional rare earths and heavy metals, and perhaps of boron and phosphorus.

The *abundance of the non-metals*, and especially of hydrogen, is difficult to estimate from the few lines which are available. *Oxygen* appears to be about as abundant by weight as all the metals together. The abundance of *hydrogen* may be found with the aid of Menzel's observations of the flash spectrum. It is finally estimated that the *solar atmosphere contains* 60 *parts of hydrogen (by volume), 2 of helium, 2 of oxygen, 1 of metallic vapors, and 0.8 of free electrons*, practically all of which come from ionization of the metals. This great abundance of hydrogen helps to explain a number of previously puzzling astrophysical facts. The temperature of the reversing layer is finally estimated at 5,600° and the pressure at its base as 0.005 atm.

A *letter from Professor Eddington* suggesting that the *departure from the thermodynamic equilibrium* noticed by Adams and the writer is due to a *deficiency* of the number of atoms in the higher excited states is quoted and discussed.

In the first two sections Russell discusses his method for obtaining the relative abundances of the elements in the sun. Tables of the lines found in solar spectra are given, together with tables of the ionization potentials and the electronic transitions of the elements giving rise to the observed lines.

The Abundance of the Elements in the Sun

By using Unsöld's[1] analysis of line contours, Russell obtains an absolute calibration of Rowland's[2] intensity scale. It takes 6×10^{12} atoms cm^{-2} to produce a line of intensity zero. The effective temperature of the reversing layer is taken to be 5,730°, and the temperature of most of the radiating material is taken to be 4,980°.[3] Russell also adopts Milne's[4] conclusion that the pressure of the reversing layer is uniform. The quantum weight of the state giving rise to the transition is then discussed,[5] and attention is called to the uncertainty arising from possible departures from local thermodynamic equilibrium.[6] The level of ionization in the sun is taken to be $I_0 = 8.26 \pm 0.08$ v, which corresponds to an electron pressure of 3.1×10^{-6} atm or 3.1 dyn cm^{-2}.[7,8,1]

Abundance of the Elements in the Sun's Atmosphere

With the constants thus determined, we may now compute the number, M, of atoms which are in any given energy state, neutral or ionized, by the equations

$$\left.\begin{aligned} \log M &= \log A_0 + \log W - 0.85E \text{ (neutral atom)},\\ \log M &= \log A_1 + \log W - 0.85E \text{ (ionized atom)},\\ \log A_1 &= \log A_0 + I_0 - I. \end{aligned}\right\} \quad (11)$$

Each observational value of $\log M$ gives one of $\log A_0$ or $\log A_1$.

Here the A's are constants that depend upon the abundance of the element; W is the total quantum weight of the state or the corresponding spectroscopic term; and E is the excitation potential in volts.

Mean values of these for each element are given in table 39.3. There are many elements for which the determination rests only on a few faint lines, and in these cases the tabular values are only an indication of the order of magnitude of the results and are denoted in the table by a colon. The remaining columns of table 39.3 give quantities related to the abundance of the atoms in the solar atmosphere. S_0 is the whole number of neutral atoms per unit area, derived from the sum of the values of M for the different energy levels, and S_1, the number of ionized atoms. In a few cases such as *Ru* II and *Rh* II, the calculation of S_1 involves assumptions regarding the nature of the lowest terms in spectra which have not yet been analyzed; but the uncertainty thus arising is much less than that of the observational data for these elements. The next column gives log T, where T is the total number of atoms of the element considered in both stages of ionization. Doubly ionized atoms may safely be neglected. They should be relatively most numerous for *Ba*, for which $\log S_2 = 1.3$, but this is only 1 per cent of the number of singly ionized atoms. For the rare earths the values of S_1 have been directly calculated by an approximate process. The ionization potentials are unknown, and S_0 cannot be given;

Table 39.3 Abundance of elements and compounds in the sun[a]

El.	log A_0	log A_1	log S_0	log S_1	log T	log Q
H	11.2::	5.7::	11.5::	5.7::	11.5::	11.5::
Li	−1.2:	1.7:	−0.9:	2.0:	2.0:	2.8:
Be	1.5	0.5	1.8	0.8	1.8	2.8
C	6.5:	3.6:	7.4:	4.4:	7.4:	8.5:
N	7.0?	0.8?	7.6?	1.8?	7.6?	8.7?
O	8.0:	2.7:	9.0:	3.3:	9.0:	10.2:
Na	3.7	6.9	4.0	7.2	7.2	8.6
Mg	6.7	7.4	7.0	7.7	7.8	9.2
Al	3.8	6.1	4.6	6.4	6.4	7.8
Si	6.0	6.2	7.0	7.0	7.3	8.8
S	4.8:	2.8:	5.7:	3.4:	5.7:	7.2:
K	2.5:	6.5:	2.8:	6.8:	6.8:	8.4:
Ca	4.2	6.4	4.6	6.7	6.7	8.3
Sc	0.6	2.3	1.9	3.6	3.6	5.3
Ti	1.9	3.4	3.6	5.2	5.2	6.9
V	1.8	3.4	1.9	5.0	5.0	6.7
Cr	3.2	4.8	4.4	5.7	5.7	7.4
Mn	4.0	4.9	5.1	5.8	5.9	7.6
Fe	5.0	5.5	6.7	7.1	7.2	9
Co	3.4	3.9	5.1	5.4	5.6	7.4
Ni	4.0	4.7	5.7	5.7	6.0	7.8
Cu	4.0	4.6	4.3	4.9	5.0	6.8
Zn	4.6	3.5	4.9	3.8	4.9	6.7
Ga	−0.6:	1.7:	0.2:	2.0:	2.0:	3.8:
Ge	1.6	2.0	2.5	2.8	3.0	4.9
As	0.0?	−1.3?	0.6?	−0.7?	0.6?	2.5?
Rb	−2.8:	1.4:	−2.5:	1.7:	1.7:	3.6:
Sr	0.3	3.0	0.6	3.3	3.3	5.2
Yt	−0.5	1.3	0.8	2.6	2.6	4.5
Zr	−0.7	0.8	0.9	2.5	2.5	4.5
Cb	−1.8:	−0.6:	−0.2:	1.0:	1.0:	3.0:
Mo	−0.5	0.5	0.5	1.4	1.4	3.4

El.	log A_0	log A_1	log S_0	log S_1	log T	log Q
Ru	−0.5	0.1	1.0	1.6	1.7	3.7
Rh	−1.5	−0.9	−0.3	0.5	0.5	2.5
Pd	0.1	0.1	0.6	0.9	1.1	3.1
Ag	−0.3	0.7	0.0	1.0	1.0	3.0
Cd	1.8:	1.3:	2.1:	1.6:	2.2:	4.2:
In	−2.8:	−0.3:	−2.0:	0.0:	0.0:	2.1:
Sn	−0.5?	0.4?	0.3?	1.2?	1.2?	3.3?
Sb	−0.3:	−0.3	0.4:	0.7:	0.8:	2.9:
Ba	−0.5	2.6	−0.2	3.3	3.3	5.4
La	−2.0:	0.3	−0.7:	1.8	1.8	3.9
Ce	—	—	—	2.4	2.4	4.6
Pr	—	—	—	0.6:	0.6:	2.8:
Nd	—	—	—	2.0	2.0	4.2
Sa	—	—	—	1.5	1.5	3.7
Eu	—	—	—	1.4:	1.4:	3.6:
Gd	—	—	—	1.1:	1.1:	3.3:
Dy	—	—	—	1.6:	1.6:	3.8:
Er	—	—	—	0.1:	0.1:	2.3:
Hf	—	—	—	0.4	0.4	2.6
W	−1.5	−1.5	−0.1	−0.1	0.2	2.5
Ir	−1.5?	−1.5?	−0.5?	−0.5?	−0.2?	2.1?
Pt	0.4	0.0	1.5	1.0	1.6	3.9
Tl	−1.2?	1.1?	−0.8?	1.4?	1.4?	3.7?
Pb	0.0	0.9	0.2	1.2	1.2	3.5
CN	—	—	3.2	—	3.2	4.6
C_2	—	—	1.3	—	1.3	2.7
CH	—	—	3.0	—	3.0	4.1
NH	—	—	2.1	—	2.1	3.3
OH	—	—	3.0	—	3.0	4.2
BO	—	—	1.4	—	1.4	2.8

a. A: indicates that the determination was made from only a few lines, whereas a ? indicates that the origin of the lines is doubtful.

there is no doubt, however, that it is small, and that no serious error is committed by neglecting it, as has been done here. For the band spectra, S_0 gives the number of neutral molecules. Finally, Q represents the total mass of the atoms or molecules of the substance per unit area of the sun's surface. The tabular values are obtained by multiplying T by the atomic or molecular weight. We have already found that the unit of T represents 6×10^{12} atoms per square centimeter. That of Q corresponds, therefore, to 1.0×10^{-11} g/cm^2.

Russell then gives an extensive discussion of his results. The main source of uncertainty is the unknown correction for departures from thermodynamic equilibrium, a correction that is small for the metals. Elements with even atomic numbers are more abundant than those with odd atomic numbers. Nonmetals provide only 1/200 of the ions or electrons in the sun, whereas the metals provide 6.3×10^{20} free ions or electrons cm^{-2} of the sun's surface. According to Russell, neutral atoms provide about 1.6×10^{20} atoms cm^{-2}. A comparison with Cecilia Payne's results (see the first half of this selection) shows general agreement. The abundance of the elements in the sun is then compared with that in the earth's crust, the major difference being the overabundance of hydrogen in the sun.

ASTROPHYSICAL CONSIDERATIONS

THE ABUNDANCE OF HYDROGEN AND ITS CONSEQUENCES The results of the present investigations leave some puzzles to be solved.

a) The calculated abundance of hydrogen in the sun's atmosphere is almost incredibly great.

b) The electron pressures calculated from the degree of ionization and from the numbers of metallic atoms and ions are discordant.

The former method gave a mean electron pressure P of 3.1 dynes/cm^2. Milne has shown that the pressure P at the bottom of the effective atmosphere should be twice this, or 6.2 dynes/cm^2. We found that the total mass of metallic elements above this level is 42 mg/cm^2; that the mean atomic weight is 32; and that 80 per cent of the atoms are ionized. At the sun's surface $g = 2.74 \times 10^4$ cm/sec.^{2}, so that the pressure due to the metallic constituents of the atmosphere is 1,160 dynes/cm^2. If no other constituents were present, the electron pressure should be 4/9 of this, or 510 dynes/cm^2. One of these values is eighty times the other. A similar discrepancy occurs between Unsöld's value determined from the ionization of Ca and that calculated from his figures for its abundance,[9] and has also been detected in the redder stars by Miss Payne and Mr. Hogg.[8]

The assumption of a higher temperature for the solar atmosphere would diminish this discordance, but by no means remove it. The rather extreme assumption $T = 5{,}740°$ (the effective temperature of the photosphere) gives from ionization $P = 50$ dynes/cm^2. At the degree of ionization in the atmosphere determined from the arc and spark lines it would be little affected by this change. A discordance by a factor of ten still remains.

c) We may add a difficulty arising from other solar observations. The calculated rate of increase of density with depth in the reversing layer is much more rapid than that indicated by observations of the flash spectrum.

It does not seem to have been noticed that the theoretical difficulties (*b*) and (*c*) can be greatly alleviated, if not removed, by the assumption that the difficulty (*a*) does not exist—in other words, the solar atmosphere really does consist mainly of hydrogen. The solution of (*c*) is obvious. The depth for which the density in an isothermal atmosphere is doubled varies inversely as the mean molecular weight $\bar{m}$. At the sun's surface (with $T = 5{,}000°$) it is $105/\bar{m}$ km. A small molecular weight accords much better with the eclipse observations. Menzel, for example, has made a brief report[10] on a determination of the law of decrease of density in the reversing layer by a study of lines in multiplets, which ought to give reliable values.

The writer is greatly indebted to Professor Menzel for permission to state that his unpublished reduction of the Lick Observatory eclipse spectra shows that, in the lower part of the reversing layer, the logarithmic decrement of density, if a temperature of 5,000° is assumed, corresponds to a mean molecular weight of about 2.

If only hydrogen and the metals were present, this would determine the abundance of the former. Take the number of metallic atoms per square centimeter as unit, and let H be the number of hydrogen atoms. Since the hydrogen is not appreciably ionized, the whole number of atoms and electrons is then $H + 1.8$, and we have $\bar{m} = (H + 32)/(H + 1.8)$, or for $\bar{m} = 2$, $H = 28.4$; hydrogen would therefore constitute more than 96 per cent of the atmosphere by volume, and nearly one-half by weight. Now let O and He be the numbers of the corresponding atoms. We have then

$$\bar{m} = (H + 4He + 16O + 32)/(H + He + O + 1.8),$$

and if $\bar{m} = 2$,

$$H = 28.4 + 2He + 14O.$$

According to table 39.3, oxygen is four times as abundant by weight and eight times by volume as all the metals together. Miss Payne makes it 1.5 times as abundant by volume as all the metals. This determination is probably better than ours, which suffers from difficulties in calibration, and we will assume $O = 2$, whence $H = 56 + 2He$. The abundance of helium is very hard to estimate, even in the stars, for its lines appear to be abnormally strong like those of hydrogen, though to a less degree. It is probably conservative to guess that it is at least as abundant as oxygen. Setting $He = 2$, we have the results shown in table 39.4. Here log T and log Q are on the scale of table 39.3. The other non-metals, C, N, S, etc., would probably add but little to the total. These estimates are provisional and will require revision when the density gradient in the flash spectrum is more accurately determined, when the oxygen lines in the deep red are better calibrated, and when—or if—the correction C for departure from thermodynamic equilibrium can be found with less uncertainty.

As they stand, these estimates go a long way toward resolving difficulty (*b*). The whole pressure due to the solar atmosphere as here contemplated is 4.1 times that due to the metals alone, or 4,800 dynes/cm^2, which is almost 1/200 of an atmosphere, or 3.6 "mm" on the familiar barometric scale; but the electronic pressure is only 0.8/65.8 of this, or 58 dynes/cm^2. To obtain this value from the observed ionization we must assume $T = 5{,}800°$, which seems improbably high; but none of the factors involved is determined with any great accuracy, and a tolerable adjustment could be made with a lower assumed temperature. It appears probable, however, that the temperature so far assumed for the reversing layer (5,040°) is too low and should be increased to about 5,600°. This would raise the computed level of ionization I_0 to 8.5 volts. The calculated ratio of the numbers of neutral and ionized atoms would be the same as before for $I = 6.3$, the average for the elements used in finding I_0. For larger values of I (Mg, Fe, Si), the computed degree of ionization would be slightly increased, but as the changes would affect log A by 0.2 at most, there is no need to revise table 39.3. With the higher assumed temperature, the correction C for departure from thermodynamic equilibrium, determined from the enhanced lines as before, would be diminished from 1.2 to 0.4 when $E = 8.2$, that is, from $0.15E$ to $0.05E$, and the computed abundances for the non-metals would be practically the same as before. With regard to the intensity of the hydrogen lines, with log $T = 9.9$, and $T = 5{,}600°$, equation (4) gives, for the two-quantum state, log $M = 1.4$, and theoretically for $H\alpha$, $H\beta$, $H\gamma$, and $H\delta$, log $N = 1.2$, 0.5, 0.0, and -0.3. The values of log N derived from Unsöld's line contours are 4.1, 3.9, 4.0, and 4.1. To obtain these contours the theoretical number of atoms must be increased by factors ranging from 800 for $H\alpha$ to 25,000 for $H\delta$. Effects peculiar to the hydrogen lines and increasing with the serial number evidently account for a great part of this discrepancy, but a good deal may remain to be set to the account of some more general departure from thermodynamic equilibrium.

Table 39.4 Probable composition of the sun's atmosphere

Element	By volume	By weight	log T	log Q
Hydrogen	60 parts	60	9.9	9.9
Helium	2 ?	8?	8.4?	9.0?
Oxygen	2	32	8.4	9.6
Metals	1	32	8.1	9.6
Free electrons	0.8	0	8.0	—
Total	65.8	132	—	—

Russell then discusses departures from thermodynamic equilibrium.

APPLICATIONS TO STARS The assumption of an atmosphere composed mainly of hydrogen serves also to resolve some difficulties which appeared in the study of stellar spectra made last year by Adams and the writer.[11] The electronic pressures, computed from the relative strength of the arc and enhanced lines, came out about 10 times greater in Procyon and 60 times greater in Sirius than in the sun, while the amounts of metallic vapor above equal areas of surface were 0.6 and 0.05 times as great. Allowance for double ionization in Sirius would increase the last figure, but could hardly double it. It was then suggested that a great abundance of hydrogen in Sirius might explain these facts, but the full effect was not realized. At the temperature of an A star, hydrogen must be heavily ionized. If the hydrogen atoms are as abundant as has been suggested for the sun, there are dozens of them for every metallic atom, and, when a considerable fraction of these are ionized, the electronic pressure may be many times that which would arise from the ionization of the metallic atoms alone. At the same time, these electrons and the hydrogen ions contribute to the general opacity, so that the photosphere is raised and the total quantity of gas above it is much diminished, and the metallic lines are thus weakened.

Hydrogen must be extremely abundant in the atmosphere of the red giants, for its lines are stronger in their spectra than in that of the sun. With any reasonable allowance for the effect of the lower temperature in diminishing the proportion of excited atoms, the relative abundance of hydrogen, compared with the metals, comes out hundreds of times greater than in the sun. If this is true, the outer portions of these stars must be almost pure hydrogen, with hardly more than a smell of metallic vapors in it.

The theory of such an atmosphere presents an interesting problem, for quantities which are ordinarily neglected may have to be considered—for example, scattering by the unexcited neutral atoms. The effect of hydrogen in reducing the electronic pressure in the sun appears to be already near its limiting value, and it cannot be invoked further to account for the extraordinary discrepancy in these stars between the degree of ionization indicated by the enhanced lines and the pressure calculated from the extent of the atmospheres and the surface gravity. Discussion of these matters, however, cannot be undertaken in the present paper.

Numerous references to the sources for spectroscopic data are then given.

1. *Zeitschrift für Physik 46*, 765, 778 (1928).

2. Russell, Adams, and Moore, *Mt. Wilson Contr.*, no. 358; *Astrophysical Journal 68*, 1 (1928).

3. Eddington, *The Internal Constitution of the Stars*, pp. 332 and 335.

4. *Monthly Notices, R.A.S. 89*, 3, 17, (1928).

5. W. Heisenberg, *Zeitschrift für Physik 38*, 411 (1926). For the application to more complex spectra, the writer is indebted to a conversation with Professor Sommerfeld.

6. Adams and Russell, *Mt. Wilson Contr*., no. 359; *Astrophysical Journal 68*, 11 (1928).

7. *Monthly Notices, R.A.S. 89*, 35 (1928).

8. *Harvard Circular*, no. 334 (1928).

9. Milne, *Nature 121*, 1017 (1928).

10. *Popular Astronomy 36*, 603 (1928).

11. *Mt. Wilson Contr*., no. 359; *Astrophysical Journal 68*, 9 (1928).

40. On the Rotation of Stars

Grigory Ambramovich Shajn and Otto Struve

(*Monthly Notices of the Royal Astronomical Society 89*, 222–239 [1929])

It was Johann Christian Doppler who, in 1842, first argued that the motion of a star might affect its color, much as the pitch of sound from a moving source is heard to vary as the source moves toward and away from the listener.[1] Doppler argued that this phenomenon would cause yellow stars to become blue stars when they moved toward us and to become red stars when they moved away. Because the stars emit a continuum of radiation at all wavelengths, however, relative motion will simply present a different portion of this continuum to the observer, and the color of stars cannot be governed by this effect. In a talk given in Paris in 1848, however, Hippolyte Louis Fizeau pointed out that the Doppler effect would cause an observable change in the wavelength of spectral lines.[2] Twenty years after Fizeau's arguments, in 1868, William Huggins presented measurements of the radial velocities of a number of bright stars; in 1890 Edward C. Pickering and Hermann Vogel showed that the orbital motion of a double-lined binary star system causes an observable, periodic change in the separation of the system's spectral lines.[3]

Although the rotation of stars themselves could in principle be measured from Doppler-widened spectral lines, only with this paper by G. A. Shajn and Otto Struve did astronomers learn that there were stars with sufficiently large rotational velocities for such measurements to be meaningful. Shajn and Struve correlate the observed widths of the spectral lines of spectroscopic binaries with their periods and conclude that the shorter-period binaries have equatorial rotational velocities ranging up to 100 km sec^{-1} at the surface. Struve himself has written,[4] "It may seem surprising that stellar rotation was not discovered long ago. The reason for the delay may be found in the enormous weight carried by the opinion of H. C. Vogel who for many years dominated all thought in this field of astrophysics. In 1877 Abney had suggested that axial rotation might be responsible for the great widths of certain stellar absorption lines [but] his paper was criticized by Vogel . . . what an irony that Vogel in a later paper re-discovered the phenomenon of stellar rotation – only to see it forgotten almost immediately."

1. J. C. Doppler, *Abhandlungen d.k. Böhmischen Gessellschaft der Wissenschaften 2*, 467 (1842), and *Annalen der Physik und Chemie 68*, 1 (1846).

2. H. L. Fizeau, Paper read before the Société Philomathique de Paris, December 23, 1848, and first published in *Annales de chimie et de physique 19*, 217 (1870).

3. W. Huggins, *Philosophical Transactions of the Royal Society 158*, 529 (1868); E. C. Pickering, *American Journal of Science 39*, 46 (1889), and *Monthly Notices of the Royal Astronomical Society 50*, 296 (1890); H. C. Vogel, *Sitzungsberichte der Akademie der Wissenschaften zu Berlin 401*, (1890).

4. In Otto Struve and Velta Zebergs, *Astronomy of the 20th Century* (New York: Macmillan Co., 1962), p. 484.

1

Captain W. de W. Abney[1] was, we believe, the first to express (in 1877) the idea that the axial rotation of the stars could be determined from measurements of the widths of spectral lines. This opinion met with severe criticism on the part of H. C. Vogel,[2] who pointed out that the great width of the hydrogen lines in certain stars could not possibly be due to rotation, since other lines in the same spectra usually appear narrow. It is now known, however, that in many stars all lines are broad, much broader in fact than the limiting width imposed by the resolving power of the spectrograph.

Professor Frank Schlesinger[3] was the first to actually observe the rotation of stars. This he did by measuring the limb-effect in the eclipsing variables δ Libræ and λ Tauri. The same problem was later discussed by G. Forbes.[4]

In 1922 J. Hellerich[5] published an important paper on the rotational effects as observed during the time of eclipse in a number of Algol-type variables.

A complete investigation of the rotational effect in β Lyræ and in Algol was carried out by R. A. Rossiter[6] and Dean B. McLaughlin.[7] More recently J. S. Plaskett[8] has shown that the effect is also present in 21 Cassiopeiæ, and McLaughlin[9] has rediscussed the eclipsing variable λ Tauri on the basis of Ann Arbor spectrograms.

All the investigations on eclipsing binaries mentioned above prove that the absorption lines are actually widened by rotation. Adams and Joy[10] have successfully predicted in a number of cases that spectroscopic binaries with very wide and diffuse lines have very short periods. Since it is probable that the rotational periods in spectroscopic binaries are equal to their orbital periods, it is clear that very short periods should be associated with wide and diffuse lines. A similar effect was also noticed by Miss Antonia C. Maury[11] in the spectroscopic binaries μ^1 Scorpii and V Puppis.

The contour of spectral lines in a rotating or pulsating star was investigated by H. Shapley and S. B. Nicholson[12] in connection with Cepheid variables.

The important theoretical discussion by J. A. Carroll[13] on the form of an absorption line in the spectrum of a rotating or expanding star unfortunately arrived only after most of our own work had been completed and prepared for publication. Both papers show that the intensities of spectral lines form a sensitive criterion for the detection of rotation. Additional sides of the problem in question are considered in this paper; we have particularly emphasized certain practical applications.

J. H. Jeans[14] has recently shown that the flow of radiation, carrying with it momentum from one part of the star to another, seriously affects the internal motions of the stars, and in the first place their axial rotations. This factor of radiative viscosity, so called because of its analogy with the ordinary viscosity of gases, leads to the interesting result that the stars do not rotate as solid bodies, but that each spherical shell has its own angular velocity roughly proportional to r^{-2}. Only the innermost portions of the stars deviate markedly from this law. The period of rotation of the outer layer is approximately nine times longer than that of the inner layers. One would, therefore, expect that the effect of axial rotation upon the spectrum would be slight. This, however, may not be true for close double stars. Immediately after fission (if double stars really originate through fission) the periods of axial rotation and of orbital revolution must be identical.

There is also observational evidence in support of this equality. Thus in the β Lyræ stars the light-variation is continuous, indicating that the components are nearly in contact. A difference between the two periods would result in marked changes in the light curve from one revolution to the next, this being caused by the elliptical shape of the components. No such changes have been observed.

One of the stars in which Adams and Joy[15] suspect rotational widening of the lines is the well-known variable W Ursæ Majoris. Being a dwarf of spectral type F8p, the radii of its components cannot be very large. Adams and Joy found that the greatest semi-axis of each component is 540,000 km. or 0.78 in terms of the Sun's radius. The total mass of this system is very nearly that of the Sun, and the period is one of the shortest yet discovered, $0^d.334$. The lines are described as "so widened and weakened that measures for velocity and estimates of line-intensity can be made only with the greatest difficulty." It is suggested that "the unusual character of the spectral lines is due mainly to the rotational effect in each star, which may cause a difference of velocity in the line of sight of as much as 240 km. between the two limbs of the star." In this connection an interesting attempt was made by J. Schilt[16] to consider W Ursæ Majoris as a rotating star just preceding fission.

It can be shown that if the rotational velocity of W Ursæ Majoris produces appreciable widening of the lines and decrease in their depth (or central intensity), a similar effect should exist in binaries of early spectral type and of somewhat longer period.[17]

Let ω be the angular velocity of the star, let the radius of the star, in km.'s, be designated as r (we neglect here any possible deformations due to tidal effects), and let the inclination of the orbit be i. The actual rotational velocity will then be

$$V_{\text{rot.}} = r\frac{d\theta}{dt} = r\omega = \frac{r}{P}\cdot \text{const.},$$

and the observed component is

$$V_0 = \frac{r}{P}\cdot \sin i \cdot \text{const.} \qquad (1)$$

We now make the assumption that in close spectroscopic binaries the periods of rotation and of revolution are actually identical, so that the value of P in our formula is given by the observations.

Consider the specific case of a spectroscopic binary of type B0 having an absolute magnitude of -3.0 for the brighter component and an orbital period of $1^{d}.5$. The approximate diameter of the brighter component is, according to Russell's tables,[18] about eight times that of the Sun, or ten times that of the diameter of each component in W Ursæ Majoris. Consequently the term r/P is about 2.2 times larger in our hypothetical star than in W Ursæ Majoris. Provided that $\sin i$ is in both cases the same, the spectral lines of our hypothetical star[19] should be much more affected by the rotation than those of W Ursæ Majoris, and this difference may be even larger if we consider that, according to Stebbins,[20] "it seems plausible that there is progressive darkening at the limb in the spectral types ranging from almost nothing in the B stars to extreme effects in classes K and M."

The value of the term $\sin i$ is not known unless the star is an eclipsing binary, and it is not possible, in general, to compute the individual values of V_0 for each star. The inclination depends upon the orbital elements in the following manner:

$$\sin i = \mathrm{const} \cdot \mathrm{KP}^{1/3}(1 - e^2)^{1/2} \frac{(m_1 + m_2)^{2/3}}{m_2}. \qquad (2)$$

If we consider stars of only one spectral class, we may put

$$m_1 + m_2 = \mathrm{const.},$$
$$m_2/m_1 = \mathrm{const.},$$

and consequently

$$\sin i = \mathrm{const} \cdot \mathrm{KP}^{1/3}(1 - e^2)^{1/2}. \qquad (3)$$

Finally we may assume for spectroscopic binaries, where e is nearly always small,

$$(1 - e^2)^{1/2} = \mathrm{const.}$$

Then

$$\sin i = \mathrm{const} \cdot \mathrm{KP}^{1/3}.$$

Substituting this in equation (1) we have

$$\mathrm{V}_0 = \mathrm{const} \cdot \frac{r\mathrm{K}}{\mathrm{P}^{2/3}}. \qquad (4)$$

We see that in a spectroscopic binary the rotational velocity is directly proportional to K and inversely proportional to $\mathrm{P}^{2/3}$. Accordingly we may expect that the effects produced by rotation will be particularly strong in binaries with short periods and large amplitudes of their velocity curves. In section 3 we shall show that this is actually the case.

2. INFLUENCE OF ROTATION UPON LINE-CONTOUR

There are a number of factors that influence the width of a spectral line and the distribution of intensities in it:[21] (1) Disturbing effects of neighbouring atoms which interfere with the perfect periodicity of the absorbing atom so that its quantum states are not entirely sharp; (2) Doppler effect due to the temperature motion of the particles; (3) Doppler effect due to ascending and descending currents of matter; (4) Doppler effect due to rotation; (5) Compton scattering by free electrons having different velocities; (6) Rayleigh scattering; etc.

The distribution of intensity in a spectral line, $\mathrm{F}(\lambda)$, is of importance in our problem. The theory of the contour of spectral lines has been developed chiefly through the researches of Stewart[22] and of Unsöld,[23] based upon the remarkable pioneering work of Schuster.[24] The influence of the spectrograph itself must also be considered.

Shajn and Struve then take the $\mathrm{F}(\lambda)$ actually recorded by a Koch microphotometer in lines of the moon's spectrum and compute the modifications of the observed line shape induced by rotation. Under the assumption that the periods of rotation and revolution of spectroscopic binaries are equal, the introduction of rotation is shown to widen the line and to produce a decrease in its depth. Deep lines are more greatly affected by rotation than shallow ones.

In actual practice the effect of rotation can be studied in two different ways. Consider a single spectral line. Its width and central intensity are functions of many factors, rotation being one of them. In a given number of stars there will be various intensities and line-widths, and it is not *a priori* possible to distinguish the influence of rotation from that of the other factors. We may, however, subdivide our stars into a number of groups, such that the expected rotations differ systematically from one group to the next, while the influences of the other factors remain, on the average, constant. In that case rotation will stand out and can be investigated separately. This method is purely statistical, and we have followed it in section 3.

Since stellar spectra contain usually many lines, wide and narrow, intense and weak, it should, theoretically, be possible to investigate the effect of rotation for each star individually. Suppose we know, from the solar spectrum, the initial intensities and widths of certain lines. If the spectrum of a star of the solar type shows the narrow lines appreciably too weak, while the wide lines remain about the same, we may conclude that the weakening is due to rotation (provided, of course, that the particular line is not sensitive to changes in absolute magnitude).

An entirely different method can be used in eclipsing variables. We have mentioned before that the effect of rotation has actually been measured from the small variations in radial velocity at the epochs just preceding and following the middle of eclipse. The line-contours observed during the same epochs will also be subject to rotational changes. The intensity distribution will be asymmetric and the degree of asymmetry will change with the phase. It should also be noted that during eclipse the central depths of the lines are less in a rapidly rotating star than in one having slow rotation. We have computed the contours of a line of the primary component of Y Cygni, for various phases. We have assumed that (1) the star is spherical, (2) there is no darkening at the limb, (3) the spectrum of the principal star is not affected by that of the fainter. Our last assumption is invalid in the case of Y Cygni, so that, strictly speaking, our computations do not refer to that star. There are, however, many other eclipsing variables that sufficiently closely resemble this hypothetical object.

3. Observational Results for Spectroscopic Binaries

We have shown theoretically in section (1) that the rotational effects, if they are at all measurable, should predominate in binaries with short periods and large velocity amplitudes. We shall now show that this is actually the case, and that indeed the effects are so pronounced that they can easily be recognized in a statistical discussion.

Here Shajn and Struve present an extensive table of the line widths of O, B, and A stars that are also spectroscopic binaries. The results are illustrated in figure 40.1 and in another figure, not reproduced, that shows that spectroscopic binaries of type A have rotational velocities ranging between 0 and 100 km sec^{-1}.

The stars occupy an area[25] limited at the bottom by the x-axis, at the left by a line that seems to be roughly straight and vertical, and at the top by a curve of the equation $K = CP^{-1/2}$. The conditions for finding a strong rotational widening of the lines (large K and small P) are fulfilled in the upper left portion of the occupied area. A glance at the figure shows that it is in these parts of the diagram that only wide lines are observed[26].

It is interesting to note that a few wide-line stars have large P and small K. If widening of spectral lines were due to rotation only this would indicate that the components of

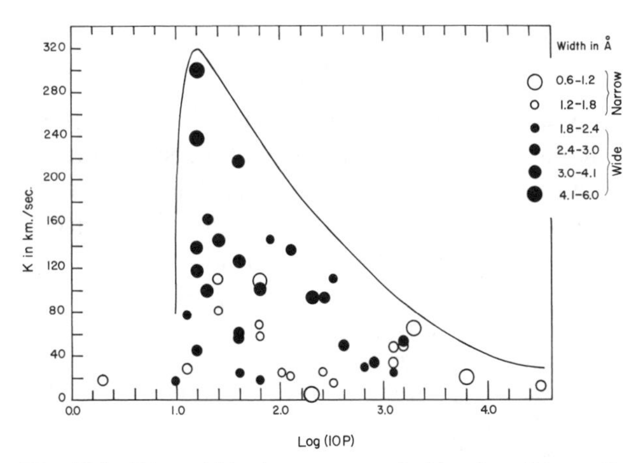

Fig. 40.1 Line widths in spectroscopic binaries of types O and B plotted as a function of orbital period P in days. The rotational velocity, K, is also given, under the assumption that the line broadening is caused by stellar rotation at a period equal to the orbital period. The two highest stars in the diagram are V Puppis and μ^1 Scorpii, for which Maury finds a width of 5.8 Å and 4.9 Å, respectively.

long-period spectroscopic binaries may occasionally have a fast rotation. But it is, of course, quite possible that the broadening depends on other factors. We conclude that the location of a spectroscopic binary in the upper left corner of the diagrams is a sufficient condition for observing wide lines in their spectra. However, if a star is located to the right or to the bottom of the diagrams the lines are not necessarily narrow.

The predominance of fast rotational velocities in short-period spectroscopic binaries can be brought out also in a different way. It is a well-known fact that stellar spectra of the same type greatly differ from one another in the appearance of their lines. This is especially true in the earliest spectral types, for which Adams and Joy have introduced the designations "sharp" and "nebulous."

Short-period binaries are shown to exhibit a distinct tendency to be associated with "nebulous" lines.

The existence of fast rotation in spectroscopic binaries of short period seems thus to be established beyond any doubt. The line-widths are strongly correlated with central intensity, so that the effect of widening is associated with one of decrease in the residual intensity of the lines, as is demanded by the theory of section 2. We hope in future to apply some of the differential methods of section 2 to individual stars, in order to get more quantitative results regarding the actual speeds of rotation.

4. The Rotational Effect in the Spectrum of Jupiter

Spectra of Jupiter taken with the slit perpendicular and parallel to its equator show no detectable rotational broadening. The conclusion of this experiment is that a rotational velocity of 25 km sec^{-1} is at the limit of detectability.

1. *M.N. 37*, 278 (1877).
2. *A.N. 90*, 71 (1877).
3. *Publ. Allegheny Observatory 1*, 134 (1909), *3*, 28; *M.N. 71*, 719 (1911).
4. *M.N. 71*, 578 (1911).
5. *A.N. 216*, 277 (1922).
6. *Ap. J. 60*, 15 (1924).
7. *Ap. J. 60*, 22 (1924); *Pop. Astr. 33*, 295 (1925).
8. *Publ. Dom. Astroph. Observatory 3*, 247 (1926).
9. *Pop. Astr. 34*, 624 (1926).
10. Adams, Joy, and Sanford, *Publ. A.S.P. 36*, 137 (1924); Joy, *Ap. J. 64*, 287 (1926); Schilt, *Ap. J. 64*, 215 (1926); Adams, Joy, Strömberg, and Burwell, *Ap. J. 53*, 94 (1921).
11. *Annals of Harvard Coll. Obs. 84*, 157 (1920).
12. *Communic. Nat. Ac. Sc. Mt. Wilson Obs.*, no. *63* (1919).
13. *M.N. 88*, 548 (1928).
14. *M.N. 86*, 328, 444 (1926).
15. *Ap. J. 49*, 189 (1919).
16. *Publ. A.S.P. 39*, 160 (1927).
17. It has previously been shown that all real double stars of spectral type B have periods longer than $1^{d}.3$ (*M.N. 86*, 63 [1925], *Ap. J. 66*, 117 [1927]).
18. *Publ. A.S.P. 32*, 315 (1920).
19. The eclipsing variable V Puppis (Sp. B1p; P = $1^{d}.454$; K = 302 km./sec.) closely resembles our hypothetical B star. According to Hellerich, the rotational velocity for the brighter component of this star is 276 km./sec., and the corresponding line-width should be 7.9 Å. Miss Maury (*Harvard Annals 84*, 186 [1920]) has measured the widths of several lines in this star and has found a mean of 5.78 Å.—an unusually high value for a B-type star.
20. *Ap. J. 54*, 91 (1921).
21. Eddington, *The Internal Constitution of the Stars* (1926) p. 353.
22. *Ap. J. 59*, 30 (1924).
23. *Zeitschrift für Physik 44*, 793 (1927), *46*, 765 (1928).
24. *Ap. J. 21*, 1 (1905).
25. *M.N. 86*, 63 (1925).
26. In figure 40.1 the stars α Camelopardalis and ξ Persei, the orbits of which are based on the detached calcium lines, have been excluded. These stars are peculiar and their velocity curves may not be due to orbital motion. Both have wide lines and small values of K. They would, of course, in no way contradict the above statement, even if they had been included. On the other hand, we have included the two stars V Puppis and μ^1 Scorpii, based on Miss Maury's measurements.

41. Intensity Measurement of the Fraunhofer Lines in the Wavelength Region 5,150 to 5,270 Å

Marcel Minnaert and Gerard Mulders
TRANSLATED BY OWEN GINGERICH

(*Zeitschrift für Astrophysik 1*, 192–199 [1930])

After the Meghnad Saha theory and its extensions revealed the importance of thermal ionization, it became possible to show that most stars have the same chemical abundances. A more detailed analysis awaited an understanding of how the shape of a spectral line is related to the number of absorbing atoms and how the strengths of individual lines are related to the entire number of atoms in a given ionization state. In a fundamental paper published in 1917, Albert Einstein gave a formula for the probability of a given atomic transition, and the later development of quantum mechanics gave procedures for calculating at least some of these probabilities theoretically.

In 1927 Albrecht Unsöld attacked the problem of the shape of spectral lines by assuming that the theory of radiation damping could be applied to solar absorption lines; he showed that if the broadening of a line is caused by the damping force of the line radiation, then the total absorption of such lines will increase with the square root of the number of absorbing atoms. Although J. Q. Stewart had derived essentially the same result three years earlier, Unsöld realized the importance of the theory for determining the actual abundance of elements in the sun. By 1928 he had applied his theory to the observed contours of solar resonance lines and had thereby shown that hydrogen is the most abundant element in the sun.[1]

Within two years, however, Marcel Minnaert and Gerard Mulders, in the very careful piece of observational work selected here, showed that the relationship between the form of a line and the number of absorbers was more complex than the simple square root dependence. Minnaert and Mulders were among the first to realize that a great simplification results if the equivalent widths of Fraunhofer lines (the areas of their profiles) are used as a measure of their intensity. By plotting the equivalent widths against the number of absorbers (in a logarithmic form), they showed for the first time the now familiar "curve of growth." Their empirical curve opened the way for a renewed theoretical understanding of line formation and a fuller comparison of theory with observation, so that in the following year they could show that for solar lines the effective damping constant is 10 times the theoretically derived classical radiation constant.[2] At about the same time, Donald Menzel independently discovered the curve of growth for emission lines, a theory he subsequently worked out in considerable detail.[3]

The curve of growth has provided an invaluable technique for the analysis of stellar abundances. It was used probably most notably by Leo Goldberg, Edith Müller, and Lawrence Aller to determine the solar abundances of seventy elements.[4]

1. A. Unsöld, *Zeitschrift für Physik 44*, 793 (1927), *46*, 765 (1928). J. Q. Stewart, *Astrophysical Journal 59*, 30 (1924).
2. M. Minnaert and G. Mulders, *Zeitschrift für Astrophysik 2*, 165 (1931).
3. D. H. Menzel, *Publications of the Lick Observatory 17*, 230 (1931).
4. L. Goldberg, E. A. Müller, and L. H. Aller, *Astrophysical Journal Supplement 5*, 1 (1960).

This investigation continues the intensity measurements of the Fraunhofer lines, previously carried out in our laboratory [Physical Institute at Utrecht] for the region 4,400 to 4,550 Å.[1] For a number of well-isolated lines in the green region, the total intensity of the energy loss in the line was measured; we have used this quantity provisionally as the simplest measure of the line strength. The unit of the line strength is always the light level in the neighboring continuous spectrum within an interval of 1 Å; thus, these line strengths are immediately "equivalent widths."

Method of Measurement

The program has been carried out exactly as the one reported earlier. Moreover, the following control measurements were made:

1. For 21 lines the total intensity was measured both on a spectrogram in the first order of our grating and on a plate from the second order (dispersion: 4 Å/mm and 2 Å/mm). The two results should theoretically be equal; if we designate the ratio $i_1:i_2$ of the two experimentally found values as μ, we have for the lines measured $\bar{\mu} = 1.05$, which shows us that there is hardly any systematic difference between the two series. Furthermore, we find $\overline{|\mu - 1|} = 0.10$, which measures the precision. This appears to be satisfactory, for most of the measured lines are weak, and drawing the continuum gives rise to many unavoidable errors.

2. Moreover, 27 lines from the blue spectral region, already measured earlier, were photographed anew with an enlarging camera placed on the grating spectrograph, and the spectrum was enlarged about 6 times.[2] By this procedure, the plate grain is made very much finer than the spectral lines, and the possibility of any disturbance from the grain is greatly diminished. The comparison between the results from the second order and those of the enlarged second order give $\bar{\mu} = 1.05$; $\overline{|\mu - 1|} = 0.20$. Again, the loss of precision arises chiefly from the weak lines.

Table 41.1 Measured total intensities of lines at different wavelengths

Rowland scale	λ	True intensity = equivalent width[a]	Rowland scale	λ	True intensity = equivalent width[a]
−2	5,213.35	0.0057*			
−1	5,211.54	.0293*	2	5,243.47	.0821*
−1	5,214.72	.0123*	3	5,166.29	.137
0	5,221.77	.0243*	3	5,198.72	.098
0	5,223.19	.0261*	3	5,210.39	.0888*
1	5,155.13	.050	3	5,215.19	.156*
1	5,176.57	.059	3	5,216.28	.135*
1	5,180.07	.048	3	5,217.40	.115*
1	5,187.92	.059	3	5,250.66	.101*
1	5,218.21	.0568*	3	5,261.71	.109*
1	5,239.82	.0554*	4	5,191.47	.242
1	5,243.78	.0659*	4	5,194.95	.142
1	5,247.06	.0637*	4	5,229.86	.123*
2	5,154.08	.086	4	5,263.32	.138*
2	5,155.77	.083	5	5,162.28	.190
2	5,159.06	.082	5	5,192.36	.276
2	5,165.42	.111	5	5,206.05	.244*
2	5,173.75	.070	6	5,171.61	.164
2	5,185.91	.064	6	5,266.56	.242*
2	5,192.98	.096	7	5,232.95	.323*
2	5,225.54	.0685*	8	5,269.55	.429*
2	5,234.63	.0917*	15	5,167.33	.600*
2	5,242.50	.0932*	20	5,172.70	1.14*
2	5,247.58	.0765*	30	5,183.62	1.50*

a. Lines measured directly with a planimeter are designated by an asterisk; the others were calculated from their central intensities.

The definitive results are made in the second order without using any enlargement; for the 21 lines also measured in the first order, only the mean of both orders is given.

Results

1. In table 41.1 are found the measured total intensities (equivalent widths) for 47 lines.

2. Now we can calibrate the Rowland scale for the green region (table 41.2). The general run of the calibration curve is about the same as in the spectral region from 4,400 to 4,550, except that the true intensity i, which corresponds to what is called R in the Rowland system, is greater in the green than in the blue region by a factor of about 1.39 in the mean. Collecting the material in two groups, namely, 5,150 to 5,210 and 5,210 to 5,270, gives only a trivial difference between the calibration curves. From table 41.1 it appears that the true values for one and the same Rowland number are very scattered, so that, for example, a single line from class 3 is stronger than another from class 4. For the weakest lines, only a few were measured well; apparently they are the strongest of their class, so that the intensities for which we have assigned the Rowland classes -2 and -1 will be considerably too high; the calibration curve should run somewhat steeper at the beginning than we have found.

Table 41.2 The Rowland scale for the green region

Rowland scale	Mean true intensity	From the curve
−2	0.0057	0.0057
−1	.0208	.0160
0	.0252	.0316
1	.0572	.0562
2	.0837	.0832
3	.117	.120
4	.161	.166
5	.237	.219
6	.203	.270
7	.323	.320
8	.429	.364
15	.600	.725
20	1.50	1.05
30	1.50	1.60

Here Minnaert and Mulders present a plot (not reproduced here) of the points given in table 41.2.

3. As was done earlier for the blue region, Russell's[3] calibration of the Rowland scale was used in order to find the relation between the "number of effective resonators" N and the total intensity i. From Russell's curve $R = F(N)$ and our result $i = f(R)$, R was eliminated.

Minnaert and Mulders next present a plot (not reproduced here) of the relationship between i and N for $\lambda = 4{,}500$ and 5,200 Å.

This curve, as well as that for the blue region, is approximated by the rule $i \sim N^{0.5}$. But the good agreement between the two spectral regions surely means that the strong deviations from this rule are real and fundamental. In the whole region of the moderately strong lines of classes 1 to 6, the growth of i is much slower, so that the rule corresponds to $i \sim N^{0.31}$; for both the strong and the weak lines the growth is faster, $i \sim N^{0.7}$. The curve for the weakest lines indeed appears to be still steeper; perhaps the rule $i \sim N$ approximates it.[4]

4. An examination of Russell's calibration shows immediately that the individual lines depart rather greatly from his mean curve. As Russell suggests, this is no doubt partly a consequence of the uncertainty of Rowland's estimates, as we found earlier. But such a source of error is completely shut off by our measurements; therefore, the study of individual multiplets now becomes possible and can show us whether perhaps the scatter around the mean calibration curve arises from some entirely different, more fundamental, cause. It will now be shown that this is in fact the case.

Here Minnaert and Mulders present a table of the wavelengths, intensities, and number of resonators for the individual lines of 6 multiplets. This table is not reproduced here, because the points are illustrated in figure 41.1.

For this investigation we will consider 6 multiplets whose lines are satisfactorily isolated. The N-values are still to be multiplied by an unknown factor that is different for each multiplet; that is, the line segments can still be arbitrarily slipped in the horizontal direction. Also plotted is the function $i = f(N)$, which was derived from a great number of multiplets via Russell's calibration and in fact agrees not badly with the mean value of the various multiplets. Nevertheless we notice at the same time that the slope of these lines is very different. A similar difference is found when we plot the Rowland estimates against the N-values, but in this case we could ascribe the difference to the lack of precision in the estimates. For the area, or equivalent width, this danger does not arise. The total intensity of a Fraunhofer line is thus not a direct function of the number of resonators that form the line, not even if we limit ourselves to one and the same spectral region. Naturally, it might be possible that some other function of the line profiles would depend directly on the

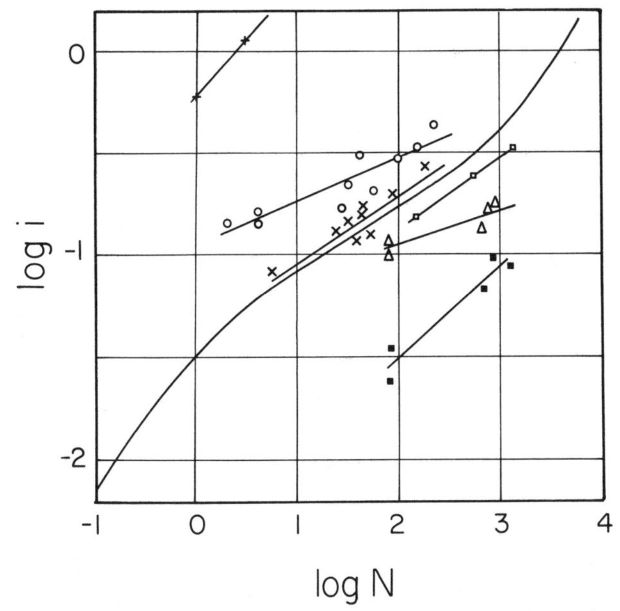

Fig. 41.1 The true line intensity, i (measured from the equivalent width), plotted as a function of the number of resonators, N, for individual multiplets. The long central curve is the mean curve from Russell's collected material.

number N, but until we have a complete theory of the Fraunhofer lines it would scarcely be possible to find such a function. The agreement between N and the area tentatively appears better than that between N and the Rowland scale.

We can gain an indication of the source of the deviations from a consideration of the first 4 multiplets of table 41.3, all from iron and belonging to the same intensity regime. We describe each of the line segments by $i = N^r$, and we place the r-value in the second column of table 41.3. The larger the r, the steeper the slope of the line segment and the weaker the self-absorption. According to the table, the r-values appear to be larger for those multiplets that have a high excitation potential and that in the chromosphere reach only a limited height.[5] If these very preliminary results are confirmed with more material, a search should be made to clarify them with a theory of Fraunhofer lines.

Table 41.3 The dependence of line intensity on the number of resonators for multiplets originating at different heights in the solar chromosphere

Multiplet	r	Excitation potential (v)	Height (km)
Fe; $1\,^3f-1\,^3f'$	0.165	1.54	420
Fe; $1\,^5f-1\,^5d'$	0.210	0.94	510
Fe; $1\,^5d'-m\,^5d$	0.331	3.24	340
Fe; $1\,^7p'-m\,^7d$	0.347	2.95	380
Ti; $1\,^3f-1\,^3f'$	0.450	0.02	380
Mg; $1\,^3p-1\,^3s$	0.570	2.70	2,000

5. In the spectral region investigated fall the three strong magnesium lines, λ5,183, 5,172, 5,167, which together form the triplet $1\,^3p-1\,^3s$. Let us use this opportunity to test whether the contours of these heavily broadened lines are connected with the description that would be required by the scattering theory: we search for that distance from the line center where the intensity is weakened by a given ratio, to see if the distances found for the components of a multiplet have the ratio $N^{0.5}$. Unfortunately, one of the magnesium lines is blended with a strong iron line, so we have only two of them left. In table 41.4 we find the width of the line, to the point where the intensity has been reduced to 90%, 80%, 70%, and so on. According to the theory, the ratio should everywhere be $(5/3)^{0.5} = 1.29$. One sees that the agreement is good for the wings of the line ($i > 40\%$); a small uncertainty arises in the correction for the weak lines at λ5,184.27 and λ5,184.56. Nearer the center, the ratio is systematically smaller. The central intensity for both these lines appear to be nearly the same.

Table 41.4 The ratio of line widths for lines whose intensities have been reduced by different percentages

Intensity in % of background continuum	Width in Ångstroms 5,183.62	5,172.70	Ratio
100	∞	∞	—
95	4.90	3.80	1.29
90	3.55	2.75	1.29
80	2.30	1.70	1.35
70	1.75	1.33	1.31
60	1.38	1.05	1.31
50	1.05	0.825	1.27
40	0.775	0.625	1.24
30	0.510	0.425	1.20
20	0.300	0.275	1.09
10	0.070	0.070	—
8.6	0	0	—

1. M. Minnaert and B. van Assenbergh, *ZS. f. Phys. 53*, 248 (1929).

2. L. S. Ornstein and M. Minnaert, *B.A.N. 5*, 175 (1930).

3. H. N. Russell, W. S. Adams, Ch. E. Moore, *Astrophys. Journ. 68*, 1 (1928).

4. To extrapolate the $N^{0.5}$ rule for the weakest lines, as we did in our paper on the blue region, is thus apparently not allowed.

5. After S. A. Mitchell, *Publ. Leander McCormick Obs. 5*, 51 (1930).

42. Electron Affinity in Astrophysics

Rupert Wildt

(*Astrophysical Journal 89*, 295–301 [1939])

Quantitative study of stellar atmospheres, that is, the outer layers of gas from which stellar absorption spectra arise, received a major impetus from the discoveries of quantum mechanics in the 1920s. Nevertheless, until this paper by Rupert Wildt, even an approximately correct understanding of the sun's continuous spectrum remained a mystery. Following the work of Karl Schwarzschild, A. S. Eddington, and Edward Milne, astronomers recognized that the sun's energy was carried through its outermost layers by a process of absorption and reemission of radiation. The detailed energy distribution of the continuum spectrum depended both on the temperature gradient of the solar atmosphere and on the particular absorption coefficients of the solar material.

By the early 1930s it had become clear that hydrogen was by far the most abundant element in the stars, and calculations made with the absorption coefficient of pure hydrogen succeeded fairly well in representing the spectra of early-type stars. For the sun, however, these coefficients predicted a Balmer discontinuity at 3,646 Å, far in excess of that actually observed. An attempt by Antonie Pannekoek[1] to consider a host of absorption edges from various metals also failed to match the observed solar spectrum.

In this paper Rupert Wildt points out the importance of the negative hydrogen ion in the atmosphere of stars later than spectral type F5. Whereas previous considerations of stellar material had only included free electrons and the positive ions, Wildt showed that collisions between neutral hydrogen atoms and free electrons could lead to the formation of the negative hydrogen ion, H^-. Because the binding energy of H^- is low, it cannot exist at high temperatures, but it can become important in the cooler atmospheres of the late-type stars. As Wildt shows here, the free-bound absorption coefficient is so great that, in spite of their low concentration, the negative hydrogen ions can provide a significant contribution to stellar opacity. This conjecture was substantiated later, when Wildt used H. S. W. Massey's detailed calculations to show that H^- is responsible for nearly all the continuous absorption in the atmospheres of late-type stars, and when S. Chandrasekhar and F. H. Breen provided even more elaborate calculations showing that the absorption coefficient of H^- is in accord with the empirically deduced values for the sun.[2]

1. A. Pannekoek, *Monthly Notices of the Royal Astronomical Society 91*, 139, 519 (1930–1931).
2. R. Wildt, *Astrophysical Journal 90*, 611 (1939); S. Chandrasekhar and F. H. Breen, *Astrophysical Journal 104*, 430 (1946).

THE CARRIERS of the emission and absorption spectra observed in celestial objects have been identified as both neutral and positively charged atoms and molecules. A number of free electrons, equivalent to the charge of the positive ions, must be present to provide for electric neutrality. Collisions between electrons and certain neutral atoms or molecules should frequently lead to attachment, a negatively charged atomic or molecular ion being formed. The existence of such negative ions has been demonstrated by a great variety of experiments, among which the analysis of ionic beams in a mass spectrograph affords the most direct evidence; and their stability is also accounted for by atomic theory.[1]

Although it has long been recognized that negative ions play an important role in the Kennelly-Heaviside layer of the terrestrial atmosphere, their presence in stellar atmospheres does not seem to have been envisaged. Several species of neutral atoms and molecules, abundant in stellar atmospheres, are able to attach electrons in stationary states. This tendency to attach electrons is generally referred to, in physical and chemical literature, as "electron affinity" of the neutral particles, and the same term is used to denote the energy of binding measured in electron volts. The bond may be severed either by a subsequent collision or by photoelectric ionization. In thermodynamic equilibrium the concentration of the negative ions is ruled by the general principles of statistical mechanics and can be evaluated numerically from the Saha equation, in which the electron affinity has to be substituted for the ionization potential. The number of reliably determined electron affinities is still very small, as will be seen from table 42.1. Many astrophysically important diatomic molecules are known to form negative ions, like H_2, C_2, O_2, NH, OH, CN, and others; but there is scarcely any information available about the respective electron affinities. On inspecting table 42.1 it will be noticed that all the electron affinities are small compared to the ionization potentials of the corresponding atoms. This circumstance counteracts the formation of negative ions in a system consisting of a single sort of atoms only, because at any temperature high enough to release electrons by ionization of the neutral atoms the negative ions would split up even more. However, in a mixture of two atoms differing greatly in ionization potential—say sodium and hydrogen—the electrons released from the sodium atoms at comparatively low temperatures may readily attach themselves to the hydrogen atoms. This is just the case realized in stellar atmospheres of medium or late spectral type. In addition, the great abundance of hydrogen will favor the formation of H^- ions. Therefore, the H/H^- equilibrium suggests itself as the most promising subject of an inquiry into the astrophysical significance of electron affinity.

If the pressure is measured in dynes per square centimeter, the Saha equation of the H/H^- equilibrium assumes the form

$$\log_{10} K = \log_{10}\left(\frac{P_H \cdot P_e}{P_{H^-}}\right)$$

$$= -\frac{0.70 \cdot 5040}{\mathrm{T}} + 2.5 \log_{10} T + 0.12,$$

the statistical weight of H^- being 1. The ratios P_{H^-}/P_H were computed as functions of the electron pressure and temperature, and for sake of comparison there are also given the ratios of the numbers of neutral hydrogen atoms in the second and first quantum state,

$$\frac{N_2}{N_1} = 4 \cdot 10^{-[(10.16 \cdot 5040)/T]}.$$

Table 2 [not reproduced] contains these quantities as functions of the temperature and the total pressure in the "Russell-mixture" of elements.[2] This mixture is rather rich in hydrogen. Unsöld's analysis of the vertical distribution of temperature and electron pressure throughout the solar atmosphere[3] is based on a smaller abundance ratio of hydrogen to metals and therefore is less favorable to the formation of H^-. Table 42.2 contains the same quantities as table 2, for several optical depths under the solar surface, and reveals the surprising fact that, in the higher levels, the negative hydrogen ions are more abundant than the neutral hydrogen atoms excited to the ground state of the Balmer series. In view of the great strength

Table 42.1 Electron affinities

Atom and ground state			I.P.	E.A.	References
1	H	^{2}S	13.53	0.70	E. A. Hylleraas, *Zs. f. Phys.* *65*, 209, (1930).
3	Li	^{2}S	5.37	0.54	Ta You-Wu, *Phil. Mag.* *22*, 837, (1936).
8	O	^{3}P	13.55	2.2	W. W. Lozier, *Phys. Rev.* *46*, 268, (1932).
9	F	^{2}P	17.34	4.13	J. E. Mayer and L. Helmholz, *Zs. f. Phys. 75*, 19, (1932).
16	S	^{3}P	10.31	2.8	J. E. Mayer and M. Maltbie, *Zs. f. Phys. 75*, 748 (1932).
17	Cl	^{2}P	12.96	3.75	Mayer and Helmholz, p. 19.
35	Br	^{2}P	11.80	3.53	Ibid.
53	I	^{2}P	10.55	3.22	Ibid.

of the Balmer series in the solar spectrum, it is obvious that the negative hydrogen ions should produce conspicuous absorption lines in the medium and late spectral types if they possessed stable excited states capable of combining with the ground state.

Table 42.2 Negative hydrogen ions in the solar atmosphere

Optical depth	5,040/T	P_{H^-}/P_H	N_2/N_1
0.12	1.0	$10^{-7.34}$	$10^{-9.56}$
0.53	0.9	$10^{-7.02}$	$10^{-8.55}$
1.25	0.8	$10^{-6.87}$	$10^{-7.53}$
2.62	0.7	$10^{-6.82}$	$10^{-6.52}$
5.40	0.6	$10^{-6.66}$	$10^{-5.50}$

Here Wildt presents a table (not reproduced) that gives the ratios P_{H^-}/P_H and N_2/N_1 for various values of electron pressure and temperature (assuming the "Russell mixture" of elements). The ratio P_{H^-}/P_H is a weak function of temperature and ranges from 10^{-7} to 10^{-13} when the total pressure ranges from 10^4 to 10^{-4} dyn cm^{-2}. The ratio N_1/N_2 ranges from $10^{5.5}$ to $10^{29.9}$ when the temperature ranges from 8,400 to 1,680 K.

At present, no definite answer can be given to the question whether or not there are stable excited states in negative ions. It has been established, as a general result, that their number could not be infinite, as in the case of an electron bound in a Coulomb field. The small binding energy of the attached electron in its ground state makes it most unlikely that any stable discrete excited states exist with appreciable binding energy, though there might be some states with very small binding energy, closely adjacent to the ionization continuum. With the coronal and certain interstellar lines defying all efforts of the spectroscopists at identification, it may not be untimely to remark upon the remote chance of negative ions being the carriers of these spectra. This has already been suggested, though rather perfunctorily, for the coronal lines by Goudsmit and Ta You-Wu.[4] This is the only reference, in astrophysical literature, to negative ions of which the writer is aware. Since the solar corona is now widely regarded, for its scattering properties, as a cloud of free electrons, it would seem to be the very place where negative ions are likely to be formed by collisions between the indigenous electrons and neutral atoms ejected from the sun during its eruptive activity. Unfortunately, there is little hope, if any, for the discovery, in the near future, of spectra of negative ions in the laboratory or for the theoretical prediction of their energy states.

The small electron affinity of hydrogen, 0.70 volt, would place any hypothetical discrete spectrum of the negative hydrogen ion far out into the infared, beyond the ionization limit of H^-, near λ17,600. The rapid decay of atomic ionization continua toward greater frequencies might make one expect that the continuous absorption coefficient of H^- is negligibly small throughout the visual spectral region and the near ultraviolet. But the negative hydrogen ions show a strikingly different behavior, according to an investigation by Massey and Smith,[5] who computed the cross-sections for the capture of electrons by hydrogen atoms. Assuming thermodynamic equilibrium, these capture coefficients can easily be transformed, by Milne's formula,[6] into cross-sections for absorption or, by a different name only, into the astrophysically relevant absorption coefficients pertaining to the bound-free transitions of the negative hydrogen ion. The absorption coefficient of H^- rises from zero, at the ionization limit, to a broad maximum at a distance of about 10 volts from the ionization limit. It then decreases gradually, extending with remarkable intensity into the region between 20 and 30 volts. Massey and Smith do not give any estimate of the accuracy of their cross-sections; but it may be hoped that their results are trustworthy for the astrophysically accessible spectral range, which is rather close to the ionization limit. Menzel and Pekeris[7] had already computed what they call the absorption resulting from the changes of kinetic energy of an electron traversing the field of a neutral hydrogen atom. From the point of view presented here, this absorption may be described as produced by the free-free transitions of the negative hydrogen ion. The absorption coefficient k_ν, resulting from the existence of negative hydrogen ions in a gas containing N_e electrons per unit volume, is, per neutral hydrogen atom, for free-free transitions

$$\frac{1}{3}\frac{e^2hN_e}{cm^3\nu^3}\left(\frac{mv_0{}^2}{2\pi kT}\right)^{3/2} v_0{}^4A(T, v_0) = a \cdot P_e,$$

and for bound-free transitions

$$\frac{1}{4}\frac{mc^2h(\nu - \nu^*)}{h^2\nu^2} Q^eN_e\left(\frac{h^2}{2\pi mkT}\right)^{3/2} e^{h\nu^*/kT} = b \cdot P_e.$$

The coefficients a have been taken from the paper of Menzel and Pekeris, and the coefficients b have been computed from the emission cross-sections Q^e published in the form of a graph by Massey and Smith, ν^* being the frequency corresponding to the electron affinity of hydrogen. As will be seen from table 42.3, the bound-free transitions produce an absorption far greater than that of the free-free transitions, especially at lower temperatures. Now the latter, as found by Menzel, is still negligible at the solar temperature, amounting to about 5 per cent of the metallic absorption. However, they would become appreciable at lower temperatures, particularly in dwarf stars. Therefore, with regard to the comparison given in table 42.3, it would seem that the contribution by negative hydrogen ions to the opacity of stellar atmospheres cannot be neglected altogether, even at the solar temperature. It is generally agreed that the present theory of the atmospheric absorption coefficient is unsatisfactory, and to take into account this new source of opacity may help to remove the discrepancies between theory and observation, at least partly. A detailed investigation is under way.

Table 42.3 Absorption coefficient of negative hydrogen ions[a]

$T°$ K		λ8,900	λ6,810	λ5,390	λ4,360	λ3,030
3,000	a	6.82	4.12	2.70	1.88	1.04
	b	150	160	250	310	370
4,000	a	5.81	3.41	2.20	1.50	0.81
	b	37	40	62	79	94
6,000	a	4.83	2.71	1.69	1.12	0.55
	b	6.7	7.2	11.3	14.4	17.1
8,000	a	4.33	2.36	1.44	0.93	0.47
	b	2.3	2.5	3.9	5.0	5.9

a. Upper line: $a \cdot 10^{28}$ free-free transitions. Lower line: $b \cdot 10^{28}$ bound-free transitions.

1. A monograph on *Negative Ions* has just been published by H. S. W. Massey (Cambridge University Press, 1938). A letter received in March, 1939, from Dr. Massey communicates preliminary results of a new calculation of the absorption coefficients of H^-. Using the Hylleraas form for the ground-state wave function instead of the former rough approximation, he finds absorption coefficients of the same order of magnitude, in the astrophysically relevant range of wave lengths, as those given by Jen (*Phys. Rev. 43*, 540[1933]), to which reference has been made in a brief note to appear soon in the *Publications of the American Astronomical Society*. Utilizing Jen's data, it has been shown that the addition of the H^- absorption to the metallic absorption produces for the solar atmosphere that desired independence of the absorption coefficient from wave length which failed to emerge from Unsöld's and Pannekoek's theory. In the opinion of the writer this result is not likely to be changed materially when Jen's data eventually will be replaced by Massey's improved data. This would mean the removal of a serious discrepancy between theory and observation and, in fact, the identification, by their continuous absorption spectrum, of the negative hydrogen ions so abundant in the solar atmosphere.
2. H. N. Russell, *Ap. J. 75*, 337 (1932).
3. A. Unsöld, *Zs. f. Ap. 8*, 262 (1934).
4. *Ap. J. 80*, 154 (1934).
5. *Proc. R.S. A 155*, 472 (1936).
6. *Phil. Mag. 47*, 209 (1924).
7. *M.N. 96*, 77 (1935).

43. The Quantitative Analysis of the B0-Star τ Scorpii, Part II

Albrecht Unsöld
TRANSLATED BY OWEN GINGERICH

(*Zeitschrift für Astrophysik 21*, 22–84 [1941])

The impact of studies of atomic structure and of quantum mechanics was nowhere more important for astronomy than in the quantitative analysis of chemical abundances. Previous selections in this chapter have illuminated some of the early steps in the understanding of physical processes in the outer layers of stars and in the formation of spectral lines. By 1938 Albrecht Unsöld had brilliantly summarized and extended these procedures in his book *Physik der Sternatmosphären*. Shortly thereafter he applied these techniques to the most detailed analysis of a single star (other than the sun) that had yet been made. By choosing a B0 star, he avoided the crowding of spectral lines and questions of the absorption coefficient that prevailed for the sun and other later-type stars. More importantly, he succeeded in deriving probably the most reliable value for the cosmic abundance of helium up to that time. He was also able to relate his derived abundances of carbon, nitrogen, and oxygen to the recently discovered source of stellar energy, the CNO cycle.

World War II delayed the appearance of Unsöld's analysis and brought the publication of further developments in this area to a virtual standstill. By the late 1940s, however, a strong resurgence of interest had emerged. Unsöld's student Erika Vitense worked out the opacities in greater detail and following earlier works of Ludwig Biermann, employed a mixing-length theory for treating atmospheric convection.[1] Subrahmanyan Chandrasekhar, in solving the problem of Rayleigh scattering in the terrestrial atmosphere, formulated an elegant mathematical framework for procedures in the field of stellar atmospheres.[2]

As early as 1931 William McCrea had calculated what he termed "model stellar atmospheres."[3] By 1950 the new work brought the physics of the models to a state improved to the extent that the formation of spectral lines could be computed throughout the successive layers of a star's atmosphere, rather than by the approximation of an isothermal gas whose temperature matched the deduced effective temperature of the star. Nevertheless, throughout the 1950s most model atmospheres were "gray"; that is, they assumed a continuous opacity independent of wavelength, although the opacity did vary with temperature and pressure. The complexity of non-gray models was so great that no appreciable progress could be made until the advent of large-scale computers. By 1964 the computational techniques had become sufficiently widespread that an international stellar atmosphere conference was convened at the Harvard and Smithsonian observatories, and workers in this field were able to compare their complex computing procedures through a standardized model.[4] In recent years, detailed quantitative analyses have yielded abundances for a variety of bright stars and for star clusters.

1. E. Vitense, *Zeitschrift für Astrophysik 28*, 81 (1951), *32*, 135 (1953).
2. S. Chandrasekhar, *Radiative Transfer* (Oxford: Clarendon Press, 1950).
3. W. McCrea, *Monthly Notices of the Royal Astronomical Society 91*, 836 (1931).
4. O. Gingerich, D. Mihalas, S. Matsushima, and S. Strom, *Astrophysical Journal 141*, 316 (1965).

1. Introduction

In the following analysis of the spectrum of τ Scorpii, I will show, using an earlier theoretical investigation as a useful approximation, that one can establish for a homogeneous stellar atmosphere a mean value for the temperature T (or $\theta = 5{,}040/T$) and for the electron pressure P_e. I will then treat the exact values as unknowns. I will also determine the chemical composition, especially the abundance of hydrogen, in the course of my investigation. My procedure will rest on the art of successive approximation, and the internal consistency of the final result will furnish the proof of its usefulness.

I will first of all use the residual intensities of the strong Fraunhofer lines and draw from them conclusions concerning the nature of the scattering process and the effective depth of the atmosphere of τ Scorpii. After a short summary of the theory of the curve of growth and some consideration of the possibility of calculating the transition probabilities or oscillator strengths for the lines in the spectrum of τ Scorpii, I will turn to my first major result: the determination of the "effective number of absorbing atoms" in the specified excited atomic levels. A procedure I previously suggested in connection with the theory of solar prominences—depending only on an order-of-magnitude choice of the temperature T—will make possible an important section on the quantitative determination of the chemical composition of the atmosphere of τ Scorpii. Examination of those elements found spectroscopically in more than one ionization state will give an insight into the ionization level of τ Scorpii. A quantitative discussion of the pressure-broadening of the Balmer lines or the number of the last visible Balmer line enables me to determine the electron density N_e and thus a definitive value of T and P_e. By a suitable combination of observations, it is also possible to measure in a purely spectroscopic way the gravitational acceleration g at the stellar surface.

In conclusion, I will consider my results in a somewhat more general manner, so that several directions for further research can be set forth.

Sections 2 through 6 of Unsöld's paper deal with the theory of line formation and with a detailed examination of the lines of each element. Unsöld himself summarizes these sections in his abstract, which follows here.

The depths R_0 or, rather, the residual intensities $1 - R_0$ of the strong Fraunhofer lines in τ Scorpii tend toward a homogeneous limiting value of $R_c = 40$–45%, regardless of their origins or transition probabilities. The lines appear to be pure absorption; R_c corresponds to the blackbody radiation for the boundary temperature T_0 of the star—in other words, to the radiation intensity of the star's surface. From R_c we calculate $\bar{\kappa}/\kappa$ as 2.3–2.5. The effective depth of the atmosphere is virtually independent of λ, particularly for the wavelength region $\lambda 4{,}860$–$3{,}710$ Å. For $\lambda < 3{,}710$ Å, the effective depth is diminished by a factor of 0.68 because of the addition of the continuous Balmer absorption.

Then the curve of growth theory, already given in a convenient form by me and in part also by D. H. Menzel, will again be presented briefly. Next, the theoretical line intensities or oscillator strengths f can be compared within multiplets, supermultiplets, and transition arrays from the known formulas and tables of Henry Norris Russell, A. Sommerfeld and Hönl, R. Kronig, Leo Goldberg, and others. An absolute calibration is possible for transitions between two hydrogenic terms (the tables of Hans Bethe et al.) or, approximately, for "strong transitions," such as $3s$–$3p$, $3p$–$3d$ . . . , for which one can expect a total $f \approx 1$ (with a limit between 0.5 and 2). Nearly all the lines observed in τ Scorpii belong to one type or the other.

After this preparation, the spectra of individual elements will be discussed in detail, and in particular the number of atoms (per cm^2 of the stellar surface) in the quantum state s of the rth ionization stage will be deduced in the form log $N_{r,s} \cdot H$. From this, I will establish two more important results: both the quantum number $n = 14$ or 15 of the last separately observed Balmer lines and the equivalent widths A_λ of Hβ, Hγ, and Hδ determine the electron density log N_e, which equals 14.48 ± 0.25. The curve of growth of the O II transition array $3s$–$3p$ (one especially suitable for this) shows that there is almost no turbulence in τ Scorpii; the turbulent velocity ξ_t is surely less than 2 km sec^{-1}.

The ionization equilibria C^{+++}:C^{++}, N^{+++}:N^{++}, O^{+++}:O^{++}, and Si^{++++}:Si^{+++} in conjunction with the electron density determined from the hydrogen lines yields a mean temperature of the atmosphere of $T = 28{,}150 \pm 750$ K and a mean electron pressure of log $P_e = 3.07 \pm 0.25$ dyn cm^{-2}.

7. Cosmic Abundances of the Light Elements and Gravitational Acceleration at the Stellar Surface

We can now proceed from the resulting numbers of atoms for the determined ionization states (or log $NH \cdot P_e$) to the total number for the elements examined, summed over all possible ionization states. The adopted values for T and P_e, together with the ionization partition functions u_r, according to the Saha equation,

$$\frac{N_{r+1} \cdot P_e}{N_r} = \frac{u_{r+1} \cdot 2}{u_r} \cdot \frac{(2\pi m)^{3/2}(kT)^{5/2}}{h^3} \cdot e^{-\chi_r/kT}, \qquad (39)$$

lead to the following seemingly trivial result: for carbon, nitrogen, oxygen, and silicon, we need to add only the two ionization states already found spectroscopically. For oxygen, there is only about 1.6% O^+, or a correction of 0.01 to be

Table 43.1 Total abundances of the elements in τ Scorpii according to number of atoms ($\log NH \cdot P_e$) and mass ($\log NH\mu \cdot P_e$)

Z	Element	$\log NH \cdot P_e$	$NH \cdot P_e \cdot 10^{-20}$	$\log NH\mu \cdot P_e$	$NH\mu \cdot P_e \cdot 10^{-20}$
1	H	25.34	219,000	25.34	219,000
2	He	24.59	38,900	25.19	155,000
6	C	21.58	38	22.66	457
7	N	21.92	83	23.07	1,180
8	O	22.33	214	23.53	3,400
10	Ne	22.39	245	23.69	4,900
12	Mg	21.10	13	22.49	310
13	Al	19.90	0.8_0	21.33	21_4
14	Si	21.14	14	22.59	390
26	Fe[a]	21.00	10	22.75	560

a. According to V. M. Goldschmidt and H. N. Russell.

added to $\log NH \cdot P_e$; for neon there is analogously 17.8% Ne^+, or a correction of +0.07 to $\log NH \cdot P_e$. The other elements exist almost solely in the ionization states already calculated.

Thus, we find that the total abundance of the elements considered is proportional to $NH \cdot P_e$ in table 43.1. The logarithm of the abundance differs from my previously given mean by more than ± 0.3 in only one case (neon, with +0.5). My values should have an error limit of this size at most.

It should further be emphasized that the relative abundances of the elements depend only slightly on the adopted values of T and P_e.

In figure 43.1 I show at the top my results, so that I can make a first precise statement about the cosmic abundance of light elements. For comparison, I show below the abundance distribution of the elements in meteorites from the analysis by V. M. Goldschmidt [5]. Then comes H. N. Russell's [10] well-known analysis of the sun and, finally, the considerably less exact determination in connection with the planetary nebulae made by I. S. Bowen and A. B. Wyse [2]. The even-numbered elements are connected with a solid line and the odd-numbered with a dashed line.

A comparison of the upper three curves (we can conveniently use the good agreement of all three analyses for the magnesium-aluminum-silicon group as a starting point) shows straightaway that of the most abundant elements only iron cannot be determined in our analysis of τ Scorpii. For the following calculations, I therefore complete table 43.1 by using the ratio iron:magnesium, aluminum, silicon from V. M. Goldschmidt and H. N. Russell, which yields the value of $\log NHP_e = 21.00$ for iron. A change of this value by ± 0.5 will scarcely affect the following considerations.

In order further easily to see the abundance of the elements by mass, I have included in table 43.1 also $\log NH \cdot \mu P_e$, where μ designates the atomic weight of the atoms in question.

I now use these numbers to calculate the acceleration of gravity g at the surface of τ Scorpii:

The gas pressure P_g at the base of the atmospheric layer in which the Fraunhofer line is essentially produced is simply equal to the weight (mass times gravitational acceleration g) of a square centimeter column, that is,

$$P_g = g \, \Sigma \, NH \cdot \mu m_H, \tag{43}$$

where $m_H = 1.67 \times 10^{-24}$ gm, the mass of the proton. The mean gas pressure of the atmosphere is thus $\bar{P}_g \approx (1/2)P_g$, which furthermore equals $2P_e$ (as there are almost equally many electrons and ions). I now multiply both sides of equation (43) by P_e, obtaining

$$4P_e^{\,2} = g \cdot m_H \sum NH \cdot P_e \cdot \mu. \tag{44}$$

According to table 43.1, $\log \sum NH \cdot P_e \cdot \mu$ is 25.59; therefore, $\log P_e$ is 3.07, and it follows (with a suitably cautiously estimated error limit) that

$$\log g = 4.93 \pm 0.40. \tag{45}$$

I next compare this value with that from the absolute magnitude M_v and the temperature T, using the mass-luminosity relation.

τ Scorpii is a well-known member of the Scorpius-Centaurus starstream. On these grounds, N. H. Rasmuson has determined in his first work an absolute visual magnitude of -2.10 mag and in his second (in which the total stream is divided into three parts) a value of -2.54 mag. F. Schlesinger gives as the most probable value -2.30 mag; E. G. Williams finds with a spectroscopic method -2.60 mag [19]. I will take as the most probable value -2.4 mag, which yields, with $T = 28{,}150°$ (simply from the diagram in [14], section 34),

$$\log_{10} g = 4.4, \tag{45a}$$

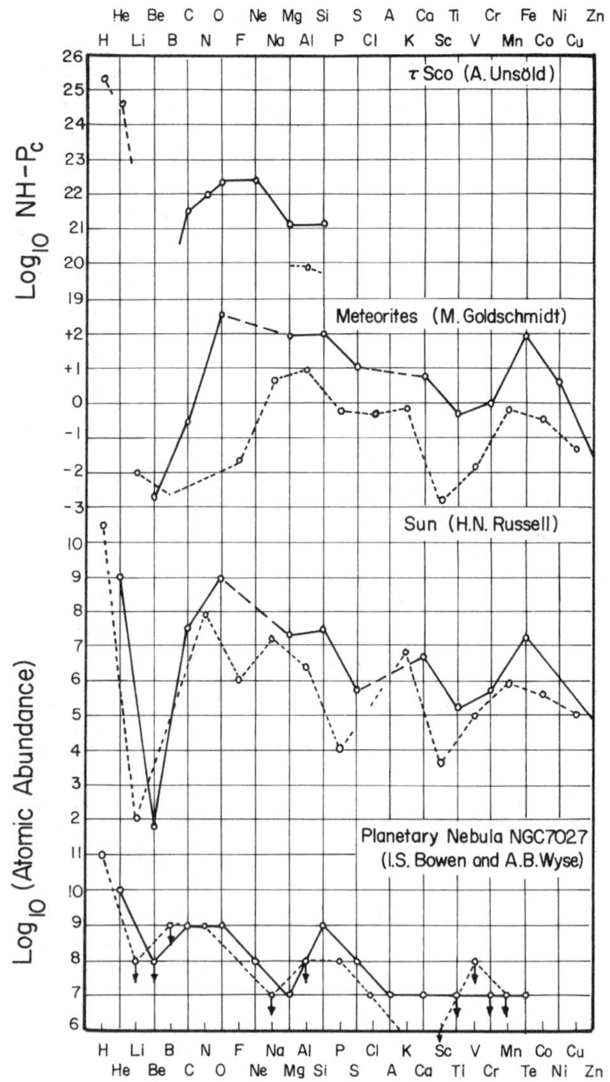

Fig. 43.1 Cosmic abundances of the elements. Elements with even atomic numbers are connected with a solid line, and those with uneven numbers are connected with a dashed line.

with an estimated error bar of ± 0.25. The agreement with (45) is very satisfactory.

Both values are essentially higher than those customarily given for the run of main-sequence B stars, $\log g = 3.90$ (see, for example, H. N. Russell, *Ap. J.* 78, 239 [1933], equation [117a]).

I have to accept this difference as real; in other words, τ Scorpii is an extreme dwarf star. Since the electron pressure P_e is connected with the temperature by approximately $\sqrt{g}$, one must obtain for a normal B0 star in comparison with τ Scorpii an electron pressure about 3 times smaller. Correspondingly, according to F. L. Mohler's version [9] of the relation from D. R. Inglis and E. Teller [7],

$$\log N = 23.26 - 7.5 \log n_m \tag{23}$$

(where n_m is the principal quantum number of the last observed Balmer line and N is the number of charged particles broadening the lines), two or three more Balmer lines should appear than in τ Scorpii. However—although, to be sure, it is not yet well worked out quantitatively—this appears to be almost exactly the case in observations made by O. Struve and me [13].

8. SOME GENERAL CONCLUSIONS FROM THE ABUNDANCE DISTRIBUTION FOUND FOR THE ELEMENTS

In conclusion, it seems appropriate to make, in an informal fashion, some general remarks naturally suggested by the abundances of the elements given in table 43.1.

THE HYDROGEN:HELIUM:OTHER ELEMENTS RATIO AND THE THEORY OF THE INTERNAL CONSTITUTION OF THE STARS According to table 43.1, the ratio of H:He:R (R = the remaining elements) is, in percent, by number of atoms,

$$\mathrm{H}:\mathrm{He}:R = 85:15:0.24, \tag{46}$$

or by mass,

$$\overline{\mathrm{H}}:\overline{\mathrm{He}}:\overline{R} = 57:40:3. \tag{47}$$

Now how exact is the ratio 0.75 for H:He or, rather, the log (H/He)? According to the remarks in table 22 [not reproduced], the error lies within ± 0.25. In a new work D. H. Menzel [8] gives the value 0.72 for log (H/He), without giving precise details about the source material. From observations of prominences, I previously [14] derived for the sun a value of 1.41 for log (H/He), using the measurements of M. Minnaert and C. Slob, or of 1.08, using those of A. Pannekoek (calculations from pp. 416 and 419 in [14], quoted here, have been improved with the more exact f-values of E. Hylleraas [6] and of L. Goldberg [4]). In hindsight, one can see that the values derived previously are surely less exact than those derived for τ Scorpii; it thus appears very unlikely that log (H/He) is greater than 1.1.

The "other elements" differ so very much from the commonly accepted distribution called the "Russell mixture" that one can say only that neon ranks highest. As far as quantitative considerations are concerned, it must always be remembered that the determination of neon is subject to an especially large error (on account of the departures of L-S coupling in the Ne II spectrum). Yet even when the most unfavorable error limits are adopted, neon always remains the second or third most abundant element after helium.

The H:metals ratio is certainly more secure. I find, for example, log (H/Mg) $\sim$ 4.24; Strömgren [12], on entirely different grounds, gets the value 4.4 for the sun, well inside the corresponding error limits.

I can now make the very plausible assumption that the chemical composition of the stellar interior agrees with that of the atmosphere of τ Scorpii, and thus my results (46) or (47) have a fundamental meaning for the theory of the internal constitutions of the stars. It now appears certain that we have not only an enormous abundance of hydrogen but also a very considerable amount of helium. In any event, it has already been very nicely established by Strömgren [11] that the right relation among mass, luminosity, and surface temperature for the sun will be obtained under the assumption that the ratio by mass is

$$\overline{\mathrm{H}}:\overline{\mathrm{He}}:\bar{R} = 60:40:2, \tag{48}$$

which agrees within our estimated limits of precision. This leads me to surmise that we have not run into any difficulties.

THE CARBON:NITROGEN RATIO AND THE THEORY OF STELLAR ENERGY GENERATION ACCORDING TO H. BETHE AND C. F. V. WEIZSÄCKER According to Bethe [1] and Weizsäcker [17], the energy generation inside a main-sequence star arises through a cyclic reaction. The end result is that four protons and two electrons combine to form an alpha particle, while carbon and nitrogen act as catalysts. The abundance of carbon 12 and nitrogen 14 must therefore be proportional to the (temperature-dependent) half-lives, just as for any radioactive equilibrium. H. Bethe calculates for the main sequence a C:N ratio of about 1:20, and we find for τ Scorpii 1:2.2 (bounded by 1:1 to 1:4).

I prefer at this time to emphasize the rough agreement rather than the difference by a factor of 10. First of all, this is an empirical determination with a factor of 2 uncertainty. Furthermore, the physical foundation of Bethe's theory is not yet known exactly. Finally, it is obvious that for Bethe's calculation assumptions made about the chemical composition of the star are significantly different from my new values. Accordingly, his assumption about the temperatures, especially in the stellar interior, must be corrected. It would undoubtedly be profitable to carry out the rather sketchy but wide-ranging calculations with the correction.

In this connection, other questions naturally arise: whether various stars might not exhibit differences in their C:N ratio (determined through the Bethe-Weizsäcker cycle) and whether in "old" stars the hydrogen might not to a great extent be already "burned" into helium. In a previous contribution [15], I have already offered some remarks on this subject. Before going back to it in detail, it would seem advantageous to have a whole series of different stars determined by the method outlined here. I hope that in the not-too-distant future I can work in this direction, using the previously mentioned material gathered with O. Struve at the McDonald Observatory.

THE COSMIC ABUNDANCE OF NEON My preceding determination of the abundance of neon has given a value between that of oxygen and magnesium; the error limits already noted would easily make neon nearly as abundant (by atoms) as oxygen. This observation fits quite well with the well-known idea that the neon at the earth's surface has for the most part escaped. It should be remarked in this connection that my investigation generally establishes for the first time a rather dependable abundance data for the group that V. M. Goldschmidt [5] has called the "atmophil" elements.

We can barely hint at the connection of the cosmic abundance of neon with the researches of W. Wefelmaier [16] concerning the nuclear binding energy and abundance of elements with atomic weights of the form 4n.

THE CONTINUOUS ABSORPTION COEFFICIENT OF STELLAR ATMOSPHERES In an earlier work, I suggested that we can understand a series of features in the behavior of stellar atmospheres only if we assume that, besides the continuous absorption of atomic hydrogen, there is for hot as well as for cool stars a second absorption mechanism.

Earlier investigations of the observations led to the assumption that the metals (especially the R elements) were particularly abundant: we should then expect that at lower temperatures the neutral metals would make an essential contribution to the continuous absorption, increasing with the increasing degree of ionization (κ changing about as Z^2).

This conception is no longer tenable; it has already been refuted by H. N. Russell and others, from the dissociation equilibrium of hydrides, and it is further refuted by the enormous abundance of hydrogen now well established through my analysis of τ Scorpii. An additional series of difficulties (height of absorption edges, size of the electron pressure, and so on) need not even be mentioned here.

R. Wildt [18] and afterward B. Strömgren [12] and P. ten Bruggencate [3] have shown a satisfactory interpretation of the observations can be given especially for the sun and later spectral types if at these temperatures the continuous absorption for an atmosphere composed primarily of hydrogen results chiefly from the negative hydrogen ion.

On the other hand, for the earlier spectral types we must seek another source of continuous absorption. I proposed previously that an important mechanism would be the Thomson scattering of free electrons (the Compton effect). In this case the continuous absorption of helium also comes into effect. I hope in some future time to carry out an exact calculation for this problem, which would necessarily lead to an interpretation of the values of $\bar{\kappa}/\kappa$ determined in table 1 [not reproduced].

CONVECTION IN STARS The large cosmic abundances found for some elements whose ionization potentials lie substantially above hydrogen (13.5 eV), namely, helium (24.5 eV) and neon (21.5 eV), lead us to expect that the calculation of the values of the specific heats C_p/C_v for the new elements will allow us to say something important about the possibility of convective currents in the stars.

The author is at this time for other reasons unfortunately not in a position to carry out these extensive numerical calculations. Perhaps it is also advisable to wait until the H:He ratio is known more exactly from observations of other stars.

A problem still not satisfactorily clarified is nonetheless recognized, that of understanding how the two separate convection mechanisms of granulation on the one hand and the 11-yr cycle (sunspots, faculae, and so on) on the other can function side by side apparently undisturbed.

I wish to express again my very hearty thanks to Dr. O. Struve and my colleagues at the Yerkes and McDonald Observatories for their complete support of this undertaking.

1. H. Bethe, *Phys. Rev. 55*, 434 (1938).

2. I. S. Bowen and A. B. Wyse, *Lick Obs. Bull. No. 495* (1939).

3. P. ten Bruggencate, *V.J.S. der Astron. Ges. 75*, 203 (1940) P. ten Bruggencate and J. Houtgast, *ZS. f. Astrophys. 20*, 149 (1940).

4. L. Goldberg, Ap. J. *90*, 414 (1939).

5. V. M. Goldschmidt, *Skrifter Norske Videnskaps-Akademi Oslo, I. Math.-Naturw. Klasse*, no. 4 (1938).

6. E. Hylleraas, *ZS. f. Phys. 106*, 395 (1937).

7. D. R. Inglis and E. Teller, *Ap. J. 90*, 439 (1939).

8. D. H. Menzel, *Popular Astronomy 46*, 47 (1938).

9. F. L. Mohler, *Ap. J. 90*, 429 (1939).

10. H. N. Russell, *Ap. J. 70*, 11 (1929); later improved in Russell, Dugan, and Stewart, *Astronomy*, 2nd ed., p. 503.

11. B. Strömgren, *Erg. d. exakt. Naturwiss. 16*, 519 (1937).

12. B. Strömgren, *On the Chemical Composition of the Solar Atmosphere: Festschrift f. E. Strömgren*, (Copenhagen: Munksgaard, 1940).

13. O. Struve and A. Unsöld, *Ap. J. 91*, 365, (1940).

14. A. Unsöld, *Physik der Sternatmosphären* (Berlin, 1938).

15. A. Unsöld, *ZS. f. techn. Phys. 21*, 301 (1940) or *Phys. ZS. 41*, 549 (1940).

16. W. Wefelmaier, *ZS. f. Phys. 107*, 332 (1937); especially see 2.

17. C. F. v. Weizsäcker, *Phys. ZS. 38*, 176 (1937), and also especially *39*, 633 (1938).

18. R. Wildt, *Ap. J. 89*, 295 (1939), *90*, 611 (1939).

19. E. G. Williams, *Ap. J. 83*, 279, 305 (1936).

CHAPTER IV

Stellar Evolution and Nucleosynthesis

44. The Equivalence of Mass and Energy

Does the Inertia of a Body Depend upon Its Energy Content?

Albert Einstein

TRANSLATED BY W. PERRETT AND G. B. JEFFERY

(*Annalen der Physik 18*, 639–641 [1905]; trans. in *The Principle of Relativity* [New York: Dodd, Mead and Co., 1923, reproduced by Dover Publications, 1953] pp. 67–71)

The Principle of the Conservation of the Motion of the Center of Gravity and the Inertia of Energy

Albert Einstein

TRANSLATED BY WOLFGANG KALKOFEN AND OWEN GINGERICH

(*Annalen der Physik 20*, 627–633 [1906])

In the first paper here Albert Einstein uses the postulates of the special theory of relativity[1] to show that energy radiated is equivalent to mass lost. He considers the conservation of energy of a radiating body in a rest system and in a system in uniform motion relative to it. Taking the difference between the two equations, he finds that the body must have lost the mass $\Delta m = \Delta E/c^2$, neglecting terms higher than $(v/c)^2$, where ΔE is the energy radiated into space and c is the velocity of light. Generalizing this result for all energy transformations, Einstein concludes for the first time that "the mass of a body is a measure of its energy content." At the end of his paper Einstein suggests that the radioactive emanations from uranium and radium might be associated with a change in the mass of the radioactive elements. Fifteen years later A. S. Eddington (see selection 45) suggested that the transformation of mass into radiated energy might be the process that makes the sun shine.

Curiously enough, Einstein's initial derivation of what was to become the most famous formula of physical science was logically fallacious and based on a circular argument. As Herbert Ives showed in 1952, Einstein essentially assumed the result in the first paper given here.[2] However, in 1906 and again in 1907, Einstein presented two further derivations, both correct. In the second part of this selection we translate from the 1906 paper Einstein's ingenious thought experiment, involving energy transport in a hollow cylinder, which leads to the inertial property of energy and the mass-energy equivalence.

In 1907 Einstein succeeded in deducing the exact expression for the equivalence of mass and energy, namely, his celebrated equation[3]

$$E = mc^2.$$

According to Einstein, every mass has an equivalent energy, just as every form of energy has an equivalent mass[4].

1. *Annalen der Physik 17*, 891 (1905).
2. See Max Jammer, *Concepts of Mass* (Cambridge: Harvard Univ. Press, 1961), pp. 176–179.
3. *Annalen der Physik 23*, 371 (1907).
4. Later in 1907 Einstein proved that a body's inertial and gravitational masses are equal to the same quantity E/c^2 and so should be considered equal exactly to each other. This result would be a stepping stone to the general theory of relativity of 1915 (see *Jahrbuch der Radioaktivität und Elektronik 4*, 411 [1907]).

A. Einstein, Does the Inertia of a Body Depend upon Its Energy Content?

THE RESULTS of the previous investigation lead to a very interesting conclusion, which is here to be deduced. I based that investigation on the Maxwell-Hertz equations for empty space, together with the Maxwellian expression for the electromagnetic energy of space, and in addition the principle that:

The laws by which the states of physical systems alter are independent of the alternative, to which of two systems of co-ordinates, in uniform motion of parallel translation relatively to each other, these alterations of state are referred (principle of relativity).

With these principles[1] as my basis I deduced *inter alia* the following result (§8):

Let a system of plane waves of light, referred to the system of co-ordinates (x, y, z), possess the energy l; let the direction of the ray (the wave-normal) make an angle ϕ with the axis of x of the system. If we introduce a new system of co-ordinates (ξ, η, ζ) moving in uniform parallel translation with respect to the system (x, y, z), and having its origin of co-ordinates in motion along the axis of x with the velocity v, then this quantity of light—measured in the system (ξ, η, ζ)—possesses the energy

$$l^* = l \frac{1 - \frac{v}{c}\cos\phi}{\sqrt{1 - v^2/c^2}},$$

where c denotes the velocity of light. We shall make use of this result in what follows.

Let there be a stationary body in the system (x, y, z), and let its energy—referred to the system (x, y, z)—be E_0. Let the energy of the body relative to the system (ξ, η, ζ), moving as above with the velocity v, be H_0.

Let this body send out, in a direction making an angle ϕ with the axis of x, plane waves of light, of energy $\frac{1}{2}L$ measured relatively to (x, y, z), and simultaneously an equal quantity of light in the opposite direction. Meanwhile the body remains at rest with respect to the system (x, y, z). The principle of energy must apply to this process, and in fact (by the principle of relativity) with respect to both systems of co-ordinates. If we call the energy of the body after the emission of light E_1 or H_1 respectively, measured relatively to the system (x, y, z) or (ξ, η, ζ) respectively, then by employing the relation given above we obtain

$$E_0 = E_1 + \tfrac{1}{2}L + \tfrac{1}{2}L,$$

$$H_0 = H_1 + \tfrac{1}{2}L \frac{1 - \frac{v}{c}\cos\phi}{\sqrt{1 - v^2/c^2}} + \tfrac{1}{2}L \frac{1 + \frac{v}{c}\cos\phi}{\sqrt{1 - v^2/c^2}}$$

$$= H_1 + \frac{L}{\sqrt{1 - v^2/c^2}}.$$

By subtraction we obtain from these equations

$$H_0 - E_0 - (H_1 - E_1) = L\left\{\frac{1}{\sqrt{1 - v^2/c^2}} - 1\right\}.$$

The two differences of the form $H - E$ occurring in this expression have simple physical significations. H and E are energy values of the same body referred to two systems of co-ordinates which are in motion relatively to each other, the body being at rest in one of the two systems (system (x, y, z)). Thus it is clear that the difference $H - E$ can differ from the kinetic energy K of the body, with respect to the other system (ξ, η, ζ), only by an additive constant C, which depends on the choice of the arbitrary additive constants of the energies H and E. Thus we may place

$$H_0 - E_0 = K_0 + C,$$
$$H_1 - E_1 = K_1 + C,$$

since C does not change during the emission of light. So we have

$$K_0 - K_1 = L\left\{\frac{1}{\sqrt{1 - v^2/c^2}} - 1\right\}.$$

The kinetic energy of the body with respect to (ξ, η, ζ) diminishes as a result of the emission of light, and the amount of diminution is independent of the properties of the body. Moreover, the difference $K_0 - K_1$, like the kinetic energy of the electron (§10), depends on the velocity.

Neglecting magnitudes of fourth and higher orders we may place

$$K_0 - K_1 = \frac{1}{2}\frac{L}{c^2}v^2.$$

From this equation it directly follows that:

If a body gives off the energy L in the form of radiation, its mass diminishes by L/c^2. The fact that the energy withdrawn from the body becomes energy of radiation evidently makes no difference, so that we are led to the more general conclusion that:

The mass of a body is a measure of its energy-content; if the energy changes by L, the mass changes in the same sense by $L/9 \times 10^{20}$, the energy being measured in ergs, and the mass in grammes.

It is not impossible that with bodies whose energy-content is variable to a high degree (e.g. with radium salts) the theory may be successfully put to the test.

If the theory corresponds to the facts, radiation conveys inertia between the emitting and absorbing bodies.

1. The principle of the constancy of the velocity of light is of course contained in Maxwell's equations.

A. Einstein, The Principle of Conservation and the Inertia of Energy

In a paper[1] published last year I have shown that the Maxwellian electromagnetic equations in combination with the relativity principle and the energy principle led to the conclusion that the mass of a body is altered by a change in its energy content, independent of the form of the energy change. It was shown that an energy change of size ΔE corresponds to a change in the mass of size $\Delta E/V^2$, where V designates the velocity of light.

In the present paper I will now show that this conclusion is the necessary and sufficient condition for the law of conservation of the motion of the center of gravity, even for systems (at least in the first approximation) in which electromagnetic as well as mechanical processes occur. Although these simple formal considerations, which are necessary to prove the assertion, are already contained in their essence in a work by H. Poincaré,[2] I shall not depend on his work in the following overview.

1. A SPECIAL CASE

Let K be a rigid hollow cylinder at rest, floating freely in space. Let A be something for sending a specified amount of radiant energy S through the cavity to B. During the sending of this radiation there will be a radiation pressure on the left inner surface of the hollow cylinder K, which will impart a certain velocity to the left side of it. If the cylinder possesses mass M, then the velocity, as can be easily proved from the law of radiation pressure, equals $S/(VM)$, where V designates the velocity of light. K maintains this velocity until the radiation complex (whose spatial extent is small compared to the cavity in K) is absorbed in B. The duration of the motion of the hollow cylinder is (apart from terms of higher order) equal to α/V, where α is the distance between A and B. After the absorption of the radiation complex in B, the body K will again be at rest. During the radiation process under consideration K will be moved to the left by the amount

$$\delta = \frac{1}{V}\frac{S}{M}\frac{\alpha}{V}.$$

In the cavity of K let there be imagined a body k (which for simplicity we consider massless) as well as a mechanism (also without mass) which allows the body k found initially at B to be moved back and forth between B and A. After the radiation of amount S has been received in B, the energy is transferred to k and then k is moved to A. Finally the energy S is again received by the cylinder K at A and k is again moved back to B. The whole system has now undergone a circular process which one may imagine to be repeated arbitrarily often.

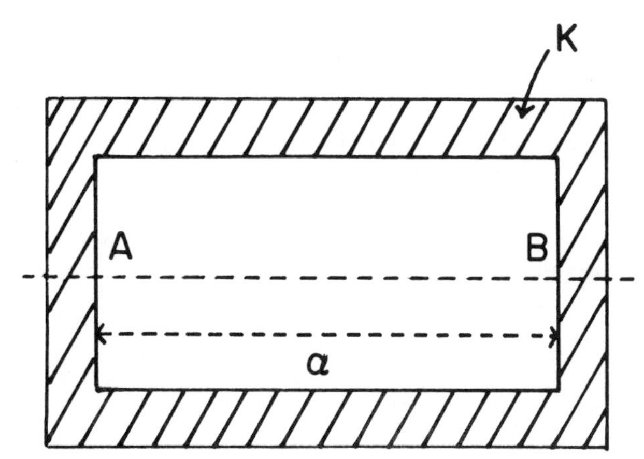

Fig. 44.1 A hollow cylinder K with an internal source which sends radiant energy from A to B along a distance α.

If we assume that the transporting body k is still massless when it has taken on the energy S, then we must assume that the return transport of the energy S is not connected with the positional change of the cylinder K. Therefore the result of the entire circular process described is the movement of the whole system to the left by the amount δ, a displacement that can be made arbitrarily large by repetition of the circular process. We thus achieve the result that a system originally at rest can change the position of its center of gravity arbitrarily without the action of any outside force and indeed without the system suffering any permanent changes.

It is clear that this result has no internal contradictions; however, it thoroughly contradicts the basic laws of mechanics, according to which a body originally at rest and on which no other bodies act cannot set out upon translational motion.

However, if we assume that to any energy E there corresponds an inertia E/V^2, the contradiction of the elements of mechanics vanishes. Thus, according to this assumption, the transporting body possesses a mass S/V^2 during the transport of the energy of amount S from B to A; and since the center of gravity *of the whole system* must remain at rest during this process according to the theorem of the center of gravity, the hollow cylinder K undergoes a displacement during this process to the right by the amount

$$\delta' = \alpha\frac{S}{V^2}\frac{1}{M}.$$

A comparison of the result found above shows that (at least in the first approximation) $\delta = \delta'$, so that the position of the system before and after the circular process is the same. Hence this removes the contradiction with the elements of mechanics.

2. Concerning the Theorem of the Conservation of the Motion of the Center of Gravity

Einstein now proceeds to derive more generally the energy content E/V^2 of a system of masses, using the Maxwell-Lorentz equations.

From the foregoing investigation it follows that one must either abandon the fundamental law of mechanics according to which a body originally at rest and not subject to external forces cannot undergo translational motion, or else that the inertia of a body depends on its energy content according to the law given here.

1. A. Einstein, *Ann. d. Phys. 18*, 639 (1905).
2. H. Poincaré, *Lorentz-Festschrift*, 1900, p. 252.

45. The Internal Constitution of the Stars

Arthur Stanley Eddington

(*Nature 106*, 14–20 [1920])

Between 1916 and 1925 A. S. Eddington published over a dozen major papers on the physical nature of stars, which were ultimately collected and extended in his book *The Internal Constitution of the Stars* (1926). During this decade his theoretical studies provided convincing evidence that stars are gaseous spheres with central energy sources and that the energy is transported to the stellar surfaces by radiation.

The problem of the source of stellar energy was brought into sharp relief by Eddington, whose calculations showed that gravitational contraction (as proposed by Hermann von Helmholtz and Lord Kelvin [William Thomson]) could suffice for only 100,000 years for the giant stars. In a highly suggestive paper published in 1919, Henry Norris Russell[1] had outlined the requirements for stellar energy generation, without, however, mentioning atomic energy in particular. Eddington promptly pointed out that nuclear processes were a likely candidate for the energy source,[2] and in this paper, a presidential address to section A of the British Association for the Advancement of Science, he even more specifically describes the possibility that hydrogen could be transformed to helium, with the resultant mass difference released as energy to power the star. This paper contains two prescient quotations: "What is possible in the Cavendish Laboratory may not be too difficult in the sun"; and "If, indeed, the sub-atomic energy in the stars is being freely used to maintain their great furnaces, it seems to bring a little nearer to fulfillment our dream of controlling this latent power for the well-being of the human race—or for its suicide."

During this decade James Jeans regularly rose in the meetings of the Royal Astronomical Society to contradict Eddington's theories; Jeans continued to support gravitational contraction as the source of stellar energy, and he denied the role of nuclear processes. Later, however, he not only changed his mind but laid claim to having suggested the idea in 1904,[3] shortly before Einstein published his famous mass-energy equivalence, $E = mc^2$. Both Jeans and Eddington had been anticipated by Thomas Chrowder Chamberlin,[4] however, who in 1899, while arguing against Kelvin's estimates for the age of the sun, had suggested that the extraordinary conditions within the centers of stars might set free some of the energy locked up in atoms themselves.

1. H. N. Russell, *Publications of the Astronomical Society of the Pacific 31*, 205 (1919).
2. A. S. Eddington, *Observatory 42*, 371 (1919).
3. J. H. Jeans, *Nature 70*, 101 (1904).
4. T. C. Chamberlin, *Science 9*, 889, *10*, 11 (1899), and *Annual Report of the Smithsonian Institution for 1899* (1900), p. 223.

LAST YEAR at Bournemouth we listened to a proposal from the President of the Association to bore a hole in the crust of the earth and discover the conditions deep down below the surface. This proposal may remind us that the most secret places of Nature are, perhaps, not 10 to the *n*th miles above our heads, but 10 miles below our feet. In the last five years the outward march of astronomical discovery has been rapid, and the most remote worlds are now scarcely safe from its inquisition. By the work of H. Shapley the globular clusters, which are found to be at distances scarcely dreamt of hitherto, have been explored, and our knowledge of them is in some respects more complete than that of the local aggregation of stars which includes the sun. Distance lends not enchantment, but precision, to the view. Moreover, theoretical researches of Einstein and Weyl make it probable that the space which remains beyond is not illimitable; not merely the material universe, but also space itself, is perhaps finite; and the explorer must one day stay his conquering march for lack of fresh realms to invade. But to-day let us turn our thoughts inwards to that other region of mystery—a region cut off by more substantial barriers, for, contrary to many anticipations, even the discovery of the fourth dimension has not enabled us to get at the inside of a body. Science has material and non-material appliances to bore into the interior, and I have chosen to devote this address to what may be described as analytical boring devices—*absit omen*!

The analytical appliance is delicate at present, and, I fear, would make little headway against the solid crust of the earth. Instead of letting it blunt itself against the rocks, let us look round for something easier to penetrate. The sun? Well, perhaps. Many have struggled to penetrate the mystery of the interior of the sun; but the difficulties are great, for its substance is denser than water. It may not be quite so bad as Biron makes out in "Love's Labour's Lost":

> The heaven's glorious sun
> That will not be deep-search'd with saucy looks:
> Small have continual plodders ever won
> Save base authority from others' books.

But it is far better if we can deal with matter in that state known as a perfect gas, which charms away difficulties as by magic. Where shall it be found?

A few years ago we should have been puzzled to say where, except perhaps in certain nebulæ; but now it is known that abundant material of this kind awaits investigation. Stars in a truly gaseous state exist in great numbers, although at first sight they are scarcely to be discriminated from dense stars like our sun. Not only so, but the gaseous stars are the most powerful light-givers, so that they force themselves on our attention. Many of the familiar stars are of this kind—Aldebaran, Canopus, Arcturus, Antares; and it would be safe to say that three-quarters of the naked-eye stars are in this diffuse state. This remarkable condition has been made known through the researches of H. N. Russell (*Nature* 93, 227, 252, 281) and E. Hertzsprung; the way in which their conclusions, which ran counter to the prevailing thought of the time, have been substantiated on all sides by overwhelming evidence is the outstanding feature of recent progress in stellar astronomy.

The diffuse gaseous stars are called *giants*, and the dense stars *dwarfs*. During the life of a star there is presumably a gradual increase of density through contraction, so that these terms distinguish the earlier and later stages of stellar history. It appears that a star begins its effective life as a giant of comparatively low temperature—a red or M-type star. As this diffuse mass of gas contracts its temperature must rise, a conclusion long ago pointed out by Homer Lane. The rise continues until the star becomes too dense, and ceases to behave as a perfect gas. A maximum temperature is attained, depending on the mass, after which the star, which has now become a dwarf, cools and further contracts. Thus each temperature-level is passed through twice, once in an ascending and once in a descending stage—once as a giant, once as a dwarf. Temperature plays so predominant a part in the usual spectral classification that the ascending and descending stars were not originally discriminated, and the customary classification led to some perplexities. The separation of the two series was discovered through their great difference in luminosity, particularly striking in the case of the red and yellow stars, where the two stages fall widely apart in the star's history. The bloated giant has a far larger surface than the compact dwarf, and gives correspondingly greater light. The distinction was also revealed by direct determinations of stellar densities, which are possible in the case of eclipsing variables like Algol. Finally, Adams and Kohlschütter have set the seal on this discussion by showing that there are actual spectral differences between the ascending and descending stars at the same temperature-level, which are conspicuous enough when they are looked for.

Perhaps we should not too hastily assume that the direction of evolution is necessarily in the order of increasing density, in view of our ignorance of the origin of a star's heat, to which I must allude later. But, at any rate, it is a great advance to have disentangled what is the true order of continuous increase of density, which was hidden by superficial resemblances.

The giant stars, representing the first half of a star's life, are taken as material for our first boring experiment. Probably, measured in time, this stage corresponds to much less than half the life, for here it is the ascent which is easy and the way down is long and slow. Let us try to picture the conditions inside a giant star. We need not dwell on the vast dimensions—a mass like that of the sun, but swollen to much greater volume on account of the low density, often below that of our own atmosphere. It is the star as a storehouse of heat which especially engages our attention. In the hot bodies familiar to us the heat consists in the energy of motion of the ultimate particles, flying at great speeds hither and thither. So, too, in the stars a great store of heat exists in this form; but a new feature arises. A large proportion, sometimes more than half the total heat, consists of imprisoned radiant energy—æther-waves travelling in all directions trying to break through the material which encages them. The star is like a sieve, which can retain them only temporarily; they are turned aside, scattered, absorbed for a moment, and flung out again in a new direction. An element of energy may thread the maze for hundreds of years before it attains the freedom of outer space. Nevertheless, the sieve leaks, and a steady stream permeates outwards, supplying the light and heat which the star radiates all round.

That some æthereal heat as well as material heat exists in any hot body would naturally be admitted; but the point on which we have here to lay stress is that in the stars, particularly in the giant stars, the æthereal portion rises to an importance which quite transcends our ordinary experience, so that we are confronted with a new type of problem. In a red-hot mass of iron the æthereal energy constitutes less than a billionth part of the whole; but in the tussle between matter and æther the æther gains a larger and larger proportion of the energy as the temperature rises. This change in proportion is rapid, the æthereal energy increasing rigorously as the fourth power of the temperature, and the material energy roughly as the first power. But even at the temperature of some millions of degrees attained inside the stars there would still remain a great disproportion; and it is the low density of material, and accordingly the reduced material energy per unit volume in the giant stars, which wipes out the last few powers of 10. In all the giant stars known to us, widely as they differ from one another, the conditions are just reached at which these two varieties of heat-energy have attained a rough equality; at any rate, one cannot be neglected compared with the other. Theoretically there could be conditions in which the disproportion was reversed and the æthereal far outweighed the material energy; but we do not find them in the stars. It is as though the stars had been measured out—that their sizes had been determined—with a view to this balance of power; and one cannot refrain from attributing to this condition a deep significance in the evolution of the cosmos into separate stars.

To recapitulate. We are acquainted with heat in two forms—the energy of motion of material atoms and the energy of æther waves. In familiar hot bodies the second form exists only in insignificant quantities. In the giant stars the two forms are present in more or less equal proportions. That is the new feature of the problem.

On account of this new aspect of the problem the first attempts to penetrate the interior of a star are now seen to need correction. In saying this we do not depreciate the great importance of the early researches of Lane, Ritter, Emden, and others, which not only pointed the way for us to follow, but also achieved conclusions of permanent value. One of the first questions they had to consider was by what means the heat radiated into space was brought up to the surface from the low level where it was stored. They imagined a bodily transfer of the hot material to the surface by currents of convection, as in our own atmosphere. But actually the problem is, not how the heat can be brought to the surface, but how the heat in the interior can be held back sufficiently—how it can be barred in and the leakage reduced to the comparatively small radiation emitted by the stars. Smaller bodies have to manufacture the radiant heat which they emit, living from hand to mouth; the giant stars merely leak radiant heat from their store. I have put that much too crudely; but perhaps it suggests the general idea.

The recognition of æthereal energy necessitates a twofold modification in the calculations. In the first place, it abolishes the supposed convection currents; and the type of equilibrium is that known as radiative instead of convective. This change was first suggested by R. A. Sampson so long ago as 1894. The detailed theory of radiative equilibrium is particularly associated with K. Schwarzschild, who applied it to the sun's atmosphere. It is perhaps still uncertain whether it holds strictly for the atmospheric layers, but the arguments for its validity in the interior of a star are far more cogent. Secondly, the outflowing stream of æthereal energy is powerful enough to exert a *direct mechanical effect* on the equilibrium of a star. It is as though a strong wind were rushing outwards. In fact, we may fairly say that the stream of radiant energy *is* a wind; for though æther waves are not usually classed as material, they have the chief mechanical properties of matter, viz. mass and momentum. This wind distends the star and relieves the pressure on the inner parts. The pressure on the gas in the interior is not the full weight of the superincumbent columns, because that weight is partially borne by the force of the escaping æther waves beating their way out. This force of radiation-pressure, as it is called, makes an important difference in the formulation of the conditions for equilibrium of a star.

Having revised the theoretical investigations in accordance with these considerations (*Astrophysical Journal 48*, 205), we are in a position to deduce some definite numerical results. On the observational side we have fairly satisfactory knowledge of the masses and densities of the stars and of the total radiation emitted by them; this knowledge is partly individual and partly statistical. The theoretical analysis connects these observational data on the one hand with the physical properties of the material inside the star on the other. We can thus find certain information as to the inner material, as though we had actually bored a hole. So far as can be judged, there are only two physical properties of the material which can concern us—always provided that it is sufficiently rarefied to behave as a perfect gas—viz. the average molecular weight and the transparency or permeability to radiant energy. In connecting these two unknowns with the quantities given directly by astronomical observation we depend entirely on the well-tried principles of conservation of momentum and the second law of thermodynamics. If any element of speculation remains in this method of investigation, I think it is no more than is inseparable from every kind of theoretical advance.

We have, then, on one side the mass, density, and output of heat, quantities as to which we have observational knowledge; on the other side, molecular weight and transparency, quantities which we want to discover.

To find the transparency of stellar material to the radiation traversing it is of particular interest, because it links on this astronomical inquiry to physical investigations now being carried on in the laboratory, and to some extent it extends those investigations to conditions unattainable on the earth. At high temperatures the æther waves are mainly of very short wave-length, and in the stars we are dealing mainly with radiation of wave-length 3 to 30 Ångström units, which might be described as very soft X-rays. It is interesting, therefore, to compare the results with the absorption of the harder X-rays dealt with by physicists. To obtain an exact measure of this absorption in the stars we have to assume a value of the molecular weight; but fortunately the extreme range possible for the molecular weight gives fairly narrow limits for the absorption. The average weight of the ultimate independent particles in a star is probably rather low, because in the conditions prevailing there the atoms would be strongly ionised; that is to say, many of the outer electrons of the system of the atom would be broken off; and as each of these free electrons counts as an independent molecule for present purposes, this brings down the average weight. In the extreme case (probably not reached in a star) when the whole of the electrons outside the nucleus are detached the average weight comes down to about 2, *whatever the material*, because the number of electrons is about half the atomic weight for all the elements (except hydrogen). We may, then, safely take 2 as the extreme lower limit. For an upper limit we might perhaps take 200; but to avoid controversy we shall be generous and merely assume that the molecular weight is not greater than—infinity. Here is the result:

For molecular weight 2, mass-coefficient of absorption = 10 C.G.S. units.
For molecular weight ∞, mass-coefficient of absorption = 130 C.G.S. units.

The true value, then, must be between 10 and 130. Partly from thermodynamical considerations, and partly from further comparisons of astronomical observation with theory, the most likely value seems to be about 35 C.G.S. units, corresponding to molecular weight 3.5.

Now this is of the same order of magnitude as the absorption of X-rays measured in the laboratory. I think the result is in itself of some interest, that in such widely different investigations we should approach the same kind of value of the opacity of matter to radiation. The penetrating power of the radiation in the star is much like that of X-rays; more than half is absorbed in a path of 20 cm. at atmospheric density. Incidentally, this very high opacity explains why a star is so nearly heat-tight, and can store vast supplies of heat with comparatively little leakage.

So far this agrees with what might have been anticipated; but there is another conclusion which physicists would probably not have foreseen. The giant series comprises stars differing widely in their densities and temperatures, those at one end of the series being on the average about ten times hotter throughout than those at the other end. By the present investigation we can compare directly the opacity of the hottest stars with that of the coolest. The rather surprising result emerges that the opacity is the same for all; at any rate, there is no difference large enough for us to detect. There seems no room for doubt that at these high temperatures the absorption-coefficient is approaching a limiting value, so that over a wide range it remains practically constant. With regard to this constancy, it is to be noted that the temperature is concerned twice over: it determines the character and wave-length of the radiation to be absorbed, as well as the physical condition of the material which is absorbing. From the experimental knowledge of X-rays we should have expected the absorption to vary very rapidly with the wave-length, and therefore with the temperature. It is surprising, therefore, to find a nearly constant value.

The result becomes a little less mysterious when we consider more closely the nature of absorption. Absorption is not a continuous process, and after an atom has absorbed its quantum it is put out of action for a time until it can recover its original state. We know very little of what determines the rate of recovery of the atom, but it seems clear that there is a limit to the amount of absorption that can be performed by an atom in a given time. When that limit is reached no increase in the intensity of the incident radiation will lead to any more absorption. There is, in fact, a saturation effect. In the laboratory experiments the radiation used is extremely weak; the atom is practically never caught unprepared, and the absorption is proportional to the incident radiation. But in the stars the radiation is very intense and the saturation effect comes in.

Even granting that the problem of absorption in the stars involves this saturation effect, which does not affect laboratory experiments, it is not very easy to understand theoretically how the various conditions combine to give a constant absorption-coefficient independent of temperature and wave-length. But the astronomical results seem conclusive. Perhaps the most hopeful suggestion is one made to me a few years ago by C. G. Barkla. He suggested that the capacity of the stars may depend mainly on *scattering* rather than on true atomic absorption. In that case the constancy has a simple explanation, for it is known that the coefficient of scattering (unlike true absorption) approaches a definite constant value for radiation of short wave-length. The value, moreover, is independent of the material. Further, scattering is a continuous process, and there is no likelihood of any saturation effect; thus for very intense streams of radiation its value is maintained, whilst the true absorption may sink to comparative insignificance. The difficulty in this suggestion is a numerical discrepancy between the known theoretical scattering and the values already given as deduced from the stars. The theoretical coefficient is only 0.2 compared with the observed value 10 to 130. Barkla further pointed out that the waves here concerned are not short enough to give the ideal coefficient; they would be scattered more powerfully, because under their influence the electrons in any atom would all vibrate in the same phase instead of in haphazard phases. This might help to bridge the gap, but not sufficiently. It must be remembered that many of the electrons have broken loose from the atom and do not contribute to the increase.[1] Making all allowances for uncertainties in the data, it seems clear that the astronomical opacity is definitely higher than the theoretical scattering. Very recently, however, a new possibility has opened up which may possibly effect a reconciliation. Later in the address I shall refer to it again.

Astronomers must watch with deep interest the investigations of these short waves, which are being pursued in the laboratory, as well as the study of the conditions of ionisation by both experimental and theoretical physics, and I am glad of the opportunity of bringing before those who deal with these problems the astronomical bearing of their work.

I can allude only very briefly to the purely astronomical results which follow from this investigation (*Monthly Notices* *77*, 16, 596, *79*, 2); it is here that the best opportunity occurs for checking the theory by comparison with observation, and for finding out in what respects it may be deficient. Unfortunately, the observational data are generally not very precise, and the test is not so stringent as we could wish. It turns out that (the opacity being constant) the total radiation of a giant star should be a function of its mass only, independent of its temperature or state of diffuseness. The total radiation (which is measured roughly by the luminosity) of any one star thus remains constant during the whole giant stage of its history. This agrees with the fundamental feature, pointed out by Russell in introducing the giant and dwarf hypothesis, that giant stars of every spectral type have nearly the same luminosity. From the range of luminosity of these stars it is now possible to find their range of mass. The masses are remarkably alike—a fact already suggested by work on double stars. Limits of mass in the ratio 3:1 would cover the great majority of the giant stars. Somewhat tentatively we are able to extend the investigation to dwarf stars, taking account of the deviations of dense gas from the ideal laws and using our own sun to supply a determination of the unknown constant involved. We can calculate the maximum temperature reached by different masses; for example, a star must have at least 1/7 the mass of the sun in order to reach the lowest spectral type, M; and in order to reach the hottest type, B, it must be at least $2\frac{1}{2}$ times as massive as the sun. Happily for the theory, no star has yet been found with a mass less than 1/7th of the sun's; and it is a well-known fact, discovered from the study of spectroscopic binaries, that the masses of the B stars are large compared with those of other types. Again, it is possible to calculate the difference of brightness of the giant and dwarf stars of type M, *i.e.* at the beginning and end of their career; the result agrees closely with the observed difference. In the case of a class of variable stars in which the light changes seem to depend on a mechanical pulsation of the star, the knowledge we have obtained of the internal conditions enables us to predict the period of pulsation within narrow limits. For example, for δ Cephei, the best-known star of this kind, the theoretical period is between four and ten days, and the actual period is $5\frac{1}{3}$ days. Corresponding agreement is found in all the other cases tested.

Our observational knowledge of the things here discussed is chiefly of a rather vague kind, and we can scarcely claim more than a general agreement of theory and observation. What we have been able to do in the way of tests is to offer the theory a considerable number of opportunities to "make a fool of itself," and so far it has not fallen into our traps. When the theory tells us that a star having the mass of the sun will at one stage in its career reach a maximum effective temperature of 9,000° (the sun's effective temperature being 6,000°) we cannot do much in the way of checking it; but an erroneous theory might well have said that the maximum temperature was 20,000° (hotter than any known star), in which case we should have detected its error. If we cannot feel confident that the answers of the theory are true, it must be admitted that it has shown some discretion in lying without being found out.

It would not be surprising if individual stars occasionally depart considerably from the calculated results, because at present no serious attempt has been made to take into account rotation, which may modify the conditions when sufficiently rapid. That appears to be the next step needed for a more exact study of the question.

Probably the greatest need of stellar astronomy at the present day, in order to make sure that our theoretical deductions are starting on the right lines, is some means of measuring the apparent angular diameters of stars. At present we can calculate them approximately from theory, but there is no observational check. We believe we know with fair accuracy the apparent surface brightness corresponding to each spectral type; then all that is necessary is to divide the total apparent brightness by this surface brightness, and the result is the angular area subtended by the star. The unknown distance is not involved, because surface brightness is independent of distance. Thus the estimation of the angular diameter of any star seems to be a very simple matter. For instance, the star with the greatest apparent diameter is almost certainly Betelgeuse, diameter 0.051″. Next to it comes Antares, 0.043″. Other examples are Aldebaran 0.022″, Arcturus 0.020″, Pollux 0.013″. Sirius comes rather low down with diameter 0.007″. The following table may be of interest as showing the angular diameters in seconds of arc expected for stars of various types and visual magnitudes:

Vis. Mag. m	A ″	F ″	G ″	K ″	M ″
0.0	0.0034	0.0054	0.0098	0.0219	0.0859
2.0	0.0014	0.0022	0.0039	0.0087	0.0342
4.0	0.0005	0.0009	0.0016	0.0035	0.0136

However confidently we may believe in these values, it would be an immense advantage to have this first step in our deductions placed beyond doubt. If the direct measurement of these diameters could be made with any accuracy it would make a wonderfully rapid advance in our knowledge. The prospects of accomplishing some part of this task are now quite hopeful. We have learnt with great interest this year that work is being carried out by interferometer methods with the 100-in. reflector at Mount Wilson, and the results are most promising. At present the method has been applied only to measuring the separation of close double stars, but there seems to be no doubt that an angular diameter of 0.05″ is well within reach. Although the great mirror is used for convenience, the interferometer method does not in principle require great apertures, but rather two small apertures widely separated, as in a range-finder. Prof. Hale has stated, moreover, that successful results were obtained on nights of poor seeing. Perhaps it would be unsafe to assume that "poor seeing" at Mount Wilson means quite the same thing as it does for us, and I anticipate that atmospheric disturbance will ultimately set the limit to what can be accomplished. But even if we have to send special expeditions to the top of one of the highest mountains in the world, the attack on this far-reaching problem must not be allowed to languish.

I spoke earlier of the radiation-pressure exerted by the outflowing heat, which has an important effect on the equilibrium of a star. It is quite easy to calculate what proportion of the weight of the material is supported in this way; it depends on neither the density nor the opacity, but solely on the star's total mass and on the molecular weight. No astronomical data are needed; the calculation involves only fundamental physical constants found by laboratory researches. Here are the figures, first for average molecular weight 3.0:

For mass $\frac{1}{2} \times$ sun, fraction of weight supported by radiation-pressure = 0.044.
For mass $5 \times$ sun, fraction of weight supported by radiation-pressure = 0.457.

For molecular weight 5.0 the corresponding fractions are 0.182 and 0.645.

The molecular weight can scarcely go beyond this range,[2] and for the conclusions I am about to draw it does not much matter which limit we take. Probably 90 per cent. of the giant stars have masses between $\frac{1}{2}$ and 5 times the sun's, and we see that this is just the range in which radiation-pressure rises from unimportance to importance. It seems clear that a globe of gas of larger mass, in which radiation-pressure and gravitation are nearly balancing, would be likely to be unstable. The condition may not be strictly unstable in itself, but a small rotation or perturbation would make it so. It may therefore be conjectured that, if nebulous material began to concentrate into a mass much greater than five times the sun's, it would probably break up, and continue to redivide until more stable masses resulted. Above the upper limit the chances of survival are small; when the lower limit is approached the danger has practically disappeared, and there is little likelihood of any further breaking-up. Thus the final masses are left distributed almost entirely between the limits given. To put the matter slightly differently, we are able to predict from general principles that the material of the stellar universe will aggregate primarily into masses chiefly lying between 10^{33} and 10^{34} grams; and this is just the magnitude of the masses of the stars according to astronomical observation.[3]

This study of the radiation and internal conditions of a star brings forward very pressingly a problem often debated in this Section: What is the source of the heat which the sun and stars are continually squandering? The answer given is almost unanimous—that it is obtained from the gravitational energy converted as the star steadily contracts. But almost as unanimously this answer is ignored in its practical consequences. Lord Kelvin showed that this hypothesis, due to Helmholtz, necessarily dates the birth of the sun about 20,000,000 years ago; and he made strenuous efforts to induce geologists and biologists to accommodate their demands to this time-scale. I do not think they proved altogether tractable. But it is among his own colleagues, physicists and astronomers, that the most outrageous violations of this limit have prevailed. I need only refer to Sir George Darwin's theory of the earth-moon system, to the present Lord Rayleigh's determination of the age of terrestrial rocks from occluded helium, and to all modern discussions of the statistical equilibrium of the stellar system. No one seems to have any hesitation, if it suits him, in carrying back the history of the earth long before the supposed date of formation of the solar system; and, in some cases at least, this appears to be justified by experimental evidence which it is difficult to dispute. Lord Kelvin's date of the creation of the sun is treated with no more respect than Archbishop Ussher's.

The serious consequences of this contraction hypothesis are particularly prominent in the case of giant stars, for the giants are prodigal with their heat and radiate at least a hundred times as fast as the sun. The supply of energy which suffices to maintain the sun for 10,000,000 years would be squandered by a giant star in less than 100,000 years. The whole evolution in the giant stage would have to be very rapid. In 18,000 years at the most a typical star must pass from the initial M stage to type G. In 80,000 years it has reached type A, near the top of the scale, and is about to start on the downward path. Even these figures are probably very much over-estimated.[4] Most of the naked-eye stars are still in the giant stage. Dare we believe that they were all formed within the last 80,000 years? The telescope reveals to us objects remote not only in distance, but also in time. We can turn it on a globular cluster and behold what was passing 20,000, 50,000, even 200,000 years ago unfortunately not all in the same cluster, but in different clusters representing different epochs of the past. As Shapley has pointed out, the verdict appears to be "no change." This is perhaps not conclusive, because it does not follow that individual stars have suffered no change in the interval; but it is difficult to resist the impression that the evolution of the stellar universe proceeds at a slow, majestic pace, with respect to which these periods of time are insignificant.

There is another line of astronomical evidence which appears to show more definitely that the evolution of the stars proceeds far more slowly than the contraction hypothesis allows; and perhaps it may ultimately enable us to measure the true rate of progress. There are certain stars, known as Cepheid variables, which undergo a regular fluctuation of light of a characteristic kind, generally with a period of a few days. This light change is *not* due to eclipse. Moreover, the colour quality of the light changes between maximum and minimum, evidently pointing to a periodic change in the physical condition of the star. Although these objects were formerly thought to be double stars, it now seems clear that this was a misinterpretation of the spectroscopic evidence. There is, in fact, no room for the hypothetical companion star; the orbit is so small that we should have to place it inside the principal star. Everything points to the period of the light pulsation being something intrinsic in the star; and the hypothesis advocated by Shapley, that it represents a mechanical pulsation of the star, seems to be the most plausible. I have already mentioned that the observed period does, in fact, agree with the calculated period of mechanical pulsation, so that the pulsation explanation survives one fairly stringent test. But whatever the cause of the variability, whether pulsation or rotation, provided only that it is intrinsic in the star, and not forced from outside, the density must be the leading factor in determining the period. If the star is contracting so that its density changes appreciably, the period cannot remain constant. Now, on the contraction hypothesis the change of density must amount to at least 1 per cent. in forty years. (I give the figures for δ Cephei, the best-known variable of this class.) The corresponding change of period should be very easily detectable. For δ Cephei the period ought to decrease 40 seconds annually.

Now δ Cephei has been under careful observation since 1785, and it is known that the change of period, if any, must be very small. S. Chandler found a decrease of period of 1/20 second per annum, and in a recent investigation E. Hertzsprung has found a decrease of 1/10 second per annum. The evidence that there is any decrease at all rests almost entirely on the earliest observations made before 1800, so that it is not very certain; but in any case the evolution is proceeding at not more than 1/400 of the rate required by the contraction hypothesis. There must at this stage of the evolution of the star be some other source of energy which prolongs the life of the star 400-fold. The time-scale so enlarged would suffice for practically all reasonable demands.

I hope the dilemma is plain. Either we must admit that whilst the density changes 1 per cent. a certain period intrinsic in the star can change no more than 1/800 of 1 per cent., or we must give up the contraction hypothesis.

If the contraction theory were proposed to-day as a novel hypothesis I do not think it would stand the smallest chance of acceptance. From all sides—biology, geology, physics, astronomy—it would be objected that the suggested source of energy was hopelessly inadequate to provide the heat spent during the necessary time of evolution; and, so far as it is possible to interpret observational evidence confidently, the theory would be held to be negatived definitely. Only the inertia of tradition keeps the contraction hypothesis alive—or, rather, not alive, but an unburied corpse. But if we decide to inter the corpse, let us frankly recognise the position in which we are left. A star is drawing on some vast reservoir of energy by means unknown to us. This reservoir can scarcely be other than the sub-atomic energy which, it is known, exists abundantly in all matter; we sometimes dream that man will one day learn how to release it and use it for his service. The store is well-nigh inexhaustible, if only it could be tapped. There is sufficient energy in the sun to maintain its output of heat for 15 billion years.

Certain physical investigations in the past year, which I hope we may hear about at this meeting, make it probable to my mind that some portion of this sub-atomic energy is actually being set free in the stars. F. W. Aston's experiments seem to leave no room for doubt that all the elements are constituted out of hydrogen atoms bound together with negative electrons. The nucleus of the helium atom, for example, consists of four hydrogen atoms bound with two electrons. But Aston has further shown conclusively that the mass of the helium atom is less than the sum of the masses of the four hydrogen atoms which enter into it; and in this, at any rate, the chemists agree with him. There is a loss of mass in the synthesis amounting to about 1 part in 120, the atomic weight of hydrogen being 1.008 and that of helium just 4. I will not dwell on his beautiful proof of this, as you will no doubt, be able to hear it from himself. Now mass cannot be annihilated, and the deficit can only represent the mass of the electrical energy set free in the transmutation. We can therefore at once calculate the quantity of energy liberated when helium is made out of hydrogen. If 5 percent of a star's mass consists initially of hydrogen atoms, which are gradually being combined to form more complex elements, the total heat liberated will more than suffice for our demands, and we need look no further for the source of a star's energy.

But is it possible to admit that such a transmutation is occurring? It is difficult to assert, but perhaps more difficult to deny, that this is going on. Sir Ernest Rutherford has recently been breaking down the atoms of oxygen and nitrogen, driving out an isotope of helium from them; and what is possible in the Cavendish Laboratory may not be too difficult in the sun. I think that the suspicion has been generally entertained that the stars are the crucibles in which the lighter atoms which abound in the nebulæ are compounded into more complex elements. In the stars matter has its preliminary brewing to prepare the greater variety of elements which are needed for a world of life. The radio-active elements must have been formed at no very distant date; and their synthesis, unlike the generation of helium from hydrogen, is endothermic. If combinations requiring the addition of energy can occur in the stars, combinations which liberate energy ought not to be impossible.

We need not bind ourselves to the formation of helium from hydrogen as the sole reaction which supplies the energy, although it would seem that the further stages in building up the elements involve much less liberation, and sometimes even absorption, of energy. It is a question of accurate measurement of the deviations of atomic weights from integers, and up to the present hydrogen is the only element for which Dr. Aston has been able to detect the deviation. No doubt we shall learn more about the possibilities in due time. The position may be summarised in these terms: the atoms of all elements are built of hydrogen atoms bound together, and presumably have at one time been formed from hydrogen; the interior of a star seems as likely a place as any for the evolution to have occurred; whenever it did occur a great amount of energy must have been set free; in a star a vast quantity of energy is being set free which is hitherto unaccounted for. You may draw a conclusion if you like.

If, indeed, the sub-atomic energy in the stars is being freely used to maintain their great furnaces, it seems to bring a little nearer to fulfilment our dream of controlling this latent power for the well-being of the human race—or for its suicide.

So far as the immediate needs of astronomy are concerned, it is not of any great consequence whether in this suggestion we have actually laid a finger on the true source of the heat. It is sufficient if the discussion opens our eyes to the wider possibilities. We can get rid of the obsession that there is no other conceivable supply besides contraction, but we need not again cramp ourselves by adopting prematurely what is perhaps a still wilder guess. Rather we should admit that the source is not certainly known, and seek for any possible astronomical evidence which may help to define its necessary character. One piece of evidence of this kind may be worth mentioning. It seems clear that it must be the high temperature inside the stars which determines the liberation of energy, as H. N. Russell has pointed out (*Pubns. Ast. Soc. Pacific*, August, 1919). If so, the supply may come mainly from the hottest region at the centre. I have already stated that the general uniformity of the opacity of the stars is much more easily intelligible if it depends on scattering rather than on true absorption; but it did not seem possible to reconcile the deduced stellar opacity with the theoretical scattering coefficient. Within reasonable limits it makes no great difference in our calculations at what parts of the star the heat energy is supplied, and it was assumed that it comes more or less evenly from all parts, as would be the case on the contraction theory. The possibility was scarcely contemplated that the energy is supplied entirely in a restricted region round the centre. Now, the more concentrated the supply, the lower is the opacity requisite to account for the observed radiation. I have not made any detailed calculations, but it seems possible that for a sufficiently concentrated source the deduced and the theoretical coefficients could be made to agree, and there does not seem to be any other way of accomplishing this. Conversely, we might perhaps argue that the present discrepancy of the coefficients shows that the energy supply is not spread out in the way required by the contraction hypothesis, but belongs to some new source available only at the hottest, central part of the star.

I should not be surprised if it is whispered that this address has at times verged on being a little bit speculative; perhaps some outspoken friend may bluntly say that it has been highly speculative from beginning to end. I wonder what is the touchstone by which we may test the legitimate development of scientific theory and reject the idly speculative. We all know of theories which the scientific mind instinctively rejects as fruitless guesses; but it is difficult to specify their exact defect or to supply a rule which will show us when we ourselves do err. It is often supposed that to speculate and to make hypotheses are the same thing; but more often they are opposed. It is when we let our thoughts stray outside venerable, but sometimes insecure, hypotheses that we are said to speculate. Hypothesis limits speculation. Moreover, distrust of speculation often serves as a cover for loose thinking; wild ideas take anchorage in our minds and influence our outlook; whilst it is considered too speculative to subject them to the scientific scrutiny which would exorcise them.

If we are not content with the dull accumulation of experimental facts, if we make any deductions or generalisations, if we seek for any theory to guide us, some degree of speculation cannot be avoided. Some will prefer to take the interpretation which seems to be indicated most immediately and at once adopt that as an hypothesis; others will rather seek to explore and classify the widest possibilities which are not definitely inconsistent with the facts. Either choice has it dangers: the first may be too narrow a view and lead progress into a cul-de-sac; the second may be so broad that it is useless as a guide, and diverges indefinitely from experimental knowledge. When this last case happens, it must be concluded that the knowledge is not yet ripe for theoretical treatment and that speculation is premature. The time when speculative theory and observational research may profitably go hand in hand is when the possibilities, or at any rate the probabilities, can be narrowed down by experiment, and the theory can indicate the tests by which the remaining wrong paths may be blocked up one by one.

The mathematical physicist is in a position of peculiar difficulty. He may work out the behavior of an ideal model of material with specifically defined properties, obeying mathematically exact laws, and so far his work is unimpeachable. It is no more speculative than the binomial theorem. But when he claims a serious interest for his toy, when he suggests that his model is like something going on in Nature, he inevitably begins to speculate. Is the actual body really like the ideal model? May not other unknown conditions intervene? He cannot be sure, but he cannot suppress the comparison; for it is by looking continually to Nature that he is guided in his choice of a subject. A common fault, to which he must often plead guilty, is to use for the comparison data over which the more experienced observer shakes his head; they are too insecure to build extensively upon. Yet even in this, theory may help observation by showing the kind of data which it is especially important to improve.

I think that the more idle kinds of speculation will be avoided if the investigation is conducted from the right point of view. When the properties of an ideal model have been worked out by rigorous mathematics, all the underlying assumptions being clearly understood, then it becomes possible to say that such-and-such properties and laws lead precisely to such-and-such effects. If any other disregarded factors are present, they should now betray themselves when a comparison is made with Nature. There is no need for disappointment at the failure of the model to give perfect agreement with observation; it has served its purpose, for it has distinguished what are the features of the actual phenomena which require new conditions for their explanation. A general preliminary agreement with observation is necessary, otherwise the model is hopeless; not that it is necessarily wrong so far as it goes, but it has evidently put the less essential properties foremost. We have been pulling at the wrong end of the tangle, which has to be unravelled by a different approach. But after a general agreement with observation is established, and the tangle begins to loosen, we should always make ready for the next knot. I suppose that the applied mathematician whose theory has just passed one still more stringent test by observation ought not to feel satisfaction, but rather disappointment—"Foiled again! This time I *had* hoped to find a discordance which would throw light on the points where my model could be improved." Perhaps that is a counsel of perfection; I own that I have never felt very keenly a disappointment of this kind.

Our model of Nature should not be like a building—a handsome structure for the populace to admire, until in the course of time someone takes away a corner-stone and the edifice comes toppling down. It should be like an engine with movable parts. We need not fix the position of any one lever; that is to be adjusted from time to time as the latest observations indicate. The aim of the theorist is to know the train of wheels which the lever sets in motion—that binding of the parts which is the soul of the engine.

In ancient days two aviators procured to themselves wings. Dædalus flew safely through the middle air across the sea, and was duly honoured on his landing. Young Icarus soared upwards towards the sun until the wax which bound his wings melted, and his flight ended in fiasco. In weighing their achievements perhaps there is something to be said for Icarus. The classic authorities tell us that he was only "doing a stunt," but I prefer to think of him as the man who certainly brought to light a constructional defect in the flying-machines of his day. So, too, in science. Cautious Dædalus will apply his theories where he feels most confident they will safely go; but by his excess of caution their hidden weaknesses cannot be brought to light. Icarus will strain his theories to the breaking-point until the weak joints gape. For a spectacular stunt? Perhaps partly; he is often very human. But if he is not yet destined to reach the sun and solve for all time the riddle of its constitution, yet he may hope to learn from his journey some hints to build a better machine.

1. *E.g.* for iron non-ionised the theoretical scattering is 5.2, against an astronomical value 120. If 16 electrons (2 rings) are broken off, the theoretical coefficient is 0.9, against an astronomical value 35. For different assumptions as to ionisation the values chase one another, but cannot be brought within reasonable range.

2. As an illustration of these limits, iron has 26 outer electrons; if 10 break away the average molecular weight is 5; if 18 break away the molecular weight is 3. Eggert (*Phys. Zeits.*, *20*, 570 [1919]) has suggested by thermodynamical

reasoning that in most cases the two outer rings (16 electrons) would break away in the stars. The comparison of theory and observation for the dwarf stars also points to a molecular weight a little greater than 3.

3. By admitting plausible assumptions closer limits could be drawn. Taking the molecular weight as 3.5, and assuming that the most critical condition is when 1/3 of gravitation is counterbalanced (by analogy with the case of rotating spheroids, in which centrifugal force opposes gravitation and creates instability), we find that the critical mass is just twice that of the sun, and stellar masses may be expected to cluster closely round this value.

4. I have taken the ratio of specific heats at the extreme possible value, 5/3; that is to say, no allowance has been made for the energy needed for ionisation and internal vibrations of the atoms, which makes a further call on the scanty supply available.

46. The Mass-Luminosity Relation for Stars

On the Relation between the Masses and Luminosities of the Stars

Arthur Stanley Eddington

(*Monthly Notices of the Royal Astronomical Society 84*, 308–332 [1924])

The Relationship between the Masses and Luminosities of the Stars

Heinrich Vogt

TRANSLATED BY VINCENT ICKE

(*Astronomische Nachrichten 226*, 301–304 [1926])

It was Jacob Halm who in 1911 first argued that the masses of stars are correlated with spectral type and therefore with their luminosities.[1] By 1919 Ejnar Hertzsprung[2] had shown that the main-sequence stars exhibit an empirical relationship between their absolute visual magnitudes and their masses. Hertzsprung's mass-luminosity relation is remarkably similar to the presently accepted relation, but according to his results, the luminosity had to increase with the seventh power of the mass; subsequent analysis of better data shows that the luminosities of the majority of stars increase approximately with the fourth power of the mass.

Arthur Eddington was probably influenced by these observational results when he established a mass-luminosity relation from theoretical considerations, although he included his own careful analysis of the observational data in support of his theory. This result was a tour de force of his physical description of stars, achieved by treating stars as gaseous spheres at a time when his major rival, James Jeans, was arguing that stars were liquid. Ironically, although Eddington had previously assumed that his theory applied only to the giant stars, which have substantial radiation pressure, he unexpectedly found that his theory describes the mass-luminosity law of the dwarf stars as well. The lower luminosity of the dwarf stars is therefore not caused by departures from the perfect gas law, as Eddington had previously thought. Accordingly, all stars can be described as gaseous spheres in hydrostatic and radiative equilibrium, essentially because the high temperature in stars like the sun, whose mass densities are actually close to that of water, cause high ionizations that make the effective size of the atoms very small.

Because the radiation pressure is negligible in the main-sequence stars, their mass is supported by gas pressure, and their internal temperatures scale inversely with their radius and directly with their mass. From the assumptions that the internal heat is transported by radiation and that the stellar opacity is relatively insensitive to mass, it follows that the luminosity must increase roughly with the third power of the mass. A fully theoretical mass-luminosity relation can be derived only if the law of energy generation is known, and so Eddington used empirical data from Capella to calibrate his mass-luminosity relation.

A closely related theorem asserts that stars of the same composition (with therefore the same laws of opacity and energy generation) have radii, luminosities, effective temperatures, and mean densities that are determined solely by the stars' masses. Heinrich Vogt first derived this theorem in the second paper given here. Henry Norris Russell included it in his textbook,[3] apparently having derived it independently, and hence it is generally called the Russell-Vogt theorem. This theorem implies that a star is uniquely described by the equation of hydrostatic equilibrium, the perfect gas law, the equation of energy balance, and the equation of radiation energy transfer, and also that a star of a given mass, age, and initial composition occupies a unique position, related to the star's evolutionary history, on the Hertzsprung-Russell diagram.

1. J. Halm, *Monthly Notices of the Royal Astronomical Society 71*, 610 (1911).
2. E. Hertzsprung, *Astronomische Nachrichten 208*, 89 (1919).
3. H. N. Russell, R. S. Dugan, and J. W. Stewart, *Astronomy II: Astrophysics and Stellar Astronomy* (Boston: Ginn and Co., 1927), section 975, pp. 909–911.

Eddington, The Masses and Luminosities of the Stars

1.

A THEORY of the stellar absorption-coefficient should, if successful, lead to formulæ determining the absolute magnitude of any giant star of which the mass and effective temperature are known. I have hitherto laid most stress on whether the theory will predict the absolute magnitude of Capella. The present position of that problem was summarised in my last paper;[1] although there appears to have been some measure of success, the final conclusion is not yet certain.

In this paper we shall consider the differential instead of the absolute results of the theory. We are not yet certain what should be the form of the absolute factor occurring in the formula connecting total radiation and mass; but apart from this factor, the form of the law seems to be fixed within narrow limits. Instead of constructing the absolute factor from physical constants we shall be content to determine its value from the observational data for Capella; and then it ought to be possible to calculate the luminosity of any other giant star, the result depending differentially on Capella.

Using the constant determined from Capella, we shall find that the formulæ of the theory appear to predict correctly the absolute magnitudes of all other ordinary stars available for the test, *regardless of whether they are giants or dwarfs.*

The evidence for this statement is shown graphically in figure 46.1.

According to the giant and dwarf theory the absolute magnitude is a double-valued function of mass and effective temperature; thus a star of mass 1 and temperature 5,860° has two possible magnitudes: (1) that of the Sun at present; (2) that of the Sun when it passed through the same temperature on the upgrade with a much larger surface area than now. It is the latter magnitude that the theory attempts to predict; but the former magnitude is actually situated on the theoretical curve.

If the theory gives the right magnitudes of the wrong stars, it is presumably wrong; if so, the question of its absolute

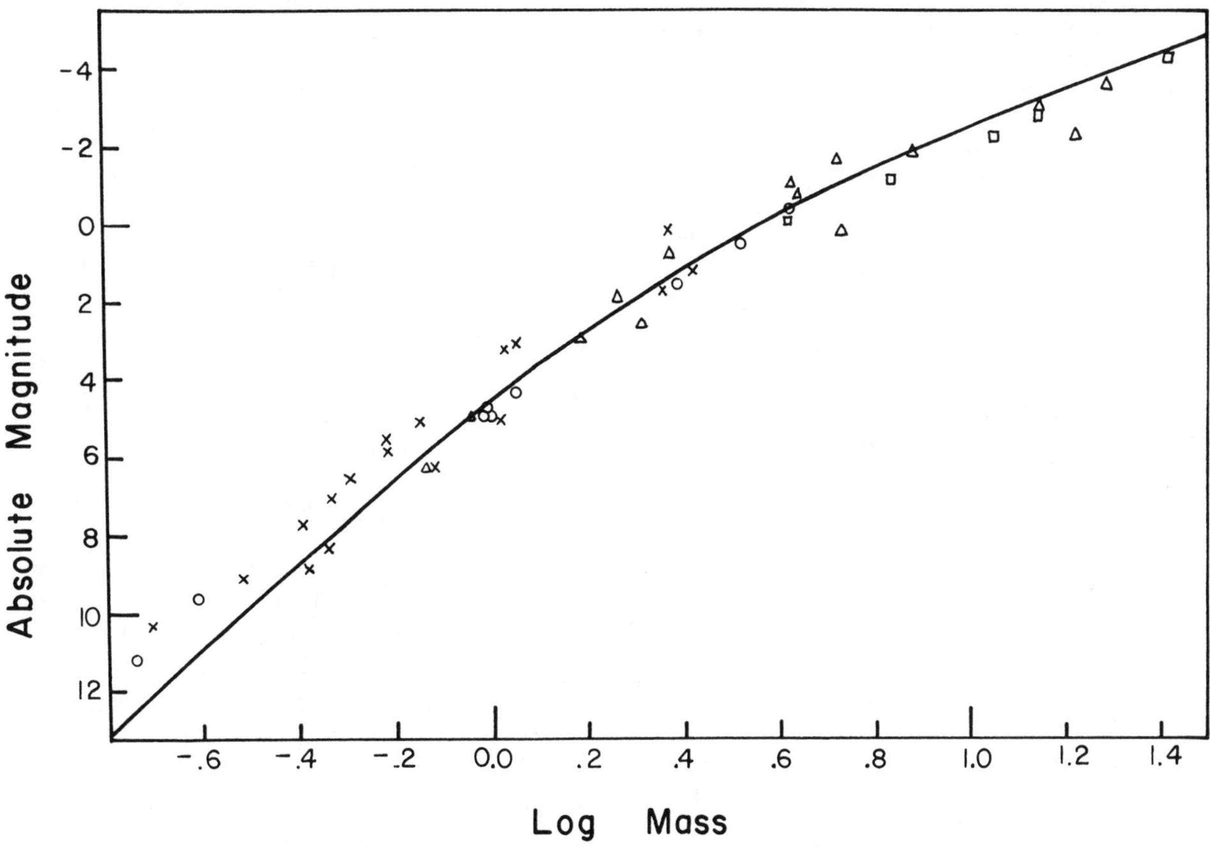

Fig. 46.1 The observed mass-luminosity relation for first-class stars (*circles*), second-class stars (*crosses*), Cepheid variable stars (*squares*), and eclipsing variable stars (*triangles*). The solid line denotes Eddington's theoretical mass-luminosity relation (equation [3] of this paper).

agreement for Capella becomes of minor importance. But it would be surprising if the accordance shown in figure 46.1 arose from mere accident, and we must face the question whether the stars there shown are really the "wrong" stars.

The suggestion is that even the dense stars like the sun are in the condition of a perfect gas, and will rise in temperature if they contract. In short, all ordinary stars are "giants" according to the usual implication of the term. In the course of this paper theoretical reasons will be given for believing that under stellar conditions matter should be able to contract to an enormously high density before deviations from the laws of a perfect gas become appreciable.

The present results come into conflict with the Lane-Ritter theory of stellar evolution as incorporated in the giant and dwarf theory at present almost universally accepted. Strong initial opposition to the results in this paper will doubtless be felt on that account; a discussion of the nature and extent of the conflict is given in §12.

Two recent investigations[2] have appeared containing diagrams which may be compared with figure 46.1. Both indicate that when absolute magnitude is plotted against mass the results tend to fall into a continuous line—a conclusion difficult to reconcile with the giant and dwarf theory. The diagram of Russell, Adams, and Joy embodies a great quantity of statistical data and gives mainly independent support to figure 46.1. Hertzsprung's results are not independent, since his data have been used by me *en bloc*, along with other evidence; owing to various details of the treatment (use of bolometric instead of visual magnitude, reduction to uniform effective temperature, determination of the mass-ratio of components of double stars) there are differences between the diagrams, and it will be seen that the scatter of his points has become considerably reduced.

2.

The determination of the absolute magnitude of a gaseous star of given mass, according to the theory of radiative equilibrium, involves the calculation of an auxiliary quantity β from the equation

$$1 - \beta = 0.00309\ M^2\mu^4\beta^4 \qquad (1)$$

where M is the mass in terms of the sun's mass, and μ the average molecular weight. If account is taken of the slight increase of μ from the centre outwards, the correct value to adopt is that corresponding to a point where the temperature is 2/3 of the central temperature.[3]

The total radiation of the star is proportional to $M(1-\beta)/k$ where k is the coefficient of absorption. In my earlier papers[4] k was believed to be approximately independent of temperature and density, so that the total radiation was considered to be proportional to $M(1-\beta)$. Absolute magnitudes calculated on this basis have been used in several researches.

On introducing the modern conception that absorption is an essentially discontinuous process, it was found that k could be more directly calculated as an emission-coefficient. All theories agree that the main part of the variation of k with temperature and density is given by $k \propto \rho/\mu T^3$. This leads to the result that the total radiation is proportional to $M(1-\beta)^2/\beta$, giving a considerably more rapid change of luminosity with mass.[5]

Detailed theories of emission supplement this by an additional factor of variation of k with temperature; but the additional factor introduces only a small correction in practical applications, and the possible doubt as to its precise form is not a serious hindrance. For the theory of nuclear capture the factor is $T^{-1/2}$; for Kramers' theory of capture it is the same, together with a logarithmic term of which the variations are usually less important. For capture of electrons by radiation under acceleration according to the classical theory, the factor is $T^{-0.9}$; and for the emission at encounters of negative electrons there is no additional factor.[6] The last two are examples of theories proved to be untenable, but they illustrate the difficulty of disturbing seriously the law proposed. We shall accept the additional factor $T^{-1/2}$, so that

$$k \propto \rho/\mu T^{7/2}. \qquad (2)$$

According to E. A. Milne[7] this gives the right order of magnitude of the optical absorption in the outer layers of the star—an extrapolation far beyond the range here required.

Introducing the effective temperature T_e, this leads to the following formula for the total radiation L.[8]

$$L = \text{const.} \times M^{7/5}(1-\beta)^{3/2}\mu^{4/5}T_e^{4/5}. \qquad (3)$$

The constant will be determined from the observational data for Capella. I have shown in the previous paper that μ is likely to be nearly 2.2 for Capella, and there will not be much change in its value for any ordinary star.[9] In the following calculations the value $\mu = 2.11$ is used.

3 and 4. Determinations

Observational data for double stars with well-known parallaxes are used to deduce masses and absolute bolometric magnitudes for these stars (first-class data). The effective temperatures were used in reducing visual to bolometric magnitudes according to table III of the author's article in the *Monthly Notices* (*77*, 605). Similar data were obtained for double stars with tolerably accurate observations (second-class data) taken from an article by Ejnar Hertzsprung (*Bulletin of the Astronomical Institutes of the Netherlands No. 43*).

In each case the theoretical magnitudes are deduced from the solid line in figure 46.1 with an addition of $-2\log(T_e/5{,}200)$ to correct for the effective temperature, T_e, of the star. The opposite correction is applied to the observed data before inserting them as points on figure 46.1. The residdual differences (O − C) between the observed and theoretical absolute magnitudes are much smaller than those expected if the residuals measure the reduction in brightness of dense stars, in contrast to diffuse stars of the same mass and effective temperature. For this reason Eddington abandons his previous assertion that the dense dwarf stars represent departures from the perfect gas law. In the following sections, we omit tables giving the star names, spectral types, masses, bolometric magnitudes, pulsation periods (for the Cepheids), orbital elements (for the eclipsing binaries), and the residuals, O − C, for the stars included in figure 46.1.

5. THE HYADES

There are 6 double stars of known period in the Hyades[10] all of much the same brightness. Although the orbits are not very accurate individually, the mean will afford a good comparison with the theory. Their mean parallax is known with considerable accuracy from the geometry of the moving cluster, viz. 0″.027.

To avoid difficulties in averaging it appears best to invert our previous procedure, and determine theoretical masses corresponding to the observed absolute magnitudes. The predicted mean mass of a system is thus 1.82. Hertzsprung found for the mean hypothetical parallax (for mass 2) the value 0″.026, which corresponds to a mean mass 1.79. If we average the masses instead of the hypothetical parallaxes of the individual systems, the mean mass is 1.93; in either case the agreement with the prediction is very good.

This has been represented as a first-class determination in figure 46.1, at the point corresponding to the mean mass of the 12 components.

6. CEPHEID VARIABLES

The Cepheid variables can be used for testing the curve if the pulsation theory is accepted.[11] The method is as follows:

Assume for trial an arbitrary value of the mass with the absolute magnitude corresponding to it. From the effective temperature (inferred from the observed spectral type) and the absolute magnitude the radius can be deduced. From the mass and radius the mean density is obtained. The central density is 54.25 times greater.[12] The period P is then obtained by the equation

$$P\sqrt{\rho_c} = 0.29(\gamma a)^{-1/2} \tag{4}$$

where $(\gamma a)^{1/2}$ is taken from *M.N.* *79*, 15, Table V.

Thus to each point of the theoretical curve will correspond a theoretical period; and we must repeat the process until we find the point which gives the period actually observed. This will give us a *calculated* mass and absolute magnitude. The observed absolute magnitude to be compared with the theoretical prediction is taken from Shapley's discussion.[13] His determination, although somewhat indirect, is entirely independent of the physical theory.

7. ECLIPSING VARIABLES

When spectroscopic orbits have been determined for *both* components of an eclipsing variable, the information is sufficient to furnish another kind of test of the curve. Owing largely to J. S. Plaskett's spectroscopic researches[14] thirteen eclipsing variables can be used in this manner.

The spectroscopic orbits determine the quantities $M \sin^3 i$ and $a \sin i$, where a is the semi-axis of the relative orbit. The discussion of the light-curve determines i and r/a, where r is the radius of the star. Hence by the combination we find M and r. Inferring the effective temperature from the spectral type, we can compute the absolute bolometric magnitude, since the total radiation is proportional to $r^2 T_e^4$. The formula is

$$m = 4.9 - 5\log_{10}(r/r_s) - 10\log_{10}(T_e/T_s),$$

r_s and T_s being the radius and effective temperature of the sun.

This result can be compared with the value m (calc.) obtained from the mass. It may be noted that the visual magnitude does not intervene in this comparison.

The results refer to the brighter component, which has been identified with the more massive component except in β Lyrae. The values of $\cos i$ and r/a are taken directly from Shapley's investigation,[15] the "darkened" solution being preferred. When the star is given as ellipsoidal the geometric mean of the three radii is used.

8

The average of all our residuals is $\pm 0^{m}.56$, most of which might fairly be attributed to errors of the observational data. Certain refinements of the nature of a second approximation will be required before the definitive theoretical curve is obtained; for example, we have used a constant value of μ throughout. But it would seem that the first approximation is already very nearly correct, except possibly at the extreme left of the curve.

Granting that the gas-laws hold for all ordinary stars, whether dense or diffuse, are we to expect that each star will have the precise luminosity deducible from its mass and effective temperature? In other words, will the theory be accurate individually, or only statistically? It is difficult to see how residual differences could arise, except from abnormal composition or abnormal rotation. As regards composition, an unduly large proportion of hydrogen would make the star fainter; apart from this not much effect is likely to be produced. As regards rotation, E. A. Milne[16] has found that a rapid rotation makes the star slightly fainter; but the effect is very small until the speed is sufficient to deform the star greatly. I think that what is most to be feared is that peculiar radiating conditions may arise, such that the observed spectrum misleads us as to the true effective temperature; but if this happened it would be a failure of the test rather than of the theory. It may be noted that an unsuspected binary should betray itself by having a magnitude fainter than that predicted from its (combined) mass.

9. Theoretical Considerations

We must now consider whether it is physically likely that a dense star, such as the sun, can obey the laws of a perfect gas.

The failure of the ordinary gas-laws at high densities is due to the finite size of the molecules which behave approximately as rigid spheres with radii of the order 10^{-8} cm. Compression proceeds with increasing difficulty until these spheres are packed tightly; the density is then of the order characteristic of solids and liquids. The idea underlying the giant and dwarf theory is that the maximum density of ordinary matter (say 10–20 gm. per c.c.) is applicable to the stars, and that the deviations from the gas-laws first begin to have serious effect when the density comes within sight of this limit.[17]

But the atoms in a star are very much smaller than ordinary atoms. Several layers of electrons have been stripped away, and the gas-laws ought therefore to hold up to far greater densities. It appears that in the interior of a star the atoms of moderate atomic weight are stripped down to the K level, and have radii of the order 10^{-10} cm.; lighter elements, such as carbon and oxygen, are reduced to the bare nucleus. The maximum density, corresponding to contact of these reduced atomic spheres, must be at least 100,000, and any star with mean density below 1,000 ought to behave as a perfect gas.

It may be asked: Does the removal of outer electrons necessarily reduce the effective size of the atom? Perhaps it is only the boundary-stone, not the boundary, that disappears. The answer seems to be given clearly by physical experiment. An α particle is a helium atom which has lost its "boundary stones," and it appears that it thereby loses its former boundary. It can now enter other atoms, and behaves in every way as a simple charged nucleus with no trace of that resisting boundary which prevents neutral helium gas from being compressed beyond a certain density. It seems clear that the effective size of the atom is determined by the existing peripheral electrons—as we should expect theoretically.

A further question arises as to the effect of the charges of the ions and electrons. It seems almost paradoxical that we should be able to force atoms closer together by ionising them, and so making them repel one another. Will not the repulsion of the ion establish a region which other ions are unable to enter, so that the volume of this region constitutes an effective size of the ion? It is very difficult to calculate the effect of these electrical forces; they are not obviously insignificant, at any rate in the stars of small mass. But it is quite easy to see that the effect does not increase when the star contracts; it is just as large when the star is diffuse as when the star is condensed, so that there is no evolution from gaseous to non-gaseous (giant to dwarf) condition.

It has often been pointed out in atomic theory that if inverse-square forces alone are acting no definite scale of size can be obtained. Thus inverse-square electrical forces will not alter the result for inverse-square gravitational forces, viz. that there is no definite scale of size for a giant star of given mass—it is equally comfortable with any radius. Stars of the same mass and different radii form a perfectly homologous series, which can only be disturbed when other than inverse-square forces begin to play an appreciable part. According to current theory this happens when the compression is great enough to bring into importance the inter-atomic forces at impact, which do not follow the inverse-square law. The star then passes into a dwarf equilibrium not homologous with its previous progress. But we have just seen that this change will not occur until the star reaches a density of at least 1,000; and electrical forces between the charged atoms and electrons do not lead us to modify this conclusion, because, being inverse-square forces, they cannot produce a breach of homology. This can be seen in more detail as follows:

Take any star A in equilibrium, and let B be a precise replica of A with lengths altered in the ratio p and velocities in the ratio $p^{-1/2}$. Hence temperature is altered in the ratio p^{-1} and density in the ratio p^{-3}, giving the familiar law $\rho \propto T^3$. Kinetic energy and potential energy (whether electrical or gravitational) will be altered in the same ratio p^{-1}. Moreover, radiant energy and material energy will be changed in the same ratio, the density of the former being proportional to T^4 and the latter to ρT; these both change in the ratio p^{-4}. Electrical-pressures, gas-pressures, and radiation-pressures are thus all altered in the same ratio, and the replica B will accordingly be in equilibrium.[18]

To give a simple example, at any moment in a certain star let there be n pairs of positive ions which have approached to a distance r and have a mutual potential energy E. Now let the star contract to 8 times its former density; according to the theory of equilibrium its temperature must increase twofold. There will now be n pairs at a distance apart $(1/2)r$, and the mutual potential energy will thus be 2E. But by the rise of temperature the kinetic energy is also increased two-fold, so that the same proportion of potential to kinetic energy is maintained.

Since the least massive stars have the greatest density for given temperature, the effect of the electrical forces, if appreciable at all, should be shown by them. We shall consider Krueger 60 which has a temperature $2.81 \cdot 10^7$ at a density 360. If it is made of iron, there will be $3.8 \cdot 10^{24}$ atoms per c.c., giving an average separation of $0.64 \cdot 10^{-8}$ cm. The charge of an ion (retaining 3 electrons) will be $23e$, and two such ions will have a mutual potential energy equal to the average kinetic energy of four free molecules when their separation is $0.61 \cdot 10^{-8}$ cm.—practically the average separation. At average separation this potential energy merely cancels the negative potential energy of the ions due to the corresponding free electrons which are at large in the neighbourhood; but if the two ions approach to $0.4 \cdot 10^{-8}$ cm., their mutual energy is increased 50 per cent., whereas there can be no appreciable change in the cancelling term—the average distribution of 46 free electrons giving a practically steady field. Hence the approach can only be made at the expense of their own kinetic energy, which will be just used up if they originally possessed the average velocity. It appears then that two ions cannot on the average approach nearer than 2/3 their normal distance; effectively, an ion is barred out from 1/3 of the whole volume.

It does not follow that exclusion by this method has the same effect as exclusion by finite size of the molecules at ordinary temperature;[19] but we can perhaps estimate the possible order of magnitude of the effect by assuming the analogy to hold. In an ordinary gas if 1/3 of the whole volume is barred to a molecule, the pressure is increased in the ratio 6/5, or 20 per cent. It must be remembered, however, that in the present case only 1/24 of the whole pressure is due to the ions, the remainder being due to the free electrons which are not barred out of these regions.[20] Thus the increase of pressure is only 1 per cent, and is therefore inappreciable even in the least massive stars.

We scarcely pretend to know all about the inter-atomic forces at present, and it would not surprise me much if some unknown circumstance were to vitiate the argument of this section. But what I wish to bring out is, that in applying the Lane-Ritter theory to stellar evolution we have been influenced by a false analogy between the mutilated stellar atoms and ordinary atoms; we can at least see now that this analogy is unfounded and approach the problem again free from this bias. There is no theoretical reason to expect a change in stellar conditions as the star reaches a density 0.1 – 1; and the indications here found, that the condition of a perfect gas persists up to higher densities, are not to be dismissed as incredible. It would have been more puzzling if we had found that it did not persist.

10. White Dwarfs

The white dwarfs Sirius (*comes*) and o_2 Eridani (bright component) should not be used for our comparisons. They have long presented a difficult problem.

The spectrum of Sirius *comes* is now given as F0. If this is to be taken literally as indicating an effective temperature of 8,000°, it gives in conjunction with the absolute magnitude 11.3 a radius of 19,600 km.—much less than Uranus. Determinations of the mass range from 0.75 to 0.95; adopting 0.85, the density is 53,000.

According to the views here reached, such a density is not absurd, and we should accept it without demur if the evidence were sufficient. The alternative is that at a very low effective temperature, probably below M*d*, the star is able in some way to produce an imitation of leading features of the F spectrum sufficiently close to satisfy the expert observer. It seems unnecessary to debate these alternatives at length, because, as several writers have pointed out, the question could probably be settled by measuring the Einstein shift of the spectrum, which should amount to about 20 km. per sec. if the high density is correct.

When compared with the curve the companion of Sirius is much too faint for its mass; this would be anticipated on either hypothesis. If the density is really 53,000, entirely new considerations enter into the calculation of k, since the electrons are in the capture zone of two or more nuclei simultaneously; moreover, the deviation from the gas-laws may be important at this high density. If the density is moderate, the effective temperature must be in the domain below 3,000°, where the visual magnitude diverges steeply from the bolometric.

It is rather surprising to find o_2 Eridani, presumably a similar star, but of smaller mass, agreeing quite well with the curve.

The companion of Procyon is also very faint for its mass, but since the spectrum is unknown there is at present no reason to suspect a discrepancy. It may be in the region below M*d* where any degree of visual faintness is possible. I suppose it is only mythological association which suggests the idea that it may be another white dwarf.

11.

The absolute magnitudes have all been calculated by using a value of 2.11 for the molecular weight. Alterations of a few tenths in the molecular weight would have no serious effects on the results. Eddington gives a table showing the degrees of ionization of atoms in stars of different mass; small corrections in the mass absorption coefficient, k, and the absolute magnitudes are indicated.

The conclusion from this section is that our first approximation, $k \propto \rho/T^{7/2}\mu = \text{const.}$, although at first sight rather crude, is likely to give absolute magnitudes very close to the true values. Further, making the best estimate we can of the effects of a more refined approximation, the slight systematic deviation between theory and observation for the stars of very small mass tends to close up.

12. STELLAR EVOLUTION

We have been led to views which in some respects are opposed to the giant and dwarf theory of evolution; and since the latter theory is considered to be strongly supported on most points by observational evidence, it is necessary to make clear the nature and extent of the conflict.

In its observational aspects the giant and dwarf theory is summed up in figure 46.2. The leading phenomenon is that the observed distribution of absolute magnitude and spectral type is concentrated along the lines PQ, QR with comparatively small scattering. The evidence for the reality of this dual concentration is extremely strong, and we shall accept it without question. Further, the observational evidence shows that the stars near the line PQ are diffuse with densities comparable to air, and those along QR have densities comparable to water; near the junction the density is of order 0.1. This difference of density is theoretically necessary unless the stars differ enormously in mass.

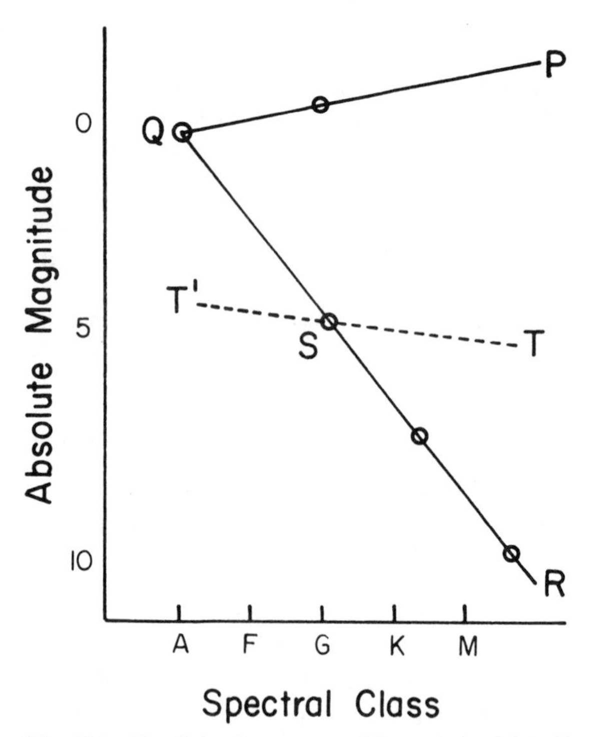

Fig. 46.2 The "giant" sequence, PQ, and the "dwarf" or "main" sequence, QR, of the Hertzsprung-Russell plot of absolute luminosity against spectral class or temperature. In this paper, Eddington argues that a main-sequence star must evolve along the line TT′ and that the giant and dwarf sequences represent a locus of equilibrium points for stars of different mass.

The discovery of this important statistical distribution appears to be a permanent contribution to our knowledge. We pass on to the evolutionary aspect of the giant and dwarf theory. It is supposed that the line PQR indicates the course of evolution of a single star. If, as usual, it is assumed that the mass of the star remains sensibly constant during its lifetime, this is in conflict with our results; for we have found that the stars at R are essentially less massive than those at P (the difference of magnitude having arisen from this mass-difference) so that no star can pass from P to R.

Considering a star at S, our results give a nearly horizontal line TST′ as its course of evolution. If the energy is derived from contraction, the direction must be from T to T′; since, however, the energy is due to some other source, the direction T′T is not definitely excluded but it seems unlikely. In any case S must be a point of quasi-equilibrium, since we seldom if ever find stars on this track except near S. Accordingly QR, instead of being a line of evolution, becomes, according to our view, a locus of equilibrium points—unless we are prepared to admit that the mass of the evolving star changes.

It is in this connection that strong opposition to the conclusions here reached is likely to be felt. The giant and dwarf theory gave an extremely plausible explanation of the distribution PQR, on the view that the turning at Q was due to the density having reached the point at which the star ceases to behave as a gas. We find it necessary to disbelieve this precise explanation, and as regards this particular feature of stellar distribution can offer only vague suggestions in its place. We make the following defence:

Firstly, the giant and dwarf explanation is essentially bound up with the view that the influence of mass on luminosity is comparatively small; for otherwise the statistics of stars of systematically different masses cannot outline the evolutionary course of a star of constant mass. It was recognised from the beginning—as a minor detail—that the stars along QR were statistically less massive than along PQ. I suppose that the explanation most generally held was that the lighter stars traversed their course more quickly—although all theoretical arguments tend to show that the heavier stars would reach R first. However that may be, there is much evidence apart from the present investigation to show that the systematic difference of mass is not a minor detail, and the correction to be applied before PQR can give the course of an individual star is very considerable. When the correction is applied, the "individual" curve PQR will make a much more acute angle than the "statistical" PQR; and the magnitude difference of the ascending and descending states is at any rate greatly exaggerated in the statistical diagram. We have attempted to determine theoretically the actual amount of the correction for mass-difference, and according to our result the corrected angle PQR is not merely more acute but closes up altogether leaving a single line like TT′.

Secondly, we have shown in §9 that the assumption of the giant and dwarf theory that a deviation from the gas-laws sets in at density 0.1 is based on a false analogy; if it occurs, it requires inter-atomic forces which have not yet appeared in experimental or theoretical physics.

Thirdly, although we can offer no serious attempt at an alternative explanation, it does not seem that explanation is impossible. The difficulty is merely that we have no knowledge of the conditions under which the star's unknown source of energy is liberated; and in these circumstances speculation is unlimited and chiefly unprofitable.

If the radiant energy of a star arises from the annihilation of matter, it may be possible to account for the relative numbers of stars of different brightness, the brighter stars having shorter lifetimes. If a star diminishes in mass during its evolution, the sequence would be along PQR in figure 46.2, the circles of which denote a progressive diminution in mass from Capella to Sirius to the sun to η Cassiopeiae *f* and on to Krueger 60.

Here Eddington takes account of the gradual increase in molecular weight from the center to the boundary of a star and shows that the mean temperature for which μ is to be calculated when used in equation (1) is between one-half and three-quarters of the central temperature.

13.

This section deals with a point of detail relating to the exactness of the fundamental equation (1). It is not concerned with the main argument of the paper.

We omit these details here.

14. Summary

1. Assuming on the evidence of previous investigations that the absorption-coefficient is proportional to $\rho/T^{7/2}$, it is possible to calculate the difference of absolute magnitude of any two gaseous (giant) stars of known mass and effective temperature. Hence, using the observed data for Capella, the absolute magnitudes of other stars can be determined differentially.

2. Collecting all suitable data 36 stars furnish comparisons between theory and observation. The average residual is $\pm 0^m.56$, and the maximum discordance is $1^m.7$. The probable errors of the observational data would account for a great part of this difference.

   The only stars omitted in the comparison are the two "white dwarfs." For these the internal conditions must (if the observations are not at fault) be so different from those of a normal star that the theoretical calculations are not expected to apply without modification.

3. More than half the stars used in the comparison are dwarf stars. The agreement of their absolute magnitudes with the predicted magnitudes for gaseous stars is in conflict with the current view that they are too dense to follow the laws of a perfect gas, and that their low luminosity is attributable to deviation from the gas-laws. According to the present results their low luminosity is fully accounted for by their comparatively small mass without appeal to any other physical difference.

4. The current expectation that between density 0.1 and 1 the compressibility of a star will fall off rapidly, as compared with the compressibility of a perfect gas, appears to rest on a false analogy between stellar ions and atoms at ordinary temperature. Owing to the high ionisation, stellar atoms have only about 1/100,000 of the bulk of ordinary atoms, and failure of the laws of a perfect gas is not to be expected till a density 100,000 times higher is reached.

   The effect of the high electric charges of the ionised atoms has been considered, but it appears that it would not appreciably affect the compressibility of any of the stars considered.

5. Notwithstanding a wide range of physical condition in the interior of the stars discussed, the ionisation level is not very different in any of them. The assumption that the same molecular weight can be used for all of them is thus closely justified. Attempting a second approximation by taking account of the small variations of molecular weight and of a slowly varying factor in the absorption-coefficient (predicted by Kramers' theory and probable on general grounds), the theoretical curve is scarcely changed for masses greater than 1/2 and is brought into rather better agreement with observation for the small stars.

6. The extent of the conflict between the present results and the current theory of stellar evolution depends on whether we admit that the mass of a star diminishes to an important extent or not by radiation of energy during its lifetime.

If the mass of the star remains sensibly constant, the statistical diagram of absolute magnitude and spectral type (the "compass-legged" diagram) cannot be in terpreted as indicating the course of evolution of a star. Instead, it indicates the locus of equilibrium points reached by stars of different initial mass.

If the star gradually burns itself away in liberating sub-atomic energy, the statistical diagram probably indicates its track of evolution as current theory supposes. In that case the divergence between the present theory and the giant and dwarf theory is narrowed down to the single point, that the diminishing brightness in the dwarf sequence is due to decreasing mass and not to a falling off of compressibility. The conception of an ascending and descending series (judged by effective temperature) is thus retained; although as judged by internal temperature there is probably a continuous ascent.

7. By way of appendix, a discussion is given of the fundamental quartic equation of the theory of radiative equilibrium in which account is taken of the gradual increase of molecular weight from the centre to the boundary of the star.

1. *M.N. 84*, 104.
2. E. Hertzsprung, *Bull. Ast. Inst. Netherlands*, no. 43. Russell, Adams, and Joy, *Pub. Ast. Soc. Pac. 35*, 189.
3. See §13.
4. *M.N. 77*, 16, 596
5. *Zeits. für Physik. 7*, 389.
6. *M.N. 83*, 40, 44, *84*, 118.
7. *Phil. Mag. 47*, 226.
8. *M.N. 83*, 107.
9. A possible refinement taking account of the slight variation of μ from star to star is considered in §11.
10. E. Hertzsprung, *B.A.N.*, no. 16.
11. *M.N. 79*, 2, 177.
12. According to §13, this factor is somewhat too large, but we must continue to use it here, since it corresponds to the theory underlying equation (4). The period will depend on the mean density rather than on the density at a particular point, so that in (4) ρ_c is to be regarded as a symbol for 54·25 ρ_m, whatever the actual central density may be.
13. *Ap. J. 48*, 282; *Mount Wilson Contrib.*, no. 153.
14. *Publications Dominion Astrophysical Obs.*, *Victoria*, *1*, *2*.
15. *Princeton Contributions*, *no. 3*, p. 82.
16. *M.N. 83*, 118.
17. Since the central density in a giant star is about 20 times greater than the mean density, we expect the deviations to be already appreciable at a mean density of 0.2.
18. The ionisation and the absorption-coefficient involve atomic constants, and are not governed by the same considerations.
19. I doubt whether the effect has even the same sign.
20. The volume from which the free electrons exclude one another is about $(23)^{-5}$ of the whole.

Vogt, The Masses and Luminosities of the Stars

The relationship between the masses and absolute luminosities of the stars, which has been derived by Eddington, is based on the assumption that for the entire stellar interior (that is, for every distance r from the center of the star) the product of the mass absorption coefficient, k, and Q is a constant, where

$$Q = \frac{4\pi}{M(r)} \int_0^r 4\pi\varepsilon\rho r^2\, dr$$

is the average value of the amount of energy generated per unit mass per unit time within a sphere of radius r; or, at least, on the assumption that kQ is sufficiently nearly a constant for the stars to be regarded as homologous systems for which the equation of state $P = \chi\rho^{4/3}$ is valid. Here $4\pi\varepsilon$ is the energy generated per unit mass per unit time, $M(r)$ is the mass within a sphere of radius r, the total pressure is P, the mass density is ρ, and χ is a constant that depends only on the mass of the star. Likewise, the relationship recently given by Jeans[1] is generally valid only when the stars are approximately homologous systems with the equation of state $P = \chi\rho^{4/3}$. Jeans starts with the assumption that Q is a constant throughout the stellar interior and that k is proportional to $\rho/(\mu T^{3+n})$. Here T is the temperature and μ is the mean molecular weight. But the approximation that he subsequently uses amounts to the assumption that, but for a constant factor, the quantity $aT^4/3$ can be taken to be the same function of the radial distance, r, in units of the stellar radius.[2] However, this can only be assumed if n equals zero or at most is very small, that is, if the stars are at least approximately the homologous systems presumed by Eddington.

An attempt will presently be made to consider the general case where the energy generated per unit mass per unit time and the mass absorption coefficient in the stellar interior change according to an arbitrary rule. The velocity of light is denoted by c, the Stefan radiation constant is denoted by a, the universal gas constant $R = 8.26 \times 10^7$, and the mean molecular weight is denoted by μ. The total gas pressure $P = R\rho T/\mu + aT^4/3$ is the sum of the gas pressure and the radiation pressure. The amount of energy flowing per unit time through an equilibrium surface at a distance r from the stellar center is given by

$$L(r) = \frac{-16\pi acr^2}{3k\rho T^3}\frac{dT}{dr}, \tag{1}$$

(assuming spherical symmetry for the star), and

$$\frac{dP}{dr} = \frac{-G\rho M(r)}{r^2} = -g\rho, \qquad (2)$$

where G is the gravitational constant, $M(r)$ is the mass within radius r, and g is the gravitational acceleration. From equations (1) and (2) it follows that

$$L(r) = \frac{4\pi acgr^2}{3k}\frac{dT^4}{dP}$$

$$= \frac{4\pi cGM(r)[1-\beta]}{k}\left[1 + \frac{P}{(1-\beta)}\cdot\frac{d(1-\beta)}{dP}\right], \qquad (3)$$

where the ratio between the radiation pressure, $aT^4/3$, and the total pressure, P, has been designated $1-\beta$. Substitution of the total stellar mass, M, for $M(r)$ and of the values corresponding to a sphere near the stellar surface for P, $1-\beta$, $d(1-\beta)/dP$, and k yields with sufficient accuracy the total luminosity, L, of the star. For although equation (1) and hence equation (3) are not valid right up to the stellar surface, the outermost layers for which equation (1) is no longer applicable contribute but little to the total energy generated because of their small mass (correspondingly, there are fewer energy sources concentrated toward the stellar center).

Using the relation among $1-\beta$, M, and μ that is valid for homologous stars, equation (3) can be cast in a somewhat different form. When unsubscripted quantities refer to a star of total mass M, radius R, and mean molecular weight μ, and quantities with subscript 1 refer to a homologous star with mass M_1, radius R_1, and mean molecular weight μ_1, then

$$\frac{T^4}{T_1{}^4} = \frac{(1-\beta)}{(1-\beta_1)}\frac{M^2}{R^4}$$

and

$$\frac{T}{T_1} = \frac{\beta}{\beta_1}\frac{M\mu}{R}.$$

By elimination of T from the latter two equations, we obtain

$$\frac{(1-\beta)}{\beta^4} = \frac{(1-\beta_1)}{\beta_1{}^4}M^2\mu^4 = \phi(\bar{r})M^2\mu^4, \qquad (4)$$

where $\phi(\bar{r})$ is a function of $\bar{r}$ only, and $\bar{r}$ is the distance from the star's center in units of the total radius. From equation (4) it follows that

$$\frac{1}{(1-\beta)}\frac{d(1-\beta)}{d\bar{r}} = \frac{\beta}{(4-3\beta)}\frac{1}{\phi(\bar{r})}\frac{d\phi(\bar{r})}{d\bar{r}},$$

and hence the total luminosity of a star is given by

$$L = \frac{4\pi cG}{k}M(1-\beta)\left[1 + \frac{A\beta}{(4-3\beta)}\right], \qquad (5)$$

where

$$\frac{1}{A} = \frac{1}{P}\frac{dP}{dr}\bigg/\frac{1}{\phi(\bar{r})}\frac{d\phi(\bar{r})}{d\bar{r}} = \frac{1}{P_1}\frac{dP_1}{dr}\bigg/\frac{1}{\phi(\bar{r})}\frac{d\phi(\bar{r})}{d\bar{r}}. \qquad (6)$$

If kQ is a constant for the entire stellar interior, as in Eddington's approximation (that is to say, when stars are homologous systems with a polytropic equation of state $P = \chi\rho^{4/3}$), then $\phi(\bar{r})$ is likewise a constant and A equals zero. [In this case L = constant $\times\, M^3$.]

If the stars are homologous systems of some different kind, the $\phi(\bar{r})$ is the same function of $\bar{r}$ for all stars and A is a constant, in the sense that A is positive when $1-\beta$ increases toward the star's center and negative when $1-\beta$ decreases. Because the parameter $1-\beta$, which enters in the expression for L, corresponds to homologous level surfaces or the same $\bar{r}$ in all stars (at least to a very high order of approximation), it obeys the relation $1-\beta$ = constant $\times\, M^2\mu^4\beta^4$, just as in Eddington's theory, with the distinction that the constant has a different value. The absorption coefficient k, which enters in equation (5), likewise corresponds to homologous level surfaces, and therefore its dependence on the total mass and the effective temperature of the star can be easily determined. For example, if k is proportional to $\rho^\lambda T^\mu$ (that is, arbitrary powers of density and temperature), then it can be cast in the form

$$k = \text{constant} \times \frac{M^{\lambda+\mu/2}}{R^{3\lambda+\mu}} \times (1-\beta)^{\mu/4},$$

as in Eddington's theory, and in this expression the effective temperature, T_e, can again be introduced by means of equation (5) and the relation $L = 4\pi acR^2T_e{}^4/4$. It should be noted that β steadily decreases with increasing mass and that therefore, in equation (5), the second part of the term in square brackets correspondingly loses in importance.

In reality, of course, kQ will not even be such a function of P, T, and ρ; the stars will not form homologous systems of any kind; and hence A and $\phi(\bar{r})$ will vary with the stellar mass. The extent of this variation from star to star follows from the dependence of the interior density distribution on the stellar mass.

The question if, and within what bounds, A and $\phi(\bar{r})$ can vary for stars of the same mass amounts to the question if, and within what bounds, the internal structure of a star with a given mass can vary at all. And this question, precisely, at least in my opinion, is one not generally treated with sufficient rigor. When one assumes that the properties of stellar matter are on the average the same for all stars (disregarding differences that, like ionization, depend on the pressure, the temperature, and the density), then the mass absorption coefficient, k, and the amount of energy liberated per unit mass per unit time, $4\pi\varepsilon$, can only be functions (explicit or implicit, but well determined) of pressure, temperature, and density. For the same combination of pressure, temperature, and density, k and $4\pi\varepsilon$ must be introduced into the equations of stellar structure as definite functions of P, T, and ρ (not merely as proportional to a function of them). The equations determining the internal structure of a star are

$$\frac{d}{dr}(P + \tfrac{1}{3}aT^4) = -g\rho,$$

$$P = \frac{R\rho T}{\mu},$$

and

$$\int_0^r 4\pi\varepsilon\rho r^2\, dr = \frac{4acr^2T^3}{3k\rho}\frac{dT}{dr}.$$

When k and $4\pi\varepsilon$ have been given as specific functions of P, T, and ρ, then in general these three equations determine only one star with a given total mass and surface boundary conditions—a star with a unique mean density, effective temperature, and luminosity. In other words, we must assume that the mean density, the effective temperature, and the luminosity of a star depend exclusively on its total mass. Among stars of the same mass, deviations in the mean density, effective temperature, and luminosity can be expected to occur only when the properties of stellar matter (even when the changes caused by pressure, temperature, and density are disregarded) are not uniform from one star to another. This could, for example, be a consequence of the transmutation of mass into energy inside stars at a different rate for different elements.

When the mass of a star changes slowly in the course of time, as a consequence of the radiation of energy, then its mean density, effective temperature, and luminosity must change accordingly.

1. *Monthly Notices of the Royal Astronomical Society 85*, 397 (1925).

2. Because of this, equation (81) of *MN 85*, 397 (1925), agrees with equation (53) of *MN 85*, 206 (1925). For (53) is valid only for stars with masses so small that their total pressure may simply be taken as equal to the gas pressure, and under this condition $aT^4/3$ is, but for a constant factor, the same function of r for all stars of very small mass and arbitrary n.

47. Atomic Synthesis and Stellar Energy I, II

Robert d'Escourt Atkinson

(*Astrophysical Journal* 73, 250–295, 308–347 [1931])

During the 1920s the theoretical work of A. S. Eddington had convinced astronomers that subatomic (that is, nuclear) energy must power the stars, but the details were lacking. In particular, even at the enormous interior stellar temperatures, the mean velocities of the particles were far smaller than those needed to penetrate the Coulomb field of a nucleus. The development of quantum mechanics provided a resolution of this difficulty. In 1928 George Gamow developed a quantum mechanical theory for the probability of the penetration of a nucleus by a fast-moving alpha particle. One year later Robert Atkinson and Fritz G. Houtermans applied and extended this theory to the nuclei within stars.[1] By combining Gamow's penetration probability with the Maxwellian distribution of velocities, they were able to provide the first attempt at a theory of nuclear energy generation within stars. It was immediately clear that the most effective nuclear interactions were those involving light nuclei with low charge and that only a few particles in the high-energy tail of the Maxwellian distribution would be able to penetrate nuclei. For this reason, nuclear reactions proceed slowly in the sun, thereby satisfying a criterion for the then unknown stellar energy source described by Henry Norris Russell in the previous decade. At the same time Atkinson and Houtermans satisfied Russell's additional criterion by showing that the rate of nuclear reactions is a sensitive function of temperature with a rate which increases steadily with the star's central temperature.

Two years later Atkinson addressed this problem in far greater detail. (Here we give little more than the abstract of this lengthy and far-reaching contribution.) By 1931 the great stellar abundance of hydrogen had just been established (see selection 39 and the references in its introduction), and Atkinson hit upon the idea that the observed relative abundances of the elements might be explained by the synthesis of heavy nuclei from hydrogen and helium by successive proton captures. Because he thought that the direct combination of four protons to form one helium nucleus (alpha particle) was very improbable, Atkinson argued that successive protons would be absorbed by nuclei until they became unstable and ejected alpha particles. Such a process, the CNO cycle, was found by Hans Bethe and by C. F. von Weizsäcker before the end of the decade (see selections 48 and 49).

Atkinson's preliminary attempt to explain the abundances of the elements failed, because he tried to produce the entire range of elements in one or two equilibrium processes acting under a single set of conditions. Nevertheless, by 1936 Atkinson had shown that the most likely nuclear reaction within stars is the collision of two protons to form a deuteron and a positron.[2] This simple reaction provides the starting point for a subsequent chain of reactions leading to the synthesis of helium and other heavier elements in stars.

1. R. d'E. Atkinson and F. G. Houtermans, *Zeitschrift für Physik 54*, 656 (1929).
2. R. d'E. Atkinson, *Astrophysical Journal 84*, 73 (1936).

Abstract—A *synthesis theory of stellar energy and of the origin of the elements* is developed, in which the various chemical elements are built up step by step from lighter ones in stellar interiors, by the successive incorporation of protons and electrons one at a time. The essential feature is that *helium*, which cannot well be formed in this way, is supposed to be *produced entirely indirectly*, by the spontaneous *disintegration* of unstable nuclei which must first themselves be formed.

A formula for the *probability of penetration of nuclei by protons* is derived from the *wave-mechanics;* it is a correction of one previously given, but agrees with that in showing that with any approach at all to the observed absolute amounts of the lighter elements *very high temperatures in normal stellar interiors are impossible.* The *model* must be approximately that of *Eddington*, but with Jeans's modification to take account of Kramers' formula for the absorption coefficient. Russell's value for the hydrogen content reconciles this formula with astrophysical data.

The low temperature and the relative amounts of the heavier elements make a *second synthesis process for protons* unavoidable; this is assumed to exist, and to possess arbitrary but extremely simple properties with no special features at any particular element. The possibility of *incorporation of electrons*, when "room" has become available, is taken for granted.

The elements whose *disintegration* supplies helium are in all cases either *those known* (or strongly suspected) to be unstable, or those isotopes, unknown now, *whose instability follows from Gamow's theory* of mass defects.

The *rate of formation of helium* must be nearly *constant*, in any one star, throughout most of its lifetime. The law of mass action demands then that all stars, after a marked *initial contraction and rebound*, should spend the greater part of their lives very *slowly expanding.* The constancy of the helium supply can be guaranteed, in the main sequence, if the average life of oxygen, until further synthesis, is about equal to the past lifetime of the star. This leads to *central temperatures* which can be calculated for the sun and estimated for other stars. The figure for the sun is 16,000,000°, which is in *good agreement with* that calculated for *a polytrope* of index about 3 and constitution somewhat over 50 per cent hydrogen. The increase necessary for heavier stars can be covered by a small systematic change in the polytropic index. The main sequence is thus accounted for.

The *relative proportions of the elements* in stars of the main sequence follow from the theory, in *excellent qualitative agreement* with Russell's figures for the sun. The scarcity of the lightest elements, the principal maximum at a fairly early point, a minimum before the iron group, a maximum in it, a scarcity of all elements above it, and minor maxima in the barium and lead regions all follow without any special assumptions, from Gamow's theory of nuclear stability, owing to the peculiarities of the Aston mass-defect curve.

For the *low-density giants* some *earlier source of helium* must be operative. This is taken to be Be^8, whose instability was already assumed by F. G. Houtermans and the writer, and has since acquired almost the status of observational fact. It must be long-lived, since it is found on the earth, and this *accounts for the Hertzsprung gap* and its continuation between the Cepheids and B stars, and for the fact that Be^8 cannot supply helium in the main sequence. In giants it can supply enough helium, and even the brightest red supergiants can exist on the *He-Li* synthesis given by the wave-mechanics alone, with *central temperatures as low as* 4,000,000°. This figure can be reached on Eddington's theory by a systematic change in the polytropic index which is considerable but not prohibitive.

All *very diffuse giants* should be *variable.* The occurrence of *R- and N-type stars* among giants is accounted for, and also their absence among dwarfs. For the *Cepheids* an *explanation of Eddington's* has become plausible, owing to the adoption of the varying polytropic index. There should be *more very bright G and M stars* than K ones, and the *concentration of the M stars round magnitudes* −5 *and* 0 can be accounted for. The occurrence of *coeval pairs* in which the *brighter is the cooler* is in entire agreement with the theory, as is the fact that the brightest stars in *clusters* are usually *all red or all blue.* The fact that *stars of small mass are all on the main sequence* (unless they are white dwarfs) can be explained. Various minor observations also seem to fit the theory.

White dwarfs follow from a *theory of Rosseland;* electrostatic forces will drive the hydrogen from the center of a star if it is not very abundant. They thus represent the final stage. The absence of hydrogen prevents generation of energy and they contract to a high density and temperature. Their energy may be purely gravitational, but need not be so. The *faint O stars* in planetary nebulae should all be "white dwarfs" generically.

The synthesis process predicted by the wave-mechanics is very sensitive to temperature; it is shown that *main-sequence stars* will nevertheless probably not be "*overstable.*"

A number of *arguments for and against older theories* are discussed. Jeans's theory of the eccentricities of binaries need not demand a long time-scale if the galaxy is expanding as fast as the universe in general is.

A *new notation* is introduced for rapid approximate calculations. It is convenient to use and cheap to print.

Introduction

WHILE THE BELIEF has now been current for more than a decade[1] that some form either of transmutation or of annihilation of matter must occur in stars, in order to provide the energy which they continually radiate, there has been little attempt as yet to investigate in detail what processes might actually occur. Treatments from the astrophysical side have in some cases made broad assumptions concerning the dependence of the rate of generation of energy on temperature and pressure, but beyond this point have generally been compelled either to introduce a large amount of quite arbitrary hypotheses or to remain so indefinite that either transmutation or annihilation could be supposed to be the fundamental process behind their equations. Direct treatment from the more purely physical side has not, until comparatively recently, been possible at all, since there was no theory of either disintegration, synthesis, or destruction of matter to start from.

Here Atkinson argues that practically nothing is known about the annihilation of matter but a good deal of theoretical knowledge has accrued about the process of synthesis. The study of the synthesis of the elements in stars should be able to predict the observed relative abundances of the elements and the types of stars that are capable of long-continued existence. It is these predictions that he sets out to derive.

The Theory of Regenerative Synthesis

It is well known that the maximum possible life of the universe is much shorter if stellar energy is due to transmutation of the elements than if it is due to total annihilation. In fact, it is definitely too short to fit some theories of stellar dynamics, associated mainly with the name of Jeans, which, to say the least, have never been disproved. We shall return to this question later. For the majority of other views, synthesis, provided hydrogen is consumed in it, furnishes an adequate, if scarcely a very liberal, time-scale,[2] if the element consumed is not hydrogen, the scale will be shortened from six to ten times at the least, and much more in most cases, since all other possible transformations upward or downward from known elements involve so much less percentage change of mass.

Russell has recently shown that the percentage of hydrogen in stars is probably very much greater even at the present time than had generally been supposed; in the sun's atmosphere, for example, sixty out of every sixty-five atoms are hydrogen. Since in addition the hydrogen nucleus is probably much simpler than any other, it seems very reasonable to assume that in its initial state any star, or indeed the entire universe, was composed solely of hydrogen; the small amount of angular momentum possessed by individual stars indicates that if this hydrogen was originally diffuse there was probably also very little small-scale motion in it; and the assumption that it was also cold is, though unimportant, at least as attractive as any other. The initial state of the universe thus becomes one of very remarkable simplicity, and we hope to show that the present complexity both of stars and of chemical elements could develop from this state by a self-regulating process.

At the outset there is one obvious difficulty that has been noticed by nearly every student of the question; the simplest direct synthesis from hydrogen to any other element is the formation of helium according to the reaction $4H^{+} + 2e = He^{++} + h\nu$, and this is almost certainly so improbable a process, and depends in any case so extremely on the density, that we cannot regard it as playing an important part in supplying stellar energy at all. For the same reasons a direct synthesis of any other element is even more objectionable.

If, however, helium can be supplied otherwise, progressive synthesis from this element onward is much less difficult to imagine. There is an element of every mass from 6 to 37 at least (and apparently no isobars, or duplicates, in this range)[3]; and we may suppose protons can penetrate light nuclei and remain inside, and that approximately every other time this happens the nucleus becomes able to take up, and sooner or later does take up, a free electron as well. Not even three-body collisions seem to be necessary; and the theory of the penetration of nuclei by protons has in fact already been investigated and found suitable in some cases.[4]

Atkinson then argues that unstable elements might be formed from hydrogen and helium and that these unstable elements would emit alpha particles. The unstable nucleus would thereby act as both a trap and an emitter of helium nuclei.

Synthesis Processes

A. The Gamow Process: Relative Amounts of the Lighter Elements We will consider first the synthesis processes that may be operative. The theory of the penetration of nuclei by α-particles was first given by Gamow,[5] and his formula for the probability of penetration can easily be modified so as to make it applicable to any other positively charged particles, in particular to protons. For small energy values of the impinging particles the formula is practically independent of the law of force in the interior of the nucleus struck; this will almost certainly be true when Z, its atomic number, is large, and may be assumed generally for the present. The probability falls off very rapidly however with decreasing velocity of the particles, and with increasing charge on either the nucleus or the impinging particle. The question whether the mere temperature movements of protons and other nuclei in stellar interiors would be violent enough for an appreciable number of nuclei to be penetrated was investigated on this basis by F. G. Houtermans and the writer[6]; it was found that for small values of Z, and when the impinging particles were protons, this was actually so, but in no other case. The outlines of the theory are repeated here, although in fact it contains a rather serious error; this is corrected in a later section.[7]

The dependence on the velocity of the protons is so great that it is not the protons having nearly the average agitational energy corresponding to the temperatures in stellar interiors (assumed to be about 40,000,000°), but the exceptional protons having ten or twenty times this energy, which, in spite of their rareness, most frequently succeed in penetrating other nuclei. This obviously makes the probability of penetration per gram per second very extremely sensitive to temperature. The formula obtained by combining the Gamow formula for penetration probability as a function of velocity with the Maxwell formula for the distribution of velocities was

$$W \approx 2\xi^3 e^{-3\xi^2}, \qquad (1)$$

where W is the average probability per nuclear collision (all collisions, including the slowest, being counted) that penetration will occur. ξ is the ratio of that velocity at which penetrations do most frequently occur, to the most probable velocity $v_0 = \sqrt{(2kT/m)}$ of the Maxwell formula for the distribution at the temperature T; its value is given, for protons and nuclei of charge Ze, by

$$\xi^3 = \frac{2\pi^2 Ze^2}{hv_0} \approx \frac{53{,}400Z}{\sqrt{T}}, \qquad (2)$$

and in all practically interesting cases it lies between $3\frac{1}{4}$ and $4\frac{3}{4}$, so that $\log_{10} W$ lies between say -12 and -28.

To obtain the actual "synthesis constant,"[8] λ, of an element of atomic number Z at a temperature T, we must multiply W by the total number of proton collisions suffered per second by a nucleus of this element, which is $N\pi\sigma^2\bar{v}$ if N is the number of protons/cc and $\bar{v}$ their mean velocity. We have thus to make an assumption about the proton density, and also about the effective "radius," σ, of the nucleus in question. The second factor is not very well known; since, however, $3\xi^2$ ranges in practice from about 30 to 70, small uncertainties in the multiplier of W play obviously a very minor rôle. In the paper referred to the nuclear radius was put at $10^{-12.4}$ cm, the temperature at $10^{7.6}$, and the density at 10^{23} protons/cc; on this basis the formula gave average lives (until further synthesis) varying from 8 seconds for *He* up to 10^9 years for *Ne*. Since the atomic number appears in the exponent in (1), the average life increases very rapidly with each step up the periodic system.

The dependence on temperature is also very marked; thus logarithmic differentiation of (1) and (2) shows that W varies approximately as T^{ξ^2}; the exponent is normally between 16 and 22, being largest when Z is large and T small, but it may fall as low as 7 for the lightest elements at the highest temperatures considered.

Roughly speaking, for every proton which enters and remains within a nucleus, energy corresponding to .007 units of atomic weight will be liberated. This number varies a little according as the mass defects of the element formed and that destroyed differ in one direction or the other; broadly speaking, however, the amount of energy is always nearly the same.

Here Atkinson shows that the relative abundances, N, of the elements observed in the sun[9] are directly related to the average lifetimes, $\tau = 1/\lambda$, calculated from equation (1).[4] (See table 47.1). This situation is analogous to the well-known law of radioactivity, in which the amounts of the various members of a series in the state of equilibrium are directly proportional to their average lives. The striking scarcity of lithium, beryllium, and perhaps boron appears to be a natural result of the application of wave mechanics, which shows that these elements are too easily destroyed to become plentiful.

Table 47.1 Relative abundances, N, and average lifetimes, τ, of elements in the sun

	Li	*Be*	*B*	*C*	*N*	*O*
$\log N$	5.0:	4.8	8.5::	10.4:	10.6?	11.4
$\log \tau$	3.3	5.5	7.6	9.5	11.4	13.2

Atkinson argues that Milne's estimate for the central temperature of stars[10] must be too high: if any appreciable fraction of a star's mass were at this temperature, the star would explode. Eddington's lower estimate for the central temperature is therefore considered more appropriate. A second synthesis process is then postulated, in order to account for the presence of elements heavier than oxygen. A discussion of Gamow's criterion for the stability of nuclei ends in the conclusion that alpha unstable elements will be fairly uncommon among those reached by a synthesis process active above xeon (Z = 54).

The Relative Amounts of the Elements

Here Atkinson gives a long account attributing the observed relative abundances of the elements to one process that results from the penetration of nuclei by protons and to another "unknown" process that depends on atomic weight and temperature in such a way that the observed abundances of the heavier elements are reproduced. Only the solar abundance of helium seems to be unexplained. A similar difficulty presents itself in trying to account for the variety of energy-generation rates of different stars on the main sequence. In Atkinson's view, some helium-producing element in the oxygen region must have a lifetime as long as that of the star itself.

The remaining forty-four pages of this paper are summarized in the abstract.

1. The idea seems first to have been suggested by Jeans. A summary of the conditions that subatomic sources must satisfy was given by H. N. Russell, *Publications of the Astronomical Society of the Pacific 31*, 205 (1919). Cf. also Eddington, *The Internal Constitution of the Stars*, chap. xi (1926); Russell, Dugan, and Stewart, *Astronomy*, chap. xxvi and pp. 589–92 (1927); Jeans, *Monthly Notices of the Royal Astronomical Society 87*, 36, 400 (1926), and *Nature* (Suppl.) *118*, 29 (1926).

2. Eddington, p. 293.

3. On the absence of C^{14} see Menzel, *Publications of the Astronomical Society of the Pacific, 42*, 34, 1930.

4. Atkinson and Houtermans, *Zeitschrift für Physik 54*, 656 (1929). The idea of an indirect production of helium was also put forward in this paper.

5. *Zeitschrift für Physik 52*, 510 (1928).

6. *Zeitschrift für Physik 54*, 656 (1929).

7. §6, C, vi. As this paper goes to press, a recalculation by A. H. Wilson is noted (*Monthly Notices of the Royal Astronomical Society 91*, 283 [1931]) which takes account of the radiation emitted in such processes. His formulae differ from those used here, and are probably more correct in principle, though a number of very rough numerical assumptions are involved. They retain the all-important exponential factor of the present and the earlier paper, and no formula which does this can lead to very seriously different temperatures; the actual values with Wilson's formula (with some changes in his numerical assumptions) seem to lie between those of Atkinson and Houtermans and those of this paper, but somewhat nearer the former. This would involve a systematic increase in the temperatures calculated, which might amount to 50 per cent, but can scarcely be more. Wilson's statement that synthesis has only a negligible probability is difficult to maintain in face of this, and the present theory is unaffected in its general outline.

8. This is the exact analogue of the "decay constant" in radioactivity, and its reciprocal, τ, will often be called the "average life" in what follows, where there is no danger of confusion.

9. *Astrophysical Journal 70*, 11 (1929).

10. *Observatory*, August 1930, and references; *Nature 127*, 16 (1931). Larmor, *Observatory*, June 1930, and references; *Monthly Notices of the Royal Astronomical Society 91*, 4 (1930).

48. Element Transformation inside Stars. II

Carl Friedrich von Weizsäcker
TRANSLATED BY RICHARD H. MILBURN

(*Physikalische Zeitschrift 39*, 633–646 [1938])

In two papers appearing in the *Physikalische Zeitschrift* in 1937 and 1938, C. F. von Weizsäcker examined the questions of what thermonuclear reactions occur within stars and what significance these reactions have on the observed relative abundances of the elements.[1] In his first paper, Weizsäcker restated Robert Atkinson's views that the relative abundances of the elements can be accounted for by thermonuclear reactions in stars and that the proton-proton reaction, $H^1 + H^1 = H^2 + e^+$, must have started the process.[2] (Here H^1, H^2, and e^+ denote, respectively, a proton, a deuteron, and a positron). In his second paper, which we reproduce in translation here, Weizsäcker renounced his previous views and proposed that the chemical elements were already formed before the formation of stars as we know them today. By assuming that all the elements in a star are present from the beginning, Weizsäcker was no longer limited to reactions that begin from hydrogen and helium. This led him to the important discovery of the cyclic CNO chain of reactions, in which carbon acts as a catalyst in the synthesis of helium from hydrogen. Weizsäcker argued that this process provided the principal source of energy for the main-sequence stars. Subsequent theoretical investigations and experimental measurements mentioned in the next introduction have shown that main-sequence stars heavier than 1.5 solar masses do shine by the CNO cycle, although main-sequence stars lighter than 1.5 solar masses shine by the proton-proton reaction.

In addition to setting forth the two major processes that provide stellar energy, Weizsäcker also presented two fundamental approaches to account for the observed abundances of the elements. In his first paper, Weizsäcker argued that the heavier elements are synthesized from the lighter elements within stars (a view he abandoned in the paper given here). We now believe that most of the elements are, in fact, synthesized within stars through different nuclear reactions occurring at different times and under different physical conditions. In the paper given here Weizsäcker shows that, to account for the abundance of the heavier elements, nuclear densities and temperatures on the order of 10^{11} degrees are required, and he argues that these conditions could be achieved in a primeval "fireball" explosion out of which the expanding universe arose. In the last decade, the discovery of the 3 degree microwave background has confirmed the existence of a hot primeval fireball, and the observed abundances of deuterium and helium appear to be consistent with an origin in this prestellar state.

1. C. F. von Weizsäcker, *Physikalische Zeitschrift 38*, 176 (1937), *39*, 633 (1938).
2. R. d'E. Atkinson, *Astrophysical Journal 73*, 250, 308 (1931), *84*, 73 (1936).

Part I. Difficulties for the Synthesis Hypothesis

1. THE MEANING OF THE "SYNTHESIS HYPOTHESIS" An earlier work[1] attempted to determine what transformations of atomic nuclei took place inside stars and what significance these transformations had for the structure and development of stars. Further investigations have shown that some of the hypothetical assumptions there utilized cannot be preserved. This present work still provides no quantitative execution of the theory; it is limited to a renewed qualitative discussion of the problem under altered assumptions.

It will be demanded of the theory, first of all, at least in certain simple cases, that it predict which nuclear reactions in a body of material of given physical and chemical constitution can occur spontaneously. But this does not finish the task. Since we cannot obtain direct observations of the physical and chemical conditions obtaining in stellar interiors, the theory must first determine them itself. At this point a hypothesis is required, since we do not know *a priori* whether, in addition to the element-building by nuclear processes, whose quantitative description is the goal of the theory, still other so far unknown or not considered effects may alter these conditions. In [1] it was assumed that such effects were not present, and this assumption was designated the "synthesis hypothesis."

This form of the hypothesis still contains an uncertainty. Specifically, the nuclear reactions exert simultaneously two different actions: they change the physical state of matter through the release of energy, and they alter the chemical composition through the transmutation of elements. The consideration of energy generation is the unproblematical part of the theory: for the explanation of the radiation of stars, nuclear reactions or effects of similar energy yield are essential, and the synthesis hypothesis is synonymous with the assumption that the nuclear processes alone are sufficient for this radiation. On the other hand, the transmutation of the elements is to some extent a result of nuclear reactions of whose significance for the life history of stars nothing is known from the outset. Now, the empirical frequency distribution of the chemical elements shows characteristic regularities, seemingly uniformly valid throughout the universe, which compel us to seek an explanation through the assumption of a uniform evolutionary process. An obvious place to seek this process is in the element transmutations necessarily connected with energy production in stars. To begin with, we cannot exclude the possibility that the chemical elements were already formed before the formation of the stars as we know them today, by another process, and that the present energy-producing reactions have brought about only a limited alteration of the original frequency distribution. We must therefore distinguish between a narrow statement of the synthesis hypothesis, in which the role of the presently occurring nuclear reactions in stars is limited to energy production, and a wider form that, in addition to the concomitant element production, takes into consideration no element-formation process anywhere else in the history of the universe.

In [1] the wider form of the hypothesis was taken as a basis, as the simplest of the possible assumptions. Moreover, it was shown that it is impossible to make the same process directly responsible for both energy production and element formation, because the accumulation of hydrogen necessary for energy production does not lead to the formation of heavier elements. There was constructed a causal connection between both processes, through the assumption that the energy-producing reaction creates neutrons as a by-product, which then take over the further building up of the elements. An attempt at quantitative execution brings this assumption up against a series of difficulties that appear scarcely surmountable. Firstly, it is uncertain whether neutrons are produced at all in perceptible numbers; if they are so produced, it appears certain that at the same time helium must be formed in a quantity incompatible with astrophysical data on its frequency. Secondly, the formation of a considerable quantity of uranium and thorium out of very short-lived radioactive intermediaries cannot apparently be explained by the mechanisms developed in [1]. Finally, the hypothesis of formation by neutrons provides no satisfactory explanation for the empirical parallel between binding energy and frequency of the various nuclei.

We are therefore well-nigh compelled to renounce the extended form of the synthesis hypothesis, but there is no empirical argument against a limitation to the narrower interpretation. It is quite possible that the formation of the elements took place before the origin of the stars, in a state of the universe significantly different from today's. To be sure, the energy-producing process necessarily leads to an alteration of the element distribution in the course of stellar evolution. Probably stars are still so young that they in their evolution have hardly had time to alter their chemical composition significantly. The hydrogen supply of an originally pure hydrogen sun is sufficient to provide its present radiation for 3×10^{11} years. On the other hand, geological and astronomical data do not allow us to assign the sun a greater age than about 3×10^9 years; and if one may interpret the redshift in the spectra of spiral nebulae as a Doppler effect, the backward extrapolation of this explosive motion provides a concrete reason to attribute a physical state significantly different from today's to the world of about 3×10^9 years ago. As a consequence, the sun would, up to now, have transformed only 1% of its mass. It is interesting that this rejection of the extended form of the synthesis hypothesis leads to a new, independent age determination that corresponds well with known data. The radioactive elements still present today must, if they were not produced continuously in the stars, have originated at a time not significantly further back than their half-life. Quantitative estimations[2] have yielded an age of the contemporary element distribution of about 5×10^9 years.

The present paper deals in the first part with the evidence contradicting the extended form of the synthesis hypothesis. The second part of the paper establishes and develops the narrower form. Finally, the third part tries to put together the conclusions one can draw about the state of the world at the time of the origin of the elements.

In section 2 Weizsäcker notes that the neutron-producing reaction needed for the formation of heavy elements must produce alpha particles as well. On the average, two helium nuclei are produced with each neutron; the proportion of helium to the number of atoms of heavier elements will therefore be as $[2(A-4)-1]:1$, where A is the mean atomic weight of the heavier elements. Observational data for stars[3] indicate that the helium to metal ratio is at most 3:1, which is a factor of 10 below the lowest theoretical limit. In section 3 Weizsäcker calls attention to the need for a central energy-producing core if uranium and thorium are built up by neutron capture. Considerations of convection[4] make the central cores discussed in [1] untenable. An alternative possibility is synthesis in small, repetitive explosive reactions,[5] but it is not clear how an explosion can stop once it is initiated. In section 4 Weizsäcker calls attention to Harkins' rule, which states that nuclei with an even number of nucleons are more energetically stable and also more abundant than those with an odd number. Considerations of the neutron interaction cross section[6,7] lead to the conclusion that Harkins' rule is not a natural consequence of synthesis by repetitive neutron capture.

Part II. The Mechanism of Energy Production

5. Review of the known energy sources We deal now with energy production independently of the question of element formation. It appears advisable to examine once more the foundations of the theory, through a compilation of the energy sources that come in question in the present state of physics. Four principal sorts of energy production come to mind:

1. Contraction without alteration of the chemical constitution of the star. The released energy is gravitational potential energy.
2. Buildup of elements. The released energy is nuclear energy.
3. Contraction resulting from conversion of a part of the matter into densely packed neutrons.[8] The released energy is again gravitational energy as well as the excess, released by the transformation, of the zero-point energy of the degenerate electron gas over that of the resulting neutron gas; for this process nuclear energy must be expended, which conforms to the mass excess of the neutrons over that of the previously present atoms. At high densities this energy balance can be favorable.
4. Complete conversion of matter into radiation. The released energy is the rest energy of the material.

In the following we shall consider only the second energy source and must therefore establish that the other three sources have no importance. Pure contraction is eliminated, at least for stars of the sun's type, on account of its limited yield. On the other hand, the energy sources named under (3) and (4) above, if they can actually occur at all, are more productive than element formation.

The complete conversion to radiation, say by the annihilation of an electron with a proton, has a forlorn probability since physics, to date, has discovered no cause that would be in a position to bring it about. According to the discovery of neutrons and positrons, it seems that in the balancing of positive and negative charges only the electron mass is converted into radiation energy and the proton mass remains conserved. That the complete conversion into radiation has never been observed is also a strong argument against its occurrence in stars. Because in the interior of planets and still smaller bodies this energy source is not present, the temperature of the stellar interior would have to be a significant factor in its taking place. On the other hand, we can today in the laboratory at least expose individual particles to the effects of energies an order of magnitude higher than those that in all probability are present in stars, and we find no conversion into radiation. If we consider that the assumption of conversion to radiation is unnecessary to explain the empirical energy production, it seems entirely reasonable to drop it altogether.

In contrast, the possibility of the third above mentioned energy source follows directly from modern physics. It can be excluded only if the conditions under which it would be effective do not occur in known stars. In fact, the usual approximations of the stellar interior yield a density lying far below the critical density at which this energy source sets in.

But we can also, perhaps, make ourselves independent of these approximations through a genetic argument. If a star begins its evolution as a ball of gas of small density, then it becomes with increasing density an energy source first of type 1, then of type 2, and finally of type 3. If the star during contraction passes through a sequence of equilibrium states, its density and with it its temperature will never rise above the level at which the energy released can be radiated away. The energy source of type 2 becomes effective at a rather sharply defined temperature, and the star should therefore remain for a long time in the vicinity of $10^{7\circ}$. To be consistent, it is necessary experimentally to assign temperatures of this order of magnitude throughout the stars of the main sequence. Because this energy source now suffices for a long interval, perhaps a hundred times longer than the probable age of the sun, one should assume that the sun (as well as the entire main sequence) is not yet old enough to have reached the necessary density for the type 3 energy source.

The only possibility of attaining this density quickly would occur in a stellar development that does not proceed through a sequence of equilibrium states; yet it appears that a development of this sort would have to lead to the explosion of the star. Since a temperature increase is coupled with a density increase, this development leads to an accelerated release of nuclear energy, and the star is at least led back into an equilibrium; if the deviation from equilibrium has already become too large for this, it can only explode, either directly or in the course of an "unstable" pulsation. The argument may also be phrased as follows: at the high density of matter, which has to be achieved before the formation of the "neutron core," and at the corresponding temperature, there must appear forthwith a thermodynamic equilibrium among the nuclear reactions. In equilibrium the mixing ratio of the elements is defined by the physical conditions. Any star that does not in advance have the correct mixing ratio must achieve it by contraction, and if this process is executed in a shorter time than about 10^{11} years, the energy released thereby destroys the cohesion of the star.

6. The course of the energy-producing nuclear reactions Which special nuclear reactions are in fact responsible for energy production could not be decided in [1]. The progress in nuclear physics to date still permits no sure answer. The problem is moved into a new light through the assumption that the elements were substantially formed before the development of the present state of the stars. This assumption is made because all known stable nuclei are now available as starting materials of reaction chains, and, in addition, the properties of neutron production and of autocatalysis need no longer be demanded of these chains; it does not matter much for the assumptions of the theory which reaction is finally proved to be the most important.

Concerning the question which of the so far proposed reactions are possible, modern nuclear physics provides the following information:

The model cycle of [1] can occur only if nuclei of mass 5 can exist. According to experimental work appearing in the meantime, nuclei of mass 5 appear not to be possible. They have not been found stable, and on the basis of nuclear reactions, one experiment has assigned to helium 5 a mass greater than the sum of the masses of an alpha particle and a neutron.[9]

Most of the buildup paths, which appear possible by way of mass 5, lead into the nucleus beryllium 8, but this nucleus also appears incapable of existence. It must, at least, be associated with a lifetime against alpha-decay that is too short for an additional charged particle to be absorbed on it, with appreciable probability, under the conditions obtaining in a star.[10]

Reaction chains leading by means of the absorption of hydrogen or of helium itself to the buildup of higher nuclei or to the increase of the quantity of helium appear therefore not to be possible through two-body collisions; and three-body collisions should, under normal stellar densities, play no role.

On the other hand, we must reckon with the possibility of a direct reaction of two protons with one another. The deuterium nucleus, which could be produced first in this way, would certainly, according to present knowledge of the forces between two protons, no longer be stable. Yet this unstable nucleus can emit a positron during the brief time of its existence, so that overall the process $H + H \rightarrow D + e^+$ occurs. This process was proposed as an energy source by Atkinson[11] and was recently investigated quantitatively by Bethe.[12] At present, all that can be said is that our knowledge of nuclear physics is insufficient to exclude this proton-proton reaction as an energy source. The very limited *a priori* probability of beta-decay in it will be balanced by the number of collisions between proton pairs in a star, but it is very hard to find a reliable estimate for the beta-decay probability. Astrophysically, the assumption of this energy source would provide considerable difficulties, because it has, on account of the small Coulomb field between two protons, only a weak temperature dependence. The nearly constant central temperature in the main sequence, in spite of the very different demands that stars of different masses make on their energy sources, would be hard to explain by this means.[13]

At this point, mention should be made of Dopel's proposal[14] that reactions, which cannot occur with particles of thermal energies, could be released by particles accelerated through stellar electric fields. My reply to this proposal is that electric fields can probably only be maintained in the outer atmospheric layers of stars, since the stellar interior, because of the high density of free electrons, must be a perfect electrical conductor.

We assume now that all elements in a star are present from the beginning, so that we are no longer limited to reactions beginning from hydrogen or helium. Concerning the behavior of the immediately higher elements in a star, we can predict on the basis of laboratory experiments the following:

All the known stable isotopes of lithium, beryllium, and boron together will be broken down by proton accretion and will lead thereby finally to the formation of helium. On the other hand, if a reaction cycle should be associated with the nucleus carbon 12, in the course of which helium would in any case be produced, the initial nucleus would remain unaltered and would serve only as a catalyst.[15] This is the cycle:

$$C^{12} + H = N^{13}; \quad N^{13} = C^{13} + e^{+}; \quad C^{13} + H = N^{14};$$
$$N^{14} + H = O^{15}; \quad O^{15} = N^{15} + e^{+}; \quad N^{15} + H = C^{12} + He^{4}.$$

The energy source of a star would thus consist first in the buildup of the elements below carbon and then in the above cycle. If, by auxiliary reactions, the quantity of carbon also decreases, then an analogously proceeding oxygen cycle would be available.

7. CONCLUSIONS FOR THE EVOLUTION OF STARS One can now estimate the change that has so far taken place in the distribution of elements in the sun through the proposed processes. In order to have simple numbers, we assume that the sun was originally composed of equal parts by weight of hydrogen and heavier elements and that the latter were uniformly distributed over all atomic weights from 1 to 50, so that one heavier atom (on the average) occurred for each hydrogen atom. The sun has up to now converted about 1% of its mass from hydrogen into heavier elements and hence has lost about one-fiftieth of its hydrogen. On the average, hydrogen atoms will react two at a time with the same heavier nucleus and its reaction products, respectively, before it is converted into helium (that is, two protons are needed for lithium 6, one for lithium 7, three for beryllium 9, and so on); hence, a fourth of all heavy nuclei should be affected by the transformation, if no heavy nucleus reacts more than once (thus excluding cyclic reactions). The lightest nuclei would be involved first, and in this way, during the evolution of the sun to the present, the element distribution should be built up all the way to carbon. At carbon the buildup stops, through the appearance of the (carbon) cycle. It is known that lithium, beryllium, and boron are especially rare in the sun,[16] not only in comparison with other elements but also in comparison to the distribution on the earth,[17] where the element transmutation ceased long ago.

The frequency of helium now comes into satisfactory agreement with experience. About as many helium nuclei should be generated as the number of protons lost. That yields a ratio of atomic numbers of about:

$$H:He = 50:1.$$

These considerations can, with suitable changes, be applied to the remaining stars of the main sequence. However, the red giants present a difficulty, because of their high luminosity with a central temperature some 10 times lower than in the main sequence. It is certain that the giants and the main-sequence stars cannot have the same energy source. On the other hand, on the assumption that the giants draw their energy purely from contraction, such a short time span is available that the pulsation periods of Cepheids would have had to change markedly as a result of contraction during historical times, in contradiction to a series of observations.

Hence, one must probably seek two different nuclear reactions for the red giants and for the main-sequence stars. The considerations above offer for this purpose two different possibilities. As Bethe has observed,[18] we may assume that the giants are still breaking down lithium, beryllium, and boron, while the main sequence has already initiated the carbon cycle. We can alternatively assume that in the red giants there may be large quantities of heavy-hydrogen isotopes present that would deliver their energy, while the breakdown of lithium and its neighbors would either be limited to a quickly transient intermediate stage or would be associated with a star already very similar to one in the main sequence. Since the energy yield is a substantial function of Z^2/T, the first assumption would reduce the required central temperature of the red giants, compared to the main sequence, by about a factor of 4 and the second by a factor that is in any case greater than 10. The first assumption may appear natural; the second perhaps expresses more clearly the characteristically wide gap between giants and dwarfs. Which of the two is correct depends upon the presently unknown primordial frequency distribution of the lightest elements. Independently of this particular question, the synthesis hypothesis in any case suggests that we attach the main sequence to a stationary working reaction cycle and the red giants to the breaking down of those nuclei that are lighter than the nuclei participating in the cycle and that must therefore be transformed before the temperature can climb high enough to stimulate the cycle.

This assumption has a few consequences that can perhaps be tested experimentally. It dictates that the giants must be very young stars. According to the above estimates, the breakdown in the sun of a nuclear variety of average frequency lasts about 10^8 years. Since the red giants possess a higher specific luminosity than the sun, this time is reduced for them to 10^7 years or less. Since globular clusters do not simultaneously contain both B stars and yellow or red stars, this suggests that we must assign to the giants a definite age (within some limits). But perhaps there are sharper tests. Moreover, the giants must contain light elements already very rare in the main sequence, in fact, either lithium and the next heavier elements or heavy hydrogen (this last, to be sure, in a quantity that, in the presence of light hydrogen, is probably not spectroscopically demonstrable).

Against the assumption that the giants also have nuclear reactions as energy sources, Gamow[19] has raised the objection that they then should be arranged in a line approximately parallel to the main sequence, whereas, in fact, the giant branch is almost perpendicular to the main sequence. The giant branch is very diffuse, however, and actually occupies an extended area above the main sequence. Now, according to our assumption, the energy source of the main sequence is approximately constant in time, and that of the giants is used up in time. Therefore, the luminosity of a giant star can alter in the course of its evolution, so that the observed giants, which vary in mass and age, should indeed fill out an area of the Russell diagram. The line of greatest star density in the diagram, normally designated the giant branch, does not then need to be a line of constant age along which the stars are distributed according to their masses; it could, to the contrary, represent the evolutionary path of the most frequently occurring masses among the giants. Perhaps there holds for the giants, because of their predominantly convective constitution, a different mass-luminosity relation than for the dwarfs.

Gamow[20] has pointed out another difficulty for the overall synthesis hypothesis. He has shown that the mass-luminosity correlation of a star depends upon its hydrogen content, and if this changes during stellar evolution the empirical existence of a universal mass-luminosity relation cannot be explained. The assumption of a resonance in the energy-producing process, introduced by Gamow himself for the removal of the difficulty, has a small *a priori* probability and appears for this purpose to lessen the difficulty but not to remove it.[21] On the other hand, perhaps there will suffice for this purpose Gamow's[22] recent observation that a large excursion from the normal mass-luminosity relation is limited to a short fraction of the lifetime of a star. However, the entire difficulty disappears if we abandon the extended synthesis hypothesis and assume that the original chemical composition of the stars has not significantly changed. For stars whose luminosity per gram is less than 10 to 100 times that of the sun this observation suffices. The universe must therefore be still so young that stellar evolution, apart from the passage through the giant stage, has hardly yet begun. The only difficulty for this concept is provided by the existence of the white dwarfs, whose small luminosity the synthesis hypothesis can hardly explain, other than by the assumption that their total hydrogen content is already used up.

These considerations of evolutionary history are still in a speculative stage, but the ideas that they apply derive from a part of a physics whose present foundations can be considered as already explained. We must therefore hope that, through some experimental advances and a close collaboration of astrophysics and nuclear physics, they can be built up in the near future into a closed theory.

Part III. The Origin of the Elements

8. Necessary physical conditions If we do not wish to give up altogether in understanding the origin of the elements, we must, in renouncing the extended form of the synthesis hypothesis, try to draw from the frequency of distribution of the elements conclusions about an earlier state of the universe in which this distribution might have originated. Obviously, such conclusions rest upon very much more uncertain foundations than the theory of energy production, since they presuppose, like the latter, the permanence in both space and time of our laws of nature. That the natural laws retain their form in the passage from the terrestrial laboratory to planetary space, and from there to galactic space, has been confirmed for at least a series of special cases. But how far we can describe with these laws the nature, unknown to us, of a substantially different and earlier state of the universe is *a priori* completely uncertain, and it concerns in essence a time about which we can, as soon as our memory no longer suffices, obtain no direct experience, but which depends upon inferences from documents. There remains no other choice than to presuppose, first of all hypothetically, the permanence in time of our natural laws and therein to be prepared for the appearance of error when this theory is compared with historical documents still available today.

The relation between mass defect and frequency urges the assumption that, in the formation of the elements, kinetic energies of reaction partners of the order of nuclear binding energies are available. At such high energies a thermodynamic equilibrium of nuclear reactions must appear quickly. For, firstly, the Coulomb repulsion, at least for the lighter nuclei, plays hardly any role; and secondly, the conversion of free protons into neutrons now becomes a very frequent process, so that for the building up and breaking down of heavier nuclei neutrons in arbitrary quantities are available.

This assumption, of course, contradicts the first impression made by the frequency distribution of the elements. But we must also consider that, after the production of the main features of the element distribution, nuclear processes have continued to take place in the universe. Hence, the theory has already accomplished what one can hope of it, if it shows that the deviation of the modern distribution from an equilibrium may be explained through nuclear processes, whose presence after the first act of creation we also, in any case, must demand on physical or astronomical grounds. Thus the abnormal rarity of the lightest elements is a consequence of the energy-producing reactions in the modern stars. We shall discuss another deviation at the end of section 9.

If we take for the original distribution a thermodynamic equilibrium, then we can try, from the empirical frequency of the elements, to calculate what temperature and what density must have obtained at the time of their formation. To the frequency ratio of two neighboring nuclei we must then apply the Saha formula.[23] We consider, say, the absorption of a neutron. Let n_A represent the number of nuclei of atomic weight A per cm^3—in particular, n_1 is the number of neutrons—and let E_A represent the energy that must be applied in order to tear out a neutron from a nucleus of mass A. Thus

$$\frac{n_{A-1} \cdot n_1}{n_A} = g_A \cdot e^{-E_A/kT} \tag{1}$$

holds with

$$g_A = \frac{G_{A-1}}{G_A} \cdot \frac{2(2\pi MkT)^{3/2}}{h^3}. \tag{2}$$

G_A is the statistical weight of a nucleus of mass A and hence a number of the order of magnitude 1. The second factor in (2) contains the statistical weight of free neutrons; it has the dimension of a reciprocal volume, and indeed it signifies the number of neutrons per unit volume for which at the given temperature "phase cells" are available.

We can now apply equation (1) to two successive neutron absorptions and obtain the following equations for the temperature and neutron density:

$$kT = \frac{E_A - E_{A-1}}{\ln \dfrac{n_{A-2} n_A}{n_{A-1}^2} \dfrac{G_{A-1}^2}{G_{A-2} G_A}} \tag{3}$$

and

$$\frac{n_1}{g_A} = \frac{n_A}{n_{A-1}} e^{-E_A/kT}. \tag{4}$$

We apply these formulae to the reactions

$$\begin{aligned} O^{16} + n^1 &\rightarrow O^{17} \\ O^{17} + n^1 &\rightarrow O^{18}. \end{aligned} \tag{5}$$

If we set n_{16} equal to 10^4, then empirically $n_{17} = 4$ and $n_{18} = 20$. Moreover, $E_{17} = 4.5$ AMU and $E_{18} = 9.8$ AMU. G_{16} and G_{18} may be set equal to 1. The spin of O^{17} is unknown; if we set it equal to 1/2, then it follows that $G_{17} = 2$. There follows

$$kT = 0.44 \text{ AMU} = 0.41 \text{ MeV},$$

or

$$T = 4.7 \times 10^{9\circ}.$$

In addition, $n_1/g_{18} = 1.2 \times 10^{-9}$.

Using the temperature calculated above,

$$1/g_{18} = (5 \times 10^{-12} \text{ cm})^3;$$

hence, $n_1/g_{18} = 1$ would already represent a neutron density comparable to the density in the atomic nucleus. The actual number is about 1,000 times larger than the average distance between two neutrons. It gives, specifically, $n_1 = 10^{25}$, that is, about 10 times the density of water for the neutrons alone.

These numbers are still very inexact. Above all, the density determination, because of the exponential dependence upon the energy, may still be wrong by several powers of 10. Moreover, through the inaccuracy of the mass defects and frequencies used, the temperature distribution is also uncertain by a factor of the order of magnitude 2. The next duty of the theory would therefore be first to test, through analogous consideration of several other reactions, whether nearly unique values for temperature and density can be determined at all from the entire frequency distribution of the elements, and, if this determination can be made, to determine the numerical values as exactly as possible. Perhaps in this way it might finally be possible to utilize the empirical frequencies of individual nuclei directly as quantitative measures of their mass defects.

Unfortunately, we possess the exact knowledge of mass defects and frequencies necessary for this test only in a preliminary form for nuclei below oxygen, whose abundance is completely altered by secondary processes. Some data for heavier nuclei give numbers of the same order of magnitude as those calculated here. We here desist from continuing the quantitative investigation and consider only a qualitative outline of the empirically more established distribution.

9. THE CONDITIONS FOR THE ORIGIN OF ALL NUCLEAR VARIETIES IN COMPARABLE QUANTITIES The empirical frequency distribution of nuclei exhibits, along with strong isolated fluctuations, a conspicuous overall uniformity. A nucleus is often 10^3 times more frequent than its neighbor; nonetheless, the mean frequency of nuclear species from oxygen to lead varies only by a factor of about 10^6. To calculate the frequency of lead, beginning with that of oxygen, one must apply equation (1) 200 times (wherein partly neutron- and partly proton-absorption is assumed); for this, so that the ratio $n_O/n_{Pb} \sim 10^6$ will result, the factor

$$g_A e^{-E_A/kT}/n_1,$$

which gives the frequency ratio of any two separate nuclei, must have an average value, in spite of very large individual fluctuations, almost exactly equal to 1. If we call this mean value f, then $f^{200} \sim 10^6$ must hold, and hence $f = 10^{0.03} = 1.07$. Such a striking relation between temperature and neutron density cannot be an accident; rather, we must demand of the theory that it explain the facts of the case physically, and in fact, this relation is proved as an almost direct consequence of the saturation of nuclear forces.

Let us permit the density of matter to vary at a fixed temperature. At small density, f is larger than 1 (the statistical weight of free protons is so large that its influence dominates the energetic preference for the bound state); therefore, the lightest nuclei are present practically by themselves. With increasing density, the equilibrium shifts in favor of composite nuclei. But if now the binding energy is proportional to the number of particles, that is, if E_A is on the average independent of A, then this shift never leads to a favoritism of a nuclear species over the next lighter but yields in the limiting case of very large densities just that distribution in which nuclei of each mass are exactly equally frequent (hence $f = 1$). The assumption made here is only that the density always remains small compared to the density of matter in the interior of nuclei; that is, that the formation of separated nuclei is in general still possible.

I base this assertion upon a simplified model, considering first only the accumulation of a unique particle species of concentration n_1 and mass M, disregarding the distinction between neutrons and protons. This procedure is harmless, because at slightly higher temperatures and densities, neutrons and protons are almost equally frequent. I assume, moreover, complete saturation of the nuclear forces and so replace E_A by the constant E. In addition, I initially disregard the top end of the periodic system (which is certainly only limited by the departure of the heaviest nuclei from saturation because of the Coulomb force) and hence assume that A can run from 1 to infinity. Finally, I set all weight factors $G_A = 1$; then g_A assumes a constant value g. From (1) it will then hold that

$$\frac{n_{A-1}}{n_A} = \frac{g}{n_1} e^{-E/kT} = f; \tag{6}$$

f is independent of A, and there results

$$n_A = n_1 f^{-(A-1)}. \tag{7}$$

If now the mass density ρ of the material is given, then n_1 (and therewith f) is determined from (6) and the additional condition

$$\rho = \sum_A n_A \cdot AM = n_1 M f \sum_A A f^{-A} = \frac{n_1 M}{(1 - 1/f)^2}. \tag{8}$$

The density will already become infinite for $f = 1$, since infinitely many nuclear species will result with equal frequency. Therefore, $f < 1$ cannot occur at all.

This result can also be expressed physically, in examination of the condensation of a neutron gas. Since $n_{A-1} = n_A$, that is, since "drops" of any size can occur with equal probability, it must hold that

$$n_1 = g e^{-E/kT}. \tag{9}$$

This is just the vapor pressure equation for the neutron liquid, which gives the concentration of the neutron gas standing in equilibrium with its condensate. But this condensate, whose presence provides the validity of equation (9), consists of the original nuclei themselves.

The question remains whether the above hypotheses prove true for actual nuclei. Only the assumption of complete saturation is doubtful. The finiteness of the periodic table represents no difficulty, because the elements above lead are indeed radioactive but capable of existence, and their frequency decreases exponentially with mass according to (7); if they disintegrate in addition, the frequency of lead, compared with its neighbor elements, will easily be raised by a factor of 10 to 100, which fits well with observations.

In order to compensate for the empirical departure of the binding energies from saturation (and indeed not only the fluctuations but also the systematic process), the temperature must be sufficiently high. This yields a substantially higher value than in the previous paragraphs. If one varies by f the binding energy of a neutron over the 200 mass numbers from oxygen to lead and applies these variations linearly, then even for very large densities, ρ, oxygen must be more plentiful than lead by a factor of about $e^{100f/kT}$. With $f = 3$ MeV and $n_O/n_{Pb} = 10^6$ this yields

$$kT = 20 \text{ MeV}; \quad T = 2.3 \times 10^{11\,\circ}.$$

This very crude approximation implies in any case that one must assume a temperature of an order of magnitude of that which would come about through the complete transformation of the nuclear binding energy into heat. The accompanying density is likewise already in the neighborhood of the density in the nucleus.

That the comparison of directly adjacent nuclei yielded a very much more modest temperature is not surprising if we consider the strong fluctuations in local frequency distribution. Physically, the temperature must have decreased steadily, even if very rapidly, following the origin of the elements. During this process, nuclear reactions must have run down. Above all, if, simultaneously, the density decreased and thence the lighter nuclei became favored in the equilibrium distribution, it is possible that the energy of the gas would simply no longer suffice to produce the distribution corresponding to the new temperature but would only allow each nucleus to run through a couple of reactions and so to adapt the fine structure of the distribution to the lower temperature. Perhaps we should be able to read from the modern distribution in the large the highest temperature achieved and from the distribution in the small the lowest temperature at which reactions still occurred. The latter temperature, because of the above calculated value, might be the temperature at which the production of neutrons from free protons and electrons ceases.

Now, how can one test these conceptions further? On the practical side, what is needed above all is accurate knowledge of the mass defects of extended isotope sequences, like the oxygen isotopes applied above, through which further quantitative application of equation (1) would be possible. We can demand of the theory that it draw up testable conjectures about when and where in the history of the universe the required temperatures and densities could be realized.

10. THE REALIZATION OF THE CONDITIONS IN THE HISTORY OF THE UNIVERSE Insofar as we know, the universe in its present state contains no region with the required temperature. It is equally hard to imagine an earlier state of the universe out of which the modern stars might have developed continuously, that is, passing through a sequence of equilibrium states, and in which such temperatures would have obtained; for, according to section 5, a star can reach a very high temperature without a subsequent explosion only if it has practically used up its hydrogen. Hence, if anywhere in the universe heavy elements should have originated in this course of continuous evolution, they could not in any case be present in modern stars, because these stars still contain their hydrogen (the hydrogen content presently associated with stars lies far above the equilibrium quantity corresponding to the distribution of heavy elements and can only be understood through the mixing in of matter that has not taken part in a thermal equilibrium with the heavy elements). Therefore, we must seek a possible state of matter before the formation of modern stars.

This state must in any case have been a starlike aggregation of matter: we could not understand the required high density without the cooperation of gravity. Now, all empirically known star types are stable and therefore unsuitable for attaining high temperature; hence, we must seek an experimentally unknown but possible type of star, perhaps a star whose mass lies above the empirical upper limit of stellar masses. In [1] it was conjectured that these stars would be unstable with pulsations of increasing amplitude. The argument given there wa. probably incorrect in details, as it underestimated the actual stability of stars, without delayed energy production.[24] Nonetheless, the assumption of the instability of these stars is probably justified, both because they should otherwise be found experimentally in at least a few examples and because a star should become all the more defenseless against an explosive development of nuclear reactions—such as might also take place singly—the more slowly it reacts to interior changes of state. And, as the dependence between pulsation period and luminosity already demonstrates, the reactivity rate of a star will in any case increase with a growing mass.

One may therefore presuppose a great primeval aggregation of matter perhaps consisting of pure hydrogen. As it collapsed into itself under the influence of gravity and thereby raised its central temperature, it came finally into a state in which nuclear reactions took place in its interior. If its mass was sufficiently small, it could remain stable as a star; if the mass was too large, then the nuclear reactions ran explosively and blew up the star, which then either was lost as a diffuse cloud in space or gathered itself together into a new and smaller star. The process repeated until only stable stars remained. In the explosions, if completed in a fixed volume, temperatures of the order of magnitude required above would be attained, because then the entire energy content of the nuclei would be transiently transformed into heat.

How large should one imagine the first aggregation to have been? Theory sets no upper limit, and our fancy has the freedom to imagine not only the Milky Way system but also the entire universe as known to us combined in it. For this speculation we can even bring up an empirical fact: the energy released in nuclear reactions is about 1% of the rest energy of matter and imparts to the nuclei on the average a velocity of the order of magnitude a tenth the velocity of light. At approximately this speed the fragments of the star should fly apart. If we ask where today speeds of this order of magnitude may be observed, we find them only in the recessional motion of the spiral nebulae. Therefore, we ought at least to reckon with the possibility that this motion has its cause in a primeval catastrophe of the sort considered above. Milne[25] a few years ago showed that an escaping "gas" of spiral nebulae in empty space must obey the Hubble Relation between distance and radial velocity as soon as its extent has become larger than its original volume. This picture fits in with my proposal. In comparison, my proposal differs from Milne's current theory in that it proposes a specific cause for the expansion and therefore foregoes the assumption of an infinite and uniform distribution of spiral nebulae. From the standpoint of cosmological speculation, Milne's theory or one of the older theories of the expanding universe may be more elegant, but it appears to me rather a superiority of the new proposal that it makes superfluous the difficult-to-prove assumption that the presently known part of the universe has qualitatively the same structure as the whole.

An empirical test of the proposal would be possible if the diminution in frequency of the spiral nebulae, which ought to occur above a certain critical radial speed, should lie within the observing range of modern telescopes. The crude approximation according to which this critical velocity is about a tenth the velocity of light is surely wrong by a factor of the order of magnitude 2. In order to obtain a more exact prediction, we would have to be able to describe the explosion process quantitatively.

SUMMARY

PART I

1. The assumption is abandoned that all known chemical elements have originated and still originate in stars existing today; on the following grounds:
2. The heavier elements must, according to [1], have been built up by neutrons whose production is necessarily coupled to the formation of helium. The quantitative investigation of this mechanism leads to the establishment of a lower bound to the abundance of helium in a star, which is incompatible with observation.
3. Uranium and thorium must have been built up from rapidly decaying intermediate nuclei on the way. In contradiction to this assumption of [1], the spatial concentration of energy sources is insufficient to give to the buildup process the rate necessary for this.
4. The explanation of Harkins' rule given in [1] cannot be justified through nuclear physics. The general relation between mass defect and frequency compels the assumption of the origin of the elements at a temperature where the breakdown and buildup processes could be set in equilibrium.

PART II

5. The energy production in stars must depend solely upon reactions of light nuclei.
6. Which reactions are the effective ones cannot yet be decided. If the elements in aggregate did not originate in the modern stars, no autocatalytic reaction cycle need be required. The model cycle of [1] is probably impossible according to nuclear physics. Most probable is a cycle in which carbon as a catalyst brings about the formation of helium.
7. The frequency distribution of the lightest elements that follows from this cycle agrees with observation. For giant and dwarf stars, we must probably assume two different types of reaction, say for the dwarfs a fixed cycle and for the giants the breakdown of nuclei lighter than those participating in the cycle.

PART III

8. I assume that the elements were produced in a thermodynamic equilibrium of nuclear reactions. For the exact verification of the proposed formulae, accurate knowledge of the mass defects of several sequential isotopes would be required. The corroboration of the formulae would justify the drawing of quantitative conclusions about their mass defects from the frequency of nuclear species.
9. From the average uniform frequency distribution of the elements follows an initial temperature of origin of about 2×10^{11}°, which corresponds well with the overall energy released through nuclear reactions. The fine structure of the distribution appears to have been determined through a lower temperature, finally at about 5×10^{9}°.
10. In a star of very large mass, the required temperatures could probably have originated transiently but would have led to its explosion. The relation of this proposal to the recessional motion of the spiral nebulae is discussed.

1. C. F. v. Weizsäcker, *Physik. Zeitschr. 38*, 176 (1973); in the text this reference is denoted [1].

2. St. Meyer, *Sitzungsber. d. Akad. Wiss. Wien. Abt. IIa, 146*, 175 (1937); W. Wefelmeier and M. Hayden, *Naturwiss. 26*, 612 (1938).

3. A. Unsöld, *Physik. der Sternatmosphären* (Berlin, 1938).

4. Mr. Biermann has shown me this; cf. Cowling, *Monthly Not. Roy. Astr. Soc.* (London) *94*, 768 (1934).

5. Cf., say, M. Bodenstein, *Naturwiss. 38*, 609 (1937), and the reports and discussion in *Zeitschr. f. Elektrochemie 42*, 439ff. (1936).

6. L. Landau, *Sow. Phys. 11*. 556 (1937).

7. W. Wefelmeier, *Naturwiss. 25*, 525 (1937), *Zeitschr. f. Phys. 107*, 332 (1937), and a shortly appearing continuation of this latter work.

8. L. Landau, *Sow. Phys. 1*, 285 (1932), *Nature 141*, 333 (1938); F. Hund, *Erg. exact. Naturwiss. 15*, 189 (1936); O. Anderson, *Veröff. d. Univ. Sternw. Dorpat.*; cf., for the following, G. Gamow and E. Teller, *Phys. Rev. 53*, 929 (1938).

9. J. H. Williams, W. G. Shepherd and R. O. Haxby, *Phys. Rev. 52*, 390 (1937). As Mr. Gamow briefly communicated to me shortly before the conclusion of this paper, He^5 and Li^5 are, according to the recent results of Joliot, nonetheless stable. Then one would, after all, consider the model cycle as operative and would have to consider heavy hydrogen the energy source of only the (red) giants.

10. Cf. M. S. Livingston and H. A. Bethe, *Rev. Mod. Phys. 9*, 245 (1937).

11. R. d'E. Atkinson, *Astrophys. J. 84*, 73 (1936); cf. Döpel (below, n. 14).

12. According to an oral communication from Mr. Gamow.

13. Mr. Biermann has demonstrated this to me.

14. R. and K. Döpel, *Zeitschr. f. Astrophys. 14*, 139 (1937).

15. From Mr. Gamow I have learned that Bethe has recently investigated the same cycle.

16. The meaning of the rarity of these elements was probably first emphasized by V. M. Goldschmidt (cf. *Gerlands Beiträge zur Geophysik 15*, 38 [1926]).

17. At least lithium and beryllium; cf. V. M. Goldschmidt, *Geochemische Verteilungsgesetze der Elemente*, vol. 9 (Oslo, 1938). Mr. Wefelmeier has informed me of these facts.

18. According to oral communication from Mr. Gamow.

19. G. Gamow, *Phys. Rev. 53*, 907 (1938).

20. Ibid., p. 595.

21. G. Gamow and E. Teller, *Phys. Rev. 53*, 608 (1938).

22. G. Gamow, *Phys. Rev. 53*, 907 (1938).

23. This test has already been often undertaken. The following publications are known to me: L. Farkas and P. Harteck, *Naturwiss. 19*, 705 (1931); T. E. Sterne, *Monthly Not. 93*, 736 (1933); K. Guggenheimer, *J. de Phys. (A) 5*, 475 (1934).

24. Cf. Cowling (n. 4, above).

25. E. A. Milne, *Zeitschr. f. Astrophys. 6*, 1 (1933).

49. Energy Production in Stars

Hans Albrecht Bethe

(*Physical Review* 55, 434–456 [1939])

The discoveries of the positron and of the neutron, measurements of nuclear reaction cross sections at stellar energies, and the revision of the theory of nuclear penetration for the conditions of stellar interiors were among the advances in the 1930s that set the stage for a renewed attack on stellar energy generation. Although Hans Bethe was one of the few young nuclear physicists familiar with these developments, he was unacquainted with the astrophysical problems of stellar energy sources until April 1938, when George Gamow organized a conference in Washington, D.C., to bring astronomers and physicists together to discuss this problem. Bethe was so stimulated by the meeting that he acquired the necessary astrophysical knowledge and presented this paper within six months.

As Robert Atkinson and C. F. von Weizsäcker had previously shown, the simplest of all nuclear reactions takes place whenever two protons collide to form a deuteron, a positron, and a neutrino.[1] At the time of Gamow's conference, however, astrophysicists believed that this proton-proton reaction was too rare to provide any significant release of energy but thought that this process must serve as the starting point in the origin of the chemical elements. By using Enrico Fermi's theory for the probability of positron emission, together with the Gamow-Teller theory for the probability of nuclear penetration within stars,[2] Bethe and C. L. Critchfield demonstrated that the proton-proton reaction gives an energy evolution of the correct order of magnitude for the sun.[3]

Although the proton-proton reaction predicts the correct energy production for the sun, it depends only weakly on temperature. Because main-sequence stars exhibit a wide variation in luminosity with a small variation in central temperature, it seemed unlikely that the proton-proton reaction would provide the fuel for all of the main-sequence stars. Furthermore, because the luminosity of main-sequence stars increases rapidly with mass, it seemed that the required temperature-dependent reaction must involve the heavier nuclei. Bethe hit upon the CNO cycle, in which carbon acts as a catalyst in the conversion of four protons and two electrons into an alpha particle, as did also C. F. von Weizsäcker independently (see selection 48). It was Bethe, in this paper, who showed that this chain provides the temperature-dependent energy generation that accounts for the luminosity of the more massive main-sequence stars. At the time, Bethe speculated that the CNO cycle is predominant in main-sequence stars more massive than the sun and that the proton-proton reaction predominates for the main-sequence stars less massive than the sun. By the early

1950s however, theoretical models had been used to show that the proton-proton reaction is dominant in the sun,[4] and careful considerations of the proton-proton reaction and the CNO cycle led to the conclusion that the crossover point between the two reactions is 1.5 solar masses.[5] As illustrated in the final selection of this chapter, however, the detective work is as yet unfinished, for the measured level of neutrinos incident on the earth is lower than that expected from both the CNO cycle and the proton-proton reaction.

1. R. d'E. Atkinson, *Astrophysical Journal 84*, 73 (1936).
2. C. F. von Weizsäcker, *Physikalische Zeitschrift 38*, 176 (1937); E. Fermi, *Zeitschrift für Physik 88*, 161 (1934); G. Gamow and E. Teller, *Physical Review 49*, 895 (1936), *53*, 608 (1938).
3. H. A. Bethe and C. L. Critchfield, *Physical Review 54*, 248 (1938).
4. I. Epstein, *Astrophysical Journal 112*, 207 (1950), *114*, 438 (1951); J. B Oke, *Journal of the Royal Astronomical Society of Canada*, *44*, 135 (1950).
5. E. E. Salpeter, *Physical Review 88*, 547 (1952); W. A. Fowler, *Mémoires de la société royale des sciences de Liège IV*, *13*, 88 (1953) *V*, *3*, 207 (1959).

Abstract—It is shown that the *most important source of energy in ordinary stars is the reactions of carbon and nitrogen with protons*. These reactions form a cycle in which the original nucleus is reproduced, *viz.* $C^{12} + H = N^{13}$, $N^{13} = C^{13} + \varepsilon^+$, $C^{13} + H = N^{14}$, $N^{14} + H = O^{15}$, $O^{15} = N^{15} + \varepsilon^+$, $N^{15} + H = C^{12} + He^4$. Thus carbon and nitrogen merely serve as catalysts for the combination of four protons (and two electrons) into an α-particle (§7).

The carbon-nitrogen reactions are unique in their cyclical character (§8). For all nuclei lighter than carbon, reaction with protons will lead to the emission of an α-particle so that the original nucleus is permanently destroyed. For all nuclei heavier than fluorine, only radiative capture of the protons occurs, also destroying the original nucleus. Oxygen and fluorine reactions mostly lead back to nitrogen. Besides, these heavier nuclei react much more slowly than C and N and are therefore unimportant for the energy production.

The agreement of the carbon-nitrogen reactions with observational data (§7, 9) is excellent. In order to give the correct energy evolution in the sun, the central temperature of the sun would have to be 18.5 million degrees while integration of the Eddington equations gives 19. For the brilliant star Y Cygni the corresponding figures are 30 and 32. This good agreement holds for all bright stars of the main sequence, but, of course, not for giants.

For fainter stars, with lower central temperatures, the reaction $H + H = D + \varepsilon^+$ and the reactions following it, are believed to be mainly responsible for the energy production. (§10).

It is shown further (§5–6) that *no elements heavier than* He^4 *can be built up in ordinary stars*. This is due to the fact, mentioned above, that all elements up to boron are disintegrated by proton bombardment (α-emission!) rather than built up (by radiative capture). The instability of Be^8 reduces the formation of heavier elements still further. The production of neutrons in stars is likewise negligible. The heavier elements found in stars must therefore have existed already when the star was formed.

Finally, the suggested mechanism of energy production is used to draw conclusions about astrophysical problems, such as the mass-luminosity relation (§10), the stability against temperature changes (§11), and stellar evolution (§12).

1. INTRODUCTION

THE PROGRESS of nuclear physics in the last few years makes it possible to decide rather definitely which processes can and which cannot occur in the interior of stars. Such decisions will be attempted in the present paper, the discussion being restricted primarily to main sequence stars. The results will be at variance with some current hypotheses.

The first main result is that, under present conditions, no elements heavier than helium can be built up to any appreciable extent. Therefore we must assume that the heavier elements were built up *before* the stars reached their present state of temperature and density. No attempt will be made at speculations about this previous state of stellar matter.

The energy production of stars is then due entirely to the combination of four protons and two electrons into an α-particle. This simplifies the discussion of stellar evolution inasmuch as the amount of heavy matter, and therefore the opacity, does not change with time.

The combination of four protons and two electrons can occur essentially only in two ways. The first mechanism starts with the combination of two protons to form a deuteron with positron emission, *viz.*

$$H + H = D + \varepsilon^+. \qquad (1)$$

The deuteron is then transformed into He^4 by further capture of protons; these captures occur very rapidly compared with process (1). The second mechanism uses carbon and nitrogen as catalysts, according to the chain reaction

$$\begin{aligned} &C^{12} + H = N^{13} + \gamma, \qquad && N^{13} = C^{13} + \varepsilon^+ \\ &C^{13} + H = N^{14} + \gamma, && \\ &N^{14} + H = O^{15} + \gamma, && O^{15} = N^{15} + \varepsilon^+ \\ &N^{15} + H = C^{12} + He^4. && \end{aligned} \qquad (2)$$

The catalyst C^{12} is reproduced in all cases except about one in 10,000, therefore the abundance of carbon and nitrogen remains practically unchanged (in comparison with the change of the number of protons). The two reactions (1) and (2) are about equally probable at a temperature of $16 \cdot 10^6$ degrees which is close to the central temperature of the sun ($19 \cdot 10^6$ degrees[1]). At lower temperatures (1) will predominate, at higher temperatures, (2).

No reaction other than (1) or (2) will give an appreciable contribution to the energy production at temperatures around $20 \cdot 10^6$ degrees such as are found in the interior of ordinary stars. The lighter elements (Li, Be, B) would "burn" in a very short time and are not replaced as is carbon in the cycle (2), whereas the heavier elements (O, F, etc.) react too slowly. Helium, which is abundant, does not react with protons because the product, Li^5, does not exist; in fact, the energy evolution in stars can be used as a strong additional argument against the existence of He^5 and Li^5 (§3).

Reaction (2) is sufficient to explain the energy production in very luminous stars of the main sequence as Y Cygni (although there are difficulties because of the quick exhaustion of the energy supply in such stars which would occur on any theory, §9). Neither of the reactions (1) or (2) is capable of accounting for the energy production in giants; if nuclear reactions are at all responsible for the energy production in these stars it seems that the only ones which could give sufficient energy are

$$\begin{aligned} H^2 + H &= He^3 \\ Li^{6,7} + H &= He^{3,4} + He^4. \end{aligned} \tag{3}$$

It seems, however, doubtful whether the energy production in giants is due to nuclear reactions at all.[2]

We shall first calculate the energy production by nuclear reactions (§2, 4). Then we shall prove the impossibility of building up heavier elements under existing conditions (§5–6). Next we shall discuss the reactions available for energy production (§5,7) and the results will be compared with available material on stellar temperatures and densities (§8, 9). Finally, we shall discuss the astrophysical problems of the mass-luminosity relation (§10), the stability of stars against temperature changes (§11) and stellar evolution (§12).

2. FORMULA FOR ENERGY PRODUCTION

The probability of a nuclear reaction in a gas with a Maxwellian velocity distribution was first calculated by Atkinson and Houtermans.[3] Recently, an improved formula was derived by Gamow and Teller.[4] The total number of processes per gram per second is[4]

$$p = \frac{4}{3^{5/2}} \frac{\rho x_1 x_2}{m_1 m_2} \frac{\Gamma}{\hbar} a R^2 e^{4(2R/a)^{1/2}} \tau^2 e^{-\tau}. \tag{4}$$

Here ρ is the density of the gas, $x_1 x_2$ the concentrations (by weight) of the two reacting types of nuclei, $m_1 m_2$ their masses, $Z_1 e$ and $Z_2 e$ their charges, $m = m_1 m_2/(m_1 + m_2)$ the reduced mass, R the combined radius,

$$a = \hbar^2/me^2 Z_1 Z_2 \tag{5}$$

the "Bohr radius" for the system, $\Gamma/\hbar$ the probability of the nuclear reaction, in sec.$^{-1}$, after penetration, and

$$\tau = 3\left(\frac{\pi^2 m e^4 Z_1{}^2 Z_2{}^2}{2\hbar^2 kT}\right)^{1/3}. \tag{6}$$

If we measure ρ in g/cm^3, Γ in volts and T in units of 10^6 degrees, we have

$$p = 5.3 \cdot 10^{25} \rho x_1 x_2 \Gamma \phi(Z_1, Z_2) \tau^2 e^{-\tau} \quad \text{g}^{-1}\ \text{sec.}^{-1}, \tag{7}$$

$$\tau = 42.7 (Z_1 Z_2)^{2/3} (A/T)^{1/3}, \tag{8}$$

$$\phi = \frac{1}{A_1 A_2 (Z_1 Z_2 A)^3} \left(\frac{8R}{a}\right)^2 e^{2(8R/a)^{1/2}}. \tag{9}$$

where $A_1 A_2$ are the atomic weights of the reacting nuclei ($A_i = m_i/m_H$), $A = m/m_H$, m_H = mass of hydrogen. For the combined radius of nuclei 1 and 2 we put

$$R = 1.6 \cdot 10^{-13} (A_1 + A_2)^{1/3} \text{ cm}. \tag{10}$$

Then we obtain for ϕ the values given in table 49.1. The values of ϕ for isotopes of the same element differ only very slightly.

Table 49.1 Values of ϕ for various nuclear reactions

Reaction	R (10^{-13} cm)	ϕ
$H^2 + H$	2.3	0.38
$H^3 + H$	2.5_5	0.48
$He^4 + H$	2.7_5	0.81
$Li^7 + H$	3.2	0.91
$Be^9 + H$	3.4_5	1.16
$B^{11} + H$	3.6_5	1.52
$C^{12} + H$	3.7_5	2.00
$N^{14} + H$	3.9_5	2.78
$O^{16} + H$	4.1	3.80
$F^{19} + H$	4.3_5	5.5
$Ne^{22} + H$	4.5_5	7.7
$Mg^{26} + H$	4.8	13.2
$Si^{30} + H$	5.0	29.3
$Cl^{37} + H$	5.4	75
$H^2 + H^2$	2.5_5	0.67
$Be^7 + H^2$	3.3	1.18
$Be^7 + He^3$	3.4_5	7.9
$He^4 + H^2$	2.9	0.57
$He^4 + He^3$	3.0_5	1.09
$He^4 + He^4$	3.2	1.29
$Li^7 + He^4$	3.5_5	4.9
$Be^7 + He^4$	3.5_5	13.2
$Be^8 + He^4$	3.6_5	16.2
$C^{12} + He^4$	4.0	230

The values of Γ for reactions giving particles can be deduced from the *observed* cross sections of such reactions with the use of the formula (cf. reference 4, Eq. (2))

$$\sigma = \frac{\pi R^2}{2E} \frac{A_1 + A_2}{A_2} \Gamma \exp\left[\left(\frac{32R}{a}\right)^{1/2} - \frac{2\pi e^2}{\hbar v} Z_1 Z_2\right], \tag{11}$$

where E is the absolute energy of the incident particle (particle 1). Table 49.2 gives the experimental results for some of the better investigated reactions. In each case, experiments with low energy particles were chosen in order to make the conditions as similar as possible to those in stars where the greatest number of nuclear reactions is due to particles of about 20

Table 49.2 Cross sections and widths for some nuclear reactions giving particles

Reaction	Ref.	E kv	σ cm^2	R cm	Γ
$H^2 + H^2 = He^3 + n^1$	7	100	$1.7\cdot10^{-26}$	$2.6\cdot10^{-13}$	$3\cdot10^5$
$Li^7 + H^1 = 2He^4$	8	42	$1.7\cdot10^{-30}$	$3.2\cdot10^{-13}$	$4\cdot10^4$
$Li^6 + H^1 = He^4 + He^3$	yield ~ same as $Li^7 + H$ in natural Li target				$5\cdot10^5$
$Li^5 + H^2 = \begin{cases} 2He^4 \\ Li^7 + H \end{cases}$	9	212	$1.9\cdot10^{-26}$	$3.2\cdot10^{-13}$	$4\cdot10^6$
$Li^7 + H^2 = 2He^4 + n$	9	212	$5.5\cdot10^{-26}$	$3.3\cdot10^{-13}$	10^7
$Be^9 + H^1 = \begin{cases} Li^6 + He^4 \\ Be^8 + H^2 \end{cases}$	10	212	$1.1\cdot10^{-25}$	$3.5\cdot10^{-13}$	$1.7\cdot10^7$
$Be^9 + H^2 = \begin{cases} Li^7 + He^4 \\ Be^8 + H^3 \\ Be^{10} + H^1 \end{cases}$	10	212	$5\cdot10^{-28}$	$3.6\cdot10^{-13}$	$6\cdot10^5$
$B^{11} + H^1 = 3He^4$	11	212	$6\cdot10^{-28}$	$3.7\cdot10^{-13}$	$2\cdot10^6$

kilovolts energy. The cross sections were in each case calculated from the thick target yield with the help of the range-energy relation of Herb, Bellamy, Parkinson and Hudson.[5] The widths obtained (last column of table 49.2) are mostly between $3\cdot10^5$ and $2\cdot10^7$ ev, with the exception of the reaction $Li^7 + H = 2He^4$ which is known to be "improbable."[6]

The γ-ray widths Γ_γ can be obtained from observed resonance capture of protons. Table 49.3 gives the experimental results. Two of the older data were taken from table XXXIX of reference 17; all the others are from more recent experiments on proton[12–15] and neutron[16] capture. Although the results of different investigators differ considerably (e.g., for $Li^7 + H^1 = Be^8$, Γ is between 4 and 40 volts the latter value being more likely) they seem to lie generally between about 1/2 and 40 volts. Ordinarily, the width is somewhat larger for the more energetic γ-rays, as is expected theoretically. A not too bad approximation to the experiments is obtained by using the theoretical formula for dipole radiation (reference 17, Eq. (711b)) with an oscillator strength of 1/50. This gives

$$\Gamma_\gamma \sim 0.1E_\gamma^{\,2}, \tag{12}$$

where E_γ is the γ-ray energy in mMU (milli-mass-units), and Γ_γ the γ-ray width in volts. For quadrupole radiation, theory gives about

$$\Gamma_\gamma \sim 5\cdot10^{-4}E_\gamma^{\,4} \quad \text{(quadrupole)}. \tag{12a}$$

Formulae (12), and (12a) will be used in the calculations where experimental data are not available; they may, in any individual case, be in error by a factor 10 or more but such a factor is not of great importance compared with other uncertainties.

It should be noted that quite generally radiative processes are rare compared with particle emission. According to the figures given in tables 49.2 and 49.3, the ratio of probabilities is 10^4–10^5 in favor of particle reactions.

Table 49.3 γ-ray widths of nuclear levels

Reaction	Reference	Width (volts)	γ-ray energy (Mev)
$Li^7 + H^1 = Be^8 + \gamma$	17	4	17
	12	40	
$B^{11} + H^1 = C^{12} + \gamma$	12	0.6	12, 16
$C^{12} + H^1 = N^{13} + \gamma$	13	0.6	2
$C^{13} + H^1 = N^{14} + \gamma$	14, 15	30	4, 8
$F^{19} + H^1 = Ne^{20} + \gamma$	17	0.6, 8, 18	6
$C^{12} + n^1 = C^{13} + \gamma$	16	<2.5	5
$O^{16} + n^1 = O^{17} + \gamma$	16	<2.5	4

In a number of cases, the reaction of a nucleus A with a heavy particle (proton, alpha-) must compete with natural β-radioactivity of A or with electron capture. In those cases where the lifetime of radioactive nuclei is not known experimentally, we use the Fermi theory. According to this theory, the decay constant for β-emission is[18]

$$\beta = 0.9\cdot10^{-4}f(W)|G|^2 \text{ sec.}^{-1}. \tag{13}$$

The matrix element G is about unity for strongly allowed transitions, and

$$f(W) = (W^2-1)^{1/2}\left(\tfrac{1}{30}W^4 - \tfrac{3}{20}W^2 - \tfrac{2}{15}\right) + \tfrac{1}{4}W\log\{W + (W^2-1)^{1/2}\}, \tag{13a}$$

where W is the maximum energy of the β-particle, including its rest mass, in units of mc^2 (m = electron mass).

The probability of electron capture is

$$\beta_C = 0.9 \cdot 10^{-4} \pi^2 N (\hbar/mc)^3 W^2 |G|^2 \text{ sec.}^{-1}, \tag{14}$$

where W is the energy of the emitted neutrino in units of mc^2 and N the number of electrons per unit volume. If the hydrogen concentration is x_H, we have (reference 1, p. 482) $N = 6 \times 10^{23}\ \rho \cdot (1/2)(1 + x_H)$ (ρ the density), and

$$\beta_C = 1.5 \cdot 10^{-11} \rho (1 + x_H) W^2 |G|^2 \text{ sec.}^{-1}. \tag{14a}$$

3. STABILITY OF UNKNOWN ISOTOPES

For the discussion of nuclear reactions it is essential to know whether or not certain isotopes exist (such as Li^4, Li^5, Be^6, Be^8, B^8, B^9, C^{10}, etc.). The criterion for the existence of a nucleus is its energetic stability against spontaneous disintegration into heavy particles (emission of a neutron, proton or alpha-particle). Whenever a light nucleus is energetically unstable against heavy particle emission, its life will be a very small fraction of a second (usually $\sim 10^{-20}$ sec.) even if the instability is slight (e.g., Be^8 will have a life of 10^{-13} sec. if it is by 50 kv heavier than two α-particles[19]).

For the question of the lifetime of radioactive nuclei, it is also necessary to know the mass difference between isobars. Similar information is required for estimating the γ-ray width in capture reactions (cf. Eqs. (12), (12a)).

Here Bethe summarizes the stability of various isotopes discussed in references 20 through 33. The nucleus He^3 is more stable than H^3 by 0.06 MeV; the nucleus H^4 is possibly stable and Li^4 might be unstable; He^5 and Li^5 are unstable and do not exist. The nucleus Be^6 is unstable against disintegration into $He^4 + 2H$, and Be^8 is unstable against disintegration into two alpha particles. Binding energies of doubtful nuclei are given in table 49.4.

Table 49.4 Corrected and additional nuclear masses, and binding energies

Nucleus	Mass	Binding energy (mMU)	Reference
n^1	1.008 93	—	21
He^3	3.016 99	5.87	20
H^4	4.025 4	0.6 ± 1	—
He^4	4.003 86	—	33
Li^4	4.026 9	−1 ± 1	—
He^5	5.013 7	−0.9 ± 0.2	25
Li^5	5.013 6	−1.6 ± 0.3	—
Be^6	6.021 9	−1.8 ± 0.8	23
Be^7	7.019 28	5.7	29
Be^8	8.007 80	−0.08 ± 0.04	31
B^8	8.027 4	0.0 ± 0.4	23
B^9	9.016 4	−0.5 ± 0.2	23
C^{10}	10.020 2	3.8	23
N^{12}	12.022 5 − 24 3	0.0 ± 0.9	23
N^{13}	13.010 08	2.03	21
O^{14}	14.013 1	5.1	23

4. REACTION RATES AT $2 \cdot 10^7$ DEGREES

We are now prepared to actually calculate the rate of nuclear reactions under the conditions prevailing in stars. We choose a temperature of twenty million degrees, close to the temperature at the center of the sun. In order to have a figure independent of density and chemical composition, we calculate (cf. 7)

$$P = (m_2/x_2) p / \rho x_1, \tag{16}$$

$P \rho x_1$ gives the probability (per second) that a given nucleus of kind 2 undergoes a reaction with any nucleus of kind 1. If there are no other reactions destroying or producing nuclei of kind 2, $1/P\ \rho x_1$ will be the mean life of nuclei of kind 2 in the star.

Table 49.5 gives the results of the calculations, based on Eqs. (7) to (9). In the first column, the nuclear reactions are listed. All reactions which seemed of importance in the interior of stars were considered; in addition, some reactions with heavier elements (O^{16} to Cl^{37}) were included in order to show the manner in which the reaction rate decreases. Moreover, seven reactions were listed in spite of the fact that their products or reactants are believed to be (§3) unstable (starred) or doubtful (question mark); these reactions are included in order to discuss the consequences if they did occur.

The second column gives the energy evolution Q in the reaction, calculated from the masses (reference 25, table LXXIII, and this paper, table 49.4). In the third column, the width Γ determining the reaction rate (cf. §2) is tabulated. Wherever possible, this was taken from experiments (table 49.2 and 49.3) or from the "empirical formulae" (12), (12a) for the radiation width. For the radiative combination of two nuclei of equal specific charge ($H^2 + He^4$, $He^4 + He^4$, $C^{12} + He^4$) quadrupole radiation was assumed, otherwise dipole radiation.[34] For almost equal specific charge (e.g., $Be^7 + He^4$), the dipole formula with an appropriate reduction was used. In some instances, the width was estimated by analogy (e.g., $N^{15} + H = C^{12} + He^4$) or from approximate theoretical calculations ($H^2 + H = He^3$).[35] The way in which Γ was obtained was indicated by a letter in each instance.

Table 49.5 Probability of nuclear reactions at $2 \cdot 10^7$ degrees

Reaction	Q (mMU)	Γ (ev)	τ	P (sec.$^{-1}$)	Life, for $\rho x_1 = 30$
$H + H = H^2 + \varepsilon^+$	1.53	Ref. 18	12.5	$8.5 \cdot 10^{-21}$	$1.2 \cdot 10^{11}$ yr.
$H^2 + H = He^3$	5.9	1 E	13.8	$1.3 \cdot 10^{-2}$	2 sec.
$H^3 + H = He^4$	21.3	10 E	14.3	$1.7 \cdot 10^{-1}$	0.2 sec.
$He^3 + H = Li^{4*}$	(0.5)	0.02 D	22.7	$3 \cdot 10^{-7}$	1 day
$He^4 + H = Li^{5*}$	(0.2)	0.005 D	23.2	$6 \cdot 10^{-8}$	6 days
$Li^6 + H = He^4 + He^3$	4.1	$5 \cdot 10^5$ X	31.1	$7 \cdot 10^{-3}$	5 sec.
$Li^7 + H = 2\ He^4$	18.6	$4 \cdot 10^4$ X	31.3	$6 \cdot 10^{-4}$	1 min.
$Be^7 + H = B^8$?	(0.5)	0.02 D	38.1	$6 \cdot 10^{-13}$	2000 yr.
$Be^9 + H = Li^6 + He^4$	2.4	10^6 X	38.1	$4 \cdot 10^{-5}$	15 min.
$B^9 + H = C^{10*}$	3.5	2 D	44.6	$2 \cdot 10^{-13}$	5000 yr.
$B^{10} + H = C^{11}$	9.2	10 D	44.6	10^{-12}	1000 yr.
$B^{11} + H = 3\ He^4$	9.4	10^6 E	44.6	$1.2 \cdot 10^{-7}$	3 days
$C^{11} + H = N^{12}$	(0.4)	0.02 D	50.6	10^{-17}	10^8 yr.
$C^{12} + H = N^{13}$	2.0	0.6 X	50.6	$4 \cdot 10^{-16}$	$2.5 \cdot 10^6$ yr.
$C^{13} + H = N^{14}$	8.2	30 X	50.6	$2 \cdot 10^{-14}$	$5 \cdot 10^4$ yr.
$N^{14} + H = O^{15}$	7.8	5 D	56.3	$2 \cdot 10^{-17}$	$5 \cdot 10^7$ yr.
$N^{15} + H = C^{12} + He^4$	5.2	10^5 E	56.3	$5 \cdot 10^{-13}$	2000 yr.
$O^{16} + H = F^{17}$	0.5	0.02 D	61.6	$8 \cdot 10^{-22}$	10^{12} yr.
$F^{19} + H = O^{16} + He^4$	8.8	10^5 E	66.9	$4 \cdot 10^{-17}$	$3 \cdot 10^7$ yr.
$Ne^{22} + H = Na^{23}$	10.7	10 D	71.7	$5 \cdot 10^{-23}$	$2 \cdot 10^{13}$ yr.
$Mg^{26} + H = Al^{27}$	8.0	10 D	81.3	10^{-26}	10^{17} yr.
$Si^{30} + H = P^{31}$	7.0	10 D	90.4	$4 \cdot 10^{-30}$	$3 \cdot 10^{20}$ yr.
$Cl^{37} + H = A^{38}$	12.0	10 D	103.1	$5 \cdot 10^{-35}$	$2 \cdot 10^{25}$ yr.
$H^2 + H^2 = He^3 + n$	3.5	$3 \cdot 10^5$ X	15.7	10^3	—
$Be^7 + H^2 = B^{9*}$	18.5	10 D'	45.9	$2 \cdot 10^{-13}$	—
$Be^7 + H^3 = B^9 + n^*$	11.9	10^6 E	50.7	$2 \cdot 10^{-10}$	—
$Be^7 + He^3 = C^{10}$	16.2	1 D'	80.5	$3 \cdot 10^{-28}$	—
$H^2 + He^4 = Li^6$	1.7	$4 \cdot 10^{-3}$ Q	27.5	$3 \cdot 10^{-10}$	—
$He^3 + He^4 = Be^7$	1.6	0.02 D'	47.3	$3 \cdot 10^{-17}$	$3 \cdot 10^7$ yr.
$He^4 + He^4 = Be^{8*}$	(0.05)	$5 \cdot 10^{-9}$ Q	50.0	10^{-24}	—
$Li^7 + He^4 = B^{11}$	9.1	1 D'	71.0	$2.5 \cdot 10^{-24}$	—
$Be^7 + He^4 = C^{11}$	8.0	1 D'	86	$3 \cdot 10^{-30}$	$3 \cdot 10^{20}$ yr.
$C^{12} + He^4 = O^{16}$	7.8	1 Q'	119	$7 \cdot 10^{-43}$	—

a. The letters in the column giving the level width mean: X = experimental value; D = calculated for dipole radiation, from Eq. (12); D' = dipole radiation with small specific charge, 1/4 to 1/20 of Eq. (12); Q = quadrupole radiation, Eq. (12*a*); and E = estimate.

* These reactions are not believed to occur, since their product or one of the reactants is unstable. They are listed merely for the sake of discussion.

The fourth column contains τ, as calculated from (8), the fifth P from (16). The wide variation of P is evident, also the smallness of P for α-particle as compared with proton reactions. E.g., the reaction between He^4 and a nucleus as light as Be is as improbable as between a proton and Si. This arises, of course, from the greater charge and mass of the α-particle both of which factors reduce its penetrability. The reaction $He^4 + He^4 = Be^8$ has an exceedingly small probability because of the small frequency and the quadrupole character of the emitted γ-rays. Thus this reaction would not be important even if Be^8 were stable. On the other hand, the reaction $He^4 + H = Li^5$ would be extremely probable if Li^5 existed. The helium in the sun would be "burnt up" completely in about six days, even if rather unfavorable assumptions are made about the probability of the reaction. Similarly, if the energy evolution per process is 0.2 mMU $= 3 \cdot 10^{-7}$ ergs, the energy produced per gram of the star would be

$$(6\cdot 10^{23}/4)\rho x_{\mathrm{H}} x_{\mathrm{He}}\cdot 3\cdot 10^{-7}\cdot 6\cdot 10^{-8}.$$

With $\rho = 80$, $x_{\mathrm{H}} = 0.35$ and $x_{\mathrm{He}} = 0.1$, this would give about 10^{10} ergs/g sec. as against 2 ergs/g sec. observed. This is a very strong additional argument against the existence of Li^5.

In the last column of table 49.5, the mean life is calculated for the various nuclei reacting with protons, by assuming a density $\rho = 80$ and hydrogen content $x_1 = 35$ percent, which correspond to the values at the center of the sun.[1] It is seen that, with the exception of H, the lifetimes of all nuclei up to boron are quite short, ranging from a fraction of a second for H^3 to 1,000 years for B^{10}. (The life of B^{10} may actually be slightly shorter because of the reaction $\mathrm{B}^{10} + \mathrm{H} = \mathrm{Be}^7 + \mathrm{He}^4$. See §6.) Of the two lives longer than 1,000 years listed, one refers to B^9 which probably does not exist (§3), the other to Be^7 which decays by positron emission with a half-life of 43 days.[29] We must conclude that *all the nuclei between* H *and* C, *notably* H^2, H^3, Li^6, Li^7, Be^9, B^{10}, B^{11}, *can exist in the interior of stars only to the extent to which they are continuously re-formed by nuclear reactions.* This conclusion does not apply to He^4 because Li^5 does not exist. To He^3 it probably applies whether Li^4 exists or not, because He^3 will also be destroyed by combination with He^4 into Be^7, although with a considerably longer period ($3\cdot 10^7$ years instead of the 1 day for the reaction giving Li^4).

The actual lifetime of carbon and nitrogen is much longer than it would appear from the table because these nuclei are reproduced by the nuclear reactions themselves (§7). This makes their actual lifetime of the order of 10^{12} (or even 10^{20}, cf. §7) years, i.e., long compared with the age of the universe ($\sim 2\cdot 10^9$ years). Protons, and all nuclei heavier than nitrogen, also have lives long compared with astronomical times.

5. THE REACTIONS FOLLOWING PROTON COMBINATION

In the last section, it has been shown that all elements lighter than carbon, with the exception of H^1 and He^4, have an exceedingly short life in the interior of stars. Such elements can therefore only be present to the extent to which they are continuously produced in nuclear reactions from elements of longer life. This is in accord with the small abundance of all these elements both in stars and on earth.

Of the two more stable nuclei, He^4 is too inert to play an important rôle. It combines neither with a proton nor with another α-particle since the product would in both cases be an unstable nucleus. The only way in which He^4 can react at all, is by triple collisions. These will be discussed in the next section and will be shown to be very rare, as is to be expected.

As the only primary reaction between elements lighter than carbon, there remains therefore the reaction between two protons,

$$\mathrm{H}^1 + \mathrm{H}^1 = \mathrm{H}^2 + \varepsilon^+. \tag{1}$$

According to Critchfield and Bethe,[18] this process gives an energy evolution of 2.2 ergs/g sec. under "standard stellar conditions" ($2\cdot 10^7$ degrees, $\rho = 80$, hydrogen content 35 percent). The reaction rate under these conditions is (cf. table 49.5) $2.5\cdot 10^{-19}$ sec.$^{-1}$, corresponding to a mean life of $1.2\cdot 10^{11}$ years for the hydrogen in the sun. This lifetime is about 70 times the age of the universe as obtained from the red shift of nebulae.

According to the foregoing, any building up of elements out of hydrogen will have to start with reaction (1). The deuteron will capture another proton,

$$\mathrm{H}^2 + \mathrm{H}^1 = \mathrm{He}^3. \tag{17}$$

This reaction follows almost instantaneously upon (1), with a delay of only 2 sec. (table 49.5). There is, therefore, always statistical equilibrium between protons and deuterons, the concentrations (by number of atoms) being in the ratio of the respective lifetimes. This makes the concentration of deuterons (by weight) equal to

$$x(\mathrm{H}^2) = x(\mathrm{H}^1)\cdot \frac{2\cdot 8.5\cdot 10^{-21}}{1.3\cdot 10^{-2}} = 1.3\cdot 10^{-18} x(\mathrm{H}^1) \tag{18}$$

(cf. table 49.5). The relative probability of the reaction

$$\mathrm{H}^2 + \mathrm{H}^2 = \mathrm{He}^3 + n^1 \tag{19}$$

as compared with (17), is then

$$p_n = \frac{1}{4}\cdot \frac{P(\mathrm{H}^2 + \mathrm{H}^2 = \mathrm{He}^3 + n^1)}{P(\mathrm{H}^2 + \mathrm{H} = \mathrm{He}^3)}\cdot \frac{x(\mathrm{H}^2)}{x(\mathrm{H}^1)}$$

$$= \frac{10^3\cdot 1.3\cdot 10^{-18}}{4\cdot 1.3\cdot 10^{-2}} = 2\cdot 10^{-14}. \tag{19a}$$

(One factor 1/2 comes from the fact that in (19) two nuclei of the same kind interact; another is the atomic weight of H^2.) Thus one neutron is produced for about $5\cdot 10^{13}$ proton combinations.

The further development of the He^3 produced according to (17) depends on the question of the stability of Li^4 and of the relative stability of H^3 and He^3.

ASSUMPTION A: LI^4 STABLE In this case, the He^3 will capture another proton, *viz.*:

$$\mathrm{He}^3 + \mathrm{H}^1 = \mathrm{Li}^4. \tag{20}$$

With the assumptions made in table 49.5, the mean life of He^3 would be 1 day. The Li^4 would then emit a positron:

$$\mathrm{Li}^4 = \mathrm{He}^4 + \varepsilon^+. \tag{20a}$$

With an assumed stability of Li^4 of 0.5 mMU compared with $He^3 + H^1$, the maximum energy of the positrons in (20a) would be 20.8 mMU = 19.4 Mev (including rest mass) which would be by far the highest β-ray energy known. The lifetime of Li^4 may accordingly be expected to be a small fraction of a second (half-life = 1/500 sec. for an allowed transition in the Fermi theory).

The most important consequence of the stability of Li^4 would be that only a fraction of the mass difference between four protons and an α-particle would appear as usable energy. For in the β-emission (20a) the larger part of the energy is, on the average, given to the neutrino which will in general leave the star without giving up any of its energy (see below). According to the Konopinski-Uhlenbeck theory, which is in good agreement with the observed energy distribution in β-spectra, the neutrino receives on the average 5/8 of the total available energy if the latter is large. In our case, this would be 13.0 mMU. Adding 0.2 mMU for the neutrino emitted in process (1), we find that altogether 13.2 mMU energy is lost to neutrinos, of a total of 28.7 mMU developed in the formation of an α-particle out of four protons and two electrons. Thus the observable energy evolution is only 15.5 mMU, i.e., 54 percent of the total. Therefore, if Li^4 is stable, process (1) would give only 1.2 ergs/g sec. instead of 2.2 (under "standard" conditions).[18]

The neutrinos emitted will have some chance of producing neutrons in the outer layers of the star. It seems reasonable to assume that a neutrino has no other interaction with matter than that implied in the β-theory. Then a free neutrino (v) will cause only "reverse β-processes"[36] of which the simplest and most probable is

$$H + v = n^1 + \varepsilon^+. \tag{21}$$

This process is endoergic with 1.9 mMU and is therefore caused only by fast neutrinos such as those from Li^4. The cross section is according to the Fermi theory

$$\begin{aligned}\sigma &= \pi^2(\hbar/mc)^3 \cdot 0.9 \cdot 10^{-4} c^{-1} W(W^2 - 1)^{1/2} \\ &= 1.7 \cdot 10^{-45} W(W^2 - 1)^{1/2} \text{ cm}^2,\end{aligned} \tag{22}$$

where W is the energy of the emitted positron, including rest mass, in units of mc^2. In reaction (21), this is the neutrino energy minus 1.35 mMU. For the Li^4 neutrinos, the average cross section comes out to be

$$\sigma_{Av} = 2.5 \cdot 10^{-43} \text{ cm}^2 \tag{22a}$$

per proton, and the probability of process (21) for a neutrino starting from the center of the star

$$p = 6 \cdot 10^{23} \cdot \sigma_{Av} x_H \int_0^R \rho\, dr = 1.5 \cdot 10^{-19} x_H \int_0^R \rho\, dr, \tag{22b}$$

where $\rho(r)$ is the density (in g/cm^3) at the distance r from the center of the star. For the sun, $p = 1.6 \cdot 10^{-7}$. This means that $1.6 \cdot 10^{-7}$ of the neutrinos emitted will cause reaction (21) before leaving the sun, and that the number of neutrons formed is $1.6 \cdot 10^{-7}$ times the number of proton combinations (1).

A further consequence of (20, 20a) would be that ordinarily no nuclei heavier than 4 mass units are formed at all, even as intermediate products. Such nuclei would only be produced in the rare cases when H^2 or He^3 capture an α-particle rather than a proton, according to the reactions

$$H^2 + He^4 = Li^6 \tag{23}$$

and

$$He^3 + He^4 = Be^7. \tag{24}$$

Under the favorable assumption that the concentration of He^4 is the same as of H^1 (by weight), the fraction of H^2 forming Li^6 is (cf. table 49.5)

$$p(Li^6) = 3 \cdot 10^{-10}/1.3 \cdot 10^{-2} = 2 \cdot 10^{-8} \tag{23a}$$

the fraction of He^3 giving Be^7 is

$$\rho(Be^7) = 3 \cdot 10^{-17}/3 \cdot 10^{-7} = 10^{-10}. \tag{24a}$$

Most of the Li^6 will give rise to the well-known reaction

$$Li^6 + H = He^3 + He^4 \tag{23b}$$

and most of the Be^7 will go over into Li^7 which in turn reacts with a proton to give two α-particles. Only occasionally, Li^6, Be^7 or Li^7 will capture an α-particle and thus form heavier nuclei. It can be shown (cf. assumption B) that Li^7 is the most efficient nucleus in this respect. Therefore, the amount of heavier elements formed is determined by Be^7, the mother substance of Li^7, and is thus 10^{-10} times the amount formed with assumption B.

Assumption B: Li^4 unstable, He^3 more stable than H^3

This assumption seems to be the most likely according to available evidence. The *only* reaction which the He^3 can undergo, is then (24), i.e., each proton combination leads to the formation of a Be^7 nucleus. The most probable mode of decay of this nucleus is by electron capture, leading to Li^7. The lifetime of Be^7 (half-life) is 43 days[29] in the complete atom, and 10 months at the center of the sun (cf. 14, 14a). This makes the mean life = 14 months and the reaction rate $2.8 \cdot 10^{-8}$ sec.$^{-1}$. The capture of a proton by Be^7 would, even if the product B^8 is stable, be 2,000 times slower (table 49.5). Each electron capture by Be^7 is accompanied by the emission of a neutrino of energy $\sim 2mc^2 = 1.1$ mMU (when Li^7 is left in its excited state, which happens rather rarely, the neutrino receives only 0.6 mMU). The total energy lost to neutrinos (including process 1) will therefore be very small in this case ($\sim$1.3 mMU per α-particle formed, i.e., $4\frac{1}{2}$ percent of the total energy evolution) and practically the full mass energy will be transformed into heat radiation. The Li^7 formed by electron capture of Be^7 will cause the well-known reaction

$$Li^7 + H = 2\,He^4 \tag{25}$$

and have (table 49.5) a mean life of only 1 minute at $2\cdot 10^7$ degrees.

The reaction chain described leads, as in the case of assumption A, to the building up of one α-particle out of four protons and two electrons, for each process (1). No nuclei heavier than He^4 are formed permanently. Such nuclei can be produced only by branch reactions alternative to the main chain described. These will be discussed in the following.

a. Reactions with protons.—When Li^7 reacts with slow protons, the result is not always two α-particles, but, in one case out of about[37] 5,000, radiative capture, giving Be^8. However, Be^8 will disintegrate again into two α-particles (§3), and during its life of about 10^{-13} sec., the probability of its reacting with another particle (e.g., capture of another proton) is exceedingly small ($\sim 10^{-24}$). Similarly, Be^7 will, in one out of about 2,000 cases (see above) capture a proton and form B^8 if that nucleus exists. However, B^8 will again go over into Be^8, by positron emission, and two alphas will again be the final result.

At this place, obviously, stability of Be^8 would increase the yield of heavy nuclei. Then one stable Be^8 would be formed for 5,000 proton combinations; and, if B^9 is also assumed to be stable, every Be^8 goes over into B^9. Since about one out of $3\cdot 10^8$ B^9 gives a C^{12} (§6), the number of heavy nuclei (C^{12}) formed would be $\sim 10^{-12}$ per α-particle. This would be the highest yield obtainable. However, Be^8 is known to be unstable (§3).

b. Reactions with α-particles.—The only abundant light nucleus other than the proton is He^4. The only reaction possible between an α-particle and Li^7 or Be^7 is radiative capture, *viz.*

$$Li^7 + He^4 = B^{11}, \tag{26}$$

$$Be^7 + He^4 = C^{11}. \tag{26a}$$

The probability of formation of B^{11} and C^{11} is (table 49.5)

$$p(B^{11}) = \frac{P(Li^7 + He^4)}{P(Li^7 + H)} = \frac{2.5\cdot 10^{-24}}{6\cdot 10^{-4}} = 4\cdot 10^{-21}, \tag{26b}$$

$$p(C^{11}) = \frac{P(Be^7 + He^4)}{P(Be^7 + \varepsilon = Li^7)} = \frac{14 \text{ months}}{3\cdot 10^{20}\text{ yr.}} = 4\cdot 10^{-21}. \tag{26c}$$

Thus the formation of B^{11} is about as probable as that of C^{11}; the effect of the lower potential barrier of Li^7 for α-particles is compensated by its shorter life. The C^{11} will, of course, also give B^{11} by positron emission.

The B^{11} will react with protons in two ways, *viz.*

$$B^{11} + H^1 = 3\,He^4, \tag{27}$$

$$B^{11} + H^1 = C^{12}. \tag{27a}$$

The branching ratio is about 10^4:1 in favor of (27) (calculated from experimental data). Thus there will be one C^{12} nucleus formed for about 10^{24} α-particles. The building up of heavier nuclei, even in this most favorable case, is therefore exceedingly improbable.

c. Reactions with He^3.—Since He^3 has a rather long life $3\cdot 10^7$ years, table 49.5) and penetrates more easily through the potential barrier than the heavier He^4, it may be considered as an alternative possibility. However, the probability of formation of C^{10} from $Be^7 + He^3$ is only 100 times greater than that of C^{11} from $Be^7 + He^4$ (table 49.5) if the concentrations of He^3 and He^4 are equal. Actually, that of He^3 is only about $3\cdot 10^{-4}$ (life of He^3 divided by life of protons) so that this process is 1/30 as probable as (26a). For $Li^7 + He^3$, the situation is even less favorable.

d. Reactions with H^2.—Deuteron capture by Be^7 would lead to B^9 whose existence is very doubtful. The probability per second would be (cf. table 49.5 and Eq. (18))

$$2\cdot 10^{-13}\rho x(H^2) = 2\cdot 10^{-13}\cdot 30\cdot 1.3\cdot 10^{-18} = 10^{-29}, \tag{28}$$

which is only 1/10 of the probability of (26a) (table 49.5). Moreover, most of the B^9 formed reverts to He^4 (§6) so that the contribution of this process is negligible.

e. Reactions with He^4 in statu nascendi.—The process (25) produces continuously fast α-particles which need not penetrate through potential barriers. These α-particles have a range of 8 cm each in standard air, corresponding to 16 cm = 0.02 g/cm^2 for both. In stars, with their large hydrogen content, a somewhat smaller figure must be used since hydrogen has a greater stopping power per gram; we take 0.01 g/cm^2. The cross section for fast particles is about

$$\sigma = \pi R^2 \frac{\Gamma}{\hbar^2/MR^2}. \tag{29}$$

With $\Gamma = 1$ volt (table 49.5) and $R = 3.6\cdot 10^{-13}$ cm (Eq. (10)), this gives $\sigma = 1.3\cdot 10^{-31}$ cm^2. The number of Be^7 atoms per gram is

$$6\cdot 10^{23}x(Be^7) = 6\cdot 10^{23}\cdot \frac{14 \text{ months}}{1.2\cdot 10^{11}\text{ yr}}x(H) = 2\cdot 10^{12} \tag{29a}$$

with $x(\mathrm{H}) = 0.35$. This gives for the number of processes (26a) per proton combination:

$$0.01 \cdot 2 \cdot 10^{12} \cdot 1.3 \cdot 10^{-31} = 2.5 \cdot 10^{-21}, \qquad (29b)$$

which is about the same as the formation of C^{11} or B^{11} by capture of *slow* alphas (cf. 26b, c).

Returning to the main reaction chain in the case of our assumption B, we note that the formation of Be^7 (Eq. 24)) is a very slow reaction, requiring $3 \cdot 10^7$ years at "standard" conditions ($2 \cdot 10^7$ degrees). At lower temperatures, the reaction will be still slower and, finally, it will take longer than the past life of the universe ($\sim 2 \cdot 10^9$ years). In this case, the amount of He^3 present will be much smaller than its equilibrium value (provided there was no He^3 "in the beginning") and the energy production due to reactions (24), (25) will be reduced accordingly. Ultimately, at very low temperatures ($< 12 \cdot 10^6$ degrees), the reaction H + H will lead only to He^3, and will therefore give an energy production of only 7.2 mMU, i.e., only one-quarter of the high temperature value, 27.4 mMU.

ASSUMPTION C: H^3 MORE STABLE THAN HE^3 In this case, He^3 will be able to capture an electron,

$$He^3 + \varepsilon^- = H^3. \qquad (30)$$

Under the assumption that a difference in mass of 0.1 electron mass exists between He^3 and H^3, the probability of (30) is, according to (14a),

$$p(H^3) = 1.5 \cdot 10^{-11}\ \mathrm{sec.}^{-1} \qquad (30a)$$

for a density $\rho = 80$ and 35 percent hydrogen content. This corresponds to a mean life of $\sim$2,000 years. The electron capture is therefore about 10^4 times more probable than the formation of Be^7 according to (24). This ratio will be reversed at temperatures $> 4 \cdot 10^7$ since (30) is independent of T and the probability of (24) increases as T^{15}.

H^3 will capture a proton and form He^4,

$$H^3 + H = He^4$$

with a mean life of about 0.2 seconds. This way of formation of He^4 from reaction (1) is probably the most direct of all. As in *B*, practically no energy is lost to neutrinos.

The formation of heavier elements goes as in *B*, but now there is only one Be^7 formed for 10^4 proton combination processes. This reduces the probability of formation of C^{12} by another factor 10^4, to one C^{12} in 10^{28} alphas.

The H^3 itself does not contribute appreciably to the building up of elements. It is true that the reaction $Be^7 + H^3 = B^9 + n^1$ is about 100 times as probable as $Be^7 + H^2 = B^9$ (considering the shorter life of H^3), and therefore 10 times as probable as (26a). However, most of the B^9 reverts to He^4 (cf. §6) so that (26) and (26a) remain the most efficient processes for building up C^{12}.

Summarizing, we find that the formation of nuclei heavier than He^4 can occur only in negligible amounts. One C^{12} in 10^{24} α-particles and one neutron in 10^{14} α-particles are the yields when Li^4 is unstable, one C^{12} in 10^{34} alphas and one neutron in 10^7 alphas when Li^4 is stable. The reason for the small probability of formation of C^{12} is twofold: First, any nonradioactive nucleus between He and C, i.e., $Li^{6,7}$, Be^9, $B^{10,11}$, reacting with protons will give α-particle emission rather than radiative capture so that a disintegration takes place rather than a building up. This will no longer be the case for heavier nuclei so that for these a building up is actually possible. Second, the instability of Be^8 causes a gap in the list of stable nuclei which is the harder to bridge because Be^8 is very easily formed in nuclear reactions (small mass excess). On the other hand, the instability of He^5 and Li^5 is of no influence because Be^7 and Li^7 are stages in the ordinary chain of nuclear reactions.

6. Triple Collisions of Alpha-Particles

In the preceding section, we have shown that collisions with protons alone lead practically always to the formation of α-particles. In order that heavier nuclei be formed, use must therefore certainly be made of the α-particles themselves. However, collisions of an α-particle with one other particle, proton or alpha, do not lead to stable nuclei. Therefore we must assume triple collisions, of which three types are conceivable:

$$He^4 + 2H = Be^6, \qquad (31)$$

$$2He^4 + H = B^9, \qquad (32)$$

$$3He^4 = C^{12}. \qquad (33)$$

The first of these reactions leads to a nucleus which is certainly unstable (Be^6). Even if it were stable, it would not offer any advantages over Be^7 which is formed as a consequence of the proton combination (1). The second reaction leads to B^9 which is probably also unstable. However, since this is not absolutely certain, we shall discuss this process in the following. The last process leads directly to C^{12}, but since it involves a rather large potential barrier for the last α-particle, it is very improbable at $2 \cdot 10^7$ degrees (see below).

THE FORMATION OF B^9 The probability of this process is enhanced by the well-known resonance level of Be^8, which corresponds to a kinetic energy of about $E = 50$ kev of two α-particles. The formation of B^9 occurs in two stages,

$$2He^4 = Be^8, \quad Be^8 + H = B^9 \qquad (32a)$$

with a time interval of about 10^{-13} sec. (life of Be^8). The process can be treated with the usual formalism for resonance disintegrations, the compound nucleus being Be^8. This nucleus can "disintegrate" in two ways (a) into two α-particles, (b) with proton capture. We denote the respective widths of the Be^8 level by Γ_α and Γ_H; the latter is given by the ordinary theory of thermonuclear processes, i.e.,

$$\Gamma_H = \hbar p m_2/x_2, \tag{34}$$

where p is given by (4), and the subscripts 1 and 2 denote H^1 and Be^8, respectively.

The cross section of the resonance disintegration becomes then

$$\sigma = \pi\lambda^2 \frac{\Gamma_\alpha \Gamma_H}{(E - E_r)^2 + \frac{1}{4}(\Gamma_\alpha + \Gamma_H)^2}. \tag{35}$$

E_r is the resonance energy. Γ_α is much larger than Γ_H (corresponding to about 10^{13} and 10^{-11} sec.$^{-1}$, respectively) but very small compared with E_r (about 10^{-2} against $5 \cdot 10^4$ volts). The resonance is thus very sharp, and, integrating over the energy, we obtain for the total number of processes per cm^3 per sec. simply

$$\rho p = B(E_r) v_r \pi \lambda_r^2 2\pi \Gamma_H. \tag{36}$$

Here $B(E)\,dE$ is the number of pairs of α-particles with relative kinetic energy between E and $E + dE$ per cm^3, *viz.*

$$B(E) = \frac{1}{2}\left(\frac{\rho x_\alpha}{m_\alpha}\right)^2 \frac{2}{\pi^{1/2}} \frac{E^{1/2}}{(kT)^{3/2}} e^{-E/kT} \tag{36a}$$

(x_α = concentration of He^4 by weight). Combining (36), (36a) and (34), (4), we find

$$p(B^9) = \frac{16\pi^{3/2}}{3^{5/2}} \frac{\rho^2 x_\alpha^2 x_H}{m_\alpha^{7/2} m_H} \frac{\hbar^2 \Gamma_{rad}}{(kT)^{3/2}} aR^2\tau^2 \times \exp\{4(2R/a)^{1/2} - \tau - E_r/kT\}, \tag{37}$$

where Γ_{rad} is the radiation width for the process $Be^8 + H = B^9$. Numerically, (37) gives for the decay constant of hydrogen the value

$$p m_H / x_H = 1.00 \cdot 10^{-4} (\rho x_\alpha)^2 \Gamma \phi T^{-3/2} e^{-11.7 E_r/T} \tau^2 e^{-\tau}, \tag{37a}$$

where T is measured in millions of degrees, the resonance energy E_r of Be^8 in kilo-electron-volts and Γ in ev. The quantities Γ, ϕ and τ refer to the process $Be^8 + H = B^9$. If there were no resonance level of Be^8, (37) would be replaced by

$$p(B^9) = \frac{8}{243} \frac{\rho^2 x_\alpha^2 x_H}{m_\alpha^2 m_H} \frac{\Gamma_{rad}}{\hbar} aR^2 a' R'^2 \tau^2 \tau'^2 \times \exp\{(32R/a)^{1/2} + (32R'/a')^{1/2} - \tau - \tau'\}, \tag{38}$$

where the primed quantities refer to the reaction $2He^4 = Be^8$, the unprimed ones to $Be^8 + H = B^9$. Numerically, (38) gives

$$p m_H / x_H = 2.0 \cdot 10^{-11} (\rho x_\alpha)^2 \Gamma \phi \tau^2 e^{-\tau} \phi' \tau'^2 e^{-\tau'}. \tag{38a}$$

Assuming $T = 20$, $\rho = 80$, $x_\alpha = 0.25$, $\Gamma = 0.02$ electron-volts, we obtain for the probability of formation of B^9 per proton per second:

$p m_H / x_H = 2 \cdot 10^{-25}$ for resonance, $E_r = 25$ kev
10^{-31} for resonance, $E_r = 50$ kev
$4 \cdot 10^{-38}$ for resonance, $E_r = 75$ kev
$2 \cdot 10^{-44}$ for resonance, $E_r = 100$ kev
$5 \cdot 10^{-42}$ for nonresonance.

The value 25 kev for the resonance level must probably be excluded on the basis of the experiments of Kirchner and Neuert.[31] But even for this low value of the resonance energy, the probability of formation of B^9 is only 10^{-6} times that of the proton combination $H + H = H^2 + \varepsilon^+$ (table 49.5, $\rho x_1 = 30$). With $E_r = 50$ kev which seems a likely value, the ratio becomes $4 \cdot 10^{-13}$. On the other hand, the building up of B^9 (if this nucleus exists) would still be the most efficient process for obtaining heavier elements (see below).

REACTIONS OF B^9 It can easily be seen[23] that B^9 cannot be positron-active but can only capture electrons if it exists at all. If B^9 is stable by 0.3 mMU, the energy evolution in electron capture would be just one electron mass. The decay constant of B^9 (for β-capture) is then, according to (14a), $1.5 \cdot 10^{-9}$ sec.$^{-1}$ ($\rho = 80$, $x_H = 0.35$) corresponding to a lifetime of about 20 years. On the other hand, the lifetime with respect to proton capture (table 49.5) is 5,000 years. Therefore, ordinarily B^9 will go over into Be^9. This nucleus, in turn, will in general undergo one of the two well-known reactions:

$$Be^9 + H = Be^8 + H^2, \tag{39}$$

$$Be^9 + H = Li^6 + He^4. \tag{39a}$$

Only in one out of about 10^5 cases, B^{10} will be formed by radiative proton capture. Therefore the more efficient way for building up heavier elements will be the direct proton capture by B^9, leading to C^{10}, which occurs in one out of about 300 cases.

The C^{10} produced will go over into B^{10} by positron emission. B^{10} may react in either of the following ways:

$$B^{10} + H = C^{11}, \tag{40}$$

$$B^{10} + H = Be^7 + He^4. \tag{40a}$$

The reaction energy of (40a) is (cf. table 49.7, §8) 1.2 mMU; the penetrability of the outgoing alphas about 1/40 (same table), therefore the probability of the particle reaction (40a) will be about 100 times that of the capture reaction (40).

The C^{11} from (40) will emit another positron. The resulting B^{11} reacts with protons as follows:

$$B^{11} + H = C^{12}, \tag{41}$$

$$B^{11} + H = 3He^4. \tag{41a}$$

Both reactions are well-known experimentally. Reaction (41) has a resonance at 160 kev. From the width of this resonance and the experimental yields, the probability of (41) with low energy protons is about 1 in 10,000 (i.e., the same as for nonresonance). Altogether, about one B^9 in $3 \cdot 10^8$ will transform into C^{12}.

With a resonance energy of Be^8 of 50 kev, and $2 \cdot 10^7$ degrees, there will thus be about one C^{12} formed for 10^{21} α-particles if B^9 is stable. This is better than any other process but still negligibly small.

At higher temperatures, the formation of B^9 will become more probable and will, for $T > 10^8$, exceed the probability of the proton combination. At these temperatures (actually already for $T > 3 \cdot 10^7$) the B^9 will rather capture a proton (giving C^{10}). Even then, there remain the unfavorable branching ratios in reactions (40), (40a) and (41), (41a),[38] so that there will still be only one C^{12} formed in 10^6 alphas. Thus even with B^9 stable and granting the excessively high temperature, the amount of heavy nuclei formed is extremely small.

DIRECT FORMATION OF C^{12} C^{12} may be formed directly in a collision between 3 α-particles. The calculation of the probability is exactly the same as for the formation of B^9. The nonresonance process gives about the same probability as a resonance of Be^8 at 50 kev. With $\rho = 80$, $x_\alpha = 1/4$, $\Gamma = 0.1$ electron-volt, $T = 2 \cdot 10^7$ degrees, the probability is 10^{-56} per α-particle, i.e., about 10^{-37} of the proton combination reaction (1). This gives an even smaller yield of C^{12} than the chains described in this and the preceding section. The process is strongly temperature-dependent, but it requires temperatures of $\sim 10^9$ degrees to make it as probable as the proton combination (1).

The considerations of the last two sections show that there is no way in which nuclei heavier than helium can be produced permanently in the interior of stars under present conditions. We can therefore drop the discussion of the building up of elements entirely and can confine ourselves to the energy production which is, in fact, the only observable process in stars.

7. THE CARBON-NITROGEN GROUP

In contrast to lighter nuclei, C^{12} is not permanently destroyed when it reacts with protons; instead the following chain of reactions occurs

$$C^{12} + H^1 = N^{13}, \tag{42}$$

$$N^{13} = C^{13} + \varepsilon^+, \tag{42a}$$

$$C^{13} + H^1 = N^{14}, \tag{42b}$$

$$N^{14} + H^1 = O^{15}, \tag{42c}$$

$$O^{15} = N^{15} + \varepsilon^+, \tag{42d}$$

$$N^{15} + H^1 = C^{12} + He^4. \tag{42e}$$

Thus the C^{12} nucleus is reproduced. The reason is that the alternative reactions producing α-particles, *viz.*

$$C^{12} + H^1 = B^9 + He^4, \tag{43}$$

$$C^{13} + H^1 = B^{10} + He^4, \tag{43a}$$

$$N^{14} + H^1 = C^{11} + He^4, \tag{43b}$$

are all strongly forbidden energetically. This in turn is due to the much greater stability of the nuclei in the carbon-nitrogen group as compared with the beryllium-boron group, and is in contrast to the reactions of Li, Be and B with protons which all lead to emission of α-particles.

The cyclical nature of the chain (42) means that practically no carbon will be consumed. Only in about 1 out of 10^4 cases, N^{15} will capture a proton rather than react according to (42e). In this case, O^{16} is formed:

$$N^{15} + H^1 = O^{16}. \tag{44}$$

However, even then the C^{12} is not permanently destroyed, because except in about one out of $5 \cdot 10^7$ cases, O^{16} will again return to C^{12} (cf. §8). Thus there is less than one C^{12} permanently consumed for 10^{12} protons. Since the concentration of carbon and nitrogen, according to the evidence from stellar spectra, is certainly greater than 10^{-12} this concentration does not change noticeably during the evolution of a star. *Carbon and nitrogen are true catalysts; what really takes place is the combination of four protons and two electrons into an α-particle.*

A given C^{12} nucleus will, at the center of the sun, capture a proton once in $2.5 \cdot 10^6$ years (table 49.5), a given N^{14} once in $5 \cdot 10^7$ years. These times are short compared with the age of the sun. Therefore the cycle (42) will have repeated itself many times in the history of the sun, so that statistical equilibrium has been established between all the nuclei occurring in the cycle, *viz.* $C^{12}C^{13}N^{13}N^{14}N^{15}O^{15}$. In statistical equilibrium, the concentration of each nucleus is proportional to its lifetime. Therefore N^{14} should have the greatest concentration, C^{12} less, and $C^{13}N^{15}$ still less. (The concentration of the radioactive nuclei N^{13} and O^{15} is, of course, very small, about 10^{-12} of N^{14}). A comparison of the observed abundances of C and N is not very conclusive, because of the very different chemical properties. However, a comparison of the isotopes of each element should be significant.

In this respect, the result for the carbon isotopes is quite satisfactory. C^{13} captures slow protons about 70 times as easily as C^{12} (experimental value!), therefore C^{12} should be 70 times as abundant. The actual abundance ratio is 94:1. The same fact can be expressed in a more "experimental" way: In equilibrium, the number of reactions (42) per second should be the same as of (42b). Therefore, if a natural sample of terrestrial carbon (which is presumed to reproduce the solar equilibrium) is bombarded with protons, equally many captures should occur due to each carbon isotope. This is what is actually found experimentally;[14] the equality of the γ-ray intensities from C^{12} and C^{13} is, therefore, not accidental.[39]

The greater abundance of C^{12} is thus due to the smaller probability of proton capture which in turn appears to be due to the smaller $h\nu$ of the capture γ-ray. Thus the great energetic stability of C^{12} actually makes this nucleus abundant. However, it is not because of a Boltzmann factor as has been believed in the past, but rather because of the small energy evolution of the proton capture reaction.

In nitrogen, the situation is different. Here N^{14} is energetically less stable (has higher mass excess) than N^{15} but is more abundant in spite of it (abundance ratio $\sim$500:1). This must be due to the fact that N^{15} can give a $p-\alpha$ reaction while N^{14} can only capture a proton; particle reactions are always much more probable than radiative capture. Thus the greater abundance of N^{14} is due not to its own small mass excess but to the large mass excess of C^{11} which would be the product of the $p-\alpha$ reaction (43b).

Quantitative data on the nitrogen reactions (42c), (42e) are not available, the figures in table 49.5 are merely estimates. If our theory about the abundance of the nitrogen isotopes is correct, the ratio of the reaction rates should be 500:1, i.e., either $N^{14} + H = O^{15}$ must be more probable[40] or $N^{15} + H = C^{12} + He^4$ less probable than assumed in table 49.5. Experimental investigations would be desirable.

Turning now to the energy evolution, we notice that the cycle (42) contains two radioactive processes (N^{13} and O^{15}) giving positrons of 1.3_5 and 1.8_5 mMU maximum energy, respectively. If we assume again that $\frac{5}{8}$ of the energy is, on the average, given to neutrinos, this makes 2.0 mMU neutrino energy per cycle, which is 7 percent of the total energy evolution (28.7 mMU). There are therefore $4.0\cdot 10^{-5}$ ergs available from each cycle. (It may be mentioned that the neutrinos emitted have too low energy to cause the transformation of protons into neutrons according to (21)).

The duration of one cycle (42) is equal to the sum of the lifetimes of all nuclei concerned, i.e., practically to the lifetime of N^{14}. Thus each N^{14} nucleus will produce $4.0\cdot 10^{-5}$ erg every $5\cdot 10^7$ years, or $3\cdot 10^{-20}$ erg per second. Under the assumption of a N^{14} concentration of 10 percent by weight, this gives an energy evolution of

$$\frac{6\cdot 10^{23}\cdot 0.1}{14}\cdot 3\cdot 10^{-20} \approx 100 \text{ ergs/g sec.} \tag{45}$$

at "standard stellar conditions," i.e., $T = 2\cdot 10^7$, $\rho = 80$, hydrogen concentration 35 percent.

This result is just about what is necessary to explain the observed luminosity of the sun. Since the nitrogen reaction depends strongly on the temperature (as T^{18}) and the temperature, as well as the density, decrease rapidly from the center of the sun outwards, the average energy production will be only a fraction, perhaps 1/10 to 1/20, of the production at the center.[41] This means that the average production is 5 to 10 ergs/g sec., in excellent agreement with the observed luminosity of 2 ergs/g sec.

Thus we see that the reaction between nitrogen and protons which we have recognized as the logical reaction for energy production from the point of view of nuclear physics, also agrees perfectly with the observed energy production in the sun. This result can be viewed from another angle: We may ignore, for the moment, all our nuclear considerations and ask simply which nucleus will give us the right energy evolution in the sun? Or conversely: Given an energy evolution of 20 ergs/g sec. at the center of the sun, which nuclear reaction will give us the right central temperature ($\sim$ 19.10^6 degrees)?

Table 49.6 Central temperatures necessary for giving observed energy production in sun, with various nuclear reactions

Reaction	T (million degrees)
$H^2 + H = He^3$	0.36
$He^4 + H = Li^5$	2.1
$Li^7 + H = 2He^4$	2.2
$Be^9 + H = Li^6 + He^4$	3.3
$B^{10} + H = C^{11}$	9.2
$B^{11} + H = 3He^4$	5.5
$C^{12} + H = N^{13}$	15.5
$N^{14} + H = O^{15}$	18.3
$O^{16} + H = F^{17}$	32
$Ne^{22} + H = Na^{23}$	37

This calculation has been carried out in table 49.6. It has been assumed that the density is 80, the hydrogen concentration 35 percent and the concentration of the other reactant

10 percent by weight. The "widths" were assumed the same as in table 49.5. Given are the necessary temperatures for an energy production of 20 ergs/g sec. It is seen that all nuclei up to boron require extremely low temperatures in order not to give too much energy production; these temperatures ($<10^7$ degrees) are quite irreconcilable with the equations of hydrostatic and radiation equilibrium. On the other hand, oxygen and neon would require much too high temperatures. Only carbon and nitrogen require nearly, and nitrogen in fact exactly, the central temperature obtained from the Eddington integrations ($19 \cdot 10^6$ degrees). Thus from stellar data alone we could have predicted that the capture of protons by N^{14} is the process responsible for the energy production.

8. Reactions with Heavier Nuclei

Mainly for the sake of completeness, we shall discuss briefly the reactions of nuclei heavier than nitrogen. For the energy production, these reactions are obviously of no importance because the higher potential barrier of the heavier nuclei makes their reactions much less probable than those of the carbon-nitrogen group.

The most important point for a qualitative discussion is the question whether a $p-\alpha$ reaction is energetically possible for a particular nucleus, and, if possible, whether it is impeded by the potential barrier.

Here a table of the energy evolution, the reaction energy, and the penetrability for the $p-\alpha$ reactions of all nonradioactive nuclei up to chlorine is given. The emission of alpha particles is strongly preferred for nuclei up to boron, although for most heavier nuclei it is not preferred. These considerations demonstrate the uniqueness of the carbon-nitrogen cycle.

9. Agreement with Observations

In table 49.7, we have made a comparison of our theory (carbon-nitrogen reaction) with observational data for five stars for which such data are given by Strömgren.[1] The first five columns are taken from his table, the last contains the necessary central temperature to give the correct energy evolution with the carbon-nitrogen reactions (cf. table 49.6). As in §7, we have assumed a N^{14} content of 10 percent, and an energy production at the center of ten times the average energy production (listed in the second column).

The result is highly satisfactory: The temperatures necessary to give the correct energy evolution (last column) agree very closely with the temperatures obtained from the Eddington integration (second last column). The only exception from this agreement is the giant Capella: This is not surprising because this star has greater luminosity than the sun at smaller density and temperature; such a behavior cannot possibly be explained by the same mechanism which accounts for the main sequence. We shall come back to the problem of energy production in giants at the end of this section.

For the main sequence we observe that the small increase of central temperature from the sun to Y Cygni (19 to $33 \cdot 10^6$ degrees) is sufficient to explain the much greater energy production (10^4 times) of the latter. The reason for this is, of course, the *strong temperature dependence of our reactions* ($\sim T^{18}$, cf. §10). We may say that *astrophysical data themselves would demand such a strong dependence* even if we did not know that the source of energy are nuclear reactions. The small deviations in table 49.7 can, of course, easily be attributed to fluctuations in the nitrogen content, opacity, etc.

In judging the agreement obtained, it should be noted that the "observational" data in table 49.7 were obtained by integration of an Eddington mode,[1,42] i.e., the energy production was assumed to be almost constant throughout the star. Since our processes are strongly temperature dependent, the "point source" model should be a much better approximation. How-

Table 49.7 Comparison of the carbon-nitrogen reaction with observations

Star	Luminosity (erg/g sec.)	Central density	H content (percent)	Central temperature (Million degrees)	
				Integration	Energy Production
Sun	2.0	76	35	19	18.5
Sirius A	30	41	35	26	22
Capella	50	0.16	35	6	32
U Ophiuchi (bright)	180	12	50	25	26
Y Cygni (bright)	1200	6.5	80	32	30

ever, it seems that the results of the two models are not very different so that the Eddington model may suffice until accurate integrations with the point source model are available.[43]

Since our theory gives a definite mechanism of energy production, it permits decisions on questions which have been left unanswered by astrophysicists for lack of such a mechanism. The first is the question of the "model," which is answered in favor of one approximating a point source model. The second is the problem of chemical composition. The equilibrium conditions permit for the sun a hydrogen content of either 35 or 99.5 percent when there is no helium, and intermediate values when there is helium. The central temperature varies from $19 \cdot 10^8$ to $9.5 \cdot 10^6$ when the hydrogen content increases from 35 to 99.5 percent. It is obvious that the latter value can be definitely excluded on the basis of our theory: The energy production due to the carbon-nitrogen reaction would be reduced by a factor of about 10^8 (100 for nitrogen concentration, 10^6 for temperature). The proton combination (1) would still supply about 5 percent of the observed luminosity; but apart from the fact that a factor 20 is missing, the proton combination does not depend sufficiently on temperature to explain the larger energy production in brighter stars of the main sequence. Thus it seems that only a small range of hydrogen concentrations around 35 percent is permitted; what this range is, depends to some extent on the N^{14} concentration and also requires a more accurate determination of the distribution of temperature and density.

Next, we want to point out a rather well-known difficulty about the energy production of very heavy stars such as Y Cygni. With an energy production of 1,200 ergs/g sec., and an available energy of $1.0 \cdot 10^{-5}$ erg per proton (formation of α-particles!), *all* the energy will be consumed in $1.7 \cdot 10^8$ years. Since at present Y Cygni still has a hydrogen content of 80 percent, its past life should be less than $3.5 \cdot 10^7$ years. We must therefore conclude that Y Cygni and other heavy stars were "born" comparatively recently—by what process, we cannot say. This difficulty, however, is not peculiar to our theory of stellar energy production but is inherent in the well-founded assumption that nuclear reactions are responsible for the energy production.[44]

Finally, we want to come back to the problem of stars outside the main sequence. The white dwarfs presumably offer no great difficulty. The internal temperature of these stars is probably rather low,[1,45] because of the low opacity (degeneracy!) so that the small energy production may be understandable. Quantitative calculations are, of course, necessary. For the giants, on the other hand, it seems to be rather difficult to account for the large energy production by nuclear reactions. If the Eddington (or the point source) model is used, the central temperatures and densities are exceedingly low, e.g., for Capella $T = 6 \cdot 10^6$, $\rho = 0.16$. Only a nuclear reaction going at very low temperature is therefore at all possible; $Li^7 + H = 2He^4$ would be just sufficient. But it seems difficult to conceive how the Li^7 should have originated in *all* the giants in the first place, and why it was not burned up long ago. The only other source of energy known is gravitation, which would require a core model[46] for giants.[47] However, any core model seems to give small rather than large stellar radii.

10. THE MASS-LUMINOSITY RELATION

In this section, we shall use our theory of energy production to derive the relation between mass and luminosity of a star. For this purpose, we shall employ the well-known homology relations (reference 1, p. 492). This is justified because we assume that all stars have the same mechanism of energy evolution and therefore follow the same model. Further, it is assumed that the matter throughout the star is nondegenerate which seems to be true for all stars except the white dwarfs. (For all considerations in this and the following two sections, cf. reference 1.)

We shall consider the mass of the star M, the mean molecular weight μ, the concentration of "Russell Mixture" y and the product of the concentrations of hydrogen and nitrogen, z, as independent variables. In addition, we introduce for the moment the radius R which, however, will be eliminated later. Then, obviously, we have for the density (at each point)

$$\rho \sim M/R^3. \tag{46}$$

From the equation of hydrostatic equilibrium

$$dp/dr = -GM_r\rho/r^2 \tag{47}$$

(G = gravitational constant) and the gas equation

$$p = RT\rho/\mu \tag{47a}$$

(R the gas constant, radiation pressure neglected), we find

$$T \sim M\mu/R. \tag{48}$$

Finally, we must use the equation of radiative equilibrium:

$$aT^3 \frac{dT}{dr} = -\frac{3}{4}\frac{k}{c}\rho \frac{L_r}{4\pi r^2}, \tag{49}$$

where a is the Stefan-Boltzmann constant, c the velocity of light, k the opacity, and L_r the luminosity (energy flux) at distance r from the center.

For the opacity, we assume

$$k \sim y\rho^\alpha T^{-\beta} \tag{50}$$

(y concentration of heavy elements). Usually, α is taken as 1 and $\beta = 3.5$ (Kramers' formula). However, the Kramers formula must be divided by the "guillotine factor" τ which was calculated from quantum mechanics by Strömgren.[48] For densities between 10 and 100, and temperatures between 10^7 and $3 \cdot 10^7$, Strömgren's numerical results can be fairly well represented by taking $\tau \sim \rho^{1/2} T^{-3/4}$. Therefore we adopt $\alpha = 1/2$, $\beta = 2.75$ in (50).

The luminosity may be written

$$L \sim M\rho z T^{\gamma}. \tag{51}$$

That the energy production per unit mass, L/M, contains a factor ρ follows from the fact that it is due to two-body nuclear reactions; this factor is apparent from all our formulae, e.g., (4). $z = x_1 x_2$ is the product of the concentrations of the reacting nuclei (N^{14} and H). For γ we obtain from (4), (6)

$$\gamma = \frac{d \log(\tau^2 e^{-\tau})}{d \log T} = \tfrac{1}{3}(\tau - 2). \tag{52}$$

For $N^{14} + H$ and $T = 2 \cdot 10^7$, this gives $\gamma = 18$. For $T = 3.2 \cdot 10^7$ (Y Cygni), $\gamma = 15.5$; generally, $\gamma \sim T^{-1/2}$. $\gamma = 18$ will be a good approximation over most of the main sequence.

Inserting (50), (51) in (49), we have

$$T^{4+\beta-\gamma} \sim yzM\rho^{2+\alpha}R^{-1}. \tag{53}$$

Combining this with (46), (48), and introducing the abbreviation

$$\delta = \gamma + 3 + 3\alpha - \beta \tag{54}$$

we find

$$R \sim M^{1-2(2+\alpha)/\delta}\mu^{1-(7+3\alpha)/\delta}(yz)^{1/\delta}, \tag{55}$$

$$T \sim M^{2(2+\alpha)/\delta}\mu^{(7+3\alpha)/\delta}(yz)^{-1/\delta}, \tag{56}$$

$$\rho \sim M^{-2+6(2+\alpha)/\delta}\mu^{-3+3(7+3\alpha)/\delta}(yz)^{-3/\delta}, \tag{57}$$

$$L \sim M^{3+2\alpha+2(2+\alpha)(\beta-3\alpha)/\delta}\mu^{4+3\alpha+(7+3\alpha)(\beta-3\alpha)/\delta} \times y^{-1}(yz)^{-(\beta-3\alpha)/\delta}. \tag{58}$$

The most important result is that the central *temperature* depends only slightly on the mass of the star, *viz.* as $M^{0.25}$ and $M^{0.30}$ for $\gamma = 18$ and 15. The reason for this is the strong temperature dependence of the reaction rate: The exponent of M in (56) is inversely proportional to δ which is mainly determined (cf. (54)) by the exponent γ in formula (51) for the temperature dependence of the reaction rate. The integration of the Eddington equations with the use of observed luminosities, radii, etc., gives, in fact, only a small dependence of the central temperature on the mass. This can *only* be explained by a strong temperature dependence of the source of stellar energy, a fact which has not been sufficiently realized in the past. Theoretically, the central temperature increases somewhat with increasing mass of the star, more strongly with the mean molecular weight, and is practically independent of the chemical composition, i.e., of y and z.

The *radius* of the star is larger for heavy stars and for high molecular weight. The density behaves, of course, in the opposite way. Both these results are in qualitative agreement with observation. The product of mass and density which occurs in Eq. (51) for the luminosity, is almost independent of the mass; therefore, for constant concentrations z, the luminosity is determined by the central temperature alone. Both radius and density are almost independent of the chemical composition, except insofar as it affects μ.

The *luminosity* increases slightly faster than the fourth power of the mass and the sixth power of the mean molecular weight. This increase is considerably less than that usually given ($M^{5.5}\mu^{7.5}$) and agrees better with observation. The difference from the usual formula is mainly due to the different dependence of the opacity on density and temperature.

The energy productions caused by the proton combination (H + H) and by the carbon-nitrogen reactions (N + H) are then compared as a function of the central temperature of the star. For temperatures below about $1.5 \times 10^{7\circ}$, the H + H predominates; for higher temperatures, N + H predominates.

11. Stability Against Temperature Changes

A brief discussion of stellar stability following Cowling[49] is presented, and it is noted that most of the observed stars are stable. The condition for stability seems to hold for both the proton combination and the carbon-nitrogen reactions, provided that assumption B of section 5 holds for the proton combination.

12. Stellar Evolution[50]

We have shown that the concentrations of heavy nuclei (Russell mixture) and, therefore, also of nitrogen, cannot change appreciably during the life of a star. The only process that occurs is the transformation of hydrogen into helium, regardless of the detailed mechanism. The state of a star is thus described by the hydrogen concentration x, and by a *fixed* parameter, y, giving the concentration of Russell mixture. The rest, $1 - x - y$, is the helium concentration. Without loss of generality, we may fix the zero of time so that the helium concentration is zero. (Then the actual "birth" of the star may occur at $t > 0$).

It has been shown that the luminosity depends on the chemical composition practically only[51] through the mean molecular weight μ. This quantity is given by

$$1/\mu = 2x + \tfrac{3}{4}(1 - x - y) + \tfrac{1}{2}y = (5/4)(x + a), \quad (62)$$

$$a = 0.6 - 0.2y, \quad (62a)$$

taking for the molecular weights of hydrogen, helium and Russell mixture the values $\frac{1}{2}$, 4/3 and 2, respectively.[1]

Now the rate of decrease of the hydrogen concentration is proportional to the luminosity, which we put proportional to μ^n (n is about 6). Then

$$dx/dt \sim -(x + a)^{-n}. \quad (63)$$

Integration gives

$$(x + a)^{n+1} = A(t_0 - t), \quad (64)$$

where A is a constant depending on the mass and other characteristics of the star. Since $x = 1 - y$ at $t = 0$, we have

$$At_0 = (1.6 - 1.2y)^{n+1}. \quad (64a)$$

It is obvious from (63) and (64) that the hydrogen concentration decreases slowly at first, then more and more rapidly. E.g., when the concentration of heavy elements is $y = 1/2$, the first half of the hydrogen in the star will be consumed in 87 percent of its life, the second half in the remaining 13 percent. If the concentration of Russell mixture is small, the result will be even more extreme: For $y = 0$, it takes 92 percent of the life of the star to burn up the first half of the hydrogen. Consequently, very few stars will actually be found near the end of their lives even if the age of the stars is comparable with their total lifespan t_0 (cf. 64a). In reality, the lifespan of all stars, except the most brilliant ones, is long compared with the age of the universe as deduced from the red-shift ($\sim 2 \cdot 10^9$ years): E.g., for the sun, only one percent of the total mass transforms from hydrogen into helium every 10^9 years so that there would be only 2 percent He in the sun now, provided there was none "in the beginning." The prospective future life of the sun should according to this be $12 \cdot 10^9$ years.

It seems to us that this comparative youth of the stars is one important reason for the existence of a *mass*-luminosity relation—if the chemical composition, and especially the hydrogen content, could vary absolutely at random we should find a greater variability of the luminosity for a *given* mass.

It is very interesting to ask what will happen to a star when its hydrogen is almost exhausted. Then, obviously, the energy production can no longer keep pace with the requirements of equilibrium so that the star will begin to contract. Gravitational attraction will then supply a large part of the energy. The contraction will continue until a new equilibrium is reached. For "light" stars of mass less than $6\mu^{-2}$ sun masses (reference 1, p. 507), the electron gas in the star will become degenerate and a white dwarf will result. In the white dwarf state, the necessary energy production is extremely small so that such a star will have an almost unlimited life. This evolution was already suggested by Strömgren.[1]

For heavy stars, it seems that the contraction can only stop when a neutron core is formed. The difficulties encountered with such a core[52] may not be insuperable in our case because most of the hydrogen has already been transformed into heavier and more stable elements so that the energy evolution at the surface of the core will be by gravitation rather than by nuclear reactions. However, these questions obviously require much further investigation.

1. B. Strömgren, *Ergebn. d. Exakt. Naturwiss. 16*, 465 (1937).

2. G. Gamow, private communication.

3. R. d'E. Atkinson and F. G. Houtermans, *Zeits. f. Physik 54*, 656 (1929).

4. G. Gamow and E. Teller, *Phys. Rev. 53*, 608 (1938), Eq. (3).

5. D. B. Parkinson, R. G. Herb, J. C. Bellamy, and C. M. Hudson, *Phys. Rev. 52*, 75 (1937).

6. M. Goldhaber, *Proc. Camb. Phil. Soc. 30*, 560 (1934).

7. R. Ladenburg and M. H. Kanner, *Phys. Rev. 52*, 911 (1937).

8. H. D. Doolittle, *Phys. Rev. 49*, 779 (1936).

9. J. H. Williams, W. G. Shepherd, and R. O. Haxby, *Phys. Rev. 52*, 390 (1937).

10. J. H. Williams, R. O. Haxby, and W. G. Shepherd, *Phys. Rev. 52*, 1031 (1937).

11. J. H. Williams, W. H. Wells, J. T. Tate, and E. J. Hill, *Phys. Rev. 51*, 434 (1937).

12. W. A. Fowler, E. R. Gaerttner, and C. C. Lauritsen, *Phys. Rev. 53*, 628 (1938).

13. R. B. Roberts and N. P. Heydenburg, *Phys. Rev. 53*, 374 (1938).

14. P. I. Dee, S. C. Curran, and V. Petržilka, *Nature 141*, 642 (1938). The γ-rays from C^{13} give about equally as many counts as those from C^{12}. The efficiency of the counter for C^{13} γ-rays is about twice that for C^{12}, because the cross section for production of Compton and pair electrons is smaller by a factor 2/3 while the range of these electrons is about 3 times longer. With an abundance of C^{13} of about 1 percent, the γ-width for this nucleus becomes 50 times that of C^{12}. I am indebted to Dr. Rose for these calculations.

15. M. E. Rose, *Phys. Rev. 53*, 844 (1938).

16. O. R. Frisch, H. v. Halban, and J. Koch, *Nature 140*, 895 (1937). *Danish Acad. Sci. 15*, 10 (1937).

17. H. A. Bethe, *Rev. Mod. Phys. 9*, 71 (1937).

18. C. L. Critchfield and H. A. Bethe, *Phys. Rev. 54*, 248, 862 (L) (1938).

19. H. A. Bethe, *Rev. Mod. Phys. 9*, 167 (1937).

20. T. W. Bonner, *Phys. Rev. 52*, 685 (1938).

21. H. A. Bethe, *Phys. Rev. 53*, 313 (1938).

22. W. Bothe and W. Gentner, *Naturwiss, 24*, 17 (1936).

23. H. A. Bethe, *Phys. Rev. 54*, 436, 955 (1938).

24. J. H. Williams, W. G. Shepherd, and R. O. Haxby, *Phys. Rev. 51*, 888 (1937).

25. M. S. Livingston and H. A. Bethe, *Rev. Mod. Phys. 9*, 247 (1937).

26. With the possible exception of a doublet structure of the ground state, similar to Li^7. However, the doublet separation should probably be much smaller than in Li^7 because of the loose binding of He^5, and presumably both components of the doublet are already contained in the rather broad α-particle group observed by Williams, Shepherd, and Haxby.

27. R. d'E. Atkinson, *Phys. Rev. 48*, 382 (1935).

28. *Note added in proof:* Recently, F. Joliot and I. Zlotowski (*J. de phys. et rad. 9*, 403 [1938]) reported the formation of stable He^5 from the reaction $He^4 + H^2 = He^5 + H^1$. The evidence is based upon the emission of singly charged particles of long range when heavy paraffin is bombarded by α-particles. However, the number of such particles observed was exceedingly small (only 6 out of a total of 126 tracks). Furthermore, the mass given for He^5 by Joliot and Zlotowski (5.0106) is irreconcilable with the stability (against neutron emission) of the well-known nucleus He^6.

29. R. B. Roberts, N. P. Heydenburg, and G. L. Locher, *Phys. Rev. 53*, 1016 (1938).

30. F. A. Paneth and E. Glückauf, *Nature 139*, 712 (1937).

31. F. Kirchner and H. Neuert, *Naturwiss. 25*, 48 (1937).

32. *Note added in proof:* These conclusions are compatible with the new measurements of S. K. Allison, E. R. Graves, L. S. Skaggs, and N. M. Smith, Jr. (*Phys. Rev. 55*, 107 [1939]) on the reaction energy of $Be^9 + H = Be^8 + H^2$.

33. K. T. Bainbridge, *Phys. Rev. 53*, 922(A) (1938).

34. If the combined initial nuclei and the final nucleus have the same parity (as may be the case, e.g., for $O^{16} + H = F^{17}$), it is still possible to have a dipole transition if only the incident particle has orbital momentum one. This does not materially affect its penetrability if $R > a$ (cf. [4], [5]), which is true in every case where the parities are expected to be the same.

35. L. I. Schiff, *Phys. Rev. 52*, 242 (1937).

36. H. A. Bethe and R. Peierls, Nature *133*, 689 (1934).

37. It was assumed that radiative capture takes place only through the reasonance level at 440 kv proton energy. The proton width of this level was taken as 11 kv, the radiation width as 40 ev.

38. The reaction $C^{11} + H = N^{12}$ becomes more probable than $C^{11} = B^{11} + \varepsilon^+$ only at $T > 3 \cdot 10^8$ degrees. The branching ratio in (40), (40a) may perhaps be slightly more favorable because the effect of the potential barrier in (40a) may be stronger.

39. It would be tempting to ascribe similar significance to the equality of intensity of the two α-groups from natural Li bombarded by protons ($Li^6 + H = He^4 + He^3$, $Li^7 + H = 2He^4$). However, the lithium isotopes do not seem to be genetically related, as are those of carbon.

40. *Note added in proof:* In this case, the life of N^{14} in the sun might actually be shorter, and its abundance smaller than that of C^{12}. Professor Russell pointed out to me that this would be in better agreement with the evidence from stellar spectra. Another consequence would be that a smaller abundance of N^{14} would be needed to explain the observed energy production.

41. *Added in proof:*—According to calculations of R. Marshak, the correct figure is about 1/30.

42. A. S. Eddington, *The Internal Constitution of the Stars* (Cambridge University Press, 1926).

43. Mr. Marshak has kindly calculated the central temperature and density of the sun for the point source model, using Strömgren's tables for which we are indebted to Professor Strömgren. With an average atomic weight $\mu = 1$, Marshak finds

for the point source model $T_c = 20.3 \cdot 10^6$, $\rho_c = 50.2$

for the Eddington model $T_c = 19.6 \cdot 10^6$, $\rho_c = 72.2$

Not only is the temperature difference very small ($3\frac{1}{2}$ percent) but it is, for the sake of the energy production, almost compensated by a density difference in the opposite direction. The product $\rho_c T_c{}^{18}$ is only 20 percent greater for the point source model.

44. Even if the most stable nuclei (Fe, etc.) are formed rather than He, the possible life will only increase by 30 percent.

45. S. Chandrasekhar, Monthly Not. *95*, 207, 226, 676 (1935).

46. L. Landau, Nature *141*, 333 (1938).

47. This suggestion was made by Gamow in a letter to the author.

48. Cf. reference 1, table 6, p. 485.

49. T. G. Cowling, *Monthly Not. 94*, 768 (1934), *96*, 42 (1935).

50. Most of these considerations have already been given by G. Gamow, *Phys. Rev. 54*, 480(L) (1938).

51. Except for the factor y^{-1}, which, however, does not change with time.

52. G. Gamow and E. Teller, *Phys. Rev. 53*, 929(A), 608(L) (1938).

50. Nuclear Reactions in Stellar Evolution

George Gamow

(*Nature 144*, 575–577 [1939])

George Gamow repeatedly brought astrophysical problems to the attention of his fellow nuclear physicists, including Hans Bethe. After Bethe had worked out the details of the CNO cycle, Gamow became the first to apply the new information specifically to stellar evolution, using a Hertzsprung-Russell diagram to illustrate his conclusions. When Henry Norris Russell had originally drawn the diagram, he had proposed an evolutionary scheme whereby the giant stars would contract to the main sequence (becoming hotter) and then move downward along this sequence as they further contracted and cooled. A. S. Eddington, after showing that contraction did not play a significant role in the evolution of main-sequence stars, pointed out that the evolutionary meaning of the diagram was completely open.

Gamow's calculations seemed to show that stars with less hydrogen were located higher on the main sequence, and thus he suggested that the stars evolved upward along the sequence as they slowly depleted their hydrogen fuel. He interpreted the mass-luminosity relation as a statistical accident arising because the low-mass stars at the lower end of the main sequence had not had enough time to evolve to more luminous states. Although on this point and others the paper erred, it was nevertheless a pioneering attempt to relate the results of nuclear physics to evolution on the H-R diagram. Much as Gamow supposed, stars do evolve through a sequence of quasi-static equilibria, of which the first stage is the hydrogen burning that characterizes the main sequence.

The question of the sources of energy of stars, and the closely connected question of stellar evolution, have presented for a long time the most important unsolved problem of theoretical astrophysics. The first step towards the understanding of the physical processes leading to the tremendous energy liberation which is responsible for the observed radiation of stars was made about ten years ago. It was shown,[1] on the basis of the quantum theory of nuclear transformations, that, at the very high temperatures existing in the interior of stars, the ordinary thermal collisions between the particles possess sufficient energy to produce artificial nuclear transformations. It was also shown that the energy liberation in such *thermo-nuclear reactions* is sufficiently high to explain the radiation of stars only in the case of collisions between the protons and the nuclei of light elements (up to about atomic number 10). It took, however, ten more years before our knowledge concerning nuclear reactions developed to such an extent that it became possible to choose particular nuclear reactions as responsible for the energy-production in different cases and to give on this basis a clear and consistent picture of stellar evolution. The exact formula for the energy production of a thermo-nuclear reaction (per gram per second) can be written in the form:[2]

$$\varepsilon = \frac{4}{3^{5/2}}\rho X_1 X_2 \frac{\hbar\Gamma r_0{}^2 Q}{m_1 m_2 m e^2 Z_1 Z_2} \exp.{}^4\left(\frac{2v_0 m e^2 Z_1 Z_2}{\hbar^2}\right)^{1/2} \cdot \tau^2 e^{-\tau} \tag{1}$$

Here ρ is the density of the gas, X_1 and X_2 the concentrations (by weight) of the two reacting types of nuclei, m_1 and m_2 their masses ($m = m_1 m_2/(m_1 + m_2)$), $Z_1 e$ and $Z_2 e$ their electric charges, r_0 the combined radius, Q the energy production per reaction, and $\Gamma/\hbar$ the probability of reaction after the penetration through the potential barrier. The dependence upon temperature is given through τ, which is defined by:

$$\tau = 3\left(\frac{\pi^2 m e^4 Z_1{}^2 Z_2{}^2}{2\hbar^2 kT}\right)^{1/3}, \tag{2}$$

T being the absolute temperature of the gas and k the Boltzmann constant.

Here Gamow gives a table of various values of Q, Γ, and ε for all possible reactions between protons and light nuclei.[3,4] Because this information may be obtained from table 49.5 (selection 49), we omit this discussion. Gamow repeats Bethe's conclusion that the energy production of the sun is caused by the carbon-nitrogen reaction cycle.

We can now answer the question concerning the characteristics of stars with the same energy-producing reaction but different masses and chemical constitution, by comparing them with the sun by means of the so-called *homology transformations* of stellar models. There are *nine* different physical quantities defining the external properties and internal structure of a star: the total mass M, the total luminosity L, the radius R (or the effective temperature defined by

$$T_{\text{eff.}} = \sqrt[4]{\frac{L}{4\pi R^2\sigma}}),$$

the average molecular weight μ, the coefficient of opacity of stellar matter κ_0, the product of concentrations of two kinds of reacting nuclei $X_1 X_2$, and the functions of ρ, p and T giving the distribution of density, pressure and temperature in the stellar interior. These quantities are connected by *five* fundamental equations: the definition of mass M through ρ and R; the equation of hydrodynamical equilibrium; the gas law; Eddington's equation for the transport of energy; and the formula giving the rate of thermo-nuclear energy production. Thus, choosing for four independent variables the M, μ, κ_0 and $X_1 X_2$, we can determine the relative changes of other variables by simple proportionality relations.

It is customary to classify the stars in the frame of the so-called Hertzsprung-Russell diagram, plotting the logarithms of absolute luminosities against the negative logarithms of effective temperatures.

Here Gamow gives formulae for the homology transformations of absolute luminosity and effective temperature.[5] They depend on chemical composition, stellar mass, and an exponent, n, that for the carbon-nitrogen cycle takes on the value $n = 18$ at temperatures around 2×10^7 °C.

If we consider stars with the same chemical constitution, the values of L and $T_{\text{eff.}}$ will depend only on M, and this dependence is shown by a straight line in figure 50.1, representing the above-mentioned Hertzsprung-Russell diagram. The numbers along the line correspond to the values of $M/M_\odot$. The shaded area represents the so-called *main sequence* in which most of the known stars are located, and the observed masses of stars are indicated by numbers in little circles. The close agreement between the observed and calculated locations brings us to the conclusion that *the main sequence should be interpreted as the collection of stars of different masses with the same chemical constitution and the same energy-producing reaction as in the sun (that is, C–N-cycles).*

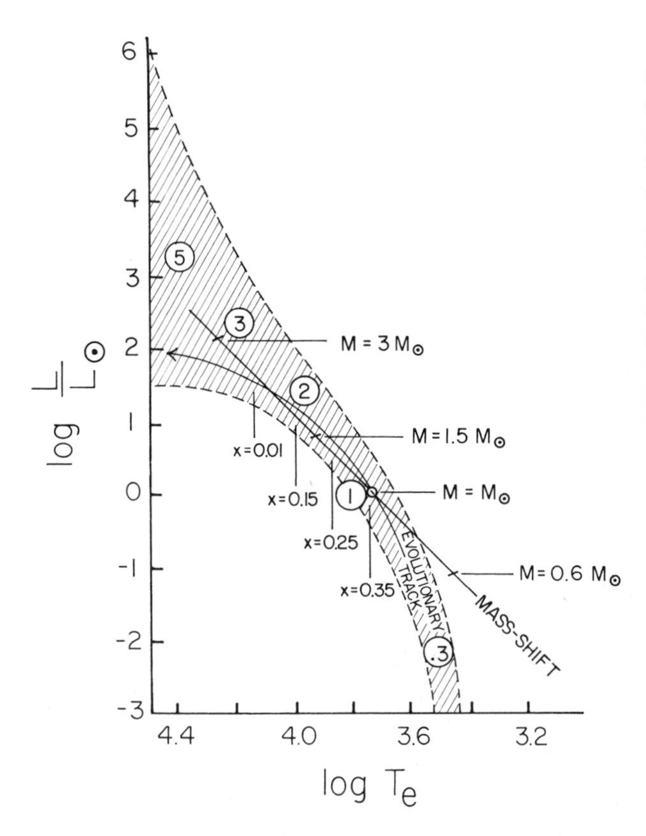

Fig. 50.1 The evolutionary track for the sun according to Gamow. The numbers along the track represent the hydrogen content at different stages of evolution. In this Hertzsprung-Russell diagram the logarithm of the absolute luminosity, L, is plotted as a function of the effective temperature, T_e. The values of x denote the hydrogen content at different stages of stellar evolution.

Although the agreement between the observed values of M, L and the theoretical relations with $n = 18$ is excellent for the stellar masses between 0.5 $M\odot$ and 2.0 $M\odot$,[6] there are quite noticeable deviations at both ends of the main sequence. The deviations for very luminous stars of large mass are, most probably, due to the effect of the high radiative pressure, which has to be neglected in the derivation of the homology transformations. On the other hand, the deviations at small luminosities might be due to the fact that for these stars the C–N-cycle is no longer the main energy-producing reaction. We have seen above that in the sun the $^1H + {}^1H$ reactions give only 1/30 of the total energy production; however, due to its comparatively small temperature dependence, this reaction becomes predominant at temperatures below 1.5×10^7 ° C., which just correspond to central temperatures of stars on the lower part of the main sequence.

We can now apply the homology transformations to the study of evolutionary changes of a star due to the continuous consumption of the hydrogen supply by the energy-producing reaction. The transmutation of hydrogen into helium increases the average molecular weight μ and the opacity κ_0 of the stellar substance and decreases the value of X_1X_2 (in which X_1 is now the constant nitrogen content, and X_2 the variable hydrogen content). This defines the changes of the luminosity and the effective temperature and enables one to draw the evolutionary track of the star in the Hertzsprung-Russell diagram. In figure 50.1 such a track for the sun is shown by the heavy line, the numbers along the track representing the hydrogen content at different stages of the evolution. We see that *during the process of the (hydrogen-) evolution the luminosity and the effective temperature of any star, and in particular of our sun, is bound to increase by quite a considerable amount.* But inasmuch as for the sun it would take 1.2×10^{10} years to burn all its present hydrogen content, the yearly changes of the luminosity and spectral class are quite negligible and could not be detected by observation.

Here, however, is an important question. If different stars at different stages of their evolution possess quite different luminosities, there should be no correlation between the luminosities and masses of stars; this is, however, in contradiction with the well-established existence of the empirical mass-luminosity relation. The solution of this difficulty is very simple.[7] The rate of consumption of hydrogen at different stages of the evolution is evidently proportional to the total luminosity of the star. Thus the star will stay a much longer time in the lower part of its evolutionary track and pass much more quickly through the stages of high luminosity. *Thus the empirical mass-luminosity relation should be considered as a statistical regularity due to the fact that most of the stars are observed in the lower part of their evolutionary track.* A more detailed survey of stellar masses must necessarily lead to the detection of stars with luminosities considerably higher than those required by the ordinary mass-luminosity relation. It should be noticed, however, that for the stars in the lower half of the main sequence, such late stages of the hydrogen evolution probably could not be observed at all, because the evolutionary life of these stars is considerably longer than the period of time ($\sim 10^9$ years) which has passed since their formation.

1. Atkinson, R. d'E., and Houtermans, F. G., *Z. Phys. 54*, 656 (1929).
2. Gamow, G., and Teller, E., *Phys. Rev. 53*, 608 (1938).
3. Bethe, H., *Phys. Rev. 55*, 434 (1939).
4. Bethe, H., and Critchfield, Ch. *Phys. Rev. 54*, 248 (1938).
5. Gamow, G., *Phys. Rev. 53*, 59 (1938), *55*, 718 (1939).
6. Gamow, G., *Astrophys. J. 89*, 130 (1939).
7. Gamow, G., *Phys. Rev. 53*, 907 (1938).

51. Stellar Structure, Source of Energy, and Evolution

Ernst Öpik

(*Publications de l'Observatoire astronomique de l'Université de Tartu 30*, No. 3, 1–115 [1938])

By the 1920s A. S. Eddington's pioneering work had established the conditions of mechanical and radiative equilibrium that determine the internal structure of stars. Robert Atkinson subsequently showed that the main-sequence stars could shine by the thermonuclear conversion of light nuclei into heavier ones in the hot, central portions of these stars. Because the giant stars have larger luminosities at lower temperatures, however, Atkinson was forced to conclude that these stars shine by some unknown, temperature-independent process.

This was the situation when Ernst Öpik presented this paper, in which he argues that both the giant and main-sequence stars shine by thermonuclear processes, following a sequence well defined by successive increases in temperature. As a star evolves, with increasing central temperature, the nuclear fuels burn from the center outward, and successively new nuclear fuels begin to burn at the center. For the giant stars, the exhaustion of the early processes begins earlier, and the central temperatures rise to open up a new source of energy not available to the main-sequence stars. Öpik argues that this rapid evolution of the giant stars is confirmed by their fewer numbers and by the absence of giants of small mass.

As Öpik realized, the key to understanding the hot central temperatures of the giant stars, which seemed to have cooler surfaces than some main-sequence stars, lay in the fact that a star generally has a convective core when nuclear burning occurs at its center. Provided that there is no mixing between the central and outer materials, hydrogen burning is initially confined to this central convective core, which is surrounded by an inert hydrogen envelope in radiative equilibrium. Once all of the core hydrogen is converted into helium, the core must undergo gravitational contraction. Öpik noticed that the rapid increase in energy generation with increasing temperature and density causes the hydrogen envelope to expand, producing the distended red giant structure. Meanwhile, the interior comprises zones of different chemical compositions, and shell energy sources are produced as well. A typical giant structure consists of a vast extended envelope of low density; an intermediate shell of about the central density of main-sequence stars, containing active sources of energy generation; and a contracting superdense core with zero hydrogen content and no energy generation.[1]

We present here parts of only the last two sections of this extensive contribution. In the first two sections, Öpik discusses gravitation and the total annihilation of matter (as opposed to atomic synthesis) as sources of stellar energy. The earlier adoption of a long time scale (10^{12} years) required a powerful source of energy to keep stars shining and led to speculations about a possible process called annihilation of matter. With a short time scale (about 10^9 years), annihilation of matter is only necessary when considering giant stars that do not have superdense cores. (This situation leads Öpik to consider that giant stars have central condensations.)

In the third section Öpik reviews the theory of atomic synthesis as a source of stellar energy. The theory was put on a sound physical basis by Atkinson, although the actual chain of processes was not well established experimentally. The only serious objection to atomic synthesis as an energy source has been the danger of pulsational instability. The work of T. G. Cowling showed that the danger of instability was exaggerated.

To account for the puzzle of giants that produce more energy at low temperatures than do dwarfs at high ones, Atkinson assumed two different processes of energy generation. In contrast, Öpik proposes that the same processes of atomic synthesis exist for giants and dwarfs and that these processes follow a sequence depending on successive increases in temperature. The high central temperatures that open up the source of energy for the high luminosity of the giants are only possible if these stars have superdense cores. Such cores are probably formed by the collapse of the central portions after the exhaustion of hydrogen in these regions. Henry Norris Russell's "giant stuff" and "dwarf stuff" hypothesis is unnecessary.

In section 4 Öpik discusses T. E. Sterne's idea that the abundance of elements is the result of thermodynamic (dissociative) equilibrium at high temperatures of the order of $3 \times 10^{9\circ}$. The high observed abundance of hydrogen and oxygen and the low abundance of iron and heavier elements in stellar atmospheres are inconsistent with Sterne's hypothesis. The composition of stellar atmospheres is not always determined by the composition of the interior, which suggests that the mixing of material may in some cases be inefficient.

1. There were two earlier papers by E. J. Öpik which set the stage for the paper given here. In 1922 Öpik concluded that the mere existence of stars in radiative equilibrium required that stellar energy production must increase with temperature, and that novae must be the result of nuclear explosions (*Publications de l'Observatoire astronomique de l'Université de Tartu 25*, no 2, 1 [1922]). In 1924 he used studies of double stars to conclude that the "dwarf" main-sequence stars have not changed much in luminosity since their origin, and that the gravitational energy source is for them negligible (*Publications de l'Observatoire astronomique de l'Université de Tartu 25*, no. 6, 1 [1924]). As a sequel to the paper given here, Öpik considered composite stellar models for unmixed stars with convective cores of increased molecular weight. In his later papers he discussed the thermonuclear burning of helium, carbon, oxygen, etc., during the evolutionary transition of low mass stars toward the white-dwarf or degenerate stage. (*Publications de l'Observatoire astronomique de l'Université de Tartu 30*, no. 4, 1 [1938], *31*, no. 1, 1 [1943], *Proceedings of the Royal Irish Academy 54A*, 49 [1951]. See also selection 52).

5. THE COMPOSITE ADIABATIC-RADIATIVE AND THE COMPLETE ADIABATIC STELLAR MODELS; GIANT AND DWARF STRUCTURE

A. TRANSFER OF HEAT BY CONVECTION

THE CONVECTIVE TRANSFER of heat (per unit of time and cross section) between two surfaces may be set equal to

$$Q_c \sim v\varrho c_p \, \Delta T, \tag{4}$$

where v is the velocity, ϱ the density, c_p the specific heat, ΔT the excess temperature of the current. The transfer by radiation is

$$Q_r \sim \frac{(T_1{}^4 - T_2{}^4)}{k\varrho x},$$

where T_1 and T_2 are the temperatures, k the coefficient of absorption, ϱ the density (supposed to be constant), x the depth. For surfaces separated by a large $k\varrho x$ the advantage of convection, as compared with radiation, is obvious; if the depth of the convection current is of the order of the radius of the star, convection is much more efficient than radiation.

Convection takes place whenever the temperature gradient exceeds the adiabatic value ξ_a. The convective region may be assumed to be built according to a polytrope of index $n = 1/(\gamma - 1)$, where γ is the ratio of specific heats.

B. THE NET FLUX OF RADIATION IN A POLYTROPE Heat that has escaped from a shell of radius r inside a star, containing a fixed mass M_r, cannot get back (because free convection cannot transport heat in the direction of the gravitational force; convection forced by rotation is too slow, and too weak, to work against the excess of the adiabatic temperature gradient required for a reversal of the transport of heat). Also, there are no subatomic processes able to absorb energy at temperatures below 10^9 K. Therefore, all the net flux of heat (radiation + convection) which has once passed outside of r, must make its way through to the surface (with the exception of a mostly small or zero fraction spent upon the heating of an expanding star). On the other hand, the temperature gradient cannot perceptibly exceed its adiabatic value ξ_a; if radiation at the maximum possible value $\xi = \xi_a$ is incapable of transporting all the heat, convection comes into play to supply the difference.

The net flux of radiation passing outwards through a shell of radius r is

$$Q_r = \frac{4\pi acr^2}{3k\varrho}\left(-\frac{dT^4}{dr}\right) \tag{5}$$

(cf.[1], p. 101); here $ac/4$ = Stefan's constant of radiation. In the case of pure radiative equilibrium, this is also equal to the net flux of the energy, L_r; in the presence of convection, however,

$$Q_r \leq L_r.$$

For Kramers' law of opacity

$$k = k_0 \varrho T^{-7/2} \tag{6}$$

(k_0 = intrinsic opacity depending upon composition, primarily upon the hydrogen content), and for a polytropic model

$$\varrho = \varrho_c u^n, \tag{7}$$

where $u = T/T_c$(ρ_c and T_c = central density and temperature).

C. CONDITION FOR CONVECTION TO START AT THE CENTRE

For an energy source $\varepsilon = \rho T^s$, convection will occur if

$$s > \frac{13}{2} - 3n, \tag{12}$$

where n is the adiabatic polytropic index for the temperature-density relation.

D. THE LUMINOSITY OF A POLYTROPE, E. THE ADIABATIC MODEL, F. THE COMPOSITE MODEL, G. REGULATION OF LUMINOSITY FOR THE ADIABATIC MODEL, H. MODEL OF NON-HOMOGENEOUS COMPOSITION. For all values of polytropic index $n < 3.25$, the net flux of radiation, Q_r, passing through a shell of radius r, increases outward to a certain maximum value at $r = r_o$, and then drops to zero at the surface. For r less than r_o, it is possible to build a star on the adiabatic model with $n = 1/(\gamma - 1)$; at $r = r_o$ the adiabatic and the radiative states of equilibrium coincide. One form of the equilibrium of a star is one with a concentrated source of energy in an adiabatic-convective core surrounded by an envelope (of considerable mass and extent) in radiative equilibrium. This is only possible when the radius of the core is smaller than r_o. If atomic synthesis is an important source of energy, convection currents always start at the center.

I. COLLAPSE OF THE EXHAUSTED CORE OF A COMPOSITE MODEL AND GIANT STRUCTURE A core devoid of hydrogen, thus presumably devoid of subatomic sources of energy, is doomed to collapse on a "Kelvin" time scale, i.e., with gravitation as the source of energy; high densities can be attained, and a super-dense core may be formed. The hydrogen-containing envelope cannot be sucked into the core as long as traces of hydrogen are present, because the corresponding immense increase of temperature and density would lead to an instantaneous release of the whole store of subatomic energy, sufficient to disperse all the envelope into space. Actually no such catastrophe happens, the contraction of the core being a gradual one; instead of blowing up, the envelope gradually expands and adjusts itself to such low values of the effective density and temperature that the release of subatomic energy remains more or less normal (it may be even less than the "normal", as the gravitational energy of the core supplies now a large fraction of the star's needs). In spite of the high gravitational force exerted by the core, the transition from the superdense core to the envelope of "normal" density and temperature is made possible by the peculiar distribution of the energy sources, and the smaller molecular weight of the envelope;[2] the presence of subatomic energy sources suddenly beginning to work outside the core creates radiation pressure that "blows away" the matter of the shell, leaving a small density of matter just sufficient for the subatomic sources to work. The conditions are similar to those in the mathematical point-source model (cf.,[1] p. 126), except that here the subatomic source of energy is not concentrated in exactly one point, and that an additional point-source of energy and a considerable point-mass complicate the problem.

Certain numerical estimates show that a star whose mass is between 1 and 10 solar masses will become convective at radii between 0.2 and 0.6 solar radii and that a superdense core with a mass density of a few grams per cm^3 may occur with a radius ranging between 0.01 and 0.0001 solar radii.

A typical giant structure results, consisting of a vast extended envelope of low density in radiative or adiabatic equilibrium, an intermediate zone in adiabatic (convectional) equilibrium, of a density about the central density of main sequence stars, containing active sources of subatomic energy, and a contracting superdense core of zero hydrogen content and no subatomic energy. The intermediate zone, with active atomic synthesis, is supposed to contain a decreased amount of hydrogen and to get in this way definitely separated from the outer envelope (cf. above); if not, the whole outer mass except the core may be stirred by convection currents (as in the purely adiabatic model), and the outer radius becomes little sensitive to eventual changes in the luminosity (corresponding to the changing mass of the core which must increase with the progress of time from the exhausted material of the shell, and decrease as the result of energy losses),[3] in which case an apparently "main sequence" star with a superdense core may result.

6. The Course of Stellar Evolution

A. PRESUMPTIONS Let us consider the course of stellar evolution determined by the most probable conditions which follow from our preceding discussion: atomic synthesis and gravitation as the only sources of stellar energy; absence of complete mixing in some stars; complete mixing for all stars (without superdense cores) in a central portion of considerable extent, without necessarily an efficient interchange of material with the outer shell; origin from condensation of diffuse matter (nebula), which also determines the original composition.

B. CONDENSATION FROM A DIFFUSE STATE The first stage of a star's life consists of a comparatively short interval of contraction from a diffuse state; the structure of the star approaches closely Eddington's radiative model (polytrope $n = 3$; $\varepsilon \sim T$), the rate of generation of gravitational energy is automatically equal to the "prescribed" loss by radiation; convection currents are practically absent (the rotational currents are too much stratified); the central temperature increases during contraction inversely to the radius, and $\rho_c = 54\bar{\rho}$ (with slight uncertainty as to the definition of the boundary of a star); the first stage may last $\sim 10^7$ years for $M = \odot$, $\sim 10^5$ years for $M \sim 10\odot$.

C. STAGE OF ATOMIC SYNTHESIS As the central temperature rises, processes of atomic transmutation come gradually into play; a second stage of the star's life starts when an outwardly steady state is reached, subatomic energy balancing the "prescribed" losses by radiation; contraction becomes extremely slow (just enough to balance exhaustion of hydrogen by an increase of the central temperature); for the sun, this stage may last for 10^{10} years, for $M \sim 10\,\odot$ perhaps 10^8 years.

D. EVOLUTION OF THE ADIABATIC MODEL If the star is of the completely adiabatic structure, with complete mixing (sufficient rotation to overcome the dead zone at $r = r_0$, cf. preceding section), it remains a "main sequence" star of more or less "normal" density; with the gradual exhaustion of hydrogen its luminosity increases (cf.[4,5]). At the same time slow changes in the radius occur which may be estimated in the following way.

Here a formula is derived for the radius of a star of constant mass and changing hydrogen content. The radius depends on the exponent, s, of the temperature in the energy source $\varepsilon = \rho T^s$ and on the relative proportion, X, of hydrogen. Öpik first chooses $s = 6.5$.

With sufficient approximation the change of radius for an adiabatic star is then $R \sim [(1 - X)X]^{0.22}$. The change is rather slow. For $X \sim 1$ per cent, the radius is about one-half of its original value at $X = 33\frac{1}{3}$ per cent. Thus for $s = 6.5$ a slow contraction proceeds during the atomic synthesis; after its exhaustion, the star starts rapid contraction, relying upon gravitational energy alone. A superdense O-type or Wolf-Rayet star results.

For $s = 19$, $R \sim [(1 - X)X]^{0.10}(1 - \beta)^{0.14}$; with increasing molecular weight $(1 - \beta)^{0.14}$ increases faster than $X^{0.10}$ decreases, and the radius starts very slowly expanding; after reaching a maximum (for the sun, at $X = 0.069$, $R = 1.2\ R_\odot$) just before exhaustion, the radius begins to decrease and ends in a collapse as described before.

E. EVOLUTION OF THE COMPOSITE MODEL If the star possessed originally a core of smaller hydrogen content, or should acquire such a core as the result of incomplete circulation and atomic synthesis, or if it originally settled into a compound radiative-adiabatic state, it will, during the second stage of its life, maintain the typical compound structure; in this stage the star is supposed to consist of a more or less extended convective core, built adiabatically according to a polytrope of $n = 1/(\gamma - 1)$ (cf. Section 5), above which an outer shell in complete or partial radiative equilibrium is placed; at first no energy is produced in the outer shell. For such stars there is little, or perhaps no interchange of matter between the inner core and the surface. With the short time scale, composite main sequence stars of solar mass and less may at present still be in this stage of evolution; if larger masses also should be found in this stage (Procyon, cf. Section 7), this could be explained by their age being less than $3 \cdot 10^9$ years.

With the exhaustion of hydrogen in the convective core the third stage of evolution for the compound model starts: the contraction of the core which gradually is transformed into a superdense nucleus. An inner core devoid of subatomic sources of energy assumes a structure very similar to an incomplete polytrope $n = 3$ (built up from the centre), generating gravitational energy according to $\varepsilon \sim T$; the persistent contraction of such a nucleus is unavoidable (the only non-collapsing form would be an isothermal structure, where the loss of energy is zero; this, however, could not maintain itself: with the first increase ΔP of the external pressure over its original equilibrium value the configuration departs from isothermity, and the net flux of energy which arises then stimulates progressive contraction and progressive departure from isothermity, until the polytrope $n = 3$ is approximately reached). The change of the radius r of the nucleus with time may be represented by

$$\frac{1}{r} = \frac{1}{r_0} + ct. \tag{29}$$

Outside the nucleus the material is not exhausted; with the progress of the central condensation the temperature of the shell adjacent to the nucleus rises, and subatomic energy is released in an intermediate shell; the rapid increase of energy generation with increasing temperature and density in the intermediate shell prevents it and the rest of the star from being drawn into the overdense nucleus; on the contrary, if the outmost shell is in radiative equilibrium, by a process described below, it is forced to expand, and a giant star is formed. (For adiabatic equilibrium in the outer shell, a giant structure is also possible.) Let figure 51.1 show the scheme of a giant star; C is the exhausted nucleus, of radius r, mass M_c, and a net output of gravitational energy L_1 transported by radiation; A is the region of release of the subatomic energy and of convective circulation (at least in its outer portion), of radius R_1, mass $M_1 - M_c$, and a net output of energy L_2, transported to the top of the shell partly by convection; B is the region of undisturbed radiative equilibrium, with a temperature T_1 at the bottom, extending to the surface

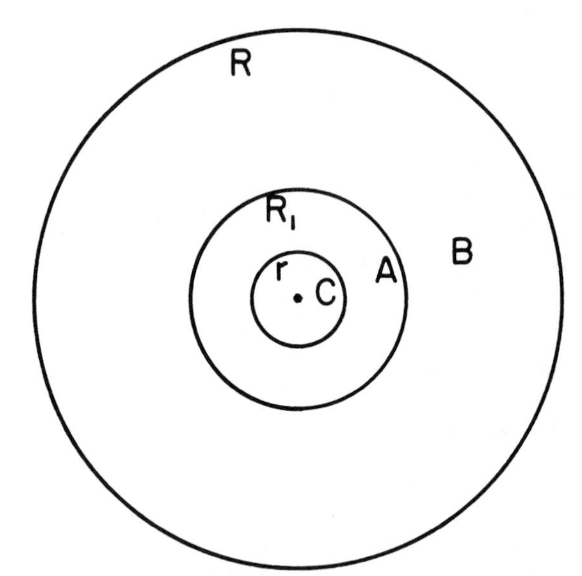

Fig. 51.1 Öpik's model for the structure of a giant star. Hydrogen has been exhausted in the contracting convective core, C, of radius r, whereas hydrogen is being converted into helium in the inner energy-burning shell, A, of outer radius R_1. The low mass densities of giant stars are accounted for by their extensive outer shells, B, which are in radiative equilibrium and in which no thermonuclear reactions take place.

of the star of radius R and mass M; no energy is generated in B (gravitational energy at eventual changes of radius being there negligible).

The condition for secular stability is

$$L_1 + L_2 = L, \tag{30}$$

where L is the luminosity. The violation of this condition, leading to an increase or decrease of the energy content, affects primarily the outer shell, B. The nucleus liberates automatically the practically fixed amount which it spends, and it is doomed to gradual collapse: there is no secular stability for the nucleus; but, if r/R is small, changes in r do not much reflect directly upon R, and the star may keep the outer appearance of being unchanged, which we describe as secular stability. The intermediate shell A, with convectional transport of heat, transports, under all circumstances, all the heat $L_1 + L_2$ to its top (cf. Section 5A,B), and no accumulation is possible there. The shell B in a given fixed state, however, with its radiative transfer of energy, is able to transport, and actually does transport, a fixed amount, equal to L, and unless L is apt to vary with the radius, secular stability cannot take place. The expansion of B, of course, causes mechanically the expansion of A, and in a minor degree of C.[6] For homologous changes of structure, L can indeed change but slowly with the radius, the change being of a non-stabilizing character ($\sim R^{-1/2}$, thus opposite to the required direction, cf.[1]), so that secular stability can be attained only by an automatic adaptation of the energy sources. For our model, L_1 is not much apt to vary; as to L_2, although an increase of it is able to prevent collapse in the case $L_1 + L_2 < L$, it is unable to prevent expansion if alone $L_1 > L$. In spite of that, as shown below, secular stability may be attained, because the changes in our complex model are not homologous, and L may vary so as to fit almost an arbitrary amount of energy generated in the interior.

To derive the luminosity we apply a simple method sufficient for our qualitative purposes. The flux of radiation between two spherical surfaces R_1, M_1, T_1 and R_2, M_2, T_2, may be represented with sufficient approximation by

$$L \sim \frac{(T_1{}^4 - T_2{}^4)R_1R_2}{k\rho(R_2 - R_1)} \tag{31}$$

This follows from the equation of radiative transfer with $k\rho = \text{const.}$, when sources of energy inside the shell $R_2 - R_1$ are absent; when such sources are present, the proportionality remains valid, with a variable proportionality factor depending upon M_1/M and M_2/M (relative internal masses), and upon the relative amount of energy developed inside the shell. For k, the coefficient of absorption, and ρ, the density, certain mean, or effective values for the given shell must be assumed.

For stars of a homologous series, when M_1/M and M_2/M are kept constant, we have: $T_1 \sim T_2 \sim (\bar{\beta}\mu M)/R$ (cf.[1]); $R_1 \sim R_2 \sim R$; $k \sim \rho T^{-7/2}\mu^{-1}$ (Kramers); $\rho \sim R^{-3}$; $1 - \bar{\beta} \sim M^2\mu^4\beta^4$ (Eddington's quartic equation for the radiation pressure); $L \sim R^2 T_e{}^4$; with these proportionalities, (31) is easily transformed into

$$L \sim M^{7/5}(1 - \bar{\beta})^{3/2} T_e{}^{4/5} \mu^{4/5},$$

which is exactly Eddington's mass-luminosity relation,[7] derived of course on the assumption $1 - \beta = \text{const.}$ throughout the star. This may be considered as a check of reliability of formula (31).

With the more general law of absorption

$$k \sim \rho T^{-3+\alpha}, \tag{32}$$

the mass-luminosity relation for stars of a homologous series (not necessarily polytropic) becomes (the influence of the molecular weight upon k and luminosity is not considered now, as it is a more complicated function of the composition, cf.[4,5] and Section 5; it cannot be represented by simple proportionality):

$$L \sim \beta^{7-\alpha} M^{5.5} R^{\alpha},$$

where β is an effective value for the whole star, depending upon its mass. For the Kramers formula $\alpha = -1/2$, for relativistic non-degenerate matter $\alpha = 0$. To degenerate matter the formula does not apply.

The "giant" model of figure. 51.1 cannot undergo homologous changes; the nucleus, an incomplete polytrope ($n \sim 3$), when having reached a sufficient degree of compression, is practically independent of changes occurring in the outer shells.

The radius of the core is $r \approx R^{0.05}$, where R is the radius of the star, and from the standpoint of external changes the nucleus is almost incompressible.

On the other hand, for regions near the outer boundary of the star the changes of radius in a shell must follow closely the changes in R. Generally, for an intermediate shell like A, we may write the intermediate formula

$$R_1 \sim R^{1-p}, \tag{33}$$

with $(0.05) < 1 - p < 1$.

We get for the dependence of luminosity upon the radius:

$$L \sim R^{6.5p+\alpha}. \tag{34}$$

A comparatively small deviation p of the exponent in (33) from unity suffices to make our model secularly stabilized by changes in the radius; (34) must for this purpose give increasing luminosity with increasing radius, thus the stabilizing condition is

$$6.5p > -\alpha.$$

Thus, for a small value of p, or for a slight deviation from proportionality of R_1 and R, the outer radius may be extremely sensitive to changes in the energy generation; the doubling of the internal source of energy may produce a typical giant, of any degree of diffusion, from an originally dense star. Now, a progressive increase of the internal generation of heat, above its original normal (original regulated) value may be actually expected, when the exhausted central core starts its collapse. This core, practically a complete polytrope as far as mass and rate of heat generation are concerned, being devoid of hydrogen, radiates more energy than if hydrogen in the normal proportion were present (cf. [4,5]) (up to 100 times more for a solar mass; the difference, however, is greatly reduced for large masses and high central temperatures, on account of electron scattering).

If L_0 is the "prescribed" luminosity of the original "main sequence" star, no great expansion of the radius can start before L_1 is a considerable fraction of L_0 (cf. formula [30]), because a moderate expansion reduces the subatomic energy L_2 and makes a balance; but when the energy output of the nucleus L_1 approaches, or even exceeds, L_0, the expansion of R must be large, and the star enters the giant stage (especially because L_2 can never drop to zero, or even to a very low value; at the boundary of the nucleus a rather peculiar zone exists where the exhausted material of the adjacent outer shell continually is driven into the nucleus, adding the released gravitational energy to L_2; the nucleus thus increases steadily, probably until a certain equilibrium size is reached, when the outwards directed resultant of the radiation pressure, being large on account of the sudden increase of the energy sources outwards, produces at the boundary of the central core a sufficiently small material density, so that the amount of inflowing material becomes equal to the radiation losses of the nuclear mass).

Reference is made here to Eddington's book[1] for estimates of luminosity, and certain details of collapse, degeneracy, and the composite model are given. One interesting conclusion is that the chief reason for the absence of diffuse stars among small masses seems to be a question of the speed of evolution. During 3×10^9 years the hydrogen in dwarfs did not get exhausted; in giants it did. Öpik also presents the interesting suggestion that white dwarfs are remnant cores of composite models after nova explosions, where the greater portion of the original mass (the hydrogen-containing shell) has been thrown away.

1. A. S. Eddington, *The Internal Constitution of the Stars* (Cambridge, 1926).
2. Without admitting such peculiar conditions, the central density and temperature of a star of fixed outer dimensions cannot exceed certain "moderate" limits, cf. A. S. Eddington, "Upper Limits to the Central Temperature and Density of a Star" *Monthly Notices 91*, 444 (1931), "Second Paper," ibid. *93*, 320 (1933).
3. Some kind of equilibrium for the mass of the core may result: increasing mass leads to rapidly increasing energy output and radiation pressure at the boundary of the core, which resists the flow of exhausted material inwards.
4. A. S. Eddington, "The Hydrogen Content of the Stars," *Monthly Notices 92*, 471 (1932).
5. B. Strömgren, "The Opacity of Stellar Matter and the Hydrogen Content of Stars," *Zeitschr. f. Astrophysik 4*, 118 (1932).
6. For C, only a slowing down of the contraction is actually imaginable.
7. A. S. Eddington, "On the Relation between the Masses and Luminosities of the Stars," *Monthly Notices 84*, 308 (1924).

52. Nuclear Reactions in Stars without Hydrogen

Edwin E. Salpeter

(*Astrophysical Journal 115*, 326–328 [1952])

By the late 1940s two key problems in nuclear astrophysics remained unresolved. Where did the heavy elements come from? What specific reactions powered the giant stars? At temperatures of 15 to 20 million degrees, chain reactions such as the proton-proton reaction or the CNO cycle could synthesize helium from hydrogen, and, under the enormously hotter conditions of a billion degrees, heavier nuclei could interact in reversible reactions to set up a thermal equilibrium of elements. In this case the relative abundances of the elements can be worked out from the equations of statistical mechanics.

Subrahmanyan Chandrasekhar and Louis R. Henrich and also Fred Hoyle noticed that the interior of a collapsing star will reach the high temperature and density appropriate for thermal equilibrium.[1] Soon thereafter, Ralph Alpher, Hans Bethe, and George Gamow noted that a similar nonequilibrium situation would prevail during the initial stages of the big bang (see selection 131). Nevertheless, one serious difficulty stood in the way of a satisfactory understanding of the origin of the heavy elements. There is no stable nucleus of atomic weight 5, and this gap seemed to provide an impenetrable barrier for the synthesis of heavier elements from hydrogen and helium.

Ernst Öpik first pointed out that other powerful nuclear reactions will occur at temperatures far below those at which considerations of the thermodynamic equilibrium of nuclei are applicable.[2] He realized that after the temperature in the contracting core of a giant star reaches about 400 million degrees, all the helium will be converted into carbon by triple collisions of helium nuclei, thus circumventing the mass 5 barrier.

In this paper, Edwin Salpeter, unaware of Öpik's work, presents in greater detail the arguments for the formation of carbon by helium burning. Two helium nuclei can combine to form beryllium 8, but because the beryllium is unstable only a tiny fraction remains at any instant. As Salpeter notices, a beryllium nucleus can occasionally combine with a third helium nucleus to form carbon 12. Nevertheless, because beryllium is so extremely rare, it must have a large cross section for helium capture if any substantial amount of carbon is to be produced. In the parlance of nuclear physics, there must be a resonance reaction. As Fred Hoyle subsequently pointed out, this means that the carbon must go through an excited state. A few years later William Fowler and his colleagues showed that the required excited state of carbon does in fact exist.[3] The formation of carbon from helium is thus enhanced enormously by two facts: the existence of beryllium 8 (itself a kind of resonance) and the existence of the excited state of carbon.

Although helium burning, which depends strongly on density, works well in the dense interiors of stars, it does not work at the lower densities following the big-bang explosion that gave rise to the expanding universe. By the time the expanding universe became sufficiently cool to allow nuclei not to be destroyed by radiation, it also became low enough in density that three-body reactions could only occur infrequently. Only the very light elements like deuterium and helium were produced by two-body reactions during the big bang, whereas carbon and the heavier elements are synthesized in the interiors of stars. Both Öpik and Salpeter considered qualitatively the stellar synthesis of heavier elements like oxygen 16, neon 20, and so forth, by the capture of additional helium nuclei at temperatures of about a billion degrees. Thus, the Salpeter and Öpik papers are generally considered the starting point for the modern understanding of the nucleosynthesis of the heavy elements and the specific nuclear reactions that occur in giant stars (see selection 55).

1. S. Chandrasekhar and L. R. Henrich, *Astrophysical Journal 95*, 288 (1942); F. Hoyle, *Monthly Notices of the Royal Astronomical Society 106*, 343 (1946).

2. E. J. Öpik, *Proceedings of the Royal Irish Academy 54*, 49 (1951), and *Contributions of the Armagh Observatory 1*, no. 3 (1951).

3. F. Hoyle, *Astrophysical Journal Supplement 1*, 121 (1954); C. W. Cook, W. A. Fowler, C. C. Lauritsen, and T. Lauritsen, *Physical Review 107*, 508 (1957).

THE MORE LUMINOUS MAIN-SEQUENCE STARS (O and B) exhaust their hydrogen supply in times of the order of magnitude of 10^9 years or less, the bulk of the hydrogen being converted into helium by means of the carbon-nitrogen cycle. When the energy supply of the carbon-nitrogen cycle has been exhausted, the star undergoes gravitational contraction, and its temperature increases. Various nuclear processes[1,2,3] have been suggested for such a contracting star, all of which require temperatures of well over 10^{9}° K. The main aim of this note is to point out that there is one nuclear process which takes place at a much lower temperature of about 2×10^{8}° K, namely, the conversion of three helium nuclei into one carbon nucleus.

We take as an example a main-sequence star of mass $5M_\odot$ (B8 star), central density $\rho = 25$ gm/cm^3, and central temperature $T = 2 \times 10^7$° K. The average energy radiated by the star $\bar{\varepsilon}$, is about 60 erg/gm sec, and most of the hydrogen is exhausted in about 10^9 years. We assume that in the ensuing gravitational contraction the central temperature and density are given by

$$T \propto R^{-1}, \quad \rho \propto R^{-3}, \tag{1}$$

where R is the radius of the star. Gravitational energy is the only source of energy during the contraction until temperatures over 10^{8}° K are reached in a few million years. At these temperatures the following nuclear reaction sets in:

$$\begin{aligned} He^4 + He^4 + 95 \text{ kev} &\rightarrow Be^8 + \gamma, \\ He^4 + Be^8 &\rightarrow C^{12} + \gamma + 7.4 \text{ mev.} \end{aligned} \tag{2}$$

The nucleus Be^8 is unstable to disintegration into two He^4 nuclei. But, since an energy of only (95 ± 5) kev, comparable with thermal energies at temperatures over 10^{8}° K, is required for its formation, a fraction of about 1 in 10^{10} of the material of the star is kept in the form of Be^8 in a state of dynamic equilibrium. The Be^8 present then easily absorbs a helium nucleus. Once carbon has been produced, the following reactions also become possible

$$C^{12} + He^4 \rightarrow O^{16} + \gamma + 7.1 \text{ mev}, \tag{3a}$$
$$O^{16} + He^4 \rightarrow Ne^{20} + \gamma + 4.7 \text{ mev}, \tag{3b}$$

and so on. Owing to the increasing Coulomb barrier, the reaction rates decrease with increasing atomic number. Assuming the absence of γ-ray resonances, the rates for reactions (2) and (3*b*) are of the same order of magnitude. Hence the helium is probably converted mainly into C^{12}, O^{16}, and Ne^{20} and into decreasing amounts of Mg^{24}, Si^{28}, S^{32}, A^{36}, and Ca^{40}.

Energies of 3–4 mev per helium nucleus are produced in these reactions (about one-seventh the production in the carbon-nitrogen cycle). At temperatures T in the neighborhood of 2×10^{8}° K, the rate of energy production ε is given by

$$\varepsilon = 10^3 \left(\frac{\rho}{2.5 \times 10^4}\right)^2 \left(\frac{T}{2 \times 10^{8}\,^\circ\text{K}}\right)^{18} X_\alpha^{\,3} \text{ erg/gm sec}, \tag{4}$$

where ρ is the density in gm/cm^3 and X_α is the concentration by weight of helium. No detailed calculations have as yet been carried out with specific stellar models, but a temperature of slightly more than 2×10^{8}° K should be sufficient for the energy generation (4) to supply the radiative-energy loss. In deriving equation (4), the nuclear γ-ray width for the formation of C^{12} (but *not* the one for Be^8) is required. This width has not yet been measured, and the position of resonance levels is not yet known accurately enough, and an estimate of 0.1 e.v. was used for this width. Hence the correct production rate could be smaller than equation (4) by a factor of as much as 10, or larger than equation (4) by as much as 1000. However, even a factor of 1000 in the production rate alters the temperature necessary by a factor of only 1.33. If we assume an average radiative-energy loss of about 200 erg/gm sec, the energy content of reactions (2) and (3) is sufficient to maintain the star at an almost constant temperature and radius (central density ρ about 2.5×10^4 gm/cm^3) for about 5×10^7 years, after which time the bulk of the helium will have been converted into carbon and heavier nuclei.

We are thus led to the conclusion that *a few per cent of all visible stars of mass $5M_\odot$ or larger are converting helium into heavier nuclei, the central temperature being about ten times larger (and radius ten times smaller) than that of a main-sequence star of the same mass.* It is hoped that some connection will be found between stars undergoing this process, on the one hand, and carbon-rich and high-temperature stars (Wolf-Rayet, nuclei of planetary nebulae), on the other (and possibly even with some of the variable stars and novae).

The fate of stars which have exhausted their helium supply and consist mainly of carbon, oxygen, and neon is much more controversial, but some tentative qualitative conclusions can be drawn about the various competing nuclear reactions. The star again contracts gravitationally until temperatures of about 10^{9}° K are reached. At these temperatures collisions between two C^{12} nuclei, giving Mg^{24} or $(Na^{23} + H^1)$, become possible and at slightly higher temperatures collisions involving O^{16} and Ne^{20}. At temperatures of (2 to 4) $\times 10^{9}$° K the dissociation of the lighter nuclei into helium nuclei and into protons becomes important. These helium nuclei and protons can be absorbed by heavier nuclei, with the result that a significant fraction of the nuclei in the star might be converted into a variety of different nuclei of atomic weight A up to about 40 or 60. The relative concentrations of the various nuclear species might be expected to be similar to that obtained on Hoyle's[1] theory of thermodynamic equilibrium. At (1 to 4) $\times 10^{9}$° K, however, the Urca processes of Gamow and Schoenberg[2] become important, which extract energy from the star in the form of escaping neutrinos. This energy loss results in a very rapid contraction of the star without any increase in temperature, as soon as the energy supply from the conversion into the very stable nuclei (A about 40–60) is exhausted. This contraction continues (unless the star becomes unstable because of its rotational momentum) until densities of more than 10^{10} gm/cm^3 are reached. The electron gas is then highly degenerate, and fairly large concentrations of beta-active nuclei are built up because of the high kinetic energies of the degenerate electron gas. More detailed calculations will be necessary to determine whether enough time is available during this collapse to build up the very heavy nuclei (up to uranium), as was suggested by Hoyle.[1] If the star becomes unstable during the collapse and becomes a supernova, one would expect the various beta-active nuclei to be expelled and to decay in the envelope of the supernova. These considerations lead to difficulties for Borst's[3] hypothesis that the energy generation in envelopes of supernovae of type I is due, to a large extent, to the beta decay of one *single* nucleus, Be^7, obtained from the reaction $He^4 + He^4 \rightarrow Be^7 + n$. It may, however, be possible that this reaction predominates over the others discussed in this note, if in a supernova of type I convection sets in suddenly (with velocities comparable to those of free fall), so that He^4 from the cooler outer layers of the star is *suddenly* brought into the central regions at a temperature of about 4×10^{9} °K. It should be emphasized again that the remarks in *this* paragraph are quite tentative and speculative.

1. F. Hoyle, *M. N. 106*, 343 (1946).
2. G. Gamow and M. Schoenberg, *Phys. Rev. 59*, 539 (1941).
3. L. Borst, *Phys. Rev. 78*, 807 (1950).

53. Inhomogeneous Stellar Models II: Models with Exhausted Cores in Gravitational Contraction

Allan R. Sandage and Martin Schwarzschild

(*Astrophysical Journal 116*, 463–476 [1952])

Shortly after C. F. von Weizsäcker and Hans Bethe described the CNO cycle for the thermonuclear conversion of hydrogen into helium as a stellar energy source, George Gamow considered the evolution of a star whose energy is supplied by this cycle (see selections 48, 49, and 50). He showed that the effective temperature and luminosity of the star should increase with time and that the star therefore evolves upward along the main sequence until all its hydrogen is exhausted. Gamow supposed that, following the complete exhaustion of hydrogen, the star would enter the final contractional stages. However, he overlooked the fact that, because the CNO cycle requires a high temperature, the nuclear burning must be restricted to the stellar core.

Mario Schönberg and Subrahmanyan Chandrasekhar criticized some aspects of Gamow's theory of stellar evolution and instead considered models in which a radiative envelope surrounds either a convective core of higher molecular weight, still generating all the star's energy, or an exhausted isothermal core of higher molecular weight, with the energy being generated in a thin shell at the interface between the envelope and the core.[1] They found it impossible to construct models in which more than about 10% of the mass of the star is included in its exhausted core. This meant that the assumption of uniform composition used earlier by Bengt Strömgren had to be abandoned and that the lifetime of a star on the main sequence is limited to the time it takes to convert 10% of its hydrogen into helium. After the star has reached this Schönberg-Chandrasekhar limit, the core contracts under its own gravity. The heating up of the core makes the envelope overlying the core generate too much energy, and the envelope expands, just as Ernst Öpik had suggested in 1938 (see selection 51). For stars of low mass the contracting exhausted core becomes degenerate; and, as first pointed out by Gamow, this is another cause for envelope expansion.

In this paper, Allan Sandage and Martin Schwarzschild discuss what happens to a star after it has burned so much of its hydrogen that, according to Schönberg and Chandrasekhar, its core must collapse. They show that the star evolves rapidly from a dwarf main-sequence star into a giant. At the time this paper was written in Princeton, Sandage was a graduate student at the California Institute of Technology, working under Walter Baade on the Hertzsprung-Russell diagram of the globular clusters. The key point of the thesis work of Sandage was that the diagram for one globular cluster included faint stars which connected the red giant branch with the main sequence. This was a decisive step at the time, for the theoreticians already more or less understood the main sequence, and they now acquired an observational basis for understanding the process of a star's evolution into the red giant phase. The theoretical calculations given here are a deliberate attempt to account for Sandage's observational data. Moreover, by comparing the evolutionary tracks of stars of different masses to the H-R diagram of a globular cluster, Sandage and Schwarzschild are able to derive a cluster age of about three billion years.

In 1955 Fred Hoyle and Schwarzschild attempted a theoretical description of all the kinks and bends of the giant branches in the H-R diagram for a globular cluster.[2] They found that stars of 1.1 to 1.2 solar masses can describe the observed data if electron degeneracy in the core is taken into account, and if the globular cluster stars have a lower proportion of metals than the Population I stars of the Milky Way. They also showed that soon after a globular cluster star leaves the main sequence, it must develop an outer convection zone whose extent is determined by the condition that the radius of the star must increase at nearly constant temperature. As Schwarzschild and Richard Härm subsequently showed, helium eventually begins to burn abruptly in the degenerate core of these stars; after this helium flash the core expands, the envelope shrinks, and the surface temperature rises again. In the meantime, Sandage had used the evolutionary information contained in the observed luminosity functions and color-magnitude diagrams of clusters to obtain semiempirical evolution tracks for the individual stars in the subgiant and giant regions of the H-R diagrams of these clusters. We represent this work by figure 53.3 appended to this selection.[3]

1. M. Schönberg and S. Chandrasekhar, *Astrophysical Journal 96*, 161 (1942).

2. F. Hoyle and M. Schwarzschild, *Astrophysical Journal Supplement 2*, 1 (1955); M. Schwarzschild and R. Härm, *Astrophysical Journal 136*, 158 (1962).

3. A. R. Sandage, *Astrophysical Journal 125*, 435, *126*, 326 (1957).

Abstract—Seven shell-source models with exhausted, gravitationally contracting cores have been computed in detail. The models form an evolutionary sequence from a configuration whose isothermal core contains the Schönberg-Chandrasekhar limitating mass. It is found that, as the cores contract, the envelopes greatly expand. Thus from the initial configuration, which is near the main sequence, the stars evolve rapidly to the right in the H-R diagram, amply covering the giant region. In this evolution the gravitational contraction contributes less than 4 per cent to the total luminosity.

A comparison of this theoretical evolution with the observed H-R diagram for globular clusters appears to explain the sudden turnoff from the main sequence to the giant region at about $M_{bol} = +3.5$.

I. INTRODUCTION

THE IDENTIFICATION in 1938 of the source of stellar energy with nuclear processes created the problem of finding equilibrium models for giant stars with the high central temperatures required for the nuclear reactions. In recent years a number of investigations[1] have indicated that the solution probably lies with chemically inhomogeneous models. These investigations have shown that certain types of inhomogeneities do lead to large radii with high central temperatures. The various models proposed have differed in the way the inhomogeneities were introduced, according to a variety of assumptions regarding the degrees of mixing and the evolutionary process involved (nuclear transmutations or accretion).

This paper considers a model whose inhomogeneity arises from the following evolutionary process. An initially homogeneous star with a convective core and radiative envelope (Cowling model) which experiences no mixing between the core and envelope starts exhausting its hydrogen supply in the core. The subsequent early stages of the evolution follow those computed by Schönberg and Chandrasekhar,[2] with the core finally exhausted of hydrogen and therefore isothermal. The nuclear-energy production is then confined to a shell between the exhausted core and the radiative envelope. The assumption of no mixing creates a chemical discontinuity between the core and the shell. When the shell has burned outward until it reaches the Schönberg-Chandrasekhar limit for an isothermal core, a new evolutionary process must take place, which is most likely a gravitational contraction of the core. This paper is concerned with the quasi-equilibrium states through which an unmixed model passes after reaching the Schönberg-Chandrasekhar limit.

The main feature differentiating the present models from earlier ones is the occurrence of an additional energy source by gravitational contraction. The distribution of this energy source throughout the exhausted core has here been treated only approximately (assumption 3 in Sec. II), in accordance with the general result that the details of the energy-source distribution have little effect on the model. However, the total amount of energy released by the contraction of the core has been accounted for explicitly (Sec. III).

II. ASSUMPTIONS, DEFINITIONS, AND EQUATIONS

The models here considered consist of a hydrogen-rich envelope in radiative equilibrium and a contracting hydrogen-exhausted core, also in radiative equilibrium.

The following simplifying assumptions have been made in the computations: (1) All the sources for nuclear energy are confined to a shell idealized to an infinitely thin sheet located immediately outside the exhausted core. (2) The temperature at the shell, necessary for the carbon cycle, is $3 \times 10^{7}\,^{\circ}$K throughout the evolutionary stages considered. (3) The distribution of the gravitational energy source in the core is assumed as $\varepsilon =$ constant, where ε is the energy liberated per gram. If L_g is the total flux due to the gravitational source in the core, the flux crossing a shell of radius r within the core is $L(r) = L_g M_r/M_1$, where M_1 is the total mass within the core. (4) The evolutionary changes are sufficiently slow that dynamical terms in the equilibrium equations are negligible. (5) The mean molecular weight is discontinuous across the interface between the envelope and the core, with an arbitrary jump of a factor of 2 consistent with the assumed abundances of $X_i = 0$, $Y_i = 0.98$, $Z_i = 0.02$ in the core, and $X_e = 0.596$, $Y_e = 0.384$, and $Z_e = 0.02$ in the envelope. (6) The switch from the Kramers opacity law to electron scattering in the envelope is abrupt and occurs at the same place for stars of different masses. (7) Radiation pressure and degeneracy are negligible.

The following definitions and equations were used for constructing the models. Subscripts: i = Region interior to the shell; e = envelope; c = center; 1 = interface between envelope and core at the shell; and s = point of switch from Kramers' opacity to electron scattering.

Nondimensional variables:

$$P = p\frac{GM^2}{4\pi R^4}, \quad T = t\frac{\mu_e H}{k}\frac{GM}{R}, \quad M_r = qM, \quad r = xR. \tag{1}$$

Transformation for core equations:

$$p = p_c p^*, \quad t = t_c t^*, \quad q = \frac{t_c^{\,2}}{l_i^{\,2}\sqrt{p_c}}\,q^*, \quad x = \frac{t_c}{\sqrt{p_c}\,l_i}\,x^*. \tag{2}$$

Absorption coefficient:

Kramers region:

$$\kappa = \kappa_0\frac{\rho}{T^{3.5}} \quad \text{with} \quad \kappa_0 = \frac{4 \times 10^{25}}{(t/\bar{g})}Z(1 + X). \tag{3}$$

Electron-scattering region:

$$\kappa = \kappa_0 \frac{\rho_s}{T_s^{3.5}}. \quad (4)$$

Eigenvalue parameters:

Envelope:

$$C = \frac{3\kappa_0}{4ac}\left(\frac{k}{\mu_e HG}\right)^{7.5}\left(\frac{1}{4\pi}\right)^3 \frac{LR^{0.5}}{M^{5.5}}; \quad (5)$$

Core:

$$C^* = \frac{C}{q_1} j_1 \frac{L_g}{L} \frac{p_s}{t_s^{4.5}} \frac{p_c}{t_c^4}. \quad (6)$$

Composition parameters:

$$l_i = \frac{\mu_i}{\mu_e} = \frac{2X_e + \frac{3}{4}Y_e + \frac{1}{2}Z_e}{2X_i + \frac{3}{4}Y_i + \frac{1}{2}Z_i} = 2.0,$$

$$j_i = \frac{Z_i(1+X_i)\mu_i}{Z_e(1+X_e)\mu_e} = 1.253. \quad (7)$$

Differential equations:

Radiative envelope:

$$\frac{dp}{dx} = -\frac{pq}{tx^2}; \quad \frac{dq}{dx} = \frac{x^2 p}{t}; \quad (8)$$

$$\frac{dt}{dx} = -\frac{Cp^2}{t^{8.5}x^2} \text{ in Kramers' region}; \quad (9)$$

$$\frac{dt}{dx} = -C\frac{p_s}{t_s^{4.5}}\frac{p}{t^4 x^2} \text{ in electron-scattering region.} \quad (10)$$

Contracting core:

$$\frac{dp^*}{dx^*} = -\frac{p^*q^*}{t^*x^{*2}}; \quad \frac{dq^*}{dx^*} = \frac{x^{*2}p^*}{t^*};$$

$$\frac{dt^*}{dx^*} = -\frac{C^*q^*p^*}{t^{*4}x^{*2}}. \quad (11)$$

Homology invariants:

$$U = \frac{lpx^3}{tq}, \quad V = \frac{lq}{tx}, \quad (12)$$

where $l_e = 1$ and $l_i = 2$;

$$(n+1)_e = \frac{t_s^{4.5}}{p_s}\frac{qt^4}{Cp}; \quad (13)$$

$$(n+1)_i = \frac{t^{*4}}{C^*p^*} = \frac{t_s^{4.5}}{p_s}\frac{q_1}{Cj_i}\frac{L}{L_g}\frac{t^4}{p}.$$

Fitting conditions at the discontinuity:

$$\frac{U_{1i}}{U_{1e}} = \frac{V_{1i}}{V_{1e}} = l_i = 2.0, \quad (14)$$

$$\frac{(n+1)_{1i}}{(n+1)_{1e}} = \frac{1}{j_i}\frac{L}{L_g}. \quad (15)$$

Density in nondimensional variables:

$$\rho = \frac{\mu}{\mu_e}\frac{p}{t}\frac{M}{4\pi R^3}. \quad (16)$$

III. Gravitational Energy Release and the Evolutionary Fitting Condition

To follow the quasi-static equilibrium states during the evolution, it is necessary to know how much energy is released by the gravitational contraction, since we must be able to apply the fitting condition equation (15). The ratio of the total luminosity to the gravitational luminosity is derived in this section.

The energy released by the core contraction in each step of the evolution is composed of three parts: (1) the change in the gravitational energy of the core between the initial and final configuration; (2) the change in the internal heat content of the core; and (3) the work, PdV, done on the core. An additional apparent source of energy must also be considered. Since the amount of mass in the core is continuously increasing as the shell exhausts its hydrogen and burns outward, additional mass is present in the final contracted state which was not present in the initial state. The additional energy due to this increment of mass, δM_1, carried into the core through the expanding interface must be subtracted from the total energy difference between the initial and final states to give the net contractional energy release. Accordingly, the net released energy as a result of contraction is

$$\bar{L}_g\delta\tau = \delta\left[-\int_0^{r_1}\left(C_vT - \frac{GM_r}{r}\right)4\pi r^2\rho\, dr\right] - 4\pi r_1{}^2 P_1\left(\delta r_1 - \left.\frac{dr}{dM_r}\right|_{r_1}\delta M_{r_1}\right) + \left(C_vT_1 - \frac{GM_{r_1}}{r_1}\right)\delta M_1. \quad (17)$$

The specific heat, C_v, is given by $C_v = k/(\gamma - 1)\mu H$. The value of γ is assumed to be 5/3. To apply equation (17) it is convenient to express it in the nondimensional variables of equation (1). After some reduction, the result is

$$\bar{L}_g\delta\tau = \frac{MkT_1}{\mu_i H}\left\{\delta q_1\left(\frac{3\gamma-4}{\gamma-1}B_1^* + \frac{\gamma}{\gamma-1} - U_1^* - V_1^*\right) + q_1\left(\frac{3\gamma-4}{\gamma-1}\delta B_1^* - \frac{1}{x_1^*}\delta[U_1^*x_1^*]\right)\right\}, \quad (18)$$

where

$$B_1^* = \frac{1}{t_1^* q_1^*} \int_0^{x_1^*} x^{*2} p^* \, dx^*. \tag{19}$$

The increment of mass δM_1 carried through the expanding interface is the mass which was burned as the shell moved outward. It then follows that the nuclear energy, released simultaneously with the gravitational energy, is

$$\bar{L}_n \delta\tau = 0.007c^2 X_e \delta M_1 = 0.007c^2 X_e M \, \delta q_1. \tag{20}$$

The ratio of equation (18) to equation (20) gives the quantity needed for the fitting condition on $n + 1$ of equation (15). With $L = L_g + L_n$, one obtains

$$\bar{L}/\bar{L}_g = 1 + 0.007c^2 X_e/(kT_1/\mu_i H)\{1.25B_1^* + 2.5 - U_1^* - V_1^* + (q_1/\delta q_1)(1.5\,\delta[B_1^*] - [1/x_1^*]\,\delta[U_1^* x_1^*])\}. \tag{21}$$

Hence, for any proposed evolutionary change from an initial stellar state with a known temperature, pressure, and mass distribution to a known final state, the value of the right side of equation (21) may be computed. The value of $\bar{L}/\bar{L}_g$ thus obtained must agree with the mean value of L/L_g obtained from equation (15) for the initial and final states of the evolution step considered. This condition makes it possible—within the present approximations—to determine uniquely the transition states in the evolution from a given initial configuration. In the next section this evolutionary track is followed for a star whose initial configuration has an isothermal core which contains the Schönberg-Chandrasekhar limiting mass.

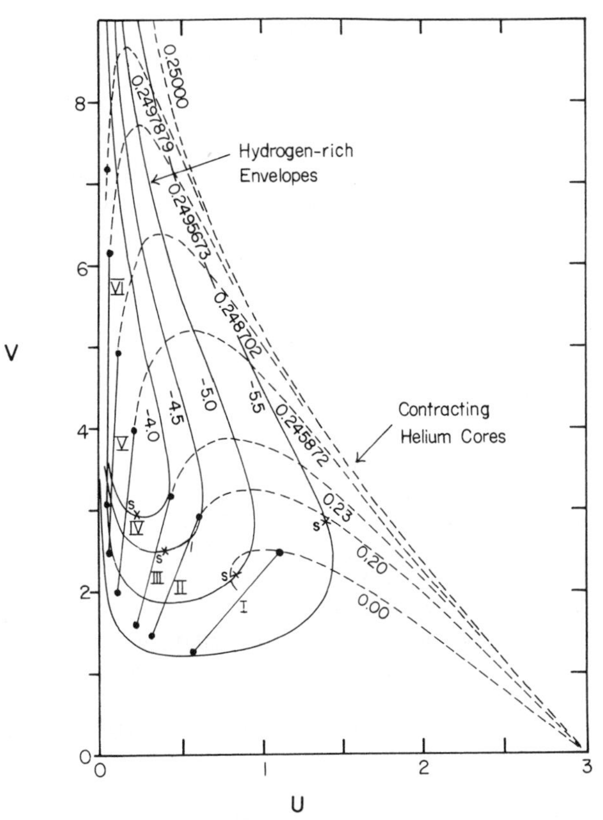

Fig. 53.1 The core and envelope solutions in the U, V plane for stellar models with nondegenerate, contracting helium cores. The solid curves are envelope solutions, with their values of log C marked. Crosses marked S on the envelope solutions show the switch point from Kramers' opacity to electron scattering. The dashed curves are core solutions, with their C^* values marked. The fitting points between envelope and core for the assembled models are indicated by dots. Roman numerals on the straight connecting lines show the model numbers.

IV. THE CONSTRUCTION OF THE MODELS

To construct the models, a family of solutions of equations (8)-(10) for the envelope and a family of solutions of equations (11) for the core are needed. Both sets of solutions are one-parameter families in terms of the eigenvalues C and C^* as long as the appropriate absorption law is given.

The solutions to the envelope equations were obtained as follows. From Paper I of this series[1] envelope solutions of equations (8) and (9), valid for Kramers' law of opacity, were available. On the basis of trial models it became apparent, however, that, for the circumstances of interest here, electron scattering provides the main source of opacity in the inner portions of the envelopes. Accordingly, with the help of the trial models, the point where electron scattering becomes dominant was estimated for each envelope solution, and the solutions were carried from these points inward with equations (8) and (10) by numerical integration.[3] The Kramers envelope with the electron-scattering extensions are shown as heavy solid curves in figure 53.1 plotted with the homology invariants U and V as co-ordinates. The electron-scattering extensions converge to the point $U = 0$, $V = 4$, which is unlike the complete Kramers envelopes that curl about $U = 1/11$, $V = 42/11$.

The one-parameter family of solutions for equation (11) of the core was obtained by numerical integration.[3] A power-series solution of equations (11) valid near the center provided starting values for the integrations. The core solutions are shown as broken curves in figure 53.1.

To find the evolutionary sequence here considered, it remains only to assemble the individual models by fitting each core solution to an appropriate envelope solution by equations (14), (15), and (21). To apply equation (21), the value of B_1^* must be known. This was found along each of the seven core solutions by an elementary quadrature of equation (19).

Since equation (21) connects two neighboring models in the evolutionary sequence, the individual models cannot be determined separately but must be derived successively—in the same sequence in which the star evolves through them.

To start with, the initial state has to be chosen, which—in accordance with the discussion in Section I—was here taken to be the Schönberg-Chandrasekhar limiting case for a model with an isothermal core. This initial state, represented by model I, is shown in the U–V plane of figure 53.1 by the straight line (marked by I) showing the jump from (U_{1e}, V_{1e}) to (U_{1i}, V_{1i}) at the envelope-core interface.

To derive in a definite manner the next state in the evolution (to be represented by model II), a time interval for the evolution from model I to model II has to be chosen, not too long so that the changes are not too radical for the difference equation (21) to hold. Instead of choosing explicitly the time interval, the condition was here chosen so that the core of model II should correspond to $C^* = 0.200$, a value for which the core solution (see figure 53.1) does not lie too far from the core solution of model I ($C^* = 0$). Correspondingly, the point (U_{1i}, V_{1i}) for model II must lie on the core solution for $C^* = 0.200$. Choosing an arbitrary point on this solution as a trial for (U_{1i}, V_{1i}), one can read from the numerical tabulation of this solution $(n + 1)_{1i}$ needed for equation (15) and all the asterisked quantities needed for equation (21). Similarly, the point (U_{1e}, V_{1e}) corresponding to the trial for (U_{1i}, V_{1i}) can be obtained from conditions (14), and, at this point, $(n + 1)_{1e}$ and q_1 can be found by interpolation between the envelope solutions. Now all the necessary quantities are known to test the compatibility of equations (15) and (21). From equation (15) L/L_g for model II is obtained; since this quantity is already known for model I, the mean (geometrical) can be formed. From equation (21) one gets $\overline{L}/\overline{L}_g$ by using for δq_1, δB_1^*, and $\delta(U_1^* x_1^*)$ the corresponding differences between models I and II and by using for T_1, X_e, and μ_i the values assumed in Section II. If the results from the two equations do not agree, the trial point (U_{1i}, V_{1i}) must be changed—along the definite core solution—until agreement is reached. Thus unique values for U_{1i} and V_{1i} (and, in consequence, for all the other nondimensional characteristics) are found for model II. The same procedure leads from model II to model III, and so on through the entire evolution considered.

The model sequence thus obtained is precisely defined by the assumptions made in Section II. It must be remembered, however, that some of these assumptions were quite arbitrary, such as the value of 2.0 for the jump in the mean molecular weight, the assumption for the distribution of the gravitational energy sources, and the assumption of no mixing within the star.

The mathematical characteristics of the seven models shown in figure 53.1 are exhibited in table 53.1.

V. Physical Properties of the Models

Table 53.1 includes several quantities of physical interest which are independent of the mass or luminosity of the stars to which the models may be applied. The first of these quantities is log C. This parameter increases somewhat from model I to model VII but never differs greatly from its value for the corresponding Cowling model (log $C = -6.0$). Since C is, by its definition (5), the numerical coefficient of the mass-luminosity law, it follows that stars built according to the present

Table 53.1 Mathematical characteristics of the models

	I	II	III	IV	V	VI	VII
U_e	0.558	0.303	0.213	0.098	0.050	0.025	0.016
V_e	1.245	1.453	1.597	1.986	2.465	3.078	3.599
C^*	0.00	0.200	0.230	0.245872	0.248702	0.2495673	0.2497879
log C	−5.490	−5.358	−5.307	−5.240	−5.172	−5.121	−5.058
log p_1	2.590	2.960	3.182	3.705	4.370	5.460	7.332
log t_1	0.182	0.294	0.360	0.508	0.680	0.964	1.430
log t_c	0.182	0.466	0.630	0.949	1.246	1.646	2.184
log p_c	3.884	5.059	5.701	6.865	8.008	9.557	11.692
$(n+1)_e$	2.52	2.95	3.12	3.53	3.80	3.95	3.98
$(n+1)_i$	∞	148	118	101	95.1	93.4	88.0
B_1^*	1.000	1.119	1.256	1.676	2.181	2.824	3.329
L/L_g	∞	62.7	47.4	36.0	31.4	29.6	27.7
$X_1 = r_1/R$	0.0643	0.0462	0.0341	0.0200	0.0109	0.0047	0.0014
$q_1 = M_1/M$	0.1200	0.1232	0.1244	0.1260	0.1278	0.1298	0.1313
log $\rho_c/\overline{\rho}$	3.526	4.418	4.893	5.739	6.587	7.734	9.333

models follow closely the mass-luminosity relation of main-sequence stars—contrary to previous red-giant models, which gave overluminous stars compared to main-sequence stars of the same mass.

Next, the values of $\rho_c/\bar{\rho}$ given in table 53.1 show a very large increase from model I to model VII, which represents an exceedingly great change in the structure of the star. The same is shown by the steeply decreasing values of r_1/R, which gives the fraction of the radius occupied by the exhausted core.

Finally, the large values of L/L_g indicate that, throughout the evolution here considered, at most 4 per cent of the total luminosity is provided by gravitational-energy release. Hence, even a rather small gravitational-energy source, if situated in an exhausted core, may thoroughly alter the stellar structure.

To derive the other physical characteristics, such as the radius, bolometric magnitude, and effective temperatures, the models must be applied to stars with definite values of the mass and the guillotine factor. The seven models, together with the Cowling main-sequence model,[4] were applied to stars with 1, 2, and 4 solar masses. The guillotine factor was determined by fitting Morse's[5] opacity values with the Russell mixture to the run of temperature and density of model V for a mass of $2M_\odot$. The coefficient κ_0 in equation (3) was adopted as $\kappa_0 = 7 \times 10^{23}$ from this fit. This corresponds to a guillotine factor of $t/\bar{g} = 1.8$.

With these values for the mass and opacity coefficient, the physical characteristics shown in table 53.2 were computed in the following manner.

First, the radius of the star was obtained from the second of equations (1) by applying the equation at the shell with $T_1 = 3 \times 10^{7\circ}$ K as assumed in Section II and t_1 given by the sixth row of table 53.1. The absolute bolometric magnitude was next found by solving equation (5) for the luminosity L, using the value of log C of the fourth row in table 53.1. The effective temperature follows from its definition in terms of the radius and luminosity.

The central temperature was next computed by noting from the second of equations (1) that $T_c = T_1(t_c/t_1)$. The sixth and seventh rows of table 53.1 then give the central temperature. Log ρ_{1i} and log ρ_c follow immediately from equation

Table 53.2 Physical characteristics of the models

	Main Sequence	I	II	III	IV	V	VI	VII
				$M/M_\odot = 1.0$				
$R/R_\odot$	0.465	0.785	1.017	1.184	1.665	2.474	4.756	13.92
M_{bol}	5.22	4.27	4.09	4.04	4.06	4.10	4.33	4.75
T_e	7,350	7,060	6,460	6,050	5,090	4,130	2,830	1,500
$T_c \times 10^{-6}$	30	30	45	56	83	110	144	170
log ρ_c	2.699	3.991	4.545	4.825	5.226	5.556	5.854	6.051
log ρ_{1i}	2.544	2.697	2.618	2.576	2.506	2.484	2.439	2.445
				$M/M_\odot = 2$				
$R/R_\odot$	0.931	1.570	2.035	2.368	3.330	4.949	9.512	27.84
M_{bol}	1.46	0.51	0.33	0.28	0.29	0.34	0.56	0.99
T_e° K	12,300	11,900	10,900	10,200	8,570	6,950	4,760	2,520
$T_c \times 10^{-6\circ}$ K	30	30	45	56	83	110	144	170
log ρ_c	2.097	3.389	3.942	4.223	4.624	4.953	5.252	5.449
log ρ_{1i}	1.942	2.095	2.015	1.974	1.904	1.881	1.837	1.843
				$M/M_\odot = 4$				
$R/R_\odot$	1.861	3.140	4.070	4.736	6.660	9.898	19.02	55.67
M_{bol}	−2.30	−3.26	−3.44	−3.48	−3.47	−3.43	−3.20	−2.78
T_e	20,700	20,000	18,300	17,100	14,400	11,700	8,020	4,240
$T_c \times 10^{-6}$	30	30	45	56	83	110	144	170
log ρ_c	1.495	2.787	3.340	3.622	4.024	4.352	4.650	4.847
log ρ_{1i}	1.340	1.493	1.413	1.373	1.304	1.280	1.235	1.241

(16) and the values in the fifth, sixth, seventh, and eighth rows of table 53.1.

The results are shown in table 53.2. First, it is seen that large radii characteristic of giant stars are obtained simultaneously with internal temperatures high enough for the nuclear processes to exist. Indeed, if the evolution is followed beyond model VII on the present assumptions, excessively large radii are reached.

Second, it appears to be characteristic of stars evolving along this sequence that as the core contracts, the envelope greatly expands. The details of this core contraction and envelope expansion can indeed be computed, since for each model the integrations give the fractional mass and radius distribution.

Here Sandage and Schwarzschild present a figure (not reproduced) that illustrates the motion of mass shells during core contraction for a mass of 2 solar masses.

Table 53.3 March of the physical variables in models I and VII for $M/M_{\odot} = 2$

	Model I			Model VII		
r/R	$M_{(r)}/M$	$\log T$	$\log \rho$	M_r/M	$\log T$	$\log \rho$
0.000	0.000	7.48	+3.39	0.000	8.23	+5.45
.005	0.001	7.48	+3.38	0.135	7.00	+0.15
.010	0.004	7.48	+3.30	0.140	6.77	−0.50
.015	0.009	7.48	+3.20	0.145	6.66	−0.82
.02	0.019	7.48	+3.09	0.150	6.59	−1.00
.03	0.040	7.48	+2.85	0.160	6.49	−1.22
.04	0.064	7.48	+2.61	0.172	6.42	−1.39
.05	0.084	7.48	+2.40	0.185	6.37	−1.50
.06	0.102	7.47	+2.14	0.200	6.32	−1.61
.07	0.120	7.45	+1.78	0.220	6.28	−1.70
.08	0.138	7.43	+1.72	0.240	6.25	−1.79
.09	0.156	7.41	+1.69	0.260	6.22	−1.84
.10	0.175	7.38	+1.63	0.290	6.18	−1.92
.15	0.285	7.27	+1.46	0.445	6.07	−2.30
.20	0.417	7.18	+1.25	0.580	5.97	−2.58
.25	0.555	7.08	+1.02	0.704	5.87	−2.88
.30	0.686	7.00	+0.80	0.800	5.78	−3.12
.40	0.849	6.83	+0.31	0.910	5.59	−3.60
.50	0.935	6.67	−0.19	0.975	5.42	−4.15
.60	0.975	6.49	−0.74	1.00	5.24	−4.75
.70	0.992	6.30	−1.40	1.00	5.07	−5.39
.80	0.998	6.06	−2.12	1.00	4.82	−6.12
.90	1.00	5.70	−3.31	1.00	4.47	−7.30

Third, table 53.2 shows a sharp rise in the central temperature from 3×10^{7} °K in model I to 1.7×10^{8} °K in model VII as the core contracts. The reason, of course, is that only part of the contractional energy is released as radiation flux, while the remainder goes to increase the internal energy. To illustrate the very great difference in the internal structure of models I and VII, table 53.3 shows their distribution of mass, density, and temperature.

Sandage and Schwarzschild then show that their assumptions regarding shell temperature, the opacity formulae (3) and (4), and the neglect of both radiation pressure and degeneracy are valid, in the sense that the models presented are consistent with the assumptions used in deriving them.

VI. The Evolutionary Track in the H-R Diagram and the Age of Globular Clusters

As Schönberg and Chandrasekhar[2] first pointed out, the star increases slightly in luminosity in going from the main sequence to model I, becoming about 1 mag. brighter when the 12 per cent limit for the exhausted mass is reached at model I. At this point $R = 1.7\ R_{\text{Cowling}}$, while the effective temperature has remained nearly constant. Thus, from the Cowling model to the limiting isothermal core model, the stars remain in the vicinity of the main sequence. The contraction of the core—with its consequent envelope expansion—begins with model I, and the stars rapidly move away from the main sequence into the giant region.

Sandage and Schwarzschild use the data in table 53.2 to derive evolutionary tracks in the Hertzsprung-Russell diagram. We do not reproduce these data here, because they are represented by figure 53.2.

Figure 53.2 suggests the following phenomenon for the Hertzsprung-Russell diagram for old stellar systems—such as globular clusters—in which all stars are presumably of the same age. The fainter stars of such a system will not yet have burned up 12 per cent of their mass and will therefore be on or near the main sequence. The brighter stars will have burned up more than 12 per cent of their mass and will therefore—under the present assumptions—have moved to the right in the H-R diagram. The evolution to the right sets in rather sharply for any given star. Hence one should expect a fairly well-defined turnoff point in the H-R diagrams of these systems, as has indeed been observed.[6] The stars at the turnoff point should then be identified with those which have

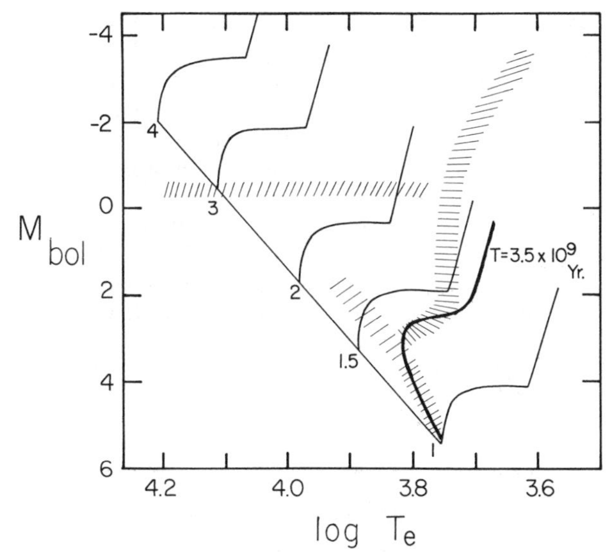

Fig. 53.2 Speculative evolutionary tracks for stars of various masses with an assumed temperature of 1.1×10^8 K for the helium burning. The schematic Hertzsprung-Russell diagram is shown by hatched markings for comparison. The heavy line is the theoretical appearance of the diagram 3.5×10^9 years after the formation of the stars.

just reached the Schönberg-Chandrasekhar limit, i.e., have just burned out 12 per cent of their mass.

If this identification is correct, one may theoretically compute the absolute magnitude at which the turnoff point should occur in a system of an age of, say, $t = 3 \times 10^9$ years. For stars which have burned out q_1 per cent of their mass, one has

$$L\tau = 0.007c^2 X_e q_1 M. \qquad (22)$$

Combining this with the appropriate mass-luminosity relation and using, as before, $X_e = 0.596$ and $q_1 = 12$ per cent, one gets a mass of 1.3 solar masses and a luminosity corresponding to $M_{\text{bol}} \approx +3.3$. (This result is fairly independent of the assumed hydrogen content, X_e, since, for varying X_e also, the jumps of the molecular weight at the discontinuity varies. In consequence, the core-mass fraction q_1 of the Schönberg-Chandrasekhar limit varies[7] in such a way that the product $X_e q_1$ occurring in eq. [22] remains nearly constant.)

On the observational side, the turnoff point in M 92 and M 3 is found at $M_{\text{bol}} = +3.6$. The agreement between the computed and observed values appears so good that it may fairly be taken as a confirmation of the above interpretation of the observed turnoff point.

Inverting the argument, one may also take this agreement as indicating the correctness of the assumed age of the globular clusters, at least within a factor of 2.

VII. Speculation on the Brighter Stars in Globular Clusters

In the previous section we have discussed the early evolutionary phases of unmixed stars, from the Cowling model through the Schönberg-Chandrasekhar limit to the beginning of the core contraction with simultaneous envelope expansion. It is tempting now to speculate on the subsequent evolutionary phases which presumably are represented by the brighter globular cluster stars which may have gone through the earlier phases at a relatively faster rate, owing to a somewhat larger mass. However, for these subsequent phases the present models are soon found quite inadequate, and only some qualitative estimates seem possible until further integrations are made.

The first difficulty arises when the rate is computed with which a star evolves from model I to model VII. By using equation (22) (but replacing L by L_n and q_1 by δq_1), one finds that a star evolves from model I to model VII in about a twentieth of the time it takes to evolve from the Cowling model I. Hence one should expect in the H-R diagram of a globular cluster a good deal fewer stars just beyond the Schönberg-Chandrasekhar limit (model I) than just below this limit contrary to observation. Preliminary tests, however, seem to indicate that the evolution rate during the core contraction may possibly be appreciably modified by moderate changes in the present assumptions, so that this difficulty, though not solved, does not seem too serious.

A second difficulty arises when the extent of the envelope expansion is considered. Under the present assumptions there is no reason why the envelope expansion should stop at or before model VII, while the observed H-R diagrams of globular clusters seem to indicate that the expansion should essentially stop about at model V, and then mainly a brightening (increase in C) and only little further expansion (little increase in t_1) should occur. One may speculate that around model V a physical process not included in the present computations should start to play an essential role.

As a first hypothesis for the needed process, one may think of the transmutation of helium into heavier elements. The central temperature reached in model V (1.1×10^{8} °K) is rather lower than the temperature needed for helium burning (2×10^8 °K) as derived by Salpeter,[8] but still just within the limits of the uncertainty of this derivation. Following this hypothesis, preliminary estimates were made for the subsequent models which should consist of a hydrogen-rich envelope, a shell in which hydrogen burns, a helium-rich intermediate zone, a shell in which helium burns, and an inert core of heavy elements. The estimated evolution through these phases is indicated in figure 53.2 in the vertical upper ends of the heavy lines, with the assumption of 1.1×10^{8} °K for the temperature of the helium burning. Whenever the evolution tracks for individual stars are given in the H-R diagram, one can derive a curve crossing the tracks and connecting all points reached by stars of various masses at the same time. Such a curve, computed for a time of 3.5×10^9 years, is shown by the heavy line in figure 53.2. This curve seems to fit well the main observed feature of the lower portions of the H-R diagram of globular clusters. Nevertheless, it seems far from certain whether this hypothesis regarding the early termination of the envelope expansion is correct, since the necessary temperature for the helium burning seems rather low.

As a second hypothesis for the process causing the termination of the expansion, one might think of moderate mixing of the layers near the shell caused by rotation. This would decrease the effect of the chemical inhomogeneity and thus reduce the expansion. The mixing by rotation should affect all evolutionary phases but possibly the later ones particularly strongly. It then seems plausible that the observed feature of the H-R diagram of globular clusters might be explained by changing the present models by introducing a moderate amount of mixing. To follow this, however, further integrations are necessary.

VIII. Summary

The application of the evolutionary fitting condition of Section III has made it possible to follow from an initial state the evolution of stars built on a shell-source model with a chemical discontinuity between the envelope and a contracting core. The detailed computations show that as the core contracts, liberating gravitational energy, the envelope greatly expands, giving giant stars with internal temperatures high enough for the nuclear processes to provide the required luminosities.

A theoretical Hertzsprung-Russell diagram based on the derived models was compared with the observed diagrams for globular clusters. Good agreement was found for the fainter stars, i.e., the earlier evolution phases. However, the present models were found inadequate to explain the brighter features of the diagram.

1. See Paper I of this series for references, Oke and Schwarzschild, *Ap. J. 116*, 317 (1952).

2. *Ap. J. 96*, 161 (1942).

3. M. Schwarzschild and R. Härm, *Ap. J. Supp. 1*, 319 (1954).

4. T. G. Cowling, *M.N. 96*, 42 (1935).

5. P. M. Morse, *Ap. J. 92*, 27 (1940).

6. Arp. Baum, and Sandage, *A.J., 57*, 4 (1952).

7. Harrison, *Ap. J. 105*, 322 (1947).

8. E. Salpeter, *Ap. J. 115*, 326 (1952).

APPENDED FIGURES

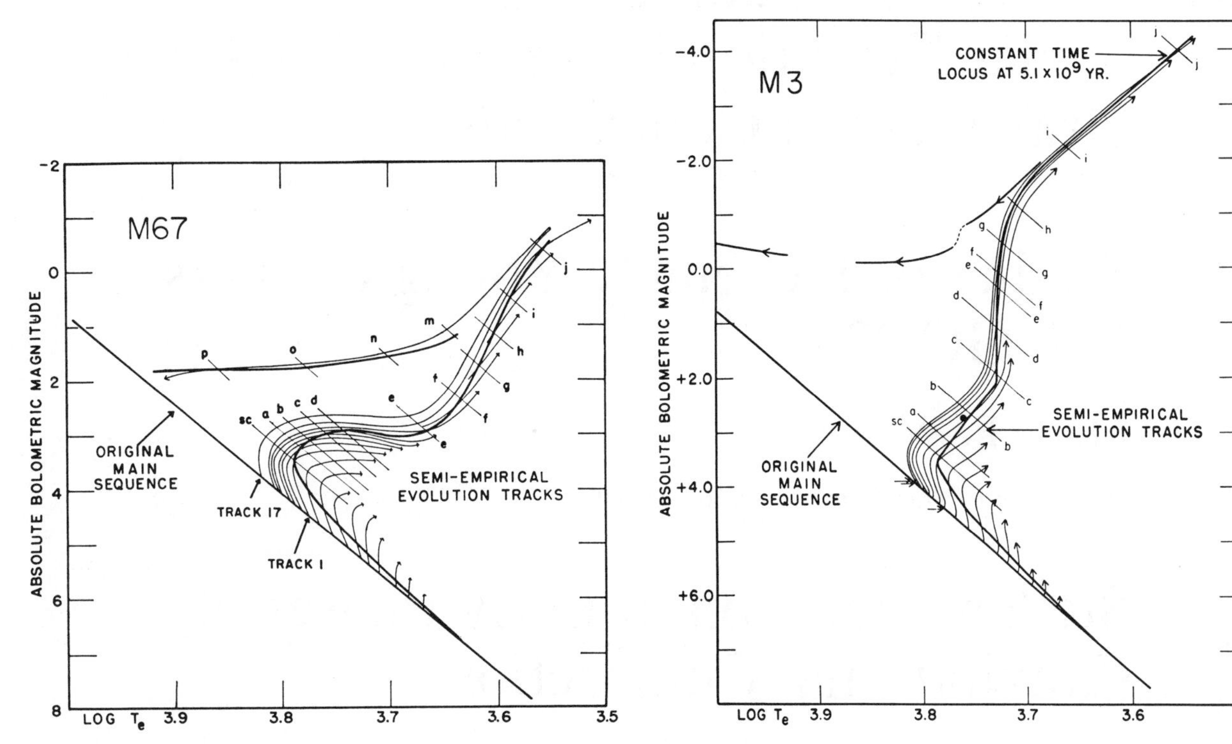

Fig. 53.3 Semiempirical tracks of evolution for stars in the globular clusters M 67 and M 3. The various mapping points are shown as lines labeled *sc*, *a*, *b*, . . . , *p*. The observed color-magnitude diagrams for the two clusters transformed to the M_{bol}, log T_e plane are shown by the heavy lines cutting across the evolutionary tracks. (From A. R. Sandage, *Astrophysical Journal 126*, 326 [1957].)

54. Studies of Young Clusters and Stellar Evolution in the Early Phases of Gravitational Contraction

Studies of Extremely Young Clusters I: NGC 2264

Merle F. Walker

(*Astrophysical Journal Supplement 2*, 365–387 [1956])

Stellar Evolution in Early Phases of Gravitational Contraction

Chushiro Hayashi

(*Publications of the Astronomical Society of Japan 13*, 450–452 [1961])

The understanding of stellar evolution involves a delicate interplay between observations (together with laboratory experiments on nuclear reactions) and theory (including computer modeling). In this selection we pair a paper on observations of pre-main-sequence stars with a paper on theoretical interpretations of these stars.

Earlier studies of stellar evolution (ca. 1940–1955) reached the general conclusion that a star spends most of its life on the main sequence, that the ages of the stars in globular clusters are on the order of 5 billion years, and that only bright, massive stars burn their hydrogen fast enough to enter the giant stage during the lifetime of our galaxy. Because brighter stars deplete their nuclear fuel comparatively quickly, such luminous stars will be found only in very young clusters. As time goes on, the bright O and B stars in young clusters will evolve into the giant stage, and the top of the main sequence will steadily disappear. It follows that clusters of O and B stars, which have main-sequence lifetimes of only a few million years, should indicate conditions in the early stages of stellar evolution.

Pavel Parenago, in studying the stars in the Orion nebula association, first showed that the more massive stars of this association fall on the main sequence and the cooler, less massive stars fall distinctly above the main sequence.[1] Three years later, Merle Walker began an extensive study of young clusters that confirmed Parenago's observation.[2] As Walker illustrates in the first of the papers given here, less massive, cooler members of NGC 2264 have not yet reached the main sequence; using the computational work of Louis Henyey et al., he explains that these stars must still be undergoing gravitational contraction from the prestellar medium and have not yet become hot enough to ignite thermonuclear reactions.

In 1956, when Walker published his Hertzsprung-Russell diagram of NGC 2264, astronomers assumed that stars slowly brightened as they contracted to the main sequence. Not until the work reported by Chushiro Hayashi in the second paper given here did they realize that strong convection would enable a contracting star to release its energy much faster than by radiation and that, consequently, pre-main-sequence stars would be considerably more luminous than their counterparts already on the main sequences of older clusters. As Hayashi shows, the evolutionary track of a contracting star moves downward on the H-R diagram until the convection zone retreats toward the stellar surface. When the surface convection zone becomes thin, the star's evolutionary track turns leftward until it reaches the main sequence. In this way, all the stars (the luminous main-sequence stars as well as the less massive contracting stars) of NGC 2264 are found to have nearly the same ages.

1. P. P. Parenago, *Astronomicheskii Zhurnal 30*, 249 (1953).
2. M. F. Walker, *Astrophysical Journal*, *125*, 636 (1957). *130*, 57 (1959), *133*, 438 (1961)

Walker, Studies of Extremely Young Clusters

Abstract—Three-color photoelectric and photographic observations of NGC 2264 have been obtained to $V = 17$, in order to investigate the color-magnitude diagram of an extremely young cluster of stars. The diagram indicates that the cluster possesses a normal main sequence extending from O7 to A0, below which the stars fall above the main sequence. The reality of this effect has been confirmed by spectroscopic observations. The shape of the color-magnitude diagram agrees approximately with that predicted theoretically for clusters which are so young that the fainter stars are still in the process of contracting gravitationally from the prestellar medium and have not yet reached the main sequence. The age of the cluster given by the point where the cluster stars depart from the main sequence is about 3×10^6 years.

In addition to luminosity effects, the spectra of the stars which lie above the main sequence show a number of peculiarities. One of the most striking is that later than F8 the rotational velocity of the stars appears to increase, suggesting that these stars have not yet had time to dissipate the original angular momentum of the material from which they have condensed.

The occurrence of T Tauri stars in the cluster and their location in the color-magnitude diagram indicate that stars of this type are extremely young and are still in the gravitational contraction state. Whether the T Tauri phenomenon represents a necessary stage in the gravitational contraction of all newly formed stars cannot be determined from the present observations. A number of new T Tauri variables was discovered during this investigation.

The luminosity function of the cluster is completed to $M_v = +8$ and appears to show a maximum at $M_v = +6$. Brighter than $M_v = +3$, the observed function agrees with the luminosity function which would be observed in the solar neighborhood if no evolution of stars away from the main sequence had taken place.

I. Introduction

While considerable observational work has now been done upon the problem of the later evolution of stars away from the main sequence, almost no observational investigations of the very early stages of stellar evolution have been made. Recent theoretical work by Salpeter (1953) and by Henyey, LeLevier, and Levée (1955) has indicated that in a group of stars which is only a few million years old the fainter members of the group will still be in the process of contracting gravitationally from the prestellar medium. The rate of contraction of an individual star depends upon its mass. As the configuration contracts, the decrease in the radiating surface is offset by the increasing temperature of the surface, so that the luminosity of the star remains nearly constant. As a result, the color-magnitude diagram of an extremely young cluster of stars should consist of a normal main sequence for the brightest stars, whose masses are greater than some critical value which depends upon the age of the cluster. Below this limit the cluster stars will fall above the main sequence and be displaced from it in the direction of lower temperatures.

The O and early B stars are known, from considerations of the rate of consumption of nuclear fuel, to have maximum lifetimes of only a few million years. Thus, to investigate the early stages of stellar evolution, we must determine the color-magnitude diagrams of clusters containing O or early B stars. At the present time, only one such group of stars has been investigated to a sufficiently faint absolute magnitude to show the divergence of the faint stars just discussed. This is the investigation of the Orion aggregate of stars by Parenago (1953). Unfortunately, Parenago did not have adequate observational material at his disposal to be able to prepare a really satisfactory color-magnitude diagram; the colors and magnitudes of his fainter stars had to be determined by measuring the images of these stars on published photographs of the field. However, his results do suggest that the fainter stars in the Orion aggregate depart from the main sequence in the direction predicted by the theory. His diagram consists of a normal main sequence extending from O to about A5. Below that, the main sequence appears to end, and one encounters a large cloud of subgiants.

This paper is the first of a series which will describe the photometric results for a number of extremely young clusters obtained during the last few years at the Mount Wilson Observatory. In order to investigate the fainter members of one of these clusters, it is necessary that the following five conditions be fulfilled as nearly as possible: (1) that the cluster contain both bright OB stars and faint stars; (2) that the cluster be relatively near, so that the intrinsically faint members will be accessible for observation; (3) that the cluster be in some manner recognizable as a group even to the faint stars, since most of the clusters are so far away that reliable proper motions cannot be obtained and since even for the few nearby groups, such as the Orion association, the proper-motion data do not extend to a sufficiently faint magnitude limit to enable one to determine the membership of the faint stars on this basis; (4) that the amount of emission nebulosity in the cluster be a minimum, since practical experience has shown that the observation of the faint stars in nebulous clusters is extremely difficult; and (5) that the amount of interstellar reddening should be small and uniform over the entire area of the cluster.

The Orion association is the nearest of these clusters of extremely young stars, and it possesses both bright and faint stars. However, it is extremely nebulous, so that the accurate observation of the faint members would be almost impossible. NGC 2264, on the other hand, is probably the most favorable of the young clusters for the purpose of investigating the faint members. It shows a very definite concentration of stars to the center of the cluster, and it is seen projected upon an extensive dark cloud (Herbig 1954), which effectively eliminates all or nearly all background stars from the central region of the cluster. While the cluster is farther away than Orion, the amount and intensity of emission nebulosity found in NGC 2264 is much smaller than is usually encountered in such objects, so that it is possible to observe accurately stars which are intrinsically much fainter than could be examined in the Orion association. Moreover, the reddening of the cluster is only 0.082 mag. in $B - V$ and appears to be uniform over the entire area of the cluster. The cluster contains a large number of faint stars which are either variable stars (Kukarkin and Parenago 1955; Wenzel 1955*a*, *b*) or which have been shown by Herbig (1954) to have bright Hα, or both. These are presumably T Tauri stars.

II. PHOTOMETRIC OBSERVATIONS

Here Walker states that photoelectric observations of NGC 2264 have been taken, using the *U-B-V* system of Johnson and Morgan (1953) and usually using the equipment described by Walker (1954). Omitted here are tables of the instruments used and the values of *V*, *B-V*, and *U-B* for the nonvariable and variable stars; also omitted are photographs of NGC 2264 numbering the observed stars.

III. BLINK SURVEY FOR VARIABLE STARS

We omit a table and photograph that give the numbers and locations respectively of 79 variable stars in NGC 2264. Walker discovered 53 of these by using a blink comparison of four plates taken between December, 1953 and April, 1954. Forty-two per cent of the variable stars in NGC 2264 show strong Hα emission. This compares favorably with the value of 47 per cent found in the Orion Nebula (Herbig, 1954), suggesting that the variable and bright Hα stars in the two clusters have similar characteristics.

IV. DISCUSSION OF PHOTOMETRIC OBSERVATIONS

Next, Walker presents the color-magnitude diagram for NGC 2264 shown in figure 54.1. For stars earlier than A0, membership in the cluster was unambiguously determined from the

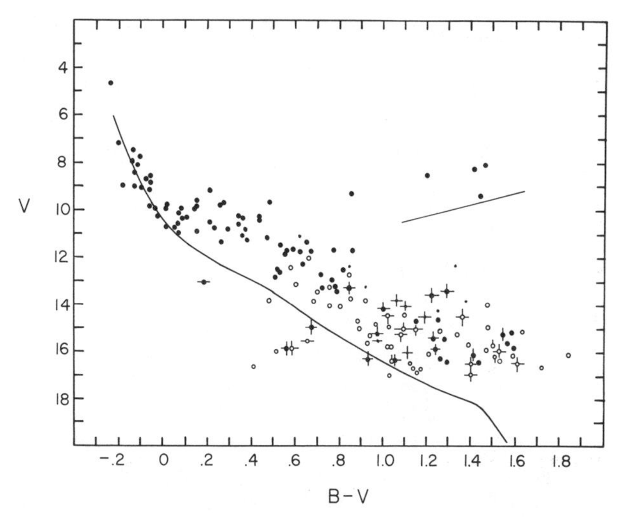

Fig. 54.1 Color-magnitude diagram of NGC 2264. Filled circles represent photoelectric, and open circles photographic, observations. Vertical lines indicate known light variables; horizontal lines indicate stars having bright Hα. Small symbols show observations of lower weight. Observed values of the magnitudes and colors have been plotted. The lines represent the standard main sequence and giant branch of Johnson and Morgan (1953), corrected for the uniform reddening of the cluster.

two-color diagram. Of later stars, only those have been included (1) that lie well within the boundaries of the dark cloud against which the cluster is seen projected, or (2) for which spectral types and luminosities have been obtained that appear to confirm cluster membership. We omit the *U-B*, *B-V* diagram for NGC 2264. A comparison of this diagram with that of unreddened, main-sequence stars shows that reddening of the cluster is 0.082 mag in *B-V*.

It will be seen that the color-magnitude diagram is quite unusual. The diagram indicates that there is a continuous sequence of stars connecting the OB stars with the T Tauri stars, the later appearing abruptly at $V = 13$. However, below A0 the sequence departs from the normal main sequence by first rising by 1.5 mag. and then extending as a broad band, 2 mag. wide, parallel to and about 2 mag. above the main sequence. Below $V = 13$, the *maximum* departure of the stars from the main sequence increases, reaching about 5.2 mag. at $B - V = 1.6$. Whether the mean line also diverges is not clear, owing to the incompleteness of the observations. Note that the cluster appears to contain five yellow giants; these will be discussed in a later section.

The distance of NGC 2264 was first determined by fitting the main sequence of the cluster above A0 to the standard main sequence of Johnson and Morgan (1953). In this manner the apparent distance modulus was found to be 10.2 mag. However, Johnson and Hiltner (1956) have recently shown that, for young groups of stars like NGC 2264, a correction of −0.5 mag. must be applied, at spectral type A0, to the distance moduli derived for such groups, using Johnson and Morgan's standard main sequence. This correction is necessary, since, on the average, stars of spectral type about A0 observed in the general field are sufficiently old that they have already begun to evolve away from the initial or "zero-age" main sequence. With this correction, the apparent distance modulus of NGC 2264 is 9.7 mag. Using the ratio of total to selective absorption of 3.0 given by Johnson and Morgan (1953), we find the true distance modulus of the cluster to be 9.5 mag., or about 800 pc. Since Johnson and Hiltner's paper appeared after the diagrams for the present paper had been completed, the main sequence shown in figure 54.1 has been left uncorrected and it is the standard main sequence of Johnson and Morgan. The only errors resulting from this procedure are that the position of the giant branch should be 0.5 mag. higher than it is shown and that there are some minor changes in the shape of the main sequence which do not affect the present discussion. All the data discussed in the text and presented in the tables have been derived on the basis of the final distance modulus of 9.5 mag.

There are three types of objects which might produce a spurious color-magnitude diagram which would resemble the one which we observe for NGC 2264. These are (1) background stars of high luminosity, (2) highly reddened early-type members of the cluster, and (3) foreground stars.

Walker concludes that these effects are not important and that the color-magnitude diagram derived for NGC 2264 is not greatly influenced by the presence of nonmembers or reddened early-type stars.

Perhaps the outstanding feature of the two-color diagram is that, while the bright stars in the cluster fall along a sequence displaced slightly from the normal relationship between $U - B$ and $B - V$, many of the fainter stars depart widely from this normal relationship. This departure is in a direction which could indicate either that these faint stars are in reality very early O- and B-type stars which are heavily reddened or that the stars are intrinsically very much brighter in the ultraviolet than normal stars having the same $B - V$ color. We have already seen that, owing to the great opacity of the dark cloud, we would expect that very few heavily reddened O and B stars would be observed in the cluster. Since most of these stars are T Tauri stars—either known light-variables or emission Hα stars or both—it is evident that at least the great majority of these stars are actually abnormally bright in the ultraviolet. This unusual brightness in the ultraviolet has already been pointed out by Haro and Herbig (1955), who detected this anomaly in several of the stars in NGC 2264 and in the Orion Nebula. So far, no interpretation of this phenomenon has been found. There also appears to be a tendency for stars between $B - V = +0.4$ and $B - V = +0.6$ to be slightly brighter than normal in the ultraviolet; perhaps it is with these stars that the ultraviolet anomaly which becomes so pronounced for the fainter members of the cluster first sets in. It is to be noted that the very intensity of the ultraviolet excess in one star, No. 79 = LHα 22, shows that the effect must be intrinsic; if the position of this star in the two-color diagram were the result of reddening, then the temperature of the star would have to be far greater than has ever been observed for any other object.

The location of the T Tauri stars in the color-magnitude diagram of NGC 2264 confirms very clearly the conclusion of Herbig (1952) that T Tauri stars in the Taurus-Aurigae dark clouds tend to lie above the main sequence. Herbig found that, beginning at about spectral class K4 or K5, the T Tauri stars begin to fall more and more above the main sequence, their departure from the main sequence reaching about 2.5 mag. at type M2 or M3. The G-type T Tauri stars, on the other hand, tend to lie below the main sequence. The same trend is evident in figure 54.1. With the possible exception of No. 90, the earliest of the variables have color classes of about F6–G0 and lie below the main sequence. At about $B - V = +0.8$ (color class about G8) the mean line of the variables crosses the main sequence, and the later variables lie progressively farther above the main sequence, reaching about 4 mag. by color class M3. In NGC 2264 the point at which the variables cross over the main sequence is earlier, and the average departure of the later variables from the main sequence is more rapid with color class than that found by Herbig in Taurus-Aurigae. This may result from a difference in the stars found in the two regions, but it is more probable that it is caused by the fact that the photometric observations at NGC 2264 extend only to $M_v = +7.5$. Also the entire light-ranges of most of the variables discussed by Herbig were available, so that he was able to plot mean light-positions, whereas in NGC 2264 it has been necessary to use "instantaneous" values of the brightness of the variables. The photometric observations of the variable stars in NGC 2264 are not sufficiently numerous to permit us to draw any conclusions regarding the nature of their light-curves. Neither are they sufficiently exact to provide much indication of how the colors of the stars change with their brightness.

V. Spectroscopic Observations

Following the discovery of the anomalous color-magnitude diagram of NGC 2264, spectroscopic observations were secured for as many as possible of the stars which appear to depart from the main sequence, in order to test the reality of this effect and to determine whether these stars exhibit any spectroscopic peculiarities.

Here we omit a description of the method of observing spectra and a figure giving the spectral class and luminosity of all the stars observed spectroscopically. In general, the spectra confirm the peculiar structure of the color-magnitude diagram (figure 54.1), and the luminosities of the cluster stars derived from their spectra agree fairly well with the luminosities determined from the color-magnitude diagram. Because the observed spectral types tend to agree with those predicted from the colors, the stars later than A0 share in the reddening of the cluster and are not unreddened foreground stars. An unusual aspect of the stellar spectra is the rotational broadening of the lines of stars of spectral type later than F8. Spectra of the five stars that fall in the yellow giant region of the color-magnitude diagram show that the stars are giants.

VI. The Luminosity Function

Walker presents a table (omitted here) that compares the luminosity function of NGC 2264 with that of the stars near the sun (van Rhijn [1936]) and with that of the "original" pre-evolution stars (Salpeter [1955]). These comparisons are discussed in the next section.

VII. Interpretation from the Standpoint of Stellar Evolution

In the preceding sections we have seen that the color-magnitude diagram for NGC 2264 consists of an apparently normal main sequence extending from O7 to A0. For stars later than A0, the sequence departs from the normal main sequence by first rising by 1.5 mag. and then extending as a broad band of points 2 mag. wide parallel to and about 2 mag. above the main sequence. The reality of the 1.5-mag. rise has been checked spectroscopically, as has the membership of a number of the fainter stars. From the situation of the cluster against a dark cloud and from consideration of the number of foreground stars to be expected, we have concluded that the color-magnitude diagram is not badly affected by non-members and therefore represents with fair accuracy the stellar content of the cluster. As pointed out in the introduction, theoretical investigations predict that in an extremely young cluster of stars the fainter stars will still be in the process of contracting from the prestellar medium, since the contraction time depends upon the mass of the star. Consequently, the color-magnitude diagram of such a cluster should consist of a main sequence extending down to some point which depends upon the age of the cluster, below which the stars will lie to the right of the main sequence, departing more and more from it as one goes to fainter stars. Thus we see that, at least in a general way, the color-magnitude diagram of NGC 2264 agrees with the diagram predicted theoretically for an extremely young cluster of stars.

Salpeter (1953) has calculated that about 10^6 years are required for an A0 star to reach the main sequence, while the more refined calculations of Henyey *et al.* (1955) give 3×10^6 years as the time required for an A0 star to reach the main sequence by gravitational contraction from a starting configuration having a surface temperature of 4,000°. It is evident that if the theory of gravitational contraction time is correct and if the turnoff of the cluster sequence is actually the result of the age of the stars, then the age of the cluster inferred from the maximum possible lifetime of the brightest star in the cluster should be greater than, or equal to, the age given by the turnoff. The age of the brightest star in NGC 2264, the O7 star S Mon, has been estimated from the rate of conversion of hydrogen to helium. If we assume that the star has a mass of $15\mathfrak{M}_\odot$ and that it has maintained its present luminosity throughout its lifetime, the time required for it to burn 10 per cent of its hydrogen is 2.1×10^6 years. However, it is possible that S Mon is somewhat older than this and has already begun to evolve away from the main sequence. An upper limit to the age of the star can be obtained by assuming that S Mon began its career on the main sequence with an absolute magnitude that was 1 mag. fainter than its present luminosity and that the star has just now reached the point where, having burned 10 per cent of its hydrogen, it is beginning to evolve away from the main sequence. In this manner the maximum possible age of S Mon is found to be 5.3×10^6 years. The true age of the star probably lies somewhere between 2 and 5×10^6 years and agrees well with the age inferred from the turnoff of the main sequence. Owing to the distance of NGC 2264, the proper-motion data available at the present time (van Maanen 1930) are not sufficiently accurate to permit the age of the cluster to be determined from any possible motions of expansion.

An interesting feature of the calculation by Henyey *et al.* (1955) of the evolutionary tracks of gravitationally contracting stars is that they indicate that the brightness of the star decreases just before it reaches the main sequence. For an A0 star, the amount of the decrease is about 0.6 mag. Thus, although the size and shape of the "hump" in the cluster sequence at A0 is not the same as predicted by the theory, it may be that this feature of the color-magnitude diagram is related to the predicted decrease in brightness.

Unfortunately, the theoretical work which has been done to date does not give us a good picture of the shape of the sequence in the cluster below the turnoff from the main sequence. However, it would appear that the sequence should continually diverge from the main sequence rather than run more or less parallel to it, as it is observed to do in NGC 2264. This disagreement is probably not too serious, since the theoretical work is still in a rather preliminary state. At the present time the most significant comparisons of the observations with the theory are probably the age determined from the point of turnoff and the *direction* of the divergence of the fainter stars.

At the same time it is worth noting that if the influence of foreground stars upon the color-magnitude diagram of NGC 2264 has been underestimated, the effect of these stars would be to make the cluster sequence appear to run more nearly parallel to the main sequence than it actually does. In addition, we have tacitly assumed that the ages of all the stars in the cluster are identical; if this is not the case, then this difference in age will also have the effect of making the cluster sequence run more nearly parallel to the main sequence than would a sequence composed of stars of the same age. Finally, we must remember that a large number of the fainter stars are variable, and, since "instantaneous" rather than mean positions of the variables have been used in the diagram, the shape of the cluster sequence may have been affected.

One feature of the color-magnitude diagram for which the present theory does not account is the occurrence of several stars *below* the position of the main sequence, discussed in Section IV. Another unexplained feature is the apparent existence of five yellow giants in the cluster. As pointed out in Section V, the membership of these stars is still not entirely certain. If they are members, this will be the first cluster in which yellow giants have been found together with OB stars. It will also differ from other galactic clusters, in that the giants in NGC 2264 are several magnitudes fainter than the brightest stars on the main sequence. According to our current ideas of stellar evolution, the giants in the somewhat older galactic clusters are stars which were once on the main sequence but have now evolved along almost horizontal tracks into the giant region (Johnson and Sandage 1955). In the case of NGC 2264, there has not been time for the observed giants to have been produced by stars evolving from the main sequence unless we suppose that these stars were formed much earlier than the other stars in the cluster. If these five stars are members of the cluster, then one possible explanation of their existence might be that they, too, are young stars which are in the process of contracting gravitationally to the main sequence.

It was pointed out in Section V that a rapid increase in the rotational velocity of stars belonging to the cluster sequence occurs between F8 and K0. As we indicated, this is the exact opposite of the variation of rotational velocity with spectral type found for ordinary stars. However, such a variation of rotation with spectral type might be expected in a group of stars such as this where the stars are in the process of contracting from the prestellar medium. While we still do not know very much about the manner in which the mutual gravitational attraction of a cloud of material becomes the dominant force acting upon it, it would seem reasonable to suppose that such a mass of material, when condensing to form a star, would possess some angular momentum. As the protostar continues to contract, the angular velocity will increase, and, if some mechanism did not act to diminish this rotational velocity, it is very probable that the star would exceed the limit of rotational stability long before it had reached the main sequence. The problem of the loss of angular momentum has been investigated by Lüst and Schlüter (1955). Making plausible assumptions regarding the density of interstellar material in the immediate neighborhood of a newly formed star, they find that if the star possesses a magnetic field of 100 gauss at its surface, the effect of magnetic braking is sufficiently great to cause a star rotating with a velocity at the limit of rotational stability to lose all its angular momentum in less than 10^6 years. It is difficult to determine the length of time required for the loss of angular momentum from the color-magnitude diagram of NGC 2264, since we cannot be sure to what extent the observed stars represent a homogeneous age or mass group. If we interpret those stars which show rotation as being stars of similar mass and slightly different age, then the evolutionary tracks computed by Henyey *et al.* (1955) indicate that the angular momentum of these stars is being lost in about 10^6 years, in agreement with the order of the length of time predicted theoretically.

The occurrence of the T Tauri stars in NGC 2264 and their location in the color-magnitude diagram seem to indicate that stars of this type are extremely young and are in the process of contracting gravitationally to the main sequence. It is not clear from the observations of NGC 2264 whether the T Tauri phenomenon represents a particular stage in the gravitational contraction of all newly formed stars or whether it is only stars fainter than a certain absolute magnitude which show the effect. The fact that there is a rather sharp upper limit to the T Tauri stars in NGC 2264 at $M_v = +3.5$ gives us no information on this point, since the rate of evolution is a function of mass or luminosity, so that the stars brighter than this limit may all have passed through the T Tauri stage. Even for the stars fainter than $M_v = +3.5$, it is not clear whether the T Tauri phenomenon is a necessary result of the contraction process or whether, as is commonly suggested, the phenomenon results from the interaction of these stars with surrounding nebulosity, although the fact that the reddening of the cluster is small would seem to militate against the interaction theory. It is certain that the study of the T

Tauri stars will be of great importance to our understanding of these early stages of stellar evolution.

Salpeter (1955) has computed an "original" luminosity function from the observed luminosity function for stars in the solar neighborhood and the known rates of evolution of stars away from the main sequence. This "original" luminosity function should represent the appearance of the luminosity function in an extremely young group of stars where no evolution away from the main sequence has yet occurred, provided that the distribution of masses in the group is the same as for stars in the vicinity of the sun. Plotting the evolutionary tracks calculated by Henyey *et al.* (1955) in terms of V and $B - V$, it turns out that, to a first approximation, the V magnitude of a gravitationally contracting star remains constant. Thus the luminosity function for a cluster containing gravitationally contracting stars can be compared with the luminosity function for main-sequence stars. A comparison of Salpeter's luminosity function with the observed luminosity function of NGC 2264 is facilitated by the observed ratio of the number of cluster stars to stars in the vicinity of the sun, reduced on the assumption that the density of stars in the cluster is twenty times the density in the solar neighborhood. It will be seen that the luminosity function for NGC 2264 agrees well with Salpeter's luminosity function for stars brighter than $M_v = +3$, although the comparison is not too exact because of the small number of bright stars in the cluster. Relative to the number of brighter stars, there is an excess of stars in the cluster fainter than $M_v = +3$, compared to the solar neighborhood. It is interesting to note that this is just the portion of the luminosity function populated by the T Tauri stars. The interpretation of the variations in the luminosity functions of different young clusters of stars and between those of the clusters and the solar neighborhood will be discussed in a later paper in this series.

Since the V magnitude of the contracting stars remains roughly constant, we may use the mass-luminosity law in connection with the observed luminosity function to estimate the total mass of the cluster. In this way the total mass of NGC 2264 is found to be about $450\mathfrak{M}\odot$. Assuming that the volume of the cluster is 300 pc^3, the mass density of the cluster is $1.5\mathfrak{M}\odot/pc^3$. The density required for stable configurations has been estimated (Bok 1934) to be $0.1\mathfrak{M}\odot/pc^3$.

Blanco, V. M. 1956, *Ap. J. 123*, 64.

Bok, B. J. 1934, *Harvard Circ.*, no. 384.

Haro, G., and Herbig, G. H. 1955, *Bol. Obs. Tonantzintla y Tacubaya*, no. 12, p. 33.

Haro, G., and Morgan, W. W. 1953, *Ap. J. 118*, 16.

Henyey, L. G., LeLevier, R., and Levée, R. D. 1955, *Pub. A.S.P. 67*, 154.

Herbig, G. H. 1952, *J.R.A.S. Canada 46*, 222, *Dom. Ap. Obs. Contr.*, no. 27 (pt. 3).

———. 1954, *Ap. J. 119*, 483.

Herbig, G. H., and Spalding, J. F. 1955, *Ap. J. 121*, 118.

Hoffmeister, C. 1949, *Ergebn. A.N. 12*, A10.

Johnson, H. L. 1956, private communication.

Johnson, H. L., and Hiltner, W. A. 1956, *Ap. J. 123*, 267.

Johnson, H. L., and Morgan, W. W. 1953, *Ap. J. 117*, 313.

Johnson, H. L., and Sandage, A. R. 1955, *Ap. J. 121*, 616

Kukarkin, B. V., and Parenago, P. P. 1955, Seventh Suppl. to *General Catalogue of Variable Stars* (Moscow).

Lüst, R., and Schlüter, A. 1955, *Zs. f. Ap. 38*, 190.

McCuskey, S. W. 1949, *Ap. J. 109*, 139.

———. 1954, ibid. *120*, 139.

Parenago, P. P. 1953, *Astr. Zhur. 30*, 249.

Salpeter, E. E. 1953, *Symposium on Astrophysics* (Ann Arbor: University of Michigan).

———. 1955, *Ap. J. 121*, 161.

Trumpler, R. J. 1930, *Lick Obs. Bull. 14*, 154.

———. 1956, private communication.

van Maanen, A. 1930, *Mt. W. Contr.*, no. 405.

van Rhijn, P. 1936, *Groningen Pub.*, no. 47, p. 17.

Walker, M. F. 1954, *Pub. A.S.P. 66*, 71.

Wenzel, W. 1955*a*, *Mitt. veränderliche Sterne*, no. 190.

———. 1955*b*, ibid. no. 193.

Wolf, M. 1924, *A.N. 221*, 379.

Hayashi, Stellar Evolution in Early Phases of Gravitational Contraction

Abstract—The surface condition for red giant stars worked out in the previous paper indicates that stars which lie in the low luminosity and low temperature region of the *H-R* diagram cannot be in equilibrium so that the evolutional path of contracting stars in this region will be different from that calculated by Henyey *et al.* The age of these stars along the loci of quasi-static solutions is calculated. The result seems to explain well the *H-R* diagram of a young cluster NGC 2264.

THE GRAVITATIONAL CONTRACTION of stars in their very early stage of evolution were calculated by Levée,[1] Salpeter[2] and Henyey *et al.*[3] under the assumption that the stars were wholly in radiative equilibrium. It is now clear, however, that the late-type stars have hydrogen convection zones and the effect of these zones to the whole internal structure cannot be neglected. Previously, the author[4] studied the structure of the outer envelope of late-type giant stars and calculated the locus E = constant in the *H-R* diagram, where $E = 4\pi G^{3/2}(\mu H/k)^{5/2}M^{1/2}R^{3/2}P/T^{5/2}$ is the characteristic value which determines the degree of the central condensation of the solutions with polytropic index 3/2. Now, 45 is a maximum value of E beyond which there exist no quasi-static solutions. In figure 54.2, APB shows the curve $E = 45$ for a given mass and a chemical composition. The right side region is forbidden for quasi-static solutions. The curve CPD shows the evolutional path calculated by the above authors. When a star is born in the forbidden region,

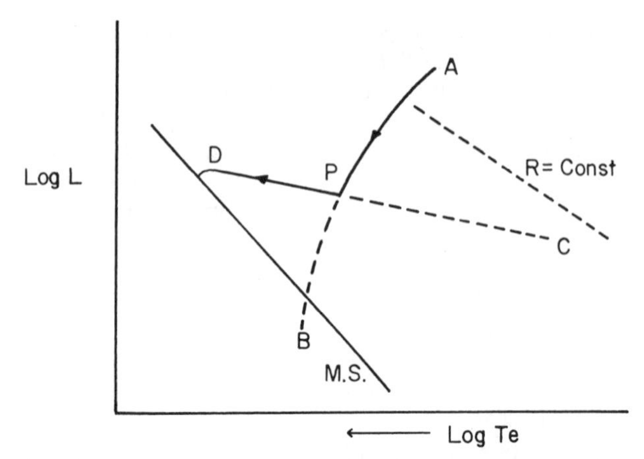

Fig. 54.2 Schematic track in the Hertzsprung-Russell diagram for contracting stars with given mass and chemical composition. The curve *CPD* shows the track calculated by Levee, Salpeter, and Henyey et al., and *APB* shows a curve with $E = 45$, the right region of which is forbidden for the existence of the quasi-static solutions.

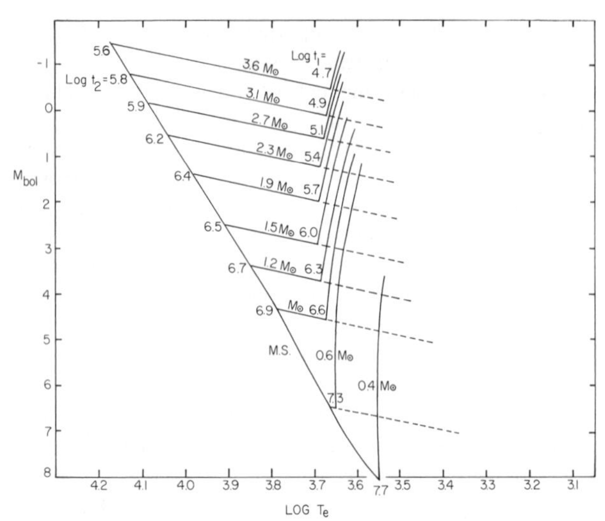

Fig. 54.3 Evolutionary tracks and ages of stars with different masses in gravitational contraction; t_1 and t_2 denote the ages (in years) at the turning point and on the main sequence, respectively.

it will adjust its internal structure in a relatively short time in such a way that it reaches to a point on the curve APB after passing through, for instance, the dotted line $R =$ constant. Then, the star moves along the curve AP downward and turns at the point P in the direction PD. Since the region which is on the left side of AP and above PD corresponds to solutions which are centrally condensed type, the stars with a uniform chemical composition and with a nearly uniform energy source cannot stay in a long time scale in a region far apart from the curve AP and the line PD.

The life time of the contraction is calculated for Population I stars using the expressions

$$\frac{dE}{dt} = -L, \qquad E = -\frac{3\gamma - 4}{3(\gamma - 1)} \frac{3}{5 - n} \frac{GM^2}{R}. \tag{1}$$

For the sake of simplicity we take $\gamma = 5/3$ and $n = 2$. For the curve APB the previous results[4] for $E = 40$ are used, and for the line CPD $L \sim M^{4.5} R^{-0.5}$ is assumed and the normalization constant is taken from the results by Henyey *et al.*[3] If $L \sim R^{-\alpha}$ along the path, the age of a star from the time when $R = \infty$ is given by

$$t = C/(1 - \alpha),$$

$$C = GM^2/2RL = 10^{7.20} \left(\frac{M}{M_\odot}\right)^2 \frac{R_\odot}{R} \frac{L_\odot}{L} \text{ years}, \tag{2}$$

where R, L are the present values. The age at a point P is approximately given by $2C/5$ by putting $\alpha = -3/2$, as compared with the value $2C$ ($\alpha = 1/2$) along the path CP.

The results for the evolutional paths and the ages are shown in figure 54.3 and figure 54.4. The comparison is made with the *H-R* diagram of a very young cluster NGC 2264 as observed by Walker.[5,6] The age of the late-type stars is greatly reduced as compared with the previous results since these stars evolve nearly vertically in the *H-R* diagram. The position of these stars in the *H-R* diagram shows that they will correspond to the *T*-Tauri stars. The radius of these stars is decreasing rapidly so that their angular momentum will have to be lost in a relatively short time scale. We have not

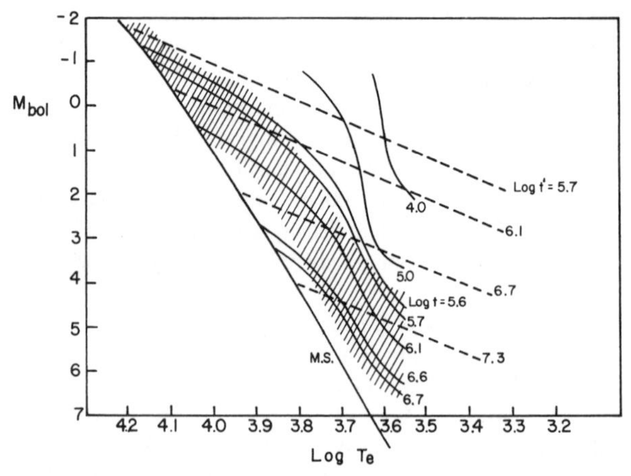

Fig. 54.4 Curves for constant ages for stars with different masses, compared with the H-R diagram of NGC 2264. Solid curves (t in years) and dashed lines (t in years) correspond to the calculation presented in this paper and to the results of Salpeter and Henyey et al., respectively.

taken into account the effect of the angular momentum to the stellar structure. The existence of the centrifugal force will correspond to the reduction of the gravitation constant G in an approximate sense and this will reduce the age somewhat more according to the Eq. (2).

1. R. D. Levée, *Ap. J. 117*, 200 (1953).
2. E. E. Salpeter, *Mem. Soc. Roy. Sci. Liège 14*, 116 (1954).
3. L. G. Henyey, R. LeLevier, and R. D. Levée, *Publ. Astronom. Soc. Pacific 67*, 154 (1955).
4. C. Hayashi and R. Hoshi, *P.A.S.J. 13*, 442 (1961).
5. M. F. Walker, *Ap. J. Suppl. 2*, 365 (1956).
6. A. R. Sandage, *Proc. Vatican Conference on Stellar Population* (1957), p. 149.

55. Synthesis of the Elements in Stars

E. Margaret Burbidge, Geoffrey R. Burbidge
William A. Fowler, and Fred Hoyle

(*Reviews of Modern Physics 29*, 547–650 [1957])

One of the most intriguing astronomical inquiries has been the origin of the chemical elements of which the material world is composed. William D. Harkins was one of the first to find an important clue to this mystery when he noticed that elements of low atomic weight are more abundant than those of high atomic weight and that, on the average, the elements with even atomic numbers are about 10 times more abundant than those with odd atomic numbers of similar value. These features led Harkins to conjecture that the relative abundances of the elements depend on nuclear rather than chemical properties and that the heavy elements must have been synthesized from the light ones.[1] Almost forty years later Hans E. Suess and Harold Clayton Urey, following previous work by Henry Norris Russell, V. M. Goldschmidt, Harrison Brown, and others, provided a detailed discussion of the element and isotopic abundances in the sun and similar Population I stars.[2] This discussion, which called attention to the many fluctuations that appear in the general trend of an exponential decline of abundance with increasing atomic weight, served as the major stimulus for current ideas concerning the synthesis of elements in stars.

During the first few decades of the twentieth century, physicists had shown that all nuclei consist of two fundamental building blocks, the proton and the neutron, and that some of the systematic properties of the abundance data are intimately related to the stability properties of nuclei. This situation led many investigators to study the possibility of element formation in a state of thermodynamic equilibrium among nuclei. It soon became clear, however, that extraordinarily high temperatures and densities were required and that no single set of physical conditions would simultaneously account for the observed relative abundances of all the elements.[3]

This failure of the early equilibrium theories led George Gamow to reason that the elements were formed in the nonequilibrium big-bang explosion that gave rise to the expanding universe. He called attention to the facts that the universe must have been remarkably hot and dense during the early stages of its expansion and that a wide range of physical conditions would have been available during its subsequent expansion and cooling. Gamow and his student Ralph Alpher next noticed the excellent inverse correlation between neutron capture cross sections and the abundances of both light and heavy elements, and reasoned that all the elements were formed by neutron capture processes during the first few minutes of the big bang (see selection 131). It is now known that two neutron capture processes, the *s*-process (*s* for slow) in red giants and the *r*-process (*r* for rapid) in supernovae do account for the abundance of most of the heavy elements above iron.

To many workers in the 1940s the attractive feature of the neutron capture theory of element formation was its apparent ability to predict the general trend of the observed relative abundance data by using only one free parameter, the density of matter during the formation process. Nevertheless, just as the equilibrium theories failed to account for the production of all the elements with just one set of physical conditions, the neutron capture theory also ultimately failed to account for the production of all the elements during the big bang.

During the time that Gamow was developing his ideas, Hans Bethe and C. F. von Weizsäcker had shown that helium 4 (alpha particles) can be synthesized from hydrogen nuclei (protons) by thermonuclear reactions in stars (see selections 48 and 49); this development led Enrico Fermi and A. Turkevich to examine all the thermonuclear reactions that might lead to element formation during the early stages of the expanding universe.[4] They considered reactions among neutrons, protons, and alpha particles and in 1950 concluded that no elements heavier than helium could have been produced in the big bang. The difficulty was that no stable nucleus of atomic mass 5 or 8 exists; therefore, the simple scheme of successive neutron captures will stop at helium 4 with a very small leakage through to lithium 7. The only reasonable mechanism for bridging this gap is through many-body reactions involving the nearly simultaneous collisions of several nuclei, and, as Fermi and Turkevich pointed out, these many-body reactions are not expected to occur during the nonequilibrium big bang.

In the early 1950s there was a concerted attack on the problem of synthesizing the heavier elements, which culminated in the paper given here. By 1951 both Ernst Öpik and Edwin Salpeter, following pioneering efforts of Hans Bethe in 1939, had shown that under suitable temperatures carbon can be produced by triple helium collisions through the intermediary of unstable beryllium 8, thus bypassing the mass gaps at 5 and 8 (see selections 51 and 52). The beryllium 8 formed momentarily by the collision of two alpha particles can capture a third one before breaking up into two alpha particles, and thus carbon 12 can be synthesized.

The rate of this triple alpha process per second per gram of interacting helium is proportional to the square of the density at a given temperature. In Gamow's big bang, synthesis cannot begin until the temperature has dropped to 1 billion degrees, since at higher temperatures photodisintegration occurs as fast as synthesis. At this temperature the big-bang density was at most a few times 10^{-4} gm cm^{-3}. This density was sufficient to produce up to 30% helium 4 by mass, plus small amounts of deuterium, helium 3, and lithium 7, but much too low by many orders of magnitude for appreciable operation of the triple alpha process.[5]

However, advocates of the steady state cosmology, such as Fred Hoyle, were unwilling to accept nucleosynthesis in a big bang in any case and actively sought alternative processes in stars. In 1954, Hoyle showed that the triple alpha process occurs at a rapid enough rate in red giant cores when the temperature is slightly over 100 million degrees K and the density is about 10,000 gm cm^{-3}.[6] The carbon 12 produced in the triple alpha process could also capture alpha particles to form oxygen 16. Hoyle went on to give the first comprehensive treatment of nucleosynthesis beyond carbon when he argued that carbon would begin to react with itself at temperatures of about a billion degrees and that oxygen would also react with itself at similar temperatures. He had also previously argued that the anomalously high abundance of the iron group of elements was due to its formation in an equilibrium process rather than a nonequilibrium chain of nuclear reactions. Because any nuclear reaction involving the iron group must absorb energy rather than release it, these nuclei cannot serve as fuel in further chains of thermonuclear reactions. Elements heavier than iron must be produced by neutron capture reactions that start with the iron group nuclei.

The thermonuclear reaction that accounts for the formation of carbon occurs when a star has used up its supply of hydrogen and evolved into the giant state, whereas the reactions accounting for the formation of heavier elements occur at even hotter, denser, later stages of stellar life. It seemed natural, then, to suppose that all the elements have been formed over long time intervals during successive static burning stages in stars and that the exponential decline in abundance is explained by the rarity of stars that have evolved to later stages of life. Moreover, as Hoyle had advocated nearly a decade earlier, the equilibrium process accounting for the iron group of elements can occur in the extraordinarily dense conditions caused by the gravitational collapse of stars that have reached the end point of stellar evolution, and this collapse might well be related to supernova explosions that can spew out the heavier elements into interstellar space.

By 1956 the detailed abundance data of Suess and Urey[2] served as an inspiration to both Alastair Cameron and Hoyle and his colleagues, who independently invoked a varied mixture of stellar equilibrium and nonequilibrium processes in order to account for the myriad details of the observed abundance distribution.[7] In 1956 Hoyle, William Fowler, Geoffrey Burbidge, and Margaret Burbidge discussed hydrogen and helium burning, rapid and slow neutron capture, and proton capture as the major element-building reactions. In the following year they considered the eight element-producing reactions that are given in this paper (which is often called "B^2FH" after the first initials of the last names of the authors). At the same time Alastair Cameron independently discussed most of the same topics and emphasized strongly that nuclear fuels might burn explosively during supernova explosions. Nevertheless, the B^2FH review generally dealt with the reactions in the greatest detail, and it was published in a more accessible journal. Consequently, it is this work that provided the fundamental framework on which virtually all subsequent studies of stellar nucleosynthesis are based.

Although the general flow of the observed relative abundances of the elements can be accounted for by successive static burning stages within stars, it is now thought that many of the details modulating the general flow are explained by fast reactions that occur during supernova explosions, just as Cameron imagined.[8] Moreover, considerations of stellar nucleosynthesis have not completed the scenario, for it is now believed that both the deuterium and the helium presently observed must have been produced during the big-bang explosion that gave rise to the expansion of the universe, and many of the other light elements are probably produced by spallation reactions in which cosmic rays strip off the components of heavy nuclei to form light ones.

1. W. D. Harkins, *Journal of the American Chemical Society 39*, 856 (1917), and *Physical Review 38*, 1270 (1931). Also see G. Oddo, *Zeitschrift für anorganische Chemie 87*, 253 (1914).

2. H. E. Suess, and H. C. Urey, *Reviews of Modern Physics 28*, 53 (1956). See also selection 39.

3. H. C. Urey and C. A. Bradley, *Physical Review 38*, 718 (1931); G. I. Pokrowski, *Physikalische Zeitschrift 32*, 374 (1931).

4. The work of E. Fermi and A. Turkevich was discussed by R. A. Alpher and R. C. Herman in the *Reviews of Modern Physics 22*, 153 (1950).

5. See R. V. Wagoner, W. A. Fowler, and F. Hoyle, *Astrophysical Journal 148*, 3 (1967), for details. Also see selections 13 and 131.

6. F. Hoyle, *Astrophysical Journal Supplement 1*, 121 (1954). See also F. Hoyle, *Monthly Notices of the Royal Astronomical Society 106*, 333 (1946).

7. F. Hoyle, W. A. Fowler, G. R. Burbidge, and E. M. Burbidge, *Science 124*, 611 (1956). The 1957 *Reviews of Modern Physics* article that constitutes this selection is reproduced in its entirety by the American Institute of Physics as Resource Letter OE-1 on the Origin of the Elements. See also A. G. W. Cameron, *Publications of the Astronomical Society of the Pacific 69*, 201 (1957); *Chalk River Report CRL-41* (1957) and *Atomic Energy of Canada Limited No. 454*, 1 (1957).

8. W. D. Arnett and D. D. Clayton, *Nature 227*, 780 (1970).

I. INTRODUCTION

ELEMENT ABUNDANCES AND NUCLEAR STRUCTURE

MAN INHABITS A UNIVERSE composed of a great variety of elements and their isotopes. Ninety elements are found terrestrially and one more, technetium, is found in stars; only promethium has not been found in nature. Some 272 stable and 55 naturally radioactive isotopes occur on the earth. In addition, man has been able to produce artificially the neutron, technetium, promethium, and ten transuranic elements. The number of radioactive isotopes he has produced now numbers 871 and this number is gradually increasing.

Each isotopic form of an element contains a nucleus with its own characteristic nuclear properties which are different from those of all other nuclei. Thus the total of known nuclear species is almost 1,200, with some 327 of this number known to occur in nature. In spite of this, the situation is not as complex as it might seem. Research in "classical" nuclear physics since 1932 has shown that all nuclei consist of two fundamental building blocks. These are the proton and the neutron which are called nucleons in this context. As long as energies below the meson production threshold are not exceeded, all "prompt" nuclear processes can be described as the shuffling and reshuffling of protons and neutrons into the variety of nucleonic packs called nuclei. Only in the slow beta-decay processes is there any interchange between protons and neutrons at low energies, and even there, as in the prompt reactions, the number of nucleons remains constant. Only at very high energies can nucleons be produced or annihilated. Prompt nuclear processes plus the slow beta reactions make it possible in principle to transmute any one type of nuclear material into any other even at low energies of interaction.

With this relatively simple picture of the structure and interactions of the nuclei of the elements in mind, it is natural to attempt to explain their origin by a synthesis or buildup starting with one or the other or both of the fundamental building blocks. The following question can be asked: What has been the history of the matter, on which we can make observations, which produced the elements and isotopes of that matter in the abundance distribution which observation yields? This history is hidden in the abundance distribution of the elements. To attempt to understand the sequence of events leading to the formation of the elements it is necessary to study the so-called universal or cosmic abundance curve.

Whether or not this abundance curve is universal is not the point here under discussion. It is the distribution for the matter on which we have been able to make observations. We can ask for the history of that particular matter. We can also seek the history of the peculiar and abnormal abundances, observed in some stars. We can finally approach the problem of the universal or cosmic abundances. To avoid any implication that the abundance curve is universal, when such an implication is irrelevant, we commonly refer to the number distribution of the atomic species as a function of atomic weight simply as the atomic abundance distribution. In graphical form, we call it the atomic abundance curve.

Here Burbidge et al. present a table of the atomic abundances of the elements produced by the different processes. We do not reproduce this table here but illustrate the general behavior with figure 55.1. The processes accounting for the different stable isotopes are given in an appendix that we do not reproduce.

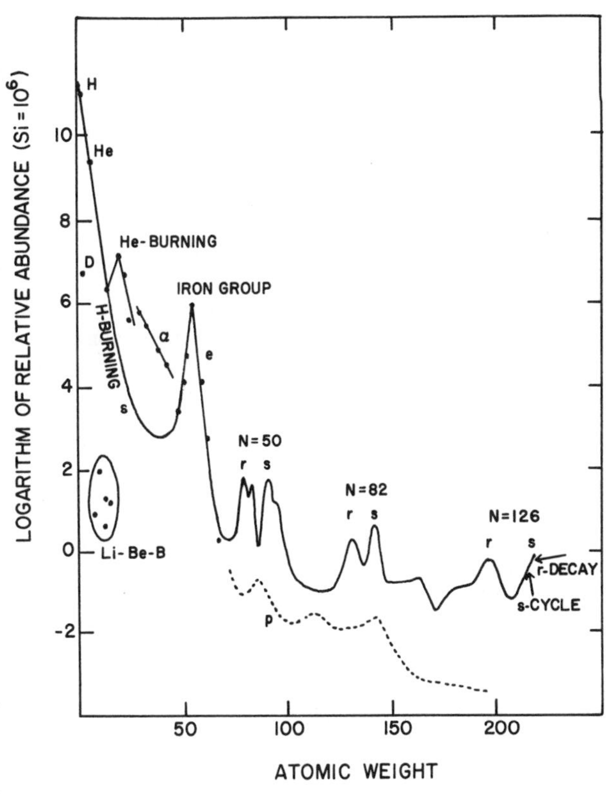

Fig. 55.1 Schematic curve of atomic abundances as a function of atomic weight based on the data of Suess and Urey (Su56). Suess and Urey have employed relative isotopic abundances to determine the slope and general trend of the curve. There is still considerable spread of the individual abundances about the curve illustrated, but the general features shown are now fairly well established. These features are outlined in Table 55.1. Note the overabundances relative to their neighbors of the alpha-particle nuclei $A = 16, 20, \ldots 40$, the peak at the iron group nuclei, and the twin peaks at $A = 80$ and 90, at 130 and 138, and at 194 and 208.

The first attempt to construct such an abundance curve was made by Goldschmidt (Go37). An improved curve was given by Brown (Br49) and more recently Suess and Urey (Su56) have used the latest available data to give the most comprehensive curve so far available. These curves are derived mainly from terrestrial, meteoritic, and solar data, and in some cases from other astronomical sources. Abundance determinations for the sun were first derived by Russell (Ru29) and the most recent work is due to Goldberg, Aller, and Müller (Go57). Accurate relative isotopic abundances are available from mass spectroscopic data, and powerful use was made of these by Suess and Urey in compiling their abundance table. This table, together with some solar values given by Goldberg *et al.*, forms the basic data for this paper.

It seems probable that the elements all evolved from hydrogen, since the proton is stable while the neutron is not. Moreover, hydrogen is the most abundant element, and helium, which is the immediate product of hydrogen burning by the *pp* chain and the CN cycle, is the next most abundant element. The packing-fraction curve shows that the greatest stability is reached at iron and nickel. However, it seems probable that iron and nickel comprise less than 1% of the total mass of the galaxy. It is clear that although nuclei are tending to evolve to the configurations of greatest stability, they are still a long way from reaching this situation.

It has been generally stated that the atomic abundance curve has an exponential decline to $A \sim 100$ and is approximately constant thereafter. Although this is very roughly true it ignores many details which are important clues to our understanding of element synthesis. These details are shown schematically in figure 55.1 and are outlined in the left-hand column of table 55.1.

It is also necessary to provide an explanation of the origin of the naturally radioactive elements. Further, the existence of the shielded isobars presents a special problem.

Table 55.1 Features of the abundance curve

Feature	Cause
Exponential decrease from hydrogen to $A \sim 100$	Increasing rarity of synthesis for increasing A, reflecting that stellar evolution to advanced stages necessary to build high A is not common.
Fairly abrupt change to small slope for $A > 100$	Constant $\sigma(n, \gamma)$ in s process. Cycling in r process.
Rarity of D, Li, Be, B as compared with their neighbors H, He, C, N, O	Inefficient production, also consumed in stellar interiors even at relatively low temperatures.
High abundances of alpha-particle nuclei, such as O^{16}, $Ne^{20} \cdots Ca^{40}$, Ti^{48}, compared to their neighbors	He burning and α process more productive than H burning and s process in this region.
Strongly-marked peak in abundance curve centered on Fe^{56}	e process; stellar evolution to advanced stage where maximum energy is released (Fe^{56} lies near minimum of packing-fraction curve).
Double peaks: $A = 80, 130, 196$	Neutron capture in r process (magic $N = 50, 82, 126$ for progenitors).
Double peaks: $A = 90, 138, 208$	Neutron capture in s process (magic $N = 50, 82, 126$ for stable nuclei).
Rarity of proton-rich heavy nuclei	Not produced in main line of r or s processes; produced in rare p process.

FOUR THEORIES OF THE ORIGIN OF THE ELEMENTS Any completely satisfactory theory of element formation must explain in quantitative detail all of the features of the atomic abundance curve. Of the theories so far developed, three assume that the elements were built in a primordial state of the universe. These are the nonequilibrium theory of Gamow, Alpher, and Herman [see (Al50)], together with the recent modifications by Hayashi and Nishida (Ha56), the polyneutron theory of Mayer and Teller (Ma49), and the equilibrium theory developed by Klein, Beskow, and Treffenberg (Kl47). A detailed review of the history and development of these theories was given by Alpher and Herman (Al53).

Each of these theories possesses some attractive features, but none succeeds in meeting all of the requirements. It is our view that these are mainly satisfied by the fourth theory in which it is proposed that the stars are the seat of origin of the elements. In contrast with the other theories which demand matter in a particular primordial state for which we have no evidence, this latter theory is intimately related to the known fact that nuclear transformations are currently taking place inside stars. This is a strong argument, since the primordial theories depend on very special initial conditions for the universe. Another general argument in favor of the stellar theory is as follows.

It is required that the elements, however they were formed, are distributed on a cosmic scale. Stars do this by ejecting material, the most efficient mechanisms being probably the explosive ejection of material in supernovae, the less energetic but more frequent novae, and the less rapid and less violent ejection from stars in the giant stages of evolution and from planetary nebulae. Primordial theories certainly distribute material on a cosmic scale but a difficulty is that the distribution ought to have been spatially uniform and independent of time once the initial phases of the universe were past. This disagrees with observation. There are certainly differences in composition between stars of different ages, and also stars at particular evolutionary stages have abnormalities such as the presence of technetium in the *S*-type stars and Cf^{254} in supernovae.

It is not known for certain at the present time whether all of the atomic species heavier than hydrogen have been produced in stars without the necessity of element synthesis in a primordial explosive stage of the universe. Without attempting to give a definite answer to this problem we intend in this paper to restrict ourselves to element synthesis in stars and to lay the groundwork for future experimental, observational, and theoretical work which may ultimately provide conclusive evidence for the origin of the elements in stars. However, from the standpoint of the nuclear physics alone it is clear that our conclusions will be equally valid for a primordial synthesis in which the initial and later evolving conditions of temperature and density are similar to those found in the interiors of stars.

GENERAL FEATURES OF STELLAR SYNTHESIS Except at catastrophic phases a star possesses a self-governing mechanism in which the temperature is adjusted so that the outflow of energy through the star is balanced by nuclear energy generation. The temperature required to give this adjustment depends on the particular nuclear fuel available. Hydrogen requires a lower temperature than helium; helium requires a lower temperature than carbon, and so on, the increasing temperature sequence ending at iron since energy generation by fusion processes ends here. If hydrogen is present the temperature is adjusted to hydrogen as a fuel, and is comparatively low. But if hydrogen becomes exhausted as stellar evolution proceeds, the temperature rises until helium becomes effective as a fuel. When helium becomes exhausted the temperature rises still further until the next nuclear fuel comes into operation, and so on. The automatic temperature rise is brought about in each case by the conversion of gravitational energy into thermal energy.

In this way, one set of reactions after another is brought into operation, the sequence always being accompanied by rising temperature. Since penetrations of Coulomb barriers occur more readily as the temperature rises it can be anticipated that the sequence will be one in which reactions take place between nuclei with greater and greater nuclear charges. As it becomes possible to penetrate larger and larger barriers the nuclei will evolve towards configurations of greater and greater stability, so that heavier and heavier nuclei will be synthesized until iron is reached. Thus there must be a progressive conversion of light nuclei into heavier ones as the temperature rises.

There are a number of complicating factors which are superposed on these general trends. These include the following.

The details of the rising temperature and the barrier effects of nuclear reactions at low temperatures must be considered.

The temperature is not everywhere the same inside a star, so that the nuclear evolution is most advanced in the central regions and least or not at all advanced near the surface. Thus the composition of the star cannot be expected to be uniform throughout. A stellar explosion does not accordingly lead to the ejection of material of one definite composition, but instead a whole range of compositions may be expected.

Mixing within a star, whereby the central material is mixed outward, or the outer material inward, produces special effects.

Material ejected from one star may subsequently become condensed in another star. This again produces special nuclear effects.

All of these complications show that the stellar theory cannot be simple, and this may be a point in favor of the theory, since the abundance curve which we are trying to explain is also not simple. Our view is that the elements have evolved, and are evolving, by a whole series of processes. These are explained in the following sections, and illustrated in figure 55.1.

II. Physical Processes Involved in Stellar Synthesis, Their Place of Occurrence, and the Time-Scales Associated with Them

MODES OF ELEMENT SYNTHESIS As was previously described in an introductory paper on this subject by Hoyle, Fowler, Burbidge, and Burbidge (Ho56), it appears that in order to explain all of the features of the abundance curve, at least eight different types of synthesizing processes are demanded, if we believe that only hydrogen is primeval. In order to clarify the later discussion we give an outline of these processes here (see also Ho54, Fo56).

Hydrogen Burning Hydrogen burning is responsible for the majority of the energy production in the stars. By hydrogen burning in element synthesis we shall mean the cycles which synthesize helium from hydrogen and which synthesize the isotopes of carbon, nitrogen, oxygen, fluorine, neon, and sodium which are not produced by helium burning and the α process.

Helium Burning These processes are responsible for the synthesis of carbon from helium, and by further α-particle addition for the production of O^{16}, Ne^{20}, and perhaps Mg^{24}.

α Process These processes include the reactions in which α particles are successively added to Ne^{20} to synthesize the four-structure nuclei Mg^{24}, Si^{28}, S^{32}, A^{36}, Ca^{40}, and probably Ca^{44} and Ti^{48}. The source of the α particles is different in the α process than in helium burning.

e Process This is the so-called equilibrium process previously discussed by Hoyle (Ho46, Ho54) in which under conditions of very high temperature and density the elements comprising the iron peak in the abundance curve (vanadium, chromium, manganese, iron, cobalt, and nickel) are synthesized.

s Process This is the process of neutron capture with the emission of gamma radiation (n, γ) which takes place on a long time-scale, ranging from ~ 100 years to $\sim 10^5$ years for each neutron capture. The neutron captures occur at a *slow* (*s*) rate compared to the intervening beta decays. This mode of synthesis is responsible for the production of the majority of the isotopes in the range $23 \leqslant A \leqslant 46$ (excluding those synthesized predominantly by the α process), and for a considerable proportion of the isotopes in the range $63 \leqslant A \leqslant 209$. Estimates of the time-scales in different regions of the neutron-capture chain in the *s* process will be considered later in this section. The *s* process produces the abundance peaks at $A = 90$, 138, and 208.

r Process This is the process of neutron capture on a very short time-scale, ~ 0.01–10 sec for the beta-decay processes interspersed between the neutron captures. The neutron captures occur at a *rapid* (*r*) rate compared to the beta decays. This mode of synthesis is responsible for production of a large number of isotopes in the range $70 \leqslant A \leqslant 209$, and also for synthesis of uranium and thorium. This process may also be responsible for some light element synthesis, e.g., S^{36}, Ca^{46}, Ca^{48}, and perhaps Ti^{47}, Ti^{49}, and Ti^{50}. The *r* process produces the abundance peaks at $A = 80$, 130, and 194.

p Process This is the process of proton capture with the emission of gamma radiation (p, γ), or the emission of a neutron following gamma-ray absorption (γ, n), which is responsible for the synthesis of a number of proton-rich isotopes having low abundances as compared with the nearby normal and neutron-rich isotopes.

x Process This process is responsible for the synthesis of deuterium, lithium, beryllium, and boron. More than one type of process may be demanded here (described collectively as the *x* process), but the characteristic of all of these elements is that they are very unstable at the temperatures of stellar interiors, so that it appears probable that they have been produced in regions of low density and temperature.

Here we omit the detailed specification of which of these eight processes are responsible for the synthesis of the different isotopes of the elements.

TIME-SCALES FOR DIFFERENT MODES OF SYNTHESIS Here Burbidge et al. discuss the internal stellar temperatures required for different modes of nucleosynthesis, together with the duration of each process. These very approximate considerations may be summarized as follows:

Hydrogen burning	$10^{7\circ}$	10^6–10^{10} yr
Helium burning	$10^{8\circ}$	10^7–10^8 yr
α process	$10^{9\circ}$	10^2–10^4 yr
s process	$3 \times 10^{8\circ}$	10^2–10^5 yr
e process	$10^{9\circ}$	Seconds to years
r and *p* processes	$10^{10\circ}$	10–1,000 sec
x process	$\ll 10^{7\circ}$	Variable function of process

[The α and *e* processes are now lumped together in what is called carbon and silicon burning.]

III. Hydrogen Burning, Helium Burning, the α Process, and Neutron Production

This section and the sections to follow are devoted to detailed elaboration and discussion of the different physical processes introduced in Sec. II. These sections treat quantitatively experimental and theoretical evaluations of the cross sections and reaction rates of the nuclear processes involved in energy generation and element synthesis in stars. The material supplements and extends that published in a series of articles in 1954 (Fo54, Bo54, Ho54), in 1955 (Fo55, Fo55a), and in 1956 (Bu56). In the first part of this section we give a discussion of the relations between nuclear cross sections and nuclear reaction rates in stellar interiors and of the notation used in this and the following sections.

CROSS-SECTION FACTOR AND REACTION RATES The experimental results to be discussed will be used to derive the numerical value of the nuclear cross-section factor for a charged particle reaction defined by

$$S = \sigma(E)E \exp(31.28 Z_1 Z_0 A^{1/2} E^{-1/2}) \text{ kev barns},$$

where $\sigma(E)$ is the cross section in barns (10^{-24} cm^2) measured at the center-of-mass energy E in kev. The charges of the interacting particles are Z_1 and Z_0 in units of the proton charge and $A = A_1 A_0/(A_1 + A_0)$ is their reduced mass in atomic mass units. S is measured in the center-of-mass system. From measurements made in the laboratory system with incident particle energy E_1 and with target nuclei at rest, the quantity S is given by

$$S = \sigma(E_1)E_1 \frac{A_0}{A_1 + A_0} \exp(31.28 Z_1 Z_0 A_1{}^{1/2} E_1{}^{-1/2}) \text{ kev barns.}$$

For a nonresonant or off-resonant reaction S is a slowly varying function of the energy E. Methods for extrapolating to the effective thermal energy E_o in stellar interiors have been given by numerous authors (Sa52a, Fo54, Sa55, Ma57). The effective thermal energy at temperature T is

$$E_o = 1.220(Z_1{}^2 Z_0{}^2 A T_6{}^2)^{1/3} \text{ kev}$$

where T_6 is the temperature measured in units of 10^6 °K. The width of the effective range of thermal energy is

$$\Delta E_o = 0.75(Z_1{}^2 Z_0{}^2 A T_6{}^5)^{1/6} \text{ kev.}$$

The mean reaction rate of a thermonuclear process may be expressed as

$$P = \rho r = n_1 n_0 \langle \sigma v \rangle_{Av} = 3.63 \times 10^{47} \rho^2 \frac{x_1 x_0}{A_1 A_0} \langle \sigma v \rangle_{Av} \quad \text{reactions cm}^{-3} \text{ sec}^{-1}$$

where n_1 and n_0 are the number densities of the interacting particles per cm^3 and $\langle \sigma v \rangle_{Av}$ is the average of the cross section multiplied by the velocity in cm^3 sec^{-1}. The quantity r is the reaction rate per gram per second. The quantities x_1 and x_0 are the amounts of the interacting nuclei expressed as fractions by weight. In terms of $S_o = S(E_o)$ kev barns, it is found for a nonresonant process that

$$P = 7.20 \times 10^{-19} n_1 n_0 f_o S_o (A Z_1 Z_0)^{-1} \tau^2 e^{-\tau} \quad \text{reactions cm}^{-3} \text{ sec}^{-1}$$

$$= 2.62 \times 10^{29} \rho^2 \frac{x_1 x_0}{A_1 A_0} f_o S_o (A Z_1 Z_0)^{-1} \tau^2 e^{-\tau} \quad \text{reactions cm}^{-3} \text{ sec}^{-1},$$

where S_o is in kev barns and

$$\tau = 42.48 \left(Z_1{}^2 Z_0{}^2 \frac{A}{T_6} \right)^{1/3}$$

(this τ is not to be confused with the mean lifetime of the interacting particles which will always be accompanied by appropriate subscripts, etc.). The term f_o is the electron screening or shielding factor discussed by Salpeter (Sa54), evaluated at E_o. The cross-section factor S_o as customarily calculated does not include allowance for electron screening.

The mean lifetime of the nuclei of type 0 for the interaction with nuclei of type 1 is given by

$$\frac{1}{\tau_1(0)} = p_1(0) = \nu_0 P / n_0$$

$$= 4.34 \times 10^5 \frac{\rho x_1}{A_1} \nu_0 S_o f_o (A Z_1 Z_0)^{-1} \tau^2 e^{-\tau} \text{ sec}^{-1}$$

$$= 7.83 \times 10^8 \frac{\rho x_1}{A_1} \nu_0 S_o f_o \left(\frac{Z_1 Z_0}{A T_6{}^2} \right)^{1/3} e^{-\tau} \text{ sec}^{-1},$$

where ν_0 is the number of nuclei of type 0 consumed in each reaction. The quantity $p_1(0)$ is the mean reaction rate per nucleus of type 0. If nuclei of type 0 are regenerated in a cycle of reactions then $\tau_1(0)$ becomes the mean cycle time for nuclei of type 0.

The most satisfactory procedure for determining S_o is to make experimental observations on cross sections over a range of energies not too large compared to E_o. The cross-section factor, S, can then be plotted as a function of E and an appropriate extrapolation to find S_o can be made. This is not always possible and computational procedures for several frequently occurring cases will now be given.

Here we omit detailed formulae for the calculation of cross sections from experimentally determined parameters of a resonance that falls outside and within the range of $E_0 \pm 2\Delta E_0$. We also omit formulae for cross sections that are averages over several resonances and for the case of light nuclei, where the interaction energy may fall in the flat minimum between resonances. [Consult W. A. Fowler, C. R. Caughlan, and B. A. Zimmerman, *Annual Reviews of Astronomy and Astrophysics 5*, 525 (1967), *13*, 69 (1975)].

PURE HYDROGEN BURNING The point of view that element synthesis begins with pure hydrogen (primordial or continuously produced) condensed in stars is based on the existence of the so-called direct *pp* chain of reactions by which hydrogen is converted into helium. This chain is initiated by the direct *pp* reaction

$$p + p \rightarrow d + \beta^+ + \nu_+ + 0.421 \text{ Mev},$$

which has good theoretical foundations but which has not yet been observed experimentally in the laboratory because of its extremely low cross section even at relatively high interaction energies. In the above equation and in what follows we use ν_+ for neutrinos emitted with positrons, β^+, and ν_- for antineutrinos emitted with electrons, β^-. *We use nuclear rather than atomic mass differences in expressing the Q values of all reactions.* There is of course practically no difference in atomic and nuclear Q values when positrons or electrons are not involved.

The calculated cross section for the *pp* reaction is 10^{-47} cm$^2 = 10^{-23}$ barn at 1-Mev laboratory energy. This is much too small for detection with currently available techniques.

Details of the reaction rate for the *pp* reaction are given for a new value of the beta-decay constant, and the results indicate that the energy generation of the proton-proton reaction is larger than that of the carbon-nitrogen cycle for central stellar temperatures less than $1.7 \times 10^{7\circ}$ (with the sun at $1.5 \times 10^{7\circ}$); the opposite is true at higher temperatures.

PURE HELIUM BURNING When hydrogen burning in a star's main-sequence stage leads eventually to hydrogen exhaustion, a helium core remains at the star's center. It has been suggested (Sa52, Sa53, Op51, Op54) that the fusion of helium plays an important role in energy generation and element synthesis in the red-giant stage of the star's evolution. The fusion occurs through the processes

$$He^4 + He^4 \rightleftharpoons Be^8$$
$$Be^8 + He^4 \rightleftharpoons C^{12*} \rightarrow C^{12} + \gamma$$

or, in a more condensed notation, through

$$3He^4 \rightleftharpoons C^{12*} \rightarrow C^{12} + \gamma.$$

We refer to this as the 3α reaction. These processes are believed to occur at a late stage of the red giant evolution in which the hydrogen in the central core has been largely converted into helium, and in which gravitational contraction (Ho55) has raised the central temperature to $\sim 10^8$ degrees, and the density to $\sim 10^5$ g/cc. Under these conditions, as shown by Salpeter, an equilibrium ratio of Be^8 to He^4 nuclei equal to $\sim 10^{-9}$ is established. This conclusion followed from experimental measurements (He48, He49, To49, Wh41) which established the fact that Be^8 was unstable to disintegration into alpha particles but only by 95 kev with an uncertainty of about 5 kev.

Rates for the triple alpha reaction and the alpha gamma reactions of C^{12} and O^{16} are given.

α PROCESS As has just been discussed, helium-burning synthesizes C^{12}, O^{16}, Ne^{20}, and perhaps a little Mg^{24}. This occurs at temperatures between 10^8 and 2×10^8 degrees and results in the exhaustion of the helium produced in hydrogen-burning. An inner core of C^{12}, O^{16}, and Ne^{20} develops in the star and eventually undergoes gravitational contraction and heating just as occurred previously in the case of the helium core. Calculations of stellar evolutionary tracks have not yet been carried to this stage, but it is a reasonable extrapolation of current ideas concerning the cause of evolution into the giant stage. Gravitation is a "built-in" mechanism in stars which leads to the development of high temperature in the ashes of exhausted nuclear fuel. Gravitation takes over whenever nuclear generation stops; it raises the temperature to the point where the ashes of the previous processes begin to burn. Implicit in this argument is the assumption that mixing of core and surrounding zones does not occur.

No important reactions occur among C^{12}, O^{16}, and Ne^{20} until significantly higher temperatures, of the order of 10^9 degrees, are attained. Two effects then arise. The γ rays present in the thermal assembly become energetic enough to promote $Ne^{20}(\gamma, \alpha)O^{16}$. The resulting transformation is strongly exothermic, *viz.*:

$$Ne^{20} + \gamma \rightarrow O^{16} + He^4 - 4.75 \text{ Mev}$$
$$Ne^{20} + He^4 \rightarrow Mg^{24} + \gamma + 9.31 \text{ Mev}.$$

Combining these two equations yields:

$$2Ne^{20} \rightarrow O^{16} + Mg^{24} + 4.56 \text{ Mev}.$$

Additional alpha gamma reactions that build up heavier elements are discussed; these elements stand out in abundance above other neighboring nuclei.

SUCCESSION OF NUCLEAR FUELS IN AN EVOLVING STAR Starting with primeval hydrogen condensed into stars, pure hydrogen burning, pure helium burning, and finally the α process successively take place at the stellar center and then move outward in reaction zones or shells. When the star first contracts the generation of energy by hydrogen burning develops internal pressures which oppose gravitational contraction, and the star is stabilized on the main sequence at the point appropriate to its mass. Similarly, the generation of energy in helium burning should lead to a period of relative stability during the red-giant stage of evolution. It is assumed that mixing does not occur. This is substantiated by the fact that, as hydrogen becomes exhausted in the interior, the star evolves off the main sequence which is the location of stars with homogenous interiors.

At the end of helium burning most of the nuclear binding energy has been abstracted, and indeed the cycle of contraction, burning, contraction ... must eventually end when the available energy is exhausted, that is, when the most stable nuclei at the minimum in the packing fraction curve are reached, near Fe^{56}. If a star which condensed originally out of pure hydrogen remains stable, it eventually forms the iron-group elements at its center, and this "iron core" continues to grow with time until gravitation, unopposed by further energy generation, leads eventually to a violent instability. At this point it suffices only to emphasize that the instability may result in the ejection of at least part of the "iron core" and its thin surrounding shells of lighter elements into the interstellar medium and that in this way a reasonable picture of the production of the abundance peak at the iron-group elements can be formulated. Production of the iron group of elements requires temperatures near 4×10^9 degrees, at which statistical equilibrium is reached (the *e* process).

BURNING OF HYDROGEN AND HELIUM WITH MIXTURES OF OTHER ELEMENTS; STELLAR NEUTRON SOURCES In the previous discussion we considered the effect of heating hydrogen and its reaction products to very high temperatures. First, the hydrogen is converted to helium, and the resulting helium is converted to C^{12}, O^{16}, and Ne^{20}. Then α particles released by (γ, α) reactions on the Ne^{20} build the α-particle nuclei Mg^{24}, Si^{28}, S^{32}, A^{36}, Ca^{40}, and also Ca^{44} and Ti^{48}. Finally, at very high temperatures, the latter nuclei are converted into the iron group. Further heating of the iron group, although of astrophysical importance, does not lead directly to any further synthesis. Thus all the remaining elements and isotopes must be provided for otherwise than by a cooking of pure hydrogen.

Very much more complicated reactions arise when we consider the cooking of hydrogen and helium mixed with small concentrations of the elements already provided for, e.g., C^{12}, Ne^{20}, Fe^{56}. It is easy to see how such admixtures can arise. Since stars eject the products of nuclear synthesis into the interstellar gas it seems highly probable that only the "first" stars can have consisted of pure hydrogen. The results of hydrogen cooking in such stars would follow the lines described above. But once the interstellar gas was contaminated by this first cooking, nuclear processes would operate on hydrogen which contains impurities. The eventual hydrogen exhaustion will lead to helium burning with impurities. As we shall see later, the presence of other nuclei can lead to highly important effects.

In addition, hydrogen and helium may in some cases become adulterated with impurities even in the "first" stars. For example C^{12}, Ne^{20} built in the inner central regions of such stars may be circulated into the outer hydrogen envelopes.

CN cycle When C^{12} produced in helium burning is mixed with hydrogen at high enough temperatures, hydrogen is converted to He^4 by the CN cycle in addition to the *pp* chain previously considered. The implications for energy generation in hot main-sequence stars have been considered by numerous authors since Bethe (Be39) and von Weizsäcker (We38). The reactions of the CN cycle are

$$\begin{aligned}&C^{12}(p, \gamma)N^{13}(\beta^+ \nu_+)C^{13}\\&C^{13}(p, \gamma)N^{14}\\&N^{14}(p, \gamma)O^{15}(\beta^+ \nu_+)N^{15}\\&N^{15}(p, \alpha)C^{12}.\end{aligned}$$

These four reactions produce C^{13}, the heavier stable isotope of carbon, and the two stable forms of nitrogen. For C^{12}, C^{13}, and N^{14} the (p, α) reaction is not exothermic and only the (p, γ) reaction occurs. At N^{15}, the (p, α) reaction becomes exothermic and much more rapid than the (p, γ) reaction, which serves only as a small leak of material to O^{16}. The $N^{15}(p, \alpha)$ reaction reproduces the original C^{12} and a true cycle of reactions is established. Not only does this give rise to the catalytic conversion of hydrogen into helium until hydrogen is exhausted, but it also results in the carbon and nitrogen isotopes *not* being consumed in hydrogen burning.

The cross sections of the CN-cycle reactions have been under experimental investigation in the Kellogg Radiation Laboratory for some years and a review of the reaction rates in stars as known up to 1954 is included in the earlier paper by Fowler (Fo54), with numerical computations by Bosman-Crespin *et al.* (Bo54). New measurements of the CN-cycle reactions in the 100-kev range of interaction energies are now underway.

A detailed discussion of the CN reaction rates follows. This section is followed by a discussion of the burning of other nuclei in hydrogen and of stellar neutron sources.

IV, V, VI, VII, VIII, IX, X. We omit here a 43-page discussion of the myriad details of the e, s, r, p and x processes.

XI. Variations in Chemical Composition among Stars, and Their Bearing on the Various Synthesizing Processes

Different processes of element synthesis take place at different epochs in the life-history of a star. Thus the problem of element synthesis is closely allied to the problem of stellar evolution. In the last few years work from both the observational and theoretical sides has led to a considerable advance in our knowledge of stellar evolution. Theoretical work by Hoyle and Schwarzschild (Ho55) has been repeatedly mentioned since it affords estimates of temperatures in the helium core and in the hydrogen-burning shell of a star in the red-giant stage. However, the calculated model is for a star of mass 1.2 solar masses and a low metal content ($\sim 1/20$ of the abundances given by Suess and Urey), and is thus intended to apply to Population II stars.

From the observational side, the attack has been through photometric observations of clusters of stars, from which their luminosities and surface temperatures have been plotted in color-magnitude or Hertzsprung-Russell (HR) diagrams, by many workers. We refer here only to two studies, by Johnson (Jo54) and by Sandage (Sa57a). A cluster may be assumed to consist of stars of approximately the same age and initial composition, and hence its HR diagram represents a "snapshot" of the stage to which evolution has carried its more massive members, once they have started to evolve fairly rapidly off the main sequence after their helium cores have grown to contain 10 to 30% of the stellar mass. Clusters of different ages will have main sequences extending upwards to stars of different luminosity, and the point at which its main sequence ends can be used to date a cluster; the rate of use of nuclear fuel, given by the luminosity, depends on the mass raised to a fairly high power (3 or 4).

Figure 55.2 is due to Sandage (Sa57a), and is a composite HR diagram of a number of galactic (Population I) clusters together with one globular cluster, M3 (Population II). The right-hand ordinate gives the ages corresponding to main sequences extending upwards to given luminosities. This diagram gives an idea of the way in which stars of different masses (which determine their position on the main sequence) evolve into the red-giant region. The most massive stars evolve into red supergiants (e.g., the cluster h and χ Persei); the difference between the red giants belonging to the Population I cluster M67 and the Population II cluster M3 should also be noted. Since stars evolve quite rapidly compared with the time they spend on the main sequence, the observed HR diagram may be taken as nearly representing the actual evolutionary tracks in the luminosity-surface-temperature plane.

A feature of Population II stellar systems is that their HR diagrams contain a "horizontal branch" [see, for example, the work by Arp, Baum, and Sandage (Ar53)] which probably represents their evolutionary path subsequent to the red-giant stage. This feature is not included in the diagram of M3 in figure 55.2, although it is well represented by that cluster, since this diagram is intended to show just the red-giant branches of clusters. The old Population I cluster M67 has a sparse distribution of stars which may lie on the Population I analogy of the horizontal branch; the other Population I clusters in figure 55.2 do not show such a feature. Presumably the evolutionary history of a more massive Population I star subsequent to its existence as a red giant is more rapid.

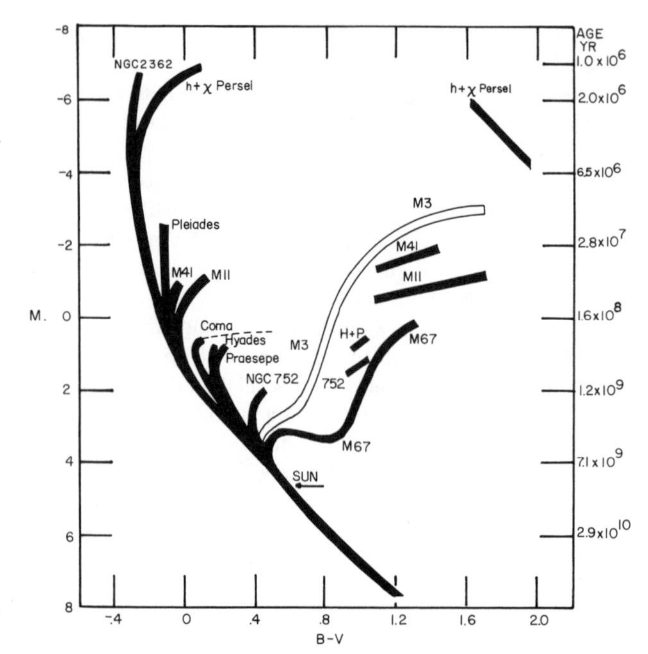

Fig. 55.2 Composite Hertzsprung-Russell diagram of a number of galactic (Population I) star clusters, together with one globular cluster, M 3 (Population II), by Sandage (Sa57a). The abscissa measures the color on the $B - V$ system and defines surface temperature (increasing from right to left). The left-hand ordinate gives the absolute visual magnitude, M_v, of the stars (showing that luminosity increases upward). The heavy black bands (Population I) and unfilled band (Population II) represent the regions in the temperature-luminosity plane that are occupied by stars. The name of each cluster is shown alongside the appropriate band. The right-hand ordinate gives the ages of the clusters, corresponding to main sequences extending upward to given luminosities. Note that the clusters all have a common main sequence below about $M_v = +3.5$ and also that the red giants have luminosities (defined by their masses) different from those they had while on the main sequence.

Whenever reference is made in different parts of this paper to particular epochs in a star's evolutionary life, we are referring to a schematic evolutionary diagram for the star which has the same general characteristics as the HR diagrams in figure 55.2. With this background in mind, we turn now to astrophysical observations which provide many indications of element synthesis in stars. This is either taking place at the present time or else it has occurred over a time-scale spanned by the ages of nearby Population II stars.

A discussion of the N:C ratio during different stages of evolution follows. Certain stars exhibit

Table 55.2 A comparison between the chemical compositions of evolved stars and young stars

	Percentage mass				
		Main-sequence young stars			
Element	Suess and Urey (Su56)	Traving	Aller	HZ 44 (Range)	HD 160641
H	75.5	58.6	[58.7]	9.0 – 9.1	0
He	23.3	39.8	[39.8]	88.7 –83.8	95.4
C	0.080	0.17	0.03	0.0047– 0.014	0.31
N	0.17	0.19	0.15	1.66 – 4.85	0.47
O	0.65	0.55	0.40	0.12 – 0.34	0.75
Ne	0.32	0.62	0.84	0.47 – 1.38	3.00
Si	0.053	0.10	0.07	0.17 – 0.48	0.065

abundances different from the "normal" abundances given by Suess and Urey (Su56). These stars include the O or B star HD 160641 (A154), the hot subdwarf HZ 44 (Mu58), and the young stars τ Scorpii and 10 Lacertae (Tr55, Tr57). These abnormal abundances are given in table 55.2.

XII. General Astrophysics

EJECTION OF MATERIAL FROM STARS AND THE ENRICHMENT OF THE GALAXY IN HEAVY ELEMENTS Here Burbidge *et al.* argue that the helium and heavier elements present at the time of formation of Population I stars must have been ejected into interstellar space by red giants, supergiants, or supernovae. Elements synthesized in processes requiring temperatures of the order of 10^9 or $10^{8\circ}$ are ejected, respectively, from supernovae or the red giants and supergiants. The estimates of the total mass of material ejected from supernovae were made by assuming that an average supernova ejects 1.4 solar masses, that supernovae occur once every 300 years (Type I) or every 50 years (Type II), and that the age of the Galaxy is 6×10^9 years. The exponential-type light curve of Type I supernovae is taken to indicate the rapid production and capture of neutrons (*r* process). The estimate of the ejection of mass by red giants and supergiants was made by assuming that these stars eject most of their material and that stars that have gone through their whole evolution make up 45% of the total mass of stars (Sa55a). Assuming that 16% and 29% lie, respectively, in the form of white dwarfs and ejected gas, Burbidge et al. conclude that some 2×10^{10} solar masses have been ejected into space by red giants and supergiants and about 2×10^8 solar masses have been ejected into space by supernovae.

The estimates are not accurate enough to establish whether or not some helium was present in the original matter of the Galaxy. Deuterium could be produced through the capture of neutrons by hydrogen in the expanding envelopes of Type II supernovae or, perhaps, in the primeval big-bang explosion of the expanding universe. Because the Galaxy was originally all gas, and because it now contains about 10% by mass of neutral hydrogen (Hu56), star formation and element synthesis must have occurred at a greater rate early in the life of the Galaxy. The present value for the expansion rate of our galaxy (Hu56a) leads to an expansion age of 5.4×10^9 years, which according to Burbidge et al. is in conflict with the greater ages of some star clusters in the Galaxy.

SUPERNOVA OUTBURSTS In the following discussion we consider the course of evolution that in our view leads to the outbursts of supernovae. According to a well-known calculation by Chandrasekhar the pressure balance in a star cannot be wholly maintained by degeneracy for masses greater than a certain critical mass. For pure He^4 this critical mass turns out to be 1.44 $M_\odot$, while for pure iron the value is 1.24 $M_\odot$.

It follows that stars with masses greater than the critical value cannot be in mechanical support unless there is an appreciable temperature contribution to their internal pressures. Mechanical support therefore demands high internal temperatures in such stars.

Our arguments depend on these considerations. We are concerned with stars above the Chandrasekhar limit, and assume that mechanical support is initially operative to a high degree of approximation. This is not a restriction on the discussion, since our eventual aim will be to show that mechanical support ceases to be operative. A discussion of catastrophic stars would indeed be trivial if a lack of mechanical support were assumed from the outset.

The next step is to realize that a star (with mass greater than the critical value) must go on shrinking indefinitely unless there is some process by which it can eject material into space. The argument for this startling conclusion is very simple. Because of the high internal temperature, energy leaks outwards from the interior to the surface of the star, whence it is radiated into space. This loss of energy can be made good either by a slow shrinkage of the whole mass of the star (the shrinkage being "slow" means that mechanical support is still operative to a high degree of approximation), or by a corresponding gain of energy from nuclear processes. But no nuclear fuel can last indefinitely, so that a balance between nuclear energy generation and the loss from the surface of the star can only be temporary. For stars of small mass the permissible period of balance exceeds the present age of the galaxy, but this is not so for the stars of larger mass now under consideration. Hence for these stars shrinkage must occur, a shrinkage that is interrupted, but only temporarily, whenever some nuclear fuel happens for a time to make good the steady outflow of energy into space.

Shrinkage implies a rising internal temperature, since mechanical support demands an increasing thermal pressure as shrinkage proceeds. It follows that the internal temperature must continue to rise so long as the critical mass is exceeded, and so long as mechanical support is maintained. The nuclear processes consequent on the rise of temperature are, first, production of the α-particle nuclei at temperatures from $1-3 \times 10^9$ degrees and, second, production of the nuclei of the iron peak at temperatures in excess of 3×10^9 degrees. It is important in this connection that the temperature is not uniform inside a star. Thus near the center, where the temperature is highest, nuclei of the iron peak may be formed, while outside the immediate central regions would be the α-particle nuclei, formed at a somewhat lower temperature, and still further outwards would be the light nuclei together with the products of the *s* process, formed at a much lower temperature. Indeed some hydrogen may still be present in the outermost regions of the star near the surface. The operation of the *r* process turns out to depend on such hydrogen being present in low concentration. The *p* process depends, on the other hand, on the outer hydrogen being present in high concentration.

Turning now to the nuclei of the iron peak, the statistical equations show that the peak is very narrow at $T \sim 3 \times 10^9$ degrees, the calculated abundances falling away sharply as the atomic weight either decreases or increases from 56 by a few units. At higher temperatures the peak becomes somewhat wider, ranging to copper on the upper side and down to vanadium on the lower side. Beyond these limits the abundances still fall rapidly away to negligible values with one exception, the case of He^4.

The statistical equation relating the abundances of Fe^{56} and He^4 is given here.

To convert 1 g of iron into helium demands an energy supply of 1.65×10^{18} ergs. This may be compared with the total thermal energy of 1 gram of material at 8×10^9 degrees which amounts to only 3×10^{17} ergs. Evidently the conversion of iron into helium demands a supply of energy much greater than the thermal content of the material. The supply must come from gravitation, from a shrinkage of the star, and clearly the shrinkage must be very considerable in order that sufficient energy becomes available. This whole energy supply must go into the conversion and hence into nuclear energy, so that very little energy is available to increase the thermal content of the material. However, instantaneous mechanical stability in such a contraction would demand a very large increase in the internal temperature. Thus we conclude that in this contraction, in which the thermal energy is, by hypothesis, only increased by a few percent, there is no mechanical stability, so that the contraction takes place by free fall inward of the central parts of the star. At a density of 10^8 g/cm^3, this implies an implosion of the central regions in a time of the order of 1/5 of a second ($\tau \simeq (4/3\pi G\rho)^{-1/2}$ and in our view it is just this catastrophic implosion that triggers the outburst of a supernova.

Here Burbidge *et al.* conclude that the loss of energy by the neutrino emission of the Urca process is ineffective, compared with the refrigerating action of the conversion of Fe^{56} to He^4.

The last question to be discussed is the relationship between the *implosion* of the central regions of a star and the *explosion* of the outer regions. Two factors contribute to produce explosion in the outer regions. The temperature in the outer regions is very much lower than the central temperature. Because of this the outer material does not experience the same extensive nuclear evolution that the central material does. Particularly, the outer material retains elements that are capable of giving a large energy yield if they become subject to sudden heating, e.g., C^{12}, O^{16}, Ne^{20}, Ne^{21}, He^4, and perhaps even hydrogen. The second point concerns the possibility of the outer material experiencing a sudden heating. Because under normal conditions the surface temperature of a star is much smaller than the central temperature, material in the outer regions normally possesses a thermal energy per unit mass that is small compared with the gravitational potential energy per unit mass. Hence any abnormal process that causes the thermal energy suddenly to become comparable with the gravitational energy must lead to a sudden heating of the outer material. This is precisely the effect of an implosion of the central regions of a star. Consequent on implosion there is a large-scale conversion of gravitational energy into dynamical and thermal energy in the outer zones of the star.

One last point remains. Will the gravitational energy thus released be sufficient to trigger a thermonuclear explosion in the outer parts of the star? The answer plainly depends on the value of the gravitational potential. Explosion must occur if the gravitational potential is large enough. In a former paper (Bu56) it was estimated that a sudden heating to 10^8 degrees would be sufficient to trigger an explosion. This corresponds to a thermal energy $\sim 10^{16}$ ergs per gram. If the gravitational potential per gram at the surface of an imploding star is appreciably greater than this value, explosion is almost certain to take place. For a star of mass 1.5 $M_\odot$, for instance, the gravitational potential energy per unit mass appreciably exceeds 10^{16} ergs per gram at the surface if the radius of the star is less than 10^{10} cm. At the highly advanced evolutionary state at present under consideration it seems most probable that this condition on the radius of the star is well satisfied. Hence it would appear as if implosion of the central regions of such stars must imply explosion of the outer regions.

Two cases may be distinguished, leading to the occurrence of the *r* and *p* processes. A star with mass only slightly greater than Chandrasekhar's limit can evolve in the manner described above only after almost all nuclear fuels are exhausted. Hence any hydrogen present in the outer material must comprise at most only a small proportion of the total mass. This is the case of hydrogen deficiency that we associate with Type I supernovae, and with the operation of the *r* process. In much more massive stars, however, the central regions may be expected to exhaust all nuclear fuels and proceed to the point of implosion while much hydrogen still remains in the outer regions. Indeed the "central region" is to be defined in this connection as an innermost region containing a mass that exceeds Chandrasekhar's limit. For massive stars the central region need be only a moderate fraction of the total mass, so that it is possible for a considerable proportion of the original hydrogen to survive to the stage where central implosion takes place. This is the case of hydrogen excess that in the former paper we associated with the Type II supernovae, where the *p* process may occur. However, the rarity of the *p*-process isotopes, and hence the small amount of material which must be processed to synthesize them, suggests that if Type II supernovae are responsible, the *p* process is a comparatively rare occurrence even among them. On the other hand, in any supernova in which a large flux of neutrons was produced, a small fraction of those having very high energies might escape to the outer parts of the envelope, and after decaying there to protons, might interact with the envelope material and produce the *p*-process isotopes.

Energy from the explosive thermonuclear reactions, perhaps as much as fifty percent of it, will be carried inwards, causing a heating of the material of the central regions. It is to this heating that we attribute the emission of the elements of the iron peak during the explosion of supernovae.

Supernova light curves are then given (see selection 70).

A150 R. A. Alpher and R. C. Herman, *Revs. Modern Phys. 22*, 153 (1950).

A153 R. A. Alpher and R. C. Herman, *Ann. Rev. Nuclear Sci. 2*, 1 (1953).

A154 L. H. Aller, *Mem. soc. roy. sci. Liège 14*, 337 (1954).

Ar53 Arp, Baum, and Sandage, *Astron. J. 58*, 4 (1953).

Be39 H. A. Bethe, *Phys. Rev. 55*, 103, 434 (1939).

Bo54 Bosman-Crespin, Fowler, and Humblet, *Bull. soc. sci. Liège 9*, 327 (1954).

Br49 H. Brown, *Revs. Modern Phys. 21*, 625 (1949).

Bu56 Burbidge, Hoyle, Burbidge, Christy, and Fowler, *Phys. Rev. 103*, 1145 (1956).

Fo54 W. A. Fowler, *Mem. soc. roy. sci. Liège 14*, 88 (1954).

Fo55 Fowler, Burbidge, and Burbidge, *Astrophys. J. 122*, 271 (1955).

Fo55a Fowler, Burbidge, and Burbidge, *Astrophys. J. Suppl. 2*, 167 (1955).

Fo56 W. A. Fowler and J. L. Greenstein, *Proc. Natl. Acad. Sci. U.S. 42*, 173 (1956).

Go37 V. M. Goldschmidt, *Skrifter Norske Videnskaps-Acad. Oslo. I. Mat.-Naturv. Kl. No. 4* (1937).

Go57 L. Goldberg, E. A. Müller and L. H. Aller, *Astrophys. J. Suppl. 5*, 1 (1960).

Ha56 C. Hayashi and M. Nishida, *Progr. Theoret. Phys. 16*, 613 (1956).

He48 A. Hemmendinger, *Phys. Rev. 73*, 806 (1948).

He49 A. Hemmendinger, *Phys. Rev. 75*, 1267 (1949).

Ho46 F. Hoyle, *Monthly Notices Roy. Astron. Soc. 106*, 343 (1946).

Ho54 F. Hoyle, *Astrophys. J. Suppl. 1*, 121 (1954).

Ho55 F. Hoyle and M. Schwarzschild, *Astrophys. J. Suppl. 2*, 1 (1955).
Ho56 Hoyle, Fowler, Burbidge, and Burbidge, *Science 124*, 611 (1956).
Hu56 H. C. van de Hulst, *Verslag Gewone Vergader. Afdel. Natuvrk. Koninkl. Ned. Akad. Wetenschap. 65*, no. 10, 157 (1956).
Hu56a Humason, Mayall, and Sandage, *Astron. J. 61*, 97 (1956)
Jo54 H. L. Johnson, *Astrophys. J. 120*, 325 (1954).
K147 O. Klein, *Arkiv mat. astron. fysik 34A*, no. 19 (1947); F. Beskow and L. Treffenberg, *Arkiv mat. astron. fysik 34A*, nos. 13, 17 (1947).
Ma49 M. G. Mayer and E. Teller, *Phys. Rev. 76*, 1226 (1949).
Ma57 J. B. Marion and W. A. Fowler, *Astrophys. J. 125*, 221 (1957).
Mu58 G. Münch, *Astrophys. J. 127*, 642 (1958).
Op51 E. J. Öpik, *Proc. Roy. Irish Acad. A54*, 49 (1951).
Op54 E. J. Öpik, *Mem. soc. roy. sci. Liège 14*, 131 (1954).
Ru29 H. N. Russell, *Astrophys. J. 70*, 11 (1929).
Sa52 E. E. Salpeter, *Astrophys. J. 115*, 326 (1952).
Sa52a E. E. Salpeter, *Phys. Rev. 88*, 547 (1952).
Sa53 E. E. Salpeter, *Ann. Rev. Nuclear Sci. 2*, 41 (1953).
Sa54 E. E. Salpeter, *Australian J. Phys. 7*, 373 (1954).
Sa55 E. E. Salpeter, *Phys. Rev. 97*, 1237 (1955).
Sa55a E. E. Salpeter, *Astrophys. J. 121*, 161 (1955).
Sa57a A. R. Sandage, *Astrophys. J. 125*, 435 (1957).
Su56 H. E. Suess and H. C. Urey, *Revs. Modern Phys. 28*, 53 (1956).
To49 Tollestrup, Fowler, and Lauritsen, *Phys. Rev. 76*, 428 (1949).
Tr55 G. Traving, *Z. Astrophys. 36*, 1 (1955).
Tr57 G. Traving, *Z. Astrophys. 41*, 215 (1957)
We38 C. F. von Weizsäcker, *Physik. Z. 39*, 633 (1938).
Wh41 J. A. Wheeler, *Phys. Rev. 59*, 27 (1941).

56. Neutrinos from the Sun

Solar Neutrinos I: Theoretical

John N. Bahcall

(*Physical Review Letters 12*, 300–302 [1964])

Search for Neutrinos from the Sun

Raymond Davis, Don S. Harmer,
and Kenneth C. Hoffman

(*Physical Review Letters 20*, 1205–09 [1968])

Although most astrophysicists believe that the sun's heat is produced by the thermonuclear conversion of hydrogen into helium in the solar interior, no direct evidence for this transformation is available. A hot stellar interior is inferred from the argument that the outer layers of the sun's atmosphere must be supported by the thermal motion of the interior gas, and nuclear physicists have shown that the inferred temperature is sufficient to ignite thermonuclear reactions. Conventional astronomical instruments, however, cannot be used to confirm directly the existence of nuclear reactions hidden in stellar interiors. The hypothetical nuclear reactions do emit neutrinos; because these neutrinos interact weakly with matter, they can penetrate the stellar atmosphere and escape into space. The detection of these neutrinos would provide direct observational confirmation of the theory that the sun shines by the thermonuclear conversion of hydrogen into helium.

As early as 1955 Raymond Davis had considered the suggestion that solar neutrinos emitted in the CNO cycle might be detected by measuring the argon produced whenever chlorine absorbs a neutrino.[1] His calculations showed that a practical solar neutrino experiment would require a detector consisting of a large volume of a chlorine-containing liquid. However, solar model calculations performed by I. Epstein and J. B. Oke indicated that the sun shines by the proton-proton reaction rather than the CNO cycle.[2] Although the first step in the proton-proton reaction includes the formation of a neutrino, these neutrinos have a maximum energy of 0.42 MeV, below the threshold for detection by reactions with chlorine. The detection of solar neutrinos seemed very difficult until 1958, when Alastair Cameron and William Fowler showed that one possible by-product of successive proton collisions is boron 8 and that the neutrinos emitted in the decay of boron 8 in the sun can have energies as large as 14 MeV.[3] John Bahcall then performed the detailed calculations of the solar neutrino emission given in the first of these two papers; here he also demonstrates the feasibility of detecting both the beryllium 7 and the boron 8 neutrinos by chlorine absorption.

Subsequently, Davis and his colleagues placed 100,000 gallons of tetrachloroethylene deep within the interior of the Homestake Mine in South Dakota. The mile of rock above the detector absorbs the subnuclear particles produced by cosmic ray collisions in the atmosphere and thereby diminishes the possibility of false signals. By 1964 Davis and his colleagues had made a preliminary search for the solar neutrinos. Four years later sufficient data had been collected to show that the neutrino flux incident on the earth is smaller than that expected from solar boron. This astonishing result, given in the second paper reproduced here, casts doubt in the minds of some on the entire idea that the sun currently shines by the thermonuclear conversion of hydrogen into helium. Additional experiments have not eradicated this doubt, and the present detection limit is a factor of 5 below theoretical expectations if the sun shines by the proton-proton chain.

Is something wrong with the theoretical calculations of the solar neutrino flux? The calculated boron 8 reaction depends critically on the assumed composition of the sun. If the heavy elements are confined to the outer layers of the sun, the resulting reduction in the solar opacity would lower the central temperature sufficiently to bring approximate agreement with the observations. Another possibility is that the present solar central temperature lies below its average value. Because the neutrinos we might now detect were generated 8 minutes ago they could reflect this lower temperature, whereas the current surface luminosity reflects the central temperature of the sun 10^7 years ago, the average time required for photons to diffuse through the solar layers. Nevertheless, these alternatives seem a little contrived, and the lack of solar neutrinos remains one of the astronomical enigmas of the 1970s.[4]

1. R. Davis, *Physical Review 97*, 766 (1955).
2. I. Epstein, *Astrophysical Journal 112*, 207 (1950), *114*, 438 (1951); J. B. Oke, *Journal of the Royal Astronomical Society of Canada 44*, 135 (1950). Also see W. A. Fowler, *Mémoires de la societé royale scientifique de Liège IV*, *13*, 88 (1954), for laboratory measurements of nuclear cross sections that helped confirm these conclusions.
3. A. G. W. Cameron, *Annual Review of Nuclear Science 8*, 299 (1958); W. A. Fowler, *Astrophysical Journal 127*, 551 (1958).
4. J. N. Bahcall and R. Davis, *Science 191*, 264 (1976).

Bahcall, Solar Neutrinos

THE PRINCIPAL ENERGY SOURCE for main-sequence stars like the sun is believed to be the fusion, in the deep interior of the star, of four protons to form an alpha particle.[1] The fusion reactions are thought to be initiated by the sequence $^1H(p, e^+\nu)^2H(p, \gamma)^3He$ and terminated by the following sequences: (i) $^3He(^3He, 2p)^4He$; (ii) $^3He(\alpha, \gamma)^7Be$ $(e^-\nu)^7Li(p, \alpha)^4He$; and (iii) $^3He(\alpha, \gamma)^7Be(p, \gamma)^8B(e^+\nu)^8Be^*(\alpha)$ ^4He. No *direct* evidence for the existence of nuclear reactions in the interiors of stars has yet been obtained because the mean free path for photons emitted in the center of a star is typically less then 10^{-10} of the radius of the star. Only neutrinos, with their extremely small interaction cross sections, can enable us to *see into the interior of a star* and thus verify directly the hypothesis of nuclear energy generation in stars.

The most promising method[2] for detecting solar neutrinos based upon the endothermic reaction ($Q = -0.81$ MeV) $^{37}Cl(\nu_{solar}, e^-)^{37}Ar$, which was first discussed as a possible means of detecting neutrinos by Pontecorvo[3] and Alvarez.[4] In this note, we predict the number of absorptions of solar neutrinos per terrestrial ^{37}Cl atom by combining results of recent theoretical investigations[5-7] of the solar neutrino fluxes with calculations[8] of the relevant neutrino absorption cross sections on ^{37}Cl. The result of a preliminary experiment by Davis[2] is then used to set an upper limit on the central temperature of the sun and also to give information about the structure of ^4Li and its role in the proton-proton chain.

The neutrino fluxes from the hydrogen-burning reactions described in the first paragraph have recently been calculated using detailed models of the sun[5,6] and the effects of uncertainties in nuclear cross sections, as well as solar composition, opacity, and age, have been determined by Sears.[7] The most important predictions are these (uncertainties estimated from the work of Sears[7]): $\phi_\nu(^7Be) = (1.2 \pm 0.5) \times 10^{+10}$ neutrinos per cm^2 per sec and $\phi_\nu(^8B) = (2.5 \pm 1) \times 10^{+7}$ neutrinos per cm^2 per sec, at the earth's surface.

The cross sections for ^7Be and ^8B neutrinos to produce transitions from the ground state of ^{37}Cl to the ground state of ^{37}Ar can readily be calculated from known quantities; the results are[8] $\sigma_g(^7Be) = 1.5\sigma_0$ and $\bar{\sigma}_g(^8B) = 3.9 \times 10^{+2}\sigma_0$, where $\sigma_0 = 1.91 \times 10^{-46}$ cm^2 is a convenient combination of ground-state parameters and $\bar{\sigma}(^8B)$ has been averaged over the ^8B neutrino spectrum. Three excited states[9] in ^{37}Ar also have large matrix elements for neutrino absorption by the ground state of ^{37}Cl (which is a $d_{3/2}{}^3$, $J = 3/2^+$, $T = 3/2$ state); the three excited states of importance in ^{37}Ar are (with their expected energies) (i) $J = 1/2^+$, $T = 1/2$ (1.4 MeV); (ii) $J = 5/2^+$, $T = 1/2$ (1.6 MeV); and (iii) $J = 3/2^+$, $T = 3/2$ (5.1 MeV). The $J = 3/2^+$, $T = 3/2$ excited state of ^{37}Ar is the analog state of the ground state of ^{37}Cl; hence the transition from the ground state of ^{37}Cl to the 5.1-MeV excited state of ^{37}Ar is *superallowed* and has a large matrix element for neutrino absorption. The calculated absorption cross sections[8] averaged over the ^8B neutrino spectrum[10] are, in order of increasing excitation energy, $\bar{\sigma}(^8B)/\sigma_0 = 0.96 \times 10^{+3}$, $1.3 \times 10^{+3}$, and $4.4 \times 10^{+3}$. The net uncertainty in the magnitude of the sum of the above cross sections is estimated to be about 25%.[8,11]

The total predicted number of absorptions per terrestrial ^{37}Cl atom per second, using the above estimates for fluxes and cross sections, is found to be

$$\sum \phi_\nu(\text{solar})\sigma_{\text{abs}} = (4 \pm 2) \times 10^{-35} \text{ sec}^{-1}. \quad (1)$$

Only about 10% of the predicted number of absorptions is due to ^7Be neutrinos, although the ^7Be neutrino flux is predicted to be approximately 500 times the ^8B neutrino flux.[12] The solar value of $\sum \phi\bar{\sigma}$ given by Eq. (1) is at least several orders of magnitude greater than one would expect from cosmic neutrinos[13] or from neutrinos produced in the earth's atmosphere by the decay of cosmic ray secondaries.[13,14]

The ^8B neutrino flux is extremely sensitive[5,12] to the central temperature of the sun because of the large Coulomb barrier, compared to solar thermal energies, for the reaction $^7Be(p, \gamma)^8B$ of sequence (iii). An upper limit on the central temperature of the sun can therefore be derived by combining the experimental upper limit already obtained by Davis,[2] on the number of solar neutrinos captured per terrestrial ^{37}Cl atom, with Eq. (1) and the known temperature dependence of the $^7Be(p, \gamma)^8B$ reaction. In this way we find that the central temperature of the sun is less than 20 million degrees[5] and that a measurement of the ^8B neutrino flux accurate to $\pm 50\%$ would determine the central temperature to better than $\pm 10\%$.

The role of ^4Li in the proton-proton chain has long been recognized as an important astrophysical problem,[1,15] but one that has not yet been solved by direct nuclear physics experiments. The upper limit obtained by Davis[2] on the number of solar neutrinos captured per terrestrial ^{37}Cl atom can be used, however, to show that ^4Li does *not* play a significant role in the proton-proton chain in the sun. The relevant cross section for neutrino absorption (with $q_\nu^{\max} =$ 20 MeV) is[8] $\bar{\sigma}(^4Li) = 2 \times 10^{-42}$ cm^2 and hence $\phi_\nu(^4Li) \leqslant$ $2 \times 10^{+8}$ neutrinos per cm^2 per sec. The fraction of terminations of the proton-proton chain that occur via ^4Li can be calculated[16] as a function of the energy, E_γ, by which the mass of the ground state of ^4Li exceeds the mass of ^3He plus a proton. One can also calculate an upper limit on the fraction of terminations that occur via $^3He(p, \gamma)^4Li(\beta^+\nu)^4He$ by comparing the above upper limit on $\phi_\nu(^4Li)$ (multiplied by 17 MeV, the thermal energy release in such a termination) with the observed solar constant ($8.7 \times 10^{+11}$ MeV cm^{-2} sec^{-1}). In this way we find that $E_\gamma \geqslant 20$ keV[17] and conclude that ^4Li

participates in at most 0.2% of the proton-proton terminations in the sun.

1. See, for example, W. A. Fowler, *Mem. Soc. Roy. Sci. Liege 3*, 207 (1960). The CNO cycle is responsible for only a few percent of the energy generation in the sun and the relatively low-energy neutrinos produced by this cycle are unimportant for solar neutrino detection with ^{37}Cl.

2. R. Davis, Jr., *Phys. Rev. Letters 12*, 303 (1964). See also R. Davis, Jr., *Phys. Rev. 97*, 766 (1955).

3. B. Pontecorvo, National Research Council of Canada Report No. P.D. 205, 1946 (unpublished), reissued by the U.S. Atomic Energy Commission as document 200-18787.

4. L. W. Alvarez, University of California Radiation Laboratory Report No. UCRL-328, 1949 (unpublished).

5. J. N. Bahcall, W. A. Fowler, I. Iben, Jr., and R. L. Sears, *Astrophys. J. 137*, 344 (1963). The central temperature of the sun for the theoretical model used in this paper (developed by Sears) is 16.2 million degrees. See also R. L. Sears, *Mem. Soc. Roy. Sci. Liege 3*, 479 (1960).

6. P. Pochoda and H. Reeves, *Planet, and Space Sci. 12*, 119 (1964).

7. R. L. Sears, *Astrophys. J. 140*, 477 (1964).

8. J. N. Bahcall, *Phys. Rev. 135*, B137 (1964). This reference will contain an extensive discussion of neutrino absorption cross sections that are relevant to the detection of solar neutrinos. A variety of experimental tests of the assumptions used to calculate the excited-state neutrino absorption cross sections for $^{37}\mathrm{Cl}(\nu, e^-)^{37}\mathrm{Ar}$ will also be discussed.

9. I am grateful to Professor B. R. Mottelson and Professor M. A. Preston for comments that sparked the investigation of excited-state transitions.

10. The proton-proton ($q_\nu^{\max} = 0.42$ MeV) and ^{7}Be electron-capture ($q_\nu = 0.86$ MeV) neutrinos do not have sufficient energy to induce transitions to excited states in ^{37}Ar.

11. The assumptions made in calculating the ^{37}Cl neutrino absorption cross sections could be directly checked by measuring the *ft* values for the $^{37}\mathrm{Ca} \rightarrow {}^{37}\mathrm{K}$ decays, one of whose branches is also superallowed. Two other experiments that would be useful in testing the assumptions made in the cross-section calculations are (i) a measurement of the branching ratios in the $^{37}\mathrm{K} \rightarrow {}^{37}\mathrm{Ar}$ decay, and (ii) a measurement with improved accuracy of the *ft* values in the $^{35}\mathrm{Ar} \rightarrow {}^{35}\mathrm{Cl}$ decay. Predictions for the lifetimes and energies of all branches involved in the above decays are available upon request and will appear in reference 8.

12. The possible importance of ^{8}B solar neutrinos was first pointed out by W. A. Fowler, *Astrophys. J. 127*, 551 (1958); A. G. W. Cameron, *Ann. Rev. Nucl. Sci. 8*, 299 (1958).

13. See, for example, H. Greisen, *Proceedings of International Conference for Instrumentation in High Energy Physics, Berkeley, California, September 1960* (New York: Interscience Publishers, 1961), p. 209; F. Reines, *Ann. Rev. Nucl. Sci. 10*, 1 (1960); B. Pontecorvo and Ya. Smorodinskii, *Zh. Eksperim. i Teor. Fiz. 41*, 239 (1961) [translation: *Soviet Phys.—JETP 14*, 173 (1962)]. The preliminary experiment of Davis (reference 2) implies that the energy density of 1-MeV cosmic neutrinos is less than 5 MeV/cm^3; however, the galactic energy density of starlight is only about 1 eV/cm^3. Thus the Davis experiment does not furnish a very stringent upper limit on the energy density of low-energy cosmic neutrinos.

14. G. T. Zatsepin and V. A. Kuz'min, *Zh. Eksperim. i Teor. Fiz. 41*, 1818 (1961) [translation: *Soviet Phys.–JETP 14*, 1294 (1962)]; M. A. Markov and I. M. Zheleznykh, *Nucl. Phys. 27*, 385 (1961); T. D. Lee, H. Robinson, M. Schwartz, and R. Cool, *Phys. Rev. 132*, 1297 (1963).

15. H. A. Bethe, *Phys. Rev. 55*, 434 (1939); H. Reeves, *Phys. Rev. Letters 2*, 423 (1959); S. Bashkin, R. W. Kavanagh, and P. D. Parker, *Phys. Rev. Letters 3*, 518 (1959). The possibility of terminating the proton-proton chain through a particle-unstable but thermally populated ground state of ^{4}Li was apparently overlooked.

16. Details of this calculation will appear in a paper by P. D. Parker, J. N. Bahcall, and W. A. Fowler *Astrophys. J. 139*, 602 (1964). I am especially grateful to Dr. Parker for valuable collaboration on this point. Note that if ^{4}Li were particle stable, all proton-proton terminations in the sun would occur via $^3\mathrm{He}(p, \gamma)^4\mathrm{Li}(\beta^+ \nu)^4\mathrm{He}$ because of the relatively low Coulomb barrier for the $^3\mathrm{He}(p, \gamma)^4\mathrm{Li}$ reaction and the high abundance of protons.

17. This result implies that there are no $T = 1$ alpha-particle bound states below 19 MeV.

Davis, Harmer, and Hoffman, Neutrinos from the Sun

Abstract—A search was made for solar neutrinos with a detector based upon the reaction $\mathrm{Cl}^{37}(\nu, e^-)\mathrm{Ar}^{37}$. The upper limit of the product of the neutrino flux and the cross sections for all sources of neutrinos was 3×10^{-36} sec^{-1} per Cl^{37} atom. It was concluded specifically that the flux of neutrinos from B^8 decay in the sun was equal to or less than 2×10^6 cm^{-2} sec^{-1} at the earth, and that less than 9% of the sun's energy is produced by the carbon-nitrogen cycle.

RECENT SOLAR-MODEL CALCULATIONS have indicated that the sun is emitting a measurable flux of neutrinos from decay of B^8 in the interior.[1–8] The possibility of observing these energetic neutrinos has stimulated the construction of four separate neutrino detectors.[9] This paper will present the results of initial measurements with a detection system based upon the neutrino capture reaction $\mathrm{Cl}^{37}(\nu, e^-)\mathrm{Ar}^{37}$. It was pointed out by Bahcall[10] that the energetic neutrinos from B^8 would feed the analog state of Ar^{37} (a superallowed transition) that lies 5.15 MeV above the ground state. The importance of the contribution of the B^8 neutrino flux is readily seen from the neutrino-capture cross sections and the solar neutrino fluxes given in table 56.1. The

Table 56.1 Solar neutrino fluxes and cross sections for the reaction $Cl^{37}(\nu,e^-)Ar^{37}$

Neutrino source	Cross section[a,b] (cm^2)	Neutrino flux[c] at the earth ($cm^{-2}\ sec^{-1}$)	$10^{35}\ \sigma\phi$ (sec^{-1})
$H + H + e^- \rightarrow D + \nu$	1.72×10^{-45}	1.7×10^8	0.03
Be^7 decay	2.9×10^{-46}	3.9×10^9	0.11
B^8 decay	1.35×10^{-42}	$1.3(1 \pm 0.6) \times 10^7$	$1.8(1 \pm 0.6)$
N^{13} decay	2.1×10^{-46}	1.0×10^9	0.02
O^{15} decay	7.8×10^{-46}	1.0×10^9	0.08
			$\sum\phi\sigma = 2.0(1 \pm 0.6) \times 10^{-35}\ sec^{-1}$

a. Ref. 4.
b. Ref. 10.
c. Ref. 8.

tabulated fluxes were taken from the calculations of Bahcall and Shaviv,[8] who studied the effect of errors in the parameters—solar composition, luminosity, opacity, and nuclear reaction cross sections. These authors have placed a probable error of 60% on the calculated B^8 flux. Their predicted B^8 flux for mean values of the various parameters agrees well with the independent calculations of Ezer and Cameron.[5] On the basis of these predictions, the total solar-neutrino-capture rate in 520 metric tons of chlorine would be in the range of 2 to 7 per day.

THE DETECTOR DESIGN

A detection system that contains 390,000 liters (520 tons chlorine) of liquid tetrachloroethylene, C_2Cl_4, in a horizontal cylindrical tank was built along the lines proposed earlier.[11] The system is located 4,850 ft underground [4,400 m (w.e.)] in the Homestake gold mine at Lead, South Dakota. It is essential to place the detector underground to reduce the production of Ar^{37} from (p, n) reactions by protons formed in cosmic-ray muon interactions. The rate of Ar^{37} production in the liquid by cosmic-ray muons at this location is estimated to be 0.1 Ar^{37} atom per day.[11] Background effects from internal α contaminations and fast neutrons from the surrounding rock wall are low. The total Ar^{37} production from all background processes is less than 0.2 Ar^{37} atom per day, which is well below the rate expected from solar neutrinos.

Neutrino detection depends upon removing the Ar^{37} from a large volume of liquid contained in a sealed tank, and observing the decay of Ar^{37} (35-day half-life) in a small proportional counter (0.5 cm^3). It is therefore necessary to have an efficient method of removing a fraction of a cubic centimeter of argon from 390,000 liters of C_2Cl_4. The Ar^{37} activity is removed by purging with helium gas. Liquid is pumped uniformly from the bottom of the tank and returned to the tank through a series of 40 eductors arranged along two horizontal header pipes inside the tank. The eductors aspirate the helium from the gas space (2,000 liters) above the liquid, and mix it as small bubbles with the liquid in the tank. The pump and eductor system passes helium through the liquid at a total rate of 9,000 liters per minute maintaining an effective equilibrium between the argon dissolved in the liquid and the argon in the gas phase.

Argon is extracted by circulating the helium from the tank through an argon extraction system. Gas flow is again achieved by a pair of eductors in the tank system, and they maintain a flow rate of 310 liters per minute through the argon extraction system. The tetrachloroethylene vapor is removed by a condenser at -40° C followed by a bed of molecular sieve adsorber at room temperature. The helium then passes through a charcoal bed at 77° K to adsorb the argon, and is finally returned to the tank.

Here we omit a schematic diagram of the solar neutrino detector.

The argon sample adsorbed on the charcoal trap is removed by warming the charcoal while a current of helium is passed through it. The argon and other rare gases from the effluent gas stream are collected on a small liquid-nitrogen-cooled charcoal trap (1 cm diam by 10 cm long). Finally, the gases from this trap are desorbed and heated over titanium metal at 1,000° C to remove all traces of chemically reactive gases. The resulting rare gas contains krypton and xenon in addition to argon. These higher rare gases were dissolved from the atmosphere during exposure of the liquid during the various manufacturing, storage, and transfer operations. Krypton and xenon are much more soluble in tetrachloroethylene than argon, and, therefore, they are more slowly removed from the liquid by sweeping with helium. Since the volume of krypton and xenon in an experimental run is comparable with or exceeds the volume of argon, it is necessary to remove these higher rare gases from the sample. A more important consideration is that atmospheric krypton contains the 10.8-yr fission product Kr^{85}. The rare gases recovered from the tank are therefore separated by gas chromatography. To insure complete removal of krypton from the argon sample, a second gas chromatographic separation is made of the argon fraction. Experience has shown that these two successive chromatographic separations reduce krypton concentration in the argon sample to less than 10^{-8} parts per volume. The entire purified argon sample is counted in a small proportional counter that will be described later.

Argon Recovery Tests

After the air and air argon had been removed from the system by prolonged sweeping with helium, the argon recovery efficiency of the system was measured by an isotope dilution method. A measured volume of 99.9% Ar^{36} was introduced into the tank and dissolved in the liquid with the eductor system. It was then recovered by six separate purging operations. The Ar^{36} recovered from each purge was determined by a volumetric and argon mass-ratio measurement. It was found that the volume of Ar^{36} in the tank dropped exponentially with the volume of helium circulated according to

$$v(\mathrm{Ar})/v_0(\mathrm{Ar}) = e^{-7.21 \times 10^{-6} V(\mathrm{He})}$$

where $v_o(\mathrm{Ar})$ is the initial volume of Ar^{36} and $v(\mathrm{Ar})$ is the volume remaining after $V(\mathrm{He})$ liters of helium have passed through the extraction system. This test showed that a 95% recovery of argon from the tank can be achieved by circulating 0.42 million liters of helium through the extraction system, which requires a period of 22 h.

Another test of the argon recovery from the tank was performed with Ar^{37} activity produced in the tank by a fast-neutron irradiation. A Ra-Be neutron source with a total neutron emission rate of 7.38×10^4 neutrons sec^{-1} was inserted in a re-entrant iron pipe that reaches to the center of the tank. The liquid was irradiated with this source for 0.703 days producing Ar^{37} in the liquid by the reaction $Cl^{37}(p,n)Ar^{37}$ from the protons produced in the liquid principally by the reaction $Cl^{35}(n,p)S^{35}$. Carrier Ar^{36} was introduced (1.18 std cc) and the tank was swept three successive times with helium in which the volumes passed were, respectively, 0.35, 0.26, and 0.34 millions of liters of helium. The recovered argon was purified and counted following the procedures given below. The Ar^{37} activities in the three separate purges were found to be 63.4 ± 3.6, 2.3 ± 1.1, and 0.7 ± 0.5 disintegrations per day at the end of the neutron irradiation. The total Ar^{37} production rate observed in this experiment was $(7.5 \pm 0.4) \times 10^{-7}$ Ar^{37} atom per neutron. This production rate compared favorably with similar measurements in containers of smaller diameters (29 and 120 cm) which gave yields of 3.0×10^{-7} and 6.4×10^{-7} Ar^{37} atom per neutron, respectively. The Ar^{36} recoveries from each of three successive purges were 90.6, 6.2, and 0.7%, matching closely the Ar^{37} recoveries.

One might question whether Ar^{37} produced by the (ν, e^-) reaction would also be removed efficiently, since it would initially have a lower recoil energy than Ar^{37} produced by the (p,n) reaction. The Ar^{37} recoil energy resulting from neutrino capture ranges from 11 to over 1,000 eV for neutrino energies of 1 to 10 MeV. These recoil energies are sufficient to assure that the Ar^{37} ion formed would be free of the parent molecule, and, therefore, it would be expected to behave chemically similarly to an Ar^{37} atom produced by the (p,n) reaction. Once an Ar^{37} atom exists as a free atom it will mix with the carrier Ar^{36} present in the liquid (10^{10} atoms cm^{-3}) and be removed by the helium purge.

Counting

The argon sample is counted in a small proportional counter with an active volume 3 cm long and 0.5 cm in diameter. A small amount of methane is added to the argon to improve the counting characteristics of the gas. The counter cathode was constructed of zone-refined iron and the exterior envelope is made of silica glass. A thin window in the envelope is located at the end of the counter to facilitate energy calibration of the counter with Fe^{55} x rays. The counter is shielded from external radiations by a cylindrical iron shield 30 cm thick lined with a ring of 5-cm-diam proportional counters for registering cosmic-ray muons. The argon counter is held in the well of a 12.5- by 12.5-cm sodium-iodide scintillation counter located inside the ring counters. Events in anticoincidence with both the ring counters and the scintillation counter are recorded on a 100-channel pulse-height analyzer.[12] The pulse-height and time distribution of the events are recorded on paper tape. Each anticoincidence pulse is displayed on a storage oscilloscope and photographed to allow examination of each pulse shape to insure that it has a proper shape and is not caused by electrical noise. The counter had a 28% resolution (full width at half-maximum) for the 2.8-keV Auger electrons from the Ar^{37} decay. The operating voltage and amplifier gain are adjusted to place the center of the 2.8-keV Ar^{37} peak at channel 50 in the spectrum. The background counting rate in the 14 channels centered around channel 50 is 0.3 count per day. The efficiency of the counter was determined by filling with argon containing a known amount of Ar^{37}. Its efficiency for Ar^{37} is 51% for the 14 channels centered about channel 50.

Results and Discussion

Two experimental runs have been performed. In both experiments a measured volume of Ar^{36} was introduced into the tank at the start of the period of exposure, and mixed into the liquid for a period of approximately two hours with the eductor system. During the period of exposure the pumps were not operated. A positive pressure of helium of approximately 250 mm of Hg exists in the tank at all times.

The first exposure was 48 days. The tank was purged with 0.50 million liters of helium. A volume of 1.27 std cc of argon was recovered from the tank, and this volume contained 94% of the carrier Ar^{36} introduced at the start of the exposure. It was counted for 39 days and the total number of counts observed in the Ar^{37} peak position (full width at half-maximum) in the pulse-height spectrum was 22 counts. This rate is to be compared with a background rate of 31 ± 10 counts for this period. The neutrino-capture rate in the tank deduced from the exposure, counter efficiency, and argon recovery from this experiment was (-1.1 ± 1.4) per day.

Here a figure is omitted that shows the pulse height spectra for 35 days and 71 days of observations together with the background counting rate for 29 days. None of the spectra show any noticeable features in the region where Ar^{37} should appear.

A second exposure was made for 110 days from 23 June to 11 October 1967. The tank was purged with 0.53 million liters of helium yielding 0.62 cm^3 of argon with a 95% recovery of the added carrier Ar^{36}. It may be seen from the pulse-height spectrum for the first 35 days of counting that 11 ± 3 counts were observed in the 14 channels where Ar^{37} should appear. The counter background for this period of time corresponded to 12 ± 4 counts. Thus, there is no increase in counts from the sample over that expected from background counting rate of the counter. One would deduce from these rates that the neutrino-capture rate in 610 tons of tetrachloroethylene was equal to or less than 0.5 per day based upon one standard deviation. A similar limit can be obtained if one examines the shape of the pulse-height spectrum for extra counts in the 14 channels centered about channel 50 in the first 35-day count.

This limit, expressed as

$$\sum \phi\sigma \leqslant 0.3 \times 10^{-35}\ \mathrm{sec}^{-1} \text{ per } Cl^{37} \text{ atom},$$

can be compared with the predicted value of $(2.0 \pm 1.2) \times 10^{-35}$ sec^{-1} per Cl^{37} atom (table 56.1). It may be seen that this limit is approximately a factor of 7 below that expected from these solar-model calculations. From this limit and the cross section for B^8 neutrinos given in table 56.1, it may be concluded that the flux of B^8 neutrinos at the earth is equal to or less than 2×10^6 cm^{-2} sec^{-1}. It may be pointed out that if one accepted all of the 11 counts in the spectrum for the 35-day count as real events, making no allowance for background, then the flux-cross-section product limit would be 0.6×10^{-35} sec^{-1} per Cl^{37} atom.

The solar-model calculation of the flux of B^8 neutrinos is dependent upon the nuclear cross sections, solar composition, solar age and luminosity, and the opacity of solar material. The effect of each of these parameters has been studied, and the present results show that the solar B^8 neutrino flux is outside the present error limits if the uncertainties are treated as probable errors.[6–8] In the following article[13] Bahcall, Bahcall, and Shaviv have re-evaluated the solar neutrino fluxes taking into account a new value for the heavy element composition of the sun, and a new rate for the reaction $H(H, e^+\nu)D$.

Since this experiment is the first one with sufficient sensitivity to detect solar neutrinos from the carbon-nitrogen cycle, it is interesting to draw a conclusion about this energy cycle. Bahcall[4] has calculated the total flux-cross-section product for the carbon-nitrogen cycle to be 3.5×10^{-35} sec^{-1} per Cl^{37} atom, based on this cycle being the only source of the sun's energy. With the limit given above one can conclude that less than 9% of the sun's energy is produced by the carbon-nitrogen cycle.

It is possible to improve the sensitivity of the present experiment by reducing the background of the counter. However, background effects from cosmic-ray muons will eventually limit the detection sensitivity of the experiment at its present location. Detailed studies of the cosmic-ray background are in progress.

1. J. N. Bahcall, W. A. Fowler, I. Iben, Jr., and R. L. Sears, *Astrophys. J. 137*, 344 (1963).

2. R. L. Sears, *Astrophys. J. 140*, 153 (1964).

3. P. Pochoda and H. Reeves, *Planetary Space Sci. 12*, 119 (1964).

4. J. N. Bahcall, *Phys. Rev. Letters 12*, 300 (1964), *17*, 398 (1966).

5. D. Ezer and A. G. W. Cameron, *Can. J. Phys. 43*, 1497 (1965), *44*, 593 (1966), and private communication.

6. J. N. Bahcall, N. Cooper, and P. Demarque, *Astrophys. J. 150*, 723 (1967); G. Shaviv, J. N. Bahcall, and W. A. Fowler, *Astrophys. J. 150*, 725 (1967).

7. J. N. Bahcall, N. Bahcall, W. A. Fowler, and G. Shaviv, *Phys. Letters 26B*, 359 (1968).

8. J. N. Bahcall and G. Shaviv, *Astrophys. J. 153*, 113 (1958).

9. For recent summary see F. Reines, *Proc. Roy. Soc.* (London) *310A*, 104 (1967).

10. J. N. Bahcall, *Phys. Rev. 135*, B137 (1964).

11. R. Davis, Jr., *Phys. Rev. Letters 12*, 303 (1964); R. Davis, Jr., and D. S. Harmer, CERN Report no. CERN 65–32, 1965 (unpublished).

12. The circuit used in this work was designed by Mr. R. L. Chase and Mr. Lee Rogers of Brookhaven National Laboratory.

13. J. N. Bahcall, N. A. Bahcall, and G. Shaviv, *Phys. Rev. Letters 20*, 1209 (1968).

CHAPTER V

Variable Stars and Dying Stars

57. Periods of Twenty-five Variable Stars in the Small Magellanic Cloud

Henrietta S. Leavitt

(*Harvard College Observatory Circular No. 173*, 1–3 [1912])

At the end of the nineteenth century, Harvard College established an observatory at Arequipa, Peru, in order to extend its photographic surveys to the entire sky. Between 1893 and 1906 the two Magellanic clouds were systematically recorded with Harvard's 24-in refractor, and from these plates Henrietta Leavitt, a researcher in Cambridge, found an extraordinary total of 1,777 variable stars. By 1908 she had derived periods for a few of these Cepheid variables, and she reported that the brighter stars tended to have the longer cycles of variation.[1] Because the extent of the Magellanic clouds is small compared to their total distance, the relation of period to apparent brightness implied also a real connection with absolute brightness or luminosity. Four years later, when she prepared the paper given here, Leavitt had obtained apparent magnitude and period data for 25 variable stars in the Small Magellanic Cloud, thereby establishing the important period-luminosity relation for the Cepheid variables.

Once this relation is suitably calibrated, observation of the period of a variable star leads to determination of its absolute magnitude; then, using the observed apparent magnitudes, the distance of the star can be calculated. Ejnar Hertzsprung was the first to calibrate Leavitt's relation, although his application for the distance of the Magellanic clouds was apparently marred by a numerical error.[2] In 1918 Harlow Shapley arrived at a similar zero point for the period-luminosity relation, and he used this calibration to establish the vast extent of the Milky Way.[3] Partly through the neglect of interstellar absorption and partly through what in retrospect proved to be an oversimplified classification of variable stars, Shapley erred in his distance determination. The calibration was dramatically corrected in 1952 by Walter Baade, who showed at the same time that at least two classes of variable stars exist, with different period-luminosity relations (see selection 110).

Although calibrated erroneously, the period-luminosity relation led to Shapley's discovery of the tremendous size of our galaxy and subsequently to Hubble's discovery of the extragalactic nature of spiral nebulae. Today, distances established with the period-luminosity relation provide an essential step in the cosmic distance scale.

1. H. Leavitt, *Annals of the Harvard College Observatory 60*, 87–108 (1908).
2. E. Hertzsprung, *Astronomische Nachrichten 196*, 201 (1913).
3. H. Shapley, *Astrophysical Journal 48*, 89 (1918).

A CATALOGUE of 1,777 variable stars in the two Magellanic Clouds is given in *H.A. 60*, No. 4. The measurement and discussion of these objects present problems of unusual difficulty, on account of the large area covered by the two regions, the extremely crowded distribution of the stars contained in them, the faintness of the variables, and the shortness of their periods. As many of them never become brighter than the fifteenth magnitude, while very few exceed the thirteenth magnitude at maximum, long exposures are necessary, and the number of available photographs is small. The determination of absolute magnitudes for widely separated sequences of comparison stars of this degree of faintness may not be satisfactorily completed for some time to come. With the adoption of an absolute scale of magnitudes for stars in the North Polar Sequence, however, the way is open for such a determination.

Fifty-nine of the variables in the Small Magellanic Cloud were measured in 1904, using a provisional scale of magnitudes, and the periods of seventeen of them were published in *H.A. 60*, No. 4, table VI. They resemble the variables found in globular clusters, diminishing slowly in brightness, remaining near minimum for the greater part of the time, and increasing very rapidly to a brief maximum. Table 57.1 gives all the periods which have been determined thus far, 25 in number, arranged in the order of their length. The first five columns contain the Harvard Number, the brightness at maximum and at minimum as read from the light curve, the epoch expressed in days following J.D. 2,410,000, and the length of the period expressed in days. The Harvard Numbers in the first column are placed in italics, when the period has not been published hitherto. A remarkable relation between the brightness of these variables and the length of their periods will be noticed. In *H.A. 60*, No. 4, attention was called to the fact that the brighter variables have the longer periods, but at that time it was felt that the number was too small to warrant the drawing of general conclusions. The periods of 8 additional variables which have been determined since that time, however, conform to the same law.

Table 57.1 Periods of variable stars in the small Magellanic Cloud

H.	Max.	Min.	Epoch d	Period d	Res. *M*	Res. *m*.
1,505	14.8	16.1	0.02	1.25336	−0.6	−0.5
1,436	14.8	16.4	0.02	1.6637	−0.3	+0.1
1,446	14.8	16.4	1.38	1.7620	−0.3	+0.1
1,506	15.1	16.3	1.08	1.87502	+0.1	+0.1
1,413	14.7	15.6	0.35	2.17352	−0.2	−0.5
1,460	14.4	15.7	0.00	2.913	−0.3	−0.1
1,422	14.7	15.9	0.6	3.501	+0.2	+0.2
842	14.6	16.1	2.61	4.2897	+0.3	+0.6
1,425	14.3	15.3	2.8	4.547	0.0	−0.1
1,742	14.3	15.5	0.95	4.9866	+0.1	+0.2
1,646	14.4	15.4	4.30	5.311	+0.3	+0.1
1,649	14.3	15.2	5.05	5.323	+0.2	−0.1
1,492	13.8	14.8	0.6	6.2926	−0.2	−0.4
1,400	14.1	14.8	4.0	6.650	+0.2	−0.3
1,355	14.0	14.8	4.8	7.483	+0.2	−0.2
1,374	13.9	15.2	6.0	8.397	+0.2	−0.3
818	13.6	14.7	4.0	10.336	0.0	0.0
1,610	13.4	14.6	11.0	11.645	0.0	0.0
1,365	13.8	14.8	9.6	12.417	+0.4	+0.2
1,351	13.4	14.4	4.0	13.08	+0.1	−0.1
827	13.4	14.3	11.6	13.47	+0.1	−0.2
822	13.0	14.6	13.0	16.75	−0.1	+0.3
823	12.2	14.1	2.9	31.94	−0.3	+0.4
824	11.4	12.8	4.0	65.8	−0.4	−0.2
821	11.2	12.1	97.0	127.0	−0.1	−0.4

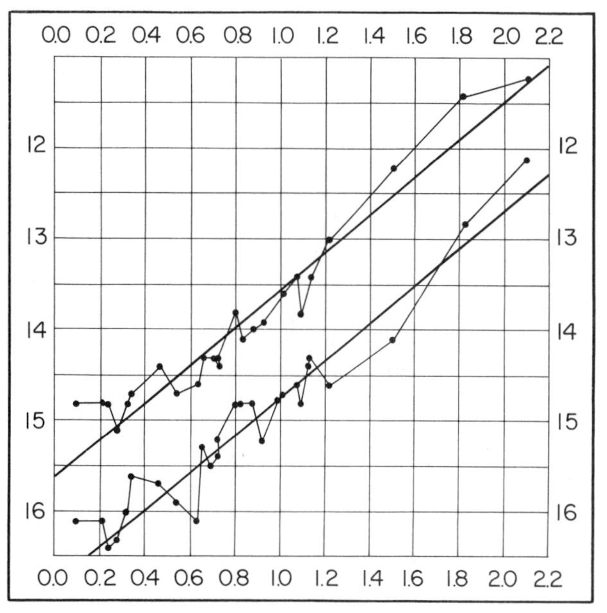

Fig. 57.1 A plot of the logarithm of the periods in days (abscissa) against the apparent magnitudes (ordinates) of twenty-five variable stars in the Small Magellanic Cloud. The two straight lines are drawn through the maximum and minimum magnitudes, and the apparent linearity of the plots reflects the period-luminosity relation, in which the absolute luminosity is a linear function of the period.

Here Leavitt presents a plot of apparent magnitude against period, omitted here because the same data appear in figure 57.1.

In figure 57.1, the abscissas are equal to the logarithms of the periods, and the ordinates to the corresponding magnitudes. A straight line can readily be drawn among each of the two series of points corresponding to maxima and minima, thus showing that there is a simple relation between the brightness of the variables and their periods. The logarithm of the period increases by about 0.48 for each increase of one magnitude in brightness. The residuals of the maximum and minimum of each star from the lines in figure 57.1. are given in the sixth and seventh columns of table 57.1. It is possible that the deviations from a straight line may become smaller when an absolute scale of magnitudes is used, and they may even indicate the corrections that need to be applied to the provisional scale. It should be noticed that the average range, for bright and faint variables alike, is about 1.2 magnitudes. Since the variables are probably at nearly the same distance from the Earth, their periods are apparently associated with their actual emission of light, as determined by their mass, density, and surface brightness.

The faintness of the variables in the Magellanic Clouds seems to preclude the study of their spectra, with our present facilities. A number of brighter variables have similar light curves, as UY Cygni, and should repay careful study. The class of spectrum ought to be determined for as many such objects as possible. It is to be hoped, also, that the parallaxes of some variables of this type may be measured. Two fundamental questions upon which light may be thrown by such inquiries are whether there are definite limits to the mass of variable stars of the cluster type, and if the spectra of such variables having long periods differ from those of variables whose periods are short.

The facts known with regard to these 25 variables suggest many other questions with regard to distribution, relations to star clusters and nebulae, differences in the forms of the light curves, and the extreme range of the length of the periods. It is hoped that a systematic study of the light changes of all the variables, nearly two thousand in number, in the two Magellanic Clouds may soon be undertaken at this Observatory.

58. On the Pulsations of a Gaseous Star and the Problem of the Cepheid Variables

Arthur Stanley Eddington

(*Monthly Notices of the Royal Astronomical Society 79*, 2–22, 177–188 [1918–1919])

At the beginning of this century Cepheid variables were commonly interpreted as eclipsing binary systems. The discovery of the period-luminosity relation posed a serious difficulty for this interpretation, because it was not clear how the period of a binary system could systematically increase with the luminosity of its brightest member. Moreover, examinations of the periodic displacement of the Cepheid's spectral lines showed that the time of their maximum brightness is very near the time of their maximum velocity of recession, and this coincidence could only be explained by rather contrived models in which the binary system was assumed to be enveloped in a resisting medium.[1] This situation led Henry Plummer to suggest in 1913 that the periodic displacement of spectral lines might be caused by the radial pulsations in the atmosphere of one star.[2] By this time Ejnar Hertzsprung and Henry Norris Russell had independently shown that the Cepheids had large intrinsic brightnesses and hence were giant stars. Harlow Shapley, who had just completed a brilliant doctoral thesis on eclipsing binaries under Russell's supervision, promptly showed that if the Cepheids were spectroscopic binaries, the giant stars had to move in orbits whose radii were less than a tenth of the sizes of the stars themselves.[3] As an alternative, Shapley independently suggested that the Cepheid variations might arise from pulsations of isolated individual stars.

Even before the variations in velocity were observed, August Ritter had argued that stellar variability might be caused by adiabatic, radial oscillations and that in this case the period of oscillation would vary inversely with the square root of the mass density of the star.[4] Such a simple interpretation depends upon the assumption that the interior conditions of a star are determined by the condition of hydrostatic equilibrium in which the gas pressure of the atoms moving within a star is just sufficient to balance the inward force of gravity; but by 1917 A. S. Eddington had shown that the assumption of radiative equilibrium is equally important in describing stellar interiors.

In the paper given here, Eddington applies his theory of radiative equilibrium to the Cepheid variables. The theory predicts, as he demonstrates, that the period of oscillations should vary inversely with the square root of the mass density of the star. His calculations also suggest that the pulsations must be determined primarily by conditions in the envelope of the star, rather than by those in the central regions where most of the mass is located.

Eddington is uncertain of the actual cause of the pulsation, but he does argue that the material in the outer atmosphere must act as a heat engine by absorbing heat when it is hottest and most compressed and releasing heat when it is coolest and most expanded; therefore, the stellar material must become more opaque when it is compressed. At the time that Eddington wrote this paper, the actual composition of stars, and therefore their sources of opacity, were unknown. By 1941, however, astronomers recognized that most stars are composed primarily of hydrogen and helium, and Eddington argued that stellar pulsation originates in an outer, convective zone where hydrogen is alternatively ionized and neutral during the course of pulsation. Because the pulsation depends upon a critical stage of ionization, it was easy to see why there is only a narrow range of mass densities for stars that become unstable to pulsation.[5]

Over a decade later, however, S. A. Zhevakin showed that Eddington's convective ionized-hydrogen zone cannot maintain pulsations because it does not absorb sufficient energy during the contraction of the star. Zhevakin showed that an outer region of doubly ionized helium can act, instead, as a special valve for the radiant flux of the star, in exactly the way Eddington had imagined.[6] The ionized helium zone absorbs the outward flow of energy from the center of the star during contraction and returns it during expansion.

1. J. C. Duncan, *Publications of the Astronomical Society of the Pacific 21*, 123 (1909).

2. H. C. Plummer, *Monthly Notices of the Royal Astronomical Society 73*, 661 (1913), *74*, 660 (1914). Forest Ray Moulton (*Astrophysical Journal 29*, 257 [1909]) had previously argued that stellar variations might be the result of an oscillation of a spherical star from an oblate to a prolate form.

3. H. Shapley, *Astrophysical Journal 40*, 448 (1914).

4. A. Ritter, *Wiedemann Annalen 5–20* (1878–1883).

5. A. S. Eddington, *Monthly Notices of the Royal Astronomical Society 101*, 182 (1941), *102*, 154 (1942).

6. S. A. Zhevakin, *Astronomicheskii zhurnal 30*, 161 (1953), *31*, 41 (1954). Also see J. P. Cox and C. Whitney, *Astrophysical Journal 127*, 561 (1958), and J. P. Cox, *Astrophysical Journal 138*, 487 (1963).

1

Although variable stars of the Cepheid type show a periodic change of radial velocity, it is improbable that they are binary stars. The theory which now appears most plausible attributes the light-changes to the pulsation of a single star;[1] and accordingly the varying radial velocity measures the approach and recession of the surface in the course of the pulsation. In order to throw light, if possible, on the phenomena of these variables, I have investigated the theory of a pulsating mass of gas. A complete solution of this problem would be very difficult; but it seems to be possible to determine the general character of the oscillation and to obtain results which may be compared with observation.

The type of pulsation here considered is symmetrical about the centre; that is to say, the star remains spherical, but expands and contracts. It is possible that the actual oscillation may be an elliptical deformation; but I think that a symmetrical oscillation is more probable in a star of low density, and it is much simpler to investigate.

It may be useful to summarise some of the leading results of observation with regard to these variables—

(1) The light-curve and the velocity-curve are closely similar. The correspondence is the more marked because both curves are usually very unsymmetrical. Maximum light corresponds to maximum velocity of approach.

(2) The light-variation is generally marked by a rapid rise to maximum and a slow decline. The velocity-curve shows a corresponding feature, which is usually expressed by saying that the periastron of the "orbit" points directly away from the earth.

(3) The period is a function of the absolute magnitude. For periods from three days upwards, the relation between log-period and absolute magnitude is practically linear; for shorter periods the relation is given by a curve. It appears to be possible to determine the absolute magnitude from the period with a probable error of less than a quarter of a magnitude.

(4) The Cepheids are giant stars, and are much more luminous than the average giants of their type.

(5) The spectral type tends to advance (towards M) as the period increases.

2

In order to gain some general knowledge of these stars and their pulsations, I have obtained the data given in tables 58.1 and 58.2 for those Cepheids which have been sufficiently investigated. The results in table 58.2 involve my theory of the radiative equilibrium of the stars;[2] those in table 58.1 are independent of the theory.

Here we omit the sources for the data given in table 58.1.

From the spectral type we obtain the effective temperature, and hence the radiation per unit area of surface. The absolute bolometric magnitude gives the total radiation. Hence, by

Table 58.1 Observed properties of Cepheid variable stars

Star	Mean type	Period (P) d	Abs. Mag. m	Light range. m	Radius (R) km.	Semi-amplitude (δR) km.	$\frac{\delta R}{R}$	$\frac{\delta T_e}{T_e}$	e	ω
Y Ophiuchi	F9	17.113	−3.99	0.46	42,200,000	1,999,000	0.047	0.054	0.10	209
ζ Geminorum	F2½	10.154	−3.18	.56	19,400,000	1,798,000	.093	.066	.22	333
S Sagittæ	F8½	8.382	−2.88	.63	24,600,000	2,000,000	.081	.075	.35	70
W Sagittarii	F5	7.595	−2.72	.85	18,400,000	1,930,000	.105	.103	.32	70
η Aquilæ	F6½	7.176	−2.64	.79	19,400,000	1,899,000	.098	.095	.47	64
X Sagittarii	F8	7.012	−2.60	.67	20,900,000	1,334,000	.064	.080	.40	94
Y Sagittarii	F9	5.773	−2.30	.74	19,400,000	1,485,000	.077	.089	.16	32
δ Cephei	F6	5.366	−2.19	.88	15,300,000	1,371,000	.090	.107	.36	83
T Vulpeculæ	F5	4.436	−2.00	.60	13,200,000	969,000	.073	.072	.43	111
SU Cygni	F1½	3.846	−1.78	.80	9,570,000	1,350,000	.141	.096	.21	346
RTAurigæ	F4	3.728	−1.73	.80	10,900,000	856,000	.078	..096	.37	95
SZ Tauri	F3	3.149	−1.58	.5	9,580,000	460,000	.048	.059	.24	77
SU Cassiopeiæ	F1½	1.950	−1.17	.4	7,230,000	295,000	.041	.047	.0	—
RR Lyræ	A5½	0.567	−0.52	.9	3,700,000	166,500	.045	.109	.27	97
Mean	—	—	—	—	—	—	.077	.082	—	—

division, we obtain the area of the surface. We thus find the radius R. For the relation between type and effective temperature, I have assumed that $\log_{10}$ T changes .0135 for each step in the spectrum (one-tenth of a type). This gives $A_0 = 11{,}200°$, $F_0 = 8{,}200°$, $G_0 = 6{,}000°$, $K_0 = 4{,}400°$—values which must be fairly near the truth. The resulting values of the radius are, of course, subject to several sources of uncertainty, but I think they must be substantially correct.

The amplitude δR of the oscillation of radius is the element $a \sin i$ of the supposed orbit. The *major* axis is usually in the line of sight, and consequently the eccentricity does not introduce any correction. In the succeeding column the relative amplitude δR/R is tabulated. Perhaps the true values of δR are somewhat greater than those here given, because the spectroscopic measures refer to the integrated light of a hemisphere. If there were no darkening at the limb, and if we make the doubtful assumption that the measures relate to the centre of gravity of the absorption, the rate of change of the star's radius would be 3/2 of the observed radial velocity. But the limb darkening will eliminate a great part of the surface, and the correction will be smaller.

In so far as the light of the star depends on the temperature, the radiation emitted varies as T^4. It is therefore convenient to take a temperature T_e proportional to the fourth-root of the luminosity, and to tabulate its relative amplitude $\delta T_e/T_e$ in table 58.1. For the moment, this is best regarded as a conventional way of measuring the light range.

On the pulsation-theory the element e of the orbit is not to be taken literally as an eccentricity; but it gives a measure of the deviation from simple-harmonic oscillation. The longitude of periastron ω should be 90° for a typical Cepheid in which the periastron points away from the earth. This is approximately true in most cases; for Y Ophiuchi, e is small and the discordance of ω is not very important; but SU Cygni (for which the orbit is uncertain) and ζ Geminorum violate the condition. Some of the other tabulated quantities run more smoothly if these two stars are cut out.

Coming to table 58.2 the mass is deduced by the formulæ in my second paper (*loc. cit.*, p. 604). The mass is simply a function of the absolute magnitude. Similarly the ratio of the force of radiation-pressure to gravitation, $1 - \beta$, is deduced.

From the mass and radius we at once obtain the mean density. I give in the table fifty-four times the mean density, which on my theory gives the central density (for a perfect gas). The central temperature, which should be proportional to $M\beta/R$, is obtained on the same theory.

The remaining columns will be of use later. n is equal to $2\pi/P$, and g is the surface-value of gravity. The last column contains a_1, the semi-axis of the orbit of a hypothetical planet which would revolve round the star of mass already given in the period P.

Determinations of spectral type based on the appearance of high dispersion spectra give a result 4 steps lower on the average. Omitted is a short comparison table based on these lower values of spectral type.

Table 58.2 Properties of Cepheid variable stars inferred from the theory of radiative equilibrium

Star	Mass (M) (sun = 1)	$1 - \beta$	Central density (ρ_c) gm./cm.^{3}	Central temperature (T_c) °	T_c/β $10^6 \times$	$P\sqrt{\rho_c}$	n^2R/g	a_1 km.
Y Ophiuchi	13.0	0.614	0.0043	3,130,000	8.10	1.12	0.8	45,100,000
ζ Geminorum	7.4	.509	.0251	4,920,000	10.00	1.61	0.4	26,400,000
S Sagittæ	6.1	.471	.0102	3,460,000	6.53	0.85	1.4	21,800,000
W Sagittarii	5.5	.451	.022	4,310,000	7.85	1.12	0.8	19,700,000
η Aquilæ	5.2	.440	.0176	3,950,000	7.05	0.95	1.1	18,600,000
X Sagittarii	5.1	.435	.0138	3,620,000	6.41	0.82	1.5	18,200,000
Y Sagittarii	4.3	.396	.0146	3,520,000	5.83	0.70	2.1	15,100,000
δ Cephei	4.0	.382	.028	4,250,000	6.88	0.89	1.3	14,100,000
T Vulpeculæ	3.6	.357	.039	4,620,000	7.19	0.88	1.3	12,000,000
SU Cygni	3.1	.330	.088	5,710,000	8.52	1.14	0.8	10,300,000
RT Aurigæ	3.0	.324	.057	4,880,000	7.22	0.89	1.3	10,000,000
SZ Tauri	2.8	.305	.079	5,340,000	7.69	0.88	1.3	8,800,000
SU Cassiopeiæ	2.3	.265	.151	6,150,000)	8.37	0.76	1.8	6,000,000
RR Lyræ	1.7	.197	.833	(9,700,000)	(12.1)	0.52	3.8	2,400,000

3

We notice first that the maximum (central) density is less than 0.1, except for SU Cassiopeiæ and RR Lyræ, and consequently the stars are in a truly gaseous condition throughout. For RR Lyræ, and perhaps SU Cassiopeiæ, the deviations from the laws of a perfect gas will be important near the centre. RR Lyræ is typical of a large class of short-period Cepheids (cluster-variables), and we must expect to find the behaviour of these somewhat different from that of the truly gaseous stars in the table. It is probably significant that the linear relation between log period and absolute magnitude holds for the truly gaseous stars, but changes to a curve when the period is less than three days, this being the point at which the central density begins to be too high.

The relative amplitudes of the pulsations, $\delta R/R$, are of much the same magnitude in the different stars, and are roughly equal to the temperature amplitudes $\delta T_e/T_e$. The light-ranges are usually not well determined, and $\delta T_e/T_e$ is probably subject to considerable accidental error. There is some indication that the two amplitudes are strictly proportional, and perhaps equal, especially if we omit ζ Geminorum, SU Cygni, and RR Lyræ for the reasons already given. The average amplitude is about one-twelfth, and probably the maximum possible is about one-ninth; the absence of smaller amplitudes may be due to selection.

The table also gives striking evidence against the binary hypothesis. Although the line of argument has been given already by Shapley,[1] the objections are perhaps made plainer when illustrated by actual numbers. We see that the radius (δR) of the supposed orbit of the principal star is on the average one-twelfth of the radius (R) of the star.[3] This would only be possible if the secondary had a very small mass, which can therefore be neglected in comparison with the mass of the primary. Thus, knowing the mass and period of the system, we can calculate the mean distance between the centres of the two components. This is tabulated under a_1 in table 58.2. Comparing R and a_1, we see that in most cases the secondary is just inside the primary; and, since the eccentricity is large, it will dip further into the primary at periastron. The binary interpretation thus seems to be inadmissible.

It will also be noticed that $P\sqrt{\rho_c}$ (table 58.2) is approximately constant, the range (again omitting the same stars) being 0.70 to 1.12. On the binary hypothesis it seems incredible that the period should have anything to do with the density of the primary, whereas a relation of this kind is predicted by almost any pulsatory theory.

4

In the course of the pulsation, the volume of the star is in many cases nearly doubled, and the consequent changes of temperature in the interior must be prodigious. It is at least worth investigating whether such a vibration could continue. Would not its energy be at once dissipated by the flow of heat tending to level the wave of temperature?

The equations for the oscillations of a gas, taking into account the flow of heat, lead to a differential equation of the fourth order, which can only be treated by quadratures. There are four boundary conditions to bring in, two at the centre and two at the boundary. Moreover, the quadrature near the boundary becomes impracticable owing to the vanishing of the density and pressure. In these circumstances there was a difficulty in finding a starting-point for attacking the problem.

This difficulty disappeared when it was realised that owing to the high opacity inside the star the vibrations near the centre must be very approximately adiabatic. In a former paper I have shown that more than half the radiation is absorbed in a layer $1\frac{1}{2}$ metres thick at atmospheric density. The explanation is that the high-temperature radiation within the stars resembles X-rays rather than ordinary light, and is absorbed on a correspondingly higher scale. This opacity tends to conserve the distribution of heat in the long waves here considered.

We start then by considering the adiabatic oscillations of a gas. We can afterwards calculate the flow of heat which would result, and determine the point in the star at which it becomes so great as to render our approximation invalid. Beyond this point the conditions would have to be treated by a different method.

5. Adiabatic Oscillations of a Star

Let p, ρ, T be the pressure, density, and temperature at a point distant ξ from the centre; and let g be the acceleration of gravity at this point. We fix attention on a particular element of matter, so that ξ oscillates in the course of a pulsation. Let $p_0, \rho_0, T_0, \xi_0, g_0$ be the values in the undisturbed state, and δp, $\delta\rho$, etc., be the deviations from the undisturbed values.

We write

$$\delta p/p_0 = p_1,$$

or

$$p = p_0(1 + p_1) \qquad \rho = \rho_0(1 + \rho_1) \qquad \xi = \xi_0(1 + \xi_1) \quad (1)$$

and similarly for all the other variables. If the period of the oscillation is $2\pi/n$, $p_1, \rho_1, \ldots$ will contain a factor $\cos nt$. We consider a small oscillation, and neglect the square of the amplitude.

For adiabatic oscillations the pressure and density of a *particular element*[4] are connected by

$$p = k\rho^{\gamma}, \quad (2)$$

where γ is the ratio of specific heats, so that

$$\frac{\delta p}{p_0} = \gamma \frac{\delta\rho}{\rho_0},$$

i.e.

$$p_1 = \gamma\rho_1. \tag{3}$$

The matter which is in the spherical shell ξ to $\xi + d\xi$ occupies in the undisturbed state the shell ξ_0 to $\xi_0 + d\xi_0$. Hence, equating the mass,

$$\rho\xi^2 d\xi = \rho_0 {\xi_0}^2 d\xi_0. \tag{4}$$

Differentiating, this gives

$$\frac{\delta\rho}{\rho_0} + \frac{2\delta\xi}{\xi_0} + \frac{d\delta\xi}{d\xi_0} = 0.$$

Hence by (1)

$$\rho_1 = -2\xi_1 - \frac{d(\xi_0\xi_1)}{d\xi_0} = -3\xi_1 - \xi_0\frac{d\xi_1}{d\xi_0}. \tag{5}$$

The ordinary hydrodynamical equation is

$$\begin{aligned}\frac{1}{\rho}\frac{dp}{d\xi} &= -g - \frac{d^2\xi}{dt^2}. \\ &= -g + n^2\xi_0\xi_1.\end{aligned} \tag{6}$$

Hence, using (4)

$$\frac{1}{\rho_0{\xi_0}^2}\frac{dp}{d\xi_0} = -\frac{g}{\xi^2} + \frac{n^2\xi_0}{\xi^2}\xi_1. \tag{7}$$

Now $g/\xi^2 = \mathrm{M}/\xi^4$ where M is the mass interior to ξ, which remains the same as the star oscillates. Hence

$$\delta\left(\frac{g}{\xi^2}\right) = \delta\left(\frac{\mathrm{M}}{\xi^4}\right) = -\frac{4\mathrm{M}}{{\xi_0}^5}\delta\xi = -\frac{4g_0}{{\xi_0}^2}\xi_1,$$

so that

$$\frac{g}{\xi^2} = \frac{g_0}{{\xi_0}^2} - \frac{4g_0}{{\xi_0}^2}\xi_1.$$

Hence (7) gives

$$\frac{1}{\rho_0{\xi_0}^2}\frac{d}{d\xi_0}(p_0 + p_0p_1) = -\frac{g_0}{{\xi_0}^2} + \frac{4g_0}{{\xi_0}^2}\xi_1 + \frac{n^2}{\xi_0}\xi_1,$$

which, if we recall that p_1 and ξ_1 contain a factor cos nt, separates into the equilibrium formula

$$\frac{dp_0}{d\xi_0} = -g_0\rho_0 \tag{8}$$

and the formula for the variation from equilibrium values

$$\frac{d(p_0p_1)}{d\xi_0} = \rho_0(4g_0 + n^2\xi_0)\xi_1. \tag{9}$$

Using (8), (9) becomes

$$p_0\frac{dp_1}{d\xi_0} - g_0\rho_0p_1 = (4g_0 + n^2\xi_0)\rho_0\xi_1. \tag{10}$$

But from (3) and (5)

$$p_1 = -\gamma\left(3\xi_1 + \xi_0\frac{d\xi_1}{d\xi_0}\right). \tag{11}$$

Eliminating p_1 from (10) and (11), we have finally,

$$\frac{d^2\xi_1}{d{\xi_0}^2} + \frac{4-\mu}{\xi_0}\frac{d\xi_1}{d\xi_0} + \left(\frac{n^2\rho_0}{\gamma p_0} - \left(3 - \frac{4}{\gamma}\right)\frac{\mu}{{\xi_0}^2}\right)\xi_1 = 0, \tag{12}$$

where $\mu = g_0\rho_0\xi_0/p_0$.

6

Numerical solutions to equation (12) are here omitted.

In equation (12) the factor

$$\frac{n^2}{\gamma}\frac{\rho_0}{p_0}$$

may be written

$$\frac{1}{u}\frac{n^2}{\gamma}\left(\frac{\rho_0}{p_0}\right)_c,$$

the suffix c indicating values at the centre. If we write

$$\omega^2 = \frac{n^2}{\gamma}\left(\frac{\rho_0}{p_0}\right)_c \tag{13}$$

and

$$\alpha = 3 - 4/\gamma, \tag{14}$$

equation (12) becomes

$$\xi_1'' + \frac{4-\mu}{\xi_0}\xi_1' + \left(\frac{\omega^2}{u} - \frac{\alpha\mu}{\xi_0{}^2}\right)\xi_1 = 0. \quad (15)$$

Here the only uncertain factor is α, which depends on the ratio of specific heats. Its greatest possible value (for a gaseous star) is 0.6, corresponding to $\gamma = 5/3$; this is only attained if the whole internal energy of the star is the translational motion of the molecules. The minimum value of α is 0, corresponding to $\gamma = 4/3$; it is well known that for lower values of γ a spherical mass of perfect gas is unstable.

As an example I have taken $\alpha = 0.2$. It is then necessary to try various values of ω^2, *i.e.* try various periods, until we find the value which gives a wave that fits the star and so determines the natural period of oscillation. For the fundamental oscillation, the first node (or place of constant pressure and density) will fall at the boundary of the star.

Here Eddington uses his theory of radiative equilibrium to determine the density distribution within the stellar interior and shows that the parameter, $\mu = g_0\rho_0\xi_0/P_0 = -4\xi du/ud\xi$, decreases from 287 to 20 when the distance from the center of the star decreases from R to 0.8R. He then provides numerical solutions to equation (12) and concludes that $\omega^2 \approx 0.3\alpha$. This result is used in section 8 to calculate the theoretical pulsation periods. By dividing equation (9) by equation (8), the boundary condition $P_1 = -(4 + n^2\xi_0/g_0)\xi_1$ is derived.

7. The Ratio of Specific Heats

When radiation pressure is taken into account, the quantity γ defined by equation (2) is given by

$$\frac{\gamma - \frac{4}{3}}{\Gamma - \frac{4}{3}} = \frac{4 - 3\beta}{1 + 12(\Gamma - 1)(1 - \beta)/\beta},$$

where Γ is the ratio of specific heats for matter alone, and $(1 - \beta)$ is the ratio of radiation pressure to the whole pressure.

8. Period of the Pulsation

Using $n = 2\pi/P$, where P is the pulsation period, and $\omega^2 = 0.3\alpha$, Eddington obtains the relation

$$P\sqrt{\rho_c} = C(\gamma\alpha)^{-1/2},$$

where the constant C = 0.290 if the period is in days and the mass density is in units of gm cm^{-3}. This theoretical value of C is consistent, within a factor of 2, with that determined from observational data. Values of $(\gamma\alpha)^{1/2}$ depend upon $(1 - \beta)$ and Γ and take on values between 0.2 and 0.6.

9. The Flow of Heat

An estimate of the outward heat flow justifies the assumption that the oscillations are adiabatic.

10. Change of Period

If a star's heat is provided by gravitational contraction, the decrease in radius during contraction will cause an increase in mass density. Because the period of pulsation varies inversely with the square root of the mass density, the period should be observed to decrease as time goes on. To provide the 10^{44} ergs of energy radiated by δ Cephei each year, the radius must decrease by 1 part in 12,000 and the period must increase by about 40 sec annually. Because these changes have not been observed, the heat radiated by a variable star cannot be provided by gravitational contraction.

11

Here Eddington uses Shapley's empirical law connecting period and luminosity, with Γ equal to 1.47, to compute the central mass densities, radii, effective temperatures, and spectral types for stars of different absolute visual magnitudes and periods. The general conclusion is that stars with longer pulsation periods have lower effective temperatures and later (redder) spectral types.

12. Dissipation of Energy

We shall now examine the rate at which oscillations of a gaseous star decay owing to thermal dissipation. When a portion of a gas is compressed, the temperature is raised, and heat tends to leak away; consequently the pressure falls, and the whole work of the compression is not recovered in the rebound. There will be other sources of dissipation, such as viscosity; but this leakage of heat seems at first sight the most threatening, and when the pulsatory theory was first suggested I found it difficult to conceive that the pulsation could last for more than a few periods. The discovery of the very high opacity within a star, however, gives a different aspect to the problem; the leakage of heat is very much smaller than would occur if we were dealing with low-temperature radiation for which gases are comparatively transparent.

Under the assumption that each element of a star acts as a reversible engine that does not dissipate heat, the heat dissipated by leakage between different parts of the star is computed. If the kinetic energy of pulsation is dissipated as heat, the total amount of heat dissipated by δ Cephei is 4.7×10^{41} ergs per year. The rate of dissipation of pulsation energy is only about one two-hundredth the total luminosity of the star, and the energy of pulsation would last about 1,500 years at the present rate of dissipation.

13. Conditions for Cepheid Variation

The uniform relations between the period, luminosity, and spectral type, which have been found by observation, show that the variation only occurs during a strictly limited stage in the development of a star. This seems to rule out the suggestion that it is provoked by some external accident such as a collision. Two possible explanations suggest themselves—

(1) During a certain stage the conditions are such that an oscillation having the appropriate period would tend to increase; thus the pulsation would start automatically.

(2) At a certain stage there is a sudden change in the state of stable equilibrium, and the collapse to the new state throws the star into a pulsation, which, according to §11, could last for a period of the order 1,000 years.

In either case it seems likely that every star of sufficient mass will become a Cepheid for a brief part of its life; this conclusion is, however, not inevitable, since the speed of rotation, for example, may make a difference in the behavior of otherwise similar bodies.

I do not think that the first explanation contradicts any thermodynamical principles, but it does not obtain much encouragement from our analytical investigation.

Because the first explanation involves a change in temperature in the interior regions where mechanical energy is dissipated, the required process would be very inefficient and unlikely.

According to either explanation the pulsation occurs at some critical state of the star. In the approximate theory of stellar equilibrium we assume that the giant stars differ from one another only in scale, so that the solution for one star is obtained from that of another by multiplying ξ, ρ, T, etc., by suitable constants. So long as the physical properties of the material—opacity, specific heat, molecular weight, etc.—are uniform throughout the star, we can find no reason for the discontinuity of condition between Cepheids and non-variable stars. We must therefore take into account a variation of these properties with temperature, or possibly with density. For example, it is not unlikely that the specific heat may change with temperature, being abnormally high for temperatures at which ionisation occurs rapidly.

It suggests itself that the condition for Cepheid variation may be a particular distribution of internal temperature; but, if our calculations in table 58.2 are correct, this cannot be strictly true. The central temperature seems to vary fairly systematically with the mass to an extent by no means negligible from the present point of view.

According to the theory of radiative equilibrium, this non-correspondence of temperature-distribution can be proved more precisely. From *Monthly Notices 77*, 600, equations (42), (43), (44), we have

$$T_e^4 \propto g(1-\beta) \propto M^{1/3}\rho^{2/3}(1-\beta)$$
$$T_c \propto \rho^{1/3}M^{2/3}\beta$$
$$1-\beta \propto M^2\beta^4.$$

Whence, eliminating M and ρ,

$$T_e \propto T_c(1-\beta)^{1/8}.$$

Thus if the central temperature T_c were the same for all Cepheids, we should have the effective temperature varying as the eighth root of $(1-\beta)$, and accordingly increasing slowly with the absolute luminosity. Observation, however, shows that the change is in the other direction, the type becoming redder as the luminosity increases.

If, however, the critical condition determining the variation occurs near the boundary, the case is somewhat altered. Owing to the smaller absorption of low-temperature radiation, radiation-pressure must become very small near the boundary. The result is that the temperatures in the outer layers depend on T_c/β instead of on T_c. From table 58.2 it appears that T_c/β is fairly constant for the Cepheids, the variations being apparently accidental. This can be seen better by taking two theoretical stars. We have

Period 4.5 days; $T_c = 3{,}750{,}000°$; $T_c/\beta = 5{,}830{,}000°$.

Period 30.8 days; $T_c = 1{,}970{,}000°$; $T_c/\beta = 5{,}810{,}000°$.

This suggests that the occurrence of variation depends on the conditions in the outer layers.

14, 15, 16

In the next three sections Eddington argues that the greatest flow of heat will occur when the opacity is diminishing most rapidly and the velocity of expansion is greatest, that the asymmetric shapes of the velocity and light curves can be explained by second-order terms of the wave equations describing the oscillation, and that the spectral lines might be expected to be sharpest at maximum light.

17. Conclusions

1. In agreement with earlier investigators, it is concluded that the binary hypothesis of Cepheids must be ruled out, because (*a*) the distance of the centres of the components would have to be less than the radius of one of them, (*b*) because there is a uniform relation between the period and density which seems to point to a cause intrinsic in the star (§3).
2. The hypothesis that the variation is due to a symmetrical dilatation and contraction of a single star leads to results in agreement with observation in the following respects:
   (*a*) The absolute value of the period, which can be determined theoretically with a factor of uncertainty not greater than 2 (§8).
   (*b*) The advance of spectral type towards the red with increasing luminosity (§11).
   (*c*) The asymmetric form of the velocity-curve (§15).
3. Owing to the high opacity, the vibrations throughout the greater part of the interior are very nearly adiabatic (§9). For the same reason, the dissipation of energy due to periodic flow of heat is comparatively small; the time of decay of a pulsation from this cause is of the order 1,500 years (§12).
4. In the interior the outward flow of heat is a maximum at the time of greatest compression. In the non-adiabatic outer layers the maximum must occur later; but it has not been possible to obtain mathematically the observed law that at the surface the delay amounts to a quarter-period—which is one of the best-known characteristics of the Cepheids (§14).
5. In the Cepheids examined the amplitude of the pulsation ranges from about 5 per cent to 11 per cent of the radius. The light-range is roughly proportional to the relative change of radius (§3).
6. Calculations of the densities of particular Cepheids show that the product $P\sqrt{\rho}$ is approximately constant. This relation should be approximately true on almost any pulsatory theory. On the present theory there should be a slight increase of $P\sqrt{\rho}$ with increasing mass, which is perhaps confirmed by the calculations based on Adams' and Joy's determinations of spectral type (§8).
7. If the star's energy is derived solely from contraction, the change of period due to increasing density should be easily measurable. Since the observed change is much too small, it appears to follow that the star must have some other source of energy (§10).
8. For stars of period greater than three days the central density is less than 1/10 water, and the condition approximates to that of a perfect gas throughout. For cluster-variables with periods less than a day the deviations from a perfect gas will have an important influence. (The effect is to increase γ in the equation $p_1 = \gamma\rho_1$, and consequently the period is shorter than if the properties of a gas were retained) (§3).
9. The suggestion that Cepheid variation may occur in all stars at the time when the central temperature reaches a certain value (and consequently the temperatures at other corresponding points have definite values) appears to be untenable. The numerical relations suggest that the occurrence of variation is determined by conditions in the outside layers of the star (§13).

1. H. Shapley, *Ap. J. 40*, 448 (1914); H. C. Plummer, *Monthly Notices 75*, 573 (1915).
2. *Monthly Notices 77*, 16, 596 (1916).
3. In particular cases, sin *i* might be small and the radius of the orbit much larger than here stated. But this could scarcely happen for all fourteen stars.
4. The undisturbed state is not supposed to be that of adiabatic equilibrium.

59. T Tauri Variable Stars

Alfred H. Joy

(*Astrophysical Journal 102*, 168–195 [1945])

In the paper given here, Alfred Joy defines for the first time a distinct class of stars that are imbedded in nebulous clouds of gas and dust and that exhibit irregular and unpredictable changes in light. Although this type of object had been known since the early 1860s, Joy demonstrates that the spectra of these stars always exhibit a distinctive pattern of emission lines. These emission line peculiarities were subsequently used to identify several hundred T Tauri stars, which are almost always found in or near nebulosity and which almost always have irregular light variations.[1]

As Joy points out, these emission lines are usually superposed upon a bright continuous spectrum, which often displays the absorption line spectrum of a late-type star. Perhaps because these stars seem to be normal main-sequence stars imbedded in nebular clouds of gas, the T Tauri stars were at first assumed to be normal stars passing through interstellar clouds. However, Viktor Ambartsumian argued that the T Tauri stars are actually very young stars in the process of gravitational collapse from the diffuse gas and dust clouds in which they are found.[2]

The verification of Ambartsumian's interpretation of the T Tauri stars had to await interpretations of the Hertzsprung-Russell diagrams of star clusters. As Joy points out in this paper, there is some uncertainty in his interpretation of the T Tauri stars as main-sequence dwarf stars. The difference in apparent magnitudes of the two components of UX Tau, for example, is about 3 mag, whereas the difference implied by the spectral types is 5 mag if the companions have the types assigned by Joy—*d*G5 and *d*M2. Therefore, both components cannot be dwarf stars, as Joy suggests.

The discrepancy between the observed differences in magnitude between the components of double T Tauri stars and those inferred from their spectral types was explained by George Herbig who showed that late type K and M stars in regions containing T Tauri stars are too bright for their spectral type.[3] Herbig's suggestion that the T Tauri stars actually lie above the main sequence in the H-R diagram was verified by Merle Walker for the T Tauri stars lying within clusters of O and B stars.[4] Walker showed that the hot, massive O and B stars in clusters like NGC 2264 lie on the main sequence and have just begun to shine by thermonuclear reactions, whereas the less massive T Tauri stars are still undergoing gravitational contraction and lie above the main sequence in the H-R diagram. It is reasonable to suppose that the T Tauri stars found in the same regions as the bright O and B stars are also young stars, with ages of about a million years, and that the relatively low-mass T Tauri stars will eventually reach the main sequence in type F or later. Nevertheless, the T Tauri stars are also found in dark cloud complexes where, because no other objects of recognized age are observed, independent age estimates may be impossible; hence, the presence of T Tauri stars does not necessarily signify regions of unusual youth.

A detailed study of the T Tauri stars has recently been published by A. Eric Rydgren, Stephen E. Strom, and Karen M. Strom.[5] Their basic conclusions include the following:

1. The emission features (Balmer continuum, Paschen continuum, infrared excess, Balmer lines) arise in a hot ($T \approx 2 \times 10^4$), dense ($n \approx 10^9 - 10^{12}$ atoms cm^{-2}), circumstellar envelope.
2. The optical variability arises from changes in the emission measure, indicating changes in density or physical size of the envelope.
3. Mass outflows or stellar winds characterize most, if not all, of these objects.
4. The largest mass outflow rates and strongest emission features are found in the youngest T Tauri stars, which are still confined within a dark cloud or within placental material. The mass outflow rates decrease as the stars approach the main sequence.
5. The large majority of T Tauri stars have relatively low mass (masses less than 2 solar masses) and are approaching the main sequence along quasi-static equilibrium tracks.

1. G. H. Herbig and N. Kameswara Rao, *Astrophysical Journal 174*, 401 (1972).
2. V. A. Ambartsumian, *Stellar Evolution and Astrophysics* (Erevan: Armenian Academy of Sciences, 1947); *Astronomischeskii zhurnal 26*, 3 (1951).
3. G. H. Herbig, *Journal of the Royal Astronomical Society of Canada 46*, 222 (1952).
4. M. F. Walker, selection 54, and *Astrophysical Journal 125*, 636 (1957), *130*, 57 (1959).
5. A. E. Rydgren, S. E. Strom, and K. M. Strom, *Astrophysical Journal Supplement 30*, 307 (1976).

Abstract—Eleven irregular variable stars have been observed whose physical characteristics seem much alike and yet are sufficiently different from other known classes of variables to warrant the recognition of a new type of variable stars whose prototype is T Tauri. The distinctive characteristics are: (1) irregular light-variations of about 3 mag., (2) spectral type F5–G5 with emission lines resembling the solar chromosphere, (3) low luminosity, and (4) association with dark or bright nebulosity. The stars included are RW Aur, UY Aur, R CrA, S CrA, RU Lup, R Mon, T Tau, RY Tau, UX Tau, UZ Tau, and XZ Tau. They are situated in or near the Milky Way dark clouds in the direction either of the center or of the anticenter of the galaxy.

The light-curves—The total light-changes are about 3 mag., the variations being extremely irregular as to range and time. The light-curves are not unique.

The individual variables—The spectrographic observations of the variables are described and the significant features of the spectra pointed out.

Discussion of the spectra—The spectral types of the T Tauri stars are estimated to be between F5 and G5, although for many of them the absorption lines generally used in classification are lacking. A small variation of type with phase was found for T Tau and RY Tau. Bright hydrogen has been found in all stars of the group, and bright *Ca* II (H and K) in all except R CrA. Most of the stars show an emission spectrum composed of many bright lines of low excitation. The strongest lines are those of *Ca* II, *H*, *Fe* II, *Ca* I, *Sr* II, *Fe* I, and *Ti* II. The identification and relative maximum intensities of 160 lines of the different stars are shown. The intensity of the emission spectrum varies greatly from time to time in each star, the bright lines becoming more prominent at maximum light of the variable.

The lines λ 4,063 and λ 4,132 of the $a^3F–y^3F^0$ multiplet of *Fe* I are greatly enhanced in strength in the stars showing strong emission spectra. This distortion is probably the result of fluorescent effects.

The marked similarity of the bright-line spectrum of the T Tauri stars to that of the upper solar chromosphere is shown in table 59.4.

Absolute magnitudes and color indices—Spectroscopic absolute magnitudes of three stars of the group, together with meager indirect evidence, indicate that the T Tauri stars are dwarfs of the main sequence. Color indices for five stars show some color excess, which is probably the result of selective absorption by surrounding nebulosity.

Radial velocities—Radial-velocity measures from absorption lines are difficult when the emission spectrum is present. Lack of agreement in the measures of both absorption and emission spectra indicates irregular atmospheric motions. In the mean the emission lines are displaced toward the violet with respect to the absorption lines.

The criteria distinguishing variable stars included in the T Tauri class are: (1) rapid irregular light-variations of about 3 mag.; (2) spectral type F5–G5, with emission lines resembling those of the solar chromosphere, particularly in the great strength of H and K of calcium; (3) low luminosity; (4) association with dark or bright nebulosity. Although little or nothing is known as to the cause of the light-variation, characteristics 1–4 are probably physically interrelated and together form a distinct stellar type which may be readily recognized. These stars differ from other known variables, especially in their low luminosity and the high intensity of bright H and K in their spectra. The well-known variable T Tauri is one of the brightest stars of the group and may properly be considered as the prototype, although there are marked differences among the stars and no two are exactly alike.

Eleven stars (table 59.1) make up the list of variables whose characteristics are sufficiently alike to entitle them to membership in this class. Heretofore, they have generally been classed as "irregular" variables, and their unpredictable light-changes indicate that they well deserve the title. Four stars (R CrA, R Mon, T Tau, and RY Tau) are definitely involved in surrounding nebulosity, and all stars of the group are situated in or near large areas of heavy obscuration by Milky Way clouds. In the course of this investigation, five others (RW Aur, UY Aur, S CrA, UX Tau, and UZ Tau) were discovered[1] to be visual double stars, leaving only two of the group (XZ Tau and RU Lup) which appear to be uninvolved either with local nebulosity or with near-by companion stars.

The galactic co-ordinates are in the fifth and sixth columns of table 59.1. The peculiar distribution of these stars is evident. Three are in the Milky Way clouds near the direction of the center of the galaxy, and the others form a group about the anticenter in the opposite direction; but, unlike the distant Cepheids, they do not lie closely along the galactic equator as seen from the sun. The mean latitude (disregarding signs) is 14°. The mean median apparent magnitude of these two groups is practically the same, indicating equal distances from the earth if all the stars are of equal absolute magnitude and suffer equal amounts of obscuration.

The Light Curves

The variations in light of the T Tauri stars are so irregular and unpredictable that classification by means of their light-curves is practically impossible. Thus far, observations have been insufficient to determine definite sequences of light-changes which are uniquely characteristic of the group. The light-curves vary greatly from year to year and from star to star.

Here we omit light curves of RW Aurigae taken by Enebo[2] in 1907 and by Zinner[3] in 1913. The

Table 59.1 T Tauri variables

Star	α (1900)	δ (1900)	Magnitude range	Galactic l	Galactic b	Remarks
RW Aur	$5^h\ 1^m\!.4$	+30°16′	9.0–12.0	142°	− 6°	Double
UY Aur	4 45.4	+30 37	11.6–14.0	139	− 8	Double
R CrA	18 55.1	−37 6	9.7–13.5	328	−19	Nucleus of variable comet-like nebula, NGC 6729
S CrA	18 54.4	−37 5	9.5–13	328	−19	Double
RU Lup	15 50.1	−37 32	9.0–11.0	307	+11	—
R Mon	6 33.7	+ 8 50	9.3–14.0	171	+ 3	Nucleus of variable comet-like nebula, NGC 2261
T Tau	4 16.2	+19 18	9.0–12.8	148	−22	Near Hind's variable nebula, NGC 1555, and surrounded by a small shell
RY Tau	4 15.6	+28 12	8.8–11.1	136	−14	Nucleus of a fan nebula
UX Tau	4 24.2	+18 0	10.5–13.4	146	−19	Double
UZ Tau	4 26.6	+25 40	9.2–<13	140	−14	Double
XZ Tau	4 25.9	+18 1	10.4–13.5	146	−19	—

light variations range between 0.5 and 2.5 mag on irregular time scales between 1 and 25 days.

Other stars such as RR Tau and certain variables in Orion have similar light-changes, yet their spectra are very different. The color index for several of the stars has been determined, but changes in color have not yet been studied. If the fluctuations in light result from the interposition of obscuring clouds, changes in color due to scattering or selective absorption might be expected.

The magnitude ranges in table 59.1 are from Schneller's Catalogue.[4] The total variation is large, averaging over 3 mag. The brightness of RU Lup and RY Tau at the time of some of the spectrographic observations was certainly a magnitude or more fainter than the minimum given by Schneller.

The Individual Variables

Here Joy embarks upon a fifteen-page discussion of the light variations, emission and absorption lines, spectral types, and radial velocities of each star listed in table 59.1. We omit the plates showing the spectra of each star but provide a summary of the discussion for each star. The flavor of Joy's individual discussions is represented by the inclusion of his text on RU Lupi, R Monocerotis, and T Tauri.

RW AURIGAE 050130 Irregular light variations were discovered by Ceraski[5] in 1906. The bright emission lines of hydrogen and ionized calcium (H and K) are distorted by broad, deep central reversals. The spectral type estimated from weak absorption lines is G5. The absorption lines are superposed on the emission lines and originate in an outerlying level of the stellar atmosphere. The outer layer seems to be expanding, and variations in the intensities and positions of the absorption lines suggest an unstable atmosphere.

UY AURIGAE 044530 Irregular light variations were discovered by Ceraski[6] in 1913. The spectrum, first observed by Joy[7] in 1932, shows strong emission lines of hydrogen, helium, calcium, and the metallic elements. The metallic lines exhibit variable intensity when the other emission lines stay at roughly the same intensity. The spectral type estimated from weak absorption lines is G5. The weighted means of the velocities inferred from emission and absorption lines are, respectively, −3 and +30 km sec^{-1}.

R CORONAE AUSTRALIS 1855*37* Irregular light variations of the star and the associated nebula, NGC 6729, were discovered by J. F. J. Schmidt and confirmed photographically by Innes,[8] Knox-Shaw,[9] and Gaposchkin.[10] Slipher[11] first photographed the spectrum in 1917, and Hubble[12] showed that the star exhibits unsymmetrically reversed hydrogen emission lines and iron emission lines superposed on an absorption spectrum of type G. The H and K emission lines of calcium are absent.

S CORONAE AUSTRALIS 1854*37* Irregular light variations were discovered by J. F. J. Schmidt in 1866. Strong emission lines of hydrogen, He I (λ4,026 and λ4,471) and He II (λ4,685), Ca I and Ca II (H and K), Fe I, and Sr II are present, but no absorption lines are present. This object is located 43 sec of time preceding and 0.3 min of arc south of R Coronae Australis.

RU LUPI 1550*37* The variation in brightness of this star (HD 142560, CoD − 37°10602) was detected by Miss J. C. Mackie[13] on Harvard photographs after Miss Cannon[14] had noted bright lines of hydrogen and calcium in its spectrum. The survey plates show irregular changes in brightness from 1893 to 1912 with a magnitude range from 9 to 11. Probably on account of its southern declination, it has been neglected by variable-star observers for nearly thirty years. No further observations are recorded until 1940, when P. W. Merrill[15] photographed its spectrum and called attention to the presence of strong emission of *Fe* II and *Ti* II, as well as of *H* and *Ca* II. He also found weak lines corresponding to *Fe* I, *Mg* II, *Cr* II, and *Sc* II. The appearance of bright lines of *Fe* I and strong H and K of *Ca* II indicated at once that the spectrum was related to that of T Tauri. Dr. Merrill has been kind enough to place at my disposal two spectrograms of RU Lup, together with his line identifications and velocity measures for discussion in this paper.

With the dispersion used, no absorption lines can be distinguished, although the distribution of light in the continuous spectrum suggests a temperature lower than G0. No titanium oxide bands are present. The lack of absorption lines may be accounted for, in part, by the great number of emission lines of neutral, as well as ionized, atoms. Probably many lines having insufficient emission to show above the continuous spectrum have enough radiation to fill up the absorption in the lines. On one plate, however, a faint central absorption line appears within the emission of both H and K.

The bright lines of RU Lup as well as of other stars of the T Tauri group are much wider than those of the Me variables or of stars like Z And, and the accuracy of measurement is correspondingly decreased. The apparent width of *H* and *Ca* II lines is about 4.5 A. Lines of other elements are, perhaps, one-half as wide, but none can be called sharp.

In the spectral region $\lambda\lambda$ 3,900–6,500, 120 emission lines have been measured and identified. The 15 elements represented, together with the number of lines, are shown in table 59.2 in order of their line intensities. Many other blended lines are probably present in RU Lup, but they are not resolved on account of the width of the lines and the low dispersion. This list, however, will be sufficient to give a general idea of the characteristics of the emission spectrum of the T Tauri stars. The metallic lines were greatly weakened in March and April, 1942, when the light of the star diminished.

Table 59.2 Elements represented in RU Lupi

Atom	No. lines	Atom	No. lines	Atom	No. lines
H	5	*Ti* II	35	*Ca* I	1
Ca II	5	*Sr* II	2	*Sc* II	5
Fe II	28	*Fe* I	24	*Cr* II	6
He I	1	*Ni* II	1	*Na* I	2
Cr I	3	*Mg* II	1	*He* II	1

Table 59.3 Radial velocities of elements in RU Lupi

Element	Vel. km/sec	No. lines	Element	Vel. km/sec	No. lines
H	−10.7	4	*Sc* II	− 3.6	5
Ca I	−12.0	1	*Ti* II	− 8.5	35
Ca II	− 3.0	2	*Cr* I	− 3.0	2
Cr II	+ 4.0	6	*Ni* II	− 7.2	1
Fe I	− 6.7	22	*Sr* II	+11.5	2
Fe II	−11.0	28			

The radial velocities from the emission lines were practically constant during the period of observation. The weighted mean velocity is −6.4 km/sec. Likewise, the range in velocity among the different elements is small, as indicated by table 59.3, in which the elements are arranged in order of atomic weight. For most of the elements the agreement of the individual lines is satisfactory, but for *Ca* II the measures indicate a discrepancy of 36 km/sec between H and K. The effect is systematic and is seen on all plates measured, K being displaced 21 km/sec toward the red and H 15 km/sec toward the violet with reference to the mean of all lines. Both lines appear to be quite symmetrical. The faint emission of $H\varepsilon$ would have little influence on the position of *H*, but it is possible that absorption in the hydrogen line might decrease the intensity of H on the red edge and tend to shift the line toward the violet. No explanation for the displacement of K, however, seems evident.

RU Lup is located in one of the dark lanes of the southern Milky Way but is not known to have any companion or to be enveloped in local nebulosity.

R MONOCEROTIS 063308 The variable star R Mon (BD + 8°1427) is the nucleus of the comet-shaped variable nebula, NGC 2261. Its light-changes were detected by J. F. J. Schmidt[16] in Athens in 1861, and later observations indicated that the variations were irregular, with a range of at least 4 mag. It is difficult to make precise estimates of brightness on account of the surrounding nebulosity. With high magnification the star resembles the nucleus of a comet near the sun.

From direct photographs taken with the 24-inch reflector of the Yerkes Observatory, E. Hubble[17] in 1916 discovered that marked changes in the outline of the nebula occurred within a period of a few months. In size and general form it is similar to NGC 6729, except that the outer parts of NGC 2261 are longer, extending to a distance of 2.5 minutes of arc.

The spectrum of the star and that of the nebula were photographed by V. M. Slipher[18] at the Lowell Observatory in 1917 and found to be identical, indicating that the light of the nebula is reflected radiation from the star. He suggested that the emission spectrum resembled that of Nova Aurigae. The lines that can be identified from his measures are those of H, *Ti* II, and *Fe* II, which do not necessarily pertain to novae, although, soon after maximum, bright lines of ionized metals are prominent in most novae and at minimum they may appear in such stars as RS Oph.[19] The high-excitation features characteristic of the spectra of novae are entirely lacking in R Mon. On the other hand, the Mount Wilson spectra of R Mon and probably the Lowell spectrum, as well, correspond closely with the spectra of the T Tauri variables. The bright hydrogen lines are accompanied by moderately strong dark lines on the violet side as in T Tau, and the decrement of the series toward the violet is steep. The H and K lines are present in emission but are not nearly so strong as in T Tau. Helium and the nebular lines are definitely lacking. The normal absorption spectrum is extremely weak.

By attributing the formation of cometary nebulae to the action of radiation pressure from the star upon the surrounding material particles, R. Minkowski[20] finds that the radiation of dwarf stars is insufficient to produce nebular forms such as NGC 2261 or NGC 6729 from surrounding material and concludes that if the spectrum is G-type, as appears from the distribution of light in its spectrum and from analogy with other stars of the group, the stars must be giants with absolute magnitudes of +0.7 or brighter. Since the absorption lines are poorly shown in the spectra, the spectroscopic absolute magnitudes cannot be well determined. A. van Maanen[21] estimates the trigonometric parallax of R Mon to be smaller than 0″.005, which, without allowance for obscuration, would indicate an absolute magnitude brighter than +2.8 at maximum and +7.5 at minimum.

The K line is three or four times as strong as H, and both are considerably weaker than in any of the other stars of the group except R CrA and UX Tau. None of the observations, however, were made at maximum light.

Although R Mon is deeply imbedded in nebulosity, its spectrum shows no marked effect of selective absorption, as far as can be estimated by visual inspection of the distribution of the continuous spectrum.

T TAURI 041619 The variable T Tau (BD + 19°706) has been chosen as the typical star of this group of peculiar variable stars because it is the best known, is among the brightest, and represents the group with respect to both emission and absorption spectra. It has an extended history, its variation in light having been discovered by Hind in 1852. It is 2.3 seconds of time following, 37 seconds of arc north of, the faint irregular patch of nebulosity well known as Hind's variable nebula (NGC 1555). The star itself is surrounded by a faint variable shell, 4″ in diameter, observed by S. W. Burnham[22] in 1890 and by H. D. Curtis[23] in 1899. On Dr. Baade's remarkable photograph of T Tau, this shell appears as a semicircular protuberance between the two diffraction rays on the right-hand side.

The brightness of T Tau changes erratically from magnitude 9.0 to magnitude 12.8. It may vary rather quickly in the course of a few weeks, or the fluctuations may be delayed for months. In the last three years the brightness, at the time of observation, has remained near maximum.

The spectrograms show emission and absorption lines, both of which lack the sharpness needed for accurate measurement. The spectral type is about G5, and the spectroscopic absolute magnitude based on the Mount Wilson system of 1935 is +5.0.

Omitted here is a table giving velocity measurements between 1915 and 1944. The weighted means for the velocities determined from the emission and absorption lines are, respectively, 19.6 and 24.6 km sec^{-1}.

The first known slit-spectrogram of T Tau was obtained in 1915 by F. G. Pease, and many of the interesting features of the spectrum were described by Adams and Pease.[24] On account of the low dispersion used, the published radial velocity and some of the identifications of lines have little weight. Fifteen emission lines were measured, H and K being the most intense. The *Fe* I lines, λ 4,063 (probably blended with λ 4,068 [*S* II]) and λ 4,132, which they did not identify, are stronger

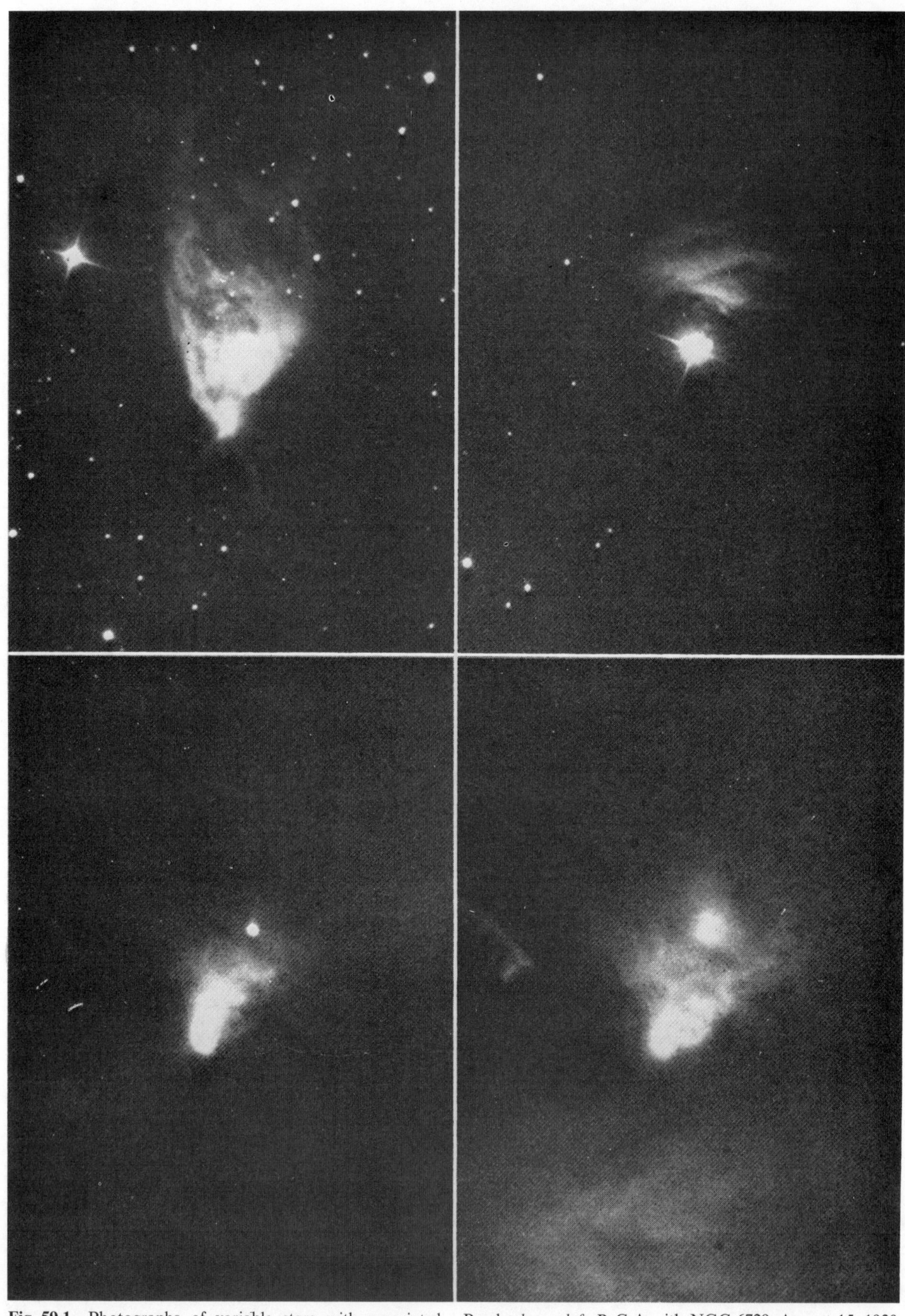

Fig. 59.1 Photographs of variable stars with associated variable nebulae. *Upper left*, R Mon with NGC 2261, January 26, 1920, photographed by E. Hubble; *upper right*, T Tau with NGC 1555, September 30, 1940, photographed by W. Baade; *lower left*, R CrA with NGC 6729, August 15, 1920, and *lower right*, August 17, 1941, photographed by E. Hubble. 100-in reflector. (Courtesy Hale Observatories.)

than the lines of *Fe* II. Lines of *Ti* II are too weak for measurement. The metallic lines are extremely weak on the second plate taken ten months later, and only a few lines of *Fe* II were measured in addition to those of *H* and *Ca* II. Dr. Sanford's low-dispersion spectrogram of 1918 is very similar to that of 1915. Sanford[25] measured 11 bright and 35 absorption lines, which he estimated to be 3–4 A in width. The metallic bright lines measured were without exception the strongest lines of *Fe* II.

The region of the D lines is well known. Sodium is doubtless present in emission in T Tau, although the strength of the lines is not much greater than that of the adjacent continuous background. Bright D_1 and D_2 show, according to Dr. Sanford's measures, approximately normal velocities, but they are accompanied by strong absorption, which is displaced 165 km/sec to the violet, as compared with the mean of the bright lines. Similar violet displacements of absorption adjacent to the strong emission lines of hydrogen and *Ca* II (H and K) appear on plates showing the violet regions of the spectrum. On one plate, the bright lines of *Fe* II are distinctly weak, and the hydrogen lines $H\beta$ and $H\alpha$, only moderately strong as compared with the continuous spectrum.

The infrared region, $\lambda\lambda$ 6,400–8,700, of the spectrum shows that the $3^2D—4^2P^0$ triplet of *Ca* II ($\lambda\lambda$ 8,498, 8,542, 8,662) cannot be present in any great strength either in emission or in absorption.

In general, the *Ca* II (H and K) emission lines of T Tau appear with an intensity unsurpassed by any star of the group; the hydrogen lines are strong, $H\gamma$ being only slightly fainter than *H* (*Ca* II), but the metallic lines are weaker than those of most of the other stars of the group. The metallic bright lines are not prominent features of the spectrum of T Tau on most of our spectrograms.

Although the absorption lines are wide, they are not greatly affected by the emission spectrum, and measures of radial velocity based on a considerable number of lines should give reliable results. The mean velocities of the emission and absorption lines agree closely, but the range in both is more than can reasonably be attributed to errors of measurement. The scatter among the mean velocities of the various elements is comparatively small and probably is not significant.

RY TAURI 041528 Variation of the star brightness was discovered by Leavitt[26] in 1907. The star is at the head of a fan-shaped nebula spreading out toward the dark nebula[27,28] Barnard 214. E. Hubble[29] used the absorption spectrum of the star to specify its spectral type as F8. The four lower members of the hydrogen series and the calcium lines H and K are the only emission lines found in this object. The weighted mean of the velocities obtained from the emission and absorption lines are, respectively, -8 and $+26.2$ km $\sec^{-1}$.

UX TAURI 042418 The variation of the star brightness was discovered by Locke[30] in 1917 and confirmed photographically in the late 1930s.[31,32,33] The absorption spectrum indicates a spectral type of *d*G5. Hydrogen and calcium (H and K) are seen in emission.

UZ TAURI 042625 Light changes of this star were first noticed by Bohlin[34] in 1921 and later confirmed by several others.[35,36,37,38] Two stars of spectral class M are involved.[39,40] Emission lines of hydrogen, calcium (H and K), Fe II, and λ4,068 S II are present, and the relative intensities of the different lines are time variable.

XZ TAURI 042518 The variability of this star was discovered by Schajn[41] in 1928 and confirmed by Esch[42] and Rugemer.[43] At minimum light, 105 bright emission lines were measured. The star lies in the heavy obscuring cloud surrounding UX Tau. The weighted mean velocities obtained from emission and absorption lines are, respectively, 32 and 109 km $\sec^{-1}$.

General Discussion of the Spectra

Only three of the T Tauri stars (T Tau, RY Tau, and UX Tau) have absorption spectra sufficiently well defined to permit a precise determination of the spectral type. For RW Aur, UY Aur, and R CrA, rough estimates may be made with the aid of such lines as are available. A variation of three- or four-tenths of a class was found for T Tau and RY Tau. The remaining stars, in which the bright-line spectrum is dominant, have few absorption lines; but, from the distribution of light and the general appearance of the spectrum, we conclude that the physical conditions correspond to those of about type G5. For some stars of the group, selective absorption by surrounding nebulosity may introduce uncertainty. Peculiar features of the spectra such as are mentioned in the preceding descriptions of the individual stars are often so prominent that the usual criteria of classification cannot be employed.

Here Joy arranges the stars according to the increasing prominence of the emission spectrum. The order is UX Tau, RY Tau, R CrA, T Tau, R Mon, UY Aur, S CrA, RW Aur, XZ Tau, RU

Lup, and UZ Tau. We omit extensive tables of the emission lines of each object.

In addition to *H* and *He*, 17 different metallic atoms are represented by more than 160 lines which arise from low levels of excitation. Iron contributes about half the total number of lines. Both neutral and ionized atoms are represented, and a trace of the strongest forbidden lines seems reasonably certain. A large number of additional lines would certainly be found by the use of higher dispersion and by an extension of the search to other spectral regions of shorter or longer wave length.

The line λ 4,068.62 [S II] is remarkably persistent in the stars with conspicuous emission spectra. Its changes in intensity do not appear to be correlated with the variations of the other bright lines. At times λ 4,068 remains practically undiminished when most of the metallic lines have disappeared. The measures support the identification with [S II]. The companion line at λ 4,076.22 is probably blended with λ 4,077.71 *Sr* II and has not been separately observed. These sulphur lines arise from a low metastable level and are frequently found in novae and occasionally in nebulae. None of the other characteristic nebular lines have been observed in these stars.

DISTORTED MULTIPLETS OF *Fe* I When the emission spectrum appears in the T Tauri stars, λ 4,063 and λ 4,132 of *Fe* I are much strengthened as compared with the other members of the multiplet a^3F—y^3F^0 and are probably affected by some such fluorescent mechanism as that proposed for λ 4,202 and λ 4,307 of the $a^3F - z^3G^0$ multiplet in Me variable stars.[44] Under the most favorable conditions, λ 4,063 may rival λ 4,233 *Fe* II in strength. This phenomenon is unique in the stars of this group and does not occur in the solar chromosphere. The two lines λ 4,063 and λ 4,132, together with λ 3,969, arise from the same upper level of the atom. The latter line has not been observed in emission, possibly because its identity is lost in the complex structure of the region between *H* and *H*ε. All the remaining members of the multiplet have been observed as bright lines of relatively low intensity in the T Tauri stars.[45]

COMPARISON WITH THE SOLAR CHROMOSPHERE The most significant characteristic of the spectra of the T Tauri stars is the low-excitation bright-line spectrum which is prominent in most of the stars at certain phases. It occurs with varying intensity in all the stars of the group except RY Tau and UX Tau, which may be considered as rudimentary members of the T Tauri class as far as the spectrum is concerned, showing only *H*, *K*, *H*α, and *H*β in emission.

This bright-line spectrum is quite different from other stellar emission spectra, but its resemblance to that of the solar chromosphere is sufficient to invite a detailed comparison.

Table 59.4 Comparison of emission lines in T Tauri stars and solar chromosphere

	Chromosphere			T Tau stars	
Element	No. lines	Mean intensity	Mean height (km)	No. lines	Mean max. intensity
Ca II	2	200	14,000	2	50
H	6	160	8,900	6	40[a]
Sr II	2	70	6,000	2	4
Ca I	1	40	5,000	1	5
Mg I	6	46	4,170	6	1
He I	11	20	3,840	3	2
He II	1	2	3,500	1	1
Ti II	21	29	2,410	19	2
Sc II	4	24	2,220	4	2
Al I	2	30	2,000	2	2
Fe II	7	25	1,700	7	12
Ba II	4	29	1,550	0	—
Cr I	3	21	1,500	3	2
Na I	2	22	1,500	2	1
Cr II	1	15	1,500	1	2
Fe I	25	13	1,470	21	3

a. Mean of 4 lines (*H*α–*H*δ) only.

The mean heights (>1,000 km) and intensities[46] of chromospheric lines between λ 3,750 and λ 6,600 are arranged in table 59.4 according to height. The numbers of these same chromospheric lines found in the T Tauri stars and their estimated mean maximum intensities (on quite a different scale) are shown in the last two columns of the table. A marked similarity between the emission spectra of the stars and that of the higher levels of the chromosphere is apparent. The bright lines of calcium (H and K) and of hydrogen are outstanding in both. In the stars, as well as in the sun, *Fe* I and *Ti* II have the greatest number of lines of high intensity. The low-excitation *Fe* I lines are more generally represented in emission in these stars than in any other known stellar source, and their maximum strength is relatively greater than in the chromosphere. The *He* I lines are few and decidedly weak in the stars. Only the triplet members, λλ 4,206, 4,471, and 5,875, are certainly present; λ 4,685 *He* II was measured in the spectra of three stars, but it is faint; and *Mg* I and *Sc* II are weaker than in the sun. On the other hand, *Fe* II is relatively stronger, and its lines are surpassed in strength only by those of *H* and *Ca* II. The lines λ 4,077 and λ 4,215 of *Sr* II are well shown but do not have the excessive intensity found in the sun.

The stellar emission lines are much broader than those of the chromosphere, and the strong lines of *H* and *Ca* II often suffer distortion by deep reversals which are not observed in the sun, yet the general resemblance of the spectra leads to the conclusion that they must have their origin under similar physical circumstances. The excitation seems to be somewhat lower in the T Tauri stars than in the solar chromosphere, while in the stars with combination spectra, such as Z And, the excitation is higher.[47]

Absolute Magnitudes and Color Indices

The spectroscopic absolute magnitudes of T Tau, RY Tau, and UX Tau may be estimated from their absorption lines. The results indicate that these three stars are dwarfs of the main sequence. For UX Tau and UZ Tau additional evidence may be sought from their visual companions, assuming that they are physically related. The spectral types of both companions are dM3e, and the absolute magnitudes are estimated to be +10.5. If we use 13.3 as the estimated apparent visual magnitude of both companions, $m - M = +2.8$. Applying this modulus to the visual magnitudes (9.9 and 9.2) at maximum of the two variable stars, we obtain values of 7.1 and 6.4, respectively, for their absolute magnitudes, 1.9 and 1.2 mag. fainter than that determined from the spectral lines of UX Tau. On the basis of these data, the variables must be subdwarfs—an inference which does not seem justified. The matter can be harmonized only by concluding that the companions have peculiar spectra, that the stars are not physical pairs, or that general or local obscuration has a greater effect in reducing the visual luminosity of the variable star than that of its fainter and probably redder companion. Both stars are in lanes of heavy obscuration, but no surrounding shells have been observed.

Here we omit a table showing that the spectroscopic absolute magnitudes and spectral types of T Tau, RY Tau, and UX Tau are, respectively, 5.0 and *d*G5, 4.3 and *d*G0, and 5.2 and *d*G5. Although values of absolute magnitude and distance are made uncertain by the presence of absorbing clouds, the available evidence indicates that the T Tauri stars belong to the main sequence and that their distance is of the order of 100–150 pc.[48,49,50]

Radial-Velocity Measures

The measures of radial velocity from both emission and absorption lines show that in the mean the emission lines are displaced toward the violet with respect to the absorption lines. This result agrees in direction with that found for the irregular Me variables[51] and the Mira stars,[52] but the measured displacement is somewhat greater for the T Tauri stars.

Lack of agreement in the measures of different plates of the same star indicates that the radial velocities may be variable, with a range of 30–40 km/sec. It seems probable that both the emission and the absorption strata are subject to irregular motions but that, in general, the emission layers are rising with respect to the absorbing clouds.

Concluding Remarks

Only 11 stars have been found which have the qualifications necessary for inclusion in the T Tauri group. The spectral criteria were given high weight. Probably other stars which have similar characteristics exist and will be recognized when competent objective-prism observations can be extended to the faint stars of the Milky Way. However, stars of this class may be expected to be comparatively few because of the peculiar circumstances of their location and their relationship to the galactic absorbing clouds.

Solely on the basis of their light-curves, RR Tau, WW Vul, and perhaps some of the irregular variables in Orion might have been included, but their spectra are of earlier type (A3–F8) and without extensive emission. Other variables, such as Z And and CI Cyg, are similar in some respects but of higher luminosity. Their emitting atmospheres have bright *Ca* II[53] and some faint *Fe* I[54] lines, although, in general, their spectra show lines requiring much higher excitation than that of the T Tauri stars. Their numerous sharp forbidden lines, doubtless, originate in an extended tenuous atmosphere which favors high ionization, while, on the contrary, the presence, in general, of lines of neutral and easily ionized atoms in the T Tauri stars points to a source of considerably lower excitation and higher density. Also, the high-luminosity variables with combination spectra showing strong 4,686 *He* II lines are not associated with dark lanes of the Milky Way or with reflection nebulae.

The changes which occur from time to time in the spectra of the T Tauri variables merit more detailed study than can be given in this general survey. They should be definitely correlated with the variation in brightness of the stars. As a result of the higher density of the gases and the smaller dimensions involved, the changes are quite different in nature from those observed in other variable stars. With further observations, the effect of near-by companion stars or of surrounding shells may be investigated. Maximum brightness, which, for some of the stars of the group, bears a resemblance to the outburst of a nova, has not been observed spectrographically. Such extreme activity might furnish a valuable clew to the interpretation of the problem.

1. A. H. Joy and G. van Biesbroeck, *Pub. A.S.P. 56*, 123 (1944).

2. *A.N. 175*, 205 (1907).

3. *A.N. 195*, 456 (1913).
4. *Kleinere Veröff. Sternwarte Berlin-Babelsberg*, no. 21 (1939).
5. *A.N. 170*, 339 (1906).
6. *A.N. 193*, 439 (1913).
7. A. H. Joy, *Pub. A.S.P. 44*, 385 (1932).
8. *Union Obs. Circ.*, no. 36, p. 282 (1916).
9. *Bull. Helwan Obs. 1*, 141 (1920).
10. S. Gaposchkin, *Harvard Ann. 105*, 514 (1937).
11. *Bull. Lowell Obs. 3*, 66 (1918).
12. *Mt. W. Contr.*, no. 241; *Ap. J. 56*, 181 (1922).
13. *Harvard Circ.*, no. 196 (1916).
14. *Harvard Circ.*, no. 201; *A.N. 207*, 215 (1918).
15. *Pub. A.S.P. 53*, 342 (1941).
16. *A.N. 55*, 91 (1861).
17. *Proc. Nat. Acad. 2*, 230 (1916); *Ap. J. 44*, 190 (1916), *45*, 351 (1917).
18. *Bull. Lowell Obs. 3*, 63 (1918).
19. Adams, Joy, and Humason, *Pub. A.S.P. 39*, 366 (1927).
20. *Pub. A.S.P. 54*, 190 (1942).
21. *Mt. W. Contr.*, no. 237 (1922).
22. *M.N. 51*, 94 (1890).
23. *Pub. A.S.P. 27*, 242 (1915).
24. *Pub. A.S.P. 27*, 132 (1915).
25. *Pub. A.S.P. 32*, 59 (1920).
26. *Harvard Circ.*, no. 130 (1907).
27. *Atlas of the Milky Way*, pl. 5 (1927).
28. *Ap. J. 86*, 543 (1937).
29. *Mt. W. Contr.*, no. 241; *Ap. J. 56*, 182 (1922).
30. *Harvard Circ.*, no. 201 (1918).
31. *Nishni-Novgorod, Veränderliche Sterne 4*, 154 (1933).
32. *A.N. 255*, 180 (1935).
33. *Valkenburg Obs. Veröff.*, no. 7, p. 13 (1937).
34. *A.N. Beob.-Zirk. 3*, no. 32, 55 (1921).
35. *A.N. Beob.-Zirk. 6*, no. 2, 5 (1924).
36. *Nishni-Novgorod, Veränderliche Sterne 4*, 228 (1934).
37. *A.N. 270*, 185 (1940).
38. *Pub. A.S.P. 54*, 33 (1942).
39. A. H. Joy, *Pub. A.S.P. 54*, 33 (1942).
40. A. H. Joy, *Pub. A.S.P. 55*, 38 (1943).
41. *A.N. 234*, 41 (1928).
42. *A.N. Beob.-Zirk. 11*, no. 4, 10, no. 9, 23 (1929).
43. *A.N. 255*, 180 (1935).
44. Thackeray and Merrill, *Pub. A.S.P. 48*, 331 (1936).
45. Since this paragraph was written, George H. Herbig (*Pub. A.S.P. 57*, 166 [1945]) has reported his observations of this phenomenon in the spectrum of RW Aur and attributed the enhancement of λ 4,063 and λ 4,132 to populating the y^3F level through the action of λ 3,968.47 *Ca* II upon λ 3,969.26. The width of the calcium line in the T Tauri stars is more than sufficient to cover the difference in wave length.
46. S. A. Mitchell, *Ap. J. 71*, 11 (1930).
47. H. H. Plaskett, *Pub. Dom. Ap. Obs. 4*, 152 (1928).
48. *Mount Wilson Obs., Annual Reports*, p. 252 (1921).
49. *Harvard Bull.*, no. 904 (1936).
50. *Mt. W. Contr.*, no. 237, p. 14, (1922).
51. A. H. Joy, *Mt. W. Contr.*, no.668; *Ap. J. 96*, 165 (1942).
52. P. W. Merrill, *Mt. W. Contr.*, no. 644; *Ap. J. 93*, 380 (1941).
53. P. W. Merrill, *Mt. W. Contr.*, no. 688; *Ap. J. 99*, 15 (1944).
54. Swings and Struve, *Ap. J. 97*, 206 (1943).

60. Binary Stars among Cataclysmic Variables III: Ten Old Novae

Robert P. Kraft

(*Astrophysical Journal 139*, 457–472 [1964])

That "novae" are neither new stars nor temporary was established by the systematic photographic surveys of the heavens begun toward the end of the last century. By the middle of our own century a nova was known to be a star that abruptly increases in brightness by as much as 100,000 times while ejecting about a thousandth of a solar mass in a shell that expands outwards at speeds of several hundred kilometers per second. At the same time, theoretical explanations for the cause of nova outbursts centered on descriptions of the structural properties and thermonuclear processes characteristic of stars occupying the nova region of the Hertzsprung-Russell diagram.[1]

A major new approach to the understanding of novae came when Merle F. Walker and Alfred H. Joy began to examine the total light and spectra of ex-novae, after the intense light of the nova outburst had faded to a relatively weak level. Twenty years after the 1934 eruption of Nova DQ Herculis, for example, Walker used photoelectric photometry to show that this nova is actually an eclipsing binary system with the remarkably short period of 4.6 hr.[2] As suggested by Walter Baade and noted by Walker in his 1954 paper, the shortness of the period of the DQ Herculis system indicates that the two stars are very close together, and if one of them has passed through the red giant phase the companion star would have had to revolve deep within the expanded atmosphere of the nova component. Even more startling conclusions came from Joy's studies of the absorption and emission lines of the dwarf nova SS Cygni.[3] Although these studies were begun in 1943, not until 1956 could Joy show that this explosive star is a spectroscopic binary with the short period of 6 hr 38 min. The emission lines originate from a blue star similar to those that seem to become novae, whereas the absorption lines come from a red star about the same size as the distance to the center of mass of the two-star system. In other words, mass could spill from the red star to the blue star, and the nova process might be related to this mass flow.

At the time that Joy and Walker made their discoveries, many astronomers were beginning to change their opinions about the conditions necessary for a star to become a nova, but it was Robert Kraft who played the leading role in showing that membership in a binary system is a necessary condition for a star to become a nova. By 1962 Kraft had shown that five dwarf novae of the U Geminorum type are binaries with periods of less than 9 hr, and two years later he presented this paper, which shows that many ex-novae are also short period binary systems. The short periods mean that the two stars are very close together; as Kraft shows, one of the component stars is usually a blue white dwarf star and the other is usually a red star of about the same mass.

Apparently, as the large, cool red star evolves, it expands into a region where the gravity of the small hot white dwarf star predominates. As a result some of the hydrogen-rich material of the red star flows onto the white dwarf star. Sumner Starrfield, Warren Sparks, and James Truran have used theoretical models to show that this hydrogen-rich envelope is gradually compressed and heated up until thermonuclear reactions occur. A runaway thermonuclear explosion then occurs, and the envelope of the white dwarf star is thrown off, producing a nova outburst.[4]

1. C. Payne-Gaposhkin, *The Galactic Novae* (Amsterdam: North Holland Publishing Co. 1957).
2. M. F. Walker, *Publications of the Astronomical Society of the Pacific 66*, 230 (1954).
3. A. H. Joy, *Publications of the Astronomical Society of the Pacific 55*, 283 (1943); *Astrophysical Journal 124*, 317 (1956).
4. S. Starrfield, W. M. Sparks, and J. W. Truran, *Astrophysical Journal Supplement 28*, 247 (1974), *Astrophysical Journal 192*, 647 (1974), *176*, 169 (1972).

Abstract—Some of the spectroscopic and photometric properties of ten old novae of the spring, summer, and early autumn sky are considered. To the four binaries previously known (T CrB, DQ Her, T Aur, WZ Sge) are added GK Per (1901), V603 Aql (1918), and V1017 Sgr (1901, 1919). Orbits are given for the first two of these three; the last is judged a binary from its composite spectrum. Two other novae, which have unusually narrow emission lines and show no certain velocity variations at the dispersions employed, are shown very probably to be pole-on objects.

Since seven of the ten objects studied are certainly binaries, the hypothesis is advanced that membership in a certain type of close-binary system is a necessary condition for a star to become a nova. A binary model similar to that advanced earlier to explain the U Gem systems (Kraft 1962) is invoked for novae; some difficulties with its application to the case of WZ Sge are cited.

A relationship, having a width of about ± 1.5 mag., is shown to exist between the absolute magnitude of the red component (at minimum light) and the orbital period, in the sense that the brighter is the minimal magnitude of the red star, the longer is the period. With decreasing period comes increasing faintness of the red star relative to the blue. There is a tendency for the masses of novae to increase with increasing period and higher luminosity.

The masses so far found for novae range from $\sim 0.1\ \mathfrak{M}_\odot$ to $\sim 3\ \mathfrak{M}_\odot$, and the luminosities from $\sim +10$ or fainter to ~ -1. Any universal mechanism advanced to explain the outbursts of novae cannot depend strongly on mass or position in the HR diagram.

It is concluded that novae can belong to any population of stars, provided that population contains some close binaries of the appropriate kind. Most of them will therefore be found in stellar populations containing many binaries, such as the galactic disk, but their occasional presence in Type II systems is distinctly possible. It is suggested that the frequency of novae per unit mass in a stellar system is a measure of angular momentum within that system.

I. Introduction

SINCE THE TIME of Merle Walker's (1954) remarkable discovery that Nova DQ Her (1934) is a binary star, considerable speculation has developed as to whether all novae might in fact be double (cf., e.g., Schatzman 1958; Kopal 1959). The period, spectrum, and physical properties of Nova DQ Her have been found (Greenstein and Kraft 1959; Kraft 1959) not to differ greatly from those of a typical dwarf nova (U Gem star). In the first paper of this series (Kraft 1962; hereinafter referred to as "Paper I"), it was demonstrated that most of the brighter U Gem variables are double, and the presumption is strong that all, in fact, are close binaries of rather short period. The present paper describes an attempt to detect binary motion among several of the brightest old novae of the spring, summer, and early autumn sky, making use of the spectroscopic facilities of the 200-inch reflector.

Because the period of DQ Her is only 4^h39^m, the problem of time resolution was thought to be of critical importance. A limit of 40 minutes was set on all exposures; using the nebular (prime-focus) spectrograph, one is able to reach magnitude 15.5 pg in this time with the following specifications: baked Eastman II*a*-O emulsions, dispersion 180 Å/mm, widening 1/3 mm at the plate. Subject to these limitations, nineteen old novae can be reached effectively from Palomar. Three of these are bright enough that coudé exposures with both the 100- and 200-inch telescopes are possible. Altogether, ten old novae will be discussed in some detail here. Since some conclusions can now be drawn from a consideration of the observational data derived from all ten stars, a very brief review of some previously obtained results appears also to be in order.

Here we omit representative spectra of old novae and the dwarf nova U Gem.

The emission-line spectra of these stars differ from those of U Gem variables in two important respects, in spite of a certain superficial similarity. First, there is a larger variety among old novae—they do not form so compact a spectroscopic group as U Gem stars (Kraft 1962.) Even if we allow for those emission lines of old novae resulting from the pure nebular material of the most recent outburst, we are still left with a considerable variety. Some, such as DQ Her and GK Per, have strong emission lines; DI Lac has shallow absorption lines with weak, narrow emission; HR Lyr is nearly continuous. (Further details can be found in a paper by Greenstein [1960] and in the early work of Humason [1938].) Some novae, for example, GK Per, show, as do certain U Gem stars, a composite spectrum with the additional absorption lines of a late-type star; individual cases will be discussed in detail in Section II.

The other important difference between the spectra of the two kinds of stars is the higher excitation of the emission lines of most old novae. Almost all novae show He II (λ 4,686), often strong, in emission; U Gem stars rarely show it, there being a faint trace, however, in U Gem itself (Kraft 1962). Presumably the stars responsible for the production of emission lines in most old novae are hotter than their counterparts in U Gem-type systems.

II. The Physical Characteristics of Some Old Novae at Minimum Light

GK PER (1901) The spectrum is composite, consisting of a blue continuum with emission lines of H, He I, He II, [O II], etc., and the absorption lines of a late-type star. Palomar prime-focus spectrograms yield the orbit of a double-line

Table 60.1 Basic physical parameters for novae

Star	Type	Spectrum	M_V(red)[a] (minimal)	Period	Eclipses	γ (km/sec)	a_{red} $\times 10^{-10}$ cm	a_{bl} $\times 10^{-10}$ cm	$\mathfrak{M}_{red}$ ($\odot$)	$\mathfrak{M}_{bl}$ ($\odot$)	References
T CrB	RN	sdBe + gM3 + Q	+ 0.2	$227^d.6$	Grazing?	−27	$\geqslant$810	$\geqslant$1130	$\geqslant$3.7	$\geqslant$2.6	Kraft 1958
GK Per	N	sdBe + K21Vp + Q	+ 4.5	$1^d.904$	No?	+26	$\geqslant$ 38	$\geqslant$ 17	$\geqslant$0.56	$\geqslant$1.29	This paper
T Aur	N	unknown	$\geqslant$ + 5.0	$4^h\ 54^m$	Yes	—	—	—	—	—	Walker 1962
DQ Her	N	sdBe + Q	$\geqslant$ + 9	$4^h\ 39^m$	Yes	−20	~2.4	~4.0	~0.20	~0.12	Greenstein & Kraft 1959 and this paper
V603 Aql	N	sdBe	$\geqslant$ + 5	$3^h\ 19^m.5$	Prob. no	−23	—	$\geqslant$0.72	—	—	This paper
WZ Sge	RN	sdBe	$\gtrsim$ +11	$1^h\ 22^m$	Yes	−30	?	?	?	?	Krzeminski 1962 and this paper
V841 Oph	N	sdBe	—	—	Prob. nearly pole-on	~ −96	—	—	—	—	This paper
DI Lac	N	sdBe	$\geqslant$ + 3.2	—	Prob. nearly pole-on	+ 1	—	—	—	—	This paper

a. From application of life-luminosity relation and appearance of minimal spectrum. For WZ Sge, to which the relation is inapplicable, we used the value of Greenstein (1957). The lifetime of V841 Oph (1848) is not known with precision.

spectroscopic binary; the leading physical elements are listed in table 60.1. The velocity-curves are shown in figure 60.1. The γ-velocity listed in table 60.1 is that of the red star—for which fairly precise measurements of radial velocity are possible; the γ-velocity for the blue star appears to be about 15 km/sec more positive, but it must be admitted that the diffuseness of the emission lines makes the velocity measurements very difficult. Only H9 and He II (λ 4,686) are used to form the velocity curve for the blue star. There is some evidence that the lower Balmer members are contaminated by emission from the nebular shell of the 1901 outburst, now lying principally southeast of the nova. The star is not known to eclipse (Mumford 1964) and the actual masses of the components may be considerably larger than those listed in table 60.1. The high orbital eccentricity of 0.4 is matched only by the orbit of the dwarf nova RX And (Kraft 1962) among cataclysmic variables.

The spectrum of the late-type star can be classified on the MK system from the 90 Å/mm spectrograms. It is not normal: the strength of λ 4,077 of Sr II varies erratically from plate to plate and the star changes from Class V to Class III. The G-band and λ 4,266 of Ca I are consistent with spectral type K0 to K5, but the ratio λ 4,254/λ 4,260 looks more like G8. We adopt K2 IVp as a rough average type.

T AUR (1891) No spectrograms have been obtained, but Walker (1962) has found the star to be an eclipsing binary of period 4^h54^m. The star is currently considerably fainter than the $V = 14.9$ reported by Walker in 1957; spectroscopic observations are planned.

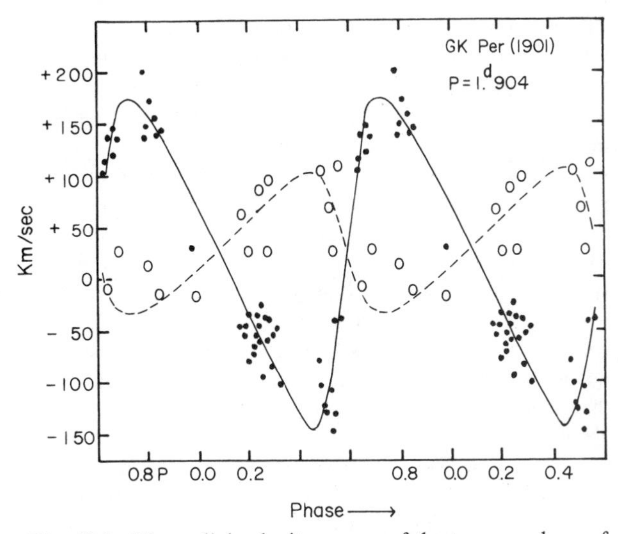

Fig. 60.1 The radial velocity curves of the two members of the binary system that gave rise to the nova GK Per (1901). The absorption lines of a red star of spectral type K are denoted by black dots; the open circles represent the means of two or more velocities from the emission lines H9 and He II (λ4,686) of the blue star.

T CRB (1866, 1946) Sanford (1949) examined the spectrum of recurrent nova T CrB after its 1946 outburst. It is the brightest of all old novae at minimum light (~10th to 11th mag.), and can therefore be reached in long coudé exposures with the 100-inch. The spectrum had been recognized as composite, that is, gM3 plus bright lines of H, He I, He II, O III, [O III], etc., for some years prior to the 1946 outburst. Sanford showed that the gM3-type spectrum varied in velocity with a period of 230 days and with K = 21 km/sec. No variation of the emission-line velocities was found, but, at the time of Sanford's study, the emission was hopelessly confused with the lines produced by the expanding nebular shell of the 1946 outburst.

By 1956–57, the writer was able to detect the motion of Hα and Hβ in emission. The velocity-curves are shown elsewhere (Kraft 1958); the improved period is 227.6 days. Principal elements are listed in table 60.1. The two leading conclusions are these:

1. The minimum masses of the blue and red components are large, about 2.1 $\mathfrak{M}_\odot$ and 2.9 $\mathfrak{M}_\odot$, respectively. No reasonable error in the velocity-curve of the blue star can change the conclusion that its mass is quite large.
2. The spectrum of the red component is completely normal. Unless i is very small, the dimensions of the orbit of the red star, viz., $a \sin i = 7.5 \times 10^{12}$ cm, are the same as its own radius, about 9×10^{12} cm. Actually, there is some evidence from spectrograms obtained by A. J. Deutsch (private communication) that a grazing eclipse occurs similar to that of ε Aur If the red star really fills its lobe of the critical zero-velocity surface defined by the restricted problem of three bodies, then $i \cong 68°$, and the masses for the blue and red components are 2.6 $\mathfrak{M}_\odot$ and 3.7 $\mathfrak{M}_\odot$, respectively.

V841 OPH (1848) The emission lines are quite narrow (Greenstein 1960) and feeble relative to the continuum. Radial velocities have been obtained from plates of 180 Å/mm dispersion taken by Greenstein and by the writer; no significant variation of velocity has been found, though two spectra were obtained in immediate succession separated by 29 minutes. The narrowness of the emission lines may well be related to the failure to detect a velocity variation (see Sec. III). An attempt to photograph the spectrum with the 200-inch coudé spectrograph at 38 Å/mm, with sufficient time-resolution, failed.

RS OPH (1898, 1933, 1958) The present state of our knowledge of the spectrum of this object at or near minimum light has been reviewed by Wallerstein (1963), to whose paper the reader is referred for details. While an absorption spectrum is present in addition to a bewildering variety of emission lines, the former appears to be that of a shell. It is not possible to argue from the spectrum alone that the object is a binary similar to T CrB, which is also recurrent. If, however, the object were a double, dynamically similar to T CrB, a small velocity range would be expected, and it would not be surprising to find the orbital velocity variations lost in the erratic photometric and spectroscopic activity characteristic of this star (see also Sec. IV).

DQ HER (1934) The photometric (Walker 1954, 1956, 1957, 1958, 1961) and spectroscopic (Greenstein and Kraft 1959; Kraft 1959) data are quite difficult to interpret, and there are many puzzling details. The leading observational features and their interpretation are these:

1. The object is a single-line binary. He II (λ 4,686) and the higher members of the Balmer series show the velocity variation; the lower Balmer lines and lines of [O II] and [S II] are stationary. Thus there are two regions producing emission lines—the expanding, low-density nebular shell and a higher-density ring, disk, or shell being carried with the blue star in its orbit. The latter has a slower decrement than the former, and the two "cross over" at about Hγ.
2. No trace of the spectrum of the secondary has been found. All mass estimates are therefore based on indirect arguments.
3. Superimposed on the photometric activity ("flickering") outside eclipse is a strictly periodic oscillation having $P = 71$ sec. and a yellow amplitude of 0.05 mag. This oscillation disappears during eclipse.
4. The He II velocity-curve shows a classical rotational disturbance that begins and ends with the eclipse. Thus the eclipse is not an ordinary stellar eclipse at all, but rather that of a semi-transparent ring or disk surrounding the hot star. This ring fluoresces ultraviolet quanta of the hot star, and free-free and free-bound hydrogen emission are responsible for most of the blue and visual light of the system.

The masses of the components can be estimated in two independent ways. Both the expansion parallax and the life-luminosity relation indicate that the hot star is a rather bright white dwarf with $M_V = \sim +8.5$. If, following Walker, we interpret the 71-sec. oscillations as a pulsation, we can get the mass because the $P\sqrt{\bar{\rho}} = Q$ relation gives us one condition on the mass and radius, and the mass-radius relation for white dwarfs gives another. According to Schatzman (1961), however, the calculations given by M. Sauvenier-Goffin (1949) of the pulsational properties of the models are incorrect; these were used earlier (Kraft 1959). The corrected calculations show that the object is very nearly a polytrope of index 3/2 (non-relativistic degeneracy), and we obtain $\mathfrak{M} =$ 0.12 $\mathfrak{M}_\odot$; the mass of the red star from the orbit as then 0.20 $\mathfrak{M}_\odot$.

A second estimate of the mass depends on an application of Jacobi's integral and will be discussed in some detail in Section III.

V1017 SGR (1901, 1919) A spectrogram of 180 Å/mm dispersion obtained by Greenstein shows the well-developed absorption spectrum of a late-type star with superimposed feeble, wide emission at Hβ, Hγ and Hδ, and with He II (λ 4,686) marginally present. The great strength of the absorption spectrum relative to the blue continuum is matched only by the recent spectroscopic appearance of T CrB. A set of three 90 Å/mm spectrograms obtained by the writer shows the emission lines almost absent; the spectral type from λ 4,226 and λ 4,383 looks later than that derived from the G-band We suggest a type of G5 IIIp.

V603 AQL (1918) Spectrograms of this object are omitted, as is the radial velocity curve. The period obtained from hydrogen emission lines is 3^h 19.5^m, and no absorption spectrum has been found in the photographic region.

WZ SGE (1913, 1946) Greenstein (1957) pointed out the existence of absorption lines characteristic of a white dwarf and the double hydrogen emission components. With continuous trailing over a long slit, the writer (1961) found an S-wave superimposed on the double emission; this S-wave has a period of about 80 minutes and $2K = 1{,}300$ km/sec. There are now nine spectrograms taken by Greenstein and the writer, all showing the wave. Krzeminski (1962) found that WZ Sge is also an eclipsing binary of period $81\frac{1}{2}$ minutes; the light-curve is of the W UMa type and exhibits erratic photometric activity.

The absolute magnitude of WZ Sge is probably quite low, perhaps $+10$ or fainter (Greenstein 1957). The mean radial velocity of the S-wave is -30 km/sec; the proper motion is well determined at 0.″080 ± 0.″004 (van Maanen 1926). If the star, for example, were as bright as $+7$, its tangential velocity would be over 200 km/sec, a value quite inconsistent both with the known velocities of white dwarfs and other old novae. Actually, WZ Sge is probably closer to a U Gem star than a nova. The small outburst range (7 to 8 mag.), faint absolute magnitude, and absence of He II in the spectrum all point to a close relationship with the U Gem group.

At present it is not possible to give a satisfactory model for this star. Two examples of its anomalous behavior will be cited. From the S-wave one can get an estimate of the masses of the stars for any assumed mass ratio (Kraft, Mathews, and Greenstein 1962). The white-dwarf absorption lines do not appear to move, but detection of their motion would be exceedingly difficult. The smallest mass for the white dwarf would be obtained if we assume that the S-wave comes from a relatively massless rotating ring with a "lump" in it. This gives $\mathfrak{M} = 1.62\ \mathfrak{M}_\odot$. A mass ratio of unity, at the other extreme, gives $6.5\ \mathfrak{M}_\odot$ for both stars. All these values seem unreasonable because they exceed the theoretical $1.2\ \mathfrak{M}_\odot$-limit for white dwarfs. If we try to avoid the difficulty by assuming the star's atmosphere has a high pressure but does not yet surround a completely electron-degenerate core, we shall have a brighter star and, therefore, trouble with the space motion.

It is of interest to apply in this connection a recent suggestion by Burbidge (1963). Suppose the "white dwarf" is really a "neutron configuration"—larger observable masses, up to $2\ \mathfrak{M}_\odot$, are then possible (corresponding to a "proper" mass of $3\ \mathfrak{M}_\odot$) (Cameron 1959). For a primary mass of (say) $2\ \mathfrak{M}_\odot$, the secondary would have a mass of $0.22\ \mathfrak{M}_\odot$; thus the largest value of the observed mass ratio would be about 0.11. Since the light-curve is of the W UMa type, an extensive envelope containing the residual internal energy would surround the neutron core, the dimensions of which are only about 8 km, and, in addition, the envelope would have a radius comparable with that of the radius of the less massive companion. Significant photometric ellipticity would also be needed. The envelope radius required is therefore about 5×10^9 cm, if the secondary fills its lobe in the inner zero-velocity surface, that is, $R_{\text{env}} \sim 10^4\ R_{\text{neutron core}}$. In that case, the gravitational redshift of the spectral lines (in absorption) would be negligible, in harmony with observation. The interpretation would lead to the radiation of gravitational energy (Kraft *et al.* 1962) at a prodigious rate. However, the possibility that the upper mass limit for neutron stars is larger than that for white dwarfs has been recently criticized by Saakyan (1963), and the entire matter remains problematical.

Even if the preceding speculations were correct, the interpretation would founder on a second complication. Simultaneous photometry and spectroscopy have been carried out by Krzeminski at Lowell and the writer at Palomar. The phasing shows that the light- and velocity-curves are almost 90° out of phase, i.e., primary eclipse occurs at maximum velocity of recession. This must mean that the mass of gas responsible for the S-wave is at some detached point in the system not directly associated with either star. The writer has attempted to interpret this observation, but so far without success.

DI LAC (1910) Superimposed on rather wide, shallow absorption lines (total width $\sim$25 Å for Hγ) of hydrogen are narrow, centrally placed emission components. The absorption lines are not so wide as those of WZ Sge (total width $\sim$90 Å for Hγ); the emission components, with half-half widths of 2.3 Å, are the narrowest so far found by the writer in any old nova (cf. also Sec. III). An apparent small variation in radial velocity with $K \sim 50$ km/sec found on 180 Å/mm spectrograms taken in the period August 7–9, 1962 (U.T.), was not confirmed from a run of 90 Å/mm plates taken in the period July 25–27, 1963 (U.T.). The object probably has no real radial velocity variation that can be measured with the low dispersions employed.

III. A MODEL FOR OLD NOVAE

Of the ten stars considered here, six have definitely been shown to be binaries (GK Per, T Aur, T CrB, DQ Her, V603 Aql, WZ Sge) and one is definitely composite (V1017 Sgr). Thus seven of the nineteen old novae, or 37 per cent, that can be reached from Palomar (subject to the restrictions set forth in Sec. I) have been shown to be binaries, or 70 per cent of all that have been extensively studied (the ten considered here). At our lowest dispersion, a limit of $K \sim 75$ km/sec $\cong 5\mu$ must be set, below which binary motion would probably go undetected. However, because of the large range of possible masses and periods associated with the novae so far demonstrated to be binaries, it is not possible to state how many "pole-on" binaries may go undetected. In some other cases, such as HR Lyr and Q Cyg, the spectra are so nearly continuous that detection of radial velocity variations would be very difficult. At the present time, a useful working hypothesis is the statement that a *necessary condition for a star to become a nova is membership in a binary system* of a kind presently to be specified in more detail.

In studies of T CrB (Kraft 1958) and DQ Her (Kraft 1959), as well as the U Gem stars (Kraft 1962), evidence has been presented in support of the view that one component of the system is a red star which overflows its lobe of the inner zero-velocity surface defined by the restricted problem of three bodies. Gaseous material flows from the red star toward the blue through the inner Lagrangian point, L_1, and takes up an orbit surrounding the blue star. This material, in the form of a ring or a disk, is responsible for the emission lines observed to be carried with the blue star in its orbit. Beyond the difficulties already cited concerning the interpretation of WZ Sge, there is nothing in the observations of GK Per, V603 Aql, or V1017 Sgr that is contrary to this picture; we will assume it is applicable to all old novae in what follows.

Two observational tests of the nova model are considered. The assumption of conservation of angular momentum for material falling onto the blue star (Kuiper, 1941) leads to an expression relating orbital elements; under certain assumptions this expression is consistent with observed parameters for U Gem, Z Cam, and DQ Her. A second test for internal consistency shows that narrow emission lines will be associated with systems seen at high inclination to the line of sight. Objects with narrow emission lines will therefore show small or nonexistent velocity variations, and this probably explains the failure to detect these variations in V841 Oph and DI Lac.

IV. A Period-Luminosity Relation for Cataclysmic Variables

If, as the preceding discussion appears to support, the picture of a nova is that of a close binary consisting of a large red star and a small blue one, then it is possible to derive a period-luminosity relation for such objects. Here we are concerned with the *orbital* period and the absolute magnitude of the red component.

Data from table 60.1 and a prior knowledge of the masses of the U Gem stars are used, together with the assumption that the red star overflows its inner lobe, to derive the expression

$$M_V(\text{red}) = -3.33 \log P - 2.5 \log k + 8.6,$$

where M_V is the absolute visual magnitude of the red star (with an intrinsic uncertainty of ± 1.5 mag), the orbital period in hours is P, and the constant k is the amount by which the observed luminosity of the red star falls below that inferred from its spectrum and the Stefan-Boltzmann law. Observed data indicate that on the average $k = 0.1$. The brighter the nova at minimum, the longer the period, and the diminution of light with decreasing period explains the difficulty in detecting the spectra of the short-period systems. The magnitudes derived from the magnitude-period relation are consistent with those derived from the life-luminosity relation (Arp, 1956, Payne-Gaposchkin, 1957). Objects whose periods and magnitudes fall outside the range given above are not expected to become novae.

V. Causes of the Outbursts

Referring to table 60.1, we see that the old novae are found over an extraordinary range of masses and periods. The blue components of T CrB and GK Per are not far from the positions in the HR diagram that one would expect for stars burning helium (Cox and Giuli 1961). The blue component of DQ Her is probably a bright white dwarf. It would appear that any mechanism advanced to explain why outbursts occur in such systems cannot depend strongly on mass or on position in the HR diagram, except for the apparent necessity for a blue-red combination. It is as if any close binary, with components which have started anywhere on the main sequence, has the potential of becoming a nova if the components can evolve into the appropriate blue-red combination.

Two theories that have been advanced to explain the outbursts will be briefly mentioned. Both assume that the blue star is the seat of the disturbance, but it must be admitted that direct observational evidence as to which of the stars is responsible is lacking. The first, to which the writer has subscribed (Kraft 1962) but which he now thinks must be quite wrong, supposes that the blue component is an electron-degenerate stellar configuration. (If it is too bright, i.e., exceeds the brightness limit for white dwarfs, then perhaps its luminosity is maintained by accretion heating or by fluorescence of ultraviolet quanta in the ring.) In any case, the core of the star is electron degenerate and contains no hydrogen except for a possible non-degenerate atmosphere which is about 100 km thick. Hydrogen-rich material from the red companion pours onto the blue star, and gradually some of this hydrogen is forced down into the degenerate core, presumed to be hot enough to start hydrogen burning. As the hydrogen burns, it heats up the material. Now a property of degenerate matter is that the equation of state does not involve the temperature. Thus the material does not expand to cool in response to the increase in the temperature. Instead, the reaction rate goes faster until, after some characteristic induction time, the degeneracy is removed, the star expands suddenly, and the nova phenomenon occurs.

The most attractive feature of this approach is that it requires the transfer of hydrogen-rich material from one star to the other, which seems to be in harmony with the observations. However, the hypothesis fails on two counts. First, the electron conduction in the core is so efficient that any locally produced heat is conducted away and distributed over the whole core. Thus one must wait for the whole core to be heated, and, as Mestel (1952) showed some years ago, when the outburst occurs, it will be on the scale of a supernova—the available energy is simply too large. Furthermore, the blue components of GK Per and T CrB have masses too big for white dwarfs. They are not likely to be electron-degenerate configurations. If, however, a process of this kind might be invoked in connection with nuclear degeneracy (cf. Burbidge 1963), the subject could possibly be reopened.

The other theory, which has been advanced by Schatzman (1958), may have fewer objectionable features. Some years ago, Cowling (1941) considered the problem of inducing non-radial oscillations in one component of a binary by the lack of synchronism between its rotation and the revolution of the other star. He showed that in certain types of polytropic configurations long-period modes of non-radial motion could show up which might come into resonance with the period of the orbit. Schatzman has extended these ideas and suggests that the blue star may have just such long-period non-radial oscillations. The theory leads naturally to the prediction of an "induction time" between outbursts and to the ejection of material in cones, rather than spherical shells—results in harmony with observation (cf. Weaver 1955). The Schatzman theory does not, however, seem to require a red companion that overflows L_1, any kind of secondary would do equally well—a conclusion that does not seem to fit in well with the observational results.

A more extensive review of these problems does not, however, appear in order in a paper devoted primarily to observational matters.

VI. NOVAE AND STELLAR POPULATIONS

Novae are usually found in the arms of spiral galaxies (Arp, 1956), but some have been found in globular clusters and dwarf elliptical galaxies. The rarity of novae in these systems and uncertainty about the number of binary systems present in them make conclusions about the binary hypothesis difficult.

Arp, H. C. 1956, *A.J. 61*, 15.
Baade, W., and Swope, H. 1963, *A.J. 68*, 435.
Burbidge, G. R. 1963, *Ap. J. 137*, 995.
Cameron, A. 1959, *Ap. J. 130*, 884.
Cowling, T. G. 1941, *M.N. 101*, 368.
Cox, J. P., and Giuli, R. T. 1961, *Ap. J. 133*, 755.
Greenstein, J. L. 1957, *Ap. J. 126*, 23.
——— 1960, "Spectra of Stars below the Main Sequence," in *Stars and Stellar Systems*, vol. 6, *Stellar Atmospheres*, ed. J. L. Greenstein (Chicago: University of Chicago Press), p. 676.
Greenstein, J. L., and Kraft, R. P. 1959, *Ap. J. 130*, 99.
Humason, M. L. 1938, *Ap. J. 88*, 228.
Kopal, Z. 1959, *Close Binary Systems* (London: Chapman & Hall), p. 7.
Kraft, R. P. 1958, *Ap. J. 127*, 625.
——— 1959, ibid. *130*, 110.
——— 1961, *Science 134*, 1433.
——— 1962, *Ap. J, 135*, 408 (Paper I).
Kraft, R. P., Mathews, J., and Greenstein, J. L. 1962, *Ap. J. 136*, 312 (Paper II).
Krzeminski, W. 1962, *Pub. A.S.P. 74*, 66.
Kuiper, G. P. 1935, *Pub. A.S.P. 47*, 121.
——— 1941, *Ap. J. 93*, 133.
Mestel, L. 1952, *M.N. 112*, 598.
Mumford, G. 1964, *A.J. 69*, 146, 553.
Payne-Gaposchkin, C. 1957, *The Galactic Novae* (Amsterdam: North-Holland Publishing Co.), pp. 22–23.
Saakyan, G. S. 1963, *Russian Astr. J. 40*, 82; English trans., *7*, 60.
Sanford, R. F. 1949, *Ap. J. 109*, 81.
Sauvenier-Goffin, E. 1949, *Ann. d'ap. 12*, 39.
Schatzman, E. 1958, *Ann. d'ap. 21*, 1.
——— 1961, ibid. *24*, 237.
van Maanen, A. 1926, *Pub. A.S.P. 38*, 325.
Walker, M. F. 1954, *Pub. A.S.P. 66*, 230.
——— 1956, *Ap. J. 123*, 68.
——— 1957, *I.A.U. Symposium No. 3*, ed. G. H. Herbig (Cambridge: Cambridge University Press), p. 46.
——— 1958, *Ap. J. 127*, 319.
——— 1961, ibid. *134*, 171.
——— 1962, *Konkoly Info. Bull., Variable Stars*, no. 2.
——— 1963, *Ap. J. 137*, 485.
Wallerstein, G. 1963, *Pub. A.S.P. 75*, 26.
Weaver, H. F. 1955, quoted by C. Payne-Gaposchkin 1957 (see above).

61. The Discovery of White Dwarf Stars

An A-Type Star of Very Low Luminosity

Walter S. Adams

(*Publications of the Astronomical Society of the Pacific 26*, 198 [1914])

The Spectrum of the Companion of Sirius

Walter S. Adams

(*Publications of the Astronomical Society of the Pacific 27*, 236–237 [1915])

By reporting the unexpected A0 spectral classification of the low luminosity companions of o Eridani and of Sirius, Walter Adams here draws attention to objects that turned out to be representatives of one of the most extraordinary groups of stars known to astronomers. In the first of these two notes, Adams announces that o Eridani B has an A0 spectrum; A-type stars have surface temperatures around 10,000°, and it is surprising that such a hot star should exhibit such a low luminosity. Nearly four years previously, Henry Norris Russell had discussed the apparent anomaly of o Eridani's luminosity and spectrum,[1] but by 1914 he omitted this discussion while including the star as a single isolated point in the lower left-hand corner of his Hertzsprung-Russell diagram (see selection 35). The perplexing behavior of o Eridani was accentuated in the next year, when Adams showed that another low-luminosity star, Sirius B, has a spectrum only slightly different from the A0-type spectrum of Sirius A.

What Adams did not point out explicitly was that the high surface temperature in combination with the low total luminosity meant that the stars had to be very small—only about the size of the earth. Furthermore, the rather ordinary mass of Sirius B meant that the density of the star had to be enormous—about 100,000 gm cm^{-3}. It was certainly Arthur Stanley Eddington as much as anyone who brought these remarkable properties to the attention of the astronomical world. Eventually he showed that there was nothing inherently absurd about the high densities of white dwarfs.[2] He argued that in the hot stellar interiors all the electrons would be stripped away from their nuclei. In this case there was nothing to prevent the free electrons from becoming closely packed with the bare nuclei, and even nuclear densities as high as a million million grams per cubic centimeter were imaginable. Using the formula from Einstein's general theory of relativity, Eddington predicted that the gravitational red shift between the two components of Sirius would be equivalent to about 20 km sec^{-1}. Within a year Adams had carried out the necessary measurements and found that, after allowance was made for the relative orbital motion of the two stars, the observed displacement is 19 km sec^{-1}.[3] Recently, Jesse Greenstein, J. Beverly Oke, and Harry L. Shipman have measured the gravitational red shift of Sirius B, finding a displacement of 89 km sec^{-1}, which is in excellent agreement with that predicted from its radius of 0.008 solar radii and its gravity.[4]

1. In a colloquium given at Princeton University Observatory in 1954, Russell recalled his visit with E. C. Pickering, then director of Harvard Observatory, which led to the discovery of the anomalous spectrum of o Eridani B. (See *In Memory of Henry Norris Russell*, ed. A. G. Davis Philip and D. H. de Vorkin, Dudley Observatory Report no. 13 [1977], pp. 90, 107). Russell thought it would be a good idea to obtain the spectra of certain parallax stars; as he recalled it: "Pickering said 'Well, name one of these stars.' Well, said I, for example, the faint component of Omicron Eridani. So Pickering said, 'Well, we make rather a speciality of being able to answer questions like that.' And so we telephoned down to the office of Mrs. Fleming and Mrs. Fleming said, yes, she'd look it up. In half an hour she came up and said I've got it here, unquestionably spectral type A. I knew enough, even then, to know what that meant. I was flabbergasted. I was really baffled trying to make out what it meant. Then Pickering thought for a moment and then said with a kindly smile, 'I wouldn't worry. It's just these things which we can't explain that lead to advances in our knowledge.' Well, at that moment, Pickering, Mrs. Fleming and I were the only people in the world who knew of the existence of white dwarfs." Letters between Pickering and Russell between December 1910 and June 1911 discuss the apparent anomaly of o Eridani's luminosity and spectrum.

2. A. S. Eddington, *Monthly Notices of the Royal Astronomical Society 84*, 308 (1924).

3. W. S. Adams, *Proceedings of the National Academy of Sciences 11*, 382 (1925).

4. J. L. Greenstein, J. B. Oke, and H. L. Shipman, *Astrophysical Journal 169*, 563 (1971).

Adams, An A-Type Star of Very Low Luminosity

IT HAS BEEN SUGGESTED by Hertzsprung that there is no such range in absolute brightness among the A-type stars as among those of types F to M, and, in fact, it is doubtful whether hitherto any certain case of a very faint A-type star has been found. A recent observation of the ninth-magnitude companion of o Eridani shows, however, that this star must be considered as such. The companion is at a distance of 83″ from the principal star and shares in its immense proper-motion of 4″.08 annually. Its parallax, therefore, may be assumed to be that of the bright star which is 0″.17. This would make the absolute magnitude of the companion 10.3, the Sun being taken as 5.5. The spectrum of the star is A0.

Adams, The Spectrum of the Companion of Sirius

WE HAVE MADE several attempts during the past two years to secure a spectrum of the companion of Sirius. Its position is favorable, the distance, according to Professor Barnard's recent measures, being more than 10″ in a position angle of about 70°. The great mass of the star, equal to that of the Sun and about one-half that of Sirius, and its low luminosity, one one-hundredth part of that of the Sun and one ten-thousandth part of that of Sirius, make the character of its spectrum a matter of exceptional interest.

Most of the spectrum photographs have been taken at the 80-foot focus of the 60-inch reflector with the Cassegrain combination of mirrors. At this focus the distance of the companion from Sirius is 1.2^{mm}. The rays from Sirius, due to the supports of the auxiliary mirrors, are very prominent, but form angles of about 45° with the line joining Sirius with the companion, and so do not reach the slit unless the images begin to blur badly. The main difficulty in securing satisfactory photographs is, of course, the strong general illumination of the field and the presence of subsidiary rays which contribute more or less light to the slit as the seeing varies. During the exposures Sirius has been kept on the black metal screen in which is cut the opening forming the star window, while the companion is held in a position slightly to one side of the center of this window. Accordingly it is possible to compare on the photographs the spectrum of the point at which the companion is maintained with the spectrum due to the general illumination of Sirius. The exposure times given have been those normal for a star of 8.5 magnitude.

Two or three photographs obtained in this way showed a decided maximum in the spectrum at the point at which the companion was kept during the exposure. Still there was no distinct line of separation from the general spectrum due to Sirius. A photograph taken on October 18th under exceptionally good conditions of seeing does show such a demarcation, however, there being a narrow spectrum corresponding to the point on the slit at which the companion was held, which is separated by a distinct break from the intense spectrum of Sirius near the edge of the star window. It is difficult to avoid the conclusion that this is the spectrum of the companion. There was no ray from Sirius near this point of the slit and during the entire exposure the companion was well visible and accurate guiding was easily maintained.

The line spectrum of the companion is identical with that of Sirius in all respects so far as can be judged from a close comparison of the spectra, but there appears to be a slight tendency for the continuous spectrum of the companion to fade off more rapidly in the violet region. The suggestion has been made by several astronomers that at least a portion of the light of the companion is due to light reflected from Sirius. It is, however, by no means necessary to have recourse to this explanation, since in the case of the companion of o_2 Eridani, where there can be no question of reflected light, we know of a similar case of a star of very low intrinsic brightness which has a spectrum of type A_0.

Direct photographs taken by Dr. van Maanen with and without the use of a yellow color screen agree with the spectrographic results in indicating that the companion of Sirius has a color index not appreciably different from that of the principal star.

62. On Dense Stars

Ralph Howard Fowler

(*Monthly Notices of the Royal Astronomical Society 87*, 114–122 [1926])

Immediately after the discovery of the white dwarf stars, it became clear that their low luminosities, high temperatures, and rather ordinary masses implied small radii and large densities. Because the matter in the hot interiors of stars is ionized, there is nothing to prevent the free electrons from being closely packed with the bare nuclei to form a very dense gas; but at first the physics of such a state was not at all clear.

In this paper, one of the first major applications of the new quantum physics to an astronomical problem, Ralph Fowler uses the statistical description of atoms published the previous year by Enrico Fermi, which was based on the newly enunciated exclusion principle of Wolfgang Pauli. Fowler here explains the fundamental discovery that the electron assembly in the white dwarf stars must be completely degenerate in the sense of Fermi's statistical description, thereby resolving the paradox described earlier by A. S. Eddington. According to Eddington, if the dense white dwarf star continued to radiate energy, it should reach a stage where the very condensed stellar material would have less energy than atoms at ordinary density and absolute zero temperature. Fowler shows that the correct relation between energy and temperature is found from quantum statistical mechanics; the individual electrons at absolute zero still have a kinetic energy comparable with the thermal energy of particles in an expanded gas whose temperature is as large as 10 million degrees. Fowler also shows that the pressure of the white dwarf gas is unaffected by its temperature. Therefore, any local heating by thermonuclear reactions will bring a higher temperature, which will increase the reaction rate, and consequently the star will explode. Hence, we must conclude that white dwarf stars do not shine by thermonuclear reactions and that their light must come from the slow leakage of the heat contained in the nondegenerate nuclei. Eventually, the nuclei themselves will become degenerate, and the white dwarf will fade into a gigantic black molecule in which all the nuclei and electrons are in their lowest quantum state.

By showing that the pressure of a degenerate gas of mass density ρ is proportional to $\rho^{5/3}$, Fowler set the stage for subsequent theoretical speculations on the internal constitution of white dwarfs. By 1929 Wilhelm Anderson had demonstrated that the electrons in the centers of degenerate masses begin to attain velocities on the order of the velocity of light and that in this case the variation of the electron mass with velocity must be taken into account by using the equations of special relativity. For a relativistic degenerate electron gas the pressure at mass density ρ is proportional to $\rho^{4/3}$.

Both Anderson and Edmund C. Stoner soon showed that a star can contract only until the decrease in gravitational potential energy becomes insufficient to balance the increase of the kinetic energy of the electrons; accordingly, for stellar masses larger than about 1 solar mass, there can be no equilibrium white dwarf configurations.[1]

Within a year of Anderson and Stoner's work, Subrahmanyan Chandrasekhar derived the equation of state of a degenerate gas in the extreme relativistic limit and found that in this limit the mass of the white dwarf is uniquely determined. When the mean molecular weight per electron is denoted by μ, the critical mass is 5.84 μ^{-2} solar masses; because μ takes on the value 2 for all elements other than hydrogen, the mass limit is quite generally 1.46 solar masses. For stars of mass greater than 1.46 solar masses, no equilibrium white dwarf configurations are possible. Although both Anderson and Stoner had previously called attention to the existence of this upper mass limit, it is called the Chandrasekhar limit, because he was the first to derive the detailed equilibrium configurations in which degenerate electron gases support their own gravity.[2]

1. E. C. Stoner, *Philosophical Magazine 7*, 63 (1929), *9*, 944 (1930); W. Anderson, *Zeitschrift für Physik 54*, 433 (1929).
2. S. Chandrasekhar, *Astrophysical Journal 74*, 81 (1931), *Monthly Notices of the Royal Astronomical Society 95*, 207, 226, 676 (1935), and *An Introduction to Stellar Structure* (Chicago: University of Chicago Press, 1938).

1. Introduction

The accepted density of matter in stars such as the companion of Sirius is of the order of 10^5 gm./c.c. This large density has already given rise to most interesting theoretical considerations, largely due to Eddington. We recognise now that matter can exist in such a dense state if it has sufficient *energy*, so that the electrons are not bound in their ordinary atomic orbits of atomic dimensions, but are in the main free—with sufficient energy to escape from any nucleus they may be near. The density of such "energetic" matter is then only limited *a priori* by the "sizes" of electrons and atomic nuclei. The "volumes" of these are perhaps 10^{-14} of the volume of the corresponding atoms, so that densities up to 10^{14} times that of terrestrial material may not be impossible. Since the greatest stellar densities are of an altogether lower order of magnitude, the limitations imposed by the "sizes" of the nuclei and electrons can be ignored in discussions of stellar densities, and the structural particles of stellar matter can be treated as massive charged points.

Eddington has recently[1] pointed out a difficulty in the theory of such matter. Assuming it to behave more or less like a perfect gas, modified by its electrostatic forces and the sizes of such atomic structures as remain undissolved, there is a perfectly definite relation between the energy and the temperature, which depends on the density only to a minor degree. This assumption even here is not so unreasonable as appears at first sight. But even without it we naturally expect a perfectly definite relation between energy and temperature, in which there is a close correlation between large energies and large temperatures, small energies and small temperatures. The emission of energy by the star will proceed in the usual way at a rate depending on the surface temperature, and the internal temperatures must provide the gradient necessary to drive the radiation out. So long as the star contains matter at a high *temperature*, radiation of energy must presumably go on. But then, according to Eddington, there may come a time when a very curious state of affairs is set up. The stellar material will have radiated so much energy that it has less energy than the same matter in normal atoms expanded at the absolute zero of temperature. If part of it were removed from the star, and the pressure taken off, what could it do?

The present note is devoted to a further consideration of this paradox. It is clear that the crucial point is the connection between the energy and the temperature. In a sense the temperature measures the "looseness" of the system, the number of possible configurations which it can assume, and therefore its radiation. These depend directly on the temperature, and only on the energy in so far as the energy determines the temperature. The excessive densities involved suggest that the most exact form of statistical mechanics must be used to discuss the relationship between the energy, temperature, and density of the material. This is a form suggested by the properties of atoms and the new quantum mechanics, which has been already applied to simple gases by Fermi and Dirac.[2] It may be accepted now as certain that classical statistical mechanics is not applicable at extreme densities, even to ideal material composed of extensionless mass-points, and that the form used here is fairly certainly the correct substitute. Its essential feature is a principle of exclusion which prevents two mass-points ever occupying exactly the same cell of extension h^3 of the six-dimensional phase-space of the mass-points. When this form of statistical mechanics is adopted, it at once appears that the suggested difficulty resolves itself, and there is really no difficulty at all. The apparent difficulty was due to the use of a wrong correlation between energy and temperature, suggested by classical statistical mechanics. When the correct relation is substituted, it is found that the limiting state of such dense stellar matter is one which the *energy* is still, as it must be, excessively great, but the *temperature* is zero! Since the temperature determines the radiation, radiation stops when the dense matter has still ample energy to expand and form normal matter if the pressure happened to be removed. As the dense matter radiates its energy away, the number of its possible configurations rapidly falls, and therewith the temperature. The absolutely final state is one in which there is only one possible configuration left. Temperature then ceases to have any meaning, for the star is strictly analogous to one gigantic molecule in its lowest quantum state. We may call the temperature then zero.

Whether or no some such explanation may not be equally possible using other forms of statistical mechanics (perhaps the classical) I am not prepared to say. The new form used here seems for entirely independent reasons so satisfactory that its applicability need not be questioned. On application it clears up Eddington's question in a convincing manner, and I am content to leave the matter so.

2. The Equilibrium State of Dense Matter

It is obviously reasonable to consider dense stellar material as an assembly of free electrons and bare nuclei of net charge zero. It is, however, necessary to idealise the problem a little further and ignore the electrostatic charges, so that the whole energy of the assembly is the kinetic energy of translation of the various particles. We shall later see reason to believe that this idealisation does not vitiate the results, or at least not their qualitative form, for we are really mainly concerned with calculations of amounts of phase-space, and these are independent of potential energy.

Consider a volume V of stellar material consisting of N/q bare nuclei of mass m_q and N free electrons of mass m, with total (kinetic) energy E. We will assume as a first approximation, as we have said, that the electrostatic forces may be ignored, and then that the statistics of assemblies of practically independent systems is applicable. Then[3] the total number of complexions C representing any state of the assembly is

$$C = \left(\frac{1}{2\pi i}\right)^3 \iiint \frac{dx}{x^{N+1}} \frac{dy}{y^{N/q+1}} \frac{dz}{z^{E+1}} \Pi_\sigma(1 + xz^{\varepsilon_\sigma})^2 \Pi_\sigma(1 + yz^{\eta_\sigma}). \tag{1}$$

In this equation $\varepsilon_\sigma(\sigma = 1, 2, \ldots)$ are the possible energies of an independent electron in a volume V, and η_σ those of an independent nucleus. The power 2 enters because the electron has a structure (axis of spin) which gives it two possible orientations in a magnetic field, that is, a statistical weight 2. The nucleus probably also has a weight greater than unity, but the nuclear terms turn out to be unimportant, so that we need not spend time over this refinement. Students of classical statistical mechanics will recall that the classical formula differs from this only in having $\exp\{\sum_\sigma 2xz^{\varepsilon_\sigma}\}$ in place of $\Pi_\sigma(1 + xz^{\varepsilon_\sigma})^2$.

All the mean values associated with the equilibrium state of the assembly can be derived from C, and the value of C can be obtained by "steepest descents." Writing

$$G = G(x, z) = \textstyle\sum_\sigma 2 \log(1 + xz^{\varepsilon_\sigma}),$$
$$G_q = G_q(y, z) = \textstyle\sum_\sigma \log(1 + yz^{\eta_\sigma}), \tag{2}$$

the critical value of the integrand is determined by the unique root λ, μ, θ of the equations

$$N = x\frac{\partial G}{\partial x}, \quad \frac{N}{q} = y\frac{\partial G_q}{\partial y}, \quad E = z\frac{\partial G}{\partial z} + z\frac{\partial G_q}{\partial z}. \tag{3}$$

With these special values of x, y, z we have, among other equations,

$$N = \lambda\frac{\partial G}{\partial \lambda}, \quad E_\varepsilon = \theta\frac{\partial G}{\partial \theta}, \tag{4}$$

where E_ε is the average energy of the free electrons. Also

$$S = k \log C = k\left[G + G_q - N \log \lambda - \frac{N}{q} \log \mu - E \log \theta\right], \tag{5}$$

$$\theta = e^{-1/kT} \tag{6}$$

The number of possible energy values for an electron of mass m in a volume V which lie between τ and $\tau + d\tau$ is[4]

$$\frac{2\pi V}{h^3}(2m)^{3/2}\tau^{1/2}\,d\tau. \tag{7}$$

Therefore

$$G = \frac{4\pi(2m)^{3/2}}{h^3} V \textstyle\sum \log(1 + \lambda\theta^\tau)\tau^{1/2}\,d\tau.$$

The values of ε_σ are extremely closely spaced, so that for a fixed value of θ we can replace the sum by an integral with negligible error and find

$$G = \frac{4\pi(2m)^{3/2}}{h^3} V \int_0^\infty \log(1 + \lambda\theta^\tau)\tau^{1/2}\,d\tau. \tag{8}$$

Strictly we shall want to make $\theta \to 0$ ($T \to 0$). The form (8) might then fail, but as we shall really not be interested in the differences between $T = 0°$, $T = 100°$, or even $T = 1000°$, this does not matter, and we shall assume the general validity of (8). The validity of some applications of (8) can be checked in the limit by direct calculation. N and E_ε are obtained from this by differentiation. Similar forms and equations hold for G_q, N/q, and E_q for the nuclei. The assembly is a thermodynamic system.

Our chief interest is now to trace the series of equilibrium states of such an assembly as its energy is slowly radiated away. A loss of energy by radiation means necessarily a loss of entropy, and therefore a fall of temperature. We have therefore to trace the behavior of the assembly as θ decreases—presumably as $\theta \to 0$. But by (4) and (8) N depends on a function of the form

$$\int_0^\infty \frac{\lambda\theta^\tau}{1 + \lambda\theta^\tau}\tau^{1/2}\,d\tau.$$

If λ has an upper bound b, this is less than

$$b\int_0^\infty \theta^\tau\tau^{1/2}\,d\tau = b\Gamma(\tfrac{3}{2})/(\log 1/\theta)^{3/2},$$

and therefore tends to 0 as $\theta \to 0$. This would make $N \to 0$, which is impossible, as the number of particles is fixed. Since λ cannot have an upper bound, it is reasonable to assume that $\lambda \to \infty$. We shall see shortly that this is a necessary consequence of $\theta \to 0$.

Let us therefore write $\theta = e^{-\alpha}$, $\lambda e^{-\alpha\tau_0} = 1$, defining τ_0, and assume that λ and therefore $\alpha\tau_0$ are large. Then

$$\frac{h^3 G}{4\pi(2m)^{3/2}V} = \int_0^{\tau_0} \{\log(\lambda e^{-\alpha\tau}) + \log(1 + e^{\alpha\tau}/\lambda)\}\tau^{1/2}\,d\tau$$
$$+ \int_{\tau_0}^\infty \log(1 + \lambda e^{-\alpha\tau})\tau^{1/2}\,d\tau,$$
$$= \tfrac{2}{3}(\log\lambda)\tau_0^{3/2} - \tfrac{2}{5}\alpha\tau_0^{5/2}$$
$$+ \sum_{j=1}^\infty \frac{(-)^{j-1}}{j}\int_0^{\tau_0} e^{-j\alpha y}(\tau_0 - y)^{1/2}\,dy$$
$$+ \sum_{j=1}^\infty \frac{(-)^{j-1}}{j}\int_0^\infty e^{-j\alpha y}(\tau_0 + y)^{1/2}\,dy.$$

In transforming the integrals we have used the fact that $e^{\alpha\tau_0} = \lambda$. Since $\alpha\tau_0$, and *a fortiori* $j\alpha\tau_0$, are large, we have approximately

$$\int_0^{\tau_0} e^{-j\alpha y}(\tau_0 - y)^{1/2}\,dy \sim \frac{{\tau_0}^{1/2}}{j\alpha},$$

$$\int_0^{\infty} e^{-j\alpha y}(\tau_0 + y)^{1/2}\,dy \sim \frac{{\tau_0}^{1/2}}{j\alpha}.$$

Therefore finally (and approximately)

$$\frac{h^3 G}{4\pi(2m)^{3/2}V} = \frac{4}{15}\frac{(\log\lambda)^{5/2}}{(\log 1/\theta)^{3/2}} + \frac{\pi^2}{6}\frac{(\log\lambda)^{1/2}}{(\log 1/\theta)^{3/2}},$$

an equation whose real value can be expressed in the form

$$\frac{h^3 G}{4\pi(2m)^{3/2}V} = \frac{4}{15}\frac{(\log\lambda)^{5/2}}{(\log 1/\theta)^{3/2}} + O\left\{\frac{(\log\lambda)^{1/2}}{(\log 1/\theta)^{3/2}}\right\}. \tag{9}$$

This equation can be differentiated. Therefore

$$\frac{h^3 N}{4\pi(2m)^{3/2}V} = \frac{2}{3}\frac{(\log\lambda)^{3/2}}{(\log 1/\theta)^{3/2}} + O\left\{\frac{1}{(\log\lambda)^{1/2}(\log 1/\theta)^{3/2}}\right\} \tag{10}$$

$$\frac{h^3 E_\varepsilon}{4\pi(2m)^{3/2}V} = \frac{2}{5}\frac{(\log\lambda)^{5/2}}{(\log 1/\theta)^{5/2}} + O\left\{\frac{(\log\lambda)^{1/2}}{(\log 1/\theta)^{5/2}}\right\}. \tag{11}$$

It follows from (10) that $\theta \to 0$ implies $\lambda \to \infty$, and therefore from (10) and (11) combined that as $\theta \to 0$

$$\frac{h^3 E_\varepsilon}{4\pi(2m)^{3/2}V} \sim \frac{2}{5}\left(\frac{3}{2}\frac{h^3 N}{4\pi(2m)^{3/2}V}\right)^{5/3}. \tag{12}$$

The temperature can (and presumably would) fall to zero, still leaving, however, the free electrons with kinetic energy E_ε given by (12). Similar relations hold for the nuclei. In the limit, however, $E_q = E_\varepsilon(m/qm_q)$, and is therefore negligible compared with E_ε.

It is wise to check up the behaviour of the entropy meanwhile. The contribution S_ε by the electrons is given by

$$S_\varepsilon = k[G - N\log\lambda + E_\varepsilon \log 1/\theta],$$

$$= k\frac{4\pi(2m)^{3/2}V}{h^3}\left[\frac{(\log\lambda)^{5/2}}{(\log 1/\theta)^{3/2}}\{\tfrac{4}{15} - \tfrac{2}{3} + \tfrac{2}{5}\} + 0\left\{\frac{(\log\lambda)^{1/2}}{(\log 1/\theta)^{3/2}}\right\}\right],$$

$$\to 0.$$

The contribution by the nuclei behaves in a similar way. This is correct.

We may note that the relation (12) can be confirmed by a direct calculation of the tight-packed phase-space required to house the N electrons. If the number of cells is the number of electrons, we find from (7)

$$\frac{N}{V} = \frac{4\pi(2m)^{3/2}}{h^3}\int_0^{\tau_0} \tau^{1/2}\,d\tau,$$

while at the same time

$$\frac{E_\varepsilon}{V} = \frac{4\pi(2m)^{3/2}}{h^3}\int_0^{\tau_0} \tau^{3/2}\,d\tau,$$

which together give (12).

3. Numerical Calculations

Let us assume that the density of the stellar matter is 10^x gm./c.c., and that it is composed on the average of iron atoms, at

$$\frac{6.06 \times 10^{23}}{56} \text{ per gm.,}$$

or

$$\frac{6.06 \times 10^{23+x}}{56} \text{ per c.c.}$$

Then for the free electrons,

$$\frac{N}{V} = \frac{6.06 \times 26}{56} 10^{23+x} = 2.81 \times 10^{23+x},$$

$$\log_{10}\frac{E_\varepsilon}{V} = -27 + 0.539 + \tfrac{5}{3}\log_{10}\frac{N}{V} = 12.621 + \tfrac{5}{3}x.$$

For density 10^5 gm./c.c.

$$E_\varepsilon/V = 9.0 \times 10^{20} \text{ ergs.}$$

Thus the average kinetic energy of an electron in this state is $9.0 \times 10^{20}/2.81 \times 10^{28}$ ergs. This is the same as the average kinetic energy of an *electron* in an expanded gas at a temperature T, where

$$\tfrac{3}{2}kT = \frac{9.0}{2.81} \times 10^{-8},$$

or

$$T = 1.56 \times 10^8\ {}^\circ\text{K}.$$

It is not, however, possible to rest quite content with this result, since the electrostatic energy terms are far from negligible, and in fact of the same order of magnitude. We observe first that the equation (12) is based entirely on considerations of phase-space. There must be enough cells of extension h^3 in six-dimensional phase-space to accommodate all the electrons. The phase-space depends only on velocities, or kinetic energy of the electrons, and is independent of the potential energy terms. The relation (12), therefore, should continue to give the necessary kinetic energy of the electrons, at least reasonably accurately, even when the effects of potential energy are allowed for. We have therefore merely to attempt to estimate the negative potential energy in order to compare the total energy of the condensed matter with its total energy in an expanded form.

The negative potential energy of a normal atom can be estimated roughly but easily if we recall that the average negative potential energy of an orbit in an inverse square field is twice the negative (total) energy of the orbit. The total negative energy of an iron atom is[5] 17,800 + 12,400 + 2,700 + 600 volts for the K, L, M, and N electrons respectively, or 33,500 volts in all. The negative potential energy of one atom may therefore be taken to be

$$67{,}000 \text{ volts} \quad \text{or} \quad 1.065 \times 10^{-7} \text{ ergs.}$$

The negative potential energy of the atoms equivalent to 1 c.c. of density 10^x is therefore

$$1.15 \times 10^{15+x} \text{ ergs.}$$

For the standard case of density 10^5 the negative energy of the normal atoms is, therefore, say 10^{20} ergs.

The negative potential energy in the condensed form is far more difficult to estimate with any pretence to accuracy. If we simply say that at a density 10^x everything will on the whole be about $(10^{x-1})^{1/3}$ closer than it could be packed before in normal atoms, we might argue that the negative potential energy would be increased about $10^{(x-1)/3}$ times, and so become

$$5.3 \times 10^{14+4x/3} \text{ ergs.,}$$

or in the standard case 2.5×10^{21}. This, however, must grossly over-estimate the increase. For half the negative energy comes from the two K electrons normally at a distance of 2×10^{-10} cm. from the nucleus. The average distance of uniformly distributed electrons at a density 10^5 is as much as 6×10^{-10} cm., and there is no reason for two electrons on the average to sink into about $2 \times 10^{-10}/10^{4/3}$, that is 10^{-11} cm. They might be expected to get hardly any closer, and the L-electrons not to experience the full shrinkage. If the contribution of the two electrons corresponding to the K electrons is unaltered, the potential energy increase, and therefore roughly the potential energy, is halved, and the incomplete shrinkage of the L group would halve the energy again. A negative potential energy of 5×10^{20} is all that can be expected.

An alternative estimate may be made as follows. The average space assigned to each nucleus is a sphere of radius r_0, where

$$1/(\tfrac{4}{3}\pi r_0^{\,3}) = \frac{2.81}{2.6} \times 10^{22+x},$$

$$r_0 = 1.30 \times 10^{-(23+x)/3}.$$

Let us estimate the electrostatic energy by distributing 26 electrons uniformly between o and r_0 for each nucleus, assume that the field of a single point charge $26e$ acts on each and ignore their own repulsions. Such an average distribution is not far wrong, for it is known that there is comparatively little crowding of bound electrons owing to the quantum restrictions, or of free electrons owing to the energy conditions.[6] Thus the negative energy per atom is

$$(26e)^2\overline{\left(\frac{1}{r}\right)}, \quad \overline{\left(\frac{1}{r}\right)} = \frac{\int_0^{r_0} r\,dr}{\int_0^{r_0} r^2\,dr} = \frac{3}{2}\frac{1}{r_0}.$$

The negative potential energy per c.c. of condensed matter is therefore

$$8.9 \times 10^{13+(4x/3)},$$

or in the standard case 4.1×10^{20}.

All this is very uncertain; the best we can do is to conclude that the negative potential energy of the condensed form is about

$$10^{14+(4x/3)},$$

or in the standard case 4.6×10^{20} ergs, compared with 10^{20} ergs when expanded in normal atoms. As a result the kinetic energy is perhaps twice the negative potential energy for a density 10^5, and the expanded form could become a perfect gas of normal *atoms* at about $1.56 \times 10^8 \times 1/2 \times 26$ say 2×10^9 °K.

Whether or no by our rough calculations we can show that the total energy of the condensed form is large and positive for a density of 10^5, we can certainly conclude that this must be true for still greater densities, for the kinetic energy per unit volume varies as $10^{(5x/3)}$, and on almost any view the negative potential energy must vary ultimately like $10^{(4x/3)}$. There seems no reason to doubt that this inequality, kinetic energy greater than negative potential energy, is true over wider conditions than we can show by these arguments. For in condensed forms at somewhat lower densities the tight packing is presumably achieved at the expense of shifting only the more lightly bound electrons, and our estimates of increase in the negative potential energy will be greatly in excess.

4. COGNATE SPECULATIONS

If we return to the condensed assembly of mass-points, electrostatic forces ignored, we may note that the pressure is given by the thermodynamic equation

$$p = T\frac{\partial G}{\partial V}.$$

This is strictly the pressure exerted by the electrons, and obeys the usual perfect gas equation

$$p = \frac{2}{3}\left(\frac{E_\varepsilon}{V}\right);$$

the contribution by the nuclei is trivial by comparison. Thus in the condensed limit

$$p = \frac{4}{15}\left(\frac{3}{2}\right)^{5/3} \frac{h^2}{(4\pi)^{2/3}2m}\left(\frac{N}{V}\right)^{5/3}.$$

If the volume is a sphere of radius R, then

$$p \propto \frac{1}{R^5}.$$

This pressure represents an interaction of the assembly with the outside world which is in a sense due to the quantum restraints to which the assembly is subject.

Now an atom, especially an atom of an inert gas, is really condensed matter of just the same general type. The assembly, whether atom or stellar material, contains just so many cells of phase-space available for electrons and every one of them is full. It is tempting to speculate as to whether the apparent forces of repulsion, which prevent the too close approach of atoms, molecules, or ions with closed configurations of electrons, may not be due in just the same way purely to the quantum constraints. The number of cells of the phase-space must be unaltered in an encounter, and the apparent force of repulsion represents just the work which must be done to make up by an increase in the velocity (or kinetic energy) space for a loss in the extent of the "space"-space. These forces of repulsion between condensed atomic configurations of electrons seem empirically to vary like $1/R^{10}$, or thereabouts, for a large variety of atoms. It would be too much to expect the pressure of the stellar material to follow exactly the same law. But the general forms of the two types of system are sufficiently alike, and the resemblance of their laws of interaction close enough to lead one to believe that the origin of this important part of inter-atomic forces is to be sought in this direction, in the quasi-thermodynamic consequences of the existence of the quantum constraints embodied in Pauli's principle.

1. Eddington, *The Internal Constitution of the Stars* (Cambridge Univ. Press, 1926), §117.

2. Fermi, *Rend. Acc. Lincei*, Ser. 6, *3*, 145 (1926); Dirac, *Proc. Roy. Soc. A 112*, 661 (1926). See also Fowler, *Proc. Roy. Soc. A 113*, 432 (1926), for a full account with other references.

3. The following formulæ will be found in detail in Fowler.

4. Dirac, p. 671.

5. Hartree, *Proc. Camb. Phil. Soc. 22*, 464 (1924).

6. For the latter see Eddington, §183.

63. Neutrino Theory of Stellar Collapse

George Gamow and Mario Schönberg

(*Physical Review 59*, 539–547 [1941])

The neutrino made its debut as a theoretical entity in 1930, when Wolfgang Pauli tried to account for the perplexing laboratory studies of the electrons emitted from certain radioactive nuclei during the so-called beta-decay process. Although these electrons always originate from a transition between two nuclear states with well-defined energies, they are found to have a variety of kinetic energies. Pauli reasoned that if the law of conservation of energy holds during beta-decay, some hitherto undetected agent had to be spiriting away energy. As Enrico Fermi soon showed, the missing neutral particle remained undetected in the laboratory because it was massless, traveled at the speed of light, and interacted only very weakly with matter. By 1934 Fermi had described how this weak interaction could account for the observed energies of the beta-decay electrons, and an inverse process soon became apparent in which a nucleus could capture an electron while emitting a neutral particle.[1] We now call the mysterious neutral particle emitted during electron capture a neutrino, or "little neutral one," and the particle emitted during the beta-decay emission of an electron is called an antineutrino.

George Gamow and Mario Schönberg soon realized that under some circumstances *both* beta-decay and electron capture might occur and that the centers of dying stars might provide the right conditions. In the paper given here, they explore the situation occurring when the temperature of the stellar material is on the order of one thousand million degrees; then a nucleus can alternately capture and release an electron while emitting a neutrino and an antineutrino. After one cycle the nucleus has the same composition, but it has lost the energy associated with the two neutrinos. Gamow accordingly named this nuclear transformation the Urca process, after its similarity to the gambling operations at the Casino de Urca near Rio de Janeiro. There also, no matter how you played the game, you always seemed to lose.

Today we know that neutrinos may be emitted during nuclear processes other than the Urca process. Furthermore, during the Urca process, it is the neutrino loss in degenerate stellar cores, rather than the ideal nondegenerate stellar cores considered by Gamow and Schönberg, that plays an important role in shaping the evolution of massive carbon-burning stars.[2]

Besides calling attention to the importance of energy losses by neutrino processes during the final stages of stellar evolution, Gamow and Schönberg also focused attention on the equally important role of neutrino processes in the onset of catastrophic stellar explosions. Although the neutrinos had not yet been detected, Gamow and Schönberg realized that the interaction cross section for neutrinos with matter had to be so small that those produced would easily escape from stars, robbing the hot interiors of their energy forever. As they pointed out, the central interior must collapse and heat up in order to compensate for this energy loss, and because the heat that is produced cannot escape, the outer stellar envelope must expand to produce an apparent stellar explosion.

Nearly twenty years later, William Fowler and Fred Hoyle showed that the photodisintegration of the iron group nuclei into alpha particles and neutrons is more efficient than neutrino energy losses in producing the implosion of massive stellar cores. Futhermore, the catastrophic energy released during stellar explosions seems to be caused by the explosive instability of nuclear reactions involving light nuclei at high temperatures.[3] Nevertheless, Gamow and Schönberg correctly reasoned that the explosion of a very massive star must be the result of rapid temperature rises accompanying the catastrophic implosion of its core, and it is the neutrino energy losses that play the important role in the explosive ignition of the nuclear fuels of stars of intermediate mass.

1. E. Fermi, *Zeitschrift für Physik 88*, 161 (1934).

2. S. Tsuruta and A. G. W. Cameron, *Canadian Journal of Physics 43*, 2056 (1965), and *Astrophysics and Space Science 7*, 374 (1970); B. E. Paczyński, *Astrophysical Letters 11*, 53 (1972).

3. F. Hoyle and W. A. Fowler, *Astrophysical Journal 132*, 565 (1960); W. A. Fowler and F. Hoyle, *Astrophysical Journal Supplement 9*, 201 (1964).

Abstract—At the very high temperatures and densities which must exist in the interior of contracting stars during the later stages of their evolution, one must expect a special type of nuclear processes accompanied by *the emission of a large number of neutrinos.* These neutrinos penetrating almost without difficulty the body of the star, must carry away very large amounts of energy and prevent the central temperature from rising above a certain limit. This must cause *a rapid contraction of the stellar body* ultimately resulting in a *catastrophic collapse.* It is shown that energy losses through the neutrinos produced in reactions between free electrons and oxygen nuclei can cause a complete collapse of the star within the time period of half an hour. Although the main energy losses in such collapses are due to neutrino emission which escapes direct observation, the heating of the body of a collapsing star must necessarily lead to the *rapid expansion of the outer layers* and the *tremendous increase of luminosity.* It is suggested that stellar collapses of this kind are responsible for the phenomena of *novae* and *supernovae,* the difference between the two being probably due to the difference of their masses.

INTRODUCTION

ONE OF THE MOST peculiar phenomena which we encounter in the evolutionary life of stars consists in vast stellar explosions known as "ordinary novae" and "supernovae." It is now well established that, although these two classes of novae possess a great many features in common, they are sharply separated insofar as their maximum luminosities are concerned. The ordinary novae, appearing at the rate of about 50 per year in our stellar system, reach at their maximum a luminosity of the order of magnitude of 10^5 suns. On the other hand, the supernovae, flaring up in any given stellar system only once in several centuries, reach luminosities exceeding that of the sun by a factor of 10^9. The intermediate luminosities have never been observed, and there seems to exist a real gap between these two classes of stellar explosions.

The common features of novae and supernovae may be briefly summarized as follows:

(1) Both ordinary novae and supernovae show a very similar form of luminosity curve (apart from the luminosity scale, of course) with a sharp rise to the maximum within a few days or weeks, and a subsequent slow decline of intensity, decreasing by a factor of two every four or five months.

(2) In both cases the spectrum shows rather high surface temperature (up to 20,000° C for ordinary novae, and probably above 30,000° C for supernovae), and the rapid expansion of the stellar atmosphere which is evidently blown up by the increasing radiative pressure. In the case of Nova Aquilae 1918, for example, the star was surrounded by a luminous gas shell expanding with a velocity of 2,000 kilometers per second, whereas the gas masses expelled by the galactic supernova of the year A.D. 1054 (observed by Chinese astronomers) form at present an extensive luminous cloud known as the "Crab-Nebula." It must be noticed here that the large surface area of this blown-up atmosphere is mainly responsible for the observed high luminosities, since the increase of the surface temperature can only account for a factor of several hundreds in the surface brightness.

(3) Whereas the "prenovae," in the rare cases when they have been observed, represent comparatively normal stars of the spectral class *A* (surface temperature about 10,000° C),[1] the "postnovae," remaining after the flare-up, possess extremely high surface temperatures (spectral class *O*) and seem to represent highly collapsed configurations, such as the stars of the "Wolf-Rayet" type.

The same evidently holds true for the case of supernovae, since the star found in the center of the "Crab-Nebula," and representing most probably the remainder of the galactic supernova A.D. 1054, shows the typical features of a very dense "white dwarf."

The change of state caused by these stellar catastrophes strongly suggests that the process involved here is not connected with any instantaneous liberation of intranuclear energy due to some explosive reaction, but rather represents a rapid collapse of the entire stellar body as was first suggested by Milne.[2]

It was suggested by Baade and Zwicky,[3] in particular application to supernovae, that such collapses may be due to the formation within the star of a large number of neutrons which would permit considerably closer "packing" in the central regions.

It is easy to see, however, that the possibility of closer packing alone is quite insufficient to explain the rapid collapse of the star, since such a collapse requires the removal of large amounts of gravitational energy produced by contraction in the interior of the star. In fact, quite independent of whether the matter in the center consists of charged nuclei or neutrons, the heat produced by contraction must pass through the entire body of the star, and the rate of energy transport, depending on the opacity of the main body, will be the same in both cases. On the other hand, if we can find some way of removing instantaneously the heat liberated in the central regions, in spite of the opacity of the stellar body, the star will collapse with a velocity comparable to that of "free fall" independent of the kind of particles existing in its interior.

The amount of gravitational energy which is liberated when a star of a mass M contracts from the original radius R_0 to the collapsed radius R_c is given approximately by

$$\Delta U \cong GM^2\left(\frac{1}{R_c} - \frac{1}{R_0}\right) \cong \frac{GM^2}{R_c}. \qquad (1)$$

Thus, for example, if a star of the mass and radius of our sun contracts to the size of the companion of Sirius ($R_c = R_0/40$),[4] the total liberation of gravitational energy will be of the order of magnitude of 10^{50} erg. On the other hand, the time of the free-fall collapse from the original radius R_0 to any small value of the radius is given approximately by

$$\Delta t \cong R_0^{3/2}/(GM)^{1/2}. \qquad (2)$$

In the case of the sun, Δt will be of the order 10^3 sec. (i.e., about half an hour), so that the mean rate of energy removal necessary for such a collapse is about 10^{14} erg/g sec. If we remember that the collapse of novae and supernovae takes place within a few days, we come to the conclusion that the rate of energy removal in these actual cases may be only several hundred times smaller than given above.

As we suggested in a recent publication,[5] this very fast removal of energy from the interior of the star can be understood on the basis of the present ideas on the role of neutrinos in nuclear transformations involving emission or absorption of β-particles. In fact, when the temperature and density in the interior of a contracting star reach certain values depending on the kind of nuclei involved, we should expect processes of the type

$$\begin{cases} {}_zN^A + e^- \rightarrow {}_{z-1}N^A + \text{antineutrino} \\ {}_{z-1}N^A \rightarrow {}_zN^A + e^- + \text{neutrino}, \end{cases} \qquad (3)$$

which we shall call, for brevity, "urca-processes." The neutrinos formed in the above processes[6] absorb a considerable part of the transformation energy (about $\frac{2}{3}$), and escape with practically no difficulty through the body of the star.

As we shall see later, these processes of absorption and reemission of free electrons by certain atomic nuclei which are abundant in stellar matter may lead to such tremendous energy losses through the neutrino emission that the collapse of the entire stellar body with an almost free-fall velocity becomes quite possible.

Before discussing in more detail the characteristic features of stellar collapse caused by the urca-processes taking place in the hot interior, we must develop the formula for the dependence of "neutrino losses" on the temperature and density of matter.

Energy Losses through Neutrino Emission

Since within a contracting star the central temperature gradually increases from comparatively low to very high values, different urca-processes will assume importance during the different stages of the contraction and collapse. In the very beginning, these processes will involve only those nuclei which can capture free electrons with the energy of few kilovolts, whereas at the height of the collapse the nuclei requiring the energy balance of several Mev may become of primary importance. Thus, it becomes necessary to have the formulas both for nonrelativistic and relativistic electrons. In the relativistic calculations we shall limit ourselves to the extreme relativistic case, since the calculations in the intermediate region are necessarily much more complicated. The values for the intermediate temperatures can be obtained with sufficient approximation by a simple interpolation.

We shall, moreover, consider only the case of an ideal electron gas, which, as we shall see later is of primary importance for the stars having large masses. It is clear that in the case of degeneracy the decreasing number of free energy levels will reduce considerably the rate of urca-processes, and that these processes will stop entirely when the electron gas becomes completely degenerated.

NONRELATIVISTIC CASE Let us consider a unit mass of stellar matter containing n_e free electrons and n_z^0 nuclei of atomic number Z, which can capture these electrons and go over into unstable nuclei of atomic number $Z - 1$. Since the matter in the stellar interior is almost completely ionized, we may write for the number of free electrons

$$n_e = \tfrac{1}{2}\rho/m_H, \qquad (4)$$

where ρ is the density of matter and m_H the mass of the hydrogen atom.[7] For the number of nuclei participating in the urca-process we have evidently

$$n_z^0 = c_z\rho/A_z \cdot m_H, \qquad (5)$$

where c_z and A_z are the concentration and atomic weight of the isotope in question.

If the electron gas is not degenerated, the number of electrons with energy between E and $(E + dE)$ is given by the Maxwellian expression

$$n_e(E)\,dE = 2\pi^{-1/2} n_e \cdot (kT)^{-3/2} \cdot e^{-E/kT} \cdot E^{1/2}\,dE. \qquad (6)$$

The average life of an electron before being captured by the nucleus through the emission of antineutrinos may be calculated on the basis of Fermi's theory of β-decay, and is given by[8]

$$\tau(E) = \pi \lg 2 \cdot \hbar^4 c^3/g^2 n_z (E - Q)^2, \qquad (7)$$

where g is the Fermi constant and Q the maximum electron energy of continuous β-spectrum ($E_{\max}$).

The number of electrons captured by the nuclei in question per unit volume per unit time is

$$N^- = \int_Q^\infty \frac{n_e(E)}{\tau(E)}\,dE. \tag{8}$$

Using (6) and (7) we get

$$N^- = \frac{2g^2 n_z n_e}{\pi^{3/2}\,\lg 2c^3\hbar^4}(kT)^2 I\left(\frac{Q}{kT}\right), \tag{10}$$

where the integral

$$I\left(\frac{Q}{kT}\right) = \int_{x_0}^\infty e^{-x}(x-x_0)x^{1/2}\,dx \tag{11}$$

is written in respect to a new variable

$$x = (E/kT)(x_0 = Q/kT). \tag{12}$$

Estimating this integral by the saddle-point method, we get

$$I(Q/kT) = I(x_0) = 2.15(x_0 + 5/2)^{1/2}\cdot e^{-x_0}. \tag{13}$$

On the other hand, the number of electrons emitted by the unstable nuclei $(Z-1)$, in the energy interval $E, (E+dE)$, is, according to Fermi,

$$dN^+ = \frac{g^2 m^{3/2} n_{z-1}}{\sqrt{2\pi^3\hbar^7c^3}}(Q-E)^2E^{1/2}\,dE \tag{14}$$

and the total number of emitted electrons per second per unit volume

$$N^+ = 0.152\frac{g^2 m^{3/2} n_{z-1}}{\sqrt{2\pi^3\hbar^7c^3}}\cdot Q^{7/2}. \tag{15}$$

In the case of equilibrium, when an equal number of electrons is absorbed and emitted, we must have $N^+ = N^-$ which leads to

$$\frac{n_{z-1}}{n_z} = \left[\frac{2g^2 n_e}{\pi^{3/2}\,\lg 2c^3\hbar^4}(kT)^2 I\left(\frac{Q}{kT}\right)\right]\left[0.152\frac{g^2m^{3/2}}{\sqrt{2\pi^3\hbar^7c^3}}Q^{7/2}\right]^{-1} \tag{16}$$

and

$$n_{z-1} = \left[1 + 6.4\times 10^{-3}\cdot\hbar^{-3}m^{3/2}(kT)^{-2}m_H\rho^{-1}\left(\frac{Q}{kT}+\frac{5}{2}\right)^{-1/2}Q^{7/2}e^{Q/kT}\right]^{-1}\times\frac{c_z\rho}{A_z m_H}. \tag{18}$$

The energy taken up by antineutrinos ejected in the process of electron capture per unit time per unit volume can be easily calculated and found to be

$$W^{(1)} = \frac{8.5g^2 n_e\cdot n_z}{\pi\,\lg 2\hbar^4c^3}\cdot\left(\frac{Q}{kT}+\frac{7}{2}\right)^{1/2}(kT)^3e^{-Q/kT}, \tag{19}$$

while the energy of neutrinos accompanying electron emission is

$$W^{(2)} = \tfrac{2}{3}Q\lambda n_{z-1}, \tag{20}$$

where λ is the decay constant of the unstable nucleus $(Z-1)$. Comparing (19) and (20), and bearing in mind the expression for λ, we get

$$W^{(1)} \cong 5.5(kT/Q)W^{(2)}, \tag{21}$$

so that for the total energy of neutrino radiation per unit time per unit volume we have

$$W = W^{(1)} + W^{(2)} \cong (1 + 5.5\cdot kT/Q)\cdot\tfrac{3}{2}Q\cdot\lambda\cdot n_{z-1}. \tag{22}$$

EXTREME RELATIVISTIC CASE When the extreme relativistic expression for $n_e(E)$ is used in equation (7) and the analysis is carried out with the relativistic expression for N^+, Gamow and Schönberg obtain

$$W = W^{(1)} + W^{(2)} \cong (1 + 8kT/Q)\frac{Q}{2}\cdot\lambda\cdot n_{z-1} \tag{35}$$

for the total energy of neutrino radiation per unit volume per unit time, where

$$n_{z-1} = [1 + 1.2\times 10^{-3}\hbar^{-3}c^{-3}m_H\rho^{-1}\times[12(kT)^2 + 6Q'kT + Q'^2]\times[1 + (1 + m_0c^2/kT)^2]Q^5e^{-Q/kT}]^{-1}\times\frac{c_Z\rho}{A_Z m_H}, \tag{31}$$

where $Q' = Q + mc^2$. Gamow and Schönberg then consider the intermediate case where the temperature of the gas is not sufficiently high for most of the particles to be relativistic, but where the electrons responsible for the Urca process are those from the relativistic Maxwellian-distribution tail.

URCA-PROCESSES WITH VARIOUS ELEMENTS

Since practically any stable nucleus leads to an unstable isobar after capturing one electron, there must be, altogether, many hundreds of possible urca-processes presenting a large display of all possible energies and decay constants. The dependence between the decay energy and the decay constant permits arranging all such processes in one row, beginning with very slow processes, at comparatively low temperatures, and ending with the very fast ones requiring extremely high thermal velocities.[9]

Since our present list of the known Fermi elements is very limited and contains only a small fraction of all possible β-active nuclei, the choice of a particular reaction, or group of reactions, which would be of special importance for the theory of stellar collapse, presents serious difficulties. This choice can be, however, somewhat simplified if we remember that the importance of any given reaction depends not only on its absolute efficiency but also, to a large extent, on the abundance of the element in question in stellar material.

Assuming that the central temperature $T_c = 10^{-15}\ M/R$ and that the central mass density $\rho_c = 13\ M/R^3$ for a star of mass M and radius R, the authors use equations (18), (22), (31), and (35) to compute the energy loss through neutrino emission in a contracting star with a mass of 5 solar masses and μ equal to 1.7. For the Urca process in iron 56 they use $E = 1.7$ MeV, $\lambda = 7.7 \times 10^{-5}\ \text{sec}^{-1}$, and a concentration of 10% to give a neutrino luminosity of 10^{12} ergs gm^{-1} sec^{-1} when the radius is smaller than one-hundredth of the solar radius. The average radiative energy loss per unit mass for this range of radii is a million times lower, but it is comparable to the neutrino energy loss for the Urca process in helium 3.

DYNAMICS OF THE COLLAPSE

Since the hydrodynamical equations describing the collapse of a stellar body, cooled from the center by neutrino emission, are necessarily very complicated, we shall limit ourselves in the present article to the entirely qualitative discussion.

It must be clear first of all that in different stages of stellar contraction the most important role is played by urca-processes with different elements. In the very beginning, when the central temperature is comparatively low, neutrino cooling can be produced only by the elements with the low value of energy balance. As we have seen above, energy losses due to such elements are rather small, even at the saturation point, so that one may expect here only a slight acceleration of normal gravitational contraction. However, as the central temperature rises, new and more powerful urca-processes are introduced, the rate of contraction increases, and there results a catastrophic collapse.

It is important to keep in mind that, since urca-processes take place in the regions of highest temperature, the energy flow towards these points, either through radiative or through convective mechanisms, is entirely excluded.

Thus, in order to maintain a central pressure necessary for the support of the outer layers, *the star must develop in its interior a rapidly growing central condensation.* The heat which will be generated in compressing the extra material into this small central region will be rapidly removed by neutrino emission, and the whole process will be somewhat analogous to the condensation of water vapor contained in a vertical vessel with the bottom cooled by liquid air.

However, whereas in the above example the vertical vapor column, unsupported from beneath, will simply fall down as a whole, gas masses forming the body of a collapsing star will be strongly heated by compression.

Since this heat cannot escape towards the center (except from the layers in the immediate vicinity of the urca-region), part of it must remain in the collapsing body and part be radiated from the surface, thus increasing the stellar luminosity. It is not difficult to see that, under such conditions, inner parts of a stellar body will continue to move towards the center, whereas *the outer regions will begin to expand.*

Such an expansion of the outer parts of the star, resulting in the decrease of their opacity and the increase of radiating surface, must necessarily lead to a rapid rise of luminosity which we observe as the "flare-up" in the nova phenomena. It must be noticed that this "flare-up" probably takes place as soon as the collapse of the interior has progressed to a high degree. At a certain stage the rapidly increasing radiative pressure overbalances the weight of stellar atmosphere, and large masses of gas begin to flow outside forming the luminous shell characterizing all nova and supernova explosions. We must stress here the point that, according to the proposed theory, the increase of luminosity during stellar collapse is an entirely secondary phenomenon, so that in the first approximation these changes of luminosity may be easily neglected.

Let us turn now to the question of the ultimate fate of a collapsing star, and the possible reasons for the difference between the ordinary novae and supernovae.

It was shown by Chandrasekhar[10] that, whereas the stars with mass smaller than $5.75M(\odot)/\mu^2$ $(=2\cdot M(\odot)$ for the molecular weight $\mu = 1.7)$ possess stable configurations with a certain minimum radius (white dwarfs), the stars having the largest masses are subject to unlimited contraction.[11] In view of this fact, it is natural to expect that, for the star masses smaller than the above-given critical value, the collapse will occur on a much smaller scale than for heavier bodies. The increasing condensation in the interior of such light collapsing stars will sooner or later lead to electron degeneracy which will slow down and finally stop all urca-processes. For these small scale collapses we should expect that the final state will be represented by a white dwarf of a mass comparable to that of the original star, and that the amount of material thrown out in the form of a gas shell will be comparatively small. This expectation seems to be in good agreement with observations of ordinary nova phenomena, where the expelled gas shell takes up only about one-hundredth of a percent of stellar mass, and is completely dissolved in space several years after the explosion.

On the other hand, the stars possessing a mass larger than the critical one will undergo a much more extensive collapse, and their ever-increasing radiation will drive away more and more material from their surface. The process will probably not stop until the expelled material brings the mass of the remaining star below the critical value. This process may be compared with the supernovae explosions, in which case the expelled gases form extensive nebulosities (as "Crab-Nebula" or "Filamentary Nebula"), apparently assuming this permanent state after the explosion. This point of view seems also to acquire some confirmation from the fact that the relative number of novae and supernovae occurring in stellar systems is of the same order of magnitude as the relative number of stars having masses smaller and larger than $2 \cdot M(\odot)$.

Conclusion

In the present article we have developed the general views regarding the role of neutrino emission in the vast stellar catastrophes known to astronomy. It must be emphasized that, while the neutrinos are still considered as highly hypothetical particles because of the failure of all efforts made to detect them, the phenomena of which we are making use in our considerations are supported by the direct experimental evidence of nuclear physics. In fact, the experiments of Ellis and Wooster[12] and of Meitner and Orthman[13] *leave no doubt that the energy balance does not hold in the processes of radioactive β-transformations*, and all later evidence on this subject strongly indicates that this "disappearance of energy" occurs in such a way that it appears to be carried away by particles of almost unlimited penetrability.

Whereas the fundamental ideas of the proposed theory are very simple and the physical part of calculations pertaining to the rate of neutrino emission can be easily carried out on the basis of existing formalism, the problem of the dynamics of the collapse represents very serious mathematical difficulties. It is to be hoped that these difficulties will be overcome by the choice of some suitable simplified model.

1. The meagerness of observational material makes it impossible to decide whether "prenovae" are located on the main sequence or to the left of it. There seems to be no doubt, however, that, as the result of explosion, the position of the star in the Hertzsprung-Russell diagram is strongly shifted towards higher surface temperatures and smaller radii.
2. E. A. Milne, *Observatory 54*, 145 (1931).
3. W. Baade and F. Zwicky, *Proc. Nat. Acad. Sci. 20*, 259 (1934).
4. The companion of Sirius possesses a mass which is almost equal to the mass of the sun, and may be considered as representing the type of the white dwarfs obtained by the collapse of the sun.
5. G. Gamow and M. Schönberg, *Phys. Rev. 58*, 1117 (1940).
6. We shall use the term "neutrino" both for ordinary neutrinos and antineutrinos involved in the reaction. since there is no noticeable difference in their behavior. It is also clear that one can *neglect the possibility of mutual annihilation of these particles within a stellar body*, since they escape from the star with practically no collisions.
7. This expression presupposes the absence of hydrogen, since the contraction of a star can start only after all the hydrogen is consumed by (C–H) and (H–H) reactions.
8. This expression differs from the formula given by H. A. Bethe and R. F. Bacher [*Rev. Mod. Phys. 8*, 83 (1936)], since the latter was obtained on the basis of the now abandoned Konopinski-Uhlenbeck form of Fermi's theory.
9. The effect of permitted and nonpermitted β-transitions impairs somewhat the above regularity, and must be investigated separately in each particular case.
10. See, for example, S. Chandrasekhar, *Introduction into the Study of Stellar Structure* (Chicago University Press, 1939), chap. 4.
11. This results from the fact that in very heavy stars the velocities of electrons in the degenerated Fermi gas approach the velocity of light, after which such a gas is unable to support the weight of the stellar body.
12. C. D. Ellis and W. A. Wooster, *Proc. Roy. Soc. A117*, 109 (1927).
13. L. Meitner and W. Orthman, *Zeits. f. Physik 60*, 143 (1930).

64. Discovery of Circularly Polarized Light from a White Dwarf Star

James C. Kemp, John B. Swedlund,
John D. Landstreet, and James R. P. Angel

(*Astrophysical Journal* (*Letters*) *161*, L77–L79 [1970])

In 1968 Franco Pacini and Thomas Gold showed, respectively, that the unusual radiation of the Crab nebula and that of the pulsars can be explained by the huge magnetic fields of rotating neutron stars (see selections 73 and 75). In 1945 Thomas Cowling had shown that for normal stars the electrical conductivity of their ionized material is sufficiently high that their magnetic fields can decay only by diffusing toward regions of the opposite polarity and being neutralized there.[1] For stellar dimensions, however, this diffusion time is longer than stellar lifetimes. Conservation of magnetic flux can therefore be assumed, and the magnetic field of a collapsing star will increase as its surface area decreases. Because normal stars have radii of about 10^{11} cm and surface magnetic field strengths on the order of 100 gauss, a neutron star with a radius of 10^6 cm will have a surface magnetic field strength on the order of 10^{12} gauss. Similarly, the white dwarf stars with radii on the order of 10^9 cm would be expected to have surface magnetic fields on the order of 10^6 gauss.

It was apparently the intense magnetic fields inferred for the newly discovered pulsars that motivated a search for magnetic white dwarfs. In the paper given here, James Kemp and his colleagues provide the first observational evidence that some white dwarfs do in fact have intense surface magnetic fields on the order of 10^7 gauss. Under the assumption that magnetic flux is conserved, this discovery provides additional evidence that white dwarfs are gravitationally collapsed normal stars with the small radii inferred from the assumption that they emit thermal radiation.

Kemp had previously predicted that the continuum radiation emitted from a hot object in the presence of an intense magnetic field would exhibit a wavelength-dependent circular polarization; this effect was demonstrated in the laboratory when metals were heated in the presence of a 10^5-gauss magnetic field. By using a system adapted from the laboratory tests of his theoretical predictions, Kemp and his colleagues were able to show that one white dwarf exhibits wavelength-dependent, circularly polarized light, believed to be caused by a surface magnetic field of 10^7 gauss. Subsequent work has shown that at least a dozen white dwarfs have surface magnetic fields of order 10^6 to 10^8 gauss.

For several of these stars values of surface field strengths between 5 and 50 megagauss are known from Zeeman patterns in the spectra of hydrogen, helium and CH.[2] Higher fields are believed to be present in other stars with unidentified spectral features and strong circular polarization. Many of the magnetic white dwarfs exhibit strong linear polarization whose origin is still controversial.[3]

1. T. G. Cowling, *Proceedings of the Royal Society* (*London*), *A183*, 453 (1945); see also M. H. Wrubel, *Astrophysical Journal 116*, 291 (1952).
2. J. R. P. Angel, *Astrophysical Journal*, *216*, 1 (1977).
3. J. C. Kemp, *Astrophysical Journal 213*, 794 (1977).

Abstract—Strong circular polarization, 1–3 percent, has been discovered in visible light from the semi-DC "peculiar" white dwarf Grw $+70°8247$. This is the first such observation on any white dwarf, and is taken as indicating a strong magnetic field. From the theory of gray-body magnetoemissivity, to which the wavelength dependence of the circular polarization would appear to conform, one estimates a mean projected B field of 1×10^7 gauss.

THIS IS TO REPORT the first detection of circularly polarized light from a white-dwarf star. The discovery follows a series of theoretical speculations regarding the properties of white dwarfs, including their probable rapid spin, and it culminates several months of fruitless search for direct evidence of strong magnetic fields in such stars (Angel and Landstreet 1970; Preston 1970). It is also the first known case of significant circular polarization in visible light from any star, apart from the polarization of narrow Zeeman-split lines in magnetic Ap stars and in sunspots (see also a forthcoming survey of circular polarization by Gehrels 1970).

The star in question is Grw $+70°8247$, visual magnitude 13.2. It is a unique, semi-DC type with very little spectral structure (Greenstein and Matthews 1957), having only a shallow (12% depth) unidentified band at λ4,135 and two even weaker bands at $\lambda\lambda$4,475 and 3,650.

We see a diffuse circular polarization, throughout the visible and near-ultraviolet spectrum and presumably extending into the infrared, of one sign and of magnitude 1–3 percent, generally increasing toward long wavelengths. The sign (sense) is such that, to an observer facing the star, the E-vector in a stationary plane rotates clockwise. The discovery was made on the 24-inch Pine Mountain telescope of the University of Oregon. A photoelastic polarization system was used which was adapted from laboratory magnetoemission experiments (Kemp, Swedlund, and Evans 1970). These measurements were then verified with more detail on a 36-inch telescope at Kitt Peak Observatory, by means of an equivalent system based on polarized photon counting (Angel and Landstreet 1970). We summarize the trend of the data in figure 64.1. Within the time span and resolution involved, the polarization was at least approximately constant in time—certainly no changes in sign were seen. Precaution against spurious results arising in the apparatus, in both sets of measurements, was taken by making readings on miscellaneous comparison stars; these showed no circular polarization on the scale of the present data. A discrepancy of rather less than a factor of 2 between the Pine Mountain and Kitt Peak data in the visible may be accounted for by calibration errors and differing sky conditions. Two Pine Mountain observations, representing about 30-minute signal averages on the nights June 14 and June 26, agreed within 10 percent. However, systematic study of long-period variation (over hours, days, etc.) remains to be done. The vertical error bars in figure 64.1 are estimates of systematic error. Short-time resolution, on the other hand, was limited purely by photon statistics: The measuring intervals required for a standard deviation σ of 10 percent in the measured fractional polarization were about 15 minutes and 5 minutes with the 24- and 36-inch telescopes, respectively; any more rapid, genuine variations are not yet detectable.

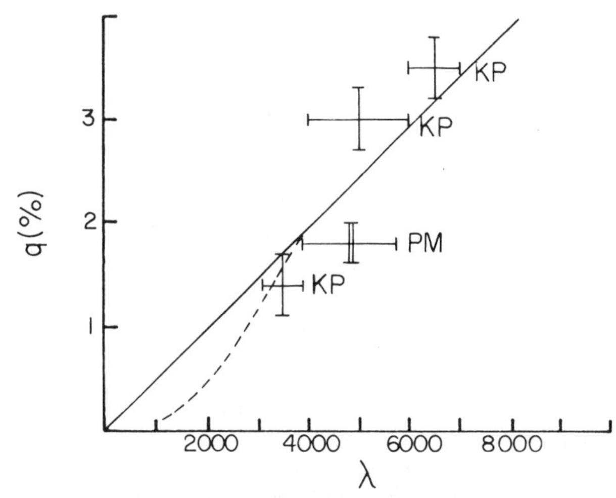

Fig. 64.1 Circular polarization, in percentages, of light from the white dwarf Grw $+ 70°8247$. The data PM refer to two measurements 12 days apart at Pine Mountain, which yielded the same result; the data KP (Kitt Peak) were taken on one occasion. Horizontal bars indicate the approximate wavelength passbands, set by the optical components and filters, where used. The straight diagonal line represents the theoretical graybody $q(\lambda)$ for $B = 1.2 \times 10^7$ gauss. The dashed line suggests the possible effect of a plasma cutoff for $\lambda_p \simeq 3{,}000$ Å.

The polarization here is a very wide-band effect. This indeed made detection feasible with telescopes of modest size, since photons from a large spectral region are collected simultaneously. The polarization data are mean values over a segment of the spectrum (*horizontal bars* in figure 64.1) set, in the Pine Mountain case, by response limits of the optics and photomultiplier alone; in the Kitt Peak work, color filters were used also, and the overall range was extended. A semiquantitative picture of the spectral polarization $q(\lambda)$ was thus obtained.

If the physical origin of the effect is indeed magnetic, we can attribute the wavelength-dependent polarization to the monotonic "graybody" magnetoemission which has been predicted theoretically (Kemp 1970) and demonstrated in the laboratory (Kemp *et al.* 1970). The first-order magnetoemission along the B field is then given by $q(\omega) = -\Omega/\omega = -\lambda\Omega/(2\pi c)$, where $\Omega = eB/(2m)$ is the Larmor frequency. Applying this to the data of figure 64.1 yields $B = (1.2 \pm 0.3) \times 10^7$ gauss—an enormous field by most standards! This must be, of course, the projection along the observer's direction of an actual stellar field B_0, of poloidal geometry for example, inclined at some angle—B_0 being still larger.

No explanation other than the gray-body magnetoemission is at hand to account for the gross features of these observations; however, even the crude data carry some suggestion of departure from the simple linearity $q \propto \lambda$. Two complications are in fact expected (Kemp *et al.* 1970). One is a "spectroscopic" term due to the λ4,135 band, superimposed on the gray-body term and proportional to the spectral derivative of the λ4,135 band shape. The band being quite shallow and narrow, this effect would scarcely affect $q(\lambda)$ under the resolution shown. Second, there is reason to expect a falloff more rapid than λ^{+1} at short λ, i.e., at $\lambda < \lambda_p$, where $\omega_p = 2\pi c/\lambda_p$ is the plasma frequency. This follows from a thermal-equilibrium model for the magnetoemissivity, and it is speculation to apply this literally to surface conditions on a white dwarf. Still, the temptation was overwhelming to sketch in such a deviation, and we have done so in figure 64.1 (*dashed line*). The indicated plasma frequency would pertain to an electron density of order 10^{21} cm^{-3}. Extensions of the data (by the use of large telescopes) into both the middle-ultraviolet and infrared are scheduled, to investigate the truth of the λ^{+1} dependence with a possible plasma correction (which might well be shifted from the location in figure 64.1).

A comment is in order on the spectrum of Grw +70°8247, which, as noted, consists only of two or three vestigial absorption bands. A B-field of 10^7 gauss would displace the central (π) components of Hγ and Hδ by the quadratic Zeeman effect (Preston 1970), to just about the locations of the observed $\lambda\lambda$4,135 and 3,650 features. Whether or not this particular assignment is correct, it is clear that a field of such magnitude would have a profound effect on any stellar spectrum.

The search for magnetic white dwarfs until now had centered on the DA types (having H-lines), with essentially null results. The quest is now shifted toward the DC and some "peculiar" types, based on the premise of diffuse or gray-body magnetic polarization. That Grw +70°8247 is a singular case would be surprising. We can at this time add only a null result on one DC star, G126—27—in which no time-averaged circular polarization as large as 0.2 percent is seen. But the accuracy and scope of the survey can be much extended.

Angel, J. R. P., and Landstreet, J. D. 1970, *Ap. J.* (*Letters*) *160*, L147.

Gehrels, T. 1970, *A.J.* (private communication).

Greenstein, J. L., and Matthews, M. S. 1957, *Ap. J. 126*, 14.

Kemp, J. C. 1970, *Ap. J. 162*, 169.

Kemp, J. C., Swedlund, J. B., and Evans, B. D. 1970, *Phys. Rev. Letters 24*, 1211.

Preston, G. W. 1970, *Ap. J.* (*Letters*) *160*, L143.

65. On the Gravitational Field of a Point Mass according to the Einsteinian Theory

Karl Schwarzschild
TRANSLATED BY BRIAN DOYLE

(*Sitzungberichte der K. Preussischen Akademie der Wissenschaften zu Berlin 1*, 189–196 [1916])

In this paper Karl Schwarzschild provides the exact solution for Einstein's field equations for the space outside a single spherically symmetric particle of constant mass. Beyond the elegance of his exact solution, Schwarzschild thought that his work was of no more importance for astronomy than the approximate solutions already worked out by Einstein.[1] If, on the other hand, the general theory of relativity was applicable to atomic physics, as Schwarzschild hoped it might be, then his exact solution, rather than Einstein's approximate solution, might be needed.[2]

The question of the singularity of the Schwarzschild metric at the now famous Schwarzschild radius $R_S = 2GM/c^2$ was not considered in Schwarzschild's paper. (Here M is the mass, G the universal constant of gravitation, and c the velocity of light.) Perhaps it was omitted because Schwarzschild's solution was for the metric outside the mass, and no known object in the universe had a physical radius anywhere near as small as the Schwarzschild radius; for example, the sun and the proton have respective radii of 7×10^{10} cm and 10^{-13} cm, whereas they have respective Schwarzschild radii of 3×10^5 cm and 10^{-52} cm. In fact, Alexandre Lemaître and Martin Kruskal later showed that the singularity in the Schwarzschild metric could be removed by a transformation of coordinates leading to a nonstatic form of the line element.[3] Nevertheless, it is entertaining to consider what might happen if a body were sufficiently massive to collapse under its own gravity to a physical radius equal to the Schwarzschild radius; this problem was considered at a later time by J. Robert Oppenheimer and Hartland Snyder (see introduction to selection 66).

1. *Sitzungberichte der K. Preussischen Akademie der Wissenschaften zu Berlin 11*, 831–839 (1915); the first English translation is selection 122 in this volume.
2. Others, such as J. Droste [*Proceedings of the Royal Academy of Sciences Amsterdam 19*, 197 (1917)], almost simultaneously worked out the exact solution.
3. A. Lemaître, *Annales de la Société scientifique de Bruxelles 53*, 51–82 (1933). M. D. Kruskal, *Physical Review 119*, 1743 (1960).

I

Albert Einstein has posed the following problem in his work on the perihelion motion of Mercury. A point mass moves according to the requirement

$$\delta \int ds = 0,$$

where (1)

$$ds = \sqrt{\sum g_{\mu\nu}\, dx_\mu\, dx_\nu} \quad \mu, \nu = 1, 2, 3, 4$$

and where $g_{\mu\nu}$ denote functions of the variable x and the variables x are held fixed for the variation at the beginning and end of the integration path. More briefly, the point moves on a geodesic line in the manifold characterized by the line element ds.

Carrying out the variation produces the equation of motion of the point

$$\frac{d^2 x_\alpha}{ds^2} = \sum_{\mu,\nu} \Gamma^\alpha_{\mu\nu} \frac{dx_\mu}{ds}\frac{dx_\nu}{ds}, \tag{2}$$

where [the Christoffel symbol of the second kind is given by]

$$\Gamma^\alpha_{\mu\nu} = -\frac{1}{2}\sum_\beta g^{\alpha\beta}\left(\frac{\partial g_{\mu\beta}}{\partial x_\nu} + \frac{\partial g_{\nu\beta}}{\partial x_\mu} - \frac{\partial g_{\mu\nu}}{\partial x_\beta}\right) \tag{3}$$

and $g^{\alpha\beta}$ denotes the normalized sub-determinant related to $g_{\alpha\beta}$ in the determinant $|g_{\mu\nu}|$.

This is now, according to Einsteinian theory, the motion of a massless point in the gravitational field of a mass located at the point $x_1 = x_2 = x_3 = 0$ if the "components of the gravitational field" Γ satisfy everywhere, except at the point $x_1 = x_2 = x_3 = 0$, the "field equations"

$$\sum_\alpha \frac{\partial \Gamma^\alpha_{\mu\nu}}{\partial x_\alpha} + \sum_{\alpha\beta} \Gamma^\alpha_{\mu\beta}\Gamma^\beta_{\nu\alpha} = 0 \tag{4}$$

and if, also, the "determinantal equation"

$$|g_{\mu\nu}| = -1 \tag{5}$$

is satisfied.

The field equations, in connection with the determinantal equation, have the fundamental property of maintaining their form under the substitution of arbitrary new variables in place of x_1, x_2, x_3, x_4, supposing only that the determinant of the substitution is equal to 1.

Let x_1, x_2, x_3 denote rectangular coordinates and x_4 the time. Moreover, let the mass be located at the origin for all times, and let the motion at infinity be uniformly rectilinear; then, according to Einstein's calculation, the following conditions are to be satisfied:

1. All components are independent of the time x_4.
2. The equations $g_{\rho 4} = g_{4\rho} = 0$ are exactly valid for $\rho = 1, 2, 3$.
3. The solution is spatially symmetric about the origin of the coordinate system in the sense that we obtain the same solution again if we subject x_1, x_2, x_3 to an orthogonal transformation (rotation).
4. The $g_{\mu\nu}$ vanish at infinity with the exception of the following four boundary values, which are unequal to zero:

$$g_{44} = 1, \quad g_{11} = g_{22} = g_{33} = -1.$$

The problem is to discover a line element with such coefficients that the field equations, the determinantal equation, and these four conditions will be satisfied.

II

Einstein has shown that this problem leads, in the first approximation, to the Newtonian law and that the second approximation correctly renders the known anomaly in the motion of the perihelion of Mercury. The following calculation yields the exact solution of the problem. It is always pleasant to have at one's disposal a rigorous solution of simple form: it is more important that the calculation produce, at the same time, the unequivocal determination of the solution. Einstein's treatment still leaves some doubt, and, as is shown below, this uniqueness could be proved only with difficulty by his method. This solution therefore manages to let Einstein's result shine through in increased purity.

III

If one names the time t and the rectangular coordinates x, y, z, the most general line element that satisfies conditions (1)–(3) is manifestly

$$ds^2 = F\, dt^2 - G(dx^2 + dy^2 + dz^2) - H(x\, dx + y\, dy + z\, dz)^2,$$

where F, G, and H are functions of $r = \sqrt{(x^2 + y^2 + z^2)}$. Condition (4) demands that, for $r = \infty$, $F = G = 1$ and $H = 0$.

If one passes over to polar coordinates, according to the relations $x = r \sin\theta \cos\phi$, $y = r\sin\theta\sin\phi$, $z = r\cos\theta$, the same line element reads

$$\begin{aligned} ds^2 &= F\, dt^2 - G(dr^2 + r^2\, d\theta^2 + r^2 \sin^2\theta\, d\phi^2) - Hr^2\, dr^2 \\ &= F\, dt^2 - (G + Hr^2)\, dr^2 - Gr^2(d\theta^2 + \sin^2\theta\, d\phi^2). \end{aligned} \tag{6}$$

However, the volume element in polar coordinates equals $r^2 \sin\theta \, dr \, d\theta \, d\phi$, and the Jacobian determinant for the transformation from old to new coordinates is $r^2 \sin\theta$, which is unequal to 1. Therefore, the field equations would not remain unaltered if polar coordinates were used; an involved transformation would be required. A simple trick sets aside this difficulty. If

$$x_1 = \frac{r^3}{3}, \quad x_2 = -\cos\theta, \quad x_3 = \phi, \tag{7}$$

then the following equation holds for the volume element:

$$r^2 \, dr \sin\theta \, d\theta \, d\phi = dx_1 \, dx_2 \, dx_3.$$

The new variables are therefore polar coordinates of determinant 1. They have the obvious merits of polar coordinates for the treatment of this problem, and at the same time, if t is still taken as equal to x_4, the field equations and the determinantal equation remain in unchanged form.

In the new polar coordinates, the line element reads

$$ds^2 = F\,dx_4{}^2 - \left(\frac{G}{r^4} + \frac{H}{r^2}\right) dx_1{}^2 - Gr^2\left[\frac{dx_2{}^2}{1 - x_2{}^2} + dx_3{}^2(1 - x_2{}^2)\right], \tag{8}$$

for which we want to write

$$ds^2 = f_4\,dx_4{}^2 - f_1\,dx_1{}^2 - f_2\frac{dx_2{}^2}{1 - x_2{}^2} - f_3\,dx_3{}^2(1 - x_2{}^2). \tag{9}$$

Then f_1, $f_2 = f_3$, f_4 are functions of x_1, which must satisfy the following conditions:

1. For $x_1 = \infty : f_1 = 1/r^4 = (3x_1)^{-4/3}, f_2 = f_3 = r^2 = (3x_1)^{2/3}, f_4 = 1$.
2. The determinantal equation in which $f_1 \cdot f_2 \cdot f_3 \cdot f_4$ is equal to 1.
3. The field equations given in equation (4).
4. f must be continuous except at $x_1 = 0$.

IV

In order to be able to set up the field equations, we must first construct the gravitational field components corresponding to the line element given in equation (9). This is most simply done by constructing the differential equations of the geodesic line through direct evaluation of the variation and by reading off the components from these equations. The differential equations of the geodesic line are produced by means of the variation directly in the form

$$0 = f_1\frac{d^2x_1}{ds^2} + \frac{1}{2}\frac{\partial f_4}{\partial x_1}\left(\frac{dx_4}{ds}\right)^2 + \frac{1}{2}\frac{\partial f_1}{\partial x_1}\left(\frac{dx_1}{ds}\right)^2 - \frac{1}{2}\frac{\partial f_2}{\partial x_1}\left[\frac{1}{1 - x_2{}^2}\left(\frac{dx_2}{ds}\right)^2 + (1 - x_2{}^2)\left(\frac{dx_3}{ds}\right)^2\right]$$

$$0 = \frac{f_2}{1 - x_2{}^2}\frac{d^2x_2}{ds^2} + \frac{\partial f_2}{\partial x_1}\frac{1}{1 - x_2{}^2}\frac{dx_1}{ds}\frac{dx_2}{ds} + \frac{f_2 x_2}{(1 - x_2{}^2)^2}\left(\frac{dx_2}{ds}\right)^2 + f_2 x_2\left(\frac{dx_3}{ds}\right)^2$$

$$0 = f_2(1 - x_2{}^2)\frac{d^2x_3}{ds^2} + \frac{\partial f_2}{\partial x_1}(1 - x_2{}^2)\frac{dx_1}{ds}\frac{dx_3}{ds} - 2f_2 x_2\frac{dx_2}{ds}\frac{dx_3}{ds}$$

$$0 = f_4\frac{d^2x_4}{ds^2} + \frac{\partial f_4}{\partial x_1}\frac{dx_1}{ds}\frac{dx_4}{ds}.$$

Comparison with equation (2) gives the components of the gravitational field:

$$\Gamma^1_{11} = -\frac{1}{2}\frac{1}{f_1}\frac{\partial f_1}{\partial x_1}, \qquad \Gamma^1_{22} = +\frac{1}{2}\frac{1}{f_1}\frac{\partial f_2}{\partial x_1}\frac{1}{1 - x_2{}^2},$$

$$\Gamma^1_{33} = +\frac{1}{2}\frac{1}{f_1}\frac{\partial f_2}{\partial x_1}(1 - x_2{}^2),$$

$$\Gamma^1_{44} = -\frac{1}{2}\frac{1}{f_1}\frac{\partial f_4}{\partial x_1},$$

$$\Gamma^2_{21} = -\frac{1}{2}\frac{1}{f_2}\frac{\partial f_2}{\partial x_1}, \qquad \Gamma^2_{22} = -\frac{x_2}{1 - x_2{}^2},$$

$$\Gamma^2_{33} = -x_2(1 - x_2{}^2),$$

$$\Gamma^3_{31} = -\frac{1}{2}\frac{1}{f_2}\frac{\partial f_2}{\partial x_1}, \qquad \Gamma^3_{32} = +\frac{x_2}{1 - x_2{}^2},$$

$$\Gamma^4_{41} = -\frac{1}{2}\frac{1}{f_4}\frac{\partial f_4}{\partial x_1}.$$

(All the others are zero.)

Because of the spherical symmetry about the origin, it suffices to construct the field equations only at the equator ($x_2 = 0$), so that $1 - x_2{}^2 = 1$ may be set from the start everywhere in the preceding expressions (because only first derivatives are taken). Then the computation of the field equations yields

a. $$\frac{\partial}{\partial x_1}\left(\frac{1}{f_1}\frac{\partial f_1}{\partial x_1}\right) = \frac{1}{2}\left(\frac{1}{f_1}\frac{\partial f_1}{\partial x_1}\right)^2 + \left(\frac{1}{f_2}\frac{\partial f_2}{\partial x_1}\right)^2 + \frac{1}{2}\left(\frac{1}{f_4}\frac{\partial f_4}{\partial x_1}\right)^2,$$

b. $$\frac{\partial}{\partial x_1}\left(\frac{1}{f_1}\frac{\partial f_2}{\partial x_1}\right) = 2 + \frac{1}{f_1 f_2}\left(\frac{\partial f_2}{\partial x_1}\right)^2,$$

c. $$\frac{\partial}{\partial x_1}\left(\frac{1}{f_1}\frac{\partial f_4}{\partial x_1}\right) = \frac{1}{f_1 f_4}\left(\frac{\partial f_4}{\partial x_1}\right)^2.$$

In addition to these three equations, the functions f_1, f_2, f_4 have still to satisfy the determinantal equation

d. $$f_1 f_2^2 f_4 = 1$$

or

$$\frac{1}{f_1}\frac{\partial f_1}{\partial x_1} + \frac{2}{f_2}\frac{\partial f_2}{\partial x_1} + \frac{1}{f_4}\frac{\partial f_4}{\partial x_1} = 0.$$

To begin with, I leave out equation (b) and determine the three functions f_1, f_2, f_4 from equations (a), (c), and (d). Then (c) can be rearranged into the form

c′. $$\frac{\partial}{\partial x_1}\left(\frac{1}{f_4}\frac{\partial f_4}{\partial x_1}\right) = \frac{1}{f_1 f_4}\frac{\partial f_1}{\partial x_1}\frac{\partial f_4}{\partial x_1}.$$

This can be immediately integrated and gives

c″. $$\frac{1}{f_4}\frac{\partial f_4}{\partial x_1} = \alpha f_1,$$

where α is an integration constant. Equations (a) and (c′) together give

$$\frac{\partial}{\partial x_1}\left(\frac{1}{f_1}\frac{\partial f_1}{\partial x_1} + \frac{1}{f_4}\frac{\partial f_4}{\partial x_1}\right) = \left(\frac{1}{f_2}\frac{\partial f_2}{\partial x_1}\right)^2 + \frac{1}{2}\left(\frac{1}{f_1}\frac{\partial f_1}{\partial x_1} + \frac{1}{f_4}\frac{\partial f_4}{\partial x_1}\right)^2,$$

which combined with equation (d) gives

$$-2\frac{\partial}{\partial x_1}\left(\frac{1}{f_2}\frac{\partial f_2}{\partial x_1}\right) = 3\left(\frac{1}{f_2}\frac{\partial f_2}{\partial x_1}\right)^2.$$

This may be integrated to

$$\frac{1}{\dfrac{1}{f_2}\dfrac{\partial f_2}{\partial x_1}} = \frac{3}{2}x_1 + \frac{\rho}{2},$$

where ρ is an integration constant, or

$$\frac{1}{f_2}\frac{\partial f_2}{\partial x_1} = \frac{2}{3x_1 + \rho}.$$

Integrating again,

$$f_2 = \lambda(3x_1 + \rho)^{2/3},$$

where λ is an integration constant. The condition at infinity requires $\lambda = 1$. Therefore,

$$f_2 = (3x_1 + \rho)^{2/3}. \tag{10}$$

With that, it follows from equations (c″) and (d) that

$$\frac{\partial f_4}{\partial x_1} = \alpha f_1 f_4 = \frac{\alpha}{f_2^2} = \frac{\alpha}{(3x_1 + \rho)^{4/3}}$$

Integrating, subject to the condition at infinity,

$$f_4 = 1 - \alpha(3x_1 + \rho)^{-1/3}. \tag{11}$$

Thus, from equation (d),

$$f_1 = \frac{(3x_1 + \rho)^{-4/3}}{1 - \alpha(3x_1 + \rho)^{-1/3}}. \tag{12}$$

The equation (b) is, as may be easily verified, satisfied by the given expressions for f_1 and f_2.

Thus, all the conditions are satisfied up to the continuity condition; and f_1 will be discontinuous when

$$1 = \alpha(3x_1 + \rho)^{-1/3}, \quad 3x_1 = \alpha^3 - \rho.$$

In order for the discontinuity to coincide with the origin, we must have

$$\rho = \alpha^3. \tag{13}$$

The continuity condition therefore couples the integration constants α and ρ in this way.

The complete solution to our problem now reads

$$f_1 = \frac{1}{R^4}\frac{1}{1 - \alpha/R}, \quad f_2 = f_3 = R^2, \quad f_4 = 1 - \alpha/R,$$

where the auxiliary quantity

$$R = (3x_1 + \rho)^{1/3} = (r^3 + \alpha^3)^{1/3}$$

has been introduced.

If one inserts these values of the function f in expression (9) for the line element and at the same time returns to the usual polar coordinates, the line element that forms the exact solution to the Einsteinian problem is produced:

$$ds^2 = (1 - \alpha/R)\,dt^2 - \frac{dR^2}{1 - \alpha/R} - R^2(d\theta^2 + \sin^2\theta\,d\phi^2),$$

$$R = (r^3 + \alpha^3)^{1/3}. \tag{14}$$

This expression contains the constant α, which depends upon the magnitude of the mass located at the origin.

For a mass M and a Newtonian gravitational constant G, we have $\alpha = 2GM/c^2$, where c is the velocity of light. Next Schwarzschild shows the difficulty of deriving his unique expression from the approximation procedures used by Einstein; he then derives the motion of a point in the gravitational field and shows that this result is identical to Einstein's previously obtained result. In general, Einstein's approximation for the orbital curve of a planet changes into the exact solution given by Schwarzschild if one introduces instead of r the quantity

$$R = (r^3 + \alpha^3)^{1/3} = r\left(1 + \frac{\alpha^3}{r^3}\right)^{1/3}.$$

Because α/r is of the order of 10^{-12} for Mercury, the two solutions are identical for all practical purposes.

In concluding, Schwarzschild derives the exact expression for the angular velocity, $n = (d\phi)/(dt)$, of an object in circular orbit about a central mass. He obtains

$$n^2 = \frac{\alpha}{2}x^3 = \frac{\alpha}{2R^3} = \frac{\alpha}{2(r^3 + \alpha^3)}.$$

Down to the solar surface, the deviation of this formula from Kepler's third law is completely imperceptible. However, for an ideal mass point, it follows that the angular velocity does not become unbounded, as in Newtonian theory, upon diminution of the orbital radius, but approaches a definite value, $n_0 = 1/(\alpha\sqrt{2})$. If similar laws prevail for molecular forces, this fact could be of interest.

66. On the Theory of Stars

Lev Davidovich Landau

(*Physicalische Zeitschrift der Sowjetunion 1*, 285–288 [1932])

The idea that the gravitational forces on the surface of a star might be so large that light could not escape from it was originally stated by Pierre Simon Laplace in 1798.[1] Laplace first noticed that a material particle can only escape a star's surface if the kinetic energy of the particle's motion exceeds the energy of gravitational attraction by the star; that is, a particle can only escape from the surface of a star of mass M and radius R if the square of the particle's velocity exceeds the limit $2GM/R$, where G is the Newtonian constant of gravity. Because light waves travel at the velocity c, Laplace was led to the conclusion that light would not escape from an object whose radius was smaller than $2GM/c^2$. This limiting radius is sometimes called the gravitational radius, and because no light can escape from an object with a radius smaller than this, such an object is called a black hole. The limiting radius derived by Laplace matches the distance at which Karl Schwarzschild's metric, which describes space-time external to a spherical, static mass, becomes "singular" (see selection 65 and its introduction). Hence, it is also often called the Schwarzschild radius.

Until recently, the densest known stars were the white dwarfs, and these stars were known to be supported against gravitational collapse by the pressure of their degenerate electron gas. For sufficiently high mass densities, however, the electrons can attain very high velocities, and the relativistic dependence of mass on velocity has to be taken into account in determining the electron gas pressure. As Lev Landau shows in the first paper given here, these relativistic effects cause the rate of change of pressure with density to decrease at high densities, and for this reason a maximum stable mass exists for stars composed of a degenerate electron gas. Landau concludes that those stars whose mass exceeds the critical mass of about 1.5 solar masses must, because of their increasing gravitational forces, collapse without limit into a point. It is Landau, then, who probably deserves the credit for first predicting theoretically in the context of modern physics the possible formation of a black hole, although it appears that Subrahmanyan Chandrasekhar had very similar ideas independently at nearly the same time. In 1931 Chandrasekhar had derived the critical mass for a degenerate electron gas, but he did not consider the fate of a more massive star. In a paper submitted for publication six months after Landau had submitted his paper, however, Chandrasekhar considered stars whose mass exceeds the critical mass and concluded that for these stars "the perfect gas equation does not break down, however high the density may become, and the matter does not become degenerate. An appeal to Fermi-Dirac statistics to avoid the central singularity cannot be made."[2]

By 1934 Chandrasekhar had worked out the exact theory for the equilibrium states of a degenerate electron gas and had pointed out that the life histories of stars of small and large mass are radically different. For a star of small mass the natural white dwarf stage is an initial stage toward complete darkness, but for a star of large mass other possibilities have to be considered. These other possibilities were delineated by A. S. Eddington, who argued that such a star will "go on radiating and radiating, and contracting and contracting until, I suppose, it gets down to a few kilometers in radius, when gravity becomes strong enough to hold in the radiation and the star can at last have peace." For Eddington, however, this situation seemed preposterous, and he stated his opinion that "there should be a law of nature to prevent a star from behaving in this absurd way."[3]

It was the physicists J. Robert Oppenheimer and Hartland Snyder who, seven years later, first focused attention on the process of gravitational collapse itself.[4] They realized that at the high mass densities in collapsing stars the gravitational effects become large enough to affect noticeably the properties of space-time and that Einstein's general theory of relativity, with its processes of time dilation, light deflection, and gravitational red shifts, was needed to describe the dynamics of collapse. By building upon the work of Karl Schwarzschild, who had shown that the space-time outside a spherical mass has a "singular" behavior at the gravitational radius, Oppenheimer and Snyder were able to show that matter collapses in a finite time within the gravitational radius and that this radius defines a trapped surface from which light cannot escape to infinity. The appearance of a collapsing star is different, however, for an observer located on the stellar surface and for an external observer. This is because we must take into account the time dilation and red shifts of signals exchanged between the star and an external observer located outside the star's gravitational field. Oppenheimer and Snyder were able to show that these effects cause a collapsing star to appear (to an external observer) to collapse in infinite time to an infinite red shift and a radius equal to the gravitational radius.

In this paper, Oppenheimer and Snyder laid the groundwork for what has been called "black hole" physics. Although they considered the pressure-free collapse of a nonrotating, chargeless, spherical mass, subsequent studies have shown that factors deriving from pressure, lack of spherical symmetry, rotation, or charge will not prevent matter from collapsing to an infinite density.[5] These considerations are rather esoteric theoretical predictions, and a practical astronomer is more interested in observational evidence for the existence of black holes. As we shall see in selection 67, black holes may be observed through their effects on nearby normal stars, and nearly thirty years after Oppenheimer and Snyder's paper, a binary X-ray star was observed that provides a candidate for a black hole.

1. P. S. Laplace, *Exposition of the System of the World* translated from the French by J. Pond (London: Printed for R. Phillips, 1809). Here Laplace asserts that a luminous body in the universe of the same density as the earth, whose diameter is 250 times larger than that of the sun, can by its attractive power prevent its light from reaching us and that consequently the largest bodies in the universe could remain invisible to us. The proof of this assertion is given in *Allgemeine geographische Ephemeriden*, 603 (May 1798), and an English translation of this proof is given by S. W. Hawking and G. F. R. Ellis in *The Large Scale Structure of Space-Time* (Cambridge: Cambridge University Press, 1973).

2. S. Chandrasekhar, *Zeitschrift für Astrophysik 5*, 321 (1932), and *Observatory 57*, 373 (1934).

3. A. S. Eddington, *Observatory 58*, 37 (1935).

4. *Physical Review 56*, 455 (1939). H. Gursky and R. Ruffini have reproduced the entire paper in *Neutron Stars, Black Holes and Binary X-Ray Sources* (Boston: D. Reidel, 1975).

5. Theoretical descriptions of black holes and the relevant references to journal articles are given by K. R. Lang in *Astrophysical Formulae* (New York: Springer-Verlag, 1974) and by W. Misner, K. S. Thorne, and J. A. Wheeler in *Gravitation* (San Francisco: W. H. Freeman, 1973). A bibliography of papers on black holes published between 1969 and 1974 appears in *Black Holes 1969–1974*, Institution of Electrical Engineers (Inspec), (London: McCorgnodale, 1974).

The astrophysical methods usually applied in attacking the problems of stellar structure are characterised by making physical assumptions chosen only for the sake of mathematical convenience. By this is characterised, for instance, Mr. Milne's proof of the impossibility of a star consisting throughout of classical ideal gas; this proof rests on the assertion that, for arbitrary L and M, the fundamental equations of a star consisting of classical ideal gas admit, in general, no regular solution. Mr. Milne seems to have overlooked the fact, that this assertion results only from the assumption of opacity being constant throughout the star, which assumption is made only for mathematical purposes and has nothing to do with reality. Only in the case of this assumption the radius R disappears from the relation between L, M and R necessary for regularity of the solution. Any reasonable assumptions about the opacity would lead to a relation between L, M and R, which relation would be quite exempt from the physical criticisms put forward against Eddington's mass-luminosity-relation.

It seems reasonable to try to attack the problem of stellar structure by methods of theoretical physics, i.e. to investigate the physical nature of stellar equilibrium. For that purpose we must at first investigate the statistical equilibrium of a given mass without generation of energy, the condition for which equilibrium being the minimum of free energy F (for given temperature). The part of free energy due to gravitation is negative and inversely proportional to some average linear dimensions of the system in question, or, in other words, it is proportional to—$\rho^{1/3}$ (ρ some average density).

The remaining inner part of free energy depends on the equation of state; for the classical ideal gas it is proportional to log ρ. In view of the fact that log ρ tends to infinity for $\rho \to \infty$ more slowly than $\rho^{1/3}$ we will always have a minimum of free energy at $\rho = \infty$. That means that, in the case of classical ideal gas, we obtain no equilibrium at all. Every part of the system would tend to a point. The state of affairs becomes quite different when we consider the quantum effects. For the non-relativistic Fermi-gas the inner free energy is changing with ρ as $\rho^{2/3}$, that means, more rapidly than the gravitational. It would lead to the existence of a stable equilibrium. The presence of sources would produce only an additional expansion due to the radiation pressure. Mr. Milne tries to escape from this conclusion by introducing a condensed inner part of the system, but he does not tell the reasons why such condensations could appear at all. The connexion of the condensed state with the normal state remains rather mysterious. It is easy to see that any equation of state leading to no discontinuous transitions (as admitted in Milne's calculations) would never make such condensations possible.

As the velocities of electrons in the Fermi-distribution rise with the density we have to apply, for sufficiently great densities, the relativistic theory. In the extreme-relativistic case the inner free energy per unit volume varies as $\rho^{1/3}$, i.e. the same power of density as the gravitational energy. The free energy F is therefore of the form $F = \alpha\rho^{1/3}$. If α is positive the system will expand in order to have F minimum, until the density becomes too small for the extreme-relativistic relation $F = \alpha\rho^{1/3}$ to be valid. If α is negative the system will have a tendency to collapse to a point. In order to find the criterion separating these two cases we have to investigate the solution of the general equation

$$\frac{1}{r^2}\frac{d}{dr}\left[r^2\frac{d\mu}{dr}\right] = -4\pi G\rho \qquad (1)$$

The chemical potential μ is in the extreme-relativistic case equal to $hc[(3\pi^2\rho)/m^4]^{1/3}$ where m is the mass per 1 electron, that means for most elements two protonic masses. The equation (1) is the $n = 3$ polytropic equation of Emden. With known substitutions it can be brought to the form

$$\frac{1}{r^2}\frac{d}{dr}\left[r^2\frac{dx}{dr}\right] = -x^3$$

with the boundary condition

$$-r^2\frac{dx}{dr} = M\left[\frac{G}{hc}\right]^{3/2} m^2\left[\frac{4}{3\pi}\right]^{1/2}$$

on the outer boundary. As Emden has shown, this equation. has a regular solution only in the case $-r^2(dx)/(dr) = 2.015$, in which case it admits arbitrary radii; that means, we will have an indifferent equilibrium corresponding to $\alpha = 0$ in the former rough treatment. Thus we get an equilibrium state only for masses [smaller] than a critical mass $M_0 = 3.1/m^2[hc/G]^{3/2} = 2.8 \cdot 10^{33}$ gr. or about $1.5\odot$ (for $m = 2$ protonic masses). For $M > M_0$ there exists in the whole quantum theory no cause preventing the system from collapsing to a point (the electrostatic forces are by great densities relatively very small). As in reality such masses exist quietly as stars and do not show any such ridiculous tendencies we must conclude that *all stars heavier than $1.5\odot$ certainly possess regions in which the laws of quantum mechanics* (and therefore of quantum statistics) *are violated.* As we have no reason to believe that stars can be divided into two physically different classes according to the condition $M >$ or $< M_0$, we may with great probability suppose that all stars possess such pathological regions. It does not contradict the above arguments, which prove only that the condition $M > M_0$ is sufficient (but not necessary) for the existence of such regions. It is very natural to think that just the presence of these regions makes stars stars. But if it is so, we have no need to suppose that the radiation of stars is due to some mysterious process of mutual annihilation of protons and electrons, which was never observed and has no special reason to occur in stars.

Indeed we have always protons and electrons in atomic nuclei very close together, and they do not annihilate themselves; and it would be very strange if the high temperature did help, only because it does something in chemistry (chain reactions!). Following a beautiful idea of Prof. Niels Bohr's we are able to believe that the stellar radiation is due simply to a violation of the law of energy, which law, as Bohr has first pointed out, is no longer valid in the relativistic quantum theory, when the laws of ordinary quantum mechanics break down (as it is experimentally proved by continuous-rays-spectra and also made probable by theoretical considerations).[1] We expect that this must occur when the density of matter becomes so great that atomic nuclei come in close contact, forming one gigantic nucleus.

On these general lines we can try to develop a theory of stellar structure. The central region of the star must consist of a core of highly condensed matter, surrounded by matter in ordinary state. If the transition between these two states were a continuous one, a mass $M < M_0$ would never form a star, because the normal equilibrium state (i.e. without pathological regions) would be quite stable. Because, as far as we know, it is not the fact, we must conclude that the condensed and non-condensed states are separated by some unstable states in the same manner as a liquid and its vapour are, a property which could be easily explained by some kind of nuclear attraction. This would lead to the existence of a nearly discontinuous boundary between the two states.

The theory of stellar structure founded on the above considerations is yet to be constructed, and only such a theory can show how far they are true.

1. L. Landau and R. Peierls, *ZS. f. Phys. 69*, 56 (1931).

67. The Discovery of a Candidate Black Hole

Cygnus X–1: A Spectroscopic Binary with a Heavy Companion?

B. Louise Webster and Paul Murdin

(*Nature 235*, 37–38 [1972])

Dimensions of the Binary System HDE 226868: Cygnus X–1

Charles Thomas Bolton

(*Nature: Physical Science 240*, 124–127 [1972])

For many years astronomers have imagined neutron stars and black holes as theoretical entities, but because no internal thermal energy sources exist in these objects, they believed that such objects would for practical purposes be invisible (see selections 66 and 69). They gradually realized, however, that the gravitational energy released during the collapse of a normal star into a neutron star will be converted into rotational energy and that this energy can be transformed into electromagnetic energy by the intense magnetic fields of the neutron stars. When pulsars were subsequently observed and interpreted as neutron stars radiating in this manner, astronomers began to believe that whatever could be imagined theoretically might actually exist and be observed. When very massive stars undergo gravitational collapse, however, there is no equilibrium state, and the star continues to collapse forever into a black hole. Neither the nuclear structure of matter nor magnetic fields are required to describe a black hole; in fact, a magnetic field cannot be anchored to a black hole, and no form of electromagnetic radiation can escape from its surface. For these reasons, it seemed that such hypothetical objects were truly black and invisible.

Nevertheless, as Iosif Shklovskii had already argued in 1962, the powerful emission of radio galaxies and quasars might arise from the release of the gravitational energy of a collapsing object during the process of mass accretion.[1] Surrounding gas would heat up and radiate as it plunged into the tremendous gravitational fields of the collapsing object. Other Russian astronomers quickly took up this idea and extended it to the case when one member of a binary star system undergoes gravitational collapse. If an invisible black hole were close enough to a large visible companion, gas would be drawn from the visible companion into the invisible black hole, where the gas might become hot enough to emit X rays. Furthermore, the presence of a black hole might be deduced from an analysis of the orbital elements of the visible companion. Examinations of catalogues of spectroscopic binary systems did not, however, lead to the discovery of an invisible companion star massive enough to be a black hole, and the only remaining hope for the detection of a black hole was through the X-ray emission of the gas it would accrete.

Although extraterrestrial X-ray sources were discovered in 1962, not until 1966 did rocket observations give positions accurate enough to identify an X-ray source with a stellar object. In 1970, the *Uhuru* satellite was launched and a number of additional X-ray stars were discovered. The most exciting discovery of the *Uhuru* satellite grew out of the detection of rapid, irregular X-ray bursts from one of the first X-ray sources to be discovered, Cygnus X–1. Because the X-ray intensity of this object varies irregularly and as rapidly as a tenth of a second, it has to be an extremely small stellar object. When radio astronomers searched for this star they found a weak, variable radio object whose accurate position led to the discovery of an optical counterpart.[2]

As illustrated in the papers given here, Cygnus X–1 is identified with a spectroscopic binary system whose period is 5.6 days. The visible companion of this system is a ninth-magnitude blue supergiant star, and the conservative estimate of about 12 solar masses for this object leads to a value larger than 3 solar masses for the invisible companion. This means that the invisible object is too massive to be a white dwarf or a neutron star and that it is in all probability a black hole. There is some uncertainty, however, in this logical chain of reasoning, and certain alternative explanations can be contrived to fit the data. For example, both the absolute luminosity and the mass of the visible companion have been determined from its optical line spectra, and in the absence of an independent luminosity estimate there is the possibility that the visible companion is a low-mass star with the same surface conditions as a blue supergiant. If this were the case, the invisible companion would have a mass characteristic of a neutron star or a white dwarf rather than a black hole. Studies of the reddening of stars in the field of Cygnus X–1 seem to rule out this possibility, but nevertheless, the controversy over this candidate black hole continues to rage.

1. I. S. Shklovskii, *Soviet Astronomy A.J. 6*, 465 (1962). Also see S. Hayakawa and M. Matsuoka, *Progress in Theoretical Physics (Japan) Supplement 30*, 204 (1964); E. E. Salpeter, *Astrophysical Journal 140*, 796 (1964); and Ya. B. Zeldovich, *Soviet Physics (Doklady) 9*, 246 (1964).

2. L. Braes and G. K. Miley, *Nature 232*, 246 (1971); R. M. Hjellming and C. M. Wade, *Astrophysical Journal (Letters) 168*, L21 (1971).

Webster and Murdin, Cygnus X–1

We have reported[1] that the spectrum, colours and interstellar features in the star HD 226868, which is coincident with the X-ray star Cygnus X-1 and a radio star,[2,3] were those of a normal B0Ib supergiant. We made measurements of the radial velocity of the star from August 1971 to October 1971 and find it to be a velocity variable, whose changes of velocity correlate with changes in the X-ray flux.

Omitted here is a plot of radial velocity as a function of phase; these data are shown in figure 67.1 of our accompanying selection.

The velocity data can be interpreted as showing that the B star (of mass M_B) is in a low eccentricity orbit of inclination, i, to the plane of the sky around an object of mass m, where the mass function of the system is

$$\frac{m^3 \sin^3 i}{(m + m_B)^2} = 0.12\, M_\odot.$$

The projected semi-major axis of the orbit of the B star is

$$a_B \sin i = 4.6 \times 10^6 \text{ km}$$
$$= 6.6\, R_\odot.$$

A lower limit to the mass of the companion is, from the mass function,

$$m \gtrsim 0.5\, M_B{}^{2/3}.$$

To have m less than 2 $M_\odot$, we must have the mass of the B0Ib star less than 6 $M_\odot$, and $\sin i = 1.0$. On this interpretation of the spectroscopic evidence alone, the system is similar to, but on the short period fringe of, other spectroscopic binaries containing blue supergiants.[4] In nearly all these systems—and especially the short period ones—the supergiant shows emission lines; emission lines are not seen in our spectra of HD 226868 (we have not examined Hα). We do not see a second spectrum or even variations in the widths of the absorption lines, and would not if the companion were a normal B0V or B8I star, or fainter.

A blue supergiant rotating synchronously in a binary system of period 5.6 days has an equatorial velocity of 9.1 R_B km/s, where R_B is the radius of the B star in solar radii. Comparing the width of the helium lines in this star with lines in spectrograms of Slettebak and Howard's stars[5] 55 Cygni (B3Ia, $v \sin i = 0$ km/s) and 6 Cephei (B3V, $v \sin i = 150$ km/s), we estimate a projected rotational velocity of

$$v \sin i = 100 \pm 20 \text{ km/s}.$$

This observation is consistent with $R_B \sim 10 \operatorname{cosec} i$, so that it is plausible to suggest that $10 \leqslant R_B \leqslant 30\, R_\odot$ (see ref. 6) implies $0.3 \leqslant \sin i \leqslant 1.0$. A much smaller value for $\sin i$ would give an equatorial speed equivalent to escape velocity and rotational break-up.

On the spectroscopic evidence alone, we cannot entirely eliminate the possibility that the velocity variations represent an expansion and contraction of the star. The fact that the projected semi-major axis of the B star orbit is smaller than the radius of a B0Ib star is reminiscent of the (incorrect) binary interpretation of the cepheid variables. The spectral type does not, however, show any of the variations with phase which one might expect if the star were changing its radius in its cycle by a factor of order two.

We have searched the X-ray data for correlations with the velocity changes. We were guided by Dolan's suggestion[7,8] that the changes in the non-pulsed intensity of the Cygnus X-1 star are attributable to its being an eclipsing binary with a period of 2.98 days. Dolan classified measurements of the X-ray flux with energy greater than 20 keV reported in the literature during the past 5 yr according to whether they were consistent with a maximum or minimum intensity. In addition, he interpreted a recent observation[9] of a sudden drop in intensity of X-rays from Cyg X-1 as the beginning of an eclipse of one of two X-ray components in the source. We have adopted *in toto* Dolan's interpretations of the X-ray flux measurements and we have computed the X-ray flux as a function of phase reduced with a period of 5.6075 days, which is also fitted by our velocity data. The phase of minimum X-ray intensity coincides with the phase of the change of velocity of the B star from positive to negative. This phase in the nth cycle occurs at times given by

$$t = \text{JD } 2{,}441{,}166.06 + 5.6075\, n.$$

Therefore, the dip in the X-radiation beam pattern (corresponding to an attenuation by a factor of about 2 over an angle of about 60°) points to Earth when the companion is between Earth and the B star.

It is this last fact which makes a simple eclipsing model for the modulation of the X-rays impossible. We can make a tentative interpretation as follows. We identify the source of the X-rays with the companion. To allow the X-radiation to pass over the B star when the companion is beyond it, the separation of the stars, d, must be such that

$$d \cos i \geqslant R_B.$$

Coupling this condition with the mass function, we find

$$\frac{(1 + m/M_B)^{2/3}}{(0.12/M_B)^{2/3}} - \frac{(1 + m/M_B)^2}{(m/M_B)^2} \geqslant R_B{}^2/(6.6)^2.$$

Using $R_B \geqslant 10$ and $10 \leqslant M_B \leqslant 30$ we find m to lie between 2.5 and 6.0 $M_\odot$. A blue supergiant with $M_B = 10\ M_\odot$ would fill the Roche lobe of such a short period orbit (see, for example, Plavec[10]) ejecting material to fall on to the companion and, presumably, causing non-isotropic X-ray emission. It is tempting to suppose that this interaction between the B star and its companion is connected with the X-ray fluctuations with 0.1 s timescales discovered in this source by Oda *et al.*[11] The small distance scale associated with fluctuations as rapid as this, and the high energy X-rays, suggest that we are not dealing with a normal main sequence or supergiant star and that the companion is a white dwarf or neutron star. The mass of the companion probably being larger than about 2 $M_\odot$, it is inevitable that we should also speculate that it might be a black hole.

1. P. Murdin and B. L. Webster, *Nature 233*, 110 (1971).
2. L. L. E. Braes and G. K. Miley, *Nature 232*, 246 (1971).
3. S. Rappaport, W. Zaumen, and R. Doxsey, *Astrophys. J. Lett. 168*, L17 (1971).
4. R. Stothers and T. Lloyd Evans, *Observatory 90*, 186 (1970).
5. A. Slettebak and R. E. Howard, *Astrophys. J. 120*, 102 (1955).
6. C. W. Allen, *Astrophysical Quantities*, 2nd ed. (London: University of London Press, 1964).
7. J. F. Dolan, *Space Sci. Rev. 10*, 830 (1971).
8. J. F. Dolan, *Nature 233*, 109 (1971).
9. P. C. Agrawal, G. S. Gokhale, V. S. Iyengar, P. K. Kunte, R. K. Manchanda, and B. V. Sreekantan, *Nature 232*, 38 (1971).
10. M. Plavec, *Bull. Astron. Inst. Czech. 18*, 253 (1967).
11. M. Oda, P. Gorenstein, H. Gursky, E. Kellogg, E. Schreier, H. Tannanbaum, and R. Giacconi, *Astrophys. J. Lett. 166* L1 (1971).

Bolton, Dimensions of the Binary System HDE 226868

It has been suggested[1] that the weak variable radio source in the position error box of Cygnus X-1 is associated with the X-ray source and that both sources coincide[2] with the star HDE 226868. Recent observations have provided strong evidence confirming this association.[3] Optical observations of HDE 226868 during the 1971 observing season showed that the star was a spectroscopic binary with a period of 5.6 day, that mass was flowing from the primary star towards the secondary, and that the secondary was too massive to be a normal white dwarf or neutron star.[4–6]

Here I combine more recent observations of HDE 226868 at the David Dunlap Observatory (DDO) with previously published data[6] from the Royal Greenwich Observatory (RGO) to derive improved orbital elements. The distance of the system, the photometric variations, and the mass exchange strongly imply that the secondary is a black hole.

The new DDO data are similar to those obtained previously.[5] All plates were taken at a dispersion of 12 Å mm^{-1}, but improved weather conditions made it possible to use smaller projected slit widths and to widen the spectra more; there are some small differences between the 1971 data and those previously published. First, a reduction error caused the originally published errors to be too large by about a factor of four. Second, a detailed study of the effect of the emission lines on the measured velocities has led me to make small changes in a few velocities for the hydrogen emission lines. These are listed here together with a qualitative estimate of the emission line strength. The principal change is the listing of the velocity of the He II λ4,686 emission line in a separate column, because recent plates indicate that the velocities for this line are in antiphase to the adsorption line velocities.

We omit a table giving the observational parameters for the star.

Orbital Elements

To compute the orbital elements the velocities were assigned subjective weights: 10 for velocities with errors of 1 to 3 km s^{-1} and 3 for velocities with errors of 10 km s^{-1}. Image tube plates and under-exposed plates were assigned somewhat lower weights than were indicated by their internal errors.

Orbital elements computed from the absorption line velocities are given in table 67.1 with their associated standard deviations. Similar orbital elements have been obtained by Kristian and Brucato.[7]

Because the total time spanned by the observations is shorter than is necessary to determine a highly accurate period, the period was included as an element in the solution. The formal error for the period that is obtained in this way is probably unrealistically small. When the period is derived by Deeming's method,[8] a period of 5.600 ± 0.003 day is obtained. The orbital elements are not significantly changed by using any period within the range allowed by Deeming's method.

Table 67.1 Orbital Elements for HDE 226868

$V_0 = -6.0 \pm 1.0$ km s^{-1}
$P = 5.5995 \pm 0.0009$ day
$K = 68.2 \pm 1.7$ km s^{-1}
$K_{(\mathrm{HeII})} = 102 \pm 6$ km s^{-1}
$e = 0.09 \pm 0.02$
$\omega = 335° \pm 14°$
$T =$ JD 2,441,214.9 ± 0.2
$a \sin i = 5.2 \pm 0.1 \times 10^6$ km $= 7.5\ R_\odot$
$f(m) = 0.182$

Separate solutions for the DDO and RGO data indicate that the centre-of-mass velocities of the two data sets were different by about three standard deviations. Accordingly, I applied a correction of -7 km s^{-1} to the RGO data to make it consistent with the DDO data. The DDO velocity system is known to be on the standard Lick Observatory velocity system for late-type stars but not enough data are available to determine corrections to the standard system for the early type stars. The corrections applied here have a significant effect on V_0 only.

SPECTRAL FEATURES

The antiphasing of the He II λ4,686 emission line velocities suggests that this line arises from a region gravitationally bound to the secondary. The broken line in figure 67.1 shows the velocity curve for the secondary for an assumed mass ratio $m_2/m_1 = 0.67$. Hutchings and Walker (private communication) have questioned this interpretation. They find that the line profile varies in a way that suggests that an absorption line is being Doppler shifted across a stationary emission line with a period of 5.6 day. Walborn and Margon (two private communications) have also seen variations in the λ4,686 line strength that seem to be consistent with this model. There are also indications that the feature's strength varies on a time scale of months. Because of the uncertain interpretation of this feature I have analysed the system as though it were a single-line binary.

The source of the high excitation He II λ4,686 feature cannot be identified readily because of the uncertain velocity behavior of the line. But the velocity behavior of the line and the spectral type of the primary seem to rule out that star as the source. Other very weak emission features appear in the λλ4,640–4,650 region of the spectrum and are probably due to either NIII or CIII. These features are too weak and distorted by underlying absorption lines to be reliably measured on my plates. It is likely that this collection of high excitation features is somehow connected with the X-ray source in the system.

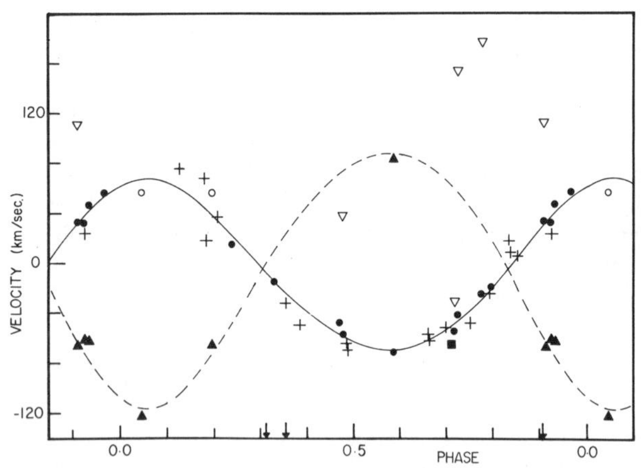

Fig. 67.1 The velocity curve of HDE 226868 = Cygnus X-1. Filled circles denote DDO velocities, and crosses denote the RGO velocities. Open circles are uncertain DDO velocities. The filled square is a velocity from the University of Arizona. He II λ4,686 velocities are denoted by filled triangles. Open triangles are hydrogen emission-line velocities. Arrows indicate phases where photometry was done by DuPhy. (Courtesy Thomas Bolton.)

The hydrogen emission lines from the gas stream seem to vary periodically. The emission is seen in the photographic spectrum only between phases 0.5 and 1.0 with the strongest emission occurring near phase 0.65. When the emission is strongest, it is seen in the hydrogen and helium lines though it never rises much about the continuum level. At other times the emission is only detectable at Hβ.

DISTANCE ESTIMATES

Previous investigations[9,10] have suggested that HDE 226868 is at least 2 kpc away. These distance estimates are based on the absolute magnitude of the primary derived from its spectral type, an estimate of the interstellar reddening, and consistency arguments involving other effects of the interstellar medium. A circumstellar shell might mimic the effects of the interstellar medium and cause the derivation of a spuriously large distance. Such a circumstellar shell would, however, be a strong infrared source, and HDE 226868 is not abnormally bright in the infrared for its spectral type and estimated distance.[11] In addition, Neckel[12] has found that HDE 226868 fits the A_v against distance relationship for this part of the sky very well at an assumed distance of 2.2 kpc.

Walborn (private communication) has recently reclassified the spectrum of HDE 226868 as O9.7 Iab which implies $M_v = -6.5$ and a distance modulus of 11.9. He describes the absorption line spectrum as "entirely normal". HDE 226868 lies less than a degree away from NGC 6871, the nuclear cluster of Cyg OB3. Walborn[13] derives a spectroscopic distance modulus of 12.0 for this OB association which is in excellent agreement with that of the star. The proper motion,[14] radial velocity[15,16] and equivalent width and velocity of the interstellar K line[16,17] of HDE 226868 are all consistent with its membership in Cyg OB3 (ref. 18). Although none of these criteria are conclusive by themselves, the evidence that the star is a member of Cyg OB3 is very strong. Thus HDE 226868 can be regarded as a "normal" OB supergiant.

MASS RATIO

The spectral type of HDE 226868 indicates that it has a mass of approximately 30 $M_\odot$ (ref. 19), and this mass will be assumed for the remainder of the discussion. The radius can be estimated from the mass and the surface gravity of an O9.7 Iab star[20] ($\log g = 3.2$) or from the distance estimate and apparent luminosity. If the effective temperature is taken to be 30,000 K, both methods yield radii of about 23 $R_\odot$.

Cherepashchuk *et al.*[21] have recently shown that HDE 226868 varies photometrically with the same period as the velocity variations. This photometric variation seems to be due to the tidal distortion of the O star by the invisible secondary. There is no evidence of eclipses and any reflexion effect due to the secondary illuminating the primary appears to be negligibly small. If the photometric variation is entirely due to tidal distortion and the primary fills its Roche lobe, the photometric and spectroscopic elements imply the mass ratio, m_2/m_1, and the inclination angle, i. I have made such a calculation using the spectroscopic elements given in table 67.1, the data given by Cherepashchuk *et al.*,[21] and the assumptions used by them in analysing their photometric data. This analysis gives $m_2/m_1 = 0.48$ and $i = 30°$. When the primary mass derived from the spectral type is combined with this mass ratio we find that $m_2 = 14.4\ M_\odot$.

The chief uncertainty in the estimate of m_2 is probably in the estimated mass m_1. If a reflexion effect is present, the derived mass ratio will be slightly underestimated. If the O star does not fill its Roche lobe, the mass ratio will be overestimated. But the size of the Roche lobe is not very sensitive to the inclination angle. For all possible inclination angles the primary fills the Roche lobe to within the accuracy of its estimated radius, so that the axis ratios for the distorted star are unlikely to be very different from those of the Roche case. The estimated uncertainty in the secondary mass is $10\ M_\odot < m_2 < 20\ M_\odot$.

THE BINARY SYSTEM AND BLACK HOLE

The derived dimensions of the binary system are given in table 67.2. The secondary mass is far too large for a white dwarf or a neutron star, yet the rapid time variations and strength of the X-ray flux seem to require that the secondary be a compact stellar object.[3] Thus it seems highly probable that the secondary is a black hole. This is true even without including the photometric data in the analysis. The improved period here excludes the phasing together of the high energy X-ray observations to show an eclipse, and there is no evidence of eclipses at other energies.[3,5] The absence of X-ray and optical eclipses requires that $m_2 \geqslant 7.4\ M_\odot$ for $m_1 = 30\ M_\odot$ and $R_1 = 23\ R_\odot$.

Table 67.2 Dimensions of HDE 226868 (Mean separation of stars $= a_1 + a_2 = 48\ R_\odot$)

$m_1 = 30\ M_\odot$ (assumed)	$R_1 = 23\ R_\odot$	$R_{\text{Roche}} = 24.5\ R_\odot$
$m_2 = 14.4\ M_\odot$	$R_2 = ?$	
$i = 30°$		

The difference between the Roche radius and the primary radius shown in table 67.2 is probably real. The emission line strengths do not seem to be as strong as would be expected if the primary filled its Roche lobe. But the primary may occasionally fill the Roche lobe. The sudden decline in X-ray intensity observed in April 1971 (ref. 3) may have been caused by the primary contracting away from the Roche surface.

The hydrogen emission lines probably arise from material flowing through the inner Lagrangian point. The variation of the emission strength would then be caused by the changing aspect of the gas stream and the eclipse of the inner Lagrangian point by the primary. Alternatively, the emission line variations may be caused by instabilities induced in the primary by the changing size of the Roche lobe in the eccentric orbit. This latter interpretation seems unlikely since the small eccentricity found could be caused by the tidal distortion of the primary star and not be due to a truly elliptical orbit.

1. R. M. Hjellming and C. M. Wade, *Astrophys, J. Lett. 168*, L21 (1971).
2. L. L. E. Braes and G. K. Miley, *Nature 232*, 246 (1971).
3. H. Tananbaum, H. Gursky, E. Kellogg, R. Giacconi, and C. Jones, *Astrophys. J. Lett. 177*, L5 (1972).
4. C. T. Bolton, *Bull. Amer. Astron. Soc. 3*, 458 (1971).
5. C. T. Bolton, *Nature 235*, 271 (1972).
6. B. L. Webster and P. Murdin, *Nature 235*, 37 (1972).
7. J. Kristian and R. Brucato, *IAU Circ. No. 2411* (1972).
8. B. W. Bopp, D. S. Evans, and J. D. Laing, *Mon. Not. Roy. Astron. Soc. 147*, 355 (1970).
9. J. Kristian, R. Brucato, N. Visvanathen, H. Lanning, and A. Sandage, *Astrophys. J. Lett. 168*, L91 (1971).
10. P. Murdin and B. L. Webster, *Nature 233*, 110 (1971).
11. E. E. Becklin, J. Kristian, G. Neugebauer, and C. G. Wynn-Williams, *Nature Physical Science 239*, 130 (1972).
12. T. Neckel, *Landessternwarte Heidelberg-Konigstuhl Veroffentlichungen 19* (1967).
13. N. R. Walborn, *Astron. J. 77*, 312 (1972).
14. Smithsonian Astrophysical Observatory *Star Catalog* (Washington, D. C.: Smithsonian Institution, 1966).
15. P. Hayford, *Lick Obs. Bull. 16*, 53 (1932).
16. R. F. Sanford, *Astrophys. J. 110*, 117 (1949).
17. P. Murdin and B. L. Webster, *Nature 233*, 110 (1971).
18. J. Ruprecht, *Trans. IAU 12B*, 350 (1966).
19. R. Stothers, *Astrophys. J. 175*, 431 (1972).
20. C. W. Allen, *Astrophysical Quantities* (Athlone Press, 1963).
21. V. M. Lyutiu, R. A. Sunyaev, and A. M. Cherepashchuk, *Soviet Astronomy 17*, 1 (1973).

68. Novae or Temporary Stars

Edwin P. Hubble

(*Astronomical Society of the Pacific Leaflet 1*, No. 14, 55–58 [1928])

For at least two thousand years astronomers have observed new or temporary stars in our galaxy that seemed to appear suddenly in the sky and then slowly fade away until they became invisible. As Edwin Hubble explains in the first part of the leaflet reprinted here, these novae are neither new nor temporary, but actually existing stars that suddenly explode. In retrospect, however, the most memorable part of Hubble's brief popularization was his identification of the Crab nebula with a nova recorded by the Chinese in the constellation Taurus in 1054 A.D.[1] Apparently his publication escaped serious notice from professional astronomers for some time; it was not until a decade later that both a systematic search of Oriental literature and a detailed astronomical discussion were undertaken. The Chinese records revealed that, like Venus, the "guest star" was seen in daylight for twenty-three days and in the nighttime sky for nearly two years.[2]

The remarkable nebula in Taurus has been known for nearly two centuries, and its name, the Crab nebula, was assigned in the 1850s. After astronomical photography superseded visual observations, John Duncan (in 1921) discovered that the nebula was expanding.[3] Around 1939–42 further measurements from plates taken over forty years apart generally confirmed the expansion age suggested by Hubble.[4] The most careful determinations show that with a uniform velocity of expansion, the original divergence of the filaments would have taken place in 1140 A.D. Because the identification of the Crab nebula with the Chinese supernova of 1054 A.D. seems quite secure, we now believe that an acceleration of the exploded material took place sometime in the history of the nebula.

1. Knut Lundmark first noticed that the "guest star" observed by the Chinese in 1054 A.D. lies in the same region of the sky as the presently observed Crab nebula [K. Lundmark, *Publications of the Astronomical Society of the Pacific 33*, 234 (1921)].
2. J. J. Duyvendak, *Publications of the Astronomical Society of the Pacific 54*, 91 (1942).
3. J. C. Duncan, *Proceedings of the National Academy of Sciences 7*, 179 (1921).
4. J. C. Duncan, *Astrophysical Journal 89*, 482 (1939); N. U. Mayall and J. H. Oort, *Publications of the Astronomical Society of the Pacific 54*, 95 (1942); Walter Baade, *Astrophysical Journal 96*, 188 (1942). Also see the article by V. Trimble in *The Crab Nebula: I.A.U. Symposium No. 46*, ed. R. D. Davies and F. G. Smith (Dordrecht: D. Reidel, 1971).

ACCOUNTS which probably refer to bright novae or temporary stars occur from time to time in ancient annals, but it was not until 1572 that scientific observation of such phenomena was undertaken. In that year Tycho Brahe observed the famous nova in Cassiopeia which rivalled Venus in its splendour and could easily be seen in broad daylight. But brilliant novae are rare. Generations pass which never witness such an event. Records are scarce until well into the last century, and not until photography had come into general use were novae discovered in considerable numbers. Our own generation is exceptional in having witnessed four novae of the first rank—Nova Persei (1901), Nova Aquilae (1918), Nova Cygni (1920) and Nova Pictoris (1925). As for the fainter novae, Bailey has estimated that in our own stellar system an average of about ten per year reach maxima brighter than the ninth magnitude (one-sixteenth the brightness of the faintest naked eye stars). These novae appear mostly along the Milky Way, perhaps because the great majority of the stars are near that plane; and they are especially numerous in the region of Sagittarius, the direction of the center of our stellar system. Probably only a few of those occurring are actually observed; the records, in fact, contain only about fifty well observed cases.

Nova Aquilae (1918) may be described as a typical example. For 30 years prior to 1918 it appeared on occasional photographs as a normal white star of the eleventh magnitude (about 100 times fainter than the faintest naked-eye star). On June 5, 1918, its appearance was still normal; but two nights later, on June 7, it had flared up, without any warning, to naked-eye visibility and was still rising at a rate which doubled its luminosity in less than an hour. By the next night it had reached the first magnitude, brighter than Antares; and later still at magnitude -1.4, it outshone all the fixed stars save Sirius alone. In three days it had blazed out to nearly 100,000 times its original brightness—as we now know from 4 to nearly 400,000 times the luminosity of the Sun. Almost at once it began to fade, rapidly at first, then more and more slowly. Eighteen days after maximum it was at the third magnitude, one-fiftieth of its maximum luminosity; eight months later it was lost to the naked eye. Now, after several years, it has faded to its pristine faintness, with only an occasional flickering to recall its great adventure.

In two respects, however, the experience has left its trace. First, before the outburst, the spectrum appears to have been the normal type A, with a surface temperature of about 10,000 degrees Centigrade, whereas the present spectrum indicates a much hotter surface temperature, from 20,000 to 30,000 degrees, and hence a greater surface brightness. The total luminosity, however, is about the same as before the outburst. This is a discrepancy as yet unexplained, unless we suppose that the size of the star has shrunk just enough to compensate for the increased surface brightness.

The second remaining trace of the outburst is a shell of nebulosity which surrounds the star. This represents material blown out at a velocity of more than a thousand miles per second, as shown by the displacement of lines in the spectrum. Since the shell was first observed by Barnard, some six months after the outburst, it has been expanding at a uniform rate; and in the year 1927 it appeared as a dim halo some 18 seconds in diameter. The angular rate of expansion combined with the linear radial velocity indicates a distance of about 1200 light years from the earth. The outburst therefore, actually occurred back in the Eighth Century, although the news did not reach us until 1918.

The histories of other novae show wide variations from this particular story, yet there is sufficient uniformity to suggest insistently that one particular type of physical causes is responsible for all the outbursts. The explanation is being sought partly in the distribution of novae among the stars and, more hopefully, in the unravelling of the bewilderingly complex and changing spectra. Among the leaders in this latter field are Wright of the Lick Observatory, and Adams and Joy of Mt. Wilson.

A nova outburst has been described thus: "A star swells up and blows off its cover"—and the prevailing opinion holds that this is not entirely wrong. The star suddenly becomes unstable and some sort of explosion results; but we do not know whether the action is spontaneous or whether it arises from some external stimulus, such for instance as a collision. Novae are so frequent, however, and the lives of stars are so long that we must suppose the outbursts to be normal episodes in the histories of stars. Probably there are preliminary indications which can be observed but as yet they have not been identified. At any moment, so far as we know, any particular star may blaze out as a nova.

Studies of the spectra indicate that outbursts are normally accompanied by the ejection of nebulous material. Only occasionally, however, is the star so near or the material in such quantity that the nebulosity can be seen or photographed. Nova Aquilae (1918) was such a case and Nova Persei (1901) as well. The Crab Nebula, Messier No. 1, is possibly a third, for it is expanding rapidly and at such a rate that it must have required about 900 years to reach its present dimensions. For, in the ancient accounts of celestial phenomena only one nova has been recorded in the region of the Crab Nebula. This account is found in the Chinese annals, the position fits as closely as it can be read, and the year was 1054! The great loop in Cygnus (NGC 6960-6922) may be a fourth case, but there the expansion is so slow that about 90,000 years would have been required for the nebula to reach its present size. Novae outbursts under exceptional conditions furnish a possible explanation of planetary nebulae in general. This statement, however, is made with the reservation that once the explosion is under way and the nebulosity is receding from the star with enormous velocity, we know of no reason why it should stop and settle down to the equilibrium that is actually observed in the planetaries.

Novae as a class are about the brightest stars in our stellar system, averaging perhaps twenty thousand times the luminosity of the Sun. Could we look back upon our system from some immense distance, from which all save the brightest stars had melted into an unresolved blur, we should still see the novae as they flash out, shine for a while and then fade from sight. In Messier 31, the great spiral in Andromeda, a vast independent stellar system nearly a million light years from our own, we actually witness these very phenomena. More than eighty novae have already been observed in that system, where they seem to occur two or three times as frequently as in our own. Novae have been observed in other systems as well—four, for instance, have been found in Messier 33—but in none do the conditions seem to be so favorable as in Messier 31. These extra-galactic novae appear to be entirely similar to novae in our own system, and investigations of them by statistical methods supplements the more intimate study of the latter stars.

69. On the Possible Existence of Neutron Stars

On Super-Novae

Walter Baade and Fritz Zwicky

(*Proceedings of the National Academy of Sciences 20*, 254–259 [1934])

On Massive Neutron Cores

J. Robert Oppenheimer and George M. Volkoff

(*Physical Review 55*, 374–381 [1939])

In 1934 Walter Baade and Fritz Zwicky communicated to the National Academy of Sciences a remarkable pair of papers on supernovae. One of these articles, reprinted as the first paper given here, demonstrated that the enormous total energy emitted in the supernova process corresponds to the annihilation of an appreciable fraction of a star's mass. The other, more speculative, paper, which we have not reproduced here, discussed their prediction that cosmic rays are produced during the supernova explosion. That paper showed that if supernovae occur once in a millenium in our galaxy, they can account for the observed flux of cosmic rays. Almost as an afterthought, Baade and Zwicky introduced under "Additional Remarks" a prescient speculation that was really a conclusion of the paper we have reproduced here:

> In addition, the new problem of developing a more detailed picture of the happenings in a super-nova now confronts us. With all reserve we advance the view that a super-nova represents the transition of an ordinary star into a *neutron star*, consisting mainly of neutrons. Such a star may possess a very small radius and an extremely high density. As neutrons can be packed much more closely than ordinary nuclei and electrons, the "gravitational packing" energy in a *cold* neutron star may become very large, and under certain circumstances, may far exceed the ordinary nuclear packing fractions. A neutron star would therefore represent the most stable configuration of matter as such. The consequences of this hypothesis will be developed in another place, where also will be mentioned some observations that tend to support the idea of stellar bodies made up mainly of neutrons.[1]

If Baade and Zwicky were right, a neutron star should be found at the center of the expanding envelope remnant of the supernova explosion. As shown in later selections of this volume, a rotating neutron star or pulsar was found in the center of the Crab nebula supernova nearly thirty-four years after Baade and Zwicky's prophetic paper postulating its existence.

By 1937 George Gamow showed that neutrons can be packed much more closely than ordinary nuclei and electrons, making possible densities of about 10^{14} gm cm^{-3} for neutron stars.[2] At these densities a normal star like the sun would have collapsed from a radius of about a million kilometers into a neutron star with a radius of only 10 km. Gamow also pointed out that the high central densities required for the formation of neutron stars can be expected only in those stages of stellar evolution subsequent to the exhaustion of the thermonuclear fuel in normal stars and that only the very massive stars would have evolved rapidly enough to reach this stage during the lifetime of our galaxy.

What is the permissible range of masses for stable neutron stars? Lev Landau first proposed an answer to this question while considering the possibility of condensed neutron cores within normal stars.[3] He estimated that the gravitational energy per neutron in the core becomes equal to the energy per particle in stable nuclei when the core mass becomes equal to 0.05 solar masses. J. Robert Oppenheimer and Robert Serber were quick to criticize this estimate, showing that if one takes only gravitational attraction into account, the critical mass is 0.3 solar masses.[4] Furthermore, they argued that this limit is unnecessarily severe, because the neutrons have additional energy when they behave like a degenerate Fermi gas at high densities.

In the second of the papers given here, Oppenheimer and George M. Volkoff show that a better estimate of the mass of a stable neutron star is obtained if one takes into account both the rapid rise of pressure at the onset of neutron degeneracy and the effect of gravitation on the properties of space-time when the mass densities become large. The equation of state of the nuclear matter was obtained in the first approximation by identifying it with a degenerate, relativistic gas of neutrons fulfilling Fermi statistics. The macroscopic structure of the star (its mass, radius, and density distribution) were determined from Einstein's theory of gravitation. Calculating equilibrium configurations along these lines, Oppenheimer and Volkoff find that a stable neutron star can exist only in a finite range of masses (one tenth to seven tenths of the sun's mass) and a finite range of densities (10^{14} to 10^{16} gm cm^{-3}).

Since the initial work of Oppenheimer and Volkoff considerable effort has gone into revising the equation of state of neutron star matter by taking into account nuclear interactions in the core as well as the lattice structure of the surface crust, but most of the conclusions reached by Oppenheimer and Volkoff have remained unchanged. It is still believed that neutron stars can only reach stable equilibrium configurations in a finite range of masses (now believed to lie in the range 0.09 to as much as two or possibly even three solar masses), and in a finite range of radii between 9 and 164 km.

1. W. Baade and F. Zwicky, *Proceedings of the National Academy of Sciences 20*, 259–263 (1934).
2. G. Gamow, *Atomic Nuclei and Nuclear Transformations* (Oxford: Oxford University Press, 1937); *Physical Review 55*, 718 (1939).
3. L. Landau, *Nature 141*, 333 (1938).
4. J. R. Oppenheimer and R. Serber, *Physical Review 54*, 540 (1938).

Baade and Zwicky, On Super-Novae

Common Novae

THE EXTENSIVE INVESTIGATIONS of extragalactic systems during recent years have brought to light the remarkable fact that there exist two well-defined types of new stars or novae which might be distinguished as *common novae* and *super-novae*. No intermediate objects have so far been observed.

Common novae seem to be a rather frequent phenomenon in certain stellar systems. Thus, according to Bailey,[1] ten to twenty novae flash up every year in our own Milky Way. A similar frequency (30 per year) has been found by Hubble in the well-known Andromeda nebula. A characteristic feature of these common novae is their absolute brightness (M) at maximum, which in the mean is -5.8 with a range of perhaps 3 to 4 mags. The maximum corresponds to 20,000 times the radiation of the sun. During maximum light the common novae therefore belong to the absolutely brightest stars in stellar systems. This is in full agreement with the fact that we have been able to discover this type of novae in other stellar systems near enough for us to reach stars of absolute magnitude -5 with our present optical equipment.

Super-Novae

The novae of the second group (super-novae) presented for a while a very curious puzzle because this type of new star was found, not only in the nearer systems, but apparently all over the accessible range of nebular distances. Moreover, these novae presented the new feature that at their maximum brightness they emit nearly as much light as the whole nebula in which they originate. Since the investigations of Hubble and others have revealed that the absolute total luminosities of extragalactic systems scatter with rather small dispersion around the mean value $M_{vis} = -14.7$, there is no doubt that we must attribute to this group of novae an individual maximum brightness of the order of $M_{vis} = -13$.

A typical specimen of these super-novae is the well-known bright nova which appeared near the center of the Andromeda nebula in 1885 and reached a maximum apparent brightness of $m = 7.5$. Since the distance modulus of the Andromeda nebula is

$$m - M = 22.2, \tag{1}$$

the absolute brightness of the nova at maximum was $M = -14.7$. An integration of the light-curve shows that practically the whole visible radiation is emitted during the 25 days of maximum brightness and that the total thus emitted is equivalent to 10^7 years of solar radiation of the present strength.

Finally, there exist good reasons for the assumption that at least one of the novae which have been observed in our Milky Way system belongs to the class of the super-novae. We refer to the abnormally bright nova of 1572 (Tycho Brahe's nova).[2]

About the final state of super-novae practically nothing is known. The bright nova of 1885 in the Andromeda nebula has faded away and must now be fainter than absolute magnitude -2. Repeated attempts to identify the nova of 1572 with one of the faint stars near its former position have so far not been very convincing.

Regarding the initial states of super-novae only the following meager facts are known. First, super-novae occur not only in the blurred central parts of nebulae but also in the spiral arms, which in certain cases are clearly resolved into individual stars. Secondly, the super-nova of 1572 in its initial stage probably was not brighter than apparent magnitude 5 as otherwise it would be registered as such in the old catalogues, which, however, is not the case.

Super-novae are a much less frequent phenomenon than common novae. So far as the present observational evidence goes, their frequency is of the order of one super-nova per stellar system (nebula) per several centuries.

We believe that on the basis of the available observations of super-novae the following assumptions are admissible:

(1) Super-novae represent a general type of phenomenon, and have appeared in all stellar systems (nebulae) at all times as far back as 10^9 years. To be conservative we shall assume for purposes of calculation that in every stellar system only one super-nova appears per thousand years.

(2) Super-novae, initially, are quite ordinary stars whose masses are not greater than 10^{33} gr. to 10^{35} gr.

(3) The super-nova of 1885 in Andromeda is a fair sample. We therefore base our calculations on the characteristics observed for this super-nova, namely:

(α) At maximum the visible radiation L_V emitted per second is equal to that of 6.3×10^7 suns. The radiation from our sun is

$$L_{\odot} = 3.78 \times 10^{33} \text{ ergs/sec.} \tag{2}$$

Therefore

$$L_V = 6.3 \times 10^7 L_{\odot} = 2.38 \times 10^{41} \text{ ergs/sec.} \tag{3}$$

The *total* visible radiation which was emitted by our super-nova represents an energy $E_V = 10^7$ years of $L_{\odot}$, that is

$$E_V = 1.19 \times 10^{48} \text{ ergs.} \tag{4}$$

(β) A common nova reaches maximum brightness in about two to three days. Indications are that a super-nova reaches maximum brightness during about the same interval.

TOTAL RADIATION FROM A SUPER-NOVA

In order to obtain an estimate of the total radiation, visible and invisible, from a super-nova, we consider two idealized limiting cases.

FIRST CASE We assume that the observed L_V corresponds to the integrated red end of a black body radiation of the effective temperature T_e, which we proceed to determine. We shall later make use of the further assumption that the total energy emitted from the super-nova is at least as great as that of the black body radiation at the temperature T_e, the red end of which is equal to that of the observed visible radiation from the super-nova. The integration over the red tail will be taken from the frequency $\nu = 0$ to the violet end ν_v of the visible spectrum; that is, to $\nu_v = 7.5 \times 10^{14}$ sec.$^{-1}$. The temperature resulting from our calculations justifies the introduction of the Rayleigh-Jeans law, namely,

$$I_\nu = \frac{2\pi k T_e \nu^2}{c^2}, \qquad (5)$$

into the integration mentioned, so that

$$L_V = S \int_0^{\nu_v} I_\nu d\nu = 1.34 \times 10^8\ S T_e, \qquad (6)$$

where S is the surface of the super-nova. If the initial stage (i) of the super-nova is a star similar to our sun, its radius would be

$$R_i = 6.95 \times 10^{10}\ \text{cm.} = R_\odot. \qquad (7)$$

At maximum brightness our super-novae will be considerably blown up and its radius R must then be considerably greater than $R_\odot$. Reasons can be advanced, however, for supposing that probably

$$R < 100\,R_\odot = R_1, \qquad (8)$$

and that almost certainly

$$R < 400\,R_\odot = R_2. \qquad (9)$$

This estimate of the radius R of a super-nova at maximum brightness is based on a comparison with common novae.[3] Roughly speaking, a common nova radiates like a black body until a radius R_m is reached such that

$$R_1 < R_m < R_2. \qquad (10)$$

For $R = R_m$ the brightness of the common nova is at its maximum. For $R > R_m$ extremely wide (50 to 100 Å) emission lines appear and gaseous shells are expelled from the nova at great speeds. Also the brightness of the nova declines; on the other hand the observed ionization of the gaseous shells indicates that a vast amount of ultra-violet radiation is emitted at this stage of the expansion. Furthermore, there are indications that R_m is the smaller the greater the speed with which the nova initially blows up. As a super-nova blows up faster than a common nova, we feel safe in assuming that R_m for a super-nova cannot be greater, and probably is smaller, than R_m for a common nova. To be conservative, we shall carry out our calculations for the two cases

$$R_m = R_1 = 7 \times 10^{12}\ \text{cm. and } R_m = R_2 = 2.8 \times 10^{13}\ \text{cm.} \qquad (11)$$

The corresponding surfaces are

$$S_1 = 6.15 \times 10^{26}\ \text{cm.}^2 \text{ and } S_2 = 9.84 \times 10^{27}\ \text{cm.}^2 \qquad (12)$$

With these figures we obtain from (6), (8) and (9) the effective surface temperature

$$\left.\begin{array}{l} T_e = L_V/1.34 \times 10^8 S = 1.78 \times 10^{33}/S, \\ T_1 = 2.89 \times 10^6 \text{ degrees}, \ T_2 = 1.81 \times 10^5 \text{ degrees.} \end{array}\right\} \qquad (13)$$

From T we obtain the total radiation per second from our sample super-nova:

$$\left.\begin{array}{l} L_T = S\dfrac{ac}{4}T_e^4 = 1.80 \times 10^{-37} L_V^4 S^{-3} \\ \qquad = 5.75 \times 10^{128} S^{-3} \\ L_{T_1} = 2.46 \times 10^{48}\ \text{ergs/sec.}, \\ L_{T_2} = 5.98 \times 10^{44}\ \text{ergs/sec.}, \end{array}\right\} \qquad (14)$$

where $a = 7.63 \times 10^{-15}$ c.g.s. is the Stefan-Boltzmann radiation constant, and c the velocity of light. The total energy emitted during the existence of the super-nova therefore is of the order of

$$\left.\begin{array}{l} E_T = L_T E_V/L_V = 5 \times 10^6\ L_T \\ E_{T_1} = 12.3 \times 10^{54}\ \text{ergs}, \quad E_{T_2} = 2.99 \times 10^{51}\ \text{ergs.} \end{array}\right\} \qquad (15)$$

These values correspond to a loss of mass

$$\Delta M = E_T/c^2, \quad \Delta M_1 = 1.37 \times 10^{34}\ \text{gr.}, \quad \Delta M_2 = 3.32 \times 10^{30}\ \text{gr.} \qquad (16)$$

In reality the mass radiated away may be several times this amount, and it therefore becomes evident that *the phenomenon of a super-nova represents the transition of an ordinary star into a body of considerably smaller mass.*

SECOND CASE It might be objected that we have far overestimated the effective temperature T_e of the super-nova, inasmuch as the surface, for some reason or other, may be so large that approximately $L_T = L_V$, in which case T_e would be approximately equal to the effective temperature of the sun. As the maximum of L_V corresponds to about 7×10^7 times the radiation from the sun, the radius R of the super-nova at maximum brightness would be

$$R = 8.36 \times 10^3 R_{\odot} = 5.81 \times 10^{14} \text{ cm.} \quad (17)$$

If the super-nova is initially an ordinary star which in about one day blows up to the radius R, the velocity of expansion of the surface will be of the order of

$$v = 6.72 \times 10^9 \text{ cm./sec.}, \quad (18)$$

which, per *proton*, gives the kinetic energy

$$E_K = m_p v^2/2 = 0.05\, m_p c^2. \quad (19)$$

It appears, therefore, that the kinetic energies E_K of the individual particles would be quite comparable with the energy of annihilation mc^2 of these particles.

Furthermore, in order to produce such energies, the radiation trapped inside the opaque surface must correspond to average temperatures T_i which are considerably higher than T_e. From this it would follow that the energy of the trapped radiation $[(4\pi/3)(R^3 a T_i^4)]$ alone is comparable with energy of total annihilation of the star (10^{54} ergs).

The above considerations seem to indicate that in any case the total energy emitted in the super-nova process represents a considerable fraction of the star's mass. We also think that our case (1) corresponds more nearly to the reality than does case (2). A more detailed discussion of the super-nova process must be postponed until accurate light-curves and high-dispersion spectra are available.

Unfortunately, at the present time only a few underexposed spectra of super-novae are available, and it has not thus far been possible to interpret them.

1. S. I. Bailey, *Pop. Astr. 29*, 554 (1921).
2. K. Lundmark, *Kungl. Svenska Vetensk. Handlingar 60*, no. 8 (1919).
3. *Handbuch d. Astrophysik*, vol. 6 (Novae).

Oppenheimer and Snyder, On Massive Neutron Cores

Abstract—It has been suggested that, when the pressure within the stellar matter becomes high enough, a new phase consisting of neutrons will be formed. In this paper we study the gravitational equilibrium of masses of neutrons, using the equation of state for a cold Fermi gas, and general relativity. For masses under $1/3 \odot$ only one equilibrium solution exists, which is approximately described by the nonrelativistic Fermi equation of state and Newtonian gravitational theory. For masses $1/3\odot < m < 3/4\odot$ two solutions exist, one stable and quasi-Newtonian, one more condensed, and unstable. For masses greater than $3/4\odot$ there are no static equilibrium solutions. These results are qualitatively confirmed by comparison with suitably chosen special cases of the analytic solutions recently discovered by Tolman. A discussion of the probable effect of deviations from the Fermi equation of state suggests that actual stellar matter after the exhaustion of thermonuclear sources of energy will, if massive enough, contract indefinitely, although more and more slowly, never reaching true equilibrium.

I. INTRODUCTION

FOR THE APPLICATION of the methods commonly used in attacking the problem of stellar structure[1] the distribution of energy sources and their dependence on the physical conditions within the star must be known. Since at the time of Eddington's original studies not much was known about the physical processes responsible for the generation of energy within a star, various mathematically convenient assumptions were made in regard to the energy sources, and these led to different star models (e.g., the Eddington model, the point source model, etc.). It was found that with a given equation of state for the stellar material many important properties of the solutions (such as the mass-luminosity law) were quite insensitive to the choice of assumptions about the distribution of energy sources, but were common to a wide range of models.

In 1932 Landau[2] proposed that instead of making arbitrary assumptions about energy sources chosen merely for mathematical convenience, one should attack the problem by first investigating the physical nature of the equilibrium of a given mass of material in which no energy is generated, and from which there is no radiation, presumably in the hope that such an investigation would afford some insight into the more general situation where the generation of energy is taken into account. Although such a model gives a good description of a white dwarf star in which most of the material is supposed to be in a degenerate state with a zero point energy high compared to thermal energies of even 10^7 degrees, and such that the pressure is determined essentially by the density only and not by the temperature, still it would fail completely to describe a normal main sequence star, in which on the basis of the Eddington model the stellar material is nondegenerate, and the existence of energy sources and of the consequent temperature and pressure gradients plays an important part in determining the equilibrium conditions.

The stability of a model in which the energy sources have to be taken into account is known to depend also on the temperature sensitivity of the energy sources and on the presence or absence of a time-lag in their response to temperature changes. However, if the view which seems plausible at present is adopted that the principal sources of stellar energy, at least in main sequence stars, are thermonuclear reactions, then the limiting case considered by Landau again becomes of interest in the discussion of what will eventually happen to a normal main sequence star after all the elements available for thermonuclear reactions are used up. Landau showed that for a model consisting of a cold degenerate Fermi gas there exist no stable equilibrium configurations for masses greater than a certain critical mass, all larger masses tending to collapse. For a mixture of electrons and nuclei in which on the average there are two protonic masses per electron Landau found the critical mass to be roughly 1.5$\odot$, and in general the critical mass is inversely proportional to the square of the mass per particle obtained by spreading out the total mass over only those particles which essentially determine the pressure of the Fermi gas.

The possibility has been suggested[3] that in sufficiently massive stars after all the thermonuclear sources of energy, at least for the central material of the star, have been exhausted a condensed neutron core would be formed. The minimum mass for which such a core would be stable has been estimated by Oppenheimer and Serber,[4] who on taking into account some effects of nuclear forces give approximately 0.1$\odot$ as a reasonable minimum mass. The gradual growth of such a core with the accompanying liberation of gravitational energy is suggested by Landau as a possible source of stellar energy.

In this connection it seems of interest to ask whether this model of the final state of a star can be right for arbitrarily heavy stars, i.e., to investigate whether there is an upper limit to the possible size of such a neutron core. Landau's original result for a cold relativistically degenerate Fermi gas quoted above gives in the case of a neutron gas an upper limit of about 6$\odot$ beyond which the core would not be stable but would tend to collapse. Two objections might be raised against this result. One is that it was obtained on the basis of Newtonian gravitational theory while for such high masses and densities general relativistic effects must be considered. The other one is that the Fermi gas was assumed to be relativistically degenerate throughout the whole core, while it might be expected that on the one hand, because of the large mass of the neutron, the nonrelativistically degenerate equation of state might be more appropriate over the greater part of the core, and on the other hand the gravitational effect of the kinetic energy of the neutrons could not be neglected. The present investigation seeks to establish what differences are introduced into the result if general relativistic gravitational theory is used instead of Newtonian, and if a more exact equation of state is used. A discussion of the general relativistic treatment of the equilibrium of spherically symmetric distributions of matter is first given, and then the special ideal case of a cold neutron gas is treated. A discussion of the results, and comparison with some results of Professor R. C. Tolman reported in an accompanying paper are given in the concluding sections.

II. Relativistic Treatment of Equilibrium

It is known[5] that the most general static line element exhibiting spherical symmetry may be expressed in the form

$$ds^2 = -e^\lambda dr^2 - r^2 d\theta^2 - r^2 \sin^2\theta\, d\phi^2 + e^\nu dt^2,$$
$$\lambda = \lambda(r), \quad \nu = \nu(r). \tag{1}$$

If the matter supports no transverse stresses and has no mass motion, then its energy momentum tensor is given by[6]

$$T_1{}^1 = T_2{}^2 = T_3{}^3 = -p, \quad T_4{}^4 = \rho \tag{2}$$

where p and ρ are respectively the pressure and the macroscopic energy density measured in proper coordinates. With these expressions for the line element and for the energy momentum tensor, and with the cosmological constant Λ taken equal to zero, Einstein's field equations reduce to:[7]

$$8\pi p = e^{-\lambda}\left(\frac{\nu'}{r} + \frac{1}{r^2}\right) - \frac{1}{r^2}, \tag{3}$$

$$8\pi \rho = e^{-\lambda}\left(\frac{\lambda'}{r} - \frac{1}{r^2}\right) + \frac{1}{r^2}, \tag{4}$$

$$\frac{dp}{dr} = -\frac{(p+\rho)}{2}\nu', \tag{5}$$

where primes denote differentiation with respect to r. These three equations together with the equation of state of the material $\rho = \rho(p)$ determine the mechanical equilibrium of the matter distribution as well as the dependence of the $g_{\mu\nu}$'s on r.

The boundary of the matter distribution is the value of $r = r_b$ for which $p = 0$, and such that for $r < r_b$, $p > 0$. For $r < r_b$ the solution depends on the equation of state of the material connecting p and ρ. For many equations of state a sharp boundary exists with a finite value of r_b.

In empty space surrounding the spherically symmetric distribution of matter $p = \rho = 0$, and Schwarzschild's exterior solution is obtained:[8]

$$e^{-\lambda(r)} = 1 + A/r, \quad e^{\nu(r)} = B(1 + A/r). \tag{6}$$

The constants A and B are fixed by the requirement that at great distances away from the matter distribution the $g_{\mu\nu}$'s must go over into their weak-field form, i.e., $B = 1$, $A = -2m$ where m is the total Newtonian mass of the matter as calculated by a distant observer.[9]

Inside the boundary Eqs. (3), (4) and (5) may be rewritten as follows. Using the equation of state $\rho = \rho(p)$ Eq. (5) may be immediately integrated.

$$\nu(r) = \nu(r_b) - \int_0^{p(r)} \frac{2dp}{p + \rho(p)},$$

$$e^{\nu(r)} = e^{\nu(r_b)} \exp\left[-\int_0^{p(r)} \frac{2dp}{p + \rho(p)}\right].$$

The constant $e^{\nu(r_b)}$ is determined by making e^ν continuous across the boundary.

$$e^{\nu(r)} = \left(1 - \frac{2m}{r_b}\right) \exp\left[-\int_0^{p(r)} \frac{2dp}{p + \rho(p)}\right]. \tag{7}$$

Thus e^ν is known as a function of r if p is known as a function of r. Further in Eq. (4) introduce a new variable

$$u(r) = \tfrac{1}{2}r(1 - e^{-\lambda}) \quad \text{or} \quad e^{-\lambda} = 1 - 2u/r. \tag{8}$$

Then Eq. (4) becomes:

$$du/dr = 4\pi\rho(p)r^2 \tag{9}$$

In Eq. (3) replace $e^{-\lambda}$ by its value from (8) and ν' by its value from (5). It becomes:

$$\frac{dp}{dr} = -\frac{p + \rho(p)}{r(r - 2u)}[4\pi p r^3 + u]. \tag{10}$$

Equations (9) and (10) form a system of two first-order equations in u and p. Starting with some initial values $u = u_0$ $p = p_0$ at $r = 0$, the two equations are integrated simultaneously to the value $r = r_b$ where $p = 0$, i.e., until the boundary of the matter distribution is reached. The value of $u = u_b$ at $r = r_b$ determines the value of $e^{\lambda(r_b)}$ at the boundary, and this is joined continuously across the boundary to the exterior solution, making

$$u_b = \frac{r_b}{2}[1 - e^{-\lambda(r_b)}] = \frac{r_b}{2}\left[1 - \left(1 - \frac{2m}{r_b}\right)\right] = m.$$

Thus the mass of this spherical distribution of matter as measured by a distant observer is given by the value u_b of u at $r = r_b$.

The following restrictions must be made on the choice of p_0 and u_0, the initial values of p and u at $r = 0$:

a. In accordance with its physical meaning as pressure, $p_0 \geqslant 0$.

b. From Eq. (8) it is seen that for all finite values of $e^{-\lambda}$, $u_0 = 0$. Since $g_{11} = -e^{\lambda}$ must never be positive, $u_0 \leqslant 0$ for infinite values of $e^{-\lambda}$ at the origin. However, it may be shown that of all the finite values of p_0 at the origin $p_0 = 0$ is the only one compatible with a negative value of u_0, and that for equations of state of the type occurring in this problem even this possibility is excluded, so that u_0 must vanish.[10]

c. A special investigation for any particular equation of state must be made to see whether solutions exist in which $0 \geqslant u_0 \geqslant -\infty$ and $p \to \infty$ as $r \to 0$.

III. PARTICULAR EQUATIONS OF STATE

Oppenheimer and Volkoff consider particles obeying Fermi statistics, with the thermal energy and all forces between them neglected. Using a parametric form for ρ and p in the equation of state,[11] they derive differential equations from (9) and (10), which are integrated numerically for several values of the time parameter t_0. These results are displayed in table 69.1. For very small values of t the equation of state reduces to $p = K\rho^{5/3}$ and $\hat{p} \propto t$, where $\hat{p}$ is the maximum momentum in the Fermi distribution and is related to the proper particle density N/V by

$$\frac{N}{V} = \frac{8\pi}{3h^3}\hat{p}^3.$$

The striking feature of the [table] is that the mass increases with increasing t_0 until a maximum is reached at about $t_0 = 3$, after which the curve drops until a value roughly $1/3\odot$ is reached for $t_0 = \infty$. In other words no static solutions at all exist for $m > 3/4\odot$, two solutions exist for all m in $3/4\odot > m > 1/3\odot$, and one solution exists for all $m < 1/3\odot$.

Some insight into this situation may be gained from the following considerations. In the non-relativistic polytrope solutions of Emden[12] the equation of state was assumed to be $p = K\rho^\gamma = K\rho^{1+1/n}$. Solutions which at first sight seem to be quite satisfactory (i.e., giving a finite mass within a finite radius) were found for values of $n < 5$ or $\gamma > 6/5$. But Landau[2] pointed out that although these solutions in every case give an equilibrium configuration, they do not in every

Table 69.1 Mass, radius and neutron density for various values of t_0

t_0	Mass In units of ⊙ for neutrons	Radius In kilometers, for neutrons	$\left(\frac{\hat{p}}{\mu_0 c}\right)_{r=0}$	$\left(\frac{N}{V}\right)_{r=0}$ Neutrons cm^3
1	0.30	21.1	0.25	0.062×10^{39}
2	0.60	13.3	0.52	0.56×10^{39}
3	0.71	9.5	0.82	2.2×10^{39}
4	0.64	6.8	1.17	6.4×10^{39}
∞	0.34	3.1	∞	∞

case give *stable* equilibrium. Thus, unless $\gamma \geqq 4/3$ the equilibrium configuration is unstable. This may be seen from the following rough calculation. The gravitational part of the free energy of the system is negative and proportional to $\rho^{1/3}$ where ρ is an appropriate average density (Newtonian gravitational theory is used). The part of the free energy caused by compression is proportional to $\int p\,dv$, and hence to $\rho^{\gamma-1}$ $(\gamma \neq 1)$. Thus

$$F = -a\rho^{1/3} + b\rho^{\gamma-1}.$$

Polytrope solutions exist for both $\gamma = 5/3(>4/3)$, i.e., for $n = 3/2$ and for $\gamma = 5/4(<4/3$, but $>6/5)$, i.e., for $n = 4$, but the former corresponds to stable equilibrium and the latter to unstable equilibrium.

Oppenheimer and Volkoff examine further the behavior of the equilibrium positions for various masses including the double equilibrium possibility for intermediate masses. Then, in section IV, they discuss the relation of their work to Tolman's solutions.

V. Discussion—Application to Stellar Matter

We have seen that for a cold neutron core there are no static solutions, and thus no equilibrium, for core masses greater than $m \sim 0.7\odot$. The corresponding maximum mass M_0 before collapse is some ten percent greater than this. Since neutron cores can hardly be stable (with respect to formation of electrons and nuclei) for masses less than $\sim 0.1\odot$, and since, even after thermonuclear sources of energy are exhausted, they will not tend to form by collapse of ordinary matter for masses under $1.5\odot$ (Landau's limit), it seems unlikely that static neutron cores can play any great part in stellar evolution;[13] and the question of what happens, after energy sources are exhausted, to stars of mass greater than $1.5\odot$ still remains unanswered. It should be observed that for the critical solution with $m \sim 0.7\odot$ the potentials $g_{\mu\nu}$ are nowhere singular, and that in particular such a core does not tend to "protect itself" from the addition of further matter by the vanishing of g_{44} at the boundary. There would then seem to be only two answers possible to the question of the "final" behavior of very massive stars: either the equation of state we have used so far fails to describe the behavior of highly condensed matter that the conclusions reached above are qualitatively misleading, or the star will continue to contract indefinitely, never reaching equilibrium. Both alternatives require serious consideration.

Can repulsive forces occur under extreme compression to prevent the collapse? Oppenheimer and Volkoff conclude that "such repulsive forces, even if they exist, will hardly make possible static solutions for arbitrarily large amounts of matter . . . From this discussion it appears probable that for an understanding of the long time behavior of actual heavy stars a consideration of non-static solutions must be essential."

[Such solutions were subsequently discussed by Oppenheimer and Hartland Snyder, *Physical Review* 56, 455–459 (1939).]

1. A. Eddington, *The Internal Constitution of the Stars* (Cambridge University Press, 1926); B. Strömgren, *Ergebn. Exakt. Naturwiss. 16*, 465 (1937); short summary in G. Gamow, *Phys. Rev. 53*, 595 (1938).

2. L. Landau, *Physik. Zeits. Sowjetunion 1*, 285 (1932).

3. G. Gamow, *Atomic Nuclei and Nuclear Transformations*, 2nd ed. (Oxford, 1936), p. 234. L. Landau, *Nature 141*, 333 (1938) and others.

4. J. R. Oppenheimer and R. Serber, *Phys. Rev. 54*, 540 (1938).

5. R. C. Tolman, *Relativity, Thermodynamics and Cosmology* (Oxford, 1934), pp. 239–241.

6. Ibid., p. 243.

7. Ibid., p. 244.

8. Ibid., p. 203.

9. Ibid., pp. 203 and 207.

10. This can be seen from the following argument. Having chosen some particular value of p_0 one may usually represent the equation of state in that pressure range by $\rho = Kp^s$ with some appropriate value of s. Using this equation of state and taking the approximate form of Eq. (10) near the origin for the case $u_0 < 0$, and finite p_0, one obtains:

$$\frac{dp}{dr} = \frac{p + \rho(p)}{2r} = \frac{p + Kp^s}{2r}.$$

Integration of this equation shows that for $s < 1$, $p_0 \gtreqless 0$ can not be satisfied, and for $s \gtreqless 1$ only the value $p_0 = 0$ is possible. For the equations of state used in this problem always $s < 1$ holds. It may also be noted that the above equation together with Eq. (7) show that $e^{\nu(r)} \to \infty$ as $r \to 0$.

11. Cf. S. Chandrasekhar, *Monthly Notices of R.A.S.* 222 (1935), but introduce *energy* density in place of his *mass* density.

12. Emden, *Gaskugeln* (1907), or cf. *Handbuch der Astrophys.* vol. 3, p. 186.

13. The mass of the shell of ordinary (but dense) matter surrounding the core must be small for cores much more massive than the lightest core stable with respect to disintegration into electrons and nuclei.

70. Spectra of Supernovae

Rudolph Minkowski

(*Publications of the Astronomical Society of the Pacific 53*, 224–225 [1941])

In this paper Rudolph Minkowski first called attention to the fact that supernovae explosions can be divided into two general categories according to the spectra of their expanding remnants. The ejected shells of supernovae of Type I have lower velocities of expansion than those of Type II, and the supernovae of Type I differ from those of Type II in having no lines of hydrogen or any other element. These two differences suggest that supernovae of Type I occur in stars that are older and less massive than the stars responsible for Type-II supernovae.

Although Minkowski does not mention it in this paper, Walter Baade and Fritz Zwicky had shown that the light curve of his prototype Type-I supernova in IC 4182 can be distinguished by a sharp peak and an exponential decline.[1] As illustrated in figure 70.1, the supernovae can also be divided into two categories according to their light curves, but it is not clear if there is a correlation between the spectra of supernovae and their light curves.

Although examinations of the light curves for the two types of supernovae show that they both emit about 10^{50} ergs of photon energy during their explosion, Minkowski has shown that there is a radical difference in the kinetic energies of their ejected shells.[2] The expanding shells of Type-I supernovae contain about one-tenth of a solar mass and expand at about 1,000 km sec^{-1} to give a kinetic energy of about 10^{48} ergs, whereas the larger masses and velocities of Type-II supernovae (several solar masses and about 6,000 km sec^{-1}) imply kinetic energies on the order of 10^{51} ergs. Furthermore, the large difference in the masses of the ejected shells suggests that the two types of supernovae originate in stars with radically different masses. Also, they seem to be associated with different stellar populations.

1. W. Baade and F. Zwicky, *Astrophysical Journal 88*, 411 (1938).
2. R. Minkowski, *Proceedings of the National Academy of Sciences 46*, 13 (1960).

SPECTROSCOPIC OBSERVATIONS indicate at least two types of supernovae. Nine objects (represented by the supernovae in IC 4182 and in NGC 4636) form an extremely homogeneous group provisionally called "type I." The remaining five objects (represented by the supernova in NGC 4725) are distinctly different; they are provisionally designated as "type II." The individual differences in this group are large; at least one object, the supernova in NGC 4559, may represent a third type or, possibly, an unusually bright ordinary nova.

Spectra of supernovae of type I have been observed from 7 days before maximum until 339 days after. Except for minor differences, the spectrograms of all objects of type I are closely comparable at corresponding times after maxima. Even at the earliest premaximum stage hitherto observed, the spectrum consists of very wide emission bands. No significant transformation of the spectrum occurs near maximum. Spectra of type II have been observed from maximum until 115 days after. Up to about a week after maximum, the spectrum is continuous and extends far into the ultraviolet, indicating a very high color temperature. Faint emission is suspected near $H\alpha$. Thereafter, the continuous spectrum fades and becomes redder. Simultaneously, absorptions and broad emission bands are developed. The spectrum as a whole resembles that of normal novae in the transition stage, although the hydrogen bands are relatively faint and forbidden lines are either extremely faint or missing. The supernova in NGC 4559, while generally similar to the other objects in this group, shows multiple absorptions of *H* and *Ca* II; the emission bands are fainter than in the other objects.

No satisfactory explanation for the spectra of type I has been proposed. Two [*O* I] bands of moderate width in the later spectra of the supernova in IC 4182 are the only features satisfactorily identified in any spectrum of type I. They are, at the same time, the only indication of the development of a nebular spectrum for any supernova. The synthetic spectra by Gaposchkin and Whipple disagree in many details with the observed spectra of type I. However, these synthetic spectra agree better with spectra of type II and provide a very satisfactory confirmation of the identifications which, in this case, are already suggested by the pronounced similarity to the spectra of ordinary novae. As compared with normal novae, supernovae of type II show a considerably earlier type of spectrum at maximum, hence a higher surface temperature (order of 40,000°), and the later spectrum indicates greater velocities of expansion (5,000 km/sec or more) and higher levels of excitation. Supernovae of type II differ from those of type I in the presence of a continuous spectrum at maximum and in the subsequent transformation to an emission spectrum whose main constituents can be readily identified. This suggests that the supernovae of type I have still higher surface temperature and higher level of excitation than either ordinary novae or supernovae of type II.

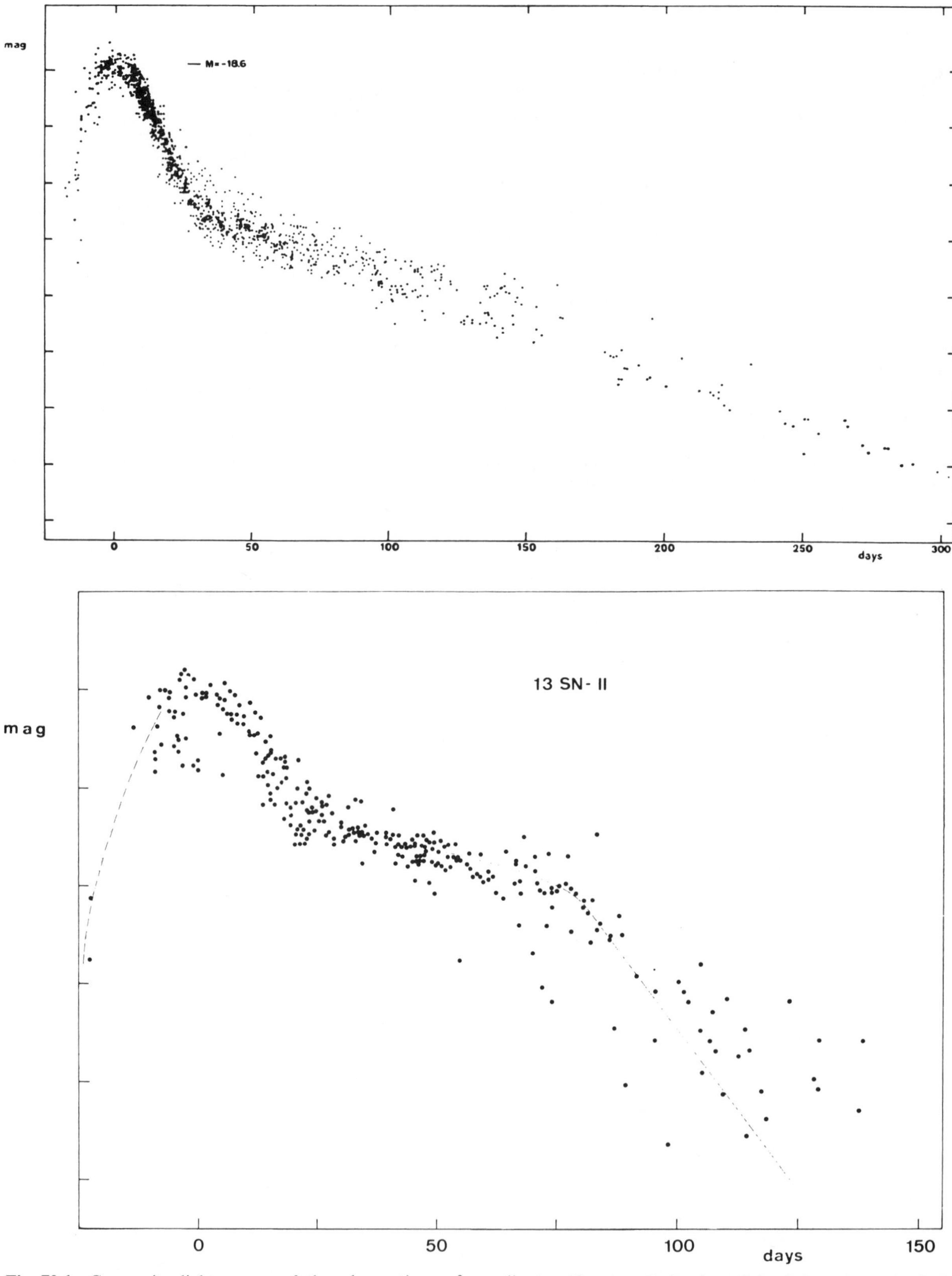

Fig. 70.1 Composite light curves of the observations of (*top*) thirty-eight Type-I supernovae and (*bottom*) thirteen Type-II supernovae. Intervals of 1 mag are marked on the ordinates. (Courtesy L. Rosino of the Asiago Astrophysical Observatory.)

71. The Crab Nebula

Rudolph Minkowski

(*Astrophysical Journal 96*, 199–213 [1942])

In the early 1900s observations of novae in spiral nebulae and the identification of these spirals as extragalactic systems similar to our own Milky Way led to the recognition of a distinct class of very intense exploding stars. These supernovae were known to flare up to a maximum luminosity of over 100 million times that of the sun, but a detailed study of their physical properties had to await the identification of the Crab nebula as the expanded debris of a supernova explosion in our galaxy.[1] Early in this century spectroscopists had noticed that the spectrum of the Crab nebula is similar to the line spectra of other gaseous emission nebulae but that it also included a continuum. In an effort to obtain further information on the continuum radiation and its origin, Walter Baade began in about 1940 to photograph the Crab nebula with the Mount Wilson 100-in refractor in selected portions of the optical wavelength region. His plates, some of the finest ever taken of the Crab, showed that the nebulosity consists of two distinct parts, a system of filaments that forms an outer envelope in which the emission lines originate and an inner amorphous region in which the continuum radiation originates.[2] In the paper given here, which appeared as a companion to Baade's paper in the *Astrophysical Journal*, Rudolph Minkowski shows that the continuum radiation contains practically all the energy emitted by the Crab nebula, and he also confirms Baade's conclusion that the nebula consists of two distinct parts. When Minkowski took spectra with a slit centered on a red filament the emission lines always appeared, but whenever the slit was centered on the inner regions only the continuum was present.

By comparing the energy received from the Crab nebula at different spectral regions with the spectra of stars of various colors, Minkowski concluded that the color temperature of the continuum decreases toward the red, which he explained in terms of free-free and free-bound radiation of electrons in a hot ionized gas. As Baade had already pointed out, however, the color of the central star implied that its luminosity was much too low to excite such a hot gas. To meet this objection, Minkowski assumed that the central star was abnormal, with a radius as small as that of a white dwarf star, but with a luminosity exceeding that of the entire Crab nebula. As we shall see in the next selection, the discovery of the nonthermal synchrotron radiation of relativistic electrons in a magnetic field led to a new explanation for the continuum radiation of the Crab nebula, and Minkowski's introduction of a new type of star became unnecessary.

Both Baade and Minkowski noticed that their candidate for the central stellar remnant of the supernova explosion did not exhibit the strong hydrogen lines expected from a normal star of its color. This fact agreed with the idea that supernovae occur at the end stages of stellar evolution, after a normal star has used up its hydrogen; but only a very massive star could have already depleted its nuclear fuel, and so massive a star should collapse into a point after the exhaustion of its fuel. As Minkowski points out in this paper, however, sufficient mass is blown off from the stellar envelope during the supernova explosion to allow the remnant star to stop its collapse at the white dwarf stage. Walter Baade and Fritz Zwicky had already used similar arguments to suggest that the remnant star of a supernova explosion would be a neutron star (see selection 69). Nearly three decades after Baade and Minkowski had pointed out the unusual spectral characteristics of one of the central stars in the Crab nebula, a series of discoveries demonstrated that it was both an optical and a radio pulsar and undoubtedly a neutron star.[2]

1. W. Baade, *Astrophysical Journal 96*, 188 (1942). Baade's photographs are reproduced as figures 72.1–72.2.

2. L. Woltjer, *Astrophysical Journal 152*, L179 (1968); D. H. Staelin and E. C. Reifenstein, *Science 162*, 1481 (1968); J. M. Comella, H. D. Craft, R. V. E. Lovelace, J. M. Sutton, and G. L. Tyler, *Nature 221*, 453 (1969); W. J. Cocke, M. J. Disney, and D. J. Taylor, *Nature 221*, 525 (1969); R. E. Nather, B. Warner, and M. Macfarlane, *Nature 221*, 527 (1969); J. S. Miller and E. J. Wampler, *Nature 221*, 1037 (1969).

Abstract—The central star of the Crab nebula is probably the south preceding of the two stars near the center of the nebula. The spectrum of this star is continuous. The nebula consists of two parts with different spectra. The filaments on the outside of the nebula produce emission lines of *H*, *He* I, *He* II, [*N* II], [*O* I], [*O* II], [*O* III], and [*S* II]. The hydrogen lines are faint. The spectrum of the main diffuse nebulosity is continuous except for a faint discontinuity at the Balmer limit. The continuous spectrum contains practically the entire energy emitted by the nebula. Its intensity distribution deviates from that of a black body; the color temperature is about 8,400° at λ 4,500 and about 6,700° at λ 6,000.

In the absence of absorption in the nebula the continuous spectrum cannot be due to scattering and must be a true emission spectrum. The only physically justified assumption is that the continuous spectrum is produced by free-free and free-bound transitions of electrons in the highly ionized gas. On this basis the observed intensity distribution in the continuous spectrum of the nebula finds a satisfactory explanation. The analysis leads to an electron density of the order 10^3 cm^{-3} and an electron temperature of the order of 50,000°. The mass contained in the nebula is about 15 solar masses. The hydrogen abundance seems to be low. For the central star a temperature of the order of 500,000° is obtained, a radius of 0.020 solar radius, and a total luminosity of 30,000 solar units.

A comparison with spectra of planetary nebulae indicates that the unusually high intensity of the [*O* II] lines in the filaments is compatible with a high temperature for the central star. The faintness of the hydrogen lines may be interpreted as due to low abundance of hydrogen.

The results as a whole support Chandrasekhar's suggestion that supernovae of type I are due to the transition of stars heavier than the limiting mass into the degenerate state. Supernovae of type II might possibly represent this transition for stars of smaller mass.

LITTLE doubt remains that the Crab nebula is the remnant of the Chinese nova of A.D. 1054.[1] This object was certainly a supernova; the records of its brightness indicate that it was a supernova of type I.[2] A study of the Crab nebula, which is the brightest object of its kind, offers the best way to obtain information on the final stage of a supernova.

An accurate determination of those spectroscopic data essential for a theoretical analysis meets considerable difficulty because of the faintness of the nebula. The observational results of the present paper have been obtained primarily with the aim of collecting data sufficient for such an analysis. The accuracy of some of them could still be improved, but better observations would not greatly affect the results of the analysis. The accuracy of the results is limited chiefly by observational and theoretical difficulties which at present cannot be overcome.

THE CENTRAL STAR

It is generally assumed that the central star of the Crab nebula—the stellar remnant of the supernova—is one of the two faint stars near the center of the nebula. The spectra of these stars are difficult to observe. The continuous spectrum of the nebula is so strong that, even with the 100-inch reflector, better than average seeing is needed to obtain sufficient contrast between star and nebulosity. So far, only one satisfactory spectrogram has been obtained. The nebular spectrograph VIB, attached at the Cassegrain focus of the 100-inch reflector, was used with one prism and a Schmidt camera of 3 inches focal length. With the slit in position angle 142°, the spectra of both the faint stars near the center of the nebula have been observed with one exposure. The features of the nebular spectrum shown on this spectrogram will be discussed later.

The north following star is distinctly redder than the south preceding. Because of insufficient contrast with the continuous spectrum of the nebula and the presence of emission lines crossing the spectrum, the spectral type is difficult to determine from the faintly visible absorption lines. Estimates between early F and early G have been given by various observers accustomed to the determination of spectral type from low dispersion spectra. The spectrum of the south preceding star is continuous without perceptible absorption lines. Compared with that of the north following star, it does not admit a color earlier than that of a late B-type star; this estimate is confirmed by Baade's measure[2] of the color index. The absence of absorption lines seems to indicate that the south preceding star is of a spectral type appropriate for the central star of a nebula with emission spectrum. On the other hand, the spectrum, which does not extend into the ultraviolet in the same way as spectra of old ordinary novae,[3] does not indicate an extremely high temperature.

The relatively late color type of the star does not necessarily indicate that the star is an object of low temperature. The color temperature of some Wolf-Rayet stars seems to be even lower; that these stars have actually a very high temperature is not in doubt. Still more in point is perhaps the example of Nova Persei (1901), for which F. H. Seares[4] has found an average color index of +0.5; it should be noted, however, that the spectrum now seems to show to some degree the extension into the ultraviolet which is usual for stars of this kind.

Taken altogether, the spectroscopic evidence admits, but does not prove, the conclusion that the south preceding star is the central star of the nebula. The measures of proper motion, which in the future may lead to a final decision, are still inconclusive.[2] The difficulties which the interpretation of the proper motion encounters if the star is not the central star, together with the spectroscopic evidence, seem to justify the assumption that the south preceding star is the central star. The following discussion will be based on this assumption; the consequences which would arise if this were not correct will be mentioned in their proper place.

The apparent photographic magnitude[2] of the south preceding star is 15.9. The distance[2] of the Crab nebula is 1,000 parsecs. From measures of the color of B-type stars in the vicinity of the Crab nebula by Stebbins, Huffer, and Whitford,[5] a value of 0.15 is obtained for the photoelectric color excess for the Crab nebula.[1] With the ratio 8.1 of interstellar absorption to reddening,[6] the photographic absorption is 1.2 mag. The absolute photographic magnitude of the star, corrected for interstellar absorption, is then, 4.8.

The Nebulosity

Baade's photographs[2] of the Crab nebula taken with various types of plates and filters show that it consists of two masses with distinctly different intensity distributions, namely, an inner diffuse nebulosity and an outlying nebulosity showing well-defined details. Observations of the spectrum by V. M. Slipher,[7] R. F. Sanford,[8] and N. U. Mayall[9] show a continuous spectrum with double emission lines superposed. The duplicity of the emission lines is a consequence of the expansion of the nebula, detected by C. O. Lampland[10] and measured by J. C. Duncan.[11] The irregular structure of the emission lines suggests that they originate mainly in the outer mass of filaments.

Here Minkowski describes his spectra of the diffuse nebulosity. The emission lines originate in the outer mass of filaments, and the emission of the inner diffuse nebulosity consists of a continuous spectrum. The intensity differences of lines is explained by emission by different features of the nebulosity (Mayall[9]). On the average, the red and violet components of lines show no systematic difference, suggesting that the nebula does not absorb light. This conclusion has been substantiated from star counts (Lundmark,[12] Baade[2]). The Hα, [N II], and [S II] lines appear with remarkably high intensity, but the contribution of the emission lines to the total luminosity is small.

The decrease of the color temperature toward the red is presented as possible evidence that the continuous spectrum is produced by free-free or free-bound transitions of electrons in a highly ionized gas. Minkowski argues that the deviation of the continuous spectrum from a blackbody spectrum demands an explanation according to the free-free and free-bound transitions, as summarized in the abstract. Scattering of light from the central star is rejected as a source of continuum emission, because the nebula is 6.9 mag brighter than the central star (nebula absolute mag = −2.2).

The central star No matter what mechanism is responsible for the emission of the continuum, the total luminosity L_s of the star must be larger than the total luminosity of the nebula and considerably larger than the luminosity L_{obs} in the limited range of the spectrum (λλ 3,700–6,600) that is accessible to observation. In this range, the sun emits 1.5×10^{33} erg/sec. If the relatively small difference between the color of the sun and of the nebula is neglected, the absolute photographic magnitude −2.2 of the nebula corresponds to a value of L_{obs} of 1.3×10^{36} erg/sec or 350 solar units. If T denotes the effective temperature of the star and R its radius,

$$L_s = 4\pi\sigma R^2 T^4. \tag{23}$$

Assuming that the star radiates as a black body, we have for its absolute photographic magnitude

$$M_{ph} = -0.72 - 5\log\frac{R}{R_\odot} + 2.5\log(e^{33,600/T} - 1). \tag{24}$$

Solving equations (23) and (24) for R and T with $M_p = 4.8$ and $L_s > 1.3 \times 10^{36}$, we obtain[13]

$$T > 120,000, \tag{25}$$

$$R < 0.042R_\odot. \tag{26}$$

If M_s is the mass of the star in solar units, the average density is

$$\rho > 19,000M_s. \tag{27}$$

It should be emphasized that the inequalities (25), (26), and (27) are quite strong. The total luminosity of the continuum may be expected to exceed by a considerable factor that contained in the observed range. Furthermore, it is not certain that the transfer of energy from star to nebula is practically complete. Thus, the luminosity, the temperature, and the density will be essentially greater and the radius essentially smaller than the limiting values. Even these values indicate that the star is a unique object.

Even the lower limit for the calculated temperature is above the upper limit (about 20,000°) indicated by the color of the star. The discrepancy shows that the star does not radiate as a black body. This result is not surprising, but it affects seriously the value of a determination of the temperature necessarily based on the assumption of black-body radiation. Whether the temperature obtained under this assumption is too high or too low depends, of course, entirely on the unknown nature of the deviation from a black-body spectrum. In any case the temperature of the star must be higher than the electron temperature of the nebula. From this point of view it appears that the color may be less significant than the temperature obtained from total luminosity and photographic magnitude.

The spread of the values for temperature and radius is surprisingly small. The values

$$L/L_{\odot} = 30{,}000, \quad T = 500{,}000^{\circ},$$

$$R/R_{\odot} = 0.020, \quad \rho = 180{,}000\, M_s,$$

are probably fair approximations. If, as seems probable, the hydrogen abundance is small, these values may even be too conservative. On the other hand, they may be extreme because of the deviation from a black-body spectrum.

At a temperature of 500,000°, the (black-body) energy distribution of the star has its maximum at a frequency of 3×10^{16} sec^{-1} corresponding to 120 V. An efficient transfer of energy from star to nebula can then be produced only by photoelectric ionization of ions with an ionization potential of about 120V. Hydrogenic ions with $Z = 3$ should thus be ionized to a high degree, and hydrogenic ions with $Z = 4$ would then be mainly responsible for the emission in the free-free and free-bound spectrum. The actual heavier atoms in the nebula can be expected to be in the fifth or higher stages of ionization.

The very high level of ionization seems to provide a reasonable explanation for the absence of emission lines from the spectrum of the diffuse nebulosity. The absence of hydrogen lines could, of course, be ascribed to low abundance of hydrogen; but the absence of such persistent lines as *He* II λ 4,686 points clearly to a level of ionization and to a temperature considerably higher than in planetary nebulae. The simplest explanation of the absence of forbidden lines at an electron density of the same order as for planetary nebulae would be provided by the absence of ions able to emit strong forbidden lines. Such ions would be absent if the level of ionization were substantially higher than 125 V, thus eliminating [*Ne* V] and [*Fe* VII], but lower than 233 V, thus insufficient to produce [*Fe* X]. A temperature of about 700,000° would meet this condition. However, the absence of [*Ne* V] and [*Fe* VII] from the observed lines is not conclusive proof for the absence of the respective ions; these lines occur in regions where their detection would be certain only if they were very strong. If these lines are present but unobserved, the level of ionization should still be higher than 55 V to exclude [*O* III] lines, which are certainly not present with noticeable intensity. A temperature of about 500,000° would then be appropriate.

As a whole, the analysis seems to lead to a very consistent picture. While owing to the uncertainties of the problem too much weight should not be given to the numerical values, the general result that the central star is an object of extremely high temperature, high density, and high total luminosity seems trustworthy.

The central star, if not the one considered here, would have to be fainter than 20.5 mag., in the absence of any other observed star near the center of the nebula. In this case the analysis would lead to temperatures of several million degrees. Perhaps the most evident difficulty would then be that even the more frequent of the heavier atoms—*C*, *N*, and *O*—would be completely ionized; it would thus become very doubtful whether the remaining ions could still provide sufficient absorption for an efficient transfer of energy from star to nebula. Difficulties of this type are absent only if the central star is bright enough to be observed. The assumption that the south preceding star is the central star thus finds some support.

THE FILAMENTS At first sight, the spectrum of the filaments seems to contradict the high temperatures derived by the analysis of the diffuse nebulosity. The general level of excitation indicated by the spectrum of the filaments is about that of an average planetary nebula, and the high relative intensity of the [*O* II] lines λ 3,726, λ 3,729 might be an indication of low excitation. Such an interpretation is, however, not supported by the results of a survey of planetary nebulae with low surface brightness.[14] Planetaries with high relative intensity of the [*O* II] lines, such as the Ring nebula in Lyra, are relatively frequent in this group. None of these objects, however, has decidedly low excitation; some definitely have high excitation, as shown by the presence of *He* II λ 4,686 and by the Zanstra temperature of the central star. Generally the [*O* II] lines appear greatly strengthened on the outside of the nebula. The relatively frequent occurrence of strong [*O* II] lines in planetaries with low surface brightness is probably to be explained as an effect of small density which should favor lines with very small transition probabilities. The strengthening of the [*O* II] lines toward the outer edge might in some nebulae have to be explained in the same way. But it appears more likely that, as is usually assumed, the gradual absorption of the exciting radiation is responsible. In any case it is evident that the high intensity of the [*O* II] lines in the Crab nebula is not inconsistent with a high temperature of the central star.

While the emission of strong [*O* II] lines by the filaments finds a parallel in the emission of these lines with increased intensity on the outside of planetary nebulae with central stars of high temperature, the mechanism of the radiative transfer is not exactly the same in both cases. In planetaries a considerable fraction of the radiation from the central star is transformed into emission lines, especially into *H* Lyman α. In the Crab nebula a considerable part of the radiation from the central star is transformed by the diffuse mass into low-frequency radiation of the continuous spectrum. In a different way in the two cases some part of the stellar radiation is thus transformed into radiation (of lower frequency than the

Lyman limit) which can escape freely and does not contribute to the ionization of the outer portion. In planetary nebulae the transfer of energy from central star to nebula occurs mainly in the Lyman continuum. If a planetary nebula is optically thick for the exciting radiation, its outer portion is ionized and excited by radiation which has been absorbed and is reradiated in the Lyman continuum and in emission lines of the heavier atoms. Estimates of the optical thickness of the diffuse mass in the Crab nebula suggest that, especially as the hydrogen abundance is low, the optical thickness in the Lyman continuum is not large and that the energy transfer occurs mainly in the corresponding absorption continua of *He* II and of the heavier ions. The nebular spectrum of the filaments can thus be excited by radiation which has been transmitted by the diffuse mass; reradiated energy may play only a minor role.

If the spectrum of the filaments is considered a low-excitation spectrum, the peculiarity is faintness of the hydrogen lines. The low intensity of $H\alpha$ compared to the [*N* II] lines might, however, be interpreted at least partly as high intensity of the [*N* II] lines. These lines have very low transition probabilities. It is not unlikely that they are strengthened by the same process as the [*O* II] lines. The observations of planetary spectra give some suggestions of such a correlation. On the other hand, the relative intensities of the hydrogen and helium lines are not to be explained in a similar way. Planetary nebulae as a whole show a high degree of correlation between the intensities of *He* I lines, *He* II λ 4,686, and the hydrogen lines. The photographic record shows *He* I λ 4,471 always considerably fainter than $H\beta$ and $H\gamma$. The relative intensity of λ 4,471 decreases as that of λ 4,686 increases, so that λ 4,471 is usually among the faintest lines when λ 4,686 reaches a photographic intensity equal to that of $H\beta$ and $H\gamma$. In the Crab nebula the image of λ 4,686 is definitely stronger than that of $H\beta$ and $H\gamma$, but λ 4,471 appears with an intensity relative to $H\beta$ and $H\gamma$ at least as great as in any planetary nebula. In other words, the hydrogen lines are considerably fainter relative to the helium lines than in any other emission nebula. The possibility that this is due to the peculiar kind of excitation in the filaments cannot be excluded, but no support for such an explanation can be found from observations in other emission nebulae. It seems, therefore, much more likely that the relative intensities of the hydrogen and helium lines in the Crab nebula indicate an abnormally low abundance of hydrogen.

CONCLUSIONS CONCERNING THE SUPERNOVA PROCESS The analysis of the Crab nebula leads directly to the conclusion that, before the outbreak, supernovae are massive stars of low hydrogen content. Stars of this type exist in the galaxy, the best-known being[15] v Sagittarii.

The relation between the present appearance and the spectrum of the nebula and the supernova spectrum will be traced in detail in a future discussion of the spectra of supernovae. A brief outline follows. The filaments offer no difficulties. They are probably to be identified with the mass in which the [*O* I] lines λ 6,300, λ 6,364 appearing in the later spectra of the supernova IC 4182 are emitted. These lines, while narrow compared to the wide bands which form the main supernova spectrum, have a width corresponding to a velocity of expansion of about 1,000 km/sec, which agrees satisfactorily with the velocity of 1,300 km/sec shown by the filaments. The low abundance of hydrogen suggested by the analysis may provide a satisfactory explanation for the appearance of the [*O* I] lines unaccompanied by *H* lines.

If the [*O* I] lines of supernovae are emitted in the matter which now forms the filamentary mass, possibly the main spectrum is emitted in the matter now forming the diffuse mass. This interpretation meets the difficulty that the width of the emission lines in the main spectrum is considerably larger than that of the [*O* I] lines, while the smaller extension of the diffuse mass indicates less rapid expansion than for the filaments. However, the width of the lines might conceivably indicate the velocity with which atoms move through a spectroscopically active zone rather than the velocity of ejection. Furthermore, unpublished observations of supernova spectra, to be discussed in a later paper, suggest a steady decrease of the velocity as a plausible explanation of the peculiar red shift of certain features. A reasonable working hypothesis is that the diffuse mass was originally the source of the main spectrum.

The analysis of the Crab nebula confirms the interpretation of the supernova outbreak proposed by Chandrasekhar in 1939 at the Paris conference on novae and white dwarfs. He suggested that the supernova outbreak might result from the inability of stars to develop degenerate cores at the center, if the mass is greater than a certain critical value

$$M_3 = 5.7\,\mu^{-2}, \tag{29}$$

where μ is the mean molecular weight. If the excess mass is blown off during the outbreak, then afterward the star will for the first time be in a position to develop a degenerate core. This core will then expand to occupy the whole star.

The values of the mass which have been obtained for the diffuse nebulosity show that the star was originally very massive and probably heavier than M_3. The following discussion of the present state of the star suggests that the present mass of the central star is of the order of one solar mass. The star has thus actually ejected the greater part of its mass.

The small radius and the high density seem to place the central star of the Crab nebula clearly among the white dwarfs. On the other hand, white dwarfs are characterized not only by high density but also by low luminosity. In this respect the central star certainly does not resemble a white dwarf. It may be assumed[16] that in this star the degenerate core does not yet occupy the whole star. The theory of degenerate configurations leads, indeed, to the conclusion that, for a star with the luminosity and radius following from the present analysis, the nondegenerate gaseous parts extend more than halfway into the star. About 70 per cent of the mass would, however, be degenerate. Thus the high luminosity of the star indicates that the development of the white dwarf is not yet concluded. It is important to note that from the stage when degeneracy sets in at the center to the stage when the star is wholly degenerate the contraction leads only to a radius of one-half to one-third the initial radius. For this reason the mass of the star can be determined—within the uncertainties of the problem—from the theoretical mass-radius relation for the white dwarfs, and a value of about 1 is thus obtained.

If the star is a still unfinished white dwarf with an extended gaseous fringe, a deviation of the intensity distribution of the stellar spectrum from that of a black body is not surprising. It is, of course, to be expected that the transformation into a degenerate configuration will become complete, possibly in the not too distant future, if 70 per cent of the mass is already degenerate. The total luminosity of the star should then decrease from the present value of the order 10,000 to about 0.01. Long before the final value is reached, the nebula will become unobservable. The limited lifetime of the remnants of supernovae as observable objects thus suggested would explain satisfactorily the scarcity of such remnants. The available observations are insufficient to decide whether the final decline of the Crab nebula is already under way.

It should finally be noted that a low abundance of hydrogen, suggested by the analysis, would fit into the picture as, according to the present theory of energy production in stars, the exhaustion of the hydrogen is the primary reason for the transition of a star into a white dwarf.

If supernovae of type I are stars of mass greater than the critical mass M_3, then it is highly suggestive to assume that supernovae of type II are stars of mass smaller than M_3. Such an assumption does not meet any contradictory observational evidence. In its favor could be cited the fact that the frequency of supernovae of type II appears to be six times as great as that of supernovae of type I. In the absence of excess mass, a supernova of type II would not necessarily have to eject a considerable fraction of its mass. The nebula surrounding a supernova of type II should thus be fainter than that around a supernova of type I. This expectation is in general agreement with the fact that any nebula surrounding Tycho's nova of 1572, which was probably a supernova of type II, is certainly much fainter than either of two nebulae connected with supernovae of type I, namely, the Crab nebula and the nebula of Kepler's nova of 1604 recently found by Baade.

1. J. J. L. Duyvendak, *Pub. A.S.P. 54*, 91 (1942); Mayall and Oort, *Pub. A.S.P. 54*, 95 (1942).

2. W. Baade, *Mt. W. Contr.*, No. 665; *Ap. J. 96*, 188 (1942).

3. M. L. Humason, *Mt. W. Contr.*, No. 596; *Ap. J. 88*, 47 (1938).

4. *Mt. W. Contr.*, No. 192; *Ap. J. 52*, 183 (1920).

5. *Mt. W. Contr.*, No. 621; *Ap. J. 91*, 61 (1940).

6. Greenstein and Henyey, *Ap. J. 93*, 327 (1941).

7. *Nature 95*, 185 (1915); *Pub. A.S.P. 28*, 192 (1916).

8. *Pub. A.S.P. 31*, 108 (1919).

9. *Pub. A.S.P. 49*, 101 (1937).

10. *Pub. A.S.P. 33*, 79 (1921).

11. *Mt. W. Comm.*, No. 76; *Proc. Nat. Acad. Sci.* 7, 179 (1921); *Mt. W. Contr.*, No. 609; *Ap. J. 89*, 482 (1939).

12. *Festskrift Tillagnad Östen Bergstrand* (Uppsala, 1938), p. 89.

13. Cf. F. Zwicky, *Rev. Mod. Phys. 12*, 66 (1940).

14. R. Minkowski, *Mt. W. Contr.*, No. 657; *Ap. J. 95*, 243 (1942).

15. J. L. Greenstein, *Ap. J. 91*, 438 (1940).

16. The author is indebted to Dr. S. Chandrasekhar for a letter of which this paragraph is an abstract.

72. On the Nature of the Luminescence of the Crab Nebula

Iosif S. Shklovskii
Translation Provided by Air Force Cambridge Research Laboratories

(*Doklady akademii nauk SSSR 90*, 983–986 [1953])

When Walter Baade and Rudolph Minkowski made the first detailed investigation of the Crab nebula supernova remnant, they showed that it consists of two distinct parts, a system of filaments forming an outer envelope in which emission lines occur and an inner amorphous region in which the continuum radiation originates. As illustrated in the previous selection, Minkowski assumed that the continuum radiation is generated thermally during collisions between electrons and ions in an extraordinarily hot gas at a temperature of nearly one hundred thousand degrees. When John Bolton and Gordon Stanley subsequently discovered that the spectral density of the flux of the Crab nebula is a thousand times more intense at radio wavelengths than at optical wavelengths, Jesse Greenstein and Minkowski rediscussed the continuum optical spectrum and concluded that it was impossible to reconcile the observed radio emission with the optical emission through purely thermal radiation at any temperature.[1]

During this time high-energy electrons were discovered to emit linearly polarized radiation when they were accelerated to high speeds in the strong magnetic fields of a synchrotron particle accelerator. This discovery stimulated Hannes Alfvén and Nicolai Herlofson to interpret the emission of "radio stars" as the synchrotron radiation of cosmic ray electrons in the stellar magnetic fields, and soon thereafter Karl Kiepenheuer reasoned that the radio frequency radiation of our Milky Way is due to the synchrotron radiation of cosmic ray electrons spiraling about the interstellar magnetic field (see selections 99 and 115). These ideas were ignored by most astronomers until Iosif Shklovskii presented this paper, in which he argues that both the radio and optical emission of the Crab nebula come from the synchrotron process. According to Shklovskii's hypothesis, electrons with extremely high energies radiate optically visible light in a weak magnetic field, whereas electrons of slightly lower energy radiate at radio wavelengths in the same magnetic field. Because the more energetic electrons lose their energy faster and also radiate at higher frequencies, the synchrotron radiation mechanism provides a natural explanation for the nonthermal spectrum of the optical continuum and for the fact that the radio emission is a thousand times more intense than the optical emission.

Although Shklovskii does not specifically mention it in this paper, the optical continuum of the Crab nebula should also be observed to be polarized if it is caused by the synchrotron radiation of relativistic electrons moving in a large-scale, well-ordered magnetic field. This polarization was independently discovered by two Soviet astronomers, V. A. Dombrovskii and M. A. Vashakidze, in 1954, and confirmed in greater detail by Walter Baade, Jan Oort, and Th. Walraven (see appended figure 72.1).[2] When Cornell Mayer and his colleagues subsequently showed that the radio emission of the Crab nebula exhibits the same direction and degree of polarization as the optical continuum,[3] Shklovskii's hypothesis for the synchrotron origin of the Crab nebula radiation was fully accepted.

Later, Shklovskii applied his theory to the young supernova remnant, Cassiopeia A, and reasoned that its rapid expansion will result in a decrease in its magnetic field strength and a corresponding diminution of its synchrotron radiation. Within a year the predicted decrease in intensity had been observed, and Shklovskii's theory was thereby confirmed by a second observational test.[4]

Shklovskii and his colleague Vitallii Ginzburg promptly applied the theory of synchrotron radiation to both the radio emission from our Milky Way and the emission from the intense extragalactic radio sources. In 1955 Shklovskii, for example, predicted that the jet observed at optical wavelengths in the central part of the radio galaxy NGC 4486 (Virgo A) was due to synchrotron emission associated with violent activity in the nucleus of its optically visible counterpart.[5] Within a year the predicted polarization of the jet's radiation was observed by Walter Baade, thus confirming Shklovskii's prediction.[6] Observations of the nonthermal spectra and polarization of the radio wavelength radiation of extragalactic sources have also confirmed that they radiate by the synchrotron process. By applying the theory of synchrotron radiation to astronomical objects, Shklovskii and Ginzburg set the stage for an entirely new picture of the universe in which magnetic fields and high energy particles, rather than the thermal radiation of stars, play the dominant role. Nevertheless, as we shall see in the next selection, there remained the fundamental difficulty of explaining exactly how the radiating particles obtain their tremendous energies in the first place.

1. J. L. Greenstein and R. Minkowski, *Astrophysical Journal 118*, 1 (1953).
2. V. A. Dombrovskii, *Doklady akademii nauk USSR 94*, 1021 (1954). M. A. Vashakidze, *Astronomical Circular* no. 147 (1954). J. H. Oort and Th. Walraven, *Bulletin of the Astronomical Institutes of the Netherlands 12*, 285 (1956).
3. C. H. Mayer, T. P. McCullough, and R. M. Sloanaker, *Astrophysical Journal 126*, 468 (1957).
4. I. Shklovskii, *Soviet Astronomy A. J. 37*, 243 (1960). J. A. Högbom and J. R. Shakeshaft, *Nature 189*, 561 (1961). (Also see W. A. Dent, H. D. Aller, and E. T. Olsen, *Astrophysical Journal* (*Letters*) *188*, L11 [1974]).
5. I. Shklovskii, *Astronomicheskii zhurnal 32*, 215 (1955).
6. W. Baade, *Astrophysical Journal 123*, 550 (1956).

THE FAMOUS NEBULA M 1 (the Crab nebula) has repeatedly drawn the attention of researchers. This object, unique of its kind, is, as has now been definitely established, the remnant of an explosion of a supernova in the year 1054. The present-day concepts about the nature of the Crab nebula are based on the important papers of Baade[1] and Minkowski,[2] who with the aid of first-class instruments obtained photographs of it and of its spectrum. According to Baade,[1] the nebula consists of two interpenetrating parts: the first is a lacelike network of filaments emitting line spectra, situated on the periphery of the nebula and expanding with a velocity of 1,000–1,300 km sec^{-1}, and the second is an amorphous mass, with a tendency to concentrate toward the center of the nebula, emitting a continuous spectrum. According to Minkowski,[2] the luminescence in the emission lines is only a few percent of that of the Crab nebula in the continuous spectrum. From the values Minkowski[2] presents for the color temperature of the nebula, it follows that the intensity of luminescence in the continuous spectrum diminishes gradually with decreased wavelength.

All current concepts concerning the physical conditions in the Crab nebula are based on the explanation of its continuous spectrum. Neither Baade[1] nor Minkowski[2] made an allowance for the possibility of some other mechanism of emission, different from that of the free-free and free-bound transitions in a strongly ionized gas, that is, in the nebular substance. This emission must be excited exclusively by a hot star (the nucleus of the nebula), a former supernova. Such a mechanism of emission is quite natural, because it is absolutely obvious that the luminescence of the Crab nebula cannot be explained by any kind of process of scattering or reflection. Other mechanisms of luminescence in the continuous spectrum of the nebulae have never been examined.[3]

However, difficulties and contradictions involved with such an assumption regarding the mechanism of emission from the nebula in the continuous spectrum cannot be overcome, it seems to us. Many of these difficulties have been known earlier, but no attempts have yet been made to analyze them critically, since no one doubted the correctness of the proposed mechanism[1,2] of emission of the continuous spectrum. We shall now briefly enumerate these difficulties.

The unavoidable results of the accepted mechanism of emission are the conclusions that in the Crab nebula: (a) the kinetic temperature is extraordinarily high, of the order of hundreds of thousands of degrees (especially if we consider that the observations do not give an indication of any rise in intensity near the boundary of the Balmer series[4] and at the boundaries of the ionized helium series); (b) the electron concentration in the nebula is about 10^3 cm^{-3}; (c) the nebular mass is about 15 $M_\odot$; (d) the temperature of the nucleus is monstrously high (500,000°); (e) the radius of a nucleus is exceedingly small (0.02 $R_\odot$).

The morphologic characteristics of the Crab nebula are simply inconceivable. One cannot understand how the filamentary network of the nebula, which has now been moving for 900 years in the amorphous, relatively dense medium, emits a continuous spectrum. We note that the density of the filaments is very low, since the most intensive emission lines are: [O II] 3,727 and [N II] $\lambda\lambda$6,548 and 6,584. But the probabilities of transitions of these lines are extraordinarily small, and they are intense only in nebulae with a low density.

Therefore, it is doubtful that the electron concentration in the filaments exceeds 300–400 cm^{-3}. Moreover, the prolonged movement of such a rarefied, relatively cold (which follows from the nature of the spectrum of the filaments) gas in a fairly dense but very hot medium is not properly understood. It is important to point out that the amorphous part of the nebula is not an expanding thin shell (similar to the filamentary network) but instead entirely fills the volume of the nebula.

Neither is it clear why the filaments have a comparatively low state of excitation. Fast electrons travelling from the diffuse part of the nebula are reaching its filamentary part, exciting and ionizing the atoms that are present there. Likewise, we cannot overlook the ionization of atoms in the filaments by the strong ultraviolet emission of the nucleus. Minkowski[2] makes an attempt to remove this difficulty, but this attempt cannot in any event be considered successful.

It is not clear why, in the spectrum of the diffuse part of the nebula, there are absent the emission lines of hydrogen and ionized helium that occur inevitably during recombinations. An attempt to explain this absence by a supposition that the Crab nebula is lacking in hydrogen and helium is unsatisfactory. In the filaments, the helium lines are comparatively intense, and hydrogen lines are also present. It was necessary to attribute to the Crab nebula some absolutely fantastic characteristics in order to explain the unusual conditions of the nebula. This is one of the weakest points in the interpretation of the continuous spectrum of M 1.

It is not known why the nebulae, which are the remainders of the galactic supernovae of the years 369,[5] 1572, and 1604 A.D., are so weak in comparison with the Crab nebula.[6] The abnormally large mass of the Crab nebula represents also a considerable difficulty.

We emphasize that all these difficulties come from the interpretation of the continuous spectrum of M 1. Because these difficulties are, as it seems to me, unsurmountable, the interpretation given above should be considered incorrect. The existing explanation of the continuous spectrum of the Crab nebula took shape when no other mechanism of emission could be proposed. At the present time, however, the situation has radically changed.

In 1949, it was learned that the Crab nebula is a fairly powerful source of radio emission.[7] Then it was reliably determined that the source of emission is not the nucleus, but the nebula itself.

The investigations have shown that the radio spectrum of the Crab nebula differs drastically from those of other discrete sources. Indeed, over an enormous range from $\lambda = 750$ cm to $\lambda = 25$ cm, encompassing about 5 octaves, the emission flux calculated for a single frequency interval remains practically constant.[8] It is known that emission of an optically thin layer of an ionized gas[5] possesses such a characteristic. Therefore, I attempted previously[5] to explain the radio emission from the Crab nebula as free-free transitions, that is, as the mechanism that, according to Baade[1] and Minkowski,[2] determines the luminescence of this nebula in the optical part of the spectrum. However, the intensity calculated for a single frequency interval in the optical portion of the spectrum must be 3 times smaller than in the region of radio frequencies; whereas observations indicate that it is 1,000 times smaller. Such a difference cannot be explained[5] by any acceptable interstellar light absorption. Therefore, the radio emission of the Crab nebula cannot be explained by the thermal radiation of gases present inside it.

Taking into account the unacceptability of the present-day concept for the mechanism of emission of the Crab nebula in the continuous spectrum and particularly in its radio spectrum, I think it is justifiable to consider that the factors causing radio emission from the Crab nebula are the causes of its optical emission with a continuous spectrum. If the intensity of radio emission calculated for a single frequency interval declines hardly at all (as is expected from observations) over a distance of 5 octaves, then across the range of the successive 18 octaves (up to the optical region), it decreases 1,000 times, which seems to be quite plausible.

According to reference 5, the mechanism of radio emission in the nebulae, which are remainders of the supernovae, is the bremsstrahlung of relativistic electrons in weak magnetic fields. The frequency at which the largest quantity of energy is emitted is expressed by the ratio

$$\nu_m = \frac{e\mathrm{H}}{4\pi mc}\left(\frac{E}{mc^2}\right)^2,$$

where E is the energy of relativistic electrons. From here it follows that if $\mathrm{H} = 10^{-4}$, then $\nu_m = 1.2 \times 10^9$ s^{-1} ($\lambda = 25$ cm) and $E = 1.1 \times 10^9$ electron volts, and with $\nu_m = 5 \times 10^{14}$ s^{-1} ($\lambda = 6{,}000$ Å), $E = 7 \times 10^{11}$ electron volts.

If in the nebula there are present, for some reason, relativistic electrons with energy of the order of 10^9 electron volts, there must also be a certain quantity of electrons with energy of the order of 10^{11}–10^{12} electron volts. Considering that the intensity of the nebula in the optical part of the spectrum is 1,000 times lower than in the radio-frequency region, with the aid of the well-known formula for the bremsstrahlung of relativistic electrons in the magnetic fields, we shall find that the concentration of electrons with energy of 7×10^{11} electron volts will be approximately 1,000 times smaller than that of electrons with energy of 10^9 electron volts and will be of the order of 10^{-9} cm^{-3}. Electrons with energy of 7×10^{11} electron volts, while emitting in the magnetic field, will be retarded. However, it is possible to demonstrate that significant losses of energy will occur only after several thousand years.

Thus, the "real nebula" of M 1 is only an expanding system of filaments, that is, a "network." It has a purely emissive spectrum. The visible stellar energy of the luminescent gases (the remnants of an explosion of the supernova in 1054 A.D.) is reduced from 9 mag to 12–13 mag, which is fairly close to the visible quantity of the remnants of the supernovae of the years 369 and 1604 A.D. (15 mag). The planetary nebulae removed from us by the same distance as M 1 usually have a stellar magnitude of about 10 mag.[6] Considering the low density of the filaments, it is doubtful whether the Crab nebula mass (consisting in our opinion of filaments only) would exceed by several hundredths that of the sun, though it apparently does exceed by several hundred times the mass of the shells of the ordinary novae. As to the mass of the diffuse nebula, which consists of relativistic particles, it is quite negligible, not more than 10^{-7} $M_\odot$. The absence of an intensive continuous spectrum in other nebulae, which are residues of supernovae of 369 and 1572 A.D., can be explained by the absence in these nebulae of a sufficient number of relativistic electrons with high energies.

Thus, the whole problem of the physics of nebulae, which are the remnants of the supernovae, and of the supernovae themselves should be reexamined.

In particular, because the mass of the shell ejected by a supernova is essentially smaller than the mass of a star itself, an explosion of the supernova would hardly induce a radical change in the structure of the star. This means that there is no qualitative difference between flare-ups of a supernova and of an ordinary nova. The difference lies only in the magnitude of the phenomenon, that is, in the quantity of energy liberated, in the mass of an ejected shell, and so on.

1. W. Baade, *Astrophys. J. 96*, 188 (1942).
2. R. Minkowski, ibid. *96*, 199 (1942).
3. An exception is the mechanism of "fractioning" the Lyman-α quantum, which Kipper has used recently to interpret the continuous spectrum of the planetary nebulae (A. Ya. Kipper, *Sborn. o razvitii sovetskoi nauki v Estonskoi SSR.* (Review of the development of Soviet science in the Estonian Sov. Soc. Republic) from 1940 to 1950 [1950]). If this mechanism were responsible for the emission from the Crab nebula in the continuous spectrum, the Crab would be emitting intensive hydrogen lines; however, this emission has not been observed. Besides, the energy distribution in the

continuous spectrum would have been different from that observed. A modification of Kipper's mechanism, that is, of "fractioning" the quantum of the helium resonance line, cannot for the very same reason explain the continuous spectrum of M 1.

4. D. Barbier, *Ann. Astrophys. 8*, 35 (1945).

5. I. S. Shklovskii, *Astron. zhurn. 30*, 15 (1953).

6. B. A. Vorontsov-Vel'yaminov, *Gazovye tumannosti i novye zvezdy* [Gaseous nebulae and supernovae] (USSR: Akad Nauk, 1948).

7. J. G. Bolton and G. J. Stanley, *Austral. J. Sci. Res. 2*, 139 (1949), G. J. Stanley and O. B. Slee, *Austral. J. Sci. Res. 3*, 234 (1950).

8. J. H. Piddington and H. C. Minnett, *Austral. J. Sci. Res. 4*, 458 (1951).

Appended Figures

Fig. 72.1 Walter Baade's photographs of the Crab nebula taken with the 200-in Palomar telescope in 1955 and published in the *Bulletin of the Astronomical Institutes of the Netherlands 12*, 312 (1956). The photograph of the outer filamentary region (above) was made in the wavelength range λ6,400–λ6,700, and the two photographs of the inner amorphous region (right) were made in the wavelength range λ5,400–λ6,400. The two photographs of the amorphous inner region were taken through a polaroid filter, and the arrows indicate the direction of the electric vector of the light recorded. In all these photographs north is on top. The southwesternmost of the two faint central stars is the star chosen by Baade and Rudolph Minkowski as the remnant neutron star of the supernova explosion. It is this star that was later found to be a pulsar. (Courtesy Hale Observatories.)

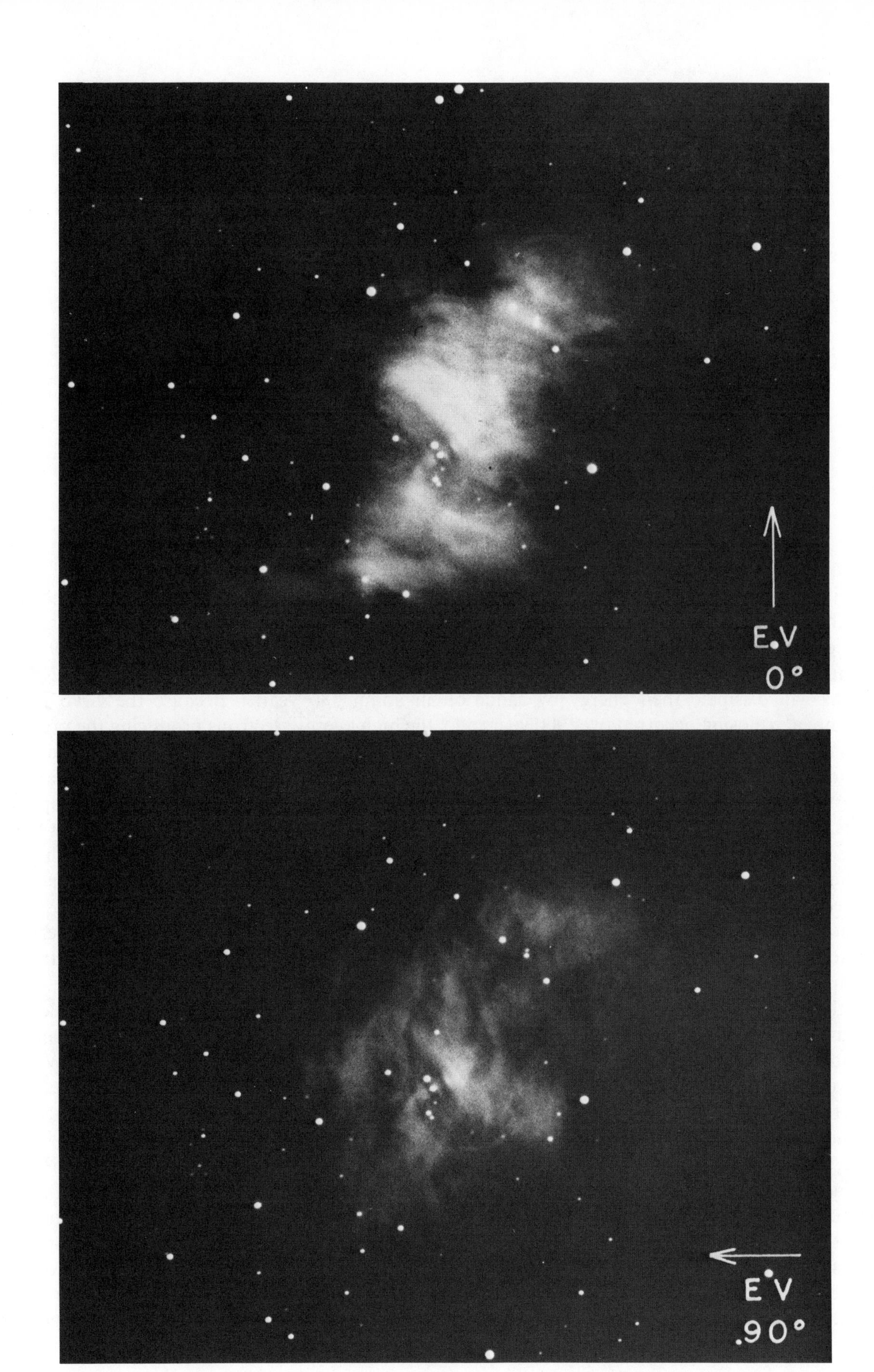
E V
0°
E V
90°

73. Energy Emission from a Neutron Star

Franco Pacini

(*Nature 216*, 567–568 [1967])

In spite of the success of the synchrotron radiation theory in explaining the nonthermal spectrum and the polarization of the radiation from the Crab nebula, accounting for the origin of the relativistic electrons that give rise to the radiation remained a fundamental difficulty. In 1956, for example, Jan Oort and Th. Walraven pointed out that the electrons radiating at optical wavelengths will dissipate their energy by radiation in about 180 years.[1] Because the Crab nebula supernova explosion occurred over 900 years ago, the high energy electrons producing the presently observed optical continuum cannot be survivors of the original supernova explosion. Instead, some unknown source must be continuously producing high-energy electrons in the Crab nebula. Oort and Walraven reminded astronomers of Carl Lampland's 1921 observation of moving wisps and knots that seemed to originate in the central double star of the Crab nebula.[2] When Walter Baade subsequently showed that these wisps and knots move through the nebula at speeds as high as one-tenth the velocity of light, it appeared that the central double star was injecting high-speed particles into the Crab nebula.

The discovery in the 1960s that the Crab nebula radiates X rays stimulated renewed interest in the source of its radiation.[3] If electrons emit synchrotron radiation at X-ray wavelengths, for example, they must have extraordinarily high energies with radiation lifetimes as short as a year, and therefore they might not last long enough to travel through the nebula. Thomas Gold long ago argued that the intense magnetic fields of collapsed stars would make them efficient emitters of radiation (see selection 116), and Fred Hoyle, Jayant Narlikar, and John Archibald Wheeler took up this idea in an effort to explain the X-ray radiation of the Crab nebula.[4] They reasoned that, because of flux conservation during collapse, the remnant neutron star will have a tremendous magnetic field of about 10 billion gauss, and that the oscillations of the neutron star will be dissipated in the form of electromagnetic waves produced during the motions of the magnetic field.

In the paper given here, Franco Pacini points out that the gravitational energy released during the collapse of a normal star will be converted into rotational energy. It follows from the conservation of angular momentum, for example, that a normal star like the sun will speed up from a rotation period of 27 days to a rotation period of much less than a second when it becomes a neutron star. The neutron star will also have a huge magnetic field; and, provided that the magnetic and rotation axes are not aligned, a rotating dipole magnetic field will convert the rotational energy into electromagnetic radiation. Using reasonable estimates for the rotation velocity and magnetic field strength of the neutron star, Pacini is able to show that this gigantic rotating magnet can provide the luminosity of the Crab nebula for its entire lifetime, and he thereby describes a tremendous new source of electromagnetic energy for dying stars.[5]

1. J. H. Oort and Th. Walraven, *Bulletin of the Astronomical Institutes of the Netherlands 12*, 285 (1956).

2. C. O. Lampland, *Publications of the Astronomical Society of the Pacific 33*, 79 (1921).

3. Selection No. 11; S. Bowyer, E. T. Byram, T. A. Chubb, and H. Friedman, *Nature*, *201*, 1307 (1964); *Science 146*, 912 (1964), *147*, 394 (1965).

4. F. Hoyle, J. V. Narlikar, and J. A. Wheeler, *Nature 203*, 914 (1964).

5. According to Iosif Shklovskii, it was N. S. Kardashev who first reasoned that "the regular magnetic field in the Crab Nebula has resulted from a twisting of the field of the neutron star in the plasma surrounding it. . . . Essentially the same prediction [as Pacini's] of a continual loss of kinetic energy by a neutron star had been made by Kardashov [sic] three years earlier. In his view the energy of the magnetic field in the Crab Nebula would be drawn from the kinetic rotational energy of the neutron star, which accordingly would tend to slow down. Pacini's service, however, was in expressing this simple concept in a clear, distinct form." [Shklovskii, *Stars: Their Birth, Life and Death* (W. H. Freeman: San Francisco, 1978), p. 319. Also see N. S. Kardashev, *Astronomicheskii Zhurnal 41*, 807 (1964), English trans. *Soviet Astronomy A. J. 8*, 643 (1965)].

Although there are still many problems concerning the supernovae, there is little doubt that a very dense stellar core has to be left behind after the explosion (at least in some cases). During the contraction of this core, inverse β reactions take place and transform most of the nuclei and electrons into neutrons. If the mass of the neutron star does not exceed a critical value of about one or two solar masses, a stable equilibrium situation can be reached with the gas pressure balancing the gravitational force.

A newly formed neutron star is an excited object. Apart from its thermal content (which will be dissipated very fast because of neutrino processes), there will also be much energy stored in vibrational and rotational form. The problem therefore arises of finding out whether the energy stored in the neutron star plays an important part in connexion with the activity observed in some supernova remnants such as the Crab Nebula.

The vibrations of the neutron star, however, do not last long enough for our purposes. The principal reason for this is that the emission of gravitational waves will damp quadrupole and higher order pulsations in a few seconds (ref. 1 and unpublished work of J. A. Wheeler and A. Zee). Moreover, because the stellar rotation will mix the radial modes of vibrations with the non-radial one, all the vibrations are going to disappear very quickly.

It seems more rewarding therefore to look for some mechanisms by which the neutron star can release either its magnetic or its rotational energy or both. In this communication I would like to outline the principal features of a possible model of this kind.

The existence of very strong magnetic fields in the neutron stars has been suggested as a consequence of the compression of an ordinary stellar field. The underlining assumption here is that the conductivity, σ, of the stellar matter is so high that the decay time for the field exceeds the collapse time. As the collapse time is of the order of seconds, this means

$$\tau_{\text{decay}} \sim \frac{4\pi\sigma R^2}{c^2} \gg 1, \tag{1}$$

where R is the radius of the neutron star (about 10^6 cm). For any conceivable value of σ this is a very weak requirement. We can therefore expect the field strength to increase as $1/R^2$ during the contraction so that fields as high as 10^{10}–10^{14} gauss can be produced.[2]

If we assume that the magnetic field is that of a dipole, the angle between the dipole and the angular momentum is likely to be arbitrary (oblique rotator). Actually, even if the two axes coincided in the pre-supernova star, the mass loss occurring during the explosion is unlikely to be perfectly symmetric, especially in the presence of strong magnetic fields. The mutual inclination of the two axes will be modified and both the rotation and the magnetic field will tend to flatten the star (but along different directions). The shape of the star is going to be rather complicated and the body will rotate about an axis which is not a principal axis of inertia. This is a non-equilibrium situation and the motion will be such that the total angular momentum is constant while the instantaneous axis of rotation precesses. Stresses and magneto-hydrodynamic waves are therefore to be expected at the surface of such a star.[3] This will dissipate energy with the final result of bringing the system into an equilibrium state, that is, the magnetic axis in coincidence with the axis of rotation. Acceleration of particles to relativistic energies is to be expected under these circumstances.

The same picture of an oblique rotator leads also to a different possibility, that is, that the neutron star might directly emit electromagnetic waves of very low frequency (in the kc/s range). This idea has been suggested by Hoyle, Narlikar and Wheeler[4] as a possible consequence of the vibrations of a magnetic neutron star. Because the rapid damping of the vibrations makes it difficult to retain this suggestion in the original form, I wish to point out that the oblique rotator model also results into an analogous emission of electromagnetic waves.

If d_0 is the projection of the dipolar moment on the plane perpendicular to the axis of rotation and Ω is the angular velocity of the star, there will be a monochromatic emission of electromagnetic waves at the frequency $\omega = \Omega$. The corresponding intensity is given[5].

$$I = \frac{2}{3}\frac{d_0{}^2\Omega^4}{c^3}. \tag{2}$$

If we take d_0 to be about $H_0 R^3 = 10^{10} \times 10^{18}$ gauss cm^3 and $\Omega = 10^4$ sec^{-1}, the intensity would be of the order of 2×10^{40} ergs/sec.

We must, however, ask ourselves the question whether this radiation will ever arise. Any variation of the magnetic field can actually be compensated by the electric currents in the surrounding matter. Hoyle, Narlikar and Wheeler[4] have stated that the strong gravitational field creates a near vacuum immediately outside the star so that in this region no propagation difficulty would arise. This equilibrium picture, however, seems unlikely for a newly formed neutron star, soon after the supernova explosion. It is then necessary to evaluate the maximum gas density which still allows the emission of electromagnetic waves. This is easily done if we note that the maximum current density in the gas is given by $j = n_e ec$, where n_e is the electron density. The Maxwell equation

$$\text{curl}\,\vec{H} = \frac{4\pi}{c}\vec{j} \tag{3}$$

gives then the maximum induced field. If we take curl $\sim 1/r$ (r is a characteristic length of the order of the size of the system) we obtain

$$H_{\max} \sim 4\pi n_e e r. \qquad (4)$$

Any variation of the magnetic field of the order of the field itself cannot therefore be compensated by the induced currents if

$$n_e < \frac{H}{4\pi e r}. \qquad (5)$$

Assuming again H to be about 10^{10} gauss and $r = R$, about 10^6 cm, we must require $n_e < 10^{13}\ \mathrm{cm}^{-3}$. This limit is certainly not very stringent and condition (5) is likely to be violated only at the beginning, that is, soon after the birth of the neutron star.

Once the electromagnetic waves are emitted and propagate in the supernova remnant, they will be reflected by the circumstellar gas if the plasma frequency exceeds the radiation frequency. For $\omega = 10^4\ \mathrm{sec}^{-1}$ this happens if $n_e > 2.5 \times 10^{-3}\ \mathrm{cm}^{-3}$ which is now a very low figure. The electromagnetic waves would therefore be unable to reach us, but by this means a large amount of energy and momentum could be pumped from the neutron star into the supernova remnant. In particular, the radiation will give an outward momentum to the nebula by being reflected and therefore accelerate its expansion. As a matter of fact, there is observational evidence that the motion of the Crab Nebula has been accelerated after the original explosion: if we divide the size of the nebula by its present expansion rate, we get too short a lifetime.[6] From a quantitative point of view, no difficulty arises because of the large amount of energy which can be stored in the neutron star under the rotational and magnetic form. As suggested by Hoyle, Narlikar and Wheeler,[4] generation of high energy electrons can be expected in the region where the electromagnetic waves are reflected and cause a rapid compression of the nebular gas.

It seems therefore that, when the oblique rotator model is realized, it can lead to a release of energy from the neutron star. It is, however, clear that the model is very idealized and requires further investigation. In particular, it would be important to evaluate the emission of gravitational waves from the star. This will depend on the mass distribution in the star as influenced by the rotation and magnetic field and will determine the ability of the gravitational waves to carry out rotational energy and angular momentum from the star.

1. W. Y. Chau, *Ap. J. 147*, 664 (1967).

2. L. Woltjer, *Ap. J. 140*, 1309 (1964).

3. L. Spitzer, *Electromagnetic Phenomena in Cosmical Physics*, Inter. Astro. Union Symposium (1958).

4. F. Hoyle, J. Narlikar, and J. A. Wheeler, *Nature 203*, 914 (1964).

5. L. D. Landau and E. M. Lifshitz, in *The Classical Theory of Fields*, 2nd ed. (Reading, Mass.: Addison Wesley, 1962).

6. L. Woltjer, in *Les Houches Lectures on High Energy Astrophysics 1*, 201 (New York: Gordon and Breach, 1967) ed. C. de Witt, E. Schatzman and P. Véron.

74. Observation of a Rapidly Pulsating Radio Source

Antony Hewish, S. Joceyln Bell, John D. H. Pilkington, Paul Frederick Scott, and Robin Ashley Collins

(*Nature 217*, 709–713 [1968])

The series of observations that eventually led to the unexpected discovery of the pulsating radio sources, or pulsars, began with Antony Hewish's pioneering studies of the twinkling of discrete radio sources caused by the motions of plasma clouds in the ionosphere and the solar corona.[1] As winds in the ionosphere blow clouds past the line of sight to a small radio source, the interference causes fluctuations in the received intensity with time scales of about 30 sec; similarly, the solar wind causes rapid fluctuations with time scales of a few tenths of a second. Hewish realized that observations of these fluctuations would provide a simple means of detecting compact radio sources. Moreover, because the distant radio galaxies and quasi-stellar objects have small angular dimensions, one method of acquiring quantitative information about the angular sizes of these sources is to study how their fluctuations change when they are observed at various angles in relation to the sun. Accordingly, Hewish and his colleagues constructed an array of 2,048 dipoles, which were spread over an area of 18,000 m^2 to provide high sensitivity and which were operated at the long wavelength of 3.7 m, because the fluctuations were known to be more prominent at the longer wavelengths.

By July 1967 the radio telescope was finished; a graduate student, Jocelyn Bell, was assigned the responsibility of using the instrument to obtain repeated observations of several radio sources at many different angles in relation to the sun. Bell diligently analyzed some 400 ft of chart recordings each week, and the radio sky was soon found to be heavily populated with compact sources. By the middle of August she had found a mysterious source that was fluctuating in the middle of the night when the effects of the solar wind should have been small. Under the assumption that this source was a flaring star, arrangements were made to install a high-speed detector to record the rapid intensity changes characteristic of solar flares. On November 28, 1967, this recorder revealed the astonishing fact that the source was emitting periodic bursts of radio noise at intervals just greater than 1 sec. Furthermore, comparisons with terrestrial clocks showed that the mysterious periodic signal kept time with an accuracy of 1 part in 1 million, and the short duration of the bursts suggested that the radiating source could not be much larger than a planet. Perhaps because objects such as this one were soon interpreted as stellar pulsations, they have been given the somewhat misleading name of pulsating radio sources, or pulsars, even though we now know that their signals come from the rotation, rather than the periodic expansion and contraction, of neutron stars.

By the time the paper reprinted here was ready for publication, Bell had scrutinized existing chart recordings for evidence of other pulsars, and within three weeks a second paper announced the discovery of three additional pulsars.[2] These papers triggered a flood of observational and theoretical papers on pulsars. In less than a year, the list of pulsars had been expanded to over two dozen, and a pulsar had been detected at the position of the very star thought to be the neutron star remnant of the Crab nebula supernova explosion.

The pulsars could probably have been discovered soon after large radio antennae were constructed, but radio astronomers were used to adding up signals over long time intervals to detect weak radiation. The long time resolutions precluded the detection of the pulsars, which are relatively weak radio sources when averaged over their periods. If time resolutions comparable to the pulsar burst durations of milliseconds had been used, the intense radio bursts would have been easily detected. It is actually because Hewish specifically designed a new type of radio telescope for a study of the rapidly changing solar wind effects that the pulsars were accidentally discovered.

1. A. Hewish, *Proceedings of the Royal Society* (*London*) *A214*, 495 (1952); A. Hewish, P. F. Scott, and D. Wills, *Nature 203*, 1214 (1964).
2. J. D. H. Pilkington, A. Hewish, S. J. Bell, and T. W. Cole, *Nature 218*, 126 (1968).

In July 1967, a large radio telescope operating at a frequency of 81.5 MHz was brought into use at the Mullard Radio Astronomy Observatory. This instrument was designed to investigate the angular structure of compact radio sources by observing the scintillation caused by the irregular structure of the interplanetary medium.[1] The initial survey includes the whole sky in the declination range $-08° < \delta < 44°$ and this area is scanned once a week. A large fraction of the sky is thus under regular surveillance. Soon after the instrument was brought into operation it was noticed that signals which appeared at first to be weak sporadic interference were repeatedly observed at a fixed declination and right ascension; this result showed that the source could not be terrestrial in origin.

Systematic investigations were started in November and high speed records showed that the signals, when present, consisted of a series of pulses each lasting ~ 0.3 s and with a repetition period of about 1.337 s which was soon found to be maintained with extreme accuracy. Further observations have shown that the true period is constant to better than 1 part in 10^7 although there is a systematic variation which can be ascribed to the orbital motion of the Earth. The impulsive nature of the recorded signals is caused by the periodic passage of a signal of descending frequency through the 1 MHz pass band of the receiver.

The remarkable nature of these signals at first suggested an origin in terms of man-made transmissions which might arise from deep space probes, planetary radar or the reflexion of terrestrial signals from the Moon. None of these interpretations can, however, be accepted because the absence of any parallax shows that the source lies far outside the solar system. A preliminary search for further pulsating sources has already revealed the presence of three others having remarkably similar properties which suggests that this type of source may be relatively common at a low flux density. A tentative explanation of these unusual sources in terms of the stable oscillations of white dwarf or neutron stars is proposed.

Position and Flux Density

The aerial consists of a rectangular array containing 2,048 full-wave dipoles arranged in sixteen rows of 128 elements. Each row is 470 m long in an E.–W. direction and the N.–S. extent of the array is 45 m. Phase-scanning is employed to direct the reception pattern in declination and four receivers are used so that four different declinations may be observed simultaneously. Phase-switching receivers are employed and the two halves of the aerial are combined as an E.–W. interferometer. Each row of dipole elements is backed by a tilted reflecting screen so that maximum sensitivity is obtained at a declination of approximately $+30°$, the overall sensitivity being reduced by more than one-half when the beam is scanned to declinations above $+90°$ and below $-5°$. The beamwidth of the array to half intensity is about $\pm\frac{1}{2}°$ in right ascension and $\pm 3°$ in declination; the phasing arrangement is designed to produce beams at roughly 3° intervals in declination. The receivers have a bandwidth of 1 MHz centred at a frequency of 81.5 MHz and routine recordings are made with a time constant of 0.1 s; the r.m.s. noise fluctuations correspond to a flux density of 0.5×10^{-26} W m^{-2} Hz^{-1}. For detailed studies of the pulsating source a time constant of 0.05 s was usually employed and the signals were displayed on a multi-channel 'Rapidgraph' pen recorder with a time constant of 0.03 s. Accurate timing of the pulses was achieved by recording second pips derived from the *MSF* Rugby time transmissions.

Figure 74.1 clearly displays the regular periodicity and also the characteristic irregular variation of pulse amplitude. On this occasion the largest pulses approached a peak flux density (averaged over the 1 MHz pass band) of 20×10^{-26} W m^{-2} Hz^{-1}, although the mean flux density integrated over one minute only amounted to approximately 1.0×10^{-26} W m^{-2} Hz^{-1}. On a more typical occasion the integrated flux density would be several times smaller than this value. It is therefore not surprising that the source has not been detected in the past, for the integrated flux density falls well below the limit of previous surveys at metre wavelengths.

The position of the source in right ascension is readily obtained from an accurate measurement of the "crossover" points of the interference pattern on those occasions when the pulses were strong throughout an interval embracing such a point. The collimation error of the instrument was determined from a similar measurement on the neighboring source 3*C* 409 which transits about 52 min later. On the routine recordings which first revealed the source the reading accuracy was only ± 10 s and the earliest record suitable for position measurement was obtained on August 13, 1967. This and all subsequent measurements agree within the error limits. The position in declination is not so well determined and relies on the relative amplitudes of the signals obtained when the reception pattern is centred on declinations of 20°, 23° and 26°. Combining the measurements yields a position

$$\alpha_{1950} = 19\text{h}\ 19\text{m}\ 38\text{s} \pm 3\text{s}$$

$$\delta_{1950} = 22°00' \pm 30'.$$

As discussed here, the measurement of the Doppler shift in the observed frequency of the pulses due to the Earth's orbital motion provides an alternative estimate of the declination. Observations throughout one year should yield an accuracy of $\pm 1'$. The value currently attained from observations during December–January is $\delta = 21°58' \pm 30'$, a figure consistent with the previous measurement.

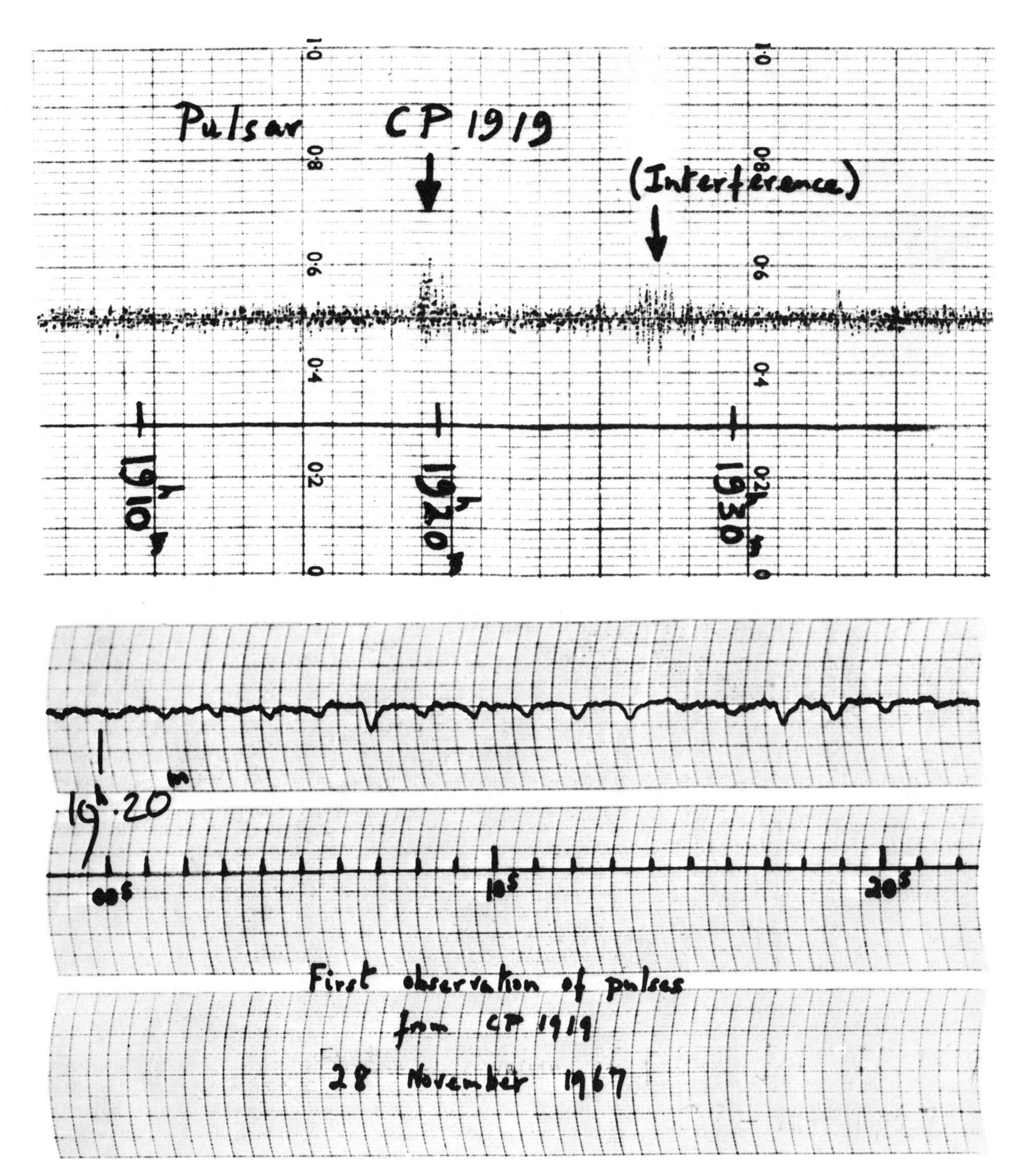

Fig. 74.1 The first observations of the pulsar CP 1919. These data were taken with an 18,000-m^2 dipole array at a wavelength of 3.7 m. The observation made on August 6, 1967 (top), shows a signal indistinguishable from terrestrial interference; the observation taken with a high-speed recorder on November 28, 1967 (bottom), shows the periodic nature of the pulsar bursts for the first time. (Courtesy Antony Hewish.)

TIME VARIATIONS

It was mentioned earlier that the signals vary considerably in strength from day to day and, typically, they are only present for about 1 min, which may occur quite randomly within the 4 min interval permitted by the reception pattern. In addition, as shown in figure 74.1, the pulse amplitude may vary considerably on a time-scale of seconds. The pulse to pulse variations may possibly be explained in terms of interplanetary scintillation,[1] but this cannot account for the minute to minute variation of mean pulse amplitude. Continuous observations over periods of 30 min have been made by tracking the source with an E.–W. phased array in a 470 m × 20 m reflector normally used for a lunar occultation programme. The peak pulse amplitude averaged over ten successive pulses for a period of 30 min suggests the possibility of periodicities of a few minutes duration, but a correlation analysis yields no significant result. If the signals were linearly polarized, Faraday rotation in the ionosphere might cause the random variations, but the form of the curve does not seem compatible with this mechanism. The day to day variations since the source was first detected are irregular and no systematic changes are clearly evident, although there is a suggestion that the source was significantly weaker during October to November. It therefore appears that, despite the regular occurrence of the pulses, the magnitude of the power emitted exhibits variations over long and short periods.

INSTANTANEOUS BANDWIDTH AND FREQUENCY DRIFT

Two different experiments have shown that the pulses are caused by a narrow-band signal of descending frequency sweeping through the 1 MHz band of the receiver. In the first, two identical receivers were used, tuned to frequencies of 80.5 MHz and 81.5 MHz. The lower frequency pulses are delayed by about 0.2 s. This corresponds to a frequency drift of ~ -5 MHz s^{-1}. In the second method a time delay was introduced into the signals reaching the receiver from one-half of the aerial by incorporating an extra cable of known length l. This cable introduces a phase shift proportional to frequency so that, for a signal the coherence length of which exceeds l, the output of the receiver will oscillate with period

$$t_0 = \frac{c}{l}\left(\frac{d\nu}{dt}\right)^{-1},$$

where $d\nu/dt$ is the rate of change of signal frequency.

For observation with $l > 450$ m the periodic oscillations were slowed down to a low frequency by an additional phase shifting device in order to prevent severe attenuation of the output signal by the time constant of the receiver. The rate of change of signal frequency has been deduced from the additional phase shift required and is $d\nu/dt = -4.9 \pm 0.5$ MHz s^{-1}. The direction of the frequency drift can be obtained from the phase of the oscillation on the record and is found to be from high to low frequency in agreement with the first result.

The instantaneous bandwidth of the signal may also be obtained because the oscillatory response as a function of delay is a measure of the autocorrelation function, and hence of the Fourier transform, of the power spectrum of the radiation. The results of the measurements indicate that the instantaneous bandwidth of the signal to exp (-1), assuming a Gaussian energy spectrum, is estimated to be 80 ± 20 kHz.

Here we omit figures illustrating measurements of the instantaneous bandwidth, frequency drift, and the day-to-day pulse arrival times of the signal. Figures showing the variation of burst intensity on time scales of minutes and days are also omitted.

PULSE RECURRENCE FREQUENCY AND DOPPLER SHIFT

By displaying the pulses and time pips from *MSF* Rugby on the same record the leading edge of a pulse of reasonable size may be timed to an accuracy of about 0.1 s. Observations over a period of 6 h taken with the tracking system mentioned earlier gave the period between pulses as $P_{obs} = 1.33733 \pm 0.00001$ s. This represents a mean value centered on December 18, 1967, at 14 h 18 m UT. A study of the systematic shift in the frequency of the pulses was obtained from daily measurements of the time interval T between a standard time and the pulse immediately following it. The standard time was chosen to be 14 h 01 m 00 s UT on December 11 (corresponding to the centre of the reception pattern) and subsequent standard times were at intervals of 23 h 56 m 04 s (approximately one sidereal day). A constant pulse recurrence frequency would show a linear increase or decrease in T if care was taken to add or subtract one period where necessary. The observations, however, show a marked curvature in the sense of a steadily increasing frequency. If we assume a Doppler shift due to the Earth alone, then the number of pulses received per day is given by

$$N = N_0\left(1 + \frac{v}{c}\cos\phi\,\sin\frac{2\pi n}{366.25}\right),$$

where N_0 is the number of pulses emitted per day at the source, v the orbital velocity of the Earth, ϕ the ecliptic latitude of the source and n an arbitrary day number obtained by putting $n = 0$ on January 17, 1968, when the Earth has zero velocity along the line of sight to the source. This relation is approximate since it assumes a circular orbit for the Earth and the origin $n = 0$ is not exact, but it serves to show that the increase of N observed can be explained by the Earth's motion alone within the accuracy currently attainable. For this purpose it is convenient to estimate the values of n for which $\delta T/\delta n = 0$, corresponding to an exactly integral value of N. These occur at $n_1 = 15.8 \pm 0.1$ and $n_2 = 28.7 \pm 0.1$, and since N is increased by exactly one pulse between these dates we have

$$1 = \frac{N_0 v}{c} \cos \phi \left[\sin \frac{2\pi n_2}{366.25} - \sin \frac{2\pi n_1}{366.25} \right].$$

This yields $\phi = 43°36' \pm 30'$ which corresponds to a declination of $21°58' \pm 30'$, a value consistent with the declination obtained directly. The true periodicity of the source, making allowance for the Doppler shift and using the integral condition to refine the calculation, is then

$$P_0 = 1.3372795 \pm 0.0000020 \text{ s}.$$

By continuing observations of the time of occurrence of the pulses for a year it should be possible to establish the constancy of N_0 to about 1 part in 3×10^8. If N_0 is indeed constant, then the declination of the source may be estimated to an accuracy of $\pm 1'$; this result will not be affected by ionospheric refraction.

It is also interesting to note the possibility of detecting a variable Doppler shift caused by the motion of the source itself. Such an effect might arise if the source formed one component of a binary system, or if the signals were associated with a planet in orbit about some parent star. For the present, the systematic increase of N is regular to about 1 part in 2×10^7 so that there is no evidence for an additional orbital motion comparable with that of the Earth.

The Nature of the Radio Source

The lack of any parallax greater than about 2′ places the source at a distance exceeding 10^3 A.U. The energy emitted by the source during a single pulse, integrated over 1 MHz at 81.5 MHz, therefore reaches a value which must exceed 10^{17} erg if the source radiates isotropically. It is also possible to derive an upper limit to the physical dimension of the source. The small instantaneous bandwidth of the signal (80 kHz) and the rate of sweep (-4.9 MHz s^{-1}) show that the duration of the emission at any given frequency does not exceed 0.016s. The source size therefore cannot exceed 4.8×10^3 km.

An upper limit to the distance of the source may be derived from the observed rate of frequency sweep since impulsive radiation, whatever its origin, will be dispersed during its passage through the ionized hydrogen in interstellar space. For a uniform plasma the frequency drift caused by dispersion is given by

$$\frac{d\nu}{dt} = -\frac{c}{L} \frac{\nu^3}{\nu_p^2},$$

where L is the path and ν_p the plasma frequency. Assuming a mean density of 0.2 electron cm^{-3} the observed frequency drift (-4.9 MHz s^{-1}) corresponds to $L \sim 65$ parsec. Some frequency dispersion may, of course, arise in the source itself; in this case the dispersion in the interstellar medium must be smaller so that the value of L is an upper limit. While the interstellar electron density in the vicinity of the Sun is not well known, this result is important in showing that the pulsating radio sources so far detected must be local objects on a galactic distance scale.

The positional accuracy so far obtained does not permit any serious attempt at optical identification. The search area, which lies close to the galactic plane, includes two twelfth magnitude stars and a large number of weaker objects. In the absence of further data, only the most tentative suggestion to account for these remarkable sources can be made.

The most significant feature to be accounted for is the extreme regularity of the pulses. This suggests an origin in terms of the pulsation of an entire star, rather than some more localized disturbance in a stellar atmosphere. In this connexion it is interesting to note that it has already been suggested[2,3] that the radial pulsation of neutron stars may play an important part in the history of supernovae and supernova remnants.

A discussion of the normal modes of radial pulsation of compact stars has recently been given by Meltzer and Thorne,[4] who calculated the periods for stars with central densities in the range 10^5 to 10^{19} g cm^{-3}. Figure 4 of their paper indicates two possibilities which might account for the observed periods of the order 1 s. At a density of 10^7 g cm^{-3}, corresponding to a white dwarf star, the fundamental mode reaches a minimum period of about 8 s; at a slightly higher density the period increases again as the system tends towards gravitational collapse to a neutron star. While the fundamental period is not small enough to account for the observations the higher order modes have periods of the correct order of magnitude. If this model is adopted it is difficult to understand why the fundamental period is not dominant; such a period would have readily been detected in the present observations and its absence cannot be ascribed to observational effects. The alternative possibility occurs at a density of 10^{13} g cm^{-3}, corresponding to a neutron star; at this density the fundamental has a period of about 1 s, while for densities in excess of 10^{13} g cm^{-3} the period rapidly decreases to about 10^{-3} s.

If the radiation is to be associated with the radial pulsation of a white dwarf or neutron star there seem to be several mechanisms which could account for the radio emission. It has been suggested that radial pulsation would generate hydromagnetic shock fronts at the stellar surface which might be accompanied by bursts of X-rays and energetic electrons.[2,3] The radiation might then be likened to radio bursts from a solar flare occurring over the entire star during each cycle of the oscillation. Such a model would be in fair agreement with the upper limit of $\sim 5 \times 10^3$ km for the dimension of the source, which compares with the mean value of 9×10^3 km quoted for white dwarf stars by Greenstein.[5] The energy requirement for this model may be roughly estimated by noting that the total energy emitted in a 1 MHz band by a type III solar burst would produce a radio flux of the right order if the source were at a distance of $\sim 10^3$ A.U. If it is assumed that the radio energy may be related to the total flare energy ($\sim 10^{32}$ erg)[6] in the same manner as for a solar flare and supposing that each pulse corresponds to one flare, the required energy would be $\sim 10^{39}$ erg yr^{-1}; at a distance of 65 pc the corresponding value would be $\sim 10^{47}$ erg yr^{-1}. It has been estimated that a neutron star may contain $\sim 10^{51}$ erg in vibrational modes so the energy requirement does not appear unreasonable, although other damping mechanisms are likely to be important when considering the lifetime of the source.[4]

The swept frequency characteristic of the radiation is reminiscent of type II and type III solar bursts, but it seems unlikely that it is caused in the same way. For a white dwarf or neutron star the scale height of any atmosphere is small and a travelling disturbance would be expected to produce a much faster frequency drift then is actually observed. As has been mentioned, a more likely possibility is that the impulsive radiation suffers dispersion during its passage through the interstellar medium.

More observational evidence is clearly needed in order to gain a better understanding of this strange new class of radio source. If the suggested origin of the radiation is confirmed further study may be expected to throw valuable light on the behavior of compact stars and also on the properties of matter at high density.

1. A. Hewish, P. F. Scott, and D. Wills, *Nature 203*, 1214 (1964).
2. A. G. W. Cameron, *Nature 205*, 787 (1965).
3. A. Finzi, *Phys. Rev. Lett. 15*, 599 (1965).
4. D. W. Meltzer and K. S. Thorne, *Ap. J. 145*, 514 (1966).
5. J. L. Greenstein, in *Handbuch der Physik 50*, 161 (1958).
6. C. E. Fichtel and F. B. McDonald, in *Annual Review of Astronomy and Astrophysics 5*, 351 (1967).

75. Rotating Neutron Stars as the Origin of the Pulsating Radio Sources

Thomas Gold

(*Nature 218*, 731–732 [1968])

As Antony Hewish and his colleagues noted in their discovery paper, the short duration of the pulsar bursts suggests that their radiation originates in a body that cannot be much larger than a planet. Furthermore, the extreme regularity of the periodic bursts immediately suggests an astronomical clock controlled by the pulsation (expansion and contraction) or rotation of a massive body. Although white dwarf stars are massive objects with planetary dimensions, they pulsate with periods longer than those of the shortest pulsars. Hewish himself sought to explain the pulsar emission in terms of the small, dense neutron stars, but all stable neutron stars have pulsation periods too short to be identified with pulsars. This left rotation as an explanation, and Jeremiah Ostriker promptly argued that the pulsar emission might come from a hot spot located on the surface of a rotating white dwarf star.[1] At about the same time, Thomas Gold presented this paper, which argues that pulsars are actually rotating neutron stars with a strong dipole magnetic field. The stress carried by this field transfers the rotational energy of the neutron star into the electromagnetic energy radiated by the lighthouse beam of the pulsar. Furthermore, as Gold successfully predicted, the loss of rotational energy should be reflected in an increase in the pulsar periods with time. He also predicted that pulsars with much shorter periods would be found.

The association of pulsars with supernova remnants became increasingly accepted when Australian radio astronomers found a pulsar with an extremely short period—89 millisec—in the center of the extended radio source Vela X, believed to mark the debris of a supernova explosion.[2] Furthermore, because white dwarf stars cannot rotate much faster than about once in a quarter of a second, this pulsar could only be a rotating neutron star. Soon afterward the Crab nebula was found to contain a pulsar with an even shorter period, 33 millisec, and the neutron star theory received further impressive support when pulsed light was observed from the very star that Walter Baade and Rudolph Minkowski had identified as the central stellar remnant of the Crab nebula supernova explosion.[3] Finally, when David Richards and John Comella showed that the period of the Crab nebula pulsar was increasing with time,[4] Gold's theory received dramatic confirmation.

The scenario was not complete, however, for Gold soon showed that the loss of rotational energy calculated from the increase in the period of the Crab nebula pulsar was exactly that required to account for the observed luminosity of the Crab nebula itself.[5] Even before the discovery of pulsars Franco Pacini had shown that the neutron star remnant of the Crab nebula might lose its rotational energy by magnetic dipole radiation (see selection 73), and the loss of rotational energy calculated from the increase in the period of the Crab nebula pulsar was also consistent with this hypothesis. Jeremiah Ostriker, James Gunn, Peter Goldreich, and William Julian then showed that the pulsars can efficiently accelerate particles to the relativistic energies that account for both the synchrotron radiation of the Crab nebula and the beamed radio waves that are observed as pulsars.[6]

1. J. Ostriker, *Nature 217*, 1227 (1968).
2. M. I. Large, A. E. Vaughan, and B. Y. Mills, *Nature 220*, 340 (1968).
3. D. H. Staelin and E. C. Reifenstein, *Science 162*, 1481 (1968); J. M. Comella, H. D. Craft, R. V. E. Lovelace, J. M. Sutton and G. L. Tyler, *Nature 221*, 453 (1969); W. J. Cocke, M. J. Disney, and D. J. Taylor, *Nature 221*, 525 (1969); R. E. Nather, B. Warner, and M. Macfarlane, *Nature 221*, 527 (1969); J. S. Miller and E. J. Wampler, *Nature 221*, 1037 (1969).
4. D. W. Richards and J. M. Comella, *Nature 222*, 551 (1969).
5. T. Gold, *Nature 221*, 25 (1969).
6. P. Goldreich and W. H. Julian, *Astrophysical Journal 157*, 869 (1969); J. P. Ostriker and J. E. Gunn, *Astrophysical Journal 157*, 1395 (1969).

Abstract—The constancy of frequency in the recently discovered pulsed radio sources can be accounted for by the rotation of a neutron star. Because of the strong magnetic fields and high rotation speeds, relativistic velocities will be set up in any plasma in the surrounding magnetosphere, leading to radiation in the pattern of a rotating beacon.

THE CASE that neutron stars are responsible for the recently discovered pulsating radio sources[1–6] appears to be a strong one. No other theoretically known astronomical object would possess such short and accurate periodicities as those observed, ranging from 1.33 to 0.25 s. Higher harmonics of a lower fundamental frequency that may be possessed by a white dwarf have been mentioned; but the detailed fine structure of several short pulses repeating in each repetition cycle makes any such explanation very unlikely. Since the distances are known approximately from interstellar dispersion of the different radio frequencies, it is clear that the emission per unit emitting volume must be very high; the size of the region emitting any one pulse can, after all, not be much larger than the distance light travels in the few milliseconds that represent the lengths of the individual pulses. No such concentrations of energy can be visualized except in the presence of an intense gravitational field.

The great precision of the constancy of the intrinsic period also suggests that we are dealing with a massive object, rather than merely with some plasma physical configuration. Accuracies of one part in 10^8 belong to the realm of celestial mechanics of massive objects, rather than to that of plasma physics.

It is a consequence of the virial theorem that the lowest mode of oscillation of a star must always have a period which is of the same order of magnitude as the period of the fastest rotation it may possess without rupture. The range of 1.5 s to 0.25 s represents periods that are all longer than the periods of the lowest modes of neutron stars. They would all be periods in which a neutron star could rotate without excessive flattening. It is doubtful that the fundamental frequency of pulsation of a neutron star could ever be so long (ref. 7 and unpublished work of A. G. W. Cameron). If the rotation period dictates the repetition rate, the fine structure of the observed pulses would represent directional beams rotating like a lighthouse beacon. The different types of fine structure observed in the different sources would then have to be attributed to the particular asymmetries of each star (the "sunspots," perhaps). In such a model, time variations in the intensity of emission will have no effect on the precise phase in the repetition period where each pulse appears; and this is indeed a striking observational fact. A fine structure of pulses could be generated within the repetition period, depending only on the distribution of emission regions around the circumference of the star. Similarly, a fine structure in polarization may be generated, for each region may produce a different polarization or be overlaid by a different Faraday-rotating medium. A single pulsating region, on the other hand, could scarcely generate a repetitive fine structure in polarization as seems to have been observed now.[8]

There are as yet not really enough clues to identify the mechanism of radio emission. It could be a process deriving its energy from some source of internal energy of the star, and thus as difficult to analyse as solar activity. But there is another possibility, namely, that the emission derives its energy from the rotational energy of the star (very likely the principal remaining energy source), and is a result of relativistic effects in a co-rotating magnetosphere.

In the vicinity of a rotating star possessing a magnetic field there would normally be a co-rotating magnetosphere. Beyond some distance, external influences would dominate, and co-rotation would cease. In the case of a fast rotating neutron star with strong surface fields, the distance out to which co-rotation would be enforced may well be close to that at which co-rotation would imply motion at the speed of light. The mechanism by which the plasma will be restrained from reaching the velocity of light will be that of radiation of the relativistically moving plasma, creating a radiation reaction adequate to overcome the magnetic force. The properties of such a relativistic magnetosphere have not yet been explored, and indeed our understanding of relativistic magnetohydrodynamics is very limited. In the present case the coupling to the electromagnetic radiation field would assume a major role in the bulk dynamical behaviour of the magnetosphere.

The evidence so far shows that pulses occupy about 1/30 of the time of each repetition period. This limits the region responsible to dimensions of the order of 1/30 of the circumference of the "velocity of light circle." In the radial direction equally, dimensions must be small; one would suspect small enough to make the pulse rise-times comparable with or larger than the flight time of light across the region that is responsible. This would imply that the radiation emanates from the plasma that is moving within 1 per cent of the velocity of light. That is the region of velocity where radiation effects would in any case be expected to become important.

The axial asymmetry that is implied needs further comment. A magnetic field of a neutron star may well have a strength of 10^{12} gauss at the surface of the 10 km object. At the "velocity of light circle," the circumference of which for the observed periods would range from 4×10^{10} to 0.75×10^{10} cm, such a field will be down to values of the order of 10^3–10^4 gauss (decreasing with distance slower than the inverse cube law of an undisturbed dipole field. A field pulled out radially by the stress of the centrifugal force of a whirling plasma would decay as an inverse square law with radius). Asymmetries in the radiation could arise either through the field or the plasma content being non-axially symmetric. A skew and non-dipole field may well result from the explosive event that gave rise to the neutron star; and the access to plasma of certain tubes of force may be dependent on surface inhomogeneities of the star where sufficiently hot or energetic plasma can be produced to lift itself away from the intense gravitational field (10–100 MeV for protons; much less for space charge neutralized electron-positron beams).

The observed distribution of amplitudes of pulses makes it very unlikely that a modulation mechanism can be responsible for the variability (unpublished results of P. A. G. Scheuer and observations made at Cornell's Arecibo Ionospheric Observatory) but rather the effect has to be understood in a variability of the emission mechanism. In that case the observed very sharp dependence of the instantaneous intensity on frequency (1 MHz change in the observation band gives a substantially different pulse amplitude) represents a very narrow-band emission mechanism, much narrower than synchrotron emission, for example. A coherent mechanism is then indicated, as is also necessary to account for the intensity of the emission per unit area that can be estimated from the lengths of the sub-pulses. Such a coherent mechanism would represent non-uniform static configurations of charges in the relativistically rotating region. Non-uniform distributions at rest in a magnetic field are more readily set up and maintained than in the case of high individual speeds of charges, and thus the configuration discussed here may be particularly favourable for the generation of a coherent radiation mechanism.

If this basic picture is the correct one it may be possible to find a slight, but steady, slowing down of the observed repetition frequencies. Also, one would then suspect that more sources exist with higher rather than lower repetition frequency, because the rotation rates of neutron stars are capable of going up to more than 100/s, and the observed periods would seem to represent the slow end of the distribution.

1. A. Hewish, S. J. Bell, J. D. H. Pilkington, P. F. Scott, and R. A. Collins, *Nature 217*, 709 (1968).
2. J. D. H. Pilkington, A. Hewish, S. J. Bell, and T. W. Cole, *Nature 218*, 126 (1968).
3. F. D. Drake, E. J. Gundermann, D. L. Jauncey, J. M. Comella, G. A. Zeissig, and H. D. Craft, Jr., *Science 160*, 503 (1968).
4. F. D. Drake, *Science 160*, 416 (1968).
5. F. D. Drake and H. D. Craft, Jr., *Science 160*, 758 (1968).
6. B. S. Tanenbaum, G. A. Zeissig, and F. D. Drake, *Science 160*, 760 (1968).
7. K. S. Thorne and J. R. Ipser, *Ap. J.* (Letters) *152*, L71, *153*, L125 (1968).
8. A. G. Lyne and F. G. Smith, *Nature 218*, 124 (1968).

CHAPTER VI

The Distribution of Stars and the Space between Them

76. Investigations of the Spectrum and Orbit of Delta Orionis

Johannes Franz Hartmann

(*Astrophysical Journal 19*, 268–286 [1904])

In this paper from the Potsdam Astrophysical Observatory, Johannes Hartmann calls attention to the comparatively narrow H and K lines of ionized calcium that do not share in the periodic displacement of the broader hydrogen and helium lines seen in the spectrum of the binary star δ Orionis. Because the calcium lines did not participate in the orbital motions of δ Orionis, he reasoned that they must be the result of absorption by a cloud of interstellar gas and that they could not originate in the atmospheres of the stars. A few years later Edwin B. Frost detected the narrow "stationary" lines of calcium in the spectra of several bright, young stars; and from observations of calcium lines in the spectra of other binary star systems, Vesto M. Slipher reaffirmed Hartmann's conclusion that they must originate in an absorbing medium outside the binary systems.[1]

Hartmann also calls attention here to the narrow, fixed D line of neutral sodium in the spectrum of Nova Persei; Slipher subsequently suggested that this sodium line might also be found in interstellar space. Nearly thirty years passed before Walter Adams and Theodore Dunham demonstrated that the atomic absorption lines of Fe I, K I, and Ti II and the molecular absorption lines of CH and CN can also be found in clouds of interstellar gas.[2]

Meanwhile, John S. Plaskett showed that the velocities of the calcium lines differ from those of their background stars by up to 50 km sec^{-1}. Moreover, the sodium and calcium velocities agree, and after correcting for solar motion, these velocities are much smaller than those of the stars in question. This evidence suggested to Plaskett that the narrow absorption lines originate in a gas pervading all of interstellar space, but it remained difficult to explain why the calcium lines do not appear in the spectra of stars later than B3. In his famous Bakerian lecture on interstellar matter, A. S. Eddington addressed this problem and argued that the small velocity differences between the stellar lines and the interstellar calcium lines are masked by the broad stellar lines of the later stars. Meanwhile, Plaskett and his colleague Joseph A. Pearce collected additional data on the velocities of the calcium lines and their background stars, and they were soon able to show that the velocities of both the stars and the calcium exhibit the sinusoidal dependence on galactic longitude characteristic of differential galactic rotation. Moreover, because the amplitudes of the radial velocity curves obtained from the calcium lines were almost exactly half those derived from the radial velocities of the stars, they concluded that the calcium must originate in interstellar clouds uniformly scattered among the stars.[3]

1. E. B. Frost, *Astrophysical Journal 29*, 233 (1909); V. M. Slipher, *Lowell Observatory Bulletin 2*, No. 51, 1 (1909).

2. T. Dunham and W. S. Adams, *Publications of the Astronomical Society of the Pacific 49*, 26 (1937), *53*, 341 (1941); W. S. Adams, *Astrophysical Journal 93*, 11 (1941).

3. J. S. Plaskett, *Monthly Notices of the Royal Astronomical Society 84*, 80 (1923); A. S. Eddington, *Proceedings of the Royal Society* (London) *A111*, 424 (1926); J. S. Plaskett and J. A. Pearce, *Publications of the Dominion Astrophysical Observatory 5*, 167 (1933).

PERIODIC VARIATIONS in the spectrum of δ Orionis were discovered by H. Deslandres in 1900, and from eleven observations he obtained a period of 1.92 days.[1] We omit here Hartmann's table of the radial velocities he observed on a total of 38 days between February 1900 and March 1903.

δ Orionis belongs to the type of Orion stars (I*b*), whose spectrum shows, besides the hydrogen lines, chiefly the lines of helium, all of which in this case are exceedingly diffuse and dim, so that their measurement is very difficult and uncertain. On account of the slight intensity of the lines, all defects of the film are very disturbing, and, in consequence of irregular distribution of the silver grains, the lines often appear crooked and unsymmetrical, sometimes indeed double. I have convinced myself by a special investigation that the indications of duplicity and unsymmetrical broadening cannot be caused by lines belonging to a second component of the stellar system; but I do not hold it to be impossible that the form of the lines is subject to small real changes, perhaps in consequence of violent motions in the gaseous envelope of the star.

Although we must, accordingly, regard δ Orionis as a binary system having one of its components "dark," in the customary phrase, I would nevertheless point out that by "dark" we here must understand only a relatively small difference of brightness. A difference of only about one magnitude would be sufficient to bring the spectrum of the fainter component to almost complete disappearance, and a difference of two magnitudes would make it impossible for even a trace of the fainter spectrum to be visible on the plate. The slight difference of magnitude necessary for the extinction of the fainter spectrum explains the fact that among the numerous spectroscopic binaries so far discovered there are very few which show the lines of the second component in the spectrum.

Here Hartmann describes his procedures for eliminating personality effects of the observer when measuring diffuse line spectra.[2] Every plate was measured twice, all the recognizable lines on each plate were measured, and the velocity curve was not drawn until after all measurements were completed. Velocities were determined mainly from the wavelengths of the lines arising from the Balmer transitions of hydrogen and ionized helium.

The calcium line at λ 3,934 exhibits a very peculiar behavior. It is distinguished from all the other lines of this spectrum, first by the fact that it always appears extraordinarily weak, but almost perfectly sharp; and it therefore attracted my attention that in computing the wavelengths for this particular line, the agreement between the results from the different plates was decidedly less than for the other, much less sharp lines. Closer study on this point now led me to the quite surprising result *that the calcium line at λ 3,934 does not share in the periodic displacements of the lines caused by the orbital motion of the star.*

Here Hartmann presents a table (not reproduced) comparing the velocities of δ Orionis deduced from the calcium line with those deduced from the other spectral lines. Whereas the calcium line velocity has a mean value of 16 km sec^{-1} with random dispersions of a few kilometers per second, the orbital velocity based on the other spectral lines varies sinusoidally with a maximum amplitude of 133 km sec^{-1}.

The fact being thus fully established that a single line of the spectrum does not participate in the oscillatory motion of the other lines, the question arises as to how it can be explained. The character of the absorption corresponding to the line makes it highly improbable that it should have originated in the Earth's atmosphere. In that case, moreover, the line in question would have to appear in every stellar spectrum, and the velocities computed from its position would come into worse agreement on applying the correction for reduction to the Sun. But the case is quite the opposite—the value does not become constant until after applying the reduction to the Sun, and thus the cosmical origin of the line is proven.

The most natural assumption, that the observed line belongs to the second component of the binary system, encounters two difficulties. We should have to assume for the fainter component a mass *at least* ten times as great as that of the brighter star. While this is itself very improbable, it is still more surprising that not a single other line should be revealed of the spectrum of the second body. The occurrence of such an isolated line would be explained by none of the spectral types hitherto known, and it points rather with a pretty fair degree of certainty to the presence of an absorbing layer of gas not in immediate connection with the star.

We are thus led to the assumption that at some point in space in the line of sight between the Sun and δ Orionis there is a cloud which produces that absorption, and which recedes with a velocity of 16 km [sec^{-1}], in case we admit the further assumption, very probable from the nature of the observed line, that the cloud consists of calcium vapor. This reasoning finds a distinct support in a quite similar phenomenon exhibited by the spectrum of Nova Persei in 1901. While the lines of hydrogen and other elements in that spectrum led us, by their enormous broadening and displacement and the continuous changing of their form, to conclude that stormy processes were going on within the gaseous envelope of the star, the two calcium lines at λ 3,934 and λ 3,969 as well as the D lines, were observed as perfectly sharp absorption lines, which yielded the constant velocity of +7 km during the whole duration of the phenomenon. I then expressed the opinion that these sharp lines probably did not have their origin in the Nova itself, but in a nebulous mass lying in the line of sight—a view which only gained in probability on the later discovery of the nebula in the neighborhood of the Nova. In the case of δ Orionis also it is not unlikely that the cloud stands in some relation to the extensive nebulous masses shown by Barnard[3] to be present in the neighborhood. The second calcium line at λ 3,969 is concealed in the spectrum of δ Orionis by the broad hydrogen line $H\varepsilon$, and therefore cannot be observed.

I would also call attention at this point to a further peculiar phenomenon. If we compute the component of the solar motion $V_{\odot}$ according to Campbell's[4] provisional elements of the apical motion, we obtain:

for δ Orionis $\quad V_{\odot} = +18.1$ km

for Nova Persei $\quad V_{\odot} = +\ 8.7$ km

These figures agree within the errors of observation with the observed velocity of the calcium clouds, which therefore in these two cases are almost completely at rest relatively to the 280 stars employed by Campbell.

The point in the line of sight at which the nebulous mass lies cannot be ascertained. In order to determine its lateral extension the spectra of neighboring stars, particularly those with variable or distinctly different velocity, should be examined for the occurrence of the calcium line. It is present in the spectrum of ε and ζ Orionis, but inasmuch as the velocities of these stars differ only slightly from the motion of the cloud given above, we cannot distinguish whether it belongs to the spectrum of the star or of the cloud.

Hartmann then presents his solution for the orbital elements of δ Orionis.

1. H. Deslandres, "Variations rapides de la vitesse radiale de l'étoile *δ Orion*," *Comptes Rendus 130*, 379 (1900).

2. *Astronomische Nachrichten 155*, 97 (1901).

3. "Diffused Nebulosities in the Heavens," *Astrophysical Journal 17*, 77 (1903).

4. "A Preliminary Determination of the Motion of the Solar System," ibid. *13*, 80 (1901).

77. Star Streaming

Jacobus Cornelius Kapteyn

(*Report of the British Association for the Advancement of Science*, 237–265 [1905])

The study of the apparent angular motions of the stars, which we now call proper motions, was initiated by Edmund Halley in 1718, when he discovered that the positions of Sirius, Aldebaran, Betelgeuse, and Arcturus were different from those given in Ptolemy's *Almagest*. At the time there was no way of knowing whether the observed motions were caused by the motion of the sun, the motion of the stars, or some combination of these motions. In 1783, however, William Herschel used the proper motions of thirteen bright stars to support his theory that the apparent motions of these stars are actually due to the sun's motion.[1]

A controversy over the exact nature of the solar motion developed during the nineteenth century, when the proper motions of many of the more distant, fainter stars were measured. In 1837 Argelander used the proper motions of 330 stars to obtain a value for the direction of solar motion similar to Herschel's value.[2] An indication that the accumulation of more data would not lead to a more accurate determination of the solar motion came in 1895, when H. Kobold used over a thousand stars to obtain a solar apex over 30° south of the generally accepted position. From this discordance Kobold concluded that the individual, peculiar motions of the stars do not have random directions and that the motions must take place in the plane of the Milky Way.

Soon after Kobold's startling announcement, Jacobus Kapteyn began in Groningen to examine the proper motions of some 2,400 stars. By dividing his data into twenty-eight separate areas, rather than averaging over the entire sky, Kapteyn not only confirmed Kobold's conclusion that the stars are not moving in a random, haphazard fashion but also discovered that they tend to move in two preferred directions. Kapteyn announced his results at the Congress of the St. Louis Exposition in 1904 and published his paper the following year; his findings were promptly confirmed by A. S. Eddington from independent data.[3] As here shown in Kapteyn's paper, the stars in each small area of the sky seem to travel in two large intermingled streams that pass through each other while moving in opposite directions in the plane of the Milky Way.

When W. W. Campbell used stellar radial velocity measurements to derive a solar motion of 20 km sec^{-1}, the relative velocities of the two star streams were set at 40 km sec^{-1}. Meanwhile, Karl Schwarzschild showed that it was unnecessary to think of two star streams and that the phenomenon of star streaming could be explained by assuming that the individual motions of the stars are distributed in an ellipsoid with the long axis in the direction of motion of Kapteyn's two star streams.[4] A decade later, when Harlow Shapley proposed a new and much larger extent for our Milky Way galaxy, it was very suggestive that his chosen center for the galaxy, in the direction of Sagittarius, lay near one of the vertices of the star streams. That star streaming is a direct consequence of the rotation of the Milky Way was first shown by Bertil Lindblad, who found that Schwarzschild's ellipsoidal distribution arises naturally from the motions of stars in the solar neighbourhood revolving under the influence of the massive central parts of our galaxy (see selections 79 and 81).

1. W. Herschel, *Philosophical Transactions of the Royal Society 15*, 402 (1783).
2. H. Argelander, *Astronomische Nachrichten 16*, 45 (1838).
3. A. S. Eddington, *Monthly Notices of the Royal Astronomical Society 68*, 588 (1908). Eddington later provided a lucid summary of the early works in his *Stellar Movements and the Structure of the Universe* (London: Macmillan, 1914).
4. W. W. Campbell, *Lick Observatory Bulletin 6, No. 196*, 125 (1910); K. Schwarzschild, *Nachrichten von der Gesellschaft der Wissenschaften in Göttingen*, 614 (1907).

IN DERIVING THE CONSTANT of precession, and in investigating the motion of the sun through space, it is usual to start from the hypothesis that the real motions of the stars, the so-called *peculiar* motions, have no preference for any particular direction.

Of late I have found anomalies in the distribution of the *apparent* proper motions, of so strongly systematic a character that I feel convinced that we are compelled to give up this hypothesis.

It will be the aim of this paper to show the nature of these anomalies, and to explain the conclusion to which they lead us.

It is only just to mention that as early as 1895 Kobold called attention to a fact which seems incompatible with a random distribution of the direction of the motion of the stars. Had Kobold been more successful in separating the systematic motions of the stars from the displacements caused by the sun's motion, he would probably have been led to conclusions similar to those which I am now about to submit to you.

In order to show clearly the anomaly in the distribution of the proper motions here alluded to, it will be necessary to call to mind how this distribution must present itself if the hypothesis of the random orientation of the motions were really satisfied.

For this purpose consider a great number of stars very near each other on the sphere, say all the stars of such a small constellation as the Southern Cross. For convenience' sake we will even assume them to be all apparently situate in the same point S (figure 77.1, *P*) of the sphere, though not in space, because their distances would be different.

The peculiar proper motions of these stars will be distributed somewhat in the manner indicated in figure 77.1, *P*.

In addition to this motion, which represents the real motion of the stars as seen projected on the sphere, they will have an *apparent* motion, the *parallactic* motion, which is due to the observer's own motion, or say the motion of the solar system, through space.

These parallactic motions, we all know, are directed away from the apex, which is the point where the sun's motion prolonged meets the sphere. For *all* the stars at *S* the parallactic motion will be directed along *Sx*.

The motions as really observed are the resultant of the peculiar and the parallactic motion.

Thus for the star whose peculiar motion is *SB*, let $S\beta$ be the parallactic motion, then the *observed* motion of that star will be *Sb*. Likewise the observed proper motion of the star having the peculiar motion *SC* will be *Sc*, and so on.

Making the composition for all the stars of figure 77.1, *P*, we get the really observable motions distributed as in figure 77.1, *Q*.

From this it must be evident that whereas, according to the hypothesis, the distribution of the *peculiar* motions would be radially symmetrical, this symmetry will be destroyed for the *observed* proper motions.

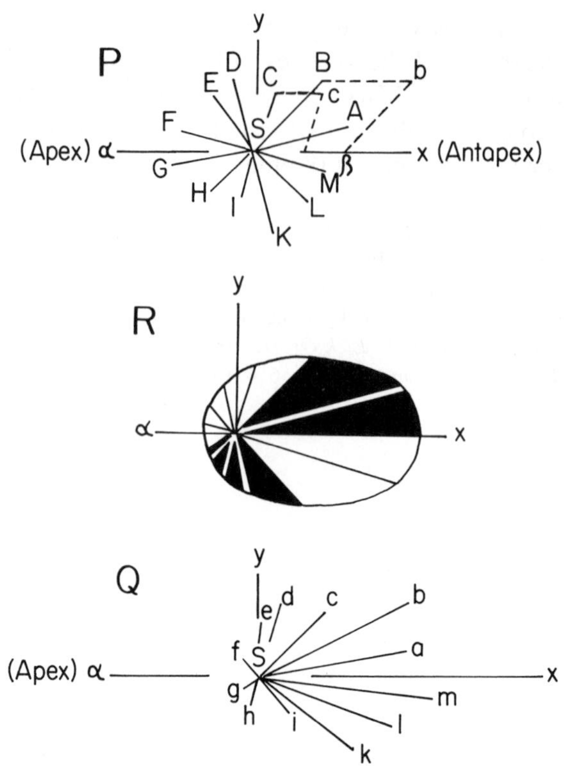

Fig. 77.1 The distribution of the peculiar proper motions with a random distribution of directions of motion is shown in *P* for a group of stars located at position *S*. Here the length of each line represents the magnitude of motion in the direction of the line. When the observer is at motion towards an apex, α, the observed motions of the stars shown in *P* will look like *Q*. Here bilateral symmetry is retained, although radial symmetry is destroyed. In *R* the lines in *Q* that make angles between 0° and −60° and between +60° and +180° have been blackened.

There will be a strong preference for motions directed towards the antapex. (See *Q*.) One thing, however, must be clear, and we want no more for what follows; it is that there will remain a *bilateral symmetry*, the line of symmetry being evidently the line αSx through the apex, the star, and the antapex.

Near to this line, on the antapex side, the proper motions will be most numerous, and they will be greater in amount.

This evident condition of bilateral symmetry would furnish probably the best means of determining the position of the apex. For if from all our data about proper motion we determine these lines of symmetry for several points of the sky and prolong them, they must all intersect in two points, which are no other than the apex and the antapex.

In trying to realize this plan we must meet with the difficulty that on account of errors of observation and the restricted number of stars included in the investigation we must be prepared to find in reality no such perfect symmetry as theory demands. For the lines of symmetry we shall thus have to substitute *lines giving the nearest approach to symmetry*. Their position will depend, at least to a certain extent, on what we choose to consider as "the nearest approach to symmetry."

If we call the required line of symmetry the axis of the x, the line at right angles thereto the axis of the y, then we may, for instance, define that position of the x-axis as the line of greatest symmetry, which makes *zero* the sum of the y's.

The lines of symmetry furnished by this definition, prolonged, will not pass through a single point; they will all cross a certain more or less extended area, the center of gravity of which might be taken as the most probable position of the apex.

Drawing great circles through this apex, we must necessarily find them diverging somewhat from the lines of best symmetry in different parts of the sky.

If, however, our hypothesis of random orientation is approximately true, the divergences will be small. The sum of the proper motions at right angles to these circles will be nearly zero in every part of the sky.

Not only that, but we will have further to expect that any other condition of symmetry will be approximately fulfilled, too, for every point of the sky. Such another condition will be, for instance, that on both sides of the great circles through the apex the total quantity of proper motion will be the same, or, again, that Σx shall be the same on both sides of these circles.

How the first of these conditions is satisfied is shown in figure 77.2. This figure summarizes the more important points in regard to the question in hand. They show in compact form the results of a complete treatment of the proper motions of all the stars observed in both co-ordinates by Bradley (over 2,400 stars). These stars are distributed over two thirds of the whole of the sky. This surface has been divided up into twenty-eight areas.

From the stars contained on each area I have derived the distribution of the proper motions corresponding to the center of the area. How this was done need not here be explained. The whole of the materials were thus embodied in twenty-eight figures, like those of figure 77.2, each of which shows at a glance the distribution of the proper motions for one particular region in the sky.

Not to overburden the plate, I have only included *ten* of the figures, for which the phenomenon to which I wish to draw your attention is most marked.

It is very suggestive that these lie all near the poles of the Milky Way.

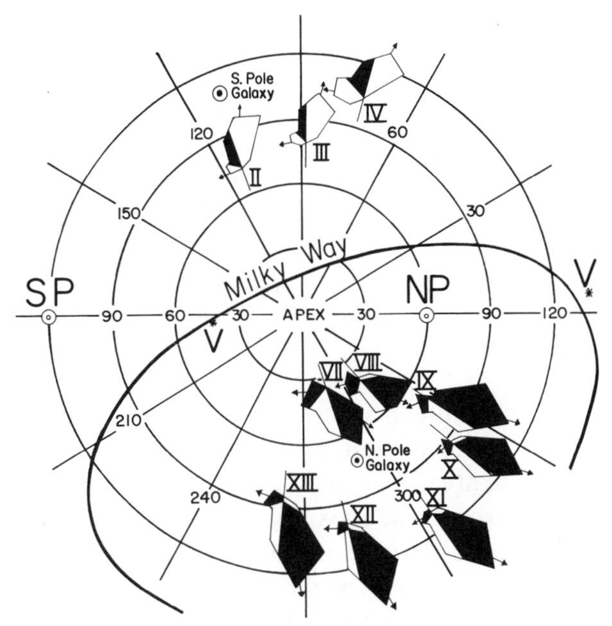

Fig. 77.2 The distribution of the proper motions of over 2,400 stars divided into twenty-eight areas (only ten of the areas are shown). For each area the diagram has been drawn with the same meanings as R in figure 77.1. If the proper motions of the stars at each point had a random distribution of directions, the areas of the black and white parts of each diagram would be equal. The stars appear to be moving in two star streams directed toward the two vertices marked V.

The figures have been constructed as follows: A line has been drawn making the angle of 15° with the great circle through the apex, the length of which represents the sum of all the proper motions making angles of between 0° and 30° with that circle.

In the same way the radius vector of 45° represents the sum of all the motions between 30° and 60°, and so on.

For the sake of uniformity all the results have been reduced to what they would have been had the total number of stars been the same for all the twenty-eight areas.

In order still better to show the important points in the distribution of the p.m. that part of the figures between the radii vectores making angles of *zero* and $-60°$, $+60°$ and $+180°$ have been blackened.

The position adopted for the apex is practically that found by a variety of methods all more or less akin to that described a moment ago.

If the random-distribution hypothesis were true, and if in consequence thereof the symmetry of our figures were complete, the blackened parts of the figures would have been equal to the corresponding lighter-tinted parts in the way of figure 77.1, *R*.

The real state of things is something quite different, and, what is all-important, we see at once that the divergences are strikingly systematic. The figures at each pole of the Milky Way show them in nearly every particular of the same character. Near the North Pole the blackened parts are invariably much greater; at the South Pole the case is reversed.

At a first glance the difference of the more extensive parts on the side of the antapex is the most striking. As a matter of fact, however, the difference between the smaller parts is by no means less important.

Here Kapteyn presents a table of the mean values of the x component of the proper motions for those areas lying on the two sides of the great circle through the apex. Significant differences of a few seconds of arc are found, and these differences should not exist if the motions only reflect the sun's motion toward its apex. Because the differences remain when the calculations are based on widely differing positions for the apex, they cannot be due to an error in the position of the apex. The character of the phenomenon also remained unchanged when Kapteyn used the lines of symmetry calculated from the condition $\Sigma y = 0$ for each of the twenty-eight areas, which means that unrelated systematic motions in each area cannot be the cause. Instead, there is an excess of proper motions in the same two favored directions for each area. The two points lie some 140° apart, one of them some 7° south of α Orionis and the other a couple of degrees south of η Sagittari.

When we see that the motions of a certain group of stars converge to a same point on the sphere we conclude either that the real motions of the stars are in reality parallel, or that the motion is only apparent and due to a motion of the observer in the opposite direction. As long as we have no fixed point of reference we cannot decide between the two.

When we see *two* groups of stars converging towards two *different* points the latter explanation fails, at least for one of the groups, because the observer can have but one motion.

From the facts set forth in what precedes we must, therefore, at once conclude that one of our sets must have a real systematic motion in respect to the other.

We can even take one further step. As early as 1843, Bravais has shown, in a paper to which sufficient attention has not been paid, that, no matter how systematic the motions of any group of stars may be, we can determine the motion of the solar system with respect to the center of gravity of that group. Therefore, if for the present we take the center of gravity of all the Bradley stars as a fixed point of reference, we can determine the direction of the sun's motion. As far as I know no extensive determination of the solar apex has as yet been made, rigorously on the basis of Bravais' theory. But for reasons, into which we cannot enter here, the result can hardly differ from the best of our modern determinations, made by other methods. This position coincides with neither of the two points found just now.

We thus get a clear indication that we have to do with two star-streams, parallel to the lines joining our solar system to the two points mentioned.

That the method is not rigorous, that, therefore, the directions here found cannot lay claim to any great accuracy, may be left out of consideration for the present. But what is important to note is: (1) that the directions are only apparent directions; that is, directions of the motion relative to the solar system. (2) That if it be true that two directions of motion predominate in the stellar world, then, if we refer all our motions to the center of gravity of the system, these two main directions of motion must be in reality diametrically opposite. Some reflection must convince you that it must be so, and I will not, therefore, stop to demonstrate it.

For the sake of brevity I will call the points of the sphere towards which the star-streams seem to be directed the *vertices* of the stellar motion.

The *apparent* vertices were thus provisionally found to lie south of α Orionis and η Sagittarii. Knowing with some approximation the velocity of the sun's motion as compared with the mean velocity of the stars, it is easy to derive from the *apparent* positions of the vertices their *true* positions, which must lie at diametrically opposite points of the sphere.

Having once got what I considered to be the clue to the systematic divergences in the proper motions, and having at the same time obtained an approximation for the position of the vertices, I have made a more rigorous solution of the problem.

I will not here enter into the details of that solution. In order to prevent misconception, however, it will be well to state expressly that the existence of two main stream-lines does not imply that the real motions of the stars are all exclusively directed to either of the two vertices; there is only a decided *preference* for these directions. In my solution I have assumed that the frequency of other directions becomes regularly smaller as the angle with the main stream becomes greater, according to the most simple law of which I could think, which makes the change dependent on a single constant.

I have as yet only finished a first approximation to the solution. The result is that one of the two vertices lies very near to ζ Orionis; the other, diametrically opposite, is not near any bright star. They have been represented by the letter V in figure 77.2. They lie almost exactly in the central line of the Milky Way. Adopting Gould's co-ordinates of the pole of this belt, I find the galactic latitude to be *two* degrees.

I will pass over the other quantities involved, but will only mention that the way in which I conducted the solution points to the conclusion that *all* the stars, without exception, belong to one of the two streams. To my regret I must pass over also the detailed comparison of theory and observation, because the detailed determination of the distribution of the proper motions from the data of our solution is such a laborious question that I have not yet made it, and would rather defer it till the real existence of the streams shall have been tested by other observations presently to be considered. I will only state that by this provisional solution the *total* amount of dissymmetry for our twenty-eight regions is reduced for the x components as well as for the y components to about *a third* of their amount in the hypothesis of random distribution of the directions. Moreover, they have lost their systematic character.

Here Kapteyn explains that he refrained from publishing these results for three years because he hoped to test his theory by using radial velocity data, which the American astronomers (notably William Campbell) were known to be accumulating. Because this data remained unpublished, Kapteyn here tests his theory using the radial velocities of the bright stars. He argues that these bright stars exhibit smaller motions than the fainter stars because the bright stars are nearby stars that participate in the sun's motion. Because the largest radial velocities of the brighter stars are not numerous near the vertices, they seem to confirm the starstream phenomenon.

Taking the evidence for what it is worth, we may say that it confirms the theory. The proof is not convincing, however, and I will conclude by giving expression to my hopes that those who are in a position to test the whole theory by more extensive and more reliable materials will not neglect to do so.

A few hundreds of stars, not pertaining to the Orion stars, and fainter than magnitude 3.5, must probably be sufficient for the purpose.

78. The Kinetic Energy of a Star Cluster

Arthur Stanley Eddington

(*Monthly Notices of the Royal Astronomical Society 76*, 525–528 [1916])

In the short paper given here, distinguished by its simplicity and elegance, A. S. Eddington demonstrates that the internal kinetic energy of a star cluster is half its potential energy. He also points out that the result could have been obtained at once from what is now known as the virial theorem, a formula whose previous use had been almost entirely restricted to gases. Eddington here extends the theorem to nonequilibrium situations by including a time-dependent term, and with this addition he considers the dissolution of the Taurus cluster (Hyades). (Since differential galactic rotation was still unimagined in 1916, Eddington could not evaluate this far more erosive effect on the Hyades.)

Eddington's applications of the virial theorem to star clusters set the stage for several important subsequent applications of this theorem. In the 1930s, for example, Fritz Zwicky assumed that the galaxies in a cluster of galaxies are in equilibrium under their mutual gravitation, and he then estimated the masses of individual galaxies from the virial theorem by using the known velocity dispersion and size of the cluster (see selection 107). Later, Subrahmanyan Chandrasekhar and Enrico Fermi used the virial theorem for nonequilibrium situations to consider the dissipation of the spiral arms of our galaxy; by adding a magnetic energy term to the theorem they were able to show that an interstellar magnetic field of only about a millionth of a gauss is needed to contain the moving interstellar gas.[1]

1. S. Chandrasekhar and E. Fermi, *Astrophysical Journal 118*, 116 (1953). See also selection 98.

1

IN ANY STAR CLUSTER *in a steady state the internal kinetic energy is one-half the exhaustion of potential energy.* Although I cannot find that this result has ever been explicitly stated, it follows almost immediately from formulae found in the theory of gases in connection with the virial.[1] It will probably be more convenient to astronomers to give in full the proof for stars, instead of quoting these formulæ. It will be seen that the result does not in any way depend on the special conditions of a gas, but is a perfectly general consequence of the law of gravitation.

Taking the centre of gravity of the cluster as origin, let

x, y, z be the co-ordinates of a star of mass m,

X, Y, Z the gravitational force on m, due to the attractions of the other stars of the cluster,

T the total kinetic energy of the motions of the stars relative to the centre of gravity,

Ω the exhaustion of potential energy of the cluster,

C the moment of inertia of the cluster about its centre of gravity.

We have

$$m\frac{d^2x}{dt^2} = \mathrm{X}.$$

From the identity

$$\frac{1}{2}\frac{d^2}{dt^2}(x^2) = \left(\frac{dx}{dt}\right)^2 + x\frac{d^2x}{dt^2}$$

we have

$$\tfrac{1}{4}m\frac{d^2}{dt^2}(x^2) = \tfrac{1}{2}m\left(\frac{dx}{dt}\right)^2 + \tfrac{1}{2}x\mathrm{X}.$$

Add this equation and the two corresponding equations for y and z, and sum for all the stars of the cluster; the result is

$$\tfrac{1}{4}\sum m\frac{d^2}{dt^2}(r^2) = \tfrac{1}{2}\sum mv^2 + \tfrac{1}{2}\sum(x\mathrm{X} + y\mathrm{Y} + z\mathrm{Z}), \tag{1}$$

where $r^2 = x^2 + y^2 + z^2$, and v is the velocity of the star.

The left-hand side

$$= \frac{1}{4}\frac{d^2}{dt^2}\sum mr^2$$

$$= \frac{1}{4}\frac{d^2\mathrm{C}}{dt^2}$$

This evidently vanishes in a steady state, because the moment of inertia cannot be altering. Equation (1) therefore reduces to

$$\mathrm{T} = -\tfrac{1}{2}\sum(x\mathrm{X} + y\mathrm{Y} + z\mathrm{Z}). \tag{2}$$

To evaluate (2), we remark that X is the sum of the attractions on a particular star of all the other stars, so that the whole expression may be split up into separate terms corresponding to every possible pair of stars. Consider a pair of stars m_1 and m_2. The contribution due to the attraction of m_2 on m_1 is

$$-\tfrac{1}{2}x_1\frac{m_1m_2(x_2 - x_1)}{r_{12}^3} - \tfrac{1}{2}y_1\frac{m_1m_2(y_2 - y_1)}{r_{12}^3} - \tfrac{1}{2}z_1\frac{m_1m_2(z_2 - z_1)}{r_{12}^3},$$

where

$$r_{12}^2 = (x_2 - x_1)^2 + (y_2 - y_1)^2 + (z_2 - z_1)^2.$$

The portion due to the attraction of m_1 on m_2 is

$$-\tfrac{1}{2}x_2\frac{m_1m_2(x_1 - x_2)}{r_{12}^3} - \tfrac{1}{2}y_2\frac{m_1m_2(y_1 - y_2)}{r_{12}^3} - \tfrac{1}{2}z_2\frac{m_1m_2(z_1 - z_2)}{r_{12}^3}.$$

Adding these together, they give

$$\frac{1}{2}\frac{m_1m_2}{r_{12}}.$$

Hence

$$-\tfrac{1}{2}\sum(x\mathrm{X} + y\mathrm{Y} + z\mathrm{Z}) = \tfrac{1}{2}\sum\frac{m_1m_2}{r_{12}},$$

the latter sum being taken over all combinations of stars in pairs.

$$= -\tfrac{1}{2}(\text{potential energy of cluster}).$$

Hence (2) gives the result enunciated, namely,

$$\mathrm{T} = \tfrac{1}{2}\Omega \tag{3}$$

The whole energy of the cluster is $\mathrm{T} - \Omega$, and the result may be expressed in the form—

Potential energy is twice the whole energy.
Kinetic energy is minus the whole energy.

2

It is of some interest to notice the general formula when the state is not steady. Equation (1) then gives

$$\frac{1}{4}\frac{d^2C}{dt^2} = T - \tfrac{1}{2}\Omega \tag{4}$$

so that the result holds for a roughly steady state, if d^2C/dt^2 is negligible.

Moreover, if $T > (1/2)\Omega$, d^2C/dt^2 is positive, so that the cluster expands (or its rate of contraction diminishes so that there is ultimately an expansion). Usually expansion diminishes Ω, and consequently diminishes $T - (1/2)\Omega$, the whole energy $T - \Omega$ remaining constant. Similarly, if $T < (1/2)\Omega$, there is a tendency for $T - (1/2)\Omega$ to increase. We see, therefore, that the value of $T - (1/2)\Omega$ is likely to be accelerated towards zero; but, since there is no absolute relation between C and Ω, we cannot draw any definite conclusion without further data as to the constitution of the cluster. Perhaps the cluster will attain the steady state by a series of rapidly damped oscillations of $T - (1/2)\Omega$.

3. RATE OF DISSOLUTION OF A MOVING CLUSTER

This example is appended to illustrate the use that can be made of the foregoing result. It has been shown by Jeans[2] that the parallelism of the motions of stars in a cluster is destroyed with exceeding slowness by perturbations due to the passage of other stars through the cluster. But although the deflections of motion are very small, they will lead to a comparatively rapid scattering of the stars in space unless counteracted by some other force such as the mutual attraction of the cluster. For example, from the present dimensions of the Taurus cluster I have shown that it could not have existed more than 57 million years unless the scattering had been counteracted.[3] We are now in a position to take this mutual attraction into account and estimate the actual rate of scattering.

The chance perturbations by external stars produce a probable deflection of the motion of any cluster star proportional to the square-root of the time, and therefore give it a probable kinetic energy (relative to the centroid of the cluster) proportional to the time elapsed. This internal energy is acquired at the expense of the translational energy of the cluster as a whole or of the stars encountered. Let M be the mass of the whole cluster, α the velocity relative to the cluster acquired by a star in unit time (mean-square value); then the cluster will be gaining internal energy at the rate $1/2M\alpha^2$. By §1 the rate of increase of potential energy of the cluster will then be $M\alpha^2$, provided an approximately steady state is maintained. (Since the cluster stars are closer and their relative motions much slower than the external stars, their mutual encounters will be more potent as a means of transferring energy; that is to say, the energy will be redistributed within the cluster more rapidly than it is acquired from outside, and a steady state should result.) Now the potential energy of the cluster is $-M^2/2c$, where c is proportional to the linear dimensions of the cluster and is of the same order of magnitude as the average radius. For Plummer's law of density, $(1 + r^2)^{-5/2}$, c is about 0.82 times the median radius of the cluster. Thus we have

$$\frac{d}{dt}\left(-\frac{M^2}{2c}\right) = M\alpha^2, \tag{5}$$

or, integrating for an interval t,

$$\frac{1}{c_0} - \frac{1}{c} = \frac{2\alpha^2 t}{M}$$

The time taken for the cluster to increase from half its present linear dimensions to its present size is thus

$$\tau = \frac{M}{2c\alpha^2}. \tag{6}$$

Taking Jeans's figures for the Taurus cluster, he finds an average deflection of 1′ in a million years, which, for a velocity of the cluster of 40 km./sec., is equivalent to a transverse velocity of 0.012 parsecs per million years. If we take for the unit of mass the sun's mass, and for the unit of time a million years, the unit of length must be 0.165 parsecs if the gravitational constant is eliminated. Hence $\alpha = 0.073$. We may take c as roughly 3 parsecs, or 18 units. Then (6) gives

$$\tau = M \times 5 \text{ million years.}$$

Forty stars are known to belong to the cluster, and there are doubtless many others. Thus τ must be several hundred million years at least, and the present concentration of the cluster does *not* indicate an unduly short period of existence.

1. Cf. Jeans, *Dynamical Theory of Gases*, 1st ed., p. 145, where the virial is found to $-(1/2)\Sigma\Sigma r\varphi(r)$. Put $\varphi(r) = -1/r^2$, and the result follows immediately. I have since learnt that the result has been given by H. Poincaré (*Hypotheses cosmogoniques*, p. 94) in a review of Ligondés' hypothesis of a swarm of meteorites. I am indebted to Mr. H. Jeffreys for this reference.

2. *Monthly Notices 74*, 109.

3. *Stellar Movements and the Structure of the Universe*, p. 254.

79. The Scale of the Universe

Part 1: Harlow Shapley

(Bulletin of the National Research Council of the National Academy of Sciences [Washington, D.C.] *2*, 171–193 [1921])

Part 2: Heber D. Curtis

(Bulletin of the National Research Council of the National Academy of Sciences, [Washington, D.C.] *2*, 194–217 [1921])

Early in 1917 Harlow Shapley, who had been working with the 60-in telescope at Mount Wilson, wrote to Jacobus Kapteyn that "the work on clusters goes on monotonously, monotonous so far as labor is concerned, but the results are continual pleasure. Give me time enough and I shall get something out of the problem yet." A year later, in January 1918, Shapley wrote to A. S. Eddington about a new breakthrough: "I have had in mind from the first that results more important to the problem of the galactic system than to any other question might be contributed by the cluster studies. Now, with startling suddenness and definiteness, they seem to have elucidated the whole sidereal structure..."

Shapley went on to explain the use of the period-luminosity law of Cepheid variation to establish the distance of globular clusters and to describe how the globular clusters outlined an immense galactic system. Later in the year Eddington wrote back, "I think it is not too much to say that this marks an epoch in the history of astronomy, when the boundary of our knowledge of the universe is rolled back to 100 times its former limit."

In 1909 the Swedish astronomer Karl Bohlin had suggested as part of his quite idiosyncratic cosmogony that the center of our galactic system is not the sun but it is within the large concentration of globular clusters lying in the direction of the constellation Sagittarius.[1] Nevertheless, it was Shapley's ingenuity in calibrating the distances to the globular clusters—a task worked out in a series of *Astrophysical Journal* papers in 1918 and 1919—that opened up a dramatically larger sidereal system.[2]

Because Shapley was oblivious to the role of interstellar absorption, he considerably overestimated the size of the Milky Way, giving a distance from the sun to the galactic center of about 50,000 l.y. At the same time, he was loathe to admit the spiral nebulae as objects of comparable dimensions, especially in light of Adrian van Maanen's measurements of internal rotations of spirals. Because Shapley's picture of our stellar system was in direct conflict with the ongoing statistical studies on stellar distributions, and because such astronomers as Heber D. Curtis of Lick Observatory strongly defended the "island universe" theory of spiral nebulae, a debate was organized between the two men at the National Academy of Sciences meeting in April 1920.

For Shapley, the topic "the scale of the universe" meant the size of the Milky Way system, since at that time he refused to accept the extragalactic distances for the spirals. Curtis's attempts to defend the smaller galactic scale of the Kapteyn universe were doomed to failure. In the second part of Curtis's argument, however, his intuition served him well when he rejected van Maanen's determinations of spiral rotations and used the novae as distance indicators. This part of Curtis's paper is not reproduced here; instead, as selection 103, we reprint his earlier and clearer 1919 address to the Washington Academy of Sciences.

The "Shapley-Curtis debate," as finally printed, is considerably different from the spoken version. The final documents were exchanged several times, with numerous modifications. There is in fact reasonable evidence that Shapley, then rather inexperienced in public speaking, was at a substantial disadvantage except for a major ad hoc response from the floor by his mentor and former thesis advisor, Henry Norris Russell.[3]

The Shapley-Curtis debate by no means settled the conflict between the sun-centered Kapteyn universe and the larger galactic system of the globular clusters centered far from the sun. Kapteyn and Pieter van Rhijn, for example, asserted that the large proper motions of the RR Lyrae stars meant that they were nearby dwarf stars, rather than distant giants as Shapley maintained;[4] and it was not at all clear at the time just how the stellar and globular cluster systems were related. Shapley had argued that the Kapteyn system was only one of many systems that make up the larger galactic system, and Bertil Lindblad subsequently adopted this idea in his explanation of star streaming. In 1927 Jan Oort used the distribution of radial velocities and proper motions of stars to show that they undergo differential rotation about a massive center in the direction of the center of the globular cluster system. It was Oort's work, combined with Lindblad's analysis, that finally overcame the resistance to Shapley's model for the galactic system (see selections 81 and 82).

1. K. Bohlin *Kungliga Svenska Vetenskapsakademiens handlingar 43*, no. 10 (1909).
2. H. Shapley, *Astrophysical Journal 48*, 89, 154 (1918), *49*, 311 (1919).
3. M. A. Hoskin, *Journal for the History of Astronomy 7*, 169 (1976); R. Berendzen, R. Hart, and D. Seeley, *Man Discovers the Galaxies* (New York: Neale Watson-Science History Publications 1976).
4. J. C. Kapteyn and P. J. van Rhijn, *Astrophysical Journal 52*, 23 (1920). Also see selection 80.

Contents—Part I: Evolution of the idea of galactic size; surveying the solar neighborhood; on the distances of globular clusters; the dimensions and arrangement of the galactic system. Part II: Dimensions and structure of the galaxy; evidence furnished by the magnitude of the stars; the spirals as external galaxies.

Shapley, Part I

Evolution of the Idea of Galactic Size

THE PHYSICAL UNIVERSE[1] was anthropocentric to primitive man. At a subsequent stage of intellectual progress it was centered in a restricted area on the surface of the earth. Still later, to Ptolemy and his school, the universe was geocentric; but since the time of Copernicus the sun, as the dominating body of the solar system, has been considered to be at or near the center of the stellar realm. With the origin of each of these successive conceptions, the system of stars has ever appeared larger than was thought before. Thus the significance of man and the earth in the sidereal scheme has dwindled with advancing knowledge of the physical world, and our conception of the dimensions of the discernible stellar universe has progressively changed. Is not further evolution of our ideas probable? In the face of great accumulations of new and relevant information can we firmly maintain our old cosmic conceptions?

As a consequence of the exceptional growth and activity of the great observatories, with their powerful methods of analyzing stars and of sounding space, we have reached an epoch, I believe, when another advance is necessary; our conception of the galactic system must be enlarged to keep in proper relationship the objects our telescopes are finding; the solar system can no longer maintain a central position. Recent studies of clusters and related subjects seem to me to leave no alternative to the belief that the galactic system is at least ten times greater in diameter—at least a thousand times greater in volume—than recently supposed.

Dr. Curtis,[2] on the other hand, maintains that the galactic system has the dimensions and arrangement formerly assigned it by students of sidereal structure—he supports the views held a decade or so ago by Newcomb, Charlier, Eddington, Hertzsprung, and other leaders in stellar astronomy. In contrast to my present estimate of a diameter of at least three hundred thousand light-years Curtis outlines his position as follows.[3]

> As to the dimensions of the galaxy indicated by our Milky Way, till recently there has been a fair degree of uniformity in the estimates of those who have investigated the subject. Practically all have deduced diameters of from 7,000 to 30,000 light-years. I shall assume a maximum galactic diameter of 30,000 light-years as representing sufficiently well this older view to which I subscribe though this is pretty certainly too large.

I think it should be pointed out that when Newcomb was writing on the subject some twenty years ago, knowledge of those special factors that bear directly on the size of the universe was extremely fragmentary compared with our information of to-day. In 1900, for instance, the radial motions of about 300 stars were known; now we know the radial velocities of thousands. Accurate distances were then on record for possibly 150 of the brightest stars, and now for more than ten times as many. Spectra were then available for less than one-tenth of the stars for which we have the types to-day. Practically nothing was known at that time of the photometric and spectroscopic methods of determining distance; nothing of the radial velocities of globular clusters or of spiral nebulae, or even of the phenomenon of star streaming.

As a further indication of the importance of examining anew the evidence on the size of stellar systems, let us consider the great globular cluster in Hercules—a vast sidereal organization concerning which we had until recently but vague ideas. Due to extensive and varied researches, carried on during the last few years at Mount Wilson and elsewhere, we now know the positions, magnitudes, and colors of all its brightest stars, and many relations between color, magnitude, distance from the center, and star density. We know some of these important correlations with greater certainty in the Hercules cluster than in the solar neighborhood. We now have the spectra of many of the individual stars, and the spectral type and radial velocity of the cluster as a whole. We know the types and periods of light variation of its variable stars, the colors and spectral types of these variables, and something also of the absolute luminosity of the brightest stars of the cluster from the appearance of their spectra. Is it surprising, therefore, that we venture to determine the distance of Messier 13 and similar systems with more confidence than was possible ten years ago when none of these facts was known, or even seriously considered in cosmic speculations?

If he were writing now, with knowledge of these relevant developments, I believe that Newcomb would not maintain his former view on the probable dimensions of the galactic system.

For instance, Professor Kapteyn has found occasion, with the progress of his elaborate studies of laws of stellar luminosity and density, to indicate larger dimensions of the galaxy than formerly accepted. In a paper just appearing as *Mount Wilson Contribution*, No. 188,[4] he finds, as a result of the research extending over some 20 years, that the density of stars along the galactic plane is quite appreciable at a distance of 40,000 light-years—giving a diameter of the galactic system, exclusive of distant star clouds of the Milky Way, about three times the value Curtis admits as a maximum for the entire galaxy. Similarly Russell, Eddington, and, I believe, Hertzsprung, now subscribe to larger values of galactic dimensions; and Charlier, in a recent lecture before the Swedish Astronomical Association, has accepted the essential features of the larger galaxy, though formerly he identified the local system of B stars with the whole galactic system and obtained distances of the clusters and dimension of the galaxy only a hundredth as large as I derive.

Surveying the Solar Neighborhood

Let us first recall that the stellar universe, as we know it, appears to be a very oblate spheroid or ellipsoid—a disk-shaped system composed mainly of stars and nebulae. The solar system is not far from the middle plane of this flattened organization which we call the galactic system. Looking away from the plane we see relatively few stars; looking along the plane, through a great depth of star-populated space, we see great numbers of sidereal objects constituting the band of light we call the Milky Way. The loosely organized star clusters, such as the Pleiades, the diffuse nebulae such as the great nebula of Orion, the planetary nebulae, of which the ring nebula in Lyra is a good example, the dark nebulosities—all these sidereal types appear to be a part of the great galactic system, and they lie almost exclusively along the plane of the Milky Way. The globular clusters, though not in the Milky Way, are also affiliated with the galactic system; the spiral nebulae appear to be distant objects mainly if not entirely outside the most populous parts of the galactic region.

This conception of the galactic system, as a flattened, watch-shaped organization of stars and nebulae, with globular clusters and spiral nebulae as external objects, is pretty generally agreed upon by students of the subject; but in the matter of the distances of the various sidereal objects—the size of the galactic system—there are, as suggested above, widely divergent opinions. We shall, therefore, first consider briefly the dimensions of that part of the stellar universe concerning which there is essential unanimity of opinion, and later discuss in more detail the larger field, where there appears to be a need for modification of the older conventional view.

Possibly the most convenient way of illustrating the scale of the sidereal universe is in terms of our measuring rods, going from terrestrial units to those of stellar systems. On the earth's surface we express distances in units such as inches, feet, or miles. On the moon, as seen in the accompanying photograph made with the 100-inch reflector, the mile is still a usable measuring unit; a scale of 100 miles is indicated on the lunar scene.

Our measuring scale must be greatly increased, however, when we consider the dimensions of a star—distances on the surface of our sun, for example. The large sun-spots shown in the illustration cannot be measured conveniently in units appropriate to earthly distance—in fact, the whole earth itself is none too large. The unit for measuring the distances from the sun to its attendant planets, is, however, 12,000 times the diameter of the earth; it is the so-called astronomical unit, the average distance from earth to sun. This unit, 93,000,000 miles in length, is ample for the distances of planets and comets. It would probably suffice to measure the distances of whatever planets and comets there may be in the vicinity of other stars; but it, in turn, becomes cumbersome in expressing the distances from one star to another, for some of them are hundreds of millions, even a thousand million, astronomical units away.

Here we omit figures showing craters on the moon with diameters on the order of 50 mi and groups of spots on the sun with total extents on the order of the earth-moon distance. Also omitted are two photographs of the diffuse nebula NGC 221, illustrating greater detail when larger exposure times are used.

This leads us to abandon the astronomical unit and to introduce the light-year as a measure for sounding the depth of stellar space. The distance light travels in a year is something less than six million million miles. The distance from the earth to the sun is, in these units, eight light-minutes. The distance to the moon is 1.2 light-seconds. In some phases of our astronomical problems (studying photographs of stellar spectra) we make direct microscopic measures of a ten-thousandth of an inch; and indirectly we measure changes in the wave-length of light a million times smaller than this; in discussing the arrangement of globular clusters in space, we must measure a hundred thousand light-years. Expressing these large and small measures with reference to the velocity of light, we have an illustration of the scale of the astronomer's universe—his measures range from the trillionth of a billionth part of one light-second, to more than a thousand light-centuries. The ratio of the greatest measure to the smallest is as 10^{33} to 1.

It is to be noticed that light plays an all-important rôle in the study of the universe; we know the physics and chemistry of stars only through their light, and their distance from us we express by means of the velocity of light. The light-year, moreover, has a double value in sidereal exploration; it is geometrical, as we have seen, and it is historical. It tells us not only how far away an object is, but also how long ago the light we examine was started on its way. You do not see the sun where it is, but where it was eight minutes ago. You do not see faint stars of the Milky Way as they are now, but more probably as they were when the pyramids of Egypt were being built; and the ancient Egyptians saw them as they were at a time still more remote. We are, therefore, chronologically far behind events when we study conditions or dynamical behavior in remote stellar systems; the motions, light-emissions, and variations now investigated in the Hercules cluster are not contemporary, but, if my value of the distance is correct, they are the phenomena of 36,000 years ago. The great age of these incoming pulses of radiant energy is, however, no disadvantage; in fact, their antiquity has been turned to good purpose in testing the speed of stellar evolution, in indicating the enormous ages of stars, in suggesting the vast extent of the universe in time as well as in space.

Taking the light-year as a satisfactory unit for expressing the dimensions of sidereal systems, let us consider the distances of neighboring stars and clusters, and briefly mention the methods of deducing their space positions. For nearby stellar objects we can make direct trigonometric measures of distance (parallax), using the earth's orbit or the sun's path through space as a base line. For many of the more distant stars spectroscopic methods are available, using the appearance of the stellar spectra and the readily measurable apparent brightness of the stars. For certain types of stars, too distant for spectroscopic data, there is still a chance of obtaining the distance by means of the photometric method. This method is particularly suited to studies of globular clusters; it consists first in determining, by some means, the real luminosity of a star, that is, its so-called absolute magnitude, and second, in measuring its apparent magnitude. Obviously, if a star of known real brightness is moved away to greater and greater distances, its apparent brightness decreases; hence, for such stars of known absolute magnitude, it is possible, using a simple formula, to determine the distance by measuring the apparent magnitude.

It appears, therefore, that although space can be explored for a distance of only a few hundred light-years by direct trigonometric methods, we are not forced, by our inability to measure still smaller angles, to extrapolate uncertainly or to make vague guesses relative to farther regions of space, for the trigonometrically determined distances can be used to calibrate the tools of newer and less restricted methods. For example, the trigonometric methods of measuring the distance to moon, sun, and nearer stars are decidedly indirect, compared with the linear measurement of distance on the surface of the earth, but they are not for that reason inexact or questionable in principle. The spectroscopic and photometric methods of measuring great stellar distance are also indirect, compared with the trigonometric measurement of small stellar distance, but they, too, are not for that reason unreliable or of doubtful value. These great distances are not extrapolations. For instance, in the spectroscopic method, the absolute magnitudes derived from trigonometrically measured distances are used to derive the curves relating spectral characteristics to absolute magnitude; and the spectroscopic parallaxes for individual stars (whether near or remote) are, almost without exception, interpolations. Thus the data for nearer stars are used for purposes of calibration, not as a basis for extrapolation.

By one method or the other, the distances of nearly 3,000 individual stars in the solar neighborhood have now been determined; only a few are within ten light-years of the sun. At a distance of about 130 light-years we find the Hyades, the well known cluster of naked eye stars; at a distance of 600 light-years, according to Kapteyn's extensive investigations, we come to the group of blue stars in Orion—another physically-organized cluster composed of giants in luminosity. At distances comparable to the above values we also find the Scorpio-Centaurus group, the Pleiades, the Ursa Major system.

These nearby clusters are specifically referred to for two reasons.

In the first place I desire to point out the prevalence throughout all the galactic system of clusters of stars, variously organized as to stellar density and total stellar content. The gravitational organization of stars is a fundamental feature in the universe—a double star is one aspect of a stellar cluster, a galactic system is another. We may indeed, trace the clustering motive from the richest of isolated globular clusters such as the system in Hercules, to the loosely organized nearby groups typified in the bright stars of Ursa Major. At one hundred times its present distance, the Orion cluster would look much like Messier 37 or Messier 11; scores of telescopic clusters have the general form and star density of the Pleiades and the Hyades. The difference between bright and faint clusters of the galactic system naturally appears to be solely a matter of distance.

In the second place I desire to emphasize the fact that the nearby stars we use as standards of luminosity, particularly the blue stars of spectral type B, are members of stellar clusters. Therein lies a most important point in the application of photometric methods. We might, perhaps, question the validity of comparing the isolated stars in the neighborhood of the sun with stars in a compact cluster; but the comparison of nearby cluster stars with remote cluster stars is entirely reasonable, since we are now so far from primitive anthropocentric notions that it is foolish to postulate that distance from the earth has anything to do with the intrinsic brightness of stars.

On the Distances of Globular Clusters

1. As stated above, astronomers agree on the distances to the nearby stars and stellar groups—the scale of the part of the universe that we may call the solar domain. But as yet there is lack of agreement relative to the distances of remote clusters, stars, and star clouds—the scale of the total galactic system. The disagreement in this last particular is not a small difference of a few percent, an argument on minor detail; it is a matter of a thousand percent or more.

Curtis maintains that the dimensions I find for the galactic system should be divided by ten or more, therefore, that galactic size does not stand in the way of interpreting spiral nebulae as comparable galaxies (a theory that he favors on other grounds but considers incompatible with the larger values of galactic dimensions). In his Washington address, however, he greatly simplified the present discussion by accepting the results of recent studies on the following significant points:

Proposition A. The globular clusters form a part of our galaxy; therefore the size of the galactic system proper is most probably not less than the size of the subordinate system of globular clusters.

Proposition B. The distances derived at Mount Wilson for globular clusters *relative to one another* are essentially correct. This implies among other things that (1) absorption of light in space has not appreciably affected the results, and (2) the globular clusters are much alike in structure and constitution, differing mainly in distance. (These relative values are based upon apparent diameters, integrated magnitudes, the magnitudes of individual giants or groups of giants, and Cepheid variables; Charlier has obtained much the same results from apparent diameters alone, and Lundmark from apparent diameters and integrated magnitudes.)

Proposition C. Stars in clusters and in distant parts of the Milky Way are not peculiar—that is, uniformity of conditions and of stellar phenomena naturally prevails throughout the galactic system.

We also share the same opinion, I believe, on the following points:

a. The galactic system is an extremely flattened stellar organization, and the appearance of a Milky Way is partly due to the existence of distinct clouds of stars, and is partly the result of depth along the galactic plane.

b. The spiral nebulae are mostly very distant objects, probably not physical members of our galactic system.

c. If our galaxy approaches the larger order of dimensions, a serious difficulty at once arises for the theory that spirals are galaxies of stars comparable in size with our own: it would be necessary to ascribe impossibly great magnitudes to the new stars that have appeared in the spiral nebulae.

2. Through approximate agreement on the above points, the way is cleared so that the outstanding difference may be clearly stated: Curtis does not believe that the numerical value of the distance I derive for any globular cluster is of the right order of magnitude.

3. The present problem may be narrowly restricted therefore, and may be formulated as follows: Show that any globular cluster is approximately as distant as derived at Mount Wilson; then the distance of other clusters will be approximately right (see Proposition B), the system of clusters and the galactic system will have dimensions of the order assigned (see Proposition A), and the "comparable galaxy" theory of spirals will have met with a serious, though perhaps not insuperable difficulty.

In other words, to maintain my position it will suffice to show that any one of the bright globular clusters has roughly the distance in light-years given in table 79.1, rather than a distance one tenth of this value or less.[5]

Table 79.1 Distances of bright globular clusters

Cluster	Distance (l.y.)	Mean photographic magnitude of brightest 25 stars	
		Apparent	Absolute
Messier 13	36,000	13.75	−1.5
Messier 3	45,000	14.23	−1.5
Messier 5	38,000	13.97	−1.4
Omega Centauri	21,000	12.3:	−1.8:

Similarly it should suffice to show that the bright objects in clusters are giants (cf. last column of table 79.1), rather than stars of solar luminosity.

4. From observation we know that some or all of these four clusters contain:

a. An interval of at least nine magnitudes (apparent and absolute) between the brightest and faintest stars.

b. A range of color-index from −0.5 to +2.0, corresponding to the whole range of color commonly found among assemblages of stars.

c. Stars of types B, A, F, G, K, M (from direct observations of spectra), and that these types are in sufficient agreement with the color classes to permit the use of the latter for ordinary statistical considerations where spectra are not yet known.

d. Cepheid and cluster variables which are certainly analogous to galactic variables of the same types, in spectrum, color change, length of period, amount of light variation, and all characters of the light-curve.

e. Irregular, red, small-range variables of the Alpha Orionis type, among the brightest stars of the cluster.

f. Many red and yellow stars of approximately the same magnitude as the blue stars, in obvious agreement with the giant star phenomena of the galactic system, and clearly in disagreement with all we know of color and magnitude relations for dwarf stars.

5. From these preliminary considerations we emphasize two special deductions:

First, a globular cluster is a pretty complete "universe" by itself, with typical and representative stellar phenomena, including several classes of stars that in the solar neighborhood are recognized as giants in luminosity.

Second, we are very fortunately situated for the study of distant clusters—outside rather than inside. Hence we obtain a comprehensive dimensional view, we can determine relative real luminosities in place of relative apparent luminosities, and we have the distinct advantage that the most luminous stars are easily isolated and the most easily studied. None of the brightest stars in a cluster escapes us. If giants or super-giants are there, they are necessarily the stars we study. We cannot deal legitimately with the average brightness of stars in globular clusters, because the faintest limits are apparently far beyond our present telescopic power. Our ordinary photographs record only the most powerful radiators—encompassing a range of but three or four magnitudes at the very top of the scale of absolute luminosity, whereas in the solar domain we have a known extreme range of 20 magnitudes in absolute brightness, and a generally studied interval of twelve magnitudes or more.

6. Let us examine some of the conditions that would exist in the Hercules Cluster (Messier 13) on the basis of the two opposing values for its distance (table 79.2).

The Blue Stars The colors of stars have long been recognized as characteristic of spectral types and as being of invaluable aid in the study of faint stars for which spectroscopic observations are difficult or impossible. The color-index, as used at Mount Wilson, is the difference between the so-called photographic (pg.) and photovisual (pv.) magnitudes—the difference between the brightness of objects in blue-violet and in yellow-green light. For a negative color-index (C. I. = pg. − pv. < 0.0) the stars are called blue and the corresponding spectral type is B; for yellow stars, like the sun (type G), the color-index is about + 0.8 mag.; for redder stars (type K, M) the color-index exceeds a magnitude.

An early result of the photographic study of Messier 13 at Mount Wilson was the discovery of large numbers of negative color-indices. Similar results were later obtained in other globular and open clusters, and among the stars of the galactic clouds. Naturally these negative color-indices in clusters have been taken without question to indicate B-type stars—a supposition that has later been verified spectroscopically with the Mount Wilson reflectors.[6]

The existence of 15th magnitude B-type stars in the Hercules cluster seems to answer decisively the question of its distance, because B stars in the solar neighborhood are invariably giants (more than a hundred times as bright as the sun, on the average), and such a giant star can appear to be of the fifteenth magnitude only if it is more than 30,000 light-years away.

We have an abundance of material on distances and absolute magnitudes of the hundreds of neighboring B's—there are direct measures of distance, as well as mean distances determined from parallactic motions, from observed luminosity curves, from stream motions, and from radial velocities combined with proper motion. Russell, Plummer, Charlier, Eddington, Kapteyn, and others have worked on these stars with the universal result of finding them giants.

Kapteyn's study of the B stars is one of the classics of modern stellar astronomy; his methods are mainly the well-tried methods generally used for studies of nearby stars. In his various lists of B's more than seventy percent are brighter than zero absolute photographic magnitude,[7] and only two out of 424 are fainter than + 3. This result should be compared with the above-mentioned requirement that the absolute magnitudes of the blue stars in Messier 13 should be + 5 or

Table 79.2 Parameters for the Hercules Cluster for two different distances

Condition	36,000 l.y.	3,600 l.y. or less
a. Mean absolute photographic magnitude of blue star (C.I. < 0.0)	0	+ 5, or fainter
b. Maximum absolute photographic magnitudes of cluster stars	Between − 1.0 and − 2.0	+ 3.2, or fainter
c. Median absolute photovisual magnitude of long-period Cepheids	− 2	+ 3, or fainter
d. Hypothetical annual proper motion	0″.004	0″.04, or greater

fainter in the mean, if the distance of the cluster is 3,600 light-years or less, and no star in the cluster should be brighter than +3.

A question might be raised as to the completeness of the material used by Kapteyn and others, for if only the apparently bright stars are studied, the mean absolute magnitudes may be too high. Kapteyn, however, entertains little doubt on this score, and an investigation[8] of the distribution of B-type stars, based on the *Henry Draper Catalogue*, shows that faint B's are not present in the Orion region studied by Kapteyn.

The census in local clusters appears to be practically complete without revealing any B stars as faint as +5. But if the Hercules cluster were not more distant than 3,600 light-years, its B stars would be about as faint as the sun, and the admitted uniformity throughout the galactic system (Proposition C) would be gainsaid: for although near the earth, whether in clusters or not, the B stars are giants, away from the earth in all directions, whether in the Milky Way clouds or in clusters, they would be dwarfs—and the anthropocentric theory could take heart again.

Let us emphasize again that the near and the distant blue stars we are intercomparing are all cluster stars, and that there appears to be no marked break in the gradation of clusters, either in total content or in distance, from Orion through the faint open clusters to Messier 13.

The Maximum Absolute Magnitude of Cluster Stars In various nearby groups and clusters the maximum absolute photographic brightness, determined from direct measures of parallax or stream motion or from both, is known to *exceed* the following values:

	M
Ursa Major system	−1.0
Moving cluster in Perseus	−0.5
Hyades	+1.0
Scorpio-Centaurus cluster	−2.5
Orion nebula cluster	−2.5
Pleiades	−1.0
61 Cygni group[9]	+1.0

No nearby physical group is known, with the possible exception of the 61 Cygni drift, in which the brightest stars are fainter than +1.0. The mean M of the above list of clusters is −0.8; yet for all distant physical groups it must be +3 or fainter (notwithstanding the certain existence within them of Cepheid variables and B-type stars), if the distance of Messier 13 is 3,600 light-years or less. Even if the distance is 8,000 light-years, as Curtis suggests in the following paper, the mean M would need to be +1.4 or fainter—a value still irreconcilable with observations on nearby clusters.

The requirement that the bright stars in a globular cluster should be in the maximum only two magnitudes brighter than our sun is equivalent to saying that in Messier 13 there is not one real giant among its thirty or more thousand stars. It is essentially equivalent, in view of Proposition B, to holding that of the two or three million stars in distant clusters (about half a million of these stars have been actually photographed) there is not one giant star brighter than absolute photographic magnitude +2. And we have just seen that direct measures show that all of our nearby clusters contain such giants; indeed some appear to be composed mainly of giants.

As a further test of the distances of globular clusters, a special device has been used with the Hooker reflector. With a thin prism placed in the converging beam shortly before the focus, we may photograph for a star (or for each of a group of stars) a small spectrum that extends not only through the blue ordinarily photographed, but also throughout the yellow and red. By using specially prepared photographic plates, sensitive in the blue and red but relatively insensitive in the green-yellow, the small spectra are divided in the middle, and the relative intensity of the blue and red parts depends, as is well known, on spectral type and absolute magnitude; giants and dwarfs, of the same type in the Harvard system of spectral classification, show markedly different spectra. The spectral types of forty or fifty of the brighter stars in the Hercules cluster are known, classified as usual on the basis of spectral lines. Using the device described above, a number of these stars have been photographed side by side on the same plate with well known giants and dwarfs of the solar neighborhood for which distances and absolute magnitudes depend on direct measures of parallax. On the basis of the smaller distance for Messier 13, the spectra of these cluster stars (being then of absolute magnitude fainter than +4) should resemble the spectra of the dwarfs. But the plates clearly show that in absolute brightness the cluster stars equal, and in many cases even exceed, the giants—a result to be expected if the distance is of the order of 36,000 light-years.

The above procedure is a variation on the method used by Adams and his associates on brighter stars where sufficient dispersion can be obtained to permit photometric intercomparison of sensitive spectral lines. So far as it has been applied to clusters, the usual spectroscopic method supports the above conclusion that the bright red and yellow stars in clusters are giants.

An argument much insisted upon by Curtis is that the average absolute magnitude of stars around the sun is equal to or fainter than solar brightness, hence, that average stars we see in clusters are also dwarfs. Or, put in a different way, he argues that since the mean spectral class of a globular cluster is of solar type and the average solar-type star near the sun is of solar luminosity, the stars photographed in globular clusters must be of solar luminosity, hence not distant. This deduction, he holds, is in compliance with proposition C—uniformity throughout the universe. But in drawing the conclusions, Curtis apparently ignores, first, the very common existence of red and yellow giant stars in stellar systems, and second, the circumstance mentioned above in Section 5 that in treating a distant external system we naturally first observe its giant stars. If the material is not mutually extensive in the solar domain and in the remote cluster (and it certainly is not for stars of all types), then the comparison of averages means practically nothing because of the obvious and vital selection of brighter stars in the cluster. The comparison should be of nearby cluster with distant cluster, or of the luminosities of the same kinds of stars in the two places.

Suppose that an observer, confined to a small area in a valley, attempts to measure the distances of surrounding mountain peaks. Because of the short base line allowed him, his trigonometric parallaxes are valueless except for the nearby hills. On the remote peaks, however, his telescope shows green foliage. First he assumes approximate botanical uniformity throughout all visible territory. Then he finds that the average height of all plants immediately around him (conifers, palms, asters, clovers, etc.) is one foot. Correlating this average with the measured angular height of plants visible against the sky-line on the distant peaks he obtains values of the distances. If, however, he had compared the foliage on the nearby, trigonometrically-measured hills with that on the remote peaks, or had used some method of distinguishing various floral types he would not have mistaken pines for asters and obtained erroneous results for the distances of the surrounding mountains. All the principles involved in the botanical parallax of a mountain peak have their analogues in the photometric parallax of a globular cluster.

Cepheid Variables Giant stars of another class, the Cepheid variables, have been used extensively in the exploration of globular clusters. After determining the period of a Cepheid, its absolute magnitude is easily found from an observationally derived period-luminosity curve, and the distance of any cluster containing such variables is determined as soon as the apparent magnitudes are measured. Galactic Cepheids and cluster Cepheids are strictly comparable by Proposition C—a deduction that is amply supported by observations at Mount Wilson and Harvard, of color, spectrum, light curves, and the brightness relative to other types of stars.

Curtis bases his strongest objections to the larger galaxy on the use I have made of the Cepheid variables, questioning the sufficiency of the data and the accuracy of the methods involved. But I believe that in the present issue there is little point in laboring over the details for Cepheids, for we are, if we choose, qualitatively quite independent of them in determining the scale of the galactic system, and it is only qualitative results that are now at issue. We could discard the Cepheids altogether, use instead either the red giant stars and spectroscopic methods, or the hundreds of B-type stars upon which the most capable stellar astronomers have worked for years, and derive much the same distance for the Hercules cluster, and for other clusters, and obtain consequently similar dimensions for the galactic system. In fact, the substantiating results from these other sources strongly fortify our belief in the assumptions and methods involved in the use of the Cepheid variables.

Since the distances of clusters as given by Cepheid variables are qualitatively in excellent agreement with the distances as given by blue stars and by yellow and red giants, discussed in the foregoing sub-sections, I shall here refer only briefly to four points bearing on the Cepheid problem, first noting that if the distances of clusters are to be divided by 10 or 15, the same divisor should be also used for the distances derived for galactic Cepheids.

(1) The average absolute magnitude of typical Cepheids, according to my discussion of proper motions and magnitude correlations, is about -2.5. The material on proper motion has also been discussed independently by Russell, Hertzsprung, Kapteyn, Strömberg, and Seares; they all accept the validity of the method, and agree in making the mean absolute magnitude much the same as that which I derive. Seares finds, moreover, from a discussion of probable errors and of possible systematic errors, that the observed motions are irreconcilable with an absolute brightness five magnitudes fainter, because in that case the mean parallactic motion of the brighter Cepheids would be of the order of $0''.160$ instead of $0''.016 \pm 0''.002$ as observed.

Both trigonometric and spectroscopic parallaxes of galactic Cepheids, as far as they have been determined, support the photometric values in demanding high luminosity; the spectroscopic and photometric methods are not wholly independent, however, since the zero point depends in both cases on parallactic motion.

(2) When parallactic motion is used to infer provisional absolute magnitudes for individual stars (a possible process only when peculiar motions are small and observations very good), the brighter galactic Cepheids indicate the correlation between luminosity and period.[10] The necessity, however, of neglecting individual peculiar motion and errors of observation for this procedure makes the correlation appear much less clearly for galactic Cepheids than for those of external systems (where proper motions are not concerned), and little importance could be attached to the period-luminosity curve if it were based on local Cepheids alone. When the additional data mentioned below are also treated in this manner, the correlation is practically obscured for galactic Cepheids, because of the larger observational errors.

On account of the probably universal uniformity of Cepheid phenomena, however, we need to know only the *mean* parallactic motion of the galactic Cepheids to determine the zero point of the curve which is based on external Cepheids; and the *individual* motions do not enter the problem at all, except, as noted above, to indicate provisionally the existence of the period-luminosity relation. It is only this *mean* parallactic motion that other investigators have used to show the exceedingly high luminosity of Cepheids. My adopted absolute magnitudes and distances for all these stars have been based upon the final period-luminosity curve, and not upon individual motions.

(3) Through the kindness of Professor Boss and Mr. Roy of the Dudley Observatory, proper motions have been submitted for 21 Cepheids in addition to the 13 in the *Preliminary General Catalogue*. The new material is of relatively low weight, but the unpublished discussion by Strömberg of that portion referring to the northern stars introduces no material alteration of the earlier result for the mean absolute brightness of Cepheids.

It should be noted that the 18 pseudo-Cepheids discussed by Adams and Joy[11] are without exception extremely bright (absolute magnitudes ranging from -1 to -4); they are thoroughly comparable with the ordinary Cepheids in galactic distribution, spectral characteristics, and motion.

(4) From unpublished results kindly communicated by van Maanen and by Adams, we have the following verification of the great distance and high luminosity of the important, high-velocity, cluster-type Cepheid RR Lyrae:

Photometric parallax	0″.003	(Shapley)
Trigonometric parallax	$+0.006 \pm 0.006$	(van Maanen)
Spectroscopic parallax	0.004	(Adams, Joy, and Burwell)

The large proper motion of this star, 0″.25 annually, led Hertzsprung some years ago to suspect that the star is not distant, and that it and its numerous congeners in clusters are dwarfs. The large proper motion, however, indicates high real velocity rather than nearness, as the above results show. More recently Hertzsprung has reconsidered the problem and, using the cluster variables, has derived a distance of the globular cluster Messier 3 in essential agreement with my value.

Hypothetical Annual Proper Motion The absence of observed proper motion for distant clusters must be an indication of their great distance because of the known high velocities in the line of sight. The average radial velocity of the globular clusters appears to be about 150 km/sec. By assuming, as usual, a random distribution of velocities, the transverse motion of Messier 13 and similar bright globular clusters should be greater than the quite appreciable value of 0″.04 a year if the distance is less than 3,600 light-years. No proper motion has been found for distant clusters; Lundmark has looked into this matter particularly for five systems and concludes that the annual proper motion is less than 0″.01.

7. Let us summarize a few of the results of accepting the restricted scale of the galactic system.

If the distances of globular clusters must be decreased to one-tenth, the light-emitting power of their stars can be only a hundredth that of local cluster stars of the same spectral and photometric types. As a consequence, I believe Russell's illuminative theory of spectral evolution would have to be largely abandoned, and Eddington's brilliant theory of gaseous giant stars would need to be greatly modified or given up entirely. Now both of these modern theories have their justification, first, in the fundamental nature of their concepts and postulates, and second, in their great success in fitting observational facts.

Similarly, the period-luminosity law of Cepheid variation would be meaningless; Kapteyn's researches on the structure of the local cluster would need new interpretation, because his luminosity laws could be applied locally but not generally; and a very serious loss to astronomy would be that of the generality of spectroscopic methods of determining star distances, for it would mean that identical spectral characteristics indicate stars differing in brightness by 100 to 1, depending only upon whether the star is in the solar neighborhood or in a distant cluster.

THE DIMENSIONS AND ARRANGEMENT OF THE GALACTIC SYSTEM

When we accept the view that the distance of the Hercules cluster is such that its stellar phenomena are harmonious with local stellar phenomena—its brightest stars typical giants, its Cepheids comparable with our own—then it follows that fainter, smaller, globular clusters are still more distant than 36,000 light-years. One-third of those now known are more distant than 100,000 light-years; the most distant is more than 200,000 light-years away, and the diameter of the whole system of globular clusters is about 300,000 light-years.

Since the affiliation of the globular clusters with the galaxy is shown by their concentration to the plane of the Milky Way and their symmetrical arrangement with respect to it, it also follows that the galactic system of stars is as large as this subordinate part. During the past year we have found Cepheid variables and other stars of high luminosity among the fifteenth magnitude stars of the galactic clouds; this can only mean that some parts of the clouds are more distant than the Hercules cluster. There seems to be good reason, therefore, to believe that the star-populated regions of the galactic system extend at least as far as the globular clusters.

One consequence of accepting the theory that clusters outline the form and extent of the galactic system, is that the sun is found to be very distant from the middle of the galaxy. It appears that we are not far from the center of a large local cluster or cloud, but that cloud is at least 50,000 light-years from the galactic center. Twenty years ago Newcomb remarked that the sun *appears* to be in the galactic plane because the Milky Way is a great circle—an encircling band of light—and that the sun also *appears* near the center of the universe because the star density falls off with distance in all directions. But he concludes as follows: "Ptolemy showed by evidence, which, from his standpoint, looked as sound as that which we have cited, that the earth was fixed in the center of the universe. May we not be the victim of some fallacy, as he was?"

Our present answer to Newcomb's question is that we have been victimized by restricted methods of measuring distance and by the chance position of the sun near the center of a subordinate system; we have been misled, by the consequent phenomena, into thinking that we are in the midst of things. In much the same way ancient man was misled by the rotation of the earth, with the consequent apparent daily motion of all heavenly bodies around the earth, into believing that even his little planet was the center of the universe, and that his earthly gods created and judged the whole.

If man had reached his present intellectual position in a later geological era, he might not have been led to these vain conceits concerning his position in the physical universe, for the solar system is rapidly receding from the galactic plane, and is moving away from the center of the local cluster. If that motion remains unaltered in direction and amount, in a hundred million years or so the Milky Way will be quite different from an encircling band of star clouds, the local cluster will be a distant object, and the star density will no longer decrease with distance from the sun in all directions.

Another consequence of the conclusion that the galactic system is of the order of 300,000 light-years in greatest diameter, is the previously mentioned difficulty it gives to the "comparable-galaxy" theory of spiral nebulae. I shall not undertake a description and discussion of this debatable problem. Since the theory probably stands or falls with the hypothesis of a small galactic system, there is little point in discussing other material on the subject, especially in view of the recently measured rotations of spiral nebulae which appear fatal to such an interpretation.

It seems to me that the evidence, other than the admittedly critical tests depending on the size of the galaxy, is opposed to the view that the spirals are galaxies of stars comparable with our own. In fact, there appears as yet no reason for modifying the tentative hypothesis that the spirals are not composed of typical stars at all, but are truly nebulous objects. Three very recent results are, I believe, distinctly serious for the theory that spiral nebulae are comparable galaxies—(1) Seares' deduction that none of the known spiral nebulae has a surface brightness as small as that of our galaxy; (2) Reynold's study of the distribution of light and color in typical spirals, from which he concludes they cannot be stellar systems; and (3) van Maanen's recent measures of rotation in the spiral M 33, corroborating his earlier work on Messier 101 and 81, and indicating that these bright spirals cannot reasonably be the excessively distant objects required by the theory.

But even if spirals fail as galactic systems, there may be elsewhere in space stellar systems equal to or greater than ours—as yet unrecognized and possibly quite beyond the power of existing optical devices and present measuring scales. The modern telescope, however, with such accessories as high-power spectroscopes and photographic intensifiers, is destined to extend the inquiries relative to the size of the universe much deeper into space, and contribute further to the problem of other galaxies.

1. The word *universe* is used in this paper in the restricted sense, as applying to the total of sidereal systems now known to exist.

2. See Part II of this article, by Heber D. Curtis.

3. Quoted from a manuscript copy of his Washington address.

4. The *Contribution* is published jointly with Dr. van Rhijn.

5. In the final draft of the following paper Curtis has qualified his acceptance of the foregoing propositions in such a manner that in some numerical details the comparisons given below are no longer accurately applicable to his arguments; I believe, however, that the comparisons do correctly contrast the present view with that generally accepted a few years ago.

6. Adams and van Maanen published several years ago the radial velocities and spectral types of a number of B stars in the double cluster in Perseus, *Ast. Jour.* (Albany, N.Y.) *27*, 187–188 (1913).

7. Stars of types B8 and B9 are customarily treated with the A type in statistical discussion; even if they are included with the B's, 64 per cent of Kapteyn's absolute magnitudes are brighter than zero and only 4 per cent are fainter than +2. No stars of types B8 or B9 fainter than +3 are in Kapteyn's lists.

8. H. Shapley, *Proceedings Nat. Acad. Sci. 5*, 434–440 (1900); a further treatment of this problem is to appear in a forthcoming *Mount Wilson Contribution.* [Ultimately published in *Astronomische Nachrichten 213*, 231 (1921) and *Harvard College Observatory Circular* no. 239 (1922).

9. The absolute visual magnitude of ε Virginis (spectrum G6) is 0.0 according to the Mount Wilson spectroscopic parallax kindly communicated by Mr. Adams.

10. Mr. Seares has called my attention to an error in plotting the provisional smoothed absolute magnitudes

against log period for the Cepheids discussed in *Mount Wilson Contribution* No. 151. The preliminary curve for the galactic Cepheids is steeper than that for the Small Magellanic Cloud, Omega Centauri, and other clusters.

11. W. S. Adams and A. H. Joy, *Publ. Ast. Soc. Pac.*, *31*, 184–186 (1919).

Curtis, Part II

DIMENSIONS AND STRUCTURE OF THE GALAXY

DEFINITION OF UNITS EMPLOYED The distance traversed by light in one year, 9.5×10^{12} km., or nearly six trillion miles, known as the light-year, has been in use for about two centuries as a means of visualizing stellar distances, and forms a convenient and easily comprehended unit. Throughout this paper the distances of the stars will be expressed in light-years.

The *absolute* magnitude of a star is frequently needed in order that we may compare the luminosities of different stars in terms of some common unit. It is that *apparent* magnitude which the star would have if viewed from the standard distance of 32.6 light-years (corresponding to a parallax of 0″.1).

Knowing the parallax, or the distance, of a star, the absolute magnitude may be computed from one of the simple equations:

$$\text{Abs. Magn.} = \text{App. Magn.} + 5 + 5 \times \log(\text{parallax in seconds of arc})$$

$$\text{Abs. Magn.} = \text{App. Magn.} + 7.6 - 5 \times \log(\text{distance in light-years}).$$

LIMITATIONS IN STUDIES OF GALACTIC DIMENSIONS By direct methods the distances of individual stars can be determined with considerable accuracy out to a distance of about two hundred light-years.

At a distance of three hundred light-years (28×10^{14} km.) the radius of the earth's orbit (1.5×10^{8} km.) subtends an angle slightly greater than 0″.01, and the probable error of the best modern photographic parallax determinations has not yet been reduced materially below this value. The spectroscopic method of determining stellar distances through the absolute magnitude probably has, at present, the same limitations as the trigonometric method upon which the spectroscopic method depends for its absolute scale.

A number of indirect methods have been employed which extend our reach into space somewhat farther for the average distances of large groups or classes of stars, but give no information as to the individual distances of the stars of the group or class. Among such methods may be noted as most important the various correlations which have been made between the proper motions of the stars and the parallactic motion due to the speed of our sun in space, or between the proper motions and the radial velocities of the stars.

The limitations of such methods of correlation depend, at present, upon the fact that accurate proper motions are known, in general, for the brighter stars only. A motion of 20 km/sec. across our line of sight will produce the following annual proper motions:

Distance	100 l.y.	500 l.y.	1,000 l.y.
Annual p. m.	0″.14	0″.03	0″.01

The average probable error of the proper motions of Boss is about 0″.006. Such correlation methods are not, moreover, a simple matter of comparison of values, but are rendered difficult and to some extent uncertain by the puzzling complexities brought in by the variation of the space motions of the stars with spectral type, stellar mass (?), stellar luminosity (?), and still imperfectly known factors of community of star drift.

It will then be evident that the base-line available in studies of the more distant regions of our galaxy is woefully short, and that in such studies we must depend largely upon investigations of the distribution and of the frequency of occurrence of stars of the different apparent magnitudes and spectral types, on the assumption that the more distant stars, when taken in large numbers, will average about the same as known nearer stars. This assumption is a reasonable one, though not necessarily correct, as we have little certain knowledge of galactic regions as distant as five hundred light-years.

Were all the stars of approximately the same absolute magnitude, or if this were true even for the stars of any particular type or class, the problem of determining the general order of the dimensions of our galaxy would be comparatively easy.

But the problem is complicated by the fact that, taking the stars of all spectral types together, the dispersion in absolute luminosity is very great. Even with the exclusion of a small number of stars which are exceptionally bright or faint, this dispersion probably reaches ten absolute magnitudes, which would correspond to a hundred-fold uncertainty in distance for a given star. However, it will be seen later that we possess moderately definite information as to the *average* absolute magnitude of the stars of the different spectral types.

DIMENSIONS OF OUR GALAXY Studies of the distribution of the stars and of the ratio between the numbers of stars of successive apparent magnitudes have led a number of investigators to the postulation of fairly accordant dimensions for the galaxy; a few may be quoted:

Wolf; about 14,000 light-years in diameter.
Eddington; about 15,000 light-years.
Shapley (1915); about 20,000 light-years.
Newcomb; not less than 7,000 light-years; later—perhaps 30,000 light-years in diameter and 5,000 light-years in thickness.
Kapteyn; about 60,000 light-years.[1]

general structure of the galaxy From the lines of investigation mentioned above there has been a similar general accord in the deduced results as to the shape and structure of the galaxy:

1. The stars are not infinite in number, nor uniform in distribution.
2. Our galaxy, delimited for us by the projected contours of the Milky Way, contains possibly a billion suns.
3. This galaxy is shaped much like a lens, or a thin watch, the thickness being probably less than one-sixth of the diameter.
4. Our Sun is located fairly close to the center of figure of the galaxy.
5. The stars are not distributed uniformly through the galaxy. A large proportion are probably actually within the ring structure suggested by the appearance of the Milky Way, or are arranged in large and irregular regions of greater star density. The writer believes that the Milky Way is at least as much a structural as a depth effect.

   A spiral structure has been suggested for our galaxy; the evidence for such a spiral structure is not very strong, except as it may be supported by the analogy of the spirals as island universes, but such a structure is neither impossible nor improbable. The position of our Sun near the center of figure of the galaxy is not a favorable one for the precise determination of the actual galactic structure.

relative paucity of galactic genera Mere size does not necessarily involve complexity; it is a remarkable fact that in a galaxy of a thousand million objects we observe, not ten thousand different types, but perhaps not more than five main classes, outside the minor phenomena of our own solar system.

1. *The stars.* The first and most important class is formed by the stars. In accordance with the type of spectrum exhibited, we may divide the stars into some eight or ten main types; even when we include the consecutive internal gradations within these spectral classes it is doubtful whether present methods will permit us to distinguish as many as a hundred separate subdivisions in all. Average space velocities vary from 10 to 30 km/sec., there being a well-marked increase in average space velocity as one proceeds from the blue to the redder stars.
2. *The globular star clusters* are greatly condensed aggregations of from ten thousand to one hundred thousand stars. Perhaps one hundred are known. Though quite irregular in grouping, they are generally regarded as definitely galactic in distribution. Space velocities are of the order of 300 km/sec.
3. *The diffuse nebulae* are enormous, tenuous, cloud-like masses; fairly numerous; always galactic in distribution. They frequently show a gaseous spectrum, though many agree approximately in spectrum with their involved stars. Space velocities are very low.
4. *The planetary nebulae* are small, round or oval, and almost always with a central star. Fewer than one hundred and fifty are known. They are galactic in distribution; spectrum is gaseous; space velocities are about 80 km/sec.
5. *The spirals.* Perhaps a million are within reach of large reflectors; the spectrum is generally like that of a star cluster. They are emphatically non-galactic in distribution, grouped about the galactic poles, spiral in form. Space velocities are of the order of 1,200 km/sec.

distribution of celestial genera With one, and only one, exception, all known genera of celestial objects show such a distribution with respect to the plane of our Milky Way, that there can be no reasonable doubt that all classes, save this one, are integral members of our galaxy. We see that all the stars, whether typical, binary, variable, or temporary, even the rarer types, show this unmistakable concentration toward the galactic plane. So also for the diffuse and the planetary nebulae and, though somewhat less definitely, for the globular star clusters.

The one exception is formed by the spirals; grouped about the poles of our galaxy, they appear to abhor the regions of greatest star density. They seem clearly a class apart. *Never* found in our Milky Way, there is no other class of celestial objects with their distinctive characteristics of form, distribution, and velocity in space.

The evidence at present available points strongly to the conclusion that the spirals are individual galaxies, or island universes comparable with our own galaxy in dimensions and in number of component units. While the island universe theory of the spirals is not a vital postulate in a theory of galactic dimensions, nevertheless, because of its indirect bearing on the question, the arguments in favor of the island universe hypothesis will be included with those which touch more directly on the probable dimensions of our own galaxy.

other theories of galactic dimensions From evidence to be referred to later Dr. Shapley has deduced very great distances for the globular star clusters, and holds that our galaxy has a diameter comparable with the distances which he has derived for the clusters, namely,—a galactic diameter of about 300,000 light-years, or at least ten times greater than formerly accepted. The postulates of the two theories may be outlined as follows:

Present Theory	Shapley's theory
Our galaxy is probably not more than 30,000 light-years in diameter, and perhaps 5,000 light-years in thickness.	The galaxy is approximately 300,000 light-years in diameter, and 30,000, or more, light-years in thickness.
The clusters, and all other types of celestial objects except the spirals, are component parts of our own galactic system.	The globular clusters are remote objects, but a part of our own galaxy. The most distant cluster is placed about 220,000 light-years away.
The spirals are a class apart, and not intra-galactic objects. As island universes, of the same order of size as our galaxy, they are distant from us 500,000 to 10,000,000, or more, light-years.	The spirals are probably of nebulous constitution, and possibly not members of our own galaxy, driven away in some manner from the regions of greatest star density.

Evidence Furnished by the Magnitude of the Stars

THE "AVERAGE" STAR It will be of advantage to consider the two theories of galactic dimensions from the standpoint of the average star. What is the "average" or most frequent type of star of our galaxy or of a globular star cluster, and if we can with some probability postulate such an average star, what bearing will the characteristics of such a star have upon the question of its average distance from us?

No adequate evidence is available that the more distant stars of our galaxy are in any way essentially different from stars of known distance nearer to us. It would seem then that we may safely make such correlations between the nearer and the more distant stars, *en masse*. In such comparisons the limitations of spectral type must be observed as rigidly as possible, and results based upon small numbers of stars must be avoided, if possible.

Many investigations, notably Shapley's studies of the colors of stars in the globular clusters, and Fath's integrated spectra of these objects and of the Milky Way, indicate that the average star of a star cluster or of the Milky Way will, in the great majority of cases, be somewhat like our Sun in spectral type, *i.e.*, such an average star will be, in general, between spectral types F and K.

CHARACTERISTICS OF F-K TYPE STARS OF KNOWN DISTANCE The distances of stars of type F–K in our own neighborhood have been determined in greater number, perhaps, than for the stars of any other spectral type, so that the average absolute magnitude of stars of this type seems fairly well determined.

There is every reason to believe, however, that our selection of stars of these or other types for direct distance determinations has not been a representative one. Our parallax programs have a tendency to select stars either of great luminosity or of great space velocity.

Kapteyn's values for the average absolute magnitudes of the stars of the various spectral types are as follows:

Type	*Average abs. magn.*
B5	+1.6
A5	+3.4
F5	+7
G5	+10
K5	+13
M	+15

The same investigator's most recent luminosity-frequency curve places the maximum of frequency of the stars in general, taking all the spectral types together, at absolute magnitude +7.7.

A recent tabulation of about five hundred modern photographically determined parallaxes places the average absolute magnitude of stars of type F–K at about +4.5.

The average absolute magnitude of five hundred stars of spectral types F to M is close to +4, as determined spectroscopically by Adams.

It seems certain that the two last values of the average absolute magnitude are too low, that is,—indicate too high an average luminosity, due to the omission from our parallax programs of the intrinsically fainter stars. The absolute magnitudes of the dwarf stars are, in general, fairly accurately determined; the absolute magnitudes of many of the giant stars depend upon small and uncertain parallaxes. In view of these facts we may somewhat tentatively take the average absolute magnitude of F–K stars of known distance as not brighter than +6; some investigators would prefer a value of +7 or +8.

COMPARISON OF MILKY WAY STARS WITH THE "AVERAGE" STARS We may take, without serious error, the distances of 10,000 and 100,000 light-years respectively, as representing the distance in the two theories from our point in space to the central line of the Milky Way structure. Then table 79.3 may be prepared.

It will be seen from table 79.3 that the stars of apparent magnitudes 16 to 20, observed in our Milky Way structure in such great numbers, and, from their spectrum, believed to be predominantly F–K in type, are of essentially the same absolute luminosity as known nearer stars of these types, if assumed to be at the average distance of 10,000 light-years. The greater value postulated for the galactic dimensions requires, on the other hand, an enormous proportion of giant stars.

PROPORTION OF GIANT STARS AMONG STARS OF KNOWN DISTANCE All existing evidence indicates that the proportion

Table 79.3 Absolute magnitudes of stars for different apparent magnitudes and distances

Apparent magnitudes	Corresponding absolute magnitudes for distances of	
	10,000 l.y.	100,000 l.y.
10	−2.4	−7.4
12	−0.4	−5.4
14	+1.6	−3.4
16	+3.6	−1.4
18	+5.6	+0.6
20	+7.6	+2.6

of giant stars in a given region of space is very small. As fairly representative of several investigations we may quote Schouten's results, in which he derives an average stellar density of 166,000 stars in a cube 500 light-years on a side, the distribution in absolute magnitude being as given in table 79.4.

Table 79.4 Distribution of absolute magnitudes among stars of known distances

Absolute magnitudes	No. of stars	Relative percentages
−5 to −2	17	0.01
−2 to +1	760	0.5
+1 to +5	26,800	16
+5 to +10	138,600	83

COMPARISON OF THE STARS OF THE GLOBULAR CLUSTERS WITH THE "AVERAGE" STAR From a somewhat cursory study of the negatives of ten representative globular clusters I estimate the average apparent visual magnitude of all the stars in these clusters as in the neighborhood of the eighteenth. More powerful instruments may eventually indicate a somewhat fainter mean value, but it does not seem probable that this value is as much as two magnitudes in error. We then have the results given in table 79.5.

Table 79.5 The absolute magnitude of an average globular cluster at two different distances

Apparent magnitude of average cluster star	Corresponding absolute magnitudes if at distances of	
	10,000 l.y.	100,000 l.y.
18	+5.6	+0.6

From table 79.5 we see again that the average F–K star of a cluster, if assumed to be at a distance of 10,000 light-years, has an average luminosity about the same as that found for known nearer stars of this type. The greater average distance of 100,000 light-years requires a proportion of giant stars enormously greater than is found in those regions of our galaxy of which we have fairly definite distance data.

While it is not impossible that the clusters are exceptional regions of space and that, with a tremendous spatial concentration of suns, there exists also a unique concentration of giant stars, the hypothesis that cluster stars are, on the whole, like those of known distance seems inherently the more probable.

It would appear, also, that galactic dimensions deduced from correlations between large numbers of what we may term average stars must take precedence over values found from small numbers of exceptional objects, and that, where deductions disagree, we have a right to demand that a theory of galactic dimensions based upon the exceptional object or class shall not fail to give an adequate explanation of the usual object or class.

THE EVIDENCE FOR GREATER GALACTIC DIMENSIONS The arguments for a much larger diameter for our galaxy than that hitherto held, and the objections which have been raised against the island universe theory of the spirals rest mainly upon the great distances which have been deduced for the globular star clusters.

I am unable to accept the thesis that the globular clusters are at distances of the order of 100,000 light-years, feeling that much more evidence is needed on this point before it will be justifiable to assume that the cluster stars are predominatingly giants rather than average stars. I am also influenced, perhaps unduly, by certain fundamental uncertainties in the data employed. The limitations of space available for the publication of this portion of the discussion unfortunately prevents a full treatment of the evidence. In calling attention to some of the uncertainties in the basal data, I must disclaim any spirit of captious criticism, and take this occasion to express my respect for Dr. Shapley's point of view, and my high appreciation of the extremely valuable work which he has done on the clusters. I am willing to accept correlations between large masses of stellar data, whether of magnitudes, radial velocities, or proper motions, but I feel that the dispersion in stellar characteristics is too large to permit the use of limited amounts of any sort of data, particularly when such data is of the same order as the probable errors of the methods of observation.

The deductions as to the very great distances of the globular clusters rest, in the final analysis, upon three lines of evidence:

1. Determination of the relative distances of the clusters on the assumption that they are objects of the same order of actual size.

2. Determination of the absolute distances of the clusters through correlations between Cepheid variable stars in the clusters and in our galaxy.

3. Determination of the absolute distances of the clusters through a comparison of their brightest stars with the intrinsically brightest stars of our galaxy.

Of these three methods, the second is given most weight by Shapley.

It seems reasonable to assume that the globular clusters are of the same order of actual size, and that from their apparent diameters the *relative* distances may be determined. The writer would not, however, place undue emphasis upon this relation. There would seem to be no good reason why there may not exist among these objects a reasonable amount of difference in actual size, say from three- to five-fold, differences which would not prevent them being regarded as of the same order of size, but which would introduce considerable uncertainty into the estimates of relative distance.

THE EVIDENCE FROM THE CEPHEID VARIABLE STARS This portion of Shapley's theory rests upon the following three hypotheses or lines of evidence:

A. That there is a close coordination between absolute magnitude and length of period for the Cepheid variables of our galaxy, similar to the relation discovered by Miss Leavitt among Cepheids of the Smaller Magellanic Cloud.

B. That, if of identical periods, Cepheids anywhere in the universe have identical absolute magnitudes.

C. This coordination of absolute magnitude and length of period for galactic Cepheids, the derivation of the absolute scale for their distances and the distances of the clusters, and, combined with A) and B), the deductions therefrom as to the much greater dimensions for our galaxy, depend almost entirely upon the sizes and the internal relationships of the proper motions of eleven Cepheid variables.

Under the first heading, it will be seen later that the actual evidence for such a coordination among galactic Cepheids is very weak. Provided that the Smaller Magellanic Cloud is not in some way a unique region of space, the behavior of the Cepheid variables in this Cloud is, through analogy, perhaps the strongest argument for postulating a similar phenomenon among the Cepheid variables of our galaxy.

Unfortunately there is a large dispersion in practically all the characteristics of the stars. That the Cepheids lack a reasonable amount of such dispersion is contrary to all experience for the stars in general. There are many who will regard the assumption made under B) above as a rather drastic one.

If we tabulate the proper motions of these eleven Cepheids, as given by Boss, and their probable errors as well, it will be seen that the average proper motion of these eleven stars is of the order of one second of arc per century in either coordinate; that the average probable error is nearly half this amount, and that the probable errors of half of these twenty-two coordinates may well be described as of the same size as the corresponding proper motions.

Illustrations bearing on the uncertainty of proper motions of the order of 0.″01 per year might be multiplied at great length. The fundamental and unavoidable errors in our star positions, the probable errors of meridian observations, the uncertainty in the adopted value of the constant of precession, the uncertainties introduced by the systematic corrections applied to different catalogues, all have comparatively little effect when use is made of proper motions as large as ten seconds of arc per century. Proper motions as small as one second of arc per century are, however, still highly uncertain quantities, entirely aside from the question of the possible existence of systematic errors. As an illustration of the differences in such minute proper motions as derived by various authorities, the proper motions of three of the best determined of this list of eleven Cepheids, as determined by Auwers, are in different quadrants from those derived by Boss.

There seems no good reason why the smaller coordinates of this list of twenty-two may not eventually prove to be different by once or twice their present magnitude, with occasional changes of sign. So small an amount of presumably uncertain data is insufficient to determine the scale of our galaxy, and many will prefer to wait for additional material before accepting such evidence as conclusive.

In view of (1) the known uncertainties of small proper motions and (2) the known magnitude of the purely random motions of the stars, the determination of *individual* parallaxes from *individual* proper motions can never give results of value, though the average distances secured by such methods of correlation from large numbers of stars are apparently trustworthy. The method can not be regarded as a valid one, and this applies whether the proper motions are very small or are of appreciable size.

As far as the galactic Cepheids are concerned, Shapley's curve of coordination between absolute magnitude and length of period, though found through the mean absolute magnitude of the group of eleven, rests in reality upon individual parallaxes determined from individual proper motions, as may be verified by comparing his values for the parallax of these eleven stars with[2] the values found directly from the upsilon component of the proper motion (namely,—that component which is parallel to the Sun's motion) and the solar motion. The differences in the two sets of values, 0.″0002 in the mean, arise from the rather elaborate system of weighting employed.

The final test of a functional relation is the agreement obtained when applied to similar data not originally employed in deducing the relation. We must be ready to allow some measure of deviation in such a test, but when a considerable proportion of other available data fails to agree within a reasonable amount, we shall be justified in withholding our decision.

If the curve of correlation deduced by Shapley for galactic Cepheids is correct in both its absolute and relative scale, and if it is possible to determine individual distances from individual proper motions, the curve of correlation, using the same method as far as the proper motions are concerned (the validity of which I do not admit), should fit fairly well with other available proper motion and parallax data. The directly determined parallaxes are known for five of this group of eleven, and for five other Cepheids. There are, in addition, twenty-six other Cepheids for which proper motions have been determined. One of these was omitted by Shapley because of irregularity of period, one for irregularity of the light curve, two because the proper motions were deemed of insufficient accuracy, two because the proper motions are anomalously large; the proper motions of the others have been recently investigated at the Dudley Observatory, but have less weight than those of the eleven Cepheids used by Shapley.

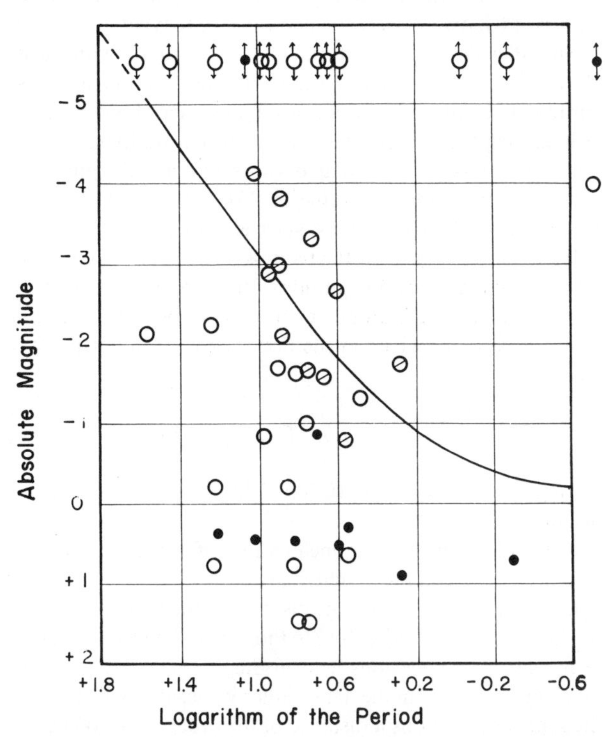

Fig. 79.1 Curtis's plot of luminosity-period data. Absolute magnitudes calculated from the upsilon component of the proper motion are indicated by circles; the eleven employed by Shapley are marked with a bar. Black dots represent directly determined parallaxes. The arrows attached to the circles at the upper edge of the diagram indicate that either the parallax or the upsilon component of the proper motion is negative and the absolute magnitude indeterminate in consequence.

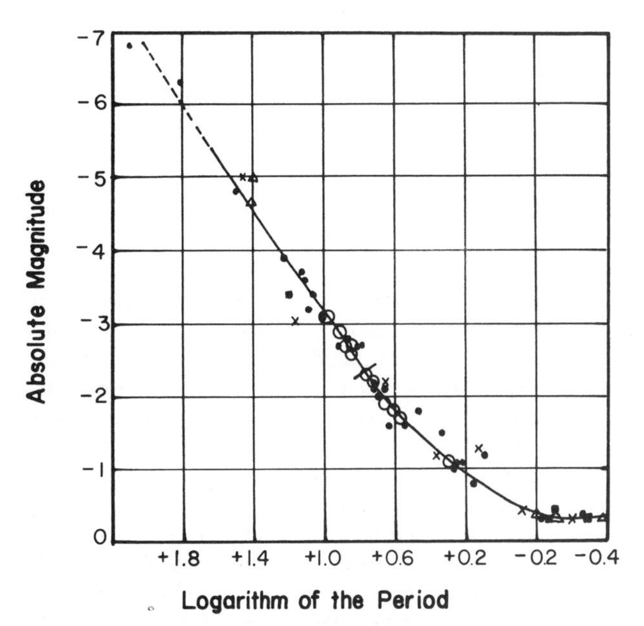

Fig. 79.2 Harlow Shapley's plot of the luminosity-period curve of Cepheid variation. Curtis did not present this plot, but we give it here for comparison. The various symbols designate variables from seven different systems. The short bisecting line at absolute magnitude −2.35, log. period 0.775, indicates the mean values for Cepheids of known proper motion. Most of the symbols for periods less than a day represent averages of about ten variables. Of the six largest deviations, four refer to values of particularly low weight. (From the *Astrophysical Journal 48*, 104 [1918].)

From the above it would seem that available observational data lend little support to the fact of a period-luminosity relation among galactic Cepheids. In view of the large discrepancies shown by other members of the group when plotted on this curve, it would seem wiser to wait for additional evidence as to proper motion, radial velocity, and, if possible, parallax, before entire confidence can be placed in the hypothesis that the Cepheids and cluster-type variables are invariably super-giants in absolute luminosity.

Argument from the intrinsically brightest stars If the luminosity-frequency law is the same for the stars of the globular clusters as for our galaxy, it should be possible to correlate the intrinsically brightest stars of both regions and thus determine cluster distances. It would seem, *a priori*, that the brighter stars of the clusters must be giants, or at least approach that type, if the stars of the clusters are like the general run of stars. Through the application of a spectroscopic method Shapley has found that the spectra of the brighter stars in clusters resemble the spectra of galactic giant stars, a method which should be exceedingly useful after sufficient tests have been made to make sure that in this phenomenon, as is unfortunately the case in practically all stellar characteristics, there is not a large dispersion, and also whether slight differences in spectral type may at all materially affect the deductions.

THE AVERAGE "GIANT" STAR Determining the distance of Messier 3 from the variable stars which it contains, Shapley then derives absolute magnitude −1.5 as the mean luminosity of the twenty-five brightest stars in this cluster. From this mean value, −1.5, he then determines the distances of other clusters. Instead, however, of determining cluster distances of the order of 100,000 light-years by means of correlations on a limited number of Cepheid variables, a small and possibly exceptional class, and from the distances thus derived deducing that the absolute magnitudes of many of the brighter stars in the clusters are as great as −3, while a large proportion are greater than −1, it would seem preferable to begin the line of reasoning with the attributes of known stars in our neighborhood, and to proceed from them to the clusters.

What is the average absolute magnitude of a galactic giant star? On this point there is room for honest difference of opinion, and there will doubtless be many who will regard the conclusions of this paper as ultra conservative. Confining ourselves to existing observational data, there is no evidence that a group of galactic giants, of average spectral type about G5, will have a mean absolute magnitude as great as −1.5; it is more probably in the neighborhood of +1.5, or three absolute magnitudes fainter, making Shapley's distances four times too large.

Russell's suggestion is worth quoting in this connection, written in 1913, when parallax data were far more limited and less reliable than at present:

> The giant stars of all the spectral classes appear to be of about the same mean brightness,—averaging a little above absolute magnitude zero, that is, about a hundred times as bright as the Sun. Since the stars of this series . . . have been selected by apparent brightness, which gives a strong preference to those of greatest luminosity, the average brightness of all the giant stars in a given region of space must be less than this, perhaps considerably so.

Some reference has already been made to the doubtful value of parallaxes of the order of 0.″010, and it is upon such small or negative parallaxes that most of the very great absolute luminosities in present lists depend. It seems clear that parallax work should aim at using as faint comparison stars as possible, and that the corrections applied to reduce relative parallaxes to absolute parallaxes should be increased very considerably over what was thought acceptable ten years ago.

From a study of the plotted absolute magnitudes by spectral type of about five hundred modern direct parallaxes, with due regard to the uncertainties of minute parallaxes, and keeping in mind that most of the giants will be of types F to M, there seems little reason for placing the average absolute magnitude of such giant stars as brighter than +2.

The average absolute magnitude for the giants in Adam's list of five hundred spectroscopic parallaxes is +1.1. The two methods differ most in the stars of type G, where the spectroscopic method shows a maximum at +0.6, which is not very evident in the trigonometric parallaxes.

In such moving star clusters as the Hyades group, we have thus far evidently observed only the giant stars of such groups.

The mean absolute magnitude of forty-four stars believed to belong to the Hyades moving cluster is +2.3. The mean absolute magnitude of the thirteen stars of types F, G, and K, is +2.4. The mean absolute magnitude of the six brightest stars is +0.8 (two A5, one G, and three of K type).

The Pleiades can not fittingly be compared with such clusters or the globular clusters; its composition appears entirely different as the brightest stars average about B5, and only among the faintest stars of the cluster are there any as late as F in type. The parallax of this group is still highly uncertain. With Schouten's value of 0.″037 the mean absolute magnitude of the six brightest stars is +1.6.

With due allowance for the redness of the giants in clusters, Shapley's mean visual magnitude of the twenty-five brightest stars in twenty-eight globular clusters is about 14.5. Then, from the equation given in the first section of this paper we have

$$+2 = 14.5 + 7.6 - 5 \times \log \text{distance},$$

or, log distance = 4.02 = 10,500 light-years as the average distance.

If we adopt instead the mean value of Adams +1.1, the distance becomes 17,800 light-years.

Either value for the average distance of the clusters may be regarded as satisfactorily close to those postulated for a galaxy of the smaller dimensions held in this paper, in view of the many uncertainties in the data. Either value, also, will give on the same assumptions a distance of the order of 30,000 light-years for a few of the faintest and apparently most distant clusters. I consider it very doubtful whether any cluster is really so distant as this, but find no difficulty in provisionally accepting it as a possibility, without thereby necessarily extending the main structure of the galaxy to such dimensions. While the clusters seem concentrated toward our galactic plane, their distribution in longitude is a most irregular one, nearly all lying in the quadrant between 270° and 0°. If the spirals are galaxies of stars, their analogy would explain the existence of frequent nodules of condensation (globular clusters?) lying well outside of and distinct from the main structure of a galaxy.

It must be admitted that the B-type stars furnish something of a dilemma in any attempt to utilize them in determining cluster distances.

From the minuteness of their proper motions, most investigators have deduced very great luminosities for such stars in our galaxy. Examining Kapteyn's values for stars of this type, it will be seen that he finds a range in absolute magnitude from +3.25 to −5.47. Dividing the 433 stars of his lists into two magnitude groups, we have:

Mean abs. magn. 249 B stars, brighter than abs. magn. 0	= −1.32
Mean abs. magn. 184 B stars, fainter than abs. magn. 0	= +0.99
Mean abs. magn. all	= −0.36

Either the value for the brighter stars, −1.32, or the mean of all, −0.36, is over a magnitude brighter than the average absolute magnitude of the giants of the other spectral types among nearer galactic stars.

Now this galactic relation is apparently *reversed* in such clusters as M. 3 or M. 13, where the B-type stars are about three magnitudes fainter than the brighter K and M stars and about a magnitude fainter than those of G type. Supposing that the present very high values for the galactic B-type stars are correct, if we assume similar luminosity for those in the clusters we must assign absolute magnitudes of −3 to −6 to the F and M stars of the clusters, for which we have no certain galactic parallel, with a distance of perhaps 100,000 light-years. On the other hand, if the F and M stars of the cluster are like the brighter stars of these types in the galaxy, the average absolute magnitude of the B-type stars will be only about +3, and too low to agree with present values for galactic B stars. I prefer to accept the latter alternative in this dilemma, and to believe that there may exist B-type stars of only two to five times the brightness of the Sun.

While I hold to a theory of galactic dimensions approximately one-tenth of that supported by Shapley, it does not follow that I maintain this ratio for any particular cluster distance. All that I have tried to do is show that 10,000 light-years is a reasonable *average* cluster distance.

There are so many assumptions and uncertainties involved that I am most hesitant in attempting to assign a given distance to a given cluster, a hesitancy which is not diminished by a consideration of the following estimates of the distance of M. 13 (The Great Cluster in Hercules).

Shapley, 1915, provisional	100,000 light-years
Charlier, 1916	170 light-years
Shapley, 1917	36,000 light-years
Schouten, 1918	4,300 light-years
Lundmark, 1920	21,700 light-years

It should be stated here that Shapley's earlier estimate was merely a provisional assumption for computational illustration, but all are based on modern material, and illustrate the fact that good evidence may frequently be interpreted in different ways.

My own estimate, based on the general considerations outlined earlier in this paper, would be about 8,000 light-years, and it would appear to me, at present, that this estimate is perhaps within fifty per cent of the truth.

THE SPIRALS AS EXTERNAL GALAXIES

Here we omit Curtis's arguments for the extra-galactic nature of spiral nebulae, because they are essentially the same as those given by him in selection 103.

1. A complete bibliography of the subject would fill many pages. Accordingly, references to authorities will in general be omitted. An excellent and nearly complete list of references may be found in Lundmark's paper "The Relations of the Globular Clusters and Spiral Nebulae to the Stellar System," in *K. Svenska Vet. Handlingar 60*, no. 8, 71 (1920).

2. *Mt. Wilson Contr.* No. 151, Table V.

80. First Attempt at a Theory of the Arrangement and Motion of the Sidereal System

Jacobus Cornelius Kapteyn

(*Astrophysical Journal 55*, 302–327 [1922])

William Herschel inaugurated the quantitative science of galactic structure at the end of the eighteenth century, when he began systematic star counts to determine the shape and size of the Milky Way. Although he could conclude that the sun is in the center of a flattened disk of stars whose diameter is 5 times its thickness, he had no way to fix an absolute distance scale to his picture.

Early in the twentieth century the slow accumulation of data on trigonometric parallaxes and proper motions made possible a renewed quantitative attack on the stellar distributions. With this research in mind, Jacobus Kapteyn drew up a Plan of Selected Areas, in which he called for a collaborative effort in extending the counts of stars to the faintest possible limits in 206 areas of the sky. The plan also called for the determination of the photographic magnitudes and proper motions of as many stars in the areas as possible. The most extensive star counts were made by Kapteyn and his colleague Pieter J. van Rhijn, and astronomers at Harvard and Mount Wilson provided a large number of the photographic magnitudes of the stars in the selected areas.[1]

By 1920 Kapteyn and van Rhijn were able to publish the first luminosity function for the stars near the sun (that is, a table of the number of stars of a given luminosity per volume of space).[2] They found that the luminosity function could be represented by a simple Gaussian curve centered at an absolute magnitude of 7.7, with a half-width of a few magnitudes. When this luminosity function was combined with the number counts of stars to very faint magnitudes, the density distribution of stars was obtained. In many respects this "Kapteyn universe" of stars is very similar to Herschel's system, for the stars are concentrated in a flattened disk with the sun at the center and with the greatest extent in the plane of the Milky Way. The Kapteyn universe is finite, with edges at about 1,500 pc in the direction perpendicular to the galactic plane and at about 8 times that distance in the direction of the plane.

Two years after Kapteyn and van Rhijn introduced the luminosity function and the Kapteyn universe, Kapteyn published the paper given here, in which he summarizes and diagrams the salient features of the stellar distribution and derives the first accurate values for the force exerted by the galaxy in the direction perpendicular to the galactic plane. The value of this force, which is essentially the same as modern determinations, can be used to determine the mass density of the galaxy in the vicinity of the sun. Furthermore, as Kapteyn points out, the inferred density of about 10^{-23} gm cm^{-3}, or 0.15 solar masses pc^{-3}, places an upper limit to the mass density of invisible interstellar matter. Larger amounts of invisible matter would produce observable gravitational effects on the motions of stars.

When published in 1922, Kapteyn's paper stood in strong contradiction to the much larger stellar system advocated by Harlow Shapley, although Shapley tried to reconcile the two by assuming that the Kapteyn universe was simply a local star cloud in the grander Milky Way structure that he envisioned. We now know that two fundamental errors undermined Kapteyn's analysis. First, he incorrectly assumed that the luminosity function obtained for the stars near the sun applies throughout the galaxy; his more serious error (and one also present in Shapley's analysis) was the neglect of the interstellar absorption of starlight. It was Robert Trumpler's studies of galactic clusters nearly a decade later that decisively demonstrated the critical role of interstellar absorption (see selection 87). The absorption meant that Kapteyn was describing a slightly foggy universe, and the diminution of stellar densities as well as the finite boundaries were immediate consequences of this fogginess.

1. E. C. Pickering and J. C. Kapteyn, *Harvard-Groningen Durchmusterung of Selected Areas, Annals of the Harvard Observatory 101, 102, 103* (1918–1924); F. H. Seares, J. C. Kapteyn, and P. J. van Rhijn, "Mount Wilson Catalogue of Photographic Magnitudes in Selected Areas 1–139," *Carnegie Institute of Washington Publication No. 402* (1930).
2. J. C. Kapteyn and P. J. van Rhijn, *Astrophysical Journal 52*, 23 (1920).

Abstract—First attempt at a general theory of the distribution of masses, forces, and velocities in the stellar system.—(1) *Distribution of stars.* Observations are fairly well represented, at least up to galactic lat. 70°, if we *assume that the equidensity surfaces are similar ellipsoids of revolution*, with axial ratio 5.1, and this enables us to compute quite readily (2) *the gravitational acceleration at various points due to such a system*, by summing up the effects of each of ten ellipsoidal shells, in terms of the acceleration due to the average star at a distance of a parsec. The *total number of stars* is taken as 47.4×10^9. (3) *Random and rotational velocities.* The nature of the equidensity surfaces is such that the stellar system cannot be in a steady state unless there is a general rotational motion around the galactic polar axis, in addition to a random motion analogous to the thermal agitation of a gas. In the neighborhood of the axis, however, there is no rotation, and the behavior is assumed to be like that of a gas at uniform temperature, but with a gravitational acceleration ($G\eta$) decreasing with the distance ρ. Therefore the density Δ is assumed to obey the barometric law: $G\eta = -\bar{u}^2(\delta\Delta/\delta\rho)/\Delta$; and taking the mean random velocity $\bar{u}$ as 10.3 km/sec., the author finds that (4) *the mean mass of the stars* decreases from 2.2 (sun = 1) for shell II to 1.4 for shell X (the outer shell), the average being close to 1.6, which is the value independently found for the average mass of both components of visual binaries. In the galactic plane the resultant acceleration—gravitational minus centrifugal—is again put equal to $-\bar{u}^2(\delta\Delta/\delta\rho)/\Delta$, $\bar{u}$ is taken to be constant and the average mass is assumed to decrease from shell to shell as in the direction of the pole. The angular velocities then come out such as to make the linear rotational velocities about constant and equal to 19.5 km/sec. beyond the third shell. If now we suppose that part of the stars are rotating one way and part the other, the relative velocity being 39 km/sec., we have a quantitative explanation of the phenomenon of star-streaming, where the relative velocity is also in the plane of the Milky Way and about 40 km/sec. It is incidentally suggested that when the theory is perfected it may be possible to determine *the amount of dark matter* from its gravitational effect. (5) The *chief defects of the theory* are: That the equidensity surfaces assumed do not agree with the actual surfaces, which tend to become spherical for the shorter distances; that the *position of the center of the system* is not the sun, as assumed, but is probably located at a point some 650 parsecs away in the direction galactic long. 77°, lat. −3°; that the average mass of the stars was assumed to be the same in all shells in deriving the formula for the variation of $G\eta$ with ρ on the basis of which the variation of average mass from shell to shell and the constancy of the rotational velocity were derived—hence either the assumption or the conclusions are wrong; and that no distinction has been made between stars of different types.

1. Equidensity Surfaces Supposed to be Similar Ellipsoids

In *Mount Wilson Contribution* No. 188[1] a provisional derivation was given of the star-density in the stellar system. The question was there raised whether the inflection appearing near the pole in the equidensity surfaces for small densities is real or not. I have since found that these inflections can be avoided without doing very serious violence to the results of observation. If this is done, the equidensity surfaces become approximately ellipsoids, and not only that, but the data can be represented without exceeding the possible limits of observation error, by assuming the equidensity surfaces to be *concentric, similar revolution ellipsoids, similarly situated.*

2. Elements of the Ellipsoids

Taking as unit of star-density that in the neighborhood of the sun, the adopted axes of the ellipsoids, which will be referred to as ellipsoids I, II, . . . , X and which correspond to the values (Δ being the density) $\log \Delta + 10 = 9.8, 9.6, \ldots, 8.0$, are as shown in table 80.1. The A-axis is directed toward the galactic Pole, the B-axis lies in the plane of the Milky Way.

Table 80.1 Equidensity ellipsoids

Ellipsoid	Log Δ	*A* (pc)	*B* (pc)	*B/A*
I	9.80 − 10	118	602	5.102
II	9.60	198	1,010	5.102
III	9.40	296	1,510	5.102
IV	9.20	413	2,106	5.102
V	9.00	553	2,820	5.102
VI	8.80	717	3,656	5.102
VII	8.60	902	4,600	5.102
VIII	8.40	1,114	5,675	5.102
IX	8.20	1,365	6,960	5.102
X	8.00	1,660	8,465	5.102

For the Milky Way and for the direction toward the Pole this table yields densities which are fairly well represented, for $\rho > 150$ parsecs, by the formulae,

$$\log \Delta = -2.135 + 2.368 \log \rho - 0.593 (\log \rho)^2 \quad \text{(M.W.)}, \quad (1)$$

$$\log \Delta = -5.356 + 4.890 \log \rho - 1.200 (\log \rho)^2 \quad \text{(Pole)}. \quad (1a)$$

A section of the equidensity-ellipsoids through the sun (which has been assumed to be the center of the system) at right angles to the plane of the Milky Way is shown in figure 80.1.

The agreement of the densities furnished by table 80.1 with those of *Contribution* No. 188 is fairly good for all galactic

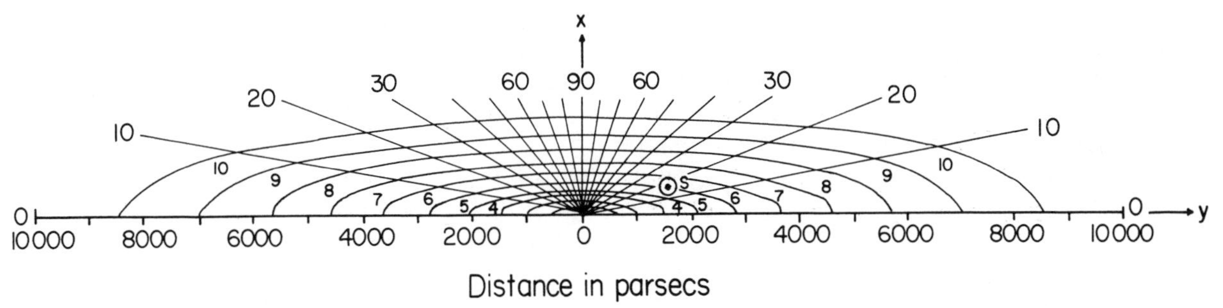

Fig. 80.1 Density distribution of the stars in a plane perpendicular to the plane of our galaxy. The curves are lines of equal density, with the star density near the sun taken to be unity. The sun, *S*, has been assumed to be near the center of the system, and the curves marked 1, 2, 3, . . . , 10 denote, respectively, relative star densities of 0.63, 0.40, 0.25, 0.19, 0.10, 0.063, 0.040, 0.025, 0.016, and 0.010.

latitudes up to 65° or 70°. For still higher latitudes it may perhaps still be called tolerable. At least the deviations hardly exceed what would be produced by an error of 0.1 mag. in the photometric scale for these regions.

In the present paper I have substituted these ellipsoids for the surfaces derived directly from observation in *Contribution* No. 188, *not* because I think they are nearer the truth, but simply because they are so enormously more convenient for further computation.

My aim in the present paper is simply to get hold of some approximate information about the real structure and motion of the system, and quantitative accuracy has been considered of secondary importance as long as we may hope that the main features are not affected. I trust that this hope will not be disappointed, notwithstanding the many defects—defects that will be duly pointed out—which still attach to the present treatment.

3. ADVANTAGE OF THE ADOPTION OF THE ELLIPSOIDS

The form of the equidensity surfaces thus adopted has the advantage that it calls attention to the possibility of determining with some precision the gravitational attraction of the *whole* of the stellar system on any point inside ellipsoid X, while at the same time it renders the computation of that attraction a relatively easy matter.

In another paper[2] van Rhijn and I have tried to show that, as soon as we possess good counts of stars for each interval of magnitude down to apparent magnitude 17 (visual), we shall know with some tolerable approximation the density of the whole region covered by figure 80.1, that is, of the whole extent of the stellar system for which the density exceeds one-hundredth of that in the neighborhood of the sun.

In the near future such counts will be available. They will be furnished by the Mount Wilson "Catalogue of the Selected Areas" (from $\delta = -15°$ to $\delta = +90°$), the discussion of which is in the hands of Seares. A few provisional counts make it probable that this work will in the main confirm the elements used for table 80.1 and figure 80.1. I will assume, therefore, that even now the densities are sufficiently well known for the whole of ellipsoid X.

The advantage just alluded to is a consequence of the well-known property that the attraction of an ellipsoidal shell of constant density, bounded by two similar and similarly situated ellipsoids, on an internal point is *zero*. For it is evident by this property that, if in all that part of the system which lies outside ellipsoid X—for which part accurate data are still wanting—the arrangement in similar ellipsoids also holds, the attraction of this outside domain on a point inside ellipsoid X would be zero. And as the distribution of density inside ellipsoid X is known, the possibility of computing the attraction of the total system on a point inside of X becomes evident. If on the contrary the same arrangement does not hold outside ellipsoid X, it still seems highly probable a priori that any change in the form of the equidensity surfaces must be gradual, that is, the equidensity surfaces in the neighborhood of X will diverge little from similar ellipsoids, and the greater changes will begin to appear only at more considerable distances. For the consecutive shells, therefore, the attraction on an internal point will begin by being very small, both on account of the near approach to similarity of these shells and their small density and greater distance from the attracted point. For more distant shells the first circumstance will probably diminish in importance with increasing distance, while, on the contrary, the second becomes more and more important. On the whole, therefore, the attraction of all of that part of the system which lies outside X will be small, and its neglect will presumably not prevent us from obtaining fairly exact ideas about the total forces.

4. COMPUTATION OF THE GRAVITATIONAL FORCES

In ellipsoid I, which for brevity I will call shell 1, and in each of the shells 2, 3, . . . , 10, between the surfaces of ellipsoids I, II, ..., X, the density varies between limiting values which are in the ratio of 1 to 1.585. In what follows I will assume for each shell a constant average density.

We include here only the ten points at a galactic latitude of 90° lying in surfaces of ellipsoids I, II, ... X.

As a unit of attraction I have used the attraction on each other of two stars of average mass separated by a distance of 1 parsec.

I first computed the attraction of the full ellipsoids I, II, ..., X on the points specified above, on the supposition that they are of a constant density such that every cubic parsec contains a single star. The formulae for this computation are given in the Appendix.

The attraction of the full ellipsoids having been found, simple subtraction gives the attraction of the separate shells 1, 2, ..., 10, all supposed to have the density corresponding to one star per cubic parsec. The actual attraction of the shells was obtained by multiplying these results by the number of stars per cubic parsec contained in each shell. For the average densities, expressed in terms of the density in the neighborhood of the sun, I adopted the values corresponding to the logarithms 9.9, 9.7, 9.5, ..., 8.1, each minus 10, multiplied by 0.0451, which according to *Contribution* No. 188 (12) is the number of stars per cubic parsec near the sun; this gives the numbers in table 80.2.

Table 80.2 Average number of stars per cubic parsec

Shell	No. stars	Shell	No. Stars
1	0.0358	6	0.00358
2	.0226	7	.00226
3	.0143	8	.00143
4	.00900	9	.000900
5	.00568	10	.000568

Having found the separate attractions, the components of the total attractions parallel to the axes can at once be determined by noting that the attraction of any shell on an internal point is *zero*, and further, by neglecting the attraction of that part of the system outside of ellipsoid X on a point inside this ellipsoid. Instead of the components I have entered in table 80.3 the total forces G and the angles that these forces make with the X-axis.

We include in table 80.3 only the forces in the direction perpendicular to the galactic plane. In this table the attraction of the whole stellar system on a body in the point I is shown to be equivalent to the attraction of 40.06 stars of average mass at a distance of 1 pc from the same body.

Table 80.3 Total attractions of the whole system

Attr. point	Log Δ	Co-ordinates attr. point (pc)	G
I, 90°	9.80 − 10	118	40.06
II, 90	9.60	198	54.26
III, 90	9.40	296	62.45
IV, 90	9.20	413	65.95
V, 90	9.00	553	66.12
VI, 90	8.80	717	64.00
VII, 90	8.60	902	60.63
VIII, 90	8.40	1,114	56.41
IX, 90	8.20	1,365	51.56
X, 90	8.00	1,660	46.52

5. Analytical Representation of G for Galactic Latitudes 0° and 90°

In trying to represent the force G by an analytical formula, I started from the consideration that, as the density is constant near the center, the attraction must be nearly proportional to the distance ρ for very small values of ρ; further, that for distances very great as compared with the dimensions of the stellar system, the attraction must be practically the same as it would be were the mass of the whole system concentrated in the center. For these latter distances, therefore, G must be proportional to $1/\rho^2$.

The following easily managed formula satisfies both conditions:

$$G = \frac{A\rho}{1 + B\rho + C\rho^2 + D\rho^3}. \tag{3}$$

In this formula A/D evidently equals the total number of stars, N, in the stellar system.

Here Kapteyn uses data from *Groningen Publication no.* 27 to show that $N = 4.74 \times 10^{10}$ stars; using the values of G given in table 80.3, he obtains $A = 0.376$, $B = 1.83 \times 10^{-3}$, $C = 3.40 \times 10^{-6}$, and $D = 7.93 \times 10^{-12}$ for a galactic latitude of 90°.

6. Application of Kinetic Theory of Gases

The results thus far obtained rest, it is true, on provisional data, which even now might be materially improved; they further depend on the supposition, not yet fully demonstrated, that, within the distances here considered, there is no appreciable extinction of light in space, but they are, nevertheless, I think, the legitimate outcome of our data.

For what follows I will now introduce some considerations borrowed from the kinetic theory of gases, the applicability of which to the stellar system might be considered doubtful. At all events I do not pretend to have demonstrated this applicability. The results which will be derived cannot lay claim to be demonstrably correct, but they seem to me to be so remarkable that, after a good deal of hesitation, I have resolved to publish them, in the hope that others, better versed in these matters, may furnish us with a more rigorous solution of the problem involved.

Even though it has been shown, in the main by unpublished investigations, that the peculiar motions[3] of the stars with some crude approximation are Maxwellian, the stellar system cannot be treated as a gas at rest; first, because of the existence of stream-motion; second, because of the form of the equidensity surfaces, which is certainly different from that of the equipotential surfaces of the gravitational force.

That they are different is proved by the fact, among others, that in general the forces are not normal to these surfaces. This is evident enough without further explanation. Moreover, it is clearly brought out by table 80.3, where the angle with the normal reaches values of more than 27°. Further it is well known that in a gas at rest under its own attraction, the equidensity surfaces are spherical.

In table 80.3 we omitted values of the density Δ and the total forces G at latitudes of 0, 30 and 57.1 degrees. Although the values of Δ decrease systematically from the inner to the outer ellipsoids at each latitude, the values of G at first increase and then decrease. This behavior is illustrated in the data given in table 80.3 for a latitude of 90 degrees.

The system cannot therefore be in a steady state unless it has a systematic motion. Since the discovery of the star-streams it is clear that such a motion really exists and that it is parallel to the plane of the Milky Way.

It seems rational, therefore, to assume that the system has a sort of rotational motion round the X-axis (see figure 80.1) which is directed toward the pole of the galaxy. The form of the equidensity surfaces found directly in *Contribution* No. 188 as well as that now adopted, strongly indicates some such motion.

This being assumed, the stars along the axis will still have no other motion than their peculiar motions, which, as was just mentioned, are Maxwellian, at least with some approximation. I venture to assume, therefore, that the stars in the immediate neighborhood of this axis are arranged as the molecules of a gas in a quiescent atmosphere.

If:

Δ be the star-density (number of stars per cubic parsec);
u one of the components of the peculiar velocity;
η the acceleration produced by the attraction of a star of average mass at a distance of one parsec, then on the above assumption

$$\overline{u^2}\,\frac{\delta\Delta}{\Delta} = -G\eta\delta\rho, \tag{8}$$

$\overline{u^2}$ being the average value of u^2.

The formula is analogous to that used for barometric determinations of altitude in an atmosphere of constant temperature throughout. On the other hand, we have found empirically formulae such as (1) and (1*a*) (see also *Contribution* No. 188, p. 13 [21]); in other words,

$$\log\Delta = -P + Q\log\rho - R(\log\rho)^2 \quad (\rho > \rho_0), \tag{9}$$

from which, by differentiation

$$\frac{\delta\Delta}{\Delta} = \frac{Q - 2R\log\rho}{\rho}\,\delta\rho. \tag{10}$$

Comparing the two expressions, (8) and (10), for $\delta\Delta/\Delta$

$$G\eta = -\overline{u^2}\left[\frac{Q - 2R\log\rho}{\rho}\right] \quad (\rho > \rho_0). \tag{11}$$

As the motions are supposed to be Maxwellian, the well-known formula used in the theory of least squares gives

$$\overline{u^2} = \frac{\pi}{2}(\overline{u})^2. \tag{12}$$

From observations of the radial velocities at the Lick Observatory, where no choice has been made on the basis of motion (*Lick Observatory Bulletin* 6, 126), I derived the value[4]

$$\overline{u} = 10.3 \text{ km/sec.} \tag{13}$$

or since

$$1 \text{ kilometer} = 3.25 \times 10^{-14} \text{ parsecs}$$
$$1 \text{ parsec} = 3.08 \times 10^{13} \text{ kilometers.} \tag{14}$$

I find, in the units parsec and second, here adopted,

$$\overline{u} = 3.35 \times 10^{-13} \tag{15}$$

$$\overline{u^2} = 1.763 \times 10^{-25}, \tag{16}$$

so that (11) becomes

$$G\eta = -1.763 \times 10^{-25} \frac{Q - 2R \log \rho}{\rho} \quad (\rho > \rho_0). \quad (17)$$

Finally, for galactic latitude 90°, we obtain from equation (1*a*) the values:

$$Q = +4.890 \quad R = +1.200 \quad (18)$$

$$\eta = \frac{(-8.620 + 4.229 \log \rho) \times 10^{-25}}{G\rho} \quad (\rho > 150). \quad (19)$$

For small values of ρ, formula (9) does not hold. According to *Contribution* No. 188, and particularly according to *Contribution* No. 229, it represents the observations excellently for values of ρ well beyond the maximum (which in the present case lies near $\rho = 110$ parsecs). For values of ρ below the maximum the density is nearly constant. The differential-quotient $\delta\Delta/\delta\rho$ thus becomes very small and $\delta\Delta/\Delta\delta\rho$ very unreliable. In the present case it will probably be well not to rely on the formula below, say, 150 parsecs. This limit was adopted in (19).

I have computed the values of η from (19) both on the supposition that G has the values found directly in table 80.3 and that it has the values yielded by formula (3). The former were adopted (table 80.4).

Table 80.4 Values of η and $\bar{m}$

Point	ρ (pc)	η form. (3)	η adopted	$\bar{m}$ (sun = 1)
II, 90°	198	11.1×10^{-30}	10.2×10^{-30}	2.2
IV, 90	413	$8.9 \quad 10^{-30}$	$9.0 \quad 10^{-30}$	2.0
VI, 90	717	$7.3 \quad 10^{-30}$	$7.5 \quad 10^{-30}$	1.7
VIII, 90	1,114	$6.6 \quad 10^{-30}$	$6.8 \quad 10^{-30}$	1.5
X, 90	1,660	$6.5 \quad 10^{-30}$	$6.5 \quad 10^{-30}$	1.4

The quantity η is, as stated above, the acceleration per second, in parsecs, produced by the attraction of a star of average mass on a body at a distance of one parsec. The acceleration which the sun would produce, expressed in the same units, is

$$\text{Acceleration by sun} = 4.53 \times 10^{-30}. \quad (20)$$

This enables us to find the average mass $\bar{m}$ of a star expressed in the mass of the sun as a unit. The values of $\bar{m}$ thus found have been inserted in the last column of table 80.4.

These values agree surprisingly well with what has been found by totally different considerations. In a recent paper[5] Jackson and Furner find for visual binary stars, as the best average

$$\frac{1}{\sqrt{m_1 + m_2}} = 0.855.$$

Consequently $m_1 + m_2 = 1.60$, which agrees with table 80.4 if we suppose that the combined mass of the two components and not that of a single component is comparable with the mass of a single star, and especially if we further consider that there are theoretical grounds for expecting that the average mass will decrease for increasing distance.[6]

Remark: Dark Matter

It is important to note that what has here been determined is the total mass within a definite volume, divided by the number of luminous stars. I will call this mass the average effective mass of the stars. It has been possible to include the luminous stars completely owing to the assumption that at present we know the luminosity-curve over so large a part of its course that further extrapolation seems allowable.

Now suppose that in a volume of space containing l luminous stars there be dark matter with an aggregate mass equal to Kl average luminous stars; then, evidently the effective mass equals $(l + K) \times$ average mass of a luminous star.

We therefore have the means of estimating the mass of dark matter in the universe. As matters stand at present it appears at once that this mass cannot be excessive. If it were otherwise, the average mass as derived from binary stars would have been very much lower than what has been found for the effective mass.

7. Angular Velocities (ω) in the Plane of the Galaxy

Ignoring for an instant the fact that the stars in the Milky Way cannot be systematically at rest and treating the stars near this plane in the same way as those near the axis, I am led by a formula analogous to (17) to values of η which are not quite half those given in table 80.4. I suppose that the difference must be wholly due to the centrifugal force induced by the rotational motions. In fact, I assume that the average mass is the same throughout the whole system, at least for points on the same equidensity surface.

8. Angular Velocity for Stars not in the Galaxy

Omitted here are Kapteyn's computations of the forces caused by rotation about an axis centered at the sun and perpendicular to the galactic plane; also omitted are tables giving estimates for the rotational velocity at different distances from the sun in different directions.

9. Explanation of Star-Streaming

The most striking feature brought out by these numbers is undoubtedly the fact that at distances from the axis exceeding 2,000 parsecs the linear velocity of the stars is nearly constant, the average being 19.5 km/sec.; that is, the great bulk of the stars must have a motion of 19.5 km in a direction parallel to the plane of the Milky Way. Observation has already proved that there really exists a systematic motion of the stars, that it is exactly parallel to the plane of the Milky Way, and that the motion takes place in two exactly opposite directions, the two streams having a relative velocity of about

$$40 \text{ km/sec.} \qquad (26)$$

Since in the preceding theory the motion is introduced simply to explain certain centrifugal forces, it is at once evident that it supposes nothing about the direction in which the motion takes place. Nothing prevents us from assuming that part of the stars circulate one way, while the rest move in the opposite direction. The relative motion of the two groups will then evidently be

$$2 \times 19.5 = 39 \text{ km/sec.} \qquad (27)$$

The motion to which our theory leads, besides being in the same plane, has therefore practically the exact value which is known from observation to exist. In fact we are led in the most direct and natural way to a complete explanation of the phenomenon of star-streaming. The circumstance that observation led us to assume two rectilinear streams, whereas we here find the motion to be circular, is probably unimportant. It is of course infinitely probable that the sun must be at a certain distance from the center of the system. If we suppose it to be at the point S (see figure 80.1) then the star-streams are derived from the observed motions of stars within a volume whose dimensions are of the order of those of the sphere around S shown in the figure. As long as the radius of this sphere is small in comparison with the distance of S from the center, the curvature of the stream-lines must be inappreciable.

When we consider that the value (27) has been obtained by a study of the arrangement of stars in space, in which the proper motions play no other part than that of a criterion of distance, while the value (26) has been obtained by a study of the motions themselves, both radial and transverse, the close agreement of the two results seems very significant. It becomes more so through the fact that both theories yield a motion exactly parallel to the plane of the Milky Way. Further, if we take into account the fact that the present theory leads to a value for the average mass of the stars which is in close accordance with what has also been found from utterly different investigations, and if we add a final point, namely, the natural explanation of the different arrangement of the stars of different spectral types, we are led irresistibly to the following conclusion:

The theory here propounded, though it may require considerable modification on account of its defectiveness both as to observational basis and mathematical treatment, is probably correct in its main features.

10, 11, 12, 13, 14 and Appendix

Here we omit discussions of the velocity of escape of stars from the system, defects in the determinations of the rotational velocities, the exact position of the sun in the stellar system, and the fact that the solutions of this paper incorrectly assume that the stars form a heterogeneous collection. Also deleted is an appendix deriving the attraction of a homogeneous revolution ellipsoid on an exterior point.

1. *Astrophysical Journal 52*, 23 (1920).
2. *Mt. Wilson Contr.*, No. 229; *Astrophysical Journal 55*, 242 (1922).
3. Peculiar velocity is defined as the motion corrected for both the solar and stream-motion. The radial and transverse velocities agree in showing a certain excess of very large motions over the Maxwellian distribution. They are both represented satisfactorily by the sum of *two* Maxwellian distributions. A thorough separate treatment of all the spectral classes is still a great desideratum.
4. There is a mistake in the derivation of this value. The true value is certainly somewhat lower. From considerations given below I have not deemed it necessary for the present paper to repeat the computations with an improved value of $\bar{u}$.
5. *Monthly Notices 81*, 4 (1920).
6. Jeans, *Problems of Cosmogony and Stellar Dynamics* (1919), p. 239.

81. Star-Streaming and the Structure of the Stellar System

Bertil Lindblad

(*Arkiv för matematik, astronomi och fysik 19A*, no. 21, 1–8 [1925])

For two decades following its discovery, star streaming remained a mysterious phenomenon. Its explanation, first proposed by Bertil Lindblad, unexpectedly involved the model of the Milky Way proposed by Harlow Shapley, in which the center lay at a considerable distance in the direction of Sagittarius.

Lindblad noticed that those rare stars with extremely high radial velocities (over 100 km sec^{-1}, compared with the more typical value of 20 km sec^{-1}) tend to have their motion in the same direction and almost exactly perpendicular to the axis of star streaming. He reasoned that if the sun and the great majority of its neighbors have a common orbital motion around a distant center, their motions would be small in relation to one another but large in relation to any stars fixed in space or revolving very slowly. In fact, these latter stars should largely appear to move in the same direction and perpendicular to the direction of the center in Sagittarius (which is also the axis of the star streaming phenomenon).

In the paper given here, Lindblad developed his hypothesis that there is a general rotation of the larger galactic system and that the asymmetric velocity distribution of the high velocity stars is related to this rotation. He assumed that the larger stellar system can be divided into subsystems that rotate at different speeds in the galactic plane about a common axis located at the center of the globular cluster system. In this picture, the asymmetric velocity distribution and the high speeds of the globular clusters are actually caused by the high rotational speed of our local system. He inferred rotational velocities of about 350 km sec^{-1} for the clouds of the Milky Way surrounding the sun, under the assumption that they are located at about 12,000 pc from the galactic center. Lindblad also derived a total galactic mass of 1.80×10^{11} solar masses, and a galactic rotational period of 200 million years for the sun.

Using a sophisticated mathematical analysis, Lindblad showed in a later paper[1] that small variations in the orbital parameters of the ordinary stars, as they revolved around the distant center, would produce in the solar neighborhood a dispersion of velocities closely resembling the ellipsoidal distribution of velocities deduced for star streaming.

Thus, the work of Lindblad, and immediately thereafter the analysis by Jan Oort, provided a powerful incentive for the acceptance of Shapley's description of the Milky Way.

1. *Arkiv för matematik, astronomi och fysik 20A*, no. 17 (1927).

IN A RECENT PAPER on the cause of star-streaming[1] I have tried to find a clue to a possible connection between the star-streaming discovered by Kapteyn, "the two star-streams," and the asymmetrical drift of high stellar velocities especially studied by Strömberg.[2] The theory outlined there rested on certain assumptions concerning the structure of the big stellar system, which were, however, rather vaguely formulated. It is the object of the present paper to show that an analysis related to that applied by Jeans[3] to Kapteyn's "local" stellar universe may help to give a stronger foundation to the ideas concerning the state of the stellar system as a whole.

We assume that the stellar system may be divided up into a series of sub-systems having rotational symmetry around one and the same axis, with different speeds of rotation at the same distance from this axis and consequently having different degrees of flattening. The inner most flattened systems have a high star-density, though decreasing with decreasing speed of rotation; the systems with the highest speed of rotation are assumed to form the Milky Way clouds. In the extreme outer systems the space-density of the individuals, stars or globular clusters, is relatively very low. The latter sub-systems show, on account of their low speed of rotation, a strong asymmetrical drift in velocity nearly at right angles to the radius vector of the big system, when the velocities of their members are measured from a star, like our sun, moving as a member of a Milky Way cloud.

The center of the stellar system should lie in the galactic plane in a direction at right angles to the direction of the asymmetrical drift. This gives nearly gal. long. 330°, gal. lat. 0°. The direction towards the center of the system of globular clusters according to Shapley[4] is gal. long. 325°, gal. lat. 0°. It seems from Strömberg's results, as if the true plane of symmetry of the stellar system may be somewhat inclined to the standard galactic plane.

Besides by a reference to the system of globular clusters, for which the strong asymmetrical drift of the velocities was first pointed out by Lundmark,[5] we may make the fundamental assumptions clearer by giving attention to another well defined class of objects, the planetary nebulæ. Their velocities are on the point of showing plainly a small asymmetrical drift; this is in harmony with their high average velocity. Though showing a considerable galactic concentration, these nebulæ are by no means typical Milky Way objects. The distribution in the sky for nebulæ of different angular diameter[6] further shows that the nebulæ of small angular diameter are very much concentrated towards that half of the sky which is so extremely favoured by the globular clusters. It is probable that the nebulæ in question define a sub-system of lower speed of rotation than the Milky Way clouds. In the same way the decrease of galactic concentration from the A stars to the late type giants appears as a consequence of the increase in the average velocity with a tendency to a slightly increased asymmetrical drift. We must conclude that the late type giants belong in the mean to a sub-system of somewhat smaller rotation and smaller general degree of flattening than the A type stars.

1

Ignoring to begin with the forces acting between individuals of the systems, we may ask for the character of the steady state of the assembly of rotating systems under the gravitational force of the sum of all the systems. If the general explanation of the asymmetrical drift of high velocities just given is correct, the analysis should give a relation between general dispersion of the velocities and asymmetrical drift of the kind found by Strömberg from observed velocities, namely a relation represented by a parabola or an elongated ellipse.

We assume the successive sub-systems to be of spheroidal form and to have the same extension in the galactic plane. The star-density is assumed to decrease from flattened to less flattened systems. Within each one of these the star-density may vary from inner to outer layers, but the surfaces of constant star-density are assumed to be spheroidal surfaces with the same diameter in the galactic plane.

Let ϖ and θ be the radius vector and position angle in the galactic plane, and z is the third coordinate parallel to the axis of the galaxy. In Jeans' system of notation we have v the star-density of a certain sub-system, Π, Z, Θ the velocities in the directions of ϖ, z, θ increasing, $p = v\overline{\Pi}^2 = v\overline{Z}^2$, $q = v\overline{\Theta}^2$, ${\Theta_0}^2 = (q-p)/v$.

p/v measures the dispersion of the random velocities at a certain point of a certain system, Θ_0 is the speed of rotation at the same point expressed in linear measure.

We then have the differential equations[7]

$$\begin{aligned}\frac{\partial p}{\partial z} &= v\frac{\partial V}{\partial z}\\ \frac{\partial p}{\partial \varpi} + \frac{p-q}{\varpi} &= v\frac{\partial V}{\partial \varpi}.\end{aligned}\tag{1}$$

We assume that every system is in a steady state considered by itself, so that every system satisfies these equations. We want to know the character of the velocity-distribution in the galactic plane for different values of ϖ. Turning our attention to a certain system we have for $z = 0$

$$p = -\int_0^{z_0} v\frac{\partial V}{\partial z}\,dz,$$

where $z = z_0$ marks the limiting surface of the sub-system in question, at which p must be zero.

z_0 is thus the value of z at the intersection between a parallel to the z-axis and the surface of one of the sub-systems. Moving from (ϖ, z) to (ϖ', z') on another of the spheroidal surfaces we note that $z'/z = z_0'/z_0$.

The component of force $\partial V/\partial z$ may be considered as a sum of the components of attraction of successive homogeneous spheroids within which the point is situated, which are proportional to z, as well as of the components of attraction due to homogeneous inner spheroids in relation to which this point is placed outside. Moving from (ϖ, z) to (ϖ', z') on the same spheroidal surface the components of attraction parallel to the z-axis due to enveloping homogeneous spheroids will be changed in the ratio $z'/z = z'_0/z_0$, and the same will hold true very nearly for the components of force due to inner spheroids, if the flattening of the sub-system in question, measured by the eccentricity of its meridional section, is great. In fact, according to the laws of attraction of spheroids, the relation in question holds for motion along a spheroidal surface confocal with the inner systems, and the difference between the actual surface considered here and the surface through (ϖ, z) confocal with one of the inner systems is then very small.

We assume for a certain value of ϖ

$$\frac{\partial V}{\partial z} = -\phi(z) = -\phi\left(\frac{z}{z_0} z_0\right).$$

For the point (ϖ', z') we have then

$$\left(\frac{\partial V}{\partial z}\right)' = -\frac{z'_0}{z_0}\phi\left(\frac{z_0}{z'_0}z'\right),$$

and then

$$p' = -\int_0^{z'_0} v\left(\frac{\partial V}{\partial z}\right)' dz' = \int_0^{z'_0} v\frac{z'_0}{z_0}\phi\left(\frac{z_0}{z'_0}z'\right)dz',$$

which gives

$$p' = \left(\frac{z'_0}{z_0}\right)^2 \int_0^{z_0} v\phi(z)dz = \left(\frac{z'_0}{z_0}\right)^2 p.$$

We may write this

$$p = Cz_0{}^2 = Cc^2\left(1 - \frac{\varpi^2}{a^2}\right), \tag{2}$$

where a and c are the semi-axes of the meridional section of the sub-system in question. Then we have

$$\frac{\partial p}{\partial \varpi} = -2Cc^2\frac{\varpi}{a^2}.$$

According to (1) we have

$$\Theta_0{}^2 = -\varpi\frac{\partial V}{\partial \varpi} + \frac{\varpi}{v}\frac{\partial p}{\partial \varpi}.$$

The first term in the right hand member is the same for all systems; we may call it $F(\varpi)$.

Thus

$$\Theta_0{}^2 = F(\varpi) - \frac{2}{v}Cc^2\frac{\varpi^2}{a^2}.$$

With the expression of $F(\varpi) = -\varpi\,\partial V/\partial\varpi$ given below this equation gives *constant angular speed of rotation*, Θ_0/ϖ, for one and the same sub-system.

Equation (2) gives

$$\overline{\Pi}^2 = \frac{1}{v}Cc^2\left(1 - \frac{\varpi^2}{a^2}\right),$$

and thus

$$\Theta_0{}^2 + 2\overline{\Pi}^2\frac{\varpi^2}{a^2 - \varpi^2} = F(\varpi). \tag{3}$$

The relation between Θ_0 and $\Pi_0 = \sqrt{\overline{\Pi}^2}$ is evidently an ellipse. In order to get an eccentricity of this ellipse corresponding to the relation between Θ_0 and Π_0 found by Strömberg,[8] ϖ/a must be near to unity, that is our point in space must be situated not far from the limiting surface of the stellar system.

$\Theta_0 =$ max. occurs for $\Pi_0 = 0$, and corresponds in our theory to the speed of rotation of the Milky Way clouds. From Strömberg's results $(\Theta_0)_{max}$ may be estimated to about 350 km. per sec. It corresponds nearly to the difference in asymmetrical drift between the extra-galactic nebulæ and the vertex of the parabola (or elongated ellipse), and gives a slight speed of rotation to the system of globular clusters. By means of this estimate of $(\Theta_0)_{max}$ we may compute the total mass of the stellar system.

We have

$$(\Theta_0)^2_{max} = -\varpi\frac{\partial V}{\partial \varpi}. \tag{4}$$

For the interior of a spheroid of uniform density ϱ we have

$$\left(\frac{\partial V}{\partial \varpi}\right)_1 = -2\pi G\varrho\varpi\frac{\sqrt{1-e^2}}{e^3}(-e\sqrt{1-e^2} + \text{arc sin } e),$$

while the mass of the spheroid is

$$M_1 = \frac{4}{3}\pi\varrho a^3\sqrt{1-e^2}.$$

Assuming e to be near to unity we may disregard the first term in the parenthesis and put arc sin $e = \pi/2$. We further put $e^3 = 1$ in the denominator. Then we get

$$\left(\frac{\partial V}{\partial \varpi}\right)_1 = -\frac{3}{4}\pi\frac{\varpi}{a^3}GM_1.$$

Taking the sum of each member for all the sub-systems and using (4), taking $\varpi/a = 0.9$, we get for the total mass of the stellar system

$$M = \frac{1.65}{\pi}\frac{1}{G}a(\Theta_0)^2_{max}. \quad (5)$$

G is the constant of gravitation $= 6.7 \cdot 10^{-8}$, $\Theta_0 = 35 \cdot 10^6$ cm per sec.

We assume further $a = 12{,}000$ parsecs $= 37 \cdot 10^{21}$ cm, adopting a scale-factor about 0.6 for Shapley's system of globular clusters.

Then we get $M = 36 \cdot 10^{43}$ gr., that is

$$M = 180 \cdot 10^9 \text{ solar masses.} \quad (6)$$

An estimation of the mass of the stellar system by equation (4) does not, of course, presuppose the special theory given here. Assuming instead a spherical shape of the main attracting mass of the universe, the mass inside a spherical surface through our point of space should be $308 \cdot 10^9$ solar masses.

Though of a higher order of magnitude than usually assumed, it cannot be claimed with assurance that the value (6) of the total mass of the stellar system is too high. If the mass were all concentrated in a spheroid with semi-axes 12,000 and 2,000 parsecs, the star-density would be in the mean one solar mass in 7 cubic parsecs.

The time of revolution of the system of Milky Way clouds corresponding to the speed 350 km. per sec. at a point $\varpi =$ 10,800 parsecs is $190 \cdot 10^6$ years.

2

In the next approximation we take into account the evident fact that the inner sub-systems, the Milky-Way clouds, show a very great richness in structure, suggesting considerable local accumulations of matter of the character of "local systems." In addition to the general field of force due to a "smoothed" stellar system we must then take into consideration the effects of local gravitational fields due to these accumulations of stars or nebulous material. In the paper referred to at the beginning I have shown that the effect of such local disturbing masses of great size would be primarily star-streaming in the galactic plane at right angles to the line of the asymmetrical drift of high velocities, and thus nearly along the radius vector of the big system. The two star-streams discovered by Kapteyn fulfill very nearly this condition. The interference between the sub-systems giving rise to these radial streamings in the galactic plane acts as a disturbance of the steady state of the big stellar system as developed above, especially for the sub-systems of only slightly smaller rotational speed than the clouds. The tendency of the interference between the systems must be to diminish the difference of rotational speed between the sub-systems. The loss in momentum of the clouds will cause these to fall slowly in the direction towards the center of the system, and thus the tendency will be to destroy the isolation of the local accumulations of matter and thereby to cause an advance towards greater homogenousness of the Milky Way structure. In this way the friction between the sub-systems may be reduced.

3

In the third approximation we should have to take into account the effects of close passages between the single individuals of different systems. The results of such passages would be that the members of the slower rotating and less flattened systems tend to be thrown out of the regions covered by the inner rapidly rotating systems. We may have an illustration to such a phenomenon in the fact that the globular clusters avoid a region next to the galactic plane. It would seem, however, that in this case the individuals of the inner systems causing the avoidance are rather to be identified with the "local systems" of the Milky Way than with single stars. When the acting cause is encounters between single stars, the time necessary for producing a considerable effect of the kind in question must be enormous. An observation bearing on the problem may be the observation by Oort[9] that the apices of high velocity vectors plotted on the sky show an avoidance of the lowest galactic latitudes of the same kind as that shown by the positions of globular clusters. Even in this case, however, the effect may be mainly caused by the star-clouds of the Milky Way, scattering the members of sub-systems of considerably lower degree of flattening and smaller speed of rotation.

1. *Astrophysical Journal 62*, 191, (1925).
2. *Mount Wilson Contr.* No. 275 (1924), *Astrophysical Journal 59*, 228; *Mount Wilson Contr.* No. 293 (1925); *Astrophysical Journal 61*, 363 (1925).
3. *Monthly Notices of the Royal Astron. Soc. 82*, 122 (1922).
4. *Mount Wilson Contr.* No. 152 (1918); *Astrophysical Journal 48*, 154 (1918).
5. *Kungl. Svenska Vetenskapsakademiens handlingar, 60*, no. 8, 33 (1920).
6. *Publications of the Lick Observatory, 13*, 60, (1918).
7. *Monthly Notices of the Royal Astron. Soc. 82*, 125 (1922).
8. The proper comparison should be with Strömberg's relation between asymmetrical drift and the velocity-dispersion in the z- or θ-direction. (Cf. §2.)
9. *Proceedings of the National Academy of Sciences 10*, 256 (1924).

82. Observational Evidence for the Rotation of Our Galaxy

Observational Evidence Confirming Lindblad's Hypothesis of a Rotation of the Galactic System

Jan H. Oort

(*Bulletin of the Astronomical Institutes of the Netherlands 3*, 275–282 [1927])

Rotation Effects, Interstellar Absorption, and Certain Dynamical Constants of the Galaxy Determined from Cepheid Variables

Alfred H. Joy

(*Astrophysical Journal 89*, 356–376 [1939])

When in the mid 1920s Jan Oort heard of Bertil Lindblad's work on the high velocity stars and star streaming, he was greatly impressed by the idea of the Milky Way's rotation. Convinced that there must be some more direct demonstration for the rotation of our galaxy, Oort guessed that the inner parts of the system should have a higher angular velocity, like the inner planets of the solar system, if the stellar rotation were strongly controlled by a distant, massive nucleus. Following up on this scheme, he promptly discovered the simple and direct observational evidence for this differential rotation of the Milky Way presented in the first paper given here.

The most obvious effect of the differential rotation is a sinusoidal variation of the mean radial velocities of stars as a function of galactic longitude. In other words, stars closer to the galaxy's nucleus will generally revolve faster than the sun, and hence those inner stars in the direction of the sun's motion will be pulling away from the sun, whereas those inner stars symmetrically opposite the direction to the nucleus will be catching up. Oort confirms this prediction in table 82.1. However, he begins with a simple mathematical analysis, which enables him to deduce the speed of solar revolution as 272 km sec^{-1} about a galactic center 5,900 pc distant (a slightly revised value was added in press at the end).

During the year following the publication of this paper, Oort used proper motions as well as radial velocities to obtain improved values of the parameters ("Oort's A and B") that described the differential rotation, and in the resulting paper he actually plotted the double sine wave of radial velocities as a function of galactic longitude.[1]

By 1933 John S. Plaskett and Joseph A. Pearce completed an exhaustive study of the radial velocities of O and B stars.[2] Not only did the rotation curves of these stars exhibit the characteristic double sine wave, but the interstellar lines seen in absorption against these bright stars also showed the sine wave, although with only half the amplitude of those of the stars. This finding meant that the interstellar gas participates in differential galactic rotation and that the gas is spread more or less continuously between the stars and the earth.

Two years after Plaskett and Pearce had completed their survey, Alfred Joy published the second paper given here, which not only confirmed the theory of differential galactic rotation but also demonstrated that interstellar gas and dust absorb the light of the distant stars. By using Cepheid variables whose distances can be specified from the period-luminosity relation, Joy showed that stars located at different distances give consistent rotation curves only if their light is being absorbed at the rate of 0.85 mag kpc^{-1} of interstellar matter.

Although Oort and Joy demonstrated conclusively that our galaxy is undergoing differential rotation, subsequent work has led to other estimates for the rotation. By observing the radial velocities of galaxies, for example, Milton L. Humason and Hugo D. Wahlquist determined that the rotational velocity of the sun about the galactic center is 250 km sec^{-1}, and Nicholas U. Mayall and Thomas D. Kinman have shown that the globular clusters may have rotational velocities of about 80 km sec^{-1} in relation to the galactic center.[3] Nevertheless, these measurements are confused by the motion of our galaxy relative to the surrounding galaxies, and we do not know whether or not the system of globular clusters has any rotation.

1. J. H. Oort, *Bulletin of the Astronomical Institutes of the Netherlands 4*, 79 (1927), *4*, 269 (1928).
2. J. S. Plaskett and J. A. Pearce, *Publications of the Dominion Astrophysical Observatory 5*, 167 (1933).
3. M. L. Humason and H. D. Wahlquist, *Astronomical Journal 60*, 254 (1955); N. U. Mayall, *Astrophysical Journal 104*, 290 (1946); T. D. Kinman, *Monthly Notices of the Royal Astronomical Society 119*, 559 (1959).

Oort, Observational Evidence Confirming Lindblad's Hypothesis

1. Introduction

It is well known that the motions of the globular clusters and RR Lyrae variables differ considerably from those of the brighter stars in our neighbourhood. The former give evidence of a systematic drift of some 200 or 300 *km/sec* with respect to the bright stars, while their peculiar velocity averages about 80 *km/sec* in one component, which is nearly six times higher than the average velocity of the bright stars.

Because the globular clusters and the bright stars seem to possess rather accurately the same plane of symmetry, we are easily led to the assumption that there exists a connection between the two. But what is the nature of the connection?

It is clear that we must not arrange the hypothetical universe in such a way that it is very far from dynamical equilibrium. Following Kapteyn[1] and Jeans[2] let us for a moment suppose that the bulk of the stars are arranged in an ellipsoidal space whose dimensions are small compared to those of the system of globular clusters as outlined by Shapley.[3] From the observed motions of the stars we can then obtain an estimate of the gravitational force and of the velocity of escape. An arrangement as supposed by Kapteyn and Jeans, which ensures a state of dynamical equilibrium for the bright stars, implies, however, that the velocities of the clusters and RR Lyrae variables are very much too high. A majority of these would be escaping from the system. As we do not notice the consequent velocity of recession it seems that this arrangement fails to represent the facts.

As a possible way out of the difficulty we might suppose[4] that the brighter stars around us are members of a local cloud which is moving at fairly high speed inside a larger galactic system, of dimensions comparable to those of the globular cluster system. We must then postulate the existence of a number of similar clouds, in order to provide a gravitational potential which is sufficiently large to keep the globular clusters from dispersing into space too rapidly. The argument that we cannot observe these large masses outside the Kapteyn-system is not at all conclusive against the supposition. There are indications that enough dark matter exists to blot out all galactic starclouds beyond the limits of the Kapteyn-system.[5]

Lindblad[6] has recently put forward an extremely suggestive hypothesis, giving a beautiful explanation of the general character of the systematic motions of the stars of high velocity. He supposes that the greater galactic system as outlined above may be divided up into sub-systems, each of which is symmetrical around the axis of symmetry of the greater system and each of which is approximately in a state of dynamical equilibrium. The sub-systems rotate[7] around their common axis, but each one has a different speed of rotation. One of these sub-systems is defined by the globular clusters for instance; this one has a very low speed of rotation. The stars of low velocity observed in our neighbourhood form part of another sub-system. As the rotational velocity of the slow moving stars is about 300 *km/sec* and the average random velocity only 30 *km/sec*, these stars can be considered as moving very nearly in circular orbits around the centre.

We may now apply an analysis similar to that used by Jeans in his discussion of the motions of the stars in a "Kapteyn-universe,"[8] the only difference being that in the present analysis we do not introduce a second system rotating in the opposite direction. Adopting some probable formula for the gravitational potential we can derive the rotational velocities for each of the sub-systems from our knowledge of the distribution of the peculiar velocities (defined as the velocities remaining after correction for the effects of rotation). The higher the average peculiar velocity in a certain sub-system the slower its rotation will be, and the less flattened it will appear in a direction perpendicular to the galactic plane. If we refer our motions to the centre of the slow moving stars in our neighbourhood, the members of a sub-system with higher internal velocities will appear to lag behind, and Lindblad has shown that in this way we can arrive at a connection between average peculiar velocity and systematic motion of the same form as that computed from observation.[9]

Lindblad's hypothesis conforms beautifully with the well-established fact that the average direction of the systematic motion of the high velocity stars is perpendicular to the direction in which the globular clusters are concentrated (galactic longitude 325°, latitude 0°). At first sight it might be hard to imagine how such a mixture of sub-systems of different angular speeds could ever come into existence; but the *possibility* cannot be denied, as is apparent from a comparison with spiral nebulae.[10]

If somewhere there existed a rapidly rotating system of stars and by some cause the internal velocities in this star-system were increased, an asymmetry in the stellar motions would necessarily result in the long run. It must be admitted, however, that the part played by the globular clusters cannot be so easily understood.

The following paper is an attempt to verify in a direct way the fundamental hypothesis underlying Lindblad's theory, namely that of the rotation of the galactic system around a point near the centre of the system of globular clusters. In a subsequent paper I hope to be able to make a more detailed comparison of the theory with the observational facts concerning the stars of high velocity collected in *Groningen Publications* No. 40.

2. Theoretical Effects of the Rotation

In the present discussion I shall altogether disregard the idea of a number of separate galactic clouds and take into consideration only the forces arising from the greater galactic system as a whole. The gravitational force, K, is consequently directed to the centre of this system and is only a function of the distance, R from this centre.

Let us now consider a group of stars at a distance r from the sun and let us suppose that r/R is so small that all terms of second or higher order in r/R can be neglected, then it is easily seen that the residual velocity caused by the rotation is equal to

$$rA \sin 2(l - l_0)$$

in radial direction, and to

$$rA \cos 2(l - l_0) + rB$$

in transverse direction, if l_0 represents the galactic longitude of the centre (about 325°), l the longitude of the stars considered, R the distance of the sun from the centre,

$$V = \sqrt{RK}$$

the circular velocity near the sun,

$$A = \frac{V}{4R}\left(1 - \frac{R}{K}\frac{\partial K}{\partial R}\right)$$

and

$$B = A - \frac{V}{R}.$$

The rotation is supposed to take place in right-hand direction as observed from a point North of the galactic plane.

If, as Lindblad tentatively supposed, the principal part of the greater galactic system is formed by an ellipsoid of constant density, the force K will be proportional to R. In this case

$$A = 0 \quad \text{and} \quad B = -\frac{V}{R},$$

the system rotates as a solid body and we shall not find any indications of rotation in the radial velocities, but the proper motions in galactic longitude should be systematically negative for stars in all longitudes.

As another extreme case we might suppose that the whole mass is concentrated in the centre and that K is inversely proportional to the square of R. We get

$$A = +\frac{3}{4}\frac{V}{R} \qquad B = -\frac{1}{4}\frac{V}{R}.$$

We shall see below that observations seem to prove that the second alternative is nearly correct. We shall then have to expect a systematic effect in the radial velocities showing maxima at 10° and 190° longitude, and minima at 100° and 280°, with a semiamplitude of $3rV/(4R)$. Now the most distant objects observed for radial velocity are at distances of about 1,000 parsecs; with $R = 10{,}000$ parsecs and $V = 300$ *km/sec* this gives a semi-amplitude of over 20 *km/sec*, which might well be verifiable. With the same assumptions the maximum effect in the proper motions in galactic longitude would be equal to $-0''.005$ per annum. The maximum will occur 90° from the direction towards the centre. In the direction of the centre and in the opposite direction the average proper motion in longitude should be equal to about $+0''.002$. The proper motion effects are, of course, independent of the distance of the objects considered.

3. DISCUSSION OF THE RADIAL VELOCITIES

Several astronomers have remarked upon instances in which the stars in different parts of the sky appeared to move differently.

The hypothesis of a rotation around a distant centre has also been put forward by Strömberg[11] on the basis of an investigation of the preferential motions of the stars. He found that the maximum peculiar radial velocity did not occur in two exactly opposite points of the sky but in directions inclined to each other. In explanation of this he suggested a rotation around a centre near 256° longitude. Later on it has become evident, however, that these results were caused by the influence of the stars of high velocity.

In a paper on the distribution of stellar velocities Gyllenberg pointed out that the so-called K-term in the radial velocities of the B type stars depended upon the galactic longitude.[12] From his drawing it is apparent that the K-term has distinct maxima somewhere around 0° and 180° longitude and minima at 90° and 270°. It is evident from the foregoing that this variation can be explained as the effect of rotation around a centre in 325° longitude, for the longitudes of the maxima are very near those expected in the case of rotation.

In 1922 Freundlich and von der Pahlen[13] have extended Gyllenberg's investigation. According to their statements they do not doubt the reality of the variation of the K-term with galactic longitude, and they propose several dynamical explanations; but none of these was considered to be very satisfactory.

It is not only the velocities of the B stars which have given evidence of systematic motions. In a statistical study of the c-stars Schilt has remarked upon the deviations from zero of the mean peculiar velocities of stars in different longitudes.[14] His table of residuals is reproduced as table 82.1.

In the last column of table 82.1 I have added the coefficient of the rotation term for which we are looking; it varies in very nearly the same manner as the average residual velocity.

A somewhat analogous variation has been found by Henroteau in a recent paper on pseudo-cepheids.[15]

Table 82.1 Schilt's table of residuals

Average longitude	Average peculiar velocity (km/sec)	mean error (km/sec)	$\sin 2(l - 325°)$
30°	+ 8	±3.5	+0.77
90	− 8	±2.7	− .94
150	0	±3.6	+ .17
210	+10	±3.9	+ .77
270	− 7	±4.3	− .94
330	0	±3.5	+ .17

The *O*-type stars have also been under suspicion of giving different systematic motions in different parts of the sky,[16] and in this case too the general character of the residuals is what we must expect if the system of stars is rotating in the way described.

Table 82.2 summarizes the results of a re-discussion of the radial velocities of all objects of which it might be hoped that they would show the effects of the rotation, if it exists. The second column gives the average apparent magnitude, for the *Md* variables the average maximum magnitude; in the case of the planetary nebulae it gives the limits of the apparent magnitude of the central stars. The third column shows the number of stars used, the fourth their average parallax and its mean error. Excepting the *Md* variables for which the mean parallaxes were estimated directly from R. E. Wilson's results,[17] all the parallaxes were computed anew from all proper motion data available, in such a way as to be uninfluenced by possible rotation terms in the proper motions (see section 4). The fifth column shows the semi-amplitude of the rotational term and its mean error. In general the stars were divided into intervals of 15° or 30° galactic longitude and stars of higher galactic latitudes were excluded as mentioned in the remarks. If the longitude of the centre of a group is called l and the average peculiar radial velocity $\bar{\rho}'$ the equations of condition are

$$\bar{r}A \sin 2(l - 325°) = \bar{\rho}',$$

the centre of rotation being assumed to lie near the centre of the system of globular clusters at 325° longitude. A positive sign of $\bar{r}A$ means that the systematic term indicates a rotation in the direction expected.

In several groups where the velocities were sufficiently rich in number or in other respects favorable for an independent determination of the longitude, l_0, of the centre of rotation, two unknowns were introduced. The computed value of l_0 is then shown in the sixth column together with its mean error. The solution gives us two opposite points; only one of these has been inserted. Unless stated otherwise in the remarks the radial velocities have been corrected for the usual value of the solar motion (20 *km/sec*). Except perhaps for the *O*-stars and planetary nebulae this value of the solar motion is very nearly equal to that found for each of the special groups of stars separately.

Here Oort discusses various sources for and the accuracy of the data given in table 82.2.

Except for the uncertain result from the early *Me* stars the values of $\bar{r}A$ in table 82.2 are all positive, indicating a rotation in the same direction as that found from our velocity relative

Table 82.2 Rotation parameter A for different stellar types

Type	$\bar{m}$	n	$\bar{\pi}$ m.e.	$\bar{r}A$ m.e.	l_0 m.e.	A m.e.
B3—B5	4.9	182	0″.0058 ±″.0004	+ 5.7 ± 1.0	322° ± 5°	+0.033 ±.006
B0—B2	4.6	86	.0037 ± .0005	+ 9.3 ± 1.5	322 ± 5	+ .034 ±.008
δ Cep variables	5.4	13	.0036 ± .0007	+11 ± 5	—	+ .040 ±.020
M6e—M8e, Se	7.0	78	.003 —	+ 2 ± 5	—	+ .006 ±.015
M1e—M5e	7.7	55	.003 —	−10 ±14	—	− .030 ±.043
c-stars	4.0	44	.0028 ± .0005	+ 9.1 ± 3.0	330 ± 8	+ .025 ±.009
Oa—Oe	7.0	8	.0028 ± .0025	+25 ±18	—	+ .070 ±.080
Oe5	6.2	27	.0020 ± .0006	+19 ± 5	—	+ .038 ±.015
N	—	18	.0017 ± .0008	+17 ± 8	—	+ .029 ±.017
c-stars	6.2	49	.0015 ± .0005	+25.1 ± 2.7	321 ± 4	+ .038 ±.014
O-star Ca clouds	—	40	—	+ 5.6 ± 2.2	—	—
Planetary nebulae	< 14	60	—	+10 ± 6	—	—
Planetary nebulae	unknown	30	—	+25 ±10	—	—
Planetary nebulae	> 14	20	—	+41 ±11	333 ±10	—

to the globular clusters. Several values are so large as to leave hardly any doubt about the reality of this sin $2(l - 325°)$-term. It is possible, of course, that the term may be explained by systematic motions arising from another cause than rotation, but these systematic motions must then bear a remarkably close resemblance to a rotation.

In general the velocities appear to be quite satisfactorily represented by the rotation term, the residuals showing no tendency to be systematic. Only in the case of the *B*-stars the average group-residual is larger than what we should have expected from the small peculiar motions of these stars.

The five determinations of the longitude of the centre are in good agreement with each other. The average $323° \pm 2\overset{\circ}{.}4$ *m.e.* is quite near the direction towards the centre of the globular cluster system as estimated by Shapley.

The objects in the first division of table 82.2 have been arranged in order of decreasing mean parallax. It appears that the better determinations give evidence that the value of $\bar{r}A$ increases proportional with the mean distance, as it should do if the term is interpreted as a rotation. The determinations with a relative mean error of less than a third are put together in table 82.3.

Table 82.3 Rotation parameter $\bar{r}A$ for different types of stars

Type	$\bar{\pi}$	$\bar{r}A$ (km/sec)	m.e.
B3—B5	0″0058	+ 5.7	±1.0
B0—B2	.0037	+ 9.3	±1.5
bright *c*-stars	.0028	+ 9.1	±3.0
Oe5	.0020	+19	±5
faint *c*-stars	.0015	±25.1	±2.7

By multiplying $\bar{r}A$ by the average parallax we have made estimates of the absolute value of *A*, as shown in the last column of table 82.2. Properly speaking these values should have been computed in a more elaborate way. For if there is considerable spreading in the distances within one group the average value of *r* will be somewhat higher than the reciprocal of the average parallax. For the present I have not tried to derive corrections, as these would of necessity be very uncertain; I do not believe that they would seriously influence the present conclusions. The total average of *A* is found to be $+0.0310 \pm 0.0037$ (*m.e.*); it represents the semi-amplitude of the rotation term for objects at a distance of one parsec.

We can now use this constant for estimating the average distance of some of the objects of unknown parallax, collected in the second division of the table. The average distance of Plaskett's Calcium clouds is thus found to be roughly 180 parsecs, which shows rather conclusively that the clouds are not connected with the *O*-stars themselves. The distances of the three groups of planetary nebulae are estimated as 320, 810 and 1,300 parsecs respectively. Though the percentage errors are large, the results confirm the serviceability of the magnitude of the central star as a criterion for relative distance. The absolute magnitudes of these central stars would seem to be at least six or seven units larger on the average than those of the *O*-stars, which they resemble in some spectral characteristics.

4. PROPER MOTION DATA

If the interpretation of the systematic term in the radial velocities as a rotation is right, a similar term should occur in the proper motions. But, as is evident from the formulae given in section 2, the rotation terms in the proper motions cannot be predicted from the radial velocity results unless we make an assumption as to the character of the general gravitational force. Now it will be shown in the next section that the radial velocity results make it very probable that a great part of this force varies inversely proportional to the square of *R*. We shall suppose that the total gravitational force in this part of the galactic system can be represented as the sum of two forces, K_1 and K_2, the first of which varies inversely proportional to R^2 and the second directly proportional to *R*. We want to determine what percentage of the total force is made up of K_1, and what of K_2.

By using the proper motions of some 600 stars,[18] Oort finds that $K_2/K_1 = 0.11$.

An empirical term in the proper motions, explicable as a rotation of the system of fixed stars, has been discussed by several authors (Anding, Seeliger, de Sitter, Woltjer, Charlier, Innes). In a memoir entitled "The motion and the distribution of the stars"[19] Charlier derives an average value of $\mu_l' = -0\overset{\prime\prime}{.}0024$, agreeing at least qualitatively with the results found above from the proper motions of distant stars.

5. CONCLUDING REMARKS

It has been shown from radial velocities that for all distant galactic objects there exist systematic motions varying with the galactic longitudes of the stars considered. The relative systematic motions are always of the same nature and they increase roughly proportional with the distance of the objects. Probably the simplest explanation is that of non-uniform rotation of the galactic system around a very distant centre. This explanation is capable of representing all the observed systematic motions within their range of uncertainty (except perhaps in the case of the *B* stars). If with this supposition we compute the position of the centre from the radial velocities, we find that it lies in the galactic plane, either at 323° longitude or at the opposite point. The first direction is in remarkably close agreement with the longitude of the centre of the system of globular clusters (325°). The observations would therefore seem to confirm Lindblad's hypothesis of a rotation of the entire galactic system around the latter centre.

The proper motions corroborate the above interpretation, at least qualitatively. They were used mainly to determine the character of the non-uniformity of the rotation. This character corresponds to a gravitational force which can sufficiently well be represented by the formula $K = c_1/R^2 + c_2R$, if R is the distance of the centre. A provisional solution gave: $c_2/c_1 = 0.11/R^3$.

Such a force would for example result if 9/10th of the total force came from mass concentrated near the centre and 1/10th from an ellipsoid of constant density large enough to contain the sun within its borders. The true character of the force will of course be more complicated.

We can derive a numerical result for R as soon as the circular velocity, V, is known. An estimate of this circular velocity may be made from the radial velocities of the globular clusters. According to Strömberg these clusters possess a systematic motion nearly perpendicular to the direction of their centre and equal to 286 *km/sec* ± 67 (*m.e.*) relatively to the sun, or 272 *km/sec* relatively to the centre of the slow moving stars. This would give us an estimate of the circular velocity if we were sure that the system of globular clusters had no rotation. From the ellipsoidal arrangement in space Lindblad derives a rotation for these objects, such that the circular velocity would be increased to 426 *km/sec.*[20] As, however, the apparent positions in the sky give no indication whatever of the system of globular clusters being flattened towards the galactic plane, it seems better to assume that they possess no rotation. In fact it seems more probable from dynamical consideration that the true circular velocity is below the value found by Strömberg than above it. Assuming $V = 272$ *km/sec* we find $R = 5{,}900$ parsecs. As the longitude of the centre of rotation agreed with that of the system of globular clusters, it is probable that the distance will agree as well. The distance of the centre of the globular cluster system is very uncertain, however, on the one hand by an uncertainty in the scale of the cluster distances and on the other hand by our incomplete knowledge of the more distant clusters. Shapley gives estimates varying from 13,000 to 25,000 parsecs. The value found above is considerably smaller. Even if we made the extreme supposition that $K_2 = 0$, or that all the mass of the galactic system were concentrated near the centre, the distance R would only be increased to 6,600 parsecs.

In order to explain the rotation there must be near the centre an attracting mass of at least 8×10^{10} times the mass of the sun. There remains the difficulty why we do not observe this large mass. Near 6,000 parsecs Kapteyn and van Rhijn find an almost negligible density, whereas it *should* be very much greater than in our neighbourhood. Part of the discrepancy may have resulted from the approximative character of their solution, in which all galactic longitudes were combined. Discussing various galactic regions separately Kreiken finds indications of a centre near 314° longitude, at a distance of 2,270 parsecs,[21] which is in the right direction, but certainly at too small a distance and too little defined.[22] The most probable explanation is that the decrease of density in the galactic plane indicated for larger distances is mainly due to obscuration by dark matter. Such a hypothesis receives considerable support from the marked avoidance of the galactic plane by the globular clusters, a phenomenon for which up to the present time no other well defensible explanation has been put forward.[23]

Lindblad suggests that the starstreaming is an indirect consequence of the rotation. In so far as his computations result in a starstreaming to and from the centre, in the same way as originally suggested by Turner, the present data do not entirely confirm the hypothesis. The true vertices according to Eddington lie at 166° and 346° longitude; it does not seem possible to admit an error of over 20°, which would be required to make vertex II coincide with the direction to the centre as derived in the present paper.

It may be remarked that the rotation offers a means of determining average distances from radial velocities, the relative accuracy of the method increasing with the distance and being independent of possible absorption of light in space. We have thus been able to derive a value for the average distance of the most distant planetary nebulae.

The question naturally arises which would be the most valuable observations that could be made in order to check the present results. Probably the most promising results could be derived from the radial velocities of some very faint *c*-stars. If one could get down to the 8th or 9th magnitude the semi-amplitude of the rotation term might become as large as 50 *km/sec*; a small number of stars would then suffice to give reliable results. There is another class of stars which deserves mentioning, viz. very faint δ Cephei variables. Several variables are known whose estimated distances are of the order 5,000 parsecs and larger, thus bringing us quite near the hypothetical centre of the galactic system. If rough values were known for the velocities of a few stars of this type favourably situated for the purpose, it might well become possible to derive reliable absolute values of the distance to this centre as well as of the circular velocity in our neighbourhood.

In conclusion I want to express my gratitude to Mr. Pels who has so ably assisted me in making the rather extensive computations required for the preparation of table 82.2.

Note Added to Proof

While this paper was going through the press a provisional correction to the constant of precession was derived from proper motions in galactic latitude, and a corresponding correction was applied to the proper motions in longitude. Both direction and amount of the angular velocity of rotation derived from the radial velocities are satisfactorily confirmed by the corrected proper motions.

The ratio K_2/K_1, which in section 4 was found to be 0.11, is changed into 0.29 by the above correction. The corresponding estimate of the distance of the centre changes from 5,900 to 5,100 parsecs.

1. *Astrophysical Journal 55*, 302 (1922); *Mt Wilson Contr.* No. 230.

2. *Monthly Notices R.A.S. 82*, 122 (1922).

3. *Astrophysical Journal 48*, 154 (1918), *49*, 311, and *50*, 107 (1919); *Mt Wilson Contributions* Nos. 152, 157, and 161.

4. Oort, *Groningen Publications* No. 40 (1926), p. 63.

5. *Cf. Hemel en Dampkring*, Jan. and Feb. 1927.

6. *Arkiv. f. mat., astr. o. fysik 19A*, nos. 21, 27, 35, *19B*, no. 7 (*Upsala Meddelanden* nos. 3, 4, 6, and 13); also *Vierteljahrschrift 61*, 265.

7. Of course the rotation considered is not generally one of constant angular velocity throughout the sub-system. In the following comparisons between the speeds of rotation these speeds are taken for stars at the same distance from the axis.

8. *Monthly Notices R.A.S. 82, 122* (1922)

9. Strömberg, *Astrophysical Journal 59*, 228 (1924); *Mt Wilson Contr.* No. 275.

10. Lindblad, *Upsala Meddel.* No. 13 (1926).

11. *Astrophysical Journal 47*, 32–34 (1918); *Mt Wilson Contr.* No. 144.

12. *Lund Meddelanden*, ser. 2, No. 13, 22–26 (1915).

13. *Astr. Nachr. 218*, 369–400 (1923).

14. *Bull. Astr. Inst. Netherlands 2*, 50 (1924).

15. *Journal R.A.S. Canada 21*, 1 (1927).

16. *Groningen Publications* No. 40, pp. 52–53 (1926).

17. *Astronomical Journal 35*, 129 (1923).

18. *Astronomical Journal 36*, 136 (1926).

19. *Memoirs of the Univ. of California 7*, 32 (1926).

20. *Upsala Meddel.* No. 4, p. 6 (1925).

21. *Monthly Notices R.A.S. 86*, 686 (1926).

22. Indications of a galactic centre near 325° longitude are also obtained from a study of the distribution of planetary nebulae and of novae.

23. *Hemel en Dampkring 25*, 67–68 (1927).

Joy, Rotation Effects from Cepheid Variables

Abstract—Data. The radial velocities of 156 variable stars of the δ Cephei type are available for a determination of the rotation of the galaxy, the coefficient of interstellar absorption, and certain galactic constants.

Distribution. The δ Cepheids are situated close to the plane of the galaxy and, as seen from the earth, they are well distributed in longitude. The known stars of this type are considerably clustered about the sun.

Interstellar absorption. The distances used are determined from apparent magnitudes and absolute magnitudes derived from the period-luminosity curve. In order that the rotation effect may show a linear relationship with distance, it is found necessary to correct the distances by postulating the presence of interstellar absorption. The coefficient of absorption is estimated to be 0.85 mag/kpc and the absorbing material is assumed to be uniformly distributed about the galactic plane in a layer having a total thickness of 0.4 kpc.

Rotational effects and galactic constants. For a study of galactic rotation the stars are divided into four distance groups. Solutions by Oort's method give for l_0, the longitude of the direction to the center of rotation, $325°.3 \pm 1°.3$, and for A, the rotation effect at one kiloparsec, 20.9 ± 0.8 km/sec. The distance to the center is estimated to be 10 kpc. By the use of Bottlinger's formula the circular orbital velocity of the sun is found to be 296 km/sec and the period 207,000,000 years.

Residual radial velocities. After taking out the effect of a solar motion of 20 km/sec and circular rotational effects, the average residual radial velocity is 10.8 km/sec.

In *Mount Wilson Contribution* No. 578 the data concerning the period, magnitude, and radial velocity of 155 stars of the δ Cephei type are compiled. The periods range from 1.5 to 45.2 days. One star, V 383 Cygni, the normal velocity of which is -20.0 km/sec, has since been added to the list. On account of their great distance, small peculiar motion, and concentration toward the plane of the galaxy, the effect of galactic rotation should be especially well shown by these stars.

The Data

Here we omit tables of the periods, galactic coordinates, apparent magnitudes, absolute magnitudes, distances, and velocities for 156 stars.

Distribution

A figure showing the galactic distribution of the observed Cepheids is omitted here.

It will be noted immediately that there is a high concentration toward the galactic equator. One hundred and thirty-six stars (87 per cent) lie within 10° and 105 (67 per cent) within 5° of the plane of the galaxy. Omitting six stars whose latitudes are numerically greater than $\pm 20°$, we find the mean latitude to be $-1°.4$. If distances are taken from the period-luminosity relationship and allowance is made for a general uniform absorption of 0.85 mag/kpc, all the stars except thirteen are found to be situated within the assumed absorbing stratum.

In longitude the distribution is fairly uniform over the region observable from northern observatories, although there is some evidence of a decrease in the number of stars in the direction of the center and in the region 115°–130°, which is nearly opposite.

A projection of the observed Cepheids within 3 kpc of the sun on the plane of the galaxy is omitted here.

The large Carina group, longitude 250°–270°, is remarkable. At longitude 90° there is a somewhat similar concentration of Cepheids in the Cassiopeia region. Taken together, these groups, which are much scattered in distance from the sun, suggest an arm of the Milky Way. A band 1 kpc in width along this arm contains 58 per cent of the known δ Cepheids. The trend of the arm appears to be more or less convex toward the center. Such a curvature would hardly be expected in the inner part of a spiral, but might be possible as a local distortion in the outer portion of an open type of galactic structure. The sun is near the axis of the arm and not in a region of low density as found by Oort[1] in a study of the distribution of faint stars in Kapteyn's Selected Areas.

Another striking feature of the diagram is the complete absence of δ Cepheids at distances greater than 1.3 kpc between longitudes 301° and 347°. Heavy obscuration doubtless prevents the discovery of distant variables at low latitudes in the general direction of the center.

Few radial velocities are available in the quadrant 230°–320°, on account of the southern declinations involved.

Table 82.4 gives the two-dimensional density of known variables and the number to be expected in zones concentric about the sun on the basis of uniform distribution in the plane of the galaxy.

Table 82.4 Distribution of Cepheids
(Distance corrected for absorption of 0.85 mag/kpc)

Zone (kpc from sun)	No.	Average density (per kpc^2)	Expected no. reduced to uniform distribution
0.0–0.4	16	31.8	16
0.4–0.8	39	25.9	48
0.8–1.2	32	12.9	80
1.2–1.6	45	12.8	112
1.6–2.0	32	7.1	144
2.0–2.4	31	5.5	176
2.4–2.8	13	2.0	208

The marked clustering of Cepheids about the sun may be attributed to one or more of three causes: (1) *local cluster*—there may actually be a local concentration of stars in the neighbourhood of the sun, possibly forming a knot in an arm of the Milky Way; (2) *insufficient correction for absorption*—a further reduction in the distances by increasing the absorption coefficient would have its greatest effect on the outlying stars and would tend toward more uniform distribution (it is doubtful, however, whether the moderate additional absorption which other considerations permit would greatly improve the distribution in this respect); (3) *the number of undiscovered variables of this type increases with distance*—although discovery lists of recent years have added surprisingly few δ Cepheids, it seems quite certain that many are yet to be detected among the fainter stars.

Galactic Rotation

Consideration of the distribution of δ Cepheid variables indicates that, except for the lack of observations in the quadrant of the Milky Way observable only from the Southern Hemisphere, they are particularly well situated for a study of galactic rotation based on radial velocities. On the other hand, for a determination of the solar motion and the location of its apex, their concentration toward the plane of the galaxy, together with the absence of observations of southern stars, gives an unbalanced solution of low weight, especially for the declination of the apex. For this reason and also because of Milne's[2] conclusions in regard to the use of the "local" solar motion, the measured radial velocities have been corrected for a standard value of 20 km/sec toward the apex $\alpha = 271°$, $\delta = +28°$, and no solar motion terms have been included in the final solutions.

One of the outstanding advantages of the use of Cepheid variables for the study of galactic rotation effects is that, although the stars are among the most distant observable for radial velocity, their distances can readily be determined by the use of apparent magnitudes and the absolute magnitudes based on Shapley's period-luminosity curve. The distances thus derived are known as "photometric" distances but, unfortunately, they are subject to considerable uncertainty on account of the effect on the apparent magnitudes of general and selective absorption in space. As a first approximation, for the purposes of this study, the absorbing material is assumed to be of a uniform structure and to lie along the plane of the galaxy in a stratum the total thickness of which is 0.4 kpc. The solutions for rotation afford a method for estimating the effective total absorption.

The general solutions for galactic rotation follow the methods of Oort,[3] and the papers of Hayford.[4] Bottlinger,[5] Plaskett and Pearce.[6] and Berman[7] have also been used extensively. The results of the preliminary solutions,[8] to which reference has been made by van Rhijn,[9] have been somewhat altered by the use of additional material obtained from later observations.

Here we omit a plot of the radial velocities of 156 Cepheids against galactic longitude, because these velocities are plotted separately in four groups in figure 82.1. Joy also gives reasons for omitting 21 stars in the following analysis.

The remaining stars, 135 in number, have been separated according to distance into five groups as shown in the accompanying tabulation.

Group	No.	Photometric Distance
1	35	0.13– 0.91 kpc
2	33	0.95– 2.40
3	34	2.51– 4.17
4	26	4.37– 9.55
5	7	10.00–22.91

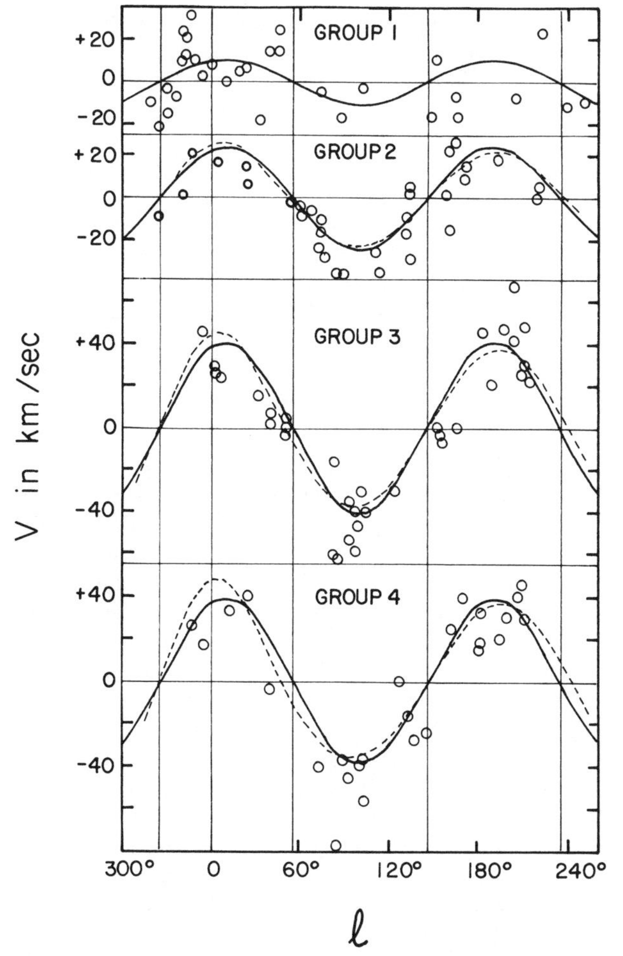

Fig. 82.1 Radial velocities, V, plotted against galactic longitude, l, for four groups of stars at successively greater distances, d, from the sun. The solid curves are the radial velocities predicted from Oort's equation $V = Ad \sin(2l)$ with $A = 25$, 23, 24.5, and 17.1 km sec^{-1} kpc^{-1} for groups 1, 2, 3, and 4, respectively. The dashed curve indicates the higher order terms according to Bottlinger's formula. Note that the four solutions yield a mean value of $A = 21$ km sec^{-1} kpc^{-1} and that the abscissa is expressed in terms of the old longitude scale l^{I}.

The first four groups were used for the solutions for galactic rotation. The stars of group 5 are too few and too scattered to be considered as a complete group.

Least-squares solutions were undertaken for each of the four groups, with the aid of the known values for the longitude, latitude, distance, and velocity. The equation of condition for each star was of the form

$$rA \sin 2(l - l_0) \cos^2 b = V',$$

where l, b are the galactic co-ordinates of the star; rA is the rotational term for a given distance r; l_0 is the galactic longitude of the center of rotation; and V' is the radial velocity of the star corrected for a solar motion of 20 km/sec. The equations were given equal weight.

The resulting values of A, the rotational term for a distance of one kpc, of $\bar{r}A$, and of l_0 are given in table 82.5.

Table 82.5 Group solutions without absorption correction

Group	No.	$\bar{r}$ (kpc)	A (km/sec)	$\bar{r}A$ (km/sec)	l_0
1	35	0.52	19.9	10.4	331°.2
2	33	1.56	14.5	22.7	323 .2
3	34	3.20	12.2	38.9	326 .5
4	26	5.56	6.8	37.7	327 .0

INTERSTELLAR ABSORPTION

The values found for the rotational constants furnish a means of estimating the effect of interstellar absorption upon the photometric distances. The results of the solution are consistent for l_0, but the values found for A do not show the agreement which would be expected if the differential rotational effect varies directly with distance in the region occupied by these stars. The discrepancy can be reasonably explained, in part at least, by assuming that the photometric distances, which depend upon observations of apparent magnitudes, are affected by considerable absorption in space.

The fact that the distant group 4 shows a rotation effect which is actually less than that of the nearer group 3 is surprising and makes the determination of an absorption coefficient difficult and uncertain. Although the observations of the stars of group 4 are mostly made with lower dispersion than those of the nearer groups, there is no reason to doubt their essential accuracy for mean results. Either there is a breaking-down of the rotation effect at great distances in a way which cannot be accounted for by the inclusion of higher-order terms in the equations or the absorption coefficient is greater for the stars of group 4. This would mean that these stars, in the mean, are situated in regions of greater obscuration than those of the nearer groups. Perhaps star counts could be made which would give information on this point.

Recent observations of interstellar calcium and sodium indicate that the gases in space are not uniformly distributed. In the existing state of our knowledge it may not be possible to set a definitive value for the absorption throughout the galaxy. The result from one group of stars may well be quite different from that given by another group, depending on their positions and distances.

Apparently the best that we can hope for at present is to adopt for our final solutions a value for the absorption coefficient which, on the assumption of constant A, will serve as a first approximation.

By means of cut-and-try methods it was found that by correcting the magnitudes for a total uniform absorption of 0.85 mag/kpc (photographic) a fair representation of the observations is obtained.

Omitted here is a plot of the rotational term rA against the distance r for the results with and without correction for absorption. When an absorption correction of 0.85 mag kpc^{-1} is used, the plot shows a linear variation of rA with distance, and the value of A is 20.9 km sec^{-1}.

Another method for estimating the absorption coefficient has been suggested by Bottlinger and Schneller.[10] In all probability the average thickness of the stratum containing Cepheid variables is the same in different parts of the galaxy. Hence, we should expect that the mean of the z co-ordinates of the stars would be the same at all distances from the sun. If, however, photometric distances are used and interstellar absorption is present, the mean value of z will increase with increasing distance. By varying the amount of absorption it is then possible to find the coefficient which gives z a constant value. The distances of the Cepheids from the plane of the galaxy have been computed and the average taken for each of four groups selected according to projected distance from the sun (x coordinate). The results indicate that, if the mean z components are to be equal at all distances from the sun, a correction corresponding to a mean total absorption of about 1.50 mag/kpc (photographic) must be applied. This value for the absorption correction may, perhaps, be considered as an upper limit. The criterion should not be applied too strictly because, in the neighborhood of the center of the system and perhaps in other regions of local clustering, we may expect that the stratum will be thicker. Group motions, unless exactly in the plane of the galaxy, tend to increase the values of the z components in certain regions. Also, for a given distance, stars with high galactic latitudes will be brighter, if absorption is limited to layers near the plane, and their variability will be more readily detected. This will increase the mean value of z, especially in the distant groups.

Solutions for Rotation Including Absorption

After introducing an absorption coefficient of 0.85 mag/kpc, new rotation solutions were made for the identical groups of the first solution with the results given in table 82.6. The distances of the stars of the outer groups are greatly reduced. For example, when allowance for absorption is made, the distance of WW Mon, the most remote star of group 4, is changed from 9.55 to 2.99 kpc. This shows conclusively why it is that variables, lying within the stratum of absorption, have not been found at greater distances. Under such conditions the brightest Cepheid, if located within the absorbing medium at a distance of 10 kpc or farther, would at median brightness appear fainter than the twentieth magnitude.

Table 82.6 Group solutions with 0.85 mag/kpc absorption

Group	$\bar{r}'$ kpc	A km/sec	$\bar{r}'A$ km/sec	l_0
1	0.42	25.1 ± 4.8	10.6	332°.2 ± 5°.4
2	1.06	22.8 ± 1.6	24.3	323.5 ± 2.0
3	1.66	24.5 ± 1.6	40.6	326.5 ± 2.3
4	2.31	17.1 ± 1.3	39.4	325.2 ± 2.6
Mean	—	20.9 ± 0.8	—	325.3 ± 1.3

The second solution shows little change in the longitude of the direction to the center. The value of 325°.3 seems to be accurately determined, as far as these stars are concerned. The value of A is not so satisfactory. In the preliminary solution of 1933 a value of 18.5 km/sec was found with the same absorption coefficient, but this is now considerably increased by the inclusion of a number of additional observations and the omission of several stars used in the preliminary discussion.

The residual term K has not been introduced in the final solutions. The lack of observations of southern variables makes its evaluation somewhat uncertain. Its mean value found from a single solution, including the stars of all groups with distances corrected for absorption, is −3.8 km/sec. Because its physical significance is doubtful, it was thought better to omit it altogether and throw the whole weight of the observations into l_0 and A. Its inclusion, however, makes very little difference in the results.

The velocities for the four groups are plotted separately in figure 82.1; the curves are those corresponding to the results found from the solutions.

Distance to the Center of Rotation

The distance to the center was determined by the simple relation[11]

$$2R_0 \cos(l_1 - l_0) = 2R_0 \cos(l_0 - l_3) = r,$$

which is based on the condition that the rotational velocity of any star at the same distance from the center as the sun has zero velocity relative to the sun. R_0 is the distance, sun-center; r is the distance, sun-star; and l_1, l_3 are the longitudes of the two points nearest l_0, the direction of the center, where the velocity-curves drawn through the observations cross the zero-axis. Thirteen stars of groups 2 and 3, located near longitude 55° where the curve is practically a straight line, were used for this purpose. Their velocities were reduced to a mean distance, and the longitude at zero-velocity was determined for each group. The values of R_0, the distance to the center, found for groups 2 and 3 are 11.7 and 7.3 kpc, respectively, and the weighted mean is 10.0 kpc. Unfortunately, there are no stars near l_3, at longitude 235°, which could be used to increase the weight of the determination.

The Sun's Orbital Velocity and the Effects of Higher Order Terms

Here Joy shows that the inclusion of higher-order terms in the relation between radial velocity and distance makes little difference in the fits of the observed data, because of the nearness of the observed Cepheids.

Group 5—The Most Distant Cepheids

Inasmuch as the stars of group 5 are few in number and would be largely affected by any uncertainties in the amount of absorption or the thickness of the stratum, they have not been included in the final results. In general, they indicate that the adopted distance to the center is of the right order.

Residual Radial Velocities

The average residual radial velocity after taking out the effects of solar motion of 20 km/sec and the circular rotation about the center is, for groups 1–4, 10.1, 7.6, 12.4, and 14.0 km/sec, respectively, with a mean of 10.8 km/sec. This includes accidental errors of observation, the effect of errors in apparent magnitude resulting from photometric estimates, and uncertainties in the absorption correction, as well as the component of the peculiar motion of the stars in the line of sight. There is little evidence for group motion among the Cepheids, except, perhaps, for the stars of the Perseus-Cassiopeia region, longitude 80°–120°, which, in general, have large negative residuals.

1. *B.A.N. 8*, 233, (1938).
2. *M.N. 95*, 564 (1935).
3. *B.A.N. 3*, 275 (1927).
4. *Lick Obs. Bull. 16*, 53 (1932).
5. *Veröff. Berlin Babelsberg 10*, no. 2 (1933).
6. *Pub. Dom. Ap. Obs. 5*, 242 (1936).
7. *Lick Obs. Bull. 18*, 57 (1937).
8. *Pub. A.S.P. 45*, 202 (1933); *Pub. A.A.S. 7*, 218 (1933).
9. *Pub. Kapteyn Astr. Lab.*, no. 47 (1936).
10. *Zs. f. Ap. 1*, 340 (1930).
11. Hayford, *Lick Obs. Bull. 16*, 73 (1932).

83. On the Dark Nebula NGC 6960

Max Wolf

Translated by Brian Doyle and Owen Gingerich

(*Astronomische Nachrichten 219*, 109–116 [1923])

The existence of dark regions in the Milky Way has been known at least since the late 1700s, when William Herschel called attention to certain dark places that appeared to be associated with bright nebulae and rich stellar regions. At the turn of the twentieth century, Edward Emerson Barnard began systematic photography of the dark and bright regions of the Milky Way. This work extended over thirty years and eventually culminated in the publication of two important catalogues of the regions he noncommittally called dark markings.[1] At first Barnard thought he was photographing dark holes in the heavens, but he soon reasoned that the close resemblence in the form and size of the dark markings to some emission nebulae meant that the dark regions were actually nebulous. Moreover, in some cases the dark and bright regions seemed to interact with each other.

Like Barnard, Max Wolf at Heidelberg also recognized the importance of photographing the Milky Way with instruments having a large field of view. While Barnard was chiefly interested in the peculiar shapes of the dark regions, Wolf was concerned with measuring their distances and absorbing powers. By counting stars in an obscured and an adjacent unobscured region, Wolf showed that the numbers of fainter stars all diminished in the obscured region as if a discrete absorbing layer were present. As shown here, the star counts could be graphically displayed in what is now called a Wolf diagram. Although subsequent work has shown that the distance, the absorption, and even the number of nebulae in the line of sight can be misjudged from such graphs.[2] Wolf's method did provide a convincing demonstration that the dark areas indeed represented dark obscuring clouds.

The region chosen by Wolf for this analysis, the so-called Cygnus Loop, has special interest because today it is recognized as a supernova remnant. The brilliant arc of bright nebulosity may be the shock front of the supernova interacting with the dark cloud measured by Wolf. His distance of 1,500 pc for the cloud is in reasonable agreement with current estimates of the distance to the supernova remnant, about 770 pc.

In Wolf's conclusion, he correctly reasons that the obscuration is caused by dust rather than gas. If the nebula were gaseous, Rayleigh scattering by the gas molecules would strongly redden the light of the more distant stars, an effect that Wolf failed to find. In fact, the fainter stars are weakly reddened in the obscured region, a phenomenon now known to result from scattering by small dust grains. Interstellar gas, while possibly producing absorption lines, is seldom present in sufficient density to cause noticeable Rayleigh scattering.

1. E. E. Barnard, *Astrophysical Journal 49*, 1 (1919), and *A Photographic Atlas of Regions of the Milky Way*, ed. E. B. Frost and M. R. Calvert (Washington, D.C.: Carnegie Institution, 1927).
2. B. J. Bok, *Distribution of Stars in Space* (Chicago: University of Chicago Press, 1937).

THE ASSERTION can be made that the dark holes in the bright Milky Way clouds are suspended at the same distance from us as the bright nebulae. If we conceive of them as dark nebulae, the light and dark nebulae adjoin one another in space.

This assertion may be proved, firstly, from the form of the border of the bright nebulae. These are regularly influenced in many places by the bounding dark nebulae.

There is an apparent brightening above the background along the edge of the great dark bay of the Orion nebula, suggesting that the dark and light regions interact with each other and that both lie at the same distance from us. Other examples in which dark nebulae break the continuity of the light nebulae with which they seem to interact include M 8, M 16, M 42, the Trifid nebula, the [North] American nebula, the nebula around 15 Monocerotis, and NGC 6611.

The luminous nebula NGC 6960 [the western part of the so-called Cygnus Loop] forms a long line through the star 52 Cygni and a bulge against a great dark cloud. Its spectrum of hydrogen, helium and "nebulium" is typical of a luminous gas mass and is essentially the same as that of the eastern part of the loop, NGC 6992.

If one examines a photograph of the nebula, a great difference in the number of stars on the two sides of the loop NGC 6960 is immediately apparent: on the western side of the boundary there are only half as many stars as on the eastern side. Thus, we conclude that the loop is the edge of a dark cloud partly catching the light of a star.

I have now succeeded not only in determining the entire star count, as before, but in carrying out for each magnitude class a count of the stars to the right and to the left of the nebula; and I have at the same time tried to determine the photographic brightness of the enumerated stars. A millimeter grid was pressed against a reflector photograph (D 371, August 31, 1908, exposure about 1 hr); this double plate was placed in the right side of the stereocomparator, while in the left side was placed another reflector plate of the same region (D 181, August 31, 1907, exposure about 70 min). The stars were counted in each square on the right plate and at the same time checked on the left plate. Each star was brought to the cross hairs of the micrometer eyepiece, and its diameter was evaluated as precisely as possible. The stellar magnitudes appropriate to the diameters so obtained were calibrated as follows: there was a third plate available, exposed equally long on the polar sequence. From a comparison of these two exposures, stellar magnitudes were determined for most of the stars. The known brightnesses for these stars gave a scale by which to evaluate the magnitudes for all the counted stars from their diameters. For the magnitude scale, the values of Seares[1] were used. The magnitudes so obtained are possibly subject to a considerable systematic error; the cause would be that the two comparison plates, which have served to ascertain the magnitudes (the polar sequence and NGC 6960), were indeed taken with similar exposures but on different emulsions. At this time I cannot ascertain the errors possibly present, because I have not yet succeeded in obtaining equivalent exposures of the two regions. The errors, however, play a secondary role in what follows.

There were 337 squares in the region of the left of the nebula, about 40 squares on the bright nebula, and 343 squares on the star-poor region west (right) of the nebula. The corresponding star counts are 1,797, 150, and 873, which give star densities of

5.332 3.750 2.563

per square millimeter (focal distance 2,818.1 mm) or

12,902 9,498 6,202

per square degree. Since the last magnitude class measured was assumed to be 17.5, it follows that if we designate by B_m the number of all stars up to those with the class m, then

$$\log B_{17.5} = 4.111 \qquad 3.978 \qquad 3.793,$$

while for the galactic latitude 0° the corresponding mean values are 4.08 (Seares) and 3.38 (Chapman).

It is immediately apparent that *the total number of stars in the darkened part is less than half that in the undarkened part.*

The count in the undarkened part corresponds almost exactly to the mean value given by Seares.

As demonstrated, the star count in the bright nebula itself is also smaller than that in the unaffected starry background. We can attribute this phenomenon to the overexposure (plate darkening) of the bright nebula, which makes many weaker stars unrecognizable. To proceed with complete safety, I have wholly excluded the stars located in the bright nebula in the following comparison.

In order to get a more certain mean value for the brighter stars (10^{mag}), whose number in such a small field is already very tiny (since one of the squares is only equal to 0.00041326 square degree), 275 more squares on the left and as many on the right were searched thoroughly for stars of 10^{mag}, and the mean count improved thereby was used in deriving the following results.

Table 83.1 gives the result of the count by magnitude class (unfortunately under the limitation of a single table). A_m denotes the number of all stars of magnitude class m in the

Table 83.1 Star count by magnitude class

A_m	B_m	$\log B_m$	m	$\log B_{m'}$	$B_{m'}$	$A_{m'}$
8,637.1	12,873	4.110	17.5	3.787	6,119	4,085.8
2,514.2	4,236	3.627	16.2	3.308	2,033	1,163.8
872.0	1,722	3.236	15.0	2.939	869	397.4
341.5	850	2.929	14.0	2.674	472	184.5
247.1	508	2.706	13.0	2.458	287	141.9
174.4	261	2.417	12.0	2.162	145	63.9
43.6	86.9	1.939	11.0	1.911	81.5	42.6
39.8	43.3	1.637	10.0	1.589	38.9	35.4
3.5	3.5	0.545	>10	0.545	3.5	3.5

unveiled part of the sky. $A_{m'}$ is the same in the veiled part (right of the nebula). B_m denotes the number of all stars from the brightest down to those with magnitude m in the unveiled region; $B_{m'}$ is the same in the veiled part. The number for the stars brighter than 10^{mag} is taken from Chapman–Melotte.

It is apparent from table 83.1 that for all fainter magnitudes the A_m are about twice as large as the $A_{m'}$; that is, a constant loss is manifest in the veiled region for all the fainter magnitude classes. This gives a very strong demonstration that we are dealing with an extinction phenomenon caused by a foreground mass.

The relationship changes very suddenly at the eleventh magnitude; the star counts A and A' become equal. The strongly absorptive effect of the nebula begins for stars of the eleventh magnitude.

Here the log B_m values are compared with those obtained by Seares and Chapman–Melotte for other regions, and the numbers tally well.

If we plot as curves the log $B_{m'}$ and log B_m, respectively, we get the graphs reproduced in figure 83.1. The solid line gives the graph of the logarithm of the star counts in the veiled part of the heavens; the dotted line, that in the unveiled part. We see the almost exactly parallel course of the curves for the fainter stars and how suddenly the curves come together at magnitude 11.

In the parallel part (that is, for the fainter stars) the displacement of the curve because of the darkening amounts to about 1 mag. Therefore, the dark nebula captures about that much light, that is, about 60%. It seems doubtful whether the difference indicated at the tenth magnitude is real. If it were, traces of the nebula would have to extend toward us.

In any case, the nebula must be located with its principal mass at approximately the distance of stars of the eleventh to twelfth magnitude. Its thickness is indicated by the fact that the curves of the stars diverge to their final separation from stars of magnitude 11 up to about those of magnitude 12.5.

If we were to take the mean parallax of 0.0022 sec of arc for the Milky Way stars of magnitude 11,[2] then the thicker nebula would begin around 1,500 l.y. (90 Siriometers) from us, and its principal mass would certainly be accepted to be about 500 l.y. (32 Siriometers) thick. Perhaps we could try to calculate more exact values according to the method of Pannekoek.[3] Because of the uncertainty in my magnitude classes and the imprecision in the numbers for the mean parallaxes, there is hardly any point to this calculation at the present time.

Of course, it is of interest to discover how the light is probably captured. Does the nebula act like a proper gas mass or

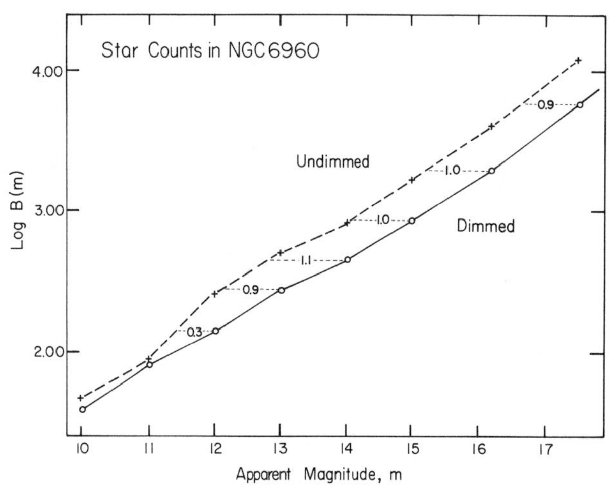

Fig. 83.1 Wolf diagram showing star counts in two regions of equal area to the west (*dimmed*) and east (*undimmed*) of NGC 6960. Here the logarithm, log $B(m)$, of the number of stars, $B(m)$, that are brighter than apparent magnitude, m, is plotted against m. Modern versions of the Wolf diagram, which look very similar to this figure, give the logarithm of the number of stars, $A(m)$, counted between apparent magnitudes $m - 0.5$ and $m + 0.5$ plotted against m for equal areas of sky in obscured and unobscured regions.

like a dust cloud? Pannekoek[4] has shown that the assumption of a gas cloud leads to such a great mass that its gravitational interaction would have to be manifest. De Sitter[5] has shown that, by assuming a dust cloud, a significantly smaller mass would suffice to produce the observed capturing action.

If the dark cloud is made of gas, then the light of stars located behind it must show a scattering in transit that, by Rayleigh's law, would remove especially the violet light and so would redden the star light that we see. In the case of dust, this reddening can hardly be present. Thus, the obscured stars would have to be redder than unobscured ones of the usual celestial background. If we determine the color indices of faint stars both inside and outside the cloudy region, then for most of the stars the mean color index must be larger in the darkened part than outside.

Wolf then compares photographs of the stars on the darkened and undarkened sides of the nebula taken with and without a yellow filter in place. Color indices of roughly 60 stars in each region are determined by comparing a photograph made with a yellow filter with photographs taken at different exposures without the filter. The color index is measured by the exposure time required to make the image diameters of these blue photographs equal to the diameter on the yellow photograph. The mean color for all stars between 11th and 15th magnitudes is the same on both sides of the nebula, which means that the dark nebula produces no detectable reddening of the starlight.

We must therefore conclude that no reddening of the stars arises from the dark cloud of the nebula, and it seems quite probable for that reason that the light-capturing cloud consists, for the most part, of a dust mass.

1. *Mt. Wilson Contr.*, No. 97 (1915).
2. *Groningen Publ.*, No. 29 (1918).
3. *Amsterdam Proc. 23*, no. 5, 707 (1920).
4. Ibid., p. 720.
5. Ibid., p. 725.

Appended Figure

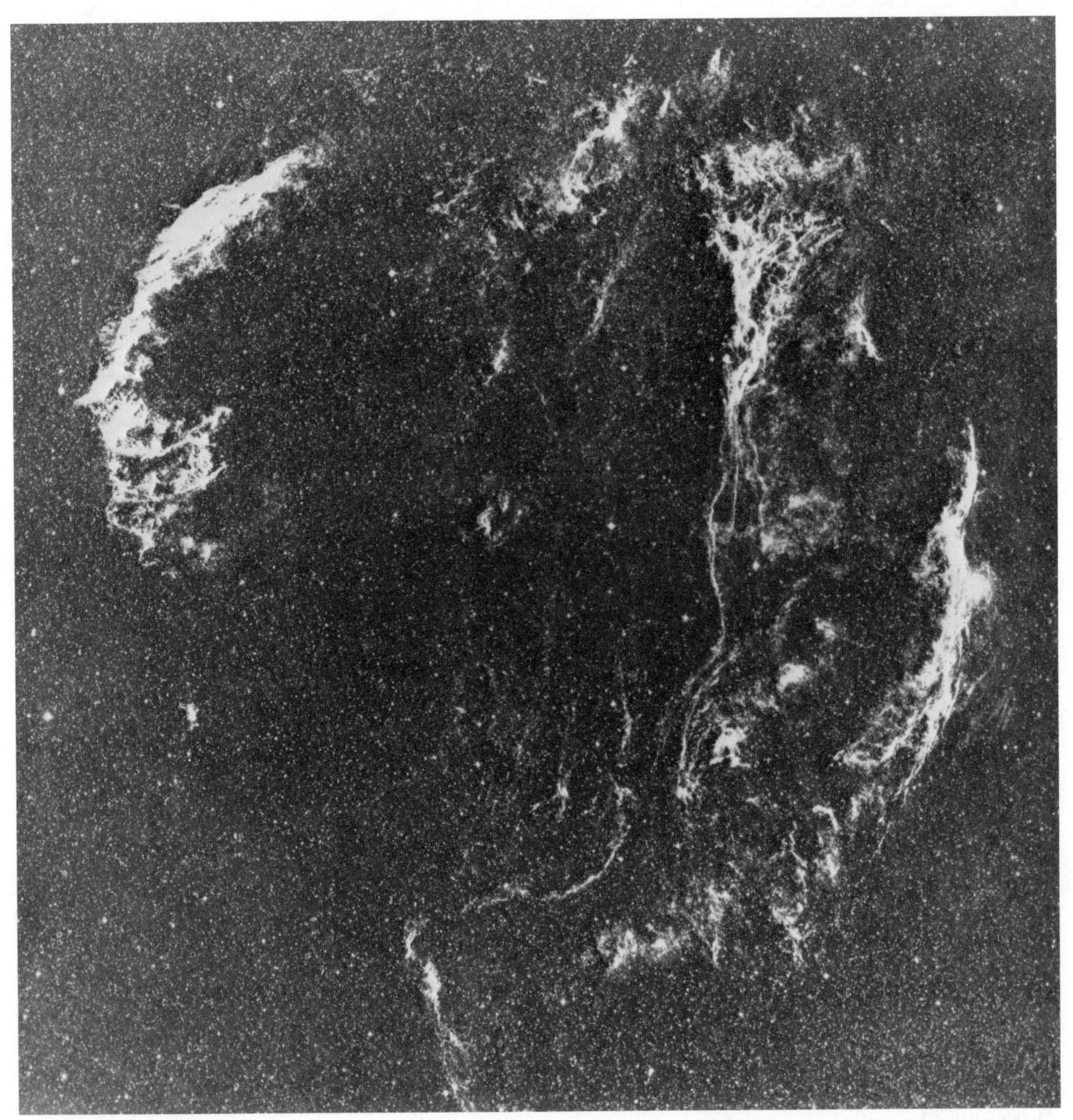

Fig. 83.2 The great nebular network known as the Cygnus Loop. The delicate filaments are thought to be the expanding remnant of a supernova explosion. The Cygnus Loop is composed of the nebular regions NGC 6960 (at the lower right) and the Veil nebula NGC 6992-6995 (upper left). (Courtesy Harvard College Observatory)

84. The Source of Luminosity of Gaseous Nebulae

The Planetary Nebulae

Donald H. Menzel

(*Publications of the Astronomical Society of the Pacific 38*, 295–312 [1926])

An Application of the Quantum Theory to the Luminosity of Diffuse Nebulae

Herman Zanstra

(*Astrophysical Journal 65*, 50–70 [1927])

When William Herschel began to "sweep the heavens" with his telescopes in the 1780s, he found many nebulous patches of light. Some, because of their uniform disks and bluish color, resembling the distant planet Uranus, he dubbed planetary nebulae; other, irregularly shaped nebulae are now called diffuse nebulae. Nearly a century later, William Huggins showed that the spectra of both the planetary and diffuse nebulae generally exhibit the same emission lines of hydrogen, nitrogen, and an unidentified substance designated "nebulum." Early in this century, V. M. Slipher reported that some diffuse nebulae do not display emission lines but a continuum marked by absorption lines similar to those of nearby stars.[1] He concluded that these objects must be "reflection" nebulae, which shine by reflected starlight.

Meanwhile, at Lick Observatory, William Hammond Wright embarked upon a thorough study of the spectra of gaseous nebulae.[2] When examining the continuum radiation of the central stars of planetary nebulae, he was surprised to find that they all radiate an extraordinary amount of ultraviolet light, just as the hot O-type stars do. The similarities among the central stars of planetary nebulae led Wright to ask whether these nebulae might be stimulated by their central stars, just as the aurora borealis seemed to be stimulated by the sun. By 1921 Henry Norris Russell had advanced the idea that nebulae might shine through the ionization of their atoms by either short-wavelength radiation or corpuscular emission from the stars imbedded within them, and one year later Edwin Hubble provided observational evidence for the idea that stellar radiation makes nebulae shine.[3] Hubble showed that the stars involved in both the diffuse emission nebulae and the planetary nebulae nearly always have spectra characteristic of hot O and B stars, whereas the stars in reflection nebulae nearly always have spectra characteristic of cooler, later-type stars. Hubble reasoned that this definite relation between the spectra of nebulae and their associated stars means that the source of nebular luminosity is radiation from stars.

By 1926 Arthur Eddington had extended the idea that the energetic photons of ultraviolet light from hot stars will ionize interstellar matter, and in his Bakerian lecture he showed that the electrons liberated by this photoionization will heat up the interstellar gas to about 10,000°.[4] A few months after Eddington's Bakerian lecture, Donald H. Menzel presented the first of the two papers given here, which raises the possibility that the Balmer emission lines of planetary nebulae are the result of photoionization of hydrogen atoms by ultraviolet starlight, followed by recombination of the free electrons and protons. Because the derived temperatures of the central stars seem unreasonably high, Menzel actually retreats from this correct description and concludes that corpuscular emission from the stars must excite the nebulae. Meanwhile, in the same year, Herman Zanstra published a short note in which he argued that the nebular emission spectra are indeed the result of recombination following ionization by the ultraviolet light of stars whose surface temperatures are about 30,000°.

In the following year Zanstra published the second paper given here, which asserts that every ultraviolet photon of starlight whose energy is greater than the ionization energy of hydrogen (radiation with wavelengths shorter than 912 Å) will ionize one hydrogen atom, producing in each case one free electron and proton. When the free electrons recombine with the protons they will not always fall into the state of lowest energy but will cascade through a series of intermediate levels, thereby producing the observed Balmer emission lines of hydrogen. In addition, the intensity of each Balmer emission line can be related to the temperature of the exciting star through the theory of the hydrogen atom and the Planck spectrum of thermal radiation.

As we shall see in selection 85, Ira Bowen subsequently interpreted the so-called nebulium lines as forbidden transitions of ionized nitrogen and oxygen excited by collisions with free electrons before recombination. Zanstra then tested this mechanism for planetary nebulae and showed that the stellar temperatures needed to produce the forbidden transitions were in approximate agreement with those required to produce the hydrogen emission-line spectrum.[5] These papers set the stage for the detailed development of the theory of both recombination and forbidden line spectra by Menzel and his colleagues, especially Lawrence H. Aller and James Baker, who in a series of papers in the *Astrophysical Journal* were able to combine observations with theory to derive the temperatures and electron densities of the nebular gas, as well as the temperatures of their associated stars.[6]

1. V. M. Slipher, *Lowell Observatory Bulletin 2*, No. 55, 26 (1912), *2*, No. 75, 155 (1916).
2. W. H. Wright, *Publications of the Lick Observatory 13*, 193 (1918).
3. H. N. Russell, *Observatory 44*, 72 (1921), and *Proceedings of the National Academy of Sciences 8*, 115 (1922); E. Hubble, *Astrophysical Journal 56*, 162, 400 (1922).
4. A. S. Eddington, *Proceedings of the Royal Society* (London) *A111*, 424 (1926).
5. H. Zanstra, *Nature 121*, 790 (1928), and *Publications of the Dominion Astrophysical Observatory 4*, 209 (1931).
6. Important early theoretical and observational papers are reproduced by D. H. Menzel in *Selected Papers on Physical Processes in Ionized Plasmas* (New York: Dover Publications, 1962).

Menzel, The Planetary Nebulae

In the first six sections of this paper Menzel reviews and references material on the measured parallaxes of planetary nebulae, the fact that the nuclear stars are of spectral class O, the masses of the nuclear stars as computed from their spectroscopic velocities of rotation, the transparency of the nebulae to visible light, the fact that the nuclear stars do not seem to adhere to the normal mass-luminosity relation for stars, and the fact that planetary nebulae seem to have no connection with either novae or Wolf-Rayet stars.

(7) THE SOURCE OF LUMINOSITY

OBSERVATION DEFINITELY LOCATES the source of excitation for the planetary nebulae in the nuclear stars, which are usually located near the geometrical center. Hubble[1] has made an extended study of the problem and finds a correlation of 0.65 between m, the magnitude of the central star, and A, the diameter of the nebula on a uniform exposure scale. The equation connecting the two was found to be $m + 5.40 \log A = 17.88$.[2] The coefficient of the second term is within the experimental error of 5.00, the value required if the inverse-square law is one of the important factors in determining the luminosity of the nebula. In many cases, the total photographic brightness of the nebula is a hundred times or more greater than that of the central star. Since the total energy radiated by the nebula cannot exceed that of the star, one naturally looks to the ultra-violet.

If we assume that, approximately, the stars radiate according to Planck's law, which is probably true as regards order of magnitude, the energy distribution curves may be drawn and the amount of light which normally affects the photographic plate compared with the ultra-violet energy beyond our vision. The former was taken from λ3,500—4,800 A. For hydrogen to emit the Balmer series requires the electron to be raised at least to the 3-quantum stage. On the simple theory, this requires the absorption of λ1,025, 972, . . . 911. It was immediately evident, no matter what temperature may be assumed for the star, that even the total energy lying between λ1,025 and λ911 falls far short of equalling that radiated by the nebula. The case did not yet appear hopeless for the absorption of λ911 corresponds to ionization of the atom, the removed electron possessing zero velocity. The ionizing power of ultra-violet radiation is well known and it is possible that radiation short of λ911 may forcibly cause the ejection of an electron from the atom, the excess energy over that necessary for ionization going into the kinetic energy of the electron in accord with Einstein's familiar law of the photo-electric effect, $1/2\ mv^2 = h\nu - I$. Let us assume the most favorable case, i.e., (a) where all the energy of the star short of λ911 is effective in ionization, (b) the surplus kinetic energy of the electrons is used in further exciting the atoms or else radiated at the head of the Balmer series, and (c) the returning electron stops only in the 2-quantum state before falling to the lowest level. This last condition favors the emission of the Balmer series, but from quantum considerations, not more than one-fourth of the energy originally absorbed can go toward the production of those lines. The remaining three-fourths produce λ1,215.

Mechanical quadratures show that for a stellar temperature of 20,000° the ratio of the energy λ3,500—λ4,800 and λ911—λ0 comes out about 10:1. For a temperature of 40,000° it is 1:22. It is evident that no reasonable temperature could be assigned to the star which would be able to explain more than a small fraction of the nebular radiation as fluorescent. To allow for possible altering of the energy curve by electron scattering in the stellar atmosphere merely serves to increase the discrepancy.

It is necessary to fall back on the theory of corpuscular excitation. Hubble,[3] Russell[4] and others have suggested it before. The hot star is emitting electrons which fly out into space in all directions causing ionization by collision. The intensity of the electron radiation will vary according to the inverse-square law. It is probable that the positive charge acquired by the star will eventually slow the electrons down and therefore decrease their effective power for doing work. The latter quantity is proportional to the electron's kinetic energy, $1/2\ mv^2$, and since the parabolic velocity depends on the square-root of the distance, the power will vary inversely as the distance of the electron from the nucleus. The total effect will, therefore, be approximately as the inverse cube, with a sudden fall in intensity when the energy of the electrons falls below the critical potential for the line in question. The assumption of parabolic velocity is justified by the fact that the star will develop a sufficient positive potential to draw back the electrons before they escape entirely.

As Russell has pointed out,[5] the sizes of the various monochromatic images in the spectra of the planetary nebulae show a tendency to arrange themselves in the order of excitation potentials for the different elements. The inverse-square law would prescribe a simple fading out of the images but the decrease in the speed of the electrons as well would require the effect noted by Russell.

In this connection the writer has measured the diameters of various images in Wright's photographs taken with the slitless spectrograph. The nebula N.G.C. 7009 was chosen because it presented measurable images of both neutral and ionized helium as well as hydrogen. The means of the longitudinal diameters of $He+$, He, and H are in ratio 1:1.20:1.56, respectively.

The ionization potentials of the three atoms are approximately in the ratio of 4:2:1, and if the ionizing force varied inversely as the distance, the ratio of the diameters would be 1:2:4. For an inverse-square law we would have 1:1.4:2 and for an inverse-cube, 1:1.26:1.58. The almost exact agreement of the observations with the figures last given is probably fortuitous and it is by no means certain that all nebulae will give the same result. A cursory examination of some of the photographs strengthens the opinion that they will not, but it appears probable that further observational material will more nearly confirm the inverse-cube than the inverse-square law. Differences in the distribution of elements would tend to introduce many irregularities and the inverse-square law would hold only if a negligible fraction of the emitted energy is absorbed. Furthermore, the atomic weights of the elements may possibly enter into the problem of the distribution.

It is possible, however, that we are avoiding one difficulty by introducing another. The energy of the expelled electrons does not come from nothing and it is just about as difficult to find this source of energy as it is to find the etherial radiation. If, as the above indicates, the total radiation, corpuscular and etherial, emitted by the star is insufficient to account for the absolute magnitude of the nebula, it would appear that there is some generation of energy within the gaseous envelope. The problem is much more complicated than has hitherto been supposed and considerably more data are needed.

1. *Ap. J. 56*, 162, 400 (1922).
2. Ibid., p. 432.
3. Ibid., p. 437.
4. *Obs. 44*, 72 (1921).
5. Ibid.

Zanstra, The Luminosity of Diffuse Nebulae

Abstract—H. N. Russell, on theoretical grounds, made the suggestion that the *emission spectrum* observed *in nebulae* is, in some way, *induced* by *radiations from neighboring stars of a high temperature. This conclusion was reached independently* by E. P. Hubble, who, by an extended study of *observational data, settled this question* beyond any reasonable doubt. He also found a number of *qualitative and quantitative relations to exist* between the *characteristics of the nebula and of the star involved in its luminosity*. In line with this idea, already widely accepted, the writer has aimed to work out a *quantitative treatment* of the *emission spectrum produced in diffuse nebulae*, using Hubble's quantitative results as a guide. The problem is simplified by assuming that the star emits radiation like a black body and that the nebula consists of hydrogen of which an appreciable amount is in the atomic form. It is assumed that the ultra-violet starlight beyond the head of the Lyman series is completely absorbed by the nebula, causing ionization of the nebular hydrogen atoms into protons and free electrons. The subsequent recombination of electrons and protons causes the re-emission by the nebula of a number of line spectra and continuous spectra, among others the Balmer series and the continuous spectrum at its head. Finally, an *approximate expression for the ratio of the photographic intensity of the nebula to the photographic intensity of the intercepted starlight* is obtained, and this ratio is calculated for different temperatures (T) of the star; it is unity for $T = 33{,}000°$. Hubble found this ratio to be approximately unity for stars of type Bo and O associated with diffuse nebulae, and hence the *theory requires for those stars an average temperature of about* 33,000°*C*. A more detailed discussion of the observational data leads to the following rough values for the *temperature of stars associated with diffuse nebulae:* B1, 21,000°; Bo, 28,000°; O, 34,000°. Stars in planetary nebulae would, in general, have a still higher temperature. These values are in fair agreement with the temperature determinations from the theory of thermal ionization of stellar atmospheres, and this agreement serves as a check for the mechanism of luminosity proposed. The fact that the continuous spectrum at the head of the Balmer series is more conspicuous in planetaries than in diffuse nebulae led to the assumption that free electrons of higher velocity are captured to a large extent by the lower levels of the hydrogen atom. This offers an explanation for the result of H. H. Plaskett in work not yet published, who finds that the first lines of the Balmer series in nebulae become relatively stronger, as the temperature of the associated star becomes higher. The assumption is also in harmony with the theoretical considerations of H. A. Kramers.

1. BRIEF REVIEW OF OBSERVATIONAL FACTS AND THEORETICAL IDEAS REGARDING GALACTIC NEBULAE

THE GALACTIC NEBULAE are usually subdivided into two classes, diffuse and planetary nebulae.

The planetary nebulae have a definite structure, appearing as ovals, rings, or disks. Their spectrum consists of a number of single lines, and frequently there is also a continuous spectrum, starting approximately at the head of the Balmer series of hydrogen and extending into the ultra-violet. Following Hubble, a spectrum of this type will be classified as an "emission" spectrum. Each planetary nebula has a star approximately at its center.

The diffuse nebulae, on the other hand, do not show a definite sharp structure. Hubble divides them into dark nebulae, obscuring the background, and luminous nebulae. The luminous nebulae have either emission spectra, similar in a general way to those of the planetaries, or continuous spectra of approximately the same type as, although less intense than, those of ordinary stars. Frequently the three types of diffuse nebulosity, "dark," "continuous," and "emission," are found in the same regions of the sky and even merge more or less into one another, as for example in the large Orion nebula.

It is generally recognized at present that both the continuous and the emission spectrum observed in nebulae are, in some way, caused by neighboring stars.

In a number of cases V. M. Slipher[1] has found that the continuous spectra of certain nebulosities are approximately the same as those of stars in their neighborhood and suggests, for this reason, that those nebulae owe their luminosity to reflected starlight. E. Hertzsprung[2] estimated that the luminosity of the Pleiades nebulae is from 3.5 to 5.6 mag. fainter than would be the case for a perfect diffuse reflection, although a more recent statistical study of a large amount of observational material by Hubble leads to a much stronger luminosity for diffuse nebulae in general.

If the nebular spectrum is of the emission type, an explanation of the luminosity by simple reflection of starlight is out of the question, since the nebular spectrum is then entirely different from that of the star. H. N. Russell[3] has suggested that such emission nebulae are excited in some way by radiation, either short wave-length or corpuscular, the most reasonable source of such radiation being a very hot body, such as a star of early type.

An extended study by E. P. Hubble[4], based on a large amount of observational material, establishes beyond any reasonable doubt that the luminosity of the different types of galactic nebulae is due to neighboring stars. In the first of his two papers a classification of nebulae is given, and he finds that with practically every luminous nebula a star is associated which can be held responsible for the nebular luminosity. There is a definite relation, which holds with very few exceptions, between the spectral characteristics of this star and of the nebulosity; stars of type B1 and later are usually associated with diffuse nebulosity of the continuous type, stars of type Bo and Oe5 with diffuse nebulosity of the emission type, and still earlier stars with planetary nebulae.

In the second paper a quantitative relation between the luminosity of a nebula and the luminosity of the associated star is derived by a statistical study of observational data.

This relation may be formulated as follows: (*a*) The light emerging from a patch of diffuse nebulosity having a continuous spectrum is the equivalent of the starlight intercepted (star type B1 and later).[5] (*b*) The light emerging from a patch of diffuse nebulosity having an emission spectrum is the equivalent of the starlight intercepted (stellar type notably Bo and Oe5). (*c*) The light emerging from a part of the nebulosity in a planetary nebula is, on the average, 4 to 5 mag. brighter than the equivalent of the starlight intercepted (type of central star earlier than Oe5).

Quantities of light are called "equivalent" if they produce the same blackening on the photographic plate under the conditions of observation.

This empirical relation will be our guide in discussing the theoretical mechanism by which the luminosity of the nebula may be caused.

2. LUMINOSITY PRODUCED IN A NEBULA CONSISTING OF HYDROGEN BY A STAR EMITTING BLACK-BODY RADIATION

Attention will be confined mainly to the production of an emission spectrum in the nebula. Apparently there are no cases known where a normal star of type B1 or later produces a nebular emission spectrum of appreciable intensity. The hotter stars of type Bo and Oe5 are able to produce a strong nebular emission spectrum. The stars in planetary nebulae, which have a higher temperature still, produce also a more intense nebular spectrum. This strongly suggests that the temperature of the associated star is the main factor in determining the intensity of the nebular emission spectrum.

In the discussion which follows the star will be replaced by a black body of surface temperature T. If the star be a sphere of radius R, the total radiation per second within the frequency interval $d\nu$ is

$$\pi R^2 c \rho_\nu \, d\nu, \tag{1a}$$

$$\rho_\nu = \frac{8\pi h}{c^3} \cdot \frac{\nu^3}{e^{h\nu/kT} - 1} \tag{1b}$$

$$h = 6.54 \times 10^{-27}, \quad k = 1.372 \times 10^{16},$$

the formula (1*b*) being Planck's formula for the energy density $\rho_\nu \, d\nu$ of black-body radiation.[6]

The total number of quanta emitted per second within $d\nu$ is obtained by dividing this by $h\nu$ and therefore becomes

$$\frac{8\pi^2 R^2 k^3}{c^2 h^3} T^3 \frac{x^2}{e^x - 1} dx, \tag{2}$$

$$x = \frac{h\nu}{kT}. \tag{2a}$$

If Ω be the solid angle subtended by a patch of nebulosity from the center of the star, the number of quanta within $d\nu$ intercepted per second by this patch is

$$\Omega \frac{2\pi R^2 k^3}{c^2 h^3} T^3 \frac{x^2}{e^x - 1} dx, \tag{3}$$

as is found by multiplying (2) by $\Omega/4\pi$.

The number of intercepted quanta which, under observational conditions, can effect the photographic plate will be taken as a rough measure of the photographic intensity of the intercepted starlight. The long wave-length limit of an ordinary photographic plate is about 5,050 A, that for the atmospheric absorption is about 3,300 A corresponding to the frequencies $\nu_1 = 5.95 \times 10^{14}$ per second and $\nu_2 = 9.10 \times 10^{14}$ per second. If a reflector is used, only quanta having frequencies between those two limits can affect the photographic plate. The number of such "photographic quanta" in the starlight intercepted by the patch of nebulosity per second is, therefore,

$$N_{ph} = \Omega \frac{2\pi R^2 k^3}{c^2 h^3} T^3 \int_{x_1}^{x_2} \frac{x^2}{e^x - 1} dx, \tag{4}$$

$$x_1 = \frac{h\nu_1}{kT}, \tag{4a}$$

$$x_2 = \frac{h\nu_2}{kT}, \tag{4b}$$

which is an approximate measure of the photographic intensity of the starlight intercepted.

The mechanism by which the radiation intercepted by a patch of nebulosity can be transformed into an emission spectrum will now be considered. The observed emission spectra, in which the Balmer series of hydrogen is very conspicuous, indicate that the diffuse nebulae are composed of certain gases, of which atomic hydrogen is the most active in producing the luminosity. Therefore the problem will be simplified by assuming the nebula to consist wholly of hydrogen, an appreciable amount of which is atomic. Such a nebula may absorb a part of the radiation which it intercepts, and for this only atoms in the normal state need to be considered.[7] The quantum theory provides two mechanisms by which this absorption may occur. In the first place, a hydrogen atom in its normal state may absorb the different frequencies of the Lyman series, and the atom thus excited will subsequently return to its normal state, either directly or in steps, the electron falling back to the different energy-levels before it reaches the lowest level. In doing this it re-emits a number of line spectra, among others the Lyman series and the Balmer series. The mechanism of ordinary excitation accounts in this way for some luminosity in the nebula. This effect, however, is much too weak to account for the total luminosity observed, and it will not be further considered.

The second, and much the more powerful, mechanism which the quantum theory provides is that of ionization and subsequent recombination. An atom in its normal state is capable of absorbing all the energy of a wave-length shorter than the head of the Lyman series, frequency $\nu_0 = 32.84 \times 10^{14}$ per second. The number of quanta per second thus absorbed from the ultra-violet starlight by a patch of nebulosity is, in view of (3),

$$N_{ul} = \Omega \frac{2\pi R^2 k^3}{c^2 h^3} T^3 \int_{x_0}^{\infty} \frac{x^2}{e^x - 1} dx, \tag{5}$$

$$x_0 = \frac{h\nu_0}{kT}, \tag{5a}$$

provided that the absorbing layer contains a sufficient number of normal atoms for the absorption to be practically complete.

For each quantum $h\nu$ thus absorbed, an electron, mass m, is knocked out of the atom and becomes a free electron with the velocity v given by the photo-electric equation

$$h\nu = W + \tfrac{1}{2}mv^2, \tag{6}$$

where $-W = -W_1 = h\nu_0$ is the energy of the first atomic level. These free electrons will recombine with the nuclei by falling back to one of the different levels, energies $-W_1$, $-W_2$, $-W_3$, etc., emitting the continuous spectrum at the head of the Lyman series, at the head of the Balmer series, at the head of the Paschen series, etc., according to the same equation (6). Subsequently the electrons on the higher levels 2, 3, 4, etc., will return to the normal level 1, either directly, under emission of the Lyman series, or in steps, emitting the other hydrogen lines, among others the Balmer series. In this way a number of continuous spectra and line spectra will be re-emitted, and this would account for the Balmer series in the nebular spectrum as well as the continuous spectrum sometimes observed at its head.

The next question is the intensity of the re-emitted spectra. To solve this completely would require a complete knowledge of the probabilities of recombination with the levels 1, 2, 3, etc., as well as of the probabilities of the transitions in the hydrogen atom, or at any rate of their ratios. Without going into details, however, a few general statements can be made. Any ionized atom will have to return, sooner or later, to its normal state, where the electron is in its first energy-level. The very last electron jump is therefore either a jump of a free electron to this first level or a jump from one of the higher levels to this first level, the corresponding frequencies emitted being those of the continuous spectrum at the head of the Lyman series and the lines of the Lyman series. Assuming a stationary state of the nebula, viz., that just as many atoms become ionized per unit time as return from the ionized to the normal state, then, since the number of electrons emitted photo-electrically from the first level is equal to the number of ultra-violet quanta absorbed, N_{ul}, the number of electrons falling back to the first level is also N_{ul}. Hence the number of quanta re-emitted by the nebula, either as the continuous spectrum at the head of the Lyman series or as lines of the Lyman series, is equal to N_{ul}, or symbolically

$$Ly_c + Ly = N_{ul}, \tag{7}$$

where Ly_c and Ly represent the numbers of quanta of the two different kinds.

Here Zanstra argues that practically all the Lyman quanta will ultimately leave the nebula as the first line of the Lyman series, $Ly\alpha$; and because the Balmer series transitions from level 2 precede the production of $Ly\alpha$; we have $Ly = Ly\alpha = Ba_c + Ba$, where Ba_c and Ba denote, respectively, the numbers of quanta emitted as the continuous spectrum at the head of the Balmer series and the lines of the Balmer series. This computation leads to

$$Ly_c + Ba_c + Ba = N_{ul}. \tag{9}$$

Therefore a number of quanta N_{ul} emerge from the patch of nebulosity, partly as Ly_c, partly as Ba_c, and partly as Ba.

Only those quanta with frequencies in the interval $\lambda_1 =$ 5,050 A to $\lambda_2 = 3{,}300$ A can affect the ordinary photographic plate under observational conditions. Hence the quanta Ly_c and the quanta of the first line of the Balmer series $H\alpha$ do not affect the photographic plate, but the remaining quanta Ba_c and the quanta of the other lines of the Balmer series $H\beta$, $H\gamma$, $H\delta$, etc., are all "photographic quanta." The number of these emerging from the nebula is therefore

$$N'_{ph} = N_{ul} - (Ly_c + H\alpha), \tag{10}$$

which is smaller than the number of absorbed ultra-violet quanta by the amount $Ly_c + H\alpha$.

Zanstra then notes that the intensity of the Hα line is comparable to the intensity of the lines Hβ and Hγ and that the quantities Ba_c and Ly_c are either negligible or on the same order as the number of quanta in the lines Hβ and Hγ. This means that, as far as order of magnitude considerations go,

$$N'_{ph} \sim N_{ul}. \tag{11}$$

Combining (11) with (4) and (5), the ratio of the number N'_{ph} of photographic quanta emerging from a patch of nebulosity to the number N_{ph} of photographic quanta in the intercepted starlight is found to be approximately

$$\frac{N'_{ph}}{N_{ph}} \sim \frac{N_{ul}}{N_{ph}} = \frac{\int_{x_0}^{\infty} \frac{x^2}{e^x - 1}\,dx}{\int_{x_1}^{x_2} \frac{x^2}{e^x - 1}\,dx}, \tag{12}$$

$$x = \frac{hv}{kT}, \quad v_0 = 32.84 \times 10^{14},$$

$$v_1 = 5.95 \times 10^{14}, \quad v_2 = 9.10 \times 10^{14}.$$

This is also a rough measure of the ratio L of the photographic intensity of the emission spectrum of a patch of nebulosity to the photographic intensity of the intercepted starlight.

If quanta of an entirely different frequency were compared, the assumption that the number of quanta between the two assigned limits is a measure of photographic intensity would not be allowed, even as an approximation. But in the final equation (12) the inaccuracy introduced by this assumption of "photographic quanta" is only due to the difference in distribution of intensity of the continuous spectrum of the star as compared with the nebular emission spectrum, both covering a large range of frequency, and here the procedure followed, although very inaccurate, should give the right order of magnitude for the luminosity ratio L.

Omitted here is a table that gives values of the integral occuring in equation (12) for different values of x.

Table 84.1 shows that the ratio L is a function of the temperature T of the associated star which increases very rapidly with T, hence if L is known from observations, the temperature of the associated star required by the mechanism of ionization and recombination may be calculated. For diffuse nebulae with emission spectra, where the stars are predominantly of the types Bo and Oe5, Hubble found empirically that the nebular luminosity is the equivalent of the intercepted starlight, and hence $L = 1$ (sec. 1, *b*). The corresponding temperature, as found from table 84.1, is approximately 33,000°. To account for the nebular luminosity observed, the theoretical

Table 84.1 Intensity ratios for different stellar temperatures*

T	$\int_{x_0}^{\infty}$	$\int_{x_1}^{x_2}$	Ratio $\int_{x_0}^{\infty} / \int_{x_1}^{x_2} \sim L$	Ratio in magnitudes
15,000°	0.0043	0.57	0.0075	5.31
20,000	0.032	.48	0.066	2.95
25,000	0.103	.38	0.271	1.42
30,000	0.216	.30	0.72	0.36
35,000	0.35	.25	1.41	−0.37
40,000	0.50	.20	2.50	−0.99
50,000	0.80	.150	5.4	−1.83
70,000	1.27	.083	15.3	−2.96
100,000	1.68	.044	38	−3.95
150,000	2.02	.020	101	−5.01
200,000	2.16	.0115	185	−5.68

* T is the temperature of the star producing the luminosity. The ratio in the fourth and fifth columns represents the approximate theoretical value for the ratio L of the photographic intensity of the nebula to the photographic intensity of the starlight intercepted.

mechanism thus requires an average temperature of 33,000° for Bo and O stars associated with diffuse nebulae. This value of the temperature seems quite reasonable, and serves as a first check of the proposed mechanism of luminosity. In the next section, a more detailed discussion of the observational data will be given.

3. Comparison with Observation: Temperature of Bo and O Stars

Here Zanstra reviews Hubble's[4] empirical estimates for L and uses average values of $L = -0.26$ and $+0.72$ mag. to obtain stellar temperatures of 34,000° and 28,000°, respectively, for O and B0 stars.

No detailed discussion of observational data for planetary nebulae will be undertaken. However, the empirical material definitely indicates that the photographic intensity of the nebula may be greatly in excess of that of the intercepted starlight. In the planetary nebulae the "nebulium" spectrum of unknown origin becomes predominant, and hence they cannot be treated satisfactorily with the approximations that have been introduced. The significant fact remains, however, that the mechanism of ionization and recombination can account for a photographic nebular luminosity greatly in excess of that of the intercepted starlight, as is shown by table 84.1. Although no quantitative statement can be made at present, there is no doubt that the mechanism requires, on the average, a temperature of the central star of planetaries very much higher than 33,000°. This is in line with present evidence, e.g., from the distribution of intensity in the continuous spectrum of the central star.[8]

Unpublished observations indicate that a higher temperature in the exciting star is associated with a strengthening of the first lines of the Balmer series, not the higher lines. This finding is interpreted as indicating a larger recapture probability for lower levels when the velocities of the free electrons increase. We omit a discussion of Kramers's[9] formulae for these recapture probabilities; they indicate that electrons with large velocities are to a great extent captured by the lower levels. Milne's[10] calculations indicate that the relative distribution of captured electrons in different levels is independent of the electron velocity, but these calculations seem to disagree with observations of nebulae.

4. Nebulosity of the Continuous Type: Temperature of a Star B1

For stars of a normal spectral type later than B1, the stellar temperatures are lower than 19,000°, and observable emission lines are not expected. The observed continuum spectrum in these cases is assumed to be starlight reflected by solid or liquid particles.

Once convinced that the consequences of the proposed mechanism fit in fairly well with the present observations and theories, the temperatures which were obtained may be considered on their own merit. The luminosity of nebulae provides a method for obtaining the temperatures of stars. An advantage of this method is that physical conditions in a nebula are very simple on account of the extremely low concentration of the nebular atoms. Certain points neglected in the present treatment may be taken into account when the physical theories of recombination and probability of transition become more reliable and more definite. The "nebulium" spectrum offers additional difficulties. The assumption that the ultra-violet starlight is practically complete may need further check. Hubble's statistical treatment of diffuse nebulae followed in this paper is based on a theoretical average, which is only approximate. The greatest source of error, however, appears to arise from the difference in distribution of intensity of nebular spectrum and star spectrum, where the number of quanta between certain limits of frequency is introduced as an approximate measure of photographic intensity. This error may be removed by the introduction of better photographic methods. It is hard to give an estimate of the accuracy of the results, but it is believed that the total error may amount to 1 or even 1 1/2 mag., which corresponds to a rather wide range in temperature, according to the first and last column of table 84.1.

1. *Publications of the Astronomical Society of the Pacific 30*, 63 (1918).
2. *Astronomische Nachrichten 195*, 448 (1913).
3. *Observatory 44*, 72 (1921); *Proceedings of the National Academy of Sciences 8*, 115 (1922).
4. *Astrophysical Journal 56*, 162, 400 (1922).
5. Hubble's interpretation implies that the light emerging from the nebula is re-emitted evenly in all directions. A different result will therefore be obtained, if the continuous spectrum is assumed to be due to reflection of the starlight by the nebula. This question will be discussed in sec. 4.
6. Well-known considerations of radiation equilibrium lead to the following result: The energy within $d\nu$ emitted by an area $d\sigma_1$ of a black body of temperature T and intercepted

by a surface element $d\sigma_2$ at distance r is $I_\nu \, d\nu \, d\sigma_{1n} \, d\sigma_{2n}/r^2$, where $I_\nu = \rho_\nu c/4\pi$ and $d\sigma_{1n} \, d\sigma_{2n}$ are the projections of $d\sigma_1$ and $d\sigma_2$ on a plane perpendicular to r. The formula (1*a*) for the radiation emitted by a spherical body is easily obtained from this by considering as σ_1 the surface of the black body and as σ_2 a large concentric sphere, through which the radiation passes outward.

7. Because the incident radiation has a very low density, it may reasonably be expected that the fraction of neutral atoms in the excited states is negligible, although it must be admitted that the present knowledge of the statistics of the atom is insufficient to decide this question.

8. See, for example, W. H. Wright, *Publications of the Lick Observatory 13*, 252, 253 (1918).

9. *Philosophical Magazine 46*, 836 (1923).

10. Ibid. *47*, 209 (1924).

85. The Origin of the Nebular Lines and the Structure of the Planetary Nebulae

Ira Sprague Bowen

(*Astrophysical Journal 67*, 1–15 [1928])

In a great procession of triumphs, nineteenth-century spectroscopists identified the celestial spectral lines, one after another, with terrestrial elements. Undoubtedly the most dramatic success came with the discovery of helium in the chromospheric spectrum of the sun; detected at an eclipse in 1868, the previously unknown element was finally confirmed in a terrestrial laboratory in 1895. Nevertheless, until 1928 two of the strongest lines in the emission spectra of nebulae, the so-called nebulium lines at 5,007 and 4,960 Å, defeated attempts to identify them with elements known on earth.

In this brilliant, lucidly presented piece of logical detective work, Ira Bowen systematically narrows down the possibilities until he can conclude that the nebulium lines must arise from doubly ionized oxygen, even though strongly differing physical conditions had prevented observation of these lines in terrestrial laboratories. Earlier Henry Norris Russell, in the Russell, Dugan, and Stewart textbook, had already argued that the unusually low densities of the nebulae might account for the nebular lines, and Bowen's solution depends on the rarity of atomic collisions in the low density nebulae. This rarification allows the occurrence of "forbidden" transitions—transitions not really forbidden, but so improbable that they seldom take place in a higher-density laboratory situation, where an atom will almost always be jostled into a different state before the forbidden radiation can be emitted.

Bowen could calculate the wavelengths of the forbidden lines from the known energy levels of the ions, and thus he could identify the green nebular lines with doubly ionized oxygen, [O III] in modern notation, where the square brackets designate forbidden transitions. In his analysis Bowen was able to match other previously unidentified nebular lines with [O II], [N II], O III, and N III. The unusual strength of the latter two groups of lines remained a mystery until seven years later, when Bowen explained them by a remarkable resonance-fluorescence pumping mechanism that depends on a series of chance coincidences in wavelengths between ionized helium and doubly ionized oxygen and nitrogen.[1]

Bowen's pioneering paper not only explained the mysterious nebulium lines in terms of known terrestrial elements but also set the stage for a similar explanation for the "coronium" lines seen during solar eclipse observations (see selection 21).

1. I. S. Bowen, *Astrophysical Journal 81*, 1(1935).

Abstract—Identification of nebular lines. Eight of the strongest nebular lines are classified as due to electron jumps from metastable states in N_{II}, O_{II} and O_{III}. Several of the weaker lines are identified with recently discovered lines in the spectrum of highly ionized oxygen and nitrogen.

Behavior of lines in nebulae. The lines thus identified are shown to behave in various nebulae in a way consistent with the foregoing classifications. A similar study of the few lines yet unknown makes it possible to estimate the stage of ionization from which they arise.

Structure of the planetary nebulae. On the basis of the foregoing identifications, the relative sizes and intensities of the monochromatic images of the planetary nebulae are explained by an extension and modification of the ideas developed by Zanstra for hydrogen in the diffuse nebulae.

Identification of Nebular Lines

In the spectrum of the gaseous nebulae several very strong lines are observed that have not been reproduced in the laboratory. At first these lines were ascribed to some unknown element "nebulium." From the character of the other elements, *H*, *He*, *C*, *N*, and *O*, known to be present in the nebulae it was quite certain that these lines must be emitted by some element of low atomic weight. The development of present ideas concerning atomic structure, however, leaves no place for an unknown element of low atomic weight. Further, Wright's studies[1] of the relative intensities of these lines in a large number of nebulae show that the various lines behave in such widely different ways that they can hardly all come from the same element.

All these considerations lead to the conclusion, expressed by H. N. Russell, that "it is now practically certain that they must be due not to atoms of unknown kinds but to atoms of known kinds shining under unfamiliar conditions."[2] This unfamiliar condition Russell suggests to be low density.[3]

One type of line, which would be possible only under conditions of very low density, is that produced by an electron jump from a metastable state. A metastable state may be considered to be one from which jumps are very improbable, i.e., one whose mean life before spontaneous emission is very long. Under laboratory conditions the mean time between impacts of a given atom with other atoms or with the walls is, even in the most extreme cases, only 1/1,000 second. Consequently, an atom in a metastable state will, in general, be dropped down to a lower state by a collision of the second kind long before it will be able to return directly with the emission of radiation. In the nebulae, however, where the mean time between impacts is variously estimated at from 10^4 to 10^7 seconds, such atoms will return spontaneously and radiate a line with the frequency corresponding to the difference in energy between the metastable state and the final state. Since the probability of emission of these lines is very small, the probability of their absorption must also be small. Thus these lines should not be observed in absorption.

As stated above, *H*, *He*, *C*, *N*, and O are the only elements known to exist in the nebulae. C_I, N_I, N_{II}, O_I, and O_{III} are the only ions of these elements that have metastable states so placed that jumps from them would give rise to lines in the region of wave-lengths that is observable in nebulae. Since the low stages of ionization of these elements are not observed in the nebulae, C_I, N_I, and O_I can at once be eliminated. In a four-valence-electron system such as N_{II} and O_{III} the normal configuration of two 2s- and two 2p-electrons is characterized by ^{3}P-, ^{1}D-, ^{1}S-terms. Of these, $^3P_0$ is the stable ground state while the rest are metastable since any jump from them involves zero change in the azimuthal quantum number. In a five-electron system such as O_{II} the normal configuration of two 2s- and three 2p-electrons has ^{4}S-, ^{2}D-, ^{2}P-terms of which ^{4}S is the stable state and ^{2}D and ^{2}P are metastable.

The only cases where the differences between these terms are accurately known[4] are ^{1}D—^{1}S of O_{III} and ^{2}D—^{2}P of O_{II}. These are 22,916 and 13,646 frequency units which correspond to wave-lengths of 4,362.54 and 7,326.2 A, respectively. Two of the strongest nebular lines are found at 4,363.21 and 7,325 A. The deviation between calculated and observed values correspond to about 3 frequency units. Since the foregoing wave-lengths were calculated from the difference in frequencies of lines in the region between 500 and 800 A, this corresponds to an error of only about .01 A in these lines. The group at 7,325 A should have three components with an extreme separation of about 10 A. As the only observations of this line were made with an instrument having a dispersion of 600 A per millimeter, it is not surprising that the line was not resolved.

The ^{4}S—^{2}D-group in O_{II} can be predicted roughly from the difference between the term values. As no intercombinations between quartets and doublets have been observed, this difference depends solely on the independent adjustment of the terms by series formulae and therefore is only approximate. This difference predicts a pair at 27,157 and 27,175 frequency units, or 3,681.25 and 3,678.81 A. The two strongest ultra-violet nebular lines are 3,728.91 and 3,726.16 A. In view of the uncertainties mentioned above, the agreement in position and separation are both satisfactory.

Two other nebular pairs are at 5,006.84, 4,958.91 and 6,583.6, 6,548.1 A. The frequency separation of these are 193 and 82.3 cm^{-1}, respectively, while the separation of $^3P_1$–$^3P_2$ in O_{III} is 192 cm^{-1} and in N_{II} 82.7 cm^{-1}. This quite certainly identifies these pairs as $^3P_2$—$^1D_2$ and $^3P_1$—$^1D_2$ of O_{III} and N_{II}. In the case of N_{II} a check on this identification is possible as A. Fowler and L. J. Freeman[5] have found inter-combinations between singlets and triplets which enable them to fix certain of the singlet levels relative to the triplet levels. The foregoing identification of the pair of nebular lines as ^{3}P—^{1}D enables one to fix the position of ^{1}D relative to these same triplet levels and then to calculate accurately the frequency of the combinations between the ^{1}D- level and certain of the singlet levels found by Fowler and Freeman. These calculations predict that the combination with the ^{1}P-level of the s^{2}p·s-configuration should give rise to a line at 746.98 A and with the ^{1}D of s^{2}p·d to a line at 582.15 A. Two strong nitrogen lines that have not previously been classified occur on plates obtained in this laboratory at 746.97 and 582.16 A.

Of the other possible jumps from the metastable state of these ions $^4S—^2P$ of O_{II}, $^3P_1—^1S$ of O_{III}, and probably $^3P_1—^1S$ of N_{II} fall below 3,300 A where they cannot be observed in the nebulae. $^1D—^1S$ of N_{II} should occur with a wave-length somewhat greater than 5,500 A where photographic observations are difficult except in the case of the strongest line. Thus all of the lines expected on this hypothesis are found, and in so doing all but two or three of the strong nebular lines are explained.[6]

These nebular lines constitute the first direct experimental evidence for the idea, expressed above, that metastable states are not absolutely metastable but are states whose mean life is long before spontaneous radiation begins, the electron being able to return from them after the proper lapse of time even under conditions where the atom cannot be affected by surrounding atoms.

In addition to the lines identified as being produced by the foregoing mechanism, certain of the remaining weak lines can also be classified by direct comparison with the lines found in recent work in the laboratory on high stages of ionization. Thus in the data of Mihul[7] on O_{III} the five strongest lines in the region above 3,300 A are 3,312.35, 3,340.78, 3,444.15, 3,754.65, and 3,759.83 A. Of these, 3,754.65 A is probably obscured by H_κ, while the remainder correspond to 3,313, 3,342, 3,445, and 3,759 in Wright's list of nebular lines. Likewise, the nebular lines at 4,097.3, 4,634.1, and 4,640.9 correspond to lines[8] that Fowler has identified in N_{III}. Strong O_{II} lines may correspond to 4,076.22 and 4,416 A. That the lines mentioned in this paragraph are regular series lines rather than lines due to jumps from metastable states is confirmed by the observation that most of these lines are also present in the spectrum of the nuclei of the planetary nebulae and in other very hot stars.

Table 85.1 is a summary of all available data on the origin, series classification, and excitation potential of all lines given by Wright.

BEHAVIOR OF LINES IN NEBULAE

Here Bowen reproduces Wright's table of the relative intensities of emission lines for different nebulae, calling attention to the fact that lines identified with ions of similar ionization potentials exhibit similar characteristics. He then notices that several of the emission lines observed by Wright remain unidentified and reasons that the unidentified emission line pairs $\lambda\lambda$3,869, 3,967 and $\lambda\lambda$3,346, 3,426 must come, respectively, from an ion of low ionization potential and an ion whose ionization potential is greater than that of O III. The two line pairs have subsequently been identified with [Ne III] and [Ne V], respectively. Other previously unidentified lines now identified include $\lambda\lambda$4,069, 4,076 from [S II] and $\lambda\lambda$4,711, 4,740 from [Ar IV].

STRUCTURE OF THE PLANETARY NEBULAE

Since the foregoing identifications determine the chief constituents of the nebulae, it now becomes possible to attempt a consideration of the mechanism of radiation and the distribution of ionization in the different parts of the planetary nebulae.

As a first approximation, consider that a star whose surface temperature is 150,000° C. is surrounded by oxygen of very low density. As Zanstra[9] has pointed out, most of the absorption of radiation in a gas is photo-electric absorption of light whose frequency is greater than the ionization frequency of the atom or ion, each quantum absorbed completely ejecting an electron from the atom. Owing to the very great absorption coefficient of any atom or ion for this radiation of frequency greater than that necessary for ionization, no atom or ion can exist long in a region where such radiation is present in appreciable strength. It is evident from figure 85.1, which represents the distribution in frequency of the quanta emitted by a black body at 150,000° C., that the oxygen in the neighborhood of the star cannot exist in a stage of ionization lower than O^{++++}.

Occasionally, however, an electron unites with an O^{++++} ion and in so doing emits the O_{IV} spectrum. It is then almost immediately ejected again by the radiation below 160 A. This ejection of electrons made necessary by their continual return to the O^{++++} ion rapidly absorbs the light below 160 A while allowing longer wave-lengths to pass unaffected. After traversing a certain distance, the radiation of wave-length

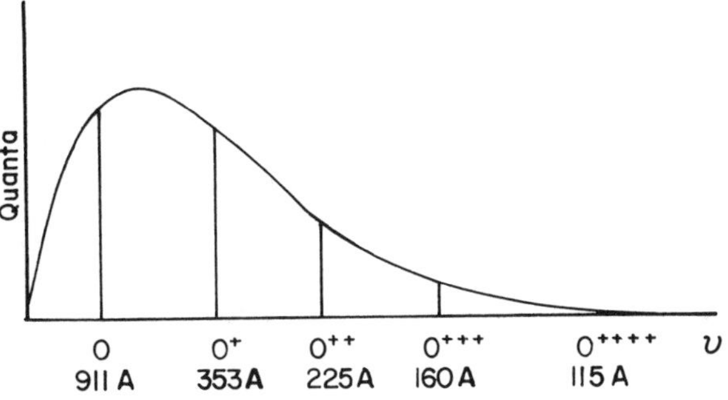

Fig. 85.1 Distribution in frequency of quanta emitted by a blackbody at 150,000 K.

Table 85.1 Series classification of nebular lines

λI.A.	Source	Series designation[a]	Excitation potential[b]
3,313	O_{III}	$3k^3P_1—3m^3S$	36.72
3,342	O_{III}	$3k^3P_2—3m^3S$	36.72
3,346	O_{IV}?	$3m^2P—3n^2D$	51.76
3,426.2	N_{IV}?	$3^3S—3^3P$	47
3,445	O_{III}	$3m^3P_2—3n^3P_2$	40.66
3,704	H_ξ, He_I	$2^3P—7^3D$	13.49, 24.22
3,712	H_ν	—	13.49
3,722	H_μ	—	13.48
3,726.12	O_{II}	$a^4S—a^2D_2$	3.31
3,728.91	O_{II}	$a^4S—a^2D_3$	3.31
3,734	H_λ	—	13.47
3,750	H_κ	—	13.45
3,759	O_{III}	$3k^3P_2—3m^3D_3$	36.19
3,771	H_ι	—	13.43
3,798	H_θ	—	13.41
3,820	He_I	$3^3P—6^3D$	24.11
3,835.5	H_η	—	13.38
3,840.2	—	—	—
3,868.74	—	—	—
3,888.96	H_ζ, He_I	$2^3S—3^3P$	13.33, 22.92
3,935	Ca_{II}?	$4^2S—4^2P_2$	3.14
3,964.8	He_I	$2^1S—4^1P$	23.65
3,967.51	—	—	—
3,970.08	H_ε	—	13.27
4,009	He_I	$2^1P—7^1D$	24.22
4,026.2	He_{II}, He_I	$2^3P—5^3D$	53.89, 23.95
4,064	—	—	—
4,068.62	—	—	—
4,076.22	O_{II}?	$3m^4D_4—3n^4F_5$	28.60
4,097.3	N_{III}	$3k^2S—3m^2P_2$	30.35
4,101.74	H_δ, N_{III}	$3k^2S—3m^2P_1$	13.17, 30.34
4,120.6	He_I	$2^3P—5^3S$	23.88
4,144.0	He_I	$2^1P—6^1D$	24.12
4,200	He_{II}	—	53.76
4,267.1	C_{II}	$3n^2D—4^2F$	20.87
4,340.46	H_γ	—	13.00
4,353	—	—	—
4,363.21	O_{III}	$a^1D—a^1S$	5.33
4,388.0	He_I	$2^1P—5^1D$	23.95
4,416	O_{II}?	$3k^2P_{1,2}—3m^2D_{2,3}$	26.18
4,471.54	He_I	$2^3P—4^3D$	23.64
4,541.4	He_{II}	—	53.54
4,571.5	—	—	—
4,634.1	N_{III}	$3m^2P_1—3n^2D_2$	33.01
4,640.9	N_{III}	$3m^2P_2—3n^2D_3$	33.01
4,649.2	C_{III}	$3^3S—3^3P$	30.1
4,658.2	—	—	—
4,685.76	He_{II}	—	50.82
4,711.4	—	—	—

Table 85.1 *(continued)*

λI.A.	Source	Series designation[a]	Excitation potential[b]
4,712.6	He_{I}	$2^3P—4^3S$	23.50
4,725.5	—	—	—
4,740.2	—	—	—
4,861.32	H_{β}	—	12.70
4,922.2	He_{I}	$2^1P—4^1D$	23.64
4,958.91	O_{III}	$a^3P_1—a^1D_2$	2.50
5,006.84	O_{III}	$a^3P_2—a^1D_2$	2.50
5,017	He_{I}	$2^1S—3^1P$	23.00
5,411.3	He_{II}	—	53.10
5,655	—	—	—
5,737	—	—	—
5,754.8	—	—	—
5,875.7	He_{I}	$2^3P—3^3D$	22.98
6,302	—	—	—
6,313	—	—	—
6,364	—	—	—
6,548.1	N_{II}	$a^3P_1—a^1D_2$	1.89
6,562.79	H_{α}	—	12.04
6,583.6	N_{II}	$a^3P_2—a^1D_2$	1.89
6,677	He_{I}	$2^1P—3^1D$	22.98
6,730	—	—	—
7,009	—	—	—
7,065	He_{I}	$2^3P—3^3S$	22.63
7,138	—	—	—
7,325	O_{II}	$a^2D—a^2P$	5.00

a. The electron configurations are indicated as follows: Terms arising from the most stable configuration of (2) 2s- and (n – 2) 2p-electrons ("n" is the number of valence electrons remaining in the atom) are indicated by "a"; from (2) 2s, (n – 3) 2p-electrons, and an excited s-electron by "k"; from (2) 2s, (n – 3) 2p-electrons, and an excited p-electron by "m"; from (2) 2s, (n – 3) 2p-electrons, and an excited d-electron by "n." The numeral preceding these letters indicates the total quantum number of the excited electron. In the case of one-and-two-valence-electron systems these letters are omitted since for all cases here considered the type of the term is the same as the type of the orbit of the excited electron.

b. The excitation potential given represents the energy necessary to place the ion or atom in a state where it can radiate the line under consideration.

below 160 A is so completely absorbed that O^{+++} can exist in the region outside this first circle. The return of electrons to this ion causes the O_{III} spectrum to be emitted in this second region. This electron is immediately ejected by the light in the 160–225 A region, but in so doing causes the rapid absorption of the remaining light below 225 A. This absorption becomes complete by the time the second circle is reached, thus enabling the O^{++} ion to exist in the third region. Returns of electrons to these ions give rise to the O_{II} spectrum, but in turn causes the light in the 225–353 A band to be completely absorbed by the time the third circle is reached. In a similar manner, the O^{+} ion can exist in the next region, and the return of an electron to it causes the O_I spectrum to be emitted but completes the absorption of all of the starlight up to 911 A.

Thus it is seen that this mechanism gives a series of concentric shells whose diameter increases with the decrease of the ionization potential of the ions present in them. This gives a qualitative explanation of the differences in the size of the monochromatic images of a planetary nebula as observed with a slitless spectrograph.

This simple theory would seem to indicate that the lines due to each stage of ionization receive the energy of the band between its ionization frequency and that of the next stage, i.e., O_I gets the energy from the 911–353 A range, O_{II} from 353–225 A, O_{III} from 225–160 A, O_{IV} from 160–115 A, etc. If this is correct, one can then follow Zanstra[10] and assume that the number of quanta emitted in the 3,300–5,100 A range for a given stage of ionization are approximately equal to those absorbed in its appropriate band in the ultra-violet. From the ratio of the intensities of these stages it should be possible to calculate the temperature of the exciting star.

Of course, in any actual nebula ions other than those of oxygen are present, but these can be grouped with the oxygen ion having an ionization potential similar to their own. Thus H_I can be classed with O_I since it has almost identically the same ionization potential, and He_{II} can be classed with O_{III}, etc.

Hydrogen exhibits somewhat different characteristics from other atoms, since it has only a single electron and consequently can be only singly ionized. Thus, whenever an electron returns it always emits the H_I spectrum and then absorbs light up to 911 A even though the hydrogen atom is in the circle nearest the star. Owing to the great increase in the absorption coefficient in the immediate neighborhood of the absorption edge, however, most of this absorption takes place from the region immediately below 911 A rather than in the region normally absorbed by the more highly ionized oxygen ions. Nevertheless, this may result in the complete absorption of the energy in the 353–911 A range before the outer region is reached, and therefore may account for the non-appearance of the O_I spectrum to be expected in that circle. It also accounts for the rather small monochromatic H-images of the nebulae.

Here we omit a table giving the relative number of photons available for the excitation of different oxygen ions for stars with different temperatures. Although the forbidden lines of [O III] are the most intense nebular emission lines, the ultraviolet radiation from stars with temperatures between 50,000° and 1,000,000° always produces more oxygen ions at ionization stages different from O III.

One possible cause of the abnormal strength of 5,007 and 4,959 and the other lines due to jumps from metastable states is that these are, in most cases, the true resonance lines of these ions and require only 1.8–5.3 volts for their excitation (see table 85.1). Thus an O^{++} ion can be excited to emit the 5,007 and 4,959 line by an electron with a velocity of only 2.5 volts. That many electrons capable of causing excitation in this way are present in the nebulae is to be expected from the fact that the band of frequencies corresponding to the difference between two successive stages of ionization has a width corresponding to about 20 volts, since in oxygen the successive ionization potentials are 13.6, 35.0, 54.8, 77.0, 107.7, etc., volts. This means that the electrons ejected photo-electrically from the ions will have an excess velocity of from 0 to 20 volts, which must, in general, be used up in exciting resonance radiation in other atoms before the electron is slowed up enough to attach itself to an ion.

This mechanism causes these resonance lines of O_{III} to be omitted when an O^{++} ion is hit by an electron and the O_{II} lines when an O^{+} ion is hit. Consequently, the resonance lines of any particular stage of ionization are emitted in the next region farther out from the star than the one where that particular stage of ionization is emitted by a returning electron. Thus the 5,007, 4,959 lines of O_{III} occur in the region where the regular series lines of O_{II} are emitted, since this O_{III} pair is emitted by the excitation of the O^{++} ion by a 2.5-volt electron, while the regular O_{II} lines are produced by the return of an electron to this same ion. This, of course, is the next region larger than the one where the regular O_{III} lines are produced.

These considerations explain the relative sizes of the various monochromatic images found in the planetaries. Thus 3,426.2 and 3,346 of N_{IV} and O_{IV} have the smallest. Next comes 4,686 of He_{II} and the 3,313, 3,342, 3,444 lines of O_{III}. These are followed by O_{III}, 5,007 and 4,959, with the 3,726 of O_{II}, the largest of all. The hydrogen lines vary somewhat from nebula to nebula, as would be expected from the differences mentioned above, but in general they are just smaller than 5,007 and 4,959.

The H-lines should also be stronger than would be expected from the absorption of the energy in the range 353–911 A, since every time an electron returns to O^{++}, O^{+++}, or O^{++++} at least one line is emitted in the region below 911 A which may be absorbed by a hydrogen atom. This enables the hydrogen lines to be excited indirectly by light leaving the star in the range below 353 A even after this light has been effective in the excitation of the appropriate O_{II}, O_{III}, or O_{IV} spectrum.

All of these secondary effects make it very difficult to perform a quantitative calculation as to the temperature of the exciting star. However, the great intensity of the He_{II} and the N_{IV} and O_{IV} lines which cannot be enhanced by any of these secondary effects require so much radiation in the range below 225 A that a temperature of at least 100,000°C must be postulated for the hottest nuclei.

The mechanisms suggested above show that in general no radiation having a frequency greater than the ionization frequency of hydrogen or a wave-length less than 911 A can leave the nebulae. All of the quanta of wave-length less than 911 A must be split up into quanta whose wave-lengths are greater than 911 A. Since a large part of the energy in the star's radiation is in the region around 250 A, this necessitates a manifold increase in the number of quanta given out by the nebula over the number received by it from the star. This helps to account for the very great brilliance of the planetaries relative to that of the central exciting star.

1. *Publications of the Lick Observatory 13*, 193 (1918); *Publications of the Astronomical Society of the Pacific 32*, 63 (1920).

2. Russell, Dugan, and Stewart, *Astronomy* (1927), p. 837.

3. Russell further states the reason why new lines might be emitted in a gas of low density as follows: "This would happen, for example, if it took a relatively long time (as atomic events go) for an atom to get into the right state to emit them, and if a collision with another atom in this interval prevented the completion of the process."

4. Bowen, *Physical Review 29*, 231 (1927).

5. *Proceedings of the Royal Society 114*, 662 (1927).

6. Since the foregoing article was written Croze and Mihul (*Comptes rendus 185*, 702 [1927]) and Russell and Meggers (*Physical Review 31*, 27 [1928]) have independently found intercombination lines in O_{II}. Russell, using his own identifications, and Fowler (*Nature 120*, 617 [1927]), the identifications of Croze and Mihul, have shown that the calculated frequencies of the ^4S–^2D lines differ from those of the nebular pair at 3,726, 3,729, by only 7 frequency units. This adds further confirmation to the identification made above.

Fowler (*Nature 120*, 582, [1927]) has also found intercombinations in O_{III} and used them to test the identification of the ^3P–^1D lines with the 5,007, 4,959 nebular pair, which was made above on the basis of their doublet separation. Using a somewhat better determination of the 374A line, viz., $\lambda = 374.03$A and $\nu = 267{,}358$ cm^{-1} than was available to Fowler, the agreement in frequency is well within the rather large experimental error in this extreme ultra-violet line.

Until a higher order of accuracy is obtainable in the extreme ultra-violet, this is as far as comparisons of the observed and calculated positions of these lines can be carried, since the identifications of all of the other lines had already been confirmed by direct comparisons given in this and also in the preceding preliminary articles (Bowen, *Physical Review 29*, 473 [1927], and *Publications of the Astronomical Society of the Pacific 39*, 295 [1927]).

7. Mihul, *Comptes rendus 183*, 1035 (1926); *184*, 89, 874, 1055 (1927).

8. Bowen, *Physical Review 29*, 231 (1927).

9. *Astrophysical Journal 65*, 1 (1927).

10. Ibid.

86. The Physical State of Interstellar Hydrogen

Bengt Strömgren

(*Astrophysical Journal 89*, 526–547 [1939])

The first calculation of the ionization of interstellar matter was carried out by A. S. Eddington in 1926, when he called attention to the fact that the ultraviolet radiation of hot stars will ionize the atoms of interstellar matter. Within a decade, Eddington had realized that interstellar hydrogen will absorb the radiation of hot stars before it has traveled very far, and as a result he changed his opinion about the transparency of interstellar matter. He then reasoned that most of the hydrogen in space will be in a neutral rather than an ionized form and that the distinction between ionized and neutral regions of hydrogen might be the criterion distinguishing the luminous and non-luminous regions of interstellar space.[1]

Soon thereafter Otto Struve and C. T. Elvey developed at the McDonald Observatory a fast, wide-field nebular spectrograph with which they detected large-scale regions of ionized hydrogen. In an attempt to reconcile this observational result with Eddington's conclusions, Bengt Strömgren in the paper presented here derives the relationship among the density of interstellar hydrogen, the temperature of the exciting star, and the radius of the sphere of ionization. He finds that although in general the hydrogen is neutral, very hot stars (particularly O and early B stars) can generate enormous but sharply bounded spheres of ionization. For some years these regions were referred to as "Strömgren spheres," but now they are almost invariably called H II regions, although their size is designated as the Strömgren radius.

The concept of the H II region played a fundamental role in tracing the spiral structure of the Milky Way (see selection 93) and has likewise been useful as a distance indicator for comparatively nearby spiral galaxies. Because the Strömgren radius is much smaller than the extent of our galaxy, the vast majority of interstellar matter consists of cold clouds of neutral hydrogen. Twelve years after Strömgren's pioneering treatment, radio astronomers used observations of the 21-cm line of neutral hydrogen to show that these implications of Strömgren's paper are correct (see selection 92).

1. A. S. Eddington, *Proceedings of the Royal Society* (London) *A111*, 424 (1926), and *Observatory 60*, 99 (1937).

Abstract—The discovery, by Struve and Elvey, of extended areas in the Milky Way in which the Balmer lines are observed in emission suggests that hydrogen exists, in the ionized state, in large regions of space. The problem of the ionization and excitation of hydrogen is first considered in a general way. An attempt is then made to arrive at a picture of the actual physical state of interstellar hydrogen. It is found that the Balmer-line emission should be limited to certain rather sharply bounded regions in space surrounding O-type stars or clusters of O-type stars. Such regions may have diameters of about 200 parsecs, which is in general agreement with the observations. Certain aspects of the problem of the ionization of other elements and of the problem of the relative abundance of the elements in interstellar space are briefly discussed. The interstellar density of hydrogen is of the order of $N = 3\ \mathrm{cm}^{-3}$. The extent of the emission regions at right angles to the galactic plane is discussed and is found to be small.

I

THE RECENT DISCOVERY, by Struve and Elvey,[1] of extended regions in the Milky Way showing hydrogen-line emission has opened up new and highly important possibilities for the study of the properties of interstellar matter. From the observed strength of $H\alpha$ in the emission regions, Struve[2] has calculated the density of interstellar hydrogen. Also, he has analyzed the problem of the ionization of interstellar calcium and sodium, taking account of the presence of ionized hydrogen.

In the present paper the problem of the ionization and excitation of interstellar hydrogen is first considered in a general way. Based upon the observed intensities of interstellar hydrogen emission, an attempt is then made to arrive at a picture of the actual physical state of interstellar hydrogen. Finally, certain aspects of the problem of the ionization of other elements and of the problem of the relative abundance of the elements in interstellar space are briefly discussed.

II

Eddington[3] has considered the problem of the ionization of interstellar hydrogen and has expressed the opinion that, in a normal region of interstellar space, hydrogen is entirely un-ionized, the reason being that the ionizing ultraviolet radiation is strongly absorbed by the interstellar hydrogen. The result of the analysis given below tends to confirm Eddington's view with the modification, however, that high-temperature stars, and especially clusters of such stars, are capable of ionizing interstellar hydrogen in regions large enough to be of importance in problems of interstellar space.

In the immediate neighborhood of a star, interstellar hydrogen will be ionized. With increasing distance from the star the proportion of neutral hydrogen atoms increases, and hence the absorption of the ionizing radiation increases. Ultimately, the ionizing radiation is so much reduced that the interstellar hydrogen is un-ionized. Our problem is to derive an expression for the extent of the ionized region as a function of the temperature and absolute magnitude of the star and the density of interstellar hydrogen. A somewhat similar problem has been considered by the author in another connection.[4]

Consider a point in interstellar space at the distance s from a star of temperature T and radius R. Let the number of neutral hydrogen atoms per unit volume be N'; the number of hydrogen ions, N''; and that of free electrons, N_e. The degree of ionization at s is determined[4] by the equation

$$\frac{N''N_e}{N'} = \frac{(2\pi m_e)^{3/2}}{h^3}\frac{2q''}{q'}(kT)^{3/2}e^{-I/kT}\cdot\sqrt{\frac{T_{\mathrm{el}}}{T}}\cdot w\,e^{-\tau_u}. \quad (1)$$

Here I is the ionization potential; q'' and q' are the statistical weights of the ion and the ground state of the atom, respectively; $\sqrt{T_{\mathrm{el}}/T}$ is a correction factor to allow for the difference between the temperature T of the exciting star and the electron temperature T_{el} at s (cf. Rosseland[5]); while $e^{-\tau_u}$ measures the reduction in the ionizing ultraviolet radiation due to absorption, τ_u being the optical depth from the ionizing star to the point s. Finally, w is the dilution factor at s, given by

$$w = \frac{R^2}{4s^2}. \quad (2)$$

We shall assume that practically all the free electrons are furnished by ionization of hydrogen, so that $N'' = N_e$. The validity of this assumption for actual interstellar space will be discussed in section V. Further, let

$$\left.\begin{aligned} N &= N' + N'', \\ N'' &= xN, \\ N' &= (1-x)N, \\ N_e &= xN, \end{aligned}\right\} \quad (3)$$

so that x is the degree of ionization of hydrogen. Introducing numerical values, equation (1) can now be written in the form

$$\frac{x^2}{1-x}N = C_1\cdot\frac{1}{s^2}\cdot e^{-\tau_u}, \quad (4)$$

with

$$\left.\begin{aligned} C_1 &= 10^{-0.51-\theta I}\cdot\frac{2q''}{q'}\sqrt{\frac{T_{\mathrm{el}}}{T}}\,T^{3/2}\cdot R^2, \\ \theta &= \frac{5040^\circ}{T}. \end{aligned}\right\} \quad (5)$$

The numerical factor of C_1 corresponds to the following

choice of units: 1 parsec $= 3.08 \cdot 10^{18}$ cm for s; the solar radius for R; and cm^{-3} for N.

Let the absorption coefficient for the ionizing radiation per neutral hydrogen atom be a_u. We simplify the problem by assuming a_u to be independent of the frequency and equal to its value at the absorption edge. The relevant range of frequency is relatively small, so that the accuracy will be sufficient for our present purpose. Then, by definition,

$$d\tau_u = (1 - x)Na_u \cdot 3.08 \cdot 10^{18}\, ds, \tag{6}$$

the factor $3.08 \cdot 10^{18}$ being derived from the choice of the parsec as the unit of s.

Equations (4) and (6) define the solution of the problem of deriving the degree of ionization as a function of the distance from the star. Solving (6) for $1 - x$ and substituting in (4), we get

$$e^{-\tau_u}\, d\tau_u = \frac{N^2}{C_1} x^2 s^2 \cdot 3.08 \cdot 10^{18} a_u\, ds. \tag{7}$$

We introduce the following new variables:

$$y = e^{-\tau_u} \qquad (1 \geqq y > 0), \tag{8}$$

$$dz = \frac{N^2}{C_1} \cdot 3.08 \cdot 10^{18} a_u \cdot s^2\, ds, \tag{9}$$

with $z = 0$ for $s = 0$, so that

$$s = \left(\frac{3C_1}{N^2 \cdot 3.08 \cdot 10^{18} a_u}\right)^{1/3} \cdot z^{1/3}. \tag{10}$$

Then (4) and (7) can be written as

$$\frac{dy}{dz} = -x^2, \tag{11}$$

$$\frac{1 - x}{x^2} = \alpha \frac{1}{y} z^{2/3}, \tag{12}$$

with

$$\alpha = \left(\frac{9}{NC_1 \cdot (3.08 \cdot 10^{18} a_u)^2}\right)^{1/3}. \tag{13}$$

Since $\tau_u = 0$ for $s = 0$, we have, according to (8) and (10),

$$y = 1 \quad \text{for} \quad z = 0. \tag{14}$$

In the cases of actual interest, α is a small quantity. When $\alpha \ll 1$, it follows from (12) that, as long as y is not small compared to 1, $1 - x \ll 1$, so that x is equal to 1, very nearly. In that case it follows from (11) and (14) that

$$y = 1 - z \quad \text{for} \quad 1 - x \ll 1. \tag{15}$$

Consequently, when α is small, ionization is strong until z is nearly equal to 1, so that (cf. [15]) y becomes a small quantity. With increasing z the degree of ionization now decreases very abruptly, so that, for z slightly greater than 1, hydrogen is practically un-ionized.

The value of s corresponding to $z = 1$, which we shall call s_0, is, according to (10),

$$s_0 = \left(\frac{3C_1}{N^2 \cdot 3.08 \cdot 10^{18} a_u}\right)^{1/3} \tag{16}$$

The result of the analysis can be stated as follows: for small α the ionization is nearly complete up to the distance from the ionizing star s_0, given by (16), while there is almost no ionization for distances greater than s_0.

Table 86.1 Degree of ionization, x, at different distances, s, from a central star

x^2	s/s_0		
	$\alpha = 0.001$	$\alpha = 0.01$	$\alpha = 0.1$
1.0	0.000	0.000	0.00
0.8	1.000	0.988	0.82
0.6	1.002	1.009	0.97
0.4	1.003	1.020	1.05
0.2	1.004	1.028	1.12

The abrupt change of the ionization can be interpreted in the following way. Once the proportion of neutral atoms begins to increase, the absorption of the ionizing radiation increases, leading to an accelerated increase of neutral atoms.

Table 86.2 Degree of ionization, x, for different distances, s, from a star ($\alpha = 0.01$)

z	s/s_0	$y = e^{-\tau_u}$	τ_u	$1 - x$	x^2
0.0	0.00	1.00	0.00	0.000	1.00
0.2	0.58	0.80	0.22	.004	0.99
0.4	0.74	0.60	0.50	.009	0.98
0.6	0.84	0.41	0.89	.017	0.97
0.8	0.93	0.22	1.52	.037	0.93
0.9	0.97	0.13	2.06	.064	0.88
1.0	1.00	0.046	3.07	.15	0.72
1.1	1.03	0.0018	6.3	.67	0.11

(For a somewhat more detailed discussion of this phenomenon see the investigation by the author quoted above.[4])

The exact dependence of the degree of ionization x upon the distance s from the ionizing star has been derived for three different values of α by numerical integration according to (11), (12), and (14). The results are shown in table 86.1. For $\alpha = 0.01$ the results are given in greater detail in table 86.2.

Finally, it may be noticed that for small α the following relation, obtained by integrating (11), (12), and (14), putting $z = 1$, holds very nearly in the transition region between almost complete ionization and negligible ionization:

$$z = \mathrm{Const} - \alpha\left\{\frac{1}{1-x} + 2ln\frac{x}{1-x}\right\}(\alpha \ll 1, |1-z| \ll 1). \quad (17)$$

Table 86.3 Degree of ionization, x, for different values of z

x^2	z + Constant Normalized to 0 for $x^2 = 0.5$	x^2	z + Constant Normalized to 0 for $x^2 = 0.5$
0.1	$+5.3\alpha$	0.6	-1.7α
.2	3.8	.7	4.2
.3	2.6	.8	8.6
.4	+1.3	.9	−20.1
.5	0.0		

Numerical values of z + constant, normalized to 0 for $x^2 = 0.5$ and calculated, according to (17), as a function of x^2, are shown in table 86.3. The table shows, for instance, that with $\alpha = 0.001$ the change from $x^2 = 0.5$ to $x^2 = 0.1$ takes place for a change in z equal to 0.0053, corresponding to a change in s of about one-sixth of 1 per cent.

For small α the relative extent of the transition region is proportional to α. From (13) and (16) it follows that αs_0 is independent of C_1, i.e., of the properties of the ionizing star. Therefore, as long as α is small, the absolute width of the transition region does not vary with the properties of the ionizing star.

Combining equations (5) and (16), we find

$$\left.\begin{aligned}\log s_0 = -6.17 + \tfrac{1}{3}\log\left(\frac{2q''}{q'}\sqrt{\frac{T_{\mathrm{el}}}{T}}\right) - \tfrac{1}{3}\log a_u - \tfrac{1}{3}\theta I \\ + \tfrac{1}{2}\log T + \tfrac{2}{3}\log R - \tfrac{2}{3}\log N.\end{aligned}\right\} \quad (18)$$

This relation holds for any element. Introducing the proper numerical values for hydrogen, viz., $q''/q' = \frac{1}{2}$, $I = 13.53$ volts, and $a_u = 6.3 \cdot 10^{-18}\ \mathrm{cm}^{-2}$, we get

$$\left.\begin{aligned}\log s_0 = -0.44 + \tfrac{1}{3}\log\left(\sqrt{\frac{T_{\mathrm{el}}}{T}}\right) - 4.51\theta + \tfrac{1}{2}\log T \\ + \tfrac{2}{3}\log R - \tfrac{2}{3}\log N.\end{aligned}\right\} \quad (19)$$

Table 86.4 gives $\log s_0$ for hydrogen for $R = 1$ and $N = 1$ as a function of T. The temperatures have been so chosen as to correspond to the spectral types from O5 to B5, according to the temperature scale recently derived by Kuiper.[6] In making the calculation the factor T_{el}/T has been put equal to 1. With $T_{\mathrm{el}}/T = 1/4$ (cf. Rosseland[5]), all values of $\log s_0$ would have to be decreased by 0.10, which would correspond to a decrease of the s_0-values of 21 per cent. Table 86.4 also gives α according to (13) for $R = 1$ and $N = 1$; α is proportional to $R^{-2/3}$ and to $N^{-1/3}$.

Table 86.4 Ionization radius, s_0, for stars of different spectral type and surface temperatures, T

Sp.	$\log T$	θ	$\log s_0$ for $R = 1$ and $N = 1$	s_0 (parsecs × $R^{2/3}N^{-2/3}$)	$\log \alpha$ for $R = 1$ and $N = 1$
O5	4.90	0.063	1.73	54	7.46—10
O6	4.80	.079	1.60	40	7.59
O7	4.70	.100	1.46	29	7.73
O8	4.60	.126	1.29	20	7.90
O9	4.50	.158	1.10	13	8.09
B0	4.40	.200	0.86	7.2	8.33
B1	4.36	.219	0.75	5.6	8.44
B2	4.31	.245	0.62	4.2	8.57
B3	4.27	.269	0.49	3.1	8.70
B4	4.23	.295	0.35	2.2	8.84
B5	4.19	.324	0.20	1.6	8.99

For any given temperature of the ionizing star, s_0 increases with the stellar radius R as $R^{2/3}$. If a region of interstellar space is ionized by a cluster of n similar stars, close together, then, s_0 has to be calculated with an equivalent R equal to $n^{1/2}R$. This follows immediately from equation (2) for the dilution factor w. Consequently, s_0 for such a cluster is equal to $n^{1/3}$ times s_0, calculated for the individual star in the cluster. This may also be expressed by saying that the volume of interstellar space ionized by a cluster of stars close together is equal to the sum of the individual volumes that would be ionized by the stars if placed so far apart in interstellar space that the volumes did not overlap.

Here we omit a table that gives values of s_0 for stars of different spectral classes whose radii have been calculated from their absolute magnitudes. The results are within a factor of 3 of those given in table 86.4, which were computed under the assumption that $R = N = 1$.

The increase in the ionized volume as one passes from low-temperature stars to high-temperature stars is so pronounced that it may be concluded that the total volume ionized by all high-temperature stars (hotter than about B2) is much larger than that ionized by all low-temperature stars, in spite of the much greater number of the latter. For instance, one O7 star ionizes a volume equal to that ionized by about two thousand B3 stars or by about five million A0 stars.

When the mean absorption coefficient is taken for all the relevant frequency range of the ionizing radiation, the value of s_0 will not differ more than 20–30% from the values derived by using the coefficient at the absorption edge. Considerations of the reemission of ionizing radiation by interstellar hydrogen will not change the estimates of s_0 by more than 15%.

Here we omit considerations of the excitation of higher energy levels of hydrogen in interstellar space, and some comparisons with observations. If the density of interstellar matter is taken to be a few hydrogen atoms per cubic centimeter, regions ionized by O and B stars have extents of a few hundred parsecs, and these extents are consistent with the observations of Struve and Elvey. We also omit considerations of the ionization of elements other than hydrogen.

1. *Ap. J. 88*, 364 (1938), *89*, 119 (1939).
2. *Proc. Nat. Acad. Sci. 25*, 36 (1939).
3. *M. N. 95*, 2 (1934); *Observatory 60*, 99 (1937).
4. G. P. Kuiper, O. Struve, B. Strömgren, *Ap. J. 86*, 570 (1937), sec. III, especially pp. 593 f. and 612.
5. *Theoretical Astrophysics* (Oxford, 1936), chap. xxii.
6. *Ap. J. 88*, 429 (1938).

87. Preliminary Results on the Distances, Dimensions, and Space Distribution of Open Star Clusters

Robert J. Trumpler

(*Lick Observatory Bulletin 14*, no. 420, 154–188 [1930])

Although this paper by Robert Trumpler is a lengthy and comprehensive study of open clusters, its most memorable impact came from the convincing demonstration of interstellar absorption. Ever since William Herschel's pioneering quantitative analysis of galactic structure, astronomers tended to take the transparency of space for granted. Jacobus Kapteyn, in the early stages of his investigation, adopted a uniform absorption of 1.6 mag kpc^{-1}, but later he rejected this idea in favor of an unobstructed view of the Milky Way.[1] Of course, the work of Max Wolf had shown the existence of discrete absorbing clouds, but apart from these special directions, students of galactic structure assumed no absorption of starlight. Harlow Shapley, writing on the eve of Trumpler's publication, admitted that the transparency of space was inferred rather than proved and conceded that "there is growing evidence of some general absorption or scattering in the direction of a few of the Milky Way star clouds."[2]

Trumpler's approach was first to observe slit spectra of individual stars in nearby open star clusters, thereby estimating their spectral types and absolute magnitudes. When distances calculated from the apparent and absolute magnitudes were compared with those inferred from the apparent angular diameters of the clusters, however, Trumpler found that either the diameters of the fainter clusters were systematically larger than those of the brighter ones, or their starlight was being absorbed by an amount depending on distance. By assuming that all clusters have the same linear diameter and that the magnitudes of the individual stars are overestimated because of the interstellar absorption of starlight, Trumpler obtained an average photographic absorption of about 0.67 mag kpc^{-1}.

An improvement on this absorption coefficient was subsequently obtained by Alfred Joy, when he noticed that the distances inferred from the period-luminosity relation for the Cepheids differed from those inferred from their velocities under the assumption that they participate in the differential rotation of our galaxy. He found that the two distances are in accord if the light from these stars is absorbed at the rate of 0.85 mag kpc^{-1} of interstellar matter (see selection 82).

These studies of interstellar absorption had major implications for studies of the shape and extent of our galaxy. For example, they meant that Shapley's estimate of the distance to the center of the globular cluster system was overestimated by a factor of almost 2, because he neglected interstellar absorption. Trumpler failed to recognize these implications, however, and he believed erroneously that his work delineated a spiral structure of the Milky Way with the sun near the center. He thereby rejected Shapley's model and the full implications of galactic rotation (see selections 79, 81, and 82). However, when photoelectric methods were subsequently used by Joel Stebbins, C. M. Huffer, and A. Whitford to determine the color excesses of O and B stars, it became clear that interstellar absorption of starlight was sufficient to account for the apparent discrepancy between the dimensions of the disk system of stars and the system of globular clusters; but their study showed at the same time that the absorption is so irregular and spotted that a constant coefficient of absorption cannot be used for any large region of interstellar space.[3]

1. J. C. Kapteyn, *Astronomical Journal 24*, 115 (1904), *Astrophysical Journal 29*, 46 (1909). See also selection 80.
2. H. Shapley, *Star Clusters* (New York: McGraw-Hill, 1930), p. 124.
3. J. Stebbins, C. M. Huffer, and A. E. Whitford, *Astrophysical Journal 91*, 20 (1940).

Although the observations of magnitudes and spectral types in open star clusters of the Milky Way undertaken by the writer are still far from being complete, it seemed of interest to utilize the data at present available for a preliminary investigation of the distances and diameters of these clusters and for a study of their space distribution.

Determination of the Distances of Clusters from Magnitudes and Spectral Types

The dimensions of most clusters are small compared with their distance from us. In any particular cluster we may thus assume that its members are at the same distance and that their absolute magnitudes M differ from their apparent magnitudes m by a constant:

$$m - M = 5 \log r - 5, \quad (1)$$

where r is the distance of the cluster in parsecs. Plotting the cluster members according to their apparent magnitudes and spectral types we obtain a diagram similar to the Hertzsprung-Russell diagram of giant and dwarf stars. Although the magnitude-spectral class diagrams of individual clusters vary considerably and although they are incomplete for the fainter stars, it is nearly always possible to decide which stars of a cluster belong to the giant or to the dwarf branch. In most clusters, moreover, only B and A type stars are observable and for these such separation is not necessary. If we assign to each cluster star the mean absolute magnitude corresponding to its spectral class and subtract it from the apparent magnitude, we obtain the distance r from formula (1).

The mean absolute magnitudes of the spectral subdivisions have been determined by various observers from trigonometric and moving cluster parallaxes or statistically from proper-motions. The values used are given in table 87.1 and are mainly based on the determinations of Adams & Joy,[1] Lundmark,[2] Malmquist,[3] and Hess:[4]

Table 87.1 Adopted mean absolute magnitudes for spectral types

Spectral type	Mean absolute magnitude: Dwarf branch, Vis.	Dwarf branch, Phtgr.	Giants, Vis.	Giants, Phtgr.
O	−4.0	−4.3	—	—
B0	−3.1	−3.4	—	—
B1	−2.5	−2.8	—	—
B2	−1.8	−2.1	—	—
B3	−1.2	−1.4	—	—
B5	−0.8	−1.0	—	—
B8	−0.2	−0.3	—	—
B9	+0.3	+0.3	—	—
A0	+0.9	+0.9	—	—
A2	+1.7	+1.7	—	—
A3	+2.0	+2.1	—	—
A5	+2.3	+2.5	—	—
F0	+2.9	+3.2	+0.5	+0.9
F2	+3.2	+3.5	—	—
F5	+3.6	+4.0	+0.5	+1.0
F8	+4.2	+4.7	—	—
G0	+4.5	+5.1	+0.5	+1.2
G5	+5.0	+5.7	+0.5	+1.4
K0	+6.2	+7.0	+0.5	+1.6

The application of this method for determining distances of clusters requires a knowledge of the spectral types and magnitudes for a number of stars in each cluster. Spectroscopic observations have so far been obtained for 57 clusters either with the 1-prism slit spectrograph attached to the 36-inch refractor for the brighter stars, or with the slitless quartz spectrograph attached to the Crossley reflector, or with both. In 43 other clusters, mostly situated too far south for observation from Mount Hamilton, the spectral types of some of the brighter stars are found in the *Henry Draper Catalogue*. This makes a total of 100 clusters for which spectroscopic data are available.

After giving details concerning apparent magnitudes, m, Trumpler lists the absolute magnitudes, M, of forty stars in the cluster Messier 36.[5,6] We omit here the detailed data, which have a mean of $m - M = 11.1$ and which, using equation (1), give a distance of 1,660 pc with a 10% error. Trumpler then notes that the ten brightest stars of M 36 give a considerably smaller value for $m - M$ than the thirty fainter ones. This behavior is noted for most clusters and is attributed to the fact that the brightest members of a cluster are of abnormally high luminosity for their spectral classes.[7] Some effect of this kind can be expected as a consequence of the luminosity dispersion of stars in any particular spectral class, but the rise always occurs at the hottest spectral types present in the cluster or among its brightest members.

The abnormally high luminosity of the brightest cluster members is probably due to large mass; it was corrected for empirically by the addition of $+1^{m}_{\cdot}0$ to the m-M for the stars in the first half-magnitude interval and of $+0^{m}_{\cdot}5$ for the stars in the second half-magnitude interval. These corrections were applied for all clusters which contain mainly stars belonging to the dwarf branch; their omission would be a serious source of error in cases where only a few of the brightest cluster stars are available for the determination of the distance. At the fainter magnitude limit to which the observations extend, the selection due to dispersion in absolute magnitude was neglected. As this dispersion seems to be small in the dwarf branch, and as the observations generally cover a considerable magnitude range, the error thus committed will be small and is partly compensated by the correction applied to the brightest stars.

The mean of the corrected values of m-M is then taken and the distance in parsecs is computed according to formula (1). When many stars covering a considerable magnitude interval were observed, as in Messier 36, the probable error of the mean m-M should be of the order of $\pm 0\overset{m}{.}2$; this includes the uncertainty of the magnitude scale and of the adopted mean absolute magnitudes (table 87.1). It corresponds to a *p.e.* of $\pm 10\%$ in the distance.

Here we omit a table, which gives the distances computed from the mean $m - M$, the linear diameters inferred from these distances, and the observed angular diameters for 100 open star clusters. The magnitude data come from references 8, 9, and 10; the apparent angular diameter data come from references 10, 12, 13, 14, and 15. The distances were calculated using equation (1), and the linear diameters, D, in parsecs were calculated from the apparent diameters, d, in minutes of arc using the relation $D = r \sin d = rd/3{,}438$.

Character of Magnitude-Spectral Class Diagram

In a former paper[11] attention has been drawn to the fact that the magnitude-spectral class diagrams of open star clusters differ considerably, although they always show some resemblance to the well known Hertzsprung-Russell diagram of giant and dwarf stars. In some clusters the giant branch is entirely missing, and the dwarf branch extends unequally far in the direction of the hotter types. A simple classification was proposed which describes, by the combination of a number and a letter, the peculiar character of the magnitude-spectral class diagram of a cluster. The number is based on the relative frequency of yellow and red giant or supergiant stars. 1: means that the giant branch is entirely missing, all cluster stars belonging to the main branch from O to M. 2: a relatively small number of stars in the giant branch. 3: the majority of the more luminous stars are yellow or red giants. The letter following the number is that of the spectral type of highest temperature reached by the main branch. In addition to the four main types 1b, 1a, 2a, 2f illustrated in the former publication two others have been introduced: 1o for clusters containing O type stars, and 3a for clusters with many red or yellow giants, but very few A type stars. Intermediate steps are also indicated; 1–2 mostly referring to cases in which it is doubtful whether the few yellow or red stars observed are physical members of the cluster.

Table 87.2 Frequency of different types of magnitude spectral class diagrams

	o	*b*	*b–a*	*a*	*a–f*	*f*	
1	7	24	5	3	—	—	39
1–2	3	15	10	3	—	—	31
2	—	1	5	18	1	1	26
2–3	—	—	—	3	—	—	3
3	—	—	—	1	—	—	1
Totals	10	40	20	28	1	1	100

Table 87.2 shows the frequency of the various types; the total number of clusters for which data are available being exactly 100, the numbers also express percentages. The most frequent types are 1b (*Pleiades*) and 2a (*Praesepe*), but there is also a well pronounced transition between them. Table 87.2 brings out plainly the peculiar feature in the distribution of these types, *i.e.*, a strong concentration along the diagonal from 1b to 2a. In other words open clusters which contain stars of highest temperature (types O and B) contain very few or no red or yellow giant stars (the few being generally supergiants); while clusters, in which types O and B are missing, most frequently contain an appreciable or even a considerable number of stars in the giant branch. The high percentage of clusters (50%) with O and B type stars is undoubtedly exaggerated by selection, as the great luminosity of these stars allows spectroscopic observations of very distant objects while only the nearer ones of type 2a are within reach of the spectroscope.

Determination of Linear Diameters (First Approximation)

Unfortunately the list of clusters which were included is very incomplete and selective. Owing to more favorable observing conditions in summer and autumn the spectroscopic and photometric observations are more complete for objects in galactic longitudes 330° to 150°, while in the opposite hemisphere practically none of the fainter and more distant clusters have been investigated. A study of the general space distribution of open star clusters should be based on a more complete number of clusters, and for most of these an estimate of the distance can at present be obtained only from the apparent diameter. A careful investigation of the linear diameters of the 100 clusters of known distance was therefore undertaken primarily for the purpose of finding a method to determine the distance of a cluster from its diameter.

The linear diameters were computed according to the procedure described previously.

In the next three sections Trumpler abandons his assumption that all open clusters have nearly the same linear dimensions and reasons that clusters of similar constitution must have the same dimensions. When the clusters are divided into subgroups according to the concentration and number of their stars, however, there is no relationship between linear diameter and subgroup classification.

Trumpler then shows that there is a pronounced correlation between the observed linear diameters and distances. This correlation is illustrated in table 87.3, which contains angular diameters, d, and average residuals, v', for clusters within different ranges of distances and angular diameters.

Still more convincing proof that our systematic error depends on the distance and not on the diameter is given in columns 6 and 9 of table 87.3 which contain the average residuals v' taken separately for two small intervals of apparent diameter; while the mean diameters (column 5 and column 8) vary very little in each column, the run of the residuals with the distance is just as pronounced as in the third column for all clusters taken together. As none of the observational errors discussed offers a plausible explanation of the observed discrepancy there are only two alternatives left; either to admit an actual change in the dimensions of open clusters with increasing distance or to assume the existence of an absorption of light within our stellar system. In favor of the first alternative it might be argued that the dimensions of clusters are possibly influenced by the star density of the space in which they are situated and that this star density on the whole decreases with greater distance from the center of the local system. If we consider, however, the amount in question it is extremely unlikely that any such influence could be powerful enough to nearly double the dimensions of remote clusters.

The assumption of an absorption of light on the other hand is not only able to give a satisfactory numerical representation of the residuals of table 87.3 but receives support also from color-index observations in clusters.

Absorption of Light in the Milky Way System

Our method of deriving cluster distances from magnitudes and spectral types was based on formula (1) which expresses the law that the apparent brightness of a star diminishes with the square of the distance from the observer. If interstellar space is not perfectly transparent this law does not hold; the apparent brightness decreases more rapidly, our distance results are too large, and the error increases with the distance of the cluster. The linear diameters computed with these distance results are then also too large, and the error also progresses with distance, just like the residuals in column 3 of table 87.3.

Let us suppose for the moment that the absorbing material is uniformly distributed throughout the galactic system. The loss of star light in magnitudes caused by absorption is then

$$\Delta m = kr,$$

where r is the star's true distance and k the absorption constant of photographically effective wavelengths. The relation between apparent and absolute magnitude of a star and its

Table 87.3 Dependence of diameter on distance

Observed distance in parsecs		Average residual		Cl. with app. diam. 10′–20′			Cl. with app. diam. 20′–40′			True distance in parsecs	Eff. of abs.	Residual	Second approxim. Average
Interval	Mean	v'	Weight	Mean d	Average v'	Weight	Mean d	Average v'	Weight	Mean	A	$v' - A$	v'
<500	294	−0.09	20	20′	−0.17	1	31′	−0.14	5	266	−0.09	0.00	−0.01
500–1,000	730	− .05	26.5	16	− .10	8.5	29	− .02	15.5	594	− .04	− .01	− .02
1,000–1,500	1,200	+ .01	13.5	15	− .01	5.5	29	+ .03	6	870	.00	+ .01	+ .01
1,500–2,000	1,620	+ .08	6	15	+ .07	4.5	22	+ .26	1	1,050	+ .03	+ .05	+ .05
2,000–3,000	2,460	+ .06	13.5	13	+ .11	3.5	27	+ .05	4	1,500	+ .10	− .04	− .03
>3,000	3,850	+ .19	8.5	15	+ .13	3	25	+ .33	0.5	1,890	+ .16	+ .03	+ .04

distance is accordingly

$$m - M = 5 \log r + kr - 5. \tag{3}$$

This equation must be substituted for equation (1) in the case of uniform absorption.

By considering the calculated and true linear diameters as well as the calculated and true distances, Trumpler is able to solve for the absorption coefficient k and to resolve the discrepancy between the two methods for finding the cluster distances. He obtains $k = 0.79$ mag kpc^{-1} and the absorption effect A for the observed linear diameters. $A = a + br$, where $a = -0.134 = k\bar{r}/5$ when $\bar{r}$ denotes the average distance, $b = 0.158 = k/5$, and r is the true distance in kiloparsecs.

It may therefore be stated that the apparent increase in the linear diameters of open star clusters is fully removed by the assumption of an absorption of light within our Milky Way system to the amount of 0.79 magnitudes per 1000 parsecs for photographically effective rays.

If the light of the stars in the more distant clusters has been dimmed by the passage through an absorbing material, it seems *à priori* likely that such absorption is selective, and varies with the wave length of the light and thus changes the color of the stars. Since the color of a star depends on its temperature a change of color by absorption can only be detected if its temperature is observable by some other means; *e. g.*, from the spectral types which measure the temperature by means of the ionization and excitation of the atoms in the stellar atmosphere. Spectral type estimates based on comparison of intensities of spectral lines are not affected by a general absorption. For the brighter and nearer stars which are little affected by absorption the relation between color indices and spectral types has been well established, but differs slightly for giant and dwarf stars. From this relation we can find the normal color index corresponding to a given spectral type. The difference between this and the color index actually observed is called the color excess of the star. The existence of large discrepancies between observed color indices and spectral types in open star clusters has been known for some time, but to test our hypothesis of an absorption of light in space it is necessary to show that the average color excesses in various clusters depend on their distance. The spectral types in open clusters determined with the 1-prism slit spectograph and the slitless quartz spectrograph made it possible to compute the average color excess for a number of clusters in which color-indices have been measured by various observers. Leaving out the few nearest clusters which are of little interest for our purpose and using only color indices determined by direct comparison with the North Polar sequence, seven clusters ranging widely in distance were available. Table 87.4 gives for each of these the observed distance r' (from magnitude and spectral types), the finally adopted distance which is corrected for absorption, the average color excess of the cluster stars, and the number of stars used.

The observed color indices given in table 87.4 come from references 16 through 23. Theoretical color excesses, E, are computed from $E = cr$, where r is the adopted corrected distance and $c = 0.32 \pm 0.03$ mag kpc^{-1}.

After discussing the possibilities of systematic error, Trumpler enunciates his principal conclusion: "We are thus led to the assumption of a general absorption of light in our stellar system by two quite independent sets of observations; by the study of the linear diameters of star clusters as well as by color-index observations in such

Table 87.4 Color-excess of open clusters

Cluster		Observed dist. r'	Adopt. corr. dist.	Color excess	Number of stars	Formula	Obs. − Form
NGC2682	Messier 67	690	740	$+0^m.26$	81	$+0^m.24$	$+0^m.02$
1647	—	800	610	+ .17	33	+ .19	− .02
				+ .19	6	+ .19	.00
2099	Messier 37	1,450	820	+ .05	25	+ .26	− .21
1960	Messier 36	1,650	980	+ .05	40	+ .31	− .26
6705	Messier 11	2,200	1,340	+ .65	46	+ .43	+ .22
7654	Messier 52	2,400	1,360	+ .49	43	+ .43	+ .06
663	—	3,500	2,170	+ .71	41	+ .69	+ .02

clusters." He then points out that the absorption may represent the average of local irregularities, and he also shows that the absorption does not depend to a great extent on galactic longitude.

It is natural to interpose here the question why such an absorption of light should not have been discovered in the discussion of the diameters of globular clusters which are much more distant, and how it is possible that we still find small color-indices in some globular clusters (such as Messier 13) despite of their great distances. There is only one way which seems to lead out of the dilemma: the hypothesis that the absorbing medium, like the open clusters, is very much concentrated toward the galactic plane. We shall see later that two-thirds of all open clusters lie within 100 parsecs of their plane of symmetry, and it is not improbable that the absorbing material has a similar distribution, thinning out very rapidly at greater distances from the galactic plane and forming so to speak a thin sheet (perhaps 200–300 parsecs thick) extending along the galactic plane to distances of at least 2,000 and perhaps 4,000 or 6,000 parsecs. It is evident that the absorption caused by a material of such distribution is practically negligible (less than $0^m.1$) for objects in high galactic latitudes. Only for objects lying within 8° of the galactic circle could the photographic absorption amount to $0^m.5$ or more and the color excess to $0^m.25$ or more. Of the relatively few globular clusters falling within these limits there are four in which variable stars or magnitudes of the brighter stars have been observed.[24] These four (NGC 6266, 6626, 6638, 6712) do indeed show some effect of absorption, that is, the distance determined from magnitudes is greater than that derived from the diameters. A more thorough investigation of the linear diameters of globular clusters is necessary, however, to draw any definite conclusions.

If our hypothesis is correct, some of the open clusters must also lie outside of this thin layer of absorbing material, and their light should be subject to absorption for part of the distance only. To show that this is the case we take the five clusters for which the distance Z' from the plane of symmetry of the clusters exceeds 200 parsecs. The absorption correction applied to the latter must have been too large; these clusters seem to be less affected by absorption than the majority of clusters at the same distance which lie close to the galactic plane.[25]

We are thus led to the conclusion that some general and selective absorption is taking place in our Milky Way system, but that this absorption is confined to a relatively thin layer extending more or less uniformly along the plane of symmetry of the system. Perhaps this absorbing material is related to interstellar calcium or to the diffuse nebulae which are also strongly concentrated to the galactic plane.

The change in the color of stars with distance exhibited by table 87.4 must be due to the fact that the absorption depends on the wave-length of light, being smaller for the longer wave length used in visual observations than for the shorter waves photographically effective. Designating by k_v the absorption constant per 1,000 parsecs of visual observations and by k_p that of photographic observations, the change of color c, previously determined, gives us the difference of the two absorption constants:

$$c = k_p - k_v = +0^m.32.$$

Combining this with the photographic absorption constant determined from the cluster diameters, we have

$$k_p = +0^m.67 \text{ per 1,000 parsecs}$$
$$k_v = +0.35.$$

The photographic absorption is nearly twice the visual one. This is not very different from the extinction in the Earth's atmosphere where the ratio of the photographic to the visual extinction is about 2.5.

Defining the absorption coefficient κ by the equation

$$I = I_0 e^{-\kappa t},$$

where I_0 is the intensity of the incident light, I that of transmitted light and t the length of the path in cm:

$$\kappa = 0.20 \times 10^{-21} \text{ for } \lambda = 4{,}300 \text{ A}$$
$$= 0.10 \times 10^{-21} \text{ for } \lambda = 5{,}600 \text{ A}.$$

Trumpler then recomputes the cluster distances, taking into account the effect of absorption; and the corrected distances are used to compute correct values for the linear diameters of the clusters. He finds that the vast majority of open clusters have diameters lying between 2 and 6 pc, that the cluster diameter increases with the number of members, and that for clusters with the same number of members the diameter decreases with increasing central concentration. He provides a new catalogue of 334 open galactic clusters, which we omit. Trumpler then considers the apparent and spatial distributions of open clusters, leading to the following treatment of the Milky Way structure, which is in sharp disagreement with the picture drawn by Shapley and with the view of a rotating system that had been emerging through the work of Lindblad and Oort.

The Structure of the Milky Way System

By the Milky Way system we wish to designate that large system which comprises practically all the stars visible to the naked eye or observable in large telescopes and which is particularly characterized by the dense accumulations of faint stars forming the conspicuous features of the Milky Way. To anybody who closely examines the beautiful Milky Way photographs of Barnard or the Franklin Adams chart it is quite apparent that the open clusters must be related to the star clouds of the Milky Way in which they often appear imbedded like condensations. We have already drawn attention to the similarity in the apparent distribution of stars and clusters except for the greater galactic concentration. Most convincing of all is perhaps the close agreement between the plane of symmetry of open clusters and that of faint stars.

In view of these facts it seems quite justifiable to make the hypothesis that the space distribution of open clusters is similar to that of stars in general and that a study of open star clusters may give us some information concerning the structure of the Milky Way system. As every test seems to indicate that our list of clusters is not much short of completeness we may expect them to represent a general outline of the Milky Way system. The features we brought out in the discussion of the open clusters should then for the most part apply also to the Milky Way system, and in fact they are generally in good agreement with the results of the statistical investigations of Seeliger,[26] Kapteyn[27] and others who describe the stellar system as a flattened lens shaped system 10,000–15,000 parsecs in diameter and 3,000–4,000 parsecs in thickness, with the stars concentrated toward the center and thinning out toward the edge. The only difference is that the clusters seem to be more strongly concentrated toward the galactic plane than the stars in general.

The close analogy between these views concerning the structure of our Milky Way system and the main features of many spiral nebulae led to the conclusion that the Milky Way system belongs to this class of objects. Our results concerning the open clusters quite support this conclusion. In some of the nearer spirals especially *M* 51 and *M* 101 we find numerous small nuclei not as well defined as star images which have all the earmarks of open star clusters and in *M* 33 some clusters are even partly resolvable into stars. The size of our Milky Way system suggested by the open clusters (10,000–12,000 parsecs in diameter) is well comparable with the dimensions of the *Andromeda* Nebula (13,000 parsecs) and Messier 33 (4,600 parsecs) according to Hubble's[28] investigations.

On the other hand it is true that the space arrangement of open clusters shows hardly any indication of spiral structure. The errors of observation in the cluster distances (p. e. 10–12%) will of course have the tendency to blur any existing spiral structure, and local obscuration by dark matter may at many places have interrupted the spiral arms. Despite of these disturbing influences it seems hardly possible to account for our diagram unless we assume that our Milky Way system is an extremely resolved spiral, even more resolved than Messier 33. Nevertheless there are some traces of spiral structure noticeable which have been drawn in figure 87.1. While they are not sufficiently prominent to prove the spiral structure of the Milky Way system, they may, once we admit the hypothesis of spiral structure, give an indication of its direction. The different fragments of branches especially between galactic longitude 60° and 180° indicate a right-handed spiral as seen from the galactic North Pole. The best marked of these branches is undoubtedly that joining the cluster groups in *Auriga*, *Perseus* and *Cassiopeia*.

It is well known that most spiral nebulae have a pronounced central nucleus, which in the case of the *Andromeda* nebula, for example, produces a nearly star like image with a short exposure. On this account there is probably some significance attached to the fact that one of the richest and most remarkable open clusters: NGC 3532 (RA = $11^{h}2^{m}.2$, Decl = $-58°8'$) falls by our distance determination quite close to the median point of the cluster system (center of local system). This cluster which is marked in figure 87.1 by an asterisk contains, according to Raab, more than 100 stars brighter than magnitude 10 (mostly of types B5–A0) and over 300 stars brighter than magnitude 12, and it is imbedded in a region which is also exceptionally rich in stars of magnitude 8–10. It thus seems not impossible that NGC 3532 and the surrounding star field represent the remainder of a central nucleus.

While our results on the space distribution of open clusters support in every way the older views concerning the structure of our Milky Way system and its similarity to a spiral nebula, they disagree entirely with the more recent conclusions by Shapley and Seares that the Sun and its surrounding star concentration (local system) are quite a secondary formation in a much larger galactic system over 100,000 parsecs in diameter, the center of which is situated in the direction of *Sagittarius* (galactic longitude 325°) at a distance of 20,000 to 40,000 parsecs. These conclusions are based mainly on three facts of observation:

1. The distribution of globular clusters (Shapley).
2. The asymmetry in the distribution of faint stars (Seares).
3. The results obtained for galactic rotation.

On the other hand there is no noticeable feature in the distribution of open star clusters which suggests a considerable extension of the Milky Way system in the direction of *Sagittarius*. A careful examination of Barnard's excellent Milky Way photographs of this region, which reach at least to the 17^{m}, did not reveal any appreciable number of small distant undiscovered star clusters. It is hardly possible that every one

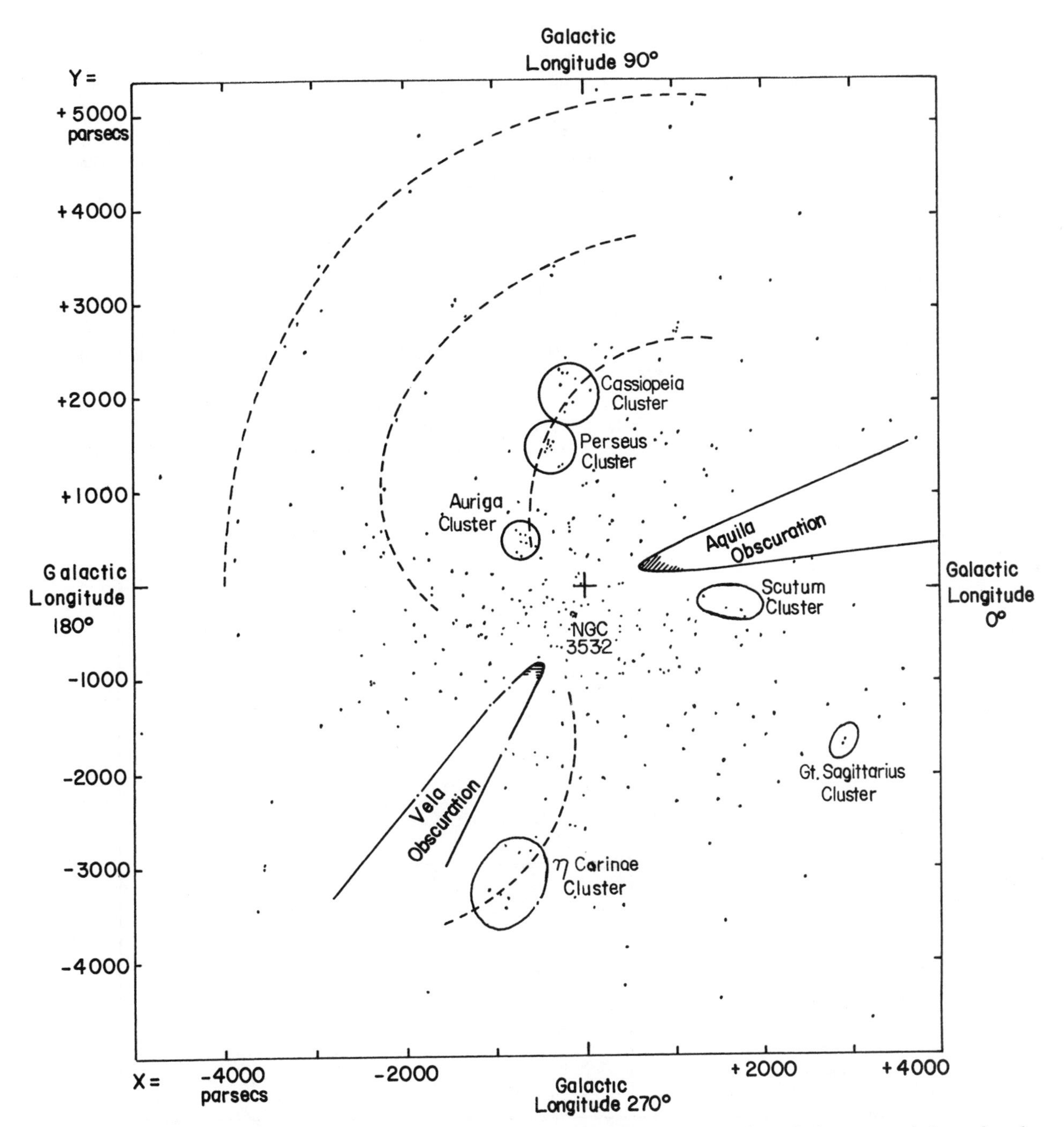

Fig. 87.1 A projection of the open clusters on the galactic plane. The position of the sun is marked by a cross, that of the median point of the system by an open circle, and the cluster NGC 3532 by an asterisk. Some traces of spiral structure are indicated by the dashed lines; the large open circles and ovals represent the probable location of cluster groups or star clouds; and the shaded areas indicate dark clouds of absorbing material with their sectors of obscuration.

of scores or hundreds of such distant clusters should be hidden from our view through absorption by dark matter. But even if we should admit such an assumption, we should still expect the visible parts of the open cluster system to show some arrangement concentric with the distant *Sagittarius* center. We should, for example, expect the limit of the cluster system in the opposite direction (galactic longitude 60°–120°), where there is not so much evidence of dark clouds, to be a segment of a circle of large radius centered in the *Sagittarius* direction; or we should expect the cluster system to widen out in the direction of the center with many distant clusters in galactic longitudes 200°–270° and 0°–70°. None of these expectations is fulfilled, and the hypothesis of a distant galactic center would leave the observed space distribution of open clusters quite unintelligible.

In examining next Seares's investigation of faint stars, Trumpler finds nothing "which should force us to assume that the galactic center is at a greater distance than 1,000 parsecs." As for galactic rotation, the observations furnish "only the direction of the center, not its distance".

Figure 87.2 shows the space distribution of globular clusters in the projection on a plane passing through the galactic pole and through galactic longitude 325°. The 93 globular clusters are entered as full dots according to Shapley's most recent data;[29] the dots, on the scale of the chart, are about twice as large as the limiting size of these clusters. The much flattened system of open clusters which according to our hypothesis outlines the Milky Way system and the two Magellanic Clouds are represented according to their dimensions. The nearest spiral nebulae (*Andromeda* Nebula, Messier 33) would be 3–4 times more distant than the diameter of the whole figure.

From the fact that the space distribution of the globular clusters is somewhat symmetrical to the galactic plane, Shapley draws the conclusion that they form an integral part of the Milky Way system and that the center of the latter should therefore be identified with the center of the globular clusters which lies at a distance of at least 20,000 parsecs in the direction of *Sagittarius*, while the dimensions of the

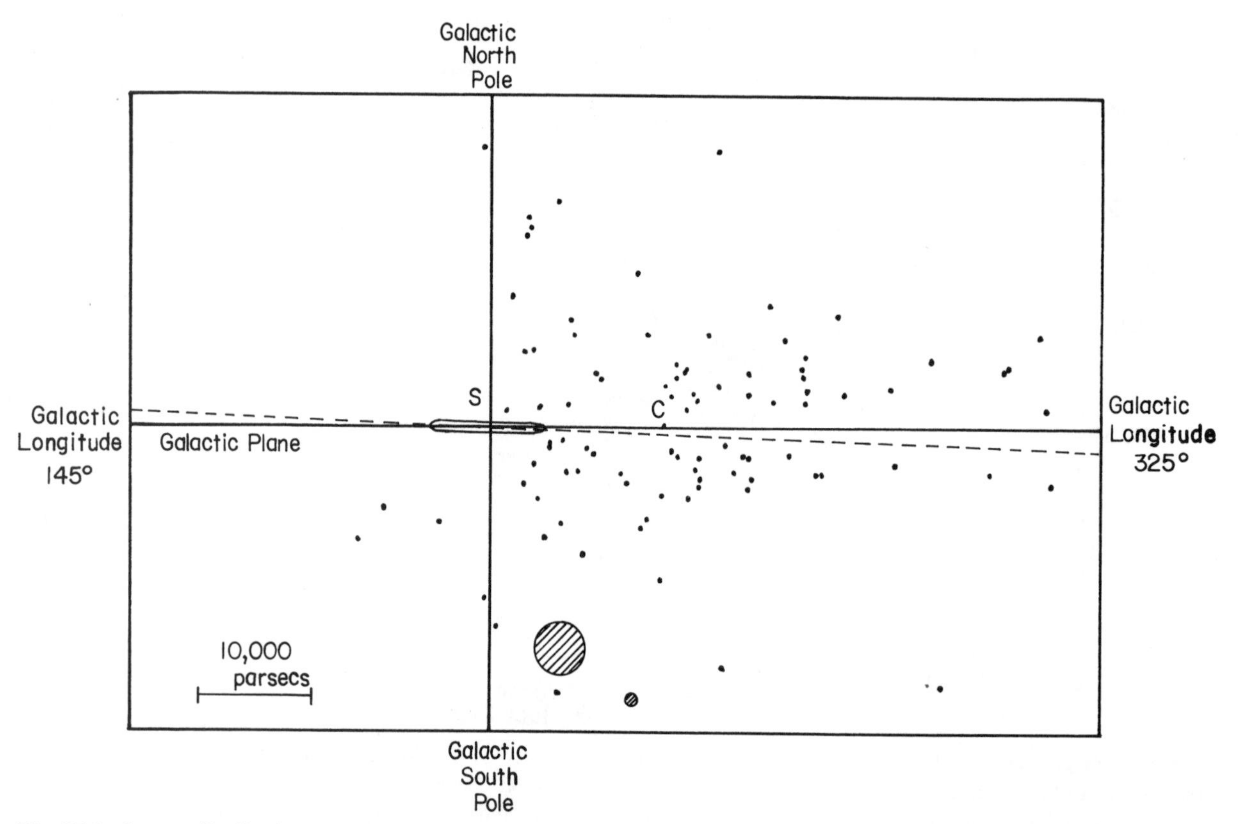

Fig. 87.2 Space distribution of open clusters, globular clusters, and the Magellanic clouds. In this figure the 93 known globular clusters are plotted as black dots in their projection on a plane passing through the galactic pole and through galactic longitude 325°. On the scale of the chart the dots are about twice as large as the limiting dimensions of the globular clusters. The system of open clusters (Milky Way system) is represented by the elongated area, and the two Magellanic clouds, by the shaded circles. The dashed line indicates the plane of symmetry of the open clusters.

galactic system should be of the order of 100,000 parsecs. It must, however, be emphasized, that the distribution of globular clusters in the sky shows practically no relationship to the general star distribution. While the globular cluster system appears nearly spherical in shape there can be no question that the Milky Way system is much flattened. On account of partial obscuration by dark matter there may be some uncertainty about the extent of the star distribution in the galactic plane. No such uncertainty, however, exists in high galactic latitudes. Statistical investigations of stellar distribution show conclusively that in high galactic latitudes the stars do not reach farther than a few thousand parsecs and that the space between the numerous globular clusters in high galactic latitudes is certainly not filled with stars. The majority of globular clusters thus lie outside of the star stratum of our Milky Way system and should in this sense be called extra-galactic systems although this does not exclude the possibility that they have some relation to it. Figure 87.2 shows a remarkable resemblance to some of the clusters of extra-galactic objects (spiral nebulae, elliptical and globular nebulae), and it seems worth while to examine the hypothesis that our Milky Way system (approximately as outlined by the open clusters) together with the two Magellanic Clouds and about a hundred globular clusters form a cluster of extra-galactic objects which we may call the "supercluster." Lundmark,[30] who quite independently came to the same conclusion as the writer, suggests also the possible existence of another large system in the *Sagittarius* region partly hidden by obscuring matter which he calls the "Hidden System" and to which the faint variable stars observed by Shapley in the direction of the "galactic center" would belong. It remains, however, to be investigated whether the faintness of these variable stars is not to some extent due to absorption of light rather than to great distance.

The striking feature of the super-cluster is the fact that it contains only one spiral system (the Milky Way system) and two large amorphous systems (the Magellanic Clouds) associated with a hundred or more globular systems of very much smaller but nearly uniform dimensions. Lundmark discusses this point and finds nothing very improbable in such an association in comparison with other clusters of extra-galactic objects.

Summary

1. The distances of 100 open clusters were determined from magnitudes and spectral types of the stars.
2. With these distances and the estimated angular diameters the linear diameters in parsecs were computed.
3. The linear diameters of open clusters vary considerably (2–16 parsecs); they depend on the constitution of the cluster.
4. The linear dimensions of an open cluster increase with the number of stars contained and with decreasing central concentration. There is a distinct but small class of clusters with exceptionally large dimensions.
5. The assumption that clusters of the same constitution have everywhere the same linear diameters leads to the conclusion that within the Milky Way system light is subject to an absorption of $0^{m}.67$ (photographic) per 1,000 parsecs.
6. The discrepancy between color-indices and spectral types observed in open clusters increases with the distance of the cluster and shows that this absorption of light is selective, the photographic absorption coefficient being about twice the visual.
7. The absorption is effective in all galactic longitudes but seems to take place mainly in a thin layer extending along the galactic plane.
8. A method is developed to determine the distance of a cluster from its angular diameter and from a classification of its constitution.
9. A catalogue of 334 open clusters is compiled, which includes 41 objects not previously listed in cluster catalogues. This catalogue should be nearly complete for all the more prominent open clusterings of our Milky Way system, except for 30–40 of the smallest or faintest ones.
10. Distances and rectangular space coordinates for the 334 clusters were derived.
11. The plane of symmetry of the open clusters is inclined $2^{\circ}.3$ to the adopted galactic plane; its pole lies at RA $12^{h}\ 50^{m}.4$ Decl: $+27^{\circ}.7$ (1900). This plane coincides very nearly with the plane of symmetry derived by Seares from the apparent distribution of faint stars.
12. A study of the space distribution of open clusters shows that they form a much flattened disk-like system about 1,000 parsecs thick with a diameter of about 10,000 parsecs.
13. This cluster system shows a strong concentration towards a point which is situated at a distance of 350 parsecs in galactic longitude 247° from the Sun. The exceptionally rich open cluster NGC 3532 falls very close to this center.
14. The Sun lies 10 parsecs north of the plane of symmetry of the open clusters.
15. The hypothesis is made that the Milky Way system is in its essential features outlined by the space

distribution of open clusters except for a greater galactic concentration of the latter.

16. This hypothesis supports the view that our Milky Way system is a highly resolved spiral nebula, a right-handed spiral as seen from the galactic North pole, of dimensions similar to those of the *Andromeda* nebula.

17. This hypothesis is not in conflict with the apparent distribution of faint stars, but it requires that the globular clusters be treated as extra-galactic objects forming, together with the Milky Way system and the two Magellanic clouds, a super-cluster of extra-galactic objects.

1. *Mount Wilson Contr.* Nos. 199, 244, 262.
2. *Publ. A. S. P. 34*, 150 (1922).
3. *Meddel. Lund*, ser. II, no. 32 (1924).
4. *Seeliger Festschrift*, p. 265.
5. *Veröffentl. Bonn*, no. 19 (1924).
6. *Meddel. Upsala*, no. 32 (1927).
7. See *Publ. A. S. P. 40*, 266 (1928).
8. *Mem. R. A. S. 49* (1888).
9. Ibid. *51*, 185 (1895), *59*, 105 (1908).
10. Ibid. *60*, 175 (1915).
11. *Publ. A. S. P. 37*, 307 (1925).
12. *H. A. 60*, 199 (1908).
13. *Lick Obs. Publ. 11*, and E. E. Barnard, *A Photographic Atlas of Selected Regions of the Milky Way*, (1927).
14. Isaac Roberts: *A Selection of Photographs of Stars, Star Clusters and Nebulae*. Vol. I (1893), Vol. II (1899).
15. *Cordoba Photographs*, by B. A. Gould, (1897).
16. *Mount Wilson Contr.* No. 117.
17. Ibid. No. 100, *Ap. J. 42*, 120 (1915).
18. Seares.
19. *Kgl. Svenska Vet. Akad. Handlingar 61*, no. 15 (1921).
20. *Meddel. Upsala* no. 32 (1927).
21. *Mount Wilson Contr.* No. 126; *Ap. J. 45*, 164 (1917).
22. *Meddel. Upsala*, no. 42 (1929).
23. *L. O. Bull. 12*, 12 (1925)
24. *H. C. O. Bull.* no. 869.
25. J. P. Van Rhijn (Derivation of the change of color with distance and apparent magnitude; Diss. Groningen 1915) from a discussion of color-indices and spectral types of the Yerkes Aktinometry also found a change of color with distance. That his value of $c(+0^{m}.15$ per 1,000 parsecs) is smaller than ours fits in well with our hypothesis that the absorbing medium is highly concentrated to the galactic plane. The stars investigated by Van Rhijn are situated in high galactic latitudes 50 to 350 parsecs distant from the galactic plane and must lie partly outside of the absorbing medium, or in the region where it is generally thinning out.
26. *Sitzungsber. d. Munch. Akad. d. Wiss.* (1920), p. 87.
27. *Mt. Wilson Contr.* No. 230 (1922).
28. Ibid. No. 310 (1926) and No. 376 (1929).
29. *H. C. O. Bull.* no. 869 (1929).
30. *Publ. A. S. P. 42*, 23 (1930).

88. The Solid Particles of Interstellar Space

Hendrik C. van de Hulst

(*Recherches astronomiques de l'Observatoire d'Utrecht 11*, pt. 2, 1–50 [1949])

The idea that interstellar matter might be composed of small solid particles as well as gas atoms had its origin in V. M. Slipher's studies of reflection nebulae and Max Wolf's studies of dark nebulae. Slipher showed that some diffuse "reflection" nebulae exhibit absorption lines similar to those of nearby stars, and he reasoned that these nebulae are composed of solid particles that reflect starlight.[1] Because Wolf could not detect any difference in the colors of stars that lie outside and behind the dark nebulae, he concluded that they must be composed of solid dust particles, whose scattering properties depend weakly on wavelength, rather than gas atoms, which scatter light much more effectively at shorter wavelengths (see selection 83).

By 1930 Robert Trumpler had published his definitive study of the diameters of open clusters, in which he conclusively showed that the interstellar medium absorbs starlight and that the absorption is only weakly dependent on wavelength (selection 87). As suggested by Trumpler's measurements and subsequently demonstrated by the photoelectric measurements of Joel Stebbins, C. M. Huffer, and A. E. Whitford, the interstellar extinction of starlight varies inversely with the wavelength in the visual wavelength region, and this variation is accurately the same for all stars, whether observed through thin or through dense clouds. Because particles larger than 10^{-3} cm will block light without scattering and because the interstellar particles cannot be much smaller than the wavelength of visual light, they have to have sizes on the order of 10^{-4} or 10^{-5} cm.

By the mid-1930s astronomers were trying to determine the composition of solid interstellar particles, which produce an inverse wavelength reddening of starlight and an extinction at the rate of about 1 mag kpc^{-1} of interstellar matter. By the late thirties, Carl Schalén had used the theory of Mie scattering by spheres whose sizes are comparable to the wavelength of the radiation to show that metallic spheres of about 10^{-4} cm in size could produce the required wavelength dependence.[2] In the meantime, Bertil Lindblad had considered the origin of solid interstellar particles and reasoned that particles of the appropriate size might grow by the sublimation of interstellar gas around particles in a process analogous to that in which nucleating particles are formed.[3] A prize competition was subsequently offered by the University of Leiden in 1942 on the subject of the origin of interstellar particles. Although many Dutch universities were almost entirely closed by the Germans, several papers were submitted for the competition, which eventually led to publications by Hendrik Kramers, Dirk ter Haar, Jan Oort, and Hendrik van de Hulst.[4] The war actually provided time for detailed theoretical considerations that might not have been carried out otherwise.

By 1946 van de Hulst was able to publish his doctoral thesis, in which he considered the extinction of light by particles of different sizes and compositions, ranging from perfectly scattering dielectric spheres to perfectly absorbing metallic spheres.[5] Within three years he had traveled to the Yerkes Observatory of the University of Chicago, where under a postdoctoral fellowship he completed his considerations, reproduced here, of the sizes and composition of interstellar particles. Comparisons of the observed extinction of light at optical wavelengths with theoretical computations of Mie scattering led him to conclude that the solid interstellar dust grains are ice particles whose sizes are about 4×10^{-5} cm and whose mass density is about equal to 10^{-24} gm cm^{-3}.

By showing that the shape of the curve of starlight extinction as a function of wavelength gives information on the size and composition of the interstellar dust grains, van de Hulst set the stage for the subsequent explanation of the polarization of starlight[6] and for nearly three decades of speculation on the exact nature of interstellar dust. When observations were subsequently extended to the near infrared and to ultraviolet wavelengths, the extinction was found to continue to grow rather than level out. This unexpected increase can naturally be attributed to scattering by smaller dust grains, whose sizes are on the order of the shorter wavelengths. As anticipated in the paper given here, the bump in the extinction curve, subsequently observed at wavelengths near 2,200 Å, can be caused by any substance with a natural resonance at this wavelength; and this bump has been attributed to absorption by small graphite particles. Because infrared absorption bands of silicates have been detected in the atmospheres of cool stars, some dust grains may also be composed of silicates. Ices, carbon, and silicates, or some combinations of these, are now viewed as the principal ingredients of grains, whose sizes may range from 10^{-4} to 5×10^{-5} cm.

1. V. M. Slipher, *Lowell Observatory Bulletin 55*, 1 (1912), *75*, 1 (1916), and *Publications of the Astronomical Society of the Pacific 30*, 63 (1918), *31*, 212 (1919).

2. C. Schalén, *Uppsala Astronomiska Observatoriums meddelanden*, no. 64, (1936), and *Uppsala Astronomiska Observatoriums annaler 1*, no. 2 (1939), *1*, no. 9 (1945).

3. B. Lindblad, *Nature 135*, 133 (1935).

4. H. A. Kramers and D. ter Haar, *Bulletin of the Astronomical Institutes of the Netherlands 10*, 137 (1946); J. H. Oort and H. C. van de Hulst, *Bulletin of the Astronomical Institutes of the Netherlands 10*, 187 (1946).

5. H. C. van de Hulst, *Recherches astronomiques de l' Observatoire d'Utrecht 11*, pt. 1 (1946). The Mie theory of the scattering and absorption of radiation by small particles had also been applied to dielectric and metallic particles by Jesse L. Greenstein in his 1937 Ph.D. thesis at Harvard University entitled *Studies of Interstellar Absorption*. Greenstein concluded that the observed color excesses of stars can be accounted for by small metallic particles or a mixture of different particles.

6. H. C. van de Hulst, *Astrophysical Journal 112*, 1 (1950).

Chapter II. Optics of Spherical Particles

THREE KINDS of observational data relating to the interstellar particles may be distinguished. First, the light of distant stars is weakened by the clouds through which it shines; the amount of this extinction and its variation with the wave length can be measured. Second, part of the light that is removed from the original beam is scattered into other directions. The amount of scattered light, its angular distribution, and its state of polarization may be measured. Third, observations of stellar motions give in an indirect way some information about the density of the interstellar clouds.

The inference which can be drawn from observations of the second and third kind has been reviewed elsewhere.[1] It appears that the scattered light and its polarization may be a promising object for further investigation but for the moment no progress has to be reported. Therefore this publication will deal with the *extinction* only.

The theory of the extinction has already been discussed by many authors; in particular the investigations of Schalén[2] and of Greenstein[3] may be mentioned. In recent years, however, both the accuracy and the wave-length range of the observations have increased, while also great progress has been made in the theoretical calculations. A new, systematic treatment of this problem seems therefore desirable.

SUMMARY OF PART I AND RECENT LITERATURE The problem of extinction and scattering by a solid particle is a problem in the field of electromagnetic light theory. Once the size and shape of the particle, its electric and magnetic properties, and the nature of the incident wave have been specified, the problem can in principle be solved. For very few cases, however, has the problem actually been solved and in all these cases a great deal of additional work is needed to convert the analytical solution into numerical results. Only for one case has this numerical work been carried out to a considerable extent, namely for the scattering of a plane wave by a homogeneous sphere. It is fortunate that the considerations of Chapter I have led us to believe that the conditions of spherical shape and homogeneous composition are at least roughly satisfied by the actual interstellar particles. The further presuppositions of the theory are that both the light source and the observer are at a large distance from the sphere[4] and also that the particles are far from one another.[5] These conditions are easily satisfied in interstellar space.

The analytic solution in a convenient form has been given by Mie. For a summary of the formulae, a compilation of numerical results, and a systematic treatment of the limiting cases of Mie's theory we refer to part I of this publication.[6] The definitions and notations, and some of the results, may be briefly repeated here.

The particle is supposed to be a smooth sphere with the radius a, consisting of a homogeneous material having the complex refractive index m. It is illuminated by a plane wave of natural light having the wave length λ. Inside the particle and immediately near it the wave motion is of a very complex form. The region of space in which the wave motion is so complex extends to a considerable distance behind the particle; it can be said to end only where the shadow cylinder behind the particle has been completely filled in by diffracted light. Outside this region the wave can be decomposed into two parts: 1. The original but slightly weakened plane wave. 2. A scattered spherical wave which seems to radiate from the scattering particle. The following formulae hold for this outer region.

Setting the unit flux equal to the flux incident on the cross-section, a^2, of the sphere, we define the following dimensionless quantities:

E = extinction = flux lost by the incident wave.

$I_1(\theta)$ = flux scattered per unit solid angle in a direction diverging by the angle θ from the original direction, and polarized with the electric vector perpendicular to the plane through the incident and scattered beams of light.

$I_2(\theta)$ = the same with the electric vector parallel to this plane.

$I(\theta) = I_1(\theta) + I_2(\theta)$ = total flux scattered per unit solid angle in the specified direction.

S = total flux scattered in all directions.

A = flux which is absorbed in the particle and transformed into heat.

From these definitions follow the identities

$$S = \int_0^\pi 2\pi I(\theta) \sin\theta \, d\theta \qquad \text{(part I; 2,1)}$$

and

$$E = S + A. \qquad \text{(part I; 2,2)}$$

All these quantities depend on m and on the parameter

$$x = 2\pi a/\lambda. \qquad (1)$$

However, it was shown in chapter 7 of part I that in the case of refractive indices near 1 the parameter

$$\varrho = 2x(m-1) \qquad (2)$$

has a more basic significance. It equals the phase-shift suffered by a light ray that passes through the particle along a diameter.

For a given value of m one can plot E against x; the curve thus obtained will be called the *extinction curve.* This curve always starts at $E = 0$ for $x = 0$, rises to a maximum where E is 2.5 to 4 and, finally, approaches the value $E = 2$ for $x = \infty$. In most cases the rising part of the curve is perfectly smooth, while beyond the first maximum further small or large fluctuations occur.

Examples of extinction curves are found in part I of van de Hulst's treatment, and we reproduce in figure 88.1 some typical curves. We omit his reproduction of extinction curves computed by La Mer.[7]

EXTINCTION BY A MIXTURE OF PARTICLES HAVING VARIOUS SIZES The definition of E, the efficiency factor for extinction of a single particle, can be extended to apply also to a cloud of particles. In that case E is equal to the cross-section for extinction of the entire cloud divided by the geometric cross-section of the entire cloud. If the individual particles are sufficiently far apart and the cloud is optically thin, the cross-sections of the individual particles have just to be added. The value of E depends then only on the quality of the mixture, not on the extent of the cloud.

In the following calculations we assume that all particles have the same refractive index.

Let in 1 cm^3 a number of $Cf(r/r_1)\,dr/r_1$ particles be found which have radii in the interval dr. The function $f(u)$ may have any form and C and r_1 are arbitrary scale factors. Introducing the definite integrals

$$A_p = \int_0^\infty f(u)u^p\,du, \tag{3}$$

we can express the following quantities.

Total number of particles (cm^{-3}):

$$N = CA_0. \tag{4}$$

Total geometrical cross-section (cm^2/cm^3):

$$\sigma = CA_2\pi r_1^{\,2}. \tag{5}$$

Total density in space (g/cm^3):

$$\varrho_s = CA_3\,\tfrac{4}{3}\pi s r_1^{\,3}, \tag{6}$$

where s is the density of each particle. The total cross-section for extinction at a given wave length is found by inserting a factor $E(\varrho)$ in the integrand of A_2, in equation (5). As a result the right hand member of (5) is multiplied by the following mean value of E:

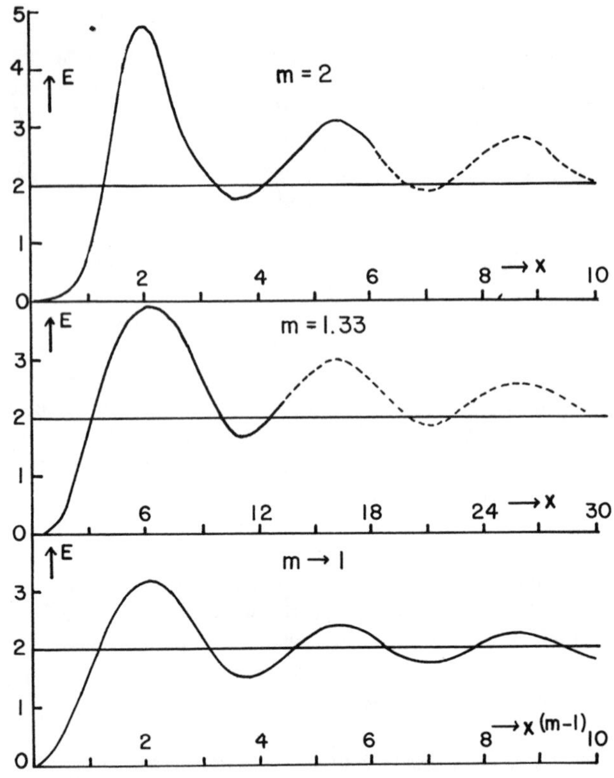

Fig. 88.1 The extinction, E, of homogeneous dielectric spheres of index of refraction m plotted against the parameter $x = 2\pi a/\lambda$, where a is the radius of the sphere and λ is the wavelength of the scattered light. The extinction begins by varying as x^{+4} for small x and gradually changes to an x^{+1} variation in the range of x from 1 to about 4. To bring out the resemblance of the extinction curves for particles of different refractive indices, the bottom curve has been plotted against $x(m-1)$. The curve for $m = 1.33$ represents that of water ice. Notice that all curves obtain an asymptotic value of 2 for large x and that the maximum value of E increases with m.

$$\bar{E}(\varrho) = \frac{1}{A_2}\int_0^\infty E(u\varrho_1)f(u)u^2\,du. \tag{7}$$

Here ϱ_1 is connected with r_1 in the same way as ϱ is with r:

$$\varrho_1 = 4\pi r_1(m-1)/\lambda.$$

The curve obtained by plotting $\bar{E}$ as a function of ϱ_1 may be called the extinction curve for the given distribution function of the radii. In the following we shall omit the bar from $\bar{E}$, writing this function simply as $E(\varrho_1)$.

The intensity of a light wave that traverses the medium over a distance L decreases by a factor $e^{-\tau}$, where the optical depth, τ, for a given wave length is

$$\tau = L\sigma E(\varrho_1).$$

The unit of L in astronomical use is 1 kpc = 3.08 10^{21} cm and the optical depth τ corresponds to a decrease of 1.086 τ magnitudes. Accordingly, the interstellar extinction coefficient becomes:

$$\alpha = 1.05\ 10^{22} C A_2 r_1{}^2 E(\varrho_1)\ \text{mag/kpc}. \quad (8)$$

It should be noted that the only way in which α depends on the wave length is by means of the factor $E(\varrho_1)$, the argument of which is proportional to λ^{-1}. Consequently, if the scales of ordinates and abscissae are properly chosen, the *extinction curve shows at once the dependence of α on λ^{-1}.*

For particles of the same size a and volume density N the absorption in magnitudes at a wavelength nearly equal to a is 1.086 $\pi a^2 NL$.

Omitted here are detailed computations of extinction curves for mixtures of particles with the same composition but different sizes. These curves have forms similar to those given in figure

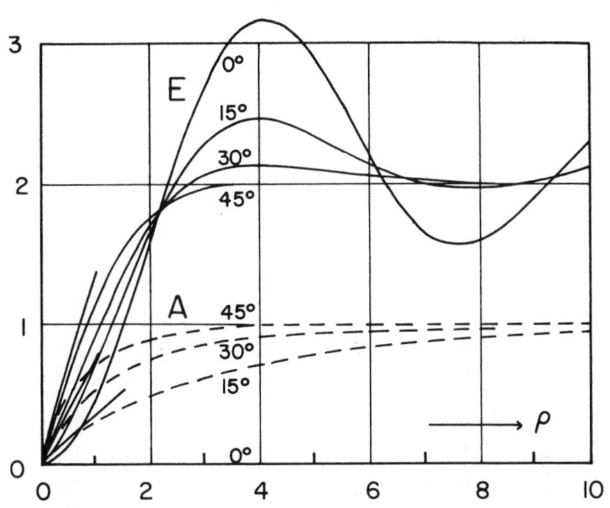

Fig. 88.2 The extinction, E, and the absorbed fraction of incident energy, A, for slightly absorbing spheres, having index of refraction $m = 1 + \varepsilon - i\varepsilon \tan \beta$, where ε is real and small. The abscissa is $\rho = 2x\varepsilon$, and values of β are written with the curves. The curved behavior of these extinction curves at low values of ρ is thought to explain optical wavelength observations of the interstellar extinction of starlight.

88.1. The main change is that the dispersion in size reduces the maximum value of extinction and tends to obliterate fluctuations in the curve beyond the maximum.

SLIGHTLY ABSORBING PARTICLES Extinction curves for absorbing particles with complex indices of refraction are illustrated in figure 88.2.

OPTICS OF PARTICLES IN THE REGION OF A SPECTRAL LINE

Van de Hulst then points out that when a particle is small compared with the wavelength, light is absorbed at the resonance wavelengths of its molecules when it is in a gaseous state. The extinction curve will exhibit a bump of apparent absorption or emission at these wavelengths according as $\rho = 2x(m - 1)$ is, respectively, greater or less than 1.

CHAPTER III. THE INTERPRETATION OF THE INTERSTELLAR EXTINCTION

INTRODUCTION The solid particles that cause the interstellar extinction[8] have been the subject of a lively discussion during the past ten years. A report on the deductive theories, relating to the growth of these particles from the interstellar gas, has been given in Chapter I. The observational data, in particular the wave-length dependence of the extinction, have also given rise to a variety of speculations on the nature of these grains. Before discussing these data in the light of the calculations reported in Chapter II, we shall briefly recall the historical development of our subject.

Several reviews of the history have been published.[9] They show that for more than a century astronomers have suspected the presence of absorbing matter in interstellar space but were unable to detect it. It was generally assumed that the space between the stars was virtually clear and most investigations on the structure of the galaxy were based on this assumption. About 1930 it became clear that this assumption led to definite inconsistencies in the picture of the galaxy. Thus the existence of an interstellar medium, with an extinction of the order of 1 magnitude per 1,000 parsec near the galactic plane, became generally recognized. The effects on astronomy at large were tremendous: the size of the galaxy had to be drastically reduced, the local system of stars proved spurious, and many other examples might be cited. At present the interstellar extinction enters into any study of galactic structure, whether local or general, and it often forms the most serious impediment to accurate results.

Only the broad features of the distribution of obscuring matter in space were needed for the first large-scale corrections. Thus, for a while, it seemed sufficient to describe the obscuring matter as being distributed in a plane-parallel slab along the galactic plane, having an extinction coefficient of 1^m/kpc inside the slab and a total extinction of $0^m.50$ across it. Exception was made for the local condensations known as dark clouds. Further studies showed that the distinction between discrete clouds at one hand and a homogeneous "stratum" at the other hand was rather artificial. The "general extinction" in interstellar space is now thought to arise from a random superposition of dark clouds; only the nearest clouds have sufficient contrast with their surroundings to be recognized as such. It should be noted, however, that the clouds are very inhomogeneous and irregular in structure.

Consequently, the general extinction coefficient is a poorly defined quantity that is highly dependent on selection effects.[10] For stars in the galactic plane at an average distance of 2,000 pc the following estimates of the average extinction may be made:

stars selected according to apparent magnitude: 1.0 mag/kpc;

stars selected according to true distance: 2.0 to 2.5 mag/kpc.

These values refer to the photographic magnitudes ($\lambda =$ 4,400 A). The accuracy by which the extinction of any one star may be computed from these coefficients may seldom be better than 50 per cent. Any attempt at a higher accuracy must be based on individual measures for the star or a small field of stars. The most important individual method is based on the measurement of the color excess.

Let $m(\lambda)$ be the apparent magnitude of a star, $M(\lambda)$ its absolute magnitude, $A(\lambda)$ the extinction of its light expressed in magnitudes, and r its distance expressed in parsecs. Then

$$m(\lambda) = M(\lambda) + 5 \log r - 5 + A(\lambda). \quad (1)$$

For a different effective wave length, λ', we have similarly

$$m(\lambda') = M(\lambda') + 5 \log r - 5 + A(\lambda'). \quad (2)$$

After defining

$$c(\lambda, \lambda') = m(\lambda) - m(\lambda') = \text{apparent color index}, \quad (2)$$

$$C(\lambda, \lambda') = M(\lambda) - M(\lambda') = \text{intrinsic color index}, \quad (3)$$

$$E(\lambda, \lambda') = A(\lambda) - A(\lambda') = \text{color excess},^{11} \quad (4)$$

we obtain by subtraction:

$$c(\lambda, \lambda') = C(\lambda, \lambda') + E(\lambda, \lambda'). \quad (5)$$

The usual convention that $\lambda < \lambda'$ makes C positive for a red star, negative for a blue one.

The color excess of a star can be determined fairly accurately by means of equation (5). For c is measured directly and C follows from the spectral type. The direct determination of $A(\lambda)$ from equation (1) is very difficult, however. For M cannot be determined very precisely from the spectrum and in addition an independent determination of r is required. Therefore it has proved useful to determine A in an indirect way by multiplying the measured value of E by a certain factor R, *the ratio of interstellar extinction to reddening*. Once this factor is known, A is known and equation (1) can be used to determine r, or M. Fine examples of this method are found in Oort's investigation of the distribution of stars at high galactic latitude[12] and in Morgan's study of the space distribution of B-stars.[13]

The chief difficulty in this method is to determine the value of the ratio R. We shall express λ in microns and specify R by

$$R = \frac{A(0.440)}{A(0.426) - A(0.477)}. \quad (6)$$

The effective wave length in the numerator corresponds to that of the photographic magnitude scale. The denominator corresponds to the photo-electric color-excesses E_1 of Stebbins, Huffer and Whitford.[14]

We omit here detailed discussions of different methods of determining R.

PREVIOUS EXPLANATIONS Early speculations on the cause of the interstellar extinction considered only Rayleigh scattering by the interstellar gas or obstruction of the light by meteoric dust. These explanations would respectively require that the extinction is proportional to λ^{-4}, or independent of λ. Later observations showed that neither hypothesis is correct.

In 1917 Halm[15] suggested that the color temperature of a star might decrease gradually on its way through space. This hypothesis would produce an interstellar extinction proportional to λ^{-1}, but the mechanism is a physical impossibility.

The extinction may therefore be written:

$$A(\lambda) = a\lambda^{-1} + b. \quad (9)$$

The constant b is unknown; it may be either positive or negative.

The suggested explanations of the λ^{-1} law have been of two kinds. The first explanation proposes that the particles are *dielectric* (i.e. non-conducting). Since the extinction changes gradually from a λ^{-4} law for particles much smaller than the wave length to a λ^{0} law for particles much larger than the wave length, it is clear that an intermediate size may be found at which the scattering is proportional to λ^{-1}. This point was brought forward by Russell[16] and by Öpik[17]

The second explanation, suggested by the high abundance of metals in meteorites, proposed metal particles. This explanation has place for a strict λ^{-1} law, for the extinction (which is pure absorption) of small metallic particles is proportional to $x = 2\pi r/\lambda$. The variation of the refractive index with λ introduces some complication. Calculations[18] have shown that an approximate λ^{-1} law is obtained for particles that are a little larger than "very small." But the essential features of the solution remain the same: the particles are fairly small and have a low albedo; b is about zero and the corresponding value of R is comparatively large.

At one time most observations seemed to confirm these features. Attempts have even been made to explain certain irregularites in the extinction curve by mixtures of various metals, to link up the diameter distribution of the absorbing particles with those of the meteors, to find a relation between meteor streams and the dark cloud in Taurus, etc. However, the evidence in favor of this explanation has faded. The observed extinction curve is better fitted by the results for dielectric particles; the observed albedo is fairly high;[19] the existence of interstellar meteors has not been confirmed by accurate velocity measurements;[20] and, finally, the deductive theory leads to particles in which the metallic atoms have a low abundance.

COMPARISON OF THE OBSERVED LAW OF REDDENING WITH THEORY We omit a detailed comparison with observations, which is represented in the following conclusion and summary.

Conclusions. The interstellar extinction from $\lambda = 0.3\ \mu$ to $\lambda = 2\ \mu$ can be explained by means of the theory for dielectric or slightly absorbing spherical particles. The scale factor, l, of the sizes in a presupposed distribution function of the sizes can be determined from the observations with an accuracy of 5 per cent. For instance, if only particles of one size are assumed we find $l = 4\pi r(m-1) = 1.25\ \mu$; with the value, $m = 1.25$, this gives $r = 0.40\ \mu$.

In order to infer from the observations something about the refractive index, or about the distribution function of the sizes, no deviations larger than $0^{m}.02$ may be permitted. It then follows that the distribution function is somewhat spread out, in particular towards the larger sizes. Further tan β in the expression.

$$m = 1 + \varepsilon - i\varepsilon \tan \beta \qquad (21)$$

is found to be small; it certainly cannot be as large as 0.27. No direct inference about the value of ε can be made.

The size function and refractive index that follow from the theory about the origin of the particles are in good agreement with the observations. The ratio, R, of extinction to reddening is 8.6. The average density of the smoke in the neighborhood of the sun is $1.4\ 10^{-26}$ gm/cm^3.

SUMMARY

The three chapters of this work form more or less separate investigations, the first two being preparations for the third one.

CHAPTER 1: OUTLINE OF THE PHYSICAL CHEMISTRY OF THE INTERSTELLAR PARTICLES The recent theories, which think the particles to originate as large molecules in the gas, to grow by accretion of more gas, and finally to evaporate by collision during an encounter of two interstellar clouds, are briefly reviewed. The gradual growth is further examined. It is made sure that the electrical charge of the particles does not change the rate of capture very much; the probable potential is about -1 volt. The accommodation coefficient of the atoms hitting the solid surface is obtained from low-temperature experiments reported in literature. Since the data for gas-covered surfaces and not those for pure surfaces are relevant, complete accommodation must take place. The energy balance is determined chiefly by radiation, with small terms from convection and latent heat. The estimated temperature, 16° to 25°, is much higher than the black-body temperature because the small particles cannot radiate effectively at long wave lengths. The vapor-pressure formula now shows that most H, He, and Ne, evaporates quickly, while the further captured atoms and ions stay. The molecules formed on the particles are probably hydrides and H_2. Thus, the particles will consist of ice with impurities. The H_2-content, roughly estimated from experimental data on multi-layer H_2 adsorption, may be about 20 per cent.

It is further noted that many of the larger particles may be the products of fusion of two particles into one during an encounter of clouds. These "secondary particles" may have somewhat different properties; they cause about one third of the interstellar extinction.

CHAPTER 2: OPTICS OF SPHERICAL PARTICLES The relevant formulae of Part I are summarized. Recently published extinction curves, computed from Mie's theory, are shown in figure 88.1. Attention is called to the small bumps on these curves, which vanish for $m \to 1$. The extinction curves for mixtures consisting of particles with different sizes are computed for seven distribution functions of the radii and for two values of the refractive index: $m = 1.33$ and $m = 1 + \varepsilon$. The theory is extended to slightly absorbing particles with $m = 1 + \varepsilon - i\,\varepsilon \tan \beta$; the resulting extinction and absorption are shown in figure 88.2. An application of this theory is found in the discussion of the extinction in the region of a spectral line. Using the ordinary dispersion theory we find that only small particles give maximal extinction at the center of the line; for larger particles the curve is asymmetric or inverted.

CHAPTER 3: THE INTERPRETATION OF THE INTERSTELLAR EXTINCTION The comparison of the observed wave-length dependence and the theoretical extinction curves is useful for two reasons. First, it provides a check on the physical theory of interstellar matter. Second, it enables us to extrapolate the observed curve to $\lambda^{-1} = 0$ and thus to determine the ratio of total extinction to reddening, which is important for investigations of galactic structure.

The reddened B-star, ζ Pers, is taken for an example of the normal reddening law. Photoelectric measurements by Stebbins and gradient measurements by other authors show the same amount of curvature. Several theoretical curves are found that represent the observations from $\lambda^{-1} = 1.0$ to $\lambda^{-1} = 3.1$ very well and also have inverse curvature in the infrared. Among them is the solution that corresponds to the size distribution and refractive index determined a priori and also takes the change of refractive index with λ into account. Extrapolation leads to $R = A_{pg}/E_1 = 8.6$. The most effective particles have $r = 0.4\ \mu$. The average space density of the solid particles is estimated at $1.4\ 10^{-24}$ gm/cm^3.

An analysis of the deviating reddening law of θ_1 Orionis indicates that the particles are about 30 percent larger, and have a stronger real absorption, than in ordinary clouds. The physical interpretation of these differences, and of further peculiarities discovered by Baade and Minkowski, is still very puzzling.

Finally, the problem of the unidentified absorption lines is reviewed. It appears impossible that lines of this regular shape are formed in the solid particles. It is suggested that they arise from free molecules or from molecules at the surface of the solid particles.

1. O. Struve, *Ann. d' astrophysique 1*, 143 (1938). H. C. van de Hulst, Harvard Symposium on interstellar matter, Dec. 1946.

2. C. Schalén, *Uppsala ann. 1*, 1 (1939), and earlier papers.

3. J. L. Greenstein, *Harvard Circular 422* (1938); *Ap. J. 104*, 403 (1946).

4. A typical case which does not satisfy this condition is the propagation of radiowaves around the earth. See B. van der Pol and H. Bremmer, *Phil. Mag. 24*, 141 and 825 (1937), *25*, 817 (1938), *27*, 261 (1939).

5. A distance of 3 diameters is large enough: W. Trincks, *Ann. der Physik*, *22*, 561 (1935).

6. See also D. Sinclair, *J.O.S.A. 37*, 475 (1947).

7. V. K. La Mer, Progress Report on "Verification of Mie Theorie. Calculations and measurements of light scattering by dielectric spherical particles"; mimeographed O.S.R.D. report No. 1857, Washington 1943.

8. The term "interstellar absorption" has mostly been used instead of extinction but appears somewhat confusing. The solid particles are also known as "grains" or "granules"; it would seem advisable to call them "dust" or "smoke" only if these terms are meant to imply a proposition about the physical origin of the grains.

9. H. Kienle, *Jahrbuch d. Radioaktivität und Elektronik 20*, 1 (1923); W. Becker, *Materie im interstellaren Raume*, *Fortschritte der Astronomie*, vol. 1 (Leipzig, 1938), F. H. Seares, *P.A.S.P. 52*, 80 (1940); O. Struve, *J. Washington Ac. Sc. 31*, 217 (1941).

10. J. Stebbins, C. Huffer, and A. E. Whitford, *Ap. J. 90*, 209 (1939).

11. We retain here the familiar notations $A(\lambda)$ and $E(\lambda, \lambda')$; they should not be confused, however, with the A and E used in Part I and in Chapter II.

12. J. H. Oort, *B. A. N. 8*, no. 308, 233 (1938).

13. W. W. Morgan, *Ap. J. 113*, 141 (1951).

14. J. Stebbins, C. M. Huffer, and A. E. Whitford, *Ap. J. 91*, 20 (1940) (1,332 B-stars).

15. J. Halm, *M.N. 77*, 243 (1917) (in particular p. 269).

16. H. N. Russell, *Proc. Nat. Ac. Sci. 8*, 115 (1922).

17. E. Öpik, *Harvard Circular 359*, 7 (1931).

18. C. Schalén, *Uppsala med. 58* (1934), *64* (1936); E. Schoenberg and B. Jung, *Breslau Veroff. 7* (1934), and *Breslau Mitt. 4* (1937).

19. L. G. Henyey and J. L. Greenstein, *Ap. J. 93*, 70 (1941); see also Harvard Observatory Monograph No. 7 (Centennial Symposium), 1948.

20. F. L. Whipple, *Proc. Am. Phil. Soc. 79*, 499 (1938) (*Harvard Reprint* no. 152).

89. The Polarization of Starlight

Polarization of Light from Distant Stars by the Interstellar Medium

William Albert Hiltner

(*Science 109*, 165 [1949])

Observations of the Polarized Light from Stars

John Scoville Hall

(*Science 109*, 166–167 [1949])

In 1946 Subrahmanyan Chandrasekhar noticed that a chief opacity source in the outer atmospheres of early-type stars should be electron scattering, and he predicted that light from the limbs of these stars would be linearly polarized. Although the limbs of these stars could not be resolved directly, in principle light from near one edge of the stellar disk could be separated out during a partial eclipse in a binary star system. Accordingly, William Hiltner and John Hall attempted to confirm the prediction during several eclipses.

As illustrated in the first of a pair of papers published together, Hiltner detected large linear polarizations, but surprisingly, the polarization did not depend on the phase of the binary motion. This meant that the polarization did not arise in stellar atmospheres themselves but was introduced by the intervening interstellar medium.

This important, accidental discovery was confirmed and extended by John Hall, who in the second paper of the pair shows that stars with highly polarized light are invariably reddened stars. This discovery suggested that the interstellar dust that causes the reddening of starlight might also produce the polarization. In the year in which Hall and Hiltner reported their discovery, Hendrik van de Hulst suggested that the polarization of starlight might be produced by elongated dust particles that are oriented in a specific direction,[1] and this speculation led to an intensive discussion of different orientation mechanisms. By 1951 Leverett Davis and Jesse Greenstein had introduced the theory that explains polarized starlight in terms of absorption and scattering by elongated dust particles aligned by an interstellar magnetic field.[2] Chandrasekhar and Enrico Fermi then noted that the interstellar magnetic field might contain and perhaps support the local spiral arm.[3] If polarized starlight is produced by dust particles aligned by a magnetic field containing the local spiral arm, we would expect the polarized light to be aligned along the galactic plane. Although Hall's early results suggested that the planes of polarization exhibit no preferential orientation, we now know that most of them do lie along the galactic plane (see figure 89.2).

Although no one had directly measured the strength of the interstellar magnetic field, Davis and Greenstein estimated that a field strength of a few microgauss could account for the observed polarization of starlight if it is caused by spinning dust particles. Chandrasekhar and Fermi showed that a similar magnetic field strength is required to maintain the local spiral arm against some of the forces tending to dissipate it.

In 1957 John Bolton and Paul Wild called attention to the possibility of actually measuring the interstellar magnetic field through the Zeeman splitting of the 21-cm transition of interstellar neutral hydrogen. After collecting nearly six hundred hours of observations over a four-year period, Gerrit Verschuur showed that several cold clouds of interstellar hydrogen do exhibit detectable Zeeman splitting caused by a magnetic field of a few microgauss.[4] This result has now been extended to the entire interstellar medium lying between the sun and the nearby pulsars. All these measurements confirm both the direction and the strength of the interstellar magnetic field inferred from observations of polarized starlight, with the assumption that the polarization is caused by scattering from interstellar dust particles aligned perpendicular to the local interstellar magnetic field. The observed field is, however, nearly a thousand times too weak to hold together and actually support the spiral arms of our galaxy.

1. H. C. van de Hulst, "Interstellar Polarization and Magneto-hydrodynamic Waves," in *Problems of Cosmical Aerodynamics*, Proceedings of IUTAM-IAU Symposium on Cosmical Gas Dynamics, Central Air Documents Office, Dayton, Ohio, 1949. See also *Astrophysical Journal 109*, 471 (1949), *112*, 1 (1950).

2. L. Davis and J. L. Greenstein, *Astrophysical Journal 114*, 206 (1951). Also see L. Spitzer and J. W. Tukey, *Astrophysical Journal 114*, 187 (1951).

3. S. Chandrasekhar and E. Fermi, *Astrophysical Journal 118*, 116 (1953). Also see selection 98.

4. J. G. Bolton and J. P. Wild, *Astrophysical Journal 125*, 296 (1957); G. L. Verschuur, *Astrophysical Journal 165*, 651 (1971).

Hiltner, Polarization of Light from Distant Stars

In the course of photoelectric observations made last summer with the 82-inch telescope of the McDonald Observatory (University of Texas) the writer found that the light from distant galactic stars is polarized. Polarizations as high as 12 percent were found. The plane of polarization appears to be close to the galactic plane in the cases examined. More recently control measures were made at the Lick Observatory, thanks to the courtesy of Director Shane and Dr. G. Kron; and during December the work at the McDonald Observatory was extended to different regions of the Milky Way.

In view of the unexpected nature of this result the circumstances leading to its discovery are recorded. Photometric observations for the detection of partially polarized radiation from eclipsing binary stars have been in progress at the Yerkes Observatory for several years with a view to establishing observationally the effect pointed out by Chandrasekhar that the continuous radiation of early-type stars should be polarized.[1,2] On the assumption that the opacity of early-type stars is due to scattering by electrons, the continuous radiation emerging from a star should be polarized with a maximum of polarization of 11 percent at the limb. Since the presence of this polarization can be detected only when the early-type star is partially eclipsed by a larger-type companion of the system, the effect is masked by radiation from this companion so that the expected maximum observable effect was only of the order of 1.2 percent in one case investigated (RY Persei).

At this stage Dr. John Hall, of Amherst College, proposed to the writer a program of collaboration whereby Dr. Hall would construct a "flicker" photometer which was to be tested jointly at the McDonald Observatory. Independently the writer was developing his own equipment which used polaroids. Dr. Hall's equipment was tested in August 1947, during a short session at the McDonald Observatory, but no dependable results were obtained and it was found that the equipment had to be remodeled. Unfortunately, Dr. Hall was unable to come for a second trial period, scheduled for August 1948.

Meanwhile the writer's own equipment was completed and put to use during the summer of 1948 and was found satisfactory. Certain Wolf Rayet stars which were known or suspected to be eclipsing binaries were examined for polarization. Fairly large polarizations were found, but *they did not appear to depend on the phase of the binary motion.* The possibility of instrumental polarization was considered, of course, but ruled out by control measures on check stars. The Wolf Rayet stars give the following results:

	Polarization	
Star	%	Position angle
CQ Cep	10.0	62°
BD 55°2721	8.0	44
WN Anon[a]	12.5	44

a. Coordinates: $22^h08^m + 57°26'$ (1945); 12.5 magnitude.

The control stars had similar color and brightness, but showed no polarization except for one object, BD 55°2723, which gave 3 percent. This star, however, is a giant and more distant than the other control stars. Similar observations made on a group of Wolf Rayet stars in Cygnus showed no appreciable polarization, while two stars in Scutum gave positive results. Other regions, such as the double cluster in Perseus, also show polarization with values ranging up to 12 percent.

We conclude from the positive and negative results quoted that the measured polarization does not arise in the atmospheres of these stars but must have been introduced by the intervening interstellar medium. If this conclusion is accepted, a new factor in the study of interstellar clouds is introduced. Further observations are in progress for relating this phenomenon with other observable characteristics of interstellar medium. As has been stated, the results already at hand indicate that the plane of polarization approximates the plane of the galaxy.

1. S. Chandrasekhar, *Astrophys. J. 103*, 365 (1946).
2. W. A. Hiltner, *Astrophys. J. 106*, 231 (1947).

Hall, Observations of the Polarized Light from Stars

Photoelectric observations of the polarization of starlight made during the period November 1948 to January 1949 with the 40-inch reflector at Washington substantiate the hypothesis of W. A. Hiltner[1] that this effect is produced by interstellar matter. Furthermore, the percentage of polarization appears to be independent of wavelength; and the plane of polarization (plane containing the magnetic vector and the line of sight) appears to have no one preferential orientation.

The observations were obtained with a photoelectric polarizing photometer[2] built at Amherst College in 1946 with the aid of a grant from the Research Corporation of New York. The light from a star is collimated and directed through a cover glass, which serves as a calibrating device, and then through a Glan-Thompson prism rotated at 15 cycles per second to a 1P21 photomultiplier. The 30-cycle voltage developed by the polarized component of the light is selectively amplified and mixed with a phasing voltage in such a way that the d-c output can be impressed as a sine wave on a Brown recorder. The amplitude of this wave is proportional to the intensity of the polarized light, and the phase of maximum defines the plane of polarization.

A record of a star showing a large percentage of polarized light is given in figure 89.1.

The vertical lines represent two-minute intervals. The trace during interval S is produced by polarized light from the star. During interval D a quartz depolarizer is placed in the light path, and C is the result when the cover glass is tilted 20° about

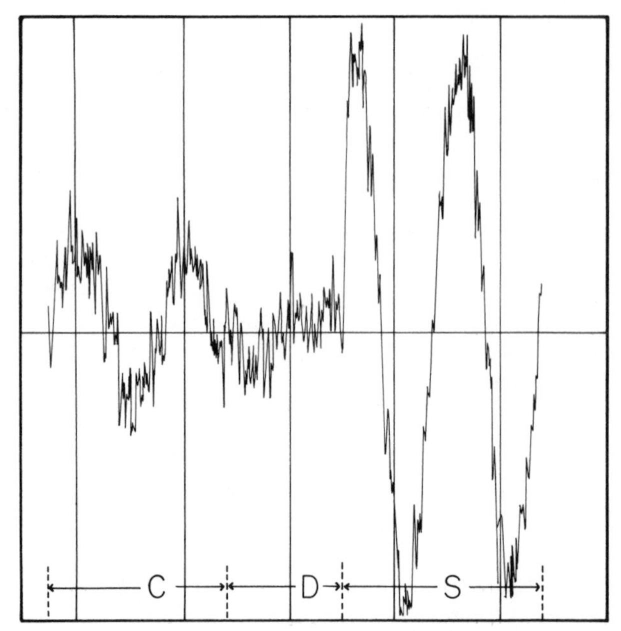

Fig. 89.1 The light from the star HD 19820 observed during 2-min intervals. The intervals *D* and *S* denote, respectively, depolarized and polarized light, and the interval *C* denotes light received during a special orientation of the cover glass. For this star the circular polarization is 5%, and the position angle of the plane of polarization is +30°.

an axis whose position angle is arbitrarily set at 94°. The starlight was already depolarized during the interval *C*. The plane of polarization is defined by the direction of the light and the axis about which the glass is tilted. A 20° tilt corresponds to 1.4% polarization.

Here we omit a figure showing that the percentage polarization increases with the color excess of the observed star and another showing that the planes of polarization for 28 stars exhibit no preferential orientation.

1. W. A. Hiltner, *Science 109*, 165 (1949).
2. John S. Hall, *Astronom. J. 54*, 39 (1948).

Appended Figure

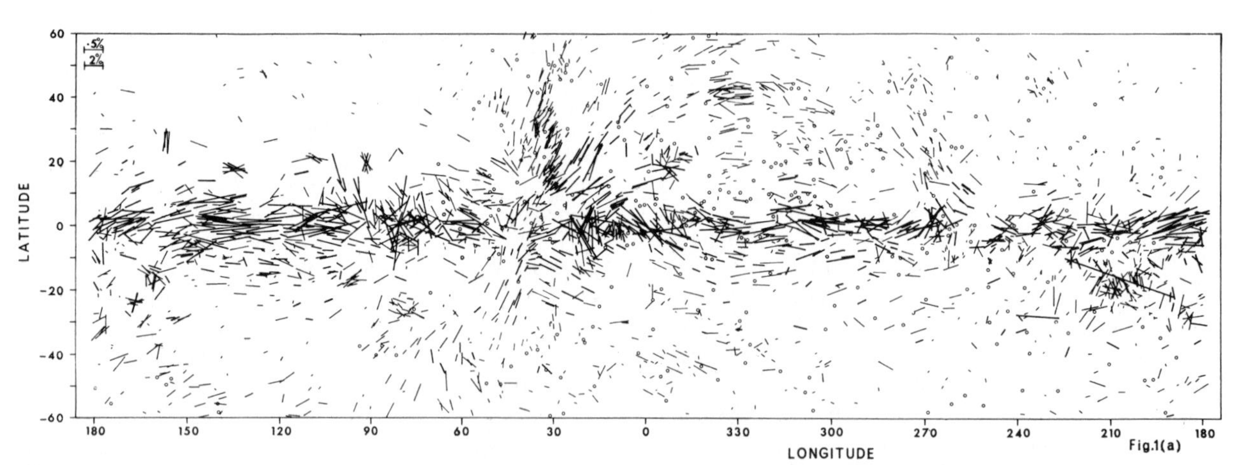

Fig. 89.2 Polarization results for nearly 7,000 stars, the majority of which lie within 1.5 kpc of the sun. The short lines indicate the strength and direction of the starlight. The results are plotted in galactic coordinates where the galactic plane runs horizontally across the middle of the figure. In most cases the direction of polarization is along the galactic plane. Where vertical projections occur, they are correlated with similar upward projections of interstellar neutral hydrogen. (Courtesy D. S. Mathewson.)

90. The Temperature of Interstellar Matter

Lyman Spitzer and Malcolm P. Savedoff

(*Astrophysical Journal 111*, 593–608 [1950])

The preceding selections have shown how astronomers of the twentieth century gradually came to appreciate the existence and role of interstellar material in our galaxy. Following the work of Bengt Strömgren (selection 86), they recognized more fully that a small fraction of the interstellar gas in the immediate vicinity of hot stars is highly ionized and at a high temperature, about 10,000°, but the temperature of the surrounding regions, in which the hydrogen is mostly neutral, remained uncertain.

In a series of papers beginning in 1948, Lyman Spitzer delineated the various processes that determine the temperatures of both the ionized (H II) and neutral (H I) regions of interstellar hydrogen. As a starting point, he adopted A. S. Eddington's conjecture that the frequent elastic collisions between interstellar particles will lead to a Maxwellian distribution of particle velocities characterized by a specific temperature. He then considered each of the processes that might heat or cool the interstellar gas and arrived at the temperatures given in this paper. For the H II regions the situation is essentially that predicted by Eddington in his Bakerian lecture of 1926, in which the heating by photoionization is balanced by cooling through the radiative recombination of electrons with ions to give a kinetic temperature of about 10,000°. In the H I regions, however, new heating and cooling processes had to be introduced, and the kinetic temperature was found to be around 60°—considerably higher than Eddington had deduced over a decade earlier. In the paper given here, Spitzer and Malcolm Savedoff introduce the ideas of cooling through the excitation of ions by electron impact and heating through cosmic ray ionization of hydrogen atoms. These processes were later considered in greater detail by Michael J. Seaton and by Spitzer.[1]

Subsequent detailed analyses have not, however, led to any changes in the general conclusions presented here. The interstellar medium is composed of two distinctly different regions with temperatures on the order of 100° and 10,000°. In fact, only a few years after this paper was published, radio astronomers were able to use measurements of the 21-cm line emission of neutral hydrogen to show that the average temperature of H I regions is about 100°.[2]

1. M. J. Seaton, *Annales d'astrophysique 18*, 188 (1955), and *Reviews of Modern Physics 30*, 979 (1958); L. Spitzer, *Diffuse Matter in Space* (New York: John Wiley, 1968); L. Spitzer and E. H. Scott, *Astrophysical Journal 158*, 161 (1969).

2. H. C. van de Hulst, C. A. Muller, and J. H. Oort, *Bulletin of the Astronomical Institutes of the Netherlands 12*, 117 (1954). See also selection 92.

Abstract—Detailed computations reveal a large difference of kinetic temperature T between H I and H II regions. In observed H II regions T probably exceeds 7,500° K, a lower limit established by observations of [O II] λ3,727, and is probably not greater than 13,000° K. The theoretical computations give actual temperatures ranging between 5,000° and 10,000° near B stars and between 7000° and 13,000° near O stars; the exact value depends on the ratio of C, N, and O ions to protons and on the cross-section for electron excitation of these ions.

In H I regions the sources of energy gain are relatively weak; the temperatures of clouds are reduced to about 60° K by three processes: electron excitation of the low-lying levels in C II and Si II, inelastic collisions between H atoms and grains, and inelastic collisions between H atoms and H_2 molecules. Even under somewhat extreme assumptions, the temperature in an H I cloud does not exceed 200° K.

Values of the time t_E required to approach the equilibrium temperature are very short if the density is high. If the density of H atoms or ions is 10^2 per cubic centimeter, t_E is only about 10^2 years in both H I and H II regions. However, if the density of H atoms or ions is taken to be 10^{-2}, a value which may represent rarefied regions between clouds, the values of t_E are about 10^6 years in H II regions and 10^9 years in H I regions.

As a result of the low temperatures found in H I clouds, the average charge on a dielectric interstellar grain is very small but may rise to several volts in H II regions.

THE PHYSICAL PROCESSES which influence the random kinetic energy of the interstellar particles have been analyzed in two earlier papers.[1] Since equipartition of kinetic energy is established, it is convenient to describe the mean kinetic energy per particle by means of the "kinetic temperature." In the present paper the previous results are used to compute kinetic temperatures to be expected under differing conditions. The computations determine essentially the kinetic temperatures at which the gain of kinetic energy in ergs per second per cubic centimeter in all types of encounters equals the corresponding rate of loss of energy. The gain of kinetic energy per free electron per second, in encounters with particles of type j, has been denoted by G_{ej}, and the corresponding loss by L_{ej}. Similarly, G_{Hj} and L_{Hj} denote the corresponding rates of gain and loss for H atoms. Since gains and losses experienced by other particles are negligible, the condition of equilibrium gives

$$n_e \sum_j G_{ej} + n_H G_{Hc} = n_e \sum_j L_{ej} + n_H \sum_j L_{Hj}. \quad (1)$$

Formulae for the different G_{ej}'s and for G_{Hc} are given in Paper I, while L_{ej} and L_{Hj} may be found from Paper II. As before, subscripts e, p, H, i, m, g, and c denote electrons, protons, H atoms, ions other than protons, H_2 molecules, grains, and cosmic rays, respectively.

Here Spitzer and Savedoff explain that they have reduced the de-excitation cross section per unit weight for a neutral atom to 2×10^{-18} cm^2, which is a factor of 10 lower than that used in Paper II.

It should be emphasized that the kinetic temperatures computed in this paper, by the use of equation (1), are those that would exist in an ideal equilibrium state, with no transfer of heat by conduction or convection. Transfer of energy from mass motion into random thermal motion, which will heat up the medium when two clouds collide, is also neglected in equation (1).

The first of the following four sections lists the various astronomical assumptions made concerning the interstellar radiation field and the characteristics of the interstellar gas and grains. Section II gives results obtained for regions of ionized hydrogen (H II regions), including temperatures, the times required to approach equilibrium conditions, and the electric charge on the grains. Section III gives comparable results for regions where the hydrogen is predominantly neutral (H I regions). In Section IV is presented a summary of the conclusions which may be drawn on the basis of these results. The appendix discusses equipartition of energy among the interstellar particles and shows that, while electrons and H atoms may in extreme cases have different kinetic temperatures, almost complete equipartition of kinetic energy is probably the more usual situation.

1. Astronomical Assumptions

A. RADIATION FIELD The energy density of interstellar radiation at different frequencies determines both the rate at which photoelectrons are emitted from atoms and the mean kinetic energy of these electrons. These effects have already been analyzed in Paper I, where an effective color temperature T_{cj} was introduced as a parameter in the equation for each G_{ej}. Values of these parameters have already been listed both for general galactic starlight and for radiation from a single luminous star. In the latter case the values of T_{cp}, in particular, were listed not for stars of a particular spectral type but rather in terms of the temperature of the equivalent black bodies. We shall here assume that the radiation emitted by stars of types O, B, and A has an actual color temperature of 40,000°, 20,000°, and 9,000° K, respectively. These values are consistent with those given in Kuiper's analysis of stellar temperatures.[2]

The intensity of stellar radiation also affects the ionization level of the interstellar gas. Here one must take into account the relatively sharp division of this gas between H I and H II regions, a phenomenon first pointed out by Strömgren.[3] It is clearly legitimate to assume that in H I regions there is virtually no radiation of wave length shorter than 912 A. Thus elements with an ionization potential higher than 13.5 volts will be considered neutral in these regions.

B. DENSITY AND COMPOSITION OF INTERSTELLAR GAS The average density of the interstellar gas in the galactic plane is usually assumed to be in the neighborhood of 1 hydrogen atom per cubic centimeter. The best evidence supporting this value is provided by the observations obtained by Struve and Elvey[4] on the galactic emission of the Balmer lines, as analyzed by B. Strömgren.[3] The occurrence of gas clouds, so beautifully demonstrated by W. S. Adams' high-dispersion spectra of the interstellar K and H lines,[5] casts some uncertainty on this earlier estimate, which was based on the assumption of a homogeneous medium. A more recent analysis by Strömgren[6] indicates that 1 hydrogen atom per cubic centimeter can still be considered consistent with the observational evidence.

It seems very likely, however, that higher densities are encountered inside the obscuring clouds, with lower densities outside. Strömgren's analysis[6] of interstellar absorption lines indicates that the electron density n_e within a cloud is about 4×10^{-3} cm^{-3} and that, if the ratio n_e/n_H is set equal to 5×10^{-4}, corresponding to the standard stellar composition, n_H is about 8. Between the clouds, values of n_H less than 0.1 are found. This value for the density within a cloud is roughly consistent with the evidence that obscuring clouds occupy about 5 per cent of the space in the galactic plane, as found by Strömgren[6] and Spitzer[7] from studies of interstellar extinction. If the average value of n_H were 1 and if n_H between the clouds were very small, this estimate would yield 20 cm^{-3} for n_H within the clouds. It is likely that somewhat higher values may be found in denser clouds. In the Orion nebula, for example, Greenstein[8] finds that the density of protons is between 10^2 and 10^3 cm^{-3}.

In the present work, densities were assumed to cover as large a range as seemed of any interest, and the values of n_H considered were 10^{-2}, 1, and 10^2. The first of these values is probably moderately representative of the more rarefied regions of the intercloud medium, while the last represents that of the denser clouds. In H II regions these same values were taken for n_p, the number of protons per cubic centimeter.

While the composition of the interstellar gas may, as a first approximation, be assumed to be the same throughout interstellar space, the level of ionization will be very different in H I and H II regions. In H II regions all the atoms may be assumed to be ionized. The value of the ratio n_i/n_p may be taken as 2×10^{-3}, in agreement with recent summaries by Unsöld[9] and Kuiper;[10] n_i is the particle density of ions other than hydrogen and helium. This value is somewhat higher than for planetary nebulae, where Aller and Menzel[11] found values of n_i/n_p ranging from 3×10^{-4} up to 2.5×10^{-3}. Since the temperatures actually computed for H II regions with the assumed abundance of ions were actually somewhat less, in general, than the lower limit established by observations of [O II] λ 3,727, computations were also made with n_i/n_p reduced by a factor of 10 to 2×10^{-4}. It is possible that the relative abundance of most atoms other than hydrogen may actually be considerably less in interstellar space than in the stars, since an appreciable fraction of these atoms may be locked up in interstellar grains. Alternatively, the collision cross-section occurring in L_{ei} may perhaps be less than the value assumed. The temperatures computed for H II regions with the low ion abundance are identical with what would be obtained with a high ion abundance but a lower cross-section. This alternative interpretation is not possible in H I regions, however, where n_e is reduced when n_i is lowered.

In H I regions, roughly three-fourths of the atoms (N and O) will be neutral, while the remaining fourth (C, Mg, etc.) will be almost completely ionized at all the densities considered; thus the "standard" values of n_a/n_H and n_i/n_H may be taken as 1.5×10^{-3} and 5×10^{-4}, respectively. As for H II regions, computations were also made for H I regions of low ion

Table 90.1 Relative abundance of interstellar particles

Region	n_e/n_H	n_g/n_H	n_i/n_H	n_a/n_H	n_m/n_H
H I					
Standard	5×10^{-4}	10^{-12}	5×10^{-4}	1.5×10^{-3}	a
Low ion density	5×10^{-5}	10^{-12}	5×10^{-5}	1.5×10^{-4}	a
Concentrated cloud	5×10^{-4}	10^{-10}	5×10^{-4}	1.5×10^{-3}	a
Region	n_e/n_p	n_g/n_p	n_i/n_p	n_a/n_p	n_m/n_p
H II					
Standard	1	10^{-12}	2×10^{-3}	0	0
Low ion density	1	10^{-12}	2×10^{-4}	0	0

a. Depends on n_g only, equaling 0.1 for $n_g = 10^{-10}$ cm^{-3} or more, and 0 for $n_g = 10^{-12}$.

density, where these ratios have one-tenth their standard values. These relative abundances are summarized in table 90.1.

In accordance with the discussion in Paper II, a substantial abundance of H_2 molecules must also be considered when n_g exceeds 10^{-12}. It has been assumed that, whenever n_g is 10^{-10} cm^{-3}, or greater, the ratio of n_m, the number of hydrogen molecules per cubic centimeter, to n_H is 0.1. Thus, when n_H is unity, only a concentrated cloud is assumed to contain H_2, while with n_H equal to 10^2, n_m/n_H is set equal to 0.1 for all compositions.

It will be noticed that no mention is made here of *He* and *Ne* atoms. Close to some very hot stars the *He* atoms will be ionized, and the electron temperatures may be somewhat higher than the values found here. More usually, however, *He* will be neutral and will have little effect on temperatures. *Ne* atoms can be expected to have even less effect.

It should be noted that the formulae adopted for L_{ei} in Paper II are somewhat approximate in *H* I regions. The low-lying electronic levels, which are important in these regions, were represented by a level at 0.015 electron volts (e.v.); half the ionized atoms were assumed to possess this level. Since *N* and *O* will be neutral in *H* I regions, the only abundant ions with levels within 0.04 e.v.of the ground state will be *C* II and *Si* II. (The excited $^2P_{3/2}$ levels in *C* II and *N* III, with excitation potentials of 0.0079 and 0.022 volts, respectively, and with values of 2 and 4 for g_A and g_B, were inadvertently omitted from table 2 of Paper II.) These excited levels have excitation potentials about half and twice, respectively, the adopted value of 0.015 e.v. It was not thought worth while to compute more precisely the contribution of *C* II and *Si* II to L_{ei}, since the error in T introduced by this approximate representation is less than other uncertainties in the temperature.

C. PROPERTIES OF INTERSTELLAR GRAINS The total cross-sectional area of interstellar grains plays an important part in the temperature of *H* I regions. If the extinction efficiency factor Q were unity (if all grains were very much larger than the wave length of visible light), an average photographic extinction of 1.6 mag. per kiloparsec in the galactic plane would yield a total absorption cross-section of 4.8×10^{-22} cm^2/cm^3. Since the radius of the grains has been arbitrarily taken as 10^{-5} cm, the corresponding value of n_g, the number of grains per cubic centimeter, would equal 1.5×10^{-12}. According to current ideas, most of the obscuring particles have dimensions comparable with the wave length and have an extinction efficiency factor of the order of magnitude of 1. Hence 10^{-12} seems a reasonable average value for the ratio of n_g to n_H, and this value has been adopted here as standard. The various particle distributions considered by van de Hulst[12] lead to values of the total geometric cross-section about the same as those used here.

According to certain theories on the formation of stars,[13,14] the interstellar clouds are produced by the radiation pressure of galactic light on the grains; this pressure pushes the grains toward one another and results in an increase in the concentration of grains relative to the gas. Hence temperatures have also been computed for concentrated clouds, with a value of 10^{-10} for n_g/n_H, an increase of 100 over the standard value of this ratio. This increased relative abundance of grains, which definitely represents an extreme case, has not been assumed for the intercloud regions, where n_H is taken to be 10^{-2}.

The internal temperature of the grains, denoted by T_g in Paper II, is an important parameter in the equation for L_{Hg}. The factors determining T_g have been analyzed in some detail by van de Hulst.[15] The results are somewhat uncertain, with temperatures ranging from 10° to 20° K for dielectric spheres and from 30° to 100° K for metal spheres. A recent theoretical explanation[16] of the polarization of light in interstellar space suggests that the percentage of the metallic particles is somewhat higher than was envisaged by van de Hulst. However, it is still likely that a larger proportion of the interstellar grains are dielectric; accordingly, T_g has been set equal to 20° K in these computations.

Finally, the photoelectric properties of the grains must be specified. Computations have been made for the three types of surfaces described in table 8 of Paper I. Surface A is equivalent in photoelectric sensitivity to the best photocells, while surface B is ordinary metal with lower efficiency. Surface C is nonmetallic and gives no photoemission. It seems likely that the actual grains are much closer to surface C than to surface B. Grains of surface A seem highly implausible but cannot be entirely excluded.

II. REGIONS OF IONIZED HYDROGEN

A. TEMPERATURES A graphical method was used to determine, under each set of assumptions, the temperature at which the sum of all the G's is equal to the sum of the L's. This process was much simplified by the fact that, in *H* II regions, only G_{ep} and L_{ei} were appreciable. The results for the different cases are shown in table 90.2. Temperatures are given for a rarefied *H* II region produced by general galactic starlight and for *H* II regions around stars of types O, A, and B. Values for general galactic starlight are given for only the lowest density, since at higher densities the *H* II region will not usually extend far enough from any one star to make the contributions from other stars of much importance.

The low temperatures found in this table are the result of the thermostatic action of the various low-lying ionic levels, an effect first analyzed in detail by Menzel and his collaborators.[17,18] For any particular ionizing radiation, the temperature depends primarily on n_i/n_p; obviously, the temperature will fall if the number of ions which cool off the gas is increased relative to the number of protons which heat up the gas.

At low densities the temperatures depend very little on the density. At the higher densities, collisional de-excitation becomes important, the population of the lower levels approaches the Boltzmann distribution, and the rate of dissipation of energy per gram of matter no longer increases with increasing density. However, G_{ep} continues to increase with increasing density, and the temperature therefore rises.

The observations of the general galactic emission by Struve and his co-workers[19] give evidence on the temperatures of H II regions. These observations show that the line [O II] λ 3,727 is, on the average, about as strong as $H\alpha$. The excitation of λ 3,727 requires 3.31 e.v. of energy, equivalent to an electron temperature of many thousand degrees. More quantitatively, from the equations given in Papers I and II, the ratio of intensities of $H\alpha$ and λ 3,727 can readily be expressed in terms of n_i/n_p, the collisional cross-section for excitation of O II, the electron temperature, and other parameters. By adopting extreme assumptions for the other quantities, a lower limit on the temperature can be obtained. In this way we find that even if $\sigma_{BA}(a)$ in equation (20) of Paper II is set equal to 10^{-17} per unit weight of the ground state, about the maximum value permitted by the conservation theorem, and if there is one O II ion to every thousand protons, the temperature is found to be about 7,500°, which we may take as a lower limit for T in the observed H II regions.

According to table 90.2, the computed temperatures for the standard composition are actually less than this lower limit in almost all cases. A small decrease in ion density from the standard composition, with a reduced value of n_i/n_p in the range 5×10^{-4} to 10^{-3}, would give good agreement between the theoretical temperatures and the lower limit found from observations. Alternatively, a further decrease by a factor of 2–4 in the cross-section for collisional de-excitation would give the same agreement, but with the standard ratio of 2×10^{-3} for n_i/n_p. It may be noted that, if the collisional cross-section had not been reduced at the beginning of this paper, the computed temperatures for n_p equal to 1 and for the standard composition would range from 210° near an A star to 1,200° near an O star.

Table 90.2 Equilibrium temperature in H II regions

Particle density (in cm^{-3})		Temperature (in °K) for ionizing radiation from:			
n_p	n_i	Galaxy	O star	B star	A star
10^{-2}	2×10^{-5}	6,000	6,800	4,800	2,800
10^{-2}	2×10^{-6}	11,000	12,000	9,500	5,800
1	2×10^{-3}	—	6,900	5,000	3,000
1	2×10^{-4}	—	12,000	9,500	5,900
10^2	2×10^{-1}	—	7,600	5,700	3,600
10^2	2×10^{-2}	—	13,000	10,000	6,600

To obtain an upper limit on T we note that the greatest temperature in table 90.2 is 13,000°. This value is already based on a relatively low abundance of ions or on a very low excitation cross-section. Further marked decreases in either of these quantities seem unlikely. Moreover, it seems most improbable that cosmic rays can increase the temperature materially; even if G_{ec} were increased by a factor of 10^5, the upper limit found in Section III, this quantity would only just equal G_{ep}. While it is always possible that unknown sources of energy may increase T, we may tentatively adopt an upper limit of about 13,000° for the kinetic temperature of most H II regions.

The contributions of photoelectric grains to the energy gain were generally quite small and have been neglected in table 90.2. Even for surface A, the value of G_{eg} is almost always less than G_{ep}, provided that T exceeds 4,000° and that the ratio n_g/n_p does not exceed the standard value 10^{-12}. For example, to make G_{eg} equal to G_{ep} with the standard value of n_g/n_p and a kinetic temperature of 4,000° requires[20] that the grains have the very high positive charge of 50 volts (γ about 100) to make collisions with electrons sufficiently frequent. With so high a charge, the wave length at the photoelectric threshold would be decreased to about 250 A! The energy density of such short-wave radiation in interstellar space is much too small to permit the maintenance of so high a charge. If n_g/n_p is increased to 10^{-10}, G_{eg} can exceed G_{ep} in a number of extreme cases, whose physical interest did not seem to warrant the computational labors required for their investigation.

B. TIME OF EQUILIBRIUM It is natural to inquire not only as to the temperature in equilibrium but also as to the rate of approach to equilibrium. We shall define the time of approach to equilibrium, designated by t_E, by the equation

$$t_E = \frac{3nk(T - T_E)}{2n_e(L_e - G_e)}, \tag{2}$$

where n is the total density of particles of all sorts, T_E is the equilibrium temperature, and G_e and L_e are the sums of all the individual G's and L's; as usual, k is the Boltzmann constant, and n_e is the number of electrons per cubic centimeter. It will be noted that t_E is simply the excess kinetic energy per cubic centimeter of the interstellar medium divided by the net rate of loss of kinetic energy per cubic centimeter per second. When T is nearly equal to T_E, the numerator and denominator in equation (2) become proportional to each other, t_E is then constant with temperature, and the approach to equilibrium temperature will become exponential. When the deviation is large, however, t_E is a function of T.

Values of t_E, computed for four different values of T, are given in table 90.3 for the standard composition and a low proton density. For different over-all densities the value of

t_E is very nearly inversely proportional to n_p. Only at the highest densities and at temperatures of 50° and 500° do the values of t_E differ by more than 10 per cent from what is found on the assumption of strict inverse proportionality to the density. For lower values of n_i/n_p, the value of t_E will be proportionally greater at the highest temperatures but will be independent of n_i/n_p at the lowest temperatures. The values in table 90.3 are computed for *H* II regions surrounding a B star, but the values obtained for other types of ionizing radiation do not usually differ by more than 20 per cent from the values in table 90.3 and are always within a factor of 2 of these values.

Table 90.3 Time for equilibrium in *H* II regions

From initial temperature (°K)	Time (in years) with $n_p = 10^{-2}$ cm^{-3} and standard composition[a]
50	1.8×10^5
500	7.5×10^5
5,000	1.4×10^6
50,000	1.7×10^6

a. For other proton densities t_E is roughly proportional to $1/n_p$.

The general behavior of the values in table 90.3 is in accord with physical expectations. As T increases, the cross-section for interaction between electrons and either protons or ions goes down as $1/T$, thus increasing t_E. At the higher temperatures this effect is partially offset by the increased number of atomic energy levels which may be excited by the electrons.

It is evident from this table that the values of t_E are not very high even when the density is low and that for higher densities equilibrium will be reached relatively rapidly. For comparison it may be noted that an *H* II region around an O star requires 10^7 years to move through a distance about equal to its radius. We may conclude that, as a hot star moves through space, the region of ionized hydrogen around it is usually at a temperature not far from its equilibrium value. This conclusion may require some modification if turbulent motions within the medium are responsible for large heating effects, but it does not seem likely that the rate of heating from such a source can exceed the rate resulting from absorption of starlight.

C. CHARGE ON THE GRAINS The charge and temperature of the grains depend upon their composition, which is unknown. Spitzer and Savedoff assume three different grain surfaces, A, B, and C, which range from a metallic to a dielectric composition. The three surfaces differ primarily in their photoelectric efficiencies, emitting, respectively, 0.01, 0.0001, and 0 electrons per absorbed photon.

Since G_{eg} was found to be unimportant in these regions, no detailed computations were made of the electrical charge on the grains. However, certain simple results are evident at once from the basic equations. For substance C the photoelectric emission vanishes, and the charge adjusts itself to equate the number of electrons captured per second to the corresponding number of positive ions captured. If the sticking probabilities ξ_e and ξ_i for electrons and positive ions, respectively, are equal to each other, then the results previously obtained[21] are applicable; the value of γ is 2.51, and the potential is $-2.2 \times (T/10{,}000°)$ volts. These potentials are quite moderate. If, however, the sticking probability ξ_i has the minimum value of 10^{-4} envisaged in Paper I and γ consequently rises to about 10, the charge now becomes $-8.6 \times (T/10{,}000°)$ volts on the basis of the equations given in Paper I. When the potential exceeds several volts, it is probable that free-field emission occurs, and thus an upper limit of about -3 volts may be taken for the charge; the importance of this free-field emission was first emphasized by van de Hulst.[12]

For surface A the photoemissivity is so high that the grains are likely to be charged to a very high positive potential, especially at close distances from stars of early type. For surface B, negative charges are to be expected for high atomic densities and weak radiation, while close to an early-type star moderate positive charges are likely; this result is similar to that obtained by Cernuschi.[22]

III. Regions of Neutral Hydrogen

A. TEMPERATURES Temperatures were computed for *H* I regions by the same methods as those used for *H* II regions. The situation was made somewhat more complicated by the increased number of processes which required consideration. In particular, the computation of G_{eg} for surfaces A and B required a simultaneous solution of equations (52), (54), and (55) of Paper I for each of a number of temperatures. The solution of these equations for γ was obtained by a series of successive approximations. Fortunately, it was found that ξ_i, the neutralization probability for ions, had no important effect on G_{eg}, and thus the large range which table 8 of Paper I exhibits for values of this quantity did not complicate the computations.

The results of these calculations are given in table 90.4. For surface C the temperatures are all low, and at high densities are only 10°–20° higher than the internal grain temperature,

T_g. In view of the importance of this result, let us examine carefully its physical basis. Two sources of energy gain, G_{Hc} and G_{ei}, and two sources of energy loss, L_{ei} and L_{Hg}, are dominant for surface C. At the lowest density, collisions of cosmic rays with neutral H atoms are largely responsible for the heating. As the density increases, G_{ei} increases, while G_{Hc} does not, and when n_H is 10^2, $n_e G_{ei}$ exceeds $n_H G_{Hc}$ in all cases. Most of the cooling results from inelastic electron-ion collisions when the composition is standard; but, for relatively low ion density and in a concentrated cloud, $n_H L_{Hg}$ is about equal to $n_e L_{ei}$. The quantity L_{ea} is less than 1 per cent of L_{ei} because of the very large cross-section for encounters between an ion and a low-speed electron. For the situations considered, L_{eH} and G_{eH} were of no importance.

Table 90.4 Equilibrium temperature in H I regions

Particle density (in cm^{-3})			Temperature (in ° K) for grains of surface		
n_H	n_i	n_g	A	B	C
10^{-2}	5×10^{-6}	10^{-14}	1,310	1,100	740
1	5×10^{-4}	10^{-12}	420	52.0	47.3
	5×10^{-5}	10^{-12}	1,200	330	230
	5×10^{-4}	10^{-10}	450	105	34.9
10^2	5×10^{-2}	10^{-10}	56	43.0	42.6
	5×10^{-3}	10^{-10}	72	37.7	35.1
	5×10^{-2}	10^{-8}	202	34.5	32.4

The increase of T with decreasing n_H, shown in table 90.4, is readily explained. As the density changes, G_{Hc} is constant, while L_{ei} and L_{Hg} both decrease proportionally with the density.

Since both L_{ei} and G_{Hc} are somewhat uncertain, one may inquire as to the variation of temperature to be expected with possible variations in these two quantities. First, we consider changes in L_{ei}. At temperatures below 100°, L_{ei} is primarily the result of electron excitation of the low-lying level of C II. To obtain an upper limit on the temperature, we may set L_{ei} equal to zero, which gives T equal to about 190° for the standard composition and n_H equal to unity. If L_{ea} is also set equal to zero, T rises to about 280°. Within an H I cloud, however, the presence of H_2 molecules may be expected to keep T below these values. For example, if n_m/n_H is set equal to 0.1, T becomes about 60° when the composition is standard, L_{ei} is neglected, and n_H is 1 or higher. One may conclude that possible errors in the adopted value of L_{ei} do not modify the conclusion that in an H I cloud T is in the neighborhood of 60°.

Possible errors in the assumed value of G_{Hc} are more serious, since the low-energy end of the cosmic-ray spectrum is highly uncertain and G_{Hc} may exceed the assumed value of 4.1×10^{-30} erg/sec by several orders of magnitude. If H_2 molecules are present, however, very large increases in G_{Hc} will be required to increase T very materially. For example, if one H_2 molecule to every ten H atoms is added to the standard composition, and n_H is taken to be unity, G_{Hc} must be increased by a factor of 10^3 to make T equal to 200°; if n_H is taken as 10^2, G_{Hc} must be increased by 10^5 to give 200° for T. Such large increases seem improbable, since they would imply a dissipation of energy by cosmic rays that would be an appreciable fraction of the total stellar radiation. If G_{Hc} is increased by a factor of 10^5, for example, the average dissipation of energy per cubic parsec, with n_H set equal to 1, becomes 3×10^{-3} $L_\odot$ per cubic parsec. It would be difficult to account for the generation within the galaxy of such intense cosmic radiation. We may conclude that cosmic rays are not likely to yield temperatures much in excess of 200° within an H I cloud.

For surface B the temperatures shown in table 90.4 tend to be somewhat higher than for surface C, especially at the lower densities. This results from the influence of G_{eg}, which becomes appreciable under these conditions. As with surface C, the temperatures tend to be greater at the lower densities. When the photoelectric efficiency is not too great and the charge on the grain relatively small as a result, the rate of photoemission of electrons from a grain will be nearly independent of the gas density. Thus G_{eg} behaves in much the same way as G_{Hc}, and T decreases with increasing n_H. In fact, when n_H is as great as 10^2, G_{eg} for surface B becomes less than $G_{ei} + n_H G_{Hc}/n_e$ and the temperatures are practically those found for surface C.

Loss of energy by molecules can be important at the temperatures reached for surface B. At the temperature of 105° shown in the fourth row of table 90.4, L_{Hm} contributes half the total loss, with L_{Hg} and L_{ei} each contributing about one-fourth.

For the highly emitting surface A the trends seen with surface B become even more marked. Now G_{eg} is the dominant energy gain under all the conditions considered, except for the lowest density, where cosmic rays contribute two-thirds of the energy gain. The change of temperature with over-all density is now less marked, since the photoelectric emission gives the grain a very high charge and the limiting factor on the photoelectric emission is the number of electrons reaching the grain per second rather than the amount of radiation available.

The only previous analysis of kinetic temperature in H I regions is apparently that by Woolley.[23] In this work Woolley considers that free-free transitions of electrons on impact with ions provide the only source of energy loss in H I regions, and he finds a temperature of 1,000°. Electron captures in excited states, which are neglected in his analysis, would raise T more nearly to the values found in H II regions. However, the neglect of the three main sources of energy loss considered here makes Woolley's results scarcely comparable to the present ones.

The depression of interstellar temperatures by H_2 molecules and by grains has already been noted by Hoyle and Lyttleton[24] and by Hoyle,[25] respectively. Excitation of the rotational states of H_2 was not considered, but formulae were given for essentially L_{em} (with an assumed spontaneous transition probability greater by a factor of 5×10^4 than the value adopted in Paper II) and for L_{Hg}. However, these authors were not primarily interested in the temperatures of the interstellar clouds observed within 1,000 parsecs of the sun and have not pointed out the probable difference of temperature between H I and H II regions.

B. TIME OF EQUILIBRIUM Equation (2) in Section IIB may also be used to determine the time of approach to equilibrium in H I regions. In this case L_{Hg}, L_{Hm}, and G_{Hc} are frequently important, and as in equation (1), these quantities must be multiplied by n_H/n_e before they are combined with the various L_{ej}'s and G_{ej}'s. The values of t_E found are given in table 90.5 for initial temperatures of 50°, 500°, and 5,000°. Results are given for grains with the surface C only. The times found for the other surfaces are usually only slightly less than the values given. At $T = 5{,}000°$, for example, the values of t_E for surface B are within 10 per cent of the values in table 90.5, although at 50° the values are in some cases only one-fourth as great. For surface A the deviation is somewhat greater, but for $T = 5{,}000°$ the times are usually within 20 per cent of the values for surface C.

It is evident from table 90.5 that, especially at low densities, the values of t_E are very long. If the region of low density was originally at a high temperature, it would scarcely have time to cool down to its equilibrium temperature during the age of the universe. At higher densities the times become much shorter and are much less than the million years required for an interstellar cloud to move through a distance about equal to its diameter. The rapid approach to equilibrium when n_H equals 10^2 is largely the result of energy dissipation by H_2 molecules. If n_m is set equal to zero, when n_H is 10^2 and n_g is 10^{-10}, then for initial temperatures between 500° and 5,000° t_E lies in the range from 10^5 to 10^6 years.

Table 90.5 Time for equilibrium in H I regions with grains of surface C

Particle density (in cm^{-3})			Time in years from initial temperature		
n_H	n_i	n_g	$T = 50°$	$T = 500°$	$T = 5{,}000°$
10^{-2}	5×10^{-6}	10^{-14}	1.1×10^9	1.2×10^9	2.1×10^9
1	5×10^{-4}	10^{-12}	1.2×10^7	1.1×10^7	1.9×10^7
	5×10^{-5}	10^{-12}	3.1×10^8	2.1×10^8	9.2×10^7
	5×10^{-4}	10^{-10}	4.3×10^6	1.3×10^5	1.8×10^4
10^2	5×10^{-2}	10^{-10}	9.2×10^4	1.8×10^3	1.8×10^2
	5×10^{-3}	10^{-10}	4.8×10^6	1.9×10^3	1.8×10^2
	5×10^{-2}	10^{-8}	4.3×10^4	1.8×10^3	1.8×10^2

C. CHARGE ON THE GRAINS As a result of the low temperatures found, the electrostatic potential of the grains becomes very low when photoelectric emission is weak. With surface C, for example, if ξ_i equals ξ_e, then γ equals 2.2; and when T equals 40°, the potential on the grain is only about 9×10^{-3} volts, corresponding to an average charge of less than 1 electron. Even if ξ_i is as low as 10^{-4}, the number of electrons computed for a grain of substance C in an H I region is only about 3. Obviously, the equations used to determine γ, which consider a uniform potential on the grain, no longer apply, except perhaps statistically, when the number of electrons per grain becomes so small.

For surfaces B and C photoelectric emission may produce an appreciable potential, especially at the low densities. Values of V, the potential of the grain in volts, were found as a by-product of the temperature calculations. These values are given in table 90.6 for the equilibrium temperatures listed in table 90.4. For comparison, values are also given for the potential of dielectric grains (surface C), computed for $\xi_i = \xi_e$. The values for surfaces A and B are mostly independent of ξ_i, the sticking probability for ions, since the metallic grains are not charged to a sufficiently great negative potential to make the number of collisions with ions as great as the number of collisions with electrons.

Table 90.6 Electrostatic potential of grains

Particle density (in cm^{-3})			Potential (in volts) for surface		
n_H	n_i	n_g	A	B	C
10^{-2}	5×10^{-6}	10^{-14}	+10.2	+2.84	−0.16
1	5×10^{-4}	10^{-12}	+ 3.5	+0.016	−0.010
	5×10^{-5}	10^{-12}	+ 8.3	+0.37	−0.050
	5×10^{-4}	10^{-10}	+ 3.4	+0.020	−0.008
10^2	5×10^{-2}	10^{-10}	+ 0.13	−0.011	−0.009
	5×10^{-3}	10^{-10}	+ 0.81	−0.002	−0.008
	5×10^{-2}	10^{-8}	+ 0.22	−0.009	−0.00

IV. Conclusions

The largely tentative nature of the present investigation should be remembered. The astronomical data on the radiation field and on the density of the interstellar medium are still quite incomplete and uncertain; numerical values of the physical parameters affecting the kinetic temperature are even more uncertain. In view of this situation, the computations summarized here have dealt with somewhat idealized and simplified situations. Evidently, the present results should be used only as an indication of general trends and not as a definite proof that certain specific kinetic temperatures may be expected in interstellar space.

Despite the uncertainty in all these numerical results, however, a large difference of temperature between *H* I and *H* II regions seems clearly evident. The observations of [*O* II] λ 3,727 in *H* II regions indicate that the temperature of these regions must be at least 7,500° and may be somewhat higher. This result can be reconciled with theory either if (*a*) the ratio of *O*, *N*, and *C* ions to protons is somewhat less than 2×10^{-3}, possibly between 5×10^{-4} and 10^{-3}, or if (*b*) the cross-section for excitation of these ions by electron impact is even smaller than the reduced value adopted at the beginning of this paper. These changes are quite within the uncertainty of our present knowledge. While one may conclude that the temperatures of *H* II regions around early-type stars probably lie between 7,500° and 13,000°, a more precise determination is not now possible.

The temperatures of *H* I regions are lower by a factor of about 100 than those of *H* II regions. When the hydrogen is neutral, cosmic rays provide most of the heating at low densities, while at higher densities capture of electrons by *C*, *Mg*, *S*, and *Fe* ions, with subsequent re-emission, makes the larger contribution. Both these sources of energy gain are relatively weak, unless cosmic rays of low energy are much more intense than has previously been assumed. On the other hand, there are three very efficient processes which tend to radiate away the kinetic energy of the assembly: electron excitation of the lowest excited levels of *C* II and *Si* II (and, at somewhat higher temperatures, of *Fe* II), inelastic collisions between *H* atoms and grains, and excitation of the low rotational levels of H_2 on impact with *H* atoms. At low densities, such as may be expected between clouds, these radiative processes are relatively ineffective, and equilibrium temperatures of 500° or more may be expected. In an *H* I cloud, where n_H is probably 10 or more, these three processes probably keep the temperature down to a value between 30° and 100°, with about 60° as the mean of the probabilities. On extreme assumptions, temperatures as high as 200° are possible, but higher values seem very improbable if, as seems likely, H_2 molecules are abundant in an *H* I cloud. On the other hand, temperatures even less than 30° are possible in *H* I regions.

The times required to approach equilibrium temperatures are relatively short for *H* II regions and for dense *H* I regions where H_2 molecules are present. In between the clouds, where the density is much less, the times of equilibrium are considerably longer, and, if the hydrogen is neutral, it is doubtful whether equilibrium will be attained. The temperature of such rarefied regions will doubtless depend on the previous history of the intercloud gas. Material torn out of a dense, cool cloud will probably remain cool until hydrogen is ionized. Gases which once formed part of an *H* II region will probably remain hot, although the hydrogen will become neutral relatively soon if the ionizing radiation is cut off.

Finally, one must consider other factors which may, in principle, affect the temperature. Conduction is readily shown to be unimportant over regions several parsecs in size, even if the density is as low as 10^{-2} *H* atoms or protons per cubic centimeter. Heating produced by the random motions of the clouds will certainly be important in localized regions, as, for example, in two clouds which are colliding; but it seems doubtful whether this influence can increase the average cloud temperatures appreciably. If, as seems likely, the cloud motions result from temperature differences in the interstellar medium, then these motions obviously cannot much increase the average cloud temperature. If, on the other hand, turbulence produced by galactic rotation is the driving force for these motions, then the kinetic energy of galactic rotation is inadequate to maintain all the *H* I clouds at temperatures of some 200° for 10^9 years, since the radiation of energy by H_2 molecules at this temperature becomes very rapid. We may therefore conclude that these other factors apparently have no great effect on average interstellar temperatures.

APPENDIX

Omitted here is an appendix showing that deviations from equipartition of energy among interstellar particles are small and that the electrons and hydrogen atoms in interstellar space can be assumed to have the same kinetic temperatures.

1. L. Spitzer, Jr., *Ap. J. 107*, 6 (1948), *109*, 337 (1949) (subsequently cited as "Paper I" and "Paper II," respectively).
2. *Ap. J. 88*, 429 (1938).
3. Ibid. *89*, 526 (1939).
4. Ibid. *88*, 364 (1938).
5. Ibid. *109*, 354 (1949).
6. Ibid. *108*, 242 (1948).
7. Ibid. p. 276.
8. Ibid. *104*, 414 (1946).
9. *Zs. f. Ap. 24*, 306 (1948).
10. *The Atmospheres of the Earth and Planets* (Chicago: University of Chicago Press, 1948), chap. xii.
11. *Ap. J. 102*, 239 (1945).
12. *Rech. Astr. Obs. Utrecht*, vol. 11, pt. 2 (1949).
13. F. L. Whipple, *Ap. J. 104* 1 (1946).

14. L. Spitzer, Jr., *Centennial Symposia, Harvard Observatory Mono.* No. 7 (Cambridge, Mass.: Harvard College Observatory, 1948), p. 87.

15. *Rech. Astr. Obs. Utrecht*, vol. 11, pt. 1 (1946).

16. L. Spitzer, Jr., and J. W. Tukey, *Science 109*, 461 (1949).

17. *Ap. J. 92*, 408 (1940).

18. D. H. Menzel and L. H. Aller, *Ap. J. 94*, 30 (1941).

19. O. Struve and C. T. Elvey, *Ap. J. 88*, 364 (1938), *89*, 119 (1939), *89*, 517 (1939); O. Struve, C. T. Elvey, and W. Linke, *Ap. J. 90*, 301 (1939).

20. See eq. (57) of Paper I; in this equation the numerical factor 10^{-11} must be corrected to 10^{-10}.

21. L. Spitzer, Jr., *Ap. J. 93*, 369 (1941).

22. *Ap. J. 105*, 241 (1947).

23. *M. N. 107*, 308 (1947).

24. *Proc. Cambridge Phil. Soc. 35*, 405 (1939), *36*, 424 (1940).

25. *M. N. 105*, 287 (1945).

91. Radio Waves from Space: Origin of Radiowaves

Hendrik C. van de Hulst
TRANSLATED BY ELSA VAN DIEN AND VINCENT ICKE

(*Nederlands tijdschrift voor natuurkunde 11*, 210–221 [1945])*

Although Karl Jansky discovered extraterrestrial radio radiation in the early 1930s (see selection 6), astronomers paid little attention to it until 1940, when Grote Reber confirmed Jansky's discovery and two astrophysicists, Louis Henyey and Philip Keenan, developed Reber's thesis that the radio emission comes from a hot interstellar gas.[1] At that time the Dutch astronomer Jan Oort was actively involved in theoretical work on galactic structure, and when Reber published his radio map of the Milky Way in 1944, Oort asked a graduate student, Hendrik van de Hulst, to study and review at a colloquium the theory behind Reber's observations. Oort also asked van de Hulst to see if there were any spectral lines in the radio spectrum.

Van de Hulst helped organize the colloquium at which he presented the paper reproduced here. His most startling prediction is the possibility of detecting a radio frequency spectral line from the neutral hydrogen (H I) regions. He realized that these regions would be cold and that most of the atoms would be in their lowest energy ground state. The electron of the hydrogen atom in this state has two possibilities for its spin, and a change from one to the other can give rise to an emission or absorption at a wavelength of 21 cm. As van de Hulst points out, these changes will occur rarely in the tenuous interstellar gas, but an observer might well detect them when looking through the vast extent of interstellar space. As we shall see in selection 92, this remarkable prediction led to the detection of interstellar neutral hydrogen and eventually to a description of the temperature, density, and motions of interstellar matter.

Van de Hulst also considered the lines emitted during the recombination of free electrons with free protons in ionized hydrogen (H II) regions. Through a numerical error in his original calculations, he incorrectly asserted that these lines will not be observable because of broadening by the Stark effect, but he was perhaps the first to call attention to the fact that these lines must occur at radio as well as optical wavelengths. In fact, subsequent theoretical considerations have shown that thermal motions dominate the broadening of these lines, and radio frequency recombination lines are now used to specify the physical conditions within H II regions.

When reviewing theoretical explanations for the continuum radio emission of the Milky Way, van de Hulst concluded that Reber's observations were consistent with radiation from a hot interstellar gas but that Jansky's observations at longer radio wavelengths were not. As Iosif Shklovskii subsequently pointed out,[2] the radio emission of the ionized hydrogen (H II) regions predominates at the shorter radio wavelengths, whereas a different mechanism, synchrotron radiation, predominates at long radio wavelengths.

In the final part of this paper, van de Hulst remarks on the tremendous importance of radio astronomy to cosmology. Here he notices that the Doppler effect, which weakens the light from distant galaxies, is not as important at radio wavelengths; therefore, deeper regions of the universe can be observed at radio wavelengths. Furthermore, by using the observations then available, van de Hulst is able to rule out the static universe in which red shift is not a Doppler effect. In this section, then, he anticipates much of the subsequent work on radio cosmology, which did not begin in earnest until over a decade later (see selection 118).

1. G. Reber, *Astrophysical Journal 91*, 621 (1940), and *Proceedings of the Institute of Radio Engineers 28*, 68 (1940); L. G. Henyey and P. C. Keenan, *Astrophysical Journal 91*, 625 (1940).

2. I. S. Shklovskii, *Astronomicheskii zhurnal 29*, 418 (1952).

* A portion of this translation, by Elsa van Dien, originally appeared in H. Shapley (ed). *Source Book in Astronomy 1900–1950* [Cambridge: Harvard University Press, 1960].

Summary—Radio waves received from any celestial object—they being the far infrared portion of its spectrum—deserve attention. Although observations of objects with small angular sizes are prevented by diffraction, the sun may be a measurable object for future instruments.

The radiation observed from our galaxy must be caused by the interstellar gas; the stars are ruled out by their small angular dimensions, and the solid smoke particles by their low temperature.

The spectral emission of a homogeneous layer of ionized hydrogen is computed. The continuous spectrum arising from free-free transitions has the intensity of blackbody radiation at wavelengths larger than 6 m and has a nearly constant intensity at wavelengths smaller than 2 m, corresponding to a large and to a small optical thickness, respectively. These intensities, shown in figure 91.1, agree with those computed by Henyey and Keenan and tally fairly well with the observations. No better accordance is to be expected, owing to the unknown electron density and extent of the interstellar gas and to unsatisfactory data about the directional sensitivity of the antenna.

Discrete lines of hydrogen are proved to escape observation. The 21.2-cm line, resulting from transitions between hyperfine structure components of the hydrogen ground level, might be observable if the lifetime of the upper level does not exceed 4×10^8 yr, which, however, is improbable.

Reber's observation of the Andromeda nebula suggests a rather high electron density. A cosmological remark concludes the article. The low background intensity caused by remote nebulae contradicts the Hubble-Tolman static model.

Astronomical Importance

Although the existence of radio waves of extraterrestrial origin has been known for about ten years, astronomers have not yet paid much attention to them. This neglect is partly due to the incomplete data at hand; not much more than an order of magnitude of the intensity and a rough dependence on direction of the radiation has been established. Hence, little is to be expected from a careful discussion of observational material.

Also, the existence of these radio waves is not very interesting from the purely theoretical point of view. The production of radio waves is by no means an essential feature of the physical condition of the interstellar matter. The amounts of energy transformed into radio radiation are so small that they are negligible in the large energy balance that starts with the ionization of interstellar atoms by the light of the stars. We cannot expect any new insight regarding the physical condition of the interstellar gas from purely theoretical considerations of the origin of these radio waves;[1] the condition of the interstellar gas is mainly characterized by its density, degree of ionization, and the distribution of electron velocities.

However, the possibility of direct observation of these waves has now made the subject attractive. For twenty centuries astronomers have obtained all their knowledge from observations in the rather narrow energy range around the visual frequencies. To this end, they built themselves powerful instruments and did not spare any trouble. Now we know that the earth's atmosphere leaves open another frequency range, near the radio frequencies. The first observations have been made, but the technique of observing is still in its infancy. Bakker has shown that it can be improved greatly with means at present available.[2]

The long wavelengths involve one difficulty. Without telescopes of enormous apertures, the radio waves will never yield a detailed picture of the sky. For the time being we shall have to be satisfied to reach a resolving power of about 1°. The sun, the Milky Way, and the brightest extragalactic nebulae would then be measurable objects. In the Milky Way the run of the intensity with latitude and longitude, and especially the distribution of H I and H II regions of the interstellar gas, could be investigated. Moreover, it is possible that the very distant extragalactic nebulae would constitute a diffuse background, which would be of special cosmological interest.

The Spectrum of the Milky Way

We shall try to derive, purely theoretically, the spectrum of the Milky Way at these frequencies. We extend our considerations over the entire width of the "radio window" in the earth's atmosphere, that is, wavelengths from 20 m to 1 cm. The observations at wavelengths 14, 16, and 1.86 m yield intensities of equal order of magnitude. Although a continuous spectrum is probable, we do not want to exclude the possibility of discrete spectral lines. (No restriction is intended by speaking about the Milky Way; the Milky Way is what we observe of our stellar system with its threefold population: stars, interstellar smoke, and interstellar gas.) The maximum intensity of the radio waves has been observed in the direction of the constellation of Sagittarius. We know indeed, from many other data, that this is the direction of the galactic center, in which we look through the deepest and densest layers of our stellar system. I shall schematize a working model[3] by assuming a homogeneous layer of constitution equal to what we know near the sun and a depth of 16,000 pc $= 5 \times 10^{22}$ cm.

Firstly, we try to establish in which group of the population of our stellar system the observed radio waves chiefly originate. We use the law of radiation of a blackbody, which in this region, since $h\nu \ll kT$, is the Rayleigh-Jeans law:

$$j_\nu = 2\nu^2 kT/c^2. \tag{1}$$

[Here j is the brightness of the blackbody in units of erg sec^{-1} cm^{-2} Hz^{-1} rad^{-2} at frequency ν.]

STARS Stars probably radiate approximately like blackbodies, with a mean temperature of $T \approx 5{,}000°$ K. The surfaces of the sun and other stars therefore have an intensity denoted by the line $T \approx 5{,}000$ in figure 91.1, which increases in proportion to ν^2. It does not seem impossible that in the future the sun might be measurable at decimeter wavelengths,[4] although Reber's method is not yet accurate enough to observe the sun. It is certain, moreover, that the stars together do not give a sky background of sufficient intensity; for the stellar discs certainly cover less than 1 part in 10^{10} of the area of the sky.

INTERSTELLAR SMOKE The solid particles of interstellar matter have a temperature of about 3° K. Even if they covered the whole sky, their contribution could be neglected. This conclusion is not changed by the greater details given by Whipple and Greenstein:[5] namely, (1) the argument that too small a particle cannot radiate at all in this frequency range because it has no "eigen-frequency," and (2) the possibility that the smoke particles near the galactic center may have a temperature as high as 30° K, because of the greater energy density of stellar radiation.

INTERSTELLAR GAS The interstellar gas has a kinetic temperature of around 10,000° K; unlike the stars, it radiates from the whole area of the sky. The question is only if the layers are optically thick enough to yield blackbody radiation, and this question cannot be answered without going in some detail into the mechanism of emission.

Here van de Hulst gives a summary of the equations that describe radiation transfer and the intensity of radiation emitted by a homogeneous layer of gas.[6,7] He reproduces the classical expression for the volume emissivity, ε_ν, of a region of ionized hydrogen with electron density N_e and temperature T. During the encounters of free electrons with free protons a bremsstrahlung (braking radiation or free-free emission) is emitted with

$$\varepsilon_\nu = 5.4 \times 10^{-39} G N_e^{\,2} T^{-1/2} \times \text{erg sec}^{-1}\ \text{cm}^{-3}\ \text{Hz}^{-1}\ \text{rad}^{-2},$$

where the Gaunt factor, G, lies between 6 and 9 for the considered case. An optically thin region of total extent l has a brightness j_ν, given by

$$j_\nu = l\varepsilon_\nu, \qquad (2)$$

whereas, because of self-absorption, an optically thick region has a brightness given by equation (1). Substituting $N_e = 1\ \text{cm}^{-3}$, $T = 10{,}000$ K, and $l = 5 \times 10^{22}$ cm into equations (1) and (2), van de Hulst obtains the curved line in figure 91.1, where

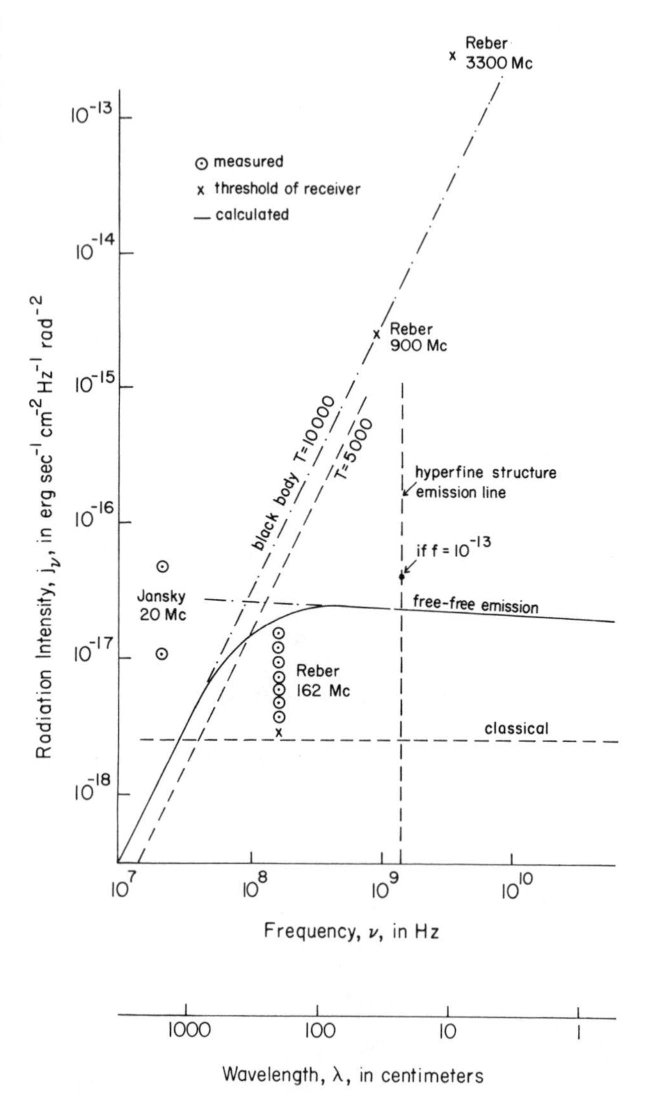

Fig. 91.1 The theoretical and observed radiation spectra at radio frequencies. The thermal radiation of stars is denoted by the straight line $T = 5{,}000°$; and the free-free emission of a hot gas with temperature $T = 10{,}000°$ K, electron density $N_e = 1\ \text{cm}^{-3}$, and extent $l = 5 \times 10^{22}$ cm is given by the solid curved line. The black dot denotes the expected intensity of the 21-cm transition of interstellar neutral hydrogen, and the dashed line denotes the classical calculations for free-free emission (which do not take into account the Gaunt factor). Reber's observations are inconsistent with a stellar origin, and Jansky's results are inconsistent with both the free-free and the stellar radiation mechanism.

the transition between optically thick and thin cases occurs at wavelengths between 1 and 10 m.

The agreement between the observed and the computed intensities is satisfactory. It is obvious why Reber has not found anything at the two short wavelengths. His sensitivity would have been just good enough for blackbody radiation, but the actual radiation is much weaker. Reber's highest intensity, at 1.8 m, agrees well with the computed intensity in the direction of the center of the Milky Way. We must not let our joy over this discovery, however, cause us to overlook the large uncertainties. The observed intensities are first of all definitely uncertain by a factor of 2 higher, which makes the intensity a factor of 4 higher.

The comparison of Jansky's observations with the theory comes off somewhat differently. Since the gas is optically thick at these wavelengths, the intensity depends only on the temperature. We must thus conclude that Jansky's measured points in figure 91.1 lie too high by a factor of 10 or more. It seems quite possible to place most of the blame for this on the insufficiently known directional characteristics [of his antenna]. We cannot reasonably gain more than a factor of 2 by assuming a higher temperature.

Are There Any Separate Spectral Lines?

We have stated that the bound-bound transitions contribute, on the average, only a negligible amount to the continuum. However, the energy liberated in these transitions is emitted in separate lines, and we might expect that within the rather narrow lines the intensity will still be appreciably higher than in the continuum.

Here van de Hulst notices that the α transitions between levels with principal quantum numbers $n+1$ and n might be seen at radio wavelengths with $n \approx 340$. He asserts that Stark broadening of these lines should make them unobservable.[8]

All these lines, like the free-free continuum, will be formed only in the H II regions. But quite a different possibility remains. The ground level of hydrogen is split by hyperfine structure into two states of distance 0.047 cm^{-1}.[9] The spins of the electron and the proton are in the one state parallel, in the other antiparallel. At the spontaneous reversal of the spin a quantum of 21.2 cm would be emitted. Of course, such a transition is forbidden. But the infrequent occurrence of this transition is compensated by the fact that in interstellar space the dilution of the radiation makes the ground state a preferred state over all others (including the free levels) by a factor of 10^{14}. In the H I regions, where the ionized state does not occur, practically all atoms are in the ground state. We suspect that these are distributed about evenly over both sublevels.

It would be of tremendous interest if this line were observable. Without self-absorption, the intensity in the line is

$$j = lNAh\nu/(4\pi\Delta\nu). \tag{3}$$

[Here A is the transition probability for spontaneous emission, l is the extent of the neutral hydrogen along the line of sight, N is the number density of the neutral hydrogen, h is Planck's constant, ν is the frequency of the hyperfine transition, and $\Delta\nu$ is the width of the line.]

The value for A can be theoretically calculated but it is as yet unknown. I therefore substitute $l = 5 \times 10^{22}$ cm, $N = 1$ cm^{-3}, and $\nu/\Delta\nu = 10^4$. The contrast between line and continuum is sufficient if j is greater than the continuum value of 2×10^{-17} erg cm^{-2}. This condition is satisfied if A is greater than 0.8×10^{-16} sec^{-1}, that is, if the lifetime of the hydrogen atom in its highest hyperfine level is less than 4×10^8 yr. That does not seem to be an impossible requirement. With the usual definition of oscillator strength, f, this requirement amounts to $f > 5 \times 10^{-14}$; with Shortley's definition[10] of magnetic dipole strength, S, this becomes $S > 0.03$. Because S is usually of order unity, the case does not seem hopeless, not even if we consider that the low sensitivity of present receivers makes it necessary to increase the requirement by a factor of 100. However, there are reasons to believe that the true S will be rather smaller. Because the rigorous calculation has not been performed, the existence of this line remains speculative.

Van de Hulst next notices that Reber's detection of radio radiation from the Andromeda nebula requires a much higher electron density in this object.

A Cosmological Remark

What do we observe of the remaining spiral nebulae? They are individually small, but together they should be able to produce a detectable background brightness. An even stronger statement can be made: if (1) the universe is perfectly homogeneous, (2) the spiral nebulae are all opaque, and (3) there exists no absorption between the nebulae, nor any other causes by which the distant nebulae are attenuated, then the entire sky must be covered with nebulae of a surface brightness equal to that of a nearby nebula. This argument has already been applied by Herschel to the stars surrounding us. The evident fact that the night sky is not everywhere as bright as the solar disk is a proof of the finite size of our galaxy. The "universe of stars" is not homogeneous.

Conclusions are not so easy to draw for the "universe of nebulae." We are now entering the domain of cosmology. Instead of distance we shall use the time T that the light from a nebula has required to reach us. Hubble's observations with the 100-in telescope on Mount Wilson reach to $T = 6 \times 10^8$ yr. Out to that distance about 1% of the sky area has already been covered, and yet no departures from homogeneity have been discovered. The extrapolation to a completely covered sky thus does not seem senseless.

The other causes by which the light from distant nebulae is weakened are also important. We see the distant nebulae in a very early stage of their development. The intensity is therefore uncertain, but we would sooner expect a higher than a lower brightness. The most important factor is the red shift. All the nebular light is shifted to longer wavelengths by an amount $\delta\lambda$ that, out to the deepest observation of $T = 2 \times 10^8$ yr, is nearly always proportional to the distance. I thank Professor Oort for this essential point in the argument. The more distant nebulae have greater red shifts and less of the ultraviolet region of their spectra is photographed. The result is a considerable apparent attenuation of the observed light from the distant nebulae. In the radio region, the radiation of (originally) shorter wavelength received from the distant nebulae is almost as intense as that from the nearby nebulae. The radio region should therefore be very suitable for tracing the influence of the remaining effects on the background brightness of the sky.

I have in particular investigated whether a decision is possible between the two models put forward for comparison by Hubble and Tolman[11] to explain the numbers of nebulae photographed at Mount Wilson. The first model is an expanding universe where the red shift is due to the Doppler effect. In the second model the red shift is ascribed to an unknown cause, and the universe is static.

Van de Hulst then calculates the ratio of the brightness of the sky background to that of a nearby nebula for objects with spectral indices $\alpha = 2.5$ and $\alpha = 0.1$ in either an expanding or a static universe.[12]

The rough calculation above indeed confirms the suggestion of Professor Oort. In the photographic region the background for both models is about 1% of the brightness of a nearby nebula. The results in the radio region, however, are entirely different. The expanding model again gives about 1% and the static model a divergent result. Therefore, we must take into account the shielding that causes a uniformly bright sky. We compare these calculations with the observations showing definitely that also in the radio region both the Milky Way and the Andromeda nebula stand out as bright objects against a dark background. Thus we may conclude that the static model is incorrect. We must bear in mind that this argument has only eliminated a certain combination of hypotheses; it is not possible to decide *which* of the hypotheses is wrong. Nevertheless, it is satisfying that these observations rule out a model that also has many theoretical objections.

1. See the review in section 2 of the Astronomical Colloquium of the Nederlandsche Astronomen-Club No. 2, *Ned. tijd. natuurkunde 10*, 243 (1943), and the literature cited therein.
2. *Ned. tijd. natuurkunde 11*, 201 (1945).
3. Chosen in agreement with L. G. Henyey and P. C. Keenan, *Ap. J. 91*, 625 (1940).
4. In the meantime, Southworth (*J. Franklin Inst. 239*, 285 [1945]) has succeeded in measuring the solar radiation in the wavelength range 1–10 cm.
5. F. L. Whipple and J. L. Greenstein, *Proc. Nat. Acad. Sci. 23*, 177 (1937).
6. Full derivation by D. H. Menzel, *Ap. J. 85*, 330 (1937).
7. Taken from D. H. Menzel and C. L. Pekeris, *M.N.R.A.S. 96*, 77 (1935).
8. Cf. D. R. Inglis and E. Teller, *Ap. J. 90*, 439 (1939).
9. H. Kopfermann, *Kernmomente* (1940), p. 15, with substitution of $\mu_p = 2.785$.
10. G. H. Shortley, *Phys. Rev. 57*, 225 (1940).
11. E. Hubble and R. C. Tolman, *Ap. J. 82*, 302 (1935).
12. O. Heckmann, *Theorien der Kosmologie* (Berlin, 1942), §17d.

92. The Radio Frequency Detection of Interstellar Hydrogen

Radiation from Galactic Hydrogen at 1,420 MHz

Harold I. Ewen and Edward M. Purcell

(*Nature 168*, 356 [1951])

The Interstellar Hydrogen Line at 1,420 MHz and an Estimate of Galactic Rotation

C. Alex Muller and Jan H. Oort

(*Nature 168*, 356–358 [1951])

Following Hendrik van de Hulst's notable prediction in 1945 that cold interstellar hydrogen might be detected at radio wavelengths through spontaneous transitions between the hyperfine components of the ground state of the hydrogen atom (see selection 91), Iosif Shklovskii investigated the problem in greater detail and in 1949 confirmed van de Hulst's speculations that this transition might be observed.[1] At about this time Harold Ewen, a graduate student at Harvard, joined an active reserve unit under the command of the Harvard astronomer Donald Menzel and became interested in the new field of radio astronomy when Menzel sent him for two weeks of active duty as a reserve officer under the direct supervision of Grote Reber. By 1949 the Harvard physicist Edward Purcell had suggested to Ewen that a search for the hydrogen transition might be undertaken despite the speculative nature of its existence as described in van de Hulst's and Shklovskii's papers. The majority of the electronic components were available to Ewen on loan from the Harvard Cyclotron Laboratory, where he was engaged in other investigations, and Purcell suggested a then novel method of signal detection based on a "switched-frequency" mode for the radiometer. In the meantime, the value of the Bohr magneton became known with sufficient accuracy to calculate the wavelength of the hyperfine transition with high accuracy, and this wavelength was confirmed in 1950 by laboratory measurements at Columbia.[2] A simple horn antenna was built of plywood and copper foil and mounted outside a laboratory window, and by March 1951 Ewen had succeeded in detecting the transition with his frequency-switched receiver.

Following the detection, Ewen met with van de Hulst, then at Harvard, to describe the results of the experiment. At this meeting the Harvard team learned for the first time that the Dutch group at Leiden had been actively trying to detect the radio wavelength hydrogen transition for several years. A description of the Harvard receiver was provided to van de Hulst to expedite the conversion of the Dutch system to the switched-frequency mode. The Dutch group had recently hired C. Alex Muller, who had done previous work on low-noise amplifiers, and within eleven weeks Muller had confirmed the discovery. Purcell also contacted the Australian radio astronomy group and provided it with the same details given the Dutch group. Although the Australians were not actively pursuing a search for the transition and did not have a detection system in operation at the transition wavelength, they were able to assemble the necessary components quickly and repeat the detection just four weeks after the Dutch confirmation. Ewen and Purcell delayed the announcement of their discovery in *Nature* until a coordinated report from all three centers was possible. The Australian report was limited to a reproduction of a cable, given here at the end of the Dutch paper. Within a year the Australians had published a more substantial account of their observations.[3]

An interesting aspect of the first observation was that the detected line was seen in emission. The equilibrium populations of the hyperfine components of the hydrogen atom are determined by frequent collisions between the atoms, which means that the observed excitation temperatures of the lines are equal to the kinetic temperature of the interstellar gas.[4] The excitation temperatures of the observed lines confirmed theoretical expectations that the neutral hydrogen in interstellar space has a temperature of about 100° (see selection 90).

When Jan Oort first suggested that van de Hulst search for a spectral line at radio wavelengths, he realized that detection of the transition would permit measurements of velocities by the Doppler effect and thus give important clues for locating the origin of the radio radiation. Oort had already used the observed patterns of stellar motions to show that the entire galaxy must rotate at large speeds about a distant center, and in the second paper given here Oort and Muller show that the effects of differential galactic rotation are seen in both the Doppler shifts and the broadening of the observed 21-cm line profiles. Within a few years the Leiden group had used these profiles to discern a spiral structure in the outer parts of the galaxy and to map out the detailed rotations of the inner parts of the galaxy (see selection 94).

1. I. S. Shklovskii, *Astronomicheskii zhurnal 26*, 10 (1949). Purcell's wife provided Ewen with an English translation of this Russian article, which stimulated Ewen's interest in detecting the hyperfine transition.

2. P. Kusch and A. G. Prodell, *Physical Review 79*, 1009 (1950). Because Ewen's original detection equipment had a tuning range of only ±200 kHz, the wavelength of the transition needed to be known with a fairly high degree of accuracy.

3. W. N. Christiansen and J. V. Hindman, *Australian Journal of Scientific Research A5*, 437 (1952).

4. G. B. Field, *Proceedings of the Institute of Radio Engineers 46*, 240 (1958).

Ewen and Purcell, Radiation from Galactic Hydrogen at 1,420 MHz

THE GROUND-STATE of the hydrogen atom is a hyperfine doublet the splitting of which, determined by the method of atomic beams, is 1,420.405 Mc/sec.[1] Transitions occur between the upper ($F = 1$) and lower ($F = 0$) components by magnetic dipole radiation or absorption. The possibility of detecting this transition in the spectrum of galactic radiation, first suggested by H. C. van de Hulst[2] has remained one of the challenging problems of radio-astronomy. In interstellar regions not too near hot stars, hydrogen atoms are relatively abundant, there being, according to the usual estimate, about one atom per cm^3. Most of these atoms should be in the ground-state. The detectability of the hyperfine transition hinges on the question whether the temperature which characterizes the distribution of population over the hyperfine doublet—which for want of a better name we shall call the hydrogen 'spin temperature'—is lower than, equal to, or greater than the temperature which characterizes the background radiation field in this part of the galactic radio spectrum. If the spin temperature is lower than the temperature of the radiation field, the hyperfine line ought to appear in absorption; if it is higher, one would expect a 'bright' line; while if the temperatures are the same no line could be detected. The total intensity within the line, per unit bandwidth, should depend only on the difference between these temperatures, providing the source is thick enough to be opaque.

We can now report success in observing this line. A microwave radiometer, built especially for the purpose, consists mainly of a double superheterodyne receiver with pass band of 17 kc., the band being shifted back and forth through 75 kc. thirty times per second. The conventional phase-sensitive detector and narrow (0.016 c./s.) filter then enable the radiometer to record the apparent radio temperature *difference* between two spectral bands 75 kc. apart. These bands are slowly swept in frequency through the region of interest. The overall noise figure of the receiver, measured by the glow-discharge method[3], is 11 db., and the mean output fluctuation at the recorder corresponds to a temperature change of 3.5°. The antenna is a pyramidal horn of about 12° half-power beam-width. It is rigidly mounted at declination −5°; scanning is effected by the earth's rotation.

The line was first detected on March 25, 1951. It appeared in emission with a width of about 80 kc., and was most intense in the direction 18 hr. right ascension. Many subsequent observations have established the following facts. At declination −5° the line is detectable, by our equipment, over a period of about six hours, during which the apparent temperature at the centre of the line rises to a maximum of 25° above background and then subsides into the background. The source appears to be an extended one approximately centred about the galactic plane. The frequency of the centre of the line, which was measured with an accuracy of ±5 kc., was displaced some 150 kc. above the laboratory value, and this shift varied during an observing period. Both the shift and its variation are reasonably well accounted for by the earth's orbital motion and the motion of the solar system toward Hercules. The period of reception shifts two hours per month, in solar time, as it ought to.

Some conclusions can already be drawn from these results. Extrapolation of radio temperature data for somewhat lower frequencies[4] suggests that the background radiation temperature near the 21-cm. line is not more than 10° K. Then the hydrogen spin temperature is not more than 35° K., if the source is 'thick'. But we can calculate the opacity of the source on the assumption of a spin temperature of 35° K. and 1 atom/$cm.^3$, using only the observed line-width and the matrix element of the transition in question, and we obtained 900 light-years for the absorption-length. As this is much smaller than galactic dimensions, we conclude that the temperature observed corresponds indeed to the spin temperature at the source. To the extent that 'self-absorption' contributes to the observed line-width, the true absorption length at the frequency of the centre may be *less* than that computed. Further evidence for relatively high opacity is the absence of large frequency-shifts, which would be expected to arise from galactic rotation were the opacity-thickness comparable to the size of the galaxy. This conclusion is contrary to the prediction of Shklovsky,[5] who has recently discussed the possibility of detecting galactic line radiation.

We have made rough theoretical estimates of the efficacy of various processes through which energy is exchanged between the hydrogen hyperfine levels and the other thermal reservoirs in the interstellar matter plus radiation complex. Of these we find exchange with the radiation field (involving spontaneous emission) and exchange with gas-kinetic energy of the hydrogen atoms (via H–H collisions) much the most important, with the latter process probably dominant. This is consistent with the observation, and if correct implies that the gas-kinetic temperature of the hydrogen exceeds, but not greatly, the spin temperature. The estimated spin relaxation time for these processes is of the order of 10^5 years.

1. P. Kusch and A. G. Prodell, *Phys. Rev. 79*, 1009 (1950).
2. H. C. van de Hulst, *Nederl. Tij. Natuurkunde 11*, 201 (1945).
3. W. W. Mumford, *Bell Syst. Tech. J. 28*, 608 (1949).
4. J. W. Herbstreit and J. R. Johler, *Nature 161*, 515 (1948).
5. I. S. Shklovsky, *Astronomicheskii Zhurnal 26*, 10 (1949) (in Russian).

Muller and Oort, The Interstellar Hydrogen Line at 1,420 MHz

FOLLOWING A SUGGESTION made by Dr. H. C. van de Hulst in 1944,[1] attempts have been made to measure the radiation at 1,420 Mc./sec. (λ21 cm.) emitted by atomic interstellar hydrogen. The first experimental evidence for the presence of this interstellar emission line was obtained by Ewen and Purcell on March 25, 1951 (see the preceding communication). In the Netherlands, the first successful measurements were made on May 11.

The receiver consists of a double superheterodyne instrument, with a crystal-controlled first local oscillator, which is switched between two frequencies 110 kc. apart thirty times per second by a reactance modulator. By varying the frequency of the second local oscillator, the two frequencies can be moved together through a 4-Mc. wide pass band at 1,420 Mc. The pass band of the second i.f. channel is 25 kc. wide. Behind the usual phase-sensitive detector a narrow-band filter with a time constant of 12 sec. is used. So the difference between two frequency bands 110 kc. apart is measured. The noise factor of the receiver has not yet been measured. The losses in the coaxial antenna cable are rather high, and an effective noise factor of the whole receiving system of 25 has been deduced from other measurements with the sun as a source of noise. Important parts of the receiver have been constructed at the Philips Laboratory in Eindhoven under the supervision of Dr. F. L. Stumpers.

The receiver has been mounted behind a movable paraboloid of 7.5 m. aperture and 1.7 m. focal-length at the radio station at Kootwijk, which was kindly put at our disposal for these measurements by the Radio Department of the Post and Telegraph Service. The beam-width at half-power is 2.8°.

The results of the measurements made on a few of the first tracings are shown in the accompanying graphs. While these tracings were being made, the instrument was left in a fixed position relative to the earth, the motion across the sky being provided by the earth's rotation. The frequency of the second local oscillator is switched every 3 min. between positions in which either of the two pass bands coincides with the spectral line. The first position gives a positive, the other a negative, deflexion on the recording meter. The curves show the intensity as function of the right-ascension; the interval between successive vertical lines is 20 min. of time. The point at which each sweep crossed the galactic equator has been marked with an arrow, accompanied by a number giving the corresponding longitude. The galactic co-ordinates at the beginning and end of each tracing have also been indicated. The curves shown may be slightly distorted because the radial components of the orbital motion of the earth, of the sun's motion relative to the nearby interstellar clouds and of differential galactic rotation are different for the various directions. However, we do not believe that these effects will have seriously affected the general shapes of the records, except for latitudes less than $1\frac{1}{2}°$, where in some directions the line is greatly widened by galactic rotation, and the measured intensities will be too small.

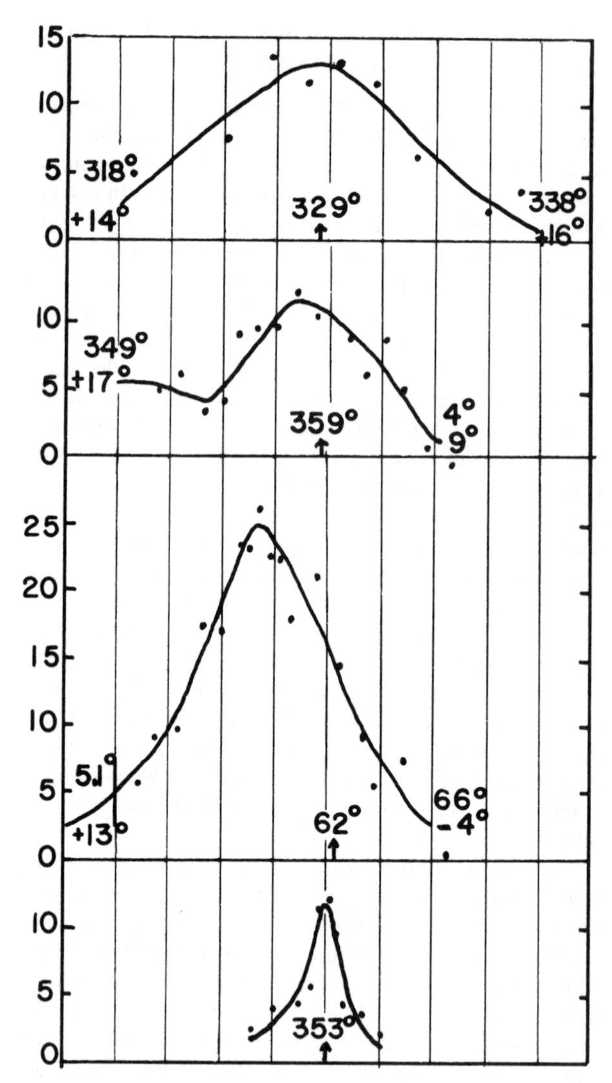

Fig. 92.1 Tracings of the intensity of radiation received at the 21 cm line of neutral hydrogen. Each tracing denotes the intensity received at different right ascensions, and each tracing crosses the galactic equator at the galactic longitude denoted by the arrow. The other numbers denote the galactic latitude and longitude at the beginning and end of each tracing, whereas the interval between successive vertical lines is 20 minutes.

It is evident from the wide spread in galactic latitude that the clouds observed must be relatively close to the earth. From the width of the emission line observed in the region of the centre of the Galaxy, where the rotation of the galactic system does not affect the line, it is found that the random velocities average about 5 km./sec. in one co-ordinate, which agrees approximately with what had been found from absorption lines in the visual region. With such small velocities it is unlikely that the gas extends to more than an average distance of about 50 parsecs from the galactic plane. With an average latitude of the order of 8°, the gas seen in the general direction of the centre cannot then be farther away than 300 or 400 parsecs. In Cygnus the distance may be twice this amount.

These small distances might be taken as an indication either that the more central parts of the Galaxy are devoid of atomic hydrogen, or else that within a distance of between 500 and 1,000 parsecs in the galactic plane the gas becomes optically thick in the wave-length of this line. Although we should like to reserve a definite judgment until more complete measures have been obtained, we believe the latter alternative is the more probable on the basis of the results so far available.

In order to test the presence of radiating hydrogen in the inner regions of the Galaxy, measures have been made across the Milky Way at $l = 355°$, 30° from the centre, where the change of frequency due to differential galactic rotation should make it possible to observe the gas up to large distances. Provisional measures indicate that the emission line in this direction has an effective width of about 330 kc./sec., or 70 km./sec., and that the rotational velocity of the galactic system at a distance from the centre equal to half the sun's distance from the centre is approximately 190 km./sec. This agrees well enough with the rotational velocity of 205 km./sec. computed from a schematic model of the galactic system.[2] (In both cases a value of 270 km./sec. has been assumed for the rotation near the sun.)

The graph at the bottom shows the results of a registration across the Milky Way at 355° longitude and at a frequency 250 kc./sec. lower than the normal frequency of the line, corresponding to a differential radial velocity of +55 km./sec. The gas clouds observed at this frequency are presumably situated at an average distance of about 8 kiloparsecs. It may be noted that the spread in latitude is quite small, and may roughly correspond to the beam-width of the instrument, as would be expected for these large distances.

The tracings reveal many irregularities in distribution. The extension to high positive latitudes in the second graph may possibly be connected with the well-known dense clouds in Ophiuchus. The maxima in the Cygnus region lie systematically about $2\frac{1}{2}°$ south of the galactic circle at longitudes between 40° and 45°, while between 56° and 62° they occur roughly 3° north of the plane. Though no weight should be attached to the exact differences in height of the various sweeps, the intensities in Cygnus appear to be definitely higher than in other regions (except in that opposite the centre). This may be ascribed to the effects of galactic rotation. The fact that the intensity is also lower in the direction towards the centre, where these effects are presumably negligible, is possibly due to the superposition of continuous background radiation, which has probably a considerable intensity in this direction.

1. H. C. van de Hulst, *Nederl. Tij. Natuurkunde 11*, 201 (1945).

2. J. H. Oort, *Bull. Astro. Inst. Netherlands 9*, 193 (1941).

Appendix

The following cable dated July 12 has been received from Sydney, N.S.W.:

Referring Prof. Purcell's letter of June 14 announcing the discovery of hyperfine structure of the hydrogen line in galactic radio spectrum, confirmation of this has been obtained by Christiansen and Hindman, of the Radio Physics Laboratory, Commonwealth Scientific and Industrial Research Organization, using narrow-beam aerial. Intensity and line-width are of same order as reported, and observations near declination 20° S. show similar extent about galactic equator.

J. L. Pawsey

93. Some Features of Galactic Structure in the Neighborhood of the Sun

William W. Morgan, Stewart Sharpless, and Donald Osterbrock

(*Astronomical Journal 57*, 3 [1951]; *Sky and Telescope 11*, 138 [1952])

Speculations that our Milky Way has a spiral structure began almost immediately after Lord Rosse announced that fourteen nebulae exhibit a spiral pattern resembling a whirlpool of light. In our own century the view that our galaxy must be a spiral gained support from observations showing that it is a highly flattened system rotating about a center located far away from the sun in the direction of Sagittarius. Nevertheless, all attempts to establish the spiral structure by counting stars failed. When Walter Baade's work on the Andromeda galaxy revealed that its spiral arms were marked out by comparatively ephemeral hot O and B stars against a rather homogeneous disk of common stars, "like the frosting on a cake," astronomers realized that the spiral arms of the Milky Way galaxy would be detected not from conspicuous density variations but from the positions of the most luminous OB stars and their associated emission nebulae.

Appropriately enough, Baade presented his findings in 1950 at the dedication of a University of Michigan telescope named in honor of Heber D. Curtis.[1] Among the participants at the symposium were two already deeply involved in studies of the luminosity criteria and space distributions of bright young stars: J. J. Nassau of Warner and Swasey Observatory and William Morgan of Yerkes Observatory. Within two years Morgan and two students, Stewart Sharpless and Donald Osterbrock, presented the paper that first traced part of the spiral pattern of our galaxy. The paper, whose abstract is reprinted as the first part of this selection, proved to be the sensation of the December 1951 Cleveland meeting of the American Astronomical Society; probably most astronomers first read of this detection in the illustrated *Sky and Telescope* report, which we also include. This latter account clearly indicates that detailed work from a large number of investigators was involved in establishing the distances to the OB stars and the associated H II regions.

By examining the distribution of emission nebulae, Morgan, Sharpless, and Osterbrock were able to delineate segments of two spiral arms, one that passes through the sun and the other at a distance of about 2,000 pc from the sun in the direction away from the galactic center. Their tentative identification of a third arm based on only one emission nebula was subsequently confirmed by Morgan and his colleagues, when they showed that associations of bright blue stars describe all three arms.[2]

The intrinsically bright O and B stars and their associated emission nebulae are actually now known to describe sections of four spiral arms: the Perseus arm, the Cygnus-Puppis or Orion arm, the Sagittarius-Carina arm, and the Norma-Centaurus arm.[3] Both of the latter two arms lie between the sun and the galactic center. Moreover, in the same year in which the optical astronomers first traced fragments of the Milky Way's spiral structure, radio astronomers first detected the 21-cm line of neutral hydrogen (see selection 92). As selection 94 shows, radio techniques soon made possible a far more extensive delineation of our galaxy's spiral arms.

1. W. Baade, *Publications of the Observatory of the University of Michigan 10*, 7 (1951).
2. W. W. Morgan, A. E. Whitford, and A. D. Code, *Astrophysical Journal 118*, 318 (1953).
3. W. Becker and G. Contopoulos, eds., *The Spiral Structure of our Galaxy: International Astronomical Union Symposium no. 38*, (Dordrecht Holland: D. Reidel 1970).

Astronomical Journal

The distribution in space of the nearer regions of ionized hydrogen has been investigated by spectroscopic parallaxes determined with the 40-inch Yerkes refractor. The regions north of $-10°$ declination occur in two long, narrow belts similar to the spiral arms observed by Baade in the Andromeda nebula. The nearer arm extends from galactic longitude 40° to 190° and passes at its nearest point about 300 parsecs distant from the sun in a direction opposite to that of the galactic center. The observed length of the arm is about 3,000 parsecs; its width is of the order of 250 parsecs. Among the constituents of this arm are the nebulosities in the neighborhood of P Cygni, the North American nebula, the ξ Persei nebulosity, the Orion nebula and loop, and the *H* II regions near λ Orionis and S Monocerotis.

A second arm can be traced from galactic longitude 70° to 140°. This arm is parallel to the first and is situated at a distance of about 2,000 parsecs from it in the anti-center direction. There is some evidence for another arm located at a distance of around 1,500 parsecs in the direction toward the galactic center. This is defined by the series of condensations of O and B stars from galactic longitude 253° in Carina to 345°, the small cloud in Sagittarius. The data are so fragmentary, however, that more observations from the southern hemisphere will be necessary before a definite conclusion can be reached.

Both arms are inclined with respect to the normal to a radius vector by approximately 25°; when this tilt is combined with the known direction of galactic rotation the arms are found to be trailing.

The dimensions of the *H* II regions are similar to those observed by Baade in the Andromeda nebula; the width of the arms is also similar, as is the frequency of *H* II regions along the arms.

The structure described above is also shown by the blue giants, O $-$ A5 stars having M_{vis} brighter than -4.0 mag. The great aggregates of early-type stars, Perseus double cluster, P Cygni region, Orion, are condensations in the arms similar to the condensations observed by Hubble in the Andromeda nebula.

Sky and Telescope

Spiral Arms of the Galaxy

Astronomers at several observatories have collaborated with Dr. W. W. Morgan, of Yerkes Observatory, in a program designed to establish the outlines of the spiral arms our Milky Way galaxy is believed to possess in the sun's general vicinity. In the symposium on the H-R diagram, Dr. Morgan presented the model of the galaxy pictured here and on the front cover. He discussed work by himself, Stewart Sharpless, and Donald Osterbrock, of Yerkes. Associated papers dealt with the collaborative work done at Warner and Swasey Observatory by Dr. J. J. Nassau, and Dr. Daniel Harris, III, of Yerkes; and by Drs. A. D. Code and A. E. Whitford, of Washburn Observatory.

For the purpose, objects easily seen and identified at relatively great distances must be used. These are the kinds of objects observed in the spiral arms of other galaxies. For instance, along the arms of M31, the Andromeda nebula, Dr. Walter Baade, of Mount Wilson and Palomar Observatories, has identified numerous regions of luminous hydrogen gas, and at Yerkes a survey of the Milky Way for similar regions has been made with the Greenstein-Henyey wide-angle camera.

The camera, which covers a field of 140°, was used with an interference filter for hydrogen-alpha light. Many of the emission nebulosities may be identified on the pictures obtained in the survey. A patch of such nebulosity is located just above the Double Cluster in Perseus. Lower is the California nebula, close to Xi Persei. The Hyades are easily identified, as are Orion and Canis Major. The Rosette nebula in Monoceros (surrounding NGC 2244) appears as the very large "star" southeast of Betelgeuse, which itself shows relatively bright because it is a red star.

In Orion and Monoceros are large nebulosities that trace the nearby spiral arm in this direction of space: the great Orion loop, surrounding the Belt and Sword; a newly discovered emission region surrounding Lambda Orionis (in the head of Orion) that has about half the apparent diameter of the Orion loop; a nebulosity associated with S Monocerotis; the Rosette; and a nebulosity connected with IC 2177, near Sirius. The positions of these five objects may be identified on the key diagram for the model; their distances from the sun are about 400, 600, 700, 1,000, and 1,100 parsecs, respectively.

These regions of hydrogen emission are ionized as a result of the ultraviolet light of hot stars associated with them. To get the distance of each nebulosity, that of its associated star has been determined. In the case of the *B* stars spectroscopic parallaxes were obtained; for *O* stars the intensity of the interstellar K line of ionized calcium was calibrated with distance by use of the *B* stars.

The cotton balls of the Morgan-Sharpless-Osterbrock model outline what are believed to be two, and possibly three, spiral arms of the galaxy. In the model, the sun is at the center of the scale, which is in galactic longitude and in thousands of parsecs (divided into hundreds for the first 500 parsecs from the sun). The cotton balls represent the approximate sizes of the nebulosities to this same scale. The galactic center is directly downward in the picture, about 8,000 parsecs, or 26,000 light-years, from the sun. The central part of a spiral arm passes through the sun's neighborhood about 300 parsecs (1,000 light-years) distant from the sun in the direction opposite to that of the center.

A second spiral arm is observed parallel to the first, at a distance of approximately 2,000 parsecs at its nearest point. Both arms can be traced over a length of about 3,000 parsecs,

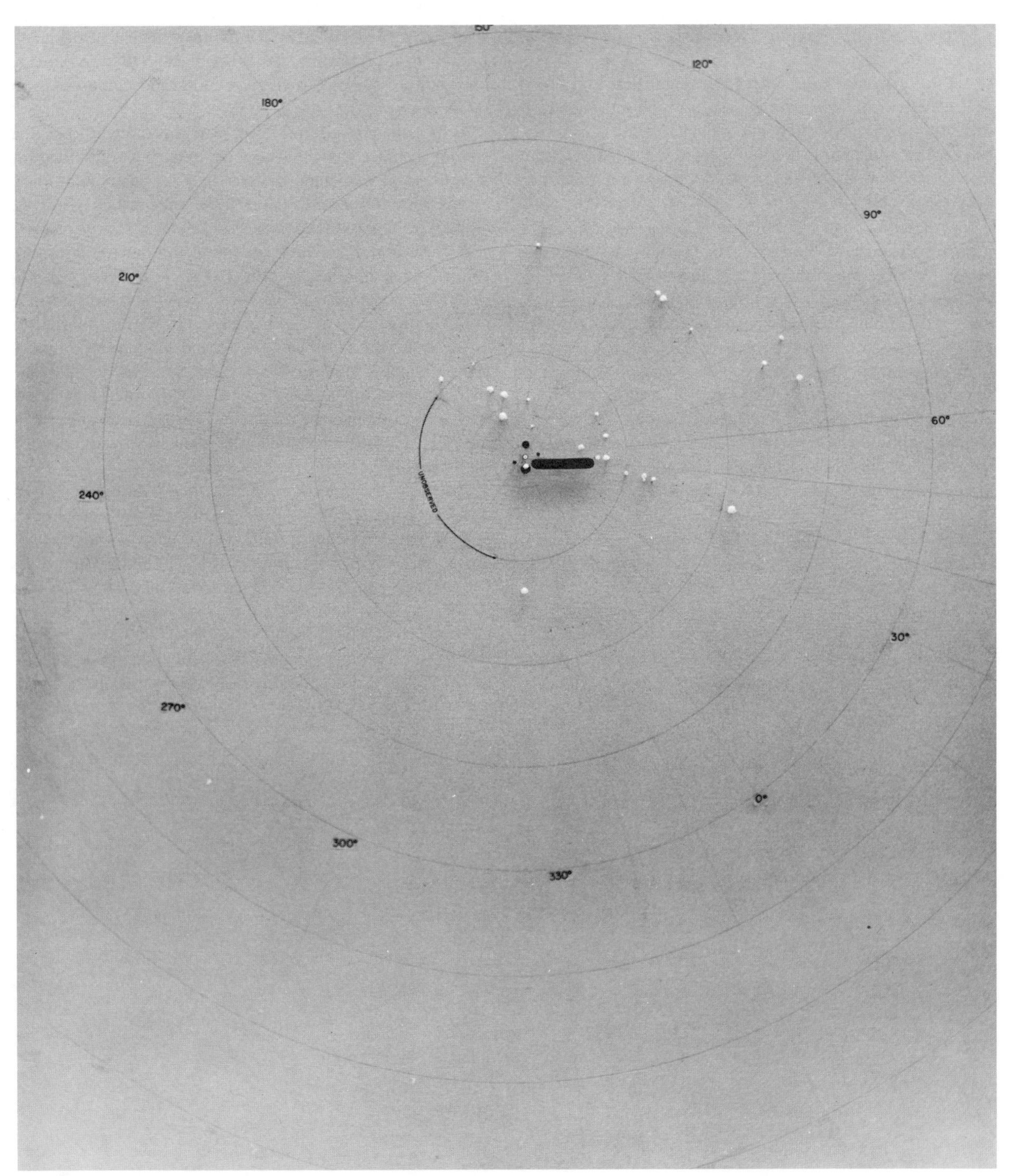

Fig. 93.1 A photograph of the Morgan-Sharpless-Osterbrock model of the galaxy in the vicinity of the sun, showing the spatial distribution of hydrogen emission (H II) regions (bright spots) that outline the nearby spiral arms. The plane of the figure lies in the plane of the Milky Way, with galactic longitudes marked by radial lines and circles centered on the sun are spaced 1 kpc apart. The dark spots denote three obscuring clouds of dust, and the Great Rift of the Milky Way is shown as the long, dark area extending from the sun to a distance of about 700 pc. (Courtesy *Sky and Telescope*.)

the inner one from Cygnus to Monoceros, the outer one from Cepheus to Auriga.

A third arm may exist nearer the center than the sun is. In the model, only one hydrogen emission object is shown in this arm, which extends from below Sagittarius to Carina, and can best be studied from Southern Hemisphere observatories, where observations of the proper kind are planned or in progress.

The direction of galactic rotation is to the right. The spiral arms are obviously inclined so that on the right they are nearer the center of the galaxy than on the left. Thus it is certain from this model that the galaxy rotates with its spiral arms trailing—it is winding up on itself. This confirms evidence of the galactic rotation obtained by analogy with other systems.

In addition to the ionized hydrogen regions, supergiant stars of spectral type *OB* were studied. These stars, with visual absolute magnitudes brighter than −3.5, outshine our sun five thousand times or more; they are very blue in color, and are found in clusters and loose groups. The spiral arms in many other galaxies are traced by them, where they are found in association with interstellar dust clouds. By analogy, they should lie along the spiral arms of our own system. Their brilliance and their association with dust make them conspicuous "skymarks" that invite investigation.

At the Warner and Swasey Observatory, a region of 89 square degrees around the star P Cygni has been examined to a limiting photographic magnitude of 12.2. The 4-degree objective prism on the 24-inch Schmidt telescope was used, by Drs. Nassau and Harris, to produce spectra on which the *OB* stars could be identified by the nature of their hydrogen and helium absorption lines. To a list of 136 *OB* stars already known in this region, 113 new stars were added, most of them rather faint.

At first, the new group of faint stars was thought to lie at a greater distance than the brighter ones, but allowance for dimming and reddening of their light by intervening dark clouds places all these *OB* stars at the same distance from us, about 5,000 light-years.

In order to obtain data for accurate photometric distances, colors and magnitudes of the *OB* stars are being photoelectrically observed with the Washburn Observatory 15-inch refractor and the 60-inch telescope on Mt. Wilson. Slit spectra for accurate classification of type and luminosity class are being obtained with the Yerkes 40-inch refractor. With the survey still incomplete. Drs. Code and Whitford have found 37 stars with color excesses (reddening) of half a magnitude or more. This more than doubles the number of such reddened stars known a dozen years ago.

Much of the obscuration of these stars is accounted for by the Great Rift in the Milky Way, which shows in the model as a long dark area extending from near the sun to a distance of about 700 parsecs. Other dark clouds that hinder our observations along the Milky Way plane are also shown in the model; they, too, are associated with the spiral arms. One star in Monoceros, in a region previously found to have very little absorption, showed a color excess of 0.42 magnitude. Its distance of 2,500–3,000 parsecs indicates that it may lie behind a remote dust cloud in an outer spiral arm in that direction.

94. The Galactic System as a Spiral Nebula

Jan H. Oort, Frank J. Kerr, and Gart Westerhout

(*Monthly Notices of the Royal Astronomical Society 118*, 379–389 [1958])

When Jan Oort asked his student Hendrik van de Hulst to see if there were any spectral lines at radio wavelengths, he realized that observations of these lines could be used to penetrate the curtains of dust hiding most of our galaxy from view at optical wavelengths. A few months after American physicists had confirmed van de Hulst's prediction that interstellar neutral hydrogen could be detected at a radio wavelength of 21 cm, both Oort's group at Leiden and the Australian radio astronomers located at the Commonwealth Scientific and Industrial Research Organization (CSIRO) in Sydney had confirmed the detection (see selection 92). These two groups embarked upon an ambitious program to map out the intensity and velocities of the 21-cm line in different directions in our galaxy.

By laboriously turning an old German radar antenna in elevation and azimuth by two small hand cranks every 2.5 min for nearly two years, the Leiden group collected observations to show that the interstellar gas seen in the northern hemisphere describes four extended armlike features.[1] In the meantime, the Sydney group used a transit antenna to discern similar features in the southern Milky Way, and the two surveys were combined to give the Leiden-Sydney map of the galaxy, which we reproduce in the review paper given here. As Oort had foreseen, neutral hydrogen can be detected out to the remotest regions of our galaxy; there elongated, nearly circular features seem to extend from the short armlike segments defined by the young stars and emission nebulae in the vicinity of the sun. Although no direct comparisons with the optical features are made in this paper, the authors note that the sun is located on the inner side of a neutral hydrogen arm. The concentrations of neutral hydrogen describe elongated features that are presumably spiral arms. Moreover, the observations summarized in this paper also provide exciting evidence for recent violent activity within the nucleus of our galaxy.

Distances of the interstellar hydrogen are not obtained directly from the line profiles and must be inferred from the dependence of rotational velocity on the distance from the galactic center. This function is derived from the velocities measured at the so-called tangential points.[2] The Leiden-Sydney map of the spiral structure of our galaxy is based on this function, under the assumption that the average motions of the interstellar gas in every direction coincide with the circular velocity of rotation at the corresponding distance from the center of the galaxy.

Detailed comparisons of the northern and southern rotation curves have subsequently shown that circular motion is only valid as a first approximation and that large-scale streaming motions exist, which might be related to the density waves that maintain the spiral structure. Hence, variations in velocity may be more important than variations in hydrogen density in determining the observed line profiles, and thus the interpretation given in this paper is actually oversimplified and uncertain. Although the general conclusion that the neutral hydrogen is concentrated in several elongated, nearly circular features remains valid today, the way in which the major neutral hydrogen features might be connected into a spiral pattern still remains a matter of controversy.[3]

1. H. C. van de Hulst, C. A. Muller, and J. H. Oort, *Bulletin of the Astronomical Institutes of the Netherlands 12*, 117 (1954).
2. M. Schmidt, *Bulletin of the Astronomical Institutes of the Netherlands 13*, 15 (1956). Also see A. Blaauw and M. Schmidt eds., *Galactic Structure: Stars and Stellar Systems V*, (Chicago: Univ. of Chicago Press, 1965).
3. W. Becker and G. Contopoulus, eds., *The Spiral Structure of our Galaxy—I.A.U. Symposium No. 38*, (New York: Springer-Verlag, 1970). Also see W. B. Burton *Astronomy and Astrophysics 10*, 76 (1971).

1. INTRODUCTION

THE VIEW that the Galaxy might have a spiral structure has been expressed almost since the first discoveries of spiral structure in nebulae. The oldest explicit reference to this seems to be in a paper by Stephen Alexander.[1] Both he and Proctor,[2] seventeen years later, tried to find support for their spiral theory and to construct to some extent the galactic spiral from the appearance of the Milky Way as inferred mainly from the observations of the Herschels.[3] In 1900, and more completely again in 1913, Easton,[4] apparently quite independently and unaware of these earlier suggestions, made a very careful study of the Milky Way as shown by visual and photographic observations, especially with a view to delineating the course of the possible spiral arms of the Galactic System; he placed the centre of the spiral in the direction of Cygnus.

Later developments have made it clear that this representation can hardly resemble the real structure of the Galaxy, the main cause of its failure and of the failure of all attempts to find spiral-like structure being the strong and uneven absorption near the galactic plane; for there is little doubt that many of the features of the apparent Milky Way structure are determined rather by the distribution of absorbing material than by that of the stars.

The extreme flatness of the Galactic System, as well as the frequency of large groupings of O and B stars, had for a long time been convincing evidence that it belonged to the class of spiral galaxies. It was not until 1951, however, that part of the galactic spiral structure was actually found. This was accomplished by using the luminosity criteria for O and B stars, as developed by Morgan, to determine the location of distant groupings of these stars. The first publication on the subject was an article by Morgan, Sharpless and Osterbrock[5] in which it was shown that regions of ionized hydrogen were arranged in long stretches which undoubtedly outlined parts of spiral arms.

It was a curious coincidence that at just about the same time systematic measures of the 21-cm line of neutral hydrogen were beginning to be made, and only two years later a much more comprehensive picture of the galactic spiral structure could be derived from these radio observations.[6,7]

The 21-cm observations brought about a revolution in the study of galactic structure. The scattering of optical radiation by small interstellar particles is so strong that in almost all directions in the galactic plane this radiation is effectively stopped in a few kiloparsecs. Although there are some places where, through accidental windows in the absorbing screens, one can observe a few stars in the galactic disk up to distances of the order of 10 kpc, these windows are so rare and so small that, but for the radio measures, it might have always remained a hopeless task to outline the general pattern of the galactic spiral.

Radiation in the range of radio waves passes without hindrance through the interstellar dust. For wave-lengths longer than a few metres absorption by ionized hydrogen becomes appreciable. With decimetre waves the most distant parts of the Galaxy can be explored. Observations in the decimetre continuum can only give the integrated radiation over the line of sight. The 21-cm line gives discrimination in distance. But although the 21-cm observations give discrimination in distance they cannot by themselves provide actual distances. The distance distribution in a given direction can only be inferred from radial velocities. For this we have to suppose that in each part of the Galactic System the average motion of the gas coincides with the circular velocity at the corresponding distance from the centre. Observations of stellar motions indicate that this condition is probably fulfilled to a fair approximation in the neighbourhood of the Sun. It seems plausible to assume that it holds as well for other parts of the Galactic System. But the possibility of deviations must certainly be kept in mind. The 21-cm observations themselves have given clear evidence of systematic divergence from circular motion in some fairly large regions. Moreover, it is certain that the hypothesis is no longer correct in the nuclear part. Within about 3 kpc from the centre the radio observations show that the large-scale motion of the gas deviates greatly from circular motion. We shall return to this below.

Beside systematic deviations from circular motion there are also the smaller-scale, internal motions. The interstellar gas is largely concentrated in clouds; the clouds have considerable random motions. These must be taken into account when computing the distance distribution of the hydrogen in a given direction from the observed velocity distribution. The difficulty is that the distribution of the random motions is very incompletely known, and that, moreover, it seems to vary from region to region.

2. ROTATION

A fundamental thing we must know in order to transform radial velocities into distances is how the circular velocity Θ_c varies with the distance R from the centre; in addition we must know the Sun's distance from the centre. We shall denote the latter by R_0.

Once we know both this distance and the circular velocity near the Sun, the 21-cm observations themselves may be used to derive Θ_c for values of R smaller than R_0. As a simple geometrical consideration shows, the circular velocities are given directly by the cut-off of the line profiles at the side of positive radial velocities for the quadrant between 328° and 58° longitude, and at the side of negative velocities for the quadrant between 238° and 328°, provided the hydrogen is distributed evenly over the disk. Actually, the derivation of Θ_c is complicated by the fact that, because of the spiral structure, the gas is very *unevenly* distributed. A second complication arises from the random motions of the clouds, as well as from *systematic* deviations from circular motion.

The rotational velocity near the Sun and the Sun's distance from the centre may be derived by combining the data just mentioned with the constants A and B of differential galactic rotation. The values of the latter constants, which rest on distances of distant stars, are more uncertain than the velocities derived from the 21-cm measures. This causes an uncertainty of about 10 per cent in the scale of the Galactic System. The distance to the centre can also be determined in a direct manner from the RR Lyrae-type variables concentrated near the centre. In this case the accuracy is limited by the uncertainty in the interstellar absorption and in the absolute magnitude of the variables.

For the regions situated farther from the centre than the Sun the rotational velocities cannot be determined from radio observations. Here we must rely on values computed from the mass densities in these outer parts, the distribution of mass density being inferred from the star density. For $R > R_0 + 2$ kpc the densities become very uncertain and this affects the calculated rotation curve. The resulting systematic uncertainty in the distances may be 10 per cent.

Here we omit a plot of the Schmidt[8] rotation curve, based upon observational data that show a fairly uniform value of circular rotation velocity of 200 ± 20 km sec^{-1} for distances of 4–8 kpc from the center of the Galaxy.[9,10] In the inner regions it is assumed that the rotation velocity drops almost linearly with decreasing distance to a value of 0 at the galactic center, but great irregularities in this region are noted. For the outer regions, where the distance lies between 8 and 12 kpc, the rotation velocity is assumed to decrease slowly with increasing distance, with high values near 200 km sec^{-1}.

3. Hydrogen Distribution. Spiral Structure

From profiles of the 21-cm line observed in various directions one can now derive the density distribution of the neutral hydrogen throughout the Galactic System. It appears that most of it is confined to a flat disk. The distance between the surfaces where the density has dropped to half the value in the central plane is about 220 pc, or only about 1/100th of the disk's diameter in the galactic plane. The arms lie for the most part extremely closely in one plane, which we may call the true galactic plane (this makes an angle of roughly $1^\circ.5$ with the standard plane as used in the Lund tables). Within a circle of 8 kpc around the centre the points of maximum density nowhere deviate more than 75 pc from the plane, except in a few small areas. Within 6 kpc from the centre the deviations are all less than 30 pc, or roughly 1/1000th of the diameter of the disk. The extreme neatness with which the gas has arranged itself into such a disk is the more remarkable when contrasted to the unevenness of the distribution *in* the plane. For $R > 10$ kpc somewhat greater deviations occur, running up to 600 and 800 pc in the outermost regions of low density. These deviations present a distinctly systematic character.[10,11]

The hydrogen distribution in the plane of the disk is shown in figure 94.1. The densities are indicated by contour lines and different types of shading. The picture is based partly on results derived in Leiden (from 340° to 220° longitude) and partly on measures made in Sydney (from 220° to 316°). There was a large region of overlap, inside which the observations agreed very well. The radio telescopes used were a 7.5-m Würzburg and a 11-m meridian telescope, respectively. The beam width to half power was $1^\circ.8 \times 2^\circ.8$ for the Dutch telescope and $1^\circ.5$ for the Australian one.

It should be emphasized that the distribution obtained depends considerably on the resolving power and on the particular assumption made regarding to velocity dispersion, temperature of the gas, circularity of the average motion, etc. We believe that the diagram gives the general pattern fairly well, but the densities must be considered very uncertain, in some cases by a factor of two or more. The relatively wide beams efface the detailed structure of the interstellar medium and are suitable for observing the large-scale features with which the present report is concerned. The zone between latitudes -10° and $+10^\circ$ was practically completely covered from 318° to 125° longitude; in the remaining longitudes the observations were spaced at somewhat larger intervals, so that roughly half of the surface between -10° and $+10^\circ$ may have been covered.

The distribution of the hydrogen evidently shows great irregularities. Nevertheless, several arms can be followed over considerable lengths. The Sun appears to be situated near the inner edge of an arm which stretches out in the direction of Cygnus and can be followed more or less continuously down to about 340° longitude. The continuation of the arm in longitudes past that of the anticentre is not so well defined; it probably passes through the big Orion association. Because of this the entire arm has been called the Orion arm. In the longitudes from 65° to 130° there is a conspicuous arm at a distance of about 10.5 kpc from the centre. This has been called the Perseus arm, after the large association around h and χ Persei which is situated close to it. The broad, almost circular structure between $R = 9$ and 12 kpc, extending from 65° down to 340°, may be a continuation of this arm. On the southern side there is no counterpart to this prominent outer arm. The pattern ends in a more

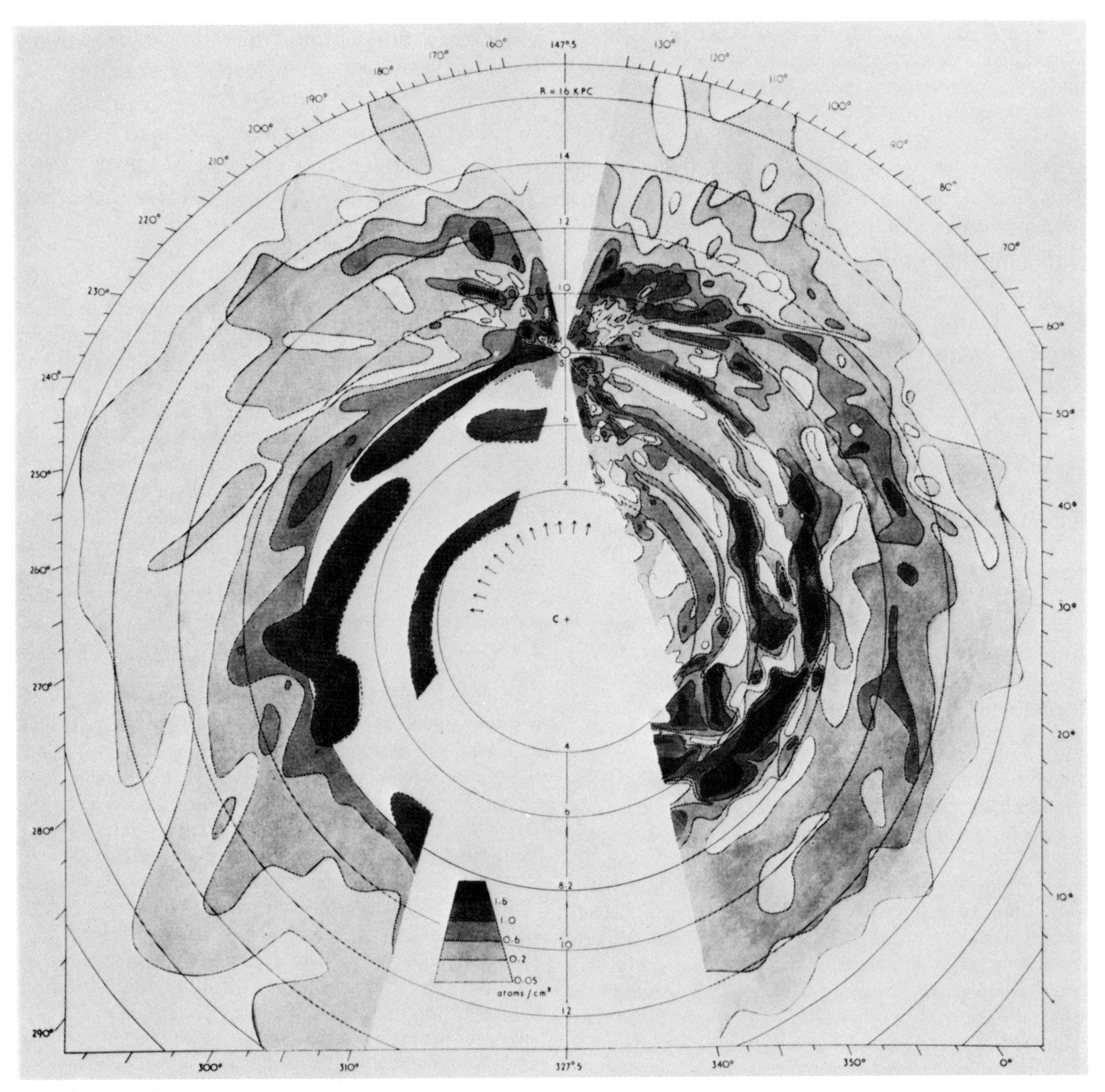

Fig. 94.1 Distribution of neutral hydrogen in the plane of our galaxy. The circles are centered at the center of our galaxy, Sagittarius A, and spaced 2 kpc apart. The sun is denoted ⊙, and the numbers around the outside of the figure denote galactic longitudes. This map is inferred from the distribution of the radial velocities and intensities of the 21-cm line along different lines of sight, with the assumption that the observed neutral hydrogen rotates with circular symmetry about the galactic center, with rotational velocities given by the Schmidt[8] rotation curve.

straggly style. Inside $R = 8$ we meet between $R = 6$ and 7 the "Sagittarius arm", which appears to move in towards the centre when we follow it in clockwise direction. At longitudes between 200° and 310° the structures may be slightly less continuous. For this reason and because of the great gap between 315° and 340°, where, except at small R, the differential rotation is too small to separate the various arms, it is not yet possible to follow the arms all around the centre.

The outer boundary of the disk appears to be about 15 kpc from the centre, but the average density diminishes very gradually, and it is quite possible that more distant bits will be discovered.

Omitted here is a figure giving the average density of neutral hydrogen in the galactic plane as a function of distance from the galactic center,[7] together with similar data for the Andromeda nebula.[12] For our galaxy, the maximum density is 1 atom cm^{-3}, and the average value between 3 and 12 kpc from the center is roughly 0.6 atoms cm^{-3}. Similar values apply to the Andromeda nebula.

There is a distinct maximum around $R = 7$ kpc. It may be noted that a similar maximum, around $R = 11$ kpc, occurs in the Andromeda nebula, but there it is much sharper. The thickness of the layer is unknown in this case; the densities have been computed on the hypothesis that it is the same as in the Galactic System.

Because of the imperfections in the observed hydrogen distribution it is still somewhat difficult to assign to the Galactic System an accurate classification among the classes of spirals. Judging from the number of continuous arms cut by a radius vector and from their spacing, it is similar to the Andromeda nebula and M 81, possibly slightly "later." It would therefore be of class Sb. It is certainly much more compact than typical Sc galaxies like M 101 and M 33, while it does not show the closely wound arms of early Sb nebulae like NGC 4594. Moreover, it clearly has a considerable central bulge of stars, but nothing comparable to that in NGC 4594. An additional indication of its place in the sequence of spiral galaxies may be obtained through a comparison of the fraction of the total mass which exists in the form of interstellar gas. For our Galaxy this is 2 per cent, for the Andromeda nebula it is 0.8 per cent, while in M 33 4 per cent of the mass is interstellar hydrogen. In elliptical galaxies, on the other hand, the interstellar component seems to be quite small. These considerations would make the Galactic System just a little "later" type than M 31.

4. EXPANSION IN NUCLEAR REGION

In figure 94.1 one conspicuous arm has been drawn in a tentative manner, viz. the arm indicated by small arrows at about $R = 3$ kpc. It can be followed from about 303° to 331° longitude. Its distance has been inferred from the fact that it appears to become tangential to the line of sight around 303°. If observed with a sufficiently narrow band and beam it stands out as a well-defined and rather sharp maximum in the line profiles.

Here we omit a plot of the radial velocity of the 3-kpc expanding arm, corrected for the motion of the sun around the galactic center, as a function of galactic longitude between 330° and 307°. The corrected radial velocity decreases linearly from -40 to -200 km sec^{-1} between these two longitude limits.

The arm can be seen in absorption against the strong source Sagittarius A, which is presumably situated at the galactic centre.[13] The absorption-line velocity agrees perfectly with the emission velocities in the surrounding points. It shows that the arm passes between the centre and us, and that, in addition to rotation, the gas of the arm has a velocity of 53 km/sec away from the centre. We shall therefore refer to it as the "3-kpc expanding arm." It is certainly not the only feature of this kind. From the line profiles in directions differing less than about 15° from 327°.7 one can see that in the part within 2 or 3 kpc from the centre large systematic deviations from circular motion must be the rule rather than the exception. Deviations up to 200 km/sec have been found. They are not of the nature of random motions, but represent mass motions of extended concentrations of matter. In the case discussed above we are clearly dealing with an *arm*. It is still unknown whether the other moving features are likewise arranged in arms or whether they are just loose bits. It is plausible to assume that, as in the 3-kpc expanding arm, the gas in these other "arms" is streaming away from the centre. Indeed, recent absorption measures in Sgr A by Rougoor show that the gas clouds for which the radial velocity (corrected for solar motion) exceeds 30 km/sec *all* move away from the centre. He has been able to verify this for radial motions between about 30 and 140 km/sec. It is worth remarking that these expanding velocities must be almost exactly in the galactic plane, as practically all of these features lie within 100 pc of the true galactic plane.

Clearly, estimates of the circular velocity in the nuclear parts are rendered very doubtful by the existence of these large deviations from circular motions, in particular because they appear to have the character of systematic velocities away from the centre. For this reason the rotation curve is quite uncertain in the part within $R = 3$.

It is possible that outward motions of smaller amount occur also at larger distances from the centre. In principle this could be investigated by special 21-cm line observations. At the time of writing these had not yet been made.

The gas motions in the central part may be strongly influenced by large-scale magnetic fields (cf. Section 5). It is unknown to what extent magnetic forces affect the motions beyond 3 kpc. But it is well to keep in mind the possibility that these may have caused appreciable errors in the gravitational forces and the mass distribution computed from the observed rotation up to about $R = 5$ kpc. However, there is good reason to believe that at distances R comparable to that of the Sun such effects are practically negligible. For, in the general vicinity of the Sun, the average motion of stars that are old enough to have become practically independent of the gas from which they originated, is nearly the same as the motion of the gas. There may be a difference of about 5 km/sec between the solar motion with respect to A stars and that with respect to the gas and the very young stars, but such an amount is of no consequence in comparison with the total velocity of rotation.

5. Evolution of Spiral Structure

What new information can the observations of the spiral structure of the Galactic System give us which could not be obtained from external galaxies, and what hope do they hold out for getting an insight into the processes causing spiral structure and maintaining it after it has originated? The information may be classed in three categories:

(*a*) The gaseous nature of spiral structure.

(*b*) Relation with the distribution of *stars.*

(*c*) The field of force.

We shall briefly consider each of these subjects.

A. Although convincing evidence that interstellar gas is concentrated in the spiral arms had been obtained previously, in particular in Baade's surveys of emission nebulae in M 31 and M 81, the 21-cm observations of our own galaxy furnish for the first time quantitative data on the concentration of gas in the arms. Although the evidence requires further substantiation, it is already clear that in many inter-arm regions the density is small compared to that in the arms.

A datum of extreme importance for the understanding of the phenomenon of spiral structure is the amount of random motion in the interstellar medium. On this point very little information can be gathered from other galaxies. For our own galaxy the large new radio telescopes will presumably enable us to collect a considerable amount of data on random velocities.

The astounding flatness of the System, which has been commented upon above, is of importance in connection with the internal motions in the medium. Not only does it provide a measure for the average random motion of the interstellar clouds, but it also indicates that there must have been a very large amount of exchange of momentum from the innermost to the outermost parts of the disk.

A fundamental question in the problem of spiral structure is whether this is essentially a phenomenon of *gas* dynamics (possibly connected with magnetic fields) or whether the stars themselves make the major contribution to the density of the arms, so that the problem would be one of *stellar* dynamics. The latter possibility has been worked out in considerable detail by Lindblad.[14] Some phenomena in external galaxies give strong indications that the spiral formation is intimately related to their content of interstellar gas, and that the presence of enough gas is an essential condition for the existence of spiral structure.[15] But it would evidently be very important to know whether in the Galactic System the arms consist mainly of gas or of stars.

B. It is clear that stars which have recently been formed from the gas must be situated in the gaseous arms. The connection between young stars and the general distribution of the gas is demonstrated in the surveys of OB associations[16] and δ Cephei variables.[17] Surveys of such supergiants, fragmentary though they are because of interstellar absorption, are extremely valuable for the investigation of the spiral structure. They can bridge gaps in regions where the differential rotation is too small to resolve the 21-cm radiation and, because the stellar distances are known, they can give information on systematic deviations of spiral arms from circular motion.

It may be estimated that stars will in general stay within the spiral arms in which they were born for several hundred million years, but that after about five hundred million years they may show appreciable differences from the distribution of the gas if the latter is supposed to be kept in its spiral pattern by other forces in addition to that of gravitation. So far, no sufficient evidence for a decisive comparison between the distribution of older stars and gas has been obtained.

C. In no other galaxy are the velocities of rotation and the gravitational field known with an accuracy comparable to that obtained in the Galactic System. Though the arms cannot yet be followed around the whole system, it is fairly clear from the parts over which they can be followed that the arms are trailing. If we suppose that like other regular Sb spirals the Galactic System has two main arms, the spacing between the large arms gives an indication of their average inclination. The observed differential rotation gives us then at once the time scale in which the present spiral structure would radically change its appearance. This scale is found to be between 100 and 200 million years. This is so short compared to the age of our galaxy that we are forced to conclude that the arms must either smooth out rapidly and then be replaced by completely new ones, or that there is some mechanism that keeps up the existing arms, notwithstanding the stretching effect of the differential rotation. A possible mechanism is that of gas transport between neighbouring arms. Rough estimates indicate that this would work in the right direction and might be of the right order. But such a mechanism, though it might preserve the arms once they are formed, would give no clue to the way in which the spiral structure could have originated, nor how the expanding arms in the nuclear region come into existence. The latter problem is acute, because, considering the speed with which they move out, these expanding features must be formed at relatively short intervals.

Though it is still entirely obscure how magnetic fields could produce these phenomena, it is tempting to think that there is some connection between large-scale interstellar magnetic fields and spiral arms. Three phenomena point in this direction. The first is that interstellar polarization of light indicates the presence of magnetic fields which, at least in some cases, seem to have their average direction along arms. The second is concerned with the continuous radio-frequency radiation. Part of this is strongly concentrated to the galactic plane. Its intensity distribution in the plane appears to be clearly connected with the spiral structure. This is well brought out by the continuous radiation at 3.5 m wave-length observed by Mills, Hill and Slee.[18] Practically all of the radiation at this wave-length, even at low latitudes, must be of non-thermal origin. This was shown by a comparison with a survey at 22 cm wave-length by Westerhout.[19] This non-thermal component may be largely "synchrotron" radiation from high-energy electrons moving in interstellar magnetic fields. If this is correct, the interstellar fields may be strongly concentrated in the arms.

The third phenomenon relevant in this connection is the radio corona of the Galactic System. We shall not deal with this in the present report, except to mention that its existence around the Galactic System as well as around the Andromeda nebula gives considerable support to the idea that magnetic fields are an essential factor in the large-scale structure of spiral nebulae. It should also be pointed out that the corona is likely to be the reservoir replenishing the central region with the gas needed to keep up the fast stream moving out of it into the larger disk.

It is clear that as yet we have made little or no progress towards an understanding of the origin of spiral structure. But the recent investigations on this structure in the Galactic System indicate at least some new roads for future research into this problem.

1. Stephen Alexander, "On the Origin of the Forms and the Present Condition of Some of the Clusters of Stars and Several of the Nebulae," *A. J. 2*, 95 a.f. (1852); see esp. pp. 101–108.

2. R. A. Proctor, "A New Theory of the Milky Way," *M.N. 30*, 30–56 (1869).

3. In 1940 Dr. Rosseland informed one of us about Alexander's and Proctor's articles. He also mentioned the possibility that Lord Rosse himself may have had the idea that the Galactic System had a spiral structure. So far, however, no direct published reference to this by Lord Rosse has been found.

4. C. Easton, "A New Theory of the Milky Way", *Ap. J. 12*, 136–158 (1900), and "A Photographic Chart of the Milky Way and the Spiral Theory of the Galactic System," ibid. *37*, 105–118 (1913).

5. W. W. Morgan, S. Sharpless, and D. Osterbrock, "Some Features of Galactic Structure in the Neighbourhood of the Sun," *A. J. 57*, 3 (1952); see also *Sky* and *Telescope 11*, 134 (1952).

6. H. C. van de Hulst, C. A. Muller, and J. H. Oort, "The Spiral Structure of the Galactic System Derived from the Hydrogen Emission at 21-cm Wave-Length," *B.A.N. 12*, 117–149 (1954).

7. C. A. Muller and G. Westerhout, "A Catalogue of 21-cm Line Profiles," *B.A.N. 13*, 151–195 (1957); A. Ollongren and H. C. van de Hulst, "Corrections of 21-cm Line Profiles," ibid., 196–200; G. Westerhout, "The Distribution of Atomic Hydrogen in the Outer Parts of the Galactic System," ibid., 201–246; M. Schmidt, "Spiral Structure in the Inner Parts of the Galactic System Derived from the Hydrogen Emission at 21-cm Wave-Length," ibid., 247–268.

8. M. Schmidt, "A Model of the Distribution of Mass in the Galactic System," *B.A.N. 13*, 15–41 (1956).

9. K. K. Kwee, C. A. Muller, and G. Westerhout, "The Rotation of the Inner Parts of the Galactic System," *B.A.N. 12*, 211–222 (1954).

10. F. J. Kerr, J. V. Hindman, and Martha Stahr Carpenter, "The Large-Scale Structure of the Galaxy," *Nature 180*, 677–679 (1957).

11. C. S. Gum and F. J. Kerr, "A 21-cm Determination of the Principal Plane of the Galaxy," Radiophysics Laboratory Report No. RPL 138 (1958).

12. H. C. van de Hulst, E. Raimond, and H. van Woerden, "Rotation and Density Distribution of the Andromeda Nebula Derived from Observations of the 21-cm Line," *B.A.N. 14*, 1–16 (1957).

13. This may be a concentration of ionized gas of about 200,000 solar masses, of the same nature as the concentrated masses of gas which are so frequently observed at the centres of elliptical galaxies; generally the latter masses are somewhat higher. See G. Westerhout, "A Survey of the Continuous Radiation from the Galactic System at a Frequency of 1,390 Mc/s," *B.A.N. 14*, 215 (1958).

14. B. Lindblad, "Contributions to the Theory of Spiral Structure," *Stockholms Obs. Ann. 19*, no. 7 (1956), and "Differential Motions in Dispersion Orbits in the Galaxy," ibid. *19*, no. 9 (1957).

15. L. Spitzer, Jr., and W. Baade, "Stellar Populations and Collisions of Galaxies," *Ap. J. 113*, 413–418 (1951).

16. W. W. Morgan, A. E. Whitford, and A. D. Code, "Studies in Galactic Structure: I. A. Preliminary Determination of the Space Distribution of the Blue Giants," *Ap. J. 118*, 318–325 (1953). Cf. also *B.A.N. 12*, 117, where the distribution of associations has been compared with that of the gas (fig. 16).

17. Th. Walraven, A. B. Muller, and P. Th. Oosterhoff, "Photoelectric Magnitudes and Colours at Maximum Brightness for 184 Cepheids," *B.A.N. 14*, 81–128 (1958).

18. B. Y. Mills, E. R. Hill, and O. B. Slee, "The Galaxy at 3.5 m," *Observatory 78*, 116–121 (1958).

19. See reference in note 13.

95. Density Waves in Disk Galaxies

Chia C. Lin and Frank H. Shu

(*International Astronomical Union Symposium No. 31*, 313–317 [1967])

In 1927 Jan Oort showed that our galaxy undergoes differential rotation in which its outer parts move at slower angular speeds than its inner parts, and this differential rate suggested that our galaxy cannot maintain a spiral form over its lifetime (see selection 82). The period of revolution of the sun about the galactic center was known to be about 2×10^8 yr, whereas the age of the galaxy was estimated to be at least 5 times and perhaps 50 times greater. Thus, the local spiral arm must have underdone at least 5 and perhaps 50 revolutions about the galactic center, and the shearing force of differential galactic rotation should have wound it up into a tight knot in the galactic center.

In 1953 Subrahmanyan Chandrasekhar and Enrico Fermi suggested that an interstellar magnetic field must contain and perhaps support the spiral arms.[1] During the 1950s and the early 1960s the prevalent opinion was that the spiral arms are magnetically bound tubes of ionized gas and that it is the interstellar magnetic field which provides the major force regulating the spiral arms. This view had to be abandoned in the late 1960s and early 1970s, when measurements of the Zeeman effect in 21-cm profiles and dispersion and Faraday rotation measurements of pulsar radiation showed that the interstellar magnetic field is not as strong as had been imagined. It is actually too weak to organize the interstellar medium into a spiral pattern. The weak magnetic fields are probably frozen into the ionized gas of the interstellar medium, so that they will be dragged along with the gas as it moves under the influence of some other force that regulates the spiral arms.

Already in the late 1920s, when Bertil Lindblad was arguing that our galaxy must rotate about a distant center, he also reasoned that it must be gravitational forces that maintain the spiral arms through the propagation of waves. During the subsequent three decades, Lindblad published over twenty papers in which he developed his theory of the spiral structure of nebulae.[2] In Lindblad's view, matter is ejected at the region of instability, where it travels out into a region where spiral orbits are possible.

Perhaps because Lindblad's papers are difficult to follow, and also because he insisted that spiral arms must lead rather than follow the direction of galactic rotation, his ideas were not widely accepted. In the mid-1960s, however, C. C. Lin and Frank Shu developed Lindblad's theory in greater detail and presented it in a simple, understandable form, while also predicting certain observational consequences. As shown in the representative paper given here, their theory allows trailing spiral arms and specifies their spatial orientation. In this theory the spiral arms are not viewed as permanent entities but rather as the loci of star concentration coinciding with a spiral component of gravitational potential. The stars and gas rotate about the galactic center at a faster speed than the spiral component, thereby continuously drifting through and replenishing the spiral arms. In fact, the stars and gas only define spiral arms because they prefer to linger in the spiral-shaped potential troughs and to speed through the interarm voids.

One important aspect of the density wave theory is that it gives predictions that can be compared with observational data. For example, large-scale streaming motions with velocities as high as 10 km sec^{-1} relative to pure circular motion are predicted. W. W. Shane, G. P. Bieger-Smith, and W. B. Burton have subsequently shown that humps in the observed curves giving the circular velocity of neutral hydrogen versus distance from the galactic center can be best understood in terms of these large-scale streaming motions.[3] Furthermore, galactic shocks associated with the spiral density waves may provide the compression to trigger the collapse of interstellar gas into young stars.[4] If this is the case hot young stars and emission nebulae will be further from the galactic center than the nearby regions of neutral hydrogen and dust. The dust lanes observed in our galaxy are on the inner edges of bright spiral arms, but the separation of stars and neutral hydrogen is very difficult to check.

The Lin-Shu density wave theory rests on the ad hoc assumption that the field of the gravitational potential of a rotating galaxy has a spiral component. The theory therefore explains only the maintenance of the spiral arms, not the origin of spiral structure. Such a mechanism, based on the existence of unstable spiral modes, was recently worked out by Lin and his collaborators Y. Y. Lau, James Mark, G. Bertin, and L. Sugiyama and was published in a series of papers.[5] It is of course possible that Lindblad's idea of nuclear ejection will eventually provide the basis for a theory of the origin of spiral arms as well.

1. S. Chandrasekhar and E. Fermi, *Astrophysical Journal 118*, 116 (1953). See also selection 98.
2. B. Lindblad, selection 81, *Arkiv för matematik, astronomi och fysik 20A*, no. 17 (1927), *Uppsala Astronomiska Observatoriums meddelanden* no. 19, (1927), and *Stockholms Observatoriums annaler 12*, no. 4 (1936), *13*, no. 5 (1940), *13*, nos. 8 and 10 (1941), *14*, no. 1 (1942), *16*, no. 1 (1950), *17*, no. 6 (1953), *21*, no. 3 (1960), *22*, no. 3 (1963). Additional references to contributions between 1925 and 1940 are given by S. Chandrasekhar in *Principles of Stellar Dynamics* (Chicago: University of Chicago Press, 1942).
3. W. W. Shane and G. P. Bieger-Smith, *Bulletin of the Astronomical Institutes of the Netherlands 18*, 263 (1966); W. B. Burton, *Astronomy and Astrophysics 10*, 76 (1971).
4. W. W. Roberts, *Astrophysical Journal 158*, 123 (1969).
5. *Proceedings of the National Academy of Sciences 73*, 1379, 3785 (1976); *74*, 4726 (1977).

Abstract—Density waves in the nature of those proposed by B. Lindblad are described by detailed mathematical analysis of collective modes in a disk-like stellar system. The treatment is centered around a hypothesis of quasi-stationary spiral structure. We examine (*a*) the mechanism for the maintenance of this spiral pattern, and (*b*) its consequences on the observable features of the galaxy.

General Observations

In discussing the spiral features of disk galaxies, it is important to note the difference between these two questions: (*a*) the organization of a spiral pattern over the whole disk, and (*b*) the detailed structure of a stretch of an individual arm, This point was clearly made by Oort (1962) in the following words:

> In systems with strong differential rotation, such as is found in all non-barred spirals, spiral features are quite natural. Every structural irregularity is likely to be drawn out into a part of a spiral. But *this* is not the phenomenon we must consider. We must consider a spiral structure extending over the whole galaxy, from the nucleus to its outermost part, and consisting of two arms starting from diametrically opposite points. Although this structure is often hopelessly irregular and broken up, the general form of the large-scale phenomenon can be recognized in many nebulae.

The primary purpose of our present theory is indeed an attempt to explain the grand design over the whole disk. We believe that the mechanism for maintaining this grand design can be described essentially in terms of *density waves* of the general nature discussed by the late Bertil Lindblad.[1] Gravitational forces are predominant, while hydromagnetic forces can at most play a secondary role. Specifically, we believe that the density waves form a quasi-stationary spiral structure.

A more detailed description of our ideas and conclusions has been published elsewhere (Lin and Shu 1964, 1966). Here we shall only indicate a few high points of the theory, and report on the latest numerical results. The density waves are primarily collective modes in a stellar system, which may be described as a 'gravitational plasma'. Numerical results reported below are based on the simplest form of the theory, but they serve well the purpose of demonstrating the relevance of these density waves to the observed spiral features. Briefly, we conclude that the principal spiral pattern in our Galaxy can be associated with a density wave propagating around the center at an angular speed of about 11 km sec^{-1} kpc^{-1} in the direction of the general galactic rotation of stars.

Self-Sustained Density Waves of a Spiral Form

Spiral patterns in the plane of a disk galaxy may be described by functions of the form

$$F(\varpi, \theta, t) = A(\varpi)\, exp\{i[\omega t - m\theta + \Phi(\varpi)]\}, \tag{1}$$

where t is the time, (ϖ, θ) are the polar coordinates in the plane, $A(\varpi)$ is a slowly varying function of ϖ, $\Phi(\varpi)$ is a slowly varying (monotonic) function multiplied by a large parameter, ω is a real constant (for neutral waves), and m is an integer. As usual, the real part of the above expression is to be taken. The lines of constant Re(F) are approximately given by the equation

$$m(\theta - \theta_0) = \Phi(\varpi) - \Phi(\varpi_0), \tag{2}$$

which represents a spiral pattern with m arms. The pattern given by (1) rotates at an angular velocity

$$\Omega_p = \omega/m. \tag{3}$$

The radial wave number of the spiral pattern is given by $|k(\varpi)|$, where

$$k(\varpi) = \Phi'(\varpi). \tag{4}$$

If the motion of the stars is in the direction of increasing θ, trailing waves correspond to $k(\varpi) < 0$ and leading waves to $k(\varpi) > 0$.

It can be shown (see Lin and Shu 1966) that a self-sustained density wave of the general nature of (1) can be obtained through an analysis of small disturbances over a symmetrical disk of stars and gas in differential rotation. These waves have the following properties.

a. They extend essentially over a range of the galactic disk where the condition

$$\Omega - \frac{\kappa}{m} < \Omega_p < \Omega + \frac{\kappa}{m} \tag{5}$$

is satisfied, where $\Omega(\varpi)$ is the angular speed of the stars, and $\kappa(\varpi)$ is the epicyclic frequency. We shall refer to this range as the *principal part* of the spiral pattern. For $m = 2$, the pattern would extend from $\varpi = 4$ kpc outwards without limit, if $\Omega_p = 11$ km sec^{-1} kpc^{-1}. For other values of m, this principal part would be quite limited in extent.

Omitted here are plots of angular velocity, Ω, and epicyclic frequency, κ, as a function of distance from the galactic center. These plots are based upon the Schmidt (1965) rotation curve. The value of Ω drops systematically from 70 to 11 km sec^{-1} kpc^{-1} as the distance increases from 4 to 20 kpc, and for a given distance κ is slightly larger than Ω.

b. The dispersion relation for such waves is expressible in terms of the wave number $\lambda = 2\pi/|k|$ and frequency

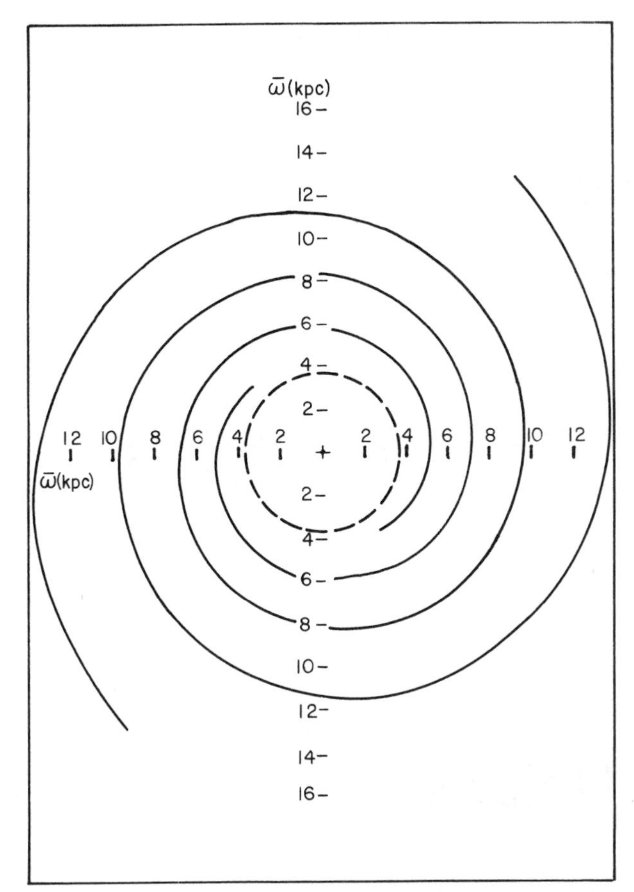

Fig. 95.1 The spiral pattern of our galaxy on the assumption that a density wave pattern rotates at an angular velocity of 11 km sec^{-1} kpc^{-1} in the general direction of rotation and that the Schmidt (1965) model for rotation velocities applies. The dispersion ring for Lindblad resonance is taken at 3.75 kpc from the center.

$v = m(\Omega_p - \Omega)/\kappa$ at which the stars see the gravitational field. This relationship includes an additional parameter, which measures the velocity dispersion of the stars. If the velocity dispersion is barely enough to stabilize the disk against gravitational collapse (Toomre 1964), the dispersion relationship is a function of $\lambda_* = 4\pi^2 G\sigma_*/\kappa^2$, G being the gravitational constant and σ_* being the projected stellar density.

We also omit a figure showing that λ/λ_* drops almost linearly from 0.55 to 0.1 as $|v|$ increases from 0 to 1.

c. Trailing patterns are preferred over leading patterns, if the velocity dispersion of stars increases toward the center. (Several other effects would also tend to prefer trailing patterns.)

An Example of a Computed Pattern

By taking $\Omega_p = 11$ km sec^{-1} kpc^{-1}, we get a pattern that starts with a dispersion ring at the Lindblad resonant point, $\bar{\omega} = 3.75$ kpc, and continues with a spiral as shown in figure 95.1. This may be compared with the spiral pattern shown by Mrs. Burbidge in her Introductory Report (W. Becker 1964, figure 4).

No criterion based on the present theory is yet available for determining the pattern frequency. The above value of Ω_p was chosen so that the location of the dispersion ring would correspond to that of the '3-kpc Arm'.[2] In general, there may be a superposition of a group of waves, and the spiral pattern becomes quasi-stationary. There may also be other spurious spiral arms superposed on the overall pattern. But density waves of the type described appear to be the most natural candidate for the explanation of the existence of a grand design.

1. For further discussions of the relationship between the present theory and that of Lindblad, see Lin and Shu (1966).

2. This is, however, not a very accurate determination, and the value of Ω_p could be as high as 20 km sec^{-1} kpc^{-1}.

Becker, W. 1964, *Z. Astrophys. 58*, 202.

Lin, C. C., Shu, F. H. 1964, *Astrophys. J. 140*, 646.

Lin, C. C., Shu, F. H. 1966, *Proc. Nat. Acad. Sci. USA 55*, 229.

Oort, J. H. 1962, in *Interstellar Matter in Galaxies*, ed. L. Woltjer (New York: W. A. Benjamin), p. 234.

Schmidt, M. 1965, in *Galactic Structure*, ed. A. Blaauw and M. Schmidt, vol. 5 in *Stars and Stellar Systems* (Chicago: Univ. of Chicago Press), p. 513.

Toomre, A. 1964, *Astrophys. J. 139*, 1217.

96. The Discovery of Protostars (?)

Small Dark Nebulae

Bart J. Bok and Edith F. Reilly

(*Astrophysical Journal 105*, 255–257 [1947])

The Spectra of Two Nebulous Objects near NGC 1999

George H. Herbig

(*Astrophysical Journal 113*, 697–699 [1951])

Herbig's Nebulous Objects near NGC 1999

Guillermo Haro

(*Astrophysical Journal 115*, 572 [1952])

Observations of an Infrared Star in the Orion Nebula

Eric E. Becklin and Gerry Neugebauer

(*Astrophysical Journal 147*, 799–802 [1967])

Because the very luminous O and B stars can be no more than about a million years old, and because these stars are found in the spiral arms of galaxies where interstellar gas and dust are most concentrated, astronomers have inferred that stars are presently being formed in dense concentrations of interstellar matter in the spiral arms of our galaxy. In the first paper given here, Bart Bok and Edith Reilly call attention to the small, dense concentrations of interstellar dust and gas that may well be the precursers of ordinary stars. Although these objects can only be seen when projected against the luminous background of intense emission nebulae, over 100 of them have now been detected in the outer peripheries of a few nebulae.[1] As Bok and Reilly show, these small globules have roughly the dimensions expected for the nebular clouds from which stars might eventually form (about 0.05 pc). Furthermore, measurements of their extinction of starlight have subsequently been used to infer a dust mass of about 0.5 solar masses.[2] If the dust furnishes only a small fraction of the mass of the globule, the rest being in gaseous form, these concentrations may be massive objects on the verge of, or in the process of, gravitational collapse leading to star formation.

The first indication that stars might actually be seen in their very early stages came when George Herbig and Guillermo Haro discovered a class of small bright nebulae now called Herbig-Haro objects. As illustrated in the next pair of papers in this selection, these nonstellar objects exhibit the strong forbidden emission lines also found in the much larger H II regions, and it is natural to suppose that, like the H II regions, the Herbig-Haro objects shine by reemission of the ultraviolet light of a hot, young star embedded deep within the nebulosity. Such surrounding nebular material may represent the material out of which the star is formed. Nevertheless, no central star has actually been seen in the Herbig-Haro objects, and the observed material might be dust reflecting the light of young, emission-line variable stars.

Although the actual process of star birth may be triggered by density waves or supernova explosions, Merle Walker's studies of young star clusters and Chushiro Hayashi's studies of stellar evolution suggest that gravitational collapse heats up concentrations of interstellar matter to the high temperatures needed to ignite the stellar thermonuclear reactions (see selection 54). When an initially cool cloud of dust and gas collapses and heats up, it should first become visible as a hot, young star embedded in an envelope of gas and dust; subsequently, the envelope will be blown off by the radiation pressure of the young star. In the last paper given here, Eric Becklin and Gerry Neugebauer report their discovery of a starlike infrared source believed to be produced by a circumstellar dust cloud that reradiates the ultraviolet radiation absorbed from a newborn central star. Although the Becklin-Neugebauer object is opaque and its central star hidden from view, similar infrared-emitting dust shells have been subsequently found surrounding T Tauri stars. These objects are almost certainly very young stars of intermediate mass in the early stages of stellar evolution, still embedded in the nebular material from which they were formed (see selection 59). If the collapsing protostars lose mass and become less obscured as time goes on, then the Becklin-Neugebauer object may represent an intermediate stage of stellar evolution linking the small dark globules and the T Tauri stars.

During the 1970s there has been a growing realization that interstellar molecules must play an important role in star formation. In the dark dust clouds in which embryonic stars are found, most of the interstellar hydrogen is in molecular form, and the mass of the embryonic stars is only a small fraction of that left behind in their molecular placenta. Moreover, in the regions in which the young OB stars are being formed, gigantic clouds of carbon monoxide are found which are hundreds of times larger than the typical spacing between the stars. Considerations of the excitation of the carbon monoxide molecules indicate that they are colliding with molecular hydrogen whose total mass is about one hundred million times that of a typical star. The massive OB stars appear to form preferentially at the edges of these giant molecular clouds in a sequential manner in which one generation of OB stars triggers the birth of the next generation.[3]

1. A. D. Thackeray, *Monthly Notices of the Royal Astronomical Society 110*, 524 (1950), *International Astronomical Union Symposium No. 20* (1964); M. V. Penston, *Monthly Notices of the Royal Astronomical Society 144*, 159 (1969).

2. B. J. Bok, *Centennial Symposium, Harvard Observatory Monograph* no. 7, 1948. See also B. T. Lynds, ed., *Dark Nebulae, Globules and Protostars* (Tuscon: University of Arizona Press, 1970).

3. T. deJong and A. Maeder, eds., *Star Formation: International Astronomical Union Symposium No. 75*, (Boston: D. Reidel, 1977).

Bok and Reilly, Small Dark Nebulae

Abstract—Attention is drawn to the small, round, dense dark nebulae with diameters varying between 5″ and 10′. We propose that these be named "globules." The region of Messier 8 abounds in globules. Published photographs show at least sixteen of them projected against the bright background of the diffuse nebula. At the derived distance of 1,260 parsecs for Messier 8, the diameters of twelve of the sixteen observed globules are between 10,000 and 35,000 A.U. It is noteworthy that there is no evidence for globules in the region of the Orion nebula.

At least twenty of the dark objects in Barnard's lists are true globules. A preference is shown for the regions of Sagittarius and Ophiuchus and for the Scutum Cloud, but some isolated examples of large globules (with diameters of the order of 100,000 A.U.) are found in the anticenter region.

The estimated *minimum* absorptions for the small globules are 2–5 mag. The larger globules are more transparent (1 mag.).

IN RECENT YEARS several authors have drawn attention to the possibility of the formation of stars from condensations in the interstellar medium.[1] It is therefore necessary to survey the evidence for the presence in our galaxy of relatively small dark nebulae, since these probably represent the evolutionary stage just preceding the formation of a star.

In the early days of astronomical photography, Barnard drew attention to the prevalence of small dark objects, and his famous two lists of dark objects in the sky are still our best source in the field.[2] Since Barnard's days there have been published many excellent photographs of bright nebulae and of Milky Way regions, notably the Lick atlas[3] and the Ross-Calvert *Atlas.*[4] The new Schmidt cameras, with their short focal ratios and excellent definition and with scales comparable to those of the standard astrographic cameras, provide new opportunities for the search for, and detection of, small dark nebulae.

In connection with the possible evolutionary process referred to above, we are primarily interested in roundish, small dark nebulae. We shall omit here from consideration the wind-blown wisps of dark nebulosity that are seen in large numbers in many parts of the sky and shall concentrate our attention on the approximately circular or oval dark objects of small size. We shall for convenience refer to these as "globules."

Our search for globules has proceeded along two lines. We have examined with care the best available photographs of some of the well-known diffuse nebulae and listed the globules that can be seen projected against the luminous background. We have, further, examined a representative number of prints of the Milky Way photographs and marked on them the objects which satisfy our definition of a globule.

The first region to be investigated was that of Messier 8, a region for which an excellent photographic print is available in the Lick atlas. To guard against defects, this photograph was compared with other published photographs of this diffuse nebula, notably a photograph by Duncan[5] and an original print of a plate taken by Struve and Elvey at the prime focus of the 82-inch reflector at McDonald Observatory.[6]

In 1908, Barnard[7] noted that there are "a number of very black, small, sharply defined spots or holes" among the markings of Messier 8. A similar observation was made by Duncan. The Lick photograph reveals the presence of twenty-three potential globules. Of these, sixteen are round and range in diameter from 6″ to 1′, and seven are irregular. The measured diameters of the sixteen true globules are distributed as shown in the accompanying tabulation. The dimensions of the irregular objects range from 6″ × 60″ and 12″ × 24″ to 60″ × 84″ and 36″ × 216″. Some of these are of an oval shape, with small dark streamers extending from them.

Approx. diam.	No.	Approx. diam.	No.
0″–10″	1	30″–40″	1
10 –20	8	40 –50	1
20 –30	4	50 –60	1

Since the globules are seen projected against the diffuse nebula, Messier 8, their distances must all be smaller than that of the diffuse nebula. The distance of Messier 8 can be determined with a fair degree of accuracy, since we have, apparently associated with the nebula, the galactic cluster NGC 6530. This region has recently been the subject of a careful study by Wallenquist,[8] who obtains $m - M = 11.00$ as the value for the distance modulus, uncorrected for absorption. This value for the distance modulus agrees well with that of 11.05 for the two stars of spectral class Oe5 and B0, which, according to Hubble,[9] are most likely to be responsible for the emission nebula.

To derive the probable distance, we must first apply a correction for absorption. Four stars in the region are in the list of Stebbins, Huffer, and Whitford,[10] and we find from the observed excesses that the total photographic absorption for $m - M = 11.0$ equals 0.6 mag. The derived distance of Messier 8 is then 1,260 parsecs, corrected for absorption.

The globules viewed against Messier 8 are, therefore, at the maximum distance of 1,260 parsecs. The *maximum* linear diameters of the globules can, therefore, be computed on the assumption that 1″ = 1,260 A.U. We find, then, that the maximum diameters of the globules viewed against Messier 8 range from 7,000 to 80,000 A.U., with twelve of the sixteen globules having diameters between 10,000 and 35,000 A.U.

The evidence for other large bright nebulae shows that the population of globules viewed against the luminous background of Messier 8 is unusually large. The Trifid nebula, which is a close neighbor of Messier 8, shows, for example, at the most, only three or four objects that could be called globules, and in other parts of the sky globules seem almost totally absent. The Orion nebula is apparently free from superimposed globules, which is all the more surprising since there is plenty of obscuring matter associated with it. Another instance of a bright nebula without globules is the delicate California nebula in Perseus.

The emission nebula near η Carinae is also rich in globules. The general appearance is very much like that of Messier 8. At least a dozen globules can be recognized on a plate taken with the 60-inch Rockefeller reflector of the Boyden Station.

In our further search for globules, we have examined the dark objects in Barnard's lists.[2] It is not a simple matter to draw the line between true globules and minor condensations in dark lanes or in regions of variable obscuration. From an inspection of the beautiful prints in Barnard's *Atlas* and in the Ross-Calvert *Atlas* we find, however, at least twenty unmistakable globules.

The regions of the star clouds in Sagittarius and Ophiuchus are rich in clearly marked globules. Plate 21 in the Barnard *Atlas* shows numerous objects that stand out against the bright stellar background. Barnard 68, 69, 70, and 255 (in the region near θ Ophiuchi) appear very similar in character to the globules in Messier 8. If one assumes that they are at the distance of the near-by dark nebulae in Ophiuchus, one finds diameters of the order of 30,000 A.U. for some of the best-defined objects. They seem to occur at the rate of one per square degree in the regions where the star density is sufficient to show them by contrast to the background.

A few exceedingly small and distinct globules are seen projected against the Scutum Cloud. It is unlikely that the linear diameter of Barnard 117 or 118 is much in excess of 20,000 A.U. About half-a-dozen globules can be recognized in this section.

The situation in Cygnus and Cepheus is of interest, since in this transition region there are indications for the presence of a few round dark holes, such as Barnard 350. The stellar background is, however, not sufficiently dense in this section to show the globules as clearly as for the Sagittarius-Ophiuchus section. A few globule-like dark spots are seen projected against the North America nebula, but these are more generally hazy and not nearly so distinct as the globules seen projected against Messier 8.

In Perseus, Auriga, and Gemini the stellar background is too thin to permit the detection of many globules. Only the larger objects cover sufficiently large areas to be detectable. Barnard 34 in Auriga, 201 in Perseus, and 227 in Gemini are among the clearest "dark holes" in the anticenter region. Their diameters are large, 10′–20′, which, in all probability, means that the corresponding linear diameters are of the order of 100,000 A.U. or more.

It is not a simple matter to measure the total photographic absorptions produced by the globules. The small ones, seen projected against diffuse nebulae, show no stars shining through them; and the best that we can do is to estimate a *minimum* absorption by comparing the surface brightness in the globule with that in the outer parts of the diffuse nebula. Because of possible overlying bright nebulosities, this represents, at best, a minimum estimate for the absorption, which, for the globules with diameters of 10,000–35,000 A.U., proves to be between 2 and 5 mag.

For some of the larger globules, absorptions can be estimated with greater accuracy, since stars shine through them in considerable numbers. Stoddard[11] has made star counts according to photographic and photovisual magnitudes for four large globules in Barnard's list, Barnard 34, 201, 226, and 227. He finds the average total photographic absorptions to be of the order of 1 mag. Star counts on plates taken with the Harvard Jewett-Schmidt telescope confirm this estimate. It is worth noting that these four objects are all in the anticenter section of the Milky Way and that their minimum diameters are of the order of 100,000 A.U.

The cosmological status of the globules was considered in a paper presented at the Harvard Observatory Centennial Symposium on Interstellar Matter (December, 1946). The globules are interesting objects, which deserve further study by the powerful Schmidt telescopes now being put into operation. Every one of them merits further careful study with the largest available reflecting telescopes. Star counts in blue, red, and infrared light should be supplemented with measurements of surface brightness for the globules and for neighboring "unobscured" areas of comparable size.

1. Spitzer, *Ap. J. 94*, 232 (1941); Whipple, *Ap. J. 104*, 1 (1946).
2. *Ap. J. 49*, 1 (1919); and Barnard's *Photographic Atlas of Selected Regions of the Milky Way* (1927), Introd., p. 18.
3. Moore, Mayall, and Chappell, *Astronomical Photographs Taken at the Lick Observatory*.
4. *Atlas of the Northern Milky Way*.
5. *Ap. J. 51*, 5 (1920).
6. See Bok and Bok, *The Milky Way*, fig. 65.
7. *A.N. 177*, 234 (1908).
8. *Uppsala Ann.*, vol. 1, no. 3 (1940).
9. *Ap. J. 56*, 184 (1922).
10. Ibid. *91*, 20 (1940).
11. Ibid. *102*, 267 (1945).

Appended Figure

Fig. 96.1 A variety of small dark globules seen against the light of an emission nebula in the southern Milky Way. These remarkable dark markings were first observed by A. D. Thackeray in 1951, and this photograph was taken on a red-sensitive emulsion at the Cerro Tololo Inter-American Observatory. (Courtesy Bart Bok.)

Herbig, The Spectra of Two Nebulous Objects near NGC 1999

On a series of direct photographs taken with the Crossley reflector in 1946 and 1947 and centered on the diffuse nebula NGC 1999, there appear several peculiar nebulous objects. The brighter of these (referred to hereafter as "No. 1") resembles, on the best plates, a slightly diffuse star with a very short, curved, nebulous "tail" extending for 5″ in p.a. 308°. Its visual magnitude was estimated at the telescope to be near 16. It lies 1′.0 west and 2′.2 south of BD −6°1253, the illuminating star of NGC 1999. Object No. 2, which is 0′.1 east and 4′.1 south of BD −6°1253, is composed of two faint stars 9″ apart, one much fainter star, and three closely associated semistellar clots of nebulosity; the entire object would be contained in a circle 20″ in diameter. It is superimposed on much fainter nebulosity in the form of a ring, and slit spectrograms indicate that still feebler emission nebulosity is present over the entire field. The two brighter stars in object No. 2 were estimated to be about visual magnitude 17.5. It is the purpose of this note to describe the spectra of objects Nos. 1 and 2 and to draw some conclusions from them.

Omitted here is a figure showing the region of NGC 1999 and the spectrum of object 1.

The spectrograms were obtained with the Crossley nebular spectrograph (dispersion 430 A/mm at $H\gamma$) with exposure times up to 2.5 hours. They show spectra composed of strong bright lines on a weak continuous background. The strongest lines are the lower members of the Balmer series, the [*O* I] lines at λ 6,300, λ 6,363, the [*O* II] doublet at λ 3,727 and the [*S* II] pairs at λ 4,068, λ 4,076 and at λ 6,717, λ 6,731. A list of emission lines is given in table 96.1, which is based on three plates of object No. 1 and one of object No. 2. The line spectra

Table 96.1 Emission lines measured in spectra of objects Nos. 1 and 2 near NGC 1999

Wave length	Identification	Estimated intensity
6,727[a]	[*S* II] λ 6,717, λ 6,731	50
6,562	*H* α	100
6,363	[*O* I]	20
6,300	[*O* I]	60
5.006	[*O* III]	3
4,958	[*O* III]	1
4,861	*H*β	15
4,657[b]	[*Fe* III]?	1
4,571[b]	*Mg* I?	1
4,452	[*Fe* II]	1
4,416[a]	[*Fe* II] λ 4,413, λ 4,416	2
4,359	[*Fe* II]	2
4,340	*H*γ	8
4,287	[*Fe* II]	2
4,249[a]	[*Fe* II] + ?	2
4,101	*H*δ	4
4,070	[*S* II] λ 4,068, λ 4,076	10
3,970	*Ca* II, *H*ε, [*Ne* III]	5
3,933	*Ca* II	2
3,888	*H*8	1
3,868[b]	[*Ne* III]	1
3,727	[*O* II] λ 3,726, λ 3,729	15

a. Measured wave length of unresolved pair or blend.

b. Measured wave length; the line was seen only on the most strongly exposed plate.

of the two are quite similar. In the case of No. 2 the slit passed through the southwest star of the pair and the clot of nebulosity immediately to the south of it, but star and nebulosity are too close together to be separately distinguishable on this plate.

These spectra are remarkable for several reasons: (*a*) the great strength of [*S* II]; (*b*) the large range in excitation energy (as represented by ionization plus excitation potentials) between such lines as those of [*O* I] (2 e.v.) to [*O* III] and [*Ne* III] (51 and 65 e.v.); and (*c*) their striking dissimilarity to the spectra of ordinary T Tauri-like stars in the same dark nebula[1] and in the Taurus-Auriga dark clouds. The explanation of reason *b* undoubtedly is that we are observing the integrated radiation from nebulous envelopes in which there exist very large variations of density.

The only object known to the writer that possesses a spectrum like that of the objects near NGC 1999 is the nebulosity close to T Tauri discovered by Burnham and by Baade.[2] In that case, owing to the proximity of T Tauri, only the lines of *H*, [*S* II], and [*O* II] could be directly observed in the nebula, but it is possible that the [*O* I] and [*Fe* II] lines found in the spectrum of the star actually originate in the surrounding nebulosity. If the source of the continuum—presumably a star—in object No. 1 were several magnitudes brighter, we should find the nebulosity near it to appear, spectroscopically, very much as does that near T Tauri, since then the weaker emission lines would be observable only with difficulty.

The generally accepted mechanism for the production of the emission lines in the gaseous nebulae involves the photoelectric ionization of the abundant lighter elements by the intense ultraviolet radiation of the very hot exciting star. The permitted lines then result from recombination or fluorescent excitation by recombination lines, while the forbidden lines are due to collisional excitation of metastable levels by the free electrons. There is no indication of the presence of a blue star, which is a vital part of this mechanism, at T Tauri. No statement can be made at the present time regarding the energy distribution or spectral type of the continuous spectra in the objects near NGC 1999, on account of both the weakness of the spectra and the interference of the numerous bright lines on the Lick plates. The absolute magnitudes inferred from the apparent distance modulus of the Orion Nebula are, however, about those to be expected for late K- or early M-type dwarfs.

One is faced with the alternative of postulating either (1) the existence of a faint blue companion near T Tauri and the presence of similar low-luminosity, high-temperature stars in the NGC 1999 objects or (2) the operation of an excitation mechanism involving the interaction of a late-type dwarf star with the nebular material. The latter seems more plausible to the writer. Certainly, at face value there is no lack of energy available to such an interaction process: the kinetic energy of, for example, a hydrogen atom falling from infinity is 1,200 e.v. at the surface of a K-type dwarf, if radiation pressure is not important. The problem is to discover a mechanism involving such infalling material that explains the observed spectrum and whose operation is demanded by the properties of the star and its environment. The relative rarity of objects such as those near NGC 1999 suggests that some special condition, not of frequent occurrence, is required for the production of this type of spectrum. Apparently, the absolute magnitude of the associated star is not critical: the distance modulus of the Taurus clouds indicates an $M_v \sim +4$ for T Tauri, while the sources of continuous spectra in the objects near NGC 1999 have $M_v \sim +9$.

The possibility should be kept in mind, in considering this situation, that the T Tauri "stars" and kindred objects in nebulae may not be normal stars. Although what can be seen of the absorption spectra resembles, with low dispersion, the spectra of late-type dwarfs, it is not certain that one can correctly infer all the physical properties of the objects from this resemblance. There is strong evidence that, in the case of the absolute magnitude, one cannot so infer in every case.

In addition to objects Nos. 1 and 2, the Crossley plates show a number of other semistellar features near NGC 1999. Most of these are very faint. The brightest, after Nos. 1 and 2, lies 3$'\!.$3 west and 0$'\!.$1 south of BD–6°1253 and resembles a very faint diffuse star with a nebulous wisp extending about 14″ in p.a. 339°. It has not been observed with a slit spectrograph, but an objective-prism plate of the field, kindly loaned by Dr. Haro, shows a bright $H\alpha$ line.

These objects define another type in the growing list of peculiar objects that occur where stars and nebular material are intimately associated.

1. This remark is based on Lick observations of the spectra of a number of the faint $H\alpha$ emission stars found in this region by G. Haro (*A.J. 55*, 72 [1950]) on objective-prism plates. In the course of that work Haro noted the bright $H\alpha$ in the two objects discussed here.

2. G. H. Herbig, *Ap. J. 111*, 11 (1950).

Haro, Herbig's Nebulous Objects near NGC 1999

IN A RECENTLY PUBLISHED NOTE G. H. Herbig[1] reports the discovery of three peculiar nebulous objects near NGC 1999 and gives the spectroscopic description of the two brightest. The writer had independently discovered these peculiar objects,[2] finding $H\alpha$ and the [*O* I] lines at λ 6,300 and λ 6,363 in emission in the spectra of the three. In the same region of the sky, covering an area of 20 square degrees, four additional, similar objects were found. It is of interest to point out that these seven peculiar objects are the only ones that were found in an investigation over Milky Way regions which cover approximately 1,000 square degrees.

The emphasis of this note, however, is upon the stars that Herbig found in his nebulous objects Nos. 1 and 2 and what he suggests about them.

Herbig is faced with the alternative of postulating either (1) the existence of faint blue, high-temperature stars in the NGC 1999 objects or (2) the interaction of a late-type dwarf star with the nebular material. The latter seems more plausible to him.

In order to prove whether the stars associated with the nebulous objects are faint blue dwarfs or whether they are late-type dwarf stars, the writer took with the Tonantzintla Schmidt camera several direct infrared plates (hypersensitized I-N plates behind a filter that cuts approximately at λ 7,200) over the NGC 1999 region. It is assumed that on the long-exposure infrared plates all stars later than type G and with photographic magnitudes as faint as 19 must appear. Thus, all the stars that are in the surroundings of the three nebulous objects and that appear on 103*a*-O plates with photographic magnitudes as faint as 19 also are easily visible on the infrared plates. However, and in spite of what has been mentioned, the stars found by Herbig in his objects Nos. 1 and 2, with visual magnitudes 16 and 17.5, do not appear on the infrared plates.

This clearly indicates that the stars associated with the three peculiar nebulous objects cannot be late K or early M-type dwarfs. If any star is associated with these nebulous objects, it must be a faint, very blue, hot star.

1. *Ap. J. 113*, 697 (1951).

2. In personal letters dated May 31, 1950, which were written to Drs. Shapley and Minkowski, the writer pointed out the peculiarity of these objects.

Becklin and Neugebauer, An Infrared Star in the Orion Nebula

IN JANUARY 1965, an intensity map of the Orion Nebula in the wavelength region from 2.0 to 2.4 μ was made using a dual-beam photometer optically similar to that described by Westphal, Murray, and Martz (1963), mounted at the Cassegrain focus of the 60-inch telescope at Mount Wilson. In this program 80 per cent of the nebula within a 2′ radius of the Trapezium was observed by slowly scanning a 13″ aperture in a raster pattern. Detectable radiation was measured throughout the region scanned, with an average measured surface intensity of $5(\pm 2) \times 10^{-9}$ W cm^{-1} μ^{-1} sterad^{-1} and a maximum intensity of approximately twice the average value. The average measured intensity corresponds to a calculated emission measure of 3.5×10^6 cm^{-6} pc if it is assumed that all of the radiation is coming from bound-free, free-free, and line hydrogen transitions. A more detailed report on the above observations as well as on further observations at 1.65 and 3.4 μ is planned for a later paper.

On the survey seven point sources were detected which could be identified positively with photographically visible stars, and one source was found which could not be identified. During the winter of 1966 photometric observations at $\lambda\lambda$1.5–1.8 μ, 2.0–2.4 μ, 3.1–3.8 μ, and 8.5–13.5 μ were made of this infrared source using the 24-inch reflector at Mount Wilson and the 200-inch Hale telescope at Mount Palomar. The photometric results are given in table 96.2.

Table 96.2 Photometric results

Effective wavelength (μ)	Wavelength pass band (μ)	Magnitude	Absolute flux (W cm^{-2} μ^{-1}) $\times 10^{16}$
1.65	0.3	9.8	0.14 ± 0.03
2.2	0.4	5.2	$3.4 \pm .2$
3.4	0.7	2.0	$13.2 \pm .8$
10.0	5.0	−1.2	$3.4 \pm .3$

Here we omit a figure comparing the normalized measurements of the flux of the infrared object with the energy distribution of a 700 K blackbody. Also omitted is a figure that shows the approximate position of the object in the Orion nebula and the infrared absorption curve for the nebula.

It is seen that the flux at the three longer wavelengths fit a black body quite well, but the flux in the region around 1.65 μ is low by a factor of 4. The absolute flux calibration was obtained by observing Johnson's (1964, 1965*a*) standard stars.

When considering the nature of this infrared object, it is important to determine whether the object is in front of, behind, or in the nebula.

If the object is in front of the nebula, it must be an extremely cool star, comparable to the coolest found in the California Institute of Technology Infrared Sky Survey (Neugebauer, Martz, and Leighton 1965; Ulrich, Neugebauer, McCammon, Leighton, Hughes, and Becklin 1966). The probability of finding such an infrared star just coinciding with the nebula is very small, and this possibility will not be discussed further.

If the object is behind the nebula, it might be an ordinary type of star that is extremely reddened. An upper limit to the probability of such a star being situated behind the nebula can be made by considering the area scanned and the surface density near the galactic plane of stars whose apparent magnitude at 10 μ is brighter than -1.2—the apparent magnitude of the object. From the work of Low and Johnson (1964), Johnson (1964), and Barnhart and Mitchell (1966) it is estimated that the number of such stars near the galactic plane is less than 100, which leads directly to a surface density less than 10^{-5} per sq. min of arc. The area scanned was about 10 sq. min of arc, so the a priori probability of the object being an ordinary star is less than 10^{-4}. If the object is a common type of star, the apparent magnitude of -1.2 at 10 μ and the minimum distance modulus of 8.5 (Markowitz 1948; Sharpless 1952; Strand 1958) requires that it be a red supergiant. If, in fact, the object is a red supergiant, extrapolation of a typical absorption curve (Johnson and Borgman 1963) shows that the total amount of absorption in the visible would be about 35 mag.

When we consider the small a priori probability of finding a common type of star in the region scanned, together with the enormous absorption that would be required, we are led to conclude that the object is probably located inside the nebula itself.

There is, in fact, some independent observational evidence that the source may be located in the nebula. Although spatial scans at the east-arm Cassegrain focus of the 200-inch telescope using a 6″ aperture indicate that the object is a point source to within 2″ at both $\lambda 2.2$ and $\lambda 10$ μ, broad wings extend 30″ from the source. This is shown by the fact that within 30″ east and west of the source the average intensity at 2.2 μ is 12.0 (± 0.5) $\times 10^{-9}$ W cm^{-2} μ^{-1}sterad^{-1}, while between 30″ and 50″ away from the source the average intensity is 7.0 (± 0.5) $\times 10^{-9}$ W cm^{-1} μ^{-1} sterad^{-1}. Wings were not observed on any stars outside the nebula. The chance that the central core is accidentally associated with these wings is small since in the 10 sq. min of arc area scanned the only other region found with a similar increase in flux was centered around the Trapezium. The probability that the coincidence is accidental is thus less than 0.1.

A correction for nebular absorption has been calculated for several amounts of total absorption typically expected in the nebula. A three-color temperature ($\lambda\lambda 2.2$, 3.4, 10 μ) of a source necessary to give the observed flux has been derived using an absorption curve obtained from the data of Johnson and Borgman (1963), Johnson (1965*b*), and infrared measurements of θ^1 (C) Orionis. Table 96.3 presents the results of these calculations, plus an approximate bolometric magnitude and a calculated radius assuming the object is at a distance of 500 pc (Markowitz 1948; Sharpless 1952; Strand 1958). The Trapezium has a visual absorption of less than 3 mag (Johnson 1965*b*); yet even if 5 times this optical depth is considered

Table 96.3 Corrected fluxes and inferred parameters

Optical depth in visible	Corrected flux (W cm^{-2} μ^{-1} $\times 10^{17}$) 1.65 μ	2.2 μ	3.4 μ	10 μ	Three-color temperature (° K)	Bolometric magnitude	Radius (a.u.)
0	1.4	34	132	34	700	-1.9	8
2.5	2.6	52	153	35	725	-2.0	8
5.0	5	79	177	36	760	-2.4	8
10.0	18	175	240	38	900	-3.0	6
15.0	65	400	320	40	1,100	-3.9	6

($\tau_v \sim 15$) the energy distribution still requires a rather low source temperature, less than 1,200° K.

From the above discussion we feel that the object is in the nebula and has a gravitationally associated cool shell. It is well known that the Orion Nebula is a very young association and that the probability of finding a star in the process of forming should be relatively high (Aller and Liller 1959). Thus an attractive interpretation of the observations is that the infrared object is a protostar. If the star has a radius R of 8 a.u., a temperature T of 700° K, and a mass M of 6 M$\odot$, the Kelvin time scale,

$$\tau = \frac{GM^2}{4\pi R^3 \sigma T^4},$$

is 10^3 years. This number is probably consistent with the approximate age of the Orion Nebula 2×10^4 years established by Vandervoort (1964) or the Trapezium expansion age of $< 10^4$ years given by Parenago (1953).

Aller, L. H., and Liller, W. 1959, *Ap. J. 130*, 45.

Barnhart, P. E., and Mitchell, W. E. 1966, *Contributions from the Perkins Observatory*, ser. 2, no. 16.

Johnson, H. L. 1964, *Bol. Tonantzintla y Tacubaya Obs. 3*, 305.

———. 1965*a*, *Com. Lunar and Planetary Lab.*, no. 53.

———. 1965*b*, *Ap. J. 141*, 923.

Johnson, H. L., and Borgman, J. 1963, *B.A.N. 17*, 115.

Low, F. J., and Johnson, H. L. 1964, *Ap. J. 139*, 1130.

Markowitz, W. M. 1948, *A.J. 54*, 111.

Neugebauer, G., Martz, D. E., and Leighton, R. B. 1965, *Ap. J. 142*, 399.

Parenago, P. P. 1953, *Astr. Zh. 30*, 249.

Sharpless, S. 1952, *Ap. J. 116*, 251.

Strand, K. 1958, *Ap. J. 128*, 14.

Ulrich, B., Neugebauer, G., McCammon, D., Leighton, R. B., Hughes, E. E., and Becklin, E. 1966, *Ap. J. 146*, 288.

Vandervoort, P. O. 1964, *Ap. J. 139*, 869.

Westphal, J. A., Murray, B. C., and Martz, D. E. 1963, *Appl. Opt. 2*, 749.

97. Radio Observations of OH in the Interstellar Medium

Sander Weinreb, Alan H. Barrett,
M. Littleton Meeks, and John C. Henry

(*Nature 200*, 829–831 [1963])

After the discovery of interstellar hydrogen at radio frequencies, astronomers began to speculate about the detection of molecules at radio wavelengths rather than in the visible spectrum. Just as the 21-cm transition of interstellar hydrogen was predicted theoretically before it was observed, the possibility of detecting the microwave emission of the OH molecule was considered by Iosif Shklovskii some fourteen years before its discovery.[1] In order for this molecule to be detected, however, the frequency of its monochromatic radio emission had to be measured accurately in the terrestrial laboratory. Furthermore, radio telescopes with surfaces accurate to a few centimeters had to be constructed for receiving the short-wavelength emission of molecules, and new methods of spectral analysis had to be developed. Not until 1953 did Charles Townes and his colleagues make the first precision laboratory measurements of the radio-frequency transitions of the OH molecule. Within a decade scientists at the Massachusetts Institute of Technology, whose report we reprint, had built the sophisticated equipment needed to obtain the first observations of interstellar OH. Here Sander Weinreb introduces his digital detection technique, which has become the basis for the spectral analysis performed at most major radio observatories.

Other radio astronomy groups soon began intensive searches for OH. Because the molecule was first detected in absorption against the strong radio source Cassiopeia A, the next observations were also made against bright radio sources. This approach accidentally led to the surprising discovery that certain bright nebulae (H II regions) exhibit emission rather than absorption lines near the OH resonance frequency.[2] Not only were these lines more intense than those expected under reasonable assumptions about the cosmic abundance of OH, but the relative line intensities showed that the molecules must be in a nonequilibrium state. Furthermore, because their telescopes did not provide high angular resolution, the early observers could not realize how compact these emission-line sources really are. Radio interferometers with intercontinental baselines have now revealed that some of these sources are as small as stars; but if they radiate like stars their temperatures must be hotter than a million million degrees. In these cases the clouds of OH seem to act like a gigantic amplifier similar to the masers built in terrestrial laboratories (maser is an acronym for the microwave amplification by stimulated emission of radiation).[3]

The radio discovery of OH did not at first stimulate a radio search for other, more complex, interstellar molecules, since at the time most astronomers thought that no more than two atoms could come together at one time in the low density regions of interstellar space. In addition, even if somewhat more complex molecules were formed, astronomers expected that they would be quickly destroyed by ultraviolet starlight and by high-energy cosmic rays. Yet as early as 1955, Charles Townes, in discussing the types of molecular spectra that might be found at radio frequencies, noted that the relatively low energies of radio photons correspond to the relatively low temperatures at which molecules are excited into rotational states. He also gave laboratory measurements of the rotational transitions of molecules that might be detected, including ammonia, water, and carbon monoxide, as well as OH.[4] Townes subsequently participated in a collaborative effort at Berkeley that led to the discovery of interstellar ammonia and water.[5]

These discoveries triggered an avalanche of molecular searches in which groups of young astronomers armed with the latest laboratory measurements engaged in an extraordinarily competitive fight to be the first to detect the next interstellar molecule. The net result has been the discovery of over 100 interstellar molecular lines ranging from those of formaldehyde to those of alcohol. These discoveries had not been anticipated even a decade earlier, because astronomers had overlooked the importance of interstellar dust grains in shielding molecules from ultraviolet light and in acting as a catalyst in forming complex molecules.

1. I. S. Shklovskii, *Astronomicheskii zhurnal 26*, 10 (1949), and *Doklady akademii nauk SSSR 92*, 25 (1953).

2. H. Weaver, D. R. W. Williams, N. H. Dieter, and W. T. Lum, *Nature 208*, 29 (1965); E. J. Gundermann, *Observations of the Interstellar Hydroxyl Radical* (Ph.D. diss., Harvard University, 1965); S. Weinreb, M. L. Meeks, J. C. Carter, A. H. Barrett and A. E. E. Rogers, *Nature 208*, 440 (1965).

3. J. M. Moran, B. F. Burke, A. H. Barrett, A. E. E. Rogers, J. C. Carter, J. A. Ball, and D. C. Cudaback, *Astronomical Journal 73*, S27, S108 (1968); M. Litvak, *Astrophysical Journal 156*, 471 (1969).

4. C. Townes, *Radio Astronomy: I.A.U. Symposium No. 4*, ed. H. C. van de Hulst (Cambridge: Cambridge University Press, 1957).

5. A. C. Cheung, D. M. Rank, C. H. Townes, D. D. Thorton, and W. J. Welch, *Physical Review Letters 21*, 1701 (1968).

In this article we wish to report the detection of 18-cm absorption lines of the hydroxyl (OH) radical in the radio absorption spectrum of Cassiopeia *A*, thereby providing positive evidence for the existence of OH in the interstellar medium. The microwave transitions of OH in the ground-state, $^2\pi_{3/2}$, $J = 3/2$, arise from two Λ-type doublet-levels, each of which is split by hyperfine interactions with the hydrogen nucleus, so that four transitions result. The two strongest lines have been previously measured in the laboratory at 1,667.34 ± 0.03 Mc/s ($F = 2 \rightarrow 2$) and 1,665.46 ± 0.10 Mc/s ($F = 1 \rightarrow 1$) with relative intensities of 9 and 5, respectively;[1] these results are in agreement with theory. The suggestion that these lines might be detected in the radio spectrum of the interstellar medium has been made by Shklovsky[2] and Townes.[3] A previous search by Barrett and Lilley,[4] in 1956, was unsuccessful, primarily because the laboratory measurements of the frequencies had not been made. A recent search for OH emission also yielded negative results.[5]

Our observations were conducted on 10 days between October 15 and October 29, 1963, using the 84-ft. parabolic antenna of the Millstone Hill Observatory of Lincoln Laboratory, Massachusetts Institute of Technology, and the spectral-line autocorrelation radiometer designed by Weinreb.[6] The receiver uses digital techniques to determine the autocorrelation function of the received signal. The resulting autocorrelation function is then coupled directly into a digital computer that performs a Fourier transformation and displays the resulting spectrum on a cathode-ray tube or a precision x–y plotter. During one integration time-interval of 2,000 sec, a 100-kc/s portion of the spectrum is determined with a frequency resolution of 7.5 kc/s. The ability to see immediately a calibrated visual display of the measured spectrum and average this result with others greatly facilitated the conduct of the experiment and eliminated almost all post-observation data handling. The system noise temperature was 420° K, of which 110° K was due to Cassiopeia *A*. System tests were performed by observing the hydrogen line.

The results obtained during the first evening of our observations showed strong evidence of the 1,667 Mc/s line in Cassiopeia *A*; the signal is visible after 2,000 sec of integration. We decided that positive identification of OH absorption lines of Cassiopeia *A* would be secured before proceeding to observations of other regions. Our results indicate that two of the three clouds showing strong H absorption,[7] namely, those at radial velocities of −0.8 km/sec and −48.2 km/sec, also give rise to OH absorption lines that we have detected at both 1,667 Mc/s and 1,665 Mc/s. The strong H absorption line at −38.1 km/sec appears to be composed of two lines at −37.4 km/sec and −42.1 km/sec when observed at the OH frequency. It is to be expected that a one-to-one correspondence between OH absorption and H absorption will not be observed because of: (*a*) greater thermal broadening of H lines; (*b*) larger optical depth of the H lines; (*c*) possible OH/H abundance variations from cloud to cloud. A typical record showing the 1,667-Mc/s line in the −0.8 km/sec cloud is shown in figure 97.1. A summary of all of our observations is presented in table 97.1.

The evidence that we are indeed detecting interstellar OH in these observations may be summarized as follows:

(1) Lines at both 1,667 Mc/s and 1,665 Mc/s have been detected with frequencies and intensity ratios that are in good agreement with the expected values.

(2) The OH absorption spectra at both frequencies show general agreement with the H absorption spectra.

(3) The absorption lines disappear when the antenna is positioned off Cassiopeia *A* by one degree in both azimuth and elevation.

(4) The lines shifted 20 kc/s between October 17 and October 29; this is the shift expected from the orbital velocity of the Earth during this time-interval.

Table 97.1 Summary of OH line absorption measurements in the Cassiopeia *A* radio source

Radial velocity (km/sec)	1,420.405 Mc/s H-line optical depth (ref. 7)	1,667.357 Mc/s OH-line optical depth	Observed line width (kc/s)	Number of OH radicals per cm^2	Abundance ratio relative to H
−0.8	1.85	0.016 ± 0.005[a]	13	~2×10^{14}	~1.5×10^{-7}
−37.2	—	.010 ± .005	13	~1.5×10^{14}	—
−42.1	—	.012 ± .005	20	~3×10^{14}	—
−48.2	4.0	.016 ± .008	25	~5×10^{14}	~1×10^{-7}

a. An optical depth of 0.010 ± 0.003 with line width of 16 kc/s was observed for the 1,665.402 Mc/s line.

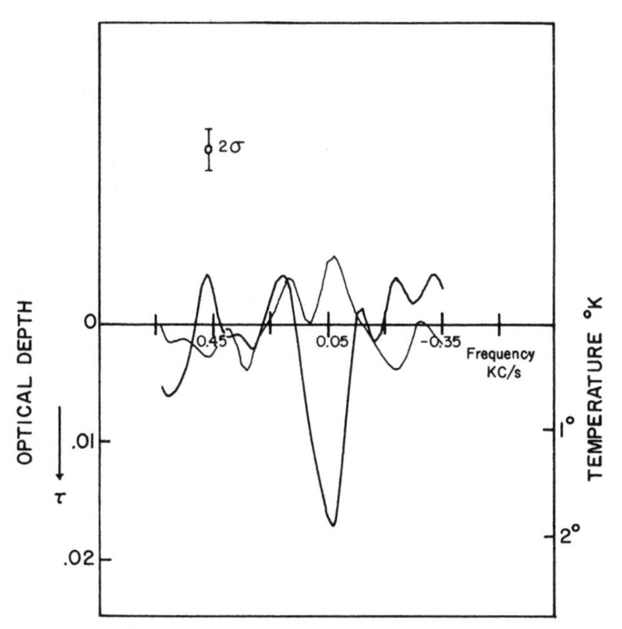

Fig. 97.1 Observed 1,667 MHz OH absorption spectrum in Cassiopeia A. The heavy line shows 8,000 sec of data taken with the antenna beam directed at Cassiopeia A, and the light line shows 6,000 sec of data taken with the beam slightly displaced from Cassiopeia A. The frequency scale is specified in kc/s with respect to the local standard of rest, assuming that the line rest frequency is 1,667,357 kc/s.

A quantity of immediate astrophysical interest which follows from our observations is the abundance ratio of OH to H. This can be obtained in the following way: the spectral change in antenna temperature ΔT_{OH} owing to the OH absorption is given by:

$$\Delta T_{OH} = \tau_{OH} T_{AC}$$

where τ_{OH} is the OH optical depth in the direction of Cassiopeia *A* and T_{AC} is the antenna temperature attributable only to Cassiopeia *A*. The maximum optical depth is given by:

$$\tau_{OH} = \frac{hc^2 A N_{OH}}{8\pi k T_s \nu_0 \Delta\nu} \frac{g_i}{\sum g_i}$$

where A is the spontaneous transition probability, T_s is the excitation temperature analogous to the H spin temperature, ν_0 is the line frequency, $\Delta\nu$ is the line width, g_i is the statistical weight of level *i*, and N_{OH} is the total number of OH radicals per unit cross-section. The statistical weight g_1 of the upper level of the 1,667-Mc/s transition is 5, and for the 1,665-Mc/s transition it is 3; the sum of the statistical weights for all levels is 16. The spontaneous transition probability A must be evaluated by using the matrix element derived from a quantum-mechanical treatment of Λ-type doubling,[8] and has the value 2.86×10^{-11} sec^{-1} for the 1,667-Mc/s transition. The OH dipole moment used in these calculations[9] is $(1.60 \pm 0.12) \times 10^{-18}$ E.S.U.

The excitation temperature T_s is the subject of considerable uncertainty because it cannot be assumed that it will be the same as that for H. The excitation temperature is defined by the relation:

$$\frac{n_i}{n_0} = \frac{g_i}{g_0} e^{(-h\nu)/(kT_s)}$$

where n_i and n_0 are the densities of radicals in the upper and lower states of the transition, respectively. For H it has been shown that T_s equals the kinetic temperature in a typical galactic gas cloud[10] because radiative transitions are relatively rare as compared with collisional transitions. For OH, however, the spontaneous transition probability is 10^4 times larger than for H, so radiative transitions play a more dominant part in establishing the equilibrium population distribution. A detailed evaluation of the processes that will be important in determining the OH excitation temperature has not been made, but a preliminary investigation shows that slow-moving positive ions may be very effective in inducing transitions between the two states, in spite of their low abundance in a H cloud.[11] This situation arises because the OH transition is of an electric dipole type, and therefore can be induced by the Coulomb field of electrons and ions which leads to large interaction radii. A similar result has been obtained by Purcell when considering the population of the fine-structure states of the $n = 2$ level of H in the solar atmosphere.[12] An upper limit on T_s can be set by our observations off Cassiopeia *A* from which one would expect to detect OH emission. From the preliminary observations we conclude that any OH emission adjacent to Cassiopeia *A* is less than 1° K; this result implies a T_s less than 50° K for an optical depth of 0.02. For purposes of computing the total number of OH radicals from the foregoing equations we have assumed a T_s of 10° K. More extensive observations for OH emission will enable a better estimate of this quantity to be made.

The values of the number of OH radicals per cm^2 in the direction of Cassiopeia *A* are shown in table 97.1. The OH/H abundance ratio can be calculated from the results of the H absorption on Cassiopeia *A*[7], and gives typical ratios of 1×10^{-7} (see table 97.1). This ratio can be compared with estimates of the CH/H abundance ratio 10^{-6} by Strömgren[13] and 2×10^{-8} by Bates and Spitzer.[14]

Our observations have enabled a more accurate determination of the frequencies of the two strongest Λ-type doublet lines of OH. The laboratory and astronomical values are shown in table 97.2. It is possible that these values will be of interest to the molecular spectroscopist for a more accurate evaluation of the hyperfine coupling constants.

Table 97.2 Rest frequency of observed lines

Transition	Laboratory measurement	Astronomical measurement
$F = 2 \rightarrow 2$	1,667,340 ± 30 kc/s	1,667,357 ± 7 kc/s
$F = 1 \rightarrow 1$	1,665,460 ± 100 kc/s	1,665,402 ± 7 kc/s

1. G. Ehrenstein, C. H. Townes, and M. J. Stevenson *Phys. Rev. Letters 3*, 40 (1959).

2. I. S. Shklovsky, *Dok. Akad. Nauk. (SSSR) 92*, 25 (1953).

3. C. H. Townes, *Intern. Astro. Union Symp.*, No. 4, ed. H. C. van de Hulst, (Cambridge: Cambridge Univ. Press, 1957) p. 92.

4. A. H. Barrett and A. E. Lilley, *Astro. J. 62*, 5 (1957).

5. A. A. Penzias, Abst. Paper *New England Radio Eng. and Manufacturers* (Nov. 4–6, 1963).

6. S. Weinreb, *Tech. Rep.* 412, *Res. Lab. Electronics, M.I.T.*, Cambridge, Mass. (Aug. 30, 1963).

7. C. A. Muller, *Symp. Radio Astronomy*, Paris, ed. R. N. Bracewell (Stanford: Stanford Univ. Press, 1959) p. 360.

8. G. C. Dousmanis, T. M. Saunders, Jr., and C. H. Townes, *Phys. Rev. 100*, 1735 (1955).

9. G. Ehrenstein, (Ph.D. diss., Columbia Univ., 1960).

10. E. M. Purcell and G. B. Field, *Astrophys. J. 124*, 542 (1955).

11. A. H. Barrett and A. E. Lilley, *Astro. J. 62*, 4 (1957).

12. E. M. Purcell, *Astrophys. J. 116*, 457 (1952).

13. B. Strömgren, *Astrophys. J. 108*, 242 (1948).

14. D. R. Bates and L. Spitzer, Jr., *Astrophys. J. 113*, 441 (1951).

98. Galactic Magnetic Fields and the Origin of Cosmic Radiation

Enrico Fermi

(*Astrophysical Journal 119*, 1–6 [1954])

On the basis of measurements made by balloon in the high atmosphere in 1912, Victor Hess discovered the energetic cosmic rays that continuously rain on the earth from some unknown, extraterrestrial source (see selection 3). For over two decades physicists adopted Hess's conclusion that the cosmic rays are uncharged γ rays. By the early 1930s, however, the discovery of the east-west asymmetry in the intensity of cosmic rays, as well as the discovery of a latitude effect, convinced physicists that the primary cosmic rays entering the terrestrial atmosphere had to be predominantly composed of positively charged particles. Some unknown extraterrestrial source had to be accelerating these particles to speeds near the velocity of light in order to account for their extraordinary energies. Plausible acceleration mechanisms were advanced in the early 1930s by William Francis Swann, who anticipated man-made particle accelerators when he argued that cosmic rays might be accelerated by changing magnetic fields in stars,[1] and by Walter Baade and Fritz Zwicky, who argued that cosmic rays are accelerated during supernova explosions (see the introduction to selection 69).

Interest in the astrophysical aspects of cosmic rays was stimulated in 1948, when balloon-borne emulsions and cloud chambers permitted the direct identification of the primary cosmic rays for the first time.[2] Not only were the protons shown to be the most abundant cosmic ray particle, but the nuclei of other cosmically abundant nuclei were also found to be present in the cosmic rays.

How do we account for the fact that cosmic rays are raining on the earth with roughly equal intensity from all directions? If supernovae are the sources of cosmic rays we might expect cosmic rays to be most intense in the directions of supernova remnants. In 1939 Hannes Alfvén argued that the motions of cosmic rays would be constrained by magnetic fields produced by ionized interstellar matter.[3] Although such a field is extraordinarily weak by terrestrial standards, it is more than adequate to produce frequent deflections of the charged cosmic rays. The continual deflection of cosmic rays by the interstellar magnetic field guarantees that, however directional they may be at their origin, they will become isotropically distributed by the time that they reach the earth.

At the time that Alfvén was developing his ideas on the containment of cosmic rays by the interstellar magnetic field, he visited the Institute for Nuclear Studies at the University of Chicago, where he discussed his ideas with Enrico Fermi. These discussions led Fermi to propose in 1949 a new cosmic ray acceleration mechanism that accounts for the cosmic ray energy spectrum, in which the number of cosmic ray particles of a given energy decreases exponentially with increasing energy. Fermi reasoned that collisions of cosmic rays with randomly moving magnetic fields will on the average neither accelerate nor decelerate most cosmic rays, but because an accelerating, head-on collision is more likely than a decelerating one, a few cosmic rays will be accelerated to extraordinarily high energies. In this paper Fermi outlines the salient features of his statistical acceleration mechanism. In addition to accounting for the declining number of high-energy cosmic rays, he also calls attention to the importance of both the cosmic rays and the interstellar magnetic field in determining the equilibrium of interstellar matter. It is not by chance that the energy densities of the cosmic rays, the interstellar magnetic field, and the moving interstellar clouds are roughly the same, for they combine to produce the outward pressures that oppose the force of gravitational attraction and keep our galaxy stable.

In the final sections of this paper, Fermi reports that his acceleration mechanism only works for particles that have already been injected into interstellar space with considerable energy; if the galaxy is to remain in equilibrium, he notes, some source is needed to supply the turbulent energy lost in accelerating cosmic rays, as well as the energy lost by the leakage and collisions of cosmic rays. Here Fermi might have mentioned Baade and Zwicky's supernova explanation for the origin of cosmic rays. Supernovae probably continuously replenish interstellar space with energetic cosmic rays through their original explosions and through the action of their remnant, rotating neutron stars.

1. W. F. G. Swann, *Physical Review 43*, 217 (1933).

2. H. L. Bradt and B. Peters, *Physical Review 74*, 1828 (1948), *80*, 943 (1950); P. Freier, E. J. Lofgren, E. P. Ney, and F. Oppenheimer, *Physical Review 74*, 1818 (1948).

3. H. Alfvén, *Physical Review 55*, 425 (1939); see also *Physical Review 75*, 1732 (1949).

I BECAME INTERESTED in the possible existence of magnetic fields extending through the volume of the galaxy in connection with a discussion on the origin of the cosmic radiation a few years ago.[1] The hypothesis was discussed then that the acceleration of cosmic-ray particles to extremely high energies was due to their interaction with a galactic magnetic field that was postulated to pervade the galactic space. According to Alfvén's ideas on magnetohydrodynamics, this field would be strongly influenced by the turbulent motions of the diffuse matter within the galaxy. Indeed, the electric conductivity of this matter is so large that any lateral shift of the magnetic lines of force with respect to the matter is effectively prevented. A strong magnetic field quenches the transverse components of the displacement due to the turbulent motion. A weak field yields to the material motions, so that its lines of force are soon bent into a very crooked pattern.

The observation by Hiltner and Hall of an appreciable polarization of the light coming to us from distant stars has been interpreted[2] as due to the orientation of nonspherical dust grains by a magnetic field. If this general type of interpretation is correct, the polarization gives us some information on the strength and the direction of the magnetic field. Hiltner's measurements[3] indicate that in the vicinity of the earth the magnetic field is approximately parallel to the direction of the spiral arm. This fact suggests that we may perhaps think that the spiral arms are magnetic tubes of force. In the following discussion we will assume that this is the case.

The direction of polarization of the stellar light indicates further that in our vicinity the magnetic lines of force show irregular deviations from parallelism of the order of 10°. This fact excludes the hypothesis that the lines of force yield completely to the turbulent motions of interstellar matter, because then they would be rapidly bent into shapes much more irregular than those observed. One is rather led to the conclusion that the field is sufficiently strong to yield only a little to the transverse component of the turbulent motion. Indeed, as was pointed out by Davis, the small deviations from parallelism of the field enable one to estimate that the intensity of the magnetic field must be of the order of 10^{-5} gauss. Recently Chandrasekhar and I[4] have re-examined this problem, considering, in particular, the balance between magnetic and gravitational effects in the spiral arm. Our conclusion is that the field intensity is about 6×10^{-6} gauss. Owing to the turbulence, the lines of force are irregularly pushed sidewise until the magnetic stress increases to the point of forcing a reversal of the material motion and of pushing back the diffuse matter, impressing on it some kind of very irregular oscillatory motion. One expects, therefore, that the lines of force sway back and forth and also that the field intensity will fluctuate along the same line of force.

A cosmic-ray particle spiraling around these moving lines of force is gradually accelerated. The acceleration mechanism was discussed in R, although the shape of the lines of force assumed then was quite different from what we now believe it to be. A cosmic-ray proton of 10 bev energy is bent in a magnetic field of 6×10^{-6} gauss in a spiral having a radius of the order of one-third the radius of the earth's orbit. This is very small on the galactic scale. The motion of a proton with this energy and also of one with much greater energy is, therefore, properly described as a very small radius spiral around a line of force. Apart from the very rapid changes due to the spiraling, the general direction of motion may change for two reasons. One is that the line of force around which the particle spirals may be curved. In R a change of direction of this kind was called a "collision of type *b*." A second type of event, called a "collision of type *a*," takes place when the particle in its spiraling encounters a region of high field strength.

Let θ be the angle between the direction of the line of force and the direction of motion of the spiraling particle. The angle θ will be called the "angle of pitch." One can prove that, in a static magnetic field, the quantity

$$q = \frac{\sin^2 \theta}{H} \tag{1}$$

is approximately constant with time. For this reason, in a static field, the particle cannot enter a region where

$$H > \frac{1}{q}. \tag{2}$$

When the particle approaches such a region, its pitch decreases until $\theta = 90°$, at which moment the particle is reflected and spirals backward in the direction whence it came.

In a variable magnetic field both *a*- and *b*-type collisions may cause changes in energy. As a rule, the energy will increase or decrease according to whether the irregularity of the field that causes the collision moves toward the particle (head-on collision) or away from it (overtaking collision). It was shown in R that, on the average, the energy tends to increase primarily because the head-on collisions are more probable than the overtaking collisions. In the present discussion the same general acceleration mechanism will be assumed. The details, however, will be quite different from those previously assumed.

It was shown in R that through this mechanism the energy of the particle increases at a rate that, for extreme relativistic particles, is proportional to their energy. The energy E, therefore, increases exponentially with time:

$$E(t) = E_0 e^{t/A}. \tag{3}$$

According to this law, the oldest particles should have the highest energy.

The time A needed for an energy increase by a factor e was estimated in R to be about 100 million years. If this estimate was correct, A would be comparable to the time B for nuclear collisions of the particle. Assuming that a nuclear collision effectively destroys the particle, the probability that a particle has the age t should be

$$e^{-t/B}\frac{dt}{B}. \tag{4}$$

Combining conditions (3) and (4), one readily finds that the probability that a particle observed now has energy E should be proportional to

$$\frac{dE}{E^n}, \tag{5}$$

with

$$n = 1 + \frac{A}{B}. \tag{6}$$

An exponent law like (5), with n of the order of 2 or 3, seems to fit fairly well the observed energy spectrum of the cosmic radiation.

Two main objections can be raised against the theory proposed in R. One is that, according to present evidence, the structure of the galactic magnetic field is much more regular than was assumed in R. We shall try to make plausible the conclusion that, in spite of this fact, the acceleration may still take place at an adequate rate. Indeed, as will be discussed below, it will be necessary to provide an acceleration process five to ten times more efficient than was previously supposed. The second difficulty arises from the fact that the protonic and the nuclear components of the cosmic radiation have very much the same energy spectrum. Heavy nuclei have a larger nuclear collision cross-section than protons. Their mean life B, therefore, should be shorter, and the exponent n given by equation (6) should be larger.

This difficulty would be removed if the process that eliminates the particles were equally effective against protons and against larger nuclei, because B would then be the same for both kinds of particle. For example, collisions against stars or planets do not differentiate between protons and nuclei. On the other hand, a simple estimate shows that the probability of collisions against such massive objects is quite negligible, even in a time equal to the age of the universe. Another means of removing cosmic-ray particles that is equally effective for protons and nuclei is diffusion outside the galaxy. Assume, for example, that the lines of force follow the spiral arms. The stretched-out length of the galactic spiral is about a million light-years, and the particles travel with a velocity very close to that of light. They could, therefore, escape in a time of the order of a million years. The escape time, of course, would be longer if the particle occasionally reversed its direction, owing, for example, to collisions of type a.

In a theory that yields essentially the same energy spectrum for cosmic-radiation protons and nuclei it will be necessary to assume that the escape time for diffusion outside the galaxy is appreciably shorter than the nuclear collision time. Then escape will be dominant with respect to nuclear collisions. We will assume that this is the case, and we will take, somewhat arbitrarily, in the numerical examples the escape time,

$$B = 10 \text{ million years}. \tag{7}$$

The escape time is then about ten times shorter than the nuclear collision time.

Assuming 2.5 as an average value of the exponent n in quantity (5), we have, from equation (6),

$$A = 1.5B = 15 \text{ million years}, \tag{8}$$

in which A is the time through which the energy of the particle increases, on the average, by the factor e. This time, 15 million years, is appreciably shorter than was estimated in R.

It therefore appears necessary to modify the acceleration mechanism of R in two ways. The mean free path must be much longer, in order to allow the escape of the particles from the galaxy in a relatively short time. And the process of acceleration must be much faster. At first sight, these two requirements seem to be contradictory, and perhaps they are. On the other hand, there is an acceleration mechanism that is potentially much more efficient than the others. This process was discussed in R and was then dismissed as of little importance for certain reasons to be mentioned shortly. I propose to criticise those reasons and to make a case in favor of this acceleration mechanism.

First of all, we observe that the knowledge that we now have of the general shape of the magnetic field makes the collisions of type b rather unimportant. We shall therefore concentrate our attention on type-a collisions. They take place, as will be remembered, when the particle encounters a region of large field strength, where condition (2) is fulfilled. A particle that finds itself between two such regions will be trapped on the stretch of line of force comprised between them. When this happens, the energy of the particle will change with time at a rate much faster than usual. It will decrease or increase according to whether the jaws of the trap move away from or toward each other.

Let H be the average value of the magnetic field and H_{max} the maximum field along the line of force that may be likely to cause a type-a reflection. If θ is the angle of pitch of the spiral where the field is average, reflections will occur only for particles having

$$\theta > \chi, \tag{9}$$

where

$$\sin \chi = \sqrt{\frac{H}{H_{max}}}. \tag{10}$$

A simple calculation shows that when a particle with $\theta > \chi$ is caught in a trap, both its energy and its angle of pitch will change with time, but the product,

$$E \sin \theta, \tag{11}$$

remains approximately constant. The particle can escape the trap only when θ has decreased to the point that condition (9) is no longer fulfilled. In this process the energy must increase by a factor

$$\frac{\sin \theta}{\sin \chi}. \tag{12}$$

This process may lead to a sizable energy gain in a relatively short time. For example, if the jaws of the trap are 10 light-years apart and move toward each other at 10 km/sec, the time needed for a 10 per cent energy increase is only a few tens of thousands of years.

To be sure, the jaws will occasionally move away from each other, causing a loss instead of a gain of energy. But in this case θ will increase, making the particle more easily caught in similar traps. The process ends only when the energy has increased to the point that θ has become less than χ, because only then will the particle be capable of passing without reflection through occasional maxima of the field intensity that it may encounter along its path.

When this condition is reached, the process of acceleration becomes exceedingly slow. Indeed, if q in equation (1) were exactly a constant of motion, the particle would always keep on spiraling in the same direction and would eventually escape from the galaxy. It is for this reason that the process of acceleration in a trap was not considered of major importance in R. The process may become important only if there is machinery whereby the angle of pitch, after having been reduced to a small value in the process of acceleration in a trap, can be increased. If this is the case, the trap mechanism again becomes operative, and the energy may be increased by a further factor.

Now one finds that q is almost exactly a constant as long as the particle is not caught in a trap and there are no sharp variations in the magnetic field either in time or in space. When I first discussed the acceleration of cosmic rays by magnetic fields, I was not aware of the possibility of sharp discontinuities of the field and for this reason did not think that the traps could be the dominant factor. I propose now to show that discontinuities in the direction of the magnetic field should not be too exceptional in the galaxy.

Recently de Hoffmann and Teller[5] have discussed the features of magnetohydrodynamic shocks. They show, in particular, that at a shock front sudden variations in direction and intensity of the field are likely to occur. One is tempted to identify the boundaries of many clouds of the galactic diffuse matter with shock fronts. If this is correct, we have a source of magnetic discontinuities. Probably many of these discontinuities will be rather small. However, either their cumulative effect or the effect of some occasional major discontinuity will tend to convert the angle of pitch that a previous trap acceleration has reduced to a small value back to a statistical distribution corresponding to isotropy of direction. At this moment the particle is ready for a new trap acceleration.

Probably our knowledge of the galactic magnetic field is still inadequate for a realistic discussion of the process here proposed. I would like, nevertheless, to list here in a purely hypothetical way a set of parameters that may be compatible with our present knowledge. We assume that a particle for part of the time has $\theta < \chi$ and that it spirals in the same direction along the line of force without being caught in traps. Let λ be the average distance of travel while the particle is in this state, measured along the line of force. After the mean path λ, the particle will change θ back to a high value and for a period of time will be frequently caught in traps, until its energy increases by a factor f and θ decreases again to $\theta < \chi$. After this the process repeats itself. Let T be the average duration of one such cycle. The time A for acceleration by a factor e, then, is

$$A = \frac{T}{\ln f}. \tag{13}$$

The escape time B can be estimated as follows. Let L be the stretched-out length of the galaxy. The motion of the particle along L can be described as a random walk, with steps of duration T. The mean displacement in a step is given in first approximation by λ, because the particle, during the acceleration phase, changes its direction very frequently and does not move far. Estimating B with the diffusion theory, one finds

$$B = \frac{T(L/\lambda)^2}{\pi^2}. \tag{14}$$

A set of parameters that yields $A = 15$ million years and $B = 10$ million years is the following.

$$L = 1.3 \times 10^{24} \text{ cm}, \qquad \lambda = 2 \times 10^{23} \text{ cm};$$
$$T = 2.3 \times 10^{6} \text{ years}, \qquad f = 1.17.$$

Naturally, these values are given here merely as an indication of possible orders of magnitude. Only a much more thorough discussion of the actual conditions in the galaxy may enable one to find reliable values of these quantities.

Two important questions should be discussed further. One is the injection mechanism that should feed into interstellar space an adequate number of particles of energy large enough for the present acceleration mechanism to take over. This problem was discussed in R, but no definite conclusion was reached there. The fact that the acceleration by the galactic magnetic field discussed here is appreciably faster than in R makes the requirements of the injection somewhat less stringent. Nevertheless, one still needs a very powerful injection mechanism. Recent evidence that cosmic-ray-like particles are emitted by the sun indicates the stars, or perhaps stars of special types, as the most likely injectors.

A second question has to do with the energy balance of the turbulence of the interstellar gas. If it is true that the cosmic radiation leaks out of the galaxy in a time of the order of 10 million years, it is necessary that its energy be replenished a few hundred times during a time equal to the age of the universe. A simple estimate shows that the energy present in the galaxy in the form of cosmic rays is comparable to the kinetic energy due to the turbulence of the interstellar gas. According to the present theory, the cosmic rays are accelerated at the expense of the turbulent energy. This last, therefore, must be continuously renewed by some very abundant source, perhaps like a small fraction of the radiation energy of the stars.

In conclusion, I should like to stress the fact that, regardless of the details of the acceleration mechanism, cosmic radiation and magnetic fields in the galaxy must be counted as very important factors in the equilibrium of interstellar gas.

1. E. Fermi, *Phys. Rev. 75*, 1169 (1949); cited hereafter as "R."

2. Davis and Greenstein, *Ap. J. 114*, 206 (1951); Spitzer and Tukey, *Ap. J. 114*, 187 (1951).

3. *Ap. J. 114*, 241 (1951).

4. Ibid. *118*, 113 (1953).

5. *Phys. Rev. 80*, 692 (1950).

99. Cosmic Rays and Radio Emission from Our Galaxy

Cosmic Rays as the Source of General Galactic Radio Emission

Karl Otto Kiepenheuer

(*Physical Review 79*, 738–739 [1950])

The Nature of Cosmic Radio Emission and the Origin of Cosmic Rays

Vitalii Lazarevich Ginzburg

(*Nuovo cimento Supplement 3*, 38–48 [1956])

The connection between extraterrestrial radio radiation and the energetic cosmic rays that bombard the earth's atmosphere was first suggested by Hannes Alfvén and Nicolai Herlofson (see introduction to selection 115 for a historical background). They argued that cosmic ray electrons originating in "radio stars" would interact with the stellar magnetic fields to emit synchrotron radiation at radio wavelengths. Their work stimulated Karl Kiepenheuer's interpretation of the radio emission of the Milky Way in terms of synchrotron radiation of cosmic ray electrons spiraling about the interstellar magnetic field. As illustrated in the first of the papers given here, Kiepenheuer showed that the extraordinarily energetic cosmic ray electrons will produce intense radiation at radio frequencies in the presence of a relatively weak interstellar magnetic field. We now believe that it is this nonthermal radiation process that accounts for the galactic radiation at the longer radio wavelengths, whereas the radio emission at shorter wavelengths is caused by the bremsstrahlung of emission nebulae, as Grote Reber suggested.

Kiepenheuer's idea received relatively little attention for some years except by the Russian astrophysicists Iosif Shklovskii and Vitalii Ginzburg, who applied the theory of synchrotron radiation to the emission of supernovae and galaxies.[1] Shklovskii, for example, first showed in 1952 that the spectrum of the galactic background emission could be explained by a combination of bremsstrahlung and synchrotron radiation. Here we represent the Russian contribution with Ginzburg's paper on radio emission from the Milky Way (Shklovskii's treatment of synchrotron radiation from the Crab nebula supernova remnant is reproduced in selection 72). Ginzburg argues that discrete radio sources other than the Milky Way can be explained by synchrotron radiation, a view that has been substantiated by the observation of both a nonthermal radiation spectrum and linearly polarized radiation in these sources.[2] Such polarization had not been detected previously because the polarization angle is rotated in a wavelength-dependent manner as the wave propagates in an ionized medium, and only observations with narrow bandwidths at short radio wavelengths are not distorted by this propagation effect. By the early 1960s, the long wavelength radio emission from our own Milky Way was shown to be linearly polarized, thus confirming its synchrotron origin.[3] It only remained to explain the origin of the relativistic electrons that give rise to the synchrotron radiation, and Ginzburg's thesis that the electrons are directly accelerated in supernovae seems to have withstood the test of time.

1. G. G. Getmantsev, *Doklady akademii nauk SSSR 83*, 557 (1952); V. L. Ginzburg, *Doklady akademii nauk SSSR 76*, 377 (1951), and *Uspekhi Fizicheskikn Nauk 51*, 343 (1953); I. S. Shklovskii, *Astronomicheskii zhurnal 30*, 15 (1953); and *Doklady akademii nauk SSSR 90*, 983 (1953), *98*, 353 (1954).

2. L. A. Moxon, *Nature 158*, 758 (1946); W. Baade, *Astrophysical Journal 123*, 550 (1956); F. F. Gardner and J. B. Whiteoak, *Physical Review Letters 9*, 197 (1962), and *Nature 197*, 1162 (1963); C. H. Mayer, T. P. McCullough, and R. M. Sloanaker, *Astrophysical Journal 139*, 248 (1964).

3. G. Westerhout, C. L. Seeger, W. N. Brouw, and J. Tinbergen, *Bulletin of the Astronomical Institutes of the Netherlands 16*, 187 (1962); D. S. Mathewson and D. K. Milne, *Nature 203*, 1273 (1964).

Kiepenheuer, Cosmic Rays as the Source of General Galactic Radio Emission

THE GALACTIC RADIO EMISSION is not a thermal free-free radiation of interstellar gas, as was first believed. The electronic temperature would have to be of the order of 100,000° in contradiction to all spectroscopic evidence which gives values around 10,000°. Stars could be considered as sources only under very artificial assumptions. The observed intensities, which must come from the outermost layers of stellar atmospheres, could not be blackbody radiation[1] and might be understood only in terms of coherent plasma oscillations of extended regions. The formation and maintenance of these oscillations is hardly possible in stellar atmospheres.[2]

It will now be shown that the general cosmic radiation of our star system is a high frequency source of sufficient power. In interstellar space, at least inside the interstellar clouds which occupy about 5 percent of space, the mean density of kinetic energy ought to be of the same order as the magnetic-field energy; therefore, fields of around 10^{-6} gauss are to be expected. An energetic electron with energy $W \gg m_0c^2$, which is circulating in this field, is radiating electromagnetic energy into a very narrow cone whose angular aperture is m_0c^2/W in the direction of motion. Therefore, an observer at rest receives very short pulses corresponding to a frequency which is very much higher than the classical Larmor frequency, ν_0. The mean spectral intensity distribution of this radiation will then be[3]

$$P(\nu) \approx (e^2/\pi R)(\nu/\nu_0)^{1/3}$$

for $\nu_0 \ll \nu < \nu_c$, where R is the radius of the electron's circular orbit and $\nu_c = \frac{2}{3}\nu_0(W/m_0c^2)^3$. If n_e is the number of electrons per cm^3 with energy W, the emissivity of high frequency radiation will be

$$\varepsilon_\nu \Delta\nu = n_e P(\nu)\,\Delta\nu = (e^3H/\pi W)n_e(\nu/\nu_0)^{1/3}\,\Delta\nu \text{ ergs/cm}^3\text{/sec.}$$

This increases steadily with frequency until $\nu = \nu_c$ and then decreases rapidly. The observed distribution[1,4] within the frequency range of 10 to 3,000 Mc seems rather to be $\propto \nu^{-0.3}$. We therefore expect to be already in a region with $\nu \geqslant \nu_c$. Also the involved interstellar magnetic field strengths might vary through space and so influence sensibly the spectral intensity distribution. If the thickness of the emitting layer is D, the intensity of the radiation becomes $I = \varepsilon_\nu \Delta\nu D$. Let us suppose that there is one electron per 100 cosmic-ray particles,[5] that is, $n_e \approx 3 \times 10^{-11}$ cm^{-3}. If we tentatively put $W = 10^8$ ev and $D = 1{,}000$ light years (1/100 of the diameter of the galaxy), assuming $\nu = 100$ Mc, and $\Delta\nu = 10$ Mc, we get a radiation intensity of $I \approx 10^{-11}$ erg/cm²/sec. This agrees in order of magnitude with the observations of Hey, Parsons, and Philips.[4]

The total radiation loss of an electron moving in a homogeneous magnetic field, H, is

$$-dW/dt = (4\pi/3)\nu_0(e^2/R)(W/m_0c^2)^4 \sim W^3H^2.$$

Since this radiation loss increases so rapidly with energy, electrons with energies greater than about 10^9 ev are not expected. It seems also possible that electrons are eliminated from cosmic rays in the vicinity of stars by collisions with thermal photons.[6] The composition of cosmic rays we observe at the outer boundary of the earth's atmosphere (being close to the sun) is therefore not an average sample of interstellar space.

The general radio emission shows some relation to the visible structure of the galaxy,[4] but does not seem to be directly correlated with stars or other galactic objects. It follows, therefore, that the sources of radio emission are more closely related to the general shape of the galaxy than to its visible components. This conclusion favors Fermi's hypothesis that the distribution of cosmic rays and of galactic matter is more or less the same, cosmic rays being created in interstellar space and not by the stars.

A relation between radio emission and cosmic rays has previously been suggested by Alfvén[7] for the special case of so-called radio stars (discrete centers of strong radio emission). This interesting suggestion, which has stimulated the above analysis, appears to the writer to be in need of re-examination.

1. A. Unsöld. *Zeits. f. Astrophysik 26*, 176 (1949).
2. M. Ryle, *Proc. Phys. Soc. London A62*, 483 (1949).
3. J. Schwinger, *Phys. Rev. 75*, 1912 (1949).
4. Hey, Parsons, and Phillips, *Proc. Roy. Soc. A192*, 425 (1947).
5. There must be a certain number of electrons in cosmic rays as a direct consequence of their ionization losses in interstellar space and of electron pair production by encounters of protons and thermal starlight photons. Also there is some indication of the existence of electrons in the primary radiation given by direct measurements (B. Rossi, *Rev. Mod. Phys. 21*, 104 [1949]) and by the east-west asymmetry (L. Jánossy, *Cosmic Rays* [London: Oxford University Press, 1948]).
6. E. Feenberg and H. Primakoff, *Phys. Rev. 73*, 449 (1948).
7. H. Alfvén and N. Herlofson, *Phys. Rev. 78*, 616 (1950).

Ginzburg, The Nature of Cosmic Radio Emission

1. Introduction

IN ORDER TO FORMULATE a theory of the origin of cosmic radiation (c.r.) it is obviously necessary to have a certain amount of information on the composition, energy spectrum and spatial distribution of the c.r. outside the earth's atmosphere, in the solar system, in the Galaxy and between galaxies. However, up to very recent times only data (and very incomplete at that) on cosmic rays in the immediate vicinity of the earth (at top of atmosphere) have been available. As a result of this, various theories of origin of the cosmic rays were founded on rather arbitrary hypotheses and even the fundamental question concerning the sources and mechanism of acceleration of c.r. particles remained obscure. One of the main aims of the present paper is to demonstrate that with the development of radio astronomy this state of affairs has radically changed. Thus, if one interprets the non-thermal cosmic radio emission as being a type of magnetic bremsstrahlung (radiation from electrons accelerated in interstellar magnetic fields) then radio astronomical data permit one to determine the spectrum and concentration of relativistic electrons of the primary c.r. in the Galaxy and beyond it. Thus the theory of origin of c.r. can rest on a reliable foundation; it now becomes a branch of astrophysics based on empirical data. The nature of cosmic radio emission will be considered in the present paper and a theory of the origin of c.r. based on radio astronomical data will be described (a more detailed account of the theory is given in references 1–3).

2. The Magnetic Bremsstrahlung Nature of Cosmic Radio Emission

Cosmic radio waves emitted with a continuous spectrum consist of radiation from discrete sources and of "general" cosmic radio emission which varies relatively slowly with the galactic coordinates. The general radiation can in turn be resolved into metagalactic and galactic components. Finally the galactic radio emission may be divided into thermal and non-thermal components. The first of these is the thermal radio emission from interstellar gas and, similarly to the gas, it is located in the plane of the Galaxy. Naturally, the corresponding maximum effective temperature of the radiation, T_{eff}, cannot exceed the kinetic temperature of the gas and therefore cannot exceed $\sim 10{,}000°$. However, in the direction of the Galaxy anticenter and pole the optical thickness of the gas is small and even for waves in the meter band the thermal radio emission intensity is weak ($T_{\text{eff}} \lesssim 1{,}000°$); the same may also be said about the metagalactic radio emission. On the other hand the total intensity of cosmic radio emission of wave length say of 16.3 m travelling from the galactic pole corresponds to an effective temperature $T_{\text{eff}} \sim 50{,}000°$. This signifies that there exists some type of non-thermal general galactic radio emission which depends weakly on the galactic coordinates and which compared with thermal and metagalactic radiation is of decisive importance in the meter band. From an analysis of the cosmic radio emission isophotes it has been concluded[4] that the non-thermal galactic radio emission comes from a quasi spherical sub-system, a sort of corona which surrounds the stellar Galaxy. The radius of this sub-system is $R \sim 3 - 5 \cdot 10^{22}$ cm. Diffuse interstellar gas with a concentration of $n \lesssim 0.1$ cm^{-3} possesses about the same spatial distribution.[5]

A convincing proof of the existence in the Galaxy of radio sources distributed in the manner described above is the discovery[6] of a similar "radio corona" around the nebula M31 in Andromeda which resembles our own Galaxy.

What is the nature of the non-thermal galactic radio emission? Even at the present time attempts are made to explain the radio emission as originating in a large number of hypothetical radio stars invisible in the optical range of the spectrum. This assumption, which for a number of reasons has always seemed to us to be improbable, becomes altogether untenable after the discovery that the discrete sources of cosmic radio waves are nebulae and not stars. There is therefore absolutely no justification in assuming the existence of an enormous number[4] of radio stars possessing some very queer properties and a very unusual spatial distribution. The "magnetic bremsstrahlung hypothesis" which associates non-thermal cosmic radio emission with the waves radiated by relativistic electrons moving in the interstellar magnetic field[7–9,1–3] seems to be more satisfactory in this respect. It will be shown in the following that not only can this hypothesis be made to agree with the probable values of the interstellar magnetic fields and with the concentration of relativistic electrons, but that it is very fruitful for radio astronomy as a whole and for the theory of origin of c.r.

As is well known, in a magnetic field H an electron moves along a helical path with a cyclic frequency

$$\omega_H = \frac{eH}{mc} \cdot \frac{mc^2}{E}, \tag{1}$$

where E is the total electron energy. In the ultra relativistic case, which is the only one of interest here, the electron radiates electromagnetic waves almost exclusively into a narrow cone whose angular aperture is $\theta \sim mc^2/E \ll 1$ in the direction of instantaneous velocity. Therefore, if the electron moves in a circle, an observer at rest will "see" radiation pulses of duration

$$\Delta t \sim \frac{r\theta}{c}\left(\frac{mc^2}{E}\right)^2 \sim \left(\frac{mc}{eH}\right)\left(\frac{mc^2}{E}\right)^2,$$

where r is the radius of the orbit. If, however, the electron moves along a helix and the angle α between velocity and field is not too small[10] then $\Delta t \sim (mc/eH_\perp)(mc^2/E)^2$, where $H_\perp$ is the component of the magnetic field perpendicular to the direction of motion. The radiation spectrum will correspondingly consist of overtones of the frequency ω_H and practically may be considered as a continuous spectrum, the maximum being at a frequency $\omega_{\text{max}} \sim 1/\Delta t \sim (eH_\perp/mc)(E/mc^2)^2$. Calculations show (see for example reference 11) that the energy radiated by an electron per second in a frequency range $d\nu$ is $P(\nu, E)\,d\nu$, where

$$\left.\begin{aligned} P(\nu, E) &\equiv P(\nu) = 2\pi P(\omega) = 16\frac{e^3 H_\perp}{mc^2}\,p\left(\frac{\omega}{\omega_m}\right) \\ &= 16\frac{e^2}{c}\,\omega_{H\perp}\left(\frac{\omega}{2\omega_{H\perp}}\right)^{1/3} Y(u) \\ \omega_{H\perp} &= \frac{eH_\perp}{mc}\cdot\frac{mc^2}{E}, \quad \omega_m = \frac{eH_\perp}{mc}\left(\frac{E}{mc^2}\right)^2, \\ u &= \left(\frac{\omega}{2\omega_m}\right)^{2/3} \quad Y(u) = \frac{p(2u^{3/2})}{u^{1/2}}. \end{aligned}\right\} \tag{2}$$

Values of the function $Y(u)$ are given in the table; in the limiting cases

$$\begin{aligned} &\frac{\omega}{\omega_m} \ll 1:\ Y = 0.256, \\ &\frac{\omega}{\omega_m} \gg 1:\ Y(u) = \frac{(2\pi)^{1/2}}{16}\,u^{1/4}\exp[-(4/3)u^{3/2}]. \end{aligned} \tag{3}$$

Table 99.1

u	$Y(u)$	u	$Y(u)$
0.0	0.256	1.8	0.0085
0.2	.204	2.0	.0050
0.4	.156	2.2	.0028
0.6	.115	2.4	.0015
0.8	.081	2.6	.0008
1.0	.055	2.8	.0004
1.2	.036	3.0	.0002
1.4	.023	3.5	.00004
1.6	.014	4.0	.000006

The function $p(\omega/\omega_m)$ has a maximum at $\omega = 0.5\omega_m$, where it equals 0.1. Thus at the maximum

$$\begin{aligned} P(\nu_{\max}) &= 1.6\frac{e^3 H_\perp}{mc^2} = 2.15\cdot 10^{-22} H_\perp \text{ erg/cycle} \\ \nu_{\max} &= 0.5\frac{\omega_m}{2\pi} = 1.4\cdot 10^6 H_\perp \left(\frac{E}{mc^2}\right)^2 \text{ cycle/s.} \end{aligned} \tag{4}$$

The intensity observed at the earth will be

$$I_\nu = \frac{1}{4\pi}\int P(\nu, E) N_e(E, \boldsymbol{r})\,dE\,dr, \tag{5}$$

where $N_e(E, \boldsymbol{r})$ is the differential electron spectrum at the point $\boldsymbol{r}$. Integration is carried out along the ray of vision and it is assumed that, due to the random distribution of the field H, the radiation on the average is isotropic (hence the factor $1/4\pi$ in (5)). This assumption means that the formulae are correct to a factor of the order of unity. Absorption and possible influence of the refraction index (which may be different from unity) have not been taken into account as they are of no consequence in the following exposition (see reference 1).

If the electron spectrum does not change along a path R and has the form

$$N_e(E) = KE^{-\gamma}, \tag{6}$$

then

$$\begin{aligned} I_\nu &= \frac{R}{4\pi}\int_0^\infty P(\nu, E) N_e(E)\mathrm{d}E \\ &= \frac{3}{\pi}(2\pi)^{(1-\gamma)/2}\frac{e^3 H_\perp}{mc^2}\left[\frac{2eH_\perp}{m^3c^5}\right]^{(\gamma-1)/2} U(\gamma)KR\nu^{(1-\gamma)/2} \\ &\simeq 1.3\cdot 10^{-22}(2.8\cdot 10^8)^{(\gamma-1)/2} U(\gamma)KRH_\perp{}^{(\gamma+1)/2}\lambda^{(\gamma-1)/2} \\ &\qquad \text{erg/cm}^2 \text{ cycle steradian,} \end{aligned} \tag{7}$$

where $U(\gamma) = \int_0^\infty Y(u)u^{(3\gamma-5)/4}\,du$ and, for example, for $\gamma = 1$, 5/3, 2, 3 and 7 the function $U(\gamma)$ has the corresponding values 0.37, 0.163, 0.128, 0.087 and 0.0153.

Data on non-thermal galactic radio emission in the band $1.5\text{ m} < \lambda < 17\text{ m}$ indicate that in a first approximation[12]

$$I_\nu \simeq a\frac{c}{\nu} = a\lambda, \quad a \sim 5\cdot 10^{-21}\text{ erg/cm}^3\text{ cycle steradian.} \tag{8}$$

From (7) and (8) we obtain

$$\gamma \cong 3, \quad I_\nu \simeq 3.1\cdot 10^{-15} H_\perp^2 KR\lambda \simeq 5\cdot 10^{-21}\lambda \tag{9}$$

and hence $KH_\perp^2 R \sim 1.6\cdot 10^{-6}$. For the reasonably highest values of $R \sim 5\cdot 10^{22}$ cm and $H_\perp \sim 10^{-5}$ oersted one obtains a minimum value $K_{\min} \sim 3\cdot 10^{-19} \simeq 10^{-5}\,(\text{eV})^2/\text{cm}^3$.

To be cautious we put $K \simeq 10^{-6}$ and thus arrive at the following electron spectrum

$$\begin{aligned} N_e(E) &\simeq \frac{10^6}{E^3}(\text{eV})^{-1}\text{ cm}^{-3}, \\ N_e(E > E_0) &= \int_{E_0}^\infty N(E)\,\mathrm{d}E \simeq \frac{10^6}{2E_0^2}\text{ cm}^{-3}, \end{aligned} \tag{10}$$

where E and E_0 are measured in eV. Hence $N_e\,(E > 10^9\text{ eV}) \simeq 5\cdot 10^{-13}$, which is not in contradiction with data on cosmic ray electrons near the earth (for more details see reference 9). On the other hand $N_e\ (E > 10^8\text{ eV}) \simeq 5\cdot 10^{-11}$, which in order of magnitude equals the concentration of all primary cosmic ray particles incident on the earth (due to high latitude cut-off $cp \geqslant 10^9$ eV). Taking into consideration the form of the function $Y(u)$, it may be shown that the spectrum (10) should refer to the region $2\cdot 10^8 < E < 3\cdot 10^9$; uncertainties in the data permit one to lower the value $N_e(E > 10^9$ eV) by several times.

Therefore, as the magnitude of the interstellar magnetic field ($\sim 10^{-5}$ oersted) required by the hypothesis of the magnetic bremsstrahlung agrees with other estimates of this quantity and this hypothesis also yields a reasonable value for the electron concentration, one may conclude that the magnetic bremsstrahlung explanation of the non-thermal general galactic radiation is probably correct. Moreover, this hypothesis automatically explains the fact that not more than 1% of the primary c.r. with $E \geqslant 10^9$ eV incident on the earth consists of electrons. The reason for this is that electrons lose energy in magnetic bremsstrahlung (which is negligible for protons and nuclei). If one also assumes (and this seems quite natural) that electrons, protons and nuclei in the energy region $E \lesssim 10^{11}$ eV are generated by the primary c.r. sources with a spectrum of the same type of (6), with $\gamma = \gamma_0 \simeq 2$, then, due to deceleration of the electrons in the magnetic field, the electron spectrum will have the same form as (6), but with $\gamma = \gamma_0 + 1 \simeq 3$ (see references 1–3, 9). According to (11) this is just the type of spectrum one would expect on the basis of radio astronomical data.[13]

The proposed magnetic bremsstrahlung mechanism may also be responsible for radio emission from powerful discrete sources.[7,8] As a matter of fact in such sources as α Cassiopeiae, α Cygni, α Tauri, $T_{\text{eff}} \gtrsim 10^7$ degrees, whereas the electron kinetic temperature is much smaller. It is impossible to explain the radiation from such galactic sources as α Cassiopeiae or α Tauri as originating from hypothetical radio stars; the magnetic bremsstrahlung hypothesis, on the other hand, does not require any impossible assumption about the magnetic fields or the electron concentration in the sources.

If the electron spectrum and the magnetic field in the source are independent of the coordinates, then the radiation flux from the source will be

$$F_\nu = I_\nu \, d\Omega = \frac{V}{4\pi R^2}\int P(\nu, E) N_e(E)\, dE, \qquad (11)$$

where V is the volume of the source and R is the distance from it. Under the optimum conditions (4)

$$F_\nu = 1.6\,\frac{e^3 H_\perp}{mc^2}\cdot\frac{N_e V}{4\pi R^2} = 1.7\cdot 10^{-23} H\,\frac{N_e V}{R^2}\ \text{erg/cm}^2\ \text{cycle}, \qquad (12)$$

where N_e is the concentration of electrons; it is considered that these possess an energy corresponding to the frequency $\nu = \nu_{\max}$ (see (4)). For α Tauri: $R \sim 5\cdot 10^{21}$ cm, $V \simeq (4\pi r^3/3) \sim 10^{56}$ cm^3, while in the meter band

$$F_\nu \cong 2\cdot 10^{-20}\ \text{erg/cm}^2\ \text{cycle}$$

and is independent of the frequency. Putting $H_\perp \sim H \sim 10^{-4}$ which seems to be allowed[14] we obtain from (12) $N_e V \sim 10^{50}$ and $N_e \sim 10^{-6}$ cm^{-3}; obviously, these values are the minimal ones. By using formulae (7) and (10) one can determine the electron spectrum, and the energy in discrete sources can be obtained. For α Tauri (Crab nebula) $\gamma \simeq 1$, $N_e V \sim 10^{51}$ for $E > 2.5\cdot 10^8$ eV, and the total energy of relativistic electrons with $E > 2.5\cdot 10^8$ is approximately $W \sim 10^{47} - 10^{49}$ erg (see reference 14 for further details). Compared with other powerful sources the source in Taurus should be considered exceptional[15] because in the majority of powerful sources (mostly extragalactic) $\gamma \simeq 3$; for Cassiopeia one also has $N_e \sim 10^{51}$ and $W \sim 10^{48}$ erg. The Crab nebula (α Tauri) is undoubtedly an expanding envelope of the supernova of 1054 A.D. According to reference 16 all other powerful galactic discrete sources are also envelopes of supernovae (thus Cassiopeia is a supernova of 369 A.D.).

We thus arrive at the conclusion that the total energy of relativistic electrons in the envelopes of supernovae is of the same magnitude as the energy of the optical radiation produced during the outburst. The radiation of powerful extragalactic discrete sources, according to the mass of collected data, should also be attributed to magnetic bremsstrahlung.[17] Thus, radio astronomy presents the possibility of detecting relativistic electrons in various parts of the universe. In particular, it is found that large amounts of electrons exist in the envelopes of supernovae and, possibly, of novae.[14,18,19] It seems difficult to overestimate the significance of this circumstance for the theory of the origin of c.r.

3. THE ORIGIN OF COSMIC RAYS

Data on the distribution of relativistic electrons in the Galaxy speak against the theory of solar origin of c.r. and also against the hypothesis of their metagalactic origin. One must therefore conclude that the main part of c.r. is generated somewhere in our Galaxy; these rays fill the Galaxy and form a quasi spherical system (mentioned in §1) with $R \lesssim 5\cdot 10^{22}$ cm. It is furthermore natural to suppose that the primary c.r. sources are supernovae and novae[1–3,14,19] the envelopes of which contain fast electrons and probably protons and nuclei. As time passes (in about 1,000 – 3,000 years) the supernovae envelopes smear out and the c.r. which previously diffused from the envelope become completely distributed throughout interstellar space. Now the particles wander through interstellar magnetic fields, diffuse towards the periphery of the Galaxy and simultaneously lose energy in nuclear collisions and, later on, in ionization collisions. Instead of nuclear collisions, electrons undergo bremsstrahlung collisions with the nuclei and with the electrons of the interstellar gas.

The electrons also lose energy in magnetic-bremsstrahlung processes discussed in §1. For a number of reasons, the acceleration of c.r. particles in the interstellar space[20] does not seem to be an effective process and even at very high energies ($E \gtrsim 10^{13}$ eV/nucleon) it probably is of no significance (see reference 1). This conclusion[2] is not affected by attempts, which seem very unconvincing to us, to retain the theory of interstellar acceleration by assuming that the life time of the particles in the Galaxy is determined by their rate of exit from it.[21,22] On the other hand Fermi's statistical acceleration mechanism[20] should be very effective[23] in turbulent envelopes of supernovae, where it is probably the main acceleration mechanism. Such is the picture one can draw of the origin of c.r. using radio astronomy data. It is impossible for us in this paper to consider the theory in detail (see references 1–3) and we will therefore discuss some of the more important points of the problem.

A most important condition which any theory of origin of c.r. must satisfy follows from energy considerations. If the cross-section for energy transfers between protons colliding in interstellar space is $\sigma \simeq 2.5 \cdot 10^{-26}$ cm^2, then on the average the life time of fast protons in the Galaxy should be $T \sim 4 \cdot 10^8$ years $\sim 10^{16}$ s (the concentration of interstellar hydrogen is taken to be on the average $n \sim 0.1$ cm^{-3}). The energy density of cosmic rays is $w \sim 1$ eV/cm^3, the total energy being $W \sim wV \sim 10^{68}$ eV, as the volume of the Galaxy $V \sim (4\pi/3)R^3 \sim 10^{68}$ cm^3 for $R \sim 3 \cdot 10^{22}$ cm. It is thus clear that the power of the sources of primary c.r. should be $W/T \sim 10^{52}$ eV/s $\simeq 10^{40}$ erg/s. This value is probably too high by one or two orders of magnitude, as most likely the accepted values of W and R are too high. Thus

$$W/T \sim 10^{38} - 10^{40} \text{ erg/s.} \qquad (13)$$

Assuming that during supernova outbursts, say 10^{48} erg appear as c.r. (see above) one finds that average power of the source should be $\lesssim 10^{38}$ erg/s as on the average supernovae outbursts take place at least once in 300 years $\sim 10^{10}$ s. It is likely that supernovae outbursts occur much more often, but because of absorption they remain invisible (this argument belongs to I. S. Šklovskij). Furthermore, according to the estimates,[1,2] the contribution from novae may reach $3 \cdot 10^{40}$ erg/s. Thus supernovae and novae can satisfactorily provide the energy balance. It should be noted that energy considerations suggesting supernovae as possible sources of c.r. were proposed previously.[24] However, these arguments acquire new significance when one associates them with radio astronomical data which confirm the generation of c.r. in the envelopes of supernovae. It should also be noted that on the average the power of the c.r. generated by the Sun is of $\sim 10^{21}$ erg/s and therefore even 10^{11} stars of the solar type would yield only 10^{32} erg/s, which is by 6–8 orders lower then the required value (see reference 25).

It has already been mentioned that in envelopes of supernovae and novae the particles are mainly accelerated by a statistical mechanism. This mechanism is capable of accelerating protons and light nuclei[26] to the highest energies observed in cosmic rays, i.e. up to $E \sim 10^{18}$ eV.[1,23] At the same time it should be emphasized that many details of the acceleration mechanism in expanding turbulent envelopes are still obscure. For this reason a reliable theoretical prediction of the spectrum of generated particles cannot be given; moreover for different sources the spectra are different (for details see reference 1). The spectra of electrons, protons and nuclei generated in the envelopes are probably of the same form and on the average, for energies $E \lesssim 10^{11}$, $\gamma \simeq 2$. Magnetic bremsstrahlung losses in interstellar space result in a softening of the electron spectrum (and therefore $\gamma \simeq 3$; see above).

The proton and nuclear spectra will remain unchanged (even when nuclear collisions are taken into account) if the energy of the secondary products (i.e. energy after the collision)

$$E_2 = \delta E_1, \qquad (14)$$

where δ is independent or depends only slightly on the primary particle energy E_1. The explanation of this is that relation (14) is a scale transformation, and collisions of this type do not lead to changes in the power exponent in spectra of the type (6). There are data which support this point of view.[2,27] Furthermore, according to the theory outlined here, one should expect that in a first approximation a stationary state should exist in the Galaxy, i.e. that the concentration of the various particles should be independent of time. On the other hand, the primary c.r. sources form a plane subsystem, lying in the plane of the Galaxy (this is exactly the way novae and probably supernovae are distributed). This leads us to the result that an approximate equilibrium number of Li, Be and B nuclei should be present in the c.r. incident on the earth; thus the concentration of these nuclei should be about four tenths the concentration of all other nuclei.[3] This is exactly the value which the latest available data yield.[28] Furthermore, according to our point of view, the high latitude cut-off in the c.r. spectrum cannot be explained by ionization losses and is probably due to the existence of a magnetic moment of the solar system.[9] This conclusion is in accord with data[29] indicating that the high-latitude cut off is of a magnetic nature.

Finally, special attention should be paid to secondary electrons and positrons produced in the Galaxy during nuclear collisions between cosmic protons and nuclei. Thus, basing on the data of reference 27, it should be expected that in each collision about 5% of the energy of the primary particles should be given to electrons and positrons.[2,3]

It can be shown[2] that in the absence of interstellar magnetic fields the amount of primary c.r. electrons and positrons with $E > 10^9$ eV should be comprised within $\sim 1 - 5\%$, even if these particles were produced exclusively in nuclear collisions. This value becomes about 10 times smaller if magnetic bremsstrahlung energy losses are taken into account. However there are reasons to believe that electrons are also produced in the primary sources (see §1), their number being larger than that of the secondary electrons. Similar considerations and also the computations carried out in §1 indicate (if no other factors must be taken into account) that the number of primary c.r. electrons near the earth with $E \geqslant 10^9$ eV should be about $0.1 - 0.5\%$ of the number of primary protons with $E \geqslant 10^9$ eV. Some of these electrons are secondary and can be separated from the primaries, as the number of secondary electrons and positrons should be about equal, while in supernovae envelopes one should expect that only electrons be accelerated. Accordingly, a study of the primary electron-positron component near the earth seems to be of special importance. Other possible means of checking the ideas outlined in this paper are discussed in reference 3. Therefore, in conclusion we confine ourselves to the statement that, as the previous discussion seems to indicate, the present data and conceptions support the picture of the origin of c.r. proposed in this paper and, at any rate, do not contradict it.

1. V. L. Ginzburg, *Usp. fiz. nauk 51*, 343 (1953); *Fortschritte der Physik* (Berlin) *1*, 659, (1954).

2. V. L. Ginzburg, *Dokl. akad. nauk SSSR 99*, 703 (1954).

3. V. L. Ginzburg, *Proceedings of 5th Conference on Cosmogony* (Moscow, 1955).

4. I. S. Šklovskij, *Astron. žu. SSSR 29*, 418 (1952).

5. S. B. Pikel'ner, *Dokl. akad. nauk SSSR 88*, 229 (1953).

6. J. E. Baldwin, *Nature 174*, 320 (1954).

7. H. Alfvén and N. Herlofson, *Phys. Rev. 78*, 616 (1950); K.O. Kiepenheuer, *79*, 738 (1950).

8. V. L. Ginzburg, *Dokl. akad. nauk SSSR 76*, 377 (1951); G. G. Getmancev, *Dokl. akad. nauk SSSR 83*, 557 (1952).

9. V. L. Ginzburg, G. G. Getmancev, and M. I. Fradkin, *Proceedings of 3rd Conference on Cosmogony* (Moscow, 1954), p. 149.

10. The radiation will be of a different nature if the angle $\alpha \lesssim \theta \sim mc^2/E$; this case is of interest for the theory of solar sporadic radio emission. See G. G. Getmancev and V. L. Ginzburg, *Dokl. akad. nauk SSSR 87*, 187 (1952).

11. V. V. Vladimirskij, *Žu. èksper. teor. fiz. 18*, 393 (1948).

12. More accurately $I_\nu = a\nu^{-\alpha}$, where $\alpha = 0.8 - 1.2$.

13. Wandering of protons and nuclei in interstellar space should not change their spectrum and therefore, near the earth, $\gamma = \gamma_0 \simeq 2$, which is the experimentally observed value for $E \sim 10^9 - 10^{11}$ eV.

14. V. L. Ginzburg, *Dokl. akad. nauk SSSR 92*, 1133 (1953).

15. The large amount of high energy electrons in α Tauri makes it probable that the magnetic bremsstrahlung mechanism is also responsible for the continuous optical radiation of the nebula. See I. S. Šklovskij, *Dokl. akad. nauk SSSR 90*, 983 (1953).

16. I. S. Šklovskij, *Astron. žu. SSSR 30*, 15 (1953).

17. I. S. Šklovskij, *Dokl. akad. nauk SSSR 98*, 353 (1953).

18. J. G. Bolton, G. J. Stanley, and O. B. Slee, *Austral. J. of Phys. 7*, 110 (1954).

19. I. S. Šklovskij, *Dokl. akad. nauk SSSR 91*, 475 (1953).

20. E. Fermi, *Phys. Rev. 75*, 1169 (1949).

21. E. Fermi, *Astrophys. Journ. 119*, 1 (1954).

22. P. Morrison, S. Olbert, and B. Rossi, *Phys. Rev. 94*, 440 (1954).

23. V. L. Ginzburg, *Dokl. akad. nauk SSSR 92*, 727 (1953).

24. D. ter Haar, *Rev. Mod. Phys. 22*, 119 (1950).

25. I. S. Šklovskij, *Dokl. akad. nauk SSSR 90*, 983 (1953).

26. In the simplest case of statistical acceleration a particle of energy Mc^2 acquires, when accelerated for a time t, energy $E = Mc^2 \exp[\alpha t]$ that is, the energy is proportional to the rest mass.

27. S. N. Vernov, N. L. Grigorov, G. T. Zacepin, and A. E. Čudakov, *Izv. akad. nauk SSSR, Ser. Fiz. 19*, 493 (1955).

28. M. F. Kaplon, J. H. Noon, and G. W. Racette, *Phys. Rev. 96*, 1408 (1954).

29. R. A. Ellis, M. B. Gottlieb, and J. A. Van Allen, *Phys. Rev. 95*, 147 (1954).

100. Expanding Stellar Associations and the Origin of the Runaway O and B Stars

Expanding Stellar Associations

Viktor Amazaspovich Ambartsumian
TRANSLATED BY ZDENKA KADLA-MICHAILOV

(Astronomicheskii zhurnal *26*, 1–9 [1949]; translated for H. Shapley (ed) *Source Book in Astronomy 1900–1950* [Cambridge: Harvard University Press, 1960])

On the Origin of the O- and B-Type Stars with High Velocities (The "Run-Away Stars") and Some Related Problems

Adriaan Blaauw

(*Bulletin of the Astronomical Institutes of the Netherlands 15*, 265–290 [1961]).

When considering the probable ages of loose, moving clusters, Bart Bok and later, Henri Mineur, pointed out the fact that the shearing force of differential galactic rotation will disrupt the clusters on time scales short compared to the age of our galaxy.[1] In fact, he showed that any star cluster in the solar neighborhood with a space density below about 0.1 solar masses pc^{-3} will be unstable and that the motions of the stars in these clusters will be governed by the effects of differential galactic rotation. Over a decade later, Viktor Ambartsumian called attention to the existence of nearby associations of bright, presumably young O and B stars, whose space densities are orders of magnitude lower than 0.1 solar masses pc^{-3} and whose roughly spherical shapes could not be maintained for long time intervals under the forces of differential galactic rotation.[2] He suggested that because these associations cannot be bound by their own gravitation, it must be their expansion and relatively young age that account for their spherical shapes. In the first paper here, Ambartsumian further develops this idea by arguing that the associations must be expanding with velocities on the order of 5 km sec^{-1} and that this expansion must have originated only about a million years ago. At the time, Ambartsumian's proposal seemed very radical to most astronomers, who found this time scale unbelievably short for such a cosmic process.

Within three years Adriaan Blaauw lent considerable support to Ambartsumian's hypothesis by measuring the expanding motions of the ζ Persei association.[3] Blaauw showed that the proper motions and the radial velocities of seventeen stars in this group indicate a group expansion at the velocity of 12 km sec^{-1}. Furthermore, at this rate the association would have expanded to its present size of 30 pc in 1.3 million years—an age comparable to the thermonuclear age of the bright, young O and B stars making up the association. This coincidence lent additional support to Ambartsumian's contention that stars and associations are being continuously formed together and that this process may account for the origin of a substantial fraction of the stars in our galaxy.

Later Ambartsumian considered the process giving rise to the expanding associations. He imagined that they must originate in very old, massive, superdense protostars that in the process of disintegration simultaneously produce the stars in the associations and the interstellar gas connected with them.[4] At this time astronomers were beginning to realize that the dust and gas in interstellar space are normally moving too fast for stars to be formed by spontaneous gravitational collapse. Some agent has to compress the clouds of interstellar matter and thereby provide the extra pressure needed to start gravitational collapse. In 1953 Ernst Öpik reasoned that the expanding remnants of supernova explosions can trigger the collapse, and one year later Jan Oort suggested that the expanding H II regions found around hot, young stars both trigger collapse and give rise to the expanding stellar associations.[5]

The expansion of H II regions cannot account for the large space velocities of the "runaway" O and B stars, however, and in the second of the two papers given here, Blaauw argues that these stars are escaped members of binary star systems in which one star has become a supernova. Blaauw notes that the runaway stars are very massive stars whose high velocities are comparable to the orbital velocities expected for massive binary star systems. Because massive stars burn their thermonuclear fuel faster than normal stars, one member of such a binary system will quickly evolve to the supernova stage and thereby release the other member as a high-velocity star. This explanation accounts for the fact that the runaway O and B stars, though single stars, seem to originate in OB associations whose component stars are normally in binary or multiple star systems. Moreover, as Blaauw points out, the expanding associations might be the result of repeated supernova explosions of this sort.[6]

1. B. J. Bok, *Harvard Circular No. 384* (1934). H. Mineur, *Annales d' Astrophysique 2*, 1 (1939).
2. V. A. Ambartsumian, *Stellar Evolution and Astrophysics* (Erevan: Armenian Academy of Sciences, 1947).
3. A. Blaauw, *Bulletin of the Astronomical Institutes of the Netherlands 11*, 405 (1952).
4. V. A. Ambartsumian, *Observatory 75*, 72 (1955).
5. E. J. Öpik, *Irish Astronomical Journal 2*, 219 (1953); J. H. Oort, *Bulletin of the Astronomical Institutes of the Netherlands 12*, 177 (1954).
6. For a later review by Blaauw, see *Annual Reviews of Astronomy and Astrophysics 2*, 213 (1964).

Ambartsumian, Expanding Stellar Associations

Omitted is Ambartsumian's introductory list of six examples of stellar associations in which stars appear to be bound together in regions of low star density. The list may be summarized as follows.

1. An association of eight T Tauri stars found in Taurus and Auriga (association diameter 25 pc).
2. An association of at least five T Tauri stars in the Aquila-Ophiuchus region (association diameter about 6°)
3. An association of O and B stars surrounding the χ and h Persei clusters (association diameter 170 pc, individual cluster diameter 10 pc).
4. An association of O and B stars surrounding the open cluster NGC 6231 (association diameter 30 pc, cluster diameter about 6 pc).
5. The NGC 1910 association of early-type supergiants in the Large Magellanic Cloud (association diameter 70 pc).
6. An association of about twenty-three faint O and B stars surrounding the open cluster NGC 381 (association diameter 100 pc, cluster diameter about 4 pc).

The Main Characteristics of Stellar Associations

From the above data the following general conclusions on stellar associations can be made:

(1) The associations are systems whose mean density is small in comparison to the density of the galactic stellar field. However, if we consider the partial concentration of stars of separate spectral types, the associations stand out sharply, owing to the relative abundance of stars of comparatively rare types. In some cases these are O and B type stars; in others, T Tauri type stars. As a result of their small density the associations cannot be in states termed, in stellar dynamics, as stationary. In distinction from globular and open clusters, the associations are nonstationary systems. Obviously the members of the association diverge in space, mixing in the course of time with stars of the field.

(2) Associations always have stars from which there is a continuous outflow of matter. In three of the six examples given above we meet with P Cyg type stars. In the fourth and sixth examples we find Wolf-Rayet stars among members of the associations. In the first two examples we meet with T Tauri type variables, the bright lines in the spectra of which have absorption components on the violet side, i.e., the same peculiarity as the bright lines in the spectra of P Cyg stars. Therefore it should be presumed that there is a continuous outflow of matter from these stars.

(3) In some cases the associations have nuclei in the form of open stellar clusters.

Stellar Associations in the Large Magellanic Cloud

It is known that the Large Magellanic Cloud is very rich in open clusters. At the same time the fact appears that the clusters of this Large Cloud have in some cases very large linear diameters (some tens of parsecs). The above-mentioned example is the most striking. The distribution of diameters of open clusters in the Large Magellanic Cloud has however a minimum, which divides all the open clusters into two groups: (*a*) clusters with diameters in excess of 20 parsecs and (*b*) clusters with diameters smaller than 20 parsecs. Already this circumstance leads us to suspect that here we have to do with objects of two different types and scales. The presence, in at least some of the clusters of the first group, of P Cyg type stars makes us think that systems with diameters in excess of 20 parsecs are objects of the type of stellar associations met with in the Galaxy, while objects of the second group are ordinary open clusters.

The following considerations make this supposition an almost authentic fact. If we observe our Galaxy from some outer system, say from the Large Magellanic Cloud, the association surrounding χ and *h* Persei will distinctly stand out on the surrounding background, because of the presence in this association of a large number of supergiant stars. If we observe this system from inside the Galaxy, we meet with the fact that the [galactic] stars of low luminosity are projected on the field of the association. These stars are at a much smaller distance than the association and have, because of this, the same apparent stellar magnitudes as the supergiants belonging to the association. Therefore, the stars of the association are lost in the general background. An observer in the Large Magellanic Cloud would, without investigating the spectra, by direct observation detect this association as a cluster of supergiants with a diameter of 170 parsecs. The χ and h Persei clusters would appear to him as only compact groups in this immense system.

On the other hand if the system NGC 1910 were transferred from the Large Magellanic Cloud to the Galaxy, to the place of χ and h Persei, it would be observed by us only as an association, i.e., it would not stand out as a noticeable condensation of stars if a separate study were not made of the distribution of early-type stars in this region of the sky.

Therefore, evidently all the giant systems in the Large Magellanic Cloud (about 15) are in reality stellar associations, the characteristic features of which were described in the previous paragraphs.

The Kinematics of Stellar Associations

As the forces of interaction of the stars in associations are small in comparison to the tidal forces of the general galactic field of force, we neglect the forces of mutual interaction for at least the peripheral members of the association.

When considering the motion of stars of the association in the galactic field of force, we should note that the differential effect of galactic rotation should lead to a mutual recession of members of the association. The velocity of mutual recession of two stars under the action of the galactic rotation effect is expressed through Oort's coefficient A by

$$v_r = Ar \sin 2(l - l_0).$$

In particular, a star at the periphery of an association with a radius R will move from the center with a velocity

$$V_r = AR \sin 2(l - l_0).$$

This velocity will also be the velocity of increase of the radius of the system in a given galactic longitude l, i.e.,

$$\frac{dR}{dt} = AR \sin 2(l - l_0).$$

From this the ratio of the radii R_1 and R_2 for two moments of time t_1 and t_2 is

$$\ln \frac{R_2}{R_1} = A(t_2 - t_1) \sin 2(l - l_0).$$

Taking into account the value of A, we see that for $l - l_0 = 45°$, the radius will be doubled in a time of the order of 40 million years.

For this deduction we in fact assumed that all the stars of the association move in circular orbits around the galactic center. Actually it cannot be said beforehand what kind of galactic orbits the different stars have. However, if large initial relative velocities in the association are not assumed, the time necessary for doubling the radius will always be of the same order.

The obtained result, irrespective of the existence of other possible additional causes for expansion, leads to the conclusion that every association originated comparatively recently and consists of stars expanding in space from an initial volume in which members of the association were formed.

However if the expansion of associations was caused only by the differential effect of galactic rotation, the dimensions of the association would increase only in the galactic plane.

With regards to the possible expansion of associations in the direction perpendicular to the galactic plane under the influence of the difference in the periods of oscillating motion along the z-coordinate, it is necessary to state that this effect will act much more slowly. The reason for this is that the periods of oscillation for small amplitudes do not depend on the magnitude of the amplitude, i.e., on the initial conditions. In fact, when a star is at a height z above the galactic plane, the value w_z of the acceleration-component is determined by the integral

$$w_z = -2\pi G \int_{-z}^{z} \rho(z)\, dz,$$

where $\rho(z)$ is the density.

If z is small, then the variations of $\rho(z)$ within the limits $-z$ to $+z$ are relatively small and therefore

$$w_z = -4\pi G \rho(0) z,$$

where $\rho(0)$ is the density in the galactic plane. We see that the acceleration is proportional to z. In other words, for small amplitudes we have to do with harmonic oscillations, the period of which does not depend on the amplitude.

As the observed associations are located in low galactic latitudes, the stars which belong to them should have approximately equal periods of oscillation along the z-coordinate. Therefore the considered effect should be very small in comparison with the differential rotation effect.

However, observations do not show especially strong flattening for the systems which were considered above as examples. This fact leads us to think that there is another cause for the expansion, which plays a much greater role than the differential effect of galactic rotation. Namely it remains to assume that the stars of the association were ejected in different directions with certain velocities from the initial volume in which they were formed.

These initial velocities should not be less than one km/sec, as otherwise, already for associations with dimensions of a few tens of parsecs, it would not be possible to disregard the differential rotation effect. On the other hand they should not exceed 10 km/sec, as otherwise when determining the radial velocities of the stars, for example in the association surrounding NGC 6231, this initial velocity would immediately be noticed.

If the initial velocity of expansion from the center is of the order of 5 km/sec, the differential effect of galactic rotation cannot predominate until the linear dimensions of the associations reach several hundred parsecs. However, such dimensions mean that the stars of the association will be completely dissolved among stars of the field, and that will be the end of the association. Consequently the flattening of the associations for such velocities will be slight, and therefore expansion velocities of the order of 5 km/sec are most plausible.

This leads to the conclusion that the stars, which compose the association of objects of the T Tauri type in Taurus-Auriga, were ejected from the above-mentioned initial volume several million years ago, and the stars composing the association surrounding Chi Persei 10 to 20 million years ago, and so on.

The moment when the association began to expand must be very close to the moment of star formation in the association itself, as the assumption that the system was in a stationary state for a long time and only later began to expand, would contradict stellar dynamics. From here we conclude that the age of the stars belonging to associations is measured by only millions, or in the extreme cases, by tens of millions of years.

This is in agreement with the fact that P Cyg type, Wolf-Rayet, or T Tauri type stars are found in associations. A star cannot be a P Cyg type star longer than 1 to 2 million years, as intense ejection of matter would lead to its complete disappearance. On the other hand P Cyg type stars, having the highest luminosities among all the known stars, have also apparently the largest masses. Even if there are other states, corresponding to larger or equal masses, their life-time must be very short, as such masses are extremely rare. P Cyg stars could not have evolved from stars of smaller mass. Therefore they must necessarily be counted as belonging to the youngest stars.

THE NUMBER OF STELLAR ASSOCIATIONS IN THE GALAXY

At present it is difficult to answer with certainty the question on the number of associations in our Galaxy. If only those associations which have early-type supergiants are considered, then these can be detected at great distances (up to 2,000–3,000 parsecs). Therefore a considerable fraction of them should be accessible. It is very probable that the number of accessible associations of this type is measured by tens. This means that the total number of such associations in the Galaxy is of the order of 100.

With regards to associations composed of T Tauri type stars and other dwarfs with bright lines in their spectra, at present only two of them are known to us. However it is very important to note that so far they are detected by us at only very close distances. In a spherical volume with a radius of the order of 100 parsecs there is one such association. This means that their total number in the Galaxy is measured by thousands.

If we adopt this number as equal to, let us say, 10,000 and take into account that associations of this type can be observed as such during a time of the order of several million years, then in order to keep up the present number of these associations in the Galaxy, not less than one association composed of T Tauri type stars must be formed on the average every thousand years.

PROBLEMS OF STAR FORMATION

Some astronomers have suggested that all the stars in our Galaxy were formed simultaneously or almost simultaneously several billions of years ago, i.e., at the epoch of formation of our Galaxy. In the light of the above-stated facts this supposition is completely refuted. The formation of stellar associations and the formation of stars in them from some other form of existence of matter takes place continuously, "almost before our eyes." The number of associations composed of T Tauri type stars, which have been formed during the life-time of the Galaxy, should be of the order of 10 million. We do not as yet know the average number of stars originating in one association, as we discern only the brightest members. However it should be assumed that this number is measured at least by hundreds. This means that at least a billion of stars in our Galaxy originated as a result of the formation of stellar associations from some other objects unknown to us.

OTHER POSSIBLE TYPES OF ASSOCIATIONS

It is very probable that the system of O and B type stars in Orion together with the Trapezium form one giant association with a diameter of more than 100 parsecs. The stars of the Trapezium and the open stellar cluster connected with it evidently form the nucleus of this association. The presence of a large diffuse nebula makes this system especially interesting. It deserves a thorough study.

The "moving cluster" in Ursa Major is a system with 32 members and a diameter of more than 200 parsecs. A group of 11 stars forms the nucleus of this system with a diameter of only 9 parsecs. However there are no direct indications showing that the stars belonging to it are young. The small number of members of this cluster is also striking. It is possible that the system is a remnant of a once rich association.

The sun is located inside this system, but as is known, it is not one of its members.

CONCLUSIONS

In the present study the presence in the Galaxy of a very large number of stellar associations is ascertained. These associations are stellar systems with a small density, unstable, and disintegrating in galactic space. Already the large role played by stellar associations in problems of stellar evolution is clear. They therefore deserve a most thorough study.

Blaauw, On the Origin of the O- and B-Type Stars with High Velocities

Abstract—A survey of presently available data on the "run-away" O- and B-type stars is given, and the following theory is considered for the explanation of these objects. It is assumed that these stars, in their earliest stage of formation, were the secondary components of proto-double stars, the more massive primaries of which have shed most of their mass in a violent process; consequently, the secondary component was released as a result of the rapidly diminishing gravitational attraction. The process of the rapid mass loss of the primary may be identified with the vibrational instability occurring in the contraction stage of stars with large masses ($> 100 \odot$ mass) or, possibly, with the supernovae of type II. The proposed hypothesis would seem to explain the two principal properties which distinguish the run-away stars from the normal ones: their lack of duplicity and the preponderance of large masses. These two properties are discussed in some detail; the discussion includes an estimation of the fraction of double and multiple stars among the normal objects, which is found to be about 75 per cent.

Numerical computations of the dynamics of a double star with rapid, but not infinitely rapid, mass loss of the primary, made by Boersma, are used as a basis for an estimation of the initial properties of the proto-double stars. The original masses of the primaries are found to be of the order of several hundreds of solar masses and the original separations of the order of several to 100 astronomical units. It is pointed out that a tendency towards formation of proto-double stars with these properties is to be expected, in view of the high incidence of duplicity among the most massive main-sequence stars known. The anomalous mass distribution of the run-away stars is related to the differences in contraction times for stars of different masses.

The required time scale for the rate of mass loss of the primary points to the identification with the supernovae of type II. Some aspects of this possible identification are discussed. The admittedly poorly-known frequency of the supernovae of type II appears to agree well with the frequency deduced from the numbers of run-away objects. The identification leads to the consideration of some special aspects of the hypothesis, such as the relation between supernovae and OB associations. We further discuss the problem of the absolute magnitudes of supernovae of type II, the question of the remnants of the proto-primaries, and a possible spread in the contraction times of the early B stars.

1. Introduction and Summary

The early-type stars with high velocities, including the so-called "run-away stars", have been briefly discussed in previous papers (Blaauw 1956a, 1958b, 1959) dealing with their individual properties and with the characteristics of the group as a whole. Their spectral types range from O5 to B5. Contrary to the generally very low space velocities for stars of these types, the velocities of the stars now under consideration amount to values up to 200 km/sec. The name "run-away" stars was assigned to those members of the group for which the direction of the space velocity indicated that the star must have originated in a known OB association. Since it is very likely that nearly all O- and B-type stars originated in such associations (Roberts 1957, Blaauw 1958b), the "run-away" stars are not to be considered as forming a distinct subdivision of the high-velocity O- and B-type stars; they merely represent those objects for which the direction of the space motion is sufficiently accurately known and for which the distance the star has travelled since it left the association is still small enough to allow the identification of its origin.

Representation of the velocity distributions of the O- and B-type stars with velocities below 30 km/sec either by means of a gaussian curve or by means of an exponential law of the type $\exp(-|v|/\eta)$, (v being the velocity, η a constant) fails to fit the numbers of higher velocities. The high-velocity objects form a distinct excess, and the investigations quoted have shown them to possess two characteristics which distinguish them from the low-velocity objects. These characteristics are:

a) the absence of double or multiple stars—a property markedly contrasting the high frequency of duplicity or multiplicity among the low-velocity objects; and
b) their different spectral distribution, which contains a ratio between the number of the O stars and that of the B stars about ten times higher than is found among the "normal" O5-B5 stars. Hence, the high-velocity group must contain a much higher proportion of large masses.

Attempts have been made to explain the expanding motions of stars in OB associations by the process of star formation in the expanding neutral hydrogen gas surrounding HII regions. This theory seems to account satisfactorily for moderate expansion velocities. The higher than average velocities may be explained by the speeding up of the motions of clouds of neutral gas by the "rocket-effect" (Oort and Spitzer 1955; Oort 1954, 1955). However, this process fails to explain the very high velocities of the stars studied in the present paper. That this is so can be readily demonstrated by means of the formula (Oort and Spitzer 1955)

$$\frac{M_0}{M} = e^{(v-v_0)/V},$$

which gives the ratio between the initial mass M_0 and the reduced mass M of a cloud by the time the cloud's velocity has increased from v_0 to v; V being the velocity with which the ionized particles escape from the cloud. It seems reasonable to suppose that for stars of about 20 solar masses to be formed within the accelerated cloud, this cloud should have had at the instant of the star's formation a total mass M of at least a few hundred solar masses, say 200 $\odot$. Taking $V = 20$ km/sec and assuming that the cloud at the beginning of the rocket acceleration may already have acquired by the general expansion a velocity $v_0 = 10$ km/sec, we find that final velocities $v = 100$ km/sec, $v = 150$ km/sec and $v = 200$ km/sec require initial masses M_0 of about 5×10^4, 2×10^5 and 3×10^6 solar masses respectively. For more refined calculations we refer to the cited papers by Oort and Spitzer; these give, however, nearly the same results. Such large masses are quite unlikely to occur for individual clouds. Moreover, as Oort and Spitzer have pointed out, the acceleration of these massive clouds would require improbably large collections of O-type stars. Ambartsumian (1955, 1958) has made the interesting suggestion that the stellar associations, including the run-away stars, were formed, simultaneously with the associated interstellar gas, out of proto-stars in a state of very high density, perhaps of the order of nuclear density. This hypothesis has attractive features, but it also meets with very serious objections. One of the most important of these latter seems to us the fact that it is hard to see how this pre-stellar matter, if it is concentrated in small super-dense bodies, could ever have obtained the extremely flat distribution in the galactic plane observed for the associations. The alternative assumption of the formation of the young stars out of the interstellar matter leads, at least qualitatively, to a natural explanation of this flat distribution since this is present in the gas, whereas the flatness of the gaseous layer may be explained by the gradual collapse of the originally more spherically shaped primordial cloud from which the Galaxy was formed. However, bodies of stellar dimension and of stellar or higher densities should not have reached this flat distribution. The observed distribution of the globular clusters confirms this expectation, and a similar spherical distribution should then also be observed for the superdense proto-stars and for the objects recently formed out of them. For this reason, among others, we prefer to adhere to the concept of stars being formed out of the interstellar medium.

In the present paper we wish to propose the following hypothesis for the origin for the high-velocity O and B stars. We suggest that these stars in their earliest stage of development—and as a rule when they were still in the later phases of the contraction—were the secondary components of proto-double stars, the more massive primaries of which have subsequently shed most of their mass, thus releasing the secondary with a velocity equal to a large fraction of its original orbital velocity. This ejection of the mass of the proto-primary must have been rather sudden, as will be shown below, and may have to be identified with the supernova phenomenon of type II. After the present investigation was concluded, it was kindly brought to our attention by Dr M. Schmidt, that several years ago Zwicky (1957) had already remarked that the supernova phenomenon in binary stars may have been the cause of observed high velocities for released companions.

We have especially in mind proto-primaries with very large masses, ranging from the largest masses actually observed to occur in spectroscopic binaries, say 90 solar masses, up to several hundred solar masses. While it is true that we do not know stars of these masses to occur as luminous and stable bodies unless the super-supergiants (Feast and Thackeray 1956, Thackeray 1958) are identified as such, it does not seem implausible that *proto*-stars with masses of this order do occur in the interstellar medium, and that they may occasionally reach advanced stages of the contraction before instability prevents its completion. (This suggestion has also been made by Gold, 1958.) Hoyle (1959) has shown that as far as the ratio between radiation pressure and gravity is concerned, stars with a hydrogen concentration of $X = 0.75$, and masses up to 1,000 solar masses may be stable; the instability with which we are dealing here is, however, one of dynamic origin. Schwarzschild and Härm (1959) find that vibrational instability must be expected to occur for stars heavier than 60 solar masses, although in the range between this and the observed upper limit of about 90 solar masses no immediate disruption may take place. For heavier objects, however, sudden disruption due to vibrational instability is indicated. Perhaps this kind of instability is to be identified with the sudden mass loss required to explain the release of the secondary component. Some aspects of the identification with the type II supernovae, an identification which allows some specific tests, are considered in Section 7.

The percentage of double and multiple systems among the O- and B-type stars is very high (see also section 4) and it is the rule rather than the exception that a massive star is formed as a binary. It is, therefore, natural to assume that the process of formation which is only partly completed by the very massive proto-stars, will as rule also tend to produce a binary or a system of still higher order. Our working hypothesis will be that the disruption of the very massive primary may sometimes take place when the system can be described as already consisting of two separate units, moving under their mutual gravitation. The contraction of the two components may yet be far from completed. All we suppose is that their sizes are of an order smaller than the separation between the two components.

For the most probable values of the distance between the two components of the proto-binary, we provisionally adopt the range of the most frequently occurring separations found for double stars in general, i.e. 10 to 100 a.u. (Kuiper 1935). This implies that the size of the massive component of the proto-binary may be as large as one a.u., i.e. about 10 times the main-sequence radius for stellar models of such large masses. (For models of masses of 121 and 218 solar masses,

Schwarzschild and Härm (1958) find main-sequence radii of 25 and 47 solar radii respectively.) It may be stressed that we are not thinking in terms of those binary systems which, after the contraction has been completed, belong to the category of the "close binaries".

Denoting the most massive component in a proto-double star and its mass by M_1, the proto-secondary (later to be identified as the high-velocity star) by M_2, and the radius of the orbit of M_2 around the common centre of gravity by a_2, the velocity of the secondary component around this centre of gravity, S_2, is $30\,[M_1/(M_1 + M_2)]\,(M_1/a_2)^{1/2}$ km/sec. For masses M_1 of about 200 solar masses with $M_2 = 20\odot$ (so that $M_1/(M_1 + M_2) = a_2/(a_1 + a_2) \sim 1$), and $a_2 = 20$ a.u., S_2 is about 100 km/sec. Generally, for masses M_1 and separations between the components in the ranges of values already mentioned, we find orbital velocities of M_2 quite comparable in amount to the speeds of the high-velocity O and B stars.

The amount of reduction of this velocity in the process of the release of M_2 by the disruption of M_1 depends on the amount of mass left of M_1 and on the interval of time during which its mass ejection takes place, or, more properly, on the ratio between this time and the orbital period in the original system. This ratio should be one half or less, according to section 6. The orbital periods of proto-binaries like the one just mentioned are of the order of a few years. The mass loss should therefore almost entirely take place within a year or so—a condition which suggests that we are dealing with the supernova phenomenon. Because of their association with Type I population, only supernovae of type II are considered. We shall show (section 7B) that the frequency of type II supernovae in the galaxy amply suffices for part of them to account for the origin of the high-velocity OB stars.

In section 2 of the present paper we present particulars about presently known high-velocity OB stars. Section 3 deals with their frequency among the various spectral subclasses and with the resulting statistics of the masses of the high-velocity stars. It appears that the incidence of high velocities among the O-type stars (masses $\geqslant 24\odot$) is about ten times higher than among the B types.

In section 4 we discuss provisionally the duplicity and multiplicity among the O to B5 stars. It is found that among the B0 to B5 stars, the occurrence of single stars must be very rare; about 75 per cent appear to be double or multiple with companions brighter than absolute magnitude + 5, and an even larger percentage would be found if fainter companions could also be taken into account. For the O-type stars, the occurrence of single objects must be even more exceptional than among the B types. The marked contrast in this respect of the high-velocity group of objects is shown in section 5; all or nearly all of these are single. This observation forms the main inducement for postulating the proto-binary hypothesis.

The proto-binary hypothesis is discussed in detail in section 6. The kinematic behaviour of a binary with a primary ejecting its mass is considered, first for the simple case of instantaneous mass loss, and next for a finite rate of loss. Results of accurate calculations for the latter case, communicated in a paper by Boersma following the present article, are used as a basis for the deduction of the initial properties of the binary systems from which the observed high-velocity objects may have originated. The proto-binary hypothesis appears to account satisfactorily for the observed properties of the high-velocity objects provided the rate of mass loss of the primary may be assumed to be as rapid as that observed in the type II supernovae, and provided it may be assumed that the contraction rates of stars of masses $10\odot$ and lower (types B2 and later) are so slow as to reduce the number of such stars which will survive the supernova explosion of the primary. Current estimates of the contraction rates support the relevant assumption.

The masses required for the proto-primaries are found to lie in the range of 100 to 1,000 solar masses as was to be expected from the foregoing considerations. A representative description of the proto-binary system at the time of its disruption would be: mass of primary: $250\odot$, mass of secondary to be released: 10 to $50\odot$, separation of the centres of the two components; 20 astronomical units, radii of the two components: 1 to 5 astronomical units. Orbital period: 5.4 years. Time during which the mass of M_1 is ejected: 3 months, remnant mass of M_1: $25\odot$ or less. Velocity with which M_2 will be encountered as a high-velocity object: 90 km/sec.

The proposed theory implies that supernovae of type II occur in OB associations at least as frequently as the high-velocity stars are produced, and probably even more frequently. Thus, the Orion, I Sco and probably also the h and χ Persei associations must have been the scene of repeated explosions. The consequences which this must have had for the distribution of the interstellar matter in the associations are briefly dealt with in section 7C. The maximum luminosity of type II supernovae should be considerably higher than currently adopted values if these supernovae occur in the regions of heavy interstellar obscuration where massive proto-stars are expected. It is then suggested that one remnant of the type II supernovae might perhaps be identified with the irregularly varying subluminous O-type star X Persei in the association II Per.

2. DATA FOR INDIVIDUAL OBJECTS

Data for nineteen high-velocity objects are assembled in table 100.1. Their origin, if known, is given in the fifth column, and in such cases the velocity in the fourth column is with respect to this origin. In the remaining cases the velocity given is the space velocity corrected for differential galactic rotation and standard solar motion, For most of the objects for which the origin is known, the visual absolute magnitude (in the 7th column) could be computed using the distance of this origin

Table 100.1 List of high-velocity O-B5 stars

name	HD	MK spectral type	space velocity (km/sec)	associated with	kinematic age (10^6 yrs)	M_v	M_{bol}	mass $\odot = 1$	l^{I}	b^{I}	distance ps
									°	°	
λ Cep	210,839	O6 f	64	I Cep	?	−5.9	−10.4	87	71.5	+ 2.5	780
ξ Per	24,912	O7	50	II Per	1.6	−5.0	− 9.3	50	128.3	−12.0	430
	157,857	O7 f	⩾ 50		?	(−5.5)	(− 9.9)	(67)	340.7	+11.8	2,400
	152,408	O7-8 fp	⩾109	I Sco	?	−7.2	−11.5	160?	311.8	+ 0.3	2,000
68 Cyg	203,064	O8	49	I Cep	5.2	−5.3	− 9.7	60	55.4	− 4.4	880
α Cam	30,614	O9.5 Ia	59	NGC 1502?	2.0	(−6.4)	(−10.5)	(90)	111.4	+14.9	1,000
AE Aur	34,078	O9.5 V	106	I Ori	2.7	−3.8	− 7.6	24	139.8	− 0.9	440
ζ Oph	149,757	O9.5	39	II Sco	1.1	−4.3	− 8.3	32	334.1	+22.1	170
μ Col	38,666	B0 V	123	I Ori	2.2	−3.6	− 7.3	21	204.5	−25.9	570
	151,397	B0.5 V	⩾180	I Sco	?	−3.2	− 6.8	17	312.1	+ 2.0	2,300
	149,363	B0.5 III	⩾115	?	?	(−4.9)	(− 8.6)	(38)	337.7	+25.2	2,600
53 Ari	19,374	B2 V	59	I Ori	4.9	−2.3	− 5.4	10	131.3	−33.0	360
	97,991	B2 V	156:	?	?	(−2.5)	(− 5.7)	(12)	231.7	+52.4	760
	197,419	B2 Ve	23:	I Lac	10:	−2.8	− 6.2	14	44.4	− 5.0	540
72 Col	41,534	B3 V	191	I Sco	14	−1.5	− 4.1	(8)	205.7	−22.2	260
	214,930	B3 V	73	?	?	(−1.5)	(− 4.1)	(8)	56.8	−30.7	440
	216,534	B3 V	82	?	?	(−1.5)	(− 4.1)	(8)	72.1	− 8.8	830
	4,142	B5 V	74	?	?	(−0.7)	(− 2.6)	(5)	89.8	−14.7	170
	201,910	B5 V	58:	I Lac	2.7:	−1.5	− 4.1	8	52.5	− 5.4	480

from the sun, and the star's proper motion and radial velocity. Assumed visual absolute magnitudes are in parentheses. Kinematic ages, given in the 6th column, are the times the stars must have travelled with the present relative speed to reach the present distances from their origin. These ages can be given only if the proper motion is known with sufficient accuracy.

The bolometric absolute magnitudes in the 8th column have been obtained by adding to the visual absolute magnitudes the bolometric corrections tentatively adopted by Limber (1960) on the basis of a compilation of available data. The next column gives the masses, derived by means of the theoretical relations between bolometric absolute magnitude and mass as computed by Schwarzschild and Härm (1958) and by Henyey, LeLevier, and Levee (1959). The masses found, while quite uncertain—especially for the very luminous objects—are usually larger than 10 solar masses and may in some cases be as much as 100 solar masses or more.

Detailed notes on the objects listed in table 100.1 are omitted. We omit section 3, which notes the greater percentage of high velocities among O types than among B types. We also omit sections 4 and 5, which show that at least 75% of B0 to B5 stars occur in double or multiple star systems, that a large fraction of O stars also occur in double or multiple star systems, but that no more than 20% of the high velocity O and B stars occur in double or multiple star systems. The details of the proto-binary hypothesis for the origin of high velocity stars are given in section 6, also omitted. See section 1 for a summary of this treatment.

7. RELATION WITH SUPERNOVAE OF TYPE II

A. IDENTIFICATION WITH SUPERNOVAE OF TYPE II According to the preceding sections, the hypothesis of the proto-binary origin of the high-velocity objects implies that as a rule the mass ejection of the proto-primary should take place within a lapse of time not exceeding the orbital period; for instance, for proto-primaries of around 200 solar masses it should happen within several months to provide that a secondary be released with a velocity of 150 km/sec. We shall consider some aspects of the possible identification of the process involved with that of the supernovae—and particularly with those of type II because of their pronounced association with Population I. Type II supernovae are found mostly in spiral arms (see, for instance, Zwicky 1958) and according to Payne–Gaposchkin (1957) all nine supernovae of type II which have been identified as such occur in spirals of types Sc, SBc, Sb or SBb. (Of 16 objects identified as type I, 9 occur in systems of these types.) The close association of supernovae type II with Population Type I fits entirely, of course, with our hypothesis which connects the phenomenon directly with the process of the star formation. Also, the rate at which the supernova phenomenon proceeds seems to meet our requirement well. We find from the composite light-curve of three well observed supernovae of type II, reproduced by Payne-Gaposchkin, that brightness within five magnitudes from the maximum lasts only for about four months. Further, it seems plausible that the bulk of the ejected material has passed beyond the orbit of the secondary within this lapse of time. Matter with ejection velocities ranging from 5,000 to 500 km/sec will have passed beyond 20 a.u. from the primary in the period between 7 and 70 days after the onset of the explosion if all these velocities are imparted to the material simultaneously. The decay time of M_1 as experienced by M_2 in the sense of the preceding kinematic considerations is then of the order $\tau_f P_0 = 70$ days or less.

B. COMPARISON WITH FREQUENCY OF SUPERNOVAE The proposed identification of the origin of the high-velocity objects with the supernovae of type II implies, among other tests, that the observed number of high-velocity objects can be related to the frequency of the occurrence of supernovae. As a basis for the estimated rate of production of high-velocity objects we take those of table 100.1 within 1,000 ps from the sun and with kinematic ages below 5 million years. For some of the objects in the table, no ages are known. An estimate may be made on the basis of the ages found for objects of similar luminosity. Even within these limits of age and distance, the table is incomplete. Objects of the lower luminosities are only partly known at distances between 500 and 1,000 ps. Further, from the statistics of the recorded ages, it appears that those between 3 and 5 million years are less completely represented than the younger ones. An, admittedly rather uncertain, factor of 4 will be tentatively adopted to take account of these two causes of incompleteness. We further notice that no objects younger than one million years occur in the table. Apparently, these do not yet show within the limits of spectral type and luminosity considered. We thus are led to the estimate that within 1,000 ps from the sun, 40 high-velocity O to B5 objects occur with kinematic ages of between 1 and 5 million years.

The origin of the high-velocity objects may be supposed to lie close to the galactic plane. The majority of them will be sufficiently close to the galactic plane within the first 5 million years to remain within the distance limit of our selection. We are therefore justified in assuming that the number of high-velocity objects which is *formed* per 4 million years within 1,000 ps is also given by the above figure. Hence the number of events which release O to B5 high-velocity objects within 1,000 ps from the sun is one per 10^5 years. In extending this to the Galaxy as a whole, we assume a factor 10^2, which leads to one such event per 10^3 years. If the event is identified with the type II supernovae, the true number of these supernovae will be larger, since only a fraction of them may lead to the release of secondaries in the range of masses encountered among the high-velocity objects. This factor is very hard to estimate, but we are inclined to tentatively estimate it to be of the order of 10—assuming a distribution of masses of the proto-secondaries similar to that of the initial luminosity function in general, and taking into account the tendency of massive objects to have more than one companion. These very rough estimates then lead to a prediction of one type II supernova in the Galaxy per 100 years, and one per 10^4 years within 1,000 ps from the sun. These figures compare reasonably well with current estimates of the actual occurrence of type II supernovae, which are about 1 per 40 years for the Galaxy as a whole (Öpik 1953, Payne-Gaposchkin 1957).

C. SUPERNOVAE IN ASSOCIATIONS If the proposed identification of the origin of the high-velocity objects with the supernovae is accepted, it leads to a number of interesting conclusions with regard to the phenomena in stellar associations. The possibility of a close relation between the structure of stellar associations and the supernova phenomenon has been discussed before. Thus, Öpik (1953, 1955) has suggested that the stars observed in the associations were formed in the interstellar matter after this was compressed by an expanding supernova shell. Shklovsky (1960), too, has suggested that type II supernovae occur in OB associations, but this author rather considers them as the product of the later evolution of these stars. The mechanism proposed by Öpik might also explain the expanding motions of the stars observed in some associations. We shall not discuss in the present context the extent to which the supernova phenomenon may have been responsible for subsequent star formation in the association in which the supernova occurs, but we shall discuss in a provisional way a few aspects of the relation between supernovae and stellar associations as inferred from the present hypothesis.

The fact that a number of high-velocity objects have originated from the same association but with different kinematic ages shows that in some associations supernovae of type II must have occurred repeatedly. In general, theories accounting for the structure of the OB associations should then take into account the fact that series of supernova explosions may have taken place, spread over an interval of time which may be as long as 10 million years and more.

Here we omit detailed discussions of the necessity of repeated supernova explosions in the Orion, I and II Scorpii, and I Persei associations.

V. A. Ambartsumian 1955, *Observatory 75*, 72.

V. A. Ambartsumian 1958, "La Structure et l'evolution de l'univers," Institut International de Physique Solvay; Onzième Conseil de Physique, Brussels, p. 241.

A. Blaauw 1956a, *P.A.S.P. 68*, 495.

A. Blaauw 1958b, *Spec. Astr. Vaticana Ric. Astr. 5*, 105; *Semaine d'etude sur le problème des populations stellaires*, ed. D. J. K. O'Connell, S. J.

A. Blaauw 1959, "The Hertzsprung-Russell Diagram," *I.A.U. Symp.* No. 10, p. 105.

M. W. Feast and A. D. Thakeray 1956 *M.N.R.A.S. 116*, 581.

T. Gold 1958, "La Structure et l'evolution de l'univers," Institut International de Physique Solvay; Onzième Conseil de Physique, Brussels, p. 276.

L. G. Henyey, R. LeLevier, and R. D. Levee 1959, *Ap. J. 129*, 2.

F. Hoyle 1959, "The Hertzsprung-Russell Diagram," *I.A.U. Symp.* No. 10, p. 83.

G. P. Kuiper 1935, *P.A.S.P. 47*, 121.

D. N. Limber 1960, *Ap. J. 131*, 168.

J. H. Oort 1954, *B.A.N. 12*, 177 (No. 455).

J. H. Oort and L. Spitzer 1955, *Ap. J. 121*, 6.

J. H. Oort 1955, "Gas Dynamics of Cosmic Clouds," *I.A.U. Symp.* No. 2, p. 147.

E. J. Öpik 1953, *Irish A.J.2*, 219.

E. J. Öpik 1955, *Mém. Soc. r. sciences Liège*, 4e Série, tome XV, 634; *Armagh Obs. Leaflet* No. 34.

C. Payne-Gaposchkin 1957, "The Galactic Novae," chaps. 9 and 11 (Amsterdam: North-Holland Publishing Company).

M. S. Roberts 1957, *P.A.S.P. 69*, 59.

M. Schwarzschild and R. Härm 1958, *Ap. J. 128*, 348.

M. Schwarzschild and R. Härm 1959, *Ap. J. 129*, 637.

I. S. Shklovsky 1960, *Astr. Journal U.S.S.R. 37*, 369.

A. D. Thackeray 1958, *Spec. Astr. Vaticana Ric. Astr. 5*, 195; *Semaine d'etude sur le problème des populations stellaires*, ed. D. J. K. O'Connell, S. J.

F. Zwicky 1957, "Morphological Astronomy," p. 258 (Berlin: Springer).

F. Zwicky 1958, *Handbuch der Physik 51*, 766.

CHAPTER VII

Normal Galaxies, Radio Galaxies, and Quasars

101. Novae in Spiral Nebulae

Heber D. Curtis

(*Publications of the Astronomical Society of the Pacific 29*, 180–182 [1917])

George W. Ritchey

(*Publications of the Astronomical Society of the Pacific 29*, 210–212 [1917])

Harlow Shapley

(*Publications of the Astronomical Society of the Pacific 29*, 213–217 [1917])

As the consummate optician who had figured the mirror for the Mount Wilson 60-in reflector (and who was at work on the 100-in mirror), George W. Ritchey received a certain amount of telescope time for taking long exposure "show plates" to demonstrate the capabilities of his instruments. These plates had research uses—for example, in the Adriaan van Maanen project to measure proper motion rotations of the spiral nebulae—and it was on such a plate, taken in July 1917, that Ritchey found a new star (or nova) not present on the earlier exposures. The telegrams announcing the discovery triggered off a search for other examples that closely paralleled the rapid discoveries of pulsars half a century later.

Meanwhile, at Lick Observatory, Heber D. Curtis had previously found three earlier novae on plates made with the Crossley reflector, but because they had already disappeared, he had seen no urgency in announcing them; only after Ritchey's telegram did he report them, in the first paper reprinted here.[1] At this same time Ritchey searched the earlier plates at Mount Wilson, finding two novae in the Andromeda nebula and four suspects in other spirals, as described in the second paper of this group. Furthermore, Francis C. Pease and Harlow Shapley each reported an additional nova from their own plates made with the 60-in reflector.

Curtis had already realized that the new discoveries had important ramifications for the island universe theory of spiral nebulae. In a follow-up note he pointed out that the mean maximum observed magnitude of the 27 novae recorded in the Milky Way was about magnitude 5.5; the novae observed in the spiral nebulae were 10 magnitudes fainter.[2] Therefore, the novae in spiral nebulae were at least 100 times as far away as the galactic novae, if they were the same type of star.

Simultaneously, Shapley drew a similar but more specific conclusion (reported in the third paper of this group), estimating a distance of a million light years for the Andromeda nebula. He also discussed two difficulties for the interpretation of the new data: the extraordinary luminosity required for the much brighter nova observed in the Andromeda nebula in 1885[3] and the problems posed by van Maanen's measurement of the internal motions in M 101. This latter problem soon took on much weight for Shapley, leading to his rejection of large extragalactic distances for the spirals and to his famous debate with Curtis (see selection 79).

1. For a historical account of Ritchey, Curtis, and the discovery of novae in spiral nebulae, see M. A. Hoskin, *Journal for the History of Astronomy 7*, 47 (1976).

2. *Publications of the Astronomical Society of the Pacific 29*, 206 (1917).

3. Frank W. Very gives an interesting account of the large distance to nova S Andromedae in *Astronomische Nachrichten 189*, 441 (1911).

Heber D. Curtis

A TELEGRAM HAS JUST BEEN RECEIVED announcing the discovery by Professor Ritchey, at Mt. Wilson, of a nova in the spiral nebula N.G.C. 6946. The new star is of about the fourteenth magnitude, and is located 105″ south and 37″ west of the nucleus of the spiral. A reproduction of this spiral may be seen in Plate 62 of Vol. VIII, *Publications of the Lick Observatory*; its position for 1900 is $\alpha = 20^h\ 32^m.8$; $\delta = +59°48'$.

In August, 1885, a new star suddenly flashed out in the Great Nebula of Andromeda, only a few seconds of arc from the nucleus; it rose to the seventh magnitude, rapidly diminished in brightness, and is now invisible.

Likewise the nova Z Centauri appeared at a distance of only 28″ from the nucleus of the spiral nebula N.G.C. 5253.

Three cases may be added to those just given. The spiral nebula N.G.C. 4527 ($\alpha = 12^h\ 29^m.0$, $\delta = +3°12'$) is a somewhat elongated spiral about five minutes of arc in total length. At some time between January 13 and March 20, 1915, a nova appeared at the edge of the inner, brighter portion of this nebula, 44″ east and 8″ north of the nucleus. It occurs on two Crossley plates of dates March 20 and April 16, and was about magnitude 14 on the earlier date. Director Pickering has kindly had a search made on the Harvard plates of this region, and Professor Barnard has placed at my disposal enlarged positives of two plates taken at Yerkes Observatory, affording, in combination with the Crossley negatives, a total of eleven plates covering the history of the object from February, 1900, to April, 1917, and serving to confirm its character as a nova. It is entirely invisible at present.

Two additional novae have been found in the spiral nebula N.G.C. 4321 (M. 100; $\alpha = 12^h\ 17^m.9$; $\delta = +16°23'$). This is a fine, approximately round, rather open spiral about five minutes of arc in diameter. At some time prior to March 17, 1901, a new star appeared in this spiral, located 110″ west and 4″ north of the nucleus; it was approximately magnitude 13.5 on March 17, 1901, and about a magnitude fainter in several images recorded on Crossley Reflector plates one month later. A second nova appeared in the same spiral at some time prior to March 2, 1914; it was about the fourteenth magnitude, and was located 24″ east and 111″ south of the nucleus. Both objects have disappeared completely. I hope later to collate photographs of this region taken at other observatories in order to give a more complete history of the objects.

It is possible that a single nova might appear, so placed in the sky as to be directly in line with a spiral nebula, tho the chances for such an occurrence would be very small. But that six new stars should happen to be thus situated in line with a nebula is manifestly beyond the bounds of probability; there can be no doubt that these novae were actually in the spiral nebulae. The occurrence of these new stars in spirals must be regarded as having a very definite bearing on the "island universe" theory of the constitution of the spiral nebulae.

George W. Ritchey

1

IN THE COURSE OF MAKING long-exposure photographs of the larger spiral nebulae with the 60-inch reflector, for the measurement of internal rotation and proper motion, fine negatives of the spiral H. IV 76 Cephei (N.G.C. 6946) were secured on the following dates:

1910, August 4, Exposure 4 hours,
1915, August 12, 13, 14, Exposure $6\frac{3}{4}$ hours,
1916, June 28, 29, 30, Exposure $9\frac{3}{4}$ hours,
1917, July 19, Exposure $4\frac{1}{2}$ hours.

Large numbers of nebulous stars or points, and groups of them, are present in the branches of this spiral, as is the case in most of the larger spirals. On the south side one branch with its nebulous stars can be traced to a distance of $4\frac{1}{2}$ minutes of arc from the nucleus, while on the north side a branch is shown to a distance of 5 minutes of arc.

In the inner and brighter parts of the spiral are two principal branches. In the southern one of these, lying centrally with respect to the width of the branch, is a small quadrilateral of four nebulous stars, approximately 105″ south and 37″ west of the nucleus. On the negative of 1917, July 19, a star of 14.6 magnitude is shown near the center of this quadrilateral, its position certainly not coinciding with that of any of the four nebulous stars. On the earlier negatives mentioned above, and on a plate made by Mr. Pease in 1912, no star is shown in this place, tho the best of these show stars down to the twenty-first magnitude.

The discovery of the Nova was announced in Harvard Bulletin No. 641. On July 27 the brightness had decreased to magnitude 15.6; this value and the magnitude at discovery were determined by Mr. Shapley. There was no opportunity to secure a spectrogram of the nova until August 16, when the photographic brightness was between 15.8 and 16.0. On that date Mr. Pease and the writer secured a small but strong spectrogram of 6 hours exposure at the 25-foot focus of the 60-inch reflector, using a slitless spectrograph of very low dispersion. The continuous spectrum is strong and is crossed by what appears to be a series of bright bands.

2

The discovery of this nova in a spiral nebula led to an examination of all the negatives of spiral nebulae made with the 60-inch reflector since 1908. The search of the Mount Wilson plates thus far has revealed the following objects which are either novae or variables.

a. In Messier 81 (N.G.C. 3031) a star of magnitude 19 in February 1910 which was of magnitude 18 in March, 1917.

b. In. H. V. 44 Camelopardalis (N.G.C. 2403) a star of magnitude 16.5 in February 1910 which is invisible on a negative of 7 hours exposure made in February, 1916, and also on one made by Mr. Pease in November, 1912.

c. In Messier 101 (N.G.C. 5457–8) a star of magnitude 18.5 in March, 1910, which was approximately of magnitude 20.5 in May, 1915.

d. In the same nebula a star of magnitude 19 in March, 1910, which was of magnitude 20.5 in May, 1915.

3

The great nebula in Andromeda is either comparatively very large or comparatively near. It is probably a highly favorable object for the detection of internal rotation and proper motion. Furthermore, the wealth of detail shown in its dark rifts and in its great streams of nebulous stars is probably typical of many or all of the spiral nebulae. During the first year of the use of the 60-inch reflector, in 1909, eleven photographs of this nebula were made for the purpose of comparison with later negatives. In the recent search for novae two stars were found near the center which are shown on many plates and merit a detailed description. They lie nearly in the same meridian, and are 118″ apart. The star *a* is approximately 191″ west and 160″ south of the nucleus; the star *b* is approximately 194″ west and 42″ south of the nucleus.

Table 101.1 gives the data for the eleven plates of 1909 and also for some earlier and later photographs. The first two were made with the 24-inch Yerkes reflector, the third at the 100-foot focus and the remainder at the 25-foot focus of the 60-inch reflector at Mount Wilson; those of August and September, 1917, were made by Mr. Pease and Mr. Shapley, respectively. The magnitudes were estimated by Mr. Shapley; but because of the non-uniform density of the nebulous background, and because of differences in exposure-time and kind of plate, estimates of brightness are necessarily only approximate.

So far as the negatives show, both stars reached a maximum at the same time, on Sept. 16, 1909. Unfortunately the evidence on this point is not complete, as no negatives were made between Sept. 16 and Oct. 11; and on the latter date both stars were faint, having decreased about two magnitudes in the interval of 25 days.

Harlow Shapley

SINCE RITCHEY'S DISCOVERY of the nova in N.G.C. 6946, the number of temporary stars known to have occurred in nebulae has increased so rapidly that an opportunity is afforded of making preliminary use of relative magnitudes

Table 101.1 Observations of two novae in the Andromeda nebula

Date	P.S.T.	Exposure	Plate	Magnitude Star *a*	Magnitude Star *b*
1901 Aug. 19	—	$2^h\ 0^m$	Cramer Crown	invisible	invisible
Sept. 18	—	4 30	Cramer Crown	invisible	invisible
1909 Aug. 13	$13^h\ 45^m$	4 0	Seed 27	invisible	invisible
Sept. 12	11 30	1 0	Cramer Inst. Iso	invisible	19.5
Sept. 12	14 38	0 45	Seed 27	18.3	18.7
Sept. 13	11 15	1 30	Seed 27	16.4	18.9
Sept. 13	14 45	1 0	Seed 27	17.8	18.8
Sept. 15	11 0	2 0	Seed 27	16.3	18.3
Sept. 16	14 45	1 30	Seed 23	16.3	17.0
Oct. 11	11 3	4 0	Seed 23	18.3	18.7
Oct. 13	9 18	2 7	Seed 23	18.3	18.5
Nov. 7	7 45	2 20	Seed Process	19.5	19.0
Nov. 7	11 27	2 20	Seed Process	19.6	19.0
1910 Aug. 4	12 10	2 30	Seed 23	invisible	invisible
1915 Aug. 9-10-11	—	9 38	Seed 23	invisible	invisible
Oct. 9-10-11	—	9 10	Seed 23	invisible	invisible
1917 Aug. 21	15 45	1 0	Seed 27	invisible	invisible
Sept. 11	10 48	0 25	Seed 27	invisible	invisible

Table 101.2 A list of temporary stars in spiral nebulae

No.	N.G.C.	Galactic latitude	Observed maximum	Date of observed maximum	Distance from nucleus	Discoverer
1	224	−20°	7^m	Aug. 1885	16″	Hartwig[a]
2	5253	+30	7	July 1895	32	Mrs. Fleming
3	4321	+77	13.5	Mar. 1901	110	Curtis
4	3147	+40	13–14	Apr. 1904	340	Mrs. Isaac Roberts
5	224	−20	16.3	Sept. 1909	249	Ritchey
6	224	−20	17.0	Sept. 1909	198	Ritchey
7	2841	+45	16	Feb. 1912	54	Pease
8	4321	+77	14	Mar. 1914	111	Curtis
9	4527	+64	14	Mar. 1915	45	Curtis
10	6946	+12	14.6	July 1917	111	Ritchey
11	224	−20	17.5	Sept. 1917	600	Shapley

a. Probably first seen by Ward.

in estimating the order of the distances of spiral systems. The results can hope only to be suggestive, but any data will be welcome that bear directly on the problem of the distances and sizes of spiral nebulae and their relation to the galactic system of stars.

A list of the novae in spirals is given in table 101.2. Six of the objects are situated in spiral arms, three in denser nuclear parts, and the other two apparently at some distance from nebulosity, but longer exposures would perhaps show connecting spiral arms. All but the last two of the eleven new stars seem to have completely vanished; but it is now safe to predict that the discovery of such objects in the future will be of frequent occurrence, and probably occasional ones will be suitable for more extensive spectroscopic studies. The first two in the table are S Andromedae and Z Centauri, respectively, which apparently are the only novae in spirals definitely recognized until two months ago. The third, seventh and eighth are found on only one plate each, and are missing from earlier and later negatives. Their authenticity can hardly be doubted, but there is little probability that a verification will ever be made. On the basis of chance, Curtis has noted the impossibility of considering these novae physically unrelated to the spirals with which they are associated. If further support of this conclusion were needed it could be obtained from the third column of the table, which gives the distance from the galaxy of the seven spirals in which new stars have appeared. All novae outside of spirals, with the exception of T Coronae, are closely confined to the Milky Way.

Thirty isolated novae are on record, half of which have been discovered by photography within the last 25 years. The magnitude at maximum ranges from −5 to +10, with an average of +5.5. It depends of course on the distances of the stars as well as their luminosities; in many cases the actual maximum was probably unobserved. Further, there is some uncertainty in the estimates, and a possibility of a marked prejudice in favor of the brighter objects.

The mean observed magnitude at maximum of the eleven temporary stars in spirals is +14, with considerable uncertainty in the individual estimates, but probably with fair accuracy in the mean. Here there has been a similar but much greater selection of novae intrinsically bright, as well as a favoring of spirals relatively near. To estimate a correction is as yet difficult, if not impossible, so the magnitudes are used as they stand.

Taking the averages as referring to objects of similar luminosity, possibly an uncertain procedure because of the great dispersion in absolute magnitude of the novae in the Andromedae nebula, and probably in the galaxy as well, we observe that the difference in apparent brightness calls for a distance at least 50 times as great for these larger spiral nebulae as for the average novae of the galactic system. Concerning the parallaxes of the latter we know practically nothing, except that they are probably very small. They must be, on the average, of the order of 0″.001 (corresponding to three or four times the distance of the B-type stars of the same apparent magnitude), or even less, if, with the above criterion, the spiral nebulae are to be well outside of our galaxy.

A few points bearing on the distance problem may be mentioned in this connection.

1. The permanent stars in the vicinity of the spirals are almost certainly not connected with the nebulae. They are generally believed to be members of the galactic system seen in projection; counts of the relative frequency of stars in and near the spirals strongly support this view. The differentiated parts of spiral

nebulae are the nebulous condensations—objects with measurable angular dimensions. If we are to believe that a spiral nebula, N.G.C. 224 for instance, is a remote stellar system, its most luminous stars must be fainter than magnitude 21, since even in its thinner parts there is no hint of resolution into distinct stars on the best of plates; nor can the nebulous condensations, with magnitudes between 15 and 20, be considered an accumulation of stars brighter than magnitude 21. This point is important, for, if in the hypothetical galactic system the brightest stars are comparable with the bright stars of our own galaxy, the minimum distance of the Andromeda Nebula must be of the order of a million light-years. At that remote distance the diameter of this largest of spirals would be about 50,000 light-years—a value that now appears most probable as a minimum for our galactic system. The same argument holds as well for other spirals, tho in some the condensations are brighter and more star-like.

2. The matter of special interest, however, in direct connection with the novae (if spirals are galactic systems of typical stars) is that these new stars attain a most extraordinary absolute brightness. The average for the eleven given in the table is at least magnitude -7, and the maximum for S Andromedae is at least -15—a luminosity nearly a hundred million times that of our Sun, the equivalent of the light emission of a million stars of zero absolute magnitude, and probably, therefore, much greater than the total light of all the stars seen with the naked eye.

This remarkable result must inevitably follow if spiral nebulae are considered external galactic systems comparable with our own in size and constituency. Yet, to parallel this luminosity in our own galaxy, it is only necessary to suppose that one of the brightest of the isolated new stars, Nova Persei, for instance, is as distant as some parts of the galactic clouds.

3. For dimensions comparable with the galactic system a spiral nebula of 5′ apparent diameter must have a parallax less than one ten-millionth of a second of arc; and, with no extinction of light in space, a star in it of the brightness of the Sun would be fainter than the thirty-fifth magnitude. Messier 101, for instance, to be comparable must not be nearer than five million light-years; and the difficulty is obvious in reconciling van Maanen's measures of internal proper motion with the hypothesis of external galactic systems. We are not prepared to accept velocities of rotation of the order of the velocity of light. If we assume the dimensions to be one four-hundredth the minimum value adopted for our system, so that measured internal velocities are about 1,000 km/sec, the distance of Messier 101 would be 32,000 light-years and any star of the absolute magnitude of the Sun or brighter could be easily photographed as a distinct stellar point. Measurable internal proper motions, therefore, can not well be harmonized with "island universes" of whatever size, if they are composed of normal stars.

4. A typical globular cluster at a distance of 30,000 light-years has an angular diameter of about 5′. Suppose it placed in a spiral nebula that is assumed to be at a distance 100 times as great. Then its integrated magnitude of 18 to 20, its diffuse outline, and its apparent diameter of 3″ would simulate very well the small nebulous condensations of spirals. The Magellanic Clouds, if located in the same nebula, would show very irregular outlines, would subtend mean diameters of approximately 30″ and 100″, and, because of their bright-line nebulae, might appear blue compared with the solar type nucleus.

102. A Spectrographic Investigation of Spiral Nebulae

Vesto M. Slipher

(*Proceedings of the American Philosophical Society 56*, 403–409 [1917])

Although the bright emission-line spectra of the diffuse gaseous nebulae had been discovered in the early days of astronomical spectroscopy, the spectral analysis of spiral nebulae proceeded slowly because they were so very faint. Heroic exposure times were required—20, 40, or even 80 hr—to obtain these elusive spectra. Today, when larger telescopes and faster emulsions are so readily available, it is difficult to comprehend how much time V. M. Slipher required with the Lowell Observatory's 24-in refractor in order to record the radial velocities of the twenty-five spirals reported in this paper. Although he necessarily used a low dispersion, the extraordinarily high velocities—sometimes exceeding 1,000 km sec^{-1}—became clear from his spectrograms.

Slipher recognized that these astonishingly high velocities, generally in recession, had significant consequences for the island universe theory of spiral nebulae. Because it was hard to believe that objects with such enormous speeds could long remain a part of the Milky Way system, these observations proved to be a major impetus for reviving this theory among Slipher's contemporaries.

Several years after Slipher reported these results, C. Wirtz tried to correlate the velocities with the distances of the spirals, but it was not until further observational material became available that Edwin Hubble could convincingly demonstrate such a connection (see selection 106).

IN ADDITION TO THE PLANETS AND COMETS of our solar system and the countless stars of our stellar system there appear on the sky many cloud-like masses—the nebulæ. These for a long time have been generally regarded as presenting an early stage in the evolution of the stars and of our solar system, and they have been carefully studied and something like 10,000 of them catalogued.

Keeler's classical investigation of the nebulæ with the Crossley reflector by photographic means revealed unknown nebulæ in great numbers. He estimated that such plates as his if they were made to cover the whole sky would contain at least 120,000 nebulæ, an estimate which later observations show to be considerably too small. He made also the surprising discovery that more than half of all nebulæ are spiral in form; and he expressed the opinion that the spiral nebulæ might prove to be of particular interest in questions concerning cosmogony.

I wish to give at this time a brief account of a spectrographic investigation of the spiral nebulæ which I have been conducting at the Lowell Observatory since 1912. Observations had been previously made, notably by Fath at the Lick and Mount Wilson Observatories, which yielded valuable information on the character of the spectra of the spiral nebulæ. These objects have since been found to be possessed of extraordinary motions and it is the observation of these that will be discussed here.

In their general features nebular spectra may for convenience be placed under two types characterized as (I.) bright-line and (II.) dark-line. The gaseous nebulæ, which include the planetary and some of the irregular nebulæ, are of the first type; while the much more numerous family of spiral nebulæ are, in the main, of the second type. But the two are not mutually exclusive and in the spirals are sometimes found both types of spectra. This is true of the nebulæ numbered 598, 1068 and 5236 of the "New General Catalogue" of nebulæ.

Some of the gaseous nebulæ are relatively bright and their spectra are especially so since their light is all concentrated in a few bright spectral lines. These have been successfully observed for a long time. Keeler in his well-known determination of the velocities of thirteen gaseous nebulæ was able to employ visually more than twenty times the dispersion usable on the spiral nebulæ.

Spiral nebulæ are intrinsically very faint. The amount of their light admitted by the narrow slit of the spectrograph is only a small fraction of the whole and when it is dispersed by the prism it forms a continuous spectrum of extreme weakness. The faintness of these spectra has discouraged their investigation until recent years. It will be only emphasizing the fact that their faintness still imposes a very serious obstacle to their spectrographic study when it is pointed out, for example, that an excellent spectrogram of the Virgo spiral N.G.C. 4594 secured with the great Mount Wilson reflector by Pease was exposed eighty hours.

Here Slipher points out that the faintness of the spiral nebulae requires 20 to 40 hr of observing time with his 24-in refractor in order to obtain adequate spectra; comparison spectra were made using an iron-vanadium spark.

The plates are measured under the Hartmann spectrocomparator in which one optically superposes the nebular plate of unknown velocity upon one of a like dark-line spectrum of known velocity, used as standard. A micrometer screw, which shifts one plate relatively to the other, is read when the dark lines of the nebula and the standard spectrum coincide; and again when the comparison lines of the two plates coincide. The difference of the two screw readings with the known dispersion of the spectrum gives the velocity of the nebula. By this method weak lines and groups of lines can be utilized that otherwise would not be available because of faintness or uncertainty of wave-length.

Table 102.1 Radial velocities of twenty-five spiral nebulæ

Nebula	Vel. km. [sec^{-1}]	Nebula	Vel. km. [sec^{-1}]
N.G.C. 221	− 300	N.G.C. 4526	+ 580
224	− 300	4565	+1,100
598	− 260	4594	+1,100
1023	+ 300	4649	+1,090
1068	+1,100	4736	+ 290
2683	+ 400	4826	+ 150
3031	− 30	5005	+ 900
3115	+ 600	5055	+ 450
3379	+ 780	5194	+ 270
3521	+ 730	5236	+ 500
3623	+ 800	5866	+ 650
3627	+ 650	7331	+ 500
4258	+ 500		

In table 102.1 are given the velocities for the twenty-five spiral nebulæ thus far observed. In the first column is the New General Catalogue number of the nebula and in the second the velocity. The plus sign denotes the nebula is receding, the minus sign that it is approaching.

Generally the value of the velocity depends upon a single plate which, in many instances, was underexposed and some of the values for these reasons may be in error by as much as 100 kilometers. This however is not so discreditable as at first it might seem to be. The arithmetic mean of the velocities is 570 km [sec^{-1}] and 100 km is hence scarcely 20 per cent. of the quantity observed. With stars the average velocity is about 20 km [sec^{-1}] and two observers with different instruments and a single observation each of an average star might differ in its velocity by 20 per cent. of the quantity measured. Thus owing to the very high magnitude of the velocity of the spiral nebulæ the percentage error in its observation is comparable with that of star velocity measures.

Since the earlier publication of my preliminary velocities for a part of this list of spiral nebulæ, observations have been made elsewhere of four objects with results in fair agreement with mine.

Here Slipher gives a table showing that other observers have obtained similar large recession velocities for some of the spiral nebulae listed in table 102.1.

Referring to the table of velocities again: the average velocity 570 km [sec^{-1}] is about thirty times the average velocity of the stars. And it is so much greater than that known of any other class of celestial bodies as to set the spiral nebulæ aside in a class to themselves. Their distribution over the sky likewise shows them to be unique—they shun the Milky Way and cluster about its poles.

The mean of the velocities with regard to sign is positive, implying the nebulæ are receding with a velocity of nearly 500 km [sec^{-1}]. This might suggest that the spiral nebulæ are scattering but their distribution on the sky is not in accord with this since they are inclined to cluster. A little later a tentative explanation of the preponderance of positive velocities will be suggested.

Slipher shows that face view, inclined, and edge view spiral nebulae have respective mean velocities of 330 km sec^{-1}, 560 km sec^{-1}, and 760 km sec^{-1}, suggesting that spiral nebulae might move edge forward.

The form of the spiral nebulæ strongly suggests rotational motion. In the spring of 1913 I obtained spectrograms of the spiral nebulæ N.G.C. 4594 the lines of which were inclined after the manner of those in the spectrum of Jupiter, and, later, spectrograms which showed rotation or internal motion in the Great Andromeda Nebula and in the two in Leo N.G.C. 3623 and 3627 and in nebulæ N.G.C. 5005 and 2683—less well in the last three. The motion in the Andromeda nebula and in 3623 is possibly more like that in the system of Saturn. It is greatest in nebula N.G.C. 4594. The rotation in this nebula has been verified at the Mt. Wilson Observatory.

Because of its bearing on the evolution of spiral nebulæ it is desirable to know the direction of rotation relative to the arms of the spirals. But this requires us to know which edge of the nebula is the nearer us, and we have not as yet by direct means succeeded in determining even the distance of the spiral nebulæ. However, indirect means, I believe, may here help us. It is well known that spiral nebulæ presenting their edge to us are commonly crossed by a dark band. This coincides with the equatorial plane and must belong to the nebula itself. It doubtless has its origin in dark or deficiently illuminated matter on our edge of the nebula, which absorbs (or occults) the light of the more brightly illumined inner part of the nebula. If now we imagine we view such a nebula from a point somewhat outside its plane the dark band would shift to the side and render the nebula unsymmetrical—the deficient edge being of course the one nearer us. This appears to be borne out by the nebulæ themselves for the inclined ones commonly show this typical dissymmetry. Thus we may infer their deficient side to be the one toward us.

When the result of this reasoning was applied to the above cases of rotation it turned out that the direction of rotation relative to the spiral arms was the same for all. (The nebula N.G.C. 4594 is unfortunately not useful in this as it is not inclined enough to show clearly the arms.) The central part—which is all of the nebulæ the spectrograms record—turns into the spiral arms as a spring turns in winding up. This agreement in direction of rotation furnishes a favorable check on the conclusion as to the nearer edge of the nebulæ, for of course we should expect that dynamically all spiral nebulæ rotate in the same direction with reference to the spiral arms. The character and rapidity of the rotation of the Virgo nebula N.G.C. 4594 suggests the possibility that it is expanding instead of contracting under the influence of gravitation, as we have been wont to think.

As noted before the majority of the nebulæ here discussed have positive velocities, and they are located in the region of sky near right ascension twelve hours which is rich in spiral nebulæ. In the opposite point of the sky some of the spiral nebulæ have negative velocities, *i.e.*, are approaching us; and it is to be expected that when more are observed there, still others will be found to have approaching motion. It is unfortunate that the twenty-five observed objects are not more uniformly distributed over the sky as then the case could be better dealt with. It calls to mind the radial velocities of the stars which, in the sky about Orion, are receding and in the opposite part of the sky are approaching. This arrangement of the star velocities is due to the motion of the solar system relative to the stars. Professor Campbell at the Lick Observatory has accumulated a vast store of star velocities and has determined the motion of our sun with reference to those stars.

We may in like manner determine our motion relative to the spiral nebulæ, when sufficient material becomes available.

A preliminary solution of the material at present available indicates that we are moving in the direction of right-ascension 22 hours and declination $-22°$ with a velocity of about 700 km. While the number of nebulæ is small and their distribution poor this result may still be considered as indicating that we have some such drift through space. For us to have such motion and the stars not show it means that our whole stellar system moves and carries us with it. It has for a long time been suggested that the spiral nebulæ are stellar systems seen at great distances. This is the so-called "island universe" theory, which regards our stellar system and the Milky Way as a great spiral nebula which we see from within. This theory, it seems to me, gains favor in the present observations.

It is beyond the scope of this paper to discuss the different theories of the spiral nebulæ in the face of these and other observed facts. However, it seems that, if our solar system evolved from a nebula as we have long believed, that nebula was probably not one of the class of spirals here dealt with.

103. Modern Theories of the Spiral Nebulae

Heber D. Curtis

(*Journal of the Washington Academy of Sciences 9*, 217–227 [1919])

As Heber Curtis remarks in this paper, the once fashionable island universe theory of spiral nebulae went into decline after the spectroscopists demonstrated the gaseous nature of diffuse nebulae. However, the extragalactic interpretation of the spirals revived its popularity after Vesto M. Slipher discovered the large radial velocities of spiral nebulae (selection 102). As a student of nebulae, Curtis subscribed to this view, and he sought to establish the nature and distances of the spirals more securely by means of the novae that he and George W. Ritchey had independently found in them (selection 101).

Meanwhile, Harlow Shapley, in delineating his novel conception of a much larger Milky Way system, became increasingly reluctant to accept the spirals as comparable systems. After Curtis presented this paper to the Washington Academy of Sciences, he became well known as the leading spokesman for the island universe theory. The two men met in a face-to-face confrontation in the now famous Shapley-Curtis debate at the April 1920 meeting of the National Academy of Sciences (see selection 79). In that encounter Shapley generally limited his discussion to his views on the structure of our own galaxy, which have proved to be essentially correct; his objections to the island universe theory were only briefly stated.

Shapley's most cogent argument against the extragalactic nature of the spiral nebulae derived from recent observations of their proper motions and rotations. In 1916 Adriaan van Maanen had found annual motions of about 0.02 sec of arc in the spiral nebula M 101;[1] if M 101 is at a great distance, these angular motions correspond to absurdly high linear speeds. In the present paper, Curtis remarks in passing that his own data failed to reveal any large proper motions, and in his 1920 rebuttal to Shapley, he voiced his skepticism concerning van Maanen's measurements. Van Maanen continued collecting data, always obtaining the same general results for a variety of spirals. In 1923 Knut Lundmark made independent measurements with the same plates at Mount Wilson Observatory;[2] he, too, found systematic motions, but an order of magnitude smaller than van Maanen reported. In 1935, however, Edwin Hubble again remeasured the plates without finding any systematic motions.[3] The source of van Maanen's spurious results is not fully explained today. But the question of the extragalactic nature of the nebulae had already been settled, at least for the insiders, by 1924, when Hubble used observations of Cepheid variables in two spiral nebulae to show that they must be outside our galaxy (selection 104).

1. A. van Maanen, *Astrophysical Journal 44*, 210 (1916), the first of a long series of papers in this journal.
2. K. Lundmark, *Studies of Anagalactic Nebulae* (Uppsala: Royal Society of Science of Uppsala, 1927); note especially plate IV.
3. E. Hubble, *Astrophysical Journal 81*, 334 (1935).

In one sense, that theory of the spiral nebulae to which many lines of recently obtained evidence are pointing, can not be said to be a modern theory. There are few modern concepts which have not been explicitly or implicitly put forward as hypotheses or suggestions long before they were actually substantiated by evidence.

The history of scientific discovery affords many instances where men with some strange gift of intuition have looked ahead from meager data, and have glimpsed or guessed truths which have been fully verified only after the lapse of decades or centuries. Herschel was such a fortunate genius. From the proper motions of a very few stars he determined the direction of the sun's movement nearly as accurately, due to a very happy selection of stars for the purpose, as far more elaborate modern investigations. He noticed that the star clusters which appeared nebulous in texture in smaller telescopes and with lower powers, were resolved into stars with larger instruments and higher powers. From this he argued that all the nebulae could be resolved into stars by the application of sufficient magnifying power, and that the nebulae were, in effect, separate universes, a theory which had been earlier suggested on purely hypothetical or philosophical grounds, by Wright, Lambert, and Kant. From their appearance in the telescope he, again with almost uncanny prescience, excepted a few as definitely gaseous and irresolvable.

This view held sway for many years; then came the results of spectroscopic analysis showing that many nebulae (those which we now classify as diffuse or planetary) are of gaseous constitution and can not be resolved into stars. The spiral nebulae, although showing a different type of spectrum, were in most theories tacitly included with the known gaseous nebulae.

We have now, as far as the spiral nebulae are concerned, come back to the standpoint of Herschel's fortunate, though not fully warranted deduction, and the theory to which much recent evidence is pointing, is that these beautiful objects are separate galaxies, or "island universes," to employ the expressive and appropriate phrase coined by Humboldt.

By means of direct observations on the nearer and brighter stars, and by the application of statistical methods to large groups of the fainter or more remote stars, the galaxy of stars which forms our own stellar universe is believed to comprise perhaps a billion suns. Our sun, a relatively inconspicuous unit, is situated near the center of figure of this galaxy. This galaxy is not even approximately spherical in contour, but shaped like a lens or thin watch; the actual dimensions are highly uncertain; Newcomb's estimate that this galactic disk is about 3,000 light-years in thickness, and 30,000 light-years in diameter, is perhaps as reliable as any other.

Of the three classes of nebulae observed, two, the diffuse nebulosities and the planetary nebulae, are typically a galactic phenomenon as regards their apparent distribution in space, and are rarely found at any distance from the plane of our Milky Way. With the exception of certain diffuse nebulosities whose light is apparently a reflection phenomenon from bright stars involved within the nebulae, both these types are of gaseous constitution, showing a characteristic bright-line spectrum.

Differing radically from the galactic gaseous nebulae in form and distribution, we find a very large number of nebulae predominantly spiral in structure. The following salient points must be taken into account in any adequate theory of the spiral nebulae.

1. In apparent size the spirals range from minute flecks, just distinguishable on the photographic plate, to enormous spirals like Messier 33 and the Great Nebula in Andromeda, the latter of which covers an area four times greater than that subtended by the full moon.

2. Prior to the application of photographic methods, fewer than ten thousand nebulae of all classes had been observed visually. One of the first results deduced by Director Keeler from the program of nebular photography which he inaugurated with the Crossley Reflector at Lick Observatory, was the fact that great numbers of small spirals are within reach of modern powerful reflecting telescopes. He estimated their total number as 120,000 early in the course of this program, and before plates of many regions were available. I have recently made a count of the small nebulae on all available regions taken at the Lick Observatory during the past twenty years[1] and from these counts estimate that there are at least 700,000 spiral nebulae accessible with large reflectors.

3. The most anomalous and inexplicable feature of the spiral nebulae is found in their peculiar distribution. They show an apparent abhorrence for our galaxy of stars, being found in greatest numbers around the poles of our galaxy. In my counts I found an approximate density of distribution as follows:

Galactic Latitude + 45° to + 90°	34 per square degree.
Galactic Latitude − 45° to − 90°	28 per square degree.
Galactic Latitude + 30° to + 45° and − 30° to − 45°	24 per square degree.
Galactic Latitude − 30° to + 30°	7 per square degree.

No spiral has as yet been found actually within the structure of the Milky Way. We have doubled and trebled our exposures in regions near the galactic plane in the hope of finding fainter spirals in such areas, but thus far without results. The outstanding feature of the space distribution of the spirals is, then, that they are found in greatest profusion where the stars are fewest, and do not occur where the stars are most numerous.

4. The spectrum of the spirals is practically the same as that given by a star cluster, showing a continuous spectrum broken by absorption lines. A few spirals show bright-line spectra in addition.

5. The space-velocities of the various classes of celestial objects are summarized in the following list:

    1. *The Diffuse Nebulae.*
       Velocities low.
    2. *The Stars.*
       Velocities vary with spectral type.
       Class B Stars: average speeds 8 miles per second.
       Class A Stars: average speeds 14 miles per second.
       Class F Stars: average speeds 18 miles per second.
       Class G Stars: average speeds 19 miles per second.
       Class K Stars: average speeds 21 miles per second.
       Class M Stars: average speeds 21 miles per second.
    3. *The Star Clusters.*
       Velocities unknown.
    4. *The Planetary Nebulae.*
       Average speeds 48 miles per second.
    5. *The Spiral Nebulae.*
       Average speeds 480 miles per second.

The peculiar variation of the space-velocity of the stars with spectral type may ultimately prove to be a function of relative mass. The radial velocities of but few spirals have been determined to date; future work may change the value given, but it seems certain that it will remain very high.

It will be seen at once that, with regard to this important criterion of space-velocity, the spiral nebulae are very distinctly in a class apart. It seems impossible to place them at any point in a coherent scheme of stellar evolution. We can not bridge the gap involved in postulating bodies of such enormous space velocities either as a point of stellar origin, or as a final evolution product.

On the older theory that the spirals are a part of our own galaxy, it is impossible to harmonize certain features of the data thus far presented. If this theory is true, their grouping near the galactic poles, inasmuch as all evidence points to a flattened or disk form for our galaxy, would indicate that they are relatively close to us. In that event, we should inevitably have detected in this class of objects proper motions of the same order of magnitude as those found for the stars at corresponding distances. Such proper motions are the more to be expected in view of the fact that the average space velocity of the spirals is about thirty times that of the stars. I have repeated all the earlier plates of the Keeler nebular program, and was able to find no certain evidence of either translation or rotation in these objects in an average time interval of thirteen years.[2] Their form, and the evidence of the spectroscope, indicate, however, that they are in rotation. Knowing that their space-velocities are high, the failure to detect any certain evidence of cross motion is an indication that these objects must be very remote.

Even if the spiral is not a stage in stellar evolution, but a class apart, is it still possible to assume that they are, notwithstanding, an integral part of our own stellar universe, sporadic manifestations of an unknown line of evolutionary development, driven off in some mysterious manner from the regions of greatest star density?

A relationship between two classes of objects may be one of avoidance just as logically as one of continuity. It has been argued that the absolute avoidance which the spirals manifest for the galaxy of the stars shows incontrovertibly that they must, by reason of this very relationship of avoidance, be an integral feature of our galaxy. This argument has proved irresistible to many, among others to so keen a thinker as Herbert Spencer, who wrote:

> In that zone of celestial space where stars are excessively abundant nebulae are rare; while in the two opposite celestial spaces that are furthest removed from this zone nebulae are abundant . . . Can this be mere coincidence? When to the fact that the general mass of the nebulae are antithetical in position to the general mass of the stars, we add the fact that local regions of nebulae are regions where stars are scarce . . . does not the proof of a physical connection become overwhelming?

It must be admitted that a distribution, which has placed three-quarters of a million objects around the poles of our galaxy, would be against all probability for a class of objects which would be expected to be arranged at random, unless it can be shown that this peculiar grouping is only apparent and due to some phenomenon in our own galaxy. This point will be reverted to later.

It has been shown that the factors of space-velocity and space-distribution separate the spirals very clearly from the stars of our galaxy; from these facts alone, and from the evidence of the spectroscope, the island universe theory is given a certain measure of credibility.

Another line of evidence has been developed within the past two years, which adds further support to the island-universe theory of the spiral nebulae.

New Stars

Within historical times some twenty-seven new stars have suddenly flashed out in the heavens. Some have been of interest only to the astronomer; others, like that of last June, have rivaled Sirius in brilliancy. All have shown the same general history, suddenly increasing in light ten thousand-fold or more, and then gradually, but still relatively rapidly, sinking into obscurity again. They are a very interesting class, nor has astronomy as yet been able to give any universally accepted explanation of these anomalous objects. Two of these novae had appeared in spiral nebulae, but this fact had not been weighed at its true value. Within the past two years over a dozen novae have been found in spiral nebulae, all of them very faint, ranging from about the fourteenth to the nineteenth magnitudes at maximum. Their life history, so far as we can tell from such faint objects, appears to be identical with that of the brighter novae. Now the brighter novae of the past, that is, those which have not appeared in spirals, have almost invariably been a galactic phenomenon, located in or close to our Milky Way, and they have very evidently been a part of our own stellar system. The cogency of the argument will, I think, be apparent to all, although the strong analogy is by no means a rigid proof. If twenty-seven novae have appeared in our own galaxy within the past three hundred years, and if about half that number are found within a few years in spiral nebulae far removed from the galactic plane, the presumption that these spirals are themselves galaxies composed of hundreds of millions of stars is a very probable one.

If, moreover, we make the reasonable assumption that the new stars in the spirals and the new stars in our own galaxy average about the same in size, mass, and absolute brightness, we can form a very good estimate of the probable distance of the spiral nebulae, regarded as island universes. Our galactic novae have averaged about the fifth magnitude. The new stars which have appeared in the spiral nebulae have averaged about the fifteenth magnitude, but it would appear probable that we must inevitably miss the fainter novae in such distant galaxies, and it is perhaps reasonable to assume that the average magnitude of the novae in spirals may be about the eighteenth, or thirteen magnitudes fainter than those in our own galaxy. They would thus be about 160,000 times fainter than our galactic novae, and on the assumption that both types of novae average the same in mass, absolute luminosity, etc., the novae in spirals should be four hundred times further away. We do not know the average distance of the new stars which have appeared in our own galaxy, but 100,000 light-years is perhaps a reasonable estimate. This would indicate a distance of the order of 4,000,000 light-years for the spiral nebulae. This is an enormous distance, but, if these objects are galaxies like our own stellar system, such a distance accords well with their apparent dimensions. Our own galaxy, at a distance of 10,000,000 light-years, would be about 10 minutes of arc in diameter, or the size of the larger spiral nebulae.

On such a theory, a spiral structure for our own galaxy would be probable. Its proportions accord well with the degree of flattening observed in the majority of the spirals. We have very little actual evidence as to a spiral structure for our galaxy; the position of our sun relatively close to the center of figure of the galaxy, and our ignorance of the distances of the remoter stars, renders such evidence very difficult to obtain. A careful study of the configurations and star densities in the Milky Way has led Professor Easton, of Amsterdam, to postulate a spiral structure for our galaxy.

Distribution of Spirals

There is still left one outstanding and unexplained problem in the island universe theory or any other theory of the spiral nebulae. Neither theory, as outlined, offers any satisfactory explanation of the remarkable distribution of the spirals. On the older theory, if a feature of our galaxy, what has driven them out to the points most remote from the regions of greatest star density? If, on the other hand, the spirals are island universes, it is against all probability that our own universe should have chanced to be situated about half way between two great groups of island universes, and that not a single object of the class happens to be located in the plane of our Milky Way.

There is one very common characteristic of the spirals which may be tentatively advanced as an explanation of the peculiar grouping of the spirals.

A very considerable proportion of the spirals show indubitable evidence of occulting matter, lying in the plane of the greatest extension of the spiral, generally outside the whorls, but occasionally between the whorls as well. This outer ring of occulting matter is most easily seen when the spiral is so oriented in space as to turn its edge toward us. But the phenomenon is also seen in spirals whose planes make a small, but appreciable angle with our line of sight, manifesting itself in such appearances as "lanes" more prominent on one side of the major axis of the elongated elliptical projection, in a greater brightness of the nebular matter on one side of this major axis, in a fan-shaped nuclear portion, or in various combinations of these effects. The phenomenon is a very common one. Illustrations of seventy-eight spirals showing evidences of occulting matter in their peripheral equatorial regions, with a more detailed discussion of the forms observed, are now being published,[3] and additional examples of the phenomenon are constantly being found.

While we have as yet no definite proof of the existence of such a ring of occulting matter lying in our galactic plane and outside of the great mass of the stars of our galaxy, there is a great deal of evidence for such occulting matter in smaller areas in our galaxy. Many such dark areas are observed around certain of the diffuse nebulosities, or seen in projection on the background furnished by such nebulosities or the denser portions of the Milky Way; these appearances seem to be actual "dark nebulae."[4] The curious "rifts" in the Milky Way may well be ascribed, at least in part, to such occulting matter.

Though we thereby run the risk of arguing in a circle, the fact that no spirals can be detected in our galactic plane, a natural result of such a ring of occulting matter, would in itself appear to lend some probability to the hypothesis. The peculiar distribution of the spiral nebulae would then be explained as due, not to an actual asymmetrical and improbable distribution in space, but to a cause within our own galaxy, assumed to be a spiral with a peripheral ring of occulting matter similar to that observed in a large proportion of the spirals. The argument that the spirals must be an integral feature of our own galaxy, based on a relationship of avoidance, would then lose its force. The explanation appears to be a possibility, even a strong probability, on the island universe theory, and I know of no other explanation, on any theory, for the observed phenomenon of nebular distribution about our galactic poles.

Summary

The Spiral Nebulae as Island Universes.

1. On this theory, it is unnecessary to attempt to coordinate the tremendous space-velocities of the spirals with the thirty-fold smaller values found for the stars. Very high velocities have been found for the Magellanic Clouds, which may possibly be very irregular spirals, relatively close to our galaxy.
2. There is some evidence for a spiral structure in our own galaxy.
3. The spectrum of the majority of the spirals is practically identical with that given by a star cluster; a spectrum of this general type is such as would be expected from vast congeries of stars.
4. If the spirals are separate universes, similar to our galaxy in extent and in number of component stars, we should observe many new stars in the spirals, closely resembling in their life history the twenty-seven novae which have appeared in our own galaxy. Over a dozen such novae in spirals have been found, and it is probable that a systematic program of repetition of nebular photographs will add greatly to this number. A comparison of the average magnitudes of the novae in spirals with those of our own galaxy indicates a distance of the order of 10,000,000 light-years for the spirals. Our own galaxy at this distance would appear 10′ in diameter, the size of the larger spirals.
5. A considerable proportion of the spirals show a peripheral equatorial ring of occulting matter. So many instances of this have been found that it appears to be a general though not universal characteristic of the spirals; the existence of such an outer ring of occulting matter in our own galaxy, regarded as a spiral, would furnish an adequate explanation of the peculiar distribution of the spirals. There is considerable evidence of such occulting matter in our galaxy.

An English physicist has cleverly said that any really good theory brings with it more problems than it removes. It is thus with the island-universe theory. It is impossible to do more than to mention a few of these problems, with no attempt to divine those which may ultimately be presented to us.

While the data are too meager as yet, several attempts have been made to deduce the velocity of our own galaxy within the super-galaxy. It would not be surprising if the space-velocity of our galaxy, like those of the spirals and the Magellanic Clouds, should prove to be very great, hundreds of miles per second.

Further, what are the laws which govern the forms assumed, and under which these spiral whorls are shaped? Are they stable structures; are the component stars moving inward or outward? A beginning has been made by Jeans and other mathematicians on the dynamical problems involved in the structure of the spirals. The field for research is, like our subject matter, practically infinite.

1. H. D. Curtis, "On the Number of Spiral Nebulae," *Proc. Amer. Phil. Soc. 57*, 513 (1918).

2. H. D. Curtis, "The Proper Motion of the Nebulae," *Publ. Astron. Soc. Pacific 27*, 214 (1915).

3. H. D. Curtis, "A Study of Occulting Matter in Spiral Nebulae," Part 2 of *Studies of the Nebulae*, *Publications of the Lick Observatory 13*, 43 (1918).

4. E. E. Barnard, "On the Dark Markings of the Sky, with a Catalogue of 182 Such Objects," *Astrophys. Journ. 49*, 1 (1919); H. D. Curtis, "Dark Nebulae," *Publ. Astron. Soc. Pacific 30*, 65 (1918).

104. Cepheids in Spiral Nebulae

Edwin P. Hubble

(*Publications of the American Astronomical Society 5*, 261–264 [1925]; reprinted in *Observatory 48*, 139–142 [1925])

On New Year's Day 1925 the great debate over the extragalactic nature of spiral nebulae was finally settled. In the following paper, presented *in absentia* to the joint meeting of the American Astronomical Society and the American Association for the Advancement of Science in Washington, D.C., Edwin Hubble announced the discovery of Cepheid variable stars in the Andromeda nebula (M 31) and the great spiral in Triangulum (M 33); using Shapley's calibration for the period-luminosity relation,[1] Hubble derived a distance of about 285,000 pc for the two nebulae. Hubble's announcement shared the prize for the best paper at the AAAS meeting, but it did not come as a surprise to the leading astronomers, who had been informed by correspondence months earlier.

John Duncan had earlier reported the discovery of three variable stars in M 33, but their nature was in doubt until Hubble took enough plates to find others and to establish that at least some were Cepheid variables following the famous period-luminosity relation. The distance Hubble found, about 285,000 pc, meant that the nebulae must lie outside our galaxy, for Shapley had already used the same calibrated period-luminosity relation to derive a maximum extent of 100,000 pc for our Milky Way (selection 79).

Although Ernst Öpik's contribution to this problem has often been overlooked, he deserves at least partial credit for the proof that spiral nebulae are extragalactic. In a paper published in 1922, Öpik used F. G. Pease's measurements of the rotation velocities of M 31 to show that its distance has to be about 480,000 pc if its mass to luminosity ratio is comparable to that of the stars in our galaxy.[2] Öpik was much more nearly correct than Hubble in estimating the distance to M 31, because, as Walter Baade showed nearly three decades later, Shapley's period-luminosity relation had been erroneously calibrated (selection 110).

1. *Astrophysical Journal 48*, 89 (1918).
2. E. Öpik, *Astrophysical Journal 55*, 406 (1922); F. G. Pease, *Proceedings of the National Academy of Sciences 4*, 21 (1918).

Messier 31 and 33, the only spirals that can be seen with the naked eye, have recently been made the subject of detailed investigations with the 100-inch and 60-inch reflectors of the Mount Wilson Observatory. Novae are a common phenomenon in M 31, and Duncan has reported three variables within the area covered by M 33[1]. With these exceptions there seems to have been no definite evidence of actual stars involved in spirals. Under good observing conditions, however, the outer regions of both spirals are resolved into dense swarms of images in no way differing from those of ordinary stars. A survey of the plates made with the blink-comparator has revealed many variables among the stars, a large proportion of which show the characteristic light-curve of the Cepheids.

Up to the present time some 47 variables, including Duncan's three, and one true nova have been found in M 33. For M 31, the numbers are 36 variables and 46 novae, including the 22 novae previously discovered by Mount Wilson observers. Periods and photographic magnitudes have been determined for 22 Cepheids in M 33 and 12 in M 31. Others of the variables are probably Cepheids, judging from their sharp rise and slow decline, but some are definitely not of this type. One in particular, Duncan's No. 2 in M 33, has been brightening fairly steadily with only minor fluctuations since about 1906. It has now reached the 15th magnitude and has a spectrum of the bright line B type.

For the determinations of periods and normal curves of the Cepheids, 65 plates are available for M 33, and 130 for M 31. The later object is too large for the area of good definition on one plate, so attention has been concentrated on three regions: around BD + 41°151, BD + 40°145, and a region some 45′ along the major axis south preceding the nucleus.

Photographic magnitudes have been determined from twelve comparisons with selected areas No. 21 and 45, made with the 100-inch using exposures from 30 to 40 minutes. This procedure seemed preferable to the much longer exposures required for direct polar comparisons with the 60-inch. It involves, however, a considerable extrapolation based on scales determined from the faintest magnitudes available for the selected areas.

Table 104.1 Cepheids in M 33

Var. no.	Period in days	Log. P	Photographic magnitudes	
			Max.	Min.
30	46.0	1.66	18.35	19.25
3	41.6	1.62	18.45	19.4
36	38.2	1.58	18.45	19.1
31	37.3	1.57	18.30	19.2
29	37.2	1.57	18.55	19.15
20	35.95	1.56	18.50	19.2
18	35.5	1.55	18.45	19.15
35	31.5	1.50	18.55	19.35
42	31.1	1.49	18.65	19.35
44	30.2	1.48	18.70	
40	26.0	1.41	19.00	
17	23.6	1.37	18.80	
11	23.4	1.37	18.85	
22	21.75	1.34	19.00	
12	21.2	1.33	18.80	
27	21.05	1.32	18.85	
43	20.8	1.32	18.95	
33	20.8	1.32	18.75	
10	19.6	1.29	18.80	
41	19.15	1.28	18.75	
37	18.05	1.26	18.95	
15	17.65	1.25	19.05	

Table 104.2 Cepheids in M 31

Var. no.	Period in days	Log. P	Photographic magnitude Max.
5	50.17	1.70	18.4
7	45.04	1.65	18.15
16	41.14	1.61	18.6
9	38	1.58	18.3
1	31.41	1.50	18.2
12	22.03	1.34	19.0
13	22	1.34	19.0
10	21.5	1.33	18.75
2	20.10	1.30	18.5
17	18.77	1.28	18.55
18	18.54	1.27	18.9
14	18	1.26	19.1

Tables 104.1 and 104.2 give the data for the Cepheids in M 33 and M 31 respectively. No magnitudes fainter than 19.5 are recorded, because of the uncertainty involved in their precise determinations. The now familiar period-luminosity relation is conspicuously present.

For more detailed investigation of the relation, the magnitudes at maxima have been plotted against the logarithm of the period in days. This procedure is necessary, not only because of the uncertainties in the fainter magnitudes, but also because most of the fainter variables at minimum are below the limiting magnitude of the plates. It assumes that there is no relation between period and range, for otherwise a systematic error in the slope of the period-luminosity curve is introduced. Among the brighter Cepheids of M 33 the assumption appears to be allowable, for the ranges show a very small dispersion about the mean value of 0.8 magnitude. The average range and the dispersion are somewhat larger in M 31, but the data are too limited for a complete investigation.

The curve for M 33 appears to be very definite. The average deviation is about 0.1 magnitude, although a considerable systematic error is allowable in the slope. For M 31 the slope is very closely the same but the dispersion is much greater, averaging about 0.2 magnitude. This is probably greater than the accidental errors of measurement.

Shapley's period-luminosity curve[2] for Cepheids, as given in his study of globular clusters, is constructed on a basis of visual magnitudes. It can be reduced to photographic magnitudes by means of his relation between period and color-index, given in the same paper, and the result represents his original data. The slope is of the order of that for the spirals, but is not precisely the same. In comparing the two, greater weight must be given the brighter portion of the curve for the spirals, because of the greater reliability of the magnitude determinations. When this is done, the resulting values of M—m are -21.8 and -21.9 for M 31 and M 33, respectively. These must be corrected by half the average ranges of the Cepheids in the two spirals, and the final values are then on the order of -22.3 for both nebulae. The corresponding distance is about 285,000 parsecs. The greater uncertainty is probably in the zero point of Shapley's curve.

The results rest on three major assumptions: (1) The variables are actually connected with the spirals. (2) There is no serious amount of absorption due to amorphous nebulosity in the spirals. (3) The nature of Cepheid variation is uniform throughout the observable portion of the universe. As for the first, besides the weighty arguments based on analogy and probability, it may be mentioned that no Cepheids have been found on the several plates of the neighboring selected areas No. 21 and 45, on a special series of plates centered on BD + 35°207, just midway between the two spirals, nor in ten other fields well distributed in galactic latitude, for which six or more long exposures are available. The second assumption is very strongly supported by the small dispersion in the period-luminosity curve for M 33. In M 31, in spite of the somewhat larger dispersion, there is no evidence of an absorption effect to be measured in magnitudes.

These two spirals are not unique. Variables have also been found in M 81, M 101 and N.G.C. 2403, although as yet sufficient plates have not been accumulated to determine the nature of their variation.

1. *Publications of the Astronomical Society of the Pacific 35*, 290 (1922).
2. *Mt. Wilson Contribution No. 151, Astrophysical Journal 48*, 89 (1918).

105. Extra-Galactic Nebulae

Edwin P. Hubble

(*Astrophysical Journal 64*, 321–369 [1926])

By exploiting the light-gathering power of the giant reflectors on Mount Wilson, Edwin Hubble was able to present this thorough summary of the extragalactic nebulae soon after he had established their extragalactic nature. He here introduces a morphological classification, based upon the forms of the photographic images, which with only small modifications still remains the standard.[1] In this classification system galaxies are divided into three broad types: ellipticals, spirals, and irregulars. The ellipticals themselves are subdivided into classes E0 to E7, in which the numbers 0 to 7 denote a progressive degree of flattening of the projected image. The spiral galaxies are assigned to types Sa, Sb, or Sc. Although Hubble also viewed his scheme as an evolutionary sequence, it now seems unlikely that spiral galaxies are a later evolutionary state of elliptical galaxies.

Hubble found that the various morphological types are uniformly distributed across the sky and that their apparent luminosities and radial velocities are of the same order. The observed numbers of extragalactic nebulae of different magnitudes agree with those computed on the assumption of a uniform distribution of extragalactic objects with a constant absolute luminosity. According to Hubble, the mean absolute magnitude is of the order of -15 and the mean nebular density of the order of 10^{-17} nebulae pc^{-3}; thus, the mean distance between nebulae was of the order of 570,000 pc.

By assuming that the masses of all nebulae have the same constant proportionality to their absolute luminosity, that they all have the same absolute luminosity, and with a mean nebular mass of 2.61×10^8 solar masses, Hubble obtained a mass density of 10^{-31} g cm^{-3}. Eight years later he revised these estimates to give a mean nebular mass of 10^9 solar masses and a mass density of 10^{-30} g cm^{-3}. The mass density of optically visible galaxies was therefore vastly smaller than the critical mass density, ρ_c, required to stop the expansion of the universe. Using Hubble's value for the recession factor of 550 km sec^{-1} Mpc^{-1}, we have $\rho_c = 3\,H_o^2/(8\pi G) = 6 \times 10^{-28}$ g cm^{-3}.

Although we now know that Hubble's constant is much smaller and that the nebulae are generally more massive (between 10^{10} and 10^{12} solar masses) than Hubble assumed, his general conclusion that there is insufficient mass in optically visible galaxies to stop the expansion of the universe seems still to hold today. In 1958, for example, Jan Oort revised Hubble's procedure for estimating the mass density of galaxies by taking into account the dispersion in their absolute luminosities.[2] Using this procedure he obtained a mass density of 3.1×10^{-31} g cm^{-3}, which is near the value of 0.5×10^{-31} g cm^{-3} currently accepted. Both values are a factor of 10 smaller than the presently accepted value of the critical mass density needed to close the universe.

1. A historical review of the Hubble classification scheme is given by R. Hart and R. Berendzen in the *Journal of the History of Astronomy 2*, 109 (1971). Photographic reproductions of Hubble's morphological sequence are found in *The Hubble Atlas of Galaxies*, ed. A. Sandage (Washington, D.C.: Carnegie Institution of Washington, 1961).

2. J. H. Oort, "Distribution of Galaxies and the Density of the Universe," in *La structure et l'évolution de l'univers*, Institut international de physique Solvay, ed. R. Stoops (Brussels: Coudenberg, 1958).

Abstract—This contribution gives the results of a statistical investigation of 400 extra-galactic nebulae for which Holetschek has determined total visual magnitudes. The list is complete for the brighter nebulae in the northern sky and is representative to 12.5 mag. or fainter.

The classification employed is based on the forms of the photographic images. About 3 per cent are irregular, but the remaining nebulae fall into a sequence of type forms characterized by rotational symmetry about dominating nuclei. The sequence is composed of two sections, the elliptical nebulae and the spirals, which merge into each other.

Luminosity relations—The distribution of magnitudes appears to be uniform throughout the sequence. For each type or stage in the sequence, the total magnitudes are related to the logarithms of the maximum diameters by the formula,

$$m_T = C - 5 \log d,$$

where C varies progressively from type to type, indicating a variation in diameter for a given magnitude or vice versa. By applying corrections to C, the nebulae can be reduced to a standard type and then a single formula expresses the relation for all nebulae from the Magellanic Clouds to the faintest that can be classified. When the minor diameter is used, the value of C is approximately constant throughout the entire sequence. The coefficient of $\log d$ corresponds with the inverse-square law, which suggests that the nebulae are all of the same order of absolute luminosity and that apparent magnitudes are measures of distance. This hypothesis is supported by similar results for the nuclear magnitudes and the magnitudes of the brightest stars involved, and by the small range in luminosities among nebulae whose distances are already known.

Distances and absolute dimensions. The mean absolute visual magnitude, as derived from the nebulae whose distances are known, is -15.2. The statistical expression for the distance in parsecs is then

$$\log D = 4.04 + 0.2\, m_T,$$

where m_T is the total apparent magnitude. This leads to mean values for absolute dimensions at various stages in the sequence of types. Masses appear to be of the order of 2.6×10^8 $\odot$.

Distribution and density of space. To apparent magnitude about 16.7, corresponding to an exposure of one hour on fast plates with the 60-inch reflector, the numbers of nebulae to various limits of total magnitude vary directly with the volumes of space represented by the limits. This indicates an approximately uniform density of space, of the order of one nebula per 10^{17} cubic parsecs or 1.5×10^{-31} in C.G.S. units. The corresponding radius of curvature of the finite universe of general relativity is of the order of 2.7×10^{10} parsecs, or about 600 times the distance at which normal nebulae can be detected with the 100-inch reflector.

RECENT STUDIES HAVE EMPHASIZED the fundamental nature of the division between galactic and extra-galactic nebulae. The relationship is not generic; it is rather that of the part to the whole. Galactic nebulae are clouds of dust and gas mingled with the stars of a particular stellar system; extra-galactic nebulae, at least the most conspicuous of them, are now recognized as systems complete in themselves, and often incorporate clouds of galactic nebulosity as component parts of their organization. Definite evidence as to distances and dimensions is restricted to six systems, including the Magellanic Clouds. The similar nature of the countless fainter nebulae has been inferred from the general principle of the uniformity of nature.

The extra-galactic nebulae form a homogeneous group in which numbers increase rapidly with diminishing apparent size and luminosity. Four are visible to the naked eye;[1] 41 are found on the Harvard "Sky Map";[2] 700 are on the Franklin-Adams plates;[3] 300,000 are estimated to be within the limits of an hour's exposure with the 60-inch reflector.[4] These data indicate a wide range in distance or in absolute dimensions. The present paper, to which is prefaced a general classification of nebulae, discusses such observational material as we now possess in an attempt to determine the relative importance of these two factors, distance and absolute dimensions, in their bearing on the appearance of extra-galactic nebulae.

The classification of these nebulae is based on structure, the individual members of a class differing only in apparent size and luminosity. It is found that for the nebulae in each class these characteristics are related in a manner which closely approximates the operation of the inverse-square law on comparable objects. The presumption is that dispersion in absolute dimensions is relatively unimportant, and hence that in a statistical sense the apparent dimensions represent relative distances. The relative distances can be reduced to absolute values with the aid of the nebulae whose distances are already known.

Part I. Classification of Nebulae

GENERAL CLASSIFICATION The classification used in the present investigation is essentially the detailed formulation of a preliminary classification published in a previous paper.[5] It was developed in 1923, from a study of photographs of several thousand nebulae, including practically all the brighter objects and a thoroughly representative collection of the fainter ones.[6] It is based primarily on the structural forms of photographic images, although the forms divide themselves naturally into two groups: those found in or near the Milky Way and those in moderate or high galactic latitudes. In so far as possible, the system is independent of the orientation of the objects in space. With minor changes in the original notation, the complete classification is as follows, although only the extra-galactic division is here discussed in detail:

	Symbol	Example
I. Galactic nebulae:		
A. Planetaries	P	N.G.C. 7662
B. Diffuse	D	—
1. Predominantly luminous	DL	N.G.C. 6618
2. Predominantly obscure	DO	Barnard 92
3. Conspicuously mixed	DLO	N.G.C. 7023
II. Extra-galactic nebulae:		
A. Regular:		
1. Elliptical ($n = 1, 2, \ldots, 7$ indicates the ellipticity of the image without the decimal point)	En	N.G.C. 3379 E0 221 E2 4621 E5 2117 E7
2. Spirals:		
a) Normal spirals	S	—
(1) Early	Sa	N.G.C. 4594
(2) Intermediate	Sb	2841
(3) Late	Sc	5457
b) Barred spirals	SB	—
(1) Early	SBa	N.G.C. 2859
(2) Intermediate	SBb	3351
(3) Late	SBc	7479
B. Irregular	Irr	N.G.C. 4449

Extra-galactic nebulae too faint to be classified are designated by the symbol "Q."

REGULAR NEBULAE The characteristic feature of extra-galactic nebulae is rotational symmetry about dominating non-stellar nuclei. About 97 per cent of these nebulae are regular in the sense that they show this feature conspicuously. The regular nebulae fall into a progressive sequence ranging from globular masses of unresolved nebulosity to widely open spirals whose arms are swarming with stars. The sequence comprises two sections, elliptical nebulae and spirals, which merge into each other.

Although deliberate effort was made to find a descriptive classification which should be entirely independent of theoretical considerations, the results are almost identical with the path of development derived by Jeans[7] from purely theoretical investigations. The agreement is very suggestive in view of the wide field covered by the data, and Jeans's theory might have been used both to interpret the observations and to guide research. It should be borne in mind, however, that the basis of the classification is descriptive and entirely independent of any theory.

Elliptical Nebulae These give images ranging from circular through flattening ellipses to a limiting lenticular figure in which the ratio of the axes is about 1 to 3 or 4. They show no evidence of resolution,[8] and the only claim to structure is that the luminosity fades smoothly from bright nuclei to indefinite edges. Diameters are functions of the nuclear brightness and the exposure times.

The only criterion available for further classification appears to be the degree of elongation. Elliptical nebulae have accordingly been designated by the symbol "E," followed by a single figure, numerically equal to the ellipticity $(a - b)/a$ with the decimal point omitted. The complete series is E0, E1, . . . , E7, the last representing a definite limiting figure which marks the junction with the spirals.

The frequency distribution of ellipticities shows more round or nearly round images than can be accounted for by the random orientation of disk-shaped objects alone. It is presumed, therefore, that the images represent nebulae ranging from globular to lenticular, oriented at random. No simple method has yet been established for differentiating the actual from the projected figure of an individual object, although refined investigation furnishes a criterion in the relation between nuclear brightness and maximum diameters. For the present, however, it must be realized that any list of nebulae having a given apparent ellipticity will include a number of tilted objects having greater actual ellipticities. The statistical average will be too low, except for E7, and the error will increase with decreasing ellipticity.

Normal Spirals All regular nebulae with ellipticities greater than about E7 are spirals, and no spirals are known with ellipticities less than this limit. At this point in the sequence, however, ellipticity becomes insensitive as a criterion and is replaced by conspicuous structural features which now become available for classification. Of these, practically speaking, there are three which fix the position of an object in the sequence of forms: (1) relative size of the unresolved nuclear region; (2) extent to which the arms are unwound; (3) degree of resolution in the arms. The form most nearly related to the elliptical nebulae has a large nuclear region similar to E7, around which are closely coiled arms of unresolved nebulosity. Then follow objects in which the arms appear to build up at the expense of the nuclear regions and unwind as they grow; in the end, the arms are wide open and the nuclei inconspicuous. Early in the series the arms begin to break up into condensations, the resolution commencing in the outer regions and working inward until in the final stages it reaches the nucleus itself. In the larger spirals where critical observations are possible, these condensations are found to be actual stars and groups of stars.

The structural transition is so smooth and continuous that the selection of division points for further classification is rather arbitrary. The ends of the series are unmistakable,

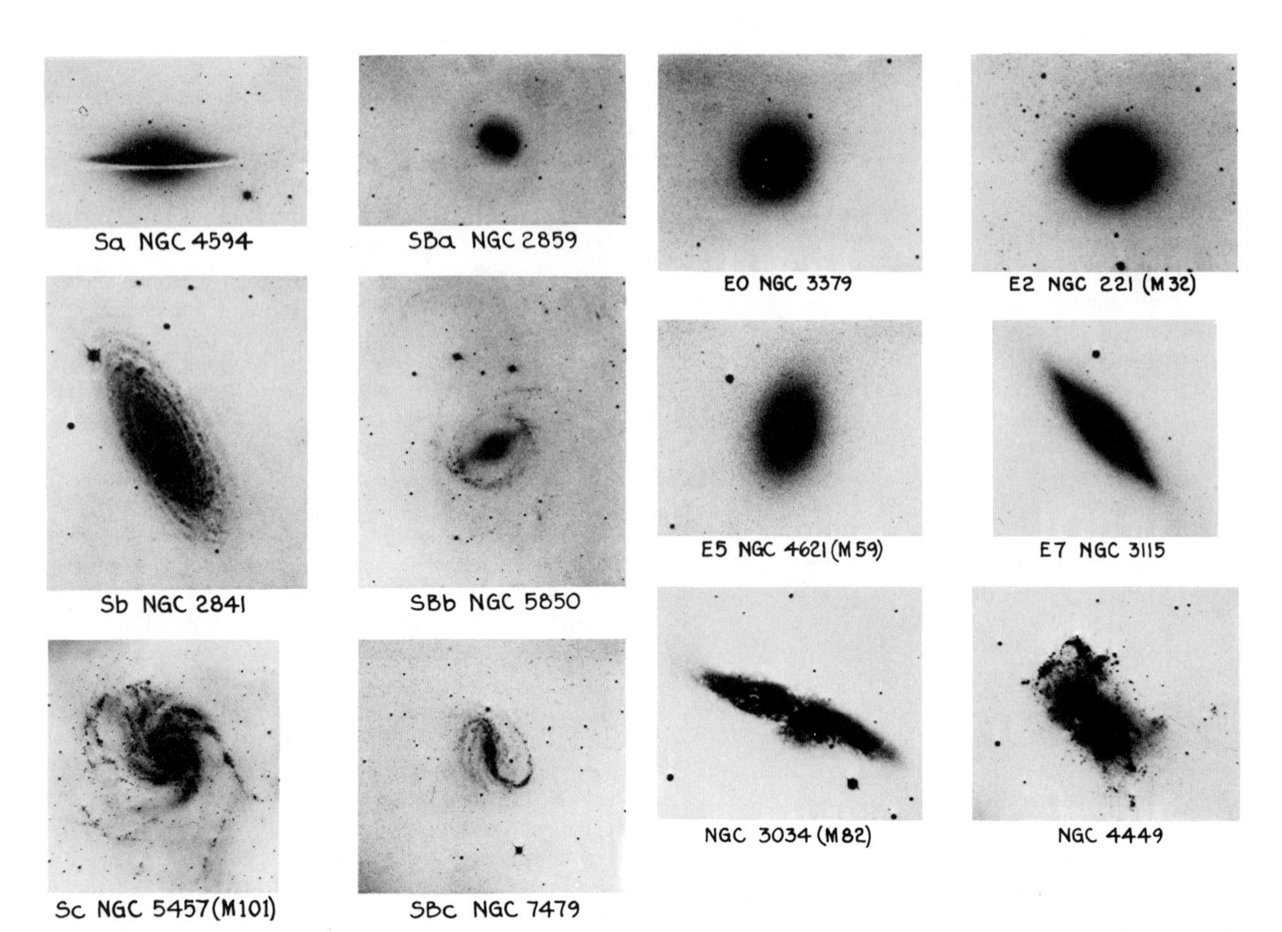

Fig. 105.1 Photographic images of normal spiral nebulae, S; barred spiral nebulae, SB; elliptical nebulae, E; and the irregular nebulae NGC 3034 and NGC 4449. The images illustrate Hubble's morphological sequence, in which the letters a, b, and c denote a progressive degree of openness in the spiral arms, with smaller nuclei, and the numbers 0 to 7 denote a progressive degree of flattening of elliptical nebulae. (Courtesy Hale Observatories.)

however, and, in a general way, it is possible to differentiate a middle group. These three groups are designated by the non-committal letters "a," "b," and "c" attached to the spiral symbols "S," and, with reference to their position in the sequence, are called "early," "intermediate," and "late" types.[9] A more precise subdivision, on a decimal scale for example, is not justified in the present state of our knowledge.

In the early types, the group Sa, most of the nebulosity is in the nuclear region and the arms are closely coiled and unresolved. N.G.C. 3368 and 4274 are among the latest of this group.

The intermediate group, Sb, includes objects having relatively large nuclear regions and thin rather open arms, as in M 81, or a smaller nuclear region with closely coiled arms, as in M 94. These two nebulae represent the lateral extension of the sequence in the intermediate section. The extension along the sequence is approximately represented by N.G.C. 4826, among the earliest of the Sb, and N.G.C. 3556 and 7331, which are among the latest. The resolution in the arms is seldom conspicuous, although in M 31, a typical Sb, it is very pronounced in the outer portions.

The characteristics of the late types, the group Sc, are more definite—an inconspicuous nucleus and highly resolved arms. Individual stars cannot be seen in the smaller nebulae of this group, but knots are conspicuous, which, in larger objects, are known to be groups and clusters of stars. The extent to which the arms are opened varies from M 33 to M 101, both typical Sc nebulae.

Barred Spirals In the normal spiral the arms emerge from two opposite points on the periphery of the nuclear region. There is, however, a smaller group, containing about 20 per cent of all spirals, in which a bar of nebulosity extends diametrically across the nucleus. In these spirals, the arms spring abruptly from the ends of this bar. These nebulae also form a sequence, which parallels that of the normal spirals, the arms apparently unwind, the nuclei dwindle, the condensations form and work inward.

H. D. Curtis[10] first called attention to these nebulae when he described several in the intermediate stages of the series and called them ϕ-type spirals. The bar, however, never extends beyond the inner spiral arms, and the structure, especially in the early portion of the sequence, is more accurately represented by the Greek letter θ. From a dynamical point of view, the distinction has considerable significance. Since Greek letters are inconvenient for cataloguing purposes, the English term, "barred spiral," is proposed, which can be contracted to the symbol "SB."

The SB series, like that of the normal spirals, is divided into three roughly equal sections, distinguished by the appended letters "a," "b," and "c." The criterion on which the division is based are similar in general to those used in the classification of the normal spirals. In the earliest forms, SBa, the arms are not differentiated, and the pattern is that of a circle crossed by a bar, or, as has been mentioned, that of the Greek letter θ. When the bar is oriented nearly in the line of sight, it appears foreshortened as a bright and definite minor axis of the elongated nebular image. Such curious forms as the images of N.G.C. 1023 and 3384 are explained in this manner. The latest group, SBc, is represented by the S-shaped spirals such as N.G.C. 7479.

IRREGULAR NEBULAE About 3 per cent of the extra-galactic nebulae lack both dominating nuclei and rotational symmetry. These form a distinct class which can be termed "irregular." The Magellanic Clouds are the most conspicuous examples, and indeed, are the nearest of all the extra-galactic nebulae. N.G.C. 6822, a curiously faithful miniature of the Clouds, serves to bridge the gap between them and the smaller objects, such as N.G.C. 4214 and 4449. In these latter, a few individual stars emerge from an unresolved background, and occasional isolated spots give the emission spectrum characteristic of diffuse nebulosity in the galactic system, in the Clouds, and in N.G.C. 6822.[11] These features are found in other irregular nebulae as well, notably in N.G.C. 1156 and 4656, and are just those to be expected in systems similar to the Clouds but situated at increasingly greater distances.

The system outlined above is primarily for the formal classification of photographic images obtained with large reflectors and portrait lenses. For each instrument, however, there is a limiting size and luminosity below which it is impossible to classify with any confidence. Except in rare instances, these small nebulae are extra-galactic, and their numbers, brightness, dimensions, and distribution are amenable to statistical investigation. For cataloguing purposes, they require a designating symbol, and the letter "Q" is suggested as convenient and not too widely used with other significations.

Part II. Statistical Study of Extra-Galactic Nebulae

In the following sections Hubble uses data for some 400 nebulae to infer a luminosity relation between the total apparent luminosities and the maximum angular diameters of the nebulae. He first notes that the various nebula types are homogeneously distributed across the sky with a roughly uniform distribution of apparent luminosities (the frequency distribution of apparent magnitudes is given in a later section). Among the nebulae of each separate type are found linear correlations between the total apparent magnitudes, m, and the logarithms of the maximum angular diameters, d. The correlations are expressed by the luminosity relation $m = C - 5 \log d$, where C changes progressively from 10.4 to 12.8 through the sequence E0 to E7, from 13.4 to 14.4 through the sequence Sa to Sc, and from 11.7 to 11.9 through the sequence SBa to SBc. A similar relation can be inferred for the nebular nuclei, whose luminosities are a constant fraction, about one-fourth, of the total luminosity of the nebulae. When the nebulae are reduced to a standard type, the luminosity relation $m = 13.0 - 5 \log d$ applies, with the total apparent luminosities varying as the square of the maximum angular diameters. The relation suggests that the mean surface brightness for each spectral type is a constant, and the variations in C suggest a progressive diminution in surface brightness from class to class throughout the entire spectral sequence. Interpreted according to Jeans's theory, the various stages of the sequence represent the time scale in the evolutionary history of an originally globular mass expanding equatorially.

Although the absolute magnitudes are available for only a few nebulae, they exhibit a small range about a mean absolute magnitude of -15.2 (inferred from individual stars whose mean absolute magnitudes are taken to be -6.2). These considerations lead to the hypothesis that the extragalactic nebulae are of the same order of absolute magnitude and of the same order of actual dimensions. In this case, the luminosity relation merely expresses the variation of the apparent luminosity with the inverse square of the distance of the nebula. For a mean absolute magnitude of -15.2 the distance, D, in parsecs is given by the relation $\log D = 0.2m + 4.04$. Using these distances with the maximum angular diameters, the maximum linear diameters in parsecs are found to range from 360 to 1,130 pc for types E0 to E7, from 1,450 to 2,500 pc for types Sa to Sc, and from 1,280 to 2,250 pc for types SBa to SBc. Spiral nebulae in the last stage of the sequence have diameters of order of 3,000 pc, with a volume of order 1.4 billion pc^3 and a mean luminosity density of order 7.7 absolute mag pc^{-3}.

MASSES OF EXTRA-GALACTIC NEBULAE Spectroscopic rotations are available for the spirals M 31[12] and N.G.C. 4594,[13] and from these it is possible to estimate the masses on the assumption of orbital rotation around the nucleus. The distances of the nebulae are involved, however, and this is known accurately only for M 31; for N.G.C. 4594 it must be estimated from the apparent luminosity.

Another method of estimating masses is that used by Öpik[14] in deriving his estimate of the distance of M 31. It is based on the assumption that luminous material in the spirals has about the same coefficient of emission as the material in the galactic system. Öpik computed the ratio of luminosity to mass for our own system in terms of the sun as unity, using Jeans's value[15] for the relative proportion of luminous to non-luminous material. The relation is

$$\text{Mass} = 2.6\, L \qquad (9)$$

The application of this method of determining orders of masses seems to be justified, at least in the case of the later-type spirals and irregular nebulae, by the many analogies with the galactic system itself. Moreover, when applied to M 31, where the distance is fairly well known, it leads to a mass of the same order as that derived from the spectrographic rotation:

Spectrographic rotation	3.5×10^9 ☉
Öpik's method	1.6×10^9

The distance of N.G.C. 4594 is unknown, but the assumption that it is a normal nebula with an absolute magnitude of -15.2 places it at 700,000 parsecs. The orders of the mass by the two methods are then

Spectrographic rotation	2.0×10^9 ☉
Öpik's method	2.6×10^8

Here again the resulting masses are of the same order. They can be made to agree as well as those for M 31 by the not unreasonable assumption that the absolute luminosity of the nebula is 2 mag. or so brighter than normal.

Öpik's method leads to values that are reasonable and fairly consistent with those obtained by the independent spectrographic method. Therefore, in the absence of other resources, its use for deriving the mass of the normal nebula appears to be permissible. The result, 2.6×10^8 ☉, corresponding to an absolute magnitude of -15.2, is probably of the right order. The two test cases suggest that this value may be slightly low, but the data are not sufficient to warrant any empirical corrections.

NUMBERS OF NEBULAE TO DIFFERENT LIMITING MAGNITUDES

Here Hubble discusses the extent and completeness of various compilations of the apparent magnitudes of nebulae. The various data, collected in table 105.1, are used to determine the numbers, N, of extragalactic nebulae to different limits of total visual magnitude, m. This data can be used to test the constancy of the density function or, on the hypothesis of uniform luminosities, to determine the distribution of nebulae in space. In table 105.1 the observed, O, numbers are compared with calculated, C, numbers under the assumption of a uniform distribution of

Table 105.1 Numbers of nebulae to various limits

m_T	Log N O	Log N C	O − C	Log D
8.5	0.85	0.65	+0.20	5.74
9.0	1.08	0.95	.13	5.84
9.5	1.45	1.25	.20	5.94
10.0	1.73	1.55	.18	6.04
10.5	1.95	1.85	.10	6.14
11.0	2.17	2.15	+ .02	6.24
11.5	2.43	2.45	− .02	6.34
12.0	2.70	2.75	.05	6.44
16.7	5.48	5.57	.09	7.38
(18.0)	—	(6.35)	—	(7.64)

nebulae having a constant absolute luminosity. The formula used for this calculation is log $N = 0.6m - 4.45$. The distances D computed from the relation $\log D = 0.2m + 4.04$ are also given in table 105.1.

The agreement between the observed and computed log N over a range of more than 8 mag. is consistent with the double assumption of uniform luminosity and uniform distribution or, more generally, indicates that the density function is independent of the distance.

The systematic decrease in the residuals O − C with decreasing luminosity is probably within the observational errors, but it may also be explained as due to a clustering of nebulae in the vicinity of the galactic system. The cluster in Virgo alone accounts for an appreciable part. This is a second-order effect in the distribution, however, and will be discussed at length in a later paper.

Distances corresponding to the different limiting magnitudes are given in the last column of table 105.1. The 300,000 nebulae estimated to the limits represented by an hour's exposure on fast plates with the 60-inch reflector appear to be the inhabitants of space out to a distance of the order of 2.4×10^7 parsecs. The 100-inch reflector, with long exposures under good conditions, will probably reach the total visual magnitude 18.0, and this, by a slight extrapolation, is estimated to represent a distance of the order of 4.4×10^7 parsecs or 1.4×10^8 light-years, within which it is expected that about two million nebulae should be found. This seems to represent the present boundaries of the observable region of space.

DENSITY OF SPACE The data are now available for deriving a value for the order of the density of space. This is accomplished by means of the formulae for the numbers of nebulae to a given limiting magnitude and for the distance in terms of the magnitude. In nebulae per cubic parsec, the density is

$$\left.\begin{aligned} \log \rho &= \log N - \log V \\ &= (0.6\, m_T - 4.45) - \log \frac{4\pi}{3} - 3(4.04 + 0.2\, m_T) \\ &= -17.19. \end{aligned}\right\} \quad (11)$$

This is a lower limit, for the absence of nebulae in the plane of the Milky Way has been ignored. The current explanation of this phenomenon in terms of obscuration by dark clouds which encircle the Milky Way is supported by the extra-galactic nature of the nebulae, their general similarity to the galactic system, and the frequency with which peripheral belts of obscuring material are encountered among the spirals. The known clouds of dark nebulosity are interior features of our system, and they do not form a continuous belt. In the regions where they are least conspicuous, however, the extra-galactic nebulae approach nearest to the plane of the Milky Way, many being found within 10°. This is consistent with the hypothesis of a peripheral belt of absorption.

The only positive objection which has been urged to this explanation has been to the effect that the nebular density is a direct function of galactic latitude. Accumulating evidence[16] has failed to confirm this view and indicates that it is largely due to the influence of the great cluster in Virgo, some 15° from the north galactic pole. There is no corresponding concentration in the neighborhood of the south pole.

If an outer belt of absorption is assumed, which, combined with the known inner clouds, obscures extra-galactic nebulae to a mean distance of 15° from the galactic plane, the value derived for the density of space must be increased by nearly 40 per cent. This will not change the order of the value previously determined and is within the uncertainty of the masses as derived by Öpik's method. The new value is then

$$\rho = 9 \times 10^{-18} \text{ nebulae per cubic parsec.} \quad (12)$$

The corresponding mean distance between nebulae is of the order of 570,000 parsecs, although in several of the clusters the distances between members appear to be a tenth of this amount or less.

The density can be reduced to absolute units by substituting the value for the mean mass of a nebula, $2.6 \times 10^8\, \odot$. Then, since the mass of the sun in grams is 2×10^{33} and 1 parsec is 3.1×10^{18} cm,

$$\rho = 1.5 \times 10^{-31} \text{ grams per cubic centimeter.} \quad (13)$$

This must be considered as a lower limit, for loose material scattered between the systems is entirely ignored. There are no means of estimating the order of the necessary correction. No positive evidence of absorption by inter-nebular material, either selective or general, has been found, nor should we expect to find it unless the amount of this material is many times that which is concentrated in the systems.

THE FINITE UNIVERSE OF GENERAL RELATIVITY The mean density of space can be used to determine the dimensions of the finite but boundless universe of general relativity. De Sitter[17] made the calculations some years ago, but used values for the density, 10^{-26} and greater, which are of an entirely different order from that indicated by the present investigations. As a consequence, the various dimensions, both for spherical and for elliptical space, were small as compared with the range of existing instruments.

For the present purpose, the simplified equations which Einstein has derived for a spherically curved space can be used.[18] When R, V, M, and ρ represent the radius of curvature, volume, mass, and density, and k and c are the gravitational constant and the velocity of light,

$$R = \frac{c}{\sqrt{4\pi k}} \cdot \frac{1}{\sqrt{\rho}}, \quad (14)$$

$$V = 2\pi^2 R^3, \quad (15)$$

$$M = \frac{\pi c^2}{2k} \cdot R. \quad (16)$$

Substituting the value found for ρ, 1.5×10^{-31}, the dimensions become

$$R = 8.5 \times 10^{28} \text{ cm} = 2.7 \times 10^{10} \text{ parsecs}, \quad (17)$$

$$V = 1.1 \times 10^{88} \text{ cm}^3 = 3.5 \times 10^{32} \text{ cubic parsecs}, \quad (18)$$

$$M = 1.8 \times 10^{57} \text{ grams} = 9 \times 10^{23} \odot. \quad (19)$$

The mass corresponds to 3.5×10^{15} normal nebulae.

The distance to which the 100-inch reflector should detect the normal nebula was found to be of the order of 4.4×10^7 parsecs, or about 1/600 the radius of curvature. Unusually bright nebulae, such as M 31, could be photographed at several times this distance, and with reasonable increases in the speed of plates and size of telescopes it may become possible to observe an appreciable fraction of the Einstein universe.

1. These are the two Magellanic Clouds, M 31, and M 33.
2. Bailey, *Harvard Annals 60* (1908).
3. Hardcastle, *Monthly Notices 74*, 699 (1914).
4. This estimate by Seares is based on a revision of Fath's counts of nebulae in Selected Areas (*Mt. Wilson Contr.* No. 297; *Astrophysical Journal, 62*, 168 [1925]).
5. "A General Study of Diffuse Galactic Nebulae," *Mt. Wilson Contr.* No. 241; *Astrophysical Journal 56*, 162 (1922).
6. The classification was presented in the form of a memorandum to the Commission on Nebulae of the International Astronomical Union in 1923. Copies of the memorandum were distributed by the chairman to all members of the Commission. The classification was discussed at the Cambridge meeting in 1925, and has been published in an account of the meeting by Mrs. Roberts in *L'Astronomie 40*, 169 (1926). Further consideration of the matter was left to a subcommittee, with a resolution that the adopted system should be as purely descriptive as possible, and free from any terms suggesting order of physical development (*Transactions of the I.A.U. 2* [1925]). Mrs. Roberts' report also indicates the preference of the Commission for the term "extragalactic" in place of the original, and then necessarily noncommittal, "non-galactic."

Meanwhile K. Lundmark, who was present at the Cambridge meeting and has since been appointed a member of the Commission, has recently published (*Arkiv för matematik, astronomi och fysik 19B*, no. 8 [1926]) a classification, which, except for nomenclature, is practically identical with that submitted by me. Dr. Lundmark makes no acknowledgments or references to the discussions of the Commission other than those for the use of the term "galactic."

7. *Problems of Cosmogony and Stellar Dynamics* (1919).
8. N.G.C. 4486 (M 87) may be an exception. On the best photographs made with the 100-inch reflector, numerous exceedingly faint images, apparently of stars, are found around the periphery. It was among these that Belanowsky's nova of 1919 appeared. The observations are described in *Publications of the Astronomical Society of the Pacific 35*, 261 (1923).
9. "Early" and "late," in spite of their temporal connotations, appear to be the most convenient adjectives available for describing relative positions in the sequence. This sequence of structural forms is an observed phenomenon. As will be shown later in the discussion, it exhibits a smooth progression in nuclear luminosity, surface brightness, degree of flattening, major diameters, resolution, and complexity. An antithetical pair of adjectives denoting relative positions in the sequence is desirable for many reasons, but none of the progressive characteristics are well adapted for the purpose. Terms which apply to series in general are available, however, and of these "early" and "late" are the most suitable. They can be assumed to express a progression from simple to complex forms.

An accepted precedent for this usage is found in the series of stellar spectral types. There also the progression is assumed to be from the simple to the complex, and in view of the great convenience of the terms "early" and "late," the temporal connotations, after a full consideration of their possible consequences, have been deliberately disregarded.

10. *Publications of the Lick Observatory 13*, 12 (1918).
11. $H\beta$ is brighter than N_2. Patches with similar spectra are often found in the arms of late-type spirals—N.G.C. 253, M 33, M 101. The typical planetary spectrum, where $H\beta$ is fainter than N_2, is found in the rare cases of apparently stellar nuclei of spirals; for instance, in N.G.C. 1068, 4051, and 4151. Here also the emission spectra are localized and do not extend over the nebulae.
12. Pease, *Mt. Wilson Comm.* No. 51; *Proceedings of the National Academy of Sciences 4*, 21 (1918).
13. Pease, *Mt. Wilson Comm.* No. 32; ibid., *2*, 517 (1916).
14. *Astrophysical Journal 55*, 406 (1922).
15. *Monthly Notices 82*, 133 (1922).
16. The latest and most reliable results bearing on the distribution of faint (hence apparently distant) nebulae are found in Seares's revision and discussion of the counts made by Fath on plates of the Selected Areas with the 60-inch reflector. When the influence of the cluster in Virgo is eliminated the density appears to be roughly uniform for all latitudes greater than about 25°.
17. *Monthly Notices 78*, 3 (1917).
18. Haas, *Introduction to Theoretical Physics 2*, 373 (1925).

106. A Relation between Distance and Radial Velocity among Extra-Galactic Nebulae

Edwin P. Hubble

(*Proceedings of the National Academy of Sciences 15*, 168–173 [1929])

As early as 1917 V. M. Slipher had shown that the majority of spiral nebulae are receding from our galaxy with tremendous velocities (selection 102). Four years later C. Wirtz looked for correlations between these velocities and other observable properties of the nebulae. Wirtz found that, when suitable averages of the available data were taken, "an approximate linear dependence of velocity and apparent magnitude is visible. This dependence is in the sense that the nearby nebulae tend to approach our galaxy whereas the distant ones move away . . . The dependence of the magnitudes indicates that the spiral nebulae nearest to us have a lower outward velocity than the distant ones."[1]

Eight years later Edwin Hubble furnished, in this selection, better quantitative data for the distances of the spiral nebulae and mentioned somewhat incidentally the linear correlation of distance, D, and radial velocity, V. Because he was primarily interested in solving for the solar motion with respect to the distant nebulae, he did not specifically state the now-famous Hubble law, $V = H_0 \times D$ (where H_0 is now called the Hubble constant). At that time one of the widely known static cosmological models, found by Wilhelm de Sitter, included the feature that distant clocks would run more slowly; hence, distant atoms, vibrating like slowed clocks, would show a redshift increasing with the square of the distance. Hubble, then unaware of the work of Friedmann and of Lemaître (selections 125 and 126), supposed that the observed velocity relation was the first approximation to the de Sitter effect.[2]

Later, after Hubble learned of the nonstatic cosmologies, he formulated his law in the form[3] $\log(V) = 0.2m + B$, where m is the apparent magnitude of the object and the constant B depends on Hubble's constant and the absolute magnitude of the object. It was later shown that the slope of 0.2 in the Hubble diagram of $\log(V)$ plotted against m is exactly the expected result for a homogeneous, isotropic expanding universe that obeys the laws of Einstein's general theory of relativity. The second form of Hubble's law is valid, within the observational errors, for the normal galaxies (see selection 111) and for the radio galaxies and the quasi-stellar objects.[4]

1. *Astronomische Nachrichten 215*, 349 (1921). Also see C. Wirtz, *Astronomische Nachrichten 222*, 21 (1924), and K. Lundmark, *Monthly Notices of the Royal Astronomical Society 84*, 747 (1924).
2. A. Sandage, in *Galaxies and the Universe*, *Stars and Stellar Systems* vol. 9 (Chicago: University of Chicago Press, 1975), pp. 761–765, gives a historical account of early efforts to explain the redshifts in terms of the de Sitter effect.
3. E. P. Hubble and M. Humason, *Proceedings of the National Academy of Sciences 20*, 264 (1934); E. P. Hubble, *The Realm of the Nebulae* (New Haven: Yale University Press, 1936; reprinted., New York: Dover Publications, 1958).
4. K. R. Lang, S. D. Lord, L. M. Johanson, and S. D. Savage, *Astrophysical Journal 202*, 583 (1975).

Determinations of the motion of the sun with respect to the extragalactic nebulae have involved a K term of several hundred kilometers which appears to be variable. Explanations of this paradox have been sought in a correlation between apparent radial velocities and distances, but so far the results have not been convincing. The present paper is a re-examination of the question, based on only those nebular distances which are believed to be fairly reliable.

Distances of extra-galactic nebulae depend ultimately upon the application of absolute-luminosity criteria to involved stars whose types can be recognized. These include, among others, Cepheid variables, novae, and blue stars involved in emission nebulosity. Numerical values depend upon the zero point of the period-luminosity relation among Cepheids, the other criteria merely check the order of the distances. This method is restricted to the few nebulae which are well resolved by existing instruments. A study of these nebulae, together with those in which any stars at all can be recognized, indicates the probability of an approximately uniform upper limit to the absolute luminosity of stars, in the late-type spirals and irregular nebulae at least, of the order of M (photographic) $= -6.3$.[1] The apparent luminosities of the brightest stars in such nebulae are thus criteria which, although rough and to be applied with caution, furnish reasonable estimates of the distances of all extra-galactic systems in which even a few stars can be detected.

Finally, the nebulae themselves appear to be of a definite order of absolute luminosity, exhibiting a range of four or five magnitudes about an average value M (visual) $= -15.2$[1] The application of this statistical average to individual cases can rarely be used to advantage, but where considerable numbers are involved, and especially in the various clusters of nebulae, mean apparent luminosities of the nebulae themselves offer reliable estimates of the mean distances.

Radial velocities of 46 extra-galactic nebulae are now available, but individual distances are estimated for only 24. For one other, N.G.C. 3521, an estimate could probably be made, but no photographs are available at Mount Wilson. The data are given in table 106.1. The first seven distances are the most reliable, depending, except for M 32 the companion of M 31, upon extensive investigations of many stars involved. The next thirteen distances, depending upon the criterion of a uniform upper limit of stellar luminosity, are subject to considerable probable errors but are believed to be the most reasonable values at present available. The last four objects appear to be in the Virgo Cluster. The distance assigned to the cluster, 2×10^6 parsecs, is derived from the distribution of nebular luminosities, together with luminosities of stars in some of the later-type spirals, and differs somewhat from the Harvard estimate of ten million light years.[2]

The data in table 106.1 indicate a linear correlation between distances and velocities, whether the latter are used directly or corrected for solar motion, according to the older solutions. This suggests a new solution for the solar motion in which the distances are introduced as coefficients of the K term, i.e., the velocities are assumed to vary directly with the distances, and hence K represents the velocity at unit distance due to this effect. The equations of condition then take the form

$$rK + X \cos\alpha \cos\delta + Y \sin\alpha \cos\delta + Z \sin\delta = v.$$

Table 106.1 Nebulae whose distances have been estimated from stars involved or from mean luminosities in a cluster

Object	m_s	r	v	m_t	M_t
S. Mag.	—	0.032	+ 170	1.5	−16.0
L. Mag	—	0.034	+ 290	0.5	17.2
N.G.C. 6822	—	0.214	− 130	9.0	12.7
598	—	0.263	− 70	7.0	15.1
221	—	0.275	− 185	8.8	13.4
224	—	0.275	− 220	5.0	17.2
5457	17.0	0.45	+ 200	9.9	13.3
4736	17.3	0.5	+ 290	8.4	15.1
5194	17.3	0.5	+ 270	7.4	16.1
4449	17.8	0.63	+ 200	9.5	14.5
4214	18.3	0.8	+ 300	11.3	13.2
3031	18.5	0.9	− 30	8.3	16.4
3627	18.5	0.9	+ 650	9.1	15.7
4826	18.5	0.9	+ 150	9.0	15.7
5236	18.5	0.9	+ 500	10.4	14.4
1068	18.7	1.0	+ 920	9.1	15.9
5055	19.0	1.1	+ 450	9.6	15.6
7331	19.0	1.1	+ 500	10.4	14.8
4258	19.5	1.4	+ 500	8.7	17.0
4151	20.0	1.7	+ 960	12.0	14.2
4382	—	2.0	+ 500	10.0	16.5
4472	—	2.0	+ 850	8.8	17.7
4486	—	2.0	+ 800	9.7	16.8
4649	—	2.0	+1,090	9.5	17.0
Mean					−15.5

m_s = photographic magnitude of brightest stars involved.
r = distance in units of 10^6 parsecs. The first two are Shapley's values.
v = measured velocities in km./sec. N.G.C. 6822, 221, 224 and 5457 are recent determinations by Humason.
m_t = Holetschek's visual magnitude as corrected by Hopmann. The first three objects were not measured by Holetschek, and the values of m_t represent estimates by the author based upon such data as are available.
M_t = total visual absolute magnitude computed from m_t and r.

Two solutions have been made, one using the 24 nebulae individually, the other combining them into 9 groups according to proximity in direction and in distance. The results are

	24 objects	9 groups
X	-65 ± 50	$+3 \pm 70$
Y	$+226 \pm 95$	$+230 \pm 120$
Z	-195 ± 40	-133 ± 70
K	$+465 \pm 50$	$+513 \pm 60$ km./sec. per 10^6 parsecs.
A	286°	269°
D	+ 40°	+ 33°
V_0	306 km./sec.	247 km./sec.

For such scanty material, so poorly distributed, the results are fairly definite. Differences between the two solutions are due largely to the four Virgo nebulae, which, being the most distant objects and all sharing the peculiar motion of the cluster, unduly influence the value of K and hence of V_0. New data on more distant objects will be required to reduce the effect of such peculiar motion. Meanwhile round numbers, intermediate between the two solutions, will represent the probable order of the values. For instance, let $A = 277°$, $D = +36°$ (Gal. long. = 32°, lat. = +18°), $V_0 = 280$ km./sec., $K = +500$ km./sec. per million parsecs. Mr. Strömberg has very kindly checked the general order of these values by independent solutions for different groupings of the data.

A constant term, introduced into the equations, was found to be small and negative. This seems to dispose of the necessity for the old constant K term. Solutions of this sort have been published by Lundmark,[3] who replaced the old K by $k + lr + mr^2$. His favored solution gave $k = 513$, as against the former value of the order of 700, and hence offered little advantage.

The residuals for the two solutions given above average 150 and 110 km./sec. and should represent the average peculiar motions of the individual nebulae and of the groups, respectively. In order to exhibit the results in a graphical form, the solar motion has been eliminated from the observed velocities and the remainders, the distance terms plus the residuals, have been plotted against the distances. The run of the residuals is about as smooth as can be expected, and in general the form of the solutions appears to be adequate.

The 22 nebulae for which distances are not available can be treated in two ways. First, the mean distance of the group derived from the mean apparent magnitudes can be compared with the mean of the velocities corrected for solar motion. The result, 745 km./sec. for a distance of 1.4×10^6 parsecs, falls between the two previous solutions and indicates a value for K of 530 as against the proposed value, 500 km./sec.

Secondly, the scatter of the individual nebulae can be examined by assuming the relation between distances and velocities as previously determined. Distances can then be calculated from the velocities corrected for solar motion, and absolute magnitudes can be derived from the apparent magnitudes. The results are given in table 106.2 and may be compared with the distribution of absolute magnitudes among the nebulae in table 106.1, whose distances are derived from other criteria. N.G.C. 404 can be excluded, since the observed velocity is so small that the peculiar motion must be large in comparison with the distance effect. The object is not necessarily an exception, however, since a distance can be assigned for which the peculiar motion and the absolute magnitude are both within the range previously determined. The two mean magnitudes, −15.3 and −15.5, the ranges, 4.9 and 5.0 mag., and the frequency distributions are closely similar for these two entirely independent sets of data; and even the slight difference in mean magnitudes can be attributed to the selected, very bright, nebulae in the Virgo Cluster. This entirely unforced agreement supports the validity of the velocity-distance relation in a very evident matter. Finally, it is worth recording that the frequency distribution of absolute magnitudes in the two tables combined is comparable with those found in the various clusters of nebulae.

The results establish a roughly linear relation between velocities and distances among nebulae for which velocities

Table 106.2 Nebulae whose distances are estimated from radial velocities

Object	v	v_s	r	m_t	M_t
N.G.C. 278	+ 650	−110	1.52	12.0	−13.9
404	− 25	− 65	—	11.1	—
584	+1,800	+ 75	3.45	10.9	16.8
936	+1,300	+115	2.37	11.1	15.7
1023	+ 300	− 10	0.62	10.2	13.8
1700	+ 800	+220	1.16	12.5	12.8
2681	+ 700	− 10	1.42	10.7	15.0
2683	+ 400	+ 65	0.67	9.9	14.3
2841	+ 600	− 20	1.24	9.4	16.1
3034	+ 290	−105	0.79	9.0	15.5
3115	+ 600	+105	1.00	9.5	15.5
3368	+ 940	+ 70	1.74	10.0	16.2
3379	+ 810	+ 65	1.49	9.4	16.4
3489	+ 600	+ 50	1.10	11.2	14.0
3521	+ 730	+ 95	1.27	10.1	15.4
3623	+ 800	+ 35	1.53	9.9	16.0
4111	+ 800	− 95	1.79	10.1	16.1
4526	+ 580	− 20	1.20	11.1	14.3
4565	+1,100	− 75	2.35	11.0	15.9
4594	+1,140	+ 25	2.23	9.1	17.6
5005	+ 900	−130	2.06	11.1	15.5
5866	+ 650	−215	1.73	11.7	−14.5
Mean				10.5	−15.3

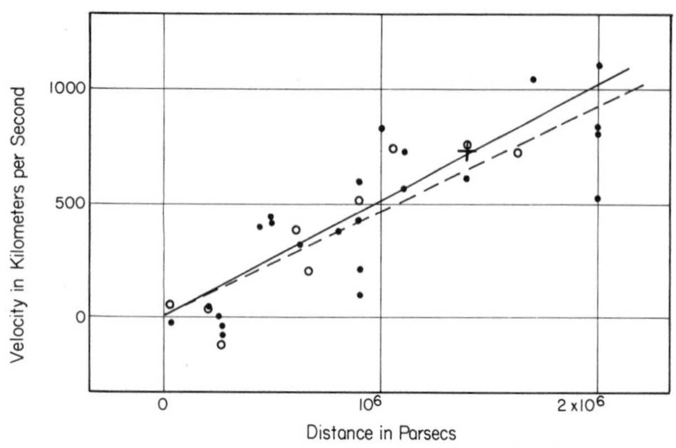

Fig. 106.1 The velocity-distance relation for extragalactic nebulae. Radial velocities, corrected for solar motion, are plotted against distances estimated from involved stars and, in the case of the Virgo cluster (represented by the four most distant nebulae), from mean luminosities of nebulae in a cluster. The filled circles and solid line represent the solution for solar motion using the nebulae individually; the open circles and dashed line represent the solution combining the nebulae into groups; the cross represents the mean velocity corresponding to the mean distance of twenty-two nebulae whose distances could not be estimated individually.

have been previously published, and the relation appears to dominate the distribution of velocities. In order to investigate the matter on a much larger scale, Mr. Humason at Mount Wilson has initiated a program of determining velocities of the most distant nebulae that can be observed with confidence. These, naturally, are the brightest nebulae in clusters of nebulae. The first definite result,[4] $v = +3{,}779$ km./sec. for N.G.C. 7619, is thoroughly consistent with the present conclusions. Corrected for the solar motion, this velocity is $+3{,}910$, which with $K = 500$, corresponds to a distance of 7.8×10^6 parsecs. Since the apparent magnitude is 11.8, the absolute magnitude at such a distance is -17.65, which is of the right order for the brightest nebulae in a cluster. A preliminary distance, derived independently from the cluster of which this nebula appears to be a member, is of the order of 7×10^6 parsecs.

New data to be expected in the near future may modify the significance of the present investigation or, if confirmatory, will lead to a solution having many times the weight. For this reason it is thought premature to discuss in detail the obvious consequences of the present results. For example, if the solar motion with respect to the clusters represents the rotation of the galactic system, this motion could be subtracted from the results for the nebulae and the remainder would represent the motion of the galactic system with respect to the extra-galactic nebulae.

The outstanding feature, however, is the possibility that the velocity-distance relation may represent the de Sitter effect, and hence that numerical data may be introduced into discussions of the general curvature of space. In the de Sitter cosmology, displacements of the spectra arise from two sources, an apparent slowing down of atomic vibrations and a general tendency of material particles to scatter. The latter involves an acceleration and hence introduces the element of time. The relative importance of these two effects should determine the form of the relation between distances and observed velocities; and in this connection it may be emphasized that the linear relation found in the present discussion is a first approximation representing a restricted range in distance.

1. *Mt. Wilson Contr.* No. 324; *Astroph. J.* (Chicago, Ill) *64*, 321 (1926).
2. *Harvard Coll. Obs. Circ.*, 294 (1926).
3. *Mon. Not. R. Astr. Soc. 85*, 865–894 (1925).
4. *Proc. Nat. Acad. Sci. 15*, 167 (1929).

107. On the Masses of Nebulae and of Clusters of Nebulae

Fritz Zwicky

(*Astrophysical Journal 86*, 217–246 [1937])

Knowledge of the average density of the universe is of prime importance for cosmology, and this knowledge in turn requires information on the masses of galaxies. In the following article Fritz Zwicky critically reviews the two previous methods of estimating the masses of galaxies and describes for the first time another important approach. The first method is based on the assumption that the mass of a nebula is a constant multiple of its absolute luminosity.[1] The second method employs the observed rotational motions with Kepler's third law to determine the mass of the galactic nucleus; Zwicky applies this procedure to NGC 4594 to obtain a nuclear mass of 2.5×10^{10} solar masses.

The virial theorem method, introduced here, uses the random motions of the galaxies in a cluster to estimate the mass of the entire cluster.[2] The method assumes that the kinetic energy of the random motions of the nebulae is just balanced by the gravitational potential energy of the cluster. Whereas the rotational curve method gives the mass within the luminous boundaries of a single nebula, the virial theorem method gives the total cluster mass, including any internebular material.

In this paper Zwicky applies the virial theorem method to the Coma cluster of galaxies to obtain an average nebular mass of 4.5×10^{10} solar masses. Because of the relatively low luminosities of the nebulae in the Coma cluster, this mass estimate leads to a mass to luminosity ratio of about 500:1, compared to a ratio of about 3:1 for our galaxy. As Zwicky points out, the rotational curve and virial theorem estimates for the masses of individual nebulae can be reconciled only if there is a substantial amount of optically invisible intergalactic gas in clusters of galaxies, or if clusters of galaxies are dynamically unstable. Even today, the mystery of the missing cluster mass remains unsolved.

1. See E. Öpik, *Astrophysical Journal 55*, 406 (1922).
2. Also see F. Zwicky, *Helvetica physica acta 6*, 110 (1933), and S. Smith, *Astrophysical Journal 83*, 499 (1936).

Abstract—Present estimates of the masses of nebulae are based on observations of the *luminosities* and *internal rotations* of nebulae. It is shown that both these methods are unreliable; that from the observed luminosities of extragalactic systems only lower limits for the values of their masses can be obtained (sec. i), and that from internal rotations alone no determination of the masses of nebulae is possible (sec. ii). The observed internal motions of nebulae can be understood on the basis of a simple mechanical model, some properties of which are discussed. The essential feature is a central core whose internal *viscosity* due to the gravitational interactions of its component masses is so high as to cause it to rotate like a solid body.

In sections iii, iv, and v three new methods for the determination of nebular masses are discussed, each of which makes use of a different fundamental principle of physics.

Method iii is based on the *virial theorem* of classical mechanics. The application of this theorem to the Coma cluster leads to a minimum value $\bar{M} = 4.5 \times 10^{10} M_{\odot}$ for the average mass of its member nebulae.

Method iv calls for the observation among nebulae of certain *gravitational lens* effects.

Section v gives a generalization of the principles of ordinary *statistical mechanics* to the whole system of nebulae, which suggests a new and powerful method which ultimately should enable us to determine the masses of all types of nebulae. This method is very flexible and is capable of many modes of application. It is proposed, in particular, to investigate the distribution of nebulae in individual great clusters.

As a first step toward the realization of the proposed program, the Coma cluster of nebulae was photographed with the new 18-inch Schmidt telescope on Mount Palomar. Counts of nebulae brighter than about $m = 16.7$ given in section vi lead to the gratifying result that the distribution of nebulae in the Coma cluster is very similar to the distribution of luminosity in globular nebulae, which, according to Hubble's investigations, coincides closely with the theoretically determined distribution of matter in isothermal gravitational gas spheres. The high central condensation of the Coma cluster, the very gradual decrease of the number of nebulae per unit volume at great distances from its center, and the hitherto unexpected enormous extension of this cluster become here apparent for the first time. These results also suggest that the current classification of nebulae into relatively few *cluster nebulae* and a majority of *field nebulae* may be fundamentally inadequate. From the preliminary counts reported here it would rather follow that practically *all* nebulae must be thought of as being grouped in clusters—a result which is in accord with the theoretical considerations of section v.

In conclusion, a comparison of the relative merits of the three new methods for the determination of nebular masses is made. It is also pointed out that an extensive investigation of great clusters of nebulae will furnish us with decisive information regarding the question whether physical conditions in the known parts of the universe are merely fluctuating around a stationary state or whether they are continually and systematically changing.

THE DETERMINATION OF THE MASSES of extragalactic nebulae constitutes at present one of the major problems in astrophysics. Masses of nebulae until recently were estimated either from the luminosities of nebulae or from their internal rotations. In this paper it will be shown that both these methods of determining nebular masses are unreliable. In addition, three new possible methods will be outlined.

I. Masses from Luminosities of Nebulae

The observed absolute luminosity of any stellar system is an indication of the approximate amount of luminous matter in such a system. In order to derive trustworthy values of the masses of nebulae from their absolute luminosities, however, detailed information on the following three points is necessary.

1. According to the mass-luminosity relation, the conversion factor from absolute luminosity to mass is different for different types of stars. The same holds true for any kind of luminous matter. In order to determine the conversion factor for a nebula as a whole, we must know, therefore, in what proportions all the possible luminous components are represented in this nebula.

2. We must know how much dark matter is incorporated in nebulae in the form of cool and cold stars, macroscopic and microscopic solid bodies, and gases.

3. Finally, we must know to what extent the apparent luminosity of a given nebula is diminished by the internal absorption of radiation because of the presence of dark matter.

Data are meager[1] on point 1. Accurate information on points 2 and 3 is almost entirely lacking. Estimates of the masses of nebulae from their observed luminosities are therefore incomplete and can at best furnish only the lowest limits for the values of these masses.

II. Masses from Internal Rotations of Nebulae

It has apparently been taken for granted by some astronomers[2] that from observations on the internal rotations good values for the masses M_N of nebulae could be derived. Values of the order of $M_N = 10^9\ M_{\odot}$ up to $M_N = 4 \times 10^{10}\ M_{\odot}$ were obtained in this way,[2] where $M_{\odot} = 2 \times 10^{33}$ gr is the mass of the sun. A closer scrutiny of the behavior of suitably chosen mechanical models of stellar systems, unfortunately, soon reveals the fact that the masses of such systems, for a given distribution of average angular velocities throughout the system, are highly indeterminate, and vice versa. This conclusion may, for instance, be derived from the consideration of two limiting models of a nebula as a mechanical system.

Here Zwicky points out that the interpretation of line of sight velocities as rotation velocities involves certain ad hoc assumptions about the distribution of mass and velocity in a nebula. The limiting cases are those in which the stars, gas, and dust in the outlying parts of the nebula have either negligible or substantial gravitational attraction. In both cases mechanical models can be constructed in which the observed angular velocities have no direct bearing on the mass of the system.

Good mechanical models of actual nebulae may presumably be constructed by combining the distinctive features of the two limiting cases described in the preceding sections. Such a combined model will possess a central, highly viscous core whose relative dimensions are not negligible but are comparable with the extension of the whole system. If the outlying, and among themselves little interacting, components of the nebula had no connection with the central core, we might, at a given instant, observe average angular velocities Ω which, as a function of the distance r from the center of rotation, would be given by

$$\Omega(r) = \Omega_0 = \text{const.} \leftrightarrow \text{ for } r < r_0, \tag{1}$$

where r_0 is the radius of the core. For $r > r_0$, the angular velocity $\Omega(r)$ would be essentially arbitrary. In reality, however, the viscosity will not drop abruptly to zero at $r = r_0$. From an inspection of the distribution of the outlying masses in many nebulae it would seem that these masses at some previous time must have formed part of the central core. They may have been ejected from this core because they acquired high kinetic energy through many close encounters, or they may be the result of a partial disruption of the core by tidal actions caused by encounters with other nebulae (Jeans). At the moment these outlying masses have passed the boundary of the core, their average tangential velocities must have been of the order $r_0\Omega_0$. Assuming that these masses in regions $r > r_0$ are essentially subject only to the gravitational attraction of a central spherical core and that the interactions among themselves may be neglected (the internal viscosity of the outlying system is equal to zero), the average angular velocity $\Omega(r)$ outside the core will be approximately given as

$$\Omega(r) = \frac{r_0^{\,2}\Omega_0}{r^2} \text{ for } r > r_0. \tag{2}$$

This relation simply expresses the fact that a mass m which, on being ejected from the core with a tangential velocity v_t relative to this core, describes an orbit whose angular momentum,

$$mr^2\omega = m[r_0 v_t + r_0^{\,2}\Omega_0] = c, \tag{3}$$

is a constant. If the average $\bar{v}_t$ for many particles leaving the core is zero, the average angular velocity $\bar{\omega} = \Omega$ of all the particles in a given point is obtained from

$$r^2\bar{\omega} = r^2\Omega(r) = r_0^{\,2}\Omega_0, \tag{4}$$

which is the same as (2). Unfortunately, relation (2) is superficially very similar to the relation obeyed by the angular velocities ω_c of a system of circular planetary orbits around a heavy central mass M.

For such an orbit we have

$$mr\omega_c^{\,2} = \frac{\Gamma mM}{r^2} \tag{5}$$

or

$$\omega_c = \frac{(\Gamma M)^{1/2}}{r^{3/2}}, \tag{6}$$

where Γ is the universal gravitational constant. The similarity of the dependence on r in (2) and (6) will in reality become still greater, since the rate of decrease of $\Omega(r)$ with increasing values of r will in actual nebulae be more gradual than that given by (2), owing to the fact that the internal viscosity will not vanish abruptly at $r = r_0$ but will disappear gradually with increasing r. The observed angular velocities in the outlying regions of nebulae actually show a dependence on r which resembles the relation (6). This relation was, therefore, sometimes erroneously used for the determination of the mass M. The preceding discussion, however, indicates that the observed angular velocities may be adequately accounted for on the basis of the considerations resulting in relation (2) rather than in relation (6). This again shows clearly that it is not possible to derive the masses of nebulae from observed rotations without the use of additional information, since the relation (2) does not contain the mass M at all.

Here Zwicky argues that the increase of surface brightness from the edge to the center of nebular cores indicates a corresponding increase in mass density and that the constancy of angular velocity in the core does not imply a constant mass density.

One further interesting problem presents itself. The distribution of matter in a stellar system which has an internal viscosity as high as we have assumed it to be in the cores of nebulae should rapidly converge toward stationary conditions. The determination of the density distribution in such a system should be analogous to the determination of the density distribution in gravitating gas spheres. R. Emden's analysis[3] of such spheres may, therefore, prove useful in the study of globular and elliptical nebulae as well as in the study of cores of spiral nebulae.

In concluding this section the question may be raised whether data on the internal rotation of nebulae, supplemented by certain additional information, make possible the determination of nebular masses. This question, in principle, must be answered in the affirmative. For instance, data on internal velocities combined with the virial theorem discussed in the next section may ultimately furnish good values for the masses of globular and elliptical nebulae as well as the cores of certain spirals. In the case of open spirals, an investigation of the geometrical structure of the spiral arms, combined with velocity data, promises to be helpful.

Zwicky then speculates that spiral arms might be caused by the tidal actions of one nebula on another. If this is the case, pulsations might ensue, and these pulsations would further complicate the interpretation of line of sight velocities as rotation velocities.

Summing up, we may say that present data on internal rotations furnish, at best, minimum values M_{min} for the masses of the nebulae. Such minimum values are obtained if we assume that nebulae are stable systems whose components have velocities v inferior to the velocity of escape v_e from the system. Therefore

$$v_e \geqslant v_{max} \geqslant r_0\Omega_0. \tag{9}$$

But

$$v_e = \left(\frac{2\Gamma M_0}{r_0}\right)^{1/2}, \tag{10}$$

where M_0, r_0, and Ω_0 are the mass, the radius, and the angular velocity of the core. Consequently, the mass M of the nebula is

$$M > M_0 \geqslant \frac{r_0{}^3\Omega_0{}^2}{2\Gamma} \tag{11}$$

For example, in the case of NGC 4594 we have $r_0 \geqslant 4.3 \times 10^{21}$ cm and $r_0\Omega_0 \cong 4 \times 10^7$ cm/sec, which gives

$$M > 5 \times 10^{43} \text{ gr} = 2.5 \times 10^{10} M_\odot . \tag{12}$$

The data used are taken from the paper on NGC 4594 by F. G. Pease,[4] who, in 1916, measured the spectroscopic rotation of this nebula. Pease also was the first to point out that the central parts of nebulae rotate like solid bodies.

III. The Virial Theorem Applied to Clusters of Nebulae

If the total masses of clusters of nebulae were known, the average masses of cluster nebulae could immediately be determined from counts of nebulae in these clusters, provided internebular material is of the same density inside and outside clusters.

As a first approximation, it is probably legitimate to assume that clusters of nebulae such as the Coma cluster are mechanically stationary systems. With this assumption, the virial theorem of classical mechanics gives the total mass of a cluster in terms of the average square of the velocities of the individual nebulae which constitute this cluster.[5] But even if we drop the assumption that clusters represent stationary configurations, the virial theorem, in conjunction with certain additional data, allows us to draw important conclusions concerning the masses of nebulae, as will now be shown.

Suppose the radius vector from a fixed point in the cluster to the nebula (σ) of mass M_σ is $\vec{r}_\sigma$. For the fixed point we conveniently chose the center of mass of the whole cluster. The fundamental law of motion of the nebula (σ) is

$$M_\sigma \frac{d^2\vec{r}_\sigma}{dt^2} = \vec{F}_\sigma, \tag{13}$$

where $\vec{F}_\sigma$ is the total force acting on M_σ. Scalar multiplication of this equation with $\vec{r}_\sigma$ gives

$$\frac{1}{2}\frac{d^2}{dt^2}(M_\sigma r_\sigma{}^2) = \vec{r}_\sigma \cdot \vec{F}_\sigma + M_\sigma\left(\frac{d\vec{r}_\sigma}{dt}\right)^2. \tag{14}$$

Summation over all the nebulae of the cluster leads to

$$\frac{1}{2}\frac{d^2\Theta}{dt^2} = Vir + 2K_T, \tag{15}$$

where $\Theta = \sum_\sigma M_\sigma r_\sigma{}^2$ is the polar moment of inertia of the cluster, $Vir = \sum_\sigma \vec{r}_\sigma \cdot \vec{F}_\sigma$ is the virial of the cluster, and K_T is the sum of the kinetic energies of translation of the individual nebulae. If the cluster under consideration is stationary, its polar moment of inertia Θ fluctuates around a constant value Θ_0, such that the time average of its derivatives with respect to time is zero. Denoting time averages by a bar, we have in this case

$$\overline{Vir} = -2\overline{K_T}. \tag{16}$$

On the assumption that Newton's inverse square law accurately describes the gravitational interactions among nebulae, it follows that

$$Vir = E_p, \tag{17}$$

where

$$E_p = -\sum_{\sigma,\nu} \frac{\Gamma M_\sigma M_\nu}{r_{\sigma\nu}} \leftrightarrow \sigma < \nu \tag{18}$$

is the total potential energy of the cluster due to the gravitational interactions of its member nebulae. Equation (16) thus takes on the well-known form

$$-\bar{E}_p = 2\overline{K_T} = \overline{\sum_\sigma M_\sigma v_\sigma^2} = \sum_\sigma M_\sigma \overline{v_\sigma^2}, \tag{19}$$

where v_σ is the velocity of the mass M_σ. In order to arrive at a quantitative estimate of the total mass $\mathcal{M}$ of a globular cluster of nebulae, we assume as a first approximation that these nebulae are, on the average, uniformly distributed inside a sphere of radius R. In this case

$$E_p = \frac{-3\Gamma \mathcal{M}^2}{5R}. \tag{20}$$

We may also write

$$\sum M_\sigma \overline{v_\sigma^2} = \mathcal{M}\overline{\overline{v^2}}, \tag{21}$$

where the double bar indicates a double average taken over time and over mass. Therefore, from (19), (20), and (21),

$$\mathcal{M} = \frac{5R\overline{\overline{v^2}}}{3\Gamma}. \tag{22}$$

Zwicky derives equation (22) by taking the time average of equation (14), which holds for an individual nebula. For the Coma cluster the nebulae are not uniformly distributed, but the potential energy is, in order of magnitude, correctly given by equation (20). A lower limit to the mass is given by equation (22) when the average line of sight velocity, v_s, is used for v [denoted as equation (33)].

From the observations of the Coma cluster so far available we have, approximately,[5]

$$\overline{\overline{v_s^2}} = 5 \times 10^{15} \text{ cm}^4 \text{ sec}^{-2} \tag{34}$$

This average has been calculated as an average of the velocity squares alone without assigning to them any mass weights, as actually should be done according to (21). It seems, however, as Sinclair Smith[6] has shown for the Virgo cluster, that the velocity dispersion for bright nebulae is about the same as that for faint nebulae. Assuming this to be true also for the Coma cluster, it follows that the mass-weighted means of v^2 and the straight means are essentially the same. Furthermore, in calculating (34) we have used velocities which belong to the bright nebulae, since only these have been measured. If brightness can be taken as a qualitative indication of mass, the error in substituting (34) for (21) cannot be great. We must, nevertheless, remember that, strictly speaking, the determination of $\mathcal{M}$ by the virial theorem is subject to the difficulty of calculating $\overline{\overline{v^2}}$ through the application of an averaging process which involves the as yet unknown masses. The mass $\mathcal{M}$, as obtained from the virial theorem, can therefore be regarded as correct only in order of magnitude.

Combining (33) and (34), we find [using $R = 2 \times 10^6$ light years].

$$\mathcal{M} > 9 \times 10^{46} \text{ gr}. \tag{35}$$

The Coma cluster contains about one thousand nebulae. The average mass of one of these nebulae is therefore

$$\bar{M} > 9 \times 10^{43} \text{ gr} = 4.5 \times 10^{10} M_\odot. \tag{36}$$

Inasmuch as we have introduced at every step of our argument inequalities which tend to depress the final value of the mass $\mathcal{M}$, the foregoing value (36) should be considered as the lowest estimate for the average mass of nebulae in the Coma cluster. This result is somewhat unexpected, in view of the fact that the luminosity of an average nebula is equal to that of about 8.5×10^7 suns. According to (36), the conversion factor γ from luminosity to mass for nebulae in the Coma cluster would be of the order

$$\gamma = 500, \tag{37}$$

as compared with about $\gamma' = 3$ for the local Kapteyn stellar system. This discrepancy is so great that a further analysis of the problem is in order. Parts of the following discussion were published several years ago, when the conclusion expressed in (36) was reached for the first time.[5]

Zwicky then discusses the case of a nonstationary cluster. In order for the discrepency between

γ and γ' to be reduced, the total energy of the Coma cluster must be positive, and it must be rapidly flying apart. After its complete dispersion from a cluster, a nebula would have a velocity nearly equal to its original velocity, but because the field nebulae have lower velocities than the cluster nebulae they cannot have originated in clusters. This conclusion suggests that clusters do not contain sufficient positive energy to reduce the discrepancy between γ and γ' substantially.

It will, nevertheless, be advisable to obtain more data on the velocities of both cluster nebulae and field nebulae in order to arrive at accurate values of the dispersion which characterizes the respective velocity distribution functions.

In addition it will be necessary to develop methods which allow us to determine the relative amounts of internebular material in clusters as well as in the general field.

It should also be noticed that the virial theorem as applied to clusters of nebulae provides for a test of the validity of the inverse square law of gravitational forces. This is of fundamental interest because of the enormous distances which separate the gravitating bodies whose motions are investigated. Since clusters of nebulae are the largest known aggregations of matter, the study of their mechanical behavior forms the last stepping-stone before we approach the investigation of the universe as a whole.

The result (36) taken at face value of course does not mean that the average masses of field nebulae must be as great as those of cluster nebulae. From the general principles discussed in a following section one would rather expect the heaviest nebulae to be favored in the process of clustering.

The distribution of nebulae in the Coma cluster rather suggests that stationary conditions prevail in this cluster. It is proposed, therefore, to study the Coma cluster in more detail. On the other hand, the virial theorem can hardly be used with much confidence in cases such as the Virgo cluster and the Pisces cluster.[7] These clusters are much more open and asymmetrical than the Coma cluster and their boundaries are thus far ill defined. Accurate values of the gravitational potentials in these clusters are difficult to determine.

In passing it should be noted that the mechanical conditions in clusters of nebulae are in some important respects different from the conditions in clusters of stars. During close encounters of stars only a minute part of their translational energy is transformed into internal energy of these stars, if the extremely rare cases of actual impacts are disregarded. Nebulae act differently. In the first place, close encounters and actual impacts in a cluster of nebulae must occur during time intervals which are not very long compared with the time of passage of one nebula through the entire system. Therefore, a considerable tendency exists toward equipartition of rotational and internal energy of nebulae with their translational energy. Star clusters in some ways are analogous to gas spheres built up of monatomic gases, whereas clusters of nebulae may be likened to gas spheres built up of polyatomic gases.

By assuming that encounters between nebulae tend to establish equipartition among the translational, rotational, and internal energies of a nebula, Zwicky is led to the conclusion that the total energy of a stationary cluster of nebulae must be positive. He explains away this contradiction by arguing that the rotational energy of a nebula cannot become equal to its translational energy, because the nebula would disintegrate.

IV. Nebulae as Gravitational Lenses

As I have shown previously,[8] the probability of the overlapping of images of nebulae is considerable. The gravitational fields of a number of "foreground" nebulae may therefore be expected to deflect the light coming to us from certain background nebulae. The observation of such gravitational lens effects promises to furnish us with the simplest and most accurate determination of nebular masses. No thorough search for these effects has as yet been undertaken. It would seem, perhaps, that if the masses of field nebulae were, on the average, as great as the masses of cluster nebulae obtained in section iii, gravitational lens effects among nebulae should have been long since discovered. Until many plates of rich nebular fields taken under excellent conditions of seeing have been carefully examined it would be dangerous, however, to draw any definite conclusions.

The mathematical analysis of the formation of images of distant nebulae through the action of the gravitational fields of nearer nebulae will be given in detail in an article to be published in the *Helvetica physica acta.*

V. Statistical Distribution in Space of Different Types of Nebulae

It will be shown elsewhere that the number of clusters of nebulae actually observed is far greater than the number that might be expected for a random distribution of noninteracting objects. This tendency of nebulae toward clustering is no doubt due to the action of gravitational forces.

By a bold extrapolation of well-known results of ordinary statistical mechanics we adopt the following working hypothesis as a tentative basis for the interpretation of future observations on the clustering of nebulae.

BASIC PRINCIPLES

1. The system of extragalactic nebulae throughout the known parts of the universe forms a statistically stationary system.

2. Every constellation of nebulae is to be endowed with a probability weight $f(\varepsilon)$ which is a function of the total energy ε of this constellation. Quantitatively the probability P of the occurrence of a certain configuration of nebulae is assumed to be of the type

$$P = A\left(\frac{V}{V_0}\right) f\left(\frac{\varepsilon}{\bar{\varepsilon}_K}\right). \tag{49}$$

Here V is the volume occupied by the configuration or cluster considered, V_0 is the volume to be allotted, on the average, to any individual nebula in the known parts of the universe, and ε is the total energy of the cluster in question, while ε_K will probably be found to be proportional to the average kinetic energy of individual nebulae. The function $A(V/V_0)$ can be determined a priori. On the other hand, $f(\varepsilon/\overline{\varepsilon_K})$ presumably will be found to be a monotonously decreasing function in $\varepsilon/\overline{\varepsilon_K}$ analogous in type to a Boltzmann factor

$$F = \text{const} \times e^{-\varepsilon/\overline{\varepsilon_K}}. \tag{50}$$

By counting the types of clusters, by assuming that nebulae of different classes have masses that are only a function of their luminosity, and by observing the velocities and separations of the nebulae in clusters, the relative masses of nebulae of different classes may be determined.

The practical application of this method necessitates a great amount of observational work. In view of this fact, it will perhaps be advantageous to apply the preceding program first in a restricted form by a consideration of the distribution of various types of nebulae in one individual great cluster. The procedure to be applied in this case is analogous to that used successfully by H. von Zeipel[9] in his determination of the masses of different types of stars in certain clusters of stars.

Since it is intended to carry out the investigation just mentioned on the Coma cluster, a few preliminary remarks concerning the distribution of nebulae in this cluster are here given.

VI. THE COMA CLUSTER OF NEBULAE

According to Hubble and Humason,[10] the Coma cluster of nebulae "consists of about 800 nebulae scattered over an area roughly 1°.7 in diameter . . . Photographic magnitudes range from about 14.1 to 19.5, with 17.0 as the most frequent value." The distance of the cluster is about 13.8 million parsecs.

In order to get some preliminary data on the distribution of nebulae in the Coma cluster, photographs of this cluster were taken on Mount Palomar with the new 18-inch Schmidt-type telescope of the California Institute of Technology. The faintest nebulae which on limiting exposures (30–60 min) with this telescope can still be clearly distinguished from stars have an apparent magnitude close to 16.5. The counts given at different distances r from the center of the Coma cluster include all the nebulae which I have been able to identify on half-hour exposures.

The nebula NGC 4874 (α 12^{h}56^m, δ 28° 20′, 1930) was taken to be the approximate center of the cluster, no effort being made to determine a mathematically accurate central point. Concentric circles were then drawn around the adopted center, with radii $r = nr_0$, where n is a whole number running from $n = 1$ to $n = 32$, and $r_0 \cong 5$ minutes of arc. The unit of area to which all counts are reduced is $s = \pi r_0^2$, or about 1/46.4 sq deg. The numbers of nebulae per unit area, n_r, are averages for the ringlike areas which lie between $r = nr_0$ and $r = (n + 1)r_0$. The first four figures in table 1, however, are averages for the full circles the radii of which are $r = r_0/5$, $r_0/3$, $r_0/2$, and r_0, respectively. The corresponding numbers N_r of nebulae per square degree are $N_r = 46.4\, n_r$.

Table 1 is omitted here, but the data are illustrated in figure 107.1.

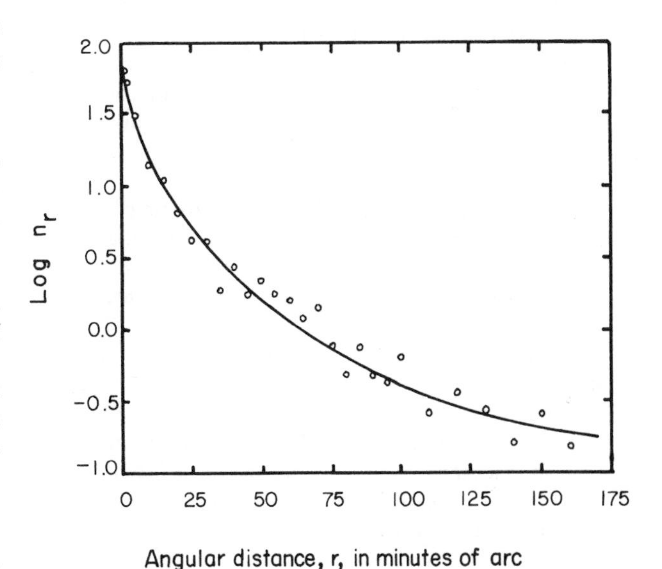

Fig. 107.1 Counts of the number of nebulae per unit area, n_r, plotted as a function of the angular distance, r, from the center of the Coma cluster of nebulae. The corresponding number of nebulae per square degree is $49.4n_r$.

In figure 107.1 values of $\log_{10} n_r$ are plotted against values of r. The full curve is drawn only approximately and does not correspond to any definite mathematical function. From the general character of this curve it is seen immediately that the Coma cluster extends to much greater distances than was originally assumed by Hubble and Humason.[10] At the edge of a circle the diameter of which is 4°.5 instead of only 1°.7, the average number of nebulae per unit area is still higher than the corresponding number in the surrounding general field. Since our counts include only the brighter nebulae, it is to be expected that counts made with more powerful telescopes will enable us to follow the extensions of the Coma cluster still farther into the general field.

The high central condensation, the very gradual decrease of the number of nebulae per unit volume at great distances from the center of a cluster, and the great extension of this cluster become here apparent for the first time. It is quite as we should expect from the considerations of section v. According to these considerations, a cluster of nebulae analogous to an isothermal gravitational gas sphere may in some cases be expected to extend indefinitely far into space, until its extension is stopped through the formation of independent clusters in the regions surrounding it.

The actual shape of the distribution curve in figure 107.1 is also of great interest. We notice at once the great similarity of this curve to the luminosity-curves of elliptical nebulae derived by Hubble.[11] According to him, the distribution of the intrinsic luminosity in globular nebulae corresponds very closely to the distribution of mass density in isothermal gas spheres as computed by R. Emden.[3] The same is approximately true for the distribution of the brighter nebulae in the Coma cluster. This result also checks the general conclusions drawn from the basic principles which, according to the discussion in section v, determine the stationary configurations of clusters of nebulae.

The total number $\mathcal{N}$ of nebulae in the Coma cluster that can be identified on photographs taken with the 18-inch Schmidt telescope is obtained as follows: The curve in figure 107.1 apparently has a horizontal asymptote which corresponds to a value of n_r not higher than $n_\infty = 0.159$. Thus, the corresponding number N of nebulae in the general field should be $N = 46.4 \times n_\infty = 7.38$. The total number of nebulae counted to the distance $r = 2°40'$ from the center is 834. From this we must subtract $22.3 \times 7.38 = 165$ nebulae, since the area in question covers 22.3 sq. deg. The Coma cluster therefore comprises a number $\mathcal{N}$ of nebulae the brightnesses of which are greater than $m = 16.5$:

$$\mathcal{N}_{16.5} = 670. \qquad (51)$$

Since the most frequent apparent magnitude is about 17, we conclude that $\mathcal{N}_{16.5}$ is less than half the total number $\mathcal{N}$ of nebulae incorporated in the Coma cluster. This number must be at least equal to $\mathcal{N} = 1{,}500$, and it may be even greater.

Here Zwicky shows that $m = 16.5$ is the limiting magnitude at which it is possible to distinguish images of average nebulae from stars on photographs taken with the 18-in Schmidt telescope.

VII. Comparison of the Three Methods

Each of the three new methods for the determination of masses of nebulae which have been described makes use of a different fundamental principle of physics. Thus, method iii is based on the virial theorem of classical mechanics; method iv takes advantage of the bending of light in gravitational fields; and method v is developed from considerations analogous to those which result in Boltzmann's principle in ordinary statistical mechanics. Applied simultaneously, these three methods promise to supplement one another and to make possible the execution of exacting tests to the results obtained.

Method iii can be applied advantageously only to clusters. Its application calls for the observation of radial velocities of cluster nebulae. The absolute dimensions of the cluster investigated also must be known.

Method iv involves the observation of gravitational lens effects. Measurements of deflecting angles combined with data on the absolute distance of the "lens nebula" from the observer suffice to determine the mass of the lens nebula. The chances for the successful application of this method grow rapidly with the size of the available telescopes. Since method iii gives only the average masses of cluster nebulae and method v furnishes only the ratios between the masses of different types of nebulae, much depends on whether or not a single image of a nebula, modified through the gravitational field of another nebula, can be found. A single good case of this kind would, so to speak, provide us with the fixed point of Archimedes in our attempt to explore the physical characteristics of nebulae.

Method v is the most powerful of all, since it enables us in principle to find the masses of all types of nebulae, provided the absolute mass of a single type of nebula is known or that we have some independent way of finding a sufficiently accurate value of $\overline{\varepsilon_K}$. Method v also results automatically in the knowledge of the statistical weight functions f which govern the distribution of nebulae. The knowledge of these functions is of interest for two reasons:

1. The weight functions derived from direct observations may be compared with those to be expected theoretically, for different "models" of the universe.

Through such a comparison it should be possible to decide whether the universe as a whole is in thermodynamic equilibrium[12,13] or is continually changing.

2. It will be of particular interest, as proposed previously,[14] to investigate the probability P of the occurrence of clusters of nebulae in our "immediate" extragalactic neighborhood, as well as at great distances. If the universe is, for instance, expanding, we should expect P to be different at different distances from the observer. The fact that a great cluster of nebulae, such as the Coma cluster, seems to represent a statistically stationary configuration suggests that a short time scale with 10^9 years as the characteristic age of the universe is hardly adequate. Considerably longer time intervals would seem to be necessary to insure the formation of a stationary distribution of nebulae in great clusters. A detailed analysis of the problem of the time scale, however, must be postponed until the distribution of nebulae in a greater number of clusters has been investigated.

1. It should, however, be mentioned that certain spiral nebulae seem to be stellar systems similar in composition to the local Kapteyn system of our galaxy. For such systems the conversion factors may with some confidence be set equal to the conversion factor of the Kapteyn system. See also E. Hubble, *Ap. J. 69*, 148 (1929).

2. E. Hubble, *The Realm of Nebulae* (New Haven: Yale University Press, 1936), p. 179; also *Ap. J. 69*, 150 (1929).

3. *Gaskugeln* (Leipzig, 1907).

4. *Proc. Nat. Acad. 2*, 517 (1916).

5. F. Zwicky, *Helv. physica acta 6*, 110 (1933).

6. *Ap. J. 83*, 499 (1936).

7. F. Zwicky, *Proc. Nat. Acad.*, May 1937.

8. F. Zwicky, *Phys. Rev. 51*, 290, 679 (1937).

9. H. von Zeipel, *Jubilaeumsnummer d. A. N.*, p. 33 (1921).

10. E. Hubble and M. L. Humason, *Ap. J. 74*, 131 (1931); see also the descriptions of the Coma cluster by M. Wolf, *Heidelberg Pub. 1*, 125 (1902), and H. Shapley, *Harvard Bull. No. 896* (1934).

11. E. Hubble, *Ap. J. 71*, 131 (1930).

12. P. S. Epstein, *Commentary on the Scientific Writing of J. W. Gibbs* (New Haven: Yale University Press, 1936), 2, 104.

13. F. Zwicky, *Proc. Nat. Acad., 14*, 592, (1928).

14. F. Zwicky, *Phys. Rev. 48*, 802 (1935).

108. Nuclear Emission in Spiral Nebulae

Carl K. Seyfert

(*Astrophysical Journal 97*, 28–40 [1943])

In this paper Carl Seyfert describes the spectra for the small, bright nuclei of a type of spiral nebula that has since been called a Seyfert galaxy. Although most spirals exhibit absorption-line spectra similar to those found in the outer layers of stars, the Seyfert galaxies exhibit "forbidden" emission lines of oxygen [O II], [O III], nitrogen [N II], neon [Ne III], and sulpher [S II], [S III]. If the widths of these emission lines are interpreted as Doppler motions, they correspond to velocities of up to 8,500 km sec^{-1}. Because the central masses derived from the rotation curves of these and other spirals are less than 10^{11} solar masses, the escape velocities of the nuclear regions are only a few hundred kilometers per second. The observed motions are far in excess of the escape velocities, and they provide the first evidence for violent, explosive events in the nuclei of galaxies.

When arguing for the extragalactic nature of discrete radio sources, Thomas Gold argued that Seyfert galaxies might produce the high-energy particles needed for intense synchrotron radio emission (see selection 116). Three years later the first radio galaxy, Cygnus A, was found to exhibit intense emission lines of [O II], [O III], [N II], and [Ne III] with velocity widths of about 1,000 km sec^{-1} (see selection 117); and at a later time the radio source 3C 84, Perseus A, was identified with the Seyfert galaxy NGC 1275. Nearly a decade later the quasar (or quasi-stellar object), 3C 48 was found to exhibit the very same unusually wide forbidden emission lines (see selection 120). Since that time most radio galaxies and quasi-stellar objects have been found to have large double or complex radio regions extending far beyond, but centered on, their optical counterparts. This situation strongly implies that the high-energy particles and magnetic fields producing the radio emission have been thrown outward and expanded from past explosions in the optical object.[1] Carl Seyfert seems to have grouped together the first examples of the nuclear activity that may well be responsible for the most luminous objects in the universe.

1. A review of the evidence for violent activity in the nuclei of Seyfert galaxies, radio galaxies, and quasi-stellar objects was given by G. R. Burbidge, E. M. Burbidge, and A. R. Sandage in *Reviews of Modern Physics 35*, 947 (1963). See also selection 112.

Abstract—Spectrograms of dispersion 37–200 A/mm have been obtained of six extragalactic nebulae with high-excitation nuclear emission lines superposed on a normal G-type spectrum. All the stronger emission lines from λ 3,727 to λ 6,731 found in planetaries like NGC 7027 appear in the spectra of the two brightest spirals observed, NGC 1068 and NGC 4151.

Apparent relative intensities of the emission lines in the six spirals were reduced to true relative intensities. Color temperatures of the continua of each spiral were determined for this purpose.

The observed relative intensities of the emission lines exhibit large variations from nebula to nebula. Profiles of the emission lines show that all the lines are broadened, presumably by Doppler motion, by amounts varying up to 8,500 km/sec for the total width of the hydrogen lines in NGC 3516 and NGC 7469. The hydrogen lines in NGC 4151 have relatively narrow cores with wide wings, 7,500 km/sec in total breadth. Similar wings are found for the Balmer lines in NGC 7469. The lines of the other ions show no evidence of wide wings. Some of the lines exhibit strong asymmetries, usually in the sense that the violet side of the line is stronger than the red.

In NGC 7469 the absorption K line of *Ca* II is shallow and 50 A wide, at least twice as wide as in normal spirals.

Absorption minima are found in six of the stronger emission lines in NGC 1068, in one line in NGC 4151, and one in NGC 7469. Evidence from measures of wave length and equivalent widths suggests that these absorption minima arise from the G-type spectra on which the emissions are superposed.

The maximum width of the Balmer emission lines seems to increase with the absolute magnitude of the nucleus and with the ratio of the light in the nucleus to the total light of the nebula. The emission lines in the brightest diffuse nebulae in other extragalactic objects do not appear to have wide emission lines similar to those found in the nuclei of emission spirals.

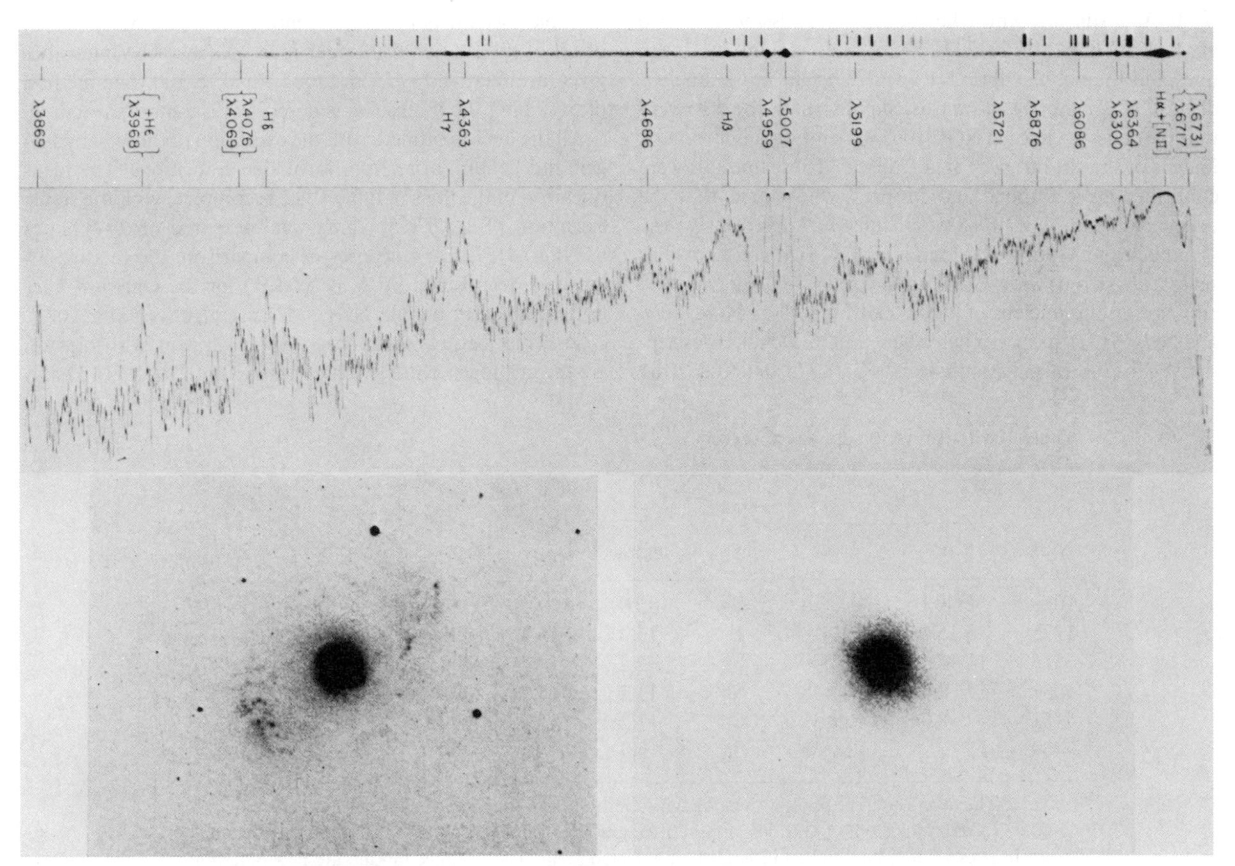

Fig. 108.1 Spectrum, microphotometer tracing, and direct photographs of NGC 4151. The spectrum is an enlargement from a 325-min exposure taken with the one-prism Cassegrain spectrograph and 10-in camera at the 60-in reflector. The photographs (enlargments from a plate taken with the 100-in reflector) show the weak, amorphous arms on the left and the semistellar nucleus on the right.

Many of the spectra of extragalactic nebulae obtained at the Mount Wilson and Lick observatories show one or more emission lines in addition to the usual absorption spectra. In particular, N. U. Mayall[1] finds that 50 percent of his spectra of spirals show the [O II] doublet λ 3,727 in emission either in the nuclear region or in the arms. However, only a very small proportion of extragalactic nebulae show spectra having many high-excitation emission lines localized in the nuclei. These emission features are similar to those found in planetary nebulae and are superposed on the characteristic solar-type absorption spectra. Twelve nebulae[2] are now known which probably belong to this unusual class of objects. Most of them are intermediate-type spirals with ill-defined amorphous arms, their most consistent characteristic being an exceedingly luminous stellar or semistellar nucleus which contains a relatively large percentage of the total light of the system. Figure 108.1 shows a photograph of NGC 4151, a typical example of this type of nebula.

Probably the earliest spectrographic observation of a member of this unusual class of objects was that of NGC 1068 by E. A. Fath,[3] in which he found five emission and two absorption lines. In 1917, V. M. Slipher[4] found hydrogen lines and the nebular lines N1 and N2 bright in the nucleus of NGC 5236. Shortly afterwards, Slipher[5] made the discovery that the emission lines in NGC 1068 were not monochromatic images of the slit but were small "discs." These findings were confirmed by Campbell and Moore,[6] who stated that the bright bands in NGC 1068 were "fully 30 Angstroms wide." Campbell and Moore[7] also found in NGC 4151 emission lines several angstroms wide. E. P. Hubble[8] refers to planetary-type emission in the nucleus of NGC 4051 (as well as NGC 1068 and NGC 4151) in a paper published in 1926. Detailed descriptions of the typical emission spirals NGC 1275 and NGC 4151 were published by M. L. Humason[9] and N. U. Mayall,[10] respectively. In addition, a private communication from Dr. Mayall indicates that NGC 3516 and NGC 7469 show broad bands of hydrogen in emission and hence belong in the group under consideration.

The Observational Material

The present investigation is an intensive study of six of the brightest extragalactic nebulae showing emission bands in their nuclei (table 108.1). Of these six, special emphasis was placed on the three having the brightest nuclei, NGC 1068, 3516, and 4151, because it was possible to observe them with higher dispersion than could be used on the fainter objects.

Here Seyfert describes some of the optical and photographic methods employed when using the 60- and 100-inch reflectors.

Identification of Lines

Table 108.2 lists the lines identified with certainty in any or all of the six nebulae under investigation. The intensities given are relative to $H\beta$ assumed equal to 100. The method of obtaining the intensities is described in a later paragraph.

All the lines identified, with the exception of those of hydrogen and helium, arise from forbidden transitions. The identification of the lines in table 108.2 is complete with the single exception of λ 5,670.5, which was measured on two plates of NGC 4151. For purposes of comparison, the relative intensities determined by A. B. Wyse[11] for the emission lines in the planetary nebula NGC 7027 are given in table 108.2.

The chief features of the spectra which appear in absorption are the cyanogen band (effective wave length, λ 3,873); H and

Table 108.1 Emission spirals observed

NGC	1950 R.A.	1950 Dec.	Type	m_{total}	$m_{\text{nuci.}}$	Spect.	Modulus	No. of Plates
1068	$2^{h}40.1$	$-0°14$	Sb	$10^{m}.0$	$13^{m}.0$	G3	$26^{m}.0$	17
1275	3 15.6	+41 18	E:	13.0	15.5	G3	30.0	4
3516	11 3.4	+72 50	Sa	12.2	13.7	G2:	28.5	6
4051	12 0.6	+44 48	Sb	11.7	14.0	G2	26.0	4
4151	12 8.0	+39 41	Sb	11.2	12.0	G2	26.0	12
7469	23 0.7	+ 8 36	Sa	13.0	14.3:	G0:	29.8	2

Note: The total apparent photographic magnitudes are from the *Shapley-Ames Catalogue of External Galaxies* (*Harv. Ann. 88*, 43, 1932). The apparent magnitudes (photographic) of the nuclei were estimated from short-exposure plates, taken in series with selected areas. The distance moduli are new determinations derived from magnitudes of resolved stars in the arms (NGC 1068), radial velocity (NGC 1068, 3516, 7469), or from association with recognized clusters or groups (NGC 1275, 4051, 4151). The plates used for determinations of nuclear magnitudes and most of the data for computing the distance moduli were supplied by E. P. Hubble. The spectral types were determined by M. L. Humason.

Table 108.2 Intensities of emission lines in six extragalactic nebulae

Atom	λ	NGC 1068	NGC 1275	NGC 3516	NGC 4051	NGC 4151 Core	NGC 4151 Wing	NGC 4151 Core+Wing	NGC 7469 Core	NGC 7469 Core+Wing	NGC 7027[a]
[*O* II]	3,726.2	80:	140:	—	—	100:	—	25:	48:	15:	8
[*O* II]	3,729.7										4
[*Ne* III]	3,869	65:[b]	35:	—	P	65:	—	15:	—	—	40
$H\zeta$	3,889.1	5:	—	—	—	—	—	—	—	—	7
[*Ne* III]	3,968	25:[b]	20:	—	—	25:	—	5:	—	—	15
$H\varepsilon$	3,970.1										8
[*S* II]	4,068.5	20	50	—	—	25	—	5	—	—	5
[*S* II]	4,076.5										2
$H\delta$	4,101.8	20	10:	25	—	20	20	20	—	35:	12
$H\gamma$	4,340.5	40	50	40[c]	40	35	30	35	—	60:[c]	20
[*O* III]	4,363.2	35	40	—	20	75	—	18	—	—	10
C IV[*Fe* III]	4,658.6	—	—	—	—	5	—	1	—	—	0.9
He II	4,685.8	40	—	—	25	25	—	5	—	—	40
[*A* IV]	4,711.4	—	—	—	—	10	—	2	—	—	3
[*A* IV]	4,740.3	—	—	—	—	10	—	2	—	—	7
$H\beta$	4,861.3	100[d]	100	100	100	100	100	100	100	100	100
[*O* III]	4,959.5	400[d]	80	15	55	375	—	90	125	35	430
[*O* III]	5,007.6	1,200[d]	270	40	190	1,150	—	275	300	80	1,190
[*Fe* VII]	5,158.3	5	—	—	—	P[e]	—	P	—	—	2
[*N* I]	5,199.2	25	—	—	—	15	—	5	—	—	3
[*Fe* VII]	5,276.1	5:	—	—	—	P:	—	P:	—	—	1.5±
—	5,670.5	—	—	—	—	P	—	P	—	—	—
[*Fe* VII]	5,720.9	10	—	—	—	20	—	5	—	—	4
[*N* II]	5,755.0	10	—	—	—	—	—	—	—	—	20
He I	5,875.6	15:	—	—	—	30	—	5	—	—	35
[*Fe* VII] [*Ca* V]	6,085.7	30	—	—	—	35	—	10	—	—	5
[*O* I]	6,300.2	85	100	—	30	150	—	35	—	—	40
[*S* III]	6,310.2										15
[*O* I]	6,363.9	30	40	—	20	50	—	10	—	—	20
[*N* II]	6,548.4				25:	100		25	—	—	90
$H\alpha$	6,562.8	1,000[f]	700	600	600	400	375	400	—	—	420
[*N* II]	6,583.9				50:	200		50	—	—	190
[*S* II]	6,717.3	140	210	—	40:	180	—	40	—	—	8
[*S* II]	6,731.5										15

a. The intensities in the planetary nebula NGC 7027 (included for comparison) are from the paper by A. B. Wyse, *Ap. J. 95*, 356, 1942.

b. The measures of λ3,869 and λ3,968 in NGC 1068 are probably affected by the strong absorptions in the underlying G-type spectrum. The true values are probably somewhat greater.

c. The G band in the underlying continuum affects the measures of $H\gamma$ in NGC 3516 and NGC 7469 (wing). The true values are probably somewhat larger.

d. The effect of the narrow central minima in $\lambda\lambda$4,861, 4,959, and 5,007 on the measures is negligibly small.

e. P indicates the presence of a line which is too weak to measure.

f. The sum of the [*N* II] intensities (λ6,548 and λ6,583) approximately equals that of $H\alpha$ in NGC 1068.

K lines of *Ca* II (λ 3,934 and λ 3,969); the G band, composed chiefly of *CH* (measured position, λ 4,303); and the D lines of *Na* (λ 5,890 and λ 5,896). Since the absorption lines are usually ill-defined and sometimes absent altogether, the spectral types in table 108.1 may be subject to fairly large uncertainties.

Relative Intensities of the Emission Lines

The conventional method of obtaining the true relative intensity of a wide emission line is to measure the intensity (corrected for the background) at each wave length in the line relative to the intensity at the corresponding wave length in a comparison star of known color temperature. These values are multiplied by the relative intensity of a black body[12] of temperature equal to that of the comparison star and summed over all wave lengths in the line. The ratio of such summations for two different emission lines, corrected for differential extinction, gives the ratio of relative intensities of the lines freed of instrumental and atmospheric absorption, i.e., the true relative intensities.

The method described above has been used for determining true relative intensities for the six objects under investigation. However, the continuum of each spiral was used instead of stars for comparison standards. Not only does this make it possible to use all the plates on which this continuum appears, but it eliminates uncertainties due to differential extinction, exposure time, changes during exposure of plate-sensitivity and sky conditions.

The method necessitates ascertaining the color temperatures of the continuum of each nebula. The values obtained were 5,250 K for NGC 1068, NGC 1275, NGC 3516, and NGC 4021; 4,750 K for NGC 4151; and 8,000 K for NGC 7469. The estimated uncertainty in the relative intensities obtained is 10%.

Profiles of the Emission Lines

All spectrograms of suitable dispersion were analyzed with the microphotometer for the purpose of measuring profiles of the emission lines. No attempt was made to allow for the instrumental profile, since only those spectrograms were used for which the instrumental widening was a negligible fraction of the width of the emission lines. The profiles of the hydrogen lines are similar when expressed in kilometers per second rather than angstroms. Most of the lines in NGC 1068 were sufficiently symmetrical to warrant combining the two halves of the line. Table 108.3 lists, for the lines measured in each object, the total widths in kilometers per second derived from the mean measured profile.

Here Seyfert exhibits line profiles and discusses the widths of different lines of different nebulae.

Table 108.3 Total widths of emission lines (in km/sec)

λ	NGC 1068	NGC 1275	NGC 3516	NGC 4051	NGC 4151	NGC 7469
3,727	2,600	—	—	—	1,050	—
3,869, 3,968	2,500:	$>$4,000	—	—	1,700	—
4,363	2,800	—	—	—	1,800	—
4,659	—	—	—	—	1,250	—
4,686	2,400	—	—	—	1,300	—
4,711, 4,740	—	—	—	—	1,150	—
4,959, 5,007	3,000	4,500	1,400	1,200	1,300	$<$1,400
6,086	3,000[a]	—	—	—	—	—
6,300	2,700	—	—	—	1,800[b]	—
6,548, 6,563	—	—	—	—	950	—
6,717 } 6,731 }	3,600[c]	—	—	—	{ 1,000: { 900:	— / —
H lines	3,600	$\geqslant$4,500	8,500	3,600	7,500	8,500
H (core of line)	—	—	—	—	1,100	2,500:

a. The true width of the line λ6,086 in NGC 1068 may be considerably smaller than that indicated, owing to the possible presence of λ6,073 (unknown origin) and λ6,102 [*K* IV].

b. In NGC 4151, since λ6,310 influences the profile, the true width of λ6,300 is probably considerably smaller than that indicated.

c. The fact that the two lines are blended in NGC 1068 contributes only a negligible amount to the width of the line.

Minima in Emission Lines

An unusual feature of six of the strongest lines in NGC 1068 and one line each in NGC 4151 and NGC 7469 is the presence of minima in the emission lines. A tentative explanation of these apparent reversals is that the broad emission lines are superposed on a G-type spectrum which contains strong absorption lines at or near the position of the emission lines.

Here Seyfert suggests identifications for the minima.

Discussion

Ten of the twelve extragalactic nebulae known to exhibit high-excitation nuclear emission are spirals of either early or intermediate type (i.e., Sa or Sb). The remaining two, NGC 1275 and NGC 3077, are objects which could be classified as either peculiar elliptical or irregular nebulae. The average absolute magnitude of the twelve emission spirals (−15.8) is somewhat brighter than the average nonemission spiral. All emission spirals have nuclei of high luminosity, which on direct photographs are scarcely distinguishable from stars—a fact first noted by Humason in his paper on NGC 1275.

The widths of the emission bands are probably correlated with the physical properties of the nucleus. The data suggest that the maximum width of the hydrogen emission lines increases with absolute magnitude and with the ratio of light in the nucleus to the total light of the nebula. No definite correlation seems to exist between the line widths and color temperatures or spectral type of the nucleus, and the number of nebulae investigated is too small to indicate any certain correlations of widths of the emission lines with emission-line intensity ratios.

Since some of the brightest emission knots in the periphery of other extragalactic nebulae have total absolute magnitudes comparable with those of the emission nuclei, the question arises whether such emission knots show spectra with wide lines. Spectra of NGC 5471 (abs. mag. −9) and NGC 604 (abs. mag. −10), the brightest emission nebulae in the spirals M 101 and M 33, revealed no evidence of widening of the lines. At the writer's request, Dr. Mayall kindly examined the available Lick Observatory spectra of the giant emission nebula 30 Doradus (abs. mag. −14) in the Large Magellanic Cloud. He reports that the emission lines on the existing plates show no conspicuous evidences of broadening. Mayall also states that the very bright knot in NGC 2366 reveals no large line width. Evidently there is no close connection between the giant diffuse nebulae and the nuclei of spirals showing nebular emission lines.

1. *Lick Obs. Bull. 19*, 33 (1939).
2. NGC 1068, 1275, 2782, 3077, 3227, 3516, 4051, 4151, 4258, 5548, 6814, and 7469.
3. *Lick Obs. Bull. 5*, 71 (1908).
4. *Pop. Astr. 25*, 36 (1917); *Proc. Amer. Phil. Soc. 56*, 403 (1917).
5. *Lowell Obs. Bull. 3*, 59 (1917).
6. *Lick Obs. Pub. 13*, 88 (1918).
7. *Lick Obs. Bull. 13*, 122 (1918).
8. *Mt. W. Contr.* No. 324; *Ap. J. 64*, 328 (1926).
9. *Pub. A.S.P. 44*, 267 (1932).
10. *Pub. A.S.P. 46*, 134 (1934).
11. *Ap. J. 95*, 356 (1942).
12. The black-body intensities may be calculated from table VII in Jahnke and Emde, *Tables of Functions* (Leipzig, 1933) p. 46.

109. The Resolution of Messier 32, NGC 205, and the Central Region of the Andromeda Nebula

Walter Baade

(*Astrophysical Journal 100*, 137–146 [1944])

In 1943, when Los Angeles and Hollywood lay subdued in a wartime blackout, Walter Baade pushed the Mount Wilson 100-in reflector to its very limits in order to achieve the resolution of the nucleus of the Andromeda galaxy and of two of its elliptical companions. The resulting publication in the *Astrophysical Journal*, reprinted here, barely suggests what a *tour de force* of observational technique was required to distinguish the individual stars in these crowded objects; however, in a later series of lectures at Harvard, Baade described in detail the painstaking preparations demanded for this breakthrough. In the course of a night, for example, the focus could vary by several millimeters, owing primarily to temperature changes at the secondary mirror. Baade taught himself to guide and to correct the focus from the comatic image of an off-axis star, at a magnification of 2,800. With ammoniated red plates and 90-min exposures, on the best of nights, he finally captured the faint red stars, the most luminous denizens of these regions.

The resolution of the red stars in the nuclear region of the Andromeda galaxy and of the two elliptical companions (M 32 and NGC 205) paved the way for a new classification of stellar populations. In this paper Baade introduces the term Type II to distinguish these collections of stars from the far more easily resolved Type I population in the spiral arms, where the most luminous stars are blue and brighter than the red nuclear stars. Baade shows that Population I is characterized by open star clusters and the highly luminous O and B stars. Population II is typified by globular clusters and, according to earlier work of Jan Oort, by high velocity stars.

Later Baade recognized the intimate connection of gas and dust with Population I, and he emphasized that the profligate O and B stars must be young, probably having emerged from a dusty womb in comparatively recent astronomical times. He envisioned Population II as a much older group of stars.

In the course of his studies Baade eventually established a further distinction in the variable stars of the two stellar populations, which led to a dramatic revision of both the extragalactic distance scale and the recession factor of the galaxies (see selection 110).

Abstract—Recent photographs on red-sensitive plates, taken with the 100-inch telescope, have for the first time resolved into stars the two companions of the Andromeda nebula—Messier 32 and NGC 205—and the central region of the Andromeda nebula itself. The brightest stars in all three systems have the photographic magnitude 21.3 and the mean color index + 1.3 mag. Since the revised distance-modulus of the group is $m - M = 22.4$, the absolute photographic magnitude of the brightest stars in these systems is $M_{pg} = -1.1$.

The Hertzsprung-Russell diagram of the stars in the early-type nebulae is shown to be closely related to, if not identical with, that of the globular clusters. This leads to the further conclusion that the stellar populations of the galaxies fall into two distinct groups, one represented by the well-known H-R diagram of the stars in our solar neighborhood (the slow-moving stars), the other by that of the globular clusters. Characteristic of the first group (type I) are highly luminous O- and B-type stars and open clusters; of the second (type II), short-period Cepheids and globular clusters. Early-type nebulae (E-Sa) seem to have populations of the pure type II. Both types seem to coexist in the intermediate and late-type nebulae.

The two types of stellar populations had been recognized among the stars of our own galaxy by Oort as early as 1926.

In contrast to the majority of the nebulae within the local group of galaxies which are easily resolved into stars on photographs with our present instruments, the two companions of the Andromeda nebula—Messier 32 and NGC 205—and the central region of the Andromeda nebula itself have always presented an entirely nebulous appearance. Since there is no reason to doubt the stellar composition of these unresolved nebulae—the high frequency with which novae occur in the central region of the Andromeda nebula could hardly be explained otherwise—we must conclude that the luminosities of their brightest stars are abnormally low, of the order of $M_{pg} = -1$ or less compared with $M_{pg} = -5$ to -6 for the brightest stars in our own galaxy and for the resolved members of the local group. Although these data contain the first clear indication that in dealing with galaxies we have to distinguish two different types of stellar populations, the peculiar characteristics of the stars in unresolved nebulae remained, in view of the vague data available, a matter of speculation; and, since all former attempts to force a resolution of these nebulae had ended in failure, the problem was considered one of those which had to be put aside until the new 200-inch telescope should come into operation.

It was therefore quite a surprise when plates of the Andromeda nebula, taken at the 100-inch reflector in the fall of 1942, revealed for the first time unmistakable signs of incipient resolution in the hitherto apparently amorphous central region—signs which left no doubt that a comparatively small additional gain in limiting magnitude, of perhaps 0.3–0.5 mag., would bring out the brightest stars in large numbers.

How to obtain these few additional tenths in limiting magnitude was another question. Certainly there was little hope for any further gain from the blue-sensitive plates hitherto used, because the limit set by the sky fog, even under the most favorable conditions, had been reached. However, the possibility of success with red-sensitive plates remained. From data accumulated in recent years it is known that the limiting red magnitude which can be reached on ammoniated red-sensitive plates at the 100-inch in reasonable exposure times is close to $m_{pr} = 20.0$, the limiting photographic magnitude being $m_{pg} = 21.0$. These figures make it clear at once that stars beyond the reach of the blue-sensitive plates can be recorded in the red only if their color indices are larger than + 1.0 mag.—the larger, the better. Now there are good reasons to believe that the brightest stars in the unresolved early-type galaxies actually have large color indices. When a few years ago the Sculptor and Fornax systems were discovered at the Harvard Observatory, Shapley introduced these members of the local group of galaxies as stellar systems of a new kind.[1] Shortly afterward, however, Hubble and the writer pointed out that in all essential characteristics, particularly the absence of highly luminous O- and B-type stars, these systems are closely related to the unresolved members of the local group.[2] It was therefore suggested that in dealing with the Sculptor and Fornax systems "we are now observing extragalactic systems which lack supergiants and are yet close enough to be resolved." Since the brightest stars in the Sculptor system, according to later observations by the present writer, have large color indices (suggesting spectral type K), it appeared probable that this would hold true for the brightest stars in the unresolved members of the Andromeda group. Altogether there was good reason to expect that the resolution of these systems could be achieved with the 100-inch reflector on fast red-sensitive plates if every precaution were taken to utilize to the fullest extent the small margin available in the present circumstances.

Since success depended so much upon a careful use of the available light-intensities, it may be surprising that the final tests were made in the light of the narrow band $\lambda\lambda$ 6,300–6,700 (on ammoniated Eastman 103E plates behind a Schott RG 2 filter). The reason is the following: It is quite true that nearly twice the speed in the red could have been obtained if a yellow filter, transmitting wave length $> \lambda$ 5,000, had been used instead of the red filter. But experience has shown that the benefits to be derived from the larger range of wave lengths are of doubtful value, particularly in long exposures, because the larger range includes two of the strongest emission lines of the night sky—the green aurora line at λ 5,577 and the red [*O* I] doublet λ 6,300, λ 6,364.

The red doublet at λ 6,300, λ 6,364 has proved to be especially troublesome for astronomical photography, partly because it falls into the region of maximum sensitivity of the E plates, partly because it displays erratic intensity changes from night to night and even in the same night. These changes are large, and it is well known that not infrequently, particularly at the times of sunspot maxima, the intensity of the red doublet surpasses that of the strong green line by a factor 2 or more. Consequently, it is impossible to predict whether on a given night the exposure time for the range $\lambda\lambda$ 5,000–6,700 has to be restricted to 1 hour or can be safely extended to several hours. To avoid any difficulties resulting from uncontrolled sky fog, which are especially serious for objects near the plate limit, it was decided to use the narrower range of wave lengths cut out by the RG 2 filter. Although this filter transmits about 24 per cent of the red doublet, no difficulties have thus far been encountered even with exposure times up to 9 hours. It may be remarked here that the plates to be discussed later are practically free from sky fog.

The minimum exposure times required with the RG 2 filter turned out to be 4 hours. Exposures of this length with a large reflector present a number of problems if critical definition is the prime requisite. That only nights with exceptionally fine definition, together with a practically perfect state of the mirror, would do hardly needs mention. Fortunately, these conditions are easily met on Mount Wilson during the fall months when the Andromeda region is in opposition. But real difficulties were presented by changes of focus during the relatively long exposures. On account of the normal drop in temperature during the night these changes are quite large under average conditions; hence repeated refocusing with the knife edge—usually once every hour—is necessary as the exposure proceeds. Although a special, precision-built plate-holder arrangement is available for such purposes, its manipulation is always somewhat risky because the change from the field to a suitable focus star and back has to be made in complete darkness. Even if such repeated manipulations are performed without mishap during a prolonged exposure, the method remains a makeshift, since between two settings the plate will gradually move out of focus. To avoid both difficulties it seemed best to use only nights on which the focus-changes at the 100-inch are very small if not entirely negligible. Such conditions are not infrequently met on Mount Wilson during the fall, when owing to a temperature inversion, the temperature stays practically constant all night. Neither was it difficult in the present case to select the proper nights. Since in the fall the Andromeda region culminates around midnight, a careful watch of the state of the mirror and of the temperature in the early evening hours permits a fair prediction of the focus-changes during the latter part of the night. Eventual small changes in focus during the exposure can then be inferred from changes in the coma of the guiding star. Although this method has fallen into disrepute because of some bad experiences of earlier observers, the writer has found it as good as the knife-edge test if the following conditions are fulfilled: (1) a nearly perfect figure of the mirror; (2) steady and crisp images; and (3) such an adjustment of the guiding eyepiece that small focus-changes produce marked changes in the coma pattern of the guiding star. All exposures discussed in the following pages have been made in this manner. As a control of the correct handling of the focus-changes, the focus was checked with the knife edge at the end of each exposure. In every case the difference between the last actually used focus and the knife-edge setting was well below 0.1 mm.

The plates of the Andromeda nebula, of Messier 32, and of NGC 205, taken in this manner at the 100-inch reflector during the fall months of 1943, led to the expected results. All three systems were resolved into stars. A description of the plates thus far obtained follows. Since the preparation of adequate reproductions would involve time-consuming experiments impossible under present conditions, illustrations will be published later. The plate of NGC 185 in the following *Contribution* will give the reader an idea how far the resolution of the hitherto unresolved systems of the local group has been successful. [See appended figure 109.2].

Here Baade discusses the details of his photographs of the Andromeda nebula, M 32, and NGC 205. Each red plate exhibits faint stellar images, and the Andromeda nebula is resolvable into stars right up to the very nucleus.

The main facts presented in the preceding descriptions can be summarized in the following four statements:

1. By using red-sensitive plates we have recorded the brightest stars in the hitherto unresolved members of the local group of galaxies.
2. The apparent magnitudes of the brightest stars are closely the same in all three systems, a result which was to be expected because the three nebulae form a triple system.
3. At the upper limit of stellar luminosity, stars appear at once in great numbers in these systems. (In what have been termed the resolvable systems, the brightest stars increase very slowly in numbers for the first 1.0–1.5 mag. below the upper limit of luminosity.)
4. With our present instruments early-type nebulae can be resolved on red-sensitive plates if their distance modulus does not exceed that of the Andromeda group.

Baade estimates that the absolute photographic magnitude of the brightest stars in the central region of the Andromeda nebula and in M 32 and NGC 205 is $M_{pg} = -1.1$.

With these data at hand we are in the position to draw an important conclusion regarding the Hertzsprung-Russell diagram of the stars in early-type nebulae. As pointed out earlier, it has been known for some time that the highly luminous stars of the main branch (O- and B-type stars), together with the supergiants of types F–M, are absent in these systems; in fact, their absence was the reason why up to now the early-type nebulae have proved to be unresolvable. But neither are the brightest stars which we find in them the common giants of the ordinary H-R diagram, because as a group they are nearly 3 mag. brighter (the average early K-type giant of the H-R diagram has the absolute photographic magnitude $M_{pg} = +1.7$, compared with $M_{pg} = -1.1$ for the mean absolute magnitude of the brightest stars in the early-type nebulae).

It is significant that the same situation is known to exist in the globular clusters. M_{25}—the mean absolute photographic magnitude of the 25 brightest stars in a globular cluster—is -1.3, compared with $M_{pg} = -1.1$ for the brightest stars in NGC 205. Now M_{25} in globular clusters and our mean value for the brightest stars in NGC 205 should be closely comparable; for although the value for NGC 205 refers to the several hundred of its brightest stars, it should define nearly the same group of stars as M_{25} in the clusters, because the population of NGC 205, according to its luminosity, exceeds that of the richest globular clusters by a factor 10 to 20. The agreement of the values quoted above is therefore as good as one could expect.

Similarly, there is perfect agreement in the color indices of the brightest stars in early-type nebulae and globular clusters. We derived $CI = +1.3$ mag. for the brightest stars in NGC 205, a value identical with that found by H. Shapley in globular clusters.[3]

We conclude, therefore, that, within the present uncertainties, absolute magnitude and color index of the brightest stars in early-type nebulae are the same as those of the brightest stars in globular clusters. However, the similarity of the stellar populations of early-type nebulae and globular clusters does not end here; for there are strong indications that another, even more unique feature of the H-R diagram of the globular clusters is shared by the stars of the early-type nebulae.

Figure 109.1 represents schematically the H-R diagrams of the stars in the neighborhood of the sun (*shaded*) and of those in globular clusters (*hatched*). To conform with the usual practice, photovisual magnitudes have been used for the absolute magnitudes; hence the brightest stars in globular clusters appear now as stars of $M_{pv} = -2.4$. Both the dispersion and the frequency of the stars have been roughly indicated to convey an idea of the distribution of the two groups of stars in the H-R plane.

As already remarked, the H-R diagram for globular clusters begins with early K-type stars of $M_{pv} = -2.4$. On its downward slope the giant branch soon splits into two separate branches, the one continuing more or less in the original direction, the other proceeding nearly horizontally from spectral type G through F and A into the early B's. For our following argument we are concerned with this horizontal branch of the cluster diagram, which is remarkable for two reasons: (1) it sweeps through the well-known Hertzsprung gap of the ordinary H-R diagram; or, to put it differently, stellar states which seem to be excluded in the ordinary H-R diagram for galactic stars in our neighborhood are quite frequent in the H-R diagram of the globular clusters; (2) the short-period Cepheids, which are such a characteristic feature of the globular clusters, are located along this horizontal branch of the cluster diagram.

In a very interesting paper M. Schwarzschild[4] has recently shown that, if the mean absolute magnitudes and the mean color indices of the short-period Cepheids in a cluster are used as co-ordinates, their domain is restricted to a well-defined, exceedingly narrow strip within the horizontal branch. More than that, Schwarzschild produces excellent evidence that any cluster star located within this strip is actually a cluster-type variable. This suggests the following interpretation: Since the short-period Cepheids are localized in a well-defined, narrow strip of the H-R plane, they can be expected in considerable numbers only in stellar populations which possess a high density in this particular region of the H-R plane. This condition is fulfilled by the H-R distribution of the stars in globular clusters.[5] It is not fulfilled by the stars

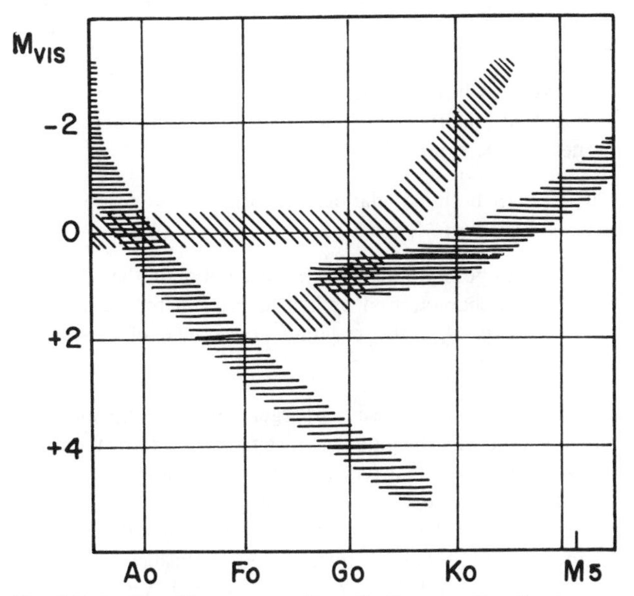

Fig. 109.1 The Hertzsprung-Russell diagram for the two stellar populations. Areas marked by short horizontal lines indicate an ordinary H-R diagram (Type I stars). Areas marked by short angled lines denote the H-R diagram of stars in globular clusters (Type II stars).

in the solar neighborhood (the slow-moving stars) because their distribution exhibits the Hertzsprung gap.[6]

Obviously, the early-type nebulae are in this respect similar to the globular clusters, for we know at least one globular nebula which, according to all indications, is rich in cluster-type variables—the Sculptor system. This extremely loose globular aggregation of stars and the similarly built Fornax system have already been mentioned in this paper. That both systems are closely related to the early-type nebulae follows at once from the fact that their brightest stars have the same luminosity as the brightest stars in globular clusters,[7] but their unusual structure made it difficult to assign them their proper places among the nebulae. It has since become clear that the Sculptor and Fornax systems are merely extreme cases of globular nebulae, because a continuous series of forms, apparently governed by decreasing stellar content, has been established between the highly concentrated objects of this class, such as Messier 32, and the Sculptor and Fornax structures.[8] As a globular nebula the Sculptor system is of particular interest because it is the only object of its kind near enough to permit a search for cluster-type variables. That they are indeed present has been shown by a preliminary test made a few years ago.[2] Although only one pair of plates were intercompared at that time, some 40 variables were found which have all the characteristics of being cluster-type variables. Undoubtedly a more thorough search will increase their number considerably. Because there is every indication that the Sculptor system is rich in cluster-type variables, we conclude that its stellar population has a high density in the Hertzsprung gap, similar to that observed in the globular clusters.

We thus have two strong arguments which indicate that the H-R diagrams of globular clusters and of early-type nebulae are similar, if not identical:

1. In both populations the brightest stars are K-type stars of $M_{pg} \sim -1.1$.
2. In both populations the distribution in the H-R plane is characterized by high density in the Hertzsprung gap, with the resulting appearance of cluster-type variables.

But we can advance a third argument which explains at the same time why the globular clusters happen to be the prototypes of this peculiar type of stellar population which we will call type II in distinction from populations defined by the ordinary H-R diagram—type I. This is the fact that, as far as the present evidence goes, globular clusters are always associated with stellar populations of type II. A good example is our own galaxy, where the globular clusters clearly have the same spatial distribution as the cluster-type variables which are representative of the stars of the second type. It is also significant that among the nebulae composed solely of stars of type II even the absolutely faintest usually have one or two globular clusters. Examples are NGC 205, the Fornax system, and the two faint globular nebulae NGC 147 and NGC 185, discussed in the following paper. This association suggests that globular clusters are properly regarded as condensations in stellar populations of the second type. Under these circumstances it is hardly surprising that their H-R diagram should be essentially identical with that of the larger populations of which they are members.[9]

Although the evidence presented in the preceding discussion is still very fragmentary, there can be no doubt that, in dealing with galaxies, we have to distinguish two types of stellar populations, one which is represented by the ordinary H-R diagram (type I), the other by the H-R diagram of the globular clusters (type II) (fig. 109.1). Characteristic of the first type are highly luminous O- and B-type stars and open clusters; of the second, globular clusters and short-period Cepheids. Early-type nebulae (E–Sa) seem to have populations of pure type II. Both types coexist, although differentiated by their spatial arrangement, in the intermediate spirals like the Andromeda nebula and our own galaxy.[10] In the late-type spirals and in most of the irregular nebulae the highly luminous stars of type I are the most conspicuous feature. It would probably be wrong, however, to conclude that we are dealing with populations of pure type I, because the occurrence of globular clusters in these late-type systems, for instance, in the Magellanic Clouds, indicates that a population of type II is present too. Altogether it seems that, whereas stars of the second type may occur alone in a galaxy, those of type I occur only in association with type II.

In conclusion it should be pointed out that these same two types of stars were recognized in our own galaxy by Oort as early as 1926.[11] Oort showed that the high-velocity stars of our galaxy (our type II) are of a kind quite different from the slow-moving stars (type I) which predominate in the solar neighborhood. Since his conclusions are based on entirely different material and since they supplement those derived in the present paper, they are worth recalling. They may be summarized as follows: (1) stars belonging to the upper main branch of the ordinary H-R diagram (highly luminous O- and B-type stars) are practically absent among the high-velocity stars; (2) the mean absolute magnitude of dwarfs of a given spectral type seems to be the same for high- and low-velocity stars; (3) the relative proportion of dwarfs to giants is much higher among the high-velocity stars then among the ordinary stars; (4) the percentage of double stars is two to three times lower among the high-velocity stars.

Conclusion 1 is in perfect agreement with the result derived in the present paper. Of special interest are conclusions 2 and 3, because they contain the first information about the dwarf branch in populations of type II. Obviously, the dwarf branch of stars of type II coincides closely with the dwarf branch of the ordinary H-R diagram. But the number of dwarfs, as we proceed to fainter absolute magnitudes, increases much faster in type II than in type I. It is very probable that this difference between the two populations is the basis for the well-known empirical criterion by which we distinguish globular clusters from open clusters and which may be stated as follows: Of two clusters with the same number of giant stars, the globular cluster has a much richer background of dwarfs than the open cluster.

Here Baade suggests that the spectral peculiarities and the H-R diagram of Population II stars should be investigated in more detail.

1. *Nature 142*, 715 (1938); *Proc. Nat. Acad. 25*, 565 (1939).
2. *Pub. A.S.P. 51*, 40 (1939).
3. *Star Clusters* ("Harvard Observatory Monographs," no. 2), p. 29 (1930).
4. *Harvard Circ.* No. 437 (1940).
5. At first sight, Messier 13, one of the richest globular clusters but exceptionally poor in short-period Cepheids, seems to present difficulties. But the reason why Messier 13 is so deficient in cluster-type variables is quite apparent from its H-R diagram if the accurate distance modulus recently derived by H. Sawyer-Hogg (*Pub. Dunlap Obs. 1*, no. 11 [1942]) is used. It turns out that the horizontal branch which contains the cluster-type variables is represented in Messier 13 by only a few scattered stars. Obviously, the strength of the horizontal branch varies from cluster to cluster, with Messier 13 and 47 Tucanae at the one extreme, Messier 3 and ω Centauri at the other. The peculiar conditions in Messier 13, therefore, only strengthen our argument.
6. We have convincing proof that the rich star clouds of the Milky Way do not contribute to the number of the cluster-type variables. In a thorough search for variables in a selected field of the Cygnus cloud (Baade, *A.N. 232*, 65 [1928]) it was found that, in contrast to all other types of variables which occur in the cloud in large numbers, the cluster-type variables brighter than magnitude 16.0 are represented in exactly the same number (1 variable per 1.6 square degrees) in which they are found in fields near the galactic north pole. The result is conclusive, because recent investigations have shown that for this particular field of the Cygnus region space absorption up to 10 kpc and more is negligible (cf. Oort and Oosterhoff, *B.A.N. 9*, 325 [1942]).
7. It should be pointed out that the value for the upper limit of luminosity in the Sculptor system, $M_{pg} = -1.8$, published in the earlier note, and the value derived in the present paper for NGC 205, $M_{pg} = -1.1$, are not contradictory. The former is an attempt to define the brightest member of the Sculptor system, the latter is the mean magnitude of the 100 or more brightest stars in NGC 205.
8. *Mt. W. Contr.* No. 697, *Ap. J. 100*, 147 (1944).
9. Similarly, we should regard the open clusters as condensations in populations of type I, an interpretation which hardly needs comment in view of the intimate association of open clusters and slow-moving stars in our own galaxy. It is the more acceptable because it would ascribe the curious variations in the composition of open clusters (Trumpler's types) to the large-scale variations in the composition of populations of type I which have been noted not only in our own galaxy but also in several of the nearer extragalactic systems.
10. The strong concentration of both globular clusters and short-period Cepheids toward the center of our galaxy indicates that the main mass of the stars of type II is located in this region, which, in turn, suggests a structure of our galaxy very similar to that of the Andromeda nebula.
11. *Groningen Pub.* No. 40 (1926).

Appended Figure

Fig. 109.2 NGC 185 (ammoniated 103E plate behind a Wratten 29F filter, $\lambda\lambda$6,000–6,700, 4-hr exposure, October 23, 1943). NGC 185, considerably brighter than NGC 147, has been classified by E. P. Hubble as Ep, the peculiarity being the abnormally slow increase in intensity toward the center. The photographic enlargment (2.7 times the original) will enable the reader to form his own judgment regarding the resolution of early-type nebulae on recent negatives. Two points may be mentioned in particular: (1) an examination of the halation rings around the brighter stars shows clearly that the grain pattern of the emulsion is very much smaller than that produced by the numerous faint stars in the nebula; and (2) the resolution decreases markedly with the distance from the optical axis. Thus the south-preceding part of the nebula, which was closer to the optical axis, is better resolved than the north-following part, where the coma begins to cause blurring. As a result of the attempt to bring out the faintest stars on the photograph, the intensity gradient in the central region of the nebula has been badly exaggerated. (Courtesy Hale Observatories, see *Astrophysical Journal 100*, 148 [1944]).

110. A Revision of the Extra-Galactic Distance Scale

Walter Baade

(*Transactions of the International Astronomical Union 8*, 397–398 [1952])

In 1950, a quarter of a century after Edwin Hubble had first set up the extragalactic distance scale using the period-luminosity relation for the Cepheid variables, the original calibration was still universally accepted (see selection 104). There emerged from this calibration, however, the uncomfortable conclusion that our Milky Way system seemed considerably larger than any other known galaxy. Also, in a curious anomaly, the globular clusters of M 31, the Andromeda galaxy, seemed systematically fainter than those of our own galaxy. Finally, the expansion age of the universe directly derived from the Hubble constant of recession yielded an age too small to be consistent with the oldest dates established from radioactive isotopes.

Hardly anyone appreciated that these three problems might be related until the dramatic announcement made by Walter Baade[1] to the Commission on Extragalactic Nebulae at the September 1952 Congress of the International Astronomical Union in Rome. As acting president of the commission, Baade rose to announce that he had recalibrated the zero point of the period-luminosity relation for classical Cepheids and that the distance to M 31 had been underestimated by a factor of 2. Consequently, M 31 was larger than had previously been believed, its globular clusters were intrinsically brighter, and the time scale of the universe was doubled.

This remarkable revision in the distance scale grew out of Baade's studies of the two stellar populations (selection 109), in which he eventually concluded that each population has its own pulsating variables, with different period-luminosity relations.[2] Over forty years earlier Harlow Shapley had unwittingly combined data from two dissimilar groups of stars, the Cepheids in the Magellanic clouds as originally given by Henrietta Leavitt (Population I) with those of the globular clusters (Population II).[3] This procedure led to incorrect estimates for the distances of extragalactic objects (based on Population I variables), but the calculations of distances of the globular clusters within our own galaxy (based on Population II variables) remained essentially correct. Hence the size of our galaxy based on estimates of the distances to globular clusters remained unaltered, but the size of its spiral structure based on the distances to Population I Cepheids was modified.

1. This report was actually transcribed by Fred Hoyle, who acted as secretary for the session.

2. A more complete discussion of the two different period-luminosity relations is given by W. Baade in *Publications of the Astronomical Society of the Pacific 68*, 5 (1956).

3. H. Shapley, *Astrophysical Journal 48*, 89 (1918).

In his opening remarks Dr Baade pointed out that although in the past instrumental opportunities for the study of extragalactic problems had been extremely limited, there was now hope that several large telescopes would soon become available. In particular, Dr Baade referred to the progress that had been made with the new reflector at the Lick Observatory.

Dr Baade then went on to describe several results of great cosmological significance. He pointed out that, in the course of his work on the two stellar populations in M 31, it had become more and more clear that either the zero-point of the classical cepheids or the zero-point of the cluster variables must be in error. Data obtained recently—Sandage's colour-magnitude diagram of M 3—supported the view that the error lay with the zero-point of the classical cepheids, not with the cluster variables. Moreover, the error must be such that our previous estimates of extragalactic distances—not distances within our own Galaxy—were too small by as much as a factor 2. Many notable implications followed immediately from the corrected distances: the globular clusters in M 31 and in our own Galaxy now come out to have closely similar luminosities; and our Galaxy may now come out to be somewhat smaller than M 31. Above all, Hubble's characteristic time scale for the Universe must now be increased from about 1.8×10^9 years to about 3.6×10^9 years.

Dr Thackeray reported that results obtained from the study of the Magellanic Clouds fitted very well with the suggestion that the old zero-point of the classical cepheids was in error by about 1 magnitude.

Work at Pretoria has led to the discovery of three cluster-type variables in NGC 121 which most probably is one of the globular clusters of the SMC. Instead of being of apparent magnitude 17.5, as one would expect according to the present zero-point for the classical cepheids (type I), they are of apparent magnitude 19. Commenting on Thackeray's result Baade stressed the point that the search for cluster-type variables in the Magellanic Clouds should be extended to, say, magnitude 20, in order to settle the question whether cluster-type variables occur in the LMC or whether they are absent. He also stated that the present discrepancy in the zero-points of the classical cepheids on the one hand and of the cluster-type variables on the other could be most easily checked in the Magellanic Clouds. If cluster-type variables could be discovered in a very few globular clusters of the Clouds (two would be sufficient), we could be certain that we are dealing with members of the Clouds and not with foreground objects. A comparison of their magnitudes with that of classical cepheids of a given period would then immediately reveal the present discrepancy in the two zero-points and would determine the error numerically.

Prof. Oort added that the distances given by the modified zero-point of the classical cepheids had also cleared up the radius discrepancy of η Aql, as found by Stebbins and Whitford. Prof. Shapley then asked Dr Baade if he could describe in a little more detail how he (Dr Baade) had arrived at the error in the zero-point of the classical cepheids. Dr Baade offered two arguments in support of his conclusion:

First argument According to the present zero-points we should expect to find the cluster-type variables of the Andromeda nebula at $m_{pg} = 22.4$ since the distance modulus of this system, derived from classical cepheids, is $m - M = 22.4$. The very first exposures on M 31, taken at the 200-inch telescope, showed at once that something was wrong. Tests had shown that we reach with this instrument, using the $f/3.7$ correcting lens, stars of $m_{pg} = 22.4$ in an exposure of 30 min. Hence we should just reach in such an exposure the cluster-type variables in M 31, at least in their maximum phases. Actually we reach only the brightest stars of population II in M 31 with such an exposure. Since, according to the latest colour-magnitude diagrams of globular clusters, the brightest stars of the population II are photographically about 1.5 mag. brighter than the cluster-type variables we must conclude that the latter are to be found in M 31 at $m_{pg} = 23.9 \pm$, and not at $m_{pg} = 22.4$ as predicted on the basis of our present zero-points.

We have also convincing proof that the brightest stars of population II in M 31 are properly identified because when they emerge above the plate limit the globular clusters of M 31 begin to be resolved into stars.

Second argument Since the observations just mentioned were made in the central lens of the Andromeda nebula one might object that the relatively dense, unresolved background could introduce serious photometric errors. In order to avoid this objection the photographic magnitude of the brightest stars in the outer parts of NGC 205, the elliptical companion of M 31, was determined. Again, the same value as in M 31 was found, i.e. $m_{pg} = 22.4 \pm$ for the brightest stars of the population II; hence, with $M_{pg} = -1.5$ for their absolute magnitude, $m_{pg} = 23.9 \pm$ for the cluster-type variables. The assumption made in this case that the population II of NGC 205 can be substituted for the population II of M 31 is fully confirmed for the area where the outlying stars of the two systems intermingle. They are completely indistinguishable from one another.

It should be emphasized that these are rough first data indicating the order of the corrections which the present constants require.

Dr Dufay pointed out that the change in the scale of extragalactic distance had an interesting application so far as the sizes of gaseous emission nebulae were concerned. Formerly there had been systematic differences of size between such nebulae in the Galaxy and both the Magellanic Clouds and NGC 6822. The changed distance scale now equalized the sizes.

Dr Dufay then went on to say that at one time it was thought that the smallest galaxies were closely comparable in luminosity with the brightest globular clusters. Then the work of Holmberg and Bigay on integrated galactic magnitudes had shown that the faintest galaxies were about 1 mag. brighter than the globular clusters. Would the new distances increase this gap still further?

Dr Baade replied that indeed the gap between the faintest galaxies and the brightest clusters is now widened to about 2 magnitudes.

111. Redshifts and Magnitudes of Extra-Galactic Nebulae

Milton L. Humason, Nicholas U. Mayall, and Allan R. Sandage

(*Astronomical Journal 61*, 97–162 [1956])

Edwin Hubble's first formulation of the velocity-distance relation (selection 106) referred only to relatively nearby galaxies (velocities less than 1,000 km sec^{-1} and distances less than 2 Mpc), but he initiated a program to explore the relation to the largest distances possible. By the early 1930s Milton Humason had extended the spectrographic observations to the limits of the 100-in telescope at Mount Wilson, and together Humason and Hubble found the velocity-distance relation to hold for velocities as large as 20,000 km sec^{-1} and for distances as large as 32 Mpc.[1] They derived a recession factor, now called the Hubble constant, of $H_o = 558$ km sec^{-1} Mpc^{-1}. Because of the impossibility of measuring the distances to the weaker nebulae, Hubble and Humason assumed that all nebulae have the same absolute magnitude and inferred a velocity-distance relation from a log redshift-apparent magnitude relation.

In the paper given here, Milton Humason, Nicholas Mayall, and Allan Sandage examine the log redshift-apparent magnitude relation for 474 nebulae, using data obtained during a 25-yr period at the Mount Wilson, Palomar, and Lick observatories. Within the accuracies of the data, the relation is linear and the slope is found to be consistent with that expected for a homogeneous, isotropic, expanding universe. Here we find that Hubble's recession factor has been revised to $H_o = 180$ km sec^{-1} Mpc^{-1} because of a revision in the estimate of extragalactic distances caused by Walter Baade's correction of the period-luminosity relation (selection 110), and by the realization that many of the objects that Hubble assumed to be stars are actually regions of ionized gas (H II regions) and in some cases clusters of stars.[2]

When compared with the initial Hubble diagram, the data given here have an extremely large scatter, presumably caused by a large spread in the absolute magnitudes of the nebulae. By choosing the brightest members of clusters of galaxies, Humason, Mayall, and Sandage were able to select nebulae with similar absolute magnitudes and thereby decrease the scatter in the log redshift-apparent magnitude relation. In this way they attempted to measure the second-order, nonlinear term in the relation. This departure from linearity is expressed as a value for the deceleration parameter, q_o, which is related to the total matter density of the universe.[3] If q_o is greater than 0.5 the universe will eventually stop its present expansion, but if q_o lies between 0.0 and 0.5 the universe will continue to expand forever. In the paper given here, a statistically significant deceleration of the universe is suggested with $q_o = 2.6 \pm 0.8$. Nevertheless, similar measurements of first-member cluster data over the past twenty years have failed to establish whether q_o is greater or less than 0.5. This uncertainty is further compounded when evolutionary effects are taken into account.[4]

1. E. Hubble and M. L. Humason, *Astrophysical Journal 74*, 43 (1931).

2. More recent revisions of the extragalactic distance scale give $H_o = 57 \pm 3$ km sec^{-1} Mpc^{-1} (A. Sandage and G. A. Tammann, *Astrophysical Journal 196*, 313 [1975]).

3. The meaning of the second-order term and the definition of q_o was first given by F. Hoyle and A. Sandage in the *Publications of the Astronomical Society of the Pacific 68*, 301 (1956).

4. A. Sandage, *Astrophysical Journal 134*, 916 (1961).

Abstract—There are three main sections to the present discussion. Part I contains redshifts of 620 extragalactic nebulae observed at Mount Wilson and Palomar. Included in these data are redshifts for 26 clusters of nebulae. Part II contains redshifts for 300 nebulae observed at Lick, together with a comparison of results for 114 nebulae in common with the Mount Wilson-Palomar lists. Part III is a discussion of these new redshift data in combination with photometric data. The redshift-apparent magnitude relation is investigated for (1) field nebulae with and without regard to nebular type, (2) isolated groups, and (3) clusters of nebulae. The principal corrections applied to the apparent magnitudes are discussed in two of the three appendices. Appendix A gives the procedure for correcting the published magnitudes for the effect of different photometer apertures. Appendix B describes the theory and computation of the correction for the selective effect of redshifts. In the final Appendix C, a provisional evaluation of the Hubble redshift parameter H is made by two independent methods.

The principal results of this study may be stated as follows. (1) For those nebulae observed in common there is a negligible mean systematic difference between the redshifts from the two sources. (2) Spectrographic coverage is 63 per cent complete to $m_{pg} = 12.9$ in the Shapley-Ames catalogue for nebulae north of $\delta = -30°$. (3) The log redshift-magnitude relation for field nebulae with and without regard to type is linear to within the accuracy of the data. (4) The log redshift-magnitude relation for the cluster data confirms the linearity for small $\Delta\lambda/\lambda_0$ and shows an apparently significant departure from linearity for shifts of the order of $\Delta\lambda/\lambda_0 = 0.2$. This non-linearity indicates deceleration of the expansion if interpretation is made by theoretical equations due to Robertson. Because of the cosmological significance of this last result, the accuracies of the various quantities that lead to it are examined. It is concluded that a deceleration should be regarded as tentative until Whitford's results are available for the spectral energy distribution of the distant nebulae, and until an adequate theory of stellar evolution is advanced to explain the Stebbins-Whitford effect. (5) The Hubble redshift parameter H is provisionally estimated from (a) the magnitudes of resolved stars in NGC 4321 that have been isolated from the emission H II regions, and (b) from the assumption that the brightest field and cluster nebulae are giants of luminosity comparable to the Andromeda nebula, with the result that $H = 180$ km/sec per 10^6 pc.

General Introduction

More than 25 years ago Hubble (1929) announced a relationship between velocities and distances of extragalactic nebulae. Since he realized that this first formulation referred to only a relatively small distance, he initiated an exploratory program to follow the relationship to the greatest distances attainable with the largest telescope. The successful outcome of that program has become widely known, especially through publication of his book, *The Realm of the Nebulae* (Hubble 1936a), and his professional lectures. The last of these (Hubble 1953) summarizes the observational basis for the early restricted velocity-distance relation, and the later far-reaching law of redshifts.

Despite the comprehensiveness of Hubble's extragalactic researches that used the 100-inch to the limits of its power for observations of faint nebulae, he regarded them in sum as a "preliminary reconnaissance." This appraisal, although a characteristic understatement, emphasized the need for many more nebular redshifts and magnitudes. These spectrographic and photometric data are now available in considerable numbers, on a systematic basis, and with improved precision, chiefly as the result of Hubble's inspiring influence on his colleagues.

Parts I and II. The Mount Wilson–Palomar and the Lick Lists of Redshifts

Here Humason, Mayall, and Sandage provide an extensive discussion of the cooperative observational program at the Mount Wilson, Palomar, and Lick observatories. The observed redshifts of 620 extragalactic nebulae and the redshifts corrected for solar motion are given in tables I, II, and III, not reproduced here. The standard deviation in this data is 39.4 km sec^{-1}. After correction for solar motion, 12 nebulae have negative velocities. Of these objects, 7 belong to the local group, 4 belong to the Virgo cluster, and 1 is the nearby nebula NGC 253. Similar redshift data are given for an additional 300 nebulae in table V, also not reproduced here. A comparison of the 114 nebulae in both sets of redshift data shows a systematic difference of only 28 km sec^{-1}. The redshift observations are essentially complete down to 11.6 mag. The measured apparent magnitudes given by Pettit (1954) and Stebbins and Whitford (1952) are corrected and given in appendix A, which is not reproduced here.

Part III. Discussion of the Spectrographic and Photometric Data

The philosophy behind the present discussion is governed by the observational approach. Two numbers, z ($\equiv \Delta\lambda/\lambda_0$) and m, are observed. Corrections are made to both quantities to free them from effects extraneous to the problem at hand. The redshifts are corrected for the solar motion with respect to the centroid of the local group. This correction is made because it appears likely that the systematic redshift does not operate within the local group (Hubble 1936a; Humason and Wahlquist 1955) and that the measured redshifts of its members reflect the motion of the sun with respect to these nebulae. The correction for solar motion is described in Parts I and II. The observed magnitudes are freed from the latitude-effect caused by obscuration in our own galaxy by the equations $\Delta P(b) = 0.25$ (csc $b - 1$) for photographic magnitudes and $\Delta V(b) = 0.18$ (csc $b - 1$) for photovisual magnitudes. These heterochromatic magnitudes are further changed to a bolometric magnitude scale by the K correction, which accounts for the effects of redshift. The theory and computation of the K correction for P and V magnitudes is given in appendix B for the case where the Stebbins-Whitford effect (1948) is zero. Discussion of the modification to the value of K due to the presence of this effect is also given. The K correction accounts only for the selective effects caused by the redshift. Other corrections to the magnitudes, such as the so-called energy and number effects, are not made, as was once the custom, since such effects are absorbed into the theoretical equations used for the interpretation of the data.

The sequel is divided into three sections. These contain the [log cz, m] relation for (1) the field nebulae, (2) selected isolated groups, and (3) the nebular clusters. Appendix C contains the calibration of these relations in terms of distance with a provisional value of the redshift parameter H.

In appendix C the Hubble expansion parameter, H, is provisionally estimated to be $H = 180$ km sec^{-1} Mpc^{-1}. We do not reproduce this appendix or appendix B, which discusses the K correction.

FIELD NEBULAE The redshift catalogues of tables I and V, together with the magnitudes in table A1, provide the data for discussion of the [log cz, m] relation for the field nebulae. Humason's redshift values in table III and the magnitudes reported in Table XII [not reproduced] provide the data for the clusters.

The Field Nebulae. For a linear redshift-distance relation of the form $cz = Hr$, with r defined by

$$\log r = [m - \Delta m(b) - K - M + 5]/5,$$

the relation between $m - \Delta m(b) - K$, called m_C in the following, and z will be of the form

$$m_C = 5 \log cz + (M - 5 - 5 \log H). \qquad (1)$$

Here all of the refinements required for a proper definition of distance are glossed over. Both Robertson (1955) and McVittie (1956) treat this problem, and their results are implicitly contained in a later equation used for the cluster data. For the relatively close field nebulae such refinement is unnecessary. Equation (1) neglects another effect. Due to the finite speed of light, we look back in time to events when light now observed was emitted from nebulae at different distances. Thus, the observed pairs [log cz, m] refer to the condition of the universe at *different* cosmic times (see e.g. Robertson (1933) for a definition of cosmic time), the difference being just the light-travel time between the source and the observer. To transform the observed "world picture" to the so-called "world map"—the condition of the universe at any given cosmic time—requires knowledge of the form of the expansion. Formulae based upon the method of Taylor series (Robertson 1955) are employed for this problem. This time effect is not important for distances such that $z \ll 1$, and this is the case for the majority of the field nebulae. Interpretation of the [log cz, m] relation for the nearby field nebulae with the simplified equation (1) is adequate for the present discussion.

The nebulae in the general field have been divided into 7 groups for analysis according to nebular type. Figures 3 to 9 [not reproduced] and figure 111.1 show the correlation between the corrected photographic magnitude $P_C \equiv P - \Delta P(b) - K$ and log $cz \equiv \log c\, \Delta\lambda/\lambda_0$ for each group. Linear relations of the form $P_C = A \log cz + B$ were fitted to the data by least squares. The linearity of the redshift-distance relation is tested by the closeness of the value of A precisely to 5. Differences in the mean absolute magnitude $\overline{M(m)}$ for the nebular types are obtained from the differences in B, on the assumption that the value of H is unique. Two solutions were made for each group. Both solutions include all the data, but Solution 1 considers A and B as unknowns, while Solution 2 adopts A as 5.000 and treats B as unknown. Table 111.1 gives the resulting solutions and probable errors. The lines drawn in figures 3 to 9 and figure 111.1 are those of Solution 2 since this case is the only one compatible with current theories. The computed probable errors are merely formal and are somewhat unrealistic, due to the nature of the scatter in the [log cz, m] pairs.

This scatter is caused by at least four effects. (1) The large spread in absolute magnitude among the nebulae appears in the correlations as a spread in apparent magnitude at a given log cz. Indeed, early attempts (Hubble 1936c) were made to derive the luminosity function for nebulae from the residuals of the [log cz, m] plot, but the results were affected by the highly selective nature of the data. (2) Redshifts represent the sum of the systematic distance effect and the random motion of the nebulae themselves. The exact size of these random motions is not known yet, but they seem to be of the order of 200 to 300 km/sec. When they are of the same size as the distance effect, unsymmetrical deviations from the [log cz, m] relation will occur if the peculiar motions themselves are

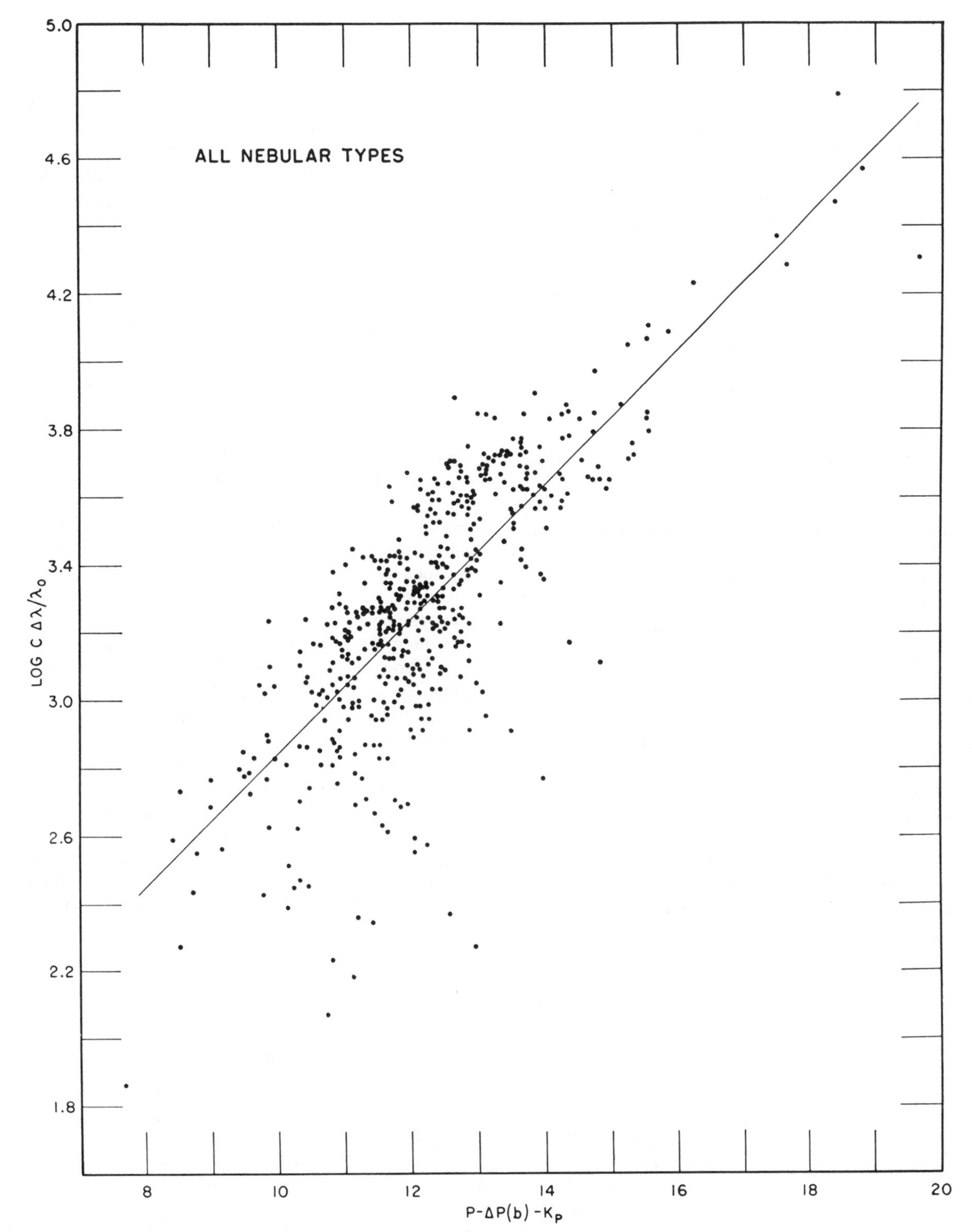

Fig. 111.1 The redshift-magnitude relation for 474 nebulae of all nebular types. The vertical axis is the logarithm of the product of the velocity of light, c, and the redshift $z = \Delta\lambda/\lambda_0$. The horizontal axis gives apparent magnitudes corrected for the dimming caused by the Doppler shift of the continuum energy spectrum of each nebulae and the absorption of light in our galaxy.

symmetrical about the distance effect. This circumstance explains part of the large scatter at log cz less than 3.0. (3) The other part of the larger scatter at small cz is explained by a selectivity effect favoring the nearer of the intrinsically faint nebulae. Objects such as the dwarf irregulars of low surface brightness are difficult to identify and observe at large distances, and hence these points are missing from the diagrams for larger redshifts than about 1,000 km/sec. (4) The values of m and z themselves contain errors of observation, but the discussion in Parts I, II, and appendix A shows these errors to be small compared with the observed scatter.

Within the total uncertainties of the solutions, all data in the first 8 groups of table 111.1 are consistent with a linear law. The solution of greatest weight, $N = 474$, gives the computed A as 5.028 ± 0.116 compared with the predicted value of 5.000.

To check the isotropy of the redshift law, correlations were made for nebulae in the north and south galactic polar regions with $|b| \geqslant 30°$. The last two solutions of table 111.1 show the result. A significant, and as yet unexplained, difference exists between the two hemispheres. The A values differ from each other, but even more serious is the difference of 0.70 mag. in B between the hemispheres for Solution 2. The southern nebulae appear to be brighter than the northern ones at the same redshift. Part of this difference is probably due to observational selection, since many nebulae in the south galactic polar cap are in high southern declinations not reachable from these latitudes. Table VIII of Part II [not reproduced] shows that the redshift catalogues are essentially complete for nebulae brighter than $m_{pg} = 11.6$ north of $\delta = -30°$. South of this declination very few redshifts are available. Comparison of the north with the south galactic hemisphere is biased, since the data for the northern hemisphere are more complete. Counts in the Shapley-Ames catalogue show that 37 nebulae brighter than $m_{pg} = 11.6$ are south of $\delta = -30°$. All of these do not satisfy $b > 30°$S but none satisfy $b > 30°$N. It would be of interest and importance to assemble [log cz, m] data for these bright southern nebulae so that an unbiased test of the isotropy could be made with the field nebulae. Observatories in the southern hemisphere could contribute significantly toward answering this fundamental question of isotropy.

The redshift-magnitude diagrams for the north and south galactic polar regions are not reproduced here, but table 111.1 illustrates the "non-cosmological" slope of the southern data.

A small part of the difference in figures 11 and 12 [not reproduced] may be due to photometric difficulties. Many nebulae south of $b = -30°$ are at high southern declinations. This is a difficult region to reach with high photometric precision from Mount Wilson due to the strong Los Angeles lights in the south and west quandrants. No check on this suggestion is possible at present because of the lack of overlap in Pettit's, Stebbins and Whitford's, and Holmberg's magnitude catalogues in the south latitudes.

Whatever the cause of the difference between figures 11 and 12, strong evidence against appreciable anisotropy of the redshift law is provided from the high degree of isotropy in the cluster data. Further work on the field nebulae is required for a satisfactory solution.

To the extent that observational selection in the present sample is comparable for the various types of nebulae, differences in their mean absolute magnitudes are reflected in the differences between the values of B from Solution 2 tabulated in table 111.1. Table X [not reproduced] exhibits these differences, normalized so that $\overline{\Delta M} = 0.00$ mag. for the solution using all data. Tabulated again are the number of nebulae N in each group; negative signs for ΔM indicate higher luminosities.

Table 111.1 Solutions for the field nebulae

	Solution 1		Solution 2		
Neb. type	A	B	A	B	N
E	$5.882 \pm .347$	$-7.400 \pm .246$	5	$-4.375 \pm .212$	117
So	$4.630 \pm .378$	$-2.843 \pm .234$	5	$-4.070 \pm .253$	67
Sa	$4.717 \pm .312$	$-3.401 \pm .229$	5	$-4.360 \pm .243$	54
Sb	$5.181 \pm .337$	$-4.974 \pm .182$	5	$-4.400 \pm .175$	76
Sc + SBc	$4.329 \pm .377$	$-1.931 \pm .307$	5	$-4.030 \pm .358$	90
SBo + SBa	$4.854 \pm .385$	$-3.466 \pm .252$	5	$-3.950 \pm .260$	36
SBb	$5.618 \pm .672$	$-6.618 \pm .261$	5	$-4.570 \pm .233$	27
All Types	$5.028 \pm .116$	$-4.324 \pm .129$	5	$-4.235 \pm .128$	474
All Types $b \geqslant +30°$	$5.102 \pm .208$	$-4.250 \pm .169$	5	$-3.895 \pm .165$	257
All Types $b \leqslant -30°$	$6.757 \pm .412$	$-10.636 \pm .283$	5	$-4.595 \pm .219$	132

According to this table, the SBb are statistically the brightest while the SBo's and SBa's are the faintest. The total range of the differences is 0.62 mag. With the SBb's excluded, for which only 27 nebulae are available, the range becomes 0.43 mag. The sizes of the probable errors for B, ranging from ± 0.21 mag. for the E nebulae to ± 0.36 mag. for the Sc plus SBc, show that most of the computed differences in $\bar{M}$ are illusory, and that the mean absolute magnitudes of nebulae along the entire sequence of classification, excluding the irregulars which show a very large dispersion, are nearly constant for this particular sample. Due to the effects of observational selection, these results may, however, be different for different samples.

ISOLATED GROUPS There exist in space several well-known isolated, physical groups of nebulae such as the local group, the nearby M81 and M101 groups, the Leo group, and Stephan's Quintet. Many of these aggregates were suspected from the geometrical aspects of the grouping before the redshift data became available. The redshift lists provide a powerful method for confirming such groups and for discovering new ones. While the general problem of the small-scale nebular distribution for nearby systems is not considered here, it is evident that steps toward its solution may now be taken with the present redshift data.

The large scatter in figures 3 to 12 is primarily due to the spread in the luminosity function for nebulae. If some *a priori* means were available for selecting nebulae with similar absolute magnitude, this scatter would become smaller and a more refined analysis of the data would be possible. It is reasonable to expect that such a homogeneous nebular sample might be found among the brightest objects in physical aggregates of moderate to large population, since such nebulae would be chosen from a definite part of the luminosity function. This expectation was tested and confirmed by analysis of 27 groups found with the redshift data. The [log cz, m] relation for the first-ranked member of each group not only has smaller scatter than figure 111.1, but it shows preference for high absolute magnitudes. The brightest member nebula for 23 of the 27 groups is at least 0.5 mag. brighter than the mean line of figure 111.1. The points for 19 of the 27 groups are at least 1.0 mag. brighter than the mean line, while 8 are more than 1.5 mag. brighter, and 5 more than 2.0 mag. brighter than this line. The increased homogeneity, gained by restricting attention to the brightest nebulae of populous aggregates, is important in the discussion of the cluster data.

The observational data for twenty-seven groups are given in table XI, not reproduced.

THE CLUSTER DATA For a given apparent magnitude, data for the brightest members of the great clusters of nebulae permit the deepest penetration into space. Further, these same nebulae provide the homogeneity of sample so important in the search for a possible second-order term in the redshift law. The [log cz, m] relation can, therefore, be carried farther and be more precisely defined with the cluster data.

Here Humason, Mayall, and Sandage discuss the schraffierkassette measurement technique for determining magnitudes of distant clusters. In this technique a jiggle-camera is used to provide square images of uniform density, with residual errors of about 0.1 mag. The photometric data for eighteen clusters are given in table XII, which we do not reproduce.

INTERPRETATION OF CLUSTER DATA Robertson (1938) has shown that, in an expanding universe, the intensity of light received at time t_0 from a source radiating at time t_1 is given by

$$l_{\text{bol}(t_0)} = \frac{L_{\text{bol}(t_1)}}{4\pi R_0{}^2\sigma^2(1+z)^2}, \tag{2}$$

where R_0 is the scale coefficient in the line element at the time of observation t_0 and σ is related to the dimensionless radial coordinate. The relation connecting m_{bol} and z is then given by Robertson (1955) as

$$m_{\text{bol}} = 5 \log cz + 1.086\left(1 + \frac{R_0\ddot{R}_0}{\dot{R}_0{}^2} - 2\mu\right)z + \text{const.} \tag{3}$$

Here, $\dot{R}_0/R_0 \equiv H$ is the Hubble redshift parameter and $\ddot{R}_0$ is the second time derivative of the metric scale factor, both evaluated at t_0. The quantity μ is related to the time rate of change of the absolute bolometric magnitude of the nebulae, plus the rate of change of that part of the K correction due to the Stebbins-Whitford effect, and plus the effect of any intergalactic obscuration. This equation accounts for the difference in the light-travel time for the nearby and distant clusters, by reducing the "world picture" to the "world map."

Following this equation, the data have been analyzed in the form $m_C = A \log cz + Bz + D$ for both P_C and V_C. Least-squares solutions were made for 3 cases: (1) A, B, and D were treated as unknowns, (2) A was considered to be precisely 5, with B and D as unknowns, and (3) A and D were considered unknowns, with $B = 0$. Table 111.2 gives the results. The goodness of fit in each case may be judged by the dispersions of the distributions of the magnitude residuals. These dispersions, σ_0, are also given in table 111.2. Case 1 fits the data best. Solutions in Cases 2 and 3 are the only ones compatible with equation (3), since in them $A = 5$ by assumption for Case 2 and to within the probable error for Case 3. Case 2 is adopted in the following discussion.

The observed magnitude dispersion of 0.32 mag is small and suggests little internebular absorption.

Table 111.2 Summary of solutions for eighteen clusters

Unknown	Case 1	Case 2	Case 3
P data			
A	5.73	5.000	5.029 ± .121
B	−5.62	−1.180 ± .875	0.00
D	−8.55	−5.81 ± .092	−6.03 ± .519
σ_0 (P) mag.	0.282	0.315	0.302
V data			
A	5.72	5.000	4.925 ± .138
B	−6.34	−1.976 ± .895	0.00
D	−9.40	−6.71 ± .094	−6.56 ± .590
σ_0 (V) mag.	0.292	0.323	0.344

The data are plotted in figure 111.2 and figure 14 [not reproduced] with the solid lines drawn from Solutions 2. The difference in the B values between the photographic and photovisual solutions is undoubtedly caused in the following way by the Stebbins-Whitford effect. The computed K corrections in appendix B are those which would be valid in the absence of the SW effect. If this effect is due to stellar evolution, K will be a function of time as well as of redshift. The correct value to be applied is $K(z, t_1)$ instead of $K(z, t_0)$ as given in appendix B. The difference between $K(z, t_1)$ and $K(z, t_0)$ is absorbed in μ of equation (3).

The term of greatest interest is B, because it describes deviations from linearity. The value of B from Solution 2 is only twice its probable error, but two uncertain elements not allowed for in the data should be emphasized. These are: (1) the aperture effect in the faint clusters, and (2) possible uniform internebular obscuration. Corrections for both effects not only preserve the negative sign of B, but they make its absolute value larger. That the aperture effect is indeed present may be seen by separate analysis of the [log cz, m] relation for the 1st and 10th brightest nebulae with the data of table XII. A larger negative B is found with the 10th ranked nebulae, due to the smaller aperture correction required for the higher-ranked cluster members.

Here μ is related to $\dot{M}$ and $\dot{K}$, the time rates of change of M and K. In the absence of interstellar absorption $\mu = 0.46\ H^{-1}(\dot{M} - \dot{K})$. The K term denotes the correction required to convert observed magnitudes to a bolometric scale, M denotes the absolute magnitude of the nebula, and $\dot{M}$ denotes its rate of change with time that is caused by stellar evolution.

We are now in a position to consider the results contained in the [log cz, m] relation of figure 111.2. This material suggests the following five major conclusions.

(1) The slope of the [log cz, m] correlation line for small z is as close to 5 as the probable errors of the determination. This conclusion rests upon (a) the small magnitude residuals of the solution for Case 2 with the slope assumed to be 5, and (b) the direct determination of the slope as 5.029 ± 0.121 and 4.925 ± 0.138 for Case 3. This result means that for small z, the redshift-distance relation is linear, on the supposition that there is no general internebular obscuration. If we postulate the existence of general uniform internebular absorption, the redshift-distance relation is non-linear. The absorption, expressed as F mag. per unit distance, must be of just the right amount to cancel the non-linearity of the redshift law so that the observed [log cz, m] relation remains linear.

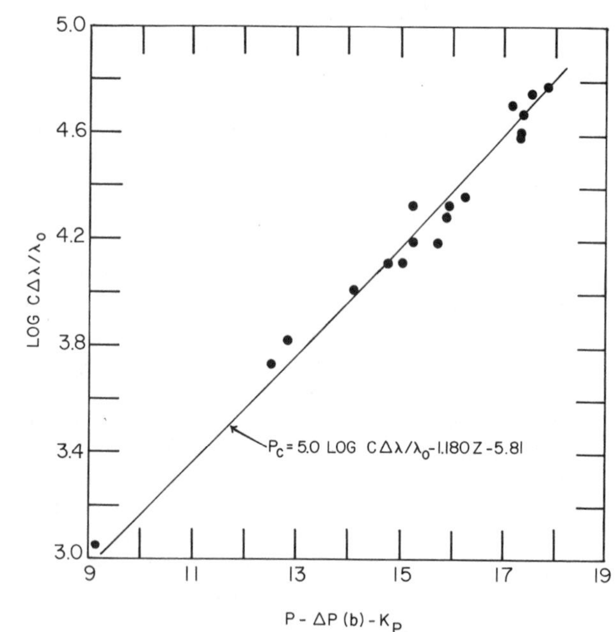

Fig. 111.2 The redshift-magnitude relation for the brightest members of clusters of nebulae. The apparent photographic magnitudes have been corrected only for the latitude effect and for the selective effect of the redshift. The "energy" and "number" corrections are not included in the data but are introduced into the theoretical equations used for the interpretation.

Such an interpretation is highly unlikely but cannot definitely be excluded.

(2) The expansion appears to be isotropic, since no separation of points occurs between the 12 clusters in north galactic latitudes and the 6 southern clusters. This is a stronger test than that for the field nebulae, since the cluster data (1) probably are less affected by observational selection and (2) show smaller scatter about the mean correlation line.

(3) The absolute magnitude of the brightest nebulae in clusters is nearly equal to the very brightest of the field nebulae. This near equality is seen if the line drawn in figure 111.2 for the clusters is transferred to figure 111.1 for the field nebulae. Such a line defines a limit above which few field nebulae occur. On this basis there appears to be an upper limit to the absolute magnitude of extragalactic nebulae close to that of the brightest cluster members.

(4) The departures from uniformity for any postulated intergalactic obscuration must be distributed with $0.30 > \sigma_F \geqslant 0$ mag.

(5) The second-order term, B, in the redshift law is negative and appears to be statistically significant. Its value is -3.0 for the photovisual data and -2.2 for the photographic data if an allowance is made for an aperture correction of 0.20 mag. at the distance of the Hydra cluster. These values, together with equation (3), give

$$\frac{R_0\ddot{R}_0}{\dot{R}_0^{\,2}} = -(3.0 \pm 0.8) + 2\mu_\mathrm{P} \qquad (5)$$

$$\frac{R_0\ddot{R}_0}{\dot{R}_0^{\,2}} = -(3.7 \pm 0.8) + 2\mu_\mathrm{V} \qquad (6)$$

where the subscripts P and V stand for photographic and photovisual wave lengths. If $2\mu_\mathrm{P} > 3.0$ or if $2\mu_\mathrm{V} > 3.7$, then $\ddot{R}_0$ is positive and the expansion is accelerating; otherwise it is decelerating.

Estimates of $\dot{M}$ can, at present, come only by appeal to some theory of stellar evolution for systems of Population II. Current ideas for such evolution stem primarily from the work of M. Schwarzschild that has appeared in a series of papers with his collaborators (Oke and Schwarzschild 1952; Sandage and Schwarzschild 1952; Härm and Schwarzschild 1955). Application of these ideas to the particular case of the globular cluster M3 (Sandage 1954b) provides a basis for an estimate of $\dot{M}$. Within the framework of this theory, the observational data show that the M3 stars were formed about 5×10^9 years ago. The theory predicts that the brightest stars in the cluster have moved from their original places on the main sequence in the H-R diagram into the giant region, and subsequently, after burning most of their fuel, have disappeared to faint luminosities. Presumably, the cluster was brighter in early times because of the presence of these bright stars. In the available time of 5×10^9 years, all stars brighter than absolute bolometric magnitude $+3.5$ have evolved from the main sequence. We know with some certainty only the evolutionary tracks for the present time t_0. If we assume that tracks for slightly different luminosities are homologous, i.e. parallel in the log T_e-M_bol plane, an evaluation of $\dot{M}$ can be made. The change in the absolute magnitude of the main-sequence break-off point in time t_1 to t_0 is, for small t_0/t_1,

$$\Delta M_\mathrm{bol} = 2.5 \log \mathfrak{M}_1/\mathfrak{M}_0 + 2.5 \log t_0/t_1, \qquad (7)$$

where $\mathfrak{M}_0$ and $\mathfrak{M}_1$ are the respective masses of the stars at the break point. We wish to compute this change in the bolometric magnitude in the last one billion years. If $t_0 = 5 \times 10^9$ yr., $t_1 = 4 \times 10^9$ yr. and with the ratio $\mathfrak{M}_0/\mathfrak{M}_1$ obtained by iteration from the mass-luminosity law, then $\Delta M_\mathrm{bol} = 0.31$ mag. For homologous evolutionary tracks this value also equals the change of the bolometric magnitude of the entire cluster if we assume that most of the light comes from stars brighter than $M_\mathrm{bol} = +3.5$. If an appreciable fraction of the total light comes from stars fainter than $M_\mathrm{bol} = +3.5$, then the ΔM_bol for the system will be less than 0.3 mag. This computation gives, therefore, an upper limit to $\dot{M}$. If the case for elliptical nebulae is similar to that of the globular clusters, then $\dot{M} \leqslant 0.3$ mag. per 10^9 yr.

The time rate of change of K is estimated to be about 0.3×10^{-9} mag. yr^{-1}. Because $\dot{M} \approx \dot{K}$ the value of μ is small ($2\mu_P = 0$ and $2\mu_V = 1.5$). If $\dot{K}$ is ignored, however, then $2\mu \approx 4$ and an accelerating universe (positive $\ddot{R}_0$) is possible.

The result that $\ddot{R}_0$ is negative has such important cosmological implications that it is well to review the steps in its evaluation and to indicate the uncertainties at each point. The basic data are the [log cz, m] pairs. Of the two, observational errors are appreciable only in the magnitudes.

In order to obtain a positive value for $\ddot{R}_0$, observational errors in photographic magnitudes must be larger than 0.6 mag, a value the authors think probably larger than can be ascribed to observational uncertainty.

Although it would be appropriate to end this paper with a definite statement of the possible cosmological models consistent with the present data, such a statement cannot be given at present for the following reason. With the field equations of general relativity, a series of mathematical models are obtained for the character of the expansion. (See, e.g., Einstein 1945 or Bondi 1952.) These models show how the function $R(t)$ depends on time, and they differ from one another according to the sign of the space curvature $(1, 0, -1)$ and of the value of the cosmological constant Λ. Three of the crucial observational items required for a choice between the models are (1) the sign of $\ddot{R}_0$, (2) the value of $1/H$, and (3) independent knowledge of the "age of the universe"—really the time since the beginning of the expansion—from say an astrophysical theory for the age of the oldest stars or from a geological age for the earth. When these three items are known, a weeding out of certain inconsistent models can be made. Unfortunately, the present uncertainty in the value of $1/H$ and the imminent rediscussion of the sign of $\ddot{R}_0$ with Whitford's anticipated results for computing $K(z, t_1)$ make such a discussion inappropriate at the present time.

Bondi, H. 1952, *Cosmology* (Cambridge University Press).

Einstein, A. 1945, *The Meaning of Relativity*, appendix for the second edition, (Princeton University Press).

Härm, R., and Schwarzschild, M. 1955, *Ap. J. 121*, 445.

Hubble, E. 1925, *Ap. J. 62*, 409.

———. 1929, *Proc. Nat. Acad. Sci. 15*, 168.

———. 1930, *Ap. J. 71*, 231.

———. 1936a, *The Realm of the Nebulae* (New Haven: Yale University Press).

———. 1936b, *Ap. J. 84*, 158.

———. 1936c, ibid., *84*, 270.

———. 1953, *M. N. 113*, 658.

Hubble, E., and Humason, M. L. 1931, *Ap. J. 74*, 43.

Humason, M. L., and Wahlquist, H. D. 1955, *A. J. 60*, 254.

McVittie, G. C. 1956, *General Relativity and Cosmology* (London: Chapman and Hill).

Oke, J. B., and Schwarzschild, M. 1952, *Ap. J. 116*, 317.

Pettit, E. 1940, *Ap. J. 91*, 159.

———. 1954, ibid., *120*, 413.

Robertson, H. P. 1933, *Rev. Mod. Phys. 5*, 62.

———. 1938, *Zs. Astroph. 15*, 69.

———. 1955, *Pub. A. S. P. 67*, 82.

Sandage, A. R. 1954a, *A. J. 59*, 180.

———. 1954b, Liege Symposium Volume, *Les Processus nucleaires dans les Astres*, p. 254.

Sandage, A. R., and Schwarzschild, M. 1952, *Ap. J. 116*, 463.

Stebbins, J., and Whitford, A. E. 1937, *Ap. J. 86*, 247.

———. 1945, ibid. *102*, 318.

———. 1948. ibid. *108*, 413.

———. 1952, ibid. *115*, 284.

112. On the Evolution of Galaxies

Viktor Amazaspovich Ambartsumian

(From *La structure et l'evolution de l'univers*, Institut International de Physique Solvay, ed. R. Stoops [Brussels: Coudenberg, 1958] pp 241–274)*

A commonly held view of galaxy formation has been that large, rotating gaseous masses collapse under their own gravity until a state of equilibrium is reached between the rotational and gravitational forces. Because no significant change in the equilibrium configuration can be produced without substantial changes in the distribution of mass and angular momentum, it was long believed that no significant departures from equilibrium would be produced during most of the lifetime of a galaxy. In an early rebuttal of this view, James Jeans noticed that it did not explain the persistence of spiral structure. Jeans argued:

> It is more and more difficult to resist a suspicion that the spiral nebulae are the seat of types of forces entirely unknown to us, forces which may possibly express novel and unsuspected properties of space. The type of conjecture which presents itself somewhat insistently, is that the centers of the nebulae are of the nature of "singular points", at which matter is poured into our universe from some other, and entirely extraneous, spatial dimension, so that, to a denizen of our universe they appear as points at which matter is being continuously created.[1]

Carl Seyfert provided some of the first specific evidence that unusual types of activity occur in the nuclei of some spiral nebulae (see selection 108). He showed that certain galaxies, subsequently named after him, have small, bright nuclei containing forbidden emission lines. When the widths of these features are interpreted as Doppler motions, the velocities are found to be well in excess of the escape velocities of the nuclei. Therefore, matter must be flowing from these nuclei out into intergalactic space.

In the paper given here, Viktor Ambartsumian provides a wide range of evidence for mass loss and other departures from equilibrium both in individual galaxies and in groups of galaxies. He shows, for example, that the measured velocities of galaxies in small groups and in clusters of galaxies exceed the escape velocity of the relevant systems. This condition, first noticed by Fritz Zwicky in 1933,[2] means that these systems are dynamically unstable. Ambartsumian goes on to argue that the newly discovered radio galaxies provide evidence for the ejection of matter from optically visible galaxies. He uses the optically visible polarized jet of the radio galaxy Virgo A as evidence for the ejection of gas and high-energy electrons. This view had already been suggested when Walter Baade and Rudolf Minkowski identified the first extragalactic radio source, Cygnus A, with an optical object that exhibits the same intense emission lines as those seen in Seyfert galaxies (see selection 117).

Additional evidence for the existence of extraordinarily energetic phenomena in the nuclei of galaxies was presented by Boris Vorontsov-Velyaminov in 1959 in his *Atlas of Interacting Galaxies*, many of which appear in chains that seem to be physically connected. Further examples of these groups have been published by Halton Arp in his *Atlas of Peculiar Galaxies*, and measurements by Margaret Burbidge and Wallace Sargent suggest that many of these systems are gravitationally unbound.[3] Even the quasars seem to confirm Ambartsumian's view that violent activity is the rule in extragalactic objects. The intense, broad emission lines of quasars suggest that these objects represent the first stages of evolution in a violent event in the nucleus of a galaxy and that at later stages the optical luminosity decreases and the radio source expands.[4]

1. J. H. Jeans, *Astronomy and Cosmogony* (Cambridge: Cambridge University Press, 1926).

2. F. Zwicky, *Helvetica physica acta 6*, 110 (1933). Also see selection 107.

3. E. M. Burbidge and W. L. W. Sargent, "Velocity Dispersions and Discrepant Redshifts in Groups of Galaxies," in *Nuclei of Galaxies*, ed. D. J. K. O'Connell (New York: American Elsevier, 1971).

4. G. R. Burbidge, E. M. Burbidge, and A. R. Sandage, *Reviews of Modern Physics 35*, 947 (1963).

* Minor stylistic changes have been made by the editors.

Introduction

Attempts to approach the solution of the question of the origin of galaxies were, until the present, based mainly on speculations connected with the remarkable fact of mutual recession of extra-galactic nebulae. In other words, these attempts have been made within the frames of the existing cosmological theories, which as a rule, are based only on some integral and averaged characteristics of the world of galaxies surrounding us.

Although the study of the nearest galaxies and also the investigation of groups and clusters of galaxies have not been advanced far enough, nevertheless rich material has been gathered, which may be possible to rely on while considering the problems of the origin and evolution of galaxies. Among the data obtained from observations those concerning multiple galaxies, groups of galaxies and clusters of galaxies deserve special attention.

In this connection it is worth while to dwell briefly on the significance that the study of multiple stars and stellar clusters have had for the problem of the origin and evolution of stars.

1. In the thirties of the present century the very existence of stellar clusters in our Galaxy, together with some statistical-mechanical considerations on the non-reversible character of the process of dissolution of clusters led to the conclusion that stars forming a cluster originate together. In other words, it was established that at least some stars in the Galaxy have been formed in groups.[1]

2. Statistical data concerning double stars have led to the conclusion that the components of each double star have a common origin.[2]

3. The mere existence of stellar associations has made it possible to come to a conclusion on the continuing process of star formation in the Galaxy.[3] The discovery of the expansion of stellar associations has allowed us to conclude that at least the majority of the stars of the disc population has also been formed within the stellar groups, now already disintegrated.[4]

4. The study of the spectrum-luminosity diagram for stellar clusters has allowed us to construct interesting schemes of the evolution of different stars. These schemes need further verification, but in any case their significance for the theory of stellar evolution is extremely great.

5. Marking out the Trapezium type multiple systems rendered it possible to establish the existence of particularly young multiple stars and thus to approach the very moment of the formation of stellar groups.

In this respect it seems to us that the situation in the world of galaxies is even more favourable. Multiple galaxies and groups of galaxies provide interesting material for the reasoning out on the group formation of galaxies. Moreover, the tendency of grouping in the world of galaxies is so strong that any study of galaxies is inevitably connected with the question of the nature of one group or another.

For instance, such nearby giant stellar systems as M 31, M 81, and M 101 are the centres of the highly interesting groups of galaxies. Our Galaxy itself has several companions of different nature.

Therefore, it is natural to think that the question of the origin of galaxies cannot be separated from the question of the origin of their groups and clusters.

Let us now turn to the fact that in multiple galaxies the periods of revolution reach a billion years and more, while in clusters, the time necessary for one revolution around the centre of the cluster is to be measured by several billions of years. However, the age of galaxies themselves is thought to reach also several billions of years. In such a case the multiple galaxies as well as clusters of galaxies in their present state must, even in configurations of their components, maintain traces of primary conditions of group formation. And this seems to mean a possibility to approach at least the kinematics of those phenomena which have led to the formation of groups.

In the present report we begin the consideration of the question with the problem of multiple galaxies and cluster galaxies. The study of some multiple systems, however, has led us to the conclusion that an intimate relation exists between the mechanism of the formation of components and the ways of formation of characteristic details in the structure of galaxies. For the time being, it is difficult to understand the exact character of this relation, but it seems to us that further study of this aspect of the question will open some prospects in solving the question of the origin of the observed structures of separate galaxies. Finally the investigation of radio-galaxies representing systems in which violent non-stable processes take place, has shown that in every such system we meet with traces of some sort of duplicity. Comparing this with other data concerning multiple galaxies we see that the narrower the double or the multiple system is the sharper the traces of non-stability become. All this emphasizes again the significance of the data concerning multiple galaxies and the tendency of galaxies to form groups for the problem of the origin and evolution of galaxies.

Confining this report to the above mentioned questions I hope that actual data referring to the stellar population of galaxies and representing interest to the problem of evolution of galaxies will be elucidated in other reports presented at this conference.

1. The Physical Meaning of Clustering Tendency

After the works of Zwicky[5] and also of Shane and Scott[6] there is a sound basis to consider that the majority of galaxies are members of clusters or of groups of galaxies, while the number of isolated galaxies in the general metagalactic field is comparatively small. In this sense, it is even difficult to speak about any general homogeneous metagalactic field, as distinguished from the condensations of galaxies. In reality we have a metagalactic field mainly composed of different clusters and groups, i.e. of inhomogeneities of different scales.

In this respect the situation in Metagalaxy strongly differs from that of the stellar systems where the general stellar field with a slowly changing density usually dominates, while clusters are comparatively rare and separate inhomogeneities that occur in this field.

It may be concluded from statistical mechanics that clusters and groups must eventually dissolve.[1] At the same time the dissolution will proceed differently and require different lengths of time, depending on whether the clusters and groups in consideration are in a steady (or in a quasi-steady) state with negative total energy, or in a state when among the cluster-members, there is a certain percentage which possesses positive energy and can immediately leave the clusters with great velocity.

Because the dissolution time scale for the first case is on the order of a hundred billion years, and because the dissolution time scale for the second case is a hundred million or a few billion years, we may conclude either that clusters dissolve because of members containing positive energy or that the dissolution proceeds so slowly that it cannot have any essential significance.

Conclusion *In the present conditions of the Metagalaxy clusters and groups may either persist or disintegrate. But they cannot be enriched at the expense of galaxies which have originated independently from them.*

2. Deviation from Dissociative Equilibrium

Attention should be paid to the fact that among the members of clusters of galaxies known to us, we meet with double and multiple galaxies. Double and multiple galaxies are met with more often in loose clusters of the Virgo type. Apparently they are seldom met with in compact clusters of the Coma type. In such relatively poor groups as the Local group of galaxies, double and multiple systems are comparatively frequent. Nevertheless, if the existence of subdwarf galaxies of the type of the objects in Sculptor and Fornax is considered, then, apparently, every one of the multiple galaxies transforms into a group consisting of approximately a dozen members. For example, our Galaxy with the Magellanic clouds forms a triple system. But it is also surrounded with several subdwarf systems of the Sculptor type. The galaxy in Andromeda is a multiple system composed of five members. But probably there are also some systems of the Sculptor type near it. That is why it seems that we should rather speak of the groups that include our Galaxy and M 31 respectively. Let us remember, however, that when speaking about the multiplicity of stars we do not take into account the possible presence of planets since the latter possess masses insignificant compared with stars. When determining the multiplicity of galaxies it is advisable not to take into account the Sculptor type systems, just as the globular clusters, which have, apparently, masses a little less than the masses of galaxies of the Sculptor type, are not taken into account.

In such a case we may admit that in our Local group, containing only a few individual galaxies (M 33, NGC 6822, IC 1613 and perhaps some others), there is one triple galaxy and one other galaxy with a higher multiplicity. We may put the question as to what the mathematical expectation of numbers of double and multiple galaxies amounts to in our Local group in case of dissociative equilibrium. It turns out that in case of dissociative equilibrium the mathematical expectation of the number of double galaxies in the Local group should be less than 0.05, and the mathematical expectation of the number of triple galaxies and galaxies of higher multiplicity even many times less.

Therefore, the fact that we have two systems of very high multiplicity in the Local group of galaxies is a very strong deviation from dissociative equilibrium. The situation is similar in many other groups and clusters. In some instances the degree of deviation from dissociative equilibrium is many times greater.

If double galaxies and multiple galaxies originate by way of mutual capture (in the course of triple encounters) or otherwise, from previously independent single galaxies, then at the primary stage of the evolution of clusters there should certainly have been a deviation in them from dissociative equilibrium. However, these deviations should have been in the opposite direction, i.e. the number of multiple galaxies should have been less than in case of dissociative equilibrium. Only after a long time the average number of multiple galaxies in clusters could have reached the theoretical value in accordance with dissociative equilibrium. The percentage of multiple galaxies with an accuracy determined by statistical fluctuations would, in this case, never surpass the indicated equilibrium value.

The fact, that the percentage of multiple systems is indeed much higher than this theoretical limit, proves our assumption that multiple galaxies are formed from individuals to be incorrect.

CONCLUSION *Components of any multiple galaxies are formed together.*

This conclusion is based on statistical considerations, and that is why it is valid for the overwhelming majority of multiple galaxies. Therefore, some exceptions are possible, and some insignificant minority of multiple galaxies could have been formed on account of mutual capture (by triple encounters or otherwise) from individual galaxies.

3. The Observed Configurations of Multiple Galaxies

During the lifetime of galaxies (a few billions of years) perturbations in the state of multiple galaxies resulting from nearby passages of outer galaxies should be insignificant. Therefore, it is possible to presume that these states still bear traces of primary conditions of formation of multiple systems. It is reasonable, therefore, to search for information about the mechanism of formation in statistical data, characterizing the bulk of double and multiple galaxies.

Among multiple galaxies the percentage of the Trapezium type significantly surpasses that of the usual type of system; among multiple stars the opposite is true. It is only among multiples containing O type stars that we observe a large percentage of Trapezium stars. By a Trapezium-type multiple system Ambartsumian means a system where it is possible to find three components, *a*, *b*, *c* whose distances *ab*, *bc*, and *ac* are of the same order of magnitude. If it is impossible to find three such components, the multiple system is called a system of the usual type.

4. The Cause of Predominance of Trapezium

The fact that the overwhelming majority of multiple stars have configurations of the usual type has been explained as follows. The configurations of the Trapezium type are, as a rule, unstable, even if the total energy of the multiple system is negative. When the motions in the system of the usual type can be approximated by a sum of a few Keplerian (i.e. periodical) motions, the movements in the system of the Trapezium type are very complex and involved. During nearby encounters of two components, occurring in the course of time, one of them may acquire enough kinetic energy to leave the system. This same mechanism operates in open stellar clusters. Calculations show that for the dissolution of a system possessing a configuration of the Trapezium type, it is necessary, on the average, that its components make a few revolutions. For most stars such an interval of time is insignificant in comparison with their age. Therefore, the overwhelming majority of the systems of the Trapezium type, originated in Galaxy, have been dissolved. This explanation gives us, at the same time, the possibility to understand the exception observed in the case of O and Bo stars. Many of these stars possess the age of the order of 10^6 years and considerably less than 10^7 years. Meanwhile the period of revolution, in observed multiple stars of the Trapezium type, must be of the order of 10^5–10^6 years. Therefore, the components of these trapeziums may have time to perform only a few revolutions around their centres of gravity. That is why these multiple systems have failed to be dissolved.

For *multiple galaxies* the situation is the same as for O–B stars. The age of multiple galaxies is measured by a few billions of years, whereas the periods of their revolutions reach the order of a billion years. Consequently, the components of multiple galaxies may have time to make only a small number of revolutions. That is the reason why multiple galaxies having configurations of the Trapezium type have not been dissolved.

Since the mechanism of the disintegration of observed multiple galaxies acting in a selected manner only on the systems of the Trapezium type should not have had time to influence most of the systems, it seems that the present distribution of configurations into types reflects that primary distribution which depends upon the laws of the formation of multiple galaxies.

CONCLUSION *The high percentage of configurations of the Trapezium type among multiple galaxies is in full agreement with the ratio of the age of galaxies to the periods of revolution in multiple systems.*

5. Multiple Systems with Positive Energy

In the previous paragraph we spoke about the "revolutions" of components in a multiple galaxy. This implies that in our systems the components, at least in their primary stage of evolution, hold one another by gravitational forces. In other words, till the present, we presumed that all multiple galaxies are systems with negative total energy.

However, in order to determine the sign of energy of a given multiple system, beside the data on the configuration of components, reliable data are necessary on the masses and velocities of components. Unfortunately, the knowledge we have on the masses of double and multiple galaxies is obtained under the assumption that these systems possess negative energies, i.e. under the assumption which it is necessary to examine.

To question the possibility of the existence of multiple systems with positive energy may seem superfluous, because, in the case of stars, all double and multiple systems, well studied until now, prove to have only negative energy. Let us, however, suppose for a moment that some multiple stars are formed in the Galaxy with negative energy while others with positive. Systems of positive energy must disintegrate in the order of 10^5 years. This time is very short if compared with the age of the overwhelming majority of stars. Only thus must we explain that multiple stars whose internal motions have so far been studied by us possess a negative energy. But as the multiple stars of the Trapezium type are young, it is impossible to state without further study that all of them possess negative energy. On the contrary some systems of the Trapezium type, met with in stellar associations, are likely to have positive energy. For example, the star ADS 13626 in the cluster IC 4996 (Cygnus association) has visible components, whose radial velocity differences are so great, that they cannot be explained by assuming a negative total energy.

Similar reasoning is true with regard to multiple galaxies, since the age of some multiple galaxies may be such (about a billion years) that the components could not recede far from one another, although they are in the process of mutual recession. However, the final solution of the question of the existence of multiple galaxies possessing positive energy is possible only on the basis of a critical study of the actual material which, indeed, is still very poor.

We shall give, at this point, some data in favour of the positive sign of total energy of some multiple galaxies.

a. If we assume that all double and multiple galaxies possess negative total energy then by observing the radial-velocity differences of their components it will be possible to come to statistical conclusions about the average masses of galaxies. Separate examination of radial-velocity differences of double galaxies and of galaxies of higher multiplicity under this assumption has brought us to the conclusion, that the average masses of galaxies in systems of higher multiplicity are approximately three times greater than the masses of components of double galaxies. But there is no reason to believe that the nature of galaxies in systems of various multiplicity is different. The only way out from this contradiction is to assume that among systems of higher multiplicity systems with positive energy are comparatively oftener met with. Admitting their energy to be negative we get an artificial increase of the probable mass value for galaxies included in those systems.

This evidence of the existence of multiple systems with positive energy has an indirect character. Therefore, in the following we shall introduce two direct facts.

b. Let us consider the group of galaxies connected with M 81. It is composed of four bright galaxies: NGC 3031 (M 81), NGC 3034 (M 82), NGC 2976 and NGC 3077 and also of some faint galaxies. The apparent integral photographic magnitudes of the four bright galaxies given above, according to Holmberg's determination,[7] are equal to 7.85, 9.20, 10.73, 10.57. If we do not want to admit too high values of mass/luminosity ratios, we must assume that the masses of all the members of the group, except those given above, are small and therefore we may consider the group as a wide quadruple system. Its configuration corresponds to the Trapezium type. That all the four given galaxies are probably members of one physical group is evident from the following considerations. Three of them (except M 82) have radial velocities close to one another. Their mean radial velocity corrected for the solar motion is equal to + 72 km/sec. But the galaxy M 82 has a radial velocity equal to + 410 km/sec. Therefore, doubt may arise whether it does or does not belong to the group. However, there is a very close physical similarity between galaxy M 82 and galaxy NGC 3077. They both belong to the class of irregular galaxies composed of the population of the second type and both possess nearly the same high surface brightness. Because the coincidence of the characteristics indicated above is very rarely met with among the relatively bright galaxies, it is very improbable that here we have an accidental projection of M 82 in the region of the sky occupied by the group. Thus, it is almost certain that all four galaxies are physically connected with one another. Then the differences in radial velocities must be explained by orbital motions.

It is natural to assume, at the beginning, that the brightest among the four galaxies, M 81, possesses the greatest mass. But its mass is determined from its rotation by Guido Münch.[8] It is about 10^{11} M⊙. The radial velocity of M 82 differs from that of M 81 by 327 km/sec. The difference of space velocities may be much more. It is not difficult to calculate that such a difference in velocities may correspond only to hyperbolic motion if the sum of the masses of galaxies M 81 and M 82 is less than $3 \cdot 10^{11}$ solar masses. So, if we assume an elliptical movement, the mass of the galaxy M 82 must, in any case, exceed $2 \cdot 10^{11}$ $M_\odot$. Thus the dominating role in the system must be placed by the galaxy M 82. If so, difficulty arises with the galaxy NGC 3077, whose radial velocity differs from the radial velocity of M 82 by 436 km/sec and whose projected distance from M 82 is approximately 55 thousand parsecs. In order to explain this difference of velocities it must be assumed that the minimum mass of M 82 is more than 10^{12} solar masses. Such an assumption will lead to an extraordinarily great value of M/L for M 82 (of the order of 500). Taking into account that real relative velocities may have considerable angles with the lines of sight we arrive at a greater mass value for M 82. The only way out of this situation is to assume that the galaxy M 82 is simply moving away from the group connected with M 81 by a velocity significantly surpassing the escape-velocity. This means that one of the members of the group has received positive energy in the process of its formation.

c. The discovery by Zwicky[9] of the group of three galaxies consisting of IC 3481, IC 3483 and of the anonymous one found between them is an interesting example. Their radial velocities are equal to + 7,011 km/sec, + 33 km/sec and 7,229 km/sec respectively. The galaxy IC 3483 is a riddle. If it is physically connected with the remaining two, which is evidenced by the filament joining all the three galaxies and by the proximity of the apparent magnitudes of IC 3481 and IC 3483, then we must directly conclude that we are dealing with a galaxy moving away from the group wherein it was formed.

But if IC 3483 is accidentally projected at the end of the filament and is actually a nearby galaxy in accordance with its radial velocity, then the absolute magnitude of this galaxy must be very low. If, for example, we suppose, that it is a member of the Virgo cluster, then we must attribute to this galaxy an absolute magnitude of about − 14.5. Such a low absolute magnitude is indeed unusual for spiral galaxies. Therefore, it is probable that the first assumption is correct.

d. Stephan's Quintet is undoubtedly a physical group. Examining the photographs of this group we particularly notice a close connection between the components NGC 7318*a* and NGC 7318*b* of this group. In spite of that the difference of the radial velocities of these two galaxies reaches about 1,000 km/sec. Since two other galaxies of this system, NGC 7317 and NGC 7319, have radial velocities differing from the radial velocity of NGC 7318*a* by not more than 100 km/sec, we conclude that the galaxy NGC 7318*b* is leaving the group with positive energy.

e. A series of narrow double galaxies are met with where it is very difficult to treat the pair as optical and yet the difference of radial velocities is great. The pair NGC 2831 and NGC 2832 may serve as an example where the distance between the components is less than 30″, which corresponds in the projection to less than 4,000 ps, while the difference of radial velocities is approximately 1,800 km/sec.[10] However, the pair under consideration is in a cluster of galaxies where the probability of casual projection may be comparatively high, while the difference of the radial velocities of the members sometimes reaches 2,000 km/sec and more. Nevertheless, it is surprising that two so closely projecting galaxies possess so great a velocity difference.

The above mentioned facts are difficult to explain if we suppose that in every physical multiple system all the components are retained owing to the force of mutual attraction.

CONCLUSION *Among multiple galaxies there are systems in which one or more components have velocities sufficient enough to leave the system.*

6. On the Sign of Energy in Large Clusters of Galaxies

Here Ambartsumian argues that the virial theorem masses of the individual galaxies of the Virgo and Coma clusters are in contradiction with their luminosities. The discrepency is only resolved if the clusters have positive total energy and are dissolving.

CONCLUSION *In some large clusters of galaxies the velocity dispersion is so great that they must represent disintegrating systems.*

7. Radio Galaxies in Perseus and Cygnus

If we admit the above conclusions that components of any multiple galaxy originate together and that the galaxies in some clusters and groups are in a state of mutual recession, then it is natural to conclude that every group, immediately after its formation, represented a narrower system than we now observe. In that case two hypotheses are possible: (*a*) galaxies of a given group or of a multiple system originate from a single amorphous mass whose diameter in the order of magnitude is not less than the diameter of an average galaxy (a few thousands of parsecs); (*b*) the primary nucleus of a galaxy for reasons unknown to us is divided into separate parts which give birth to independent galaxies that become components of the system. In this case the process of division must take place in a small volume with a diameter measured by some parsecs or tens of parsecs.

The parts of the divided nucleus must recede from one another, in their primary stage, with velocities of the order of hundreds or even of a thousand kilometers a second. Otherwise, their mutual attraction cannot be overcome and a few galaxies with coinciding centre arise which eventually join again into one galaxy.

The wide emission lines of Perseus A and Cygnus A suggest the high velocities needed for escape from the optically visible nucleus. [The optical counterpart of Cygnus A was at this time thought to represent two galaxies in collision.]

CONCLUSION *Radio galaxies Perseus A and Cygnus A are systems in which division of nuclei has taken place but the separation of galaxies is not yet complete:*

8. Radio Galaxy Virgo A = NGC 4486 = M 87

Two structural peculiarities in the optical region distinguish this radio galaxy from other elliptical galaxies: 1) the presence of a jet with condensations that emit polarized radiation, and 2) the presence of a great number of globular clusters.[11]

The fact that the jet comes from the centre leaves no doubt that here we have an ejection from the nucleus of the galaxy. On the other hand polarization of the radiation shows that the mechanism of luminosity is partly, if not wholly, similar to that of the Crab nebula. It follows that the sources of luminosity in the condensations consist not only of stars but also of diffuse matter which is in the same state as in the Crab nebula. In other words, a considerable amount of high energy electrons should be expected in these condensations. On the other hand the sources of radio emission are known to be concentrated continuously in the body of galaxy NGC 4486.

Two conjectures are likely to be made: (*a*) the high energy electrons were directly emitted from the nucleus of the galaxy, and (*b*) objects are ejected from the nucleus which are the sources of the electrons of very high energy, a considerable part of the synchrotron radiation of which is concentrated in the optical region.

One cannot confine oneself to the first hypothesis since the concentration of optical radiation in a small volume of condensations would remain unexplained. Therefore, one should believe that the sources emitting electrons of high energy are concentrated within the condensations themselves. Observations of the objects of our Galaxy reveal that various non-steady stars (supernovae, stars of the T Tauri type and others as well) constitute powerful sources of high energy electrons. Therefore, it is very likely that a large number of similar non-steady objects are present in the condensations referred to. Thus we seem to get closer to an understanding of the nature of the condensations under consideration. They represent conglomerates of clouds of relativistic electrons, gaseous clouds and non-steady stars. Such conglomerates can hardly be assumed to exist in the nuclei of galaxies. It should be inferred, therefore, that the matter ejected from the nucleus evolves, over a short period, into such conglomerates. The emission line λ 3,727 noticed in the region of the nucleus of NGC 4486 gives, apparently, some idea of the velocity of the ejection from the nucleus. Hence one can evaluate the order of the periods in the course of which such conversions take place. They turn out to be of the order of $3 \cdot 10^6$ years.

CONCLUSION *Apart from the division of the nuclei of galaxies, ejection of relatively small masses from the nuclei of galaxies may occur in nature. These ejected masses can, over a short period, turn into conglomerates made up of young non-steady stars, interstellar gas and clouds of high energy particles.*

9. Blue Ejections from the Nuclei of Elliptical Galaxies

The galaxy NGC 4486 is not the only one where an ejection of matter from the nucleus is observed. We have paid attention to some other similar cases of which galaxy NGC 3561*a* is of especial interest.[12] That galaxy is apparently spherical and has an outflow that looks like a jet. The jet ends in a condensation which is bright enough on the blue photograph and almost invisible on a red one. The colour index of the condensation in the international system is $-0^m.5$.

The ejection observed in NGC 4486 is also bluer than the main galaxy. The luminosities of the blue ejections from elliptical galaxies exceed those of common stellar associations, and this suggests that the ejections can be taken as separate galaxies.

CONCLUSION *In certain cases ejections from central parts of elliptical galaxies have very blue colour. This colour may be the result of a preponderance of a great number of highly luminous blue stars or of the strong continuous violet emission. In both cases it cannot last very long. It is, therefore, highly probable that these blue ejections and companions are very young galaxies.*

10. Bridges and Filaments Joining Galaxies

The valuable service of Zwicky[9] is connected with the investigation of many double and triple galaxies joined together by means of filaments or bridges of various thickness. Zwicky himself is inclined to believe that these bridges were formed as a result of the tidal interaction arising from the encounter of two galaxies. According to Zwicky bridges and filaments consist of stars ejected as a result of tidal action from the given galaxy. It is not hard to see that such an interpretation is at variance with the facts.

Ambartsumian argues that the thinness of the filaments and the fact that some filaments are continuations of spiral arms mean that the filaments cannot be caused by tidal interaction.

CONCLUSION *Bridges and filaments between galaxies are not the result of tidal interactions. They can be supposed to arise during the process of mutual recession of two or more galaxies that have originated from the same nucleus.*

11. Type M 51 Galaxies

The presence of the companion NGC 5195 at the end of the spiral arm of the galaxy M 51 has always seemed to us to be a strong argument in support of the concepts put forth in the foregoing paragraph. That the spiral arm discontinues almost immediately beyond NGC 5195 furnishes, in our opinion, a serious argument against the conjecture of NGC

5195 accidentally projecting on the arm of the spiral galaxy NGC 5194. It is desirable, however, to discover another case in which the connection between the spiral structure and the presence of the companion should prove more convincing. Such a case has been discovered by a student of mine—Iskoudarian—on the prints of the Palomar atlas. This refers to the double galaxy NGC 7752–7753.

The similarity between the double galaxy under consideration and M 51 is emphasized by the fact that in both cases, the curvature of the spiral arm sharply decreases in the region adjacent to the companion. Thus, formations of the M 51 type should not be considered the result of mere projections. As it was pointed out by B.A. Vorontsov-Velyaminov,[13] they represent a class of double galaxies in which the components are bound together by means of a powerful spiral arm and by a thin filament. Perhaps this is partly connected with the fact that the distance between the two companions, at least at the given stage of the evolution of the group, is small. In the case of M 51 this distance is of the order of three thousand parsecs. As the distance between the components increases the bridge becomes considerably thinner.

CONCLUSION *The existence of galaxies of the M 51 type confirms the suggestion of the connection between the process of the division of the original nucleus and the formation of spiral arms.*

12. Big Condensations in Spiral Arms

In luminosity there is no clear-cut boundary between condensations in spiral arms and companion galaxies. Examples of galaxies with bright condensations are NGC 4861 and NGC 2366.

CONCLUSION *Besides cases where a spiral arm links a given galaxy with companions consisting of population of the second type, there are other cases where a spiral arm ends in a companion which presents a large conglomerate of objects of population of the first type.*

13. Nature of Nuclei of Galaxies

Our knowledge of the nuclei of galaxies is very poor. Referring to the nuclei we mean very small formations a few parsecs in diameter with a very high surface brightness in the center of a galaxy.[14] Last year Dr. Baade was kind enough to show me a photograph of the nucleus of the galaxy M 31. This is, in fact, an amazing formation with an exceedingly bright surface luminosity. Unfortunately, in most galaxies, we are unable to pick out the nucleus from the whole central body of the stellar system.

We concluded above that the nuclei can be divided and also eject spiral arms or radial jets containing certain condensations. But the spontaneous division of a stellar system consisting only of stars is dynamically impossible. Therefore, if the nucleus were made up of only common stars we should have to give up the concepts put forth above in which the basic role in the genesis of galaxies and the formation of spiral arms is ascribed to nuclei. A grave difficulty arises from the fact that in the region of the nucleus the amount of interstellar hydrogen is small as compared to the density of the interstellar hydrogen in the outer parts of stellar systems such as the arms of our Galaxy. Meanwhile in some instances the outflow of the matter from the nucleus can be observed almost immediately. I mean not only the jet in NGC 4486 and in NGC 3561 but also the outflow from the central part of our Galaxy of interstellar hydrogen with considerable velocity as discovered by Dutch astronomers. Thus, it turns out that hydrogen flows out from where there is none. To form a fuller idea of the difficulties involved we must realize that the spiral arms of galaxies contain large masses of hydrogen and also, irrespective of any hypothesis, that there is a definite genetic bond between the arms and the central nuclei.

We must consider this as one of the greatest difficulties in astrophysics which may be solved provided a radical change in the conception of the nuclei of galaxies is adopted.

Apparently we must reject the idea that the nucleus of a galaxy is composed of common stars alone. We must admit that highly massive bodies are members of the nucleus which are capable not only of splitting into parts that move away at a great velocity but also of ejecting condensations of matter many times more massive than the Sun.

The new bodies, resulting from splitting, or ejection, move away from the volume of the original nucleus with velocities which are sufficient to overcome the attractive force of the nucleus and eject considerable masses of gases as well as some denser condensations. After some time these condensations can come to a state of quasi-stability under the influence of their own gravitation, i.e. turn into stars.

Not all the transformations referred to above should end immediately after the formation of the spiral arm or of the new galaxy. In some cases these transformations are very likely to be delayed because of the transition of a number of fragments into some kinds of meta-stable states and only then would they turn into stars and gas. It is this last transformation that we probably observe in the arms of our Galaxy as phenomena of the origin of stars and nebulae in stellar associations. This refers to both O- and T-associations. This point of view may raise some objections. One can say that it is difficult to offer, at present, a model of massive bodies with the above described properties. Even if we should not try to understand directly the concrete mechanism of the division of massive bodies located within the nuclei, nevertheless, difficulties can be encountered with the preservation of the various laws of conservation such as that of the rotational momentum. On the other hand, it is quite possible that the consideration of the common origin of two or more stellar systems might be instrumental in overcoming these difficulties.

The basic idea we want to emphasize lies in the fact that before starting the construction of theories of the origin of galaxies it is necessary to determine from observations the type and character of the processes leading to the formation of new stellar systems. Next, the problem of theoretical interpretation of the observed processes should arise.

CONCLUSION *There are evidences in favour of the formation of new galaxies and arms at the expense of matter contained in nuclei of galaxies. These nuclei are very small in size and of high density. Since these processes of the genesis of stellar systems cannot occur at the expense of the common stellar population of a nucleus we should conclude that large masses of prestellar matter are present in nuclei.*

14. On the Repetition of the Processes of Formation of Components and Arms

Here Ambartsumian argues that the complex structure of spiral galaxies suggests nuclear ejections at different times and that nuclear activity is in some way connected with the masses and luminosities of galaxies.

15. On the Role of Interstellar Gas

As the radio observations of the 21-cm line of interstellar hydrogen show, the interstellar gas comprises a considerable part of the masses of spirals and of irregular galaxies. Confronting this with the richness of these systems or associations, one usually concludes that young stars are being formed from interstellar gas.

However, the parallelism between the abundances of interstellar gas and O associations permits two different interpretations: (*a*) The formation of stars from gas, and (*b*) The common origin of stars and of interstellar gas from some protostellar bodies. Therefore, of the greatest value are the data which enable us to choose between these two interpretations.

Ambartsumian presents evidence that star formation is not directly correlated with the presence of interstellar gas. Some systems with more interstellar gas have fewer young stars and vice versa.

CONCLUSION *The facts connected with interstellar gas and associations are speaking in favour of the common formation of stars and gas from protostars rather than of formation of stars from gas.*

16. On the Origin of Population II Stars

There are observational indications that globular clusters are moving in the Galaxy in highly elongated orbits. This may serve as a direct evidence that they were some time ejected from the nucleus of our Galaxy with velocities of the order of some hundreds of km/sec. This shows that globular clusters have been formed not in a thin diffuse cloud, but in a volume of sufficiently high density.

Probably further investigation of globular clusters will give us the possibility to approach nearer to the solution of the problem of the origin of the type II population.

Here Ambartsumian argues that the mechanisms of star formation in flat and spherical subsystems are different and to a considerable degree independent of each other (see Kukarkin[15]).

17. On Double Spirals

We have admitted above that the formation of spiral arms is connected with the formation of a double galaxy. Then, the following possibility of verification of the hypothesis of the division of the primary nucleus arises. Since the primary nucleus, having a small volume, cannot have rotational momentum comparable with the rotational momenta of spiral galaxies, we may assume that during the formation of two spirals the sum of the momenta will remain small. This condition is easily satisfied when the spirals formed have momenta of opposite directions. In this case, we must expect that the directions of winding of spiral arms should be opposite to each other, that is, the angle between these directions will be near to 180°.

CONCLUSION *In the overwhelming majority of cases physical pairs of spiral galaxies have opposite directions of spiral arms.*

1. V. A. Ambartsoumian, Outchenie zapiski LGU, no. 22, *Seria matematitcheskikh naouk* (astronomia), vol. 4, 19–22 (1938).

2. V. A. Ambartsoumian, *Astronomitcheskii Journal 14*, 207–219, (1937).

3. V. A. Ambartsoumian, *Evolioutsia zvezd i astrophisika* (Erevan, 1947).

4. V. A. Ambartsoumian, *Astronomitcheskii Journal 26*, 3, (1949); A. Blaauw, BAN *11*, 405 (1952).

5. F. Zwicky, PASP *50*, 218 (1938), *64*, 247 (1952).

6. J. Neyman, E. L. Scott, and C. D. Shane, *Ap. J. 117*, 92 (1953).

7. E. Holmberg, *Meddelande Lund Observatory*, ser. II, no. 136 (1958).

8. Guido Münch, *A.J. 62*, 28 (1957); "Report of the Director, Mount Wilson and Palomar Observatories" (1955–1956), p. 49.

9. F. Zwicky, *Ergebnisse d. exakt Naturwissenschaften*, vol. 29, 344 (1956).

10. M. L. Humason, N. U. Mayall, and A. R. Sandage, *A.J. 61*, 97 (1956).

11. W. Baade and R. Minkowski, *Ap. J. 119*, 215 (1954).

12. V. A. Ambartsoumian and R. K. Chakhbazian, *DAN Armianskoi SSR 25*, 185 (1957).

13. B. A. Vorontsov-Veliaminov, *Astronomitcheskii Zhurnal 34*, 8 (1957).

14. W. Baade, *IAU symposium*, no. 5, p. 1 (Cambridge, 1958).

15. B. V. Kukarkin, "Issledovanie stroenia i razvitia zvezdnykh sistem," M.-L. (1949).

113. Fluctuations in Cosmic Radiation at Radio Frequencies

James S. Hey, S. J. Parsons, and J. W. Phillips

(*Nature 158*, 234 [1946])

The paper given here reports the discovery of a discrete, variable radio source in the direction of the constellation Cygnus. This discovery of the first discrete radio source other than the Sun had enormous implications, for it led to the use of interferometers in detecting and locating other discrete radio sources (see selection 114) and to the eventual identification of many of these sources with distant extragalactic objects (see selections 117 and 118).

This paper also provided support for the erroneous hypothesis that the discrete radio sources are actually stars. Because the variations of the radio source Cygnus A are similar to the fluctuations of the radio noise coming from sunspots, James S. Hey, S. J. Parsons, and J. W. Phillips suggest that the radio radiation of our galaxy may come from stars rather than from the interstellar medium. Moreover, they argue that any radiator like Cygnus A that exhibits short-period fluctuations in intensity must be localized in a discrete source. Because no source of radiation can move at speeds faster than the velocity of light, a source such as Cygnus A, which fluctuates with time scales of seconds, must be smaller than 100,000 km in linear extent. This conclusion suggests that Cygnus A has dimensions comparable to those of stars.

Two years later Hey, Parsons, and Phillips presented a more extensive survey of the radio emission coming from the Milky Way.[1] Except for the region in Cygnus, no fluctuations in the intensity of the radio signals were detected. The correlation of radio emission with the near and bright stars was poor, suggesting that only the more distant stars provide the observed radio emission. Furthermore, the ratio of average radio to light radiation for the sun is ten million times less than the ratio for the Milky Way. Unless the sun and nearby stars are abnormally weak in the radio spectrum, the radio star hypothesis seemed untenable.

Four years after the discovery of the radio wavelength fluctuations from Cygnus A, it was found that they are not intrinsic to the source at all. Just as stars twinkle at optical wavelengths because of moving clouds in the earth's atmosphere, the radio waves coming from Cygnus A were varying in intensity because of moving clouds in the earth's ionosphere. If the observed fluctuations were actually generated in a distant source, they should appear the same at all points on the earth's surface. When two radio antennas were separated by only a few kilometers, the variation in intensity from Cygnus A seen at the two antennas was identical; but when the antennas were separated by 210 km, the correlation between the fluctuations observed at the two antennas completely disappeared.[2] This difference meant that moving clouds of several kilometers extent had to be causing the observed fluctuations. Paradoxically, the argument that Cygnus A had to be a localized source of emission remained valid. Just as the moon does not twinkle but the stars, because of their smaller angular extent, do, Cygnus A has to be smaller than a few minutes of arc in extent if the ionosphere causes the observed fluctuations at radio wavelengths. It is this nonstellar, extended source that was later identified as a powerful radio galaxy (see selection 117).

1. J. S. Hey, S. J. Parsons and J. W. Phillips, *Proceedings of the Royal Society* (London) *A192*, 425 (1948).

2. F. Graham Smith, C. G. Little, and A. C. B. Lovell, *Nature 165*, 422 (1950). Also see G. J. Stanley and O. B. Slee, *Australian Journal of Scientific Research 3A*, 234 (1950).

In a previous publication[1] we described the results of an investigation into the spatial distribution of cosmic electromagnetic noise radiation at 5 metres wavelength. We have recently been engaged in an attempt to make a more detailed determination by using a more sensitive receiver of narrower beam-width. An interesting new feature which has emerged from these latter experiments is the occurrence of short-period irregular fluctuations which have been found to be associated with the direction of Cygnus. This region, which is a secondary peak in the cosmic noise distribution, appears to be unique in being characterized by short-period variations of marked amplitude in the intensity of power flux.

A watch on this region has been kept intermittently during the last four months. The receiving apparatus, situated in Richmond Park, has an aerial beam rotatable in bearing but fixed in elevation at an angle of 12°. The region of the fluctuations ascended and descended through the aerial beam on bearings 30° and 330° respectively. The corresponding times were 0100 hr. and 1900 hr. G.M.T. in February, when the watch was commenced, while in June they were 1800 hr. and 1200 hr. G.M.T. Care was taken to avoid including recordings taken in daylight periods when the powerful solar noise emission associated with the great sunspot in February was also present. Since the observations covered a wide range of bearings and solar times, we were able to rule out the possibilities of terrestrial or solar causes, and the interpretation of the results was consistent only with an origin in the direction of Cygnus.

It is not easy to determine the bearing of a source of irregular disturbance with a high order of accuracy unless an exceptionally narrow beam is used. The aerial of the equipment has a beam width of approximately $\pm 6°$ to half power in bearing and elevation, and the average of a large number of observations indicated a source of disturbance subtending an angle not exceeding 2°. There may be other areas of occasional fluctuation in the immediate vicinity (within 8°).

The average amplitude of the fluctuation is 15 per cent of the mean power received. If the disturbed area be assumed to extend over a circle of angular diameter 2°, then this solid angle is 1/36 of that for the equivalent acceptance cone of the aerial beam. The variations in power per unit solid angle therefore correspond to more than five times the mean power per unit solid angle for the whole beam. The centre of the region is approximately R.A. 2000 hr., Decl. + 43°. The type of fluctuation, which itself varies from day to day, is illustrated in the accompanying figure. The noise from Sagittarius would, by comparison appear as a straight line on a diagram of this scale.

Omitted here is a figure showing that the radio noise from Cygnus varies with time scales ranging between a few seconds and a minute and that similar fluctuations appear in the radio noise received from the sun.

It appears probable that such marked variations could only originate from a small number of discrete sources. This suggests at once the analogy with the radio-frequency sunspot radiation.[2,3,4] The solar radio noise from sunspots is also characterized by strong fluctuations. A recording of these solar radiations, taken on July 2, is also shown in the figure. On the other hand, Greenstein, Henyey and Keenan[5] have recently pointed out the difficulties in attempting to account for the magnitude of cosmic radiations in terms of the solar phenomena; further, they direct attention to the close agreement between experimental observations of cosmic noise intensity and their calculations of the expected interstellar radiation arising from free transitions of electrons in the field of protons. A theory in terms of widely distributed interstellar matter does not, however, appear readily to account for the localized disturbances just described. These fluctuations therefore appear of special importance in that they may prove particularly relevant to the explanation of the origin of cosmic radiations at radio frequencies.

We are indebted to the Director General of Scientific Research and Development (Defence), Ministry of Supply, for permission to publish this communication.

1. Hey, Parsons, and Phillips, *Nature 157*, 296 (1946).
2. Appleton, *Nature 156*, 534 (1945).
3. Hey, *Nature 157*, 47 (1946).
4. Pawsey, Payne-Scott, and McCready, *Nature 157*, 158 (1946).
5. Greenstein, Henyey, and Keenan, *Nature 157*, 805 (1946).

114. Positions of Three Discrete Sources of Galactic Radio-Frequency Radiation

John G. Bolton, Gordon J. Stanley, and O. B. Slee

(*Nature 164*, 101–102 [1949])

In order for the nature of the newly discovered radio sources to be understood, their positions and angular sizes had to be measured and their distances deduced. Once sufficiently accurate positions were obtained, the optical wavelength counterparts of the radio sources could perhaps be found and their distances determined by the techniques of optical astronomy. The major obstacle to obtaining accurate positions and angular sizes at radio wavelengths was the poor angular resolution of even the largest radio antennas. Because angular resolution increases linearly with wavelength, a radio telescope would have to be about a million times as large as an optical telescope to provide the same resolution.

In order to obtain adequate angular resolutions at radio wavelengths, research groups led by J. L. Pawsey in Australia and by Martin Ryle in Cambridge, England, constructed radio interferometers similar to that used by Albert Michelson at optical wavelengths. Both groups showed that the intense radio emission from the sun comes from sunspots,[1] and they then began to survey the entire sky. In the selection reproduced here, the Australian group gives the first three optical identifications for discrete radio sources: Taurus A—the Crab nebula, an expanding supernova remnant in our galaxy; Virgo A—NGC 4486 (M87), an elliptical galaxy; and Centaurus A—NGC 5128, a peculiarly distorted galaxy. These observations provided the first definitive suggestion that the radio stars might not be stars at all, since two of them were associated with extragalactic nebulae.

1. M. Ryle and D. D. Vonberg, *Nature 158*, 339 (1946); L. L. McCready, J. L. Pawsey, and R. Payne-Scott, *Proceedings of the Royal Society* (London) *A190*, 357 (1947).

In a recent communication[1] an account was given of the discovery of a number of discrete sources of galactic radio-frequency radiation. Accurate measurements of the positions of three of these sources have since been made from sites on the east and west coasts of New Zealand and on the east coast of Australia. The technique employed was to observe the sources at rising or setting, with an aerial on a high cliff overlooking the sea. These observations, when corrected for atmospheric refraction, allow the path of a source above the horizon to be plotted, and the time of its rising and setting—and hence its celestial co-ordinates—to be determined.

It is found that all three sources correspond within limits of experimental error to positions of certain nebulous objects. The positions of the sources determined and details of these objects are given in table 114.1.

The position of the source Taurus A agrees, within the limits of experimental error, with that of the Crab nebula. This is believed to be the expanding shell of the supernova of A.D. 1054. The present dimensions of the nebula are $4' \times 6'$, and photographs taken with different wave-length ranges show considerable fine structure, including $H\alpha$ filaments. The emission from the nebulosity is some 600 times that of the central star identified by Baade.[2] Minkowski[3] considers that most of the radiation from the central star is in the far ultra-violet and estimates the surface temperature as 500,000° K. He suggests a temperature of 50,000° K. for the nebulosity and explains a low ratio of $H\alpha$ to [N II] emission as being due to a deficiency of hydrogen. It should be pointed out, however, that a much higher temperature would have a similar effect on this ratio and at the same time increase the electron density. Intensity measurements we have made at 100 Mc./s., assuming a source size of 5′, give an apparent temperature of 2,000,000° K. for Taurus A. The present estimates of density and temperature in the Crab nebula would fall well short of explaining this result by strictly thermal processes. Non-thermal components resulting from the expansion of the nebula do not, however, seem unlikely.

Both N.G.C. 5128 and N.G.C. 4486 are generally classed as extra-galactic nebulæ, though Paraskevoulos[4] considers that the position of the obscuring band splitting the disk of N.G.C. 5128 into halves is unusual. Such obscuring bands are normally seen in nebulæ viewed on edge. Neither of these objects has been resolved into stars, so there is little definite evidence to decide whether they are true extra-galactic nebulæ or diffuse nebulosities within our own galaxy. If the identification of these objects with the discrete sources of radio-frequency energy can be accepted, it would tend to favour the latter alternative, for the possibility of an unusual object in our own galaxy seems greater than a large accumulation of such objects at a great distance.

Full details of the technique and observations involved in the determination of the positions of these three discrete sources will be published elsewhere.[5]

1. J. G. Bolton, *Nature 162*, 141 (1948).
2. W. Baade, *Astrophys. J. 96*, 188 (1942).
3. R. Minkowski, *Astrophys. J. 96*, 199 (1942).
4. J. S. Paraskevoulos, *Harvard Bull. 890*, 1 (1935).
5. J. G. Bolton and G. J. Stanley, *Aust. J. Sci. Res. A2*, 139 (1949).

Table 114.1 Optical identifications of discrete radio sources

	Position (Epoch 1948)		Possible associated visible object		
Source	Right ascension	Declination	Object	Spectrum	Remarks
Taurus A	$5^h31^m00^s \pm 30^s$	$+22°01' \pm 7'$	N.G.C. 1952 (Messier 1)	Continuous. Weak emission lines of H, He, forbidden lines of N, O and Si	The Crab nebula, expanding shell of an old super-nova
Virgo A	$12^h28^m06^s \pm 37^s$	$+12°41' \pm 10'$	N.G.C. 4486 (Messier 87)	Continuous	Spherical nebula—unresolved
Centaurus A	$13^h22^m20^s \pm 60^s$	$-42°37' \pm 8'$	N.G.C. 5128	Continuous. Weak emission lines, $H\beta$, $H\gamma$, $H\delta$, and λ 4,686	Unresolved nebula crossed by a marked obscuring band

115. Cosmic Radiation and Radio Stars

Hannes Alfvén and Nicolai Herlofson

(*Physical Review 78*, 616 [1950])

When Karl Jansky first discovered radio emission from the Milky Way, the similarity of the sound of the galactic radio noise to that produced by the thermal agitation of electric charges in a resistor led him to suggest that the source of the radio emission is actually hot, charged particles speeding through the interstellar medium (see selection 6). Grote Reber subsequently reasoned that radiating particles would exhibit a Planck spectrum, in which the radiation intensity increases with decreasing wavelength. He therefore searched for extraterrestrial radio emission at wavelengths shorter than those used by Jansky (9 cm, compared with 14.6 m), but when he failed to detect any signal he concluded that the Planck law does not hold for the radio frequency radiation. When he eventually detected radio radiation from the Milky Way at 187 cm, he interpreted it according to the only other radiation mechanism known at the time, bremsstrahlung, or braking radiation (see selection 6 and its references). According to this interpretation, interactions between free electrons and ions in the material surrounding hot stars produce the radio frequency radiation. Nevertheless, no radio source had been identified with a star, and the bremsstrahlung process failed to account for the intense emission detected by Jansky at the longer radio wavelengths.

Already in 1912, G. A. Schott had discussed another process for the generation of radiation,[1] which only attracted intense interest several decades later when its study became connected with electron accelerators[2] and the behavior of cosmic ray particles in the earth's magnetic field.[3] These studies showed that a charged particle moving in the presence of a magnetic field is accelerated and emits electromagnetic radiation. Electrons accelerated to high speeds in the strong magnetic fields of the General Electric Company's synchrotron were subsequently found to emit linearly polarized light,[4] and this discovery stimulated additional theoretical treatments of this type of radiation.[5] It was only after the actual observation of this synchrotron radiation in the terrestrial laboratory that such a mechanism was considered as a possible source for the radiation from extra-terrestrial objects.

In the following paper Hannes Alfvén and Nicolai Herlofson introduce the proposition that the synchrotron radiation mechanism can generate the observed radio emission from discrete radio sources. They also suggest that the sources are optically invisible radio stars located in interstellar space. We now know that our galaxy and the discrete extragalactic radio sources do radiate by the synchrotron process, but the high energy electrons which account for the diffuse radio emission from our galaxy are thought by many to be cosmic ray electrons spiraling about the interstellar magnetic field. Moreover, these galactic electrons were probably accelerated in supernovae remnants rather than invisible stars (see selection 99).

1. G. A. Schott, *Electromagnetic Radiation* (Cambridge: Cambridge University Press, 1912).

2. L. A. Artsimovich and I. Ya. Pomeranchuk, *Zhurnal Éksperimental' noĭ i Teoreticheskoĭ Fiziki 16*, 379 (1946). L. I. Schiff, *Review of Scientific Instruments 17*, 6 (1946).

3. I. Ya. Pomeranchuk, *Zhurnal Éksperimental' noĭ i Teoreticheskoĭ Fiziki 18*, 392 (1948).

4. F. R. Elder, R. V. Langmuir, and H. C. Pollock, *Physical Review 74*, 52 (1948).

5. J. Schwinger, *Physical Review 75*, 1912 (1949).

THE NORMAL RADIO WAVE EMISSION from the sun amounts to 10^{-17} of the heat radiation, and increases during bursts[1] to as much as 10^{-13}. If a radio star, e.g., the source in Cygnus, is situated at a distance of 100 light years, its radio emission is of the order of 10^{-4} of the heat radiation of our sun. It is very unlikely that the atmosphere of any star could be so different from the sun's atmosphere as to allow a radio emission which is 10^9 to 10^{13} times greater, and it seems therefore to be excluded that the source could be as small as a star. The recent discovery[2] that the intensity variations of radio stars is a "twinkling" makes it possible to assume larger dimensions.

Ryle has suggested that there should be a connection between radio stars and cosmic radiation.[3]

According to a recent development of Teller and Richtmyer's theory of cosmic radiation, the sun should be surrounded by a "trapping field" of the order of 10^{-6} to 10^{-5} gauss, which confines the cosmic rays to a region with dimensions of about 10^{17} cm (0.1 light year).[4] It is likely that almost every star has a cosmic radiation of its own, trapped in a region of similar size. We suggest that the radio star emission is produced by cosmic-ray electrons in the trapping field of a star.

Electrons with an energy $W \gg m_0c^2$ moving in a magnetic field H radiate at a rate

$$-dW/dt = (2e^2/3c)\omega_0{}^2\alpha^2, \qquad (1)$$

where[5] $\omega_0 = eH/m_0c$ is the gyro-frequency corresponding to the rest mass m_0, and $\alpha = W/m_0c^2$. Most of the energy is emitted with a frequency of the order

$$\nu = \omega_0\alpha^2/2\pi = 2.8 \times 10^6 H\alpha^2 \text{ sec.}^{-1}. \qquad (2)$$

As soon as the energy is much higher than the rest energy the emitted frequency becomes much higher than the gyro-frequency, a phenomenon which is observed in large synchrotrons, where the electron beam emits visual light.[6]

According to (2) an emission of radio waves of 100 Mc/sec. requires

$$H\alpha^2 = 36 \text{ gauss}. \qquad (3)$$

The acceleration process of cosmic radiation should accelerate electrons as well as positive particles. In the solar environment the electron component is eliminated by Compton collisions with solar light quanta as discussed by Feenberg and Primakoff.[7] In the neighborhood of a star which does not emit much light, the electrons would be accelerated until their energy is so high that they radiate. The wave-length falls in the meter band if, for example, $\alpha = 300$ ($W = 1.5 \times 10^8$ ev) and $H = 3 \times 10^{-4}$ gauss. This field is about 100 times the estimated strength of the sun's trapping field. As the strength is determined by the "interstellar wind," a radio star should be situated in an interstellar cloud moving rather rapidly relative to the star.

In order to account for the total energy emitted by a radio star we must suppose either that the radio emission is a transitory phenomenon, lasting a time which is short compared to the lifetime of cosmic rays in the trapping field (10^8 years), or that the cosmic-ray acceleration close to the star is supplemented by a Fermi process[8] further out in the trapping field.

According to the views presented here, a radio star must not emit very much light, and should be situated in an interstellar cloud. This would explain why it is so difficult to find astronomical objects associable with the radio stars.

1. J. S. Hey, *M.N.R.A.S. 109*, 179 (1949).
2. Smith, Little, and Lovell, *Nature 165*, 424 (1950).
3. M. Ryle, *Proc. Phys. Soc. London A62*, 491 (1949).
4. Alfvén, Richtmyer, and Teller, *Phys. Rev. 75*, 892 (1949); R. D. Richtmyer and E. Teller, *Phys. Rev. 75*, 1729 (1949); H. Alfvén, *Phys. Rev. 75*, 1732 (1949), *77*, 375 (1950).
5. D. Iwanenko and I. Pomeranchuk, *Phys. Rev. 65*, 343 (1944); J. Schwinger, *Phys. Rev. 75*, 1912 (1949).
6. Elder, Langmuir, and Pollock, *Phys. Rev. 74*, 52 (1948).
7. E. Feenberg and H. Primakoff, *Phys. Rev. 73*, 449 (1948).
8. E. Fermi, *Phys. Rev. 75*, 1169 (1949).

116. The Origin of Cosmic Radio Noise

Thomas Gold

WITH A DISCUSSION BY FRED HOYLE, GEORGE MCVITTIE, AND MARTIN RYLE

(*Proceedings of the Conference on Dynamics of Ionized Media*, Department of Physics, University College, London, 1951)

When radio telescopes of limited resolving power were used to scan the sky, the angular variation in received intensity seemed to be in good agreement with the distribution of stars in the Milky Way. The sun was known to exhibit intense radio fluctuations and since one discrete radio source, Cygnus A, exhibited similar fluctuations, it was widely believed that the radio sources were some kind of radio star. Because the nearer and brighter stars did not exhibit detectable radio emission, and because Cygnus A was one hundred million times more intense than the sun at radio wavelengths, Martin Ryle[1] argued for the existence of a new kind of radio star located between the optically visible stars.

At the time of the following discussion (1951), Ryle's group had used radio telescopes of high angular resolution to discover fifty discrete radio sources, isotropically distributed across the sky.[2] Although four of these objects had been identified with extragalactic nebulae, their radio emission was a thousand times weaker than that of most of the other discrete radio sources. Thomas Gold argues here, with remarkable foresight, that if stars are to radiate intense radio emission they must be collapsed stars, which would have a strong magnetic field. In close analogy with Heber Curtis's arguments for the extragalactic nature of spiral nebulae three decades earlier, Gold maintains that the isotropic distribution of the radio sources favors an extragalactic origin and that the difficult problem of accounting for the intense radio emission is less severe if the sources are placed at extragalactic distances. In spite of several additional compelling arguments for the extragalactic nature of radio sources, Martin Ryle and George McVittie, in the ensuing discussion, emphatically supported the radio star hypothesis.

1. M. Ryle, *Proceedings of the Physical Society* (London) *A62*, 491 (1949).
2. M. Ryle, F. G. Smith, and B. Elsmore, *Monthly Notices of the Royal Astronomical Society 110*, 508 (1950).

Thomas Gold

I SHALL DISCUSS here only the basis which one has at present for speculations regarding the origin of cosmic radio noise; I shall not even discuss the particular mechanisms which might be held responsible.

On the basis of the present observations, one is entitled to assume either that all radio noise comes from "radio stars", or that all of it comes from diffuse gas clouds in galaxies, or that its origin lies partly in the one and partly in the other.

In the corresponding speculation about the origin of cosmic rays, one has an important consideration which is independent of the particular mechanism which is assumed; the mean power required is so great that one can restrict the possibilities very much by confining oneself to sufficiently energetic processes. In the present problem we have no guidance of that sort. The mean power contained in radio noise in the universe is without doubt so small that there is a large variety of free energy which is ample for the purpose; the problem lies in finding the conversion process.

It is a widely held and very reasonable guess that the radio waves come entirely from objects of roughly stellar dimensions which are scattered throughout galaxies. Ryle and his collaborators have shown that the measurements are compatible with the assumption that these "radio stars" are between three and ten light years apart within the galaxy and that the majority of them make up the unresolved point sources. This statistical test is, however, a rather insensitive one, and the conclusion is by no means compelling.

Why, on this basis, does one not find any identifiable visual object where those very near radio stars are supposed to be?

Knowing that the radio emission is far too intense to be thermal, (as there are no fields of force which could hold together a gas at the required temperatures), we must see where we would expect intense electrodynamic processes to occur. Such processes are prevented almost everywhere by the high conductivity of the tenuous gas in space, which virtually implies an absence of any sort of electric field. The only regions where space behaves as an insulator, at least in some directions, are those which combine strong magnetic fields and low density. This condition can be set up only where a dense and tenuous region are closely adjacent, so that the currents responsible for the magnetic field can flow where the density is great enough, whilst the resulting magnetic field extends into the tenuous region.

The most favourable conditions of this sort would be expected in the neighbourhood of collapsed, dense stars. Their magnetic field must be stronger than it was before their collapse, merely as a result of compression in a time short compared with the decay time; and as a result of the high value of the gravitational force at the surface, any atmosphere or corona would be less extended than in the case of an ordinary star. If those conditions of "magnetic insulation" are essential for the generation of intense radio noise, then we should expect radio stars to be in practice invisible or very dim.

One would, I think, feel more confidence in such an interpretation if the radio signals showed substantial fluctuations like those associated with solar activity. Their steadiness over times which are at any rate millions of times longer than the transit time of light across the size of the source seems to run counter to such an explanation.

The alternative view which one might hold is that radio noise is generated diffusely in the galactic gas. The contribution of our own galaxy would then be merely the unresolved radiation which shows a strong galactic concentration. The point sources would then have to be other galaxies in which similar processes occur at somewhat greater intensities.

It cannot be ruled out that other galaxies may behave quite differently from our own, for it is known that there are very different types. In particular, one might mention those which show light resembling that of novae, with greatly broadened emission lines, suggesting that far more violent motions occur there than here. The necessary differences in radio emission may hence exist; the variation between galaxies required for this explanation is certainly much smaller than that which is required between stars for the previous explanation, where some stars would have to emit 10^8 times more than others. It must be added that the only point sources which have been identified are some of the nearer extragalactic nebulae; it is hence not very farfetched to consider whether the remaining ones may have a similar origin.

The fifty point sources which are known at present are distributed fairly isotropically over the sky. For the stellar interpretation this implies that they are near, at distances small compared with the galactic thickness. With the possibility of the extragalactic origin, we must hence say that these sources are either very near or very far from us; only then could one understand the absence of a galactic concentration.

Lastly, I must stress the connection between this problem and cosmic rays. The energy in cosmic rays is so vastly greater, that the radio noise could be a very small by-product of their generation or absorption. If, for example, one supposes cosmic rays to be as intense in the whole galaxy as they are here, then it would suffice if one part in 10^6 of the power they would dissipate by collisions appeared in the form of radio noise. If it were possible to treat this dissipation process theoretically with sufficient precision to discover such small effects, one might be able to show that this alone would account for the galactic noise; or one might even show that cosmic rays cannot be so general, if they would result in more than the observed power of radio waves.

Discussion

Hoyle: I want finally to mention the topic of radio stars. Is there some special type of nonvisible star giving enormously high acceleration energies, or are the so-called radio stars simply extragalactic nebulae? For my part I am inclined to prefer Gold's suggestion of the extragalactic nebulae, although I think Gold was mainly concerned to stress that both possibilities should be kept in mind. Ryle objected to the extragalactic nebula hypothesis on the grounds that we are now forced to regard the diffuse galactic noise as arising from a large number of distant stars. Then why not suppose the discrete sources to be a few exceptionally close stars? The objections to this argument are:

(i) the invisibility of the assumed stars

(ii) the assumed stars would have to emit noise of vastly greater intensity than the distant stars producing the diffuse galactic emission.

On Ryle's view the radio stars would be an astronomical mystery. In contrast with this, the lower level emission from the distant stars is only of an order that could conveniently be associated with members of the collapsed sequence.

McVittie: I would like to raise one or two small points. Firstly, about the origin of radio noise being outside our galaxy: I was rather puzzled listening to Mr. Gold this morning about two points. It is taken for granted that many galaxies emit radio noise. That surprises me, for until then I was under the impression that radio noise had been detected from the Andromeda galaxy alone.

Gold: It has been found in four galaxies.

McVittie: Well, four out of 6×10^7, which is about the number that one can see with a 100-in telescope, seems a very small percentage. And I cannot see that one can establish the existence of radio noise from all galaxies from such a small number.

Gold: Fifty stars out of all the galaxy is equally small.

McVittie: But the sources need not necessarily be stars. Then there is another point. We were told that very large velocities had been observed in certain (unspecified) galaxies. I believe that M 33 is the galaxy that has been most carefully studied from the point of view of the radial densities of emission line objects, and the velocities so found are not very different from the velocities in our own galaxy. A few more have been studied by Mayall, and there, too, the velocities were not particularly large. Now these so called radio stars obviously are not stars in the ordinary sense of the word; they are sources of radiation which presumably are in our own galaxy. At least, if they are not, it does not seem to me that an extragalactic nebula will do the trick.

Gold: Might I say something about the radiation from the galaxies? In the first place, I don't wish to be misunderstood; I merely wish to point out that so far as we know it is possible to identify radio stars with other galaxies. This is a possible point of view, though I do not necessarily wholeheartedly accept it. But since the tenability itself seems to have been challenged, I wish to add some remarks. Of the fifty sources in the sky that have been observed, it seems to be the case that one is known to be a particular very peculiar star in our own galaxy—namely the Crab nebula—and five, the only other five that have been identified as anything at all, are other galaxies. So I don't see that, with that state of affairs, one can conclude that it is inconceivable that other radio sources may also be extragalactic.

There is the additional point that Mr. Hoyle has mentioned, that in terms of stars we would have to suppose that some produce 10^8 times as much radio noise as others. In terms of galaxies we have to assume only a corresponding variability by a factor of at most 1,000.

Ryle: I think the theoreticians have misunderstood the experimental data.

First of all, I think the coincidence of one of the radio stars with the Crab nebula should not necessarily be taken too seriously. The accuracy of present position-finding observations is such that if you take the positions of fifty radio stars and compare them with fifty positions selected at random, there is roughly a 50-50 chance of one of them coinciding; whilst there may be some relation between the Crab nebula and this particular radio star, I think that the present evidence cannot be regarded as suggesting a general origin for radio stars of this type. It might be worth adding that we found no radio star near the positions of the only other recent supernovae (of 1572 and 1604) in the area surveyed.

As regards the possibility of accounting for radio stars by the emission from extragalactic nebulae, I think we already have a good deal of evidence against this and in favour of a "stellar" source within the galaxy. If we looked at the visual stars with a telescope having the same resolving power as our radio telescope we should find that among the fifty brightest sources four or five would not be stars, but the nearest of the extragalactic nebulae; in the same way we have found that four of the weaker radio stars coincide with four of the known nearest extragalactic nebulae. This result would suggest straight away that there may be approximate equality between the number of radio stars and the number of visible stars within the galaxy.

The alternative possibility of accounting for all the radio stars by extragalactic bodies presents very great difficulties, which were perhaps not fully appreciated in Mr. Hoyle's summary. There is first the necessity of providing some other mechanism to account for the total radiation from our own galaxy; as I indicated this morning the emission from the interstellar gas appears quite inadequate to account for the observations.

In order to see whether there is any evidence for the suggestion that other extragalactic nebulae might emit more intense radio waves than our own galaxy, we made an analysis to relate the known radio stars in the northern hemisphere to about 300 of the nearest extragalactic nebulae. Apart from the four sources (which have been mentioned and which produced about the right intensity, on the supposition that they were the same as our own galaxy), no significant correlation was found. We may therefore conclude that there is as yet no evidence to suggest that other extragalactic nebulae emit radio waves having a much greater intensity than our own galaxy.

If we are to account for radio stars by extragalactic bodies, we therefore have to postulate the existence of a new type of body whose visible emission is negligible; if we are to account for the observed intensities and diameters of the intense radio stars, we must conclude that these bodies have a radio emission at least 10^5 times as great as that from our own galaxy. If this emission is supposed to be due to the interstellar medium, where plasma frequencies will be too low for oscillation-type mechanisms, we should require an interstellar temperature of at least 10^8 degrees. If on the other hand we postulate the existence, in these invisible nebulae, of dense bodies capable of radio emission, they will have to radiate about 10^5 times as strongly as such bodies situated in our own galaxy.

Thus, the suggestion that radio stars are situated within the galaxy not only provides an explanation for the observed background radiation from our own galaxy (and the nearer extragalactic nebulae) but also requires less formidable requirements of their radio emission.

Hoyle: From the remarks of Professor McVittie and Mr. Ryle it might be thought that Gold and I have once more been riding a dogmatic hobbyhorse—in this instance to the effect that the discrete radio sources are extragalactic nebulae. What we said was that the extragalactic nebulae hypothesis must be kept actively in mind. The boot is really on the other foot, for Professor McVittie and Mr. Ryle have dogmatically asserted that the discrete sources cannot be of the extragalactic origin, although of the half dozen or so discrete sources that have indeed been identified, five have been found to correspond with nearby extragalactic nebulae. Presumably a discrete source ceases to be a discrete source as soon as it is identifiable as a galaxy.

117. Identification of the Radio Sources in Cassiopeia, Cygnus A, and Puppis A

Walter Baade and Rudolph Minkowski

(*Astrophysical Journal 119*, 206–214 [1954])

The most startling aspect of this paper is the identification of the "radio star" Cygnus A with a distant extragalactic object whose redshift is 0.057 (a recession velocity of 17,000 km sec^{-1}). Photography of this source with the 200-in reflector became possible only after Francis Graham Smith had supplied highly accurate positions from the Cambridge radio interferometer.[1] On the basis of his first photographs, Walter Baade deduced that the object was a pair of colliding galaxies and bet Rudolph Minkowski a bottle of whiskey that the spectrogram would confirm his hunch. According to Baade, even before the spectrogram was taken, Minkowski could see the emission lines in the detection equipment and agreed to pay up. The spectrogram showed strong forbidden emission lines similar to those found in the nuclei of Seyfert galaxies. However, the view of Cygnus A as a pair of colliding galaxies was discarded some years later when it became clear that an improbably high number of galaxy collisions would be required to account for the violent activity of so many radio sources.

Using a value of Hubble's constant of $H_0 = 540$ km s^{-1} Mpc^{-1}, Baade and Minkowski found the distance to Cygnus A to be 3.3×10^7 pc. Cygnus is the second brightest radio object in the sky, and at an extragalactic distance its absolute radio luminosity has to be enormous. The present estimate of its absolute radio luminosity is, in fact, 10^{45} erg sec^{-1}, which amounts to a total energy of 10^{60} erg radiated over a lifetime of 10^8 years. Because 10^{60} erg is the energy contained in the rest mass of a million solar masses, finding a suitable radiation mechanism has proved an ongoing problem for astrophysicists.

In a companion paper to that given here,[2] Baade and Minkowski reviewed the optical identifications of discrete radio sources. They showed that galactic radio sources are associated with supernova remnants and emission nebulae, whereas extragalactic radio sources are associated with normal spiral galaxies and peculiar radio galaxies. The normal galaxies are similar to the Andromeda nebula and have radio luminosities a million times lower than their optical luminosities;[3] the radio galaxies have radio luminosities approximately equal to their optical luminosities. The optical counterparts of the radio galaxies all seem to have peculiar features suggesting explosive activity. During the decade following Baade and Minkowski's paper, the radio galaxies were resolved at radio wavelengths and were generally found to consist of two radio sources centered on opposite sides of the optical counterpart.[4] The obvious inference is that the radio objects are produced by violent explosions within the optical object.[5]

1. F. Graham Smith, *Nature 168*, 555 (1951).
2. W. Baade and R. Minkowski, *Astrophysical Journal 119*, 215 (1954).
3. M. Ryle, F. Graham Smith, and B. Elsmore, *Monthly Notices of the Royal Astronomical Society 110*, 508 (1950); R. Hanbury Brown and C. Hazard, *Monthly Notices of the Royal Astronomical Society 111*, 357 (1951).
4. The basic double structure of the intense radio source Cygnus A was first derived by R. C. Jennison and M. K. Das Gupta [*Nature 172*, 996 (1953)].
5. In 1959 G. R. Burbidge [*Astrophysical Journal 129*, 849 (1959)] estimated that the minimum energy content of the radio lobes of an extended radio galaxy might correspond to that released by the complete annihilation of a million solar masses of material. This indicated that events may occur in galactic nuclei which release energy on scales that vastly exceed those of known stellar explosions, and that this energy is somehow channeled primarily into the form of relativistic particles and magnetic fields.

Abstract The radio sources in Cassiopeia and Puppis A are identified with a new type of galactic emission nebulosity. The outstanding features of these nebulosities are very large internal random velocities. The radio source Cygnus A is an extragalactic object, two galaxies in actual collision.

ONLY VERY FEW INDIVIDUAL SOURCES of cosmic radio emission have been identified with conspicuous astronomical objects. Although the sources in Cassiopeia[1] and Cygnus A[2] are among the brightest and earliest-known radio sources of the sky, all attempts to identify them with astronomical objects in the visible range have failed so far. In the fall of 1951, F. G. Smith, of the Cavendish Laboratory, communicated to us in advance of publication new positions for both sources which were very much more accurate than any previous data. Using an interferometric method, Smith had reduced the uncertainties in the positions to $\pm 1^s$ in right ascension and $\pm 40''$ in declination. As it turned out, the new positions were accurate enough for an unambiguous identification of both radio sources on plates taken in September, 1951, at the 200-inch telescope. The Cassiopeia source coincides with a galactic-emission nebulosity of a new type, whereas Cygnus A is an extragalactic affair, two galaxies in collision.

Baade and Minkowski first identify the radio source Cassiopeia A with a nebulosity, a coincidence that had been suggested by D. W. Dewhirst; [later studies of this source have shown it to be an expanding supernova remnant located in our galaxy]. Baade and Minkowski then show that the radio source Puppis A is identified with a galactic emission nebula.

Cygnus A

Except for the first positions by Bolton and Stanley,[3] all positions of the source given in table 117.1 are within an area of 18′ in right ascension and 10′ in declination. No bright star or any other conspicuous object is near the position of the source. Faint outlying parts of the nebulosities near γ Cygni cover the field. But interstellar absorption is low, as shown by the fact that numerous faint galaxies appear in the field.

Previous observers[4] had already noted that the region around Cygnus A must be quite transparent in spite of its low galactic latitude ($b = -4°$), because extragalactic nebulae shine through in large numbers. The 200-inch exposures clearly show the reason why, aside from the general transparency of the field, extragalactic nebulae are so frequent. The radio source lies amid a rich cluster of galaxies, the brightest members of which are of about the seventeenth photographic magnitude. In conformity with the general rule, the dominating nebular types in the cluster are E- and So-systems. The radio source coincides in position with one of the brightest members of the cluster:[5] Cyg A (radio position by F. G. Smith[6]), 1950.0, $19^h57^m45\overset{s}{.}3 \pm 1^s$, $+40°35\overset{\prime}{.}0 \pm 1'$; position of nebula, 1950.0, $19^h57^m44\overset{s}{.}49$, $+40°35'46\overset{\prime\prime}{.}3$.

In the center of this nebula are two bright condensations separated by about 2″ in position angle 115°. The bright central region of about 3″ × 5″ is surrounded by much fainter outer parts of elliptical outline, about 18″ × 30″, with the major axis in position angle 150°.

At first sight, this nebula is a very curious object which seems to defy classification. The clue to a proper interpreta-

Table 117.1 Position of Cygnus A

α (1950)	δ (1950)	Method	Observer
$19^h58^m47^s \pm 10^s$	$+41°41' \pm 7'$	Radio interferometry	Bolton and Stanley[a]
$19^h58^m14^s \pm 60^s$	$+40°36' \pm 10'$	Radio interferometry	Stanley and Slee[b]
$19^h57^m46^s \pm 5^s$	$+40°30' \pm 7'$	Radio interferometry	Ryle, Smith, and Elsmore[c]
$19^h57^m37^s \pm 6^s$	$+40°34' \pm 3'$	Radio interferometry	Mills and Thomas[d]
$19^h57^m44^s \pm 2\overset{s}{.}5$	$+40°35' \pm 1\overset{\prime}{.}5$	Radio interferometry	Mills[e]
$19^h57^m22^s \pm 25^s$	$+40°22' \pm 16'$	Radio paraboloid	Hanbury Brown and Hazard[f]
$19^h57^m45\overset{s}{.}3 \pm 1^s$	$+40°35\overset{\prime}{.}0 \pm 1'$	Radio interferometry	Smith[g]
$19^h57^m44\overset{s}{.}49$	$+40°35'46\overset{\prime\prime}{.}3$	Photographic	Baade

a. *Nature 161*, 312 (1948). This position has been superseded by the value given by Stanley and Slee.
b. *Australian J. Sci. Res. A*, *3*, 234 (1950).
c. *M.N. 110*, 508 (1950).
d. *Australian J. Sci. Res. A*, *4*, 158 (1951).
e. Ibid. *5*, 456 (1952).
f. *M.N. 111*, 576 (1951).
g. *Nature 168*, 555 (1951).

tion lies in the fact that it has two nuclei which are tidally distorted and that hence we are dealing with the superimposed images of two galaxies. Both are late-type systems, judging by the low density gradients of the two disks. Spatially they are oriented face to face, they are slightly decentered, and we look upon them at an angle not far from 45°. Actually, the two systems must be in close contact because of the strong signs of tidal distortions which the nuclei show. This suggests that we are dealing with the exceedingly rare case of two galaxies which are in actual collision. The main features of such a collision have been discussed by Spitzer and Baade.[7] On the cosmical time scale, collisions of galaxies are a rather frequent phenomenon in the rich clusters of galaxies. As far as the stars of the colliding systems are concerned, such a collision is an absolutely harmless affair. The average distance between two stars is so large that the two galaxies penetrate each other without any stellar collisions. The situation is very different for the gas and dust imbedded in the two systems. Because of the much shorter free paths of the gas and dust particles, the collision of the two galaxies means a real collision of the imbedded gas and dust, which are heated up to very high temperatures, since the collisional velocities range from hundreds to thousands of kilometers per second.

It is obvious that this behavior of the gas and dust provides a beautiful test of our hypothesis that the two galaxies which we identify with the radio source Cygnus A are actually colliding; for we should expect a very unusual spectrum for our nebula. Besides the continuous spectrum provided by the stars of the two systems, the emission spectrum of the colliding gases should be visible. In fact, this emission spectrum should be quite strong, since we are dealing with a face-on collision of the two galaxies which makes it certain that maximum amounts of the gas are in collision simultaneously. Moreover, on account of the high velocities involved in the collision of two galaxies, the resulting emission spectrum should be one of unusually high excitation and the lines themselves should show broadening. Spectra taken at the 100-inch and 200-inch telescopes largely confirmed these predictions.

Here we omit photographs of Cygnus A taken in the light of different forbidden emission lines, which suggest to Baade and Minkowski that the optical object is two galaxies in collision. We also omit the spectrum of Cygnus A showing emission lines of Hα, [O I] at λ6,300, λ6,364, [O II] at λ3,727, [O III] at λ4,959, λ5,007, [N II] at λ6,548, λ6,584, and [Ne III] at λ3,868.

Spectra of the bright nuclear region have been obtained with the nebular spectrographs at the prime focus of the 200-inch telescope and at the Newtonian focus of the 100-inch telescope. The spectrum was obtained with the grating mentioned earlier, which permits the simultaneous observation of the third-, fourth-, and fifth-order red, green, and violet, respectively. For the photographic region, a grating with 15,000 lines/inch blazed for violet has also been used with a dispersion of 435 A/mm.

The spectrograms show that the nebula is indeed a very peculiar, if not unique, object. Strong forbidden-line emission is recorded, while the continuous spectrum barely shows. Photoelectric measures by Dr. W. A. Baum indicate that more than 50 per cent of the total light of the condensations is in the emission lines. The emission lines and the nebular red shift derived from them are in table 117.2. From the red shift, the distance of the nebula is 3.3×10^7 psc, with the value 540 km/sec per 10^6 psc for the red-shift constant.

Table 117.2 Emission lines in Cygnus A

λ Observed	Identification		$c d\lambda/\lambda_0$ (km/sec)
3,619.9	3,425.86	[*Ne* V]	17,010
3,937.2	3,726.06/8.82	[*O* II]	16,930
4,087.5	3,868.77	[*Ne* III]	16,930
4,189.6	3,967.48	[*Ne* III]	16,820
5,234.3	4,958.91	[*O* III]	16,660
5,284.6	5,001.85	[*O* III]	16,955
6,642.5	6,300.27	[*O* I]	16,770
6,718.6	6,363.88	[*O* I]	16,720
6,916.6	6,548.06	[*N* II]	(16,870)[a]
6,928.0	6,562.66	$H\alpha$	(16,300)[a]
6,949.2	6,583.43	[*N* II]	16,670
Mean			16,830

a. $\lambda\lambda$6,916.6 and 6,928.0 are blended: the accuracy of the measured wave lengths therefore is low, and the values of the red shift for both lines have been excluded from the mean.

The large red shift of 16,830 km/sec places all lines in regions of the spectrum where the plate sensitivity differs from its normal values for the lines; reliable estimates of relative line intensities, therefore, cannot be given at this time. It is obvious that $H\alpha$ is relatively faint compared to the adjacent [*N* II] lines. The absence of $H\beta$ from the observed lines indicates relatively low intensity compared to the neighboring [*O* III] lines. The lines are diffuse, but the dispersion is too low to determine reliable values of the width. Individual lines seem to differ in width, but at least part of this is caused by intensity differences. The velocity spread is of the order of 400 km/sec.[8] The [*O* II] line λ 3,727 extends through the full length of the slit. The corresponding diameter of more than 30″ in position angle 90° is slightly larger than the visible size of the nebula. Inclination of the line indicates rotation of the nebula; the south-following side is receding relative to the center.

On account of the great intensity of some of the emission lines, it is possible, by selecting sufficiently narrow filter ranges, to photograph the emission regions. There are marked differences in the pictures, depending on the emission lines selected. Owing to the nebular red shift, the emission lines do not appear in their normal wave-length region.

The determinations of the size of the radio source confirm that the source is not stellar, as had been assumed originally, but has a small diameter measurable with refined interferometric methods. Measures by Mills[9] give an effective size of 1'.1. Smith[10] finds a diameter of about 3'.5. Both values should apply to the east-west direction. Measures by Hanbury Brown, Jennison, and Das Gupta[11] suggest an elliptical outline of about $1' \times 2'.1$, with the major axis in a position angle close to that of the major axis of the nebula. This is three to four times larger than the visible size, which, as is usual for faint nebulae, is only that of the main body of the nebula. The emission line λ 3,727 [*O* II] extends beyond the main body, and it is quite possible that the fainter outer parts may have a diameter much larger than the main body. If, however, a large part of the disagreement has to be explained by an essential contribution of the outer parts to the radio emission, the radio and the optical emission cannot be related. It seems improbable that this assumption entirely suffices to explain the discordance. In view of the difficulty of the radio measures, however, the lack of agreement in regard to the precise extent seems less important than the agreement as to the order of magnitude of the size.

Photoelectric magnitudes and color of Cygnus A have been measured by Dr. W. A. Baum. With a diaphragm of 11''.5 diameter, the magnitudes and colors reduced to the international system are $m_{pg} = 17.90$; $m_{pv} = 16.22$; $CI = +1.68$. To determine the absolute magnitude, corrections have to be applied for interstellar absorption and for the size of the diaphragm. By comparison with nebulae of similar appearance, a conservative estimate is that the nebula is 0.75 mag. brighter than the part admitted by the diaphragm. The interstellar absorption cannot be determined from the color index of the nebula, since the color is affected by the emission lines. But, on the assumption that the elliptical nebula 25'' north-preceding is a member of the cluster to which Cygnus A belongs and therefore at the same distance, this apparently normal nebula may be used. Its magnitudes and colors are $m_{pg} = 18.95$; $m_{pv} = 17.42$; $CI = +1.53$. The color of an elliptical nebula with a red shift of 17,000 km/sec is[12] $CI = +1.08$. The color excess of $+0.45$ mag. between this value and the measured color index is due to interstellar reddening. The ratio of photographic absorption to international color excess is somewhat less than 5. The photographic absorption, therefore, is 2.1 mag. Thus we have for Cygnus A the values in the accompanying tabulation.

	Mag.
Apparent m_{pg}	17.90
Correction for incompleteness	0.75
Photographic absorption	2.1
Corrected m_{pg}	15.05
Distance modulus $m - M$	32.6 (from red shift)
M_{pg}	-17.5

The system, therefore, is a giant system which may be expected to have a diameter of more than 10,000 psc. The major axis of 30'' of the visible main body of the nebula is only 5,000 psc. The assumption is justified that the nebula has a larger diameter than this; but a diameter of 20,000 psc, which would correspond to the radio interferometer measures, would call for strong radio emission from the outermost parts of the nebula, where the average density is exceedingly low.

Since the distance of the nebula is known, the total energy emitted in the radio region may be computed. The flux received from the nebula is of the order of 1.3×10^{-22} watts m^{-2} $(c/sec)^{-1}$ in a band of an equivalent width of 500 Mc/sec. The total flux, therefore, is of the order of 6×10^{-14} watts m^{-2} or 6×10^{-11} ergs cm^{-2} sec^{-1}. With the source at the distance of 3.3×10^7 psc, the total energy emitted in the radio region is 8×10^{42} ergs/sec. This is larger than the total optical emission of 5.6×10^{42} ergs/sec and larger than the energy contained in the line emission. The source of energy for the radio emission may be the relative kinetic energy of the colliding nebulae, which is of the order of 10^{59} ergs for a relative velocity of 500 km/sec.

1. M. Ryle and G. Smith, *Nature 162*, 462 (1948).
2. J. G. Bolton and G. J. Stanley, *Nature 161*, 312 (1948).
3. *Nature 161*, 312 (1948).
4. B. Y. Mills and A. B. Thomas, *Australian J. Sci. Res. A*, *4*, 158–171 (1951).
5. This coincidence was already noted in 1951 by Mills and Thomas (see n. 4), who had obtained for the radio position of Cyg A: 1950.0, $19^h57^m37^s \pm 6^s$, $+40°34' \pm 3'$; but it seemed unlikely at that time that a distant galaxy could be the radio emitter. Moreover, the coincidence established by them was not convincing, since, besides the nebula in question, three of the brighter members of the cluster fall into the area defined by the uncertainty of the position. Minkowski therefore wrote Mills that he did not think it was permissible to identify the source with one of the faint extragalactic nebulae in the area and emphasized that what was wanted was a more accurate radio position. The accuracy of Smith's position was needed to make the identification among the cluster members unambiguous. After Smith's position became known, the coincidence of the nebula with the radio source was also noted by D. W. Dewhirst (*Observatory 71*, 212 [1951]). B. Y. Mills

later gave the improved position: 1950.0, $19^h57^m44^s \pm 2\frac{1}{2}^s$, $+40°35' \pm 1\frac{1}{2}'$.

6. Nature *170*, 1064 (1952).

7. *Ap. J. 113*, 413 [1951].

8. The spectrum of the nebula resembles to a certain degree the spectra of those relatively rare nebulae whose nuclei show strong emission lines (C. K. Seyfert, *Ap. J. 97*, 28 [1943]; *Mt. W. Contr.* No. 671), but there are marked differences. In NGC 1068 and NGC 4151, the outstanding representatives of the emission nuclei, only 13 and 22 per cent of the light of the nucleus, respectively, is in the emission lines. The outstanding difference between Cygnus A and the nebulae with emission nuclei is that in Cygnus A the area of very strong emission has a diameter of 800 psc, whereas in the other nebulae the emission is restricted to diameters of the order of 25 psc.

9. *Nature 170*, 1063 (1952).

10. Ibid., 1064 (1952).

11. Ibid., 1060 (1952).

Added October, 1953: Interferometric measures by Jennison and Das Gupta, reported at the Symposium on Radio Astronomy at the Jodrell Bank Experimental Station in July, 1953, show that the radio emission originates in two areas separated by 82″ and 43″ long in position angle 95° with a width of 30″. No object of this description can be seen on any of our plates which cover the wave-length range from the ultraviolet to the infrared.

12. J. Stebbins and A. E. Whitford, *Ap. J. 108*, 413 (1948).

118. The Nature of Cosmic Radio Sources

Martin Ryle

(*Proceedings of the Royal Society* (London) *A248*, 289–308 [1958])

Four years after the identification of Cygnus A, Martin Ryle abandoned his radio star hypothesis and made the interesting speculation that radio galaxies might be used to test cosmological models. In the lecture presented here, his arguments portray the majority of radio sources as powerful emitters at extragalactic distances.

Because radio sources have approximately the same intensity over a wide wavelength range, in contrast to the optical spectrum of galaxies, the effect of the Doppler shift on distant sources will be much less important for radio than for optical detection. As first pointed out by Hendrick van de Hulst and Jan Oort in 1945, this difference means that radio telescopes can sample more distant regions of the universe than optical telescopes, and the radio data can therefore be used to test different cosmological models (see selection 91).

In this paper, Ryle reviews the early attempts to draw cosmological conclusions from counts of radio sources. If the nearby, brighter radio sources represent a uniform distribution of stationary sources, then the number of fainter, presumably more distant, sources should be far fewer than actually observed. This effect is opposite to that expected when the expansion of the universe is taken into account, because the large redshifts would reduce the apparent brightness of a source and cause fewer to be observed. Since distant objects are seen at an earlier stage in the development of the universe than are nearby objects, the observed excess can be interpreted as an evolutionary effect in which the more distant, younger objects are either more luminous or more concentrated than the nearby older objects. Such evolutionary effects are, of course, inconsistent with the steady state cosmology.

In the past twenty years considerable controversy has arisen over the nature of the number counts of radio sources. The Cambridge group of radio astronomers has argued that there is an excess number of radio sources present down to 4 flux units at 73-cm wavelength, and for weaker flux values it diminishes.[1] Other radio astronomy groups, using wavelengths of 6 cm and 21 cm, claim that there is virtually no detectable difference in counts from that expected from a stationary, Euclidean universe in which no evolution takes place.[2] Later work has shown that the number counts vary with wavelength.[3]

1. G. G. Pooley and M. Ryle, *Monthly Notices of the Royal Astronomical Society 139*, 515 (1968).
2. K. I. Kellermann, M. M. Davis, and I. I. K. Pauliny-Toth, *Astrophysical Journal* (*Letters*) *170*, L1 (1971); H. H. Bridle, M. M. Davis, E. B. Fomalont, and J. Lequeux, *Nature-Physical Science 235*, 123 (1972); K. I. Kellermann, *Astronomical Journal 77*, 531 (1972).
3. J. V. Wall and D. J. Cooke, *Monthly Notices of the Royal Astronomical Society 171*, 9 (1975).

1. Introduction

When a radio telescope of limited resolving power is used to scan the sky, the variation of the received power can be related to the general structure of the galaxy; the greatest intensity is received from directions near the galactic equator, particularly in the direction of the center of the system.

When observations are made with an instrument of greater resolving power, a number of localized sources, superimposed on the general radiation, can be distinguished. These sources do not simply represent brighter patches of the general radiation, since their angular extent in most cases is less than a few minutes of arc, whilst except for areas close to the galactic equator, the background radiation appears to have no features with an angular structure finer than a few degrees. The surface brightness of these localized sources, or 'radio stars' may exceed by a factor of 1,000, that of the general radiation.

The possibility that the general radiation represents the smoothed out emission from large numbers of faint radio stars belonging to our galaxy, but too weak to be detected individually, has been considered by a number of authors (Ryle 1950; Westerhout and Oort 1951; Priester 1954*b*). An alternative suggestion, developed by Shklovsky (1952) and Burbidge (1956*a*) associates the emission with the acceleration of relativistic electrons in extensive interstellar magnetic fields; under these circumstances the radio stars might contribute a negligible fraction of the total emission.

The purpose of this lecture is to collect all the available evidence on the nature of the radio stars. I shall first discuss the angular distribution of the background radiation in relation to the observations of radio stars. Attempts to identify these sources with photographic objects have met with only limited success, and the small number of such identifications (§4) includes a wide range of types of source, some of which are relatively weak sources within the galaxy, whilst others represent extremely powerful and distant extra-galactic sources.

In attempting to find the nature of the remainder, we must rely largely on a more detailed analysis of the radio data. We can study the angular distribution of the sources and their contribution to the background radiation; in many cases we can measure their angular diameters and spectra and set limits to their optical brightness. By relating all these quantities, the range of possible types of source can be limited.

Initially, it will be supposed that all the sources are of the same 'radio luminosity' and each emits a power P per unit solid angle. We shall proceed to find what ranges of P, and the corresponding values of space density ρ, are compatible with the observations. It is of course possible that more than one class, for example, both galactic and extra-galactic sources, might make roughly equal contributions to the number of sources observed in a given range of intensity; we must examine this possibility, as well as the effect of a spread in the luminosity within any one class (§8).

For the purpose of numerical computation, data have been taken for the wavelength of 3.7 m at which two of the main surveys were made. When considering extra-galactic sources we adopt a value of Hubble's constant of 180 km s^{-1} Mpc^{-1} (Humason, Mayall and Sandage 1956). Further revision of the extra-galactic distance scales is now in progress (Sandage 1958); any modification will alter the numerical values of P and ρ, but will not affect the general conclusions on the nature of the sources.

2. The Angular Distribution of the General Radiation

Surveys of the general radiation have now been made over a range of wavelengths from 30 cm to 15 m. Observations at the shorter wavelengths have provided maps of good resolution, but lack of sensitivity has restricted them to areas close to the galactic equator. Observations at the long wavelengths have covered the whole sky, but these have been limited in their resolving power.

Here Ryle discusses the 3.7-m wavelength isophotes obtained by Baldwin (1955*a*,*b*). The radio emission may be divided into a disk distribution concentrated within 150 pc of the galactic plane, a distribution similar to the distribution of mass within our galaxy, a spherical halo surrounding our galaxy with a radius of 14 kpc, and an extra-galactic component. We omit a cross-sectional diagram of our galaxy showing the different regions from which radio emission can be distinguished.

3. The Observation of Radio Stars

A large number of observations of radio stars have been made since their original recognition by Hey, Parsons and Phillips (1946). In addition to surveys aimed at establishing the positions and intensities of the sources, the angular diameters and spectra of a number of them have been measured.

Most of the observations have used interferometric methods because of the improved positional accuracy, but observations with single instruments of large resolving power have been made at centimetric wavelengths (Haddock, Mayer and Sloanaker 1954) and at longer wavelengths using a large paraboloid (Hanbury-Brown and Hazard 1953) and more recently with the method described by Mills and Little (1953).

The most detailed surveys which have been published are those made at Cambridge using an interferometer at a wavelength of 3.7 m (Shakeshaft, Ryle, Baldwin, Elsmore and Thomson 1955) and in Australia by Mills and Slee (1957) at a wavelength of 3.5 m. A second survey has recently been completed at Cambridge, using the same instrument on a wavelength of 1.9 m, in which the resolving power is improved by a factor of about 4. Some of the results of this survey have been compared with the previous two by Edge, Scheuer and Shakeshaft (1958); although the observations of the weaker sources are still unreliable, as shown by the considerable disagreements between the different surveys, there is a good measure of agreement for the more intense sources, and a number of important conclusions can be drawn.

(*a*) At small galactic latitudes, a large number of extended sources occur, having angular diameters between a few minutes of arc and 1°. The fact that these only occur close to the galactic equator indicates that they are sources situated within the galaxy.

(*b*) A few extended sources having angular diameters greater than 2° have been observed at high galactic latitudes.

(*c*) Most of the sources more than a few degrees from the galactic plane have angular diameters below the limit of resolution; their distribution over the sky appears to be isotropic and not related to the galactic structure.

It has also been found that the spectra of the latter sources differ appreciably from those of most of the sources situated near the plane; Whitfield (1957, 1958) has shown that the average spectrum of 18 sources situated within 10° of the equator can be represented by a flux density $S(\lambda) \propto \lambda^{0.78}$. The average for 63 sources at galactic latitudes greater than 10° was given by $S(\lambda) \propto \lambda^{1.2}$.

Observations at centimetric wavelengths have revealed additional sources at very low galactic latitudes, having flux densities independent of wavelength (Haddock 1955); they undoubtedly represent the thermal emission from ionized hydrogen regions, with which a number of the sources have been related by optical observations. These sources have also been detected both in emission and in absorption at metric wavelengths (Mills, Little and Sheridan 1956).

4. Identification of Radio Stars with Photographic Objects

Since the existence of radio stars was recognized, considerable efforts have been made to observe optically the objects responsible for the radio emission. About a dozen related sources were found fairly early, but subsequent progress has been remarkably slow, despite the fact that the accuracies of the source positions have improved considerably.

Although about 500 sources are known with positional accuracies of a few minutes of arc, only for some 40 are there reasonably reliable identifications, and a number of these represent the thermal radiation from known regions of ionized hydrogen. The remainder include both galactic and extragalactic sources; it is possible that none of them is representative of the majority of the radio stars.

a. SOURCES WITHIN THE GALAXY Apart from the ionized hydrogen regions, eight sources have been associated with galactic nebulosities, and their details are given in table 118.1. The radio luminosity P, the power emitted per unit solid angle, is also given for those sources whose distances are known. In addition to the three optically observed supernovae, Oort (1946) has concluded that the Cygnus loop represents the remains of a supernova outburst which occurred some 10^5 years ago. It has further been suggested that both Cassiopeia A (Shklovsky and Parenago, 1952) and IC 443 (Shklovsky 1954) are the remnants of early supernovae.

Table 118.1 Galactic sources

	Angular diameter		Distance	Radio flux density	P
	Optical	Radio	(kpc)	(10^{-26} W (c/s)$^{-1}$ m^{-2})	(W (c/s)$^{-1}$ ster^{-1})
Cygnus Loop	2.7°	—	0.3	300	2.6×10^{14}
Supernova 1572 (Tycho Brahe)	—	5′	~1	280	2.7×10^{15}
Supernova 1604 (Kepler)	~1.5′	<1′	—	110	—
Supernova 1054 (Crab Nebula)	4′ × 6′	3.5′ × 5.5′	1	1,900	1.8×10^{16}
Cassiopeia A	6.3′	3′	3(?)[a]	22,000	1.9×10^{18}
IC 443	49′	24′	—	580	—
Puppis A	50′ × 80′	33′	—	530	—
Auriga A	2°	1.4°	—	280	—

a. Radio observations; optical data uncertain.

Table 118. 2 Extragalactic sources

	Angular diameter		Distance	Radio flux density	P	
	Optical	Radio	(Mpc)	(10^{-26} W (c/s)$^{-1}$ m^{-2})	(W (c/s)$^{-1}$ ster^{-1})	
SMC	7° × 4°	—	0.047	600	1.3×10^{19}	normal irregular galaxies
NGC 55	25′ × 3′	—	~1	3.6	3.4×10^{19}	
LMC	21° × 19°	—	0.042	4,000	6.7×10^{19}	
NGC 253	24′ × 5′	—	1.3	22	3.5×10^{20}	normal spiral galaxies
NGC 224	160′ × 40′	140′	0.61	240	8×10^{20}	
NGC 300	20′ × 10′	—	—	9.6	—	
NGC 6774	15′ × 10′	—	—	6	—	
NGC 4945	14′ × 2′	—	—	24	—	
NGC 5236	12′ × 12′	—	1.5	38	8.1×10^{20}	
NGC 5194–5	~14′	—	3.3	19	2.0×10^{21}	
NGC 1068	7′	—	6.1	15	5.3×10^{21}	
NGC 1316	5′ × 3′	~40′	9.6	295	2.5×10^{23}	optically abnormal
NGC 5128	20′	3′ × 6.5′ with 2° halo	2.5(?)	6,700	3.5×10^{23}	
NGC 4486	~5′	5′ with 1° halo	6.8	1,800	7.9×10^{23}	
NGC 1275	~2′	2.4′	38	190	2.6×10^{24}	colliding galaxies
NGC 6166	~1′	—	—	155	—	
Hydra A	—	1.6′	88.5	550	4.1×10^{25}	
Hercules A	—	2.3′	142	680	1.2×10^{26}	
Cygnus A	—	2 disks each 40″ diam. separated by 85″	93.5	13,300	1.1×10^{27}	

b. SOURCES OUTSIDE THE GALAXY Besides the Magellanic clouds and the other small irregular galaxy NGC 55, a number of nearby galaxies have been identified with radio sources (Hanbury-Brown and Hazard 1952; Mills 1955). The intensities of these sources are consistent with an emission comparable with that from our own galaxy for which $P = 2 \times 10^{21}$ W (c/s)$^{-1}$ ster^{-1}; as already mentioned, the surface brightness of such sources is very much less than that of the 'point' sources.

A further eight identifications have been made with peculiar extra-galactic objects; five of these represent collisions between pairs of galaxies, and all of them show unusual optical features. The radio emission from these sources is very much greater than that from the previous group, as can be seen from the last column in table 118.2.

c. DISCUSSION OF OPTICAL IDENTIFICATIONS In addition to the thermal emission from regions of ionized hydrogen, a number of sources have been related to galactic nebulosities; most of these are believed to be the remains of supernovae outbursts. It is possible that most of the extended sources situated near to the galactic equator are of similar origin and are due to supernovae which occurred before recorded history.

We can be far less certain about the nature of the sources situated away from the galactic equator; a number of them with appreciable angular diameters have been related to nearby extra-galactic nebulae, and the radio emission from these sources is about the same as that of our own galaxy. Most of the others have angular diameters too small to measure, and their distribution appears to be isotropic. Many of these are intense sources, whose positions are known with good accuracy, and the failure to identify them with optical objects implies a very small ratio of optical/radio brightness. This result suggests that they might belong to a class of object whose optical emission is intrinsically very small. An alternative possibility is suggested by the identification of three of them with the remote extra-galactic nebulae in Hydra,

Hercules and Cygnus. Owing to the relatively slow variation of the radio emission with wavelength, in comparison with the optical emission, the effect of the red-shift will be less important for radio detection than for optical detection. Sources similar to these three nebulae situated at greater distances would be very difficult to detect photographically whilst still being readily observable with large radio telescopes.

Here we omit a figure illustrating the fact that objects with large recession velocities produce a large reduction in photographic energy and a relatively less important reduction in the energy received at a given radio wavelength.

Let us now examine the radio evidence in detail to see how far we may exclude various possible classes of source. First of all we will see within what ranges the radio luminosity P, and the spatial density ρ may lie.

5. The Number-Intensity Relation

Consider a region containing a random distribution of sources having an average spatial density ρ pc^{-3}, and suppose that each source emits a power P W (c/s)$^{-1}$ ster^{-1}.

Then the number of sources (N) per unit solid angle having a flux density greater than S is given by

$$N = \alpha S^{-3/2}, \quad \text{where} \quad \alpha = \tfrac{1}{3}\rho P^{3/2}.$$

The results of any survey of radio sources may be used to find the relationship between N and S, and hence to determine α.

In the Cambridge 3.7 m survey it was found that α was not constant, but appeared to increase with decreasing S (Ryle and Scheuer 1955). Since this effect might have been due to misinterpretation of the weaker sources, an additional method of analysis was employed, which was based on a statistical study of the record amplitude and avoided this difficulty (Scheuer 1957); the increase of α was again found.

A similar effect was found in the later survey of higher resolving power (Edge et al. 1958); the value of α found for the most intense sources is about half that found for the weaker sources. Without discussing possible explanations for this effect, it is clear that the former value, whilst being more directly established, may be subject to statistical errors associated with the smaller number of sources. In the subsequent analysis, it will be supposed that the two values derived represent the limits of the possible range of α. For the wavelength 3.7 m, expressing P in terms of W (c/s)$^{-1}$ster^{-1} and ρ in terms of ps^{-3}, the values of α are: (0.7 to 1.4) $\times 10^{14}$. $\rho P^{3/2}$ then lies in the range (2 to 4) $\times 10^{14}$.

It is convenient to represent the radio star observations in terms of a plot of $\rho P^{3/2}$ = constant, and this is shown in figure 118.1 where the width of the line corresponds to the adopted limits of α. Any point on this line represents one possible combination of P and ρ which would satisfy the observations of the number of radio stars falling within a given range of intensity. The values of P for those identified sources whose distances are known have been included in figure 118.1 as well as the approximate spatial densities of common visible stars and extra-galactic nebulae.

6. The Isotropic Distribution of Radio Stars

Except in directions close to the galactic equator, where the observations are complicated by the presence of additional sources which are clearly situated within the galaxy, the "point" sources appear to be distributed isotropically. This result is to be expected if most of the sources are extra-galactic, but if they are situated within the galaxy it allows a lower limit to be set to ρ, their average spatial density. For example, if observations in different directions reveal N sources per steradian above a given flux density then $(1/3)\rho r_1{}^3 \nless N$, where r_1 represents the smaller extent of the emitting region.

In order to obtain the most sensitive determination, without the uncertainty involved in measurements of individual sources near the limit of the resolution of the survey, the recent observations at a wavelength of 1.9 m were analyzed by the statistical method developed by Scheuer (1957). By comparing the curves of the probability distribution of the record amplitude derived for different areas of sky, it becomes possible to detect reliably quite small departures from uniformity.

In this analysis an area of sky was selected between RA 08 to 16 h and declination -8 to $+50°$. Radio stars could be associated either with the extended 'halo' or with the spiral arms. The intermediate distribution of emission like the mass-distribution is mainly confined to regions nearer to the galactic nucleus than the solar system and the emission near the solar system is hardly distinguishable from that associated with the more extended halo.

The analysis was therefore carried out to detect the sort of anisotropy associated with radio stars distributed either in the halo or throughout the local spiral arm. In each case the observations revealed no significant departure from isotropy for values of N, the number of sources per steradian, up to 500.

From the known extent of the halo and of the local spiral arm, it is then possible to establish lower limits of ρ for the two cases; the values are 3×10^{-9} ps^{-3} for the halo distribution and 5×10^{-4} ps^{-3} for radio stars distributed in the spiral arm. These limits are indicated in figure 118.1.

7. The Integrated Emission from the Radio Stars

We have seen that the majority of the radio stars may be relatively common objects distributed throughout the spiral arms with a spatial density of at least 5×10^{-4} ps^{-3}, or they may be more powerful but less numerous sources distributed

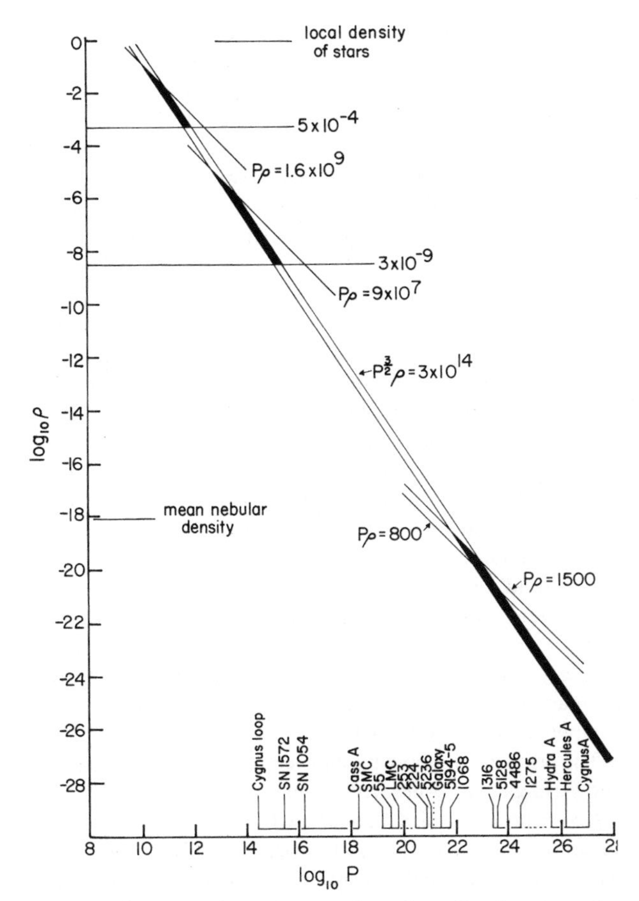

Fig. 118.1 Plot of $\log_{10}\rho$ against $\log_{10}P$, showing the possible values of spatial density, ρ, in pc^{-3} for sources that emit power, P, in W Hz^{-1} sterad^{-1} when $\rho P^{3/2} = 3 \times 10^{14}$. The limitations set by the isotropic distribution of the sources and by their maximum contribution to the general emission ($P\rho$ = constant) are shown; the solid areas represent the possible ranges of ρ and P. The values of P are also shown for those identified sources whose distances are known.

througho it the galactic halo. Alternatively, they might be sources of very much greater luminosity situated outside the galaxy.

We do not know what fraction of the background radiation is due to the integrated emission from the radio stars, but it is clear that the emission per unit volume $P\rho$ which they produce, cannot exceed that required to account for the observed distribution of sky brightness.

Baldwin has shown that at a wavelength of 3.7 m the average emission per unit volume throughout the halo is 1.8×10^8 W $(\mathrm{c/s})^{-1}$ ster^{-1} ps^{-3}; his corresponding figure for the spiral arm is 3.2×10^9 W $(\mathrm{c/s})^{-1}$ ster^{-1} ps^{-3}. The emissivity of the halo measured over the range of wavelength 1.7 to 8 m, varies as $\lambda^{0.5}$; for the spiral arm the index is less than 0.5. Over the same range of wavelength, the emission of the radio stars varies as $\lambda^{1.2}$. In the absence of reabsorption, some other source of emission must therefore contribute, and it can be shown that at a wavelength of 3.7 m, not more than half of the total radiation can be due to the radio stars. By comparing observations of the general radiation made at very low frequencies by Reber and Ellis (1956), with those at higher frequencies, Baldwin (1958) has shown that the total absorption in the halo, and normal to the spiral arm, is negligible at metre wavelengths.

We may therefore conclude that if most of the radio stars are within the spiral arm, the product $P\rho$ cannot exceed 1.6×10^9 W $(\mathrm{c/s})^{-1}$ ster^{-1} ps^{-3}, whilst if they are distributed throughout the halo, $P\rho$ cannot exceed 9×10^7 W $(\mathrm{c/s})^{-1}$ ster^{-1} ps^{-3}.

These limits are shown in figure 118.1 and it can be seen that the range of possible values of P, if the sources are within the galaxy, is restricted to two regions, both of which correspond to sources of considerably lower luminosity than any of the identified sources.

We may now apply similar arguments to the case in which the radio stars are supposed outside the galaxy. Baldwin's observations led him to conclude that the total contribution of extra-galactic sources results in a brightness temperature at 3.7 m, which cannot exceed 500° K. From this figure an upper limit may be found for the average emission per unit volume of extra-galactic space.

Owing to the comparatively slow variation of the radio emission with wavelength, the red-shift has a much smaller effect than in the optical case and the integrated intensity converges only slowly with distance; much of the integrated intensity will therefore be due to very distant sources. This problem has been considered in detail by Shakeshaft (1954) and Priester (1954*a*), who show that the relation between the emission per unit volume and the total sky brightness depends strongly on the cosmological model adopted.

In the case of a steady-state model (Bondi and Gold 1948) the maximum brightness temperature of 500° K, corresponds to an average emissivity of 1,700 W $(\mathrm{c/s})^{-1}$ ster^{-1} pc^{-3}.

For a relativistic cosmology having zero space curvature, the convergence is less rapid, but even if the limit of integration is taken as the optical limit of observation, so that marked evolutionary changes in the sources are unlikely, it is possible to establish an upper limit of 1,000 W $(\mathrm{c/s})^{-1}$ ster^{-1} pc^{-3}.

Part of the emission in extra-galactic space is due to normal nebulae, which contribute about 200 W $(\mathrm{c/s})^{-1}$ ster^{-1} pc^{-3}. It is thus possible to fix upper limits of $P\rho$ of 800 and 1,500 W $(\mathrm{c/s})^{-1}$ ster^{-1} pc^{-3}, according to which of these two models is adopted.

When these limits are added to figure 118.1 it is apparent that if most of the radio stars are extra-galactic, then their luminosity is considerably in excess of that of the ordinary galaxies, whilst the space density is correspondingly lower. We shall later examine some independent arguments leading to the same conclusion (9).

8. The Effects of a Spread in the Luminosities

So far, for simplicity, we have supposed that all the radio stars belong to a single class having a fixed value of P.

With this assumption it has been shown that the observations are compatible with any of three classes of sources:

a. Sources distributed throughout the local spiral arm having values of P between 10^{11} to 10^{12} W $(c/s)^{-1}$ $ster^{-1}$.

b. Sources distributed throughout the galactic halo, with P in the range 5×10^{13} to 10^{15} W $(c/s)^{-1}$ $ster^{-1}$.

c. Extra-galactic sources having P greater than about 10^{23} W $(c/s)^{-1}$ $ster^{-1}$.

Even in the extreme case in which all three classes contribute equally, the effect on the permissible range of P in any class is not important. It is therefore assumed that one of the three classes is mainly responsible for the observed sources. A spread in the luminosity, P, within a single class of sources is associated with an earlier appearance of anisotropy, a greater contribution to the integrated radiation per unit volume, $\sum P\rho$, and a greater likelihood of sources being identified or having measurable angular diameters. The values of P extend over the ranges 10^4:1, 10^8:1, and 10^4:1, respectively, for identified galactic sources, for identified extragalactic sources, and for seven "abnormal" galaxies. By choosing a smaller spread in luminosity, P, of 10^3:1, Ryle shows that the radio stars cannot be in our galaxy. This is essentially because a dispersion in P increases the lower limit on the effective space density, ρ_0, and the quantity $P\rho$ in figure 118.1 must be reduced if the allowed integrated emission is not to be exceeded.

9. The Source of Energy

We will now examine the problem of supplying sufficient energy to account for each of the three classes of source we have considered. From the spatial density and maximum angular sizes of the sources we can compute, for each class, a minimum volume emissivity which we can then relate to the physical mechanism of radio emission.

Only one satisfactory mechanism has been proposed to account for the 'non-thermal' emission from sources other than the sun. This mechanism, first proposed by Shklovsky (1952), is based on the acceleration of high-energy electrons in weak magnetic fields. It has been shown capable of explaining the radiation from a number of different sources, including the Galaxy (Shklovsky 1952; Burbidge 1956*a*), the colliding galaxies NGC 5128 and NGC 1316 (Burbidge and Burbidge 1957), the Crab Nebula (Oort and Walraven 1956) and NGC 4486 (Burbidge 1956*b*). The latter two sources are of particular interest, since part of their optical emission has been shown to be plane polarized, a result which implies radiation of light by the same mechanism (Shklovsky 1953*a*; Dombrovsky 1954; Oort and Walraven 1956; Baade 1956).

There is, therefore, little doubt that the mechanism is responsible for the radio emission from these particular sources, and in the absence of any alternative theory, it seems reasonable to suppose that it is responsible for the emission from other sources.

The total energy of the particles required to produce a given radio emissivity depends on the magnetic field, and there is a minimum value for the combined energies of the particles and field. This condition, which approximates to equipartition between the two forms of stored energy, can be computed (Burbidge 1956*b*).

The problem of supplying sufficient energy both for galactic and extra-galactic sources has been considered by Greenstein (1955), who concluded that apart from the special case of supernova outbursts, the most powerful supply is likely to be the translational kinetic energy of colliding gas clouds. The only other adequate source of energy which has been suggested is that associated with the annihilation of anti-matter (Burbidge and Hoyle 1956; Gruber 1957).

Here Ryle uses the equipartition argument to deduce total energies of 10^{42}, 10^{48}, and 10^{57} ergs, respectively, for the sources in the spiral arms, for the sources in the galactic halo, and for the extragalactic sources. There seems to be no plausible mechanism for providing the energy in the first two cases, but the kinetic energy of colliding galaxies seems to be sufficient to provide the energy for the extragalactic sources.

10. Evidence Derived from Measurements of Angular Diameters and from Optical Data

We have seen that if there is appreciable spread in the values of the radio luminosities, it is difficult to account for the radio stars in terms of sources situated either in the spiral arm or in the galactic halo, and additional arguments against either of these possibilities have been based on the difficulty of supplying sufficient energy.

If on the other hand, most radio stars are extra-galactic, then we have seen that their effective luminosity P must lie in the range greater than about 10^{23} W $(c/s)^{-1}$ $ster^{-1}$. Seven sources in this range of luminosity have been identified with optically observable extra-galactic nebulae (NGC 1275, 1316, 4486, 5128, Hydra A, Hercules A and Cygnus A) and in each case their physical sizes and optical luminosities have been found typical of giant systems. The diameters of the radio-emitting regions are respectively, 25, 68, 100, 87, 40, 80 and 40 kpc.

It cannot be assumed that other sources with similar radio luminosities are necessarily as large as these, but the problem of supplying sufficient energy suggests that they must have masses comparable with those of normal extra-galactic nebulae.

If we suppose that their sizes and optical luminosities are comparable with those of normal galaxies, then measurements of the angular diameter or optical brightness can be used as indicators of distance, to define more closely the possible values of P.

Under these assumptions Ryle derives an upper limit of 2×10^{-24} pc^{-3} for the space density of extragalactic radio sources.

11. COSMOLOGICAL EFFECTS

If most of the sources have luminosities comparable with that of Cygnus A, then many of those which have been detected must be at distances at which the effects of the red-shift are important. Two particular effects may occur: one is to cause an additional reduction in the flux density of distant sources, and whilst it cannot easily be measured directly unless the distances of individual sources can be found, it may have an important effect on the relationship between the observed number of sources falling in different ranges of flux density. The second effect arises if the spectral distribution of the emission cannot be represented by x = constant in the relation $S(\lambda) \propto \lambda^x$.

Here Ryle argues that the Doppler effect will cause the spectral index of a distant source to be greater than that of a nearby source when measurements are made over a fixed wavelength interval. Sources with lower flux densities are observed to have larger spectral indices. The red-shift effect will also cause a reduction in brightness over and above the inverse square law effect. Unexpectedly, an excess of weak radio sources is observed, and this observation seems to rule out the steady state cosmology.

Whether or not the present arguments are regarded as conclusive evidence in favour of an evolving cosmological model, I believe that what they have shown is that most radio stars are very powerful sources at very great distances. The ability to observe to distances at which large red-shifts occur, without the restrictions set in optical astronomy by the form of the spectral distribution, means that the predictions of different cosmological models can be tested with a directness which has been impossible photographically.

Baade, W. 1956, *Astrophys. J. 123*, 550.

Baldwin, J. E. 1955*a*, *Mon. Not. R. Astr. Soc. 115*, 684.

Baldwin, J. E. 1955*b*, *Mon. Not. R. Astr. Soc. 115*, 691.

Baldwin, J. E. 1958, *Observatory 78*, 166.

Bondi, H., and Gold, T. 1948, *Mon. Not. R. Astr. Soc. 108*, 252.

Burbidge, G. R. 1956*a*, *Astrophys. J. 123*, 1.

Burbidge, G. R. 1956*b*, *Astrophys. J. 124*, 416.

Burbidge, G. R. 1958, *Astrophys. J. 127*, 48.

Burbidge, G. R., and Burbidge, E. M. 1957, *Astrophys. J. 125*, 1.

Burbidge, G. R., and Hoyle, F. 1956, *Nuovo Cim. 4*, 558.

Dombrovsky, V. A. 1954, *Dokl. nauk., U.S.S.R. 94*, 1021.

Edge, D.O., Scheuer, P. A. G., and Shakeshaft, J. R. 1958, *Mon. Not. R. Astr. Soc. 118*, 183.

Greenstein, J. L. 1955, *I.A.U. Symposium on Radio Astronomy* (Cambridge), p. 179.

Gruber, F. 1957, *Two Kinds of Matter in the Universe* (Vienna).

Haddock, F. T. 1955, *I. A. U. Symposium no. 4: Radio Astronomy*, ed. H. C. van de Hulst (Cambridge: Cambridge University Press, 1957), p. 192.

Haddock, F. T., Mayer, C. H., and Sloanaker, R. M. 1954, *Nature* (Lond.) *174*, 176.

Hanbury-Brown, R., and Hazard, C. 1952, *Phil. Mag. 43*, 137.

Hanbury-Brown, R., and Hazard, C. 1953, *Mon. Not. R. Astr. Soc. 113*, 123.

Hey, J. S., Parsons, S. J., and Phillips, J. W. 1946, *Nature* (Lond.) *158*, 234.

Humason, M. L., Mayall, N. U., & Sandage, A. R. 1956, *Astron. J. 61*, 97.

Mills, B. Y. 1955, *Aust. J. Phys. 8*, 368.

Mills, B. Y., and Little, A. G. 1953, *Aust. J. Phys. 6*, 272.

Mills, B. Y., Little, A. G., and Sheridan, K. V. 1956, *Aust. J. Phys. 9*, 218.

Mills, B. Y., and Slee, O. B. 1957, *Aust. J. Phys. 10*, 162.

Oort, J. H. 1946, *Mon. Not. R. Astr. Soc. 106*, 159.

Oort, J. H., and Walraven, Th. 1956, *B.A.N. 12*, 285.

Priester, W. 1954*a*, *Z. Astrophys. 34*, 283.
Priester, W. 1954*b*, *Z. Astrophys. 34*, 295.
Reber, G., and Ellis, G. R. 1956, *J. Geophys. Res. 61*, 1.
Ryle, M. 1950, *Rep. Progr. Phys. 13*, 184.
Ryle, M., and Scheuer, P.A.G. 1955, *Proc. Roy. Soc. A. 230*, 448.
Sandage, A. R. 1958, *Sky and Telescope 17*, 275.
Scheuer, P.A.G. 1957, *Proc. Camb. Phil. Soc. 53*, 764.
Shakeshaft, J. R. 1954, *Phil. Mag. 45*, 1136.
Shakeshaft, J. R., Ryle, M., Baldwin, J. E., Elsmore, B., and Thomson, J. H. 1955, *Mem. R. Astr. Soc. 67*, 106.
Shklovsky, I. S. 1952, *Astr. J. Moscow 29*, 418.
Shklovsky, I.S. 1953*a*, *Dokl. nauk. U.S.S.R. 90*, 983.
Shklovsky, I.S. 1953*b*, *Astr. J. Moscow 30*, 15.
Shklovsky, I.S. 1954, *Dokl. nauk. U.S.S.R. 94*, 417.
Shklovsky, I. S., and Parenago, P. P. 1952, *Astr. tsirc. akad. nauk. 131*, 1.
Westerhout, G., and Oort, J. H. 1951, B.A.N. *11*, 323.
Whitfield, G. R. 1957, *Mon. Not. R. Astr. Soc. 117*, 680.
Whitfield, G. R. 1958, *I. A. U. Symposium no. 9: Paris Symposium on Radio Astronomy*, ed. R. N. Bracewell (Stanford: Stanford University Press, 1959), p. 297.

119. First True Radio Star?

Thomas A. Matthews, John G. Bolton, Jesse L. Greenstein, Guido Münch, and Allan R. Sandage

(*Sky and Telescope 21*, 148 [1961])

This report by the editors of *Sky and Telescope* documents the first step in the discovery of the remarkable quasi-stellar objects, or quasars. In the early 1950s, when only a handful of discrete radio sources were known, these sources were commonly called "radio stars." However, as the first optical identifications became known, astronomers realized that the radio noise generally emanated from nebulae and galaxies, rather than from stars themselves. In 1954 Walter Baade and Rudolph Minkowski had, for example, shown that the intense radio source Cygnus A coincides with a distant extragalactic object (see selection 117). When interferometric measurements showed that the intense radio source 3C 295 has an angular size more than 10 times smaller than Cygnus A, Minkowski reasoned that 3C 295 must be a distant analogue to Cygnus A. This idea was confirmed in 1960, when Minkowski used accurate positional information provided by Thomas Matthews to identify 3C 295 with the brightest member of a cluster of galaxies whose redshift of 0.46 was 8 times larger than the redshift of Cygnus A.[1]

Stimulated by the identification of 3C 295 with a remote galaxy, Matthews and Allan Sandage began a search for the optical counterparts of other intense radio sources with small angular sizes. The article reproduced here summarizes the results for the small radio source 3C 48. This paper was presented at the December 1960 meeting of the American Astronomical Society, but because the authors were mystified by the nature of the object, they never published the customary abstract of their paper; the results remained unpublished in the professional journals for two years. Even when the formal paper announcing the results was written,[2] the true nature of 3C 48 was not yet realized. It was only after the discovery of the intermediate large redshift of another "stellar" object, 3C 273, that 3C 48 was recognized to have an even larger redshift of 0.3685 (see selection 120). The object's starlike appearance was caused not by its stellar size but by its tremendous distance. The spectral lines were unidentified not because of unknown elements or physical conditions but because of their large redshift.

1. R. Minkowski, *Astrophysical Journal 132*, 908 (1960).
2. T. A. Matthews and A. R. Sandage, *Astrophysical Journal 138*, 30 (1963).

In an unscheduled paper, Allan R. Sandage, Mount Wilson and Palomar Observatories, described a 16th-magnitude object in Triangulum that appears to be the first case where strong radio emission originates from an optically observed star. It was found in the course of the survey now being conducted at the radio observatory of California Institute of Technology to obtain very precise positions of radio sources. Twin 90-foot parabolic antennas, movable to various positions along a huge cross formed by a pair of tracks, are used for this work.

Thomas A. Matthews first determined a six-times more accurate position for the source 3C-48 than had hitherto been established at Cambridge University in England. The world's largest steerable paraboloid, the Jodrell Bank 250-foot radio telescope, had set the source's diameter at less than four seconds of arc.

At 96 megacycles, the Caltech instruments gave a radio flux intensity of 21.3×10^{-26} watt per square meter per cycle per second, while in England values of 50 and 43 were found at 159 and 178 megacycles, respectively. So strong a flux from such a small area led Dr. Matthews to suggest that the 200-inch telescope be used to search for an optical object in the same place in he sky.

Dr. Sandage, in a recent 90-minute exposure, found a star at the precise position. It is accompanied by a faintly luminous nebulosity measuring about five by 12 seconds of arc. The star is at right ascension $1^{h}\ 34^{m}\ 51^{s}$, declination $+32^{\circ}\ 54'.2$ (1950 co-ordinates), just where the radio source is located, so there is practically no doubt that the two are the same.

Spectrograms taken by Jesse L. Greenstein and Guido Münch of Caltech, and by Dr. Sandage, show a combination of strong emission and absorption lines unlike that of any other star known. The star or its gaseous envelope contains ionized calcium, ionized and neutral helium, and possibly oxygen ionized many times. The spectrum shows no hydrogen, the main constituent of all normal stars.

Since the distance of 3C-48 is unknown, there is a remote possibility that it may be a very distant galaxy of stars; but there is general agreement among the astronomers concerned that it is a relatively nearby star with most peculiar properties. It could be a supernova remnant. The radio output may be intrinsically 10 million times stronger than the sun's.

If the star were surrounded by high-energy electrons traveling in a magnetic field at near the velocity of light, the resulting synchrotron radiation would produce both its light and the radio energy. Despite its yellow color, the star is unusually bright in the ultraviolet region of the spectrum, an indication of synchrotron radiation.

120. The Discovery of Quasars

Investigation of the Radio Source 3C 273 by the Method of Lunar Occultations

Cyril Hazard, M. B. Mackey, and A. J. Shimmins

(*Nature 197*, 1037–39 [1963])

3C 273: A Star-like Object with Large Red-Shift

Maarten Schmidt

(*Nature 197*, 1040 [1963])

Red-Shift of the Unusual Radio Source: 3C 48

Jesse L. Greenstein and Thomas A. Matthews

(*Nature 197*, 1041–42 [1963])

Absolute Energy Distribution in the Optical Spectrum of 3C 273

John Beverly Oke

(*Nature 197*, 1040–41 [1963])

A dramatic and rapid sequence of events in late 1962 and early 1963 led to the discovery of the nature of quasi-stellar objects, or quasars. In late 1962 the moon happened to occult the radio source 3C 273. As Cyril Hazard had already shown,[1] a careful timing of the disappearance and reappearance of an occulted source could allow the determination of radio positions accurate to 1 sec of arc, since the position of the edge of the moon is known accurately for any time. Furthermore, because the moon's edge acts as a diffracting screen, an estimate of the size and structure of a source can be obtained from the diffraction pattern observed during its occultation. Hazard and his colleagues used this method to show that 3C 273 is a double radio source, one component of which coincides with a thirteenth-magnitude blue stellar object.

This coincidence prompted Maarten Schmidt to obtain an optical spectrum for the object. Schmidt used the 200-in Hale telescope to confirm the optical identification; at the same time, he noticed a jetlike feature extending from the stellar image toward the other radio component. The most remarkable aspect of Schmidt's work was his interpretation of the optical spectrum of 3C 273. Four of the emission lines seem to coincide with the Balmer series of hydrogen red shifted by 16%. If this redshift is interpreted as a Doppler shift, the "radio star" has to be receding from our galaxy at 16% of the velocity of light; if the redshift obeys Hubble's law, then 3C 273 is at a distance of 470 Mpc (for $H_0 = 100$ km sec^{-1} Mpc^{-1}). In this case, the intrinsic optical luminosity of 3C 273 is over 100 times brighter than the nearby spiral galaxies, and its intrinsic radio luminosity is comparable to that of the brightest radio galaxies.

Perhaps as a result of Schmidt's discovery, Jesse Greenstein and Thomas Matthews reexamined the spectrum of the starlike object that coincided with the position of the radio source 3C 48. They found that the spectral lines could be interpreted as indicating an even larger redshift—37%.[2] Stimulated by the previously reported ultraviolet excess of 3C 48, Beverly Oke determined the absolute energy distribution of 3C 273 and found it to have a nearly flat, nonthermal optical spectrum.

The sequence of events leading to the discovery of these hitherto unknown objects of tremendous velocity, luminosity, and distance happened so quickly that the articles reporting the major discoveries appeared together in a six-page sequence in *Nature*, reprinted below. Within a month another extraordinary characteristic of 3C 273 was reported by Harlan Smith and Dorrit Hoffleit.[3] They examined Harvard's historical plate collection and found that the "star" coinciding with 3C 273 had fluctuated in brightness over the past 80 years. The observed variations were often as short as 1 year in duration. The corresponding size of the emitting region is a few parsecs if the disturbing effect is propagated at the speed of light and proportionately smaller if propagated more slowly. Therefore, the optical variable has an extraordinarily small linear extent and an extraordinarily high surface brightness. Nevertheless, the overall extent of 3C 273 is comparable to that of a galaxy (about 50 kpc) if the angular size of the jet is used with the distance estimate of 470 Mpc.

1. C. Hazard, *Nature 191*, 58 (1961).
2. J. L. Greenstein has called attention to the fact that he took the spectrum of a quasar (Tonantzintla 202) in 1960, listed it as a white dwarf in 1965 and with J. B. Oke in 1970 showed it is actually a quasar.
3. H. J. Smith and D. Hoffleit, *Nature 198*, 650 (1963).

Hazard, Mackey, and Shimmins, Investigation of the Radio Source 3C 273

THE OBSERVATION of lunar occultations provides the most accurate method of determining the positions of the localized radio sources, being capable of yielding a positional accuracy of the order of 1 sec of arc. It has been shown by Hazard[1] that the observations also provide diameter information down to a limit of the same order. For the sources of small angular size the diameter information is obtained from the observed diffraction effects at the Moon's limb which may be considered to act as a straight diffracting edge.

The method has so far been applied only to a study of the radio source *3C* 212 the position of which was determined to an accuracy of about 3 sec of arc.[1,2] However, *3C* 212 is a source of comparatively small flux density and although the diffraction effects at the Moon's limb were clearly visible the signal-to-noise ratio was inadequate to study the pattern in detail and hence to realize the full potentialities of the method. Here we describe the observation of a series of occultations of the intense radio source *3C* 273 in which detailed diffraction effects have been recorded for the first time permitting the position to be determined to an accuracy of better than 1″ and enabling a detailed examination to be made of the brightness distribution across the source.

The observations were carried out using the 210-ft. steerable telescope at Parkes, the method of observation being to direct the telescope to the position of the source and then to record the received power with the telescope in automatic motion following the source. Three occultations of the source have been observed, on April 15, at 410 Mc/s, on August 5 at 136 Mc/s and 410 Mc/s, and on October 26 at 410 Mc/s and 1,420 Mc/s, although in October and April only the immersion and emersion respectively were visible using the Parkes instrument. The 410 Mc/s receiver was a double-sided band receiver, the two channels, each of width 10 Mc/s, being centred on 400 Mc/s and 420 Mc/s, while the 136 Mc/s and 1,420 Mc/s receivers each had a single pass band 1.5 Mc/s and 10 Mc/s wide respectively.

The record of April 15, although of interest as it represents the first observation of detailed diffraction fringes during a lunar occultation, is disturbed by a gradient in the received power and is not suitable for accurate position and diameter measurements. Therefore, attention will be confined to the occultation curves recorded in August and October and which are reproduced in figure 120.1. It is immediately obvious from these records that *3C* 273 is a double source orientated in such a way that whereas the two components passed successively behind the Moon at both immersions, they reappeared almost simultaneously. The prominent diffraction fringes show that the angular sizes of these components must be considerably smaller than 10″, which is the order of size of a Fresnel zone at the Moon's limb.

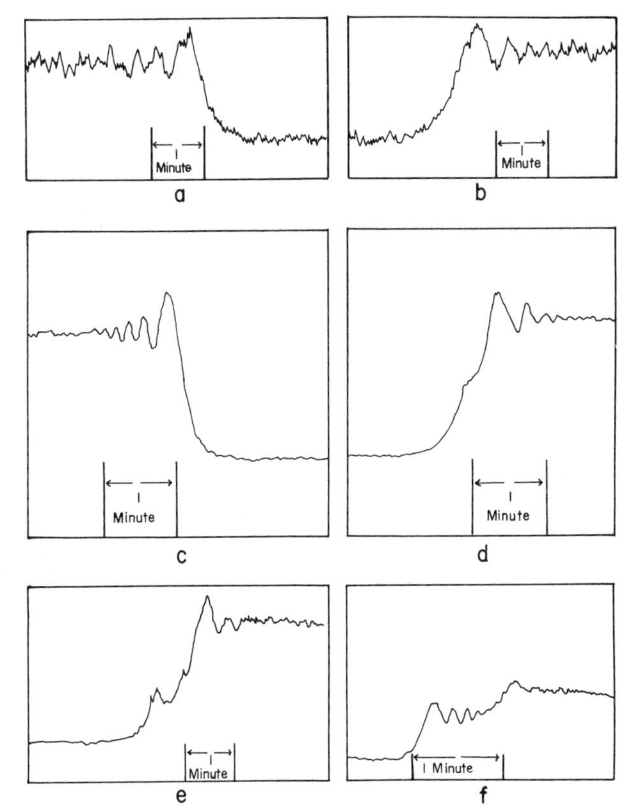

Fig. 120.1 Facsimiles of records showing occultations of 3C 273 on August 5 and October 26, 1962, at different frequencies: *a* emersion of August 5, 1962, at 136 Mc/s; *b* immersion of August 5, 1962, at 136 Mc/s; *c* emersion of August 5, 1962, at 410 Mc/s; *d* immersion of August 5, 1962, at 410 Mc/s; *e* immersion of October 26, 1962, at 410 Mc/s; *f* immersion of October 26, 1962, at 1,420 Mc/s. Abscissae, U.T.; ordinates, flux density.

The most interesting feature of figure 120.1 (e) and (f) is the change in the ratio of the flux densities of the two components with frequency. The ratio of the flux density of the south preceding source (component *A*) to that of the north following source (component *B*) is 1:0.45 at 410 Mc/s and 1:1.4 at 1,420 Mc/s, indicating a striking difference in the spectra of the two components. If it be assumed that the flux densities[3] of *3C* 273 at 410 Mc/s and 1,420 Mc/s are 60 and 35 Wm^{-2} $(c/s)^{-1}$ and that over this frequency-range the spectrum of each component may be represented by $S \propto f^n$, then the above ratios correspond to spectral indices for components *A* and *B* of -0.9 and 0.0 respectively. The spectral index of *A* is a representative value for a Class II radio source; but the flat spectrum of *B* is most unusual, no measurements of a comparable spectrum having yet been published. If the spectral indices were assumed constant down to 136 Mc/s then at this frequency component A must contribute almost 90 per cent of the total emission, a conclusion which is confirmed by a comparison of the times of immersion at 136 Mc/s and 410 Mc/s on August 5.

It has been shown by Scheuer[4] that it is possible to recover the true brightness distribution across the source from the observed diffraction pattern, the resolution being subject only to limitations imposed by the receiver bandwidth and the finite signal to noise ratio and being independent of the angular scale of the diffraction pattern. However, in this preliminary investigation we have not attempted such a detailed investigation but based the analysis on the calculated curves for uniform strip sources of different widths as published by Hazard.[1] As a first step in the investigation approximate diameters were estimated from the intensity of the first diffraction lobe and the results corresponding to the three position angles defined by the occultations and indicated in figure 120.2 are given in table 120.1.

As already indicated here, the 136-Mc/s measurements refer only to component A and hence no diameter measurements are available for B at this frequency. The 410-Mc/s observations of the August occultation are the most difficult to interpret owing to the components having both comparable flux density and small separation relative to the angular size of the first Fresnel zone. At immersion the widths were estimated by using a process of curve fitting to reproduce figure 120.1(d); at emersion (position angle 313°) the diameter of component B was assumed to be 3″ as indicated by the estimates at position angles 105° and 83°. The individual measurements at each frequency are reasonably consistent but there is a striking variation of the angular size of component A with frequency and evidence of a similar variation for component B. As at the time of the August occultation the angular separation of the Sun and the source was about 50° and hence coronal scattering of the type observed by Slee[5] at 85 Mc/s is not likely to be significant, this variation in size suggests that the model of two uniform strip sources is inadequate.

Therefore, a more detailed analysis was made of the intensity distributions of the lobe patterns given in figure 120.1(c) and (f), and it was found that in neither case can the pattern be fitted to that for a uniform strip source or a source with a gaussian brightness distribution. The 1,420-Mc/s observations of component B can be explained, however, by assuming that this source consists of a central bright core about 0.5″ wide contributing about 80 per cent of the total flux embedded in a halo of equivalent width of about 7″. Figure 120.1(b) where component A predominates, suggests that this source has a similar structure but with a core of effective width about 2″ at 410 Mc/s and a halo of width 6″. It therefore seems that the overall extent of both components are comparable but that the emission is more highly concentrated to the nucleus in B than in A. The close agreement between the halo size of A and its effective diameter at 136 Mc/s suggests that the observed variation of effective size with frequency may be due to a difference in the spectra of the halo and central regions. This would imply that the spectrum becomes steeper in the outer regions of the sources, that is, in the regions of lower emissivity. It is of interest that the integrated spectral indices of the two components show an analogous effect. Thus the spectrum of B, where most of the emission arises in a source about 0.5″ wide, is markedly flatter than that of A, where it arises in a source about 2″ wide.

Table 120.1 Effective width of equivalent strip source (sec of arc)

Frequency (Mc/s)	Component A, Position angle 106°	Component A, 313°	Component A, 84°	Component B, Position angle 105°	Component B, 314°	Component B, 83°
136	6.4	6.4	—	—	—	—
410	3.1	4.2[a] 2(6)[b]	4.2	3.1	3.0[a]	2.7
1,420	—	—	2.9	—	—	2.1 0.5(7)[b]

a. Component B assumed to have width of 3″.
b. Estimated from an analysis of the whole diffraction pattern.

The analysis is not sufficiently accurate to reach any reliable conclusions on the ellipticity of the individual components of 3C 273, but allowing for the uncertainty in the estimated widths and position angle 314°, the 410-Mc/s observations indicate that both components may be elliptical with A elongated approximately along the axis joining the two components and B elongated perpendicular to this axis.

Table 120.2 Observed occultation times of the two components of 3C 273

	Component A (U.T.)	Component B (U.T.)
Time of disappearance August 5, 1962	07h 46m 00s ± 1s	07h 46m 27.2s ± 0.5s
Time of reappearance August 5, 1962	09h 05m 45.5s ± 1s	09h 05m 45.7s ± 1.5s
Time of disappearance October 26, 1962	02h 55m 09.0s ± 1s	02h 56m 01.5s ± 0.4s

The position of each source was calculated from the observed times of disappearance and reappearance, which were estimated from the calculated flux density at the edge of the geometrical shadow and, where possible, from the positions of the diffraction lobes; these times are given in table 120.2. In estimating the values of $T_D{}^A$ and $T_R{}^A$ from the 136-Mc/s records a small correction was applied for the effects of component *B*, this correction being estimated by comparison with the 410-Mc/s records. The corresponding times for *B* were estimated from the 410-Mc/s observations using the estimated position of component *A* and the known flux density ratio of the two components. For each component the times and associated errors given in table 120.2 define three strips in each of which the source should lie; the centre lines of these strips represent the limb of the Moon at the time of observation and define in each case a triangular-shaped area. In principle, the position of the source lies in the area common to the three associated strips but it was found that for each component, and in particular for component *A*, that the size of the triangles defined by the Moon's limb was larger than would be expected from the estimated timing errors. This suggests that errors in the positions of the Moon's limb are more important than the estimated timing errors, and possibly that

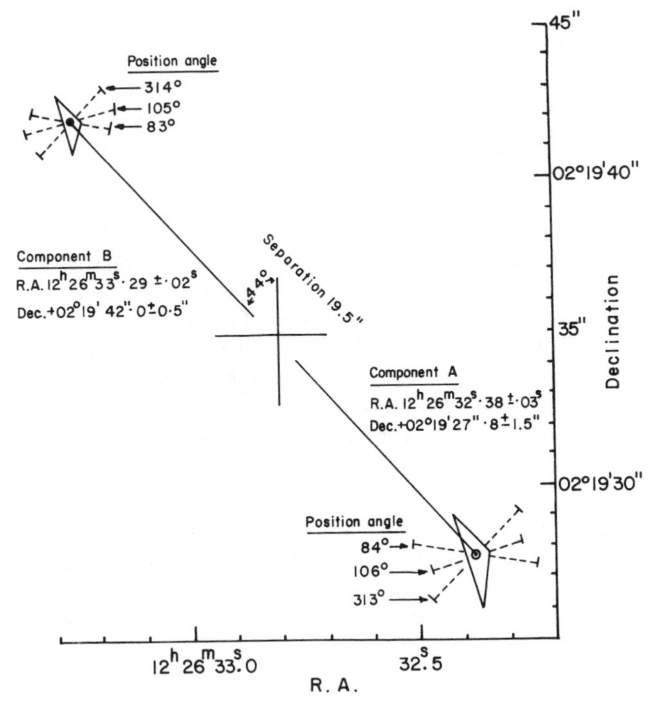

Fig. 120.2 Diagram of the radio source 3C 273. The sides of the solid line triangles represent the positions of the limb of the moon at the times of occultation. The dashed lines represent the widths of the equivalent strip source as measured at 410 Mc/s for each of three position angles indicated.

the effective position of the source varies slightly with frequency. The position of each source was therefore assumed to be given by the centre of the circle inscribed in the triangle defined by the Moon's limb at the relevant times. Dr. W. Nicholson of H.M. Nautical Almanac Office has kindly carried out these calculations and the estimated positions are as follows:

Component *A*	R.A.	12h 26m 32.38s ± 0.03s
(Epoch 1950)	Decl.	02° 19′ 27.8″ ± 1.5″
Component *B*	R.A.	12h 26m 33.29s ± 0.02s
(Epoch 1950)	Decl.	02° 19′ 42.0″ ± 0.5″

The average positions of the two sources given here represent the most accurate determination yet made of the position of a radio source. The quoted errors were estimated from the size of the triangles defined by the Moon's limb at the times of disappearance and reappearance, for the method is not subject to uncertainties introduced by refraction in the Earth's ionosphere or troposphere and is also free from the effects of confusion. A comparison of the times of disappearance at different frequencies indicates that there is also no significant source of error due to refraction in either the solar corona or a possible lunar ionosphere; any refraction appears to be less than 0.3″ even at 136 Mc/s. This may be compared with the upper limit of 2″ at 237 Mc/s and 13″ at 81 Mc/s as estimated by Hazard[1] and Elsmore[6] respectively, and allows a new limit to be set to the density of the lunar ionosphere. Thus, from his observations at 81.5 Mc/s, Elsmore has set an upper limit to the electron density of 10^3 cm^{-3}; and it follows that the present measurements set a limit of about 10^2 cm^{-3}. Similarly, Buckingham[7] has estimated that at 50 Mc/s a ray passing at 50° to the Sun would be deviated by 1″ if the electron density in the solar corona at the Earth's distance from the Sun is 100 cm^{-3}. The present observations at 136 Mc/s and 410 Mc/s on August 5 indicate that at 50 Mc/s the deviation is less than 2″ at this angle, setting an upper limit to the electron density of about 200 cm^{-3}, which may be compared with an upper limit of 120 cm^{-3}, set by Blackwell and Ingham[8] from observations of the zodiacal light.

In a preliminary examination of a print from a 200-in plate it was noted that the position of component *B* agreed closely with that of a thirteenth magnitude star. We understand that the investigations by Drs. A. Sandage and M. Schmidt of the Mount Wilson and Palomar Observatories have revealed that this star and an associated nebulosity is very probably the source of the radio emission.

1. C. Hazard, *Mon. Not. Roy. Astro. Soc. 134*, 27 (1962).
2. C. Hazard, *Nature 191*, 58 (1961).
3. J. G. Bolton, F. F. Gardner, and M. B. Mackey, *Austral. J. Phys. 17*, 340 (1964).
4. P. A. G. Scheuer, *Austral. J. Phys. 15*, 333 (1962).

5. O. B. Slee, *Mon. Not. Roy. Astro. Soc. 123*, 223 (1961).
6. B. Elsmore, *Phil. Mag. 2*, 1040 (1957).
7. M. J. Buckingham, *Nature 193*, 538 (1962).
8. D. E. Blackwell and M. F. Ingham, *Mon. Not. Roy. Astro. Soc. 122*, 129 (1961).

Schmidt, 3C 273: A Star-like Object with Large Red-Shift

THE ONLY OBJECTS seen on a 200-in. plate near the positions of the components of the radio source 3*C* 273 reported by Hazard, Mackey and Shimmins in the preceding article are a star of about thirteenth magnitude and a faint wisp or jet. The jet has a width of 1″ – 2″ and extends away from the star in position angle 43°. It is not visible within 11″ from the star and ends abruptly at 20″ from the star. The position of the star, kindly furnished by Dr. T. A. Matthews, is R.A. 12h 26m 33.35s ± 0.04s, Decl. +2° 19′ 42.0″ ± 0.5″ (1950), or 1″ east of component B of the radio source. The end of the jet is 1″ east of component *A*. The close correlation between the radio structure and the star with the jet is suggestive and intriguing.

Spectra of the star were taken with the prime-focus spectrograph at the 200-in. telescope with dispersions of 400 and 190 Å per mm. They show a number of broad emission features on a rather blue continuum. The most prominent features, which have widths around 50 Å, are, in order of strength, at 5,632, 3,239, 5,792, 5,032 Å. These and other weaker emission bands are listed in the first column of table 120.3. For three faint bands with widths of 100–200 Å the total range of wave-length is indicated.

The only explanation found for the spectrum involves a considerable red-shift. A red-shift $\Delta\lambda/\lambda_0$ of 0.158 allows identification of four emission bands as Balmer lines, as indicated in table 120.3. Their relative strengths are in agreement with this explanation. Other identifications based on the above red-shift involve the Mg II lines around 2,798 Å, thus far only found in emission in the solar chromosphere, and a forbidden line of [O III] at 5,007 Å. On this basis another [O III] line is expected at 4,959 Å with a strength one-third of that of the line 5,007 Å. Its detectability in the spectrum would be marginal. A weak emission band suspected at 5,705 Å, or 4,927 Å reduced for red-shift, does not fit the wave-length. No explanation is offered for the three very wide emission bands.

It thus appears that six emission bands with widths around 50 Å can be explained with a red-shift of 0.158. The differences between the observed and the expected wave-lengths amount to 6 Å at the most and can be entirely understood in terms of the uncertainty of the measured wave-lengths. The present explanation is supported by observations of the infra-red spectrum communicated by Oke in a following article, and by the spectrum of another star-like object associated with the radio source 3*C* 48 discussed by Greenstein and Matthews in another communication.

Table 120.3 Wavelengths and identifications

λ	$\lambda/1.158$	λ_0	
3,239	2,797	2,798	Mg II
4,595	3,968	3,970	Hε
4,753	4,104	4,102	Hδ
5,032	4,345	4,340	Hγ
5,200–5,415	4,490–4,675	—	—
5,632	4,864	4,861	Hβ
5,792	5,002	5,007	[O III]
6,005–6,190	5,186–5,345	—	—
6,400–6,510	5,527–5,622	—	—

The unprecedented identification of the spectrum of an apparently stellar object in terms of a large red-shift suggests either of the two following explanations.

(1) The stellar object is a star with a large gravitational red-shift. Its radius would then be of the order of 10 km. Preliminary considerations show that it would be extremely difficult, if not impossible, to account for the occurrence of permitted lines and a forbidden line with the same red-shift, and with widths of only 1 or 2 per cent of the wavelength.

(2) The stellar object is the nuclear region of a galaxy with a cosmological red-shift of 0.158, corresponding to an apparent velocity of 47,400 km/sec. The distance would be around 500 megaparsecs, and the diameter of the nuclear region would have to be less than 1 kiloparsec. This nuclear region would be about 100 times brighter optically than the luminous galaxies which have been identified with radio sources thus far. If the optical jet and component *A* of the radio source are associated with the galaxy, they would be at a distance of 50 kiloparsecs, implying a time-scale in excess of 10^5 years. The total energy radiated in the optical range at constant luminosity would be of the order of 10^{59} ergs.

Only the detection of an irrefutable proper motion or parallax would definitely establish 3*C* 273 as an object within our Galaxy. At the present time, however, the explanation in terms of an extragalactic origin seems most direct and least objectionable.

Greenstein and Matthews, Red-Shift of the Unusual Radio Source: 3C 48

THE RADIO SOURCE 3*C* 48 was announced to be a star[1] in our Galaxy on the basis of its extremely small radio diameter,[2] stellar appearance on direct photographs and unusual spectrum. Detailed spectroscopic study at Palomar by Greenstein during the past year gave only partially successful identifications of its weak, broad emission lines; the possibility that they might be permitted transitions in high stages of ionization could not be proved or disproved. Hydrogen was absent but several approximate coincidences with He II and O VI were suggested.

The discovery by Schmidt (a preceding article) of much broader emission lines in the apparently stellar radio source, *3C* 273, suggested a red-shift of 0.16 for *3C* 273 if the lines were interpreted as the Balmer series. In *3C* 48 no such series was apparent; measurable lines still do not coincide with the hydrogen series. However, the possibility of a very large red-shift, which had been considered many times, was re-explored successfully. *3C* 48 has a spectrum containing one very strong emission feature near $\lambda 3{,}832$ which is 35 Å wide and about 10 other weaker features near 23 Å in width. The sharper lines are listed in table 120.4 in order of decreasing intensity. Some broad lines or groups of lines between 50 and 100 Å width may be red-shifted hydrogen lines.

Table 120.4 Identifications and observed red-shifts

Wave-length			
λ^*	λ lab.	Source	λ^*/λ lab.
3,832.3	2,796, 2,803 } 2,798	Mg II	1.3697:
4,685.0	3,426	[Ne V]	1.3676
5,098	3,729, 3,726 } 3,727	[O II]	1.3679
4,575	3,346	[Ne V]	1.3673
5,289	3,868	[Ne III]	1.3671
4,065.7	2,975:	[Ne V]	1.3667:

The weighted mean red-shift $d\lambda/\lambda_0$ is 0.3675 ± 0.0003, an apparent velocity of $+110{,}200$ km/sec. The slightly discrepant value for $\lambda 2{,}975$ of [Ne V] is compatible with the uncertainty of ± 3 Å in the wave-length predicted by Bowen.[3] The Mg II permitted resonance doublet has a small additional displacement to longer wave-lengths, possibly caused by self-absorption in an expanding shell; it is the strongest emission line in the rocket-ultra-violet spectrum of the Sun. The forbidden lines are similar to those in other intense extragalactic radio sources.

So large a red-shift, second only to that of the intense radio source *3C* 295, will have important implications in cosmological speculation. A very interesting alternative, that the source is a nearby ultra-dense star of radius near 10 km containing neutrons, hyperons, etc., has been explored and seems to meet insuperable objections from the spectroscopic point of view. The small volume for the shell required by the observed small gradient of the gravitational potential is incompatible with the strength of the forbidden lines.

The distance of *3C* 48, interpreted as the central core of an explosion in a very abnormal galaxy, may be estimated as 1.10×10^9 parsecs; the visual absolute magnitude is then -24.0, or -24.5 corrected for interstellar absorption. The minimum correction for the effect of red-shift is of the order of $2\, v/c$ and a value between 4 and 5 times v/c is probable for a normal galaxy. The absolute visual magnitude of *3C* 48 is then brighter than -25.2 and possibly as bright as -26.3, 10–30 times greater than that of the brightest giant ellipticals[4] hitherto recognized, which are near -22.7 and another factor of five brighter than our own Galaxy, near -21.0.

As a radio source at a distance of 1.1×10^9 parsecs *3C* 48 is not markedly different from other known strong radio sources like *3C* 295 or Cygnus *A*. The one feature in which it does differ from most sources is in its high surface brightness. This is partially due to its extremely small radio size of $\leqslant 1$ sec of arc.[2] The optical size is comparable, being also $\leqslant 1$ sec of arc.[5] At the assumed distance such angular sizes indicate that both the optical and radio emission arise within a diameter of $\leqslant$ 5,500 parsecs. The radio diameter might even be comparable with or less than that of *3C* 71 (*NGC* 1068) the diameter of which is about 700 parsecs. However, *3C* 71 has 5 orders of magnitude less radio emission.

If we determine the integrated radio emission of *3C* 48 from the observed spectral index of the radio spectrum, and correct for the red-shift, we find that *3C* 48 is comparable with *3C* 295, emitting 4×10^{44} erg/sec of radio-frequency power. The cut-off frequencies were 7×10^7 c/s and 10^{11} c/s. The lower limit is indicated by the observed radio spectrum and the upper limit is an assumed one.

The absolute magnitudes of the galaxies connected with *3C* 295 and Cygnus *A*, corrected for interstellar absorption, are $M_v = -21.0$ and -21.6 (using a red-shift correction of $2\, v/c$) or $M_v = -22.4$ and -21.8 (using a correction of $5\, v/c$) respectively. Thus *3C* 48 radiates about 50 times more powerfully in the optical region than other more normal but intense radio galaxies. In contrast, the absolute radio luminosity of *3C* 48 is the same as that of Cygnus *A* and *3C* 295. The unusually strong optical radiation may be synchrotron radiation as suggested (for other reasons) by Matthews and Sandage.[5]

1. T. A. Matthews, J. G. Bolton, J. L. Greenstein, G. Münch, and A. R. Sandage, Amer. Astro. Soc. meeting, New York, 1960; *Sky and Telescope 21*, 148 (1961). J. L. Greenstein and G. Münch, *Ann. Rep. Dir. Mt. Wilson and Palomar Obs. 80*, (1961).

2. L. R. Allen et al., *Mon. Not. Roy. Astro. Soc. 124*, 447 (1962); B. Rowson, *Mon. Not. Roy. Astro. Soc. 125*, 177 (1962).

3. I. S. Bowen, *Astrophys. J. 132*, 1 (1960).

4. G. Abell, *Problems of Extragalactic Research*, I.A.U.

Symp. No. 15, ed. G. C. McVittie (New York: Macmillan, 1962), p. 213.
5. T. A. Matthews and A. R. Sandage, *Astrophys. J. 138*, 30 (1963).

Oke, Absolute Energy Distribution in the Optical Spectrum of 3C 273

THE RADIO SOURCE 3*C* 273 has recently been identified with a thirteenth magnitude star-like object. The details are given by M. Schmidt in the preceding communication. Since 3*C* 273 is relatively bright, photoelectric spectrophotometric observations were made with the 100-in. telescope at Mount Wilson to determine the absolute distribution of energy in the optical region of the spectrum; such observations are useful for determining if synchrotron radiation is present. In the wavelength region between 3,300 Å and 6,000 Å measurements were made in 16 selected 50-Å bands. Continuous spectral scans with a resolution of 50 Å were also made. The measurements were placed on an absolute-energy system by also observing standard stars whose absolute energy distributions were known.[1] The accuracy of the 16 selected points is approximately 2 per cent. The strong emission features found by Schmidt were readily detected; other very faint features not apparent on Schmidt's spectra may be present.

The source 3*C* 273 is considerably bluer than the other known star-like objects 3*C* 48, 3*C* 196, and 3*C* 286 which have been studied in detail.[2] The absolute energy distribution of the apparent continuum can be accurately represented by the equation

$$F_\nu \propto \nu^{+0.28}, \quad (1)$$

where F_ν is the flux per unit frequency interval and ν is the frequency. The apparent visual magnitude of 3*C* 273 is +12.6, which corresponds to an absolute flux at the Earth of 3.5×10^{-28} W m^{-2} (c/s)$^{-1}$ at 5,600 Å. At radio frequencies[3] the spectral index is −0.25 and the flux at 960 Mc/s is 5.0×10^{-25} W m^{-2} (c/s)$^{-1}$.

Between 6,000 Å and 10,250 Å, eleven 120-Å bands were measured with an accuracy of 10 per cent. These measures indicate that the relatively flat energy distribution given in the equation here applies as far as 8,400 Å. Beyond 8,400 Å the flux may increase significantly. Between 3,300 Å and 8,400 Å the energy distribution cannot be represented, even approximately, by the flux from a black-body or a normal star. At least part of the optical continuum radiation must be synchrotron radiation.

During the course of the infra-red observations a strong emission feature was found near 7,600 Å. Further observations with a 50-Å band-width placed the emission line at 7,590 Å with a possible error of about 10 Å. The emission profile was found to be similar to that of the emission line at 5,632 Å. Using this line and others in the visual spectrum Schmidt has shown that the most prominent emission features are Balmer lines and that the line at 7,590 Å is $H\alpha$. Using Schmidt's red-shift $\Delta\lambda/\lambda_0$ of 0.158, $H\alpha$ should appear at 7,599 Å; this is in satisfactory agreement with observation, when it is recalled that the atmospheric *A* band absorbs strongly beyond 7,594 Å. It is possible that the [*N* II] lines which have unshifted wavelengths of 6,548 Å and 6,583 Å can contribute to the emission feature identified as $H\alpha$. A large contribution, however, would shift the line significantly towards the red. The relative positions of $H\alpha$, $H\beta$, $H\gamma$, and $H\delta$ cannot be produced by applying a red-shift to any other hydrogen-like ion spectrum.

1. J. B. Oke, *Astrophys. J. 131*, 358 (1960).
2. T. A. Matthews and Allan Sandage, *Astrophys. J. 138*, 30 (1963).
3. D. E. Harris and J. A. Roberts, *Pub. Astro. Soc. Pacific 72*, 237 (1960).

121. The Quasi-Stellar Radio Sources 3C 48 and 3C 273

Jesse L. Greenstein and Maarten Schmidt

(*Astrophysical Journal 140*, 1–34 [1964])

Soon after the discovery of the large redshifts of the quasars, a controversy arose that lasted for nearly a decade.[1] If the redshifts are interpreted as Doppler shifts, the quasars must be receding from our galaxy at velocities up to 90% of the velocity of light. Furthermore, if the redshifts obey Hubble's law, the quasars must be extremely distant and 100 times brighter than the nearby spiral galaxies. These unusually high velocities and luminosities have led several astrophysicists to search for alternative explanations for the observed redshifts.[2] One possibility raised is that the redshifts are gravitational, occurring when a light photon escapes through the gravitational field of a massive body.[3] If the quasars are sufficiently dense, then a substantial fraction of the observed redshifts may be gravitational in origin. If this is the case, the quasars may not be moving at high velocities, and they may be relatively near our own galaxy.

In the paper given here, Jesse Greenstein and Maarten Schmidt provide two compelling structural arguments against the gravitational redshift hypothesis. They first notice that the densest known visible objects, the neutron stars, can only provide gravitational redshifts less than 10 percent. They then estimate the densities of 3C 48 and 3C 273 from their observed forbidden emission lines, and conclude that these lines cannot arise in a region of substantial gravitational potential gradient. Thus, Greenstein and Schmidt interpret the quasars as distant, superluminous objects. This cosmological interpretation of quasars as remote objects whose large redshifts and small angular sizes are a result of their large distances is now favored by the large majority of astronomers.

1. See, for example, *The Redshift Controversy*, ed. G. B. Field, H. Arp, and J. N. Bahcall (Reading, Mass: W. A. Benjamin, 1973).

2. One of the most vociferous proponents of the "noncosmological" nature of the quasar redshifts has been Geoffrey Burbidge (*Nature-Physical Science 246*, 17 [1973]).

3. The formula for the gravitational redshift is the same for both the Newtonian and the General Theory of gravitation. It was first given by A. Einstein in *Annalen der Physik 35*, 898 (1911), and an English translation may be found in *The Principle of Relativity* (New York: Dodd, Mead and Co., 1923; reproduced New York: Dover Pub., 1952).

Abstract The spectra of two quasi-stellar radio sources, 3C 48 and 3C 273, have been studied in detail. We present as full conclusions as we can derive from the redshift, luminosity, emission-line, and continuous spectra. Together with the radio-frequency data and the light variability, these indicate the presence of very large total energies in a relatively small volume of space. We deliberately have not attempted to discuss the origin of these large energies, nor do we discuss the numerous other physical problems concerned with suggested mechanisms in the quasi-stellar objects.

We first consider other explanations for the large redshifts, in particular the possibility of gravitational redshift. The presence of relatively narrow emission lines excludes objects near 1 $M_\odot$ which are stable because of the small emitting volume. The presence of forbidden lines sets an upper limit to the gas density. Together with a limit to gravitational perturbations on our Galaxy, this leads to a lower limit of 10^{11} $M_\odot$, condensed to a 10^{17}-cm radius. Whether such large masses can be even quasi-stable has not yet been demonstrated.

We then adopt the interpretation that the redshifts are cosmological in origin. The absolute visual magnitudes are about -26 for 3C 273 and -25 for 3C 48. The forbidden lines of high ionization potential are quite strong in 3C 48 relative to hydrogen. By analogy with planetary nebulae and assuming normal abundances, with astrophysical details given in the appendices, we derive the electron density, N_e, probably near to or less than 3×10^4 cm^{-3}; the electron temperature is not very high, and the mass is about 5×10^6 $M_\odot$ within a radius of 10 pc or more. The emitting volumes are obtained from N_e and the observed luminosity in Hβ and Mg II. The forbidden lines and the Balmer lines are optically thin, but Mg II is optically thick, leading to discussions in the appendices. For 3C 273, in which the forbidden lines are weaker, the surprising weakness of [O II] permits a closer estimate of N_e near 3×10^6 cm^{-3} and a mass of 6×10^5 $M_\odot$ within a radius of about 1 pc.

The light variations observed in both, with cycles of 10 years or less, suggest the presence of a source of optical continuum with a diameter of 1 pc, possibly much less. We urgently need continued observations of the absolute intensities of lines and continuum and their variations. The thermal energy supply in the H II region is small. The ionized gas must be of low density in the region in which the radio frequencies are generated, because of free-free absorption and Faraday rotation, i.e., $N_e < 10$ cm^{-3} if $R = 500$ pc for the radio source.

We explore models for synchrotron generation of radio and optical frequencies. If $R = 500$ pc, total energies required for radio emission are relatively low, about 10^{57} erg at equipartition. The lifetimes for exhausting the total energy supply are about 10^6 years. If we wish to obtain optical synchrotron from the same volume as produces radio frequencies, the equipartition energy reaches 10^{58} erg. If optical synchrotron radiation is to arise within a volume 1 pc^3, however, the total energy is small, 10^{54} erg, the life about a year, and serious problems arise, such as cosmic-ray proton collisional loss, and inverse-Compton effect electron loss. Models for the jet radio source 3C 273A offer no particular difficulty.

We review conditions under two possible age estimates of the inner components of the quasi-stellar objects. At 10^3 years, the object can be in expansion, with a velocity compatible with the emission-line width, about 1,000 km/sec. The energy supply is sufficient for the radio spectrum, and the kinetic energy of the H II region is nearly enough to maintain the optical emission. On this hypothesis, the jet in 3C 273 and the nebular wisps in 3C 48, which are 150,000 light-years in size, must have originated in a separate event.

If the age is 10^6 years, the H II region energy is much too small; in addition, its small radius and large internal motion would need to be stabilized by a gravitational mass near 10^9 $M_\odot$, inside the H II region. The radiated energy in 10^6 years is near 10^{60} erg, so that nuclear-energy sources of 10^9 $M_\odot$ are required. The simplest model of the quasi-stellar sources is one in which a small mass of 10^9 $M_\odot$ is surrounded by shells of increasing radius in which the optical continuum, the emission lines, and the radio continuum, respectively, originate. The relation of these objects to the most intense radio galaxies is unclear. The quasi-stellar sources have small optical size, a high ratio of optical to radio emission, and an optical luminosity so high that, if their age exceeds 1,000 years, continued input of energy is required from some not directly observable source. Table 121.2 gives a brief résumé of numerical results.

I. Introduction

The present paper deals with optical objects of *stellar* appearance that have been associated with radio sources. The first radio source so identified was 3C 48, for which Matthews, Bolton, Greenstein, Münch, and Sandage (1960) announced the stellar appearance of the associated optical object. Subsequently, the radio sources 3C 196 and 3C 286 were identified with similar optical objects (Matthews and Sandage 1963), as was the source 3C 147 (Schmidt and Matthews 1964). The optical spectra of these four quasi-stellar objects appeared quite dissimilar, and no satisfactory identifications of the emission features could be obtained. The identification of the radio source 3C 273 with a bright object of stellar appearance provided a clue when it was found that its spectrum could be understood on the basis of an unexpectedly large redshift (Schmidt 1963). The spectrum of 3C 48, although of a rather different nature, could be explained by an even larger redshift (Greenstein and Matthews 1963).

The present discussion is limited to the *quasi-stellar radio sources* 3C 48 and 3C 273; the observational data are given in Sections II and III. The possibility of interpreting the redshift as the gravitational effect of either very dense or very massive objects is discussed in Section IV. The finally adopted interpretation of these quasi-stellar radio sources as distant, superluminous objects in galaxies, or intergalactic objects, is discussed in the remaining sections.

II. The Source 3C 273

A remarkably detailed study of the radio source 3C 273 has been made by Hazard, Mackey, and Shimmins (1963) using lunar occultations. They found that the source is double, with a separation of 19.5″ between the components. Component A has a diameter of 4″ at 400 Mc/s, and a spectral index of 0.9. Component B has a diameter of 3″ at 400 Mc/s, with a spectral index near zero. This component has a core with a diameter of about 0.5″ which contributes about half the flux at higher frequencies (Hazard, private communication). The position of each of the components was determined with an accuracy of around 1″.

The two components coincide almost precisely with a thirteenth-magnitude star and the end of a faint jet. Figure 121.1 shows an enlarged portion of a 200-inch photograph taken by Sandage. The star is 1″ east of Component B and the end of the jet 1″ east of Component A. The jet is between 1″ and 2″ in width, begins 11″ and ends 20″ from the star. This is one of the few radio sources where one of the two components coincides with the optical object. Another case may be M87 where an asymmetric jet is also present.

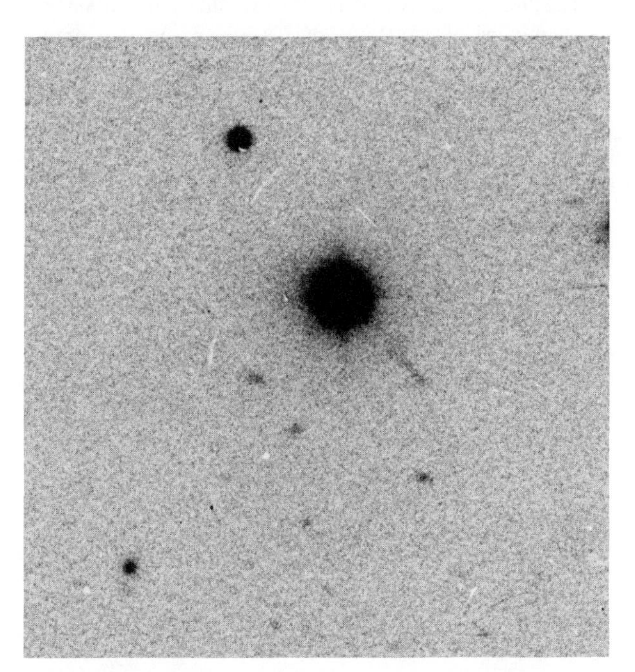

Fig. 121.1 Enlarged portion of 200-in photograph of 3C 273 (from a 103*a-D* plate taken by A. R. Sandage). The weak narrow jet visible in position angle 223° reaches to about 20 sec of arc from the quasi-stellar object. The optical object coincides with component B of the radio object, and the jet extends toward the other radio component, A. The other dark round spots in the figure are foreground stars in our galaxy. (Courtesy Hale Observatories.)

[See W. Baade, *Astrophysical Journal 123*, 550 (1956) for a description of the polarized jet in Virgo A (M 87).]

Here Greenstein and Schmidt list the emission lines in 3C 273, which are given in selection 120.

III. The Source 3C 48

Here Greenstein and Schmidt notice that the photographic image of 3C 48 is stellar, but wisps appear at an approximate distance of 50 kpc. They list the emission lines, which are given in selection 120.

IV. The Possibility of Gravitational Redshifts

Redshifts as large as those found for 3C 48 and 3C 273 have thus far only been encountered in distant galaxies. Both sources have a dominant optical component with an angular diameter of less than 1″, simulating the appearance of a star. The observed redshifts cannot be explained as velocity shifts of ordinary stars; the low upper limit to the proper motion of 3C 273 in conjunction with a transverse velocity of the same order as the observed redshift would lead to a minimum distance of 10 Mpc, and an absolute magnitude higher than −16. We shall discuss in the present section a possible interpretation in terms of gravitational redshifts. We divide the considerations into those relevant to collapsed "neutron" stars and to very massive objects.

Let us investigate the consequences of assuming that the redshift is of gravitational origin. The gravitational redshift is

$$\frac{\Delta\lambda}{\lambda_0} = \frac{GM}{Rc^2} = 1.47 \times 10^5 \frac{M/M_\odot}{R}, \quad (1)$$

where R is the mean radius in centimeters. The width w of the emission lines, though considerable, is but a small part of the redshift $\Delta\lambda$. If the line width is due solely to the variation of gravitational potential over the region containing ionized gases, they must be confined within a shell with thickness ΔR, such that

$$\Delta R/R = w/\Delta\lambda. \quad (2)$$

Table 121.1 shows some relevant data for the two sources.

Table 121.1 Interpretation involving gravitational redshift

	3C 48	3C 273
$\Delta\lambda/\lambda_0$	0.367	0.158
w (Ångstroms)	~ 30	~ 50
R (cm)	$4.0 \times 10^5\, M/M_\odot$	$9.3 \times 10^5\, M/M_\odot$
$\Delta R/R$	0.016	0.07

Greenstein and Schmidt then discuss various theoretical considerations of collapsed stellar configurations of very high density, concluding that low values of gravitational redshift ($z \leqslant 0.6$) are obtained for neutron stars.

A much more positive argument arises from evaluation of the emission from ionized hydrogen. At an electron temperature T_e and electron density N_e the emissivity in Hβ is (Aller 1956)

$$E(\mathrm{H}\beta) = 2.28 \times 10^{-19} N_e{}^2 T_e{}^{-3/2} b_4 \exp(0.98 \times 10^4/T_e), \quad (3a)$$

where b_4 measures the deviation from equilibrium population of the upper level of Hβ. We assume $T_e = 10^{4\circ}$ K, and a typical average nebular value of $b_4 = 0.16$. Then we compute

$$E(\mathrm{H}\beta) = 1.0 \times 10^{-25} N_e{}^2 \text{ erg sec}^{-1} \text{ cm}^{-3}. \quad (3b)$$

The observed brightness of Hβ in 3C 273 is 3.4×10^{-12} erg cm^{-2} sec^{-1}. Equating this to the emission from a volume, V, of ionized hydrogen at a distance, r, yields the relation

$$10^{-25}\, V N_e{}^2 = 3.4 \times 10^{-12}\, 4\pi r^2. \quad (4a)$$

Here $V = 4\pi R^2 \Delta R$, and $\Delta R = 0.07\, R$ is obtained in table 121.1 from the width of the emission lines. Finally we have

$$N_e{}^2 R^3 r^{-2} = 4 \times 10^{14} \text{ cm}^{-5}. \quad (4b)$$

Let us first consider the possibility of a collapsed body of nearly normal stellar mass. The absence of observable proper motion makes a distance less than 100 pc very unlikely. Introducing this minimum distance into equation (4b) leads to the inequality

$$N_e{}^2 R^3 \geqslant 4 \times 10^{55} \text{ cm}^{-3}. \quad (4c)$$

If the mass of the object is 1 $M_\odot$, then $R \approx 10^6$ cm, from table 121.1 and we need $N_e \geqslant 6 \times 10^{18}$ cm^{-3}. This density is very much larger than the maximum electron density of about 10^8 cm^{-3} which is imposed by the appearance of a forbidden line in the spectrum. This enormous discrepancy cannot be removed by any other choice of T_e, or of b_4, in equation (3a), and we are forced to the conclusion that the spectrum of 3C 273 cannot be explained by gravitational redshift near a star of about 1 $M_\odot$. Similar conclusions can be reached with regard to 3C 48, in which the forbidden lines are very strong, making the upper limit to N_e even lower.

The total luminosity in Hβ at $N_e = 6 \times 10^8$ cm^{-3} exceeds 10^{30} erg sec^{-1}. A black body of $R = 10^6$ cm and $T = 10^{4\circ}$ K emits 7×10^{24} erg sec^{-1}; therefore the emission-line flux exceeds that from a black body, which is impossible. In fact, so high an electron density would result in a largely neutral gas opaque to its own radiation. The monochromatic flux at Earth from an object of radius R cm at a distance of r_{pc}, using the infra-red approximation to the black body, is

$$F_\nu\, d\nu = 10^{-73} \left(\frac{R}{r_{\mathrm{pc}}}\right)^2 T \nu^2\, d\nu \text{ erg sec}^{-1} \text{ cm}^{-2} (\text{c/s})^{-1}. \quad (5)$$

With $R = 10^6$ cm, $r_{\mathrm{pc}} = 10^2$, $\nu = 6 \times 10^{14}$, we have $F_\nu \approx 4 \times 10^{-36}\, T$ erg sec^{-1} cm^{-2} (c/s)$^{-1}$ at 5,000 Å. Since the observed $F_\nu \approx 4 \times 10^{-25}$ for 3C 273, we require that $T = 10^{11\circ}$ K if the black-body radiation of a 10-km star is to produce the continuum.

The discovery of hyperdense stars would be of great interest in theoretical physics; our conclusions suggest that optical detection is hopelessly difficult. From the theoretical physics side, in addition, if only neutron degeneracy determines the pressure, the redshifts already observed are too large to be explained in this manner.

Another possible interpretation of a gravitational redshift, from Equation (1), is that it arises in an object of very great mass and moderate radius, e.g., $10^8\, M_\odot$ and $R = 10^{14}$ cm. We disregard for the moment questions regarding the stability of such a configuration. If such an object is located in or near our Galaxy, an additional limiting condition on its distance can be derived from its gravitational perturbations on the local dynamics of the Galaxy. The acceleration for stars near the Sun due to a single radio source certainly should be less than 10 per cent of that of the whole Galaxy, which gives

$$M/M_\odot \leqslant 10^{-35} r^2. \quad (6)$$

The emission in Hβ, from equation (3b), yields a flux at the Earth of

$$F(\mathrm{H}\beta) = 10^{-25} N_e{}^2 R^2\, \Delta R\, r^{-2}. \quad (7)$$

Combining equations (1), (6), and (7),

$$r^4 \geqslant N_e{}^{-2} (\Delta R/R)^{-1} (\Delta\lambda/\lambda_0)^3\, 10^{114.5} F(\mathrm{H}\beta), \quad (8a)$$

$$M/M_\odot \geqslant N_e{}^{-1} (\Delta R/R)^{-1/2} (\Delta\lambda/\lambda_0)^{3/2}\, 10^{22.2} [F(\mathrm{H}\beta)]^{1/2}. \quad (8b)$$

The Hβ fluxes are 3.4×10^{-12} and 4.4×10^{-14} erg cm^{-2} sec^{-1} for 3C 273 and 3C 48, respectively; the redshifts and $\Delta R/R$ from table 121.1 give approximately the same results for the two sources:

$$r_{pc} > 8 \times 10^6 N_e^{-1/2},$$
$$M/M_\odot > 7 \times 10^{15} N_e^{-1}. \quad (9)$$

We will see that the spectra, containing forbidden lines, are not consistent with an N_e very different from 10^7 cm^{-3} for 3C 273, and a somewhat lower value, 10^5 cm^{-3} for 3C 48. The resulting values are for 3C 273, $r_{pc} \geqslant 2{,}500$, $M/M_\odot \geqslant 7 \times 10^8$ and for 3C 48, $r_{pc} \geqslant 25{,}000$ and $M/M_\odot \geqslant 7 \times 10^{10}$. These minimum values are derived for the unlikely case of a gravitational perturbation 10 per cent that of the attraction of the Galaxy. From the structure of the equations, when we lower this upper limit to the perturbation, we increase the mass and distance in equation (9). The high latitude of 3C 273 ($b = 60°$) and the fact that the above mass and distance are minimum values make it quite safe to conclude that the interpretation of the redshift as a gravitational effect also requires an extragalactic nature for these quasi-stellar objects. We are reminded here of the gravitationally collapsing masses investigated by Hoyle and Fowler (1963; see also Fowler and Hoyle 1964) with a view to gaining some understanding of the origin of the very large energies required in the strong extragalactic radio sources. At present, there is no assurance from the theory, and it may even be viewed as unlikely, that 10^{11} $M_\odot$ within a radius of 0.01 pc are stable. If gravitational implosions should occur, it is difficult to see what train of theoretical arguments would lead to a gaseous shell of fixed radius and 10^{-4} pc thickness. Under large gravitational attraction, only a very narrow range of temperature (or explosion velocities) could produce such a scale height. It is difficult also to understand why the density distribution would be such as to make the emission lines quite symmetrical. On this hypothesis the red wing would originate in the inner part and the blue wing in the outer part of a thin shell.

We believe that the postulation of stable, or nearly stable, objects of size less than 1 pc, with masses near those of galaxies, in nearby intergalactic space is not really justified at present. Consequently, we reject gravitational redshift as a basis for explanation of the large redshifts observed in 3C 48 and 3C 273. If stable, massive configurations exist, we must re-examine this possibility. The mass-radius relation would have to be such as to give larger gravitational redshifts for fainter objects. Although the range of shifts is small (a factor of 3 with recent discoveries), if we interpret the shifts as cosmological we deduce consistent optical and radio absolute fluxes for sources at various redshifts. The unprecedented combination of redshift, apparent luminosity, and appearance of 3C 48 and 3C 273 will require an unorthodox explanation.

V. EMISSION LINES IN A SUPERLUMINOUS GALAXY

Here Greenstein and Schmidt compare the observed and predicted intensities for the forbidden lines observed in 3C 48 and 3C 273. They adopt electron densities of $N_e = 3 \times 10^4$ cm^{-3} and 3×10^6 cm^{-3}, respectively, for 3C 48 and 3C 273. Using these values of electron densities, they compute radii, R, and hydrogen masses, M, from the Hβ luminosities. They obtain $R = 11$ pc and $M = 5 \times 10^6$ $M_\odot$ for 3C 48 and $R = 1.2$ pc and $M = 6 \times 10^5$ $M_\odot$ for 3C 273.

VI. AGES, LIGHT VARIATIONS, AND ENERGY SUPPLY

The southern end of the optical wisps of 3C 48 is at a distance of about 50 kpc, in projection, from the quasi-stellar object. The end of the jet of 3C 273 is also at a projected distance of about 50 kpc from the quasi-stellar object. If these features have been ejected from the central object, this must have taken place at least 2×10^5 years ago.

The upper limits to the angular diameters of the optical, starlike components from visual inspection are 1″ for 3C 48 and 0.5″ for 3C 273, corresponding to diameters of 5 kpc and 1 kpc. The widths of the emission lines correspond to 2,000 and 3,000 km/sec, respectively. These speeds are much larger than the escape velocities of about 100 km/sec, determined from the mass of hydrogen and the radius of the emission-line region. If material is indeed escaping from a very small inner region with the observed velocities, and if we may assume that these velocities represent the mean expansion velocity in the past, we obtain from the upper limit of size a maximum age of 2×10^6 years for 3C 48 and of 3×10^5 years for 3C 273. It would be more realistic to use here the diameters of the gaseous emission regions as determined in the preceding section. Together with the above velocities, these lead to ages of 10^4 and 10^3 years for 3C 48 and 3C 273, respectively.

These ages are considerably smaller than the minimum ages mentioned above. One possibility is that several events have occurred, one 2×10^5 years earlier than the other. Alternatively, the ages derived from the size of emission regions and the line widths would be meaningless if the emission regions are *not* expanding.[1] In that case there must be a mass of about 10^9 $M_\odot$ at the center of the gas cloud. The presence of a mass of this order is also attractive in view of the energy requirements to be discussed presently.

The discovery of light variations in 3C 48 by Matthews and Sandage (1963) and in 3C 273 by Smith and Hoffleit (1963) imposes quite stringent conditions on the light-travel time across these objects. The brighter object, 3C 273, has been studied for a long period. Smith and Hoffleit (1963) and Smith (1964) give results which may be summarized as follows: (1) there are small cyclic variations with a period of about 13 years; (2) flashes lasting of the order of a week or a month may occur, during which the object is up to 1 mag. brighter; and (3) in 1929 a sharp drop in brightness of $0^m.4$ occurred. after which normal brightness was restored by 1940. The cyclic variations with a period around 13 years present no problems as far as the light-travel time is concerned; the diameter of the optical model of the gas cloud for 3C 273 found in Section V is 7 light-years.

VII. Free-Free Emission, Absorption, and Faraday Rotation

Here Greenstein and Schmidt investigate models for the optical emission of 3C 48 and 3C 273. They consider both free-free radiation and synchrotron radiation; their general results are summarized in the following discussion. The absence of free-free absorption and the existence of intrinsic polarization and moderate Faraday rotation lead them to exclude models in which the radio-frequency synchrotron radiation comes from the same volume of space as the emission lines.

VIII. Models for Field and Particle Energies from Synchrotron Emission

Here we omit formulae and tables 8 and 9, which give the particle and magnetic energies for 3C 48 and 3C 273 for different values of the cutoff frequency, ν_c, of the synchrotron radiation. Three cases, A, B, and C, are given with $=$ 6.3×10^{14} $\sec^{-1}$, 6.3×10^{10} $\sec^{-1}$, and 6.3×10^{15} $\sec^{-1}$, respectively. The total luminosities, L, of the synchrotron radiation over all frequencies and the total energies, U, (assuming equipartition) for the three cases are given in Table 121.2.

IX. The Jet, 3C 273A

Here we omit models for the synchrotron radiation of the optically visible jet extending from the source 3C 273A.

X. Discussion

We have considered the explanation of the observed redshifts in terms of (1) Doppler effect from a high-velocity star, (2) gravitational redshift, and (3) cosmological redshift.

The first two alternatives have been shown to lead to an extra-galactic nature of the object. In case 1 the small proper motion results in a distance measured in megaparsecs and a luminosity closer to the galaxies than of stars. The four quasi-stellar objects with known velocity are all receding from us (Schmidt and Matthews 1964). These facts, when combined with the problem of how to accelerate an apparently very large star (or stellar system) to velocities that are an appreciable fraction of that of light, make case 1 an exceedingly unlikely interpretation.

We have shown that alternative 2 also leads to an extragalactic nature of these objects. In fact, the mass would have to be of the order of that of a galaxy or more. It seems rather likely on the basis of current theoretical work that objects of such a mass, condensed to a diameter of less than 1 pc, cannot be stable and thus cannot exist for any length of time. The thinness of the surrounding emission-line shell would also be a problem. Altogether, we believe that it is quite unlikely (but not definitely disproven) that gravitational redshifts explain the spectrum of the quasi-stellar objects.

Accordingly, we have adopted the interpretation of the redshifts as cosmological redshifts. The ensuing lengthy astrophysical discussion of the emission spectra gave radii for the gaseous nebulae of about 1 pc for 3C 273, and 10 pc or more for 3C 48. The light variations seem to require even smaller sizes for the source of optical continuum radiation, especially for 3C 48. The non-thermal character of the radio spectrum shows that it must originate outside the gas cloud which emits the emission lines. We find it attractive to think of a model in which a small inner core produces most of the optical continuum, surrounded by a gas cloud producing the emission lines and thermal continuum. This would itself be surrounded by the radio-emitting regions.

The models for optical continuum synchrotron radiation from an inner volume small enough to admit of the light variations encounter serious difficulties. The high energy-density in a region containing a gaseous nebula produces rapid loss of cosmic-ray protons; the electrons are lost rapidly by either inverse Compton-effect or synchrotron radiation, so that it is nearly impossible to maintain the high-energy particle supply.

An important parameter in further considerations regarding the quasi-stellar objects is their age, i.e., their lifetime as objects producing large optical luminosity from an intrinsically small volume. Let us consider the consequences of two quite different estimates of their age, namely, 10^3 years and 10^6 years.

Age 10^3 years. This is the age, to an order of magnitude, that follows from the size of the gaseous nebula and the interpretation of the widths of the emission lines as caused by expansion. It also seems a lower limit to the possible age of 3C 273, because the secular decrease in optical light amounts to less than 0.1 mag. per century (Smith 1964). The total energy output would amount to 10^{57} ergs, or the rest-mass energy of about 500 suns. This amount can be supplied by nuclear fusion of $10^5\ M_\odot$ of hydrogen to helium. Such a mass is less than that of the gases producing the emission lines; the expansion of the fast-moving gas would be unimpeded. Any of the synchrotron emission models in Tables 8 and 9 [omitted here] would have sufficient energy content to maintain present radiation over 10^3 years. *Not* explained by so short an age would be the radio halo of 3C 273B, the radio source 3C 273A and the jet, or the optical wisps near 3C 48. We assume here that all these features originated in what is now the quasi-stellar object. Their existence requires either a number of separate events, or a much larger age.

Age 10^6 years. The above objections are met if we assume an age of about 10^6 years. Total energy output would amount to 10^{60} ergs, requiring gravitational collapse of a large mass or the nuclear energy for 10^8 $M_\odot$ of hydrogen. This would probably involve a total mass of some 10^9 $M_\odot$. Such a mass could stabilize the large internal motions of the observed H II region at the radius of a few parsecs, derived from the electron density and the intensity of the emission lines. Synchrotron emission models (Tables 8 and 9) with $B \approx 10^{-3}$ gauss (close to equipartition of energies) do not have sufficient energy content to last for 10^6 years at present luminosity. Either there is a steady injection of high-energy particles or a non-equipartition of energy. In the latter case we would consider fields of 0.01–0.1 gauss most likely.

We discussed above a model of the quasi-stellar objects consisting of an optical-continuum source (radius < 1 pc) surrounded by an emission region (radius ≈ 1 pc [3C 273], $\approx$ 10 pc [3C 48]) and by a radio-emitting region. If there is indeed an object with a mass of about 10^9 $M_\odot$ present in these objects, then, presumably, this would be inside the small optical-continuum source. The radius of the 10^9 $M_\odot$ object could have any value below 1 pc. Its gravitational redshift, if any exists, cannot easily be observed because the observed redshift of emission lines refers to a distance from this mass of about 1 pc, where the gravitational effect is negligible. The radius of a Schwarzschild sphere is $2GM/c^2$, about 10^{-4} pc for such a mass. It would be important to know whether continued energy and mass input from such a "collapsed" region are possible.

It is not yet possible to establish the role of these quasi-stellar radio sources in the evolution of galaxies or radio galaxies. No trace of a galaxy around these objects has been found as yet. It is not certain at present that this really excludes the possibility of a galaxy being present; seeing and scattering in the photographic emulsion make detection of a low surface-brightness galaxy, containing a stellar object 100 times brighter, very difficult.

The quasi-stellar sources might have been thought to be a precursor stage of the radio galaxies. Their radio luminosities are about equal to that of the most intense radio galaxies. However, the linear sizes of the radio-emitting regions in the quasi-stellar sources show a range quite similar to that seen in the radio galaxies (Schmidt and Matthews 1964). This does not support the idea that radio galaxies start their radio life as a quasi-stellar source. Either the quasi-stellar stage can occur at any time in the life of a radio galaxy, or the two phenomena may be completely unrelated, with the quasi-stellar objects primary intergalactic condensations.

In table 121.2 we give a final résumé of the properties of possible models. The "distance" r is obtained simply from czH^{-1}, with $H = 100$ km sec^{-1} Mpc^{-1}. The Hβ luminosity is observed; the N_e refers to the H II regions producing permitted and forbidden lines, the internal velocity, v, is deduced from the line widths. The N_e in 3C 273 is determined from the weakness of [O II]; in 3C 48 a wide range is possible, and we give nearly the largest acceptable. The angular diameter gives the sizes, but R(H II) is derived from L(Hβ) and N_e, so the masses depend on $N_e R^3$. The total luminosity is well determined for the visual region, but bolometric corrections may be large. If the radiation is of synchrotron origin, for various ν_c, the L becomes fixed, and the total L and U are given in the last columns.

Table 121.2 Résumé of data on quasi-stellar sources

Parameter	3C 48	3C 273
z	0.367	0.158
r (Mpc)	1,100	474
Luminosities (erg sec^{-1}):		
Hβ	6×10^{42}	9×10^{43}
Visual	10^{45}	4×10^{45}
Radii (R_{pc}):		
Optical	$<2{,}500$	<500
Radio	$<2{,}500$	500
H II region	$\geqslant 10$	1
H II region:		
N_e cm^{-3}	$\leqslant 3 \times 10^4$	3×10^6
$M/M_\odot$	$\geqslant 5 \times 10^6$	6×10^5
v(km/sec)	1,000	1,500
Equipartition synchrotron models*		
L (erg sec^{-1}):		
Case A	2×10^{46}	2×10^{47}
Case B	10^{45}	2×10^{44}
Case C	2×10^{46}	5×10^{46}
U (erg):		
Case A	10^{58}	2×10^{58}
Case B	2×10^{57}	3×10^{57}
Case C	4×10^{55}	4×10^{53}

* For cases A, B, and C, respectively, the luminosity, L, is at both radio and optical frequencies, only at radio frequencies, and only at optical frequencies.]

1. Electron scattering broadens emission lines when the density is high. The scattering optical depth, t, would be near 10 for 3C 273 and 1 for 3C 48. At a fixed hydrogen-line strength, t varies as $R^{-1/2}$; thus the effectiveness of scattering arises from the small size. Thermal electron velocities are $600\,(T_e/10^4)^{1/2}$ km/sec, so that an initially sharp emission line would be broadened to a width, at half-intensity, of 1,000 km/sec at 10,000° K. Multiple electron scattering slows the diffusion of quanta out of the H II region and, therefore, further slows the light variations. In both sources the emission lines have greater velocity widths, and in addition the line profiles seem to vary from line to line. For example, [O III] is sharper than Hβ in 3C 273, and [O II] is sharper than Hγ or [Ne V] in 3C 48. Only if T_e were greater than $10^{5\circ}$ K and the sizes less than a parsec would electron and Compton scattering dominate the line widths.

Aller, L. H. 1954, *Ap. J. 120*, 401.

———. 1956, *Gaseous Nebulae* (New York: John Wiley & Sons), p. 162.

Fowler, W. A., and Hoyle, F. 1964, *Quasi-stellar Sources and Gravitational Collapse*, ed. I. Robinson, A. E. Schild, and E. L. Schucking (Chicago: University of Chicago Press, 1965), p. 17.

Greenstein, J. L., and Matthews, T. A. 1963, *Nature 197*, 1041.

Hazard, C., Mackey, M. B., and Shimmins, A. J. 1963, *Nature 197*, 1037.

Hoyle, F., and Fowler, W. A. 1963, *M.N. 125*, 169.

Matthews, T. A., Bolton, J. G., Greenstein, J. L., Münch, G., and Sandage, A. R. 1960, Am. Astr. Soc. meeting, New York; 1961, *Sky and Telescope 21*, 148.

Matthews, T. A., and Sandage, A. R. 1963, *Ap. J. 138*, 30.

Schmidt, M. 1963, *Nature 197*, 1040.

Schmidt, M., and Matthews, T. A. 1964, *Ap. J. 139*, 781.

Smith, H. J. 1964, *Quasi-stellar Sources and Gravitational Collapse*, ed. I. Robinson, A. E. Schild, and E. L. Schucking (Chicago: University of Chicago Press, 1965), p. 221.

Smith, H. J., and Hoffleit, D. 1961, *Pub. A.S.P. 73*, 292.

CHAPTER VIII

Relativity and Cosmology

122. Explanation of the Perihelion Motion of Mercury by Means of the General Theory of Relativity

Albert Einstein
TRANSLATED BY BRIAN DOYLE

(*Sitzungsberichte der Preussischen Akademie der Wissenschaften zu Berlin 11*, 831–839 [1915])

In 1911 Albert Einstein used the principle of equivalence between gravitational and inertial mass to show that a grazing light ray would be bent by 0.83 sec of arc by the Newtonian gravitational field of the sun.[1] After unsuccessfully searching for evidence of this effect by examining old eclipse photographs,[2] the Babelsberg astronomer Erwin Freundlich arranged to check Einstein's predicted bending of starlight during an eclipse visible in central Russia on August 21, 1914. The outbreak of World War I intervened, and in the meantime, in this paper, Einstein corrected his first calculation by a factor of 2.

Here, by using his general theory of relativity, which is, in effect, a theory of gravitation expressed in terms of curvature of space by means of the tensors of Riemannian geometry, Einstein first derives a formula for the deflection of light by a massive body. A ray of light passing a minimum distance, Δ, from a mass, M, will be deflected from a straight line path by the amount

$$\phi = \frac{2\alpha}{\Delta} = \frac{4GM}{\Delta c^2} \text{ radians,}$$

where α is a parameter appearing in the approximation to the fundamental gravitational tensor, G is the Newtonian constant of gravitation, and c is the velocity of light. For the case of a light ray just grazing the sun's surface, $\phi = 1.75$ sec of arc, and it is this larger deflection that was confirmed by eclipse measurements on May 29, 1919 (see selection 123).

At about the same time that Freundlich was trying to confirm the earlier prediction of the bending of light, he called attention to the anomalies in the motion of Mercury.[3] This was an unsolved problem first discussed by the French celestial mechanician Urbain Jean Joseph Leverrier:[4] although the planet Mercury moves in a nearly elliptical orbit, the attraction of the other planets rotates the perihelion of Mercury by about 530 sec of arc in a century.[5] Leverrier noted that the observed advance of perihelion was greater than could be accounted for by about 40 sec of arc per century, and he suggested that a hitherto unrecognized planet revolving at somewhat less than half of Mercury's mean distance from the sun would produce the extra change in the perihelion of Mercury. However, the hypothetical planet, Vulcan, has never been detected.

In the second part of this selection, Einstein employs a second approximation to the fundamental tensor for the gravitational field to obtain a perihelion advance of

$$\varepsilon = \frac{3\pi\alpha}{a(1-e^2)} = \frac{6\pi GM}{c^2a(1-e^2)} \text{ radians per revolution}$$

where a is the semimajor axis of the planet's orbit and e is its eccentricity. Einstein was delighted to find that his prediction gave Mercury an additional perihelion advance of 43 sec of arc per century. In 1915 the observed perihelion advance for Earth and Mars seemed discrepant when compared with Einstein's predicted values, but today the available observational values compare favorably with his predictions.

1. *Annalen der Physik 35*, 898 (1911). English translation in *The Principle of Relativity* (New York: Dodd, Mead and Co., 1923; reprod. Dover Pub., 1952).
2. *Astronomische Nachrichten 193*, 369 (1913).
3. Ibid. *201*, 49 (1915).
4. *Compte rendu de l' Académie des sciences* (Paris) *49*, 379 (1859); *Annales de l' Observatoire de Paris 5*, 104 (1859).
5. About 530 seconds of arc per century with respect to the fixed stars; with respect to the rotating (precessing) earth-based coordinate system it is about 5,557 seconds of arc per century.

In a work recently published in these reports, I set up the gravitational field equations that are covariant with respect to arbitrary transformations of determinant 1. In a supplement I showed that these equations are generally covariant if the contraction of the energy tensor of "matter" vanishes, and I demonstrated that no important considerations oppose the introduction of this hypothesis, through which time and space are robbed of the last trace of objective reality.[1]

In the present work I find an important confirmation of this most fundamental theory of relativity, showing that it explains qualitatively and quantitatively the secular rotation of the orbit of Mercury (in the sense of the orbital motion itself), which was discovered by Leverrier and which amounts to 45 sec of arc per century.[2] Furthermore, I show that the theory has as a consequence a curvature of light rays due to gravitational fields twice as strong as was indicated in my earlier investigation.

The Gravitational Field

From my last two communications it follows that the gravitational field in a vacuum has to satisfy, upon properly choosing a reference frame, the equations

$$\sum_{\alpha} \frac{\partial \Gamma^{\alpha}_{\mu\nu}}{\partial x_{\alpha}} + \sum_{\alpha\beta} \Gamma^{\alpha}_{\mu\beta}\Gamma^{\beta}_{\nu\alpha} = 0, \tag{1}$$

where the $\Gamma^{\alpha}_{\mu\nu}$ are defined by the equations

$$\begin{aligned} \Gamma^{\alpha}_{\mu\nu} &= -\begin{Bmatrix} \mu\nu \\ \alpha \end{Bmatrix} = -\sum_{\beta} g^{\alpha\beta}\begin{bmatrix} \mu\nu \\ \beta \end{bmatrix} \\ &= -\frac{1}{2}\sum_{\beta} g^{\alpha\beta}\left(\frac{\partial g_{\mu\beta}}{\partial x_{\nu}} + \frac{\partial g_{\nu\beta}}{\partial x_{\mu}} - \frac{\partial g_{\mu\nu}}{\partial x_{\alpha}}\right). \end{aligned} \tag{2}$$

Let us make, moreover, the hypothesis established in the last communication, that the contraction of the energy tensor of "matter" always vanishes, so that, in addition, the determinantal condition is imposed:

$$|g_{\mu\nu}| = -1. \tag{3}$$

A point mass, the sun, is located at the origin of the coordinate system. The gravitational field this point mass produces can be calculated from these equations by means of successive approximations.

Nevertheless, we should consider that the $g_{\mu\nu}$ are still not completely determined mathematically by equations (1) and (3), because these equations are covariant with respect to arbitrary transformations of determinant 1. Yet we are justified in assuming that all these solutions can be reduced to one another by such transformations that they are distinguished (by the given boundary conditions) formally but not, however, physically, from one another. Consequently, I am satisfied for the time being with deriving here a solution, without discussing the question whether the solution might be unique.

To proceed, let the $g_{\mu\nu}$ be given in the 0th approximation by the following scheme corresponding to the original theory of relativity:

$$\left.\begin{matrix} -1 & 0 & 0 & 0 \\ 0 & -1 & 0 & 0 \\ 0 & 0 & -1 & 0 \\ 0 & 0 & 0 & +1 \end{matrix}\right\}, \tag{4}$$

or, more briefly,

$$\left.\begin{aligned} g_{\rho\sigma} &= \delta_{\rho\sigma} \\ g_{\rho 4} &= g_{4\rho} = 0 \\ g_{44} &= 1 \end{aligned}\right\}. \tag{4a}$$

Here ρ and σ signify the indices 1, 2, 3; $\delta_{\rho\sigma}$ is equal to 1 or 0 if $\rho = \sigma$ or $\rho \neq \sigma$, respectively.

I assume in what follows that the $g_{\mu\nu}$ differ from the values given in equation (4a) only by quantities small compared to unity. I treat this deviation as a small quantity of first order, whereas functions of the nth degree in these deviations are treated as quantities of the nth order. Equations (1) and (3) together with equation (4a) enable us to calculate by successive approximations the gravitational field up to quantities of nth order exactly. The approximation given in equation (4a) forms the 0th approximation.

The solution has the following properties, which determine the coordinate system:

1. All components are independent of x_4.
2. The solution is spatially symmetric about the origin of the coordinate system, in the sense that we encounter the same solution again if we subject it to a linear orthogonal spatial transformation.
3. The equations $g_{\rho 4} = g_{4\rho} = 0$ are exactly valid for $\rho = 1, 2, 3$.
4. The $g_{\mu\nu}$ possess the values given in equation (4a) at infinity.

First approximation It is easy to verify that to quantities of first order the equations (1) and (3) are satisfied for the just-named four conditions by the assumed solution

$$\left.\begin{aligned} g_{\rho\sigma} &= -\delta_{\rho\sigma} + \alpha\left(\frac{\partial^2 r}{\partial x_{\rho}\partial x_{\sigma}} - \frac{\delta_{\rho\sigma}}{r}\right) = -\delta_{\rho\sigma} - \alpha\frac{x_{\rho}x_{\sigma}}{r^3} \\ g_{44} &= 1 - \frac{\alpha}{r} \end{aligned}\right\}. \tag{4b}$$

The $g_{4\rho}$ ($g_{\rho 4}$) are determined by condition 3, the r denotes the quantity $+\sqrt{x_1{}^2 + x_2{}^2 + x_3{}^2}$, and α is a constant determined by the mass of the sun.

That condition 3 is fulfilled to terms of first order we see immediately. More simply, the field equations (1) are also fulfilled in the first approximation. We need only to consider that upon neglect of quantities of second and higher order, the left side of equation (1) can be permuted successively through

$$\sum_\alpha \frac{\partial \Gamma^\alpha_{\mu\nu}}{\partial x_\alpha}$$

$$\sum_\alpha \frac{\partial}{\partial x_\alpha}\begin{bmatrix}\mu\nu \\ \alpha\end{bmatrix},$$

where α runs over only 1–3.

As we perceive from equation (4b), my theory implies that in the case of a resting mass, the components g_{11} up to g_{33} are in quantities of first order already different from 0. Therefore, as we shall see later, no disagreement with Newton's law arises in the first approximation. This theory, however, produces an influence of the gravitational field on a light ray somewhat different from that given in my earlier work, because the velocity of light is determined by the equation

$$\sum g_{\mu\nu}\, dx_\mu\, dx_\nu = 0. \tag{5}$$

Upon the application of Huygen's principle, we find from equations (5) and (4b), after a simple calculation, that a light ray passing at a distance Δ suffers an angular deflection of magnitude $2\alpha/\Delta$, while the earlier calculation, which was not based upon the hypothesis $\sum T^\mu_\mu = 0$, had produced the value α/Δ. A light ray grazing the surface of the sun should experience a deflection of 1.7 sec of arc instead of 0.85 sec of arc. In contrast to this difference, the result concerning the shift of the spectral lines by the gravitational potential, which was confirmed by Mr. Freundlich on the fixed stars (in order of magnitude), remains unaffected, because this result depends only on g_{44}.

Since we have obtained the $g_{\mu\nu}$ in the first approximation, we can also calculate the components $T^\alpha_{\mu\nu}$ of the gravitational field to the first approximation. From equations (2) and (4b) we have

$$\Gamma^\tau_{\rho\sigma} = -\alpha\left(\delta_{\rho\sigma}\frac{x_\tau}{r^2} - \frac{3}{2}\frac{x_\rho x_\sigma x_\tau}{r^5}\right), \tag{6a}$$

where ρ, σ, τ signify any one of the indices 1, 2, 3, and

$$\Gamma^\sigma_{44} = \Gamma^4_{4\sigma} = -\frac{\alpha}{2}\frac{x_\sigma}{r^3}, \tag{6b}$$

where σ signifies the index 1, 2, or 3. Those components in which the index 4 appears once or three times vanish.

Second Approximation It will subsequently be seen that we need to determine only three components Γ^σ_{44} exactly to quantities of the second order in order to be able to determine the orbits of the planets with the appropriate degree of accuracy. For this process, the last field equation, together with the general conditions we have imposed on our solution, suffices. The last field equation,

$$\sum_\sigma \frac{\partial \Gamma^\sigma_{44}}{\partial x_\sigma} + \sum_{\sigma\tau} \Gamma^\sigma_{4\tau}\Gamma^\tau_{4\sigma} = 0,$$

becomes upon consideration of equation (6b) and upon neglect of quantities of third and higher order

$$\sum_\sigma \frac{\partial \Gamma^\sigma_{44}}{\partial x_\sigma} = \frac{\alpha^2}{2r^4}.$$

From this we deduce, upon considering equation (6b) and the symmetry properties of our solution,

$$\Gamma^\sigma_{44} = -\frac{\alpha}{2}\frac{x_\sigma}{r^3}\left(1 - \frac{\alpha}{r}\right). \tag{6c}$$

The Motion of the Planets

The equation of motion of the point mass in the gravitational field yielded by the general theory of relativity reads

$$\frac{d^2 x_\nu}{ds^2} = \sum_{\sigma\tau} \Gamma^\nu_{\sigma\tau}\frac{dx_\sigma}{ds}\frac{dx_\tau}{ds}. \tag{7}$$

From this equation we first deduce that it contains the Newtonian equations of motion as a first approximation. Of course, if the motion of the planet takes place with a velocity less than the velocity of light, then dx_1, dx_2, dx_3 are smaller than dx_4. In consequence, we get a first approximation in which we consider on the right side only the term $\sigma = \tau = 4$. Upon considering equation (6b), we obtain

$$\left.\begin{aligned} \frac{d^2 x_\nu}{ds^2} &= \Gamma^\nu_{44} = -\frac{\alpha}{2}\frac{x_\nu}{r^3}\ (\nu = 1, 2, 3) \\ \frac{d^2 x_4}{ds^2} &= 0 \end{aligned}\right\} \tag{7a}$$

These equations show that we can set $s = x_4$ for the first approximation. Then the first three equations are exactly the Newtonian equations. If we introduce polar variables in the orbital plane, then, as is well known, the energy law and the law of areas yield the equations

$$\left.\begin{aligned} \frac{1}{2}u^2 + \Phi &= A \\ r^2\frac{d\phi}{ds} &= B \end{aligned}\right\}, \tag{8}$$

where A and B signify the constants of the energy law, and where

$$\left.\begin{aligned} \Phi &= -\frac{\alpha}{2r} \\ u^2 &= \frac{dr^2 + r^2\,d\phi^2}{ds^2} \end{aligned}\right\} \tag{8a}$$

is granted.

We have now to evaluate the equations to the next order. The last of the equations (7) yields, then, together with equation (6b),

$$\frac{d^2x_4}{ds^2} = 2\sum_\sigma \Gamma^4_{\sigma 4}\frac{dx_\sigma}{ds}\frac{dx_4}{ds} = -\frac{dg_{44}}{ds}\frac{dx_4}{ds},$$

or, correct to the first order,

$$\frac{dx_4}{ds} = 1 + \frac{\alpha}{r}. \tag{9}$$

We now turn to the first of the three equations (7). The right side yields:

a) for the index combination $\sigma = \tau = 4$

$$\Gamma^\nu_{44}\left(\frac{dx_4}{ds}\right)^2,$$

or considering equations (6c) and (9), correct to the second order,

$$-\frac{\alpha}{2}\frac{x_\nu}{r^3}\left(1 + \frac{\alpha}{r}\right);$$

b) for the index combination $\sigma \neq 4$, $\tau \neq 4$ (which alone still needs to be considered), upon considering the product $(dx_\sigma/ds)(dx_\tau/ds)$, using equation (8) to first order, correct to the second order,

$$-\frac{\alpha x_\nu}{r^3}\sum\left(\delta_{\sigma\tau} - \frac{3}{2}\frac{x_\sigma x_\tau}{r^2}\right)\frac{dx_\sigma}{ds}\frac{dx_\tau}{ds}.$$

The summation gives

$$-\frac{\alpha x_\nu}{r^3}\left(u^2 - \frac{3}{2}\left(\frac{dr}{ds}\right)^2\right).$$

Using this value we obtain for the equation of motion the form, correct to the second order,

$$\frac{d^2x_\nu}{ds^2} = -\frac{\alpha}{2}\frac{x_\nu}{r^3}\left(1 + \frac{\alpha}{r} + 2u^2 - 3\left(\frac{dr}{ds}\right)^2\right), \tag{7b}$$

which together with equation (9) determines the motion of the mass point. Moreover, it should be observed that equations (7b) and (9) for the case of circular motion give no deviation from Kepler's three laws.

From equation (7b) follows, above all, the exact validity of the equation

$$r^2\frac{d\phi}{ds} = B, \tag{10}$$

where B is a constant. The law of areas is therefore valid to second order if we use the "proper time" of the planet to measure time. In order to determine the secular rotation of the orbital ellipse from equation (7b), we substitute the terms of the first order in the parentheses most advantageously by means of equation (10) and the first of the equations (8), through which procedure the terms of second order on the right side are not altered. The parentheses take on the form

$$\left(1 - 2A + \frac{3B^2}{r^2}\right).$$

Finally, if we choose $s\sqrt{(1 - 2A)}$ as the time variable, and if we redesignate it as s, we have, with a somewhat different meaning of the constant B;

$$\frac{d^2x_\nu}{ds^2} = -\frac{\partial\Phi}{\partial x_\nu}, \qquad \Phi = -\frac{\alpha}{2}\left[1 + \frac{B^2}{r^2}\right]. \tag{7c}$$

In order to determine the equation of the orbit, we now proceed exactly as in the Newtonian case. From equation (7c) we obtain first

$$\frac{dr^2 + r^2\,d\phi^2}{ds^2} = 2A - 2\Phi.$$

If we eliminate ds from this equation with the help of equation (10), we obtain

$$\left(\frac{dx}{d\phi}\right)^2 = \frac{2A}{B^2} + \frac{\alpha}{B^2}x - x^2 + \alpha x^3, \tag{11}$$

where we denote by x the quantity $1/r$. This equation differs from the corresponding one in Newtonian theory only in the last term on the right side.

The angle described by the radius vector between the perihelion and the aphelion is consequently given by the elliptic integral

$$\phi = \int_{\alpha_1}^{\alpha_2}\frac{dx}{\sqrt{\frac{2A}{B^2} + \frac{\alpha}{B^2}x - x^2 + \alpha x^3}},$$

where α_1 and α_2 signify the roots of the equation

$$\frac{2A}{B^2} + \frac{\alpha}{B^2}x - x^2 + \alpha x^3 = 0$$

and closely correspond to the neighboring roots of the equation that arises from this one by the omission of the last term.

Thus, it can be established with the precision demanded of us that

$$\phi = [1 + \alpha(\alpha_1 + \alpha_2)] \cdot \int_{\alpha_1}^{\alpha_2} \frac{dx}{\sqrt{-(x - \alpha_1)(x - \alpha_2)(1 - \alpha x)}},$$

or upon expansion of $(1 - \alpha x)^{-1/2}$,

$$\phi = [1 + \alpha(\alpha_1 + \alpha_2)] \cdot \int_{\alpha_1}^{\alpha_2} \frac{\left(1 + \frac{\alpha}{2}x\right) dx}{\sqrt{-(x - \alpha_1)(x - \alpha_2)}}.$$

The integration yields

$$\phi = \pi\left[1 + \frac{3}{4}\alpha(\alpha_1 + \alpha_2)\right],$$

or if we consider that α_1 and α_2 signify the reciprocal values of the maximum and minimum distances, respectively, from the sun,

$$\phi = \pi\left(1 + \frac{3}{2}\frac{\alpha}{a(1 - e^2)}\right). \tag{12}$$

Therefore, after a complete orbit, the perihelion advances by

$$\varepsilon = 3\pi \frac{\alpha}{a(1 - e^2)} \tag{13}$$

in the sense of the orbital motion, where the semimajor axis is denoted by a and the eccentricity by e. If we introduce the orbital period T (in seconds), we obtain

$$\varepsilon = 24\pi^3 \frac{a^2}{T^2c^2(1 - e^2)}, \tag{14}$$

where c denotes the velocity of light in units of cm $\sec^{-1}$. The calculation yields, for the planet Mercury, a perihelion advance of 43″ per century, while the astronomers assign 45″ ± 5″ per century as the unexplained difference between observations and the Newtonian theory. This theory therefore agrees completely with the observations.

For Earth and Mars, the astronomers assign, respectively, forward motions of 11″ and 9″ per century, while our formula yields, respectively, 4″ and 1″ per century. Nevertheless, a small value seems to be proper to these assignments because of the small eccentricities of the orbits of these planets. A more certain confirmation of the perihelion motion will be made by determining the product of the motion with the eccentricity. If we consider these quantities assigned by Newcomb,

	$e\frac{d\pi}{dt}$
Mercury	8.48″ ± 0.43
Venus	−0.05 ± 0.25
Earth	0.10 ± 0.13
Mars	0.75 ± 0.35,

for which I thank Dr. Freundlich, then we obtain the impression that the advance of the perihelion is, after all, demonstrated really only for Mercury. However, I prefer to relinquish a final decision to the astronomical specialists.

1. In a forthcoming communication it will be shown that this hypothesis is unnecessary. It is because such a choice of reference frame is possible that the determinant $|g_{\mu\nu}|$ takes on the value −1. The following investigation is independent of this choice.

2. E. Freundlich recently wrote a noteworthy article on the impossibility of satisfactorily explaining the anomalies in the motion of Mercury on the basis of the Newtonian theory (*Astronomische Nachrichten 201*, 49 [1915]).

123. A Determination of the Deflection of Light by the Sun's Gravitational Field, from Observations Made at the Total Eclipse of May 29, 1919

Frank Watson Dyson, Arthur Stanley Eddington, and Charles Davidson

(*Philosophical Transactions of the Royal Society* (*London*) *220*, 291–333 [1920])

By 1916 Einstein had prepared the comprehensive formulation of his general theory of relativity.[1] By then he had also proposed the following three tests of his theory:

1. The rate of a clock will depend on the gravitational field or the curvature of space-time; the frequency of an atomic transition, and hence the wavelength of a spectral line, will appear changed if the observer has a different gravitational field than the atomic source. At the time this selection was written, this first prediction had not yet been tested experimentally, but now it has been accurately confirmed by the Pound-Rebka laboratory experiment.[2]
2. The general theory predicts a hitherto unexplained advance of perihelion for Mercury; selection 122 provides Einstein's account of the residual 43 sec of arc per century that was observed and accounted for by the general theory.
3. In the same selection, Einstein showed that a light ray grazing the solar surface should be deflected by an angle of 1.75 sec of arc. This deflection is twice that predicted by using the equivalence principle and a Newtonian gravitational potential.

A complete account of these predictions was first brought to an English-speaking audience in A. S. Eddington's *Report on the Relativity Theory of Gravitation* (1918), published in the closing days of World War I. Although the theory had been proposed by a German professor, interest ran high in Britain, and funding was obtained for a joint Greenwich Observatory-Royal Society expedition to measure the deflection of starlight around the sun during the total eclipse of May 1919.

In the last evening before sailing to Principe, a small island off the coast of West Africa, E. T. Cottingham, who was to accompany Eddington, asked Sir Frank Dyson, the astronomer royal, what would happen if they found *twice* Einstein's predicted deflection. Sir Frank replied, "Then Eddington will go mad and you will have to come home alone!" Three months later, at Principe, when he had made his first plate reduction, Eddington turned to his companion and said, "Cottingham, you won't have to go home alone."[3] Eclipse conditions were favorable both at Principe and in Brazil, and a light deflection equivalent to 1.98 ± 0.16 sec of arc at the solar limb was measured.

The deflections measured by observing stars during subsequent solar eclipses have, at best, verified Einstein's theory to within 25 percent. As first suggested by Irwin Shapiro, the predicted deflection can also be tested by observing the microwave radiation from extragalactic radio sources, and this test does not need to be made during a solar eclipse.[4] The measured bending of the microwave radiation in terms of that at the solar limb is 1.775 ± 0.019 sec of arc.[5]

1. *Annalen der Physik 49*, 769 (1916). English translation in *The Theory of Relativity* (New York: Dodd, Mead and Co., 1923; reprod. Dover Pub., 1952).
2. R. V. Pound and G. A. Rebka, *Physical Review Letters 4*, 337 (1960); R. V. Pound and J. L. Snider, *Physical Review Letters 13*, 539 (1964).
3. A. Vibert Douglas, *The Life of Arthur Stanley Eddington* (London: Thomas Nelson and Sons, 1956), p. 40.
4. I. I. Shapiro, *Science 157*, 806 (1967).
5. E. B. Fomalont and R. A. Sramek, *Astrophysical Journal 199*, 749 (1975).

Purpose of the Expeditions

1. The purpose of the expeditions was to determine what effect, if any, is produced by a gravitational field on the path of a ray of light traversing it. Apart from possible surprises, there appeared to be three alternatives, which it was especially desired to discriminate between—

(1) The path is uninfluenced by gravitation.

(2) The energy or mass of light is subject to gravitation in the same way as ordinary matter. If the law of gravitation is strictly the Newtonian law, this leads to an apparent displacement of a star close to the sun's limb amounting to 0″.87 outwards.

(3) The course of a ray of light is in accordance with Einstein's generalized relativity theory. This leads to an apparent displacement of a star at the limb amounting to 1″.75 outwards.

In either of the last two cases the displacement is inversely proportional to the distance of the star from the sun's centre, the displacement under (3) being just double the displacement under (2).

It may be noted that both (2) and (3) agree in supposing that light is subject to gravitation in precisely the same way as ordinary matter. The difference is that, whereas (2) assumes the Newtonian law, (3) assumes Einstein's new law of gravitation. The slight deviation from the Newtonian law, which on Einstein's theory causes an excess motion of perihelion of Mercury, becomes magnified as the speed increases, until for the limiting velocity of light it doubles the curvature of the path.

2. The displacement (2) was first suggested by Prof. Einstein[1] in 1911, his argument being based on the Principle of Equivalence, viz., that a gravitational field is indistinguishable from a spurious field of force produced by an acceleration of the axes of reference. But apart from the validity of the general Principle of Equivalence there were reasons for expecting that the electromagnetic energy of a beam of light would be subject to gravitation, especially when it was proved that the energy of radio-activity contained in uranium was subject to gravitation. In 1915, however, Einstein found that the general Principle of Equivalence necessitates a modification of the Newtonian law of gravitation, and that the new law leads to the displacement (3).

3. The only opportunity of observing these possible deflections is afforded by a ray of light from a star passing near the sun. (The maximum deflection by Jupiter is only 0″.017.) Evidently, the observation must be made during a total eclipse of the sun.

Immediately after Einstein's first suggestion, the matter was taken up by Dr. E. Freundlich, who attempted to collect information from eclipse plates already taken; but he did not secure sufficient material. At ensuing eclipses plans were made by various observers for testing the effect, but they failed through cloud or other causes. After Einstein's second suggestion had appeared, the Lick Observatory expedition attempted to observe the effect at the eclipse of 1918. The final results are not yet published. Some account of a preliminary discussion has been given,[2] but the eclipse was an unfavorable one, and from the information published the probable accidental error is large, so that the accuracy is insufficient to discriminate between the three alternatives.

4. The results of the observations here described appear to point quite definitely to the third alternative, and confirm Einstein's generalised relativity theory. As is well-known the theory is also confirmed by the motion of the perihelion of Mercury, which exceeds the Newtonian value by 43″ per century—an amount practically identical with that deduced from Einstein's theory. On the other hand, his theory predicts a displacement to the red of the Fraunhofer lines on the sun amounting to about 0.008 Å in the violet. According to Dr. St. John this displacement is not confirmed.[3] If this disagreement is to be taken as final it necessitates considerable modifications of Einstein's theory, which it is outside our province to discuss. But, whether or not changes are needed in other parts of the theory, it appears now to be established

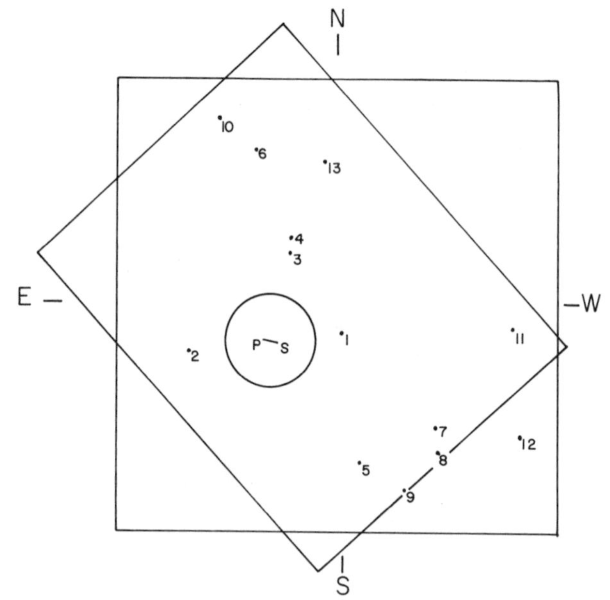

Fig. 123.1 The spatial positions of the stars near the sun during the solar eclipse of May 29, 1919. The sun moved from S to P in the interval between totality at Sobral, Brazil, and at the Island of Principe. The angular scale of the diagram can be inferred from the disk of the sun, which is 31.6 min of arc in angular diameter.

Table 123.1 Expected gravitational displacements of star positions

No.	Names	Photog. mag.	Coordinates (unit = 50′)		Gravitational displacement			
					Sobral		Principe	
			x	y	x	y	x	y
		m.			″	″	″	″
1	B.D., 21°, 641	7.0	+0.026	−0.200	−1.31	+0.20	−1.04	+0.09
2	Piazzi, IV, 82	5.8	+1.079	−0.328	+0.85	− .09	+1.02	− .16
3	κ^2 Tauri	5.5	+0.348	+0.360	−0.12	+ .87	−0.28	+ .81
4	κ^1 Tauri	4.5	+0.334	+0.472	−0.10	+ .73	−0.21	+ .70
5	Piazzi, IV, 61	6.0	−0.160	−1.107	−0.31	− .43	−0.31	− .38
6	υ Tauri	4.5	+0.587	+1.099	+0.04	+ .40	+0.01	+ .41
7	B.D., 20°, 741	7.0	−0.707	−0.864	−0.38	− .20	−0.35	− .17
8	B.D., 20°, 740	7.0	−0.727	−1.040	−0.33	− .22	−0.29	− .20
9	Piazzi, IV, 53	7.0	−0.483	−1.303	−0.26	− .30	−0.26	− .27
10	72 Tauri	5.5	+0.860	+1.321	+0.09	+ .32	+0.07	+ .34
11	66 Tauri	5.5	−1.261	−0.160	−0.32	+ .02	−0.30	+ .01
12	53 Tauri	5.5	−1.311	−0.918	−0.28	− .10	−0.26	− .09
13	B.D., 22°, 688	8.0	+0.089	+1.007	−0.17	+ .40	−0.14	+ .39

that Einstein's law of gravitation gives the true deviations from the Newtonian law both for the relatively slow-moving planet Mercury and for the fast-moving waves of light.

It seems clear that the effect here found must be attributed to the sun's gravitational field and not, for example, to refraction by coronal matter. In order to produce the observed effect by refraction, the sun must be surrounded by material of refractive index $1 + .00000414/r$, where r is the distance from the centre in terms of the sun's radius. At a height of one radius above the surface the necessary refractive index 1.00000212 corresponds to that of air at 1/140 atmosphere, hydrogen at 1/60 atmosphere, or helium at 1/20 atmospheric pressure. Clearly a density of this order is out of the question.

PREPARATIONS FOR THE EXPEDITIONS

In late 1917 it was noted that the eclipse of May 29, 1919, was especially favorable for testing Einstein's theory because of the unusual number of bright stars in the field of stars surrounding the sun. As the eclipse track ran from North Brazil across the Atlantic Ocean through the Island of Principe and then across Africa, an application was made to the English government for a grant of 1,100 pounds to cover expenses for expeditions to Sobral, North Brazil, and to the Island of Principe. The photographic magnitudes, standard coordinates, and expected gravitational displacements for the stars visible at the two sights are given in table 123.1. Here the gravitational displacements are calculated on the assumption of a radial displacement of $1.75\ r_0/r$ sec of arc, where r is the distance from the sun's center and r_0 is the radius of the sun.

Figure 123.1 shows the relative positions of the stars given in table 123.1. The square shows the limits of the plates used at Principe and the oblique rectangle the limits with the 4-in lens at Sobral. The sun moved from S to P in the 2.25-h interval between totality at the two stations, and the sun is here represented as midway between them.

THE EXPEDITION TO SOBRAL

(*Observers, Dr. A. C. D. Crommelin and Mr. C. Davidson*)

For the expedition to Sobral, the Greenwich astrographic telescope of 3.43-m focal length was stopped to 8 in, mounted in a steel tube, and fed with a 16-in coelstat. An auxilliary 4-in lens was mounted in a 19-ft-long square wooden tube in conjunction with an 8-in coelstat. A description of the moment of eclipse follows.

The morning of the eclipse day was rather more cloudy than the average, and the proportion of cloud was estimated at 9/10 at the time of first contact, when the sun was invisible; it appeared a few seconds later showing a very small encroachment of the moon, and there were various short intervals of sunshine during the partial phase which enabled us to place the sun's image at its assigned position on the ground glass, and to give a final adjustment to the rates of the driving clocks. As totality approached, the proportion of cloud diminished, and a large clear space reached the sun about one minute before second contact. Warnings were given 58s., 22s. and 12s. before second contact by observing the length of the disappearing crescent on the ground glass. When the crescent disappeared the word "go" was called and a metronome was started by Dr. Leocadio, who called out every tenth beat during totality, and the exposure times were recorded in terms of these beats. It beat 320 times in 310 seconds; allowance has been made for this rate in the recorded times. The programme arranged was carried out successfully, 19 plates being exposed in the astrographic telescope with alternate exposures of 5 and 10 seconds, and eight in the 4-inch camera with a uniform exposure of 28 seconds. The region round the sun was free from cloud, except for an interval of about a minute near the middle of totality when it was veiled by thin cloud, which prevented the photography of stars, though the inner corona remained visible to the eye and the plates exposed at this time show it and the large prominence excellently defined. The plates remained in their holders until development, which was carried out in convenient batches during the night hours of the following days, being completed by June 5.

Reference photographs of the eclipse field were made on July 11 from the same station with the same instruments as those used to photograph the eclipse. The positions of the stars on the reference and eclipse photographs were then compared to obtain a mean displacement of 0.625 sec of arc at a distance of 50 min of arc from the sun center for the photographs taken with the 4-in object glass. Because the radius of the sun at the time of the eclipse was 15.8 min of arc, the observed displacement is equivalent to a 1.98 sec of arc deflection at the limb. After making corrections for differential refraction, aberration, plate orientation, and changes of scale, the displacements of the different stars given in table 123.2 were established. In this table the calculated values are those given by Einstein's theory with the value of 1.75 sec of arc at the sun's limb.

Table 123.2 Displacements of star positions

	Displacement in right ascension		Displacement in declination	
No. of star	Observed (″)	Calculated (″)	Observed (″)	Calculated (″)
11	−0.19	−0.32	+0.16	+0.02
5	− .29	− .31	−0.46	− .43
4	− .11	− .10	+0.83	+ .74
3	− .20	− .12	+1.00	+ .87
6	+ .10	+ .04	+0.57	+ .40
10	− .08	+ .09	+0.35	+ .32
2	+ .95	+ .85	−0.27	− .09

The photographs taken with the astrographic object glass had diffuse images caused by focusing changes induced by the sun's heat. The results were of poorer quality than those taken with the 4-in refractor; they conflicted with the latter results, in that a mean deflection of 0.93 sec of arc was obtained when the observations were referred to the sun's limb.

The Expedition to Principe
(*Observers, Prof. A. S. Eddington and Mr. E. T. Cottingham*)

For the expedition to Principe, the Oxford astrographic telescope was stopped to 8 in and fed with a 16-in coelstat. A description of the moment of eclipse follows.

The days preceding the eclipse were very cloudy. On the morning of May 29 there was a very heavy thunderstorm from about 10 a.m. to 11.30 a.m.—a remarkable occurrence at that time of year. The sun then appeared for a few minutes, but the clouds gathered again. About half-an-hour before totality the crescent sun was glimpsed occasionally, and by 1.55 it could be seen continuously through drifting cloud. The calculated time of totality was from 2h. 13m. 5s. to 2h. 18m. 7s. G.M.T. Exposures were made according to the prepared programme, and 16 plates were obtained. Mr. Cottingham gave the exposures and attended to the driving mechanism, and Prof. Eddington changed the dark slides. It appears from the results that the cloud must have thinned considerably during the last third of totality, and some star images were shown on the later plates. The cloudier plates give very fine photographs of a remarkable prominence which was on the limb of the sun.

A few minutes after totality the sun was in a perfectly clear sky, but the clearance did not last long. It seems likely that the break-up of the clouds was due to the eclipse itself, as it was noticed that the sky usually cleared at sunset.

In addition to the eclipse plates, a check field including Arcturus was photographed at Oxford and at Principe. Comparison photographs of the eclipse field could not be made at nighttime for several months. The check plates were used to show that none of the displacements exhibited on the eclipse plates were present on the plates of the field containing Arcturus. The inference is that the displacements in the former case could only be attributed to the presence of the eclipsed sun in the field.

If X and W denote the eclipse plates and G, H, D, and I denote comparison plates taken at Oxford, the following results are obtained for the equivalent limb deflection.

The four determinations from the two eclipse plates are

X − G	1″.94
X − H	1″.44
W − D	1″.55
W − I	1″.67

giving a mean of 1.65 sec of arc, with a probable error of 0.30 sec of arc. They evidently agree with Einstein's predicted value 1″.75.

General Conclusions

In summarising the results of the two expeditions, the greatest weight must be attached to those obtained with the 4-inch lens at Sobral. From the superiority of the images and the larger scale of the photographs it was recognised that these would prove to be much the most trustworthy. Further, the agreement of the results derived independently from the right ascensions and declinations, and the accordance of the residuals of the individual stars provides a more satisfactory check on the results than was possible for the other instruments. These plates gave

From declinations	1″.94
From right ascensions	2″.06

The result from declinations is about twice the weight of that from right ascensions, so that the mean result is

1″.98

with a probable error of about ±0″.12.

The Principe observations were generally interfered with by cloud. The unfavorable circumstances were perhaps partly compensated by the advantage of the extremely uniform temperature of the island. The deflection obtained was

1″.61.

The probable error is about ±0″.30, so that the result has much less weight than the preceding.

Both of these point to the full deflection 1″.75 of Einstein's generalised relativity theory, the Sobral results definitely, and the Principe results perhaps with some uncertainty. There remain the Sobral astrographic plates which gave the deflection

0″.93

discordant by an amount much beyond the limits of its accidental error. For the reasons already described at length not much weight is attached to this determination.

It has been assumed that the displacement is inversely proportional to the distance from the sun's centre, since all theories agree on this, and indeed it seems clear from considerations of dimensions that a displacement, if due to gravitation, must follow this law. From the results with the 4-inch lens, some kind of test of the law is possible though it is necessarily only rough. The evidence is summarised in figure 123.2, which shows the radial displacement of the individual

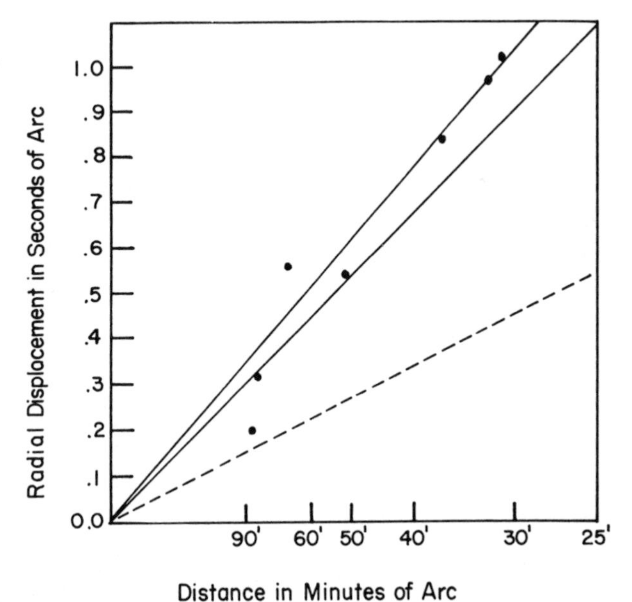

Fig. 123.2 The mean radial displacement of the positions of stars near the sun during the solar eclipse of May 29, 1919, plotted against the angular distance of each star from the center of the sun in minutes of arc. The displacement according to Einstein's theory is indicated by the middle line, that according to the Newtonian law by the dashed line, and that from these observations by the top line.

stars (mean from all the plates) plotted against the reciprocal of the distance from the centre.

Thus the results of the expeditions to Sobral and Principe can leave little doubt that a deflection of light takes place in the neighborhood of the sun and that it is of the amount demanded by Einstein's generalised theory of relativity, as attributable to the sun's gravitational field. But the observation is of such interest that it will probably be considered desirable to repeat it at future eclipses. The unusually favorable conditions of the 1919 eclipse will not recur, and it will be necessary to photograph fainter stars, and these will probably be at a greater distance from the sun. This *can* be done with such telescopes as the astrographic with the object-glass stopped down to 8 inches, if photographs of the same high quality are obtained as in regular stellar work. It will probably be best to discard the use of cœlostat mirrors. These are of great convenience for photographs of the corona and spectroscopic observations, but for work of precision of the high order required, it is undesirable to introduce complications, which can be avoided, into the optical train. It would seem that some form of equatorial mounting (such as that employed in the Eclipse Expeditions of the Lick Observatory) is desirable.

1. *Annalen der Physik 35*, 898.
2. *Observatory 42*, 298.
3. *Astrophysical Journal 46*, 249.

124. Fourth Test of General Relativity: New Radar Result

Irwin I. Shapiro, Michael E. Ash, Richard P. Ingalls, William B. Smith, Donald B. Campbell, Rolf B. Dyce, Raymond F. Jurgens, and Gordon H. Pettengill

(*Physical Review Letters 26*, 1132–35 [1971])

As early as 1915 Albert Einstein showed that the curvature of space-time accounts for the previously unexplained perihelion advance of Mercury, and he predicted that the light from distant stars would be deflected at the solar limb by 1.75 sec of arc because of the sun's gravitational field (see selection 122). This prediction was soon verified by A. S. Eddington and his colleagues, who found a solar limb deflection of 1.98 ± 0.16 sec of arc (see selection 123). Nevertheless, the margin of error in this observation was sufficiently large that it could not be used to test alternative gravitational theories predicting different curvatures of the space-time structure.

In 1964 Irwin Shapiro suggested a new test of gravitational theories, made possible through the development of modern technology.[1] This test measures the total time required for a radar signal to make a round trip between the earth and a planet. When the line of sight to the planet passes near the sun, the radar echo should have an extra time delay caused by the gravitational field of the sun. Radar signals permit the measurement of the distance from the earth to some planets to within a hundred meters, and hence Shapiro's test can be used not only to tighten the observational verification of Einstein's theory but also to evaluate various alternative gravitational theories. The paper given here provides such a criterion by presenting measurements of the residual time delay of radar signals to Mercury and Venus.

Because gravity in the solar system is a relatively weak force, the small departures from Newtonian gravitation can be expressed in any metric theory with the first few terms of a series—different metric theories then differing only in the numerical values of the coefficients.[2] The coefficient β describes the nonlinearity of the gravitational theory, whereas the coefficient γ describes the degree to which mass curves space. For the bending of light by the sun, for example, the solar limb deflection is 1.749 $(1 + \gamma)/2$ sec of arc, and Einstein's theory gives $\gamma = 1.000$. For the perihelion advance of a planet, the parameter $[2(1 + \gamma) - \beta]/3$ is measured, where for Einstein's theory β and γ are identically unity.

One of the most widely discussed alternative theories of gravity is the scaler-tensor theory developed by Carl Brans and Robert H. Dicke.[3] According to their formulation, β has a value of one, but space is curved by a lesser amount for a given mass than in Einstein's theory. As mentioned at the end of this selection, Dicke's preferred value for γ is 0.86, in disagreement with the radar measurements, which give $\gamma = 1.03 \pm 0.04$. This inconsistency was subsequently confirmed by measurements of the solar limb deflection of the microwave radiation of radio sources ($\gamma = 1.007 \pm 0.009$) and by the radar determinations of Mercury's perihelion advance $[(2 + 2\gamma - \beta)/3 = 1.005 \pm 0.007]$.[4] The radar tests of Shapiro and his colleagues further substantiate the correctness of Einstein's theory and place stringent limits on alternative theories of gravitation.

1. I. I. Shapiro, *Physical Review Letters 13*, 789 (1964).

2. These equations are derived by A. S. Eddington in *The Mathematical Theory of Relativity* (Cambridge: Cambridge University Press, 1923) and reprinted by K. R. Lang in *Astrophysical Formulae* (New York: Springer-Verlag, 1974), p. 577.

3. C. Brans and R. H. Dicke, *Physical Review 124*, 925 (1961).

4. I. I. Shapiro, G. H. Pettengill, M. E. Ash, R. P. Ingalls, D. B. Campbell, and R. B. Dyce, *Physical Review Letters 28*, 1594 (1972).

Abstract—New radar observations yield a more stringent test of the predicted relativistic increase in echo times of radio signals sent from Earth and reflected from Mercury and Venus. These "extra" delays may be characterized by a parameter λ which is unity according to general relativity and 0.93 according to recent predictions based on a scalar tensor theory of gravitation. We find that $\lambda = 1.02$. The formal standard error is 0.02, but because of the possible presence of systematic errors we consider 0.05 to be a more reliable estimate of the uncertainty in the result.

General relativity predicts that the round-trip time delay of an electromagnetic wave is influenced by the gravitational potential along the path of the radiation. A test of this prediction involving the transmission of radar signals from Earth to either Mercury or Venus and the detection of the echoes was suggested in 1964[1] These echoes are expected on the basis of general relativity to be retarded by solar gravity by an amount[2]

$$\Delta t \simeq (4r_0/c)\ln[(r_e + r_p + R)/(r_e + r_p - R)], \quad (1)$$

where Δt, expressed in harmonic coordinates, is the coordinate-time retardation, $r_0 \simeq 1.5$ km is the gravitational radius of the sun, c is the speed of light far from the sun, r_e is the Earth-sun distance, r_p is the planet-sun distance, and R is the Earth-planet distance. The quantity Δt is not an observable but is indicative of the magnitude and behavior of the measurable effect as predicted by general relativity. The operational interpretation of the effect has been discussed in detail elsewhere.[3] To test whether or not the echo time-delay data are in agreement with this theory, we may insert an *ad hoc* multiplicative parameter λ on the right side of Eq. (1) and estimate it along with the other unknown parameters that affect the data.[4]

This experiment was first performed in 1967 and yielded the result[5] $\lambda = 0.09 \pm 0.2$ which corresponds to a value of 0.8 ± 0.4 for γ, the relevant coefficient in the generalized metric for the Schwarzschild solution.[6] Over the past three years a substantial body of consistent echo time-delay and (less important) Doppler data has accumulated from radar observations of Mercury, Venus, and Mars made at the Haystack Observatory and from observations of Mercury and Venus made at the Arecibo Observatory. (The inconsistencies in the data from these two observatories, noted previously,[5] have been resolved for Venus and will be the subject of a separate publication; small differences in the Mercury data are still under investigation.) The "crucial" measurements near superior conjunctions were obtained primarily at Haystack, but only for Mercury and Venus. The Arecibo measurements were of most use in the refinement of the orbits of Earth and Venus.[7] How do these additional data improve the estimate of λ? With 24 relevant parameters[8] estimated simultaneously in a weighted least-squares analysis, we obtained

$$\lambda = 1.01_5 \pm 0.02, \quad (2)$$

or, equivalently,

$$\gamma \simeq 1.03 \pm 0.04, \quad (3)$$

where 0.02 is the formal standard error in the estimate of λ. This error reflects a uniform reduction by a factor of 0.6 of the individual measurement errors; it was made so that the root mean square of the weighted post-fit residuals for the approximately 1,700 measurements would be unity. When the known topographical effects are taken into account (see below and figure 124.1), these residuals decrease appreciably, providing a further indication that the individual measurement errors originally assigned were in most cases quite conservative.

There are two main sources of systematic error that might affect this result: (i) the solar corona, and (ii) uncharted topography on the target planets. Since the accuracy of the echo-delay measurements is so much greater at inferior than at superior conjunctions, the necessity to estimate the planetary orbits *per se* introduces no important errors-either systematic or otherwise.

The solar corona and, more generally, the interplanetary plasma cause the group echo delays to be increased by an amount inversely proportional to the square of the radar signal frequency which is 430 MHz at Arecibo and 7,840 MHz at Haystack. Because of the former's relatively low radar frequency, the interplanetary medium affects the time-delay data noticeably. To take this plasma into account approximately, we assumed the interplanetary medium to be static with a charged-particle density varying as the inverse square of the distance from the sun. This law appears from other evidence to be an adequate representation for heliocentric distances between about 20 and 200 solar radii—a range encompassing the ray paths for all of the Arecibo radar measurements. The proportionality constant ρ, normalized to yield the electron density at the Earth's orbit, was one of the estimated parameters.[8] The result

$$\rho = 7 \pm 2 \text{ electrons/cm}^3 \quad (4)$$

represents an average condition over the period 1966–1970 spanned by the most precise Arecibo Venus data. Other more sensitive measurements yield average values between about 5 and 7 electrons/cm^3 in good agreement with our estimate which cannot be expected to be nearly so accurate in view of the limited magnitude of the effect—only about 2 μsec for Arecibo Venus observations near elongation—and the possible systematic errors that might have been introduced by planetary topography.

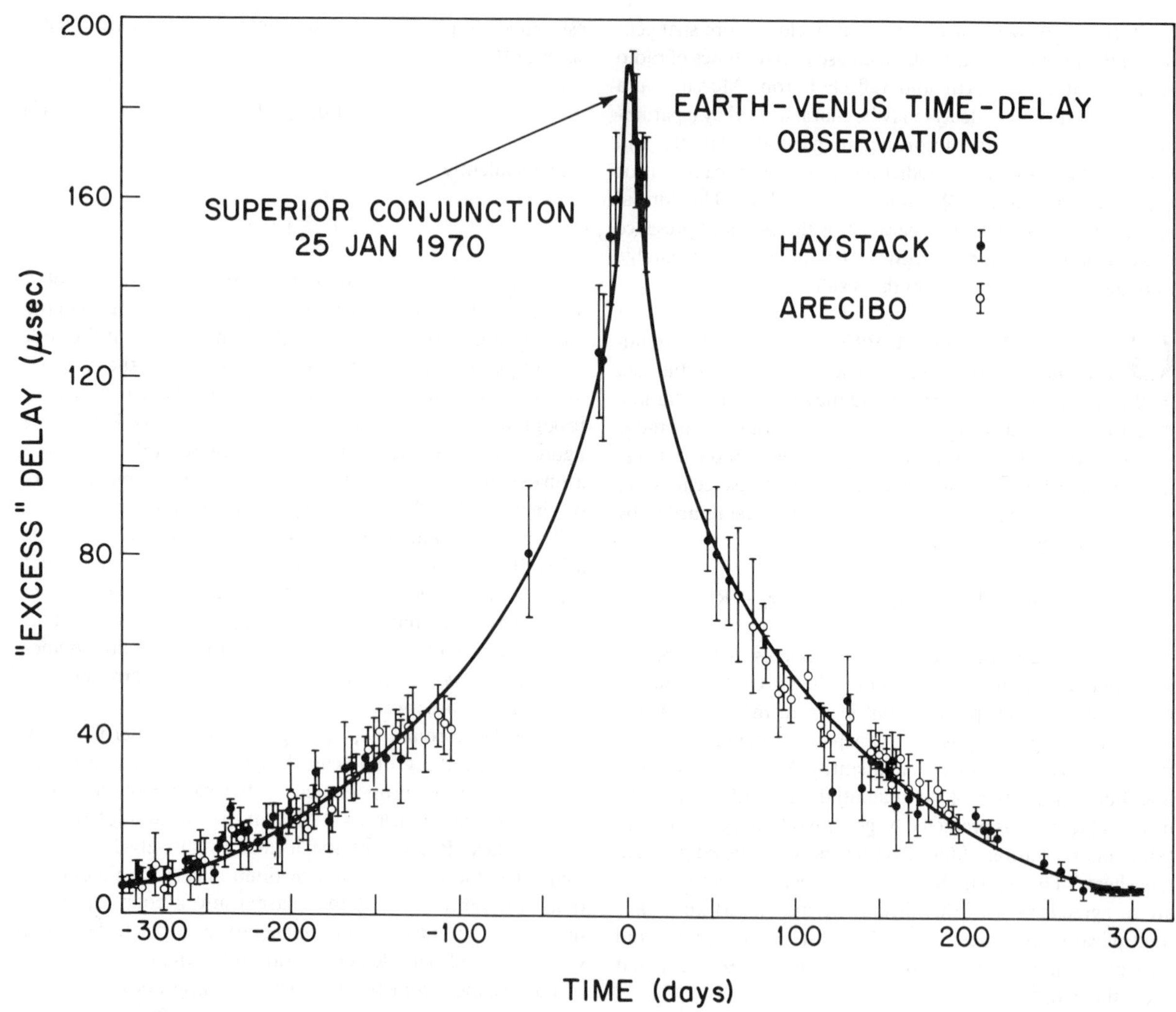

Fig. 124.1 Typical sample of post-fit residuals for Earth-Venus time-delay measurements, displayed relative to the "excess" delays predicted by general relativity. Here the delays are given in microseconds, or 10^{-6} sec, for different times before and after inferior conjunction, which is denoted by a zero on the horizontal axis. Corrections were made for known topographic trends on Venus. The bars represent the original estimates of the measurement standard errors. Note the dramatic increase in accuracy obtained with the radar-system improvements incorporated at Haystack just prior to the inferior conjunction of November 1970. (Courtesy Irwin Shapiro.)

Inside 20 solar radii, the coronal density increases more steeply with decreasing distance to the sun than in the inverse-square model. Nevertheless, for the Haystack measurements, the two-way coronal delay most likely never exceeded 3 μ sec.[9] Thus, even though the model underestimates delays near superior conjunction, the value of λ could thereby be increased spuriously by no more than a few percent.

Planetary topography affects the time delay in two ways: First, the altitude at, and in the neighborhood of, the sub-radar point on the target planet will directly influence the time delay; and, second, the radar scattering law,[10] which varies with aspect, will exert an indirect influence through the cross-correlation or "template-matching" technique that is used to estimate the delay to the subradar point.[5] In principle these effects can be determined once and for all and taken into account. There is no question of dynamics: The time scale for changes in surface height and in scattering law are undoubtedly long compared to the decade-length scale of these radar astronomy experiments. Although considerable progress has been made recently in charting surface-height and scattering-law variations[11] for the inner planets, uncertainties still remain that, for example, might contribute an error of up to 5 or 10 μ sec to the interpretation of some of the Venus echo delays. To investigate empirically the sensitivity of our result for λ to such uncertainties, we performed a number of computer experiments in which different sets of parameters were estimated and different subsets of data deleted. In some, the parameters included low-order terms in the sectoral harmonic expansions for the effective surface heights in the equatorial regions spanned by the subradar points. For more than a dozen such computer experiments, the variations in λ in all but three instances were smaller than the formal standard error. However, the full spectrum of topographic effects cannot be investigated economically by such studies, and we therefore cannot set accurate limits on the potential contribution of systematic errors. Our best judgement is that the contributions from these sources raise the uncertainty to an equivalent standard error of about 0.05 for the estimate of λ. A sample of the post-fit Earth-Venus delay residuals, displayed relative to the "excess" delays given by Eq. (1), is shown in figure 124.1 after correction for known topographic variations.[11] When the remaining topographic uncertainties are reduced sufficiently, the analysis can be repeated, and the resultant formal standard error may then be a reliable measure of the accuracy of the λ estimate.

How does this result for λ compare with the value predicted by the Brans-Dicke scalar-tensor theory of gravitation? Based primarily on his interpretation of the Princeton solar-oblateness experiment, Dicke[12] considers the most likely value of s, the fractional contribution of the scalar field, to be 0.07 which implies a value for λ of 0.93. This prediction appears to differ significantly from our determination.

Here Shapiro et al. argue that radar observations of Venus and Mercury with new bistatic radar facilities and, possibly, radio frequency communication with spacecraft will improve the accuracy of the time-delay test.

1. I. I. Shapiro, *Phys. Rev. Lett. 13*, 789 (1964).
2. See, for example, M. J. Tausner, Lincoln Laboratory Technical Report no. 425, 1966 (unpublished); D. B. Holdridge, Jet Propulsion Laboratory Space Program Summary no. 37–48, 1967 (unpublished), vol. 3.
3. I. I. Shapiro, *Phys. Rev. 141*, 1219 (1966), *145*, 1005 (1966).
4. For a description of the methods used see M. E. Ash, I. I. Shapiro, and W. B. Smith, *Astron. J. 72*, 338 (1967); and I. I. Shapiro, W. B. Smith, M. E. Ash, and S. Herrick, *Astron. J. 76*, 588 (1971).
5. I. I. Shapiro, G. H. Pettengill, M. E. Ash, M. L. Stone, W. B. Smith, R. P. Ingalls, and R. A. Brockelman, *Phys. Rev. Lett. 20*, 1265 (1968).
6. For a definition of γ see A. S. Eddington, *The Mathematical Theory of Relativity* (Cambridge: Cambridge University Press, 1922).
7. For this latter purpose, use was also made of earlier observations made with the Millstone Hill radar [see J. V. Evans, R. P. Ingalls, L. P. Rainville, and R. R. Silva, *Astron. J. 71*, 902 (1966)].
8. These parameters were (in addition to λ) the four "in-plane" orbital elements for each of the four inner planets, the mean equatorial radius of each of the three target planets, the speed of light in astronomical units, a plasma constant, the mass of Mercury, and the Earth-moon mass ratio. The theoretical model for planetary motion was based on the Schwarzschild metric, the Newtonian mutual perturbations of planetary orbits, and a zero value for asteroidal masses and for the solar gravitational quadrupole moment. The possible errors in these assumptions have a negligible effect on the accuracy of the estimate of λ.
9. This result is based on the integrated coronal electron densities found from pulse time-of-arrival measurements made daily at Arecibo during the June 1969 and 1970 occultations of the Crab pulsar (C. C. Counselman and J. M. Rankin, private communication).
10. See, for example, *Radar Astronomy*, ed. J. V. Evans and T. Hagfors (New York: McGraw-Hill, 1968).
11. W. B. Smith, R. P. Ingalls, I. I. Shapiro, and M. E. Ash, *Radio Sci. 5*, 411 (1970). Recent, far more precise results for Venus were used in figure 124.1 and are being prepared for publication.
12. R. H. Dicke, private communication.

125. On the Curvature of Space

Aleksandr Friedmann
TRANSLATED BY BRIAN DOYLE

(*Zeitschrift für Physik 10*, 377–386 [1922])

In deriving a cosmological model from his general theory of relativity, Einstein somewhat arbitrarily opted for a static universe. The mathematical consequence of this decision was a nonzero value for one of the constants of integration, the so-called cosmological constant, Λ. From a Newtonian analogue, Λ can be viewed as representing a repulsive force that increases with distance and that keeps the universe from collapsing under gravitational attraction.

In the following selection, the Russian mathematician Aleksandr Friedmann considers nonstatic models for the first time. By treating the spatial curvature of the universe as a function of time, he shows the possibility of nonstationary worlds with positive and negative curvature. These dynamic world models became especially important several years later when the universe of galaxies was found to be expanding. Although Friedmann's nonstatic cosmology was for some years overlooked by astronomers, Einstein noticed this paper and within a few months issued a one-paragraph critique in the same journal, only to retract his objection early in 1923.

Friedmann begins with the general idea that at any given instant of time the cosmological model represents a space of positive spatial curvature $R(t)$. If R is independent of time, then the stationary world models of Einstein and Wilhelm de Sitter follow. If $R(t)$ depends only on the time variable, then a variety of monotonically expanding or periodically oscillating models result, depending on the value chosen for Λ. Friedmann notes that with $\Lambda = 0$, there follows an oscillating model whose period depends on the total mass of the universe.

In a second paper[1] Friedmann considers models with negative curvature. He finds a nonstationary world with negative spatial curvature and positive matter density, but no static model. Friedmann also notes in this later contribution that Einstein's field equations do not suffice to extract a conclusion about the finiteness of space without some supplementary assumptions.

1. A. Friedmann, *Zeitschrift für Physik 21*, 326 (1924).

I

A. In their well-known works on general cosmological questions, Einstein[1] and de Sitter[2] arrive at two possible types of universe: Einstein obtains the so-called cylindrical world, in which space[3] possesses a constant curvature independent of time and in which the radius of curvature is connected with the total mass of matter existing in space. De Sitter obtains a spherical world in which not only space but also the world can be spoken of, in a certain sense, as a world of constant curvature.[4] In doing so, certain assumptions about the matter tensor are made by both Einstein and de Sitter; these correspond to the incoherence of matter and its being relatively at rest, e.g. the velocity of matter is assumed to be sufficiently small in comparison with the fundamental velocity,[5] the velocity of light.

The goal of this notice is, first, the derivation of the cylindrical and spherical worlds (as special cases) from some general assumptions and, second, the proof of the possibility of a world whose spatial curvature is constant with respect to three coordinates that are permissible spatial coordinates and that depend on time, e.g. on the fourth (time) coordinate. This new type is, as far as its remaining properties are concerned, an analogue of the Einsteinian cylindrical universe.

B. The assumptions on which we shall base our considerations break down into two classes. To the first class belong assumptions that coincide well with the assumptions of Einstein and de Sitter. They refer to the equations that the gravitational potentials satisfy and to the state and motion of matter. To the second class belong assumptions about the general, so-to-speak geometric, character of the world. From our hypothesis the cylindrical world of Einstein and the spherical world of de Sitter follow as special cases.

Class 1 The assumptions of the first class are the following:

1. The gravitational potentials satisfy the Einstein system of equations with the cosmological term, which we may also set equal to zero:

$$R_{ik} - \frac{1}{2} g_{ik}\bar{R} + \lambda g_{ik} = -\kappa T_{ik}\,(i, k = 1, 2, 3, 4). \qquad (1)$$

Here the g_{ik} are the gravitational potentials, T_{ik} is the matter tensor, κ is a constant, $\bar{R} = g^{ik}R_{ik}$, [the cosmological constant is denoted by λ] and R_{ik} is determined by the equations

$$R_{ik} = \frac{\partial^2 \log(\sqrt{g})}{\partial x_i \partial x_k} - \frac{\partial(\log\sqrt{g})}{\partial x_\sigma}\begin{Bmatrix} ik \\ \sigma \end{Bmatrix} - \frac{\partial}{\partial x_\sigma}\begin{Bmatrix} ik \\ \sigma \end{Bmatrix} + \begin{Bmatrix} i\alpha \\ \sigma \end{Bmatrix}\begin{Bmatrix} k\sigma \\ \alpha \end{Bmatrix}, \qquad (2)$$

where the x_i $(i = 1, 2, 3, 4)$ are world coordinates and $\begin{Bmatrix} ik \\ l \end{Bmatrix}$ are the Christoffel symbols of the second kind.[6]

2. The matter is incoherent and relatively at rest. Stated less strongly, the relative velocities of matter are vanishingly small in comparison with the velocity of light. In consequence of these assumptions, the matter tensor is given by the equations

$$T_{ik} = 0 \text{ for } i \text{ and } k \neq 4$$
$$T_{44} = c^2 \rho g_{44}. \qquad (3)$$

Here ρ is the density of matter and c is the fundamental velocity. Moreover, the world coordinates are divided into three spatial coordinates x_1, x_2, x_3 and the time coordinate x_4.

Class 2 The assumptions of the second class are the following:

1. After distribution of the three spatial coordinates x_1, x_2, x_3, we have a space of constant curvature, which, however, may depend on x_4, the time coordinate. The interval[7] ds, determined by $ds^2 = g_{ik}dx_i dx_k$, can be brought into the following form by the introduction of suitable spatial coordinates:

$$ds^2 = R^2(dx_1{}^2 + \sin^2 x_1\, dx_2{}^2 + \sin^2 x_1 \sin^2 x_2\, dx_3{}^2) + 2g_{14}dx_1 dx_4 + 2g_{24}dx_2 dx_4 + 2g_{34}dx_3 dx_4 + g_{44}dx_4{}^2. \qquad (4)$$

Here R depends only on x_4 and it is proportional to the radius of curvature of space, which may therefore change with time.

2. In the expression for ds^2, the g_{14}, g_{24}, g_{34} can be made to vanish by a suitable choice of the time coordinate. In brief, time is orthogonal to space. It seems to me that no physical or philosophical grounds can be given for the second assumption. It serves exclusively to simplify the calculation. One must still notice that the worlds of Einstein and de Sitter are contained in our assumptions as special cases.

In consequence of assumptions 1 and 2, ds^2 can be brought into the form

$$ds^2 = R^2(dx_1{}^2 + \sin^2 x_1\, dx_2{}^2 + \sin^2 x_1 \sin^2 x_2\, dx_3{}^2) + M^2\, dx_4{}^2, \qquad (5)$$

where R is a function of x_4 and M depends, in the general case, on all four world coordinates. The Einstein universe is obtained if one replaces R^2 by $-R^2/c^2$ in equation (5) and if one also sets $M = 1$, whereby R signifies the constant (independent of x_4) radius of curvature of space.

$$d\tau^2 = -\frac{R^2}{c^2}(dx_1{}^2 + \sin^2 x_1\, dx_2{}^2 + \sin^2 x_1 \sin^2 x_2\, dx_3{}^2) + dx_4{}^2 \qquad (6)$$

The universe of de Sitter is obtained if one replaces R^2 by $-R^2/c^2$ and M by $\cos x_1$ in equation[8] (5)

$$d\tau^2 = -\frac{R^2}{c^2}(dx_1{}^2 + \sin^2 x_1\, dx_2{}^2 + \sin^2 x_1 \sin^2 x_2\, dx_3{}^2) + \cos^2 x_1\, dx_4{}^2 \tag{7}$$

C. Now we must still strike an agreement about the boundaries within which the world coordinates are confined, e.g. what points of the 4-dimensional manifold we will treat as different. Without engaging in a more detailed motivation, we shall assume that the spatial coordinates are confined to the following intervals: x_1 in the interval $(0, \pi)$, x_2 in the interval $(0, \pi)$, and x_3 in the interval $(0, 2\pi)$. With respect to the time coordinate we make, for the present, no restricting assumptions, but we shall consider this question further below.

II

A. From equations (3) and (5) it follows, if one sets $i = 1$, 2, 3 and $k = 4$ in equation (1), that

$$R'(x_4)\frac{\partial M}{\partial x_1} = R'(x_4)\frac{\partial M}{\partial x_2} = R'(x_4)\frac{\partial M}{\partial x_3} = 0.$$

Two cases arise. (1) $R'(x_4) = 0$, R is independent of x_4. We shall designate this world as a *stationary* world. (2) $R'(x_4) \neq 0$, M depends only on x_4. This shall be called a *nonstationary* world.

We consider, first, the stationary world and write the equations (1) for $i, k = 1, 2, 3$ and moreover $i \neq k$. Then we obtain the following system of formulae:

$$\frac{\partial^2 M}{\partial x_1\, \partial x_2} - \operatorname{cotg} x_1 \frac{\partial M}{\partial x_2} = 0$$

$$\frac{\partial^2 M}{\partial x_1\, \partial x_3} - \operatorname{cotg} x_1 \frac{\partial M}{\partial x_3} = 0$$

$$\frac{\partial^2 M}{\partial x_2\, \partial x_3} - \operatorname{cotg} x_2 \frac{\partial M}{\partial x_3} = 0.$$

The integration of these equations yields the following expression for M:

$$M = A(x_3, x_4)\sin x_1 \sin x_2 + B(x_2, x_4)\sin x_1 + C(x_1, x_4), \tag{8}$$

where A, B, C are arbitrary functions of their arguments. If we solve the equations (1) for R_{ik} and eliminate the unknown density[9] ρ from the still-unused equations, we obtain, if we insert for M equation (8), the following two possibilities for M after some long, but elementary calculations:

$$M = M_0 = \text{const.} \tag{9}$$

$$M = (A_0 x_4 + B_0)\cos x_1, \tag{10}$$

where M_0, A_0 and B_0 are constants.

If M is equal to a constant, then the stationary world is the cylindrical world. Here it is advantageous to work with the gravitational potentials of equation (6). If we determine the density and the quantity λ, then the well-known result of Einstein is obtained:

$$\lambda = \frac{c^2}{R^2}, \qquad \rho = \frac{2}{\kappa R^2}, \qquad \bar{M} = \frac{4\pi^2}{\kappa}R,$$

where $\bar{M}$ denotes the total mass of space.[8]

In the second possible case, when $\bar{M}$ is given by equation (10), we get, by means of a judicious transformation[10] of x_4, the spherical world of de Sitter in which $M = \cos x_1$. With the help of equation (7) we obtain the relations of de Sitter:

$$\lambda = 3c^2/R^2, \qquad \rho = 0, \qquad \bar{M} = 0.$$

We thus have the following result: *the stationary world is either the Einstein cylindrical world or the de Sitter spherical world.*

B. We now want to consider the nonstationary world. M is now a function of x_4. By an appropriate choice of x_4 one can obtain $M = 1$, without loss of generality. In order to couple to our customary presentation, we give ds^2 a form that is analogous to equations (6) and (7):

$$d\tau^2 = \frac{-R^2(x_4)}{c^2}(dx_1{}^2 + \sin^2 x_1\, dx_2{}^2 + \sin^2 x_1 \sin^2 x_2\, dx_3{}^2) + dx_4{}^2. \tag{11}$$

Our task is now the determination of R and ρ from the equations (1). It is clear that the equations (1) with different indices yield nothing. The equations (1) for $i = k = 1, 2, 3$ give the relation

$$\frac{R'^2}{R^2} + \frac{2RR''}{R^2} + \frac{c^2}{R^2} - \lambda = 0. \tag{12}$$

The equations (1) with $i = k = 4$ yield the relation

$$\frac{3R'^2}{R^2} + \frac{3c^2}{R^2} - \lambda = \kappa c^2 \rho \tag{13}$$

with

$$R' = \frac{dR}{dx_4} \quad \text{and} \quad R'' = \frac{d^2R}{dx_4{}^2}.$$

Because $R' \neq 0$, the integration of equation (12), if we write t for x_4, gives the following equation:

$$\frac{1}{c^2}\left(\frac{dR}{dt}\right)^2 = \frac{A - R + \frac{\lambda}{3c^2}R^3}{R}, \tag{14}$$

where A is an arbitrary constant. From this equation, we obtain R through the inversion of an elliptic integral, e.g. through the solution for R of the equation

$$t = \frac{1}{c}\int_a^R \sqrt{\frac{x}{A - x + \frac{\lambda}{3c^2}x^3}}\,dx + B \tag{15}$$

in which B and a are constants. Attention must still be paid to the usual conditions about sign variation in the square root. The mass density, ρ, may be determined from equation (13):

$$\rho = \frac{3A}{\kappa R^3}. \tag{16}$$

The constant A is expressed in terms of the total mass of space $\bar{M}$ in the following way:

$$A = \frac{\kappa \bar{M}}{6\pi^2}. \tag{17}$$

If $\bar{M}$ is positive, then A is also positive.

c. We must base the consideration of the nonstationary world on equations (14) and (15). The quantity λ is not determined by these equations. We shall postulate that it can have an arbitrary value. We now determine that value of the variable x, for which the square root of equation (15) changes its sign. If we restrict our consideration to positive radii of curvature, it will suffice to consider the interval $(0, \infty)$ for x and in this interval the values of x that make the radicand equal to zero or ∞. One value of x for which the square root in equation (15) equals 0 is $x = 0$. The remaining values of x, for which the square root in equation (15) changes sign, are given by the positive roots of the equation $A - x + (\lambda/3c^2)x^3 = 0$. We denote $\lambda/3c^2$ by y and consider the system of third degree curves in the x–y plane:

$$yx^3 - x + A = 0. \tag{18}$$

Here A is the parameter of the curve, which varies over the interval $(0, \infty)$. A curve of the system cuts the x-axis at the point $x = A$, $y = 0$ and has a maximum at the point

$$x = \frac{3A}{2}, \qquad y = \frac{4}{27A^2}.$$

From figure 125.1 it is obvious that the equation $A - x + (\lambda/3c^2)x^3 = 0$ has a positive root x_0 in the interval $(0, A)$ for negative λ. If one considers x_0 as a function of λ and A, then

$$x_0 = \Theta(\lambda, A),$$

one finds that Θ is an increasing function of λ and of A. If λ is in the interval $[0, 4c^2/(9A^2)]$, the equation has two positive roots $x_0 = \Theta(\lambda, A)$ and $x_0' = \Phi(\lambda, A)$, where x_0 is the root in the interval $(A, 3A/2)$ and x_0' is in the interval $(3A/2, \infty)$. $\Theta(\lambda, A)$ is an increasing function of A and λ, whereas $\Phi(\lambda, A)$ is a decreasing function of A and λ. Finally, if λ is bigger than $4c^2/(9A^2)$, then the equation has no positive roots.

Let us now pass on to a discussion of equation (15) taking into consideration the following remark: Let the radius of curvature equal R_0 for $t = t_0$. The sign of the square root in equation (15) is, for $t = t_0$, positive or negative depending on whether the radius of curvature is increasing or decreasing for $t = t_0$. Since we can replace t by $-t$, if need be, we can always make the square root positive, e.g. by choice of the

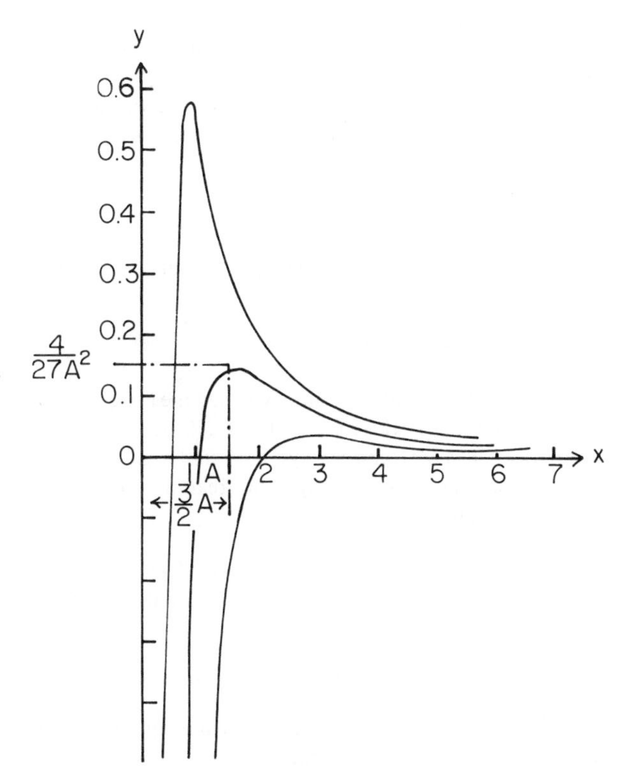

Fig. 125.1 A plot in the $x - y$ plane of the curves satisfying the non-stationary world equation $yx^3 - x + A = 0$ where $y = \lambda/3c^2$ is the reciprocal of the radius R and A is a constant which is proportional to the total mass of the universe.

time it can always be arranged such that the radius of curvature increases with increasing time at $t = t_0$.

D. We consider first the case $\lambda > 4c^2/(9A^2)$, e.g. the case in which the equation $A - x + (\lambda/3c^2)x^3 = 0$ has no positive roots. Equation (15) can then be written

$$t - t_0 = \frac{1}{c}\int_{R_0}^{R}\sqrt{\frac{x}{A - x + \frac{\lambda}{3c^2}x^3}}\,dx, \qquad (19)$$

where, in consequence of our remark, the square root is always positive. From that, it follows that R is an increasing function of t. The positive initial value R_0 is free of any restriction.

Since the radius of curvature may not be smaller than zero, it must decrease with decreasing time, t, from R_0 to the value zero at time t'. We shall call the growth time of R from 0 to R_0 the time since the creation of the world.[11] This time, t', is given by

$$t' = \frac{1}{c}\int_{0}^{R_0}\sqrt{\frac{x}{A - x + \frac{\lambda}{3c^2}x^3}}\,dx \qquad (20)$$

We denote the world under consideration as a *monotonic world of the first kind.*

The time since the creation of the monotonic world of the first kind, considered as a function of R_0, A, λ, has the following properties:

1) It increases with increasing R_0.

2) It decreases if A increases, e.g. if the mass in space is increased.

3) It decreases if λ increases.

If $A > 2R_0/3$, then for an arbitrary λ the time elapsed since the creation of the world is finite. If $A \leqslant 2R_0/3$, then a value of $\lambda = \lambda_1 = 4c^2/(9A^2)$ can always be found that as λ approaches this value, the time since the creation of the world increases without limit.

E. Now let λ lie in the interval $[0, 4c^2/(9A^2)]$; then the initial value of the radius of curvature can lie in one of the intervals

$$(0, x_0), \quad (x_0, x_0'), \quad \text{or} \quad (x_0', \infty).$$

If R_0 falls in the interval (x_0, x_0'), then the square root in equation (15) is imaginary. A space with this initial curvature is impossible.

We devote the next section to the case where R_0 lies in the interval $(0, x_0)$. Here we consider the third case: $R_0 > x_0'$ or $R_0 > \Phi(\lambda, A)$. Through considerations that are analogous to the preceding ones, it can be shown that R is an increasing function of time, whereby R can begin with the value $x_0' = \Phi(\lambda, A)$. The time that has elapsed from the moment when $R = x_0'$ to the moment that corresponds to $R = R_0$, we again call the time since the creation of the world. Let it be t'; then

$$t' = \frac{1}{c}\int_{x_0'}^{R_0}\sqrt{\frac{x}{A - x + \frac{\lambda}{3c^2}x^3}}\,dx. \qquad (21)$$

We call this world a *monotonic world of the second kind.*

F. We now consider the case that λ falls between the limits $(-\infty, 0)$. If $R_0 > x_0 = \Theta(\lambda, A)$, the square root in equation (15) becomes imaginary, and the space with this R_0 is impossible. If $R_0 < x_0$, the considered case is identical with that which we have left aside in the preceding sections. We therefore assume that λ lies in the interval $[-\infty, 4c^2/(9A^2)]$ and $R_0 < x_0$. By means of a well-known argument[12] one can now show that R becomes a periodic function of t with the period t_π, which we name the *world period*; t_π is given by the formula

$$t_\pi = \frac{2}{c}\int_{0}^{x_0}\sqrt{\frac{x}{A - x + \frac{\lambda}{3c^2}x^3}}\,dx. \qquad (22)$$

The radius of curvature varies between 0 and x_0. We shall call this universe the *periodic world.* The period of the periodic world increases if we increase λ and tends to infinity if λ tends to the value $\lambda_1 = 4c^2/(9A^2)$.

For small λ, the period is represented by the approximate formula

$$t_\pi = \pi A/c. \qquad (23)$$

With reference to the periodic world two points of view are possible: We count two events as coincident if their spatial coordinates coincide and the difference of time coordinate is an integral multiple of the period, so that the radius of curvature grows from 0 to x_0 and thereafter decreases to the value 0. The time of world existence is finite. On the other hand, if the time varies between $-\infty$ and $+\infty$ (e.g. if we consider two events as coincident only when not only their spatial but also their world coordinates coincide), we come to a real periodicity of the space curvature.

G. Our information is completely insufficient to carry out numerical calculations and to distinguish which world our universe is. It is possible that the causality problem and the problem of centrifugal force will illuminate these questions. It remains to note that the "cosmological" magnitude λ remains undetermined in our formula, because it is a superfluous constant in the problem. Possibly electrodynamic considerations can lead to its evaluation. It we set $\lambda = 0$ and

$M = 5 \times 10^{21}$ solar masses, the world period becomes of the order 10 billion years. However, these numbers are valid only as an illustration of our calculation.

1. Einstein, "Kosmologische Betrachtungen zur allegemeinen Relativitätstheorie," *Sitzungsberichte Berl. Akad. 1*, 142 (1917).

2. De Sitter, "On Einstein's theory of gravitation and its astronomical consequences," *Monthly Notices of the R. Astronom. Soc. 78*, 3 (1917).

3. By "space" we understand here a space that is described by a manifold of 3 dimensions; the "world" corresponds to a manifold of 4 dimensions.

4. Klein, "Über die Integralform der Erhaltungssätze und die Theorie der räumlich-geschlossenen Welt," *Götting. Nachr.* (1918).

5. See this name in Eddington's book *Espace, Temps et Gravitation*, 2 Partie (Paris: Hermann, 1921), p. 10.

6. The sign of R_{ik} and $\bar{R}$ differs here from the usual convention.

7. See, for example, Eddington, *Espace, Temps et Gravitation*, 2 Partie (Paris: Hermann, 1921).

8. The *ds*, which is taken to have the dimension of time, we designate $d\tau$; then the constant κ has the dimension Length/Mass and in c.g.s. units equals 1.87×10^{-27}. See Laue, *Die Relativitätstheorie*, vol. 2 (Braunschweig, 1921), p. 185.

9. The density ρ is for us an unknown function of the world coordinates x_1, x_2, x_3, and x_4.

10. This transformation is given by the formula

$$d\bar{x}_4 = \sqrt{A_0 x_4 + B_0}\, dx_4.$$

11. The time since the creation of the Universe is the time that has elapsed from the moment when space was a point ($R = 0$) to the present state ($R = R_0$): this term may also be infinite.

12. See, for example, Weierstrass, "Über eine Gattung reell periodischer Funktionen," *Monatsber. d. Königl. Akad. d. Wissensch.* (1866), and Horn, "Zur Theorie der kleinen endlichen Schwingungen," *ZS. f. Math. und Physik 47*, 400 (1902). In our case the considerations of these authors have to be altered appropriately. However, the periodicity can be established in our case by elementary considerations.

126. A Homogeneous Universe of Constant Mass and Increasing Radius accounting for the Radial Velocity of Extra-Galactic Nebulae

Georges Lemaître

(*Annales de la Société scientifique de Bruxelles A47*, 49 [1927]; *translation in Monthly Notices of the Royal Astronomical Society 91*, 483–490 [1931])

Around 1930 A. S. Eddington suggested that the apparent stalemate in cosmology arose because only static models had been considered. The Belgium astronomer Georges Lemaître promptly pointed out that he had worked out an expanding model, and consequently Eddington sponsored the English translation of his paper in the *Monthly Notices*. Neither Lemaître nor the British astronomers were aware of Aleksandr Friedmann's earlier work, and as a result Lemaître's solution was widely heralded for its novelty.

Lemaître here reviews the dilemma between the de Sitter and Einstein world models. The de Sitter model ignores the existence of matter, and it lacks symmetry by separately partitioning space and time. However, his model explains the observed recession velocities of the spiral nebulae as a simple consequence of the gravitational field. Einstein's solution allows for the presence of matter and leads to a relation between matter density and the radius of the universe. It cannot, however, explain the observed recession of the galaxies. Lemaître sought a solution of the Einstein field equations intermediate between the Einstein and de Sitter models: a solution having both material content and an explanation for the recession of the nebulae.

Lemaître assumed that the radius of curvature of the universe, R, is a function of time, t. Like Friedmann, he found that $R(t)$ increases without limit with increasing time, and he obtained nearly identical differential equations for $R(t)$. The major difference was that Lemaître assumed the conservation of energy and included radiation pressure as well as a term for matter density. By including radiation pressure, Lemaître was able to show its importance in the early stages of the expansion of the universe. In fact, Lemaître here first introduces the idea that the recessional velocities of the extragalactic nebulae are the cosmical consequence of the expansion of the universe. After he showed that Einstein's model was unstable, Eddington then proposed the static Einstein world model as the stage from which Lemaître's dynamical model began. Lemaître himself always preferred to think of the universe expanding from an initial singularity, the "primeval atom."

1. Introduction

According to the theory of relativity, a homogeneous universe may exist such that all positions in space are completely equivalent; there is no centre of gravity. The radius of space R is constant; space is elliptic, i.e. of uniform positive curvature $1/R^2$; straight lines starting from a point come back to their origin after having travelled a path of length πR; the volume of space has a finite value $\pi^2 R^3$; straight lines are closed lines going through the whole space without encountering any boundary.

Two solutions have been proposed. That of de Sitter ignores the existence of matter and supposes its density equal to zero. It leads to special difficulties of interpretation which will be referred to later, but it is of extreme interest as explaining quite naturally the observed receding velocities of extra-galactic nebulae, as a simple consequence of the properties of the gravitational field without having to suppose that we are at a point of the universe distinguished by special properties.

The other solution is that of Einstein. It pays attention to the evident fact that the density of matter is not zero, and it leads to a relation between this density and the radius of the universe. This relation forecasted the existence of masses enormously greater than any known at the time. These have since been discovered, the distances and dimensions of extra-galactic nebulae having become known. From Einstein's formulae and recent observational data, the radius of the universe is found to be some hundred times greater than the most distant objects which can be photographed by our telescopes.

Each theory has its own advantages. One is in agreement with the observed radial velocities of nebulae, the other with the existence of matter, giving a satisfactory relation between the radius and the mass of the universe. It seems desirable to find an intermediate solution which could combine the advantages of both.

At first sight, such an intermediate solution does not appear to exist. A static gravitational field for a uniform distribution of matter without internal stress has only two solutions, that of Einstein and that of de Sitter. De Sitter's universe is empty, that of Einstein has been described as "containing as much matter as it can contain." It is remarkable that the theory can provide no mean between these two extremes.

The solution of the paradox is that de Sitter's solution does not really meet all the requirements of the problem. Space is homogeneous with constant positive curvature; space-time is also homogeneous, for all events are perfectly equivalent. But the partition of space-time into space and time disturbs the homogeneity. The co-ordinates used introduce a centre. A particle at rest at the centre of space describes a geodesic of the universe; a particle at rest otherwhere than at the centre does not describe a geodesic. The co-ordinates chosen destroy the homogeneity and produce the paradoxical results which appear at the so-called "horizon" of the centre. When we use co-ordinates and a corresponding partition of space and time of such a kind as to preserve the homogeneity of the universe, the field is found to be no longer static; the universe becomes of the same form as that of Einstein, with a radius no longer constant but varying with the time according to a particular law.

In order to find a solution combining the advantages of those of Einstein and de Sitter, we are led to consider an Einstein universe where the radius of space or of the universe is allowed to vary in an arbitrary way.

2. Einstein Universe of Variable Radius. Field Equations. Conservation of Energy.

As in Einstein's solution, we liken the universe to a rarefied gas whose molecules are the extra-galactic nebulae. We suppose them so numerous that a volume small in comparison with the universe as a whole contains enough nebulae to allow us to speak of the density of matter. We ignore the possible influence of local condensations. Furthermore, we suppose that the nebulae are uniformly distributed so that the density does not depend on position. When the radius of the universe varies in an arbitrary way, the density, uniform in space, varies with time. Furthermore, there are generally interior stresses, which, in order to preserve the homogeneity, must reduce to a simple pressure, uniform in space and variable with time. The pressure, being two-thirds of the kinetic energy of the "molecules," is negligible with respect to the energy associated with matter; the same can be said of interior stresses in nebulae or in stars belonging to them. We are thus led to put $p = 0$.

Nevertheless it might be necessary to take into account the radiation-pressure of electromagnetic energy travelling through space; this energy is weak but it is evenly distributed through the whole of space and might afford a notable contribution to the mean energy. We shall thus keep the pressure p in the general equations as the mean radiation-pressure of light, but we shall write $p = 0$ when we discuss the application to astronomy.

We denote the density of total energy by ρ, the density of radiation energy by $3p$, and the density of the energy condensed in matter by $\delta = \rho - 3p$. We identify ρ and $-p$ with the components T_4^4 and $T_1^1 = T_2^2 = T_3^3$ of the material energy tensor, and δ with T. Working out the contracted Riemann tensor for a universe with a line-element given by

$$ds^2 = -R^2 d\sigma^2 + dt^2, \tag{1}$$

where $d\sigma$ is the elementary distance in a space of radius unity, and R is a function of the time t, we find that the field equations can be written

$$3\frac{R'^2}{R^2} + \frac{3}{R^2} = \lambda + \kappa\rho \tag{2}$$

and

$$2\frac{R''}{R} + \frac{R'^2}{R^2} + \frac{1}{R^2} = \lambda - \kappa p. \quad (3)$$

Accents denote derivatives with respect to t. λ is the unknown cosmological constant, and κ is the Einstein constant whose value is $1.87 \cdot 10^{-27}$ in C.G.S. units (8π in natural units).

The four identities giving the expression of the conservation of momentum and of energy reduce to

$$\frac{d\rho}{dt} + \frac{3R'}{R}(\rho + p) = 0, \quad (4)$$

which is the energy equation. This equation can replace (3). As $V = \pi^2 R^3$ it can be written

$$d(V\rho) + pdV = 0, \quad (5)$$

showing that *the variation of total energy plus the work done by radiation-pressure in the dilatation of the universe is equal to zero.*

3. Universe of Constant Mass

If $M = V\delta$ remains constant, we write, α being a constant,

$$\kappa\delta = \frac{\alpha}{R^3}. \quad (6)$$

As

$$\rho = \delta + 3p$$

we have

$$3d(pR^3) + 3pR^2 dR = 0 \quad (7)$$

and, β being a constant of integration,

$$\kappa p = \frac{\beta}{R^4}, \quad (8)$$

and therefore

$$\kappa\rho = \frac{\alpha}{R^3} + \frac{3\beta}{R^4}. \quad (9)$$

By substitution in (2) we have

$$\frac{R'^2}{R^2} = \frac{\lambda}{3} - \frac{1}{R^2} + \frac{\kappa\rho}{3} = \frac{\lambda}{3} - \frac{1}{R^2} + \frac{\alpha}{3R^3} + \frac{\beta}{R^4}, \quad (10)$$

and

$$t = \int \frac{dR}{\sqrt{\frac{\lambda R^2}{3} - 1 + \frac{\alpha}{3R} + \frac{\beta}{R^2}}}. \quad (11)$$

When α and β vanish, we obtain the de Sitter solution in Lanczos's form

$$R = \sqrt{\frac{3}{\lambda}}\cosh\sqrt{\frac{\lambda}{3}}(t - t_0). \quad (12)$$

The Einstein solution is found by making $\beta = 0$ and R constant. Writing $R' = R'' = 0$ in (2) and (3) we find

$$\frac{1}{R^2} = \lambda, \qquad \frac{3}{R^2} = \lambda + \kappa\rho, \qquad \rho = \delta$$

or

$$R = \frac{1}{\sqrt{\lambda}}, \qquad \kappa\delta = \frac{2}{R^2} \quad (13)$$

and from (6)

$$\alpha = \kappa\delta R^3 = \frac{2}{\sqrt{\lambda}}. \quad (14)$$

The Einstein solution does not result from (14) alone; it also supposes that the initial value of R' is zero. If we write

$$\lambda = \frac{1}{R_0^2}, \quad (15)$$

we have for $\beta = 0$ and $\alpha = 2R_0$

$$t = R_0\sqrt{3}\int \frac{dR}{R - R_0}\sqrt{\frac{R}{R + 2R_0}}. \quad (16)$$

For this solution the two equations (13) are of course no longer valid.

Writing

$$\kappa\delta = \frac{2}{R_E^2}, \quad (17)$$

we have from (14) and (15)

$$R^3 = R_E^2 R_0. \quad (18)$$

The value of R_E, the radius of the universe computed from the mean density by Einstein's equation (17), has been found by Hubble to be

$$R_E = 8.5 \times 10^{28} \text{ cm.} = 2.7 \times 10^{10} \text{ parsec.} \tag{19}$$

We shall see later that the value of R_0 can be computed from the radial velocities of the nebulae; R can then be found from (18).

Finally, we shall show that a serious departure from (14) would lead to consequences not easily acceptable.

4. Doppler Effect due to the Variation of the Radius of the Universe

From (1) we have for a ray of light

$$\sigma_2 - \sigma_1 = \int_{t_1}^{t_2} \frac{dt}{R}, \tag{20}$$

where σ_1 and σ_2 relate to spatial co-ordinates. We suppose that the light is emitted at the point σ_1 and observed at σ_2. A ray of light emitted slightly later starts from σ_1 at time $t_1 + \delta t_1$ and reaches σ_2 at time $t_2 + \delta t_2$. We have therefore

$$\frac{\delta t_2}{R_2} - \frac{\delta t_1}{R_1} = 0, \qquad \frac{\delta t_2}{\delta t_1} - 1 = \frac{R_2}{R_1} - 1, \tag{21}$$

where R_1 and R_2 are the values of the radius R at the time of emission t_1 and at the time of observation t_2. If δt_1 is the period of the emitted light, δt_2 is the period of the observed light. Now δt_1 is also the period of light emitted under the same conditions in the neighbourhood of the observer, because the period of light emitted under the same physical conditions has the same value everywhere when reckoned in proper time. Therefore

$$\frac{v}{c} = \frac{\delta t_2}{\delta t_1} - 1 = \frac{R_2}{R_1} - 1 \tag{22}$$

is the apparent Doppler effect due to the variation of the radius of the universe. *It equals the ratio of the radii of the universe at the instants of observation and emission, diminished by unity.*

v is that velocity of the observer which would produce the same effect. When the light source is near enough, we have the approximate formulae

$$\frac{v}{c} = \frac{R_2 - R_1}{R_1} = \frac{dR}{R} = \frac{R'}{R}dt = \frac{R'}{R}r,$$

where r is the distance of the source. We have therefore

$$\frac{R'}{R} = \frac{v}{cr}. \tag{23}$$

From a discussion of available data, we adopt

$$\frac{R'}{R} = 0.68 \times 10^{-27} \text{ cm.}^{-1} \tag{24}$$

and find from (16)

$$\frac{R'}{R} = \frac{1}{R_0\sqrt{3}}\sqrt{1 - 3y^2 + 2y^3} \tag{25}$$

where

$$y = \frac{R_0}{R}. \tag{26}$$

Now from (18) and (26)

$$R_0{}^2 = R_E{}^2 y^3, \tag{27}$$

and therefore

$$3\left(\frac{R'}{R}\right)^2 R_E{}^2 = \frac{1 - 3y^2 + 2y^3}{y^3}. \tag{28}$$

With the adopted numerical data (24) and (19), we have

$$y = 0.0465$$

giving

$$\begin{aligned} R &= R_E\sqrt{y} = 0.215 R_E \\ &= 1.83 \times 10^{28} \text{ cm.} = 6 \times 10^9 \text{ parsecs.} \\ R_0 &= Ry = R_E y^{3/2} \\ &= 8.5 \times 10^{26} \text{ cm.} = 2.7 \times 10^8 \text{ parsecs.} \\ &= 9 \times 10^8 \text{ light-years.} \end{aligned}$$

Integral (16) can easily be computed. Writing

$$x^2 = \frac{R}{R + 2R_0}, \tag{29}$$

it can be written

$$\begin{aligned} t &= R_0\sqrt{3}\int \frac{4x^2 dx}{(1 - x^2)(3x^2 - 1)} \\ &= R_0\sqrt{3}\log\frac{1 + x}{1 - x} + R_0 \log\frac{\sqrt{3}x - 1}{\sqrt{3}x + 1} + C. \end{aligned} \tag{30}$$

If σ is the fraction of the radius of the universe travelled by light during time t, we have also

$$\sigma = \int \frac{dt}{R} = \sqrt{3}\int \frac{2dx}{3x^2 - 1} = \log\frac{\sqrt{3}x - 1}{\sqrt{3}x + 1} + C'. \tag{31}$$

The following table gives values of σ and t for different values of R/R_0:

Table 126.1 Values of σ and t

$\frac{R}{R_0}$	$\frac{t}{R_0}$	σ Radians	σ Degrees	$\frac{v}{c}$
1	$-\infty$	$-\infty$	$-\infty$	19
2	-4.31	-0.889	$-51°$	9
3	-3.42	-0.521	-30	$5\frac{2}{3}$
4	-2.86	-0.359	-21	4
5	-2.45	-0.266	-15	3
10	-1.21	-0.087	-5	1
15	-0.50	-0.029	-1.7	$\frac{1}{3}$
20	0.00	0.000	0.0	0
25	0.39	0.017	1	. .
∞	∞	0.087	5	. .

The constants of integration are adjusted to make σ and t vanish for $R/R_0 = 20$ in place of 21.5. The last column gives the Doppler effect computed from (22). The approximate formula (23) would make v/c proportional to r and thus to σ. The error is only 0.005 for $v/c = 1$. The approximate formula may therefore be used within the limits of the visible spectrum.

5. THE MEANING OF EQUATION (14)

The relation (14) between the two constants λ and α has been adopted following Einstein's solution. It is the necessary condition that the quartic under the radical in (11) may have a double root R_0 giving on integration a logarithmic term. For simple roots, integration would give a square root, corresponding to a minimum of R as in de Sitter's solution (12). This minimum would generally occur at time of the order of R_0, say 10^9 years—i.e. quite recently for stellar evolution. If the positive roots were to become imaginary, the radius would vary from zero upwards, the variation slowing down in the neighbourhood of the modulus of the imaginary roots. In both cases the time of variation of R in the same sense would be of the order of R_0 if the relation between λ and α were seriously different from (14).

6. CONCLUSION

We have found a solution such that

(1) The mass of the universe is a constant related to the cosmological constant by Einstein's relation

$$\sqrt{\lambda} = \frac{2\pi^2}{\kappa M} = \frac{1}{R_0}.$$

(2) The radius of the universe increases without limit from an asymptotic value R_0 for $t = -\infty$.

(3) The receding velocities of extragalactic nebulae are a cosmical effect of the expansion of the universe. The initial radius R_0 can be computed by formulae (24) and (25) or by the approximate formula

$$R_0 = \frac{rc}{v\sqrt{3}}.$$

This solution combines the advantages of the Einstein and de Sitter solutions.

Note that the largest part of the universe is forever out of our reach. The range of the 100-inch Mount Wilson telescope is estimated by Hubble to be 5×10^7 parsecs, or about R/200. The corresponding Doppler effect is 3,000 km/sec. For a distance of 0.087R it is equal to unity, and the whole visible spectrum is displaced into the infra-red. It is impossible to see ghost-images of nebulae or suns, as even if there were no absorption these images would be displaced by several octaves into the infra-red and would not be observed.

It remains to find the cause of the expansion of the universe. We have seen that the pressure of radiation does work during the expansion. This seems to suggest that the expansion has been set up by the radiation itself. In a static universe light emitted by matter travels round space, comes back to its starting-point, and accumulates indefinitely. It seems that this may be the origin of the velocity of expansion R'/R which Einstein assumed to be zero and which in our interpretation is observed as the radial velocity of extragalactic nebulae.

1. For the different partitions of space and time in the de Sitter universe, see K. Lanczos, *Phys. Zeits. 23*, 539 (1922); H. Weyl, *Phys. Zeits. 24*, 230 (1923); P. du Val, *Phil. Mag.* 6, *47*, 930 (1924); G. Lemaître, *Journal of Math. and Phys. 4*, No. 3, (1925).

2. Equations of the universe of variable radius and constant mass have been fully discussed, without reference to the receding velocity of nebulae, by A. Friedmann, "Über die Krümmung des Raümes," *Z. f. Phys. 10*, 377 (1922); see also A. Einstein, *Z. f. Phys. 11*, 326 (1922), and *16*, 228 (1923). The universe of variable radius has been independently studied by R. C. Tolman, *P.N.A.S. 16*, 320 (1930).

3. Discussion of the theory, and recent developments are found in A. S. Eddington, *M.N. 90*, 668 (1930); W. de Sitter, *Proc. Nat. Acad. Sci. 16*, 474 (1930), and *B.A.N. 5*, No. 185, 193, and 200 (1930); G. Lemaître, *B.A.N. 5*, No. 200 (1930).

4. Popular expositions have been given by G. Lemaître, "La grandeur de l'espace," *Revue des questions scientifiques*, March 1929; W. de Sitter, "The Expanding Universe," *Scientia*, January 1931.

127. On the Relation between the Expansion and the Mean Density of the Universe

Albert Einstein and Wilhelm de Sitter

(*Proceedings of the National Academy of Sciences 18*, 213–214 [1932])

In order to produce a cosmology for a closed, static universe with material content, Albert Einstein added the "cosmological constant" to his field equations.[1] Nearly a decade later, after Edwin Hubble had demonstrated the velocity-distance relation, Einstein and Wilhelm de Sitter withdrew the cosmological constant from the field equations. Aleksandr Friedmann had already shown that nonstationary, expanding universes were possible for which the cosmological constant was zero and space was positively or negatively curved (see selection 125). In the paper given here, Einstein and de Sitter show that an expanding universe is possible without introducing space curvature. Provided that the present matter density of the universe is equal to the critical value $\rho_c = 3H^2/(8\pi G)$, space is Euclidean and the universe is open, infinite, and ever-expanding. Here G is the Newtonian gravitation constant, and Hubble's constant, H, is the rate of expansion of the universe. Einstein and de Sitter use a value of 500 km $\sec^{-1}$ Mpc^{-1} for Hubble's constant, approximately an order of magnitude larger than presently accepted. They notice that the observed matter density of optically visible galaxies, ρ_G, is on the order of ρ_c, and therefore Euclidean space appears a distinct possibility. The precise balance between ρ_G and ρ_c is still a matter of considerable controversy, although for many years the preponderance of evidence has suggested that ρ_G is less than ρ_c, and that the universe is open and ever-expanding in an infinite, hyperbolic space of negative curvature.

1. A. Einstein, *Sitzungsberichte der Preussischen Akademie der Wissenschaften zu Berlin 1*, 142 (1917); English transl. in *The Principle of Relativity* (New York: Dodd, Mead and Co., 1923; reprod. Dover Pub., 1952).

IN A RECENT NOTE in the *Göttinger Nachrichten*, Dr. O. Heckmann has pointed out that the non-static solutions of the field equations of the general theory of relativity with constant density do not necessarily imply a positive curvature of three-dimensional space, but that this curvature may also be negative or zero.

There is no direct observational evidence for the curvature, the only directly observed data being the mean density and the expansion, which latter proves that the actual universe corresponds to the non-statical case. It is therefore clear that from the direct data of observation we can derive neither the sign nor the value of the curvature, and the question arises whether it is possible to represent the observed facts without introducing a curvature at all.

Historically the term containing the "cosmological constant" λ was introduced into the field equations in order to enable us to account theoretically for the existence of a finite mean density in a static universe. It now appears that in the dynamical case this end can be reached without the introduction of λ.

If we suppose the curvature to be zero, the line-element is

$$ds^2 = -R^2(dx^2 + dy^2 + dz^2) + c^2\,dt^2, \tag{1}$$

where R is a function of t only, and c is the velocity of light. If, for the sake of simplicity, we neglect the pressure p,[1] the field equations without λ lead to two differential equations, of which we need only one, which in the case of zero curvature reduces to

$$\frac{1}{R^2}\left(\frac{dR}{c\,dt}\right)^2 = \frac{1}{3}\kappa\rho. \tag{2}$$

The observations give the coefficient of expansion and the mean density:

$$\frac{1}{R}\frac{dR}{c\,dt} = h = \frac{1}{R_B}; \qquad \rho = \frac{2}{\kappa R_A{}^2}.$$

Therefore we have, from (2), the theoretical relation

$$h^2 = \frac{1}{3}\kappa\rho \tag{3}$$

or

$$\frac{R_A{}^2}{R_B{}^2} = \frac{2}{3}. \tag{3'}$$

Taking for the coefficient of expansion

$$h = 500 \text{ km./sec. per } 10^6 \text{ parsecs}, \tag{4}$$

or

$$R_B = 2 \times 10^{27} \text{ cm.},$$

we find

$$R_A = 1.63 \times 10^{27} \text{ cm.},$$

or

$$\rho = 4 \times 10^{-28} \text{ gr. cm.}^{-3}, \tag{5}$$

which happens to coincide exactly with the upper limit for the density adopted by one of us.[2]

The determination of the coefficient of expansion h depends on the measured red-shifts, which do not introduce any appreciable uncertainty, and the distances of the extra-galactic nebulae, which are still very uncertain. The density depends on the assumed masses of these nebulae and on the scale of distance, and involves, moreover, the assumption that all the material mass in the universe is concentrated in the nebulae. It does not seem probable that this latter assumption will introduce any appreciable factor of uncertainty. Admitting it, the ratio h^2/ρ, or $R_A{}^2/R_B{}^2$ as derived from observations, becomes proportional to Δ/M, Δ being the side of a cube containing on the average one nebula, and M the average mass of the nebulae. The values adopted above would correspond to $\Delta = 10^6$ light years, $M = 2 \cdot 10^{11}\odot$, which is about Dr. Oort's estimate of the mass of our own galactic system. Although, therefore, the density (5) corresponding to the assumption of zero curvature and to the coefficient of expansion (4) may perhaps be on the high side, it certainly is of the correct order of magnitude, and we must conclude that at the present time it is possible to represent the facts without assuming a curvature of three-dimensional space. The curvature is, however, essentially determinable, and an increase in the precision of the data derived from observations will enable us in the future to fix its sign and to determine its value.

1. It seems certain that the pressure p in the actual universe is negligible as compared with the material density ρ_0. The same reasoning, however, holds good if the pressure is not neglected.
2. *Bull. Astronom. Inst. Netherlands* (Haarlem) *6*, 142 (1931).

128. The Cosmical Constants

Paul Adrien Maurice Dirac

(*Nature 139*, 323 [1937])

In this paper Paul Dirac constructs a dimensionless number that combines the fundamental constants of atomic and cosmic physics.[1] The force of electric attraction characterizes atomic physics just as the force of gravitational attraction characterizes cosmic physics. For the electron and the proton the ratio of the two forces is a dimensionless constant whose value is 10^{39}. In a remarkable coincidence, the same number, 10^{39}, is equal to the ratio of the age of the universe to the time it takes light to traverse an electron's radius.

A. S. Eddington had previously tried to unify the general theory of relativity with the quantum theory of matter[2] and in the process attributed special powers to a dimensionless number of the order of 10^{78}. As Dirac notices here, this number is equal to the ratio of the mass of the universe to the proton mass, and it is also equal to the square of 10^{39}. These coincidences suggested to Dirac that the important constants of cosmological and atomic theory are functions of the age of the universe and hence of time.

Assuming that the mass of the electron and proton do not depend on time, Dirac was led to the conclusion that the Newtonian gravitational constant, G, must be a function of time.[3] At least three plausible cosmological theories are consistent with a decrease of G with increasing time.[4] Nevertheless, Dirac's suggestion that the constant of gravitation is changing has remained rather controversial. Evidence in favor of a decrease has been adduced from the moon's motion, and evidence against any change from the structure and history of the sun (as revealed by paleontological data). Spectroscopic observations of neutral hydrogen and fine structure lines at a high red shift place very low limits to the time variation of three products of the fine structure constant, the Landé nuclear g factor of the proton, and the ratio of the electron to proton mass.[5]

1. Dirac modified and enlarged upon this contribution in his article in the *Proceedings of the Royal Society* (London) *A165*, 199 (1938).

2. A. S. Eddington, *Monthly Notices of the Royal Astronomical Society 92*, 3 (1932). *Proceedings of the Cambridge Philosophical Society 40*, 37 (1944). See also Eddington's *Fundamental Theory*, ed. E. T. Whittaker (Cambridge: Cambridge University Press, 1948).

3. At nearly the same time S. Sambursky (*Physical Review 52*, 335 [1937]) independently showed that the observed redshifts of nebulae are consistent with a static, nonexpanding universe when one assumes that the radius of the electron and the other universal lengths of atomic physics decrease with time. Instead of postulating a growing world radius, Sambursky reasoned that the so-called quantum of action, h(Planck's constant), and also G decrease with time.

4. C. Brans and R. H. Dicke, *Physical Review 124*, 925 (1961); F. Hoyle and J. V. Narlikar, *Monthly Notices of the Royal Astronomical Society 155*, 323 (1972); P. A. M. Dirac, *Proceedings of the Royal Society* (London) *A333*, 403 (1973).

5. A. M. Wolfe, R. L. Brown, and M. S. Roberts, *Physical Review Letters 37*, 179 (1976).

THE FUNDAMENTAL CONSTANTS of physics, such as c the velocity of light, h Planck's constant, e the charge and m mass of the electron, and so on, provide for us a set of absolute units for measurement of distance, time, mass, etc. There are, however, more of these constants than are necessary for this purpose, with the result that certain dimensionless numbers can be constructed from them. The significance of these numbers has excited much interest in recent times, and Eddington has set up a theory for calculating each of them purely deductively. Eddington's arguments are not always rigorous, and, while they give one the feeling that they are probably substantially correct in the case of the smaller numbers (the reciprocal fine-structure constant hc/e^2 and the ratio of the mass of the proton to that of the electron), the larger numbers, namely the ratio of the electric to the gravitational force between electron and proton, which is about 10^{39}, and the ratio of the mass of the universe to the mass of the proton, which is about 10^{78}, are so enormous as to make one think that some entirely different type of explanation is needed for them.

According to current cosmological theories, the universe had a beginning about 2×10^9 years ago, when all the spiral nebulae were shot out from a small region of space, or perhaps from a point. If we express this time, 2×10^9 years, in units provided by the atomic constants, say the unit e^2/mc^3, we obtain a number about 10^{39}. This suggests that the above-mentioned large numbers are to be regarded, not as constants, but as simple functions of our present epoch, expressed in atomic units. We may take it as a general principle that all large numbers of the order 10^{39}, 10^{78}... turning up in general physical theory are, apart from simple numerical coefficients, just equal to t, t^2..., where t is the present epoch expressed in atomic units. The simple numerical coefficients occurring here should be determinable theoretically when we have a comprehensive theory of cosmology and atomicity. In this way we avoid the need of a theory to determine numbers of the order 10^{39}.

Let us examine some of the elementary consequences of our general principle. In the first place, we see that the number of protons and neutrons in the universe must be increasing proportionally to t^2. Present-day physics, both theoretically and experimentally, provides no evidence in favour of such an increase, but is much too imperfect to be able to assert that such an increase cannot occur, as it is so small; so there is no need to condemn our theory on this account. Whether the increase is a general property of matter or occurs only in the interior of stars is a subject for future speculation.

A second consequence of our principle is that, if we adopt a scheme of units determined by atomic constants, the gravitational 'constant' must decrease with time, proportionally to t^{-1}. Let us define the gravitational power of a piece of matter to be its mass multiplied by the gravitational constant. We then have that the gravitational power of the universe, and presumably of each spiral nebula, is increasing proportionally to t. This is to some extent equivalent to Milne's cosmology[1], in which the mass remains constant and the gravitational constant increases proportionally to t. Following Milne, we may introduce a new time variable, $\tau = \log t$, and arrange for the laws of mechanics to take their usual form referred to this new time.

To understand the present theory from the point of view of general relativity, we must suppose the element of distance defined by $ds^2 = g_{\mu\nu}\, dx_\mu\, dx_\nu$ in the Riemannian geometry to be, not the same as the element of distance in terms of atomic units, but to differ from this by a certain factor. (The former corresponds to Milne's $d\tau$ and the latter to Milne's dt.) This factor must be a scalar function of position, and its gradient must determine the direction of average motion of the matter at any point.

1. Milne, *Proc. Roy. Soc. A158*, 324 (1937).

129. The Steady-State Theory of the Expanding Universe

Hermann Bondi and Thomas Gold

(*Monthly Notices of the Royal Astronomical Society 108*, 252–270 [1948])

During the early part of the twentieth century many astronomers believed that the universe satisfies the cosmological principle asserting that, except for local irregularities, the universe presents the same aspect from every point.[1] In the paper given here Hermann Bondi and Thomas Gold extend the cosmological principle to include temporal as well as spatial homogeneity. They argue that this "perfect cosmological principle" is a natural extrapolation of the fundamental axiom that the outcome of an experiment is restricted neither by the position nor the time at which it is carried out. Unless we accept this principle, Bondi and Gold argue, terrestrially obtained knowledge of the physical world cannot be applied to the universe at large. Moreover, if the universe presents an unchanging aspect in space and time, there is no need to allow for evolutionary changes in the average properties of galaxies. This unchanging, steady state aspect of Bondi and Gold's cosmology makes it the one most easily compared with observations.

Although the universe does not evolve and must remain forever in the same steady state in this theory, individual galaxies can evolve and the universe can expand. In fact, the expansion of the universe follows from the steady state theory when the thermo-dynamic disequilibrium of the universe is taken into account. As the universe expands, the density of matter apparently decreases, and new matter must be continually created in order to keep the density constant. According to Bondi and Gold, about one atom of hydrogen is created *ex nihilo* in a cubic meter of empty, intergalactic space every 300,000 years.

By building upon Thomas Gold's suggestion that matter might be continually created, Fred Hoyle soon showed that a steady state, expanding universe can follow from a modification of Einstein's equations.[2] By introducing a symmetrical tensor into the field equations in the same way that Einstein introduced the cosmological constant, Hoyle was able to obtain an automatic balance between the expansion of the universe and the origin of new matter. The resultant model corresponds to a de Sitter universe with a constant matter density that is at least an order of magnitude higher than the mean density of optically visible galaxies and it thus requires most of the mass density of the universe to be in the form of optically invisible intergalactic matter.

At the time that the steady state theory was introduced, astronomers were vexed by serious incompatibilities between various methods for estimating the age of the universe. The expansion age inferred from the reciprocal of Hubble's constant was a factor of 10 younger than the ages estimated for our solar system and for the older stars in our galaxy. Because many of the objects that Hubble assumed to be stars in other galaxies are actually regions of ionized gas (H II regions) and in some cases clusters of stars, and also because Hubble used an incorrect period-luminosity relation (see selection 110), his estimates for the distances to galaxies and the expansion age of the universe were a factor of 10 too small, and the two age estimates are now in agreement. Moreover, there is now considerable evidence for the violation of the perfect cosmological principle.

The first indication that the universe is not homogeneous in time came from the number counts of radio sources, which showed an apparent excess of weak sources (see selection 118). Although this excess is still a matter of controversy, it indicates that extragalactic radio sources must evolve with time and that the more distant, younger objects are either more luminous or more concentrated than the nearby older objects. Nearly two decades after its first formulation, one of the most outspoken proponents of the steady state theory, Fred Hoyle, recanted the theory in the face of strong observational evidence for a dense, primeval origin of the universe.[3] Most of the deuterium and helium now present in the universe, and also the microwave background radiation, must have been created during the first 3 minutes of a gigantic big-bang explosion that occurred about 10 billion years ago.[4] Nevertheless, the steady state theory has been the longest-lived alternative to the big-bang cosmology.

1. In 1926 Edwin Hubble provided some observational evidence in support of the cosmological principle, when he professed to show that the distribution of galaxies is spatially homogeneous and isotropic (see selection 105). Although some of Hubble's number counts have been subsequently discredited, the galaxies do seem to be homogeneously and isotropically distributed when sufficiently large regions are observed.

2. F. Hoyle, *Monthly Notices of the Royal Astronomical Society 108*, 372 (1948).

3. F. Hoyle, *Nature 208*, 111 (1965).

4. Selections 13 and 132. Also see F. Hoyle and R. J. Tayler, *Nature 203*, 1108 (1964), and R. V. Wagoner, *Astrophysical Journal 179*, 343 (1973).

Summary—The applicability of the laws of terrestrial physics to cosmology is examined critically. It is found that terrestrial physics can be used unambiguously only in a stationary homogeneous universe. Therefore a strict logical basis for cosmology exists only in such a universe. The implications of assuming these properties are investigated.

Considerations of local thermodynamics show as clearly as astronomical observations that the universe must be expanding. Hence, there must be continuous creation of matter in space at a rate which is, however, far too low for direct observation. The observable properties of such an expanding stationary homogeneous universe are obtained, and all the observational tests are found to give good agreement.

The physical properties of the creation process are considered in some detail, and the possible formulation of a field theory is critically discussed.

1. The Perfect Cosmological Principle

THE UNRESTRICTED REPEATABILITY of all experiments is the fundamental axiom of physical science. This implies that the outcome of an experiment is not affected by the position and the time at which it is carried out. A system of cosmology must be principally concerned with this fundamental assumption and, in turn, a suitable cosmology is required for its justification. In laboratory physics we have become accustomed to distinguish between conditions which can be varied at will and the inherent laws which are immutable.

Such a distinction between the "accidental" conditions and the "inherent" laws and constants of nature is justifiable so long as we have control over the "accidental," and can test the validity of the distinction by a further experiment. In astronomical observations we do not have this control, and we can hence never prove which is "accidental" and which "inherent." This difficulty, though logically a very real one, need not concern us in an interpretation of the dynamics of the solar system. We may be satisfied when we discover that the solar system with all its numerous orbits is accurately one of the many systems permitted by our "inherent" laws.

But when we wish to consider the behaviour of the entire universe, then the logical basis for a distinction between "inherent" laws and "accidental" conditions disappears. Any observation of the structure of the universe will give as unique a result as, for instance, a determination of the velocity of light or the constant of gravitation. And yet, if we were to contemplate a changing universe we should have to assume some such observations to represent "accidental" conditions and others "inherent" laws.

Such assumptions were in fact implied in all theories of evolution of the universe; they were necessary to specify the problem. Without them, there would be no rules and hence unlimited freedom in any extrapolation into the future or into the past. Some such sets of assumptions may be intellectually much more satisfying than others, and accordingly they are adopted. We may place much reliance in such aesthetic judgments, but we cannot claim any logical foundation for them.

1.2. Let us take an example to demonstrate the point of this criticism. Present observations indicate that the universe is expanding. This suggests that the mean density in the past has been greater than it is now. If we are now to make any statement regarding the behaviour of such a denser universe, and the way in which it is supposed to have reached its present condition, then we have to know the physical laws and constants applicable in denser universes. But we have no determinations for those. An assumption which is commonly made is to say that those physical laws which we have learnt to regard as "intrinsic" because they are unaffected by any changes of conditions which we may produce, are in fact not capable of any change. It is admitted that such a change of the density of the universe would have slight local effects; there would be more or less light arriving here from distant galaxies. But, it is assumed, there would be no change in the physical laws which an observer would deduce in a laboratory.

Such a philosophy may be intellectually very agreeable: it gives permanence to the abstract things, the physical laws, whilst it regards the present condition of the universe as merely a particular demonstration of the consequences of such laws. There are, however, grave difficulties inherent in such a view. The most striking of those is concerned with the absolute state of rotation of a body. Mach[1] examined this problem very thoroughly and all the advances in theory which have been made have not weakened the force of his argument. According to "Mach's Principle" inertia is an influence exerted by the aggregate of distant matter which determines the state of motion of the local frame of reference by means of which rotation or acceleration is measured. A particular rotational state which is described as "non-rotating" can be found by a purely local experiment (a Foucault pendulum, for instance). This rotational state is always found to coincide accurately with the rotational state of the distant stars around the observer. Mach's principle is the recognition that this coincidence must in fact be due to a causal relation. If, as is now widely agreed, we adopt Mach's principle, then we imply that the nature of any local dynamical experiment is fundamentally affected by distant matter. We can hence not contemplate a laboratory which is shielded so as to exclude all influence from outside; and for the same reason we cannot have any logical basis for choosing physical laws and constants and assigning to them an existence independent of the structure of the universe.

Any interdependence of physical laws and large-scale structure of the universe might lead to a fundamental difficulty in interpreting observations of light emitted by distant objects. For if the universe, as seen from those objects, presented a different appearance, then we should not be justified in assuming familiar processes to be responsible for the emission of the light which we analyse. This difficulty is partly removed by the "cosmological principle." According to this principle all large-scale averages of quantities derived from astronomical observations (i.e. determination of the mean density of space, average size of galaxies, ratio of condensed to uncondensed matter, etc.) would tend statistically to a similar value independent of the positions of the observer, as the range of the observation is increased; provided only that the observations from different places are carried out at equivalent times. This principle would mean that there is nothing outstanding about any place in the universe, and that those differences which do exist are only of local significance; that seen on a large scale the universe is homogeneous.

This principle is widely recognized, and the observations of distant nebulae have contributed much evidence in its favour. An analysis of these observations indicates that the region surveyed is large enough to show us a fair sample of the universe, and that this sample is homogeneous.

But in the sense in which the principle came to be adopted there is the qualification regarding the time of the observation. The universe is still presumed to be capable of altering its large-scale structure, but only in such a way as not to upset its homogeneity. The result of a large-scale observation may hence be a measure of a universal time.

We might have looked to the cosmological principle for a justification of the assumption of the general validity of physical laws; but whilst the principle supplies the justification with respect to changes of place, it still leaves the possibility of a change of physical laws with universal time. Any system of cosmology must still involve a speculation about this dependence as one of its basic assumptions. Indeed, we are not even in a position to interpret observations of very distant objects without such an assumption, for the light which we receive from them was emitted at a different instant in this scale of universal time, and accordingly the processes responsible for its emission may be unfamiliar to us.

The systems of cosmology may well be classified according to the assumption made or implied at this stage of the argument. While one school of thought considers all the results of laboratory physics to be always applicable, without regard to the state of the universe, another starts from the narrow cosmological principle and with the aid of a number of intellectually agreeable assumptions arrives at the conclusion that laboratory physics is qualitatively permanently valid, though some of its "constants" are changing. There are yet other schools of thought which attempt to distinguish the changeable from the permanent constants by their magnitudes.

We shall proceed quite differently at this point. As the physical laws cannot be assumed to be independent of the structure of the universe, and as conversely the structure of the universe depends upon the physical laws, it follows that there may be a stable position. We shall pursue the possibility that the universe is in such a stable, self-perpetuating state, without making any assumptions regarding the particular features which lead to this stability. We regard the reasons for pursuing this possibility as very compelling, for it is only in such a universe that there is any basis for the assumption that the laws of physics are constant; and without such an assumption our knowledge, derived virtually at one instant of time, must be quite inadequate for an interpretation of the universe and the dependence of its laws on its structure, and hence inadequate for any extrapolation into the future or the past.

Our course is therefore defined not only by the usual cosmological principle but by that extension of it which is obtained on assuming the universe to be not only homogeneous but also unchanging on the large scale. This combination of the usual cosmological principle and the stationary postulate we shall call the *perfect cosmological principle*, and all our arguments will be based on it. The universe is postulated to be homogeneous and stationary in its large-scale appearance as well as in its physical laws.

We do not claim that this principle must be true, but we say that if it does not hold, one's choice of the variability of the physical laws becomes so wide that cosmology is no longer a science. One can then no longer use laboratory physics without relying on some arbitrary principle for their extrapolation.

But if the perfect cosmological principle is satisfied in our universe then we can base ourselves confidently on the permanent validity of all our experiments and observations and explore the consequences of the principle. Unless and until any disagreement appears we therefore accept the principle, since this is the only assumption on the basis of which progress is possible without further hypothesis. As will be seen later many conclusions can be drawn which can be compared with experiment. It would be most fortuitous if no contradictions arose, although the perfect cosmological principle was not valid.

2. Application of the Perfect Cosmological Principle

2.1. Although the perfect cosmological principle states that the universe is homogeneous and stationary on the large scale no statement is made about spatial isotropy. This is usually assumed together with homogeneity for mathematical convenience, but in our view the position of the two assumptions is quite different. Homogeneity and the stationary property are vital to the very existence of the subject. There may be similarly strong arguments for isotropy, but those so far advanced do not seem to us to be equally compelling. Further considerations of the nature of the stable self-perpetuating state referred to above can be made, which suggest but do not prove isotropy.

We shall therefore introduce spatial isotropy as an assumption which is strongly suggested both by local and by astronomical observations and which is simplest.

We must now apply the perfect cosmological principle to laboratory physics and to astronomical observations. We regard the principle as of such fundamental importance that we shall be willing if necessary to reject theoretical extrapolations from experimental results if they conflict with the perfect cosmological principle even if the theories concerned are generally accepted. Of course we shall never disregard any direct observational or experimental evidence and we shall see that we can easily satisfy all such requirements.

2.2. For the perfect cosmological principle to apply, one might at first sight expect that the universe would have to be static, i.e. to possess no consistent large-scale motion. This, however, would conflict with the observations of distant galaxies, and it would also conflict with the thermodynamic state which we observe. For such a static universe would be very different indeed from the universe we know. A static universe would clearly reach thermodynamical equilibrium after some time. An infinitely old universe would certainly be in this state. There would be complete equilibrium between matter and radiation, and (apart possibly from some slight variations due to gravitational potentials) everything would be at one and the same temperature. There would be no evolution, no distinguishing features, no recognizable direction of time. That our universe is not of this type is clear not only from astronomical observations but from local physics and indeed from our very existence. Accordingly there must be large-scale motions in our universe. The perfect cosmological principle permits only two types of motion, viz. large-scale expansion with a velocity proportional to distance, and its reverse, large-scale contraction.

In a contracting universe there would be even more radiation compared with matter than in a static universe. Therefore we reject this possibility and confine our attention to an expanding universe.

2.3. The observations of distant galaxies, which are now capable of a more rigorous interpretation by means of the perfect cosmological principle, inform us of the notion of expansion. This motion in which the velocity is proportional to the distance (apart from a statistical scatter) is well known to be of the only type compatible with homogeneity; but the compatibility with the hypothesis of a stationary property requires investigation. If we considered that the principle of hydrodynamic continuity were valid over large regions and with perfect accuracy then it would follow that the mean density of matter was decreasing, and this would contradict the perfect cosmological principle. It is clear that an expanding universe can only be stationary if matter is continuously created within it. The required rate of creation, which follows simply from the mean density and the rate of expansion, can be estimated as at most one particle of proton mass per litre per 10^9 years. In interpreting the universe as stationary we have to assume that such a process of creation is operative; we have to infringe the principle of hydrodynamic continuity. But this principle is not capable of experimental verification to such a precision, and this infringement does not constitute a contradiction with observational evidence. It is true that hydrodynamic continuity has been regarded as an unqualified truth and not as an approximation to physical laws, but this was merely a bold simplifying extrapolation from evidence. Hydrodynamic continuity is no doubt approximately true but this does not compel us to assume that it holds without any deviation whatever. In the conflict with another principle which is much more far-reaching and capable of making many more statements about the nature of the universe and the applicability of physical laws, there is no reason for upholding the principle of continuity to an indefinite accuracy, far beyond experimental evidence.

We have referred to the principle infringed by the creation process as the principle of hydrodynamic continuity; as we shall see later it is by no means clear whether we can regard any principle of conservation of energy as infringed.

2.4. Much of the structure of the universe is closely defined by the perfect cosmological principle and the knowledge of the velocity-distance law of expansion. As we shall see later such a universe must be described by a de Sitter metric. The universe is therefore spatially as well as temporally infinite, but the effect of distant matter on an observer tends to zero as its velocity of recession approaches the velocity of light.

The motion of matter in such a universe follows a particular pattern; though in the actual case there are fluctuations, there is a preferred motion defined at each place. Any observer equipped with a sufficiently good telescope can make observations of the velocity-distance law in different directions, and from such observations he can deduce any deviation that he may have from this preferred motion in his locality. If he follows this motion without any deviation then he will see symmetry around him; although all the objects which he observes may have some deviations from the preferred motions in their localities, a statistical averaging method can serve to specify the preferred motion at the observer exactly. (It is not necessary to specify the averaging method at this point beyond stating that it must give sufficient weight to distant matter.) At any rate there is not much latitude. Such a preferred motion could be established by an observer at any place, and the resulting pattern forms a vector field in a four dimensional space-time. Much physical significance must be attached to this vector-field, for the motion of all matter appears to be closely dictated by it. What deviations there are, are statistically distributed around the preferred motion, and these deviations are quite small compared with the velocity of light.

An interesting problem connected with the preferred direction is that of the existence of a "cosmic time." In the cosmologies of general relativity the existence of a cosmic time is deduced from Weyl's postulate.[2] As observations show, there are irregularities and deviations from uniformity. These are only small, but if they are such that there are no surfaces orthogonal to the world lines of all the nebulae, then there is no cosmic time. The existence of a cosmic time in the presence of these irregularities has not been proved and seems improbable. The ambiguities in cosmic time arising from such a breakdown are only small in any region, but over the universe as a whole they completely undermine the logical structure on which a homogeneous but not stationary theory rests. Our position is very different. Owing to the stationary character of the universe, homogeneity can be defined in the absence of cosmic time. Whether or not there are surfaces orthogonal to the preferred direction is therefore of no importance for our theory. On general grounds one would not expect these surfaces to exist, so that an unambiguous definition of cosmic time is impossible. In the simplified smoothed-out model it will of course exist, but probably not in the actual universe with its irregularities.

The stationary property implies that the factor of the velocity-distance law must stay constant. This would not be fulfilled if matter moved such that the velocity between two particular masses remained constant; the velocity has to increase as the distance increases. Accordingly, there must be relative acceleration between matter.

We can now examine the requirements which the perfect cosmological principle places on the evolution of stars and galaxies. The mean ratio of condensed to uncondensed matter has to stay constant, and for this reason new galaxies have to be formed as older ones move away from each other. We take the perfect cosmological principle to imply that no feature of the universe is subject to any consistent change, and no observer hence capable of any unique definition of a universal time. This will be satisfied only if the ages of galaxies in any sufficiently large volume follow a certain statistical distribution. With such a distribution it is of course possible for any particular volume to contain a very old galaxy; but very old galaxies are thinly scattered, as they have had time to move far apart in the general expansion. The characteristic time in this age distribution must of course be related to the characteristic time of expansion of the universe. In opposition to most other theories we should hence expect to find much diversity in the appearance of galaxies, as they will be of greatly different ages; and much diversity is in fact observed. Furthermore the age distribution of galaxies in any volume will be independent of the time of observation, and it will hence be the same for distant galaxies as for near ones, although in the former case the light has taken long to reach us. This is a property of great importance, for it is in practice the only basis on which we can construct any interpretation of the observations of distant galaxies.

2.5. The next point to discuss is the thermodynamic state of the universe. It is known that this state is very far from any equilibrium. Very much more energy is in the form of matter than in radiation; and very much more energy is radiated away than is absorbed by matter.

The disequilibrium is very great indeed, and we have grown so used to it that we take its existence to be self-evident. Such statements as "A hydrogen atom in an excited state returns quickly to its ground state, since this has less energy", do not indicate any lack of faith in the principle of conservation of energy but merely an awareness of the paucity of radiation in the universe which makes the reverse process so very unlikely.

This disequilibrium is such a marked and characteristic feature of our universe that its explanation must be one of the main tasks of cosmology. The connection between this phenomenon and cosmology was noticed by Olbers[3], who pointed out that in an infinite homogeneous static Newtonian universe the mean radiation density would be as high as it is on the surface of a star! In the present case, however, this disequilibrium is the direct consequence of the motion of expansion. Any photon emitted by a star has a very large probable free path, owing to the tenuous distribution of opaque matter in the universe. The most likely career for such a photon is that it will travel on into regions where the local motion of matter is very different from the motion of the star which caused its emission. When such a photon is finally absorbed, then it will, as seen by an observer on the absorbing matter, be subject to a Doppler-shift which reduces its frequency by a large factor. The thermal energy it will supply to the intercepting matter will hence be only a small fraction of the energy it removed from the emitting star. It is not necessary to consider here whether the difference in energy can be regarded as being supplied to the expansion process of the universe; but it is clear that the universe provides in this way a sink for radiative energy, and that this sink is in fact available to the majority of photons produced at the surface of stars. The process of getting rid of both matter and radiation from any fixed volume is by pushing both across the surface bounding this volume; and both are replenished from within.

It is of course important for any theory of cosmology that it should not predict thermodynamic equilibrium, and this is possible only by supposing either that insufficient time has been available since the commencement of radiation by stars, or, and this is the view we take, by suggesting that the bulk of the energy radiated does not again become available as heat.

A thermodynamic disequilibrium is a requirement for any local thermodynamical evolution, and in this sense it is necessary in order that any physical significance should be attached to the passage of time. In a stationary universe this condition can be fulfilled only if there is a motion of expansion, and if the mean free path of a photon is at least comparable with the dimension defined by the velocity-distance law and the velocity of light. (The radius of a de Sitter universe.) Expressing the same consideration differently: the perfect cosmological principle and the knowledge of the thermodynamic disequilibrium require that the universe should be expanding. The experimental evidence of the recession of nebulae is hence a good indication that this specific requirement of the perfect cosmological principle is in fact satisfied. The narrow cosmological principle would have resulted in no such requirement, for there the disequilibrium would result simply if the luminosity of all stars was a suitable function of universal time.

Let us now consider the question of the overall conservation of mass (energy) in such a universe. An observer attempting to estimate the quantity of matter in the universe will find that his result depends on the threshold intensities and wavelengths of his apparatus. Keeping these fixed, he will always observe a finite amount of matter and this will be constant in time. At a great range matter is drifting into an unobservable state by approaching the velocity of light, and without a process of creation this would not allow any principle of conservation of energy to be applied to the sum total of all observable matter.

A further point to be mentioned in relation to the stationary property of the universe is the coincidence of numbers pointed out by Eddington.[4] Two non-dimensional numbers which can be constructed from observation are both found to be of the order 10^{39}; one is the ratio of the characteristic length defining the expansion of the universe to the classical radius of the electron (e^2/mc^2), and the other is the ratio of the electric to the gravitational force between a proton and an electron. The coincidence is striking, and though we have no causal connection to offer we feel that one exists. This and some other coincidences cannot be ignored by a system of cosmology to the extent of suggesting that they occur only in a fleeting phase. It is satisfactory that according to the theory presented here such coincidences are permanent.

3. THE OBSERVATIONAL TESTS

3.1. Our observations of the universe depend on the interactions between the rest of the universe and ourselves. These interactions are of two kinds: dynamical and electromagnetic. Dynamically, the rest of the universe affects us primarily through the long-range force of inertia, i.e. it defines inertial frames of reference in our neighbourhood. This dynamical interaction is strong and arises almost certainly from the whole of the observable universe, but it is ill-understood and is discussed in other parts of this paper. Therefore, in this section we shall confine ourselves to a discussion of the electromagnetic interactions, which are more easily interpreted.

Light reaches us in observable intensities from large numbers of extragalactic nebulae, and a considerable amount of information has been gathered. In particular the distribution over the sky and the number of nebulae of given apparent magnitude is quite well known, and the radial velocities have been determined (from the Doppler shift) for many nebulae.

In trying to interpret these data, we must first decide whether the observations extend over a sufficiently large region to have any bearing on cosmology, or whether they must be taken to refer to a purely local district.

There are two strong arguments (isotropy and homogeneity of nebular distribution and magnitude of the radial velocities), for assuming that the observations cover a fair sample of the universe and there is no evidence against it. In common with most writers on the subject we assume therefore that the observations refer to a sufficiently large region to be of relevance to cosmology.

3.2. We must now see whether our theory, at this stage of its development, can make any statements about the aspect of the universe which can be compared with observation. The mathematical apparatus we can use is strictly limited, since the field equations of the general theory of relativity with their insistence on conservation of mass are clearly not acceptable to us. In fact we must rely largely on kinematical arguments and on the generally accepted results of laboratory physics dealing with the propagation of light (special theory of relativity). The power of these kinematical arguments was demonstrated by the pioneer work of Milne and by the papers of Walker and Robertson.

A simplified model of the universe can be obtained by assuming the conditions of homogeneity and isotropy to be absolutely (instead of only on a large scale) satisfied. There seems to be good reason to believe that the features amenable to long distance observation are nearly the same in this simplified model as in the actual universe with all its irregularities. In common with most authors we shall assume that this representation of the universe is sufficient for comparison with the observations of extragalactic nebulae.

The mathematical properties of such a model have been studied in great detail by Robertson[5] and by Walker.[6] These authors adopt a narrower cosmological principle and they do not assume a stationary property. They do, however, assume that the particle paths do not intersect except that possibly they all intersect at one singular point in the past (origin of universe). Although we do not wish to make this assumption in general it is clear that in the simple model we are now discussing the assumption will be satisfied. For if there were an intersection at a point P, then the isotropy condition would show that this point had to be a focus, and the homogeneity condition would then lead us to a universe in which every point is a focus. Such a universe is evidently so different from the one we know that we need not hesitate to reject this possibility.

Accordingly, the work of Robertson and Walker is applicable to our theory, except that we have to specialize their results by the stationary condition. In this way the following theorem can be established:

In a non-static stationary universe the light-paths are the null geodesics of the metric

$$ds^2 = dt^2 - (dx^2 + dy^2 + dz^2)\exp(2t/T),$$

where the (x, y, z) coordinates of any particle partaking of the general motion of the universe are constant and t measures the proper time of any observer travelling with such a particle. T is a universal constant and the velocity of light has been put equal to unity. This metric is very well known, having been discovered by de Sitter[7] in 1917. It is the simplest of all metrics of the Robertson type, but could not be used in general relativity, since the field equations imply that it describes empty space only.

The geometrical properties of the de Sitter universe are very simple. If we change from Cartesians (x, y, z) to spherical polars (r, θ, ϕ) in the usual way, then an observer at the origin would find that the spectral lines of a source of light of luminosity L at (r, θ, ϕ) received by him at time $t = 0$ would show a red shift

$$1 + \delta = \frac{\lambda_{\text{received}}}{\lambda_{\text{emitted}}} = \exp(-t_{em}/T) = 1 + r/T, \tag{1}$$

while the intensity of light received would be

$$\frac{L}{4\pi r^2}\exp\left(\frac{2t_{em}}{T}\right) = \frac{L}{4\pi r^2(1 + r/T)^2}. \tag{2}$$

These results are well known[8] and apply equally in general relativity and in our theory. An important difference occurs, however, when we consider the third observable quantity, the number of nebulae in a given magnitude class. In general relativity it is assumed that the density of nebulae was greater in the past than it is now, owing to the expansion of the universe. The theory here presented assumed that the density was the same in the past as it is now. Accordingly, if n is the number of nebulae per unit volume, the number of nebulae between r and $r + dr$ is

$$4\pi r^2 n\,dr\exp(3t_{em}/T) = 4\pi r^2 n\,dr(1 + r/T)^{-3}. \tag{3}$$

The comparison with observation consists in checking that the interdependence of the functions (1), (2), (3) is the same in observation as in theory. This process is familiar from relativistic cosmology, but there is another difference not yet referred to. In our theory the average nebular luminosity L must be constant by virtue of our fundamental assumption. This is not so in any other cosmology, where the distant nebulae, from which light reaches us only after a long time, must be assumed to have been in an earlier stage of development when the light observed was emitted. Owing to the likely variation of nebular luminosities the measurement of distances becomes very inaccurate just at the ranges where space curvature begins to be important. The theory here presented is therefore the only one which can be compared with observation without having to make any assumptions about nebular evolution.

3.3. Here Bondi and Gold argue that uncertainties in the Doppler effect corrections to observed magnitudes allow an interpretation of the number-magnitude relation in terms of the steady state theory. Under this interpretation, the observed paucity of faint nebulae is caused by a constant density of nebulae in an expanding universe; the relativistic interpretation ascribes this paucity to positive space curvature.

3.4. A difficulty which occurs in other cosmological theories is absent in ours: the problem of the time-scale. It will be useful to discuss at some length the difficulties associated with the time-scale in other theories.

The reciprocal of Hubble's constant, which is a time (T), defined by the observations of the velocity-distance law, is between 1,800 and 2,000 million years. It is clear that this time must be of fundamental importance in any theory.

Terrestrial observations of radioactive decay in rocks indicate an age of these rocks of at least 2×10^9 years, and some samples seem to imply an age of more than 3×10^9 years. Astrophysical considerations tend to indicate an age of the stars in our galaxy of about 5×10^9 years.

How do these results affect cosmology? If the continuous creation of matter is denied then the following two possibilities exist:

(i) The rate of expansion of the universe in the past was the same as now or greater. Then the origin of the universe must have been catastrophic and took place less than 2×10^9 years ago. This clearly contradicts the terrestrial and astrophysical evidence referred to above. A distinction between a "dynamical" and a "cosmic" time, as suggested by Milne, removes only part of the difficulty. It makes some of the astrophysical evidence inadmissible but, since atoms and nuclei must be supposed to follow the cosmic time, the terrestrial evidence remains.

(ii) The rate of expansion was slower in the past than it is now. Then some disruptive force must have been responsible for the acceleration. Such a force is described by the hypothetical and much debated cosmological term in general relativity. A general criticism of any model based on such a disruptive force can be made, but it will be useful to discuss a special example, viz. the famous model due to Lemaître.[9] In this model the universe is supposed to have been for an arbitrary long time in the unstable Einstein state in which λ repulsion exactly balanced gravitational attraction. The formation of condensations led to an expansion proceeding at an exponentially increasing speed.

Much more attention has been paid to the attractive formal features of this model than to the several objections to which it is open on physical grounds.

Although its past is mathematically infinite it must be taken to be entirely featureless and void of any physical development. For any physical process such as the formation of a condensation would have upset its equilibrium. In fact in its "pre-historical" state (before the first condensation) circumstances were such that the formation of a condensation was only just possible, though very unlikely, since otherwise it would have occurred earlier. However, as soon as the physical history of the model started, expansion began and so the mean density decreased. Very early on in the expansion this decrease would have been sufficient to make the previously improbable formation of condensations quite impossible. Therefore the period of physical evolution of this model is brief in spite of its infinitely long mathematical existence.

The unsatisfactory features of this model are shared by the other accelerating models of general relativity. The theory presented here differs from all these models very profoundly. Although the universe is accelerating in the sense that the speed of each particular nebula is increasing, the mean density remains constant. Therefore conditions for the condensation of new nebulae are always the same and may be taken to be always moderately favourable. The probability of the formation of condensations is presumably determined by the important non-dimensional quantity $\gamma\rho T^2$. In our theory this is a constant, since all three factors are constant. (In general relativity it is variable, but in kinematical relativity it is constant, although all the factors vary.) Observations indicate that the value of this constant is between 10^{-3} and 1, probably nearer 1.

Under the steady state theory the ages of nebulae follow a statistical law, and there is no reason to suppose that a particular nebula is of some age rather than another. In this theory there is therefore no difficulty in taking the age of our galaxy to be anything indicated by local observations (say $5-8 \times 10^9$ years), although the reciprocal of Hubble's constant, T, may be much shorter.

Although this view of nebular evolution does not form an integral part of the theory, we should mention why we do not follow the view that clusters are evaporating rather than condensing. The time constant for gravitational capture is $(\gamma M/d^3)^{-1/2}$, where M is the mass of a nebula and d the internebular distance. This time is therefore only about 10^{10} years, compared with the period of 10^{12} years required for evaporation by dynamical friction.[10] Furthermore, the presence of even a very tenuous intergalactic gas would not allow nebulae to acquire the "accidentally" high velocities required for evaporation. Therefore we expect that in general the tendency for evaporation will be negligible compared with the gravitational clustering tendency.

If we assign to our galaxy, on the astrophysical evidence, an age of between 5×10^9 and 10^{10} years, it follows that there should be a moderate degree of clustering such as is actually observed.

4. The Physics of Creation

4.1. We shall now attempt to consider some of the details of the physical process of creation. The discussion must necessarily be somewhat speculative but some deductions can be made which limit the range of the possibilities. The average rate of creation is determined by the rate of expansion and the density of the universe[11] and is approximately 10^{-43} g. per sec. per cm.^{3}. We now must discuss how this creation rate is distributed in space. Will it be roughly the same everywhere or will it be very much greater in some regions than in others? In particular will it be as concentrated as the existing matter is (in which case it would be more appropriate to speak of creation per unit mass per unit time)?

The answer to this question is intimately connected with the theory of evolution. There are, roughly speaking, at present two schools of thought in this matter:

(i) One body of astronomers is of the opinion that the main line of development is centrifugal. For example, they consider that nebulae evaporate from clusters and so form the general field, that far more matter is ejected by stars than is accreted by them, so that interstellar space is populated primarily by matter which once was stellar, etc.

(ii) Another body of astronomers considers that centripetal development is predominant. In their view, accretion is far more important for stellar evolution than the ejection of matter from stars; they consider the gradual condensation of stars in the interstellar material (which is taken to constitute the bulk of the mass of a nebula) to be the main process of nebular evolution, etc.

Now it is clear from the discussion in Section 1, that in order to ensure the stationary aspect of the universe, the new material must be created in the form from which evolution starts. For, if evolution turns matter from state A into state B, then the proportion of matter in state A will be diminished unless the newly created matter is in state A. Therefore the created matter must start at the beginning of the evolutionary chain to replenish the reservoir that would otherwise dry up. Accordingly if we assume that creation is by unit mass, then we have to assume that the centrifugal line of thought is correct over an even wider field than its adherents claim, while if we assume the rate of creation per unit volume to be about the same everywhere, then we have to follow the centripetal school.

In our opinion the first view is untenable. For if matter were created mainly in the stellar interior it would have to boil off or be ejected in explosions to a distance and at a rate sufficient to supply intergalactic space with enough material for the required rate of formation of new galaxies. This implies that on an average each star loses its entire mass, by such a process of evaporation or explosion, in $(1/3)T \log_e 2 = 4 \times 10^8$ years! The known rate of occurrence of novae and supernovae shows that these phenomena can only make an insignificant contribution to this. This calculation given here does not allow for the necessary recondensation of the ejected material in its original galaxy, so that the period would be even shorter. This conclusion seems to be so absurd that we take it to prove that the rate of creation per unit volume per unit time is roughly constant and that evolution proceeds mainly in the centripetal direction. It must be clearly understood that some variation of the rate of creation is still permitted, it is only necessary that far more matter is created in space than in the stars.

Accordingly we assume that the rate of creation is everywhere about 10^{-43} g. per sec. per cm.^{3}, although variations by a factor of, say 100, are quite permissible. This rate corresponds to one new atom of hydrogen per cubic metre per 3×10^5 years. Accordingly we cannot expect the process to be directly observable.

4.2. We have seen that the process of creation will take place everywhere and that most of the matter will be created in the vast intergalactic spaces. The next question to engage our attention will be the initial motion of the created matter.

It is evident that this velocity associated with creation cannot be relativistically invariant, nor can invariance be introduced by assuming a random distribution of velocities, since the only invariant distribution diverges.

It follows, therefore, that there will be a preferred time-like direction at each point of space-time, and that the initial motion of newly created matter will be either along this direction (in the 4-dimensional sense) or will be distributed about it in a non-invariant way. The question of this distribution will be discussed later; for the moment it will be assumed that the matter is actually created along the preferred direction.

The adoption of a non-invariant law of creation may at first sight seem surprising; but are we entitled to demand invariance of creation when the resulting movement of matter is well known not to be invariant? Even in the cosmologies of general relativity the existence of a preferred direction is explicitly recognized in Weyl's postulate. General relativity demands that the laws of cosmology should be invariant while admitting that the one and only application is not invariant. We can see no reason why the laws of nature determining the structure of our universe should be invariant, although the universe is unique and does not bear an invariant aspect.

The identification of the preferred direction defined by the creation of matter with the preferred direction defined by the motion of matter (as discussed in 2.4) is compelling. It is most satisfactory that our theory can connect the direction defined everywhere by the motion of matter with another phenomenon, while in other theories no such connection exists.

It should be realized that the existence of this preferred state of motion not only conflicts with general relativity, but also with restricted (and even with Galilean) relativity. However, this lack of invariance is nothing new. As we have seen in Section 2 it is an immediate consequence of the law of the recession of nebulae.

4.3. We must now consider the problem of the probability distribution of the initial velocity about the preferred direction. If the spread of velocities were great, then this would lead either to large amounts of radiation from intergalactic space or to high temperatures of intergalactic matter.

The amount of radiation observed and the postulated intergalactic condensation limit the velocity spread to at most 10^7 cm./sec., corresponding to a temperature of at most 50,000 deg.

There do not seem to be any arguments setting a lower

limit to the temperature and the arguments limiting the temperature from above all depend on the absence of any effects of the temperature. If one wishes to have a definite and simple picture of the process, one can therefore assume that matter is created without random velocities (i.e. at zero temperature), although this is not a necessary conclusion.

4.4. The next question we must discuss is the electric charge of the created matter. Now it is clear that an average excess charge of either sign would lead to forces of repulsion tending to remove the excess charge to infinity. The speed with which the excess charge is repelled to infinity depends on the ratio of excess charge to mass. It can be shown[12] that if this ratio is greater than about $10^{-19} \times$ electron charge/electron mass, then this repulsion of excess charge would take place at a speed greater than the speed of expansion of the material universe. But such a rapidly "expanding universe of excess charge" cannot be present together with a slowly expanding "universe of matter" without violating the principle of homogeneity. Accordingly we can conclude that if there is any average charge excess of the created matter, the average charge-mass ratio must not exceed the very low limit mentioned above.

These results give us some indication of what the created matter may be. If we reject as far-fetched the possibility of the creation of dust, etc., and concentrate on the simplest building material, then we are left with the following possibilities:

(i) Protons and electrons, created separately but at identical (or very nearly identical) rates. The radiation arising from collisions or captures would be just within the limits laid down in 4.3.

(ii) Neutrons. The energy released by the disintegration of the neutrons would (within a few hours of creation) lead to velocities and radiation somewhat in excess of the limits mentioned. Nevertheless, this possibility is so attractive that it should not be ruled out without further examination.

(iii) Hydrogen atoms.

The last possibility seems to agree best with the view adopted in 4.3, although (i) cannot be excluded. Since (i) would, but for the necessarily small amount of radiation, also lead to the formation of hydrogen in the ground state, we can for simplicity and definiteness adopt (iii).

In agreement with the widely held view that the building-up of the elements is a complex stellar process and that the traces of elements other than hydrogen in interstellar space are due to explosion and ejection of matter from stars, we have assumed here that all matter starts its career as hydrogen. This seems to us to be very plausible, but is a separate subject not directly connected with the creation theory.

4.5. We finally have to consider where in space-time matter is created and it seems to be most reasonable to assume that it is created in a random manner. According to this view the probability of creation taking place in any particular four-dimensional element of volume (spatial volume element × element of time) is simply proportional to its (four-dimensional) volume, the factor of proportionality being a function of position.

By our argument in 4.1 this factor cannot vary very much from point to point.

Accordingly randomness would be introduced by the random origin of particles, but not necessarily by any initial random motion.

5. THE FUNCTION OF A FIELD THEORY

Here Bondi and Gold argue that a new field theory must be developed and that in their view the general theory of relativity is inadequate. They also find Hoyle's[13] adaptation of general relativity to the steady state situation unsatisfactory and unacceptable. The theory of general relativity is unacceptable because it requires conservation of mass, and both the general relativity theory and Hoyle's adaptation are unacceptable because they violate Mach's principle.

1. E. Mach, *The Science of Mechanics* (London, 1893), pp. 229–238.
2. H. Weyl, *Phys. Z. 24*, 230–232 (1923); *Phil. Mag.*, ser. 7, *9*, 936–943 (1930).
3. Olbers, *Bode's Jahrbuch 10*, 1826.
4. A. S. Eddington, *Proc. Roy. Soc. A 133*, 605 (1931); *M.N. 92*, 3 (1932).
5. H. P. Robertson, *Ap. J. 82*, 284 (1935).
6. A. G. Walker, *Proc. Lond. Math. Soc.*, ser. 2, *42*, 90 (1936).
7. W. de Sitter, *Proc. Akad. Wetensch. Amst. 19*, 1217 (1917).
8. H. P. Robertson, *Rev. Mod. Phys. 5*, 71 (1933).
9. G. Lemaître, *M.N. 91*, 483 (1931).
10. M. Tuberg, *Ap. J. 98*, 501 (1943).
11. *Cf.* E. Hubble, *The Realm of the Nebulae* (Oxford, 1936), p. 189.
12. Unpublished theorem of Newtonian Cosmology.
13. F. Hoyle, *M.N. 108*, 372 (1948).

130. The Origin of Chemical Elements

Ralph A. Alpher, Hans Bethe, and George Gamow

(*Physical Review 73*, 803–804 [1948])

The idea that the early, dense stages of the universe were hot enough to generate thermonuclear reactions was first proposed by George Gamow in the 1930s and 1940s.[1] By 1946 he had put forward the idea that the observed relative abundances of the elements were determined by nonequilibrium nucleosynthesis during the early stages of the expansion of the universe.[2] Gamow assumed that the cosmic ylem (the primordial substance from which the elements were formed) consisted solely of neutrons at high temperature (about 10 billion degrees). Protons were formed by neutron decay, and successive captures of neutrons led to the formation of the elements. When Ralph Alpher and Gamow worked out this account of element buildup, they added Hans Bethe's name in order to make a pun on the first three letters of the Greek alphabet. Their α-β-γ theory for the buildup of elements by neutron capture during the early stages of the expansion of the universe was subsequently developed in greater detail by Alpher and Robert Herman.[3]

Astronomers now believe that stellar nucleosynthesis explains the observed abundances of the heavier elements (see selection 55), whereas the observed hydrogen, helium, and deuterium must have been produced in the early stages of the universe. Although Gamow's hypothesis for the origin of the elements appears basically correct for these lighter elements, his initial assumption that matter was first composed solely of neutrons was an oversimplification. In the first few seconds of the expansion of the universe, the temperature must have exceeded 10 billion degrees, and sufficient energy was available to form neutrinos and electron-positron pairs. As Chushiro Hayashi first showed, the neutrons will interact with thermally excited positrons to form protons and antineutrinos. Because the reverse process happens just as rapidly, the relative abundance of neutrons and protons is determined by conditions of thermal equilibrium rather than by neutron decay.[4] Modern calculations including these considerations indicate that nucleosynthesis during the early stages of the expansion of the universe accounts for the observed abundance of deuterium and helium.[5]

1. G. Gamow, *Ohio Journal of Science 35*, 406 (1935), and *Journal of the Washington Academy of Sciences 32*, 353 (1942).
2. G. Gamow, *Physical Review 70*, 572 (1946).
3. R. A. Alpher and R. C. Herman, *Physical Review 74*, 1737 (1948), *84*, 60 (1951), and *Reviews of Modern Physics 22*, 153 (1950).
4. C. Hayashi, *Progress in Theoretical Physics* (Japan) *5*, 224 (1950). A discussion of this result is given, together with the first thorough analysis of the early history of the universe, by R. A. Alpher, J. W. Follin, and R. C. Herman in *Physical Review 92*, 1347 (1953).
5. R. V. Wagoner, *Astrophysical Journal 179*, 343 (1973). See also selection 13.

As POINTED OUT BY ONE OF US,[1] various nuclear species must have originated not as the result of an equilibrium corresponding to a certain temperature and density, but rather as a consequence of a continuous building-up process arrested by a rapid expansion and cooling of the primordial matter. According to this picture, we must imagine the early stage of matter as a highly compressed neutron gas (overheated neutral nuclear fluid) which started decaying into protons and electrons when the gas pressure fell down as the result of universal expansion. The radiative capture of the still remaining neutrons by the newly formed protons must have led first to the formation of deuterium nuclei, and the subsequent neutron captures resulted in the building up of heavier and heavier nuclei. It must be remembered that, due to the comparatively short time allowed for this process,[1] the building up of heavier nuclei must have proceeded just above the upper fringe of the stable elements (short-lived Fermi elements), and the present frequency distribution of various atomic species was attained only somewhat later as the result of adjustment of their electric charges by β-decay.

Thus the observed slope of the abundance curve must not be related to the temperature of the original neutron gas, but rather to the time period permitted by the expansion process. Also, the individual abundances of various nuclear species must depend not so much on their intrinsic stabilities (mass defects) as on the values of their neutron capture cross sections. The equations governing such a building-up process apparently can be written in the form

$$\frac{dn_i}{dt} = f(t)(\sigma_{i-1}n_{i-1} - \sigma_i n_i) \quad i = 1, 2, \ldots 238, \tag{1}$$

where n_i and σ_i are the relative numbers and capture cross sections for the nuclei of atomic weight i, and where $f(t)$ is a factor characterizing the decrease of the density with time.

We may remark at first that the building-up process was apparently completed when the temperature of the neutron gas was still rather high, since otherwise the observed abundances would have been strongly affected by the resonances in the region of the slow neutrons. According to Hughes,[2] the neutron capture cross sections of various elements (for neutron energies of about 1 Mev) increase exponentially with atomic number halfway up the periodic system, remaining approximately constant for heavier elements.

Using these cross sections, one finds by integrating Eqs. (1) that the relative abundances of various nuclear species decrease rapidly for the lighter elements and remain approximately constant for the elements heavier than silver. In order to fit the calculated curve with the observed abundances[3] it is necessary to assume the integral of $\rho_n\,dt$ during the building-up period is equal to 5×10^4 g sec./cm^3.

Here we omit a plot of the relative abundance of the elements as a function of atomic number, because an identical plot appears in figure 55.1. Alpher, Bethe, and Gamow's interpretation of the data is different, however, for they show that their calculations produce a smooth decrease in abundance with increasing atomic number.

On the other hand, according to the relativistic theory of the expanding universe[4] the density dependence on time is given by $\rho \cong 10^6/t^2$. Since the integral of this expression diverges at $t = 0$, it is necessary to assume that the building-up process began at a certain time t_0, satisfying the relation

$$\int_{t_0}^{\infty} (10^6/t^2)\,dt \cong 5 \times 10^4, \tag{2}$$

which gives us $t_0 \cong 20$ sec. and $\rho_0 \cong 2.5 \times 10^3$ g sec./cm^3. This result may have two meanings: (a) for the higher densities existing prior to that time the temperature of the neutron gas was so high that no aggregation was taking place, (b) the density of the universe never exceeded the value 2.5×10^3 g sec./cm^3 which can possibly be understood if we use the new type of cosmological solutions involving the angular momentum of the expanding universe (spinning universe).[5]

More detailed studies of Eqs. (1) leading to the observed abundance curve and discussion of further consequences will be published by one of us (R. A. Alpher) in due course.

1. G. Gamow, *Phys. Rev. 70*, 572 (L) (1946).
2. D. J. Hughes, *Phys. Rev. 70*, 106(A) (1946).
3. V. M. Goldschmidt, *Geochemische Verteilungsgesetz der Elemente und der Atom-Arten*. IX. (Oslo, Norway, 1938).
4. See, for example: R. C. Tolman, *Relativity, Thermodynamics and Cosmology* (Oxford, England: Clarendon Press, 1934).
5. G. Gamow, *Nature, 158*, 549 (1946).

131. The Evolution and Physics of the Expanding Universe

Evolution of the Universe

Ralph A. Alpher and Robert C. Herman

(*Nature 162*, 774–775 [1948])

The Physics of the Expanding Universe

George Gamow

(*Vistas in Astronomy 2*, 1726–32 [1956])

In the late 1940s and the early 1950s George Gamow and his colleagues developed a detailed evolutionary theory based upon Georges Lemaître's hypothesis that the universe began in an intensely hot, compressed state from which it expanded. By 1948 Gamow had examined the physical state of this primeval stage, where he believed the elements must have been synthesized (see selection 130), and he had also traced out the evolution of the expanding universe through the formation of galaxies.[1] In the same year Ralph Alpher and Robert Herman published the first paper reprinted here, in which they follow Gamow's basic ideas but reformulate the physics of the early expanding universe. They conclude for the first time that there should be a very low-temperature (5°), relict blackbody radiation produced by the redshifted light left over from the hot, prestellar stages of the universe.[2] In 1941 Andrew McKellar had interpreted observations made by Walter Adams to indicate that the rotational excitation temperature of cyanogen radicals in interstellar space was about 2.3°,[3] but the significance of this observation was not realized until two and a half decades later, when Arno Penzias and Robert Wilson accidentally detected the 2.8° microwave background radiation that proved to be the relict radiation of the primeval fireball (see selection 132).

During the late 1940's Gamow and his colleagues stressed that the very early stages of the universe were completely governed by the density of this radiation.[4, 2] In the later lucid account given as the second paper here, Gamow outlines the basic scenario of the evolving universe. In discussing the relative importance of matter and radiation at different stages in the evolution of the universe, he notes that at first radiation is more important than matter. This radiation is characterized by a temperature that decreases as the universe expands. After 100 million years the universe becomes cool enough for the matter density to exceed the radiation energy density, and only at this time can galaxies and stars begin to form.[5]

1. G. Gamow, *Nature 162*, 680 (1948).
2. Also see R. A. Alpher and R. C. Herman, *Physical Review 75*, 1089 (1949).
3. A. McKellar, *Publications of the Dominion Astrophysical Observatory* (Victoria) *7*, 251 (1941).
4. G. Gamow, *Physical Review 74*, 505 (1948), *Reviews of Modern Physics 21*, 367 (1949).
5. For a good early account of physical conditions during the expansion of the universe, also see R. A. Alpher, J. W. Follin, and R. C. Herman, *Physical Review 92*, 1347 (1953). Gamow later published a paper with Alpher and Herman which provides a more thorough account of the approximate treatment given in his *Vistas in Astronomy* paper (R. A. Alpher, G. Gamow and R. C. Herman, *Proceedings of the National Academy of Sciences 58*, 2179 [1967]).

Alpher and Herman, Evolution of the Universe

IN CHECKING THE RESULTS PRESENTED by Gamow in his recent article on "The Evolution of the Universe" [*Nature* of October 30, p. 680], we found that his expression for matter-density suffers from the following errors: (1) an error of not taking into account the magnetic moments in Eq. (7) for the capture cross-section, (2) an error in estimating the value of α by integrating the equations for deuteron formation (the use of an electronic analogue computer leads to $\alpha = 1$), and (3) an arithmetical error in evaluating ρ_0 from Eq. (9). In addition, the coefficient in Eq. (3) is 1.52 rather than 2.14. Correcting for these errors, we find

$$\rho_{\text{mat.}} = \frac{4.83 \times 10^{-4}}{t^{3/2}}.$$

The condensation-mass obtained from this corrected density comes out not much different from Gamow's original estimate. However, the intersection point $\rho_{\text{mat.}} = \rho_{\text{rad.}}$ occurs at $t = 8.6 \times 10^{17}$ sec. $\cong 3 \times 10^{10}$ years (that is, about ten times the present age of the universe). This indicates that, in finding the intersection, one should not neglect the curvature term in the general equation of the expanding universe. In other words, the formation of condensations must have taken place when the expansion was becoming linear with time.

Accordingly, we have integrated analytically the exact expression:[1]

$$\frac{dl}{dt} = \left[\frac{8\pi G}{3}\left(\frac{aT^4}{c^2} + \rho_{\text{mat.}}\right)l^2 - \frac{c^2 l_0^{\,2}}{R_0^{\,2}}\right]^{1/2}$$

with $T \propto 1/l$ and $R_0 = 1.9 \times 10^9\sqrt{-1}$ light-years. The integrated values of $\rho_{\text{mat.}}$ and $\rho_{\text{rad.}}$ intersect at a reasonable time, namely, 3.5×10^{14} sec. $\cong 10^7$ years, and the masses and radii of condensations at this time become, according to the Jeans' criterion, $M_c = 3.8 \times 10^7$ sun masses, and $R_c = 1.1 \times 10^3$ light-years. The temperature of the gas at the time of condensation was 600°K., and the temperature [radiation] in the universe at the present time is found to be about 5° K.

1. G. Gamow, *Phys. Rev. 70*, 572 (1946).

Gamow, The Physics of the Expanding Universe

INTRODUCTION

ACCORDING TO THE GENERALLY ACCEPTED point of view, the red-shift of spectral lines observed in distant galaxies must be interpreted as the consequence of a rapid expansion (or, rather, dispersal) of the far-flung system of galaxies populating the limitless space of the Universe. It follows that, once upon the time, the matter forming our Universe was strongly compressed possessing a, presumably uniform, high density and temperature. An alternative hypothesis, according to which the continuous dispersal of galaxies is exactly compensated by a continuous creation of new matter, and formation of new galaxies, was proposed several years ago by Bondi, Gold, and Hoyle.[1] However, at least in the opinion of the present writer, this hypothesis is at present neither necessary nor tenable. In fact, the old discrepancy between the figures for the age of the Universe, obtained from geological data, and from the observed recession velocities, was recently removed by new information concerning intergalactic distances. On the other hand, the excess reddening of distant elliptical galaxies, observed by Stebbins and Whitford,[2] definitely indicates the presence of evolutionary trends in the Universe at large.

IMPORTANCE OF THERMAL RADIATION IN COSMOLOGY

Returning to the picture of evolutionary expansion of the Universe from the original dense state, we may ask the question about the physical conditions which must have existed at different stages of this expansion, and about the processes which could have caused the differentiation of the originally homogeneous material into the present highly heterogeneous system of individual galaxies, stars, planets, *etc.*

In the study of physical processes which have taken place in the past history of the Universe, it is important to distinguish between the part played by ordinary matter (material particles), and the part played by thermal radiation (heat quanta). Within the limits of our ordinary physical experience, thermal radiation penetrating all material bodies (unless, of course, these are at absolute zero temperature) plays a comparatively unimportant role in their mechanical and thermal behaviour. Consider, for example, a unit volume of air at normal density and temperature; we have here a mixture of ordinary atmospheric gases with a gas formed by quanta of thermal radiation. For the mass-density ρ and heat capacity C of that system we can write

$$\rho = mn + \frac{aT^4}{c^2} \tag{1}$$

and

$$C = \frac{3}{2}kn + 4aT^3. \tag{2}$$

where m and n are the mean mass of the particles and their number per unit volume, and k and a Boltzmann's and Stefan's constants.

Substituting numerical values, we find for the mass-density and for the heat-capacity of radiation 7×10^{-26} gm per cm^3 and 8×10^{-7} erg per cm^3-degree, respectively; both figures are negligibly small as compared with those for the material component.

However, if we now consider a mixture of matter and radiation in the central regions of stars, we find the situation somewhat different. In fact, for $\rho = 100$ gm per cm^3, and $T = 2 \times 10^{7\circ}$ K the mass-density of radiation is still negligible as compared with that of matter, but its heat-capacity is now 2.4×10^8 erg per cm^3-degree, which is $2\frac{1}{2}$ per cent of the heat-capacity of matter under these conditions. In the radiation of hotter stars the heat-capacity (and pressure) becomes still larger, and must be taken into account in all thermodynamic and hydrodynamic considerations.

Let us now consider a unit volume of interstellar gas with a density of 10^{-24} gm per cm^3 and a temperature of about 100° K. The mass-density of radiation under these circumstances (0.8×10^{-28} gm per cm^3) is still very small, and negligible as compared with that of matter. But the heat capacity of radiation is now 3×10^{-8} erg per cm^3-degree, *exceeding the heat-capacity of matter by a factor of two-hundred million.* Thus in all thermodynamic considerations pertaining to such a mixture of atoms and heat quanta, thermal radiation will play the principal role. In particular, if such a mixture is adiabatically expanded or contracted, the change of temperature will be determined entirely by radiation. Since, as is well known, the temperature of adiabatically expanding radiation drops in inverse proportion to the cube root of the volume, whereas the temperature of a (monoatomic) gas changes inversely as that volume to the power of two-thirds, the material part of the system will cool much faster, and there will be a constant stream of energy from the radiation into the gas. However, owing to the much larger heat-capacity of radiation, the equilibrium temperature of the mixture will follow very closely the inverse cube-root law. It must be noticed here that, in the process of adiabatic expansion or compression, the ratio of heat capacities of radiation and matter remains constant. In fact, the expression (2) shows that they both change in inverse proportion to the volume.

The balance between matter and radiation in the Universe as a whole is rather similar to the above-discussed example of interstellar gas. In fact, if we take for the mean material density in the Universe the value of 10^{-30} gm per cm^3, and assume the temperature to be as low as 1° K, the radiation density will be ten-thousand times smaller than that of matter, whereas its heat-capacity will be two-hundred-million times larger. The heat-capacity of radiation can be made smaller than that of matter only if we assume the average temperature of the Universe to be less than one-thousandth of a degree.

Turning now to the actual physical conditions existing at present in intergalactic space, we shall first estimate its minimum temperature by making the assumption that all the present heat content of the Universe is exclusively provided by the radiation of the stars, accumulated during the five-thousand-million years of its existence. For our purpose we can take the Sun as a representative star, since, indeed, galaxies possess the same average spectral class as the Sun. Our Sun radiates 2 erg per sec. per unit mass and, in the course of its existence, must have sent out a total of 3×10^{17} ergs or 3×10^{-4} gm of radiation for each gramme of material. Since the average density of stellar matter in the Universe is about 10^{-30} gm per cm^3, the present average density of thermal radiation should be at least 3×10^{-34} gm per cm^3. Using (1), we calculate that the present temperature of the Universe must be at least a few degrees above absolute zero, so the total heat capacity of thermal radiation must exceed that of matter by a factor of several hundred million.

We can also fix a maximum value for the temperature of the Universe considered as a consequence of stellar radiation, using the energy-balance of the hydrogen-to-helium transformation which is responsible for stellar energy production. According to the data of nuclear physics, this transformation results in the liberation of 0.6 per cent of the original mass in the form of radiation. Thus the maximum mass-density of radiation which could be released by pure hydrogen uniformly distributed with density 10^{-30} gm per cm^3 is 6×10^{-33} gm per cm^3, *i.e.*, only twenty times larger than our previous estimate. The corresponding maximum temperature is only 6° absolute! Thus we see that if, instead of condensing into stars, the matter of the Universe were to remain uniformly distributed through space up to its present degree of rarefaction, it could not have been heated much above the absolute zero even if, by some miracle, the transformation of hydrogen into helium had released all available nuclear energy. Even though nuclear reactions would supply enough energy per particle to raise the "particle temperature" to many thousand-million degrees,[3] all that energy would be taken away by thermal radiation due to its enormous heat-capacity, and as a result the temperature of the mixture would hardly rise above absolute zero!

RADIATION- AND MATTER-ERAS IN THE HISTORY OF THE UNIVERSE

Apart from the heat content of the Universe produced by nuclear reactions in stars, there also could be some heat *left over* from the earlier pre-stellar stage, since, indeed, it would be rather unnatural to assume that the original homogeneous material of the young Universe was completely free from any thermal motion. To estimate this residual heat, we must consider in some detail the laws of expansion of space if the latter is filled with a mixture of gas and thermal radiation. According to the general theory of relativity the time-behaviour of a homogeneous isotropic Universe is described by the equation:[4]

$$\frac{1}{l}\frac{dl}{dt} = \sqrt{\frac{8\pi G}{3}\rho - \frac{c^2}{R^2}}, \tag{3}$$

where l is the distance between any two material points in the expanding space, ρ the total mass-density, G Newton's gravitational constant, c the velocity of light, and R the radius of curvature.

The value of ρ is given by the formula (1); we notice that, while in the process of expansion, the first term varies as l^{-3}, the second term goes as l^{-4} (because $T \sim l^{-1}$). Since the second term under the radical in (3) varies as l^{-2}, we conclude that, for sufficiently early stages of the expansion, the only important term under the radical is the one representing the mass-density of the radiation. Neglecting the other two terms, and replacing the logarithmic derivative of l by the negative logarithmic derivative of T, we obtain

$$\frac{1}{T}\frac{dT}{dt} = -\sqrt{\frac{8\pi G}{3} \times \frac{aT^4}{c^2}}, \tag{4}$$

which integrates as

$$T = \sqrt[4]{\frac{3c^2}{32\pi aG}} \times \frac{1}{t^{1/2}} = \frac{1.5 \times 10^{10}}{t^{1/2}} \,^\circ\text{K}, \tag{5}$$

where t is counted in seconds from the singular state representing the "beginning" of the Universe. For the mass-density of radiation during that period we obtain

$$\rho_{\text{rad}} = \frac{4.4 \times 10^5}{t^2} \text{ gm per cm}^3. \tag{6}$$

Let us assume for a moment that these formulae still hold for the present state of the Universe, *i.e.* that even at present the radiation-density exceeds the density of matter. Using for t the present age of the Universe, 5×10^9 years or 1.5×10^{17} sec. (as given by geological data and by the study of stellar evolution), we would get $T_{\text{rad.}}$ (present) $= 40^\circ$K, and $\rho_{\text{rad.}}$ (present) $= 2 \times 10^{-29}$ gm per cm^3 $= 20\rho_{\text{mat.}}$ (present); these are results which by themselves do not contradict known facts. However, assuming that the formula (4) still holds at the present time, and calculating the rate of expansion from the expression

$$\frac{1}{l}\frac{dl}{dt} = -\frac{1}{T}\frac{dT}{dt} = \frac{1}{2t}, \tag{7}$$

we would get for the present rate of expansion (*i.e.* the present value of Hubble's constant) a value which is only one-half of that actually observed. Thus we are forced to the conclusion that radiation gave up its priority over matter sometime in the past history of the Universe, and that in all calculations pertaining to the present era we must retain only the matter term in the expression for the total mass-density. Returning to formula (3), using the correct value 6.5×10^{-18} sec.$^{-1}$ for Hubble's constant, and taking $\rho = \rho_{\text{mat.}} = 10^{-24}$ gm per cm^3, we find that the first term under the radical can be neglected as compared with the last. Thus we have

$$\left[\frac{1}{l}\frac{dl}{dt}\right]_{\text{present}} = \sqrt{-\frac{c^2}{R^2}} = \frac{ic}{R}, \tag{8}$$

with the result that the Universe possesses an imaginary curvature (meaning that space is open and infinite), and that the present numerical value of the radius of curvature is 4×10^9 light years. The small value of the density term, as compared with the curvature term in (3), means physically that the kinetic energy of the dispersing system of galaxies is at present much larger than their mutual potential energy, so that the expansion proceeds linearly with time. Thus we write

$$\rho_{\text{matter}}(\text{present}) = \frac{3.4 \times 10^{21}}{t^3} \text{ gm per cm}^3, \tag{9}$$

where the numerical coefficient is chosen so as to give the present density (10^{-30} gm per cm^3) for the present time (1.5×10^{17} sec.).

We are now ready to calculate the demarcation point between the radiation- and matter-eras in the evolutionary history of our Universe: in fact, that point is simply the intersection of the curves (6) and (9). Writing $\rho_{\text{rad.}}(t_0) = \rho_{\text{mat.}}(t_0)$, and solving for t_0, we find

$$t_0 = 8 \times 10^{15} \text{ sec.} = 2.5 \times 10^8 \text{ years.}$$

For the temperature and density at this point of transition the equations give

$$T_0 = 170^\circ \text{ K and } \rho_{0,\,\text{rad}} = \rho_{0,\,\text{mat}} = 7.5 \times 10^{-27}.$$

In figure 131.1 the changes in time of the densities of matter and radiation are shown graphically on a logarithmic scale.

We can also easily write down the expression for the changes of the matter-density during the radiation-era, and of the radiation-density during the matter-era. In fact, during the early stages of the expansion all linear dimensions (being inversely proportional to the temperature) must have been changing in direct proportion to the square root of the time. Thus matter-density must have been changing inversely as the three-halves power of the time. Adjusting the numerical coefficient so that this curve would pass through the point $[t_0, \rho_0]$ we find

$$\rho_{\text{matter}}(\text{early}) = \frac{4 \times 10^{-3}}{t^{3/2}} \text{ gm per cm}^3. \tag{10}$$

Similarly, noticing that, in the present era, the radiation-density must vary in inverse proportion to the fourth power

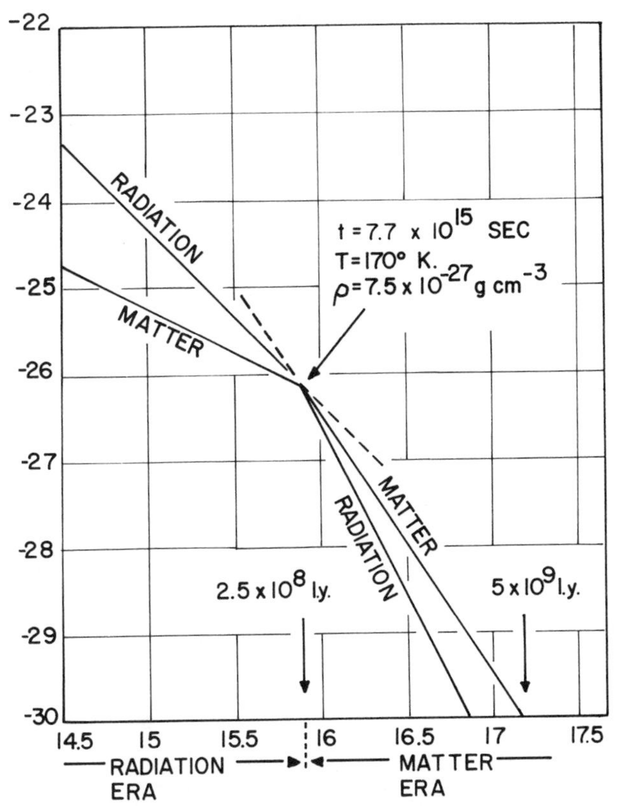

Fig. 131.1 The densities, in gm cm^{-3}, of matter and radiation (ordinates) plotted against time in seconds (abscissae); logarithmic scale.

of the time (because $\rho \sim T^4$, $T \sim l^{-1}$, and $l \sim t$), we find

$$\rho_{\text{rad.}}(\text{present}) = \frac{3.1 \times 10^{37}}{t^4} \text{ gm per cm}^3. \qquad (10)$$

For the present density of residual radiation we obtain 6×10^{-32}, corresponding to about 6° K. Thus we may conclude that the residual heat found at present in the Universe is comparable with the heat provided by nuclear transformations in stars.

FORMATION OF CHEMICAL ELEMENTS AND ORIGIN OF GALAXIES

The above considerations give us a general picture of changing physical conditions characteristic of the evolutionary history of our Universe. We will indicate here only quite briefly how this information can be used for the explanation of various characteristic properties of the Universe as we know it to-day. First of all, it may be suggested that, at least partially, the relative abundances of the atoms of various chemical elements were conditioned by thermonuclear reactions which took place at high speed during the very early stages of expansion while the temperature of the Universe was exceedingly high. And, in fact, the calculations in that direction, carried out by the present writer,[5] and later in some more detail by Fermi and Turkevich,[6] lead to a value of the H/He ratio which is in good agreement with observational data. However, there are still some difficulties to be overcome in understanding the abundances of heavier elements, and there is a possibility that the original distribution was partially modified by various processes during the later stages of the evolution.

The second interesting point is connected with the problem of the formation of galaxies. It is natural to believe that, during the radiation-era of the evolution, material particles were distributed uniformly through space, being carried around by the more abundant gas of heat quanta. However, as soon as the density of matter became higher than that of radiation Jeans' process of gravitational instability must have set to work, and broken up the homogeneous gas of material particles into separate gas clouds. According to the well-known formula of Jeans the diameter of these gravitational condensations is given by

$$D = \sqrt{\frac{5\pi T}{3Gm\rho}}, \qquad (11)$$

where m is the mass of the particles in question. Applying this formula to the ordinary H, He-mixture, and using for T and ρ the values corresponding to the transition-era ($t = t_0$), we obtain

$$D = 1 \times 10^{21} \text{ cm} = 40{,}000 \text{ light-years.}$$

For the mass we get

$$M = 4 \times 10^{41} \text{ gm} = 2 \times 10^8 \text{ sun-masses.}$$

These values are in a reasonably good agreement with the observed mean values of galactic dimensions and masses.

Thus it seems that an insight into the varying physical conditions in the expanding Universe, as described in this article, may lead to a real understanding of cosmogonic processes, and it may be hoped that further studies in this direction will help to clarify the past evolutionary history of the Universe.

1. See H. Bondi, *Cosmology* (Cambridge: Cambridge Univ. Press, 1952).
2. See A. E. Whitford, *Astron. J. 58*, 49 (1953).
3. In fact, if the total energy 3.7×10^{-5} ergs of the 4H → He reaction were distributed only between the resulting α-particle and two electrons, without being given to thermal

radiation, the particle temperature would be

$$\frac{1}{3} \times \frac{2}{3} \times \frac{3.7 \times 10^{-5}}{1.4 \times 10^{-16}} = 6 \times 10^{10\circ}\,\mathrm{K}.$$

4. See G. Gamow, *Dansk. Mat. Fys. Medd.* 27, no. 10 (1953).

5. G. Gamow, *Nature 162*, 680 (1948).

6. See R. A. Alpher and R. C. Herman, "Theory of the Origin and Relative Abundance Distribution of Elements," *Rev. Mod. Phys. 22*, no. 2, 153–212 (1950).

132. A Measurement of Excess Antenna Temperature at 4080 MHz

Arno A. Penzias and Robert W. Wilson

(*Astrophysical Journal 142*, 419–421 [1965])

One of the most important twentieth-century discoveries bearing on cosmology came about quite serendipitously. When testing the microwave-receiving system intended to be used for measurements of the high-latitude continuum radiation of our galaxy, scientists at the Bell Telephone Laboratories found a few degrees of unexpected noise temperature. Because the excess noise showed no sidereal, solar, or directional variation, they first supposed that the noise arose in the receiving antenna itself. Here Arno Penzias and Robert W. Wilson conclude that there remains an external noise contribution with an antenna temperature of $3.5 \pm 1.0°$ at the wavelength of 7.3 cm.

In a companion letter in the same issue of the *Astrophysical Journal*, Robert Dicke and his colleagues explained the excess 3° radiation as the residual temperature of the primeval explosion that initiated the expansion of the universe.[1] It was not surprising that Dicke was one of the first to realize the importance of this radiation, since nearly twenty years earlier he had used a sensitive microwave radiometer to derive an upper limit of 20° for the residual temperature of the "radiation from cosmic matter."[2] Nevertheless, the idea was not altogether new, for George Gamow, Ralph Alpher, and Robert Herman had already shown that a residual temperature of 5° results from the expansion of the 10 billion degree primeval fireball (see selection 131).

If the cosmic noise discovered by Penzias and Wilson is the relic fireball radiation, it should have a blackbody spectrum. Because of the interactions of matter and radiation, the primeval radiation will very rapidly reach thermodynamic equilibrium, and the resulting thermal spectrum will be preserved for all time as the universe expands and cools. While P. G. Roll and D. T. Wilkinson confirmed the initial measurements, G. B. Field and J. L. Hitchcock verified the expected thermal spectrum by examining the excitation of interstellar molecules at the shorter wavelength of 2.5 mm.[3] Balloon measurements were subsequently used to confirm the thermal spectrum at submillimeter wavelengths (see figure 132.1).[4] The microwave background radiation has also provided additional evidence for the isotropy of the universe as a whole. Following the pioneering observations of E. K. Conklin and R. N. Bracewell, and of Wilkinson and R. B. Partridge,[5] the background radiation has been found to be isotropic to 0.1% on all angular scales greater than 1 min of arc. Most recently, an anisotropy of a few millidegrees has been detected and attributed to the motion of the earth relative to the radiation at a velocity of 390 km sec^{-1}.[6] When this component is removed, the radiation is isotropic to 1 part in 3,000.

1. R. H. Dicke, P. J. E. Peebles, P. G. Roll, and D. T. Wilkinson, *Astrophysical Journal 142*, 414 (1965).
2. R. H. Dicke, R. Beringer, R. L. Kyhl, and A. B. Vane, *Physical Review 70*, 340 (1946).
3. P. G. Roll and D. T. Wilkinson, *Physical Review Letters 16*, 405 (1966); T. F. Howell and J. R. Shakeshaft, *Nature 210*, 1318 (1966); G. B. Field and J. L. Hitchcock, *Physical Review Letters 16*, 817 (1966).
4. D. P. Woody, J. C. Mather, N. S. Nishioka, and P. L. Richards, *Physical Review Letters 34*, 1036 (1975).
5. E. K. Conklin and R. N. Bracewell, *Physical Review Letters 18*, 614 (1967); D. T. Wilkinson and R. B. Partridge, *Nature 215*, 719 (1967).
6. G. F. Smoot, M. V. Gorenstein, and R. A. Muller, *Physical Review Letters 39*, 898 (1977).

Measurements of the effective zenith noise temperature of the 20-foot horn-reflector antenna (Crawford, Hogg, and Hunt 1961) at the Crawford Hill Laboratory, Holmdel, New Jersey, at 4,080 Mc/s have yielded a value about 3.5° K higher than expected. This excess temperature is, within the limits of our observations, isotropic, unpolarized, and free from seasonal variations (July, 1964–April, 1965). A possible explanation for the observed excess noise temperature is the one given by Dicke, Peebles, Roll, and Wilkinson (1965) in a companion letter in this issue.

The total antenna temperature measured at the zenith is 6.7° K of which 2.3° K is due to atmospheric absorption. The calculated contribution due to ohmic losses in the antenna and back-lobe response is 0.9° K.

The radiometer used in this investigation has been described elsewhere (Penzias and Wilson 1965). It employs a traveling-wave maser, a low-loss (0.027-db) comparison switch, and a liquid helium–cooled reference termination (Penzias 1965). Measurements were made by switching manually between the antenna input and the reference termination. The antenna, reference termination, and radiometer were well matched so that a round-trip return loss of more than 55 db existed throughout the measurement; thus errors in the measurement of the effective temperature due to impedance mismatch can be neglected. The estimated error in the measured value of the total antenna temperature is 0.3° K and comes largely from uncertainty in the absolute calibration of the reference termination.

The contribution to the antenna temperature due to atmospheric absorption was obtained by recording the variation in antenna temperature with elevation angle and employing the secant law. The result, 2.3° ± 0.3° K, is in good agreement with published values (Hogg 1959; DeGrasse, Hogg, Ohm, and Scovil 1959; Ohm 1961).

The contribution to the antenna temperature from ohmic losses is computed to be 0.8° ± 0.4° K. In this calculation we have divided the antenna into three parts: (1) two non-uniform tapers approximately 1 m in total length which transform between the $2\frac{1}{8}$-inch round output waveguide and the 6-inch-square antenna throat opening; (2) a double-choke rotary joint located between these two tapers; (3) the antenna itself. Care was taken to clean and align joints between these parts so that they would not significantly increase the loss in the structure. Appropriate tests were made for leakage and loss in the rotary joint with negative results.

The possibility of losses in the antenna horn due to imperfections in its seams was eliminated by means of a taping test. Taping all the seams in the section near the throat and most of the others with aluminum tape caused no observable change in antenna temperature.

The backlobe response to ground radiation is taken to be less than 0.1° K for two reasons: (1) Measurements of the response of the antenna to a small transmitter located on the ground in its vicinity indicate that the average back-lobe level is more than 30 db below isotropic response. The horn-reflector antenna was pointed to the zenith for these measurements, and complete rotations in azimuth were made with the transmitter in each of ten locations using horizontal and vertical transmitted polarization from each position. (2) Measurements on smaller horn-reflector antennas at these laboratories, using pulsed measuring sets on flat antenna ranges, have consistently shown a back-lobe level of 30 db below isotropic response. Our larger antenna would be expected to have an even lower back-lobe level.

From a combination of the above, we compute the remaining unaccounted-for antenna temperature to be 3.5° ± 1.0° K at 4,080 Mc/s. In connection with this result it should be noted that DeGrasse *et al.* (1959) and Ohm (1961) give total system temperatures at 5,650 Mc/s and 2,390 Mc/s, respectively. From these it is possible to infer upper limits to the background temperatures at these frequencies. These limits are, in both cases, of the same general magnitude as our value.

Note added in proof. The highest frequency at which the background temperature of the sky had been measured previously was 404 Mc/s (Pauliny-Toth and Shakeshaft 1962), where a minimum temperature of 16° K was observed. Combining this value with our result, we find that the average spectrum of the background radiation over this frequency range can be no steeper than $\lambda^{0.7}$. This clearly eliminates the possibility that the radiation we observe is due to radio sources of types known to exist, since in this event, the spectrum would have to be very much steeper.

Crawford, A. B., Hogg, D. C., and Hunt, L. E. 1961, *Bell System Tech. J. 40*, 1095.

DeGrasse, R. W., Hogg, D. C., Ohm, E. A., and Scovil, H. E. D. 1959, "Ultra-low Noise Receiving System for Satellite or Space Communication," *Proceedings of the National Electronics Conference 15*, 370.

Dicke, R. H., Peebles, P. J. E., Roll, P. G., and Wilkinson, D. T. 1965, *Ap. J. 142*, 414.

Hogg, D. C. 1959, *J. Appl. Phys. 30*, 1417.

Ohm, E. A. 1961, *Bell System Tech. J. 40*, 1065.

Pauliny-Toth, I. I. K., and Shakeshaft, J. R. 1962, *M.N. 124*, 61.

Penzias, A. A. 1965, *Rev. Sci. Instr. 36*, 68.

Penzias, A. A., and Wilson, R. W. 1965, *Ap. J. 142*, 1149.

APPENDED FIGURE

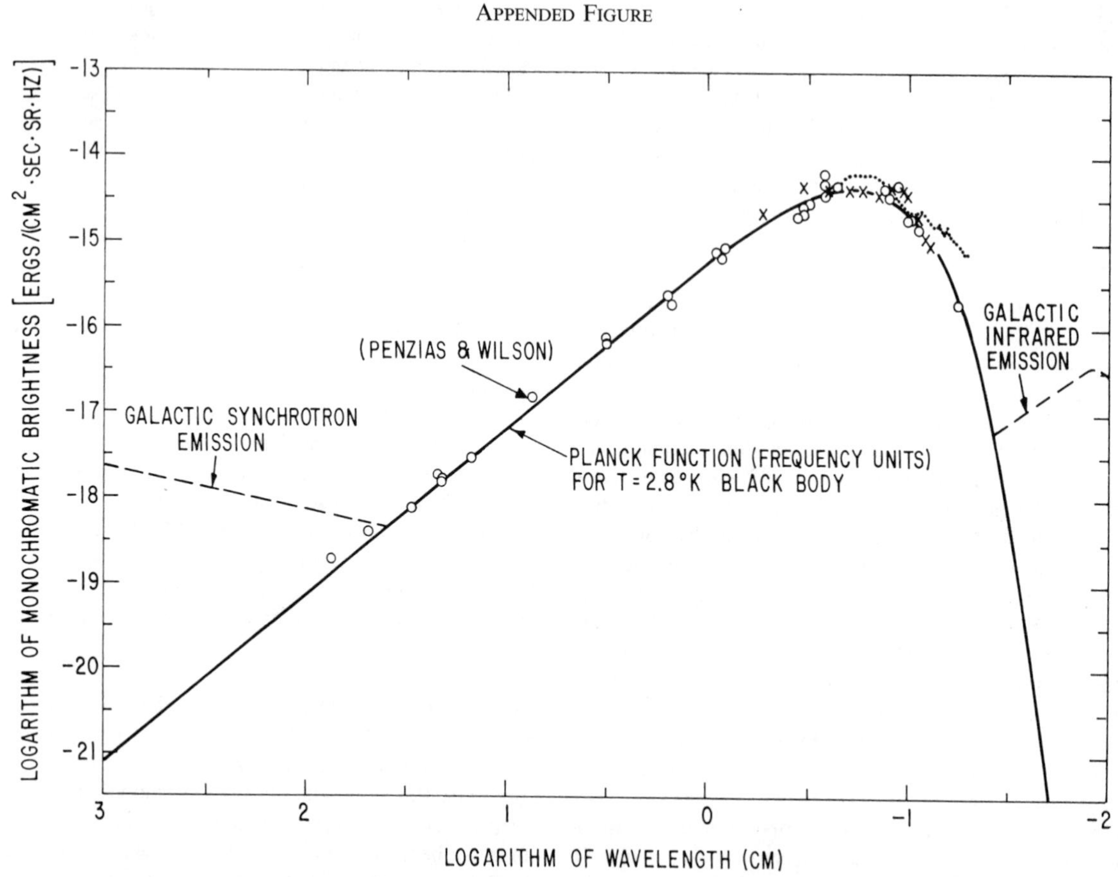

Fig. 132.1 Comparison of observations of the relict cosmic blackbody radiation with the theoretical monochromatic brightness in erg cm^{-2} sec^{-1} sterad^{-1} Hz^{-1} as a function of wavelength, according to the Planck spectral distribution function for a 2.8 °K blackbody. The maximum brightness occurs at 0.18 cm. The data point recorded by Penzias and Wilson is denoted by an arrow and the data points obtained by other radio astronomers by open circles. The data near the peak of the radiation curve were taken by balloon or rocket-borne equipment, and separate observations are denoted by a sequence of crosses and a sequence of dots. (From R. A. Alpher and R. Herman, *Proceedings of the American Philosophical Society 119*, 325 [1975]).

Indexes

Author Index

Subject Index